MICROMECHATRONICS

Modeling, Analysis, and Design with MATLAB®

Nano- and Microscience, Engineering, Technology, and Medicine Series

Series Editor
Sergey Edward Lyshevski

Logic Design of NanoICS
Svetlana Yanushkevich

MEMS and NEMS: Systems, Devices, and Structures
Sergey Edward Lyshevski

Microelectrofluidic Systems: Modeling and Simulation
Tianhao Zhang, Krishnendu Chakrabarty, and Richard B. Fair

**Micromechatronics: Modeling, Analysis, and Design
with MATLAB®, Second Edition**
Victor Giurgiutiu and Sergey Edward Lyshevski

Microdrop Generation
Eric R. Lee

Molecular Electronics, Circuits, and Processing Platforms
Sergey Edward Lyshevski

**Nano- and Micro-Electromechanical Systems: Fundamentals
of Nano- and Microengineering**
Sergey Edward Lyshevski

Nano and Molecular Electronics Handbook
Sergey Edward Lyshevski

Nanoelectromechanics in Engineering and Biology
Michael Pycraft Hughes

MICROMECHATRONICS

Modeling, Analysis, and Design with MATLAB®

SECOND EDITION

Victor Giurgiutiu

University of South Carolina
Columbia, South Carolina, U.S.A.

Sergey Edward Lyshevski

Rochester Institute of Technology
Rochester, New York, U.S.A.

CRC Press
Taylor & Francis Group
Boca Raton   London   New York

CRC Press is an imprint of the
Taylor & Francis Group, an **informa** business

First Indian Reprint, 2015

CRC Press
Taylor & Francis Group
6000 Broken Sound Parkway NW, Suite 300
Boca Raton, FL 33487-2742

© 2009 by Taylor & Francis Group, LLC
CRC Press is an imprint of Taylor & Francis Group, an Informa business

Printed and bound in India by Nutech Print Services

International Standard Book Number-13: 978-1-4200-6562-6

Library of Congress Cataloging-in-Publication Data

Giurgiutiu, Victor.
    Micromechatronics modeling, analysis, and design with MATLAB / by Victor Giurgiutiu and Sergey Edward Lyshevski. -- 2nd ed.
        p. cm. -- (Nano- and microscience, engineering, technology, and medicine series)
    Includes bibliographical references and index.
    ISBN-13: 978-1-4200-6562-6
    ISBN-10: 1-4200-6562-9
    1. Mechatronics. 2. Microelectromechanical systems. I. Lyshevski, Sergey Edward. II. Title. III. Series.

TJ163.12.G58 2009
621--dc22   2008047996

Visit the Taylor & Francis Web site at
http://www.taylorandfrancis.com

and the CRC Press Web site at
http://www.crcpress.com

For sale in India, Pakistan, Nepal, Bhutan, Bangladesh and Sri Lanka only.

Dedication

To our loving and understanding families

In memory of Nikola Tesla

This book is dedicated to the memory of Nikola Tesla (1856–1943), a brilliant inventor, engineer,

and scientist who made indispensable contributions to engineering, science, and technology

propelling progress for centuries

Contents

Preface

This book reports and covers comprehensive, consistent, and coherent studies on mechatronic and electromechanical systems. We have strived to address, deliver, and cover a spectrum of major opportunities and challenges in electromechanics, mechatronics, and electromechanical systems. We span transformative educational and research activities focused on application and utilization of recent developments in engineering science (mechanics, electromagnetics, control, electronics, and other disciplines), enabling hardware (motion devices, power electronics, microelectronics, etc.), advanced technologies (micromachining, materials, etc.), and software.

It is unlikely that one can imagine a prosperous life without electric energy and electromechanical systems. Electricity is produced by power plants and energy systems perform energy conversion. For example, synchronous generators convert the mechanical energy of various origins into electrical energy. Piezoelectric materials and devices convert mechanical energy into electrical energy. Power plants, piezoelectric devices, and other systems can be referred to as electromechanical or mechatronic systems. We use thousands of high-performance mechatronic systems every day. There are hundreds of electromechanical motion devices in each passenger car. There are numerous mechatronic systems in household appliances such as consumer electronics, computers, sound systems, fans, etc. In a stand-alone computer hard drive, there are two electromechanical systems, in particular, an electric drive with a radial-topology permanent-magnet synchronous motor and a limited-angle axial-topology actuator to properly position a magnetic head. There are many electromechanical motion devices in a CD/DVD player and computer DVD-RW drives, for example, spinning drives, repositioning servo, loading servo, etc.

We integrate recent technological developments with sound engineering. This book directly contributes to the design and analysis of high-performance mechatronic systems, energy systems, efficient energy conversion, etc. The fact that $\sim$20% of air pollution comes from power generation, while transportation and industrial processes contribute $\sim$60%, reinforces the need to improve and optimize a broad spectrum of electromechanical systems widely utilized in industrial and transportation systems. Our overall goal is to contribute, enable, create, generate, and apply new or advanced knowledge in mechatronics by advancing electromechanical and energy systems and improving energy efficiency. This book is written due to

- An enormous impact of electromechanical systems, which has been signified in recent years
- Rising awareness of energy deficiencies and urgent needs for advanced and energy-efficient electromechanical systems
- Continuous need for a sound and broad engineering education, which links the classical engineering concept with core engineering science and technology developments

There is a need for responsible, efficient, sustainable, diversified, and environmentally friendly energy utilization in electromechanical systems. The overall objective of this book is to cover mechatronic (electromechanical) systems and their components. We will examine motion devices (actuators, motors, sensors, and transducers), power electronics, control solutions, induced-strain devices, active sensors, microcontrollers, etc. A variety of enabling mechatronics systems and devices are covered.

Why micromechatronics, and what does this terminology imply? Why micromechatronics and not mechatronics? We use the term micromechatronics to expand the horizon of mechatronics to

1. Analysis, design, and integration from *top* (system) to *bottom* (device) and from *bottom* to *top* guarantying the overall coherency and soundness at the device, component, and system levels. Correspondingly, we emphasize and focus on both device- and system-level coherency, functionality, organization, and hierarchy. (This rationale does not imply the application of *system engineering* principles, we rather advocate that *systems engineering* may be exercised, if needed, with great caution and restraints.)

2. Application of advanced microelectronics, microactuation, and microsensing solutions (e.g., advanced integrated circuits, electromagnetic devices with permanent-magnets, advanced energy transduction, application of active materials, etc.).

3. Utilization of microfabrication technologies such as CMOS, micromachining, etc.

Micromechatronics focuses and integrates both conventional and "micro" solutions and technologies. From the terminology standpoints, we examine and cover only *macroscopic* systems. Micromechatronics does not imply its focus on *microscopic* (molecular, atomic, and subatomic) devices and systems that are far-reaching frontiers. One may wonder why we use the term micromechatronics rather than mechatronics. Almost all high-performance electromechanical systems utilize electronics and microelectronics, as well as conventional fabrication and microfabrication (CMOS, *surface* micromachining, *bulk* micromachining, etc.). Perhaps, it is logical to define these systems as micromechatronic because mechatronics may imply predominantly conventional technologies and solutions. These differences between mechatronics and micromechatronics may be beyond semantics and preferred terminology. In fact, one distinguishes electronics and microelectronics, fabrication and microfabrication, etc. However, terminology is not of importance and is a minor issue. One may use mechatronics, electromechanics, or any other term for that matter. This is absolutely understandable and perfectly fine.

Recent trends in engineering have increased the emphasis on integrated systems analysis and design with the ultimate objective to achieve the enhanced functionality, ensure superior capabilities, guarantee optimal performance, etc. Micromechatronic (or mechatronics) is aimed to ensure a coherent overall design at the device and system levels integrating advanced electromechanical motion devices, actuators, sensors, power electronics, integrated circuits, microprocessors, digital signal processors, and other components. Even though the basic fundamentals have been developed, some urgent areas were less emphasized and studied. This book aims to ensure descriptive features, extend the results to modern hardware–software developments, utilize enabling solutions, place the integrated perspectives in favor of sound engineering, as well as focus on the unified studies. We combine traditional engineering with the latest technologies and developments in order to stimulate new advances in the design of state-of-the-art systems. Micromechatronics integrates synergetic intellectual partnerships to strengthen the fundamentals of theoretical, applied, and experimental electrical and mechanical engineering.

The major objective of this book is to enable a deep understanding of multidisciplinary engineering underpinnings. The modern picture of electromechanics, energy conversion, electromechanical motion devices, electronics, and control is provided. This book targets the frontiers of electromechanical engineering, applying basic theory, emerging technologies, advanced software, and enabling hardware. We demonstrate the application of cornerstone fundamentals to design electromechanical systems, cover emerging software and hardware, introduce the rigorous theory, assist the designer in the design of high-performance systems, and help one to develop problem-solving skills. This book offers an in-depth presentation and adequate coverage.

To accelerate analysis and design tasks, ensure productivity and creativity, integrate advanced control algorithms, attain rapid prototyping, generate C codes, and visualize the results, MATLAB® is used. MATLAB embeds the Simulink®, Real-Time Workshop, Control, Optimization, Signal Processing, Symbolic Math, and other application-specific environments and toolboxes. The book demonstrates the MATLAB capabilities, helps one to master this viable environment, studies important practical examples, facilitates designer productivity by showing how to use the advanced software, etc. MATLAB offers a rich set of capabilities to efficiently solve a variety of complex problems. One can modify the covered application-specific problems and utilize the reported results to solve particular problems. We focus on solutions of various modeling, simulation, control, optimization, and other problems.

Engineers and students come from an exceedingly diverse background. Many have extensive electrical or mechanical engineering experience, while others have expertise in aerospace, biomedical, chemical, computer, or industrial engineering. For this reason, to avoid excessive and unnecessary superficiality, the book offers an in-depth presentation covering the subject in sufficient details. We hope that engineers and students who master this book will know what they are doing, why they are doing it, and how to do it.

By using this textbook, the following courses can be developed and taught:

1. Mechatronics

2. Electromechanics

3. Electromechanical motion devices

4. Electric machines

5. Electromechanical systems or mechatronic systems

6. Energy conversion

7. Induced-strain actuators

8. Smart structures and materials

9. Introduction to structural health monitoring

10. Microcontroller applications

Depending on coverage, this book can be utilized for courses in *mechatronics, power systems, energy systems, active materials and smart structures, solid-state actuation, structural health monitoring, and applied microcontroller* engineering. Instructional materials—including suggestions for and examples of lecture plans, laboratory instructions, a typical course package (supplemental information handouts, five laboratory notes, additional homework, software in MATLAB and Assembly languages), and a solutions manual—are available to those professors who adopt this book for their courses. We hope that readers will enjoy this book and provide us with valuable suggestions. Your feedback will be truly appreciated. It has been an inspiring project and a thrilling experience to work on this book. We significantly refocus and refine the first edition, and hope the reader will like it.

MATLAB® is a registered trademark of The MathWorks, Inc. For product information, please contact:

The MathWorks, Inc.
3 Apple Hill Drive
Natick, MA 01760–2098
UAS Tel: 508-647-7000
Fax: 508-647-7001
E-mail: info@mathworks.com
Web: www.mathworks.com

Acknowledgments

We would like to express our sincere acknowledgments and gratitude to all colleagues, peers, and students who provided suggestions. With great pleasure we acknowledge the assistance and help from the outstanding CRC Press team, especially Nora Konopka (acquisitions editor, electrical engineering), Jessica Vakili (project coordinator), Richard Tressider (project editor) Anithajohny Mariasusai (account manager), and Arunkumar Aranganathan (project manager). We gratefully acknowledge the assistance from Math-Works, Inc. for supplying the MATLAB® environment (MathWorks, Inc., 24 Prime Park Way, Natick, MA 01760-15000 http://www.mathworks.com). Many thanks to all of you.

Victor Giurgiutiu
University of South Carolina
Columbia, South Carolina

and

Sergey Edward Lyshevski
Rochester Institute of Technology
Rochester, New York

Authors

Victor Giurgiutiu was born in Bucharest, Romania. He received a BS in aeronautical engineering (1972) and a PhD in aeronautical structures (1977) from Imperial College of Science, Technology, and Medicine, London, United Kingdom. For the next 15 years, he worked in Romania on aeronautical teaching and research being involved with the building of several fixed and rotary wind aircraft under British and French licenses, as well as of Romanian design. From 1992 through 1996 he worked as a research professor at Virginia Polytechnic Institute and State University. Currently, he is professor of mechanical engineering at the University of South Carolina. He maintains a broad interest in many areas of applied mechanics and mechatronics with special interest on active materials, smart structures, microcontroller applications, structural health monitoring, nondestructive evaluation, and engineering diagnosis and prognosis. He teaches courses in microcontroller applications for students of mechanical engineering, as well as courses on active materials and smart structures, wave propagation in solids, composite materials, among others. He is a fellow of the Royal Aeronautical Society of London, United Kingdom; fellow of the American Society of Mechanical Engineers; and member in several other professional societies. He is a member of the editorial board on three journals: *Structural Health Monitoring—An International Journal*; *Aeronautical Journal* of the Royal Aeronautical Society; and *International Journal of Aerospace Engineering*. Through a competitive process, he was selected to serve for three years as program manager for structural mechanics at the Air Force Office of Scientific Research in suburban Washington DC under an IPA assignment away from his university.

Sergey Edward Lyshevski was born in Kiev, Ukraine. He received his MS (1980) and PhD (1987) in electrical engineering from Kiev Polytechnic Institute. From 1980 through 1993, Dr. Lyshevski held faculty positions at the Department of Electrical Engineering at Kiev Polytechnic Institute and the Academy of Sciences of Ukraine. From 1989 through 1993, he was the microelectronic and electromechanical systems division head at the Academy of Sciences of Ukraine. From 1993 through 2002 he was with Purdue School of Engineering as an associate professor of electrical and computer engineering. In 2002, Dr. Lyshevski joined Rochester Institute of Technology as a professor of electrical engineering.

Dr. Lyshevski served as the full professor and senior faculty fellow at the Air Force Research Laboratories, U.S. Surface and Undersea Naval Warfare Centers. He is the author of 15 books (including *MEMS and NEMS: Systems, Devices, and Structures*, CRC Press, 2002; *Nano- and Micro-Electromechanical Systems: Fundamentals of Micro- and Nanoengineering*, CRC Press, 2005; and *Electromechanical Systems and Devices*, CRC Press, 2008), and author and coauthor of more than 300 journal articles, handbook chapters, and regular conference papers. His current research activities are in the areas of high-performance electromechanical systems, nano- and micro-engineering, molecular and biomolecular processing, and systems informatics. Dr. Lyshevski has made significant contributions in design, application, analysis, and deployment of advanced aerospace, automotive, electromechanical, and naval systems. He has also made more than 30 invited and keynote presentations.

1

Introduction to Mechatronic Systems

1.1 Outline of Basic Fundamentals

By applying a micromechatronic (mechatronic) paradigm, we concentrate on a coherent analysis, design, and integration of motion devices (actuators, sensors, transducers, etc.), power electronics, integrated circuits (ICs), controllers (analog controllers, microcontrollers, digital signal processors (DSPs), etc.), and other components of electromechanical systems. Designers have always strived to design integrated high-performance electromechanical systems, which, in the last three decades, were frequently referenced as mechatronic systems [1–8]. The terms high-performance electromechanical systems and mechatronic systems can be used interchangeably.

Advanced hardware (permanent-magnet actuators, piezoelectric transducers, ICs, DSPs, etc.), enabling technologies, state-of-the-art software, and far-reaching tools have been developed utilizing multidisciplinary engineering, science, and technology. Reflecting these developments, the term "mechatronics" was introduced and widely applied. By use of the term micromechatronics, the authors emphasize:

1. Analysis, design, and integration from *top* (system) to *bottom* (device) and from *bottom* to *top* guaranteeing the overall coherency and soundness at the device, component, and system levels. Correspondingly, we focus on both device- and system-level coherency, functionality, organization, and hierarchy.

2. Application of advanced microelectronics, microactuation, and microsensing solutions.

3. Utilization of microfabrication technologies such as complimentary metal-oxide-semi-conductor (CMOS), micromachining, etc.

The current progress in microelectronics, utilization of microelectronic components, and application of microfabrication technologies likely may necessitate using the term "micro-mechatronics." The terminology applied depends on the preference and inclination. Hence, the terminology issue is very minor and may not be of great importance.

The concurrent design of electromechanical (mechatronic) systems has always aimed to provide the needed coherency integrating various electronic, electromagnetic (including electrostatic, optical, and other), and mechanical subsystems, modules, and devices. With stringent requirements on systems performance and capabilities, the designer applies advanced concepts in analysis and design. In general, mechatronic systems can be classified as

- Conventional high-performance electromechanical systems
- Minielectromechanical and microelectromechanical systems (MEMS)

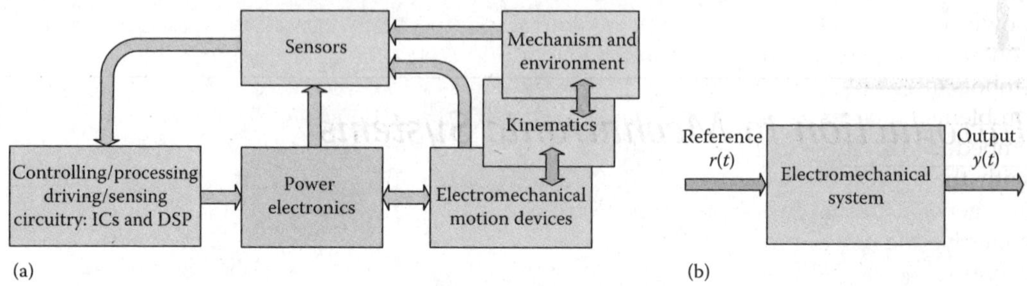

FIGURE 1.1
(a) Functional diagram of an electromechanical system; (b) electromechanical system with input reference $r(t)$ and output $y(t)$.

Figure 1.1a illustrates the functional diagram of an electromechanical (mechatronic) system that comprises different components and modules. The electromechanical systems' performance and capabilities are measured using many criteria, for example, functionality, efficiency, stability, robustness, sensitivity, transient behavior, accuracy, disturbance attenuation, noise immunity, thermodynamics, and overloading. The specifications depend on the requirements imposed in the full operating envelope. For example, the designer can examine and evaluate the input–output behavior studying the reference (command) $r(t)$, output $y(t)$, and state $x(t)$ variables as illustrated in Figure 1.1b.

An electromechanical system comprises electromechanical motion device (devices), sensors, power electronics, controlling/processing and driving/sensing circuitry (ICs and DSPs), energy sources, and mechanics (kinematics). The electromechanical system

1. Performs energy conversion, actuation, and sensing ensuring overall functionality, specifications, and requirements
2. Converts physical stimuli and events to electrical and mechanical quantities and vice versa
3. Comprises control, diagnostics, data acquisition, and other features

The functional and structural designs of high-performance electromechanical systems frequently imply the components, modules, and device selection and developments. Electromechanical motion devices, used as actuators, generators, and sensors, are one of the major components. At the motion device level, the following problems are usually emphasized:

- Design and optimization of electromechanical motion devices (actuators, transducers, etc.) according to their applications within the system kinematics and overall specifications
- Integration of high-performance electromechanical motion devices with sensors, power electronics, and ICs (one emphasizes integrity, regularity, modularity, compliance, matching, and completeness) within a system organization
- Control of electromechanical motion devices

By applying the mechatronic paradigm, one strives to achieve a synergistic combination of the most advanced electromechanical/electronic hardware (kinematics, motion devices, power electronics, ICs, DSPs, etc.), enabling optimization, and sound software ensuring coherent functional, structural, and behavioral designs. For electromechanical systems (robots, servomechanisms, pointing systems, drives, etc.), one solves various interrelated problems, such as actuation, sensing, control, optimization, etc. Actuators and sensors embedded in the kinematic hardware must be designed (or chosen) and integrated with the corresponding power electronics. The principles of matching and compliance are general design principles which require that the systems should be synthesized soundly. The matching conditions have to be determined and guaranteed, and kinematics–actuators–sensors–ICs–power electronics compliance must be satisfied. Electromechanical systems are controlled by using controllers striving to ensure best performance and enhanced capabilities. To implement analog and digital controllers, ICs, microprocessors, DSPs, and other solutions are used. In aircraft and other flight and marine vehicles, various control surfaces must be properly displaced. The functional diagram of a closed-loop electromechanical system is documented in Figure 1.2 for an aircraft emphasizing a stand-alone actuator with corresponding electronics (input–output devices, A/D and D/A converters, optocouplers, transistor drivers, etc.).

As reported in Figure 1.2, each actuator (electromechanical motion device) is regulated by using the state variables given as a vector $\mathbf{x}(t)$, as well as the difference between the desired reference input $r(t)$ and the output $y(t)$. The tracking error $e(t) = r(t) - y(t)$ is used. For robots, aircraft, submarine, and other systems, the Euler angles (θ, ϕ, and ψ) and other variables (velocity, displacement, etc.) are commonly considered as the outputs. The output vector for a vehicle or robotic arm can be given as $\mathbf{y} = [\theta \ \phi \ \psi]^T$. The reference inputs for an aircraft, depicted in Figure 1.2, are the desired Euler angles r_θ, r_ϕ, and r_ψ, yielding the reference vector $\mathbf{r} = [r_\theta \ r_\phi \ r_\psi]^T$. To control the aircraft, one deflects various control surfaces utilizing the control surface servos. By applying the voltage to the actuator windings, one changes the angular or linear displacement of actuators, thereby accomplishing control of aircraft. For each rotational or translational actuator, the desired (reference) deflection $r(t)$ is compared with the actual displacement $y(t)$ resulting in $e(t)$. The variables $r(t)$, $y(t)$, and $e(t)$ are used for an actuator, while the vectors $\mathbf{r}(t)$, $\mathbf{y}(t)$, and $\mathbf{e}(t)$ are used for an aircraft. Advanced fighter aircraft are controlled by varying the thrust as well as

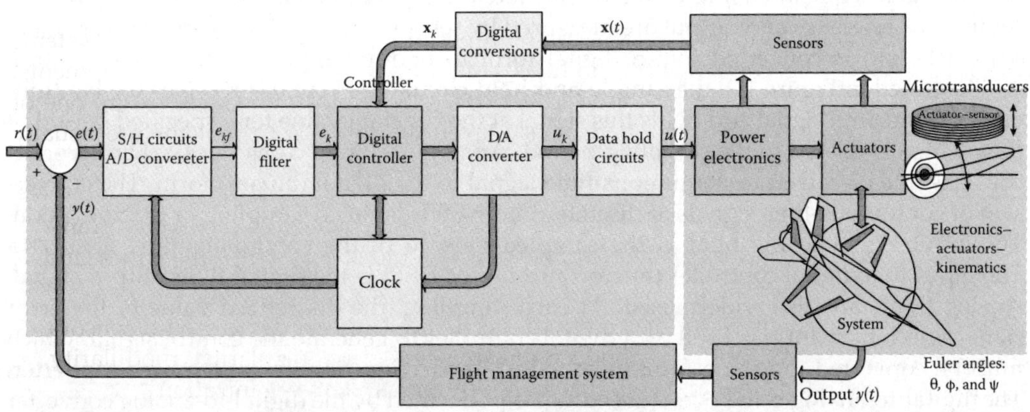

FIGURE 1.2
Functional diagram of electromechanical systems within a flight control system: Flight actuators displace control surfaces, and actuators are controlled by digital controllers. Control of aircraft is accomplished by the flight management system.

FIGURE 1.3
Underwater vehicle hull with actuators to displace fins. As illustrated, actuators are controlled by using power electronics with controllers.

displacing hundreds of rotational and translational actuators with outputs y_i. Figure 1.3 depicts an underwater vehicle hull with actuators (to displace fins) and power electronics. If one controls the underwater vehicle path, defined by $[\theta \ \phi \ \psi]^T$, using actuators, control surfaces are displaced.

Microprocessors and DSPs are widely used to control electromechanical systems. Specifically, DSPs are used to derive control signals based upon the control algorithms, perform data acquisition, implement filters, and attain decision making. Assuming that the continuous reference and output are measured by sensors, the continuous-time error signal $e(t) = r(t) - y(t)$ is converted into a digital form to perform digital filtering and control. As illustrated in Figure 1.2, the sample-and-hold circuit (S/H circuit) receives the continuous-time (analog) signal and holds this signal at the constant value for a specified period of time which is related to the sampling period. Analog-to-digital converter (A/D converter) converts this piecewise or continuous-time signal to the digital (binary) form. The conversion of continuous-time signals to discrete-time signals is called sampling or discretization. The input signal to the filter is the sampled version of the continuous-time error $e(t)$. The input to a digital controller (microcontroller or DSP) is the digital filter output signal. Analog filters are also widely used. At each sampling, the discretized value of the error signal e_k in binary form is used by a digital controller to generate the control signal, which must be converted to analog form to be fed to the driving circuitry of a power converter. The digital-to-analog conversion (decoding) is performed by the digital-to-analog converter (D/A converter) and the data hold circuit. Coding and decoding are synchronized by using a clock. This brief description illustrates that there are various signal conversions, for example, multiplexing, demultiplexing, sample and hold, analog-to-digital (quantizing and encoding), and digital-to-analog (decoding).

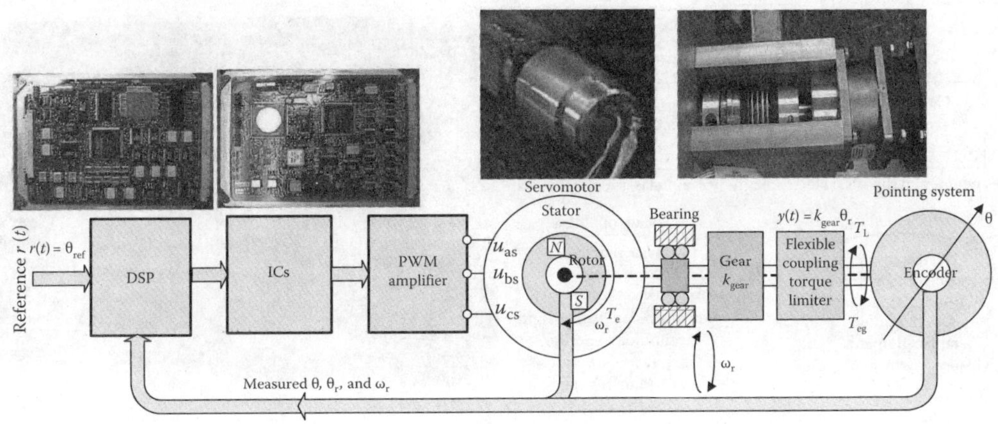

FIGURE 1.4
Electromechanical system.

In electromechanical systems, advanced ICs, microprocessors, DSPs, power electronics, sensors, actuators, and kinematics are used. Consider an electromechanical system, as shown in Figure 1.4, which consists of pointing system kinematics, a geared motor (integrated within the kinematics using the *torque limiter*), a pulse-width modulation (PWM) amplifier, ICs, and DSP. Using the reference r and actual θ angular displacements (θ is measured by the encoder), the DSP (which performs ariphmetics and processing to accomplish data filtering, control, conversions, etc.) develops PWM signals to drive high-frequency insulated-gate bipolar transistors (IGBTs) or metal-oxide-semiconductor field-effect transistors (MOSFETs). The number of the PWM outputs depends upon the converter output stage topology. A three-phase permanent-magnet synchronous motor is utilized. Usually, six PWM outputs drive six transistors to vary the phase voltages u_{as}, u_{bs}, and u_{cs}. The magnitude of the output voltage of the PWM amplifier is controlled by changing the transistors' duty cycle, and the Hall-effect sensors measure the rotor angular displacement θ_r to generate the balanced phase voltages u_{as}, u_{bs}, and u_{cs}. A fully integrated electromechanical system hardware (actuator–mechanism coupling is established utilizing a *torque limiter*) is shown in Figure 1.4.

Electromechanical motion devices are the core components of mechatronic systems. The operating principles and basic foundations of conventional and mini/microscale electromechanical motion devices are identical or similar. The overall device physics and analysis are based on classical electromagnetics and mechanics. However, the device physics (electromagnetic phenomena and effect utilized, for example, electromagnetics versus electrostatics), device topology, and fabrication technologies (including processes and materials used) can be profoundly different. Figure 1.5a illustrates those features. As an example, in the high-end applications as computer and camera hard drives, two high-performance electromechanical systems (drive and servo) are utilized, as illustrated in Figure 1.5b. In cars, there are hundreds of electromechanical systems, for example, from the starter/alternator to various solenoids, relays, fans, microphones, speakers, etc.

A phenomenal growth in electromechanical systems has been accomplished due to

1. Rising industrial and societal needs
2. Strong growing market and demands

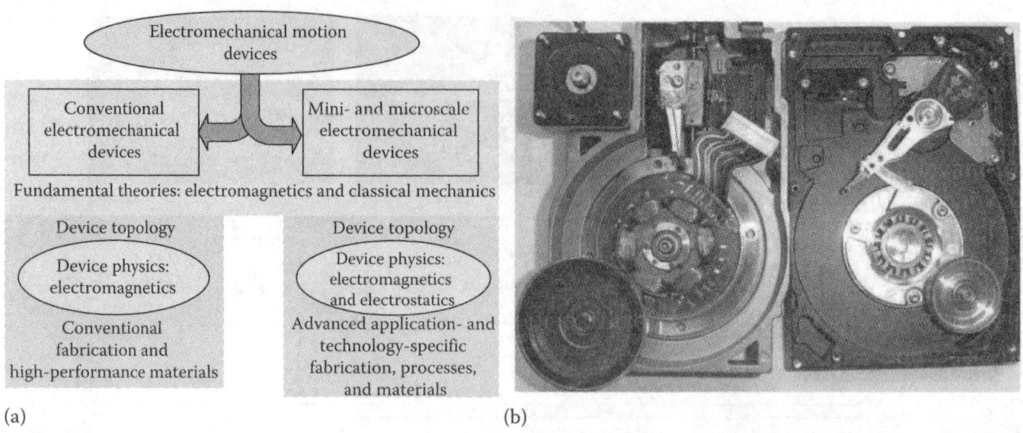

FIGURE 1.5
(a) Electromechanical motion devices; (b) two hard drives. In the hard drive on left, to displace the pointer, the rotational motion of the stepper motor (at the top left) is converted to the translational motion (displacement) by using a gear within the corresponding kinematics. The direct-drive limited-angle axial topology actuator is illustrated in the hard drive on the right. Permanent-magnet synchronous motors (at the center of hard drives), controlled by power electronics, rotate the disk. As documented, the phase windings are on the stator (stationary machine member), while the radial-segmented permanent magnets are on the rotor.

3. Affordability and overall superiority of electromechanical systems as compared to other (hydraulic, pneumatic, etc.) drives and servos

4. Rapid advances in actuators, sensors, power electronics, and ICs hardware

5. Existing software and computer-aided design (CAD) tools

6. Well-developed cost-effective fabrication technologies

Fundamental and applied developments in actuators, electric machines, sensors, and power electronics notably contributed and motivated the need to apply coherent concepts. By applying mechatronics, the design focuses on

1. Devising, advancing, and integrating leading-edge actuation, sensing, and electronic paradigms

2. Enhancing/devising device physics of electromechanical motion devices, thereby ensuring enabling performance and capabilities

3. Advancing hardware and developing novel software

4. Developing and implementing advanced fabrication technologies

5. Integrating optimization, evaluation, and other tasks

The structural and organizational complexity of electromechanical systems has increased drastically due to hardware advancements in response to stringent performance requirements. To meet the rising demands on the system performance with evolved complexity, novel solutions and concepts have been introduced and applied. In addition to a coherent choice of system components (subsystems, modules, devices, etc.), there are various issues that must be addressed in view of the constantly evolving nature of integrated

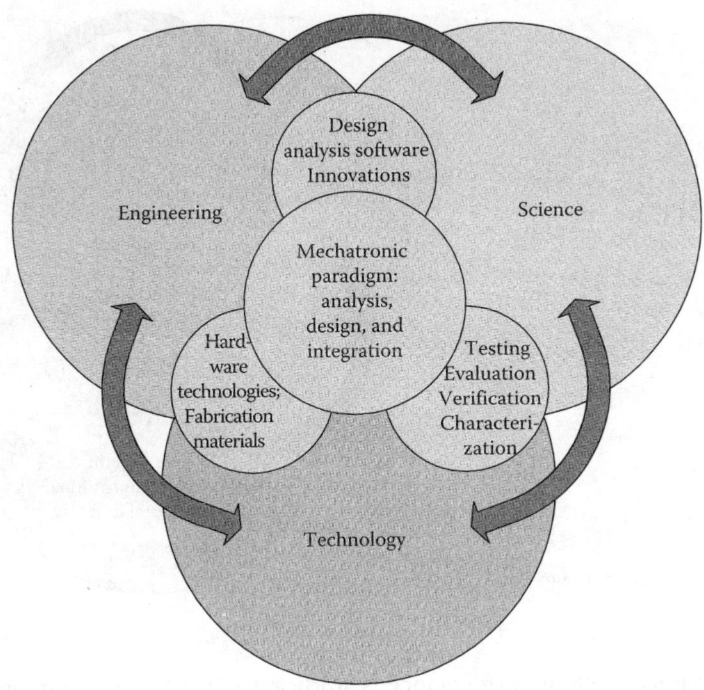

FIGURE 1.6
Integration of engineering, science, and technology.

developments in optimization, diagnostics, packaging, etc. The *optimum-performance* systems can be designed only by using advanced hardware and software. Integrated multidisciplinary features approach quickly. As documented in Figure 1.6, integration of engineering (electrical, mechanical, and computer), science, and technologies is taking place.

One of the most challenging problems in electromechanical systems design is the development and integration of advanced hardware components (actuators, sensors, power electronics, ICs, microcontrollers, DSPs, and others), device/system-level optimization, and software development. Software development includes environments, tools, and algorithms to perform design, control, sensing, data acquisition, simulation, visualization, prototyping, evaluation, and other tasks. Attempts to design high-performance electromechanical systems can be pursued through analysis of complex patterns and paradigms of enabling devices, systems, and technologies. For example, permanent-magnet actuators are enabling solutions, while digital electronics may ensure processing advantages as compared to analog ICs. However, any solution must be applied using sound engineering judgment. Recent trends in engineering have increased the emphasis on integrated design and analysis of advanced systems in order to ensure concurrency.

The design process is evolutionary in nature. It starts with a given set of requirements and specifications. High-level functional design can be performed first in order to produce specifications at the system and component levels. The physical and technological limits must be considered. Using the advanced components, the initial design is performed, and

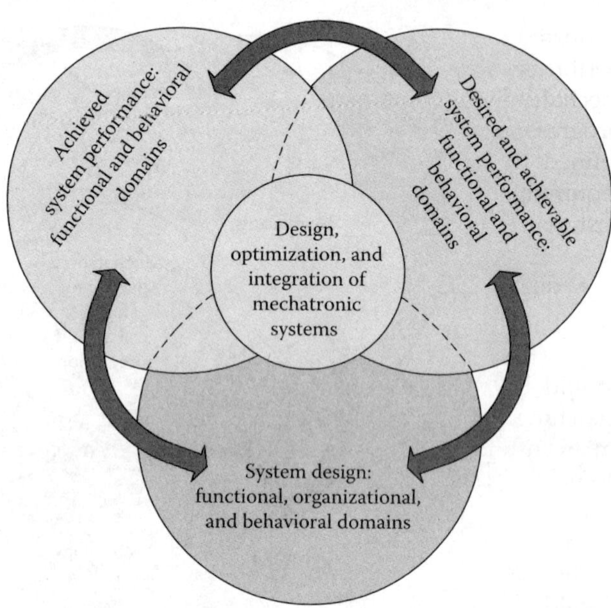

FIGURE 1.7
Mechatronics: interactive design flow.

the electromechanical system performance is studied against the requirements. If requirements and specifications are not met, the designer revisits the system organization and refines the design by devising and evaluating alternative solutions. At each level of the design hierarchy, the system performance and capabilities are used to evaluate and refine the solutions applied. Each level of the design hierarchy corresponds to a particular abstraction level and has the specified set of supporting activities and tools. Different concepts are applied and used to design actuators and ICs due to different device physics, operational principles, behavior, and performance criteria. The level of hierarchy must be defined, and there is no need to study the behavior of thousands of transistors in each IC component because electromechanical systems may integrate hundreds of ICs components and modules. The end-to-end behavior of ICs is usually evaluated because ICs are assumed to be optimized. The design flow is illustrated in Figure 1.7. Only through sound structural and behavioral designs, optimal performance can be accomplished. The requirements on electromechanical system performance cannot exceed the *achievable* performance and capabilities defined by fundamental and technological limits and constraints. Distinct performance estimates, metrics, and measures are used for integrating different specifications on the efficiency, robustness, redundancy, power/torque density, accuracy, and other requirements. The system performance and capabilities are tailored to specific hardware solutions. The so-called *systems design* should be applied with great care. To apply a high-abstraction-level methodology, one should have a great deal of expertise on the available hardware, possible device/component/system solutions, performance analysis, and capabilities. The blind application of *systems design*, without the use of fundamentals, very likely may result in catastrophic failure.

Among the most important criteria, one can emphasize force/torque requirements, velocity/acceleration specifications, and efficiency, accuracy, and stability in the full operating envelope. A wide variety of other requirements are usually imposed. Electromechanical systems are nonlinear and can be multivariable. One solves challenging problems by performing a systematic design, analysis, and integration in order to design high-performance systems within sound design flows and taxonomies. The design of electro-mechanical systems is a process which starts from the specification and requirements progressively proceeding to a functional design and optimization that are gradually refined through a sequence of steps. Specifications typically include the performance requirements derived from systems functionality, operating envelope, sizing features, cost, and hardware availability. Both

- device-to-system and system-to-device hardware integration and
- hardware–software integration

should be combined to study hierarchy, integrity, regularity, modularity, compliance, matching, and completeness. The electromechanical systems synthesis should guarantee an eventual consensus between functional, structural (organizational), and behavioral domains. The descriptive and integrative features must be ensured. The unified analyses of advanced hardware (actuators, sensors, power electronics, ICs, microprocessors, and other components) and software are carried out.

The device physics, functionality, and performance of distinct components and modules should be examined. Synthesis, optimization, modeling, and simulation are complementary activities performed to ensure sound design and analysis. Analysis starts with model developments and simulations, while synthesis starts with the behavioral (dynamic and steady-state) specifications imposed, assessment of system functionality, evaluation of performance, etc. The designer studies, analyzes, and evaluates the system behavior and capabilities using state, output, performance, control, events, disturbance, and other variables. Simulation and optimization are critical aspects for the development of advanced electromechanical systems. As a flexible high-performance environment, MATLAB® has become a standard software tool. To accelerate design and analysis tasks, apply the best knowledge and tools, facilitate significant gains in productivity and creativity, integrate control and signal processing, accelerate prototyping features, generate real-time C code, visualize the results, perform data acquisition, as well as ensure that data-intensive analysis, MATLAB is used. Within MATLAB, the following toolboxes can be applied: Real-Time Workshop, Control System, Nonlinear Control Design, Optimization, Robust Control, Signal Processing, Symbolic Math, System Identification, Partial Differential Equations, Neural Networks, etc. We will demonstrate MATLAB and Simulink® abilities by solving practical examples enhancing user competence. The MATLAB environment offers a rich set of capabilities to efficiently solve a variety of complex problems. The examples will provide the practice and educational experience within the highest degree of comprehensiveness and coverage.

By applying the mechatronic paradigm, conventional and mini/microscale electromechanical systems are studied from a unified prospective. The operating features, basic phenomena, and dominant effects are based on electromagnetics and classical mechanics. Electromechanical systems integrate various components. No matter how well an individual component (actuator, motion device, sensor, power amplifier, or DSP) performs, the overall performance can be degraded if the designer fails to soundly integrate and optimize the system. While actuators, generators, sensors, power electronics, and micro-controllers/DSPs should be analyzed, designed, and optimized, the focus also should be centered on the hardware, software, and hardware–software integration and compliance. The designer sometimes fails to grasp and understand the global picture. While the component-based *divide-and-solve* approach or *systems design* can be applicable in a preliminary design phase, it is important that the design of an integrated electromechanical system must be accomplished in the context of sound physics, general/specific objectives, specifications, requirements, limits, etc. These features are not within the reach of component confinement or *systems engineering*. Enabling hardware, advanced technologies, high-performance software, and software–hardware codesign tools must be applied. We focus on high-performance electromechanical systems covering a broad spectrum of the cornerstone problems. There is a need to further enhance the basic theory and practice, as well as provide a coherent coverage of the current state-of-the-art developments reporting most notable results.

The steady-state and dynamic behavior (evolution) of mechatronic systems can be described by a sixruple $(\mathbf{r}, \mathbf{x}, \mathbf{y}, \mathbf{u}, \mathbf{d}, \mathbf{e})$, where $\mathbf{r} \in \mathbb{R}^b$, $\mathbf{x} \in \mathbb{R}^n$, $\mathbf{y} \in \mathbb{R}^k$, $\mathbf{u} \in \mathbb{R}^m$, $\mathbf{d} \in \mathbb{R}^l$, and $\mathbf{e} \in \mathbb{R}^s$ are the reference (input), state, output, control, disturbance, and events vectors. The system evolves in a set $R \times X \times Y \times U \times D \times E$. The measured (observed or accessible) set is $M_m = \{(\mathbf{r}_m, \mathbf{x}_m, \mathbf{y}_m, \mathbf{u}_m, \mathbf{d}_m, \mathbf{e}_m) \in R_m \times X_m \times Y_m \times U_m \times D_m \times E_m, \; \forall t \in T\}$. The system performance and capabilities are examined using the system evolution sets.

There are many different concepts to synthesize the performance and capability functionals, measures, metrics, and indexes. The best performance and *achievable* capabilities can be achieved by performing various tasks reported, including solving the optimization problem to design closed-loop systems. For example, a mini–max problem can be solved by using the following functionals

$$J_p = \min_{\mathbf{r} \in R, \mathbf{x} \in X, \mathbf{y} \in Y, \mathbf{u} \in U, \mathbf{d} \in D, \mathbf{e} \in E} \max \left[W_{p1}(\mathbf{r}, \mathbf{x}, \mathbf{y}, \mathbf{u}, \mathbf{d}, \mathbf{e}) + \int_0^T W_{p2}(\mathbf{r}, \mathbf{x}, \mathbf{y}, \mathbf{u}, \mathbf{d}, \mathbf{e})dt \right],$$

$$J_p: R \times X \times Y \times U \times D \times E \rightarrow \mathbb{R},$$

$$J_c = \min_{\mathbf{r} \in R, \mathbf{x} \in X, \mathbf{y} \in Y, \mathbf{u} \in U, \mathbf{d} \in D, \mathbf{e} \in E} \max \left[W_{c1}(\mathbf{r}, \mathbf{x}, \mathbf{y}, \mathbf{u}, \mathbf{d}, \mathbf{e}) + \int_0^T W_{c2}(\mathbf{r}, \mathbf{x}, \mathbf{y}, \mathbf{u}, \mathbf{d}, \mathbf{e})dt \right],$$

$$J_c: R \times X \times Y \times U \times D \times E \rightarrow \mathbb{R},$$

where $W_i(\mathbf{r}, \mathbf{x}, \mathbf{y}, \mathbf{u}, \mathbf{d}, \mathbf{e})$ are the performance (subscript p), capability (subscript c) measures and integrands.

Using J_p and J_c, as well as the specified performance levels $\gamma_p \in \mathbb{R}$ and $\gamma_c \in \mathbb{R}$, we define the overall performance goals as

$$J_p(\mathbf{r}, \mathbf{x}, \mathbf{y}, \mathbf{u}, \mathbf{d}, \mathbf{e}) \leq \gamma_p \quad \text{and} \quad J_c(\mathbf{r}, \mathbf{x}, \mathbf{y}, \mathbf{u}, \mathbf{d}, \mathbf{e}) \leq \gamma_c,$$

$$\forall (\mathbf{r}, \mathbf{x}, \mathbf{y}, \mathbf{u}, \mathbf{d}, \mathbf{e}) \in R \times X \times Y \times U \times D \times E \quad \forall t \in T.$$

The desired performance and capability are achieved if the measured system sixruple lie in the set $J(\gamma) = \{(\mathbf{r}_m, \mathbf{x}_m, \mathbf{y}_m, \mathbf{u}_m, \mathbf{d}_m, \mathbf{e}_m) \in R_m \times X_m \times Y_m \times U_m \times D_m \times E_m | J_p \leq \gamma_p, \; J_c \leq \gamma_c\}$. By using the performance specification and requirements, one performs the design with the well-defined multiple tasks thereby ensuring optimal performance. In general, the designer strives to solve very complex nonlinear optimization problems. After simplification of these problems to control design and optimization, the results are reported in Section 8.1.

1.2 Introduction to Taxonomy of Electromechanical System Synthesis and Design

The designer may follow the taxonomy of devising (synthesis) and design which is relevant to synergetic study, assessment, and evaluation of any system. Devising electromechanical systems can be viewed as an evolutionary process of discovering and examining evolving topologies, organizations, and architectures utilizing basic physics and studying possible system evolutions based upon synergetic integration of all components in a unified functional core.

The ability to devise and optimize electromechanical systems to a large extent depends on basic physical principles, as well as available software and hardware solutions. One may hope to complement design and analysis tasks by expected high-level hierarchy, abstraction, efficiency, adaptability, functionality, integrity, compliance, robustness, flexibility, prototypeability, scalability, visualability, interactability, decision making, virtual reality, and other expected features that may be associated with the envisioned CAD tools. However, it is unlikely that one can ensure sound synthesis taxonomy, utilize *generic* solutions, and guarantee cornerstone fundamentals by applying current CADs. There is only a hope that synergetic quantitative synthesis, symbolic descriptions, *knowledge-based* libraries, *evolutionary learning*, and *intelligent* tools can be embedded and utilized, searching and evaluating possible solutions and other descriptive features while providing meaningful prospects at the system and device levels.

Biosystems provide a proof-of-concept principle for highly integrated systems. However, though engineering biomimetics may provide conceptual ideas how living (*natural*) and human-made (*engineered*) systems may function, in general, biomimetics may not be applied. In fact, though biomimetics and bioprototyping are appealing premises, one cannot utilize them due to insufficient knowledge and inadequate understanding of the basic biophysics at device and system levels. The designer may, to some extent, only typify complex living devices and systems. For example, the designer cannot utilize the knowledge learned from studies of actuation and sensing exhibited by a single-cell *Escherichia coli* (*E. coli*) bacterium (1 μm diameter, 2 μm length, and 1 pg weight), shown in Figure 1.8. This bacterium has a plasma membrane, cell wall, and capsule that contains the cytoplasm and nucleoid. Considering locomotion, sensing, and actuation exhibited by this bacterium, one finds that *E. coli* exhibits remarkable performance and capabilities. For example, bacteria's control and propulsion have not been achieved by any human-made micro-, mini-, or conventional underwater vehicles including the most advanced torpedoes and submarines. Advanced conventional torpedoes achieve a maximum speed of ~25 m/s, and speed is a function of vehicle length. The bacterium could propel with a speed of ~20 μm/s. That is, the speed/length ratio is 10. Only unconventional and most advanced supercavity torpedoes can reach this ratio. Other examples reported in Figure 1.8 illustrate the ant and the butterfly. *Engineered* micro-airvehicles most likely will never achieve agility, controllability, maneuverability, efficiency, and other capabilities exhibited by living organisms. Processing, actuation, and sensing in living systems, not to mention intelligence and learning, are far-reaching frontiers. This book reports the radial and axial topology electromechanical motion devices which to some extent typify topologies observed in bioactuators. However, the basic physics of *engineered* actuators are entirely distinct as compared to the biophysics of bioactuators.

There exist sound principles to synthesize high-performance systems and devices. Cornerstone physical laws and sound operating principles should be examined to attain

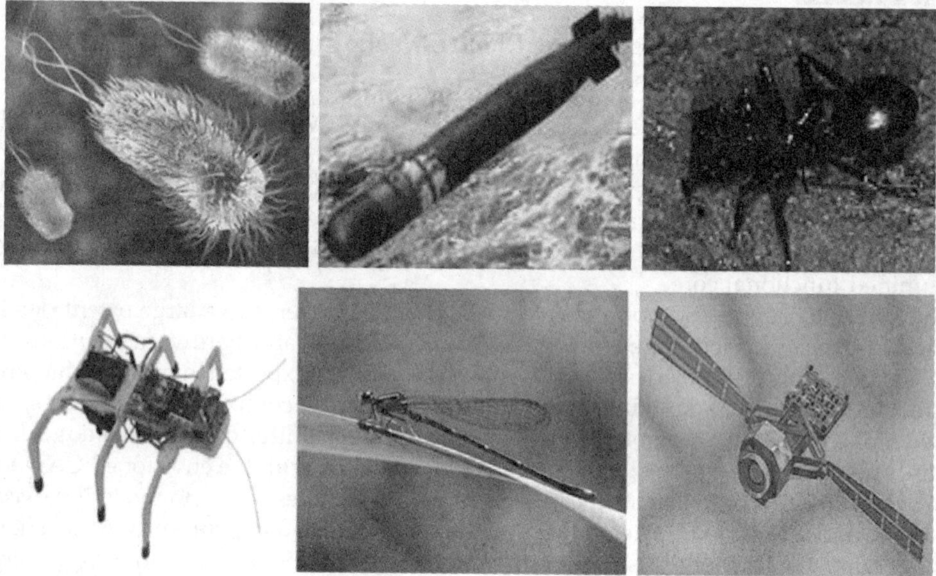

FIGURE 1.8
E. coli bacteria and torpedo, ant and minirobot (with infrared emitters and receivers), and dragonfly-and-mini-airvehicle. (From UC Berkeley. With permission.)

functionality. General and systematic approaches have been developed to analyze and design electromechanical systems radiating energy, and optical and electronic devices without applying *generic* and abstract tools. For electromechanical motion devices, based upon the device physics and restricting the number of possible solutions, the specific synthesis can be performed using electromechanical classifiers [7] utilizing, for example, electromagnetic, electrostatic, piezoelectric, and thermal actuation. Well-defined analysis and design tasks can be accomplished.

Synthesis of electromechanical systems integrates consecutive functional design, analysis, and optimization. For electromechanical devices and systems, we emphasize four distinct domains, for example, (1) devising (synthesis), (2) analysis–design, (3) optimization–refining, and (4) fabrication–evaluation. A design flow-map is reported in Figure 1.9a. The evolving synthesis of electromechanical systems is represented as a bidirectional X-flow-map. This map illustrates the synthesis flows within various sequential tasks. The designer may synthesize novel devices and systems utilizing specific physics, phenomena, effects, and solutions (topologies, organizations, architectures, etc.). The failure to verify the design in the early phases at least causes the failure to design high-performance or functional systems. The interaction between four domains allows one to guarantee interactive bidirectional design applying low-level data to high-level design and using the high-level requirements in devising and designing low-level components. The reported synthesis taxonomy guarantees a multiple hierarchy, modularity, locality, and integrity. A flowchart of the different tasks involved, starting from basic physics and up to verification, is reported in Figure 1.9b.

For electromechanical systems, various benchmarking and emerging problems can be formulated and solved utilizing a mechatronic paradigm by integrating electrical, mechanical, and computer engineering, and science and technology. The existing solutions

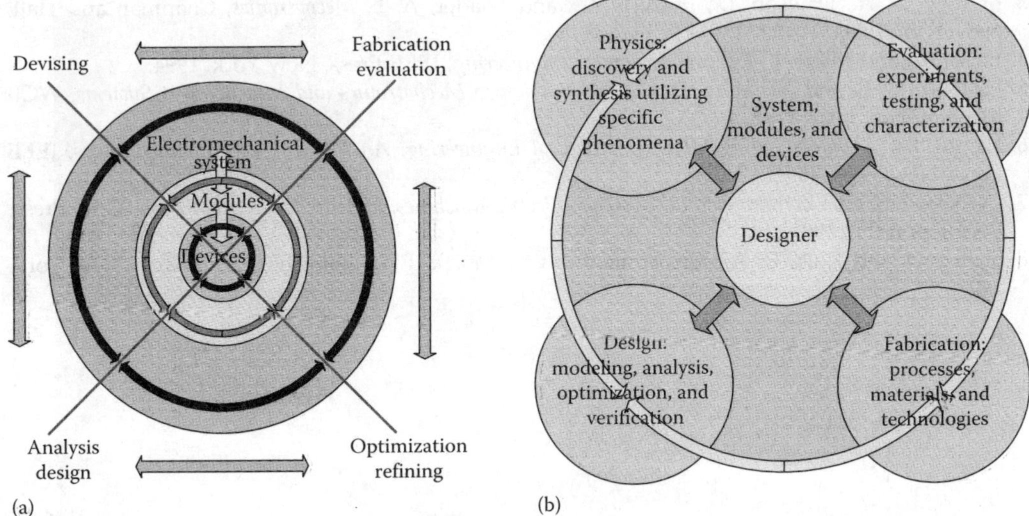

FIGURE 1.9
(a) *X*-flow-map with four domains; (b) flowchart of major tasks.

may not be optimal and may be further developed. The application of a mechatronic paradigm reflects

- Progress in fundamental, applied, and experimental research in response to long-standing challenges, needs to advance various systems, and unsolved problems
- Engineering and technological enterprises and entreaties of steady evolutionary demands
- Evolving technologies and solutions

Mechatronics focuses on the multidisciplinary synergy, and this paradigm promises

1. Improving the overall system performance
2. Enhancing and enabling system capabilities, including adaptability, controllability, flexibility, integrity, robustness, safety, functionality, operationability, and others
3. Ensuring sound, cost-effective, and affordable solutions

References

1. Auslander, D. M. and Kempf, C. J., *Mechatronics: Mechanical System Interfacing*, Prentice Hall, Upper Saddle River, NJ, 1996.
2. Bolton, W., *Mechatronics: Electronic Control Systems in Mechanical Engineering*, Addison Wesley Longman Publishing, New York, 1999.

3. Bradley, D. A., Dawson, D., Burd, N. C., and Loader, A. J., *Mechatronics*, Chapman and Hall, New York, 1996.
4. Fraser, C. and Milne, J., *Electro-Mechanical Engineering*, IEEE Press, New York, 1994.
5. Histand, M. B. and Alciatore, D. G., *Introduction to Mechatronics and Measurement Systems*, WCB McGraw-Hill, New York, 1999.
6. Kamm, J. L., *Understanding Electro-Mechanical Engineering: An Introduction to Mechatronics*, IEEE Press, New York, 1996.
7. Lyshevski, S. E., *Electromechanical Systems, Electric Machines, and Applied Mechatronics*, CRC Press, Boca Raton, FL, 1999.
8. Shetty, D. and Kolk, R. A., *Mechatronics System Design*, PWS Publishing Company, New York, 1997.

2

Electromagnetic and Electromechanics Fundamentals

2.1 Introduction to Design and Analysis

Mechatronic systems must be designed and analyzed with the ultimate objective to ensure best performance and guarantee *achievable* capabilities under physical (electromagnetic, mechanical, thermal, etc.), technological, and other limits. To ensure optimal performance and capabilities, various tasks should be performed, for example, synthesis, modeling, optimization, control, etc. These tasks, in some extent, were emphasized in Chapter 1, where the mechatronic system organization and basics were covered. In general, the terms and abbreviation design and analysis have very broad meanings and intent. For example, design may imply overall syntheses at device and system levels (devising device physics and synthesis of electromagnetic systems thereby *engineering* devices, devising system topology/ organization/architecture, etc.); closed-loop system design, which is also frequently referred to as system optimization; etc. The evolving design and analysis taxonomy, applied at the devices and systems, integrates functional, structural, and behavioral design tasks implying

1. Devising and assessing system organization and architecture to ensure overall system functionality and operationability meeting the specified requirements
2. Device synthesis (*engineering* devices) with device physics analysis, assessment, and evaluation
3. Components matching, compliance, and completeness
4. Preliminary evaluation: Device-level data-intensive electromagnetic and mechanical analyses to estimate device- and system-level *achievable* performance and capabilities, while avoiding costly and time-consuming hardware testing of an entire system
5. Development of advanced software and hardware to attain the highest degree of synergy, integration, efficiency, and performance
6. Coherent experimental evaluation to examine the system performance
7. Hardware and software testing, characterization, and evaluation with possible redesign tasks

Studying mechatronic systems, the emphasis is placed on

- Devising and design of high-performance systems by applying advanced (or discovering innovative) components and devices (actuators, power electronics, sensors, controllers, driving/sensing circuitry, ICs, etc.)

- Analysis and optimization of motion devices (actuators, motors, sensors, trans-ducers, etc.)
- Development of high-performance power electronics, signal processing, and con-trolling ICs
- Synthesis and implementation of optimal control algorithms

The quantitative and qualitative design and analyses are carried out by performing synthesis, evaluation, assessment, and other tasks. As the soundness of functional and structural designs is guaranteed, the designer focuses on various tasks. For example, performing integration, one studies different components, such as ICs, power electronics, actuators, motion devices, kinematics, etc. The functional, structural, and behavioral designs are related to the analysis problems within the following three-step taxonomy. The first step is to accomplish various preliminary design and analysis tasks. In particular,

- *Engineer*, design, examine, and analyze a mechatronic system using a multilevel hierarchy concept applying sound principles: Define the system organization and develop multivariable input–output pairs between system components, for example, study kinematics, motion, and motionless devices (actuators, sensors, transducers), electronics, ICs, controller, input/output, and other devices.
- Assess the soundness of system organization to guarantee overall functionality and operationability.
- Collect and evaluate the data and information. Develop input–output variable pairs, identify the independent and dependent control, disturbance, output, reference (command), state, performance and capability variables, and quantities.
- Making accurate assumptions, simplify the problem to estimate system and device performance and capabilities.

The second step is to carry out a coherent analytic design and analysis emphasizing the device level. In particular,

- Examine the device physics focusing on effects exhibited, phenomena utilized, controllability, etc.
- Using laws of electromagnetics and mechanics, derive equations which relate the variables and events. Define and specify the basic laws (Kirchhoff, Maxwell, Newton, Lagrange, and others) to be used. Describe and examine the electromagnetic, mechanical, thermal, vibroacoustic, and other phenomena exhibited by electromagnetic, electronic, and mechanical components. Examine mathematical models in the form of differential (or difference) and constitutive equations.
- Solve control and optimization problems designing a closed-loop system. Analytically and/or numerically solve equations and study system evolution and behavior.

The third step is testing, characterization, and validation:

- Perform testing, characterization, and evaluation of devices, components, and systems assessing their performance, capabilities, and other features of interest.
- Define the criteria to assess the functionality, operationability, practicality, adequacy, correspondence, and other features.

- Evaluate the overall fleetness and accuracy of analytic and experimental results.
- Perform overall evaluation and assessment of a mechatronic system in order to achieve the best performance and *achievable* capabilities by carrying out coherent design and analysis.

To assess the system performance and examine system capabilities, one performs testing, characterization, and evaluation at the device and system levels. The integrated multidisciplinary features approach quickly. The complexity of electromechanical systems and devices has been increased due to hardware and software advancements as well as stringent performance requirements. Answering the demands of the rising systems complexity and performance specifications, the fundamental theory is coherently applied. Design and analysis of high-performance electromechanical systems require coherent studies of electromagnetics, mechanics, and energy conversion. Analytical, numerical, and experimental studies ensure quantitative and qualitative analysis. In order to analyze and optimize steady-state and dynamic system behavior, one integrates phenomena exhibited, effects utilized, and control concepts.

Analysis may imply modeling, simulation, performance evaluation, capabilities assessment, etc. It is evident that from a control, optimization, and modeling viewpoint, complex electromagnetic, mechanical, thermodynamic, vibroacoustic, and other phenomena must be described and studied. High-fidelity and lamped-parameter mathematical models are developed by utilizing physical laws. The term modeling means the deviations of the equations of motion (governing equations) which describe the electromagnetic, mechanical, thermal, and other phenomena, effects and transitions in the studied device, component, or system. High-fidelity modeling, as three-dimensional Maxwell's equations and tensor calculus are applied, results in data-intensive analysis of electromechanical motion devices. The complexity may be relaxed by applying sound methods without loss of generality, thereby ensuring accuracy and tractability. Any descriptive models are the idealization of physical systems, phenomena, effects, processes, etc. Mathematical models are never absolutely accurate, but models must be adequate allowing the designer to coherently support various tasks.

For moderate complexity mechatronic systems, the experienced designers usually can accomplish a near-optimal design utilizing the expertise and practice. Using the cornerstone physical laws, some performance features (efficiency, force/torque and power densities, settling time, acceleration, etc.) can be estimated avoiding high-fidelity modeling and experiments. Oversimplification and oversophistication result in overall failure due to variety of reasons. From a modeling prospective, adequate and accurate mathematical models, which describe the device physics, are derived using the basic physical laws. The need for models is obvious. Many key performance measures and capabilities (stability, robustness, sensitivity, dynamic and steady-state accuracy, acceleration rate, dynamic responses, etc.) are determined by (1) examining baseline physical phenomena taking into account device physics, hardware limits, and software constraints; and (2) validating and evaluating the results by performing numerical, analytic, and experimental studies. In general, any fundamental and analytic results cannot replace practical and experimental data. The analytic and experimental results must be examined and related. Using the informative variables (measured or observed) and events, one may define the fitting and mismatch functionals and metrics. By comparing the solution (modeled input-state–output-event mapping sets) versus the experimental data (experimental input-state–output-event mapping sets), the results are verified. The steady-state and dynamic behavior can be mapped by the following behavioral modeling (m) and experimental (e) sixtuples $(\mathbf{r}_m, \mathbf{x}_m, \mathbf{y}_m, \mathbf{u}_m, \mathbf{d}_m, \mathbf{e}_m)$ and $(\mathbf{r}_e, \mathbf{x}_e, \mathbf{y}_e, \mathbf{u}_e, \mathbf{d}_e, \mathbf{e}_e)$. Here, $\mathbf{r} \in \mathbb{R}^b$, $\mathbf{x} \in \mathbb{R}^n$, $\mathbf{y} \in \mathbb{R}^k$, $\mathbf{u} \in \mathbb{R}^m$, $\mathbf{d} \in \mathbb{R}^l$, and $\mathbf{e} \in \mathbb{R}^s$ are the reference (input), state, output, control, disturbance, and events vectors. The measured sets, which map the steady-state and dynamic system behavior (evolution), are

$$M_m = \left\{ (\mathbf{r}_m, \mathbf{x}_m, \mathbf{y}_m, \mathbf{u}_m, \mathbf{d}_m, \mathbf{e}_m) \in R_m \times X_m \times Y_m \times U_m \times D_m \times E_m, \forall t \in T \right\}$$

and

$$M_e = \left\{ (\mathbf{r}_e, \mathbf{x}_e, \mathbf{y}_e, \mathbf{u}_e, \mathbf{d}_e, \mathbf{e}_e) \in R_e \times X_e \times Y_e \times U_e \times D_e \times E_e, \forall t \in T \right\}.$$

Hence, one examines the system in M_m and M_e. The strong and weak behavior matching are guaranteed if $M_m = M_e$ and $M_m \subseteq M_e$, respectively. The fitting and mismatch functionals and metrics can be examined assessing analytical and/or numerical data against experimental data and evidence. The goal is to evaluate the fitness and accuracy with the ultimate objective to ensure consistency and coherency between fundamental and experimental findings. Until overall coherency, consistency, and accuracy are not guaranteed, the designer returns to a specific task (step) within the analysis cycle.

It is virtually impractical to apply *generic* approaches, *model-free* concepts, *linguistic* models, *descriptive* techniques, and other *abstract systems design* (systems engineering) methodologies in order to perform coherent synthesis, design, and analysis. Usually, the above-mentioned approaches are not based on the underlying device physics and focus on the narrative descriptive features which lead to a limited applicability and practicality.

Mathematical models should be developed with minimum level of simplifications. At the same time, one should avoid overspecificity and unjustifiable complexity. These models must coherently describe the underlying phenomena (time-varying fields, energy conversion, torque or force development, voltage induction, energy storage, etc.), effects, and processes. The application of physical laws to derive equations of motion, which describe system behavior, are reported in this chapter by applying electromagnetics and classical mechanics. As the model is obtained, some tasks are carried out. For example, one may design control laws and examine the closed-loop system performance.

2.2 Energy Conversion and Force Production in Electromechanical Motion Devices

All components and devices, which constitute mechatronic systems, are very important. For example, one cannot achieve the specified angular displacement accuracy if the steady-state sensor resolution (number of pulses per revolution in resolver or optical encoder) is not adequate. The sensor accuracy is limited imposing the limits. As the electromechanical system is synthesized and its organization is derived, one focuses on various analysis and design activities. The overall system performance and capabilities are largely defined by chosen hardware (actuators, sensors, power electronics, and ICs). The actuation solutions frequently predefine the *achievable* system performance and capabilities. The actuators, as one of the most critical components, are predominantly emphasized in this section.

Various electromechanical motion devices have been devised and applied. For example, electromagnetic versus electrostatic, permanent-magnet versus variable reluctance- or

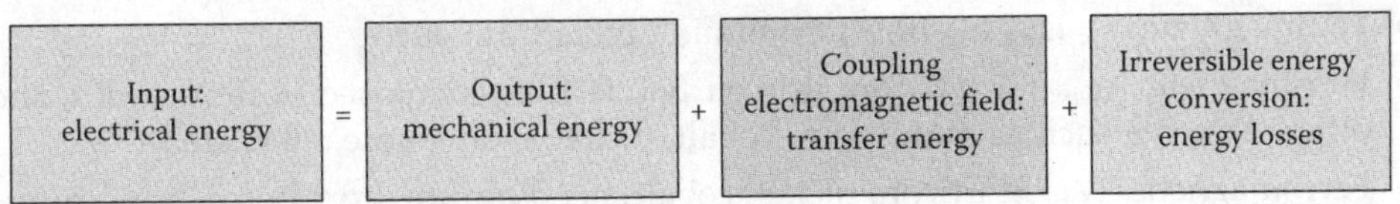

FIGURE 2.1
Energy conversion and transfer in an electromechanical device.

inductance-centered, piezoelectric versus acoustic, thermal versus hydraulic, etc. We will examine a great number of electromagnetic and electrostatic electromechanical motion devices applying classical electromagnetics and mechanics. Energy conversion takes place in electromechanical motion devices that convert electrical energy to mechanical energy and vice versa, for example, convert mechanical energy to electrical energy [1–8]. The device physics is based on the specific phenomena (effects) exhibited and utilized affecting the energy conversion and force/torque production. One examines energy conversion and force production performing various tasks with the ultimate objective to guarantee optimal overall performance. The cornerstone principle of energy conversion is formulated as follows: For a lossless electromechanical motion device (in the conservative system no energy is lost through friction, heat, or other irreversible energy conversion), the sum of the instantaneous kinetic and potential energies of the system remains constant. The energy conversion is represented by Figure 2.1.

The general equation which describes the energy conversion is

$$\underset{\text{Electrical energy input}}{\mathbf{E}_E} - \underset{\text{Ohmic losses}}{\mathbf{L}_E} - \underset{\text{Magnetic losses}}{\mathbf{L}_M} = \underset{\text{Mechanical energy}}{\mathbf{E}_M} + \underset{\text{Friction losses}}{\mathbf{L}_E} + \underset{\text{Stored energy}}{\mathbf{E}_S}.$$

For conservative (lossless) energy conversion, one yields

$$\underset{\text{Change in electrical energy input}}{\Delta\mathbf{W}_E} = \underset{\text{Change in mechanical energy}}{\Delta\mathbf{W}_M} + \underset{\text{Change in electromagnetic energy}}{\Delta\mathbf{W}_m}.$$

The electrical energy, mechanical energy, and energy losses must be defined and examined. The electromagnetic motion devices ensure superior performance and capabilities. Very high power and force densities in actuators and electric machines are achieved by utilizing permanent magnets to establish a stationary magnetic field. The total energy stored in the magnetic field is expressed using the magnetic field density B and intensity H. In particular,

$$W_m = \frac{1}{2}\int_v \vec{B}\cdot\vec{H}\,dv.$$

The material becomes magnetized in response to the external field $\vec{H}$. Using the dimensionless magnetic susceptibility χ_m or relative permeability μ_r, one has $\vec{B}=\mu\vec{H}=\mu_0(1+\chi_m)\,\vec{H}=\mu_0\mu_r\vec{H}$.

The magnetization is expressed as $\vec{M}=\chi_m\vec{H}$.

Based upon the value of the magnetic susceptibility χ_m, the materials are classified as

- Nonmagnetic, $\chi_m=0$, and, thus, $\mu_r=1$
- Diamagnetic, χ_m is $\sim-1\times10^{-5}$ ($\chi_m=-9.5\times10^{-6}$ for copper, $\chi_m=-3.2\times10^{-5}$ for gold, and $\chi_m=-2.6\times10^{-5}$ for silver)

- Paramagnetic, χ_m is $\sim 1 \times 10^{-4}$ (aluminum, palladium, etc.)
- Ferrimagnetic, $|\chi_m| > 1$ (yttrium iron garnet, ferrites composed of iron oxides, and other elements such as aluminum, cobalt, nickel, manganese, zinc, etc.)
- Ferromagnetic, $|\chi_m| \gg 1$ (iron, nickel, cobalt, neodymium–iron–boron, samarium–cobalt, and other permanent magnets). Ferromagnetic materials exhibit high magnetizability, and materials are classified as hard (alnico, rare earth elements, copper–nickel, and other alloys) and soft (iron, nickel, cobalt, and their alloys) materials.

The relative permeability μ_r of some bulk materials is reported in Table 2.1. The permeability, as well as other descriptive material parameters, strongly depends on the fabrication technologies and dimensionality (sizing). Therefore, for example, one cannot assume that the superpermalloy will always ensure μ_r to be $\sim 1,000,000$, while for the electric steel μ_r not always 5,000.

The magnetization behavior of the ferromagnetic materials is described by the magnetization curve, where H is the externally applied magnetic field and B is the total magnetic flux density in the medium. Typical $B–H$ curves for hard and soft ferromagnetic materials are shown in Figure 2.2.

Assume that initially $B_0 = 0$ and $H_0 = 0$. Let H increase from $H_0 = 0$ to $H_{\max}$. Then, B increases from $B_0 = 0$ until the maximum value $B_{\max}$ is reached. If then H decreases to $H_{\min}$, B decreases to $B_{\min}$ through the remanent value B_r (the so-called residual magnetic flux density) along the different curve as illustrated in Figure 2.2. For variations of H, $H \in [H_{\min} \; H_{\max}]$, B changes within the *hysteresis loop*, and $B \in [B_{\min} \; B_{\max}]$. Figure 2.2 reports typical curves representing the dependence of magnetic induction B on magnetic field

TABLE 2.1

Relative Permeability of Some Diamagnetic, Paramagnetic, Ferromagnetic, and Ferrimagnetic Materials

Material	Relative Permeability, μ_r
Diamagnetic	
Silver	0.9999736
Copper	0.9999905
Paramagnetic	
Aluminum	1.000021
Tungsten	1.00008
Platinum	1.0003
Manganese	1.001
	Relative permeability μ_r reaches
Ferromagnetic	
Purified iron (Fe$_{99.96\%}$)	280,000
Electric steel (Fe$_{99.6\%}$)	5,000
Permalloy (Ni$_{78.5\%}$Fe$_{21.5\%}$)	70,000
Superpermalloy (Ni$_{79\%}$Fe$_{15\%}$Mo$_{5\%}$Mn$_{0.5\%}$)	1,000,000
Ferrimagnetic	
Nickel–zinc ferrite	600–1000
Manganese–zinc ferrite	700–1500

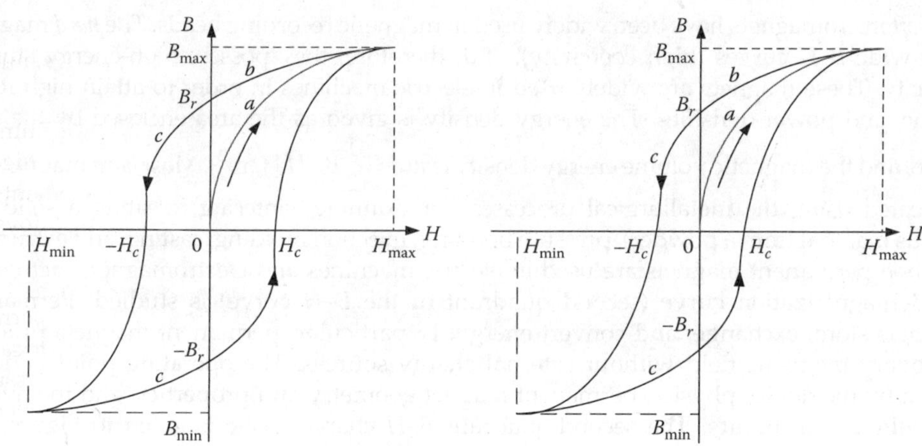

FIGURE 2.2
B–H curves for hard and soft ferromagnetic materials.

H for ferromagnetic materials. When H is first applied, B follows curve a as the favorably oriented magnetic domains grow. This curve reaches the saturation. When H is then reduced, B follows curve b, but retains a finite value (the remanence B_r) at $H = 0$. In order to demagnetize the material, a negative field $-H_c$ must be applied. Here, H_c is called the coercive field or coercivity. As H is further decreased, and, then, increased to complete the cycle (curve c), a hysteresis loop is formed.

The area within this loop is a measure of the energy loss per cycle for a unit volume of the material. The B–H curve allows one to establish the energy analysis. In the per-unit volume, the applied field energy is $W_F = \oint_B H\,dB$, while the stored energy is expressed as $W_c = \oint_H B\,dH$. The equations for field and stored energies represent the areas enclosed by the corresponding curve.

In the volume v, the expressions for the field and stored energies are

$$W_F = v\oint_B H\,dB \quad \text{and} \quad W_c = v\oint_H B\,dH.$$

In ferromagnetic materials, time-varying magnetic flux produces core losses which consist of hysteresis losses (due to the hysteresis loop of the B–H curve) and the eddy-current losses, which are proportional to the current frequency and lamination thickness. The area of the hysteresis loop is related to the hysteresis losses. Soft ferromagnetic materials have narrow hysteresis loop and they are easily magnetized and demagnetized. Therefore, the lower hysteresis losses, compared with hard ferromagnetic materials, result. Different *soft* and *hard* magnets have been utilized. The following magnets have been commonly used in electromechanical motion devices: neodymium iron boron ($Nd_2Fe_{14}B$), samarium cobalt (usually Sm_1Co_5 and Sm_2Co_{17}), ceramic (ferrite), and alnico (AlNiCo). The term *soft* is used to describe those magnets that have high saturation magnetization and a low coercivity (narrow B–H curve). Another property of these magnets is their low magnetostriction.

The *soft* micromagnets have been widely used in magnetic recording heads. The *hard* magnets have wide *B–H* curves (high coercivity), and, therefore, they possess high-energy storage capacity. These magnets are widely used in electric machines in order to attain high force, torque, and power densities. The energy density is given as the area enclosed by the *B–H* curve, and the magnetic volume energy density is $w_m = \frac{1}{2}\vec{B} \cdot \vec{H}$ [J/m^3]. Most *hard* magnets are fabricated using the metallurgical processes, for example, sintering (creating a solid but porous material from a powder), pressure bonding, injection molding, casting, and extruding.

When permanent magnets are used in electric machines and electromagnetic actuators, the demagnetization curve (second quadrant of the *B–H* curve) is studied. Permanent magnets store, exchange, and convert energy. In particular, permanent magnets produce stationary magnetic field without external energy sources. The operating point is determined by the device physics, permanent magnet geometry and properties, and the hysteresis minor loop occurs. The second-quadrant *B–H* characteristic is given in Figure 2.3a (permanent magnets operate on the demagnetization curve of the hysteresis loop). Figure 2.3a also illustrates the energy product curve for a permanent magnet. The demagnetization and energy product curves are in mutual correspondence.

The operating point is at H_d and B_d. For a given air gap, where the magnetic flux exists in machines or actuators, according to Ampere's law, under common assumption, we have $H_d l_m = H_{ag} l_{ag}$, where l_m is the length of the magnet, l_{ag} is the length of the airgap parallel to the flux lines, H_{ag} is the magnetic field intensity in the air gap. For electromechanical motion devices, the flux linkages are plotted versus current because i and ψ are commonly used as the variables rather than the field intensity and density. In actuators and electric machines almost all energy is stored in the air gap. Air is a conservative medium, implying

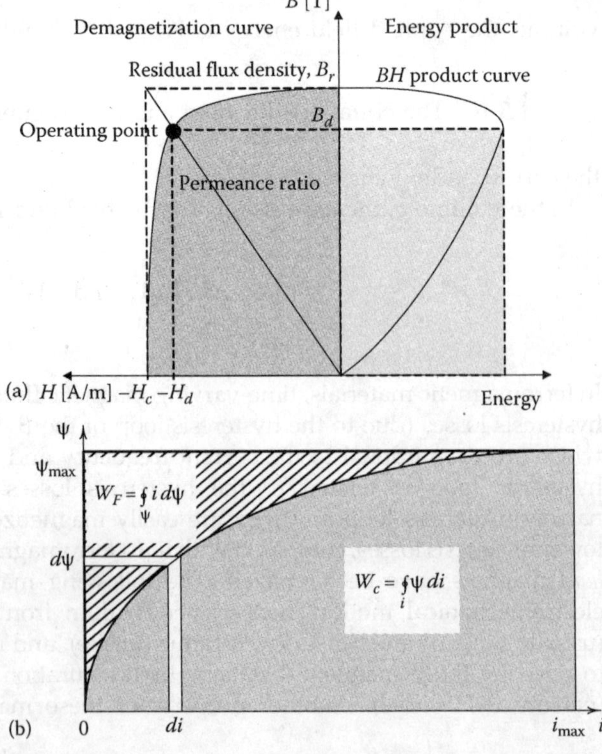

FIGURE 2.3
(a) *B–H* demagnetization and energy product curves; (b) magnetization curve and energies.

that the coupling filed is lossless. Figure 2.3b illustrates the nonlinear magnetizing characteristic (normal magnetization curve). The energy stored in the magnetic field is $W_F = \oint_{\psi} i \, d\psi$, while the coenergy is $W_c = \oint_i \psi \, di$. The total energy is given as

$$W_F + W_c = \oint_{\psi} i \, d\psi + \oint_i \psi \, di = \psi i.$$

The cross-sectional area of a magnet required to produce a specific flux density in the airgap is $A_m = \dfrac{B_{ag} A_{ag}}{B_d}$, where A_{ag} is the air gap area and B_{ag} is the flux density in the air gap. The flux linkages due to the magnet in the air gap are $\psi = N\Phi = N B_{ag} A_{ag}$, and the coenergy is $W = \int_{\psi} i \, d\psi = \int_i \psi \, di$. In a lossless system, an informative energy equation for the air gap is

$$(\text{Volume})_{ag} B_{ag} H_{ag} = A_{ag} l_{ag} A_m l_m B_d H_d / A_{ag} l_{ag} = A_m l_m B_d H_d = \psi i.$$

The flux density at position **r** can be derived. In one-dimensional case for cylindrical magnets (length l_c and radius r_m) which have *near-linear* demagnetization curves, the flux density at a distance x can be approximated as $B = \dfrac{B_r}{2}\left(\dfrac{l_c + x}{\sqrt{r_m^2 + (l_c + x)^2}} - \dfrac{x}{\sqrt{r_m^2 + x^2}} \right).$

Using the Biot–Savart law, the magnetic flux density on the axis of a uniformly magnetized circular magnet (the magnetization M is constant) is

$$\vec{B} = \frac{1}{2}\mu_0 M \left(\frac{z}{\sqrt{z^2 + r_m^2}} - \frac{z - l_c}{\sqrt{(z - l_c)^2 + r_m^2}} \right) \vec{a}_z.$$

The flux linkage is the function of the current i and position x (for translational motion) or angular displacement θ (for rotational motion). That is, $\psi = f(i, x)$ or $\psi = f(i, \theta)$. Assuming that the coupling field is lossless, the differential change in the mechanical energy (found using the differential displacement $d\vec{l}$ as $dW_{mec} = \vec{F}_m \cdot d\vec{l}$) is related to the differential change of the coenergy. For the displacement dx at constant current, one obtains $dW_{mec} = dW_c$. Hence, for a one-dimensional case, the electromagnetic force is

$$F_e(i, x) = \frac{\partial W_c(i, x)}{\partial x}.$$

For rotational motion, the electromagnetic torque is given as $T_e(i, \theta) = \dfrac{\partial W_c(i, \theta)}{\partial \theta}$.

Different material can be used to fabricate electromechanical devices. The device performance is affected by the material properties and characteristics. Table 2.2 reports the initial permeability μ_i, maximum relative permeability μ_{rmax}, coersivity (coercive force) H_c, saturation polarization J_s, hysteresis loss per cycle W_h, and Curie temperature T_C for high-permeability bulk metals and alloys. Table 2.3 reports the remanence B_r, flux coersivity H_{Fc}, intrinsic coercivity H_{Ic}, maximum energy product BH_{max}, Curie temperature, and the maximum operating temperature T_{max} for *hard* permanent magnets. As emphasized, the provided data and constants are strongly affected by the dimensions, temperature, fabrication, etc. The designer must be aware that these parameters can significantly vary for specific applications and devices.

TABLE 2.2

Magnetic Properties of High-Permeability Soft Metals and Alloys

Material	Composition	μ_i	$\mu_{r\,max}$	H_c [A/m]	J_s [T]	W_h [J/m³]	T_C [K]
Iron	$Fe_{99\%}$	200	6,000	70	2.16	500	1043
Iron	$Fe_{99.9\%}$	25,000	35,0000	0.8	2.16	60	1043
Silicon–iron	$Fe_{96\%}Si_{4\%}$	500	7,000	40	1.95	50–150	1008
Silicon–iron (110) [001]	$Fe_{97\%}Si_{3\%}$	9,000	40,000	12	2.01	35–140	1015
Silicon–iron {100} <100>	$Fe_{97\%}Si_{3\%}$		100,000	6	2.01		1015
Steel	$Fe_{99.4\%}C_{0.1\%}Si_{0.1\%}Mn_{0.4\%}$	800	1,100	200			
Hypernik	$Fe_{50\%}Ni_{50\%}$	4,000	70,000	4	1.60	22	753
Deltamax {100} <100>	$Fe_{50\%}Ni_{50\%}$	500	200,000	16	1.55		773
Isoperm {100} <100>	$Fe_{50\%}Ni_{50\%}$	90	100	480	1.60		
78 Permalloy	$Ni_{78\%}Fe_{22\%}$	4,000	100,000	4	1.05	50	651
Supermalloy	$Ni_{79\%}Fe_{16\%}Mo_{5\%}$	100,000	1,000,000	0.15	0.79	2	673
Mumetal	$Ni_{77\%}Fe_{16\%}Cu_{5\%}Cr_{2\%}$	20,000	100,000	4	0.75	20	673
Hyperco	$Fe_{64\%}Co_{35\%}Cr_{0.5\%}$	650	100,00	80	2.42	300	1243
Permendur	$Fe_{50\%}Co_{50\%}$	500	6,000	160	2.46	1200	1253
2V Permendur	$Fe_{49\%}Co_{49\%}V_{2\%}$	800	4,000	160	2.45	600	1253
Supermendur	$Fe_{49\%}Co_{49\%}V_{2\%}$		60,000	16	2.40	1150	1253
25 Perminvar	$Ni_{45\%}Fe_{30\%}Co_{25\%}$	400	2,000	100	1.55		
7 Perminvar	$Ni_{70\%}Fe_{23\%}Co_{7\%}$	850	4,000	50	1.25		
Perminvar (magnetically annealed)	$Ni_{43\%}Fe_{34\%}Co_{23\%}$		400,000	2.4	1.50		
Alfenol (Alperm)	$Fe_{84\%}Al_{16\%}$	3,000	55,000	3.2	0.8		723
Alfer	$Fe_{87\%}Al_{13\%}$	700	3,700	53	1.20		673
Aluminum–iron	$Fe_{96.5\%}Al_{3.5\%}$	500	19,000	24	1.90		
Sendust	$Fe_{85\%}Si_{10\%}Al_{5\%}$	36,000	120,000	1.6	0.89		753

Sources: Lide, D.R., *Handbook of Chemistry and Physics*, CRC Press, Boca Raton, FL, 2002; Dorf, R.C., *Handbook of Engineering Tables*, CRC Press, Boca Raton, FL, 2003; and Lyshevski, S.E., *Nano- and Micro-Electromechanical Systems: Fundamentals of Nano- and Microengineering*, CRC Press, Boca Raton, FL, 2004. With permission.

2.3 Fundamentals of Electromagnetics

The elegance and uniformity of electromagnetics arise from related fundamental laws which allow one to study the field quantities. We denote the vector of electric flux density as $\vec{D}$ [F/m] and the vector of electric field intensity as $\vec{E}$ [V/m or N/C]. Using the Gauss law, the total electric flux Φ [C] through a closed surface is equal to the total force charge enclosed by the surface. That is,

$$\Phi = \oint_s \vec{D} \cdot d\vec{s} = Q_s, \quad \vec{D} = \varepsilon\vec{E},$$

TABLE 2.3

Magnetic Properties of High-Permeability Hard Metals and Alloys

Composite and Composition	B_r [T]	H_{Fc} [A/m]	H_{Ic} [A/m]	BH_{max} [kJ/m^3]	T_C [°C]	T_{max} [°C]
Alnico1: 20Ni,12Al,5Co	0.72		35	25		
Alnico2:17Ni,10Al,12.5Co,6Cu	0.72		40–50	13–14		
Alnico3: 24–30Ni,12–14Al,0–3Cu	0.5–0.6		40–54	10		
Alnico4: 21–28Ni,11–13Al,3–5Co,2–4Cu	0.55–0.75		36–56	11–12		
Alnico5: 14Ni,8Al,24Co,3Cu	1.25	53	54	40	850	520
Alnico6: 16Ni,8Al,24Co,3Cu,2Ti	1.05		75	52		
Alnico8: 15Ni,7Al,35Co,4Cu,5Ti	0.83	1.6	160	45		
Alnico9: 5Ni,7Al,35Co,4Cu,5Ti	1.10	1.45	1.45	75	850	520
Alnico12: 3.5Ni,8Al,24.5Co,2Nb	1.20		64	76.8		
Ferroxdur: $BaFe_{12}O_{19}$	0.4	1.6	192	29	450	400
$SrFe_{12}O_{19}$	0.4	2.95	3.3	30	450	400
$LaCo_5$	0.91			164	567	
$CeCo_5$	0.77			117	380	
$PrCo_5$	1.20			286	620	
$NdCo_5$	1.22			295	637	
$SmCo_5$	1.00	7.9	696	196	700	250
$Sm(Co_{0.76}Fe_{0.10}Cu_{0.14})_{6.8}$	1.04	4.8	5	212	800	300
$Sm(Co_{0.65}Fe_{0.28}Cu_{0.05}Zr_{0.02})_{7.7}$	1.2	10	16	264	800	300
$Nd_2Fe_{14}B$ (sintered)	1.22	8.4	1120	280	300	100
Vicalloy II: Fe,52Co,14V	1.0	42		28	700	500
Fe,24Cr,15Co,3Mo (anisotropic)	1.54	67		76	630	500
Chromindur II: Fe,28Cr,10.5Co	0.98	32		16	630	500
Fe,23Cr,15Co,3V,2Ti	1.35	4		44	630	500
Fe,36Co	1.04		18	8		
Co (rare earth)	0.87		638	144		
Cunife: Cu,20Ni,20Fe	0.55	4		12	410	350
Cunico: Cu,21Ni,29Fe	0.34	0.5		8		
Pt,23Co	0.64	4		76	480	350
Mn,29.5Al,0.5C (anisotropic)	0.61	2.16	2.4	56	300	120

Sources: Lide, D.R., *Handbook of Chemistry and Physics*, CRC Press, Boca Raton, FL, 2002; Dorf, R.C., *Handbook of Engineering Tables*, CRC Press, Boca Raton, FL, 2003; and Lyshevski, S.E., *Nano- and Micro-Electromechanical Systems: Fundamentals of Nano- and Microengineering*, CRC Press, Boca Raton, FL, 2004. With permission.

where $d\vec{s}$ is the vector surface area, $d\vec{s} = ds\vec{a}_n$ $\vec{a}_n$ is the unit vector which is normal to the surface and Q_s is the total charge enclosed by the surface.

Ohm's law for circuits is $V = ir$. However, for a media, the Ohm law relates the volume charge density $\vec{J}$ and electric field intensity $\vec{E}$ using conductivity σ. In particular,

$$\vec{J} = \sigma\vec{E}.$$

TABLE 2.4

Fundamental Equations of Electrostatic and Magnetostatic Fields in Media

	Electrostatic Equations	**Magnetostatic Equations**
Governing equations	$\nabla \times \vec{E}(x,y,z) = 0$	$\nabla \times \vec{H}(x,y,z) = \vec{J}(x,y,z)$
	$\nabla \cdot \vec{D}(x,y,z) = \rho_v(x,y,z)$	$\nabla \cdot \vec{B}(x,y,z) = 0$
Constitutive equations	$\vec{D} = \varepsilon \vec{E}$	$\vec{B} = \mu \vec{H}$

For copper, the conductivity is $\sigma = 5.8 \times 10^7$ A/V-m, while for aluminum $\sigma = 3.5 \times 10^7$ A/V-m.

The resistance r of the conductor is related to the resistivity and conductivity as $r = \dfrac{\rho l}{A}$ and $r = \dfrac{l}{\sigma A}$, where l is the length and A is the cross-sectional area. The resistivity, as well as other constants and parameters, varies. For example, the resistivity depends on temperature T, and $\rho(T) = \rho_0 \left[1 + \alpha_{\rho 1}(T - T_0) + \alpha_{\rho 2}(T - T_0)^2 + \cdots \right]$, where $\alpha_{\rho i}$ are the coefficients. In the small temperature range (up to 160°C) for copper at $T_0 = 20$°C, we have $\rho(T) = 1.7 \times 10^{-8} [1 + 0.0039(T - 20)]$.

Electromagnetic theory and classical mechanics form the basis to devise and examine the device physics, study the inherent phenomena exhibited and utilized, as well as derive the equations of motion. The electrostatic and magnetostatic equations in linear isotropic media are found using the vectors of the electric field intensity $\vec{E}$, electric flux density $\vec{D}$, magnetic field intensity $\vec{H}$, and magnetic flux density $\vec{B}$. In addition, one uses the constitutive equations $\vec{D} = \varepsilon \vec{E}$ and $\vec{B} = \mu \vec{H}$. The basic equations in the Cartesian coordinate system are reported in Table 2.4.

In the static (time-invariant) fields, electric and magnetic field vectors form separate and independent pairs. That is, $\vec{E}$ and $\vec{D}$ are not related to $\vec{H}$ and $\vec{B}$, and vice versa. However, in electromechanical motion devices, the electric and magnetic fields are time-varying. The changes of magnetic field influence the electric field, and vice versa. Four Maxwell's equations in the differential form for time-varying fields are

1. Faraday's law $\nabla \times \vec{E}(x,y,z,t) = -\dfrac{\partial \vec{B}(x,y,z,t)}{\partial t}$

2. Ampere's law $\nabla \times \vec{H}(x,y,z,t) = \vec{J}(x,y,z,t) + \dfrac{\partial \vec{D}(x,y,z,t)}{\partial t}$

3. Gauss's law for electric field $\nabla \cdot \vec{D}(x, y, z, t) = \rho_v(x, y, z, t)$

4. Gauss's law for magnetic field $\nabla \cdot \vec{B}(x, y, z, t) = 0$

Here, $\vec{E}$ is the electric field intensity, and using the permittivity ε, the electric flux density is $\vec{D} = \varepsilon \vec{E}$; $\vec{H}$ is the magnetic field intensity, and using the permeability μ, the magnetic flux density is $\vec{B} = \mu \vec{H}$; $\vec{J}$ is the current density, and using the conductivity σ, we have $\vec{J} = \sigma \vec{E}$; ρ_v is the volume charge density, and the total electric flux through a closed surface is

$$\Phi = \oint_s \vec{D} \cdot d\vec{s} = \oint_v \rho_v \, dv = Q \text{ (Gauss's law)},$$ while the magnetic flux crossing surface is

$$\Phi = \oint_s \vec{B} \cdot d\vec{s}.$$

The constitutive (auxiliary) equations are given using the permittivity, permeability, and conductivity tensors ε, μ, and σ, respectively. One has

$$\vec{D} = \varepsilon\vec{E} \quad \text{or} \quad \vec{D} = \varepsilon\vec{E} + \vec{P},$$
$$\vec{B} = \mu\vec{H} \quad \text{or} \quad \vec{B} = \mu(\vec{H} + \vec{M}),$$
$$\vec{J} = \sigma\vec{E} \quad \text{or} \quad \vec{J} = \rho_v\vec{v}.$$

Examining actuators and other electromechanical motion devices, one also concentrates on the deriving the expressions for force, torque, *electromotive* and *magnetomotive* forces (*emf* and *mmf*), etc. The Lorenz force, which relates the electromagnetic and mechanical variables, is

$$\vec{F} = \rho_v(\vec{E} + \vec{v} \times \vec{B}) = \rho_v\vec{E} + \vec{J} \times \vec{B}.$$

The expressions for energies stored in electrostatic and magnetic fields in terms of field quantities should be derived. The total potential energy stored in the electrostatic field is obtained using the potential difference V as

$$W_e = \frac{1}{2}\int_v \rho_v V \, dv,$$

where the volume charge density is given as $\rho_v = \vec{\nabla} \cdot \vec{D}$, $\vec{\nabla}$ is the curl operator.

Using $\rho_v = \vec{\nabla} \cdot \vec{D}$ and $\vec{E} = -\vec{\nabla}V$, one obtains the following expression for the energy stored in the electrostatic field $W_e = \frac{1}{2}\int_v \vec{D} \cdot \vec{E} \, dv$, and the electrostatic volume energy density is $\frac{1}{2}\vec{D} \cdot \vec{E}$. For a linear isotropic medium, one finds

$$W_e = \frac{1}{2}\int_v \varepsilon|\vec{E}|^2 \, dv = \frac{1}{2}\int_v \frac{1}{\varepsilon}|\vec{D}|^2 \, dv.$$

From $W_e = \frac{1}{2}\int_v \rho_v V \, dv$, the potential energy which is stored in the electric field between two surfaces (e.g., in capacitor) is $W_e = \frac{1}{2}QV = \frac{1}{2}CV^2$.

Using the principle of virtual work, for the lossless conservative system, the differential change of the electrostatic energy dW_e is equal to the differential change of mechanical energy dW_{mec}. That is,

$$dW_e = dW_{mec}.$$

For translational motion, one has $dW_{mec} = \vec{F}_e \cdot d\vec{l}$, where $d\vec{l}$ is the differential displacement.

From $dW_e = \vec{\nabla}W_e \cdot d\vec{l}$ one concludes that the force is the gradient of the stored electrostatic energy, and

$$\vec{F}_e = \vec{\nabla}W_e.$$

In the Cartesian coordinates, we have $F_{ex} = \dfrac{\partial W_e}{\partial x}$, $F_{ey} = \dfrac{\partial W_e}{\partial y}$, and $F_{ez} = \dfrac{\partial W_e}{\partial z}$.

The stored energy in the magnetostatic field in terms of field quantities is

$$W_m = \frac{1}{2} \int_v \vec{B} \cdot \vec{H} \, dv \quad \text{or} \quad W_m = \frac{1}{2} \int_v \mu |\vec{H}|^2 \, dv = \frac{1}{2} \int_v \frac{|\vec{B}|^2}{\mu} \, dv.$$

The energy stored in the magnetic field can be expressed as $W_m = \frac{1}{2} i_i L_{ij} i_j$.

Using the current vector $\mathbf{i} = [i_1, i_2, \ldots, i_{n-1}, i_n]^T$ and the inductance mapping $\mathbf{L} \in \mathbb{R}^{n \times n}$, we have

$$W_m = \frac{1}{2} \mathbf{i}^T \mathbf{L} \mathbf{i},$$

where T denotes the transpose symbol.

The magnetic energy, stored in the inductor with a single winding is $W_m = \frac{1}{2} L i^2$.

The force is the gradient of the stored magnetic energy, and $\vec{F}_m = \vec{\nabla} W_m$. Hence, in the *xyz* coordinate system for the translational motion, we have

$$F_{mx} = \frac{\partial W_m}{\partial x}, \quad F_{my} = \frac{\partial W_m}{\partial y}, \quad \text{and} \quad F_{mz} = \frac{\partial W_m}{\partial z}.$$

For the rotational motion, the torque should be used. Using the differential change in the mechanical energy as a function of the angular displacement θ, the following formula results if the rigid body (rotor) is constrained to rotate around the *z*-axis

$$dW_{mec} = T_e \, d\theta,$$

where T_e is the *z*-component of the electromagnetic torque.

Assuming that the system is lossless, one obtains the expression for the electromagnetic torque as

$$T_e = \frac{\partial W_m}{\partial \theta}.$$

The *emf* and *mmf* are given as

$$emf = \oint_l \vec{E} \cdot \vec{dl} = \underbrace{\oint_l \left(\vec{v} \times \vec{B} \right) \cdot \vec{dl}}_{\text{Motional induction(generation)}} - \underbrace{\oint_s \frac{\partial \vec{B}}{\partial t} \, d\vec{s}}_{\text{Transformer induction}}$$

and

$$mmf = \oint_l \vec{H} \cdot \vec{dl} = \oint_s \vec{J} \cdot d\vec{s} + \oint_s \frac{\partial \vec{D}}{\partial t} \, d\vec{s}.$$

The motional *emf* is a function of the velocity and the magnetic flux density, while the *emf* induced in a stationary closed circuit is equal to the negative rate of increase of the magnetic flux (transformer induction). The induced *mmf* is the sum of the induced current and the rate of change of the flux penetrating the surface bounded by the contour.

2.4 Classical Mechanics and Its Application

Distinct electromechanical systems and motion devices can be studied applying classical mechanics. Newtonian mechanics, Lagrange's concept, and Hamilton's method provide meaningful approaches to derive the governing equations.

2.4.1 Newtonian Mechanics

2.4.1.1 Newtonian Mechanics, Energy Analysis, Generalized Coordinates, and Lagrange Equations: Translational Motion

We study the system behavior with the corresponding analysis of forces that cause motion. The equations of motion for mechanical systems can be found using Newton's second law of motion. Using the position (displacement) vector $\vec{r}$, the Newton equation in the vector form is given as

$$\sum \vec{F}(t, \mathbf{r}) = m\vec{a}, \tag{2.1}$$

where $\Sigma\vec{F}(t, \mathbf{r})$ is the vector sum of all forces (*net* force) applied to the body, $\vec{a}$ is the vector of acceleration of the body with respect to an inertial reference frame, and m is the mass of the body.

In (2.1), $m\vec{a}$ represents the magnitude and direction of the applied net force acting on the object. Hence, $m\vec{a}$ is not a force. A body is at equilibrium (the object is at rest or is moving with constant speed) if $\Sigma\vec{F} = 0$, for example, there is no acceleration if $\Sigma\vec{F} = 0$. Using (2.1), in the Cartesian system, we have the mechanical equations of motion in the xyz coordinates

$$\sum \vec{F}(t, \mathbf{r}) = m\vec{a} = m\frac{d\vec{r}^2}{dt^2} = m \begin{bmatrix} \dfrac{d\vec{x}^2}{dt^2} \\[2ex] \dfrac{d\vec{y}^2}{dt^2} \\[2ex] \dfrac{d\vec{z}^2}{dt^2} \end{bmatrix}, \quad \begin{bmatrix} \vec{a}_x \\ \vec{a}_y \\ \vec{a}_z \end{bmatrix} = \begin{bmatrix} \dfrac{d\vec{x}^2}{dt^2} \\[2ex] \dfrac{d\vec{y}^2}{dt^2} \\[2ex] \dfrac{d\vec{z}^2}{dt^2} \end{bmatrix}.$$

One obtains the second-order ordinary differential equations which are the rigid-body mechanical equations of motion. The forces to control the motion should be developed by actuators. The force can be a function of current (electromagnetic actuators), voltage (electrostatic actuators), pressure (hydraulic actuators), temperature gradient, field quantities, etc. Correspondingly, equation (2.1) must be integrated with the actuator's equations of motion.

In the Cartesian coordinate system, Newton's second law is expressed as

$$\sum \vec{F}_x = m\vec{a}_x, \quad \sum \vec{F}_y = m\vec{a}_y, \quad \sum \vec{F}_z = m\vec{a}_z.$$

Newton's second law in terms of the linear momentum $\vec{p} = m\vec{v}$, is given by

$$\sum \vec{F} = \frac{d\vec{p}}{dt} = \frac{d(m\vec{v})}{dt},$$

where $\vec{v}$ is the vector of the object velocity.

Thus, the force is equal to the rate of change of the momentum. The object or particle moves uniformly if $\dfrac{d\vec{p}}{dt} = 0$ which implies that $\vec{p} = $ constant.

Taking note of the expression for the potential energy function $\Pi(\vec{r})$, for the conservative mechanical system we have

$$\sum \vec{F}(\vec{r}) = -\nabla \Pi(\vec{r}).$$

Therefore, the work done per unit time is $\dfrac{dW}{dt} = \sum \vec{F}(\vec{r}) \dfrac{d\vec{r}}{dt} = -\nabla \Pi(\vec{r}) \dfrac{d\vec{r}}{dt} = -\dfrac{d\Pi(\vec{r})}{dt}.$

From Newton's second law, one obtains $m\vec{a} - \sum \vec{F}(\vec{r}) = 0$ or $m\dfrac{d^2\vec{r}}{dt^2} - \sum \vec{F}(\vec{r}) = 0.$

Hence, for a conservative system the following equation results:

$$m\frac{d^2\vec{r}}{dt^2} + \nabla \Pi(\vec{r}) = 0.$$

In the derived equation, the kinetic and potential energies are utilized. One recalls that the total kinetic energy of the particle is $\Gamma = \dfrac{1}{2}mv^2$. Having documented the Newtonian mechanics and the use of the kinetic and potential energies, we will cover one of the most general concepts in the modeling of dynamic systems which is based on the Lagrangian mechanics. Instead of applying the displacement, velocity, or acceleration, the generalized coordinates $(q_1, \ldots, q_n)$ and generalized velocities $\left(\dfrac{dq_1}{dt}, \ldots, \dfrac{dq_n}{dt} \right)$ are used. The total kinetic $\Gamma \left(q_1, \ldots, q_n, \dfrac{dq_1}{dt}, \ldots, \dfrac{dq_n}{dt} \right)$ and potential $\Pi(q_1, \ldots, q_n)$ energies can be found in terms of q_i and dq_i/dt. Using the expressions for the total kinetic and potential energies, Newton's second law of motion is

$$\frac{d}{dt}\left(\frac{\partial \Gamma}{\partial \dot{q}_i} \right) + \frac{\partial \Pi}{\partial q_i} = 0.$$

Example 2.1:

Consider a positioning table actuated by a motor. The work required to accelerate a 20 g payload ($m = 0.02$ kg) from $v_0 = 0$ m/s to $v_f = 1$ m/s is

$$W = \frac{1}{2}\left(mv_f^2 - mv_0^2 \right) = \frac{1}{2}20 \times 10^{-3} \times 1^2 = 0.01 \text{ J}.$$

We utilized the work–energy theorem which is formulated as following for a particle: The work done by the *net* force on a particle equals the change in the particle's kinetic energy, for example, $W_{total} = \Gamma_2 - \Gamma_1 = \Delta\Gamma$. For a varying force, one can find the total work done by the *net* force as

$$W = \int_{x_1}^{x_2} F\,dx, \text{ and, using } a = \frac{dv}{dt} = \frac{dv}{dx}\frac{dx}{dt} = v\frac{dv}{dx},$$

$$\text{we have } W = \int_{x_1}^{x_2} F\,dx = \int_{x_1}^{x_2} ma\,dx = \int_{x_1}^{x_2} mv\frac{dv}{dx}\,dx = \int_{v_1}^{v_2} mv\,dv. \qquad \blacksquare$$

Example 2.2:

Consider a body of mass m in the XY coordinate system. The free-body diagram is illustrated in Figure 2.4. The force $\vec{F}_a$ is applied in the x-direction, and let $\vec{F}_a(t,x) = x\sin(6t - 4)e^{-0.5t} + \frac{dx}{dt}t^2 + x^3\cos\left(\frac{dx}{dt}t - x^2t^4\right)$. We assume that the viscous friction force is $F_{fr} = B_v\frac{dx}{dt}$, where B_v is the viscous friction coefficient. Our goal is to find the equations of motion neglecting *Coulomb* and static friction.

The sum of the forces, acting in the y-direction, is $\sum \vec{F}_Y = \vec{F}_N - \vec{F}_g$ where $\vec{F}_g = mg$ is the gravitational force acting on the mass m, $\vec{F}_N$ is the normal force which is equal and opposite to the gravitational force.

From (2.1), the equation of motion in the y-direction is $\vec{F}_N - \vec{F}_g = ma_y = m\frac{d^2y}{dt^2}$, where a_y is the acceleration in the y-direction, $a_y = \frac{d^2y}{dt^2}$.

Making use of $\vec{F}_N = \vec{F}_g$, the resulting equation is

$$\frac{d^2y}{dt^2} = 0.$$

The sum of the forces acting in the x-direction is found using the applied force $\vec{F}_a$ and the friction force $\vec{F}_{fr}$. We have $\sum \vec{F}_X = \vec{F}_a - \vec{F}_{fr}$.

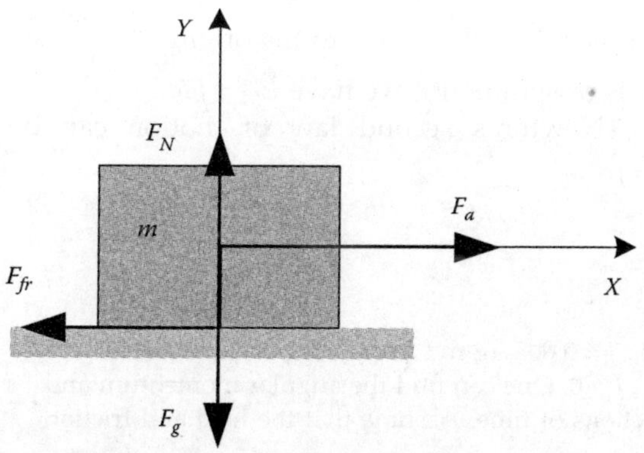

FIGURE 2.4
Free-body diagram.

The applied force can be time invariant $\vec{F}_a = $ constant or nonlinear time-varying $\vec{F}_a(t) = f(t, x, y, z)$. Using (2.1), the equation motion in the x-direction is $\vec{F}_a - \vec{F}_{fr} = ma_x = m\dfrac{d^2x}{dt^2}$, where a_x is the acceleration in the x-direction, $a_x = \dfrac{d^2x}{dt^2}$, and the velocity in the x-direction is $v = \dfrac{dx}{dt}$.

Assume that the *Coulomb* and static friction can be neglected. The friction force, as a function of the viscous friction coefficient B_v and velocity v, is $F_{fr} = B_v\dfrac{dx}{dt} = B_v v$. One obtains the second-order nonlinear differential equation to describe the rigid-body dynamics in the x-direction

$$\frac{d^2x}{dt^2} = \frac{1}{m}\left(F_a - B_v\frac{dx}{dt}\right) = \frac{1}{m}\left[x\sin(6t - 4)e^{-0.5t} + \frac{dx}{dt}t^2 + x^3\cos\left(\frac{dx}{dt}t - x^2t^4\right) - B_v\frac{dx}{dt}\right].$$

From the derived equation of motion, a set of two first-order linear differential equations results. In particular,

$$\frac{dx}{dt} = v,$$

$$\frac{dv}{dt} = \frac{1}{m}\left[x\sin(6t - 4)e^{-0.5t} + vt^2 + x^3\cos(vt - x^2t^4) - B_v v\right], \quad t \geq 0. \qquad \blacksquare$$

2.4.1.2 Newtonian Mechanics: Rotational Motion

For rotational devices, instead of linear displacement and acceleration, the angular displacement and acceleration are used. The *net* torque is considered. The rotational Newton's second law for a rigid body is

$$\sum \vec{T}(t, \vec{\theta}) = J\vec{\alpha}, \qquad (2.2)$$

where $\Sigma\vec{T}$ is the *net* torque [N-m], J is the moment of inertia (*rotational inertia*) [kg-m^2], $\vec{\alpha}$ is the angular acceleration vector [rad/s^2], $\vec{\alpha} = \dfrac{d}{dt}\dfrac{d\vec{\theta}}{dt} = \dfrac{d^2\vec{\theta}}{dt^2} = \dfrac{d\vec{\omega}}{dt}$, $\vec{\theta}$ is the angular displacement, and $\vec{\omega}$ is the angular velocity.

The angular momentum of the system $\vec{L}_M$ is $\vec{L}_M = \vec{R}\times\vec{p} = \vec{R}\times m\vec{v}$, and

$$\sum\vec{T} = \frac{d\vec{L}_M}{dt} = \vec{R}\times\vec{F},$$ where $\vec{R}$ is the position vector with respect to the origin.

For the rigid body, rotating around the axis of symmetry, we have $\vec{L}_M = J\vec{\omega}$.

For one-dimensional rotational systems, Newton's second law of motion can be expressed using the *net* moment as $\sum M = J\alpha$.

Example 2.3:

A motor has the equivalent moment of inertia $J = 0.005$ kg-m^2. Let the motor accelerates, and the angular velocity of the rotor is $\omega_r = 1000t^3$, $t \geq 0$. One can find the angular momentum and the developed electromagnetic torque as functions of time. Assume that the load and friction torques are zero. The angular momentum is $L_M = J\omega_r = 5t^3$. The developed electromagnetic torque is $T_e = \dfrac{dL_M}{dt} = 15t^2$ N-m. $\blacksquare$

From Newtonian mechanics one concludes that the applied *net* force or torque plays a key role in describing the motion and to control devices. The analysis of motion was performed using the energy and momentum quantities, which are conserved. The principle of conservation of energy states that the energy can be only converted from one form to another. Kinetic energy is associated with motion, while potential energy is associated with position. The sum of the kinetic Γ, potential Π, and dissipated D energies is called the total energy of the system Σ_T, which is conserved. The total amount of energy remains constant, for example, $\Sigma_T = \Gamma + \Pi + D = \text{constant}$.

Example 2.4:

Consider the translational motion of a body which is attached to an ideal spring that exerts the force which obeys ideal Hooke's law. Neglecting friction, one obtains the following expression for the total energy

$$\Sigma_T = \Gamma + \Pi = \frac{1}{2}\left(mv^2 + k_s x^2\right) = \text{constant}.$$

The translational kinetic energy is $\Gamma = \dfrac{1}{2}mv^2$, while the elastic potential energy of the spring is $\Pi = \dfrac{1}{2}k_s x^2$, where k_s is the force constant of the spring.

For rotational motion and torsional spring, we have

$$\Sigma_T = \Gamma + \Pi = \frac{1}{2}\left(J\omega^2 + k_s\theta^2\right) = \text{constant},$$

where the rotational kinetic energy and the elastic potential energy are, respectively, $\Gamma = \dfrac{1}{2}J\omega^2$ and $\Pi = \dfrac{1}{2}k_s\theta^2$. $\blacksquare$

The kinetic energy of a rigid body having translational and rotational components of motion is

$$\Gamma = \frac{1}{2}(mv^2 + J\omega^2).$$

That is, motion of the rigid body is represented as a combination of translational motion of the center of mass and rotational motion about the axis through the center of mass. The moment of inertia J depends upon how the mass is distributed with respect to the axis, and J is different for different axes of rotation. If the body is uniform in density, J can be calculated for regularly shaped bodies using their dimensions. For example, a rigid cylinder of mass m (which is uniformly distributed), radius R, and length l has the following horizontal and vertical moments of inertia:

$$J_{horizontal} = \frac{1}{2}mR^2 \quad \text{and} \quad J_{vertical} = \frac{1}{4}mR^2 + \frac{1}{12}ml^2.$$

The *radius of gyration* can be found for irregularly shaped objects, and the moment of inertia can be easily obtained.

In electromechanical motion devices, the force and torque are of great interest. Assuming that the body is rigid and the moment of inertia is constant, one has

$$\vec{T}\,d\vec{\theta} = J\vec{\alpha}\,d\vec{\theta} = J\frac{d\vec{\omega}}{dt}\,d\vec{\theta} = J\frac{d\vec{\theta}}{dt}\,d\vec{\omega} = J\vec{\omega}\,d\vec{\omega}.$$

The total work, as given by $W = \int_{\theta_0}^{\theta_f} \vec{T}\,d\vec{\theta} = \int_{\omega_0}^{\omega_f} J\vec{\omega}\,d\vec{\omega} = \frac{1}{2}\left(J\omega_f^2 - J\omega_0^2\right)$, represents the change

of the kinetic energy. Furthermore, $\dfrac{dW}{dt} = \vec{T}\dfrac{d\vec{\theta}}{dt} = \vec{T} \times \vec{\omega}$, and the power is $P = \vec{T} \times \vec{\omega}$.

This equation is an analog of $P = \vec{F} \times \vec{v}$, which is applied for translational motion.

Example 2.5:

Assume that the rated power and angular velocity of a motor are 1 W and 1000 rad/s, respectively. The rated electromagnetic torque is $T_e = \dfrac{P}{\omega_r} = \dfrac{1}{1000} = 1 \times 10^{-3}$ N-m.   ∎

Example 2.6:

Given a point mass m suspended by a massless unstretchable string of length l as shown in Figure 2.5. One can derive the equations of motion for a simple pendulum.

The restoring force, which is $-mg\sin\theta$, is the tangential component of the *net* force. Therefore, the sum of the moments about the pivot point O is

$$\sum M = -mgl\sin\theta + T_a,$$

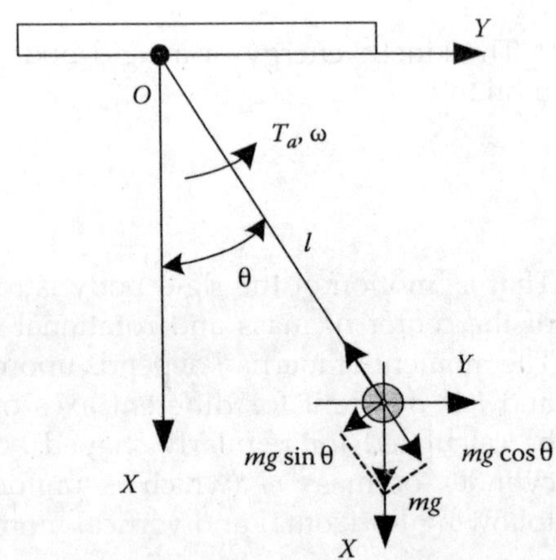

FIGURE 2.5
A simple pendulum.

where T_a is the applied torque (one cannot apply the torque if there is a string, but using a rod, one may apply T_e) and l is the length of the pendulum measured from the point of rotation. Using (2.2), one obtains the equation of motion

$$J\alpha = J\frac{d^2\theta}{dt^2} = -mgl\sin\theta + T_a,$$

where J is the moment of inertial of the mass about the point O.

Hence, the second-order differential equation is $\dfrac{d^2\theta}{dt^2} = \dfrac{1}{J}(-mgl\sin\theta + T_a)$.

Using the differential equation for the angular displacement $\dfrac{d\theta}{dt} = \omega$, one obtains a set of two first-order differential equations:

$$\frac{d\omega}{dt} = \frac{1}{J}(-mgl\sin\theta + T_a)$$

and

$$\frac{d\theta}{dt} = \omega.$$

The moment of inertia is $J = ml^2$. Hence, we have the following differential equations to describe the dynamics of a simple pendulum:

$$\frac{d\omega}{dt} = -\frac{g}{l}\sin\theta + \frac{1}{ml^2}T_a,$$
$$\frac{d\theta}{dt} = \omega.$$

■

Example 2.7:   Friction in Motion Devices

Friction is a very complex nonlinear phenomenon. The classical *Coulomb* friction is a retarding frictional force (for translational motion) or torque (for rotational motion) that changes its sign with the reversal of the direction of motion, and the amplitude of the frictional force or torque is constant. For translational and rotational motions, the *Coulomb* friction force and torque are

$$F_{Coulomb} = k_{Fc}\,\text{sgn}(v) = k_{Fc}\,\text{sgn}\left(\frac{dx}{dt}\right)$$

and

$$T_{Coulomb} = k_{Tc}\,\text{sgn}(\omega) = k_{Tc}\,\text{sgn}\left(\frac{d\theta}{dt}\right),$$

where k_{Fc} and k_{Tc} are the *Coulomb* friction coefficients.

Figure 2.6a illustrates the *Coulomb* friction.

Viscous friction is a retarding force or torque which is a linear (or nonlinear) function of linear or angular velocity. The viscous friction force and torque versus linear and angular

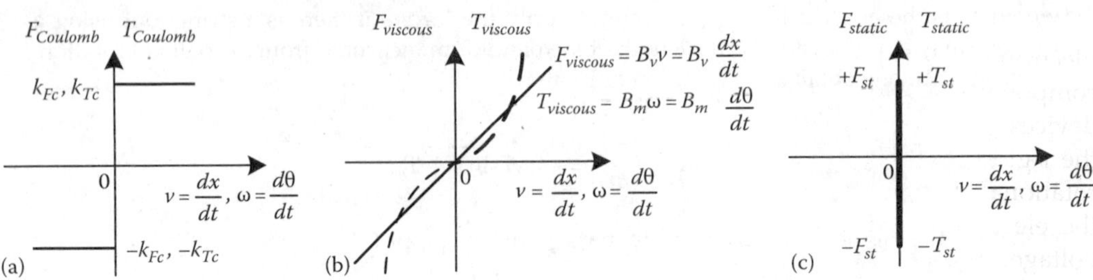

FIGURE 2.6
(a) *Coulomb* friction; (b) viscous friction; (c) static friction.

velocities are shown in Figure 2.6b. The following expressions are commonly used to describe the viscous friction for translational and rotational motions

$$F_{viscous} = B_v v = B_v \frac{dx}{dt} \quad \text{or in nonlinear form} \quad F_{viscous} = \sum_{n=1}^{\infty} B_{vi} v^{2n-1},$$

and

$$T_{viscous} = B_m \omega = B_m \frac{d\theta}{dt} \quad \text{or in nonlinear form} \quad T_{viscous} = \sum_{n=1}^{\infty} B_{mi} \omega^{2n-1},$$

where B_v and B_m are the viscous friction coefficients.

The static friction exists only when the body is stationary and vanishes as motion begins. The static friction is a force F_{static} or torque T_{static} which may be expressed as

$$F_{static} = \pm F_{st} \Big|_{v=\frac{dx}{dt}=0} \quad \text{and} \quad T_{static} = \pm T_{st} \Big|_{\omega=\frac{d\theta}{dt}=0}.$$

The static friction is a retarding force or torque which tends to prevent the initial translational or rotational motion at beginning (see Figure 2.6c).

The friction force and torque are nonlinear functions that are modeled using frictional memory, presliding conditions, etc. The formulas commonly used are $F_{fr} = \left(k_{fr1} - k_{fr2} e^{-k|v|} + k_{fr3}|v|\right)\mathrm{sgn}(v)$ and $T_{fr} = \left(k_{fr1} - k_{fr2} e^{-k|\omega|} + k_{fr3}|\omega|\right)\mathrm{sgn}(\omega)$. The typifying plots for F_{fr} and T_{fr} are shown in Figure 2.7.   ■

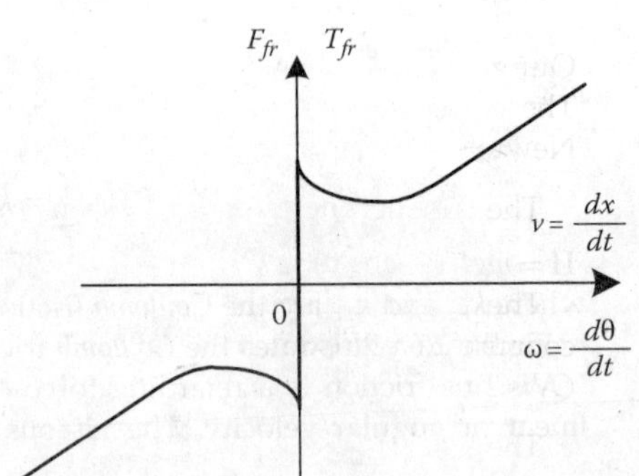

FIGURE 2.7
Friction force and torque as functions of linear and angular velocities.

2.4.2 Lagrange Equations of Motion

Electromechanical systems integrate mechanical, electromagnetic, circuit, and electronic components. Therefore, one studies mechanical, electromagnetic, circuitry, solid-state devices, and other transients. The designer may apply Newtonian dynamics deriving the rigid-body dynamics (translational and *torsional–mechanical* dynamics for linear and rotational actuators), and, then, using the coenergy concept derive the expression for the electromagnetic (or electrostatic) force or torque that are the functions of current, voltage, or electromagnetic field quantities. The energy conversion, force/torque production, and electromagnetic dynamics, as obtained by using laws of electromagnetics, must be integrated.

In contrast, the Lagrange and Hamilton concepts are based on the energy analysis of an entire system. Using the Lagrange equations, one can integrate the rigid-body (mechanical) dynamics and circuitry-electromagnetic equations of motion. Hence, the Lagrange and Hamilton concepts are more general approaches. Using the system variables, one finds the total kinetic $\Gamma\left(t, q_1, \ldots, q_n, \dfrac{dq_1}{dt}, \ldots, \dfrac{dq_n}{dt}\right)$, dissipation $D\left(t, q_1, \ldots, q_n, \dfrac{dq_1}{dt}, \ldots, \dfrac{dq_n}{dt}\right)$, and potential $\Pi(t, q_1, \ldots, q_n)$ energies. Using the total energies, the Lagrange equations of motion are

$$\frac{d}{dt}\left(\frac{\partial \Gamma}{\partial \dot{q}_i}\right) - \frac{\partial \Gamma}{\partial q_i} + \frac{\partial D}{\partial \dot{q}_i} + \frac{\partial \Pi}{\partial q_i} = Q_i. \tag{2.3}$$

Here, q_i and Q_i are the *generalized* coordinates and the *generalized* forces (applied forces and disturbances), respectively. These generalized coordinates q_i are used to derive energies Γ, D, and Π. As the generalized coordinates, one uses

- Linear or angular displacement (for translational and rotational devices)
- Charges (please pay attention that the charge is usually denoted using the symbol q)

For conservative (lossless) systems $D = 0$. One obtains the following Lagrange's equations of motion

$$\frac{d}{dt}\left(\frac{\partial \Gamma}{\partial \dot{q}_i}\right) - \frac{\partial \Gamma}{\partial q_i} + \frac{\partial \Pi}{\partial q_i} = Q_i.$$

Example 2.8: Simple Pendulum

Our goal is to derive the equations of motion for a simple pendulum as depicted in Figure 2.5. The equations of motion for the simple pendulum were derived in Example 2.6 using the Newtonian mechanics. For the studied conservative (lossless) system, $D = 0$.

The kinetic energy of the pendulum bob is $\Gamma = \dfrac{1}{2}m(l\dot{\theta})^2$, while the potential energy is $\Pi = mgl(1 - \cos\theta)$.

The angular displacement is the generalized coordinate. We have only one generalized coordinate, for example, $q_1 = \theta$. The generalized force is the applied torque T_a, and, hence $Q_1 = T_a$.

The kinetic and potential energies are $\Gamma = \dfrac{1}{2}m(l\dot{q}_1)^2$ and $\Pi = mgl(1 - \cos q_1)$.

In the governing equation of motion $\dfrac{d}{dt}\left(\dfrac{\partial\Gamma}{\partial\dot{q}_1}\right) - \dfrac{\partial\Gamma}{\partial q_1} + \dfrac{\partial\Pi}{\partial q_1} = Q_1$, one obtains the following expressions for the derivatives $\dfrac{\partial\Gamma}{\partial\dot{q}_1} = \dfrac{\partial\Gamma}{\partial\dot{\theta}} = ml^2\dot{\theta}$, $\dfrac{\partial\Gamma}{\partial q_1} = \dfrac{\partial\Gamma}{\partial\theta} = 0$, and $\dfrac{\partial\Pi}{\partial q_1} = \dfrac{\partial\Pi}{\partial\theta} = mgl\sin\theta$.

The first term of the Lagrange equation is $\dfrac{d}{dt}\left(\dfrac{\partial\Gamma}{\partial\dot{\theta}}\right) = ml^2\dfrac{d^2\theta}{dt^2} + 2ml\dfrac{dl}{dt}\dfrac{d\theta}{dt}$.

Assuming that the string (rod) is unstretchable, we have $dl/dt = 0$. If this assumption is not valid, one should use the appropriate expression for the length as a function of the generalized coordinate q. For $dl/dt = 0$, we have

$$ml^2\frac{d^2\theta}{dt^2} + mgl\sin\theta = T_a.$$

Thus, one obtains $\dfrac{d^2\theta}{dt^2} = \dfrac{1}{ml^2}(-mgl\sin\theta + T_a)$.

Recall that the equation of motion, derived by using Newtonian mechanics, is

$$\frac{d^2\theta}{dt^2} = \frac{1}{J}(-mgl\sin\theta + T_a), \quad \text{where } J = ml^2.$$

One concludes that the results are the same. The equations of motion are

$$\frac{d\omega}{dt} = -\frac{g}{l}\sin\theta + \frac{1}{ml^2}T_a \quad \text{and} \quad \frac{d\theta}{dt} = \omega.$$

The Lagrange equations of motion provide more general results. We will illustrate that electromechanical devices can be modeled using the Lagrange equations of motion. Newton's laws can be used only to model the rigid-body dynamics unless *electromechanical analogies* are applied. Furthermore, the coordinate-dependent system parameters can be accounted. For example, l can be a function of θ. $\blacksquare$

Example 2.9:   Double Pendulum

Consider a two-degree-of-freedom double pendulum with no external forces applied, see Figure 2.8. Using the Lagrange equations of motion, we should derive the differential equations.

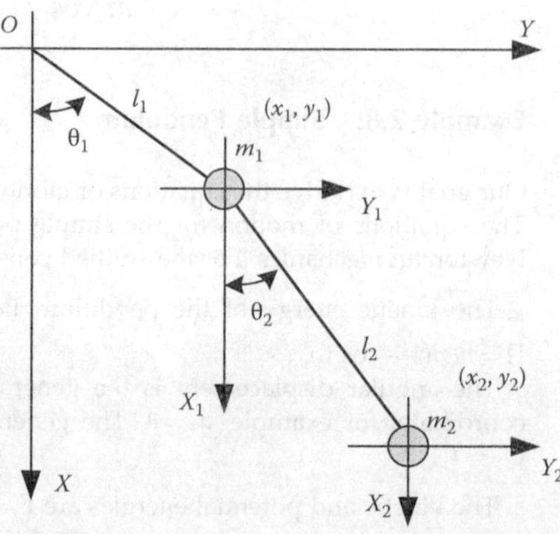

FIGURE 2.8
Double pendulum.

The angular displacement θ_1 and θ_2 are the independent generalized coordinates q_1 and q_2. In the xy plane, let (x_1, y_1) and (x_2, y_2) be the rectangular coordinates of point masses m_1 and m_2. We obtain

$$x_1 = l_1 \cos\theta_1, \quad x_2 = l_1 \cos\theta_1 + l_2 \cos\theta_2, \quad y_1 = l_1 \sin\theta_1, \quad \text{and} \quad y_2 = l_1 \sin\theta_1 + l_2 \sin\theta_2.$$

The total kinetic energy Γ is found to be a nonlinear function of the displacements, for example,

$$\Gamma = \frac{1}{2}m_1\left(\dot{x}_1^2 + \dot{y}_1^2\right) + \frac{1}{2}m_2\left(\dot{x}_2^2 + \dot{y}_2^2\right) = \frac{1}{2}(m_1 + m_2)l_1^2\dot{\theta}_1^2 + m_2 l_1 l_2 \dot{\theta}_1 \dot{\theta}_2 \cos(\theta_2 - \theta_1) + \frac{1}{2}m_2 l_2^2 \dot{\theta}_2^2.$$

One obtains

$$\frac{\partial\Gamma}{\partial\theta_1} = m_2 l_1 l_2 \sin(\theta_2 - \theta_1)\dot{\theta}_1\dot{\theta}_2, \quad \frac{\partial\Gamma}{\partial\dot{\theta}_1} = (m_1 + m_2)l_1^2\dot{\theta}_1 + m_2 l_1 l_2 \cos(\theta_2 - \theta_1)\dot{\theta}_2,$$

$$\frac{\partial\Gamma}{\partial\theta_2} = -m_2 l_1 l_2 \sin(\theta_1 - \theta_2)\dot{\theta}_1\dot{\theta}_2, \quad \frac{\partial\Gamma}{\partial\dot{\theta}_2} = m_2 l_1 l_2 \cos(\theta_2 - \theta_1)\dot{\theta}_1 + m_2 l_2^2\dot{\theta}_2.$$

The total potential energy is

$$\Pi = m_1 g x_1 + m_2 g x_2 = (m_1 + m_2)g l_1 \cos\theta_1 + m_2 g l_2 \cos\theta_2,$$

yielding $\dfrac{\partial\Pi}{\partial\theta_1} = -(m_1 + m_2)g l_1 \sin\theta_1$ and $\dfrac{\partial\Pi}{\partial\theta_2} = -m_2 g l_2 \sin\theta_2$.

The Lagrange equations of motion are

$$\frac{d}{dt}\left(\frac{\partial\Gamma}{\partial\dot{\theta}_1}\right) - \frac{\partial\Gamma}{\partial\theta_1} + \frac{\partial\Pi}{\partial\theta_1} = 0, \quad \frac{d}{dt}\left(\frac{\partial\Gamma}{\partial\dot{\theta}_2}\right) - \frac{\partial\Gamma}{\partial\theta_2} + \frac{\partial\Pi}{\partial\theta_2} = 0.$$

Hence, the differential equations which describe the motion are

$$l_1\left[(m_1 + m_2)l_1\ddot{\theta}_1 + m_2 l_2 \cos(\theta_2 - \theta_1)\ddot{\theta}_2 - m_2 l_2 \sin(\theta_2 - \theta_1)\dot{\theta}_2^2 - m_2 l_2 \sin(\theta_2 - \theta_1)\dot{\theta}_1\dot{\theta}_2\right.$$
$$\left. - (m_1 + m_2)g \sin\theta_1\right] = 0,$$

$$m_2 l_2\left[l_2\ddot{\theta}_2 + l_1 \cos(\theta_2 - \theta_1)\ddot{\theta}_1 + l_1 \sin(\theta_2 - \theta_1)\dot{\theta}_1^2 + l_1 \sin(\theta_2 - \theta_1)\dot{\theta}_1\dot{\theta}_2 - g \sin\theta_2\right] = 0.$$

If the torques T_1 and T_2 are applied to the first and second joints (two-degree-of-freedom robot), the following equations of motions result

$$l_1\left[(m_1 + m_2)l_1\ddot{\theta}_1 + m_2 l_2 \cos(\theta_2 - \theta_1)\ddot{\theta}_2 - m_2 l_2 \sin(\theta_2 - \theta_1)\dot{\theta}_2^2 - m_2 l_2 \sin(\theta_2 - \theta_1)\dot{\theta}_1\dot{\theta}_2\right.$$
$$\left. - (m_1 + m_2)g \sin\theta_1\right] = T_1,$$

$$m_2 l_2\left[l_2\ddot{\theta}_2 + l_1 \cos(\theta_2 - \theta_1)\ddot{\theta}_1 + l_1 \sin(\theta_2 - \theta_1)\dot{\theta}_1^2 + l_1 \sin(\theta_2 - \theta_1)\dot{\theta}_1\dot{\theta}_2 - g \sin\theta_2\right] = T_2.$$

Torques T_1 and T_2 are time-varying and controlled by actuators. Furthermore, the load torques, can be added to the equations derived. $\blacksquare$

Having illustrated the use of the Lagrange equations of motion to mechanical systems, the following examples demonstrate the use of Lagrangian mechanics to electric circuits.

Example 2.10: Electric Circuit

Consider an electric circuit as shown in Figure 2.9. Our goal is to derive the equations that describe the circuitry dynamics.

We use the electric charges as the generalized coordinates. That is, the electric charge in the first loop q_1 and the electric charge in the second loop q_2 are the independent generalized coordinates (variables) as shown in Figure 2.9. These generalized coordinates are related to the currents i_1 and i_2. We have $i_1 = \dot{q}_1$ and $i_2 = \dot{q}_2$, and $q_1 = i_1/s$ and $q_2 = i_2/s$. The supplied voltage u_a is the generalized force Q_1 applied, and $u_a(t) = Q_1$.

The Lagrange equations of motion are expressed using the independent coordinates, for example,

$$\frac{d}{dt}\left(\frac{\partial \Gamma}{\partial \dot{q}_1}\right) - \frac{\partial \Gamma}{\partial q_1} + \frac{\partial D}{\partial \dot{q}_1} + \frac{\partial \Pi}{\partial q_1} = Q_1, \quad \frac{d}{dt}\left(\frac{\partial \Gamma}{\partial \dot{q}_2}\right) - \frac{\partial \Gamma}{\partial q_2} + \frac{\partial D}{\partial \dot{q}_2} + \frac{\partial \Pi}{\partial q_2} = 0.$$

The total magnetic energy (kinetic energy) is $\Gamma = \frac{1}{2}L_1\dot{q}_1^2 + \frac{1}{2}L_{12}(\dot{q}_1 - \dot{q}_2)^2 + \frac{1}{2}L_2\dot{q}_2^2$.

Hence, $\dfrac{\partial \Gamma}{\partial q_1} = 0$, $\dfrac{\partial \Gamma}{\partial \dot{q}_1} = (L_1 + L_{12})\dot{q}_1 - L_{12}\dot{q}_2$, $\dfrac{\partial \Gamma}{\partial q_2} = 0$, $\dfrac{\partial \Gamma}{\partial \dot{q}_2} = -L_{12}\dot{q}_1 + (L_2 + L_{12})\dot{q}_2$.

Using the expression for the total electric energy (potential energy) $\Pi = \dfrac{1}{2}\dfrac{q_1^2}{C_1} + \dfrac{1}{2}\dfrac{q_2^2}{C_2}$, one finds

$$\frac{\partial \Pi}{\partial q_1} = \frac{q_1}{C_1} \quad \text{and} \quad \frac{\partial \Pi}{\partial q_2} = \frac{q_2}{C_2}.$$

The total heat energy dissipated is $D = \dfrac{1}{2}R_1\dot{q}_1^2 + \dfrac{1}{2}R_2\dot{q}_2^2$, yielding $\dfrac{\partial D}{\partial \dot{q}_1} = R_1\dot{q}_1$ and $\dfrac{\partial D}{\partial \dot{q}_2} = R_2\dot{q}_2$.

The differential equations for the circuit are found to be

$$(L_1 + L_{12})\ddot{q}_1 - L_{12}\ddot{q}_2 + R_1\dot{q}_1 + \frac{q_1}{C_1} = u_a$$

and

$$-L_{12}\ddot{q}_1 + (L_2 + L_{12})\ddot{q}_2 + R_2\dot{q}_2 + \frac{q_2}{C_2} = 0.$$

To assess the transient dynamics, one examines the resulting second-order differential equations

$$\ddot{q}_1 = \frac{1}{(L_1 + L_{12})}\left(-\frac{q_1}{C_1} - R_1\dot{q}_1 + L_{12}\ddot{q}_2 + u_a\right),$$

$$\ddot{q}_2 = \frac{1}{(L_2 + L_{12})}\left(L_{12}\ddot{q}_1 - \frac{q_2}{C_2} - R_2\dot{q}_2\right)$$

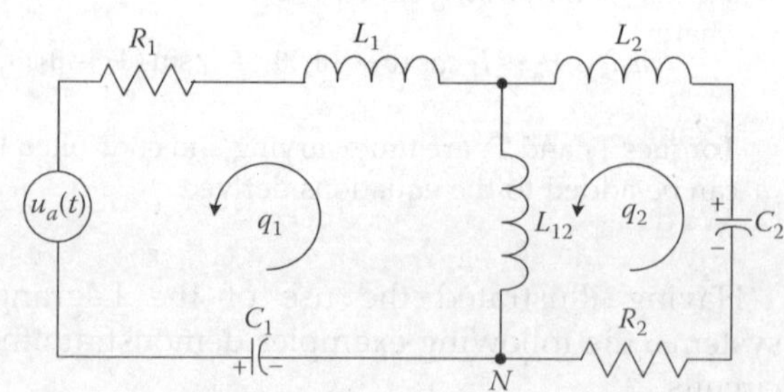

FIGURE 2.9
Electric circuit.

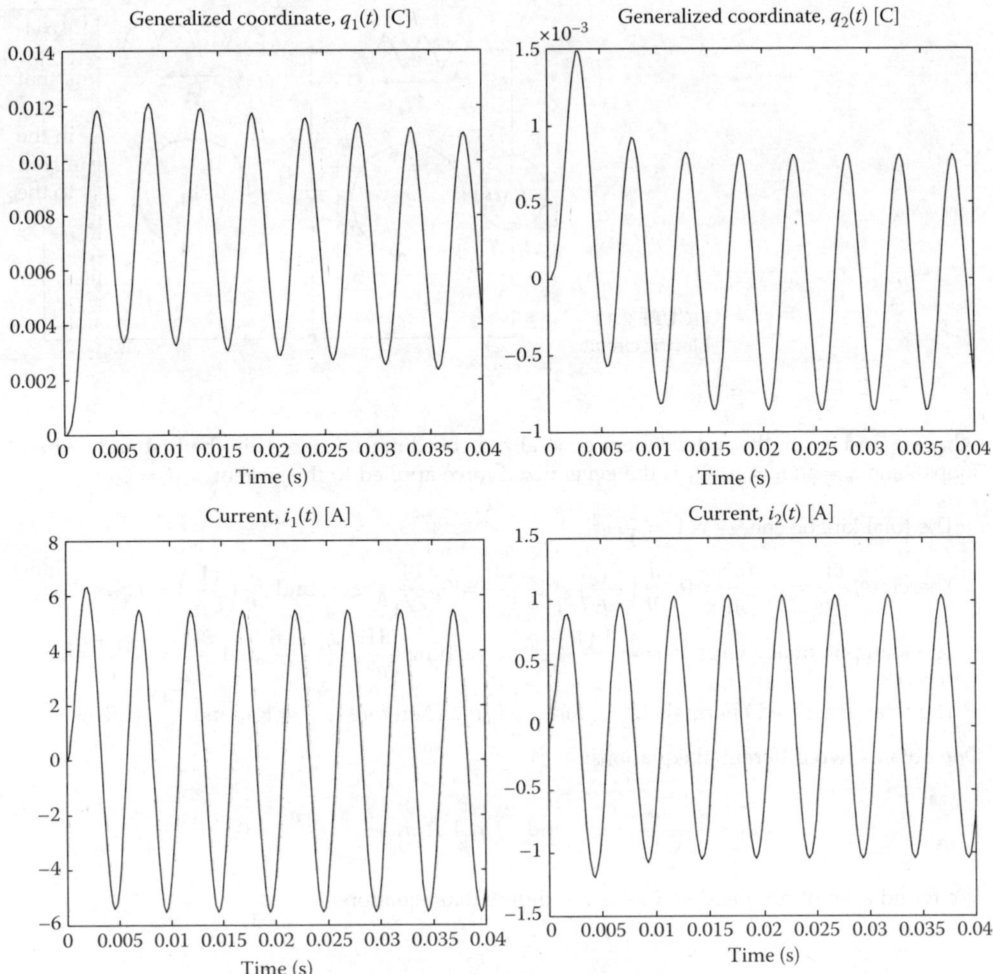

FIGURE 2.10

Circuit dynamics: transients of the generalized coordinates and currents.

which describe the circuit dynamics. To perform numerical simulations, the MATLAB® environment will be used. Let the circuitry parameters be $L_1 = 0.01$ H, $L_2 = 0.005$ H, $L_{12} = 0.0025$ H, $C_1 = 0.02$ F, $C_2 = 0.1$ F, $R_1 = 10$ ohm, and $R_2 = 5$ ohm. The applied voltage is $u_a = 100 \sin(200t)$ V. The simulation results, reported in Figure 2.10, provide the evolution of the variables of interest $q_1(t)$, $q_2(t)$, $i_1(t)$, and $i_2(t)$ in the time domain. The currents i_1 and i_2 are expressed in terms of charges as $i_1 = \dot{q}_1$ and $i_2 = \dot{q}_2$.

Example 2.11:   Electric Circuit

We derive the corresponding differential equations for the circuit depicted in the Figure 2.11. We apply the Lagrange equations of motion

$$\frac{d}{dt}\left(\frac{\partial \Gamma}{\partial \dot{q}_1}\right) - \frac{\partial \Gamma}{\partial q_1} + \frac{\partial D}{\partial \dot{q}_1} + \frac{\partial \Pi}{\partial q_1} = Q_1 \quad \text{and} \quad \frac{d}{dt}\left(\frac{\partial \Gamma}{\partial \dot{q}_2}\right) - \frac{\partial \Gamma}{\partial q_2} + \frac{\partial D}{\partial \dot{q}_2} + \frac{\partial \Pi}{\partial q_2} = 0,$$

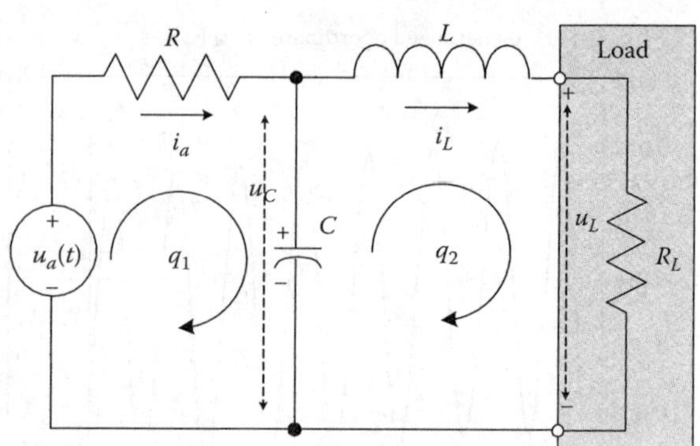

FIGURE 2.11
Electric circuit.

where q_1 and q_2 are the independent generalized coordinates (charges in the first and second loops), and $i_a = \dot{q}_1$, $i_L = \dot{q}_2$; Q_1 is the generalized force applied to the system, $u_a(t) = Q_1$.

The total kinetic energy is $\Gamma = \dfrac{1}{2} L \dot{q}_2^2$.

Therefore, $\dfrac{\partial \Gamma}{\partial q_1} = 0$, $\dfrac{\partial \Gamma}{\partial \dot{q}_1} = 0$, $\dfrac{d}{dt}\left(\dfrac{\partial \Gamma}{\partial \dot{q}_1}\right) = 0$, $\dfrac{\partial \Gamma}{\partial q_2} = 0$, $\dfrac{\partial \Gamma}{\partial \dot{q}_2} = L \dot{q}_2$, and $\dfrac{d}{dt}\left(\dfrac{\partial \Gamma}{\partial \dot{q}_2}\right) = L \ddot{q}_2$.

The total potential energy is $\Pi = \dfrac{1}{2} \dfrac{(q_1 - q_2)^2}{C}$, yielding $\dfrac{\partial \Pi}{\partial q_1} = \dfrac{q_1 - q_2}{C}$ and $\dfrac{\partial \Pi}{\partial q_2} = \dfrac{-q_1 + q_2}{C}$.

The total dissipated energy is $D = \dfrac{1}{2} R \dot{q}_1^2 + \dfrac{1}{2} R_L \dot{q}_2^2$. Therefore, $\dfrac{\partial D}{\partial \dot{q}_1} = R \dot{q}_1$ and $\dfrac{\partial D}{\partial \dot{q}_2} = R_L \dot{q}_2$.

One obtains two differential equations:

$$R \dot{q}_1 + \frac{q_1 - q_2}{C} = u_a \quad \text{and} \quad L \ddot{q}_2 + R_L \dot{q}_2 + \frac{-q_1 + q_2}{C} = 0.$$

We found a set of first- and second-order differential equations

$$\dot{q}_1 = \frac{1}{R}\left(\frac{-q_1 + q_2}{C} + u_a\right),$$

$$\ddot{q}_2 = \frac{1}{L}\left(-R_L \dot{q}_2 + \frac{q_1 - q_2}{C}\right).$$

The equations of motion derived using the Lagrange concept should be equivalent to the model developed using Kirchhoff's law. Using Kirchhoff's law, the following two differential equations result:

$$\frac{du_C}{dt} = \frac{1}{C}\left(-\frac{u_C}{R} - i_L + \frac{u_a(t)}{R}\right),$$

$$\frac{di_L}{dt} = \frac{1}{L}(u_C - R_L i_L).$$

From $i_a = \dot{q}_1$ and $i_L = \dot{q}_2$, using $C\dfrac{du_C}{dt} = i_a - i_L$, we obtain $u_C = \dfrac{q_1 - q_2}{C}$. The equivalence of the differential equations derived using the Lagrange equations of motion and Kirchhoff's law is proven.

## Example 2.12:	Electromechanical Actuator

Consider an electromechanical motion device which actuates the load (robotic arm, pointer, etc.). The actuator has two independently excited stator and rotor windings as reported in Figure 2.12. The magnetic coupling between the stator and rotor windings results in an electromagnetic torque T_e. The developed T_e is countered by the torsional spring which causes a *counterclockwise* rotation. The load torque T_L is considered. We derive the differential equations to describe the device dynamics.

The following notations are used for the actuator variables and parameters: i_s and i_r are the currents in the stator and rotor windings; u_s and u_r are the applied voltages to the stator and rotor windings; ω_r and θ_r are the rotor angular velocity and displacement; T_e and T_L are the electromagnetic and load torques; r_s and r_r are the resistances of the stator and rotor windings; L_s and L_r are the self-inductances of the stator and rotor windings; L_{sr} is the mutual inductance of the stator and rotor windings; $\Re_m$ is the reluctance of the magnetizing path; N_s and N_r are the number of turns in the stator and rotor windings; J is the equivalent moment of inertia of the rotor and attached load; B_m is the viscous friction coefficient; k_s is the spring constant.

By using the Lagrange concept, the independent generalized coordinates are q_1, q_2, and q_3, where q_1 and q_2 are the electric charges in the stator and rotor windings; q_3 is the rotor angular displacement. The generalized forces, applied to a system, are Q_1, Q_2, and Q_3, where Q_1 and Q_2 are the applied voltages to the stator and rotor windings; Q_3 is the load torque.

The first derivative of the generalized coordinates $\dot{q}_1$ and $\dot{q}_2$ represent the stator and rotor currents i_s and i_r, while $\dot{q}_3$ is the angular velocity of the rotor ω_r. We have, $q_1 = i_s/s$, $q_2 = i_r/s$, $q_3 = \theta_r$, $\dot{q}_1 = i_s$, $\dot{q}_2 = i_r$, $\dot{q}_3 = \omega_r$, $Q_1 = u_s$, $Q_2 = u_r$, and $Q_3 = -T_L$.

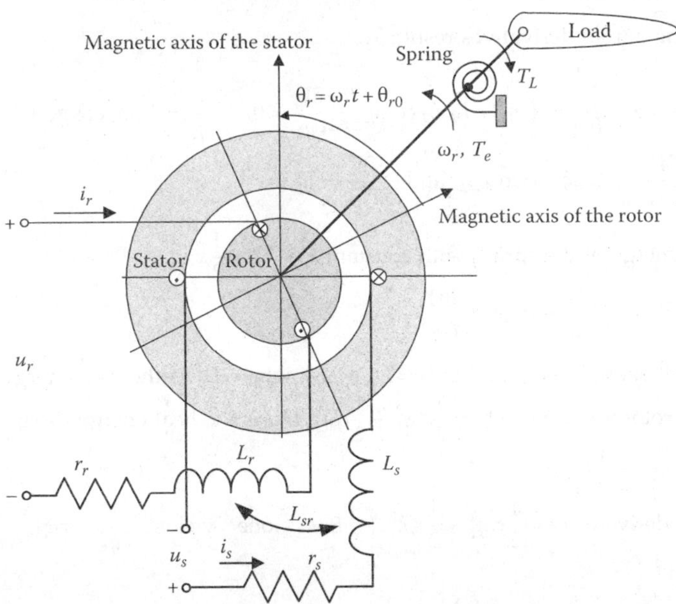

FIGURE 2.12
Actuator with stator and rotor windings.

The Lagrange equations are expressed in terms of each independent coordinate, yielding

$$\frac{d}{dt}\left(\frac{\partial \Gamma}{\partial \dot{q}_1}\right) - \frac{\partial \Gamma}{\partial q_1} + \frac{\partial D}{\partial \dot{q}_1} + \frac{\partial \Pi}{\partial q_1} = Q_1,$$

$$\frac{d}{dt}\left(\frac{\partial \Gamma}{\partial \dot{q}_2}\right) - \frac{\partial \Gamma}{\partial q_2} + \frac{\partial D}{\partial \dot{q}_2} + \frac{\partial \Pi}{\partial q_2} = Q_2,$$

$$\frac{d}{dt}\left(\frac{\partial \Gamma}{\partial \dot{q}_3}\right) - \frac{\partial \Gamma}{\partial q_3} + \frac{\partial D}{\partial \dot{q}_3} + \frac{\partial \Pi}{\partial q_3} = Q_3.$$

The total kinetic energy of electrical and mechanical systems is found as a sum of the total electromagnetic (magnetic, electrical, etc.) Γ_E and mechanical Γ_M energies. The total kinetic energy of the stator and rotor circuitry is $\Gamma_E = \frac{1}{2}L_s\dot{q}_1^2 + L_{sr}\dot{q}_1\dot{q}_2 + \frac{1}{2}L_r\dot{q}_2^2$. The total kinetic energy of the mechanical system is $\Gamma_M = \frac{1}{2}J\dot{q}_3^2$. Therefore,

$$\Gamma = \Gamma_E + \Gamma_M = \frac{1}{2}L_s\dot{q}_1^2 + L_{sr}\dot{q}_1\dot{q}_2 + \frac{1}{2}L_r\dot{q}_2^2 + \frac{1}{2}J\dot{q}_3^2.$$

The mutual inductance depends on the displacement of the rotor winding with respect to the stator winding. It is obvious that $L_{sr}(\theta_r)$ is a periodic function of the angular rotor displacement, and $L_{sr\,min} \leq L_{sr}(\theta_r) \leq L_{sr\,max}$. The amplitude of the mutual inductance between the stator and rotor windings is denoted as L_M, and $L_M = L_{sr\,max}$. If the windings are orthogonal, we have $L_{sr} = 0$. The mutual inductance L_{sr} is approximated as $L_{sr}(\theta_r) = L_M \cos \theta_r = L_M \cos q_3$.

The total kinetic energy is $\Gamma = \frac{1}{2}L_s\dot{q}_1^2 + L_M\dot{q}_1\dot{q}_2 \cos q_3 + \frac{1}{2}L_r\dot{q}_2^2 + \frac{1}{2}J\dot{q}_3^2.$

The following partial derivatives result:

$$\frac{\partial \Gamma}{\partial q_1} = 0, \quad \frac{\partial \Gamma}{\partial \dot{q}_1} = L_s\dot{q}_1 + L_M\dot{q}_2 \cos q_3, \quad \frac{\partial \Gamma}{\partial q_2} = 0, \quad \frac{\partial \Gamma}{\partial \dot{q}_2} = L_M\dot{q}_1 \cos q_3 + L_r\dot{q}_2,$$

$$\frac{\partial \Gamma}{\partial q_3} = -L_M\dot{q}_1\dot{q}_2 \sin q_3, \quad \text{and} \quad \frac{\partial \Gamma}{\partial \dot{q}_3} = J\dot{q}_3.$$

The potential energy of the spring with constant k_s is $\Pi = \frac{1}{2}k_s q_3^2$.

Therefore, $\dfrac{\partial \Pi}{\partial q_1} = 0, \ \dfrac{\partial \Pi}{\partial q_2} = 0, \ \text{and} \ \dfrac{\partial \Pi}{\partial q_3} = k_s q_3.$

The total heat energy dissipated is $D = D_E + D_M$, where D_E is the heat energy dissipated in the stator and rotor windings, $D_E = \frac{1}{2}r_s\dot{q}_1^2 + \frac{1}{2}r_r\dot{q}_2^2$; D_M is the heat energy dissipated by mechanical system, $D_M = \frac{1}{2}B_m\dot{q}_3^2$.

From the derived $D = \frac{1}{2}r_s\dot{q}_1^2 + \frac{1}{2}r_r\dot{q}_2^2 + \frac{1}{2}B_m\dot{q}_3^2$ one yields $\dfrac{\partial D}{\partial \dot{q}_1} = r_s\dot{q}_1, \ \dfrac{\partial D}{\partial \dot{q}_2} = r_r\dot{q}_2,$ and $\dfrac{\partial D}{\partial \dot{q}_3} = B_m\dot{q}_3.$

Using the following relationships between the generalized coordinates and state variables

$$q_1 = \frac{i_s}{s}, \ q_2 = \frac{i_r}{s}, \ q_3 = \theta_r, \ \dot{q}_1 = i_s, \ \dot{q}_2 = i_r, \ \dot{q}_3 = \omega_r, \ Q_1 = u_s, \ Q_2 = u_r, \ \text{and} \ Q_3 = -T_L,$$

we have three differential equations for the considered actuator. In particular,

$$L_s \frac{di_s}{dt} + L_M \cos\theta_r \frac{di_r}{dt} - L_M i_r \sin\theta_r \frac{d\theta_r}{dt} + r_s i_s = u_s,$$

$$L_r \frac{di_r}{dt} + L_M \cos\theta_r \frac{di_s}{dt} - L_M i_s \sin\theta_r \frac{d\theta_r}{dt} + r_r i_r = u_r,$$

$$J \frac{d^2\theta_r}{dt^2} + L_M i_s i_r \sin\theta_r + B_m \frac{d\theta_r}{dt} + k_s \theta_r = -T_L.$$

The last equation may be rewritten recalling that $\dfrac{d\theta_r}{dt} = \omega_r$. Using the stator and rotor currents, angular velocity, and displacement as the state variables, the nonlinear differential equations in Cauchy's form are

$$\frac{di_s}{dt} = \frac{-r_s L_r i_s - \frac{1}{2}L_M^2 i_s \omega_r \sin 2\theta_r + r_r L_M i_r \cos\theta_r + L_r L_M i_r \omega_r \sin\theta_r + L_r u_s - L_M \cos\theta_r u_r}{L_s L_r - L_M^2 \cos^2\theta_r},$$

$$\frac{di_r}{dt} = \frac{r_s L_M i_s \cos\theta_r + L_s L_M i_s \omega_r \sin\theta_r - r_r L_s i_r - \frac{1}{2}L_M^2 i_r \omega_r \sin 2\theta_r - L_M \cos\theta_r u_s + L_s u_r}{L_s L_r - L_M^2 \cos^2\theta_r},$$

$$\frac{d\omega_r}{dt} = \frac{1}{J}\left(-L_M i_s i_r \sin\theta_r - B_m \omega_r - k_s \theta_r - T_L\right),$$

$$\frac{d\theta_r}{dt} = \omega_r.$$

The developed nonlinear differential equations cannot be linearized. One must analyze the actuator and examine the system performance using a complete set of equations of motion. ∎

Example 2.13: Beam Equations of Motion

Consider an elastic beam of length l with constant cross-sectional area A and uniform weight per unit volume (density) ρ. We denote the static vertical displacement at the free end of the beam, which is illustrated in Figure 2.13, as q.

One needs to find the kinetic and potential energies. We will make some assumption to illustrate the application of the Lagrange equations of motion. Let the third-order deflection polynomial be $y(x) = \dfrac{1}{2}\left(3\dfrac{x^2}{l^2} - \dfrac{x^3}{l^3}\right)q$. Using $q(t)$, we have $y(t, x) = \dfrac{1}{2}\left(3\dfrac{x^2}{l^2} - \dfrac{x^3}{l^3}\right)q(t)$.

The kinetic energy is $\Gamma(q) = \dfrac{1}{2}\displaystyle\int_0^l \dot{y}^2 dm = \dfrac{1}{2}A\rho\int_0^l \dfrac{1}{4}\left(3\dfrac{x^2}{l^2} - \dfrac{x^3}{l^3}\right)^2 \dot{q}^2 dx = \dfrac{33}{280}A\rho l\dot{q}^2$, while the poten-

tial energy of elastic deformation is $\Pi(q) = \dfrac{1}{2}\displaystyle\int_0^l EI\left(\dfrac{\partial^2 y}{\partial x^2}\right)^2 dx = \dfrac{1}{2}EI\int_0^l \dfrac{3}{2l^3}\left(1 - \dfrac{x}{l}\right)^2 d\left(\dfrac{x}{l}\right) = \dfrac{3}{2}\dfrac{EI}{l^3}q^2$,

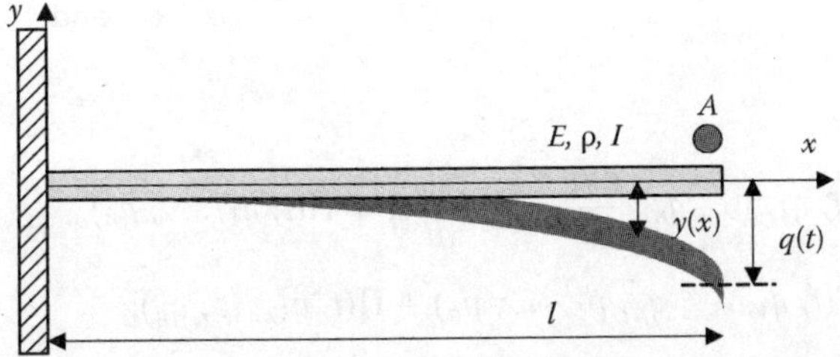

FIGURE 2.13
Beam in the xy plane.

where E is the Young' modulus of elasticity and I is the moment of inertia of the cross section about its neural axis.

From the Lagrange equation $\dfrac{d}{dt}\left(\dfrac{\partial\Gamma}{\partial\dot{q}}\right) - \dfrac{\partial\Gamma}{\partial q} + \dfrac{\partial\Pi}{\partial q} = Q$, the beam equation of motion is

$$\frac{d^2 q}{dt^2} = -12.7\frac{EI}{A\rho l^4}q + F_q(t,x).$$

The reported simplified concept results in the second-order differential equation which describes the beam dynamics. In general, the potential energy of elastic beam is $\Pi = \dfrac{1}{2}\displaystyle\int_r \sigma_{ij}\varepsilon_{ij}\,d\mathbf{r} + \int_r T(\mathbf{r})w(\mathbf{r})\,d\mathbf{r} + \int_r F(\mathbf{r})w(\mathbf{r})\,d\mathbf{r}$, where $T(\mathbf{r})$ and $F(\mathbf{r})$ are the beam surface traction and force, respectively. The term $\dfrac{1}{2}\sigma_{ij}\varepsilon_{ij}$ gives the strain energy stored. Various equations of motions were derived. For example, for the laterally distributed load $T(x)$, the equation for the beam bending is $a_b\dfrac{d^4 w}{dx^4} = T(x)$, $a_b = EI$. Other commonly used equation is $\xi EI(x)\dfrac{\partial^5 y(t,x)}{\partial x^4 \partial t} + EI(x)\dfrac{\partial^4 y(t,x)}{\partial x^4} + m_0(x)\dfrac{\partial^2 y(t,x)}{\partial t^2} + m(x)\dfrac{d^2\varphi}{dt^2} = F(t,x)$, where $F(t,x)$ is the distributed force through the beam. ∎

2.4.3 Hamilton Equations of Motion

The Hamilton concept allows one to describe the system dynamics. The differential equations are found using the generalized momenta p_i, $p_i = \dfrac{\partial L}{\partial \dot{q}_i}$. The generalized coordinates were used in the Lagrange equations of motion. The Lagrangian function $L\left(t, q_1, \ldots, q_n, \dfrac{dq_1}{dt}, \ldots, \dfrac{dq_n}{dt}\right)$ for the conservative systems is the difference between the total kinetic and potential energies. We have

$$L\left(t, q_1, \ldots, q_n, \frac{dq_1}{dt}, \ldots, \frac{dq_n}{dt}\right) = \Gamma\left(t, q_1, \ldots, q_n, \frac{dq_1}{dt}, \ldots, \frac{dq_n}{dt}\right) - \Pi(t, q_1, \ldots, q_n).$$

One concludes that $L\left(t, q_1, \ldots, q_n, \dfrac{dq_1}{dt}, \ldots, \dfrac{dq_n}{dt}\right)$ is the function of $2n$ independent variables.

The Hamiltonian function is

$$H(t, q_1, \ldots, q_n, p_1, \ldots, p_n) = -L\left(t, q_1, \ldots, q_n, \frac{dq_1}{dt}, \ldots, \frac{dq_n}{dt}\right) + \sum_{i=1}^{n} p_i\dot{q}_i,$$

and

$$H\left(t, q_1, \ldots, q_n, \frac{dq_1}{dt}, \ldots, \frac{dq_n}{dt}\right) = \Gamma\left(t, q_1, \ldots, q_n, \frac{dq_1}{dt}, \ldots, \frac{dq_n}{dt}\right) + \Pi(t, q_1, \ldots, q_n),$$

$$\text{or} \quad H(t, q_1, \ldots, q_n, p_1, \ldots, p_n) = \Gamma(t, q_1, \ldots, q_n, p_1, \ldots, p_n) + \Pi(t, q_1, \ldots, q_n).$$

Hence, the Hamiltonian, which represents the total energy, is expressed as a function of the generalized coordinates and generalized momenta. The equations of motion are governed by the following equations:

$$\dot{p}_i = -\frac{\partial H}{\partial q_i}, \quad \dot{q}_i = \frac{\partial H}{\partial p_i}. \tag{2.4}$$

These equations are called the Hamiltonian equations of motion. Using the Hamiltonian mechanics, one obtains the system of $2n$ first-order differential equations to describe the system dynamics. In contrast, using the Lagrange equations of motion, the system of n second-order differential equations results. However, the derived differential equations are equivalent.

Example 2.14:

Consider the harmonic oscillator which is formed by the sliding mass m attached to the spring assuming that there is no viscous friction. The total energy is given as the sum of the kinetic and potential energies, for example, $\Sigma_T = \Gamma + \Pi = \frac{1}{2}(mv^2 + k_s x^2)$.

One can find the equations of motion using the Lagrange and Hamilton concepts.

Recall that $q = x$. The Lagrangian function is

$$L\left(x, \frac{dx}{dt}\right) = \Gamma - \Pi = \frac{1}{2}\left(mv^2 - k_s x^2\right) = \frac{1}{2}\left(m\dot{x}^2 - k_s x^2\right).$$

From the Lagrange equations of motion $\dfrac{d}{dt}\dfrac{\partial L}{\partial \dot{x}} - \dfrac{\partial L}{\partial x} = 0$, where $q = x$, we have the following second-order differential equation:

$$m\frac{d^2 x}{dt^2} + k_s x = 0.$$

From Newton's second law, the second-order differential equation of motion is given as

$$m\frac{d^2 x}{dt^2} + k_s x = 0.$$

The Hamiltonian function is

$$H(x, p) = \Gamma + \Pi = \frac{1}{2}\left(mv^2 + k_s x^2\right) = \frac{1}{2}\left(\frac{1}{m}p^2 + k_s x^2\right).$$

From the Hamiltonian equations of motion (2.4), one obtains the following differential equations:

$$\dot{p} = -\frac{\partial H}{\partial x} = -k_s x, \quad \dot{x} = \dot{q} = \frac{\partial H}{\partial p} = \frac{p}{m}.$$

The equivalence of the resulting equations of motion is obvious.　　　　∎

2.5 Application of Electromagnetics and Classical Mechanics to Electromechanical Systems and Devices

The cornerstone laws of electromagnetics and mechanics are applied to examine device physics as well as to describe physical phenomena, effects, and processes. We derived the differential equations which describe the device behavior. It was illustrated that electromechanical devices can be described (modeled) using different concepts. Forces and torques can be found using the Maxwell stress tensor and energies. The nonlinear partial differential equations and tensor calculus, though lead to high-fidelity modeling, are very complex. Frequently, the lumped-parameter models are used. Consider the motor's rotor (bar magnet, current loop, and solenoid) in a uniform magnetic field as illustrated in Figure 2.14.

The torque tends to align the magnetic moment $\vec{m}$ with $\vec{B}$, and $\vec{T} = \vec{m} \times \vec{B}$.

Consider a magnetic bar with the length of dipole l and the magnetic pole strength Q_m. The magnetic dipole moment is $m = Q_m l$, while the force is $F = Q_m B$. In particular, the north pole (+) experiences a force $Q_m B$ to the right, while the south pole (−) exhibits a force $Q_m B$ to the left. As illustrated in Figure 2.14, the electromagnetic torque is $T = 2F\dfrac{1}{2}l\sin\alpha = Q_m l B \sin\alpha = mB\sin\alpha$. The torque tends to align the bar with the magnetic field $\vec{B}$, and it rotates clockwise.

Using the vector notations, one obtains $\vec{T} = \vec{m} \times \vec{B} = m\vec{a}_m \times \vec{B} = Q_m l \vec{a}_m \times \vec{B}$, where $\vec{a}_m$ is the unit vector in the magnetic moment direction.

For a current loop with the cross-sectional loop area A, the torque is

$$\vec{T} = \vec{m} \times \vec{B} = m\vec{a}_m \times \vec{B} = iA\vec{a}_m \times \vec{B}.$$

For a solenoid with N turns, shown in Figure 2.14, one obtains

$$\vec{T} = \vec{m} \times \vec{B} = \vec{a}_m m \times \vec{B} = iAN\vec{a}_m \times \vec{B}.$$

The expression for the electromagnetic torque developed should be integrated with the *torsional–mechanical* dynamics given by Newton's second law for the rotational motion $J\dfrac{d\omega_r}{dt} = \sum \vec{T}_\Sigma$. One recalls that $\sum \vec{T}_\Sigma$ is the *net* torque. The transient evolution of the angular displacement θ_r is $\dfrac{d\theta_r}{dt} = \omega_r$.

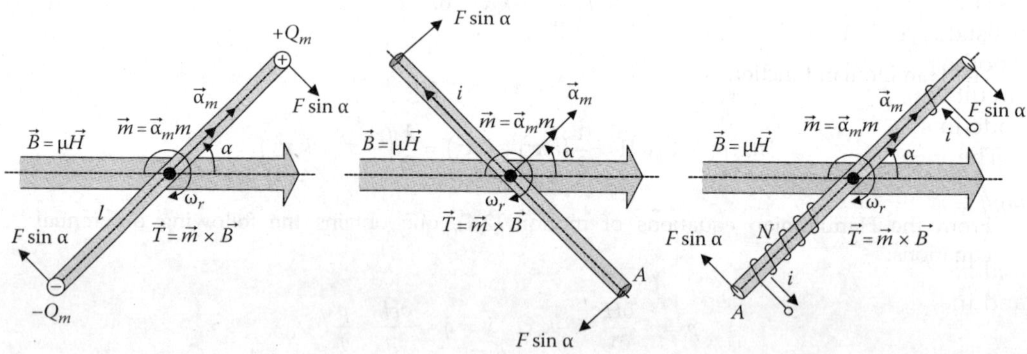

FIGURE 2.14
Clockwise rotation of magnetic bar, current loop, and solenoid.

Combining the equations for the electromagnetic torque (found in terms of the magnetic field variables $\vec{m}$ and $\vec{B}$, or current i and $\vec{B}$) and the *torsional–mechanical* dynamics (the state variables are the angular velocity ω_r and displacement θ_r), the resulting mathematical model results.

The total magnetic flux through the surface is $\Phi = \int \vec{B} \cdot d\vec{s}$. The Ampere circuital law is $\oint_l \vec{B} \cdot d\vec{l} = \mu_0 \int_s \vec{J} \cdot d\vec{s}$. For the filamentary current, Ampere's law connects the magnetic flux with the algebraic sum of the enclosed (linked) currents (*net current*) i_n, and $\oint_l \vec{B} \cdot d\vec{l} = \mu_0 i_n$.

The time-varying magnetic field produces the *emf*, denoted as $\mathscr{E}$, which induces the current in the closed circuit. Faraday's law relates the *emf* (induced voltage due to conductor motion in the magnetic field) to the rate of change of the magnetic flux Φ penetrating the loop. Lenz's law should be used to find the direction of *emf* and the current induced. In particular, the *emf* is in such a direction as to produce a current whose flux, if added to the original flux, would reduce the magnitude of the *emf*. According to Faraday's law, the induced *emf* in a closed-loop circuit is defined in terms of the rate of change of the magnetic flux Φ. One has the following equation for the induced *emf* (induced voltage)

$$\mathscr{E} = \oint_l \vec{E}(t) \cdot d\vec{l} = -\frac{d}{dt} \int_s \vec{B}(t) \cdot d\vec{s} = -N \frac{d\Phi}{dt} = -\frac{d\psi}{dt},$$

where N is the number of turns and ψ is the flux linkages.

Equation $\mathscr{E} = -\dfrac{d\psi}{dt} = -N\dfrac{d\Phi}{dt}$ represents the Faraday law of induction. The current flows in an opposite direction to the flux linkages. The unit for the *emf* is volts. The *emf* (*energy-per-unit-charge quantity*) represents a magnitude of the potential difference V in a circuit carrying a current. We have

$$V = -ir + \mathscr{E} = -ir - \frac{d\psi}{dt}.$$

The Kirchhoff voltage law states that around a closed path in an electric circuit, the algebraic sum of the *emf* is equal to the algebraic sum of the voltage drop across the resistance. This formulation will be used to examine various electromagnetic actuators. Another formulation is the algebraic sum of the voltages around any closed path in a circuit is zero. The Kirchhoff current law states that the algebraic sum of the currents at any node in a circuit is zero.

The *mmf* is the line integral of the time-varying magnetic field intensity $\vec{H}(t)$. That is, $mmf = \oint_l \vec{H}(t) \cdot d\vec{l}$. The unit for the *mmf* is amperes or ampere-turns. The duality of the *emf* and *mmf* can be observed using the following two equations given in terms of the electric and magnetic field intensity vectors

$$\mathscr{E} = \oint_l \vec{E}(t) \cdot d\vec{l} \quad \text{and} \quad mmf = \oint_l \vec{H}(t) \cdot d\vec{l}.$$

The inductance is the ratio of the total flux linkages to the current which they link, $L = \dfrac{N\Phi}{i}$. The reluctance is the ratio of the *mmf* to the total flux, $\Re = \dfrac{mmf}{\Phi}$. Hence, we have

$$\Re = \frac{\oint_l \vec{H} \cdot d\vec{l}}{\int_s \vec{B} \cdot d\vec{s}}.$$

The *emf* and *mmf* are used to find inductance and reluctance. The equation $L = \psi/i$ yields

$$\mathscr{E} = -\frac{d\psi}{dt} = -\frac{d(Li)}{dt} = -L\frac{di}{dt} - i\frac{dL}{dt}.$$

If $L = $ constant, one obtains $\mathscr{E} = -L\dfrac{di}{dt}$. That is, the self-inductance is the magnitude of the self-induced *emf* per unit rate of change of current.

The force- and torque-energy relations in electromagnetic and electrostatic actuators should be examined. The energy stored in the capacitor is $\dfrac{1}{2}CV^2$, while the energy stored in the inductor is $\dfrac{1}{2}Li^2$. The energy in the capacitor is stored in the electric field between plates, while the energy in the inductor is stored in the magnetic field within the coils.

As examples, consider the variable reluctance relays and solenoids which are studied in details in Chapter 3. The varying reluctance results in the electromagnetic force. In contract, high-performance electromagnetic motion devices utilize a coupling (magnetic interaction) between windings that are carrying currents and the stationary magnetic field developed by permanent magnets or electromagnets. In separately exited DC and induction machines, there is a magnetic coupling between windings due to their mutual inductances. To derive the electromagnetic force or torque, one applies $\vec{T} = \vec{m} \times \vec{B}$, or uses the expressions for the coenergy $W_c[i, L(x)]$ (translational motion) or $W_c[i, L(\theta)]$ (rotational motion). The developed electromagnetic force and torque are $F_e(i, x) = \dfrac{\partial W_c[i, L(x)]}{\partial x}$ and $T_e(i, x) = \dfrac{\partial W_c[i, L(\theta)]}{\partial \theta}$, or $T_e = m \times B$.

Example 2.15:

Solenoids usually integrate movable (plunger) and stationary (fixed) members made from high-permeability ferromagnetic materials. The windings wound within a helical pattern. These solenoids, as electromechanical devices, convert electrical energy to mechanical energy. Solenoids and relays operate due to the varying reluctance, and the force is generated due to the changes of reluctance. Performance of solenoids is strongly affected by the magnetic system, materials, geometry, relative permeability, windings, friction, etc. We find the inductances of a solenoid with the air-core ($\mu_r = 1$) and with a filled-core letting $\mu_r = 10,000$. The solenoid has 100 turns ($N = 100$), the length is 5 cm ($l = 0.05$ m). Let the uniform circular cross-sectional area be $A = 1 \times 10^{-4}$ m^2.

The magnetic field inside a solenoid is $B = \dfrac{\mu Ni}{l}$, where $\mu = \mu_0 \mu_r$. From $\mathscr{E} = -N\dfrac{d\Phi}{dt} = -L\dfrac{di}{dt}$, by applying $\Phi = BA = \dfrac{\mu NiA}{l}$, one obtains the following expression for the inductance

$$L = \frac{\mu N^2 A}{l}.$$

For the solenoid with air-core one obtains $L = 2.5 \times 10^{-5}$ H.

The MATLAB statement to calculate the numerical value of L is

```
mu0 = 4*pi*1e-7; mur = 1; N = 100; A = 1e-4; l = 5e-2; L = mu0*mur*N*N*A/l
```

The result displayed in the Command Window is $L = 2.5133e-005$
If the solenoid is filled with a ferromagnetic material, we have $L = 0.25$ H.　　　■

Remark.　The magnetic flux crossing surface is given as $\Phi = \oint_s \vec{B} \cdot d\vec{s}$. The expression $\Phi = BA$

must be used with a great caution considering the device physics and examining the electromagnetic system. In general, $\oint_s \vec{B} \cdot d\vec{s} \neq BA$.

Example 2.16:

We derive a formula for the self-inductance of a torroidal solenoid which has a rectangular cross section $2a \times b$ and mean radius r. The magnetic flux through a cross section is

$$\Phi = \int_{r-a}^{r+a} Bb\,dr = \int_{r-a}^{r+a} \frac{\mu Ni}{2\pi r} b\,dr = \frac{\mu Nib}{2\pi} \int_{r-a}^{r+a} \frac{1}{r}\,dr = \frac{\mu Nib}{2\pi} \ln\left(\frac{r+a}{r-a}\right)$$

yielding $L = \dfrac{N\Phi}{i} = \dfrac{\mu N^2 b}{2\pi} \ln\left(\dfrac{r+a}{r-a}\right)$.　　　■

Example 2.17:

Calculate the stored magnetic energy of the torroidal solenoid if the self-inductance is $L = 0.2$ H when the current is $i = 1 \times 10^{-3}$ A.

The stored field energy is $W_m = \dfrac{1}{2} L i^2$. Hence, $W_m = 1 \times 10^{-6}$ J.　　　■

Example 2.18:

Derive the expression for the electromagnetic force developed by the relay which is depicted in Figure 2.15.

The current i_a in N coils produces the flux Φ. We assume that the flux is constant. The displacement (the virtual displacement is denoted as dy) changes only the magnetic energy stored in the air gaps. From $W_m = \dfrac{1}{2} \int_v \mu |\vec{H}|^2 dv = \dfrac{1}{2} \int_v \mu^{-1} |\vec{B}|^2 dv$, we have

$$dW_m = dW_{m\ \text{air gap}} = 2\frac{B^2}{2\mu_0} A\,dy = \frac{\Phi^2}{\mu_0 A}\,dy,$$

where A is the cross-sectional area, $A = l_w l_d$.

If $\Phi = $ constant (the current is constant), one concludes that the increase of the air gap dy leads to increase of the stored magnetic energy. Using $F_e = \dfrac{\partial W_m}{\partial y}$, one finds the expression for

the electromagnetic force as $\vec{F}_e = -\vec{a}_y \dfrac{\Phi^2}{\mu_0 A}$. The result indicates that the force tends to reduce the airgap length, for example, to minimize the reluctance. The movable member, for which the gravitational force is mg, is attached to the springs which develop three forces in addition to the electromagnetic force.

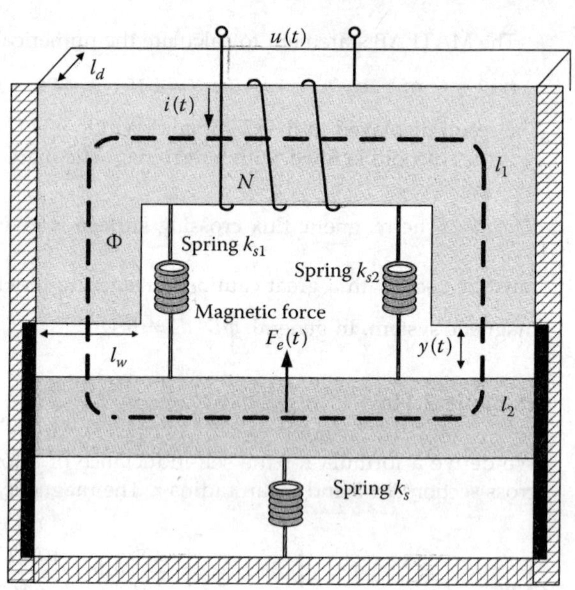

FIGURE 2.15
Relay with springs.

The airgap reluctance (two air gaps are in series) is $\Re_g = \dfrac{2y}{\mu_0 A} = \dfrac{2y}{\mu_0 l_w l_d}$.

The *fringing* effect may be integrated. The airgap reluctance can be approximated as $\Re_g = \dfrac{2y}{\mu_0\left(k_{g1}l_w l_d + k_{g2}y^2\right)}$, where k_{g1} and k_{g2} are the nonlinear functions of the ferromagnetic material, l_d/l_w ratio, B–H curve, load, etc.

The reluctances of the ferromagnetic materials of stationary and movable members $\Re_1$ and $\Re_2$ are $\Re_1 = \dfrac{l_1}{\mu_0\mu_{r1}A} = \dfrac{l_1}{\mu_0\mu_{r1}l_w l_d}$ and $\Re_2 = \dfrac{l_2}{\mu_0\mu_{r2}A} = \dfrac{l_2}{\mu_0\mu_{r2}l_w l_d}$.

The magnetizing inductance is expressed as $L(y) = \dfrac{N^2}{\Re_g(y) + \Re_1 + \Re_2}$.

The electromagnetic force is $F_e = \dfrac{1}{2}i^2\dfrac{dL(y)}{dy} = \dfrac{1}{2}i^2\dfrac{d\left(\dfrac{N^2}{\Re_g(y) + \Re_1 + \Re_2}\right)}{dy}$.

Using $\Re_g(y)$, one performs the differentiation deriving the expression for F_e. ∎

Example 2.19:

Figure 2.16a documents a cross-sectional view of a variable-reluctance actuator (relay or solenoid) which has N turns. The equivalent magnetic circuit with the reluctances of the various paths is illustrated in Figure 2.16b. The distance between the stationary and movable members is $x(t)$. The mean lengths of the stationary and movable members are l_1 and l_2, and the cross-sectional area is A. One can find the force exerted on the movable member as a function of the current $i_a(t)$ in the winding as well as the displacement $x(t)$. The permeabilities of stationary and movable members are μ_{r1} and μ_{r2}.

The electromagnetic force is $F_e = \dfrac{\partial W_m}{\partial x}$, where $W_m = \dfrac{1}{2}Li_a^2(t)$.

The magnetizing inductance is $L = \dfrac{N\Phi}{i_a(t)} = \dfrac{\psi}{i_a(t)}$, where the magnetic flux is $\Phi = \dfrac{Ni_a(t)}{\Re_1 + \Re_x + \Re_x + \Re_2}$. The reluctances of the ferromagnetic stationary and

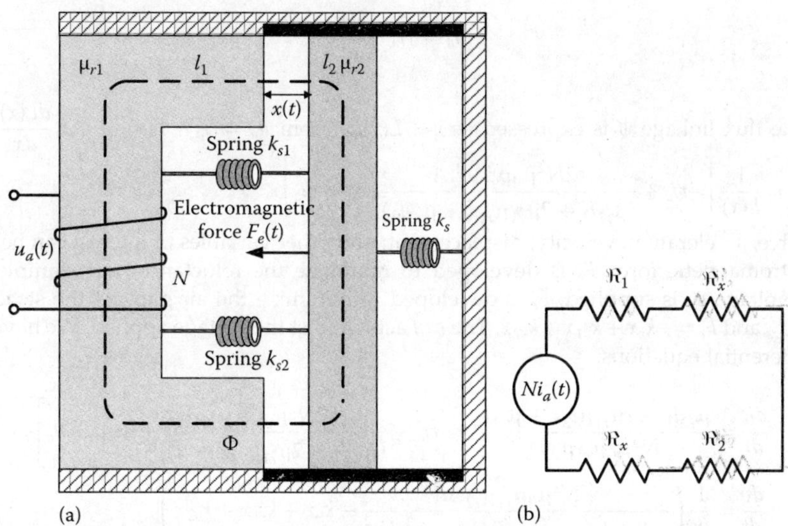

FIGURE 2.16
(a) Schematic of a variable-reluctance actuator (relay or solenoid); (b) equivalent magnetic circuit.

movable members $\Re_1$ and $\Re_2$, as well as the reluctance of the air gap $\Re_x$, are

$$\Re_1 = \frac{l_1}{\mu_0\mu_{r1}A}, \quad \Re_2 = \frac{l_2}{\mu_0\mu_{r2}A}, \quad \text{and} \quad \Re_x = \frac{x(t)}{\mu_0 A}.$$

Using the expression for reluctances of the movable and stationary members and air gap, one obtains the following formula for the flux linkages $\psi = N\Phi = \dfrac{N^2 i_a(t)}{\dfrac{l_1}{\mu_0\mu_{r1}A} + \dfrac{2x(t)}{\mu_0 A} + \dfrac{l_2}{\mu_0\mu_{r2}A}}$. The magnetizing inductance is a nonlinear function of the displacement. One has

$$L(x) = \frac{N^2}{\dfrac{l_1}{\mu_0\mu_{r1}A} + \dfrac{2x(t)}{\mu_0 A} + \dfrac{l_2}{\mu_0\mu_{r2}A}} = \frac{N^2\mu_0\mu_{r1}\mu_{r2}A}{\mu_{r2}l_1 + 2\mu_{r1}\mu_{r2}x(t) + \mu_{r1}l_2}.$$

From $F_e = \dfrac{\partial W_m}{\partial x} = \dfrac{1}{2}\dfrac{\partial\left(L(x(t))i_a^2(t)\right)}{\partial x}$, the force in the x direction is

$$F_e = -\frac{N^2\mu_0\mu_{r1}^2\mu_{r2}^2 A i_a^2}{(\mu_{r2}l_1 + 2\mu_{r1}\mu_{r2}x + \mu_{r1}l_2)^2}.$$

Using Newton's second law of motion, one obtains following nonlinear differential equations to examine the performance and analyze the steady-state and dynamic behavior

$$\frac{dx}{dt} = v,$$

$$\frac{dv}{dt} = \frac{1}{m}\left(-\frac{N^2\mu_0\mu_{r1}^2\mu_{r2}^2 A i_a^2}{(\mu_{r2}l_1 + 2\mu_{r1}\mu_{r2}x + \mu_{r1}l_2)^2} - k_s x + k_{s1}x + k_{s2}x\right).$$

This set of nonlinear differential equations describes the mechanical dynamics of a plunger. The voltage $u_a(t)$ is applied changing $i_a(t)$. The energy conversion, *emf* induction, and other electromagnetic phenomena and effects must be integrated. The Kirchhoff's voltage law gives

$$u_a = r i_a + \frac{d\psi}{dt},$$

where the flux linkage ψ is expressed as $\psi = L(x) i_a$. From $u_a = r i_a + L(x)\frac{di_a}{dt} + i_a \frac{dL(x)}{dx}\frac{dx}{dt}$, one

finds $\dfrac{di_a}{dt} = \dfrac{1}{L(x)}\left[-r i_a + \dfrac{2N^2 \mu_0 \mu_{r1}^2 \mu_{r2}^2 A}{\left(\mu_{r2} l_1 + 2\mu_{r1}\mu_{r2} x + \mu_{r1} l_2\right)^2} i_a v + u_a\right].$

The force, acceleration, velocity, displacement, and other variables of interest can be vectors. The electromagnetic force F_e is developed to minimize the reluctance, for example, as the applied voltage u_a is supplied, F_e is developed to minimize the air gap. At the steady state, $F_e = F_{springs}$ and $F_e = -k_s x + k_{s1} x + k_{s2} x$. The *emf* acts against the voltage applied. We have a set of three differential equations:

$$\frac{di_a}{dt} = \frac{\mu_{r2} l_1 + 2\mu_{r1}\mu_{r2} x + \mu_{r1} l_2}{N^2 \mu_0 \mu_{r1}\mu_{r2} A}\left[-r i_a - \frac{2N^2 \mu_0 \mu_{r1}^2 \mu_{r2}^2 A}{\left(\mu_{r2} l_1 + 2\mu_{r1}\mu_{r2} x + \mu_{r1} l_2\right)^2} i_a v + u_a\right],$$

$$\frac{dv}{dt} = \frac{1}{m}\left(-\frac{N^2 \mu_0 \mu_{r1}^2 \mu_{r2}^2 A i_a^2}{\left(\mu_{r2} l_1 + \mu_{r1}\mu_{r2} 2x(t) + \mu_{r1} l_2\right)^2} - k_s x + k_{s1} x + k_{s2} x\right),$$

$$\frac{dx}{dt} = v.$$

■

Example 2.20:

Two coils have a mutual inductance $L_{12} = 0.00005$ H. The current in the first coil is $i_1 = \sqrt{\sin 4t}$. One can find the induced *emf* in the second coil as $\mathscr{E}_2 = L_{12}\dfrac{di_1}{dt}$. By using the power rule, for the time-varying current in the first coil $i_1 = \sqrt{\sin 4t}$, we have $\dfrac{di_1}{dt} = \dfrac{2\cos 4t}{\sqrt{\sin 4t}}$. Hence, $\mathscr{E}_2 = \dfrac{0.0001\cos 4t}{\sqrt{\sin 4t}}$ V. Formally, we found di/dt and $\mathscr{E}$. However, can current, current derivative (rate of change), and *emf* be complex? From engineering prospects, the problem formulations and result are sound if one imposes the following condition $\sin 4t > 0$. ■

2.6 Simulation of Systems in MATLAB Environment

MATLAB (MATrix LABoratory) is a high-performance interacting software environment for high-efficiency engineering and scientific numerical calculations [9,10]. This environment can be applied to perform heterogeneous simulations and data-intensive analysis of electromechanical systems. MATLAB enables the users to solve a wide spectrum of control, optimization, identification, data acquisition, and other problems using matrix-based methods. Excellent interactive capabilities are attained. In addition, it allows compiling features with high-level programming languages, state-of-the-art numerical algorithms, powerful graphical and interface capabilities, etc. Due to high flexibility and versatility, the MATLAB environment has been significantly enhanced in recent years. A family of application-specific toolboxes, with a specialized collection and libraries of m-files for solving problems, guarantees comprehensiveness, and effectiveness. For example, Simulink® is a companion graphical mouse-driven interactive environment enhancing MATLAB. A great number of outstanding books and MathWorks user manuals in MATLAB, Simulink, and different MATLAB toolboxes are available. In addition, the MathWorks Inc. educational Web site can be used, for example, http://education.mathworks.com and http://www.math works. com.

This section introduces the MATLAB environment helping one to use this environment efficiently. The MATLAB environment (version 7.3) is used, and the Web site http://www. mathworks.com/access/helpdesk/help/helpdesk.shtml can assist users to master MATLAB. MATLAB documentation and user manuals (1000 pages each) are available in the Portable Document Format (pdf) using the Help Desk. This book focuses on MATLAB applications to electromechanical systems, educating one how to solve practical problems using step-by-step instructions. This section is not aimed to substitute hundreds of excellent stand-alone books on MATLAB and manuals.

Start MATLAB by double-clicking the MATLAB icon on the desktop. The MATLAB Command Window with Launch Pad and Command History appear on the screen as shown in Figure 2.17. Typing ver or demo, the available toolboxes are listed as reported in Figure 2.17.

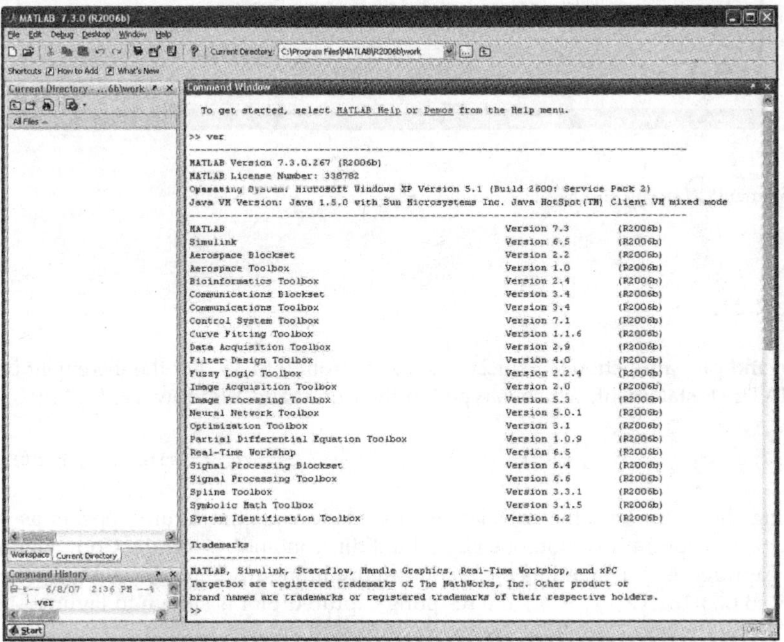

The line

```
>>
```

is the MATLAB prompt.

Typing

```
>> a = 1+2+3*4
```

and pressing the Enter (Return) key, we have the value for a. The following result is displayed

```
a = 15
```

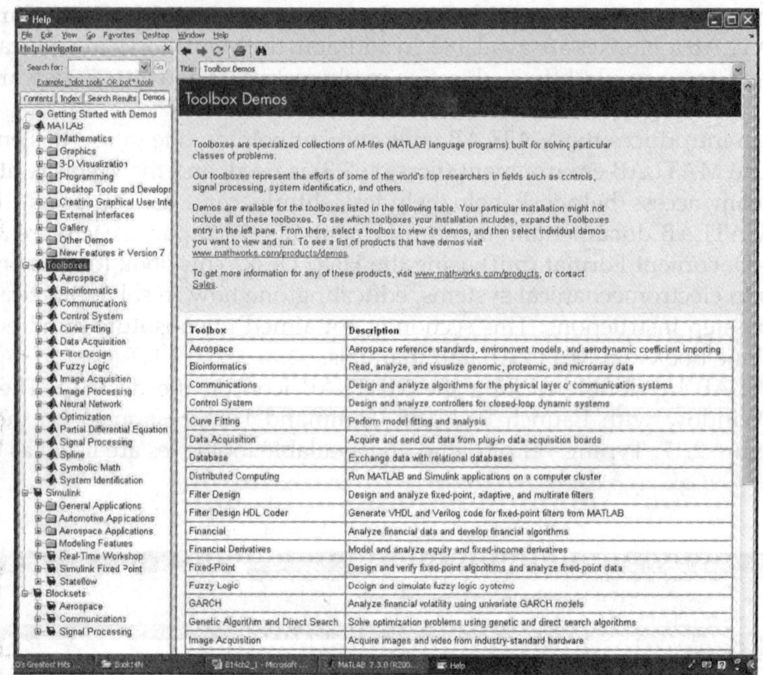

FIGURE 2.17
MATLAB Command Window and MATLAB toolboxes.

Example 2.21:

Calculate and plot a function $y = \sin 2x$, if x varies from 0 to 3π. Let the increment be 0.025π. The MATLAB statement, which is typed in the Command Window, is

```
>> x=0:0.025*pi:3*pi; y=sin(2*x); plot(x,y); title('y=sin(2x)','FontSize',14)
```

By pressing the Enter key, the calculations are made, and the figure appears as shown in Figure 2.18. To capture this plot, one clicks the Edit icon and selects the Copy Figure option. The plot captured is illustrated as the second figure in Figure 2.18. If one refines the plotting statement to be `plot(x, y, 'o')`, the resulting captured plot is shown in Figure 2.18. ∎

To master the plotting, one is advised to type in the Command Window

```
>> help plot
```

and as the Enter key is pressed, the detailed instructions are provided. At the very end, clicking on the doc plot, the Microsoft Word file with the explanations, user options, and various examples is assessable.

The nonlinear differential equations, which describe the dynamics of electromechanical systems, usually cannot be solved analytically. However, for simple equations, analytic solution can be derived using MATLAB. Considering the translational and rotational

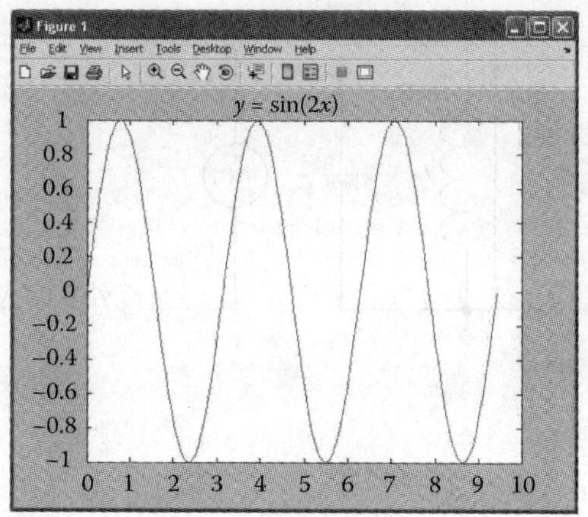

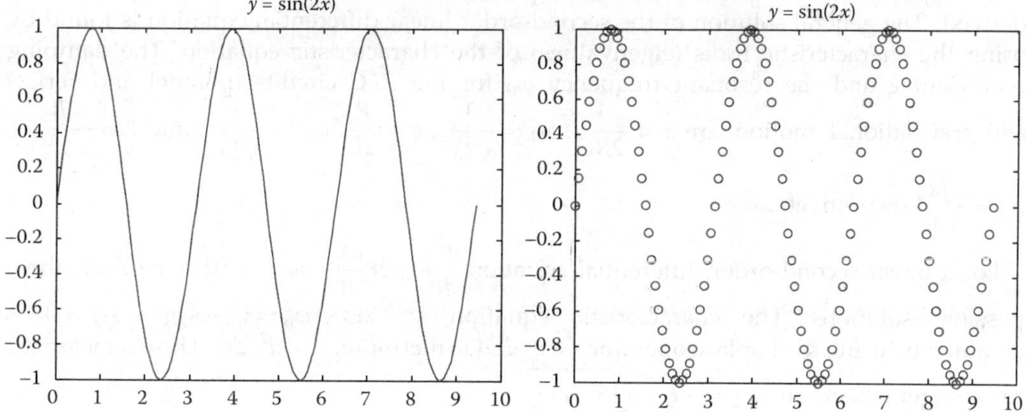

FIGURE 2.18
Plots for $y = \sin 2x$.

rigid-body one-dimensional mechanical systems, we found the second-order differential equations

$$m\frac{d^2x}{dt^2} + B_v\frac{dx}{dt} + k_s x = F_a(t) \quad \text{and} \quad J\frac{d^2\theta}{dt^2} + B_m\frac{d\theta}{dt} + k_s\theta = T_a(t),$$

where $F_a(t)$ and $T_a(t)$ are the time-varying applied force and torque.

For parallel and series *RLC* circuits, illustrated in Figure 2.19, one obtains the following equations:

$$C\frac{d^2u}{dt^2} + \frac{1}{R}\frac{du}{dt} + \frac{1}{L}u = \frac{di_a}{dt} \quad \text{or} \quad \frac{d^2u}{dt^2} + \frac{1}{RC}\frac{du}{dt} + \frac{1}{LC}u = \frac{1}{C}\frac{di_a}{dt},$$

and

$$L\frac{d^2i}{dt^2} + R\frac{di}{dt} + \frac{1}{C}i = \frac{du_a}{dt} \quad \text{or} \quad \frac{d^2i}{dt^2} + \frac{R}{L}\frac{di}{dt} + \frac{1}{LC}i = \frac{1}{L}\frac{du_a}{dt}.$$

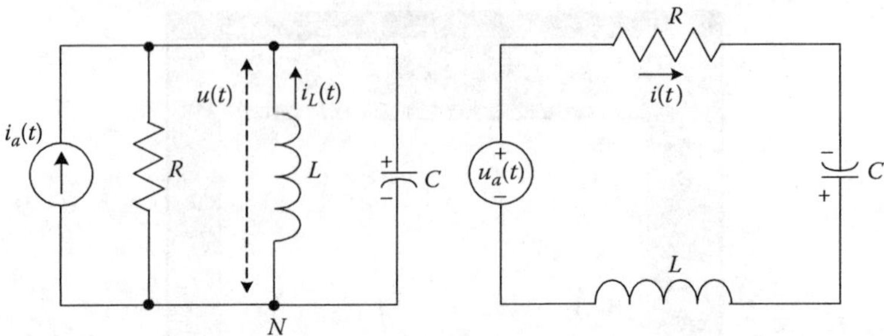

FIGURE 2.19
Parallel and series *RLC* circuits.

The analytic solution of linear differential equations with constant coefficients can be easily derived. The general solution of the second-order linear differential equation is found by using the characteristic roots (eigenvalues) of the characteristic equation. The damping coefficient ξ and the resonant frequency ω_0 for the *RLC* circuits (parallel and series) and translational motion are $\xi = \dfrac{1}{2RC}$, $\omega_0 = \dfrac{1}{\sqrt{LC}}$, $\xi = \dfrac{R}{2L}$, $\omega_0 = \dfrac{1}{\sqrt{LC}}$, and $\xi = \dfrac{B_v}{2\sqrt{k_s m}}$, $\omega_0 = \sqrt{\dfrac{k_s}{m}}$, respectively.

For a linear second-order differential equation $\dfrac{d^2x}{dt^2} + 2\xi\dfrac{dx}{dt} + \omega_0^2 x = f(t)$, one finds three possible solutions. The characteristic equation $s^2 + 2\xi s + \omega_0^2 = (s - s_1)(s - s_2) = 0$ is obtained by using the Laplace operator $s = d/dt$. Furthermore, $s^2 = d^2/dt^2$. The characteristic roots (eigenvalues) are $s_{1,2} = -\xi \pm \sqrt{\xi^2 - \omega_0^2}$.

Case 1. If $\xi^2 > \omega_0^2$, the real distinct characteristic roots s_1 and s_2 result. The general solution is $x(t) = ae^{s_1 t} + be^{s_2 t} + c_f$, where coefficients a and b are obtained using the initial conditions; c_f is the solution due to the *forcing* function f, and for the *RLC* circuits f is $i_a(t)$ or $u_a(t)$.

Case 2. For $\xi^2 = \omega_0^2$, the characteristic roots are real and identical, for example, $s_1 = s_2 = -\xi$. The solution of the second-order differential equation is given as $x(t) = (a + b)e^{-\xi t} + c_f$.

Case 3. If $\xi^2 < \omega_0^2$, the complex distinct characteristic roots are $s_{1,2} = -\xi \pm j\sqrt{\omega_0^2 - \xi^2}$. The general solution is

$$x(t) = e^{-\xi t}\left[a\cos\left(\sqrt{\omega_0^2 - \xi^2}\,t\right) + b\sin\left(\sqrt{\omega_0^2 - \xi^2}\,t\right) \right] + c_f$$

$$= e^{-\xi t}\sqrt{a^2 + b^2}\cos\left[\left(\sqrt{\omega_0^2 - \xi^2}\,t\right) + \tan^{-1}\left(\dfrac{-b}{a}\right)\right] + c_f.$$

Example 2.22:

Consider the series *RLC* circuit illustrated in Figure 2.19. Derive and plot the transient response due to the unit step input with different initial conditions. Let $R = 0.5$ ohm, $L = 1$ H, $C = 2$ F, $a = 1$ and $b = -1$.

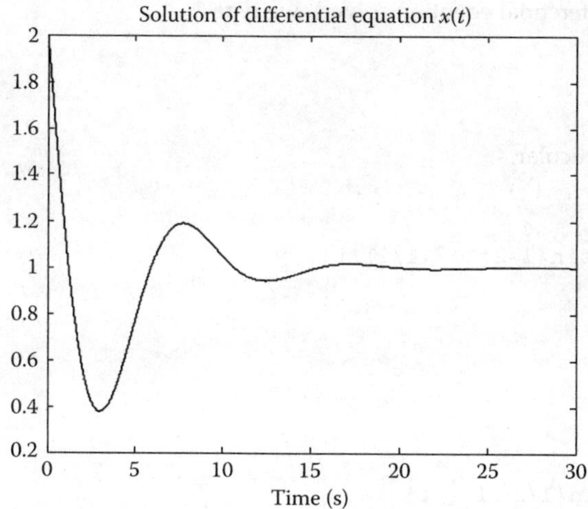

FIGURE 2.20
Dynamics due to the unit step and initial conditions.

The series RLC circuit is described by the differential equation $\dfrac{d^2i}{dt^2} + \dfrac{R}{L}\dfrac{di}{dt} + \dfrac{1}{LC}i = \dfrac{1}{L}\dfrac{du_a}{dt}$ which yields the following characteristic equation $s^2 + \dfrac{R}{L}s + \dfrac{1}{LC} = 0$. The characteristic roots are

$$s_1 = -\frac{R}{2L} - \sqrt{\left(\frac{R}{2L}\right)^2 - \frac{1}{LC}} \quad \text{and} \quad s_2 = -\frac{R}{2L} + \sqrt{\left(\frac{R}{2L}\right)^2 - \frac{1}{LC}}.$$

If $\left(\dfrac{R}{2L}\right)^2 > \dfrac{1}{LC}$, the characteristic roots are real and distinct. For $\left(\dfrac{R}{2L}\right)^2 = \dfrac{1}{LC}$, the characteristic roots are real and identical. If $\left(\dfrac{R}{2L}\right)^2 < \dfrac{1}{LC}$, the characteristic roots are complex.

For the assigned values for R, L, and C one concludes that the characteristic roots are complex, and the dynamics is underdamped because the solution is

$$x(t) = e^{-\xi t}\left[a\cos\left(\sqrt{\omega_0^2 - \xi^2}\,t\right) + b\sin\left(\sqrt{\omega_0^2 - \xi^2}\,t\right)\right] + c_f, \quad \xi = \frac{R}{2L} = 0.25, \quad \text{and} \quad \omega_0 = \frac{1}{\sqrt{LC}} = 0.71.$$

The MATLAB statements are

```
R=0.5; L=1; C=2; a=1; b=-1; cf=1; e=R/2*L; w0=1/sqrt(L*C);
t=0:.01:30;
x=exp(-e*t).*(a*cos(sqrt(w0^2-e^2)*t)+b*sin(sqrt(w0^2-e^2)*t))+cf;
plot(t,x); xlabel('Time (seconds)','FontSize',14);
title('Solution of Differential Equation x(t)','FontSize',14);
```

The resulting dynamics is documented in Figure 2.20.   ∎

Example 2.23:   Application of MATLAB to Analytically Solve Differential Equations

We demonstrate the MATLAB capabilities analytically solving the third-order differential equation $\dfrac{d^3x}{dt^3} + 2\dfrac{dx}{dt} + 3x = 10f$. We will use the Symbolic Math Toolbox.

Using the `dsolve` command (analytic differential equation solver), we type

```
x=dsolve('D3x+2*Dx+3*x=10*f')
```

The resulting solution is displayed. In particular,

```
x =
10/3*f+C1*exp(-t)+C2*exp(1/2*t)*sin(1/2*11^(1/2)*t)
+C3*exp(1/2*t)*cos(1/2*11^(1/2)*t)
```

Using the `pretty` command, we find

```
>> pretty(x)
                                        1/2
10/3 f + C1 exp(-t) + C2 exp(1/2 t) sin(1/2 11    t)
                            1/2
    + C3 exp(1/2 t) cos(1/2 11     t)
```

Hence, $x(t) = \dfrac{10}{3}f + c_1 e^{-t} + c_2 e^{0.5t} \sin\left(\dfrac{1}{2}\sqrt{11}t\right) + c_3 e^{0.5t} \cos\left(\dfrac{1}{2}\sqrt{11}t\right).$

Using the initial conditions, the unknown constants are found. As an example, we assign the following initial conditions $\left(\dfrac{d^2x}{dt^2}\right)_0 = 5$, $\left(\dfrac{dx}{dt}\right)_0 = 15$, and $x_0 = -20$. Using the statement

```
x=dsolve('D3x+2*Dx+3*x=10*f','D2x(0)=5','Dx(0)=15','x(0)=-20'); pretty(x)
```

the resulting solution, with the derived c_1, c_2, and c_3, is

```
10/3 f + (−2 f - 14) exp(-t)
                1/2                              1/2
    − 8/33 11     (f - 3) exp(1/2 t) sin(1/2 11    t)
                                        1/2
    + (- 4/3 f - 6) exp(1/2 t) cos(1/2 11    t)
```

Thus, $x(t) = \dfrac{10}{3}f + (-2f - 14)e^{-t} - \dfrac{8}{33}\sqrt{11}(f-3)e^{0.5t}\sin\left(\dfrac{1}{2}\sqrt{11}t\right) + \left(-\dfrac{4}{3}f - 6\right)e^{0.5t}\cos\left(\dfrac{1}{2}\sqrt{11}t\right).$

If the forcing function is time-varying, the analytic solution of $\dfrac{d^3x}{dt^3} + 2\dfrac{dx}{dt} + 3x = 10f(t)$ is found by using the statement

```
x=dsolve('D3x+2*Dx+3*x=10*f(t)'); pretty(x)
```

If $f(t)$ is defined, and letting, $f(t) = 5\cos(10t)$, by using

```
x=dsolve('D3x+2*Dx+3*x=10*5*cos(5*t)','D2x(0)=5','Dx(0)=15','x(0)=-20');
pretty(x)
```

we found $x(t)$ as

```
  2875                    75                  187
- ----- sin(5 t) + - ----- cos(5 t) - ---- exp(-t)
  6617                  6617                 13

       5702                        1/2    1/2
   + ---- exp(1/2 t) sin(1/2 11    t) 11
       5599

       2864                        1/2
   - ---- exp(1/2 t) cos(1/2 11    t)
       509
```

One concludes that

$$x(t) = -\frac{2875}{6617}\sin 5t + \frac{75}{6617}\cos 5t - \frac{187}{13}e^{-t} + \frac{5702}{5599}e^{0.5t}\sin\left(\frac{1}{2}\sqrt{11}t\right)\sqrt{11}$$
$$- \frac{2864}{509}e^{0.5t}\cos\left(\frac{1}{2}\sqrt{11}t\right).$$

For nonlinear differential equations, which describe the dynamics of electromechanical systems, frequently, analytic solution cannot be derived. Therefore, numerical solutions must be found. The following example illustrates the application of MATLAB to numerically solve the ordinary differential equations.

Example 2.24:

Using the MATLAB `ode45` solver (built-in `ode45` command), numerically solve a system of highly nonlinear differential equations

$$\frac{dx_1(t)}{dt} = -20x_1 + |x_2x_3| + 10x_1x_2x_3, \quad x_1(t_0) = x_{10},$$
$$\frac{dx_2(t)}{dt} = -5x_1x_2 - 10\cos x_1 - \sqrt{|x_3|}, \quad x_2(t_0) = x_{20},$$
$$\frac{dx_3(t)}{dt} = -5x_1x_2 + 50x_2\cos x_1 - 25x_3, \quad x_3(t_0) = x_{30}.$$

The initial conditions are $x_0 = \begin{bmatrix} x_{10} \\ x_{20} \\ x_{30} \end{bmatrix} = \begin{bmatrix} 2 \\ 1 \\ -2 \end{bmatrix}$.

Two m-files (`ch2_1.m` and `ch2_2.m`) are developed in order to numerically simulate this set of nonlinear differential equations. The evolution of the state variables $x_1(t)$, $x_2(t)$, and $x_3(t)$ must be plotted as the differential equations are solved. To illustrate the transient responses for $x_1(t)$, $x_2(t)$, and $x_3(t)$, the `plot` command is used. Comments, which are not executed, appear after the % symbol. These comments explain sequential steps. The MATLAB file (`ch2_1.m`) with the `ode45` solver, two-dimensional plotting statements using the `plot` command, as well as three-dimensional plotting statements using the `plot3` command, is reported below

```
echo on; clear all
t0 = 0; tfinal = 1; tspan = [t0 tfinal];    % initial and final time
y0 = [2 1 -2]';   % initial conditions for state variables
[t,y] = ode45 ('ch2_2',tspan,y0);    %ode45 MATLAB solver using ode45 solver
% Plot of the transient dynamics of the state variables solving differential equations
% These differential equations are assigned in file ch2_2.m
plot (t,y(:,1),'-',t,y(:,2),'-',t,y(:,3),':'); % plot the transient dynamics
xlabel ('Time (seconds)','FontSize',14);
ylabel ('State Variables','FontSize',14);
title ('Solution of Differential Equations: x_1(t), x_2(t) and x_3(t)','FontSize',14);
pause
% 3-D plot using x1, x2 and x3
plot3 (y(:,1),y(:,2),y(:,3));
xlabel ('x_1','FontSize',14); ylabel ('x_2','FontSize',14); zlabel ('x_3','FontSize',14);
title ('Three-Dimensional State Evolutions: x_1(t), x_2(t) and x_3(t)','FontSize',14);
text (0,-2.5,2,'0 Origin','FontSize',14);
```

The second MATLAB file (`ch2_2.m`), with the specified set of differential equations to be numerically solved, is

```
% Simulation of the third-order differential equations
function yprime = difer(t,y);
% Differential equations coefficients
a11 = -20; a12 = 1; a13 = 10; a21 = -5; a22 = -10; a31 = -5; a32 = 50; a33 = -25;
% Three differential equations: System of three first-order differential equations
yprime = [a11*y(1,:)+a12*abs(y(2,:)*y(3,:))+a13*y(1,:)*y(2,:)*y(3,:);  ...% first dif.eq.
a21*y(1,:)*y(2,:)+a22*cos(y(1,:))+sqrt(abs(y(3,:)));  ...% second differential equation
a31*y(1,:)*y(2,:)+a32*cos(y(1,:))*y(2,:)+a33*y(3,:)];% third differential equation
```

To calculate the transient dynamics and plot the transient dynamics, one types in the Command Window

```
ch2_1
```

and presses the Enter key. The resulting transient behavior (two-dimensional plot) is documented in Figure 2.21. The three-dimensional evolution of the state variables is illustrates in Figure 2.21.

Typing in the Command Window

```
who, x = [t,y]
```

the variables used and arrays are displayed. We have

```
Your variables are:
t       t0      tfinal  tspan   y       y0
```

The dynamics of $x_1(t)$, $x_2(t)$, and $x_3(t)$ were illustrated in Figure 2.21. The resulting data for x, which is displayed in the Command Window, is reported below. In particular, we have four columns for time t, as well as for three state variables $x_1(t)$, $x_2(t)$, and $x_3(t)$. That is $x = [t, x_1, x_2, x_3]$.

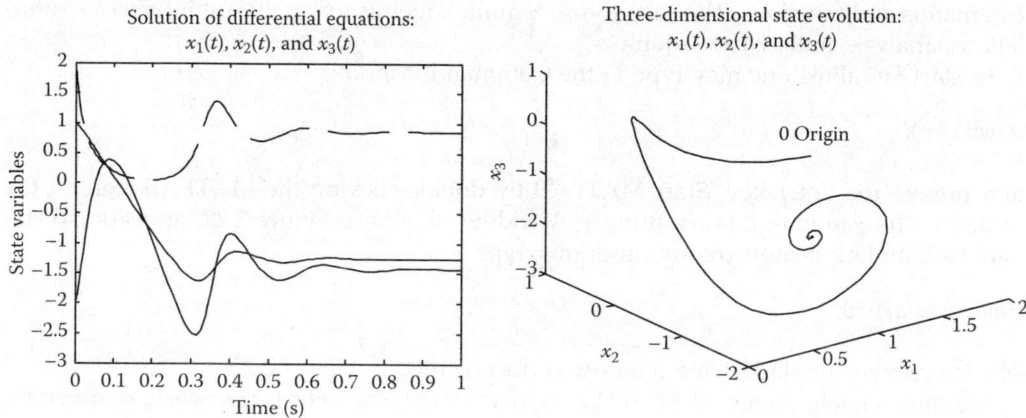

FIGURE 2.21
Dynamics and evolution of the state variables.

```
x =

          0      2.0000      1.0000     -2.0000
     0.0013      1.9025      0.9941     -1.9724
     0.0026      1.8108      0.9876     -1.9392

      . . . . . . . . . . . . . . . . . . . .

     0.9951      0.8539     -1.2543     -1.4336
     0.9976      0.8541     -1.2543     -1.4336
     1.0000      0.8543     -1.2543     -1.4336
```

One can perform plotting, data mining, filtering, and other advanced numerics. ■

Simulink, as a part of the MATLAB environment, is an interactive computing package for simulation dynamic systems. Simulink is a graphical mouse-driven program which allows one to numerically simulate and analyze systems by developing and utilizing block-diagrams. Simulink is applied to linear, nonlinear, continuous-time, discrete-time, multi-variable, multirate, and hybrid systems. Blocksets are built-in blocks in Simulink which provide a full comprehensive block library for different system components, and C-code from block diagrams is generated using the Real-time Workshop Toolbox. Using a mouse-driven block-diagram interface, the Simulink diagrams (models) can be built. These block diagrams (mdl models) represent systems which are described by differential, difference, and constitutive equations. Hybrid and discrete-even systems can be simulated and examined. The distinct advantage is that Simulink provides a graphical user interface (GUI) for building models (block diagrams) using "select-drag-connect-click" mouse-based operations.

A comprehensive library of sinks, sources, linear and nonlinear components (blocks), connectors, as well as customized blocks (S-functions) provide a great flexibility, immense interactability, superior efficiency, and excellent prototyping features. For example, complex systems can be built using high- and low-level blocks. It was illustrated that systems can be numerically simulated by solving differential equations using various MATLAB ode solvers. Different methods and various MATLAB algorithms are embedded and utilized in Simulink. However, one interacts using the Simulink menus rather than entering the

commands and functions. The easy-to-use Simulink menus uniquely suit interactive simulations, analysis, and visualization.

To start Simulink, one may type in the Command Window

```
simulink
```

and presses the Enter key. Start MATLAB by double-clicking the MATLAB icon on the
desktop. The Simulink Library Browser Window, shown in Figure 2.22, appears. To run
various Simulink demonstration programs, type

```
demo simulink
```

The interactive Simulink demo window is documented in Figure 2.22.

Simulink notably extends the MATLAB environment offering a large variety of ready-to-
use building blocks to build diagrams which represent the models derived. One can learn
and explore Simulink using the Simulink and MATLAB Demos. Different MATLAB and
Simulink releases are used. Though there are some differences, the overall coherence
between all releases is ensured. The Simulink manuals are supplied with the software, and

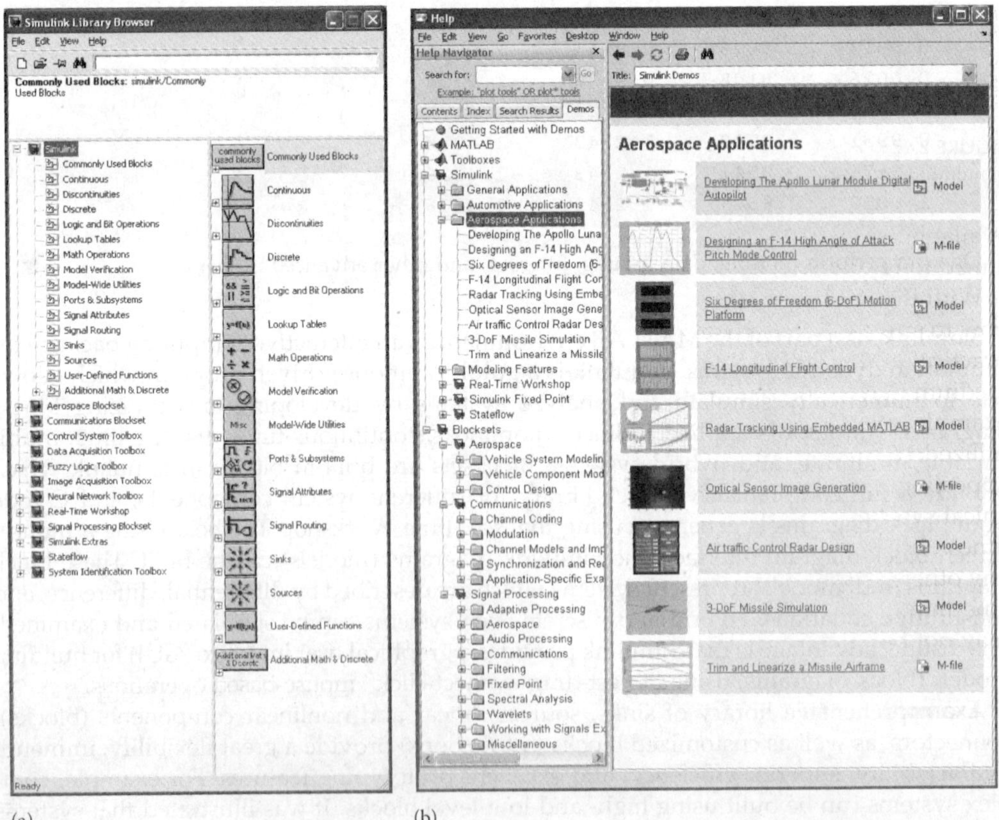

(a)   (b)

FIGURE 2.22

(a) Simulink Library Browser; (b) Simulink demo window.

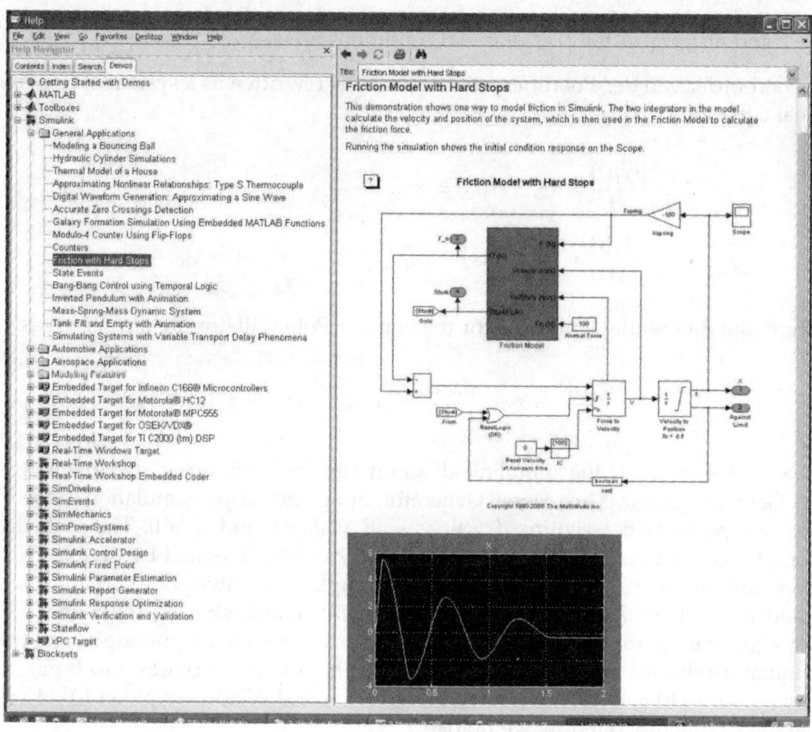

FIGURE 2.23
Simulink demo features.

available in the pdf. These user-friendly manuals can be accessed, and this section is not aimed to rewrite the excellent user manuals.

With the ultimate goal of providing supplementary coverage and educate the reader on how to solve practical problems, we introduce Simulink with step-by-step instructions and practical examples. Figure 2.23 reports the Simulink demo features with various simple, medium complexity, and advanced examples which are ready to be assessed and used. For example, the friction model was covered in Example 2.7. The Index and Search icon can be utilized. As illustrated in Figure 2.23, MATLAB offers the model of friction. Various appealing examples, from aerospace to automotive applications, from electronics to mechanical systems, and other are available. However, the designer must coherently assess the fitness, applicability, and accuracy of MATLAB (Simulink and other toolboxes), as well as any other environment, for specific problems under consideration and objectives targeted. One may find that the presumably ready-to-use files, blocks, diagrams, and other tools may not be sound or require significant refinements.

Example 2.25:   Van der Pol Differential Equations Simulations Using Simulink

The van der Pol oscillator is described by the second-order nonlinear differential equation:

$$\frac{d^2x}{dt^2} - k(1 - x^2)\frac{dx}{dt} + x = d(t),$$

where $d(t)$ is the forcing function.

Let $k = 2$ and $d(t) = d_0 \operatorname{rect}(\omega_0 t)$. The initial conditions are $x_0 = \begin{bmatrix} x_{10} \\ x_{20} \end{bmatrix} = \begin{bmatrix} 1 \\ -1 \end{bmatrix}$.

The second-order van der Pol differential equation is rewritten as a system of two first-order differential equations:

$$\frac{dx_1(t)}{dt} = x_2, \quad x_1(t_0) = x_{10},$$

$$\frac{dx_2(t)}{dt} = -x_1 + kx_2 - kx_1^2 x_2 + d(t), \quad x_2(t_0) = x_{20}.$$

In literature, the differential equations for the van der Pol oscillator are also given as

$$\frac{dx_1(t)}{dt} = x_2, \quad \frac{dx_2(t)}{dt} = \mu\big[(1 - x_1^2)x_2 - x_1\big].$$

As shown in Figure 2.24, the Simulink diagram can be built using the following blocks: Function, Gain, Integrator, Mux, Signal Generator, Sum, and Scope. Simulation of the transient dynamics was performed assigning $d_0 = 0$, as well as $d_0 \neq 0$ and $\omega_0 \neq 0$. The coefficients and initial conditions must be uploaded. The coefficient k can be assigned by double-clicking the Gain block and entering the value needed, for example, one enters 2. Alternatively, one can type k, and in the Command Window type $k = 2$. By double-clicking the Signal Generator block, we select the square function and assign the corresponding magnitude d_0 and frequency ω_0. The initial conditions are set by double-clicking the Integrator blocks and typing x10 and x20. The values for x10 and x20 are entered in the Command Window typing x10 = 1, x20 = −1. Hence, in the Command Window we upload

```
k=2; d0=0; w0=5; x10=1; x20=-1;
```

Specifying the simulation time to be 20 s (see Figure 2.24 where the simulation parameters window is illustrated), the Simulink model is run by clicking the ▦ icon. The simulation results are illustrated in Figure 2.24 which provides the behavior of two variables displayed by three Scopes.

The plotting statements can be effectively used. In the Scopes, in the General/Data History icon (the second icon to the left), one can specify the variable names. The variables x_1, x_2, and x_{12} are specified in the first, second, and third scopes and stored. By using

```
plot(x12(:,1),x12(:,2),x12(:,1),x12(:,3));
xlabel('Time (seconds)','FontSize',14);
ylabel('State Variables x_1 and x_2','FontSize',14);
title('Solution of van der Pol Equation: x_1(t) and x_2(t)','FontSize',14);
```

the plotting is performed. The resulting plot for x_1 and x_2 are illustrated in Figure 2.24.

Many illustrative and valuable examples are given in the MATLAB and Simulink demos. The van der Pol equations simulations are covered. In the MATLAB demo, the following set of differential equations $\dfrac{dx_1(t)}{dt} = x_2, \dfrac{dx_2(t)}{dt} = -x_1 + x_2 - x_1^2 x_2$ is simulated. ∎

All demo Simulink models can be modified for the specific problems. To start, stop, or pause the simulation, the Start, Stop, and Pause buttons are available in the Simulation menu (Start, Stop, and Pause buttons can be clicked using the toolbar commands as well).

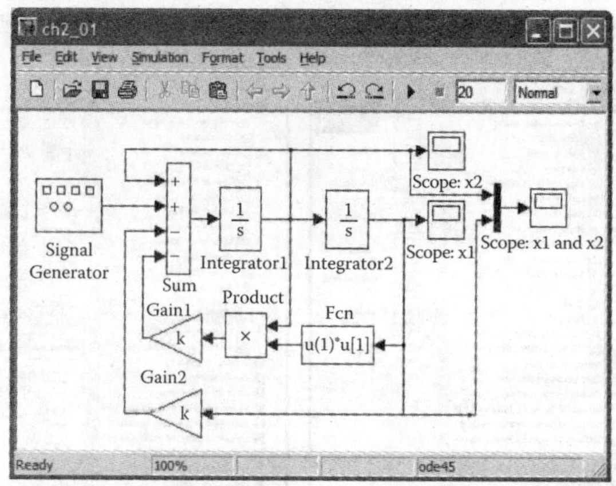

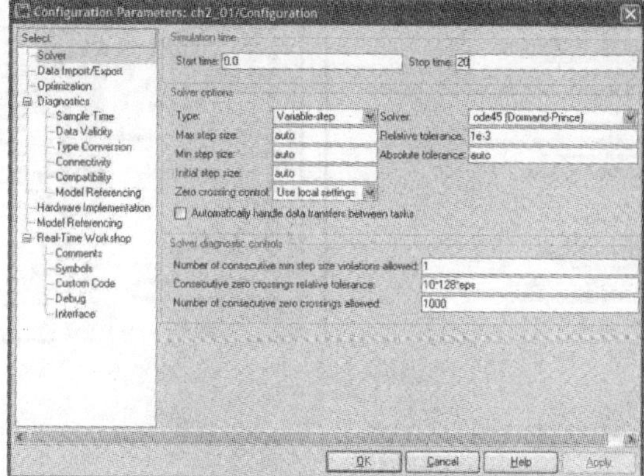

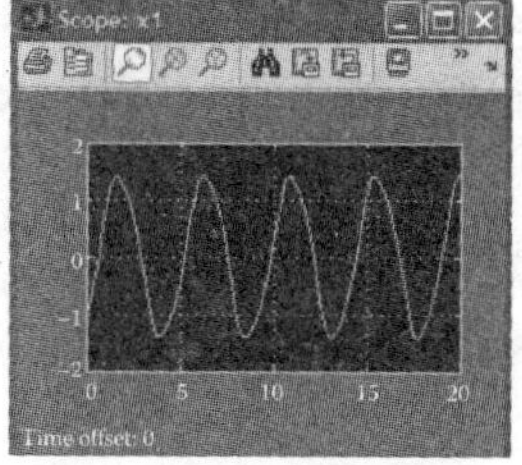

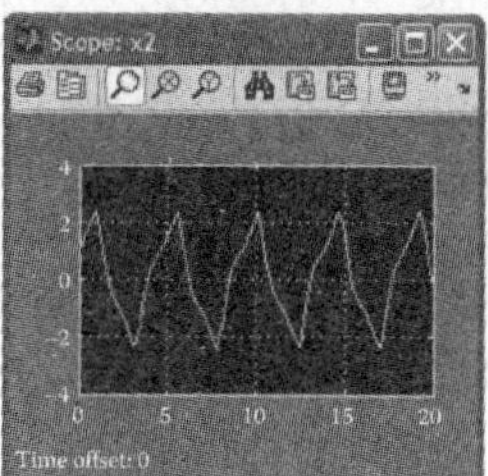

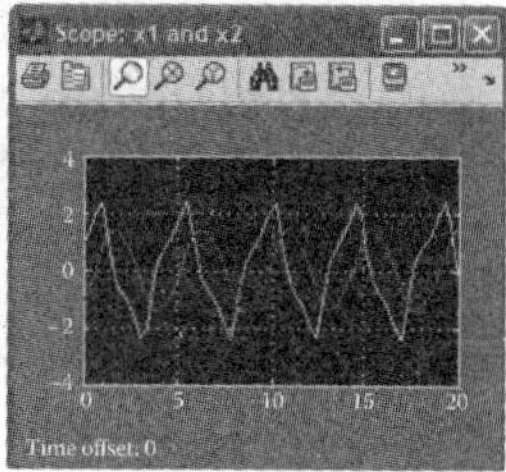

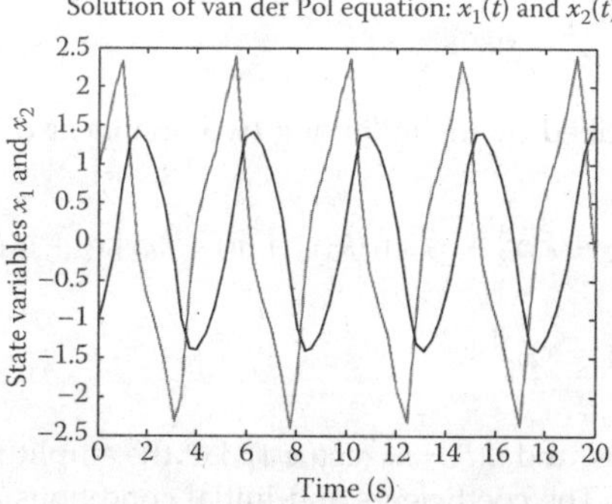

FIGURE 2.24

Simulink block diagram (`ch2_01.mdl`), simulation configuration, transient dynamics displayed in the scopes, and plot for the state variables $x_1(t)$ and $x_2(t)$.

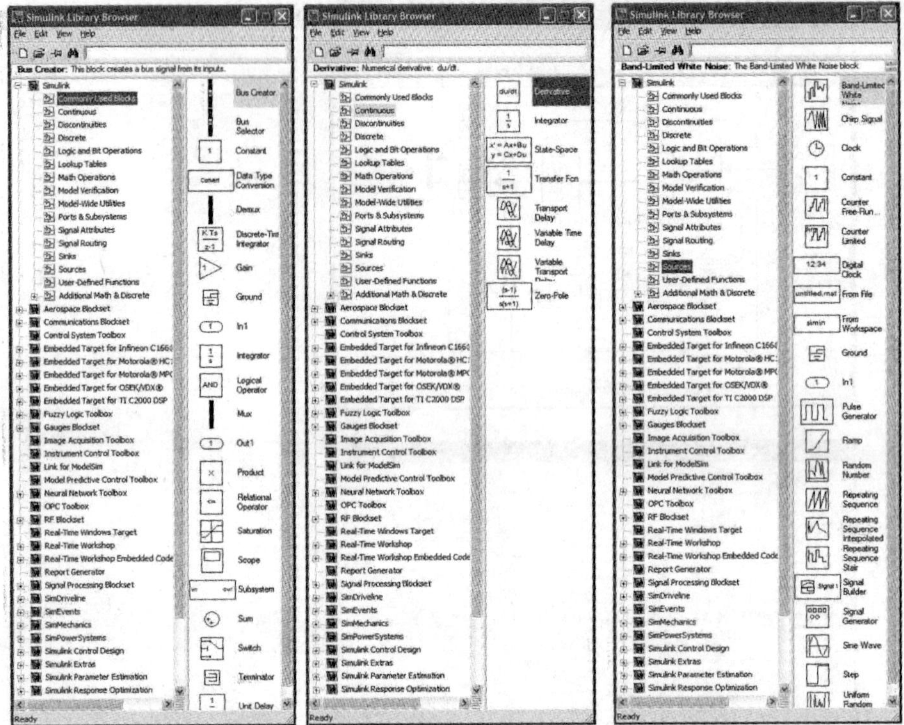

FIGURE 2.25
Commonly used Blocks, Continuous, and Sources Simulink libraries.

One can open Aerospace, Real Time Workshop, SimMechanics, and other toolboxes as documented in the Simulink demo features in Figure 2.23. The application-specific built-in blocks are available in the Simulink Library Browser. For example, the Commonly Used Blocks, Continuous, and Sources, which were used to build the Simulink diagram reported in Figure 2.24, are illustrated in Figure 2.25. The designer can utilize other libraries depending on the problem under the consideration. One selects and drags those blocks to the Simulink diagram, and, then connects the blocks in order to perform simulations.

Example 2.26:

We simulate the system described by the following two nonlinear differential equations:

$$\frac{dx_1(t)}{dt} = -k_1 x_1 - k_2 x_2 + k_3 x_2^3 + k_4 \sin(k_5 x_1 + \pi) + k_6 x_1 x_2 + u(t), \quad x_1(t_0) = x_{10},$$

$$\frac{dx_2(t)}{dt} = k_7 x_1, \quad x_2(t_0) = x_{20}.$$

The input $u(t)$ is a train of steps, and $u(t) = u_0 \, \text{rect}(\omega_0 t)$. Let the amplitude is $u_0 = 0$ or $u_0 = 4$, and the frequency is $\omega_0 = 1$ rad/s. The coefficients and initial conditions are $k_1 = 2$, $k_2 = 3$, $k_3 = -4$, $k_4 = -5$, $k_5 = 6$, $k_6 = -7$, $k_7 = 8$, $x_{10} = 2$, and $x_{20} = -2$.

We use the Signal Generator, Sum, Gain, Integrator, Function, Mux and Scope blocks. These blocks are dragged from the Simulink block libraries to the untitled mdl model, positioned (placed), and connected using the lines as shown in Figure 2.26. That is, by connecting the blocks and typing the coefficients and nonlinear term $k_3 x_2^3 + k_4 \sin(k_5 x_1 + \pi) + k_6 x_1 x_2$ in the Function block, the Simulink block diagram to be used results as shown in Figure 2.26.

The differential equations parameters and initial conditions are uploaded by typing in the Command Window

```
k1 = 2; k2 = 3; k3 = −4; k4 = −5; k5 = 6; k6 = −7; k7 = 8; u0 = 0; w0 = 1; x10 = 2; x20 = −2;
```

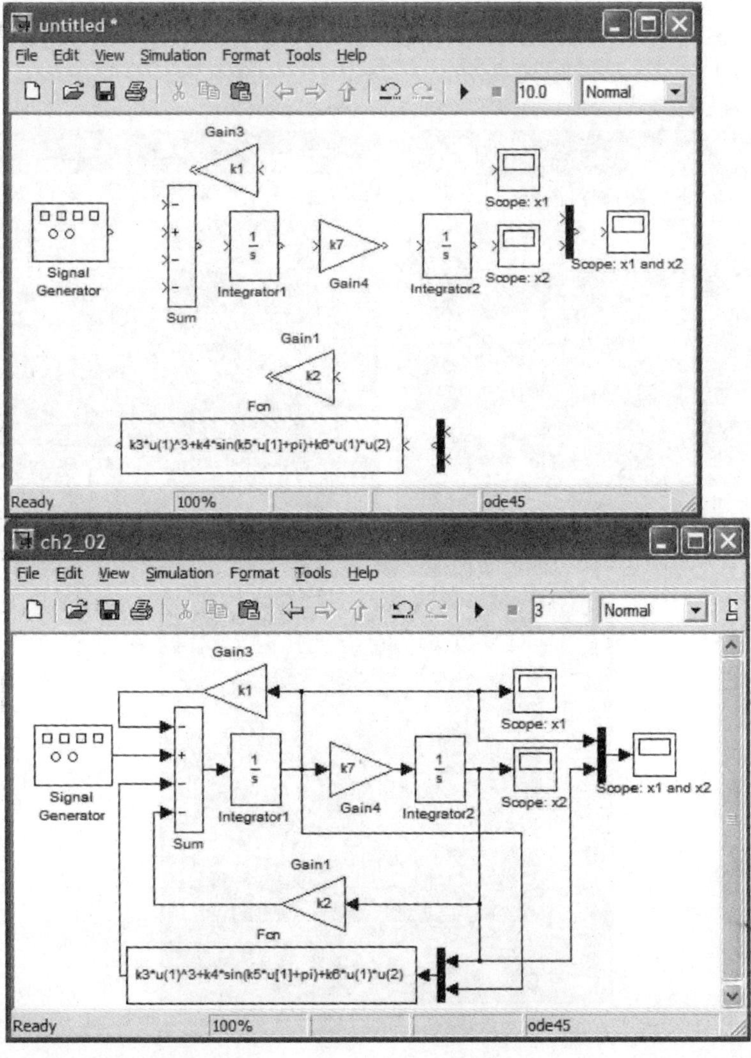

FIGURE 2.26

Simulink block diagram to simulate a set of two differential equations (ch2_02.mdl).

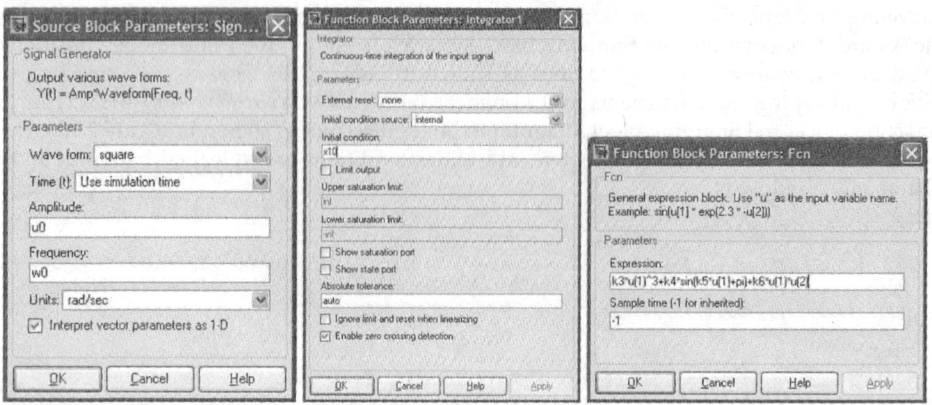

FIGURE 2.27
Signal generator, integrator, and function blocks.

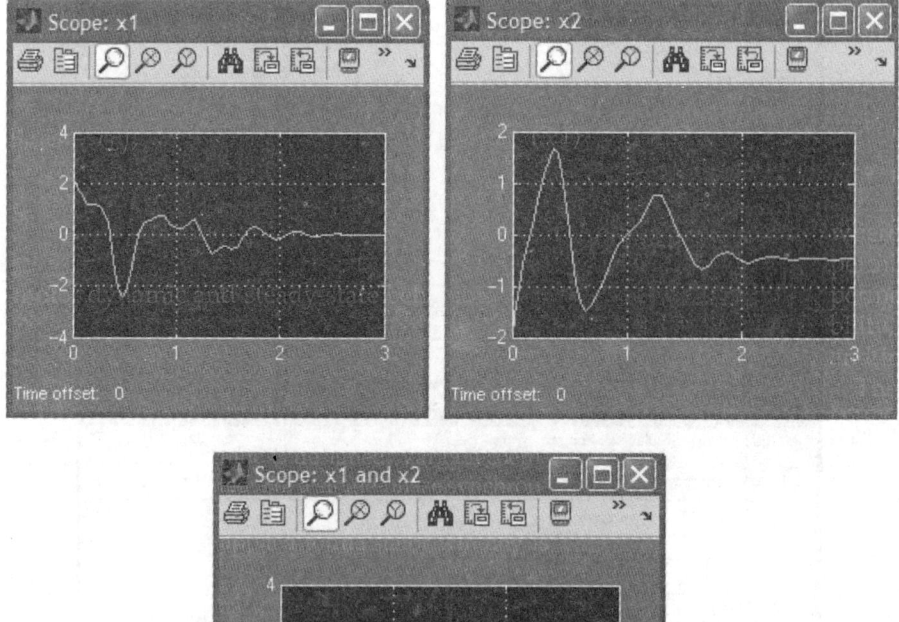

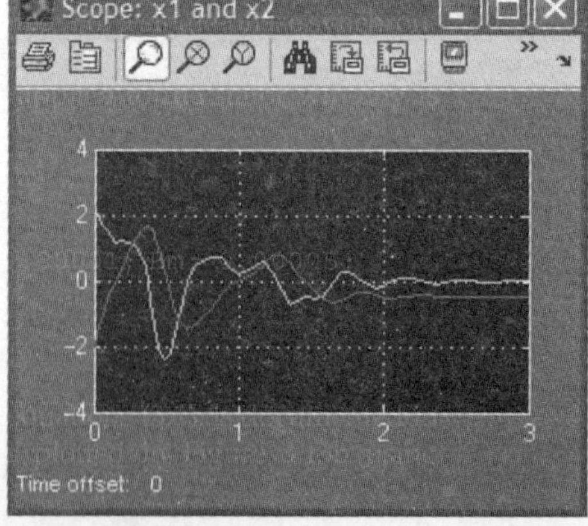

FIGURE 2.28
Simulation results displayed in the Scopes.

The Signal Generator, Integrator, and Function blocks are used to generate the input $u(t)$, set up the specified initial conditions in the integrators, as well as implement the nonlinear functions. These blocks are illustrated in Figure 2.27.

For $u_0 = 0$, specifying the simulation time to be 2 s, the transient behavior of the states $x_1(t)$ and $x_2(t)$, as provided by three scopes, are depicted in Figure 2.28.

The plotting statements can be used. In the scopes, in the General/Data History icon (the second icon from the left), specify the variable to be x1, x2, and x12. The statements to plot $x_1(t)$ are

```
plot(x1(:,1),x1(:,2)); xlabel('Time (seconds)','FontSize',14);
ylabel('State Variable x_1','FontSize',14);
title('Solution of Differential Equations: x_1(t)','FontSize',14);
```

Similar statements to plot $x_2(t)$. The resulting plots are illustrated in Figure 2.29.

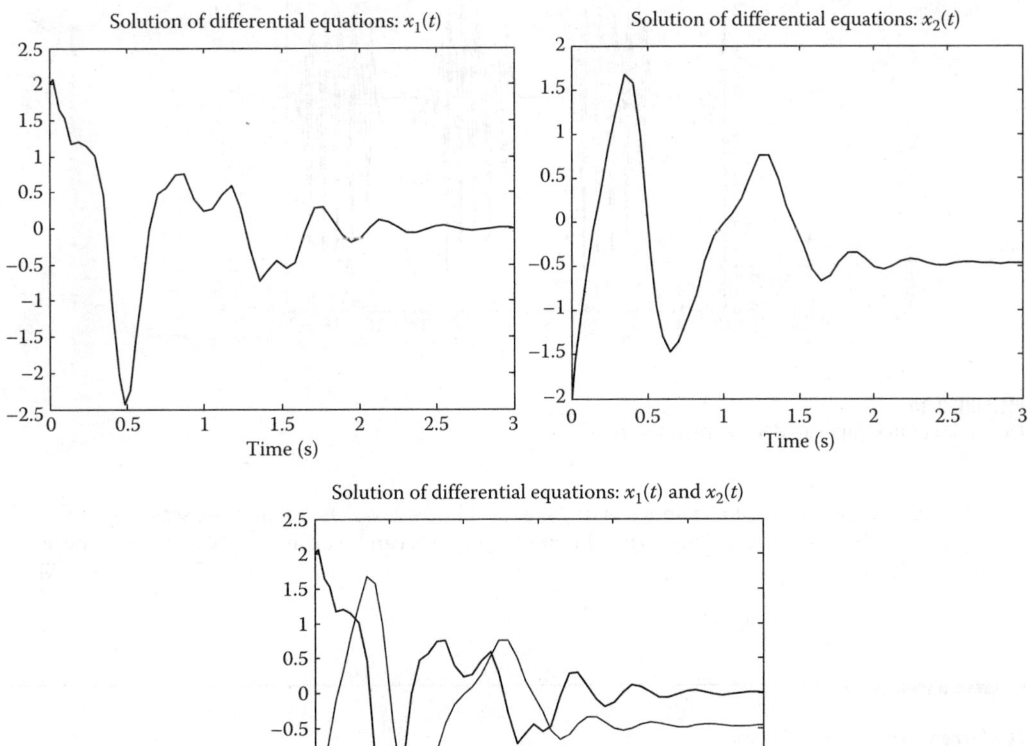

FIGURE 2.29
Dynamics of the state variables $x_1(t)$ and $x_2(t)$, $u_0 = 0$.

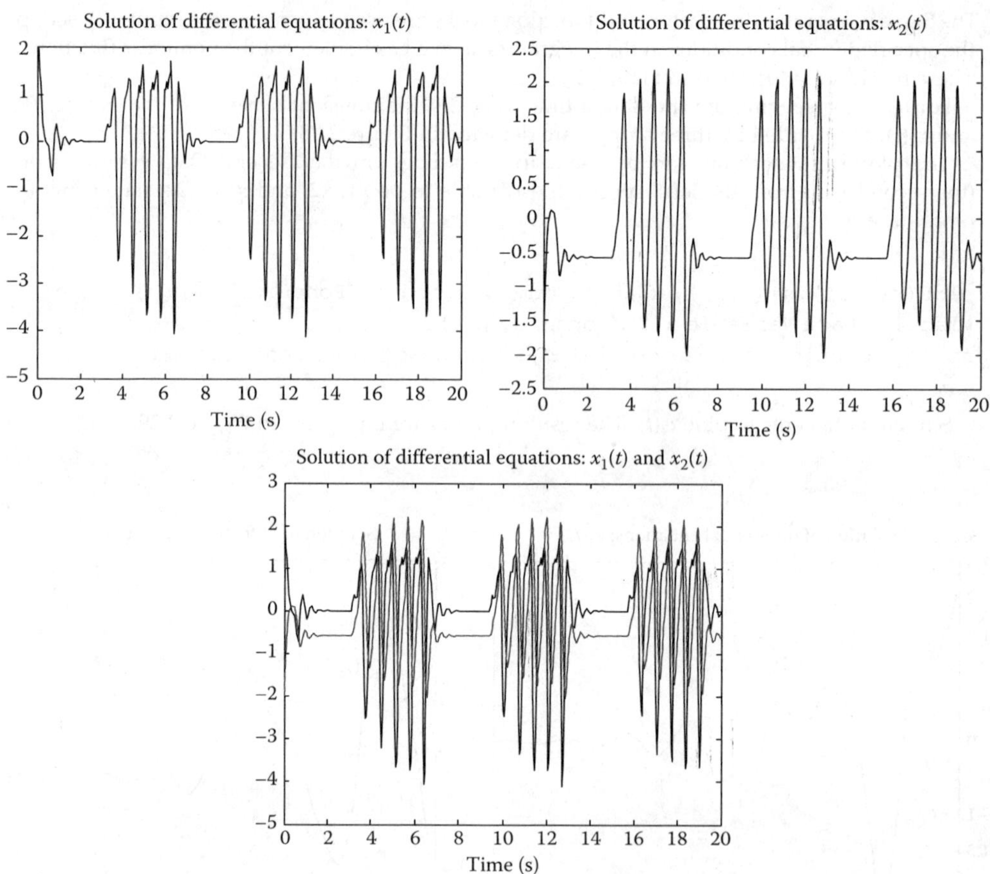

FIGURE 2.30
Dynamics of the state variables $x_1(t)$ and $x_2(t)$, $u_0 = 4$.

For $u_0 = 4$, the transient dynamics is depicted in Figure 2.30. The reader assesses the impact of input on the system transients. The obtained dynamics can be examined from the prospects of stability. ∎

References

1. Chapman, S. J., *Electric Machinery Fundamentals*, McGraw-Hill, New York, 1999.
2. Fitzgerald, A. E., Kingsley, C., and Umans, S. D., *Electric Machinery*, McGraw-Hill, New York, 1990.
3. Krause, P. C. and Wasynczuk, O., *Electromechanical Motion Devices*, McGraw-Hill, New York, 1989.
4. Krause, P. C., Wasynczuk, O., and Sudhoff, S. D., *Analysis of Electric Machinery*, IEEE Press, New York, 1995.
5. Leonhard, W., *Control of Electrical Drives*, Springer, Berlin, 1996.

6. Lyshevski, S. E., *Electromechanical Systems, Electric Machines, and Applied Mechatronics*, CRC Press, Boca Raton, FL, 1999.
7. Slemon, G. R., *Electric Machines and Drives*, Addison-Wesley Publishing Company, Reading, MA, 1992.
8. White, D. C. and Woodson, H. H., *Electromechanical Energy Conversion*, Wiley, New York, 1959.
9. Lyshevski, S. E., *Engineering and Scientific Computations Using MATLAB®*, Wiley, Hoboken, NJ, 2003.
10. *MATLAB R2006b*, CD-ROM, MathWorks, Inc., Natick, MA, 2007.

3

Electrostatic and Variable Reluctance Electromechanical Motion Devices

3.1 Introduction

Electrostatic and electromagnetic devices are among the major classes of electromechanical motion devices. The electromagnetic devices, as compared to the electrostatic devices, ensure the superior force/torque and power densities. The stored electric and magnetic volume energy densities ρ_{We} and ρ_{Wm} are

- $\rho_{We} = \frac{1}{2}\varepsilon E^2$ for electrostatic transducers
- $\rho_{Wm} = \frac{1}{2}\dfrac{B^2}{\mu} = \frac{1}{2}\mu H^2$ for magnetic (electromagnetic) transducers

where ε is the permittivity, $\varepsilon = \varepsilon_0\varepsilon_r$; ε_0 and ε_r are the permittivity of free space and relative permittivity, respectively, and $\varepsilon_0 = 8.85\times10^{-12}$ F/m, E is the electric field intensity, μ is the permeability, $\mu = \mu_0\mu_r$; μ_0 and μ_r are the permeability of free space and relative permeability, respectively, and $\mu_0 = 4\pi \times 10^{-7}$ T-m/A, and B and H are the magnetic field density and intensity.

The electrostatic actuators are commonly utilized in microelectromechanical systems (MEMS) due to the applicability of the well-established complementary metal-oxide-semiconductor (CMOS) technology that has been enhanced to fabricate MEMS using *surface* and *bulk* micromachining. The fabrication issues are covered in Chapter 13. The maximum energy density of electrostatic actuators is limited by the maximum field (voltage), which can be applied before electrostatic breakdown occurs. In mini- and microstructures, the maximum electric field is constrained resulting in the maximum energy density ρ_{Wemax}. For example, in 100×100 μm to millimeter size structures with a few micrometer airgap, E_{max} is $\sim3\times10^6$ V/m. For $\varepsilon_r = 2.5$, one estimates ρ_{Wemax} to be ~100 J/m^3. In contrast, for electromagnetic actuators, the maximum energy density ρ_{Wmmax} is limited by saturation flux density B_{sat} (B_{sat} could be ~2.5 T) and material permeability (μ_r varies from 1,000 to 1,000,000). Thus, the resulting magnetic energy density ρ_{Wmmax} could be $\sim100,000$ J/m^3, and $\rho_{We} \ll \rho_{Wm}$. The electromagnetic transducers can store energy 1000 times larger than electrostatic. The ratio $\rho_{Wm}/\rho_{We} > 100$ is guaranteed even for the most favorable microscale actuators dimensions if the soft magnets (Fe, Ni, or NiFe) with low B_{sat} are used. The application of hard magnets, in most cases, ensures ρ_{Wm}/ρ_{We} to be ~1000.

One must perform a coherent design and technology assessment integrating electronics, actuators, and mechanism kinematics. For example, electrostatic actuators may ensure favorable kinematics, fabrication, actuator–ICs integration, and packaging solutions. This may result in significant reduction in the force or torque required. Therefore, in MEMS, the use of electrostatic actuators could be a favorable solution that may ensure the desired performance. The translational and rotational electrostatic actuators are studied in this chapter. For example, $\sim 1,000,000$ micromirrors (each $\sim 10 \times 10$ μm), controlled by a DSP, are repositioned in the Texas Instruments digital light processing (DLP) module used in high-definition displays and projection systems. Each electrostatic microactuator is controlled, and the settling time is ~ 0.0001 s. To actuate any mechanism, the developed electrostatic force or torque must be greater than the load force or torque. The actuator integration, packaging, housing, and other aspects are examined. Hence, while in conventional and miniscale mechatronic systems, the electromagnetic actuators ensure the superiority, in MEMS, the electrostatic, thermal, and piezoelectric actuators may be a preferable solution.

Using the CMOS technology integrated with the *bulk* and *surface* micromachining processing, the images of fabricated structures, electromagnetic actuators (with deposited coils on ~ 30 μm silicon diaphragm), and electrostatic actuator are reported in Figure 3.1. For the micromachined electromagnetic actuators, as documented on Figure 3.1b, the deflection of the diaphragm is measured by using the variations of resistances of four polysilicon resistors that form the Wheatstone bridge.

By using $W_e = \frac{1}{2} \int_v \rho_v V dv$, the potential energy stored in the electric field between two surfaces (capacitor) is $W_e = \frac{1}{2} QV = \frac{1}{2} CV^2$. The force- and torque-energy relations were reported in Chapter 2. The energy stored in the capacitor is $\frac{1}{2} CV^2$, while the energy stored in the inductor is $\frac{1}{2} Li^2$. The energy in the capacitor is stored in the electric field between plates, while the energy in the inductor is stored in the magnetic field within the coils.

Considering device physics of electromechanical motion devices [1–6], the major torque production and energy conversion mechanisms that lead to the corresponding operation should be emphasized. In particular, the devise physics of electromagnetic devices is based on

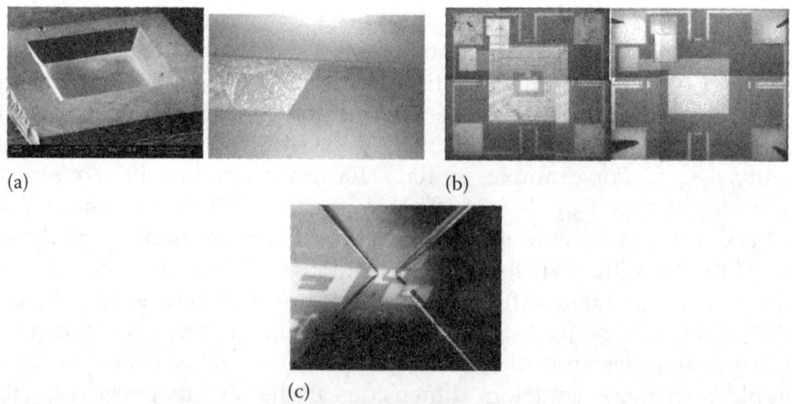

(a)   (b)

(c)

FIGURE 3.1
(a) Etched silicon structure and cross-section of a 35 μm silicon diaphragm; (b) micromachined $\sim 1 \times 1$ mm electromagnetic and thermal actuators with displacement sensors; (c) micromachined electrostatic actuator with the suspended (released) plate. Voltage is applied to displace the top plate.

- *Induction electromagnetics*: The phase voltages are induced in the rotor windings due to the time-varying stator magnetic field and motion of the rotor with respect to the stator, the electromagnetic torque (force) results due to the interaction of time-varying electromagnetic fields.

- *Synchronous electromagnetics*: The torque (force) results due to the interaction of time-varying magnetic field established by the stator windings and stationary magnetic field established by the windings or magnets on the rotor.

- *Variable reluctance electromagnetics*: The force (torque) is produced to minimize the reluctance of the electromagnetic system (solenoids, relays, etc.), and the torque is created by the magnetic system in an attempt to align the minimum-reluctance path of the rotor with the time-varying rotating airgap *mmf* (synchronous electric machine).

In general, permanent magnet electromechanical motion devices surpass other actuator solutions such as induction and variable reluctance. However, various variable reluctance motion devices are utilized in many systems. Different radial and axial topologies are covered for translational and rotational electromechanical motion devices. The following sequential procedure can be utilized to ensure a coherent design flow:

1. Define application, environmental, and other requirements
2. Specify performance specifications and capabilities
3. Devise, advance, or examine existing electromechanical motion devices examining device physics, operating principles, topologies, electromagnetic systems, etc.
4. Perform electromagnetic, energy conversion, mechanical, thermal, and sizing-dimensional estimates
5. Based upon data-intensive analysis, design a device (derive windings, determine airgap, select permanent magnets, define materials, etc.)
6. Define processes and technologies to fabricate structures (stator and rotor with windings or magnets, bearing, etc.), assemble and package them
7. Perform a coherent electromagnetic, mechanical, vibroacoustic, and thermodynamic design and optimization with performance and capabilities analysis
8. Define matching power electronics and sound control solutions (this task itself can be partitioned to many subtasks and problems related to power converter topologies, control laws design, controller implementation, circuitry design, ICs fabrication, actuator–sensor–ICs integration, etc.)
9. Integrate all modules, devices, and components
10. Test, characterize, and evaluate devices, modules, and system
11. Optimize and redesign system ensuring best performance and *achievable* capabilities

We study the electromechanical motion devices that are profoundly different as compared to biological motion devices. For example, *Escherichia coli* bacteria has biomotors made from proteins with a diameter ∼50 nm. The motors rotate flagella that vary their geometry to perform propulsion. The device physics of these motors is not comprehended. These devices unlikely utilize magnets or windings. One may expect that biosystems do

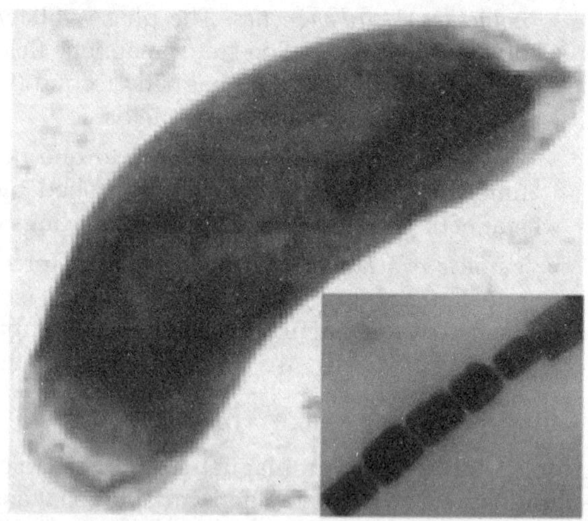

FIGURE 3.2
Magnetotactic bacterium and image of a chain of 60–100 nm diameter cylindrical magnetosome mineral magnetic particles (rectangular, octahedral, prismatic, and other shapes of 30–100 nm magnetosome particles exist).

not use magnets or magnetic materials. In 1962, Professor Heinz A. Lowenstam discovered magnetite (Fe_3O_4) biomineralization in the teeth of chitons (mollusks of the class *Polyplacophora*) demonstrating that living organisms are able to precipitate the mineral magnetite. Another intriguing finding was the discovery of the magnetotactic bacteria by Richard Blakemore in 1975. Three-billion-year-evolved magnetotactic bacteria contain magnetosomes (magnetic mineral particles) enclosed in the protein-based membranes, Figure 3.2. In most cases, the magnetosomes are arranged in a chain or chains fixed within the cell. In many magnetotactic bacteria, the magnetosome mineral particles are either 30–100 nm magnetite (Fe_3O_4) or, in marine and sulfidic environments, greigite (Fe_3S_4). These magnets interact with an external magnetic field, and bacteria migrate along the magnetic field lines. The chain of magnetosome particles constitutes a permanent magnetic dipole fixed within the bacterium. The magnetosome particles are uniformly magnetized forming permanent magnetic domains. All particles are arranged along the chain axis such that the crystallographic magnetic easy axes are aligned.

3.2 Electrostatic Actuators

We consider the translational and rotational actuators. The translation electrostatic actuators are widely used as microactuators, and high-yield micromachining technologies are utilized to achieve the affordable fabrication of high-performance MEMS. The images of MEMS are documented in Figure 3.1.

Example 3.1:

Consider the capacitor (the plates have area A and they are separated by x), which is charged to a voltage V. The permittivity of the dielectric is ε. We find the stored electrostatic energy and the force F_{ex} in the x direction.

Neglecting the fringing effect at the edges, one concludes that the electric field is uniform, and $E = V/x$. Therefore,

$$W_e = \frac{1}{2}\int_v \varepsilon\left|\vec{E}\right|^2 dv = \frac{1}{2}\int_v \varepsilon\left(\frac{V}{x}\right)^2 dv = \frac{1}{2}\varepsilon\frac{V^2}{x^2}Ax = \frac{1}{2}\varepsilon\frac{A}{x}V^2 = \frac{1}{2}C(x)V^2,$$

where $C(x) = \varepsilon A/x$.

The force is

$$F_{ex} = \frac{\partial W_e}{\partial x} = \frac{\partial\left[\frac{1}{2}C(x)V^2\right]}{\partial x} = \frac{1}{2}V^2\frac{\partial C(x)}{\partial x}.$$

Using the expression for $C(x)$, one finds

$$F_{ex} = \frac{1}{2}V^2\frac{\partial C(x)}{\partial x} = -\frac{1}{2}\varepsilon A\frac{1}{x^2}V^2.$$

The force is a nonlinear function of the voltage applied V and displacement x. ◼

Example 3.2:

Rotational electrostatic motors have been widely examined as microelectromechanical motion devices. The cross-sectional view of the electrostatic motor, as well as $\sim$500 μm diameter micromachined motor, are shown in Figure 3.3.

As the voltage V is applied to the parallel conducting rotor and stator plates, the charge is $Q = CV$, where C is the capacitance:

$$C = \varepsilon\frac{A}{g} = \varepsilon\frac{WL}{g}$$

where A is the overlapping area of the plates, $A = WL$, W and L are the width and length of the plates, respectively, ε is the permittivity of the media between the plates, $\varepsilon = \varepsilon_0\varepsilon_r$; $\varepsilon_0 = 8.85 \times 10^{-12}$ C^2/N-m$^2 = 8.85\times10^{-12}$ F/m and $\varepsilon_r = 1$, and g is the airgap between the plates.

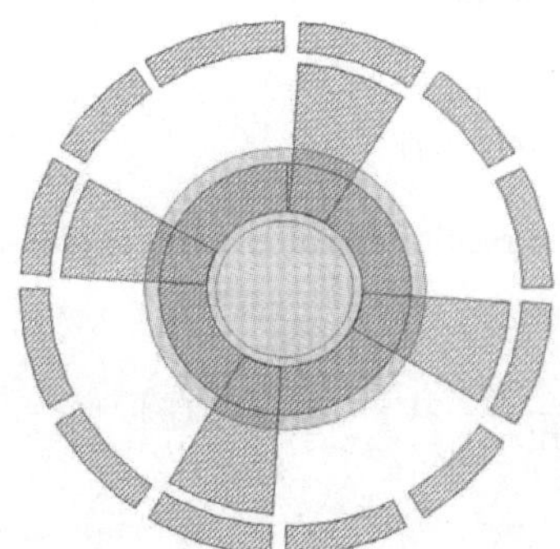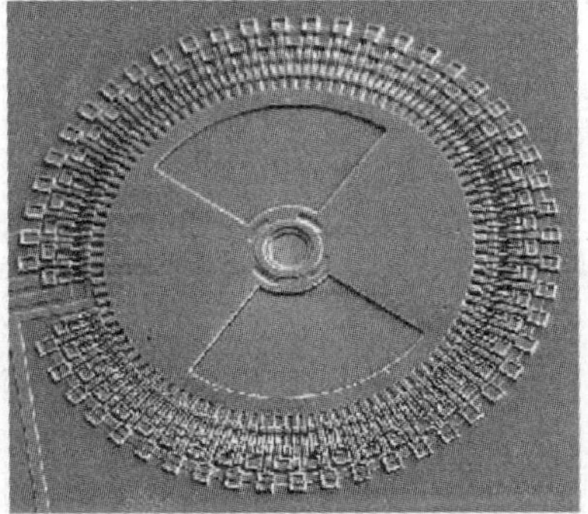

FIGURE 3.3
Electrostatic motor.

The energy associated with electric potential is $W_e = \frac{1}{2}CV^2$. The electrostatic force at each overlapping plate segment $F_{el} = \frac{\partial W_e}{\partial g} = -\frac{1}{2}\frac{\varepsilon WL}{g^2}V^2$ is balanced by the opposite segment (we assume an ideal fabrication for which W, L, and g are the same for all segments).

The tangential force due to misalignment is $F_t = \frac{\partial W_e}{\partial x} = \frac{1}{2}\frac{\varepsilon}{g}\frac{\partial (WL)}{\partial x}V^2$, where x is the direction in which misalignment potentially occurs. If the misalignment occurs in the width direction, one has $F_{t,w} = \frac{\partial W_e}{\partial x} = \frac{1}{2}\frac{\varepsilon L}{g}\frac{\partial W(x)}{\partial x}V^2$.

The capacitance of a cylindrical capacitor is found to derive the electrostatic torque that drives the rotor. The voltage between the cylinders is obtained by integrating the electric field. The electric field at a distance r from a conducting cylinder has only a radial component denoted as E_r. We have $E_r = \frac{\rho}{2\pi\varepsilon r}$, where ρ is the linear charge density, and $Q = \rho L$. Hence, the potential difference is

$$\Delta V = V_a - V_b = \int_a^b \vec{E}d\vec{l} = \int_a^b E_r dr = \frac{\rho}{2\pi\varepsilon}\int_{r_1}^{r_2} \frac{1}{r}dr = \frac{\rho}{2\pi\varepsilon}\ln\frac{r_2}{r_1},$$

where r_1 and r_2 are the radii of the rotor and stator where the plates are positioned. Thus,

$$C = \frac{Q}{\Delta V} = \frac{2\pi\varepsilon L}{\ln(r_2/r_1)}.$$

The capacitance per unit length is

$$\frac{C}{L} = \frac{\rho}{\Delta V} = \frac{2\pi\varepsilon}{\ln(r_2/r_1)}.$$

Using the stator–rotor plates overlap, for the rotational electrostatic motor, the capacitance can be expressed as a function of the angular displacement as

$$C(\theta_r) = N\frac{2\pi\varepsilon}{\ln(r_2/r_1)}\theta_r,$$

where N is the number of overlapping stator–rotor plates.

Taking note of the derived $C(\theta_r)$, the electrostatic torque developed is found to be

$$T_e = \frac{1}{2}\frac{\partial C(\theta_r)}{\partial\theta_r}V^2 = N\frac{\pi\varepsilon}{\ln(r_2/r_1)}V^2.$$

Other expressions for capacitances $C(\theta_r)$ can be found resulting in alternative expressions for T_e.

The *torsional–mechanical* equations of motion are

$$\frac{d\omega_r}{dt} = \frac{1}{J}(T_e - B_m\omega - T_L) = \frac{1}{J}\left(N\frac{\pi\varepsilon}{\ln(r_2/r_1)}V^2 - B_m\omega_r - T_L\right),$$

$$\frac{d\theta_r}{dt} = \omega_r.$$

Let us study some performance estimates and design steps. To rotate the motor, inequality $T_e > T_L + T_{\text{friction}}$ must be guaranteed. That is, motor must develop the electrostatic torque

$T_e = N \dfrac{\pi \varepsilon}{\ln(r_2/r_1)} V^2$ higher than the maximum (or rated) load torque. Estimating the load

torque T_{Lmax} and assigning the desired acceleration capabilities $\dfrac{\Delta \omega_r}{\Delta t} = \dfrac{1}{J}(T_e - B_m \omega - T_L)$, tak-

ing note of the estimated J, one obtains the desired T_e. One evaluates the following motor parameters N, r_1, r_2, and J. Taking note of those parameters, the motor sizing estimates are derived, and the fabrication technologies (processes and materials) are assessed. The applied voltage V is bounded. The fabrication technologies and processes significantly affect the motor dimensions and parameters. For example, one may attempt to minimize the airgap to attain the minimal value of $(r_2 - r_1)$ minimizing the expression $\ln(r_2/r_1)$ thereby maximizing T_e. The moment of inertia J can be minimized reducing the rotor mass using cavities, plastic materials, etc. However, there are physical limits on the maximum V and E. The technologies affect and define materials used, tolerance, achievable ratio r_2/r_1, etc. Other critical issue is the need to form a mechanical contact with the rotating rotor to apply the voltage to the rotor plates. These features reduce the overall performance of rotational electrostatic actuators limiting their applications.　　　■

3.3　Variable Reluctance Electromagnetic Actuators

We consider the translational (solenoids and relays) and rotational (variable reluctance synchronous motors) electromagnetic devices. The varying reluctance results in the electromagnetic force or torque. In contrast, high-performance electromagnetic motion devices utilize a magnetic coupling and interaction between windings that are carrying currents and the stationary magnetic field that is developed by permanent magnets or electromagnets. In separately exited DC and induction machines, there is a magnetic coupling between windings due to their mutual inductances. The device physics of electromagnetic device, energy conversion, torque production, *emf* induction, and other topics are covered in Chapter 2.

3.3.1　Solenoids and Relays

Solenoids kinematics usually integrates movable member (commonly called plunger or rotor) and stationary member. These members are made from high-permeability ferromagnetic materials. The windings wound within a helical pattern. These solenoids convert electrical energy to mechanical energy. Solenoids and relays operate due to the varying reluctance, and the force is developed due to the changes of reluctance. Performance of solenoids is strongly affected by the magnetic system, materials, geometry, relative permeability, winding resistance, friction, etc. Solenoids with movable plunger are shown in Figure 3.4. The electromagnetic system is formed by stationary and movable members forming the path for flux. When the voltage is applied to the winding, current flows in the winding, and the electromagnetic force is developed causing the plunger to move to minimize the reluctance. When the applied voltage becomes zero, the plunger resumes its equilibrium position due to the returning spring (assuming that the static and *Coulomb* frictions are negligibly small). The undesirable phenomena such as residual magnetism and friction must be minimized. Different materials for the central guide (nonmagnetic sleeve) and plunger coating (plating) should be chosen to attain minimum friction and minimize wear. Glass-filled nylon and brass (for the guide), silver, copper, aluminum,

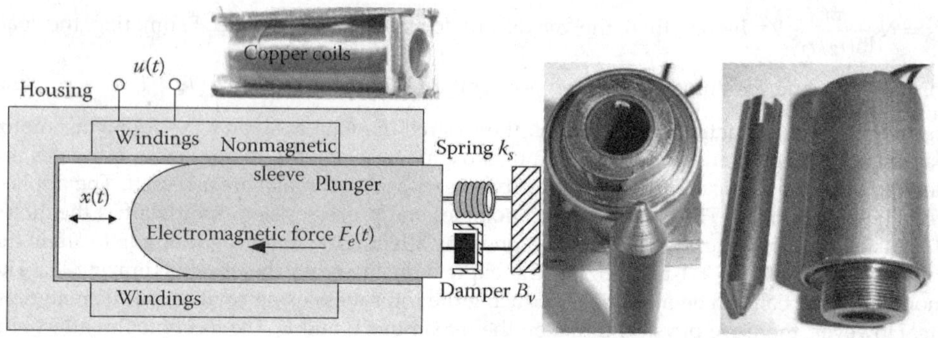

FIGURE 3.4
Schematic and image of solenoids.

tungsten, platinum plating, or other low friction coatings (for the plunger) are among possible solutions. The friction coefficients of lubricated (solid film and oil) and unlubricated for different possible materials are tungsten on tungsten 0.04–0.1 and ~0.3, copper on copper 0.04–0.08 and ~1.2, aluminum on aluminum 0.04–0.12 and ~1, platinum on platinum 0.04–0.25 and ~1.2, and titanium on titanium 0.04–0.1 and ~0.6. The design and analysis of solenoids require application of basic physics (electromagnetics, mechanics, thermodynamics, etc.), consideration of different fabrication processes and materials, etc. In many cases, it is essential to make electromagnetic, mechanical, thermal, and other trade-offs.

Figure 3.5 illustrates a solenoid (relay) with a stationary member and a movable plunger. Our goal is to examine the electromagnetic force production, derive the differential equations, and perform analysis.

Applying Newton's second law of motion, one finds the differential equation of translational motion. Let the restoring/stretching forces exerted by the springs be $F_s = k_s x$. We have

$$m\frac{d^2x}{dt^2} = F_e(t) - B_v\frac{dx}{dt} - k_s x - F_L(t).$$

Assuming that the magnetic system is linear, the coenergy is

$$W_c(i,x) = \frac{1}{2}L(x)i^2.$$

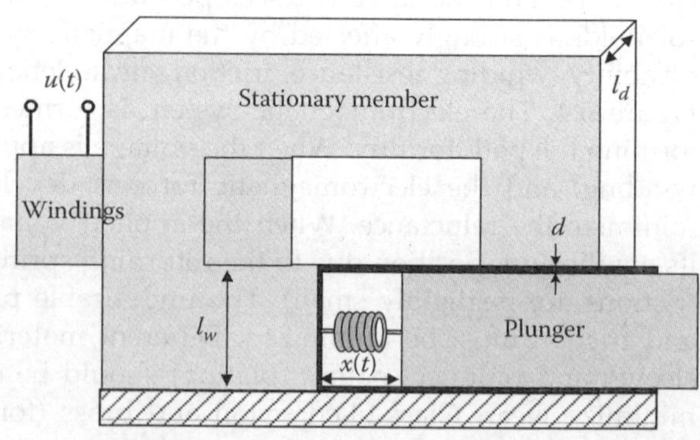

FIGURE 3.5
Solenoid schematics.

From

$$F_e(i,x) = \frac{\partial W_c(i,x)}{\partial x},$$

the electromagnetic force is

$$F_e(i,x) = \frac{1}{2}i^2\frac{dL(x)}{dx}.$$

The magnetizing inductance is

$$L(x) = \frac{N^2}{\Re_f + \Re_x} = \frac{N^2\mu_0\mu_r A_f A_x}{A_x l_f + A_f \mu_r(x+2d)},$$

where $\Re_f$ and $\Re_x$ are the reluctances of the ferromagnetic material and *effective* airgap (which integrates airgap x and thickness d of two nonmagnetic sleeves), A_f and A_x are the associated cross-section areas, l_f is the equivalent lengths of the magnetic flux in the ferromagnetic stationary member and plunger, $(x+2d)$ is the *effective* airgap, and d is the nonmagnetic sleeve thickness that is usually ~ 1 µm, and, correspondingly, one may assume $d=0$.

Though $l_{fplunger}$ varies as x changes, we assume $l_f=$ constant because $l_{fplunger}(x)$ variations do not significantly affect the results due to high μ_r. Using Kirchhoff's law $u = ri + \dfrac{d\psi}{dt}$, $\psi = L(x)i$, one obtains

$$u = ri + L(x)\frac{di}{dt} + i\frac{dL(x)}{dx}\frac{dx}{dt},$$

where

$$\frac{dL}{dx} = -\frac{N^2\mu_0\mu_r^2 A_f^2 A_x}{\left[A_x l_f + A_f\mu_r(x+2d)\right]^2}.$$

We neglect the leakage inductance, and the resulting nonlinear differential equations that describe the dynamics of the solenoid are

$$\frac{di}{dt} = -\frac{r\left[A_x l_f + A_f\mu_r(x+2d)\right]}{N^2\mu_0\mu_r A_f A_x}i - \frac{\mu_r A_f}{A_x l_f + A_f\mu_r(x+2d)}iv + \frac{A_x l_f + A_f\mu_r(x+2d)}{N^2\mu_0\mu_r A_f A_x}u,$$

$$\frac{dv}{dt} = \frac{N^2\mu_0\mu_r^2 A_f^2 A_x}{2m\left[A_x l_f + A_f\mu_r(x+2d)\right]^2}i^2 - \frac{1}{m}k_s x - \frac{B_v}{m}v - \frac{1}{m}F_L,$$

$$\frac{dx}{dt} = v.$$

The force, velocity, and displacement are vectors. Furthermore, the *back emf* opposes the applied voltage, and the friction, spring, and load forces act against the electromagnetic force. These features are integrated in the equations reported. Depending on the applications, requirements (accuracy, efficiency, power consumption, etc.), materials used, and other factors, the designer may refine the model developed enhancing its accuracy and applicability. For example, nonlinear magnetization, various friction models, and the *fringing* effect can be integrated.

Example 3.3:

For the relay, given in Figure 3.5, we perform the modeling and simulations. The restoring spring force is $F_s = k_{s1}x + k_{s2}x^3$. The relay parameters are $r_a = 8.5$ ohm, $L_l = 0.001$ H, $N = 700$, $m = 0.095$ kg, $\mu_{rs} = 4500$, $\mu_{rp} = 5000$, $l_{fp} = 0.055$ m, $l_{fs} = 0.095$ m, $A_f = A_x = 0.00025$ m^2, $B_v = 0.06$ N-s/m, $k_{s1} = 1$ N/m, and $k_{s2} = 2$ N/m^3. The subscripts p and s stand for the plunger and stationary member. Due to the fact that $\mu_r \gg \mu_0$ and $l_{fs} > l_{fp}$, we use the average magnetic plunger path length that is $\frac{1}{2}l_{fp}$. In fact, variations $l_{fp}(x)$ result in very minor overall changes while complicate the equations and obscure the expression for F_e that should be comprehended. The assumption that $l_{fp} = $ constant results in less than 0.1% error.

The magnetizing inductance is found using the reluctances, yielding

$$L(x) = \frac{N^2}{\dfrac{l_{fs}}{\mu_0 \mu_{rs} A} + \dfrac{l_{fp}}{\mu_0 \mu_{rp} A} + \dfrac{x}{\mu_0 A}} = \frac{N^2 \mu_0 \mu_{rs} \mu_{rp} A}{\mu_{rp} l_{fs} + \mu_{rs} l_{fp} + \mu_{rs} \mu_{rp} x}.$$

The electromagnetic force is

$$F_e = \frac{\partial\left[\frac{1}{2}L(x)i^2\right]}{\partial x} = -\frac{N^2 \mu_0 \mu_{rs}^2 \mu_{rp}^2 A}{2\left(\mu_{rp} l_{fs} + \mu_{rs} l_{fp} + \mu_{rs} \mu_{rp} x\right)^2} i^2.$$

Using the leakage inductance L_l, and approximating the spring force applying the nonlinear Hook law with the equilibrium x_{s0}, one finds the following nonlinear differential equations:

$$\frac{di}{dt} = \frac{1}{L(x) + L_l}\left[-ri - \frac{N^2 \mu_0 \mu_{rs}^2 \mu_{rp}^2 A}{2\left(\mu_{rp} l_{fs} + \mu_{rs} l_{fp} + \mu_{rs} \mu_{rp} x\right)^2} iv + u\right], \quad L(x) = \frac{N^2 \mu_0 \mu_{rs} \mu_{rp} A}{\mu_{rp} l_{fs} + \mu_{rs} l_{fp} + \mu_{rs} \mu_{rp} x},$$

$$\frac{dv}{dt} = \frac{1}{m}\left[-\frac{N^2 \mu_0 \mu_{rs}^2 \mu_{rp}^2 A}{2\left(\mu_{rp} l_{fs} + \mu_{rs} l_{fp} + \mu_{rs} \mu_{rp} x\right)^2} i^2 + k_{s1}(x_{s0} - x) + k_{s2}(x_{s0} - x)^3 + B_v v + F_L\right],$$

$$\frac{dx}{dt} = v.$$

The Simulink model is documented in Figure 3.6a, and parameters are uploaded as

```
ra = 8.5; Ll = 0.001; N = 700; m = 0.095; mu0 = 4*pi*1e-7; mus = 4500; mup = 5000;
ls = 0.095; lp = 0.0275; x0 = 0.05; A = 2.5e-4; Bv = 0.06; ks1 = 5; ks2 = 10;
```

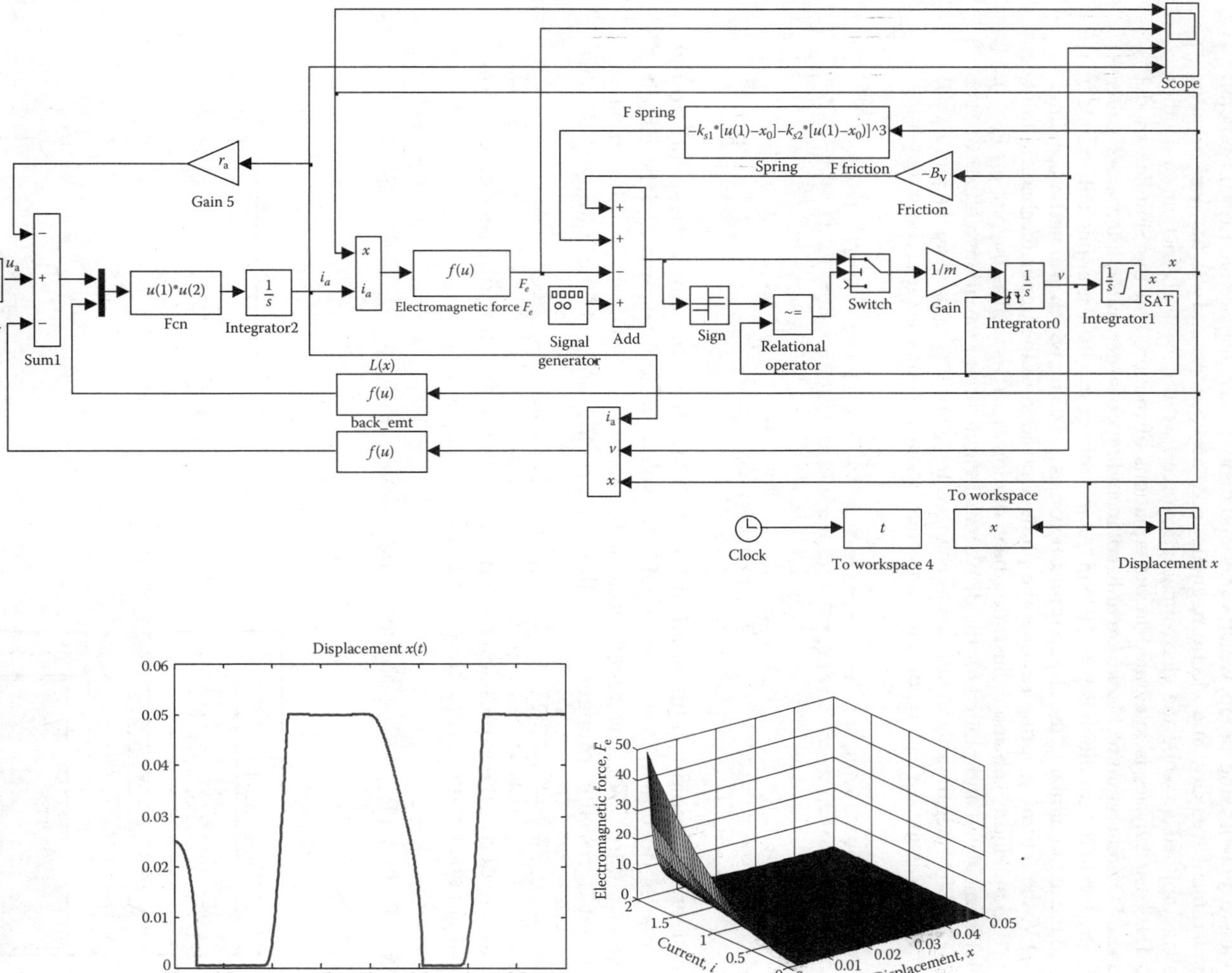

FIGURE 3.6
(a) Simulink model (ch801.mdl); (b) dynamics for $x(t)$ as $u_a = 10$ V and $u_a = 0$ V; (c) plot for $F_e(x, i)$.

The displacement of the plunger $x(t)$ is constrained by hard mechanical limits, and $x \in [0.0005$ $0.05]$ m. The limits on $x(t)$ are set in the integrator. Analyzing the relative displacement of the plunger with respect to the stationary member, we recall that the electromagnetic force is developed to minimize the airgap, which varies and denoted as a displacement $x(t)$. In general, the displacement, velocity, and electromagnetic force are vectors, and the *back emf* opposes the voltage applied u, while the electromagnetic force F_e counters load, friction, and other forces. The basic physics is integrated in the equations of motion, and the Simulink model developed. Different Simulink blocks are utilized, including pulse generator, integrators, gains, switch, etc. The initial condition is $x_0 = 0.025$ m. As the voltage $u_a = 10$ V is applied, the plunger moves to the left minimizing the airgap until x becomes 0.0005 m, which is a mechanical limit. As $u = 0$ V, the returning spring restores the plunger position to the upper mechanical limit, and x is 0.05 m. Figure 3.6b and c illustrates the evolution of $x(t)$ as well as the plot for $F_e(x, i)$. This plot is calculated and plotted for the displacement and current variations $x \in [0.01 \ 0.05]$ m and $i \in [0 \ 2]$ A. The negative sign with F_e was changed to be positive to provide the plot $F_e(x, i)$ as reported in catalogs, which report $F_e(x)$ for different currents or voltages. The statement is

```
x = linspace(0.0025,0.05,50); i = linspace(0,2,50); [X,Y] = meshgrid(x,i);
Fe = (0.5*N^2*mu0*mup^2*mus^2*A*Y.*Y)./(lp*mus+ls*mup+mup*mus*X).^2;
surf(x,i,Fe);
xlabel('Displacement, \itx','FontSize',14); ylabel('Current, \it i','FontSize',14);
zlabel('Electromagnetic Force, \itF_e','FontSize',14);
```

We consider a solenoid as illustrated in Figure 3.7a. This schematic corresponds to the solenoid images documented in Figure 3.4. The solenoid equivalent magnetic circuit is illustrated in Figure 3.7b. The reluctances of the stationary member $\Re_{fs}$, the stationary member that faces the plunger $\Re_{fsp}$, the airgap $\Re_x$, and of the plunger $\Re_{fp}$ are

$$\Re_{fs} = \frac{l_{fs}}{\mu_0 \mu_r A_1}, \ \Re_{fsp} = \frac{l_{fsp}}{\mu_0 \mu_r A_2}, \ \Re_x = \frac{x}{\mu_0 A_2}, \text{ and } \Re_{fp} = \frac{l_{fp}}{\mu_0 \mu_r A_2}.$$

Our goal is to derive the expression for F_e and obtain the resulting differential equations. From the magnetic circuit, as illustrated in Figure 3.7b, one has

$$\frac{1}{2}Ni = \Re_{fs}\Phi_1 + \left(\Re_{fsp} + \Re_x + \Re_{fp}\right)\Phi_3 \quad \text{and} \quad \frac{1}{2}Ni = \Re_{fs}\Phi_2 + \left(\Re_{fsp} + \Re_x + \Re_{fp}\right)\Phi_3.$$

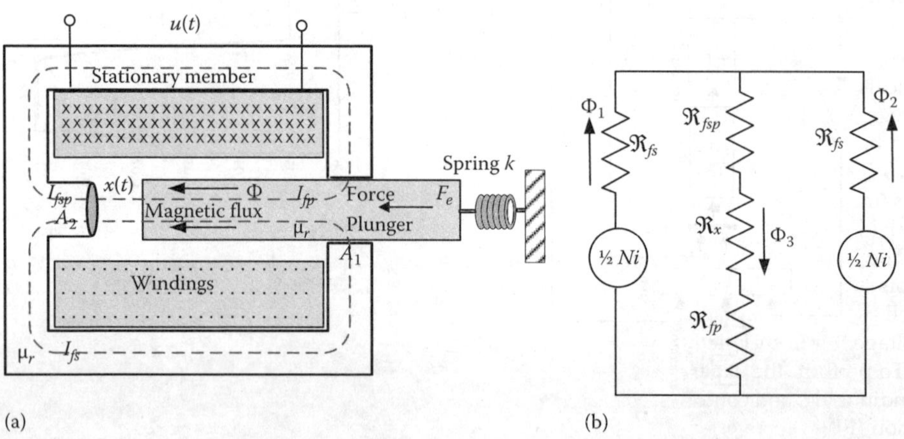

FIGURE 3.7
(a) Solenoid schematics; (b) equivalent magnetic circuit for a solenoid.

From $\quad Ni = \Re_{fs}(\Phi_1 + \Phi_2) + 2(\Re_{fsp} + \Re_x + \Re_{fp})\Phi_3, \quad$ using $\quad \Phi_1 + \Phi_2 = \Phi_3, \quad$ we $\quad$ obtain $Ni = (\Re_{fs} + 2\Re_{fsp} + 2\Re_x + 2\Re_{fp})\Phi_3$. Hence, the total magnetic flux Φ_3 and flux linkages $\psi = N\Phi_3$ are

$$\Phi_3 = \frac{Ni}{\Re_{fs} + 2\Re_{fsp} + 2\Re_x + 2\Re_{fp}} \quad \text{and} \quad \psi = \frac{N^2 i}{\Re_{fs} + 2\Re_{fsp} + 2\Re_x + 2\Re_{fp}}.$$

The magnetizing inductance $L(x)$ is

$$L(x) = \frac{N^2}{\Re_{fs} + 2\Re_{fsp} + 2\Re_x + 2\Re_{fp}} = \frac{N^2 \mu_0 \mu_r A_1 A_2}{l_{fs} A_2 + 2l_{fsp} A_1 + 2A_1 \mu_r x + 2l_{fp} A_1}.$$

Using the coenergy concept, the electromagnetic force F_e is found to be

$$F_e = \frac{\partial\left[\frac{1}{2}L(x)i^2\right]}{\partial x} = -\frac{N^2 \mu_0 \mu_r^2 A_1^2 A_2}{\left(l_{fs} A_2 + 2l_{fsp} A_1 + 2A_1 \mu_r x + 2l_{fp} A_1\right)^2} i^2.$$

From the Kirchhoff voltage law and the second Newton law, we have

$$\frac{di}{dt} = \frac{1}{L(x) + L_l}\left[-ri - \frac{2N^2 \mu_0 \mu_r^2 A_1^2 A_2}{\left(l_{fs} A_2 + 2l_{fsp} A_1 + 2A_1 \mu_r x + 2l_{fp} A_1\right)^2} iv + u\right],$$

$$\frac{dv}{dt} = \frac{1}{m}\left[\frac{N^2 \mu_0 \mu_r^2 A_1^2 A_2}{\left(l_{fs} A_2 + 2l_{fsp} A_1 + 2A_1 \mu_r x + 2l_{fp} A_1\right)^2} i^2 - k_{s1} x - B_v v - F_L\right],$$

$$\frac{dx}{dt} = v.$$

We examined translational variable reluctance electromechanical motion devices, for example, solenoids and relays. Example 3.4 documents numerical and experimental studies of an electromechanical system with a solenoid, while Example 3.7 illustrates a possible use of a single-phase synchronous reluctance motor as a limited-angle rotational relay.

Example 3.4:

We perform the numerical and experimental studies for open-loop and closed-loop electromechanical systems with a Ledex solenoid (model B11M-254), driving circuit, and proportional–integral controller. The solenoid schematics and magnetic circuit were reported in Figure 3.7a and b. The parameters are $r_a = 17.3$ ohm, $L_a = 0.0064$ H, $L_l = 0.001$ H, $N = 1780$, $m = 0.017$ kg, $\mu_{rs} = 5500$, $\mu_{rp} = 5500$, $l_{fp} = 0.02$ m, $l_{fs} = 0.048$ m, $l_{fsp} = 0.08$ m, $A_f = A_x = 2 \times 10^{-4}$ m^2, $B_v = 0.25$ N-s/m, and $k_{s1} = 3$ N/m. Recalling that $\mu_r \gg \mu_0$ and $l_{fs} > l_{fp}$, we use the average magnetic plunger path length $\frac{1}{2}l_{fp}$.

The Simulink diagram to perform the simulations is illustrated in Figure 3.8a. The resulting plots for the transients of the linear displacement $x(t)$, linear velocity $v(t)$, and current $i(t)$, as well as the evolution of the electromagnetic force $F_e(t)$, are reported in Figure 3.8b. The applied voltage 6 V is supplied at 0 s, and $u(t)$ is reduced from 6 to 3 V at 1 s.

To perform the experimental studies, a solenoid is integrated with a power electronic, signal conditioning, and control circuits: in particular, one-quadrant *step-down* (*buck*) pulse width modulation (PWM) converter, filters, proportional–integral controller, etc. The solenoid displacement $x(t)$ is measured and compared with the reference (command) displacement $r(t)$ in order to obtain the tracking error $e(t) = r(t) - x(t)$. A PWM concept implies controlling the duty cycle of the switch

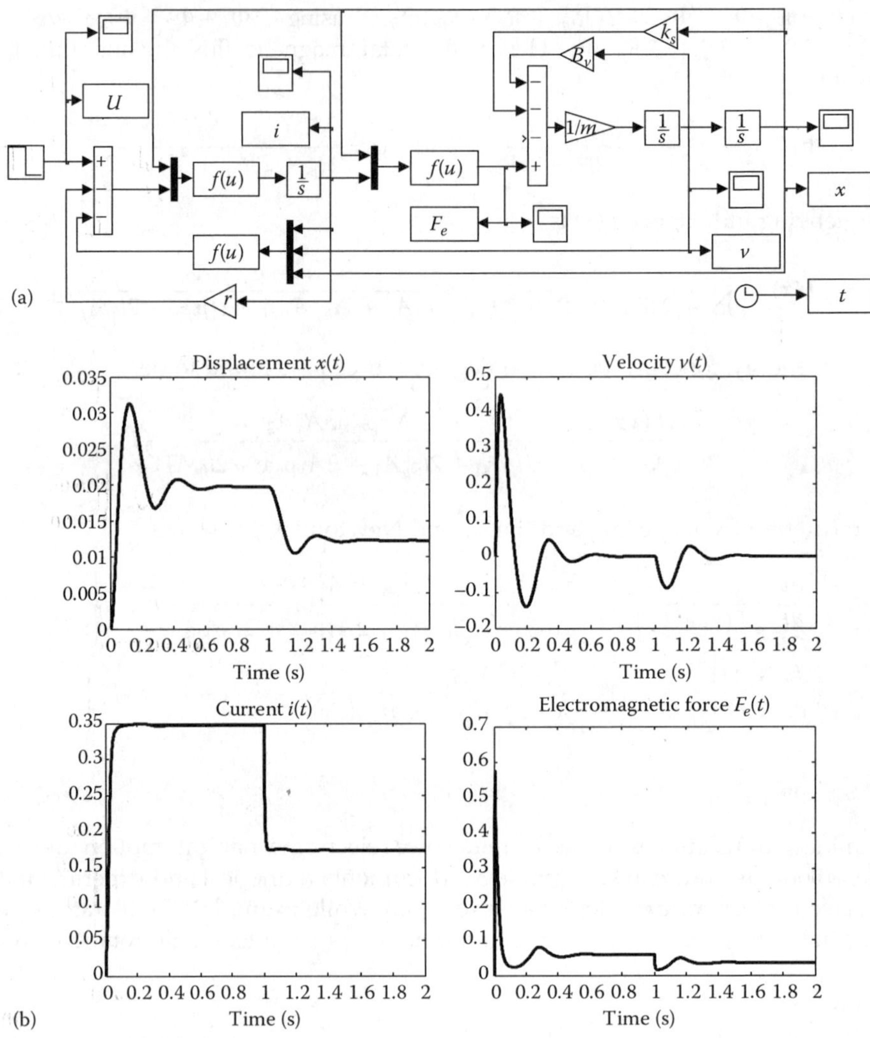

FIGURE 3.8
(a) Simulink model (ch802.mdl); (b) dynamics for $x(t)$, $v(t)$, and $i(t)$, and evolution of $F_e(t)$ if $u = 6$ and 3 V at 1 s.

(transistor) thereby changing the *average* voltage $u(t)$ applied to a winding. As reported in Chapter 7, the control voltage $u_c(t)$ is compared with a periodic (triangular or sinusoidal) signal. The circuit topology and images of the closed-loop system are depicted in Figure 3.9a and b. All definitions and imperative subcircuitry, components, and signals are labeled and accentuated correspondingly. The corresponding results in electronics and control are covered in-depth in Chapters 7 and 8.

The first component of the circuit is the error circuit to obtain $e(t)$. The error circuit adds the reference voltage signal $r(t)$ with the inverted linear potentiometers output signal that corresponds to the plunger displacement $x(t)$. Hence, we obtain the tracking error $e(t)$. The error circuit is implemented as a unit-gain instrumentation amplifier INA128.

The tracking error signal $e(t)$ is fed into the controller circuit that aims to implement a proportional–integral control law $u_c = k_p e + k_i \int edt$, where k_p and k_i are the proportional and integral feedback gains, respectively. Consider the proportional feedback of the proportional–integral controller. The proportional gain applied to the error $e(t)$ is implemented by using an operational amplifier TLC277 in an inverting configuration. The proportional gain k_p is realized by using the input and feedback resistors, and $k_p = -R_{P2}/R_{P1}$. To implement the integral feedback, an integrator is realized using an operational amplifier configured as an inverter with an input resistor R_I and a feedback capacitor C_I. The input resistor and feedback capacitor values determine the value of k_i, and $k_i = -1/R_I C_I$. The time over which the error is integrated is $\tau_I = R_I C_I$. Due to the inverting configuration for both proportional and integral feedback, the outputs must be inverted and summed to produce the control voltage u_c. To perform the inversion and summation, instrumentation amplifiers are utilized due to their robustness, tolerance, and other advantages. By fixing the positive input of the instrumentation amplifier to ground and feeding the respective signal to the negative input, a simple inverter is realized. By feeding the output of either inverting instrumentation amplifier to the reference node of the following circuitry, a summing circuit is implemented.

The developed control signal u_c is fed to a comparator circuit, which compares its two inputs to produce an output, Figure 3.9. The comparator is constructed using a single operational amplifier with the control signal u_c supplied to its inverting terminal and a periodic near-triangular waveform u_t applied to its noninverting terminal. The comparator circuit will output its positive rail voltage V_{CC} for the duration of time when its positive input is greater than its negative input, and will conversely output its negative voltage rail $-V_{CC}$ for the amount of time when its negative input is greater than its positive input. The comparator develops the PWM waveform altering the duty cycle of a square wave, whose amplitude is equal to the

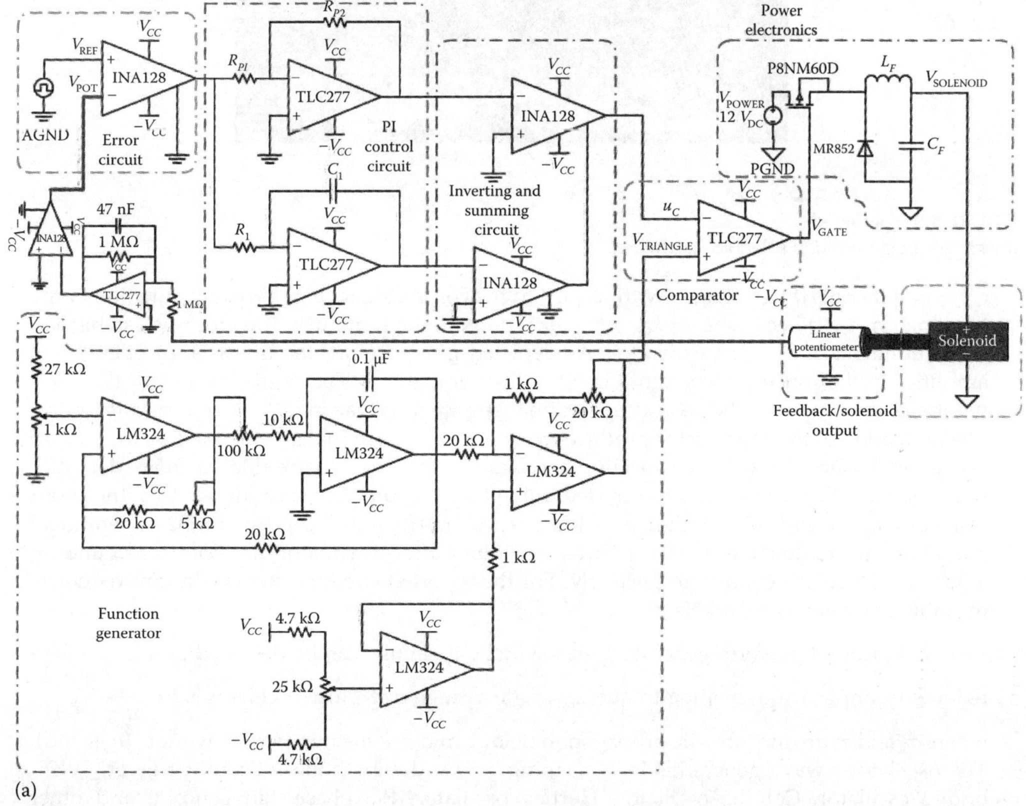

(a)

FIGURE 3.9
(a) Closed-loop system schematics: solenoid with power electronics, control, and filter circuits

(continued)

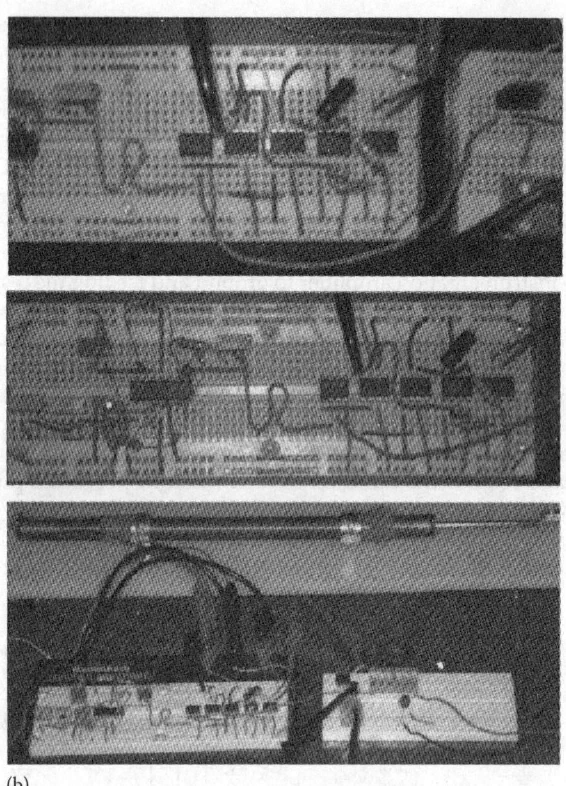

(b)

FIGURE 3.9 (continued)
(b) images of closed-loop solenoid hardware.

comparators rails (bus voltage), with respect to the control signal u_c. This u_c is a function of the tracking error $e(t)$ because $u_c = k_p e + k_i \int e \, dt$. The periodic triangular waveform is established by a rudimentary function generator. As depicted in Figure 3.9, the first LM324 operational amplifier in the function generator circuit is the comparator. The positive input of the comparator is the output of the second operational amplifier in the circuit, which is an integrator. The oscillation of the comparator produces a square waveform that is integrated by the second operational amplifier with a capacitor as negative feedback to produce a near-triangular waveform u_t. The frequency of u_t is determined by the time constant defined by the input resistance and feedback capacitance of the integrator. The remaining two LM324 operational amplifiers in the function generator circuit are the buffer/attenuator amplifier for u_t and an adjustable DC offset control, respectively. For the reported circuitry, the maximum frequency of stable operation is ∼6.8 kHz.

Remark. Various waveform generating and switching circuits can be designed and implemented. For example, the oscillating frequency of the n-stage ring pulse-oscillator is $f = \dfrac{1}{n(\tau_1 + \tau_2)}$, where τ_1 and τ_2 are the intrinsic propagation delays, and n is the number of inverters (n is odd). The oscillators, wave generation, and shaping circuit design using a Schmitt trigger, Wien-bridge oscillator, Colpitts oscillator, Hartley oscillator, RC phase-shift concept, and other solutions are well known and can be utilized.

The PWM waveform, produced by comparing the control signal u_c and generated near-triangular waveform u_t, allows one to control the *average* voltage u supplied to the solenoid.

The TLC277 operational amplifier that produces the PWM waveform can output a maximum current and voltage of $\sim$30 mA and 18 V, respectively. For the studied solenoid, the rated current and voltage are $\sim$0.6 A and 10 V respectively. The TLC277 comparator circuit is not able to deliver the necessary current and power to operate the solenoid.

The power electronics circuitry is based on the use of a one-quadrant *step-down* converter as reported in Figure 3.9. The *step-down* converter employs a power MOSFET (switch), which is controlled by applying the voltage to a gate. We connect the output of the comparator circuit to the gate to control the MOSFET switching activity. As reported in Chapter 7, a low pass filter is implemented utilizing an inductor L_F and capacitor C_F. The corner frequency of the filter $f_c = \dfrac{1}{2\pi\sqrt{L_F C_F}}$ is set to be 3.4 kHz. This implies $L_F = 500$ μH and $C_F = 4.7$ μF. The plunger of the solenoid is connected to a linear potentiometer to measure the linear displacement $x(t)$ that is fed to the error circuit to generate $e(t)$. A second-order low pass filter with two poles is implemented to condition the potentiometer output signal by filtering the noise. The low pass filter is documented in Figure 3.9 between the potentiometer and the error circuit input. Our goal is to ensure a high signal-to-noise ratio within the operating frequency envelope. The design of the unity-gain filter to achieve a 3 Hz cutoff frequency f_c is performed using the expression $f_c = \dfrac{1}{2\pi R_f C_f}$, which leads to $R_f = 1$ Mohm and $C_f = 47$ nF.

The control voltage u_c (output of the proportional–integral controller) and near-triangular signal u_t are illustrated in Figure 3.10a. One recalls that those signals are compared by the comparator to produce a signal to drive the MOSFET. The MOSFET gate voltage and the voltage applied to the solenoid are documented in Figure 3.10b.

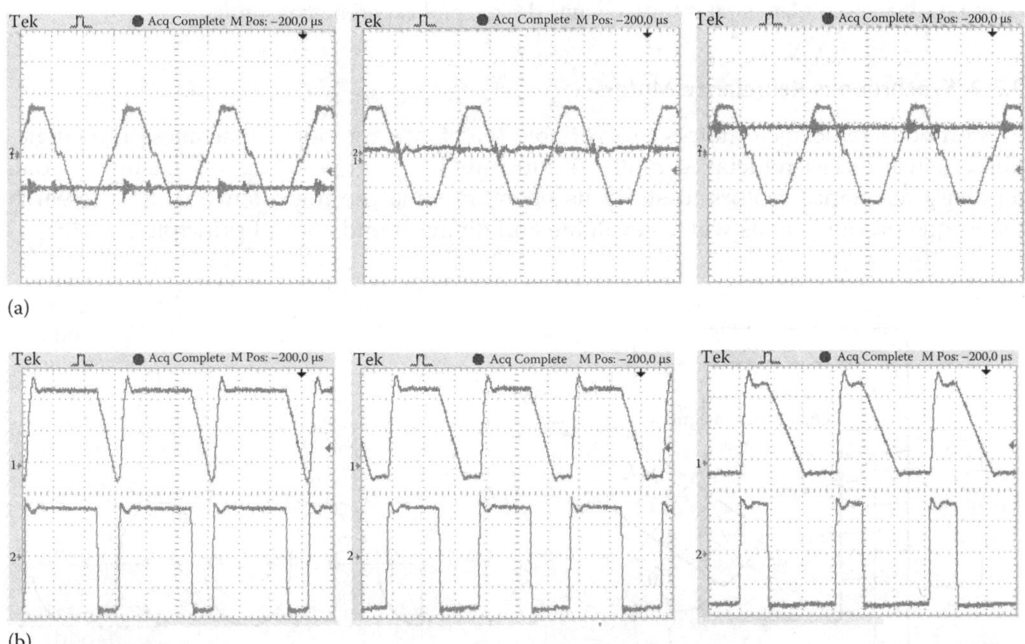

FIGURE 3.10
(a) Near-triangular waveform u_t and control voltage u_c when a plunger is inserted at $\sim$¾, ½, and ¼ near-maximum-stroke position; (b) MOSFET gate voltage and voltage applied to a solenoid winding u when a plunger is $\sim$¾ inserted (3.2 and 2.6 V), half-inserted (8.4 and 6.8 V), and at $\sim$¼ near-maximum-stroke position (12.2 and 9.8 V).

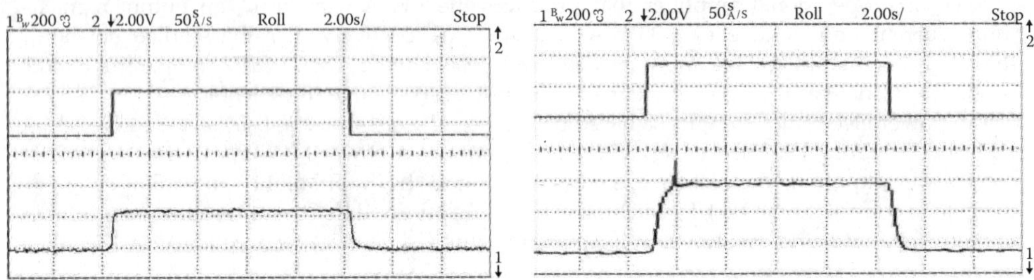

FIGURE 3.11
Reference input r (top) and plunger displacement x (bottom) when a plunger is initially at the half-inserted and near-maximum stroke positions.

The transient dynamics for the closed-loop system is reported in Figure 3.11. The comparison of the assigned reference $r(t)$ and plunger displacements $x(t)$ provide an evidence that (1) the desired vary fast nonoscillatory repositioning is accomplished, (2) stability is guaranteed, (3) the steady-state tracking error is zero (within allowed positioning sensor accuracy), (4) the disturbances are attenuated (see the attenuation of F_L when x reached the steady-state value examining the second bottom plot for x), (5) robustness to parameter variations is accomplished, etc. One recalls that due to the device physics, one-directional control of the displacement $x(t)$ can be accomplished, and the returning spring is used to restore the plunger position at the equilibrium as $u = 0$. We conclude that a sound design is accomplished and verified using coherent fundamental results, circuits design, and experimental studies.

3.3.2 Synchronous Reluctance Motors

Synchronous reluctance motors are variable reluctance rotational machines. The single-phase synchronous reluctance motor is illustrated in Figure 3.12a. We examine radial topology reluctance motors to study its operation, analyze important features, research the torque production, as well as evaluate and define sound control principles.

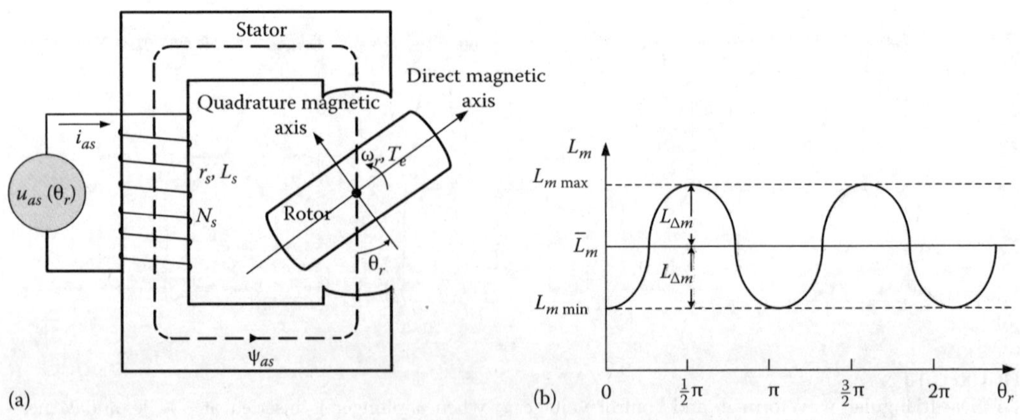

FIGURE 3.12
(a) Radial topology reluctance motor with a variable reluctance path; (b) magnetizing inductance $L_m(\theta_r)$.

The *quadrature* and *direct* magnetic axes are fixed with the rotor, which rotates with angular velocity ω_r. These magnetic axes rotate with the angular velocity ω. Under normal operation, the angular velocity of synchronous machines is equal to the synchronous angular velocity ω_e. Hence, $\omega_r = \omega_e$ and $\omega = \omega_r = \omega_e$. The angular displacements of the rotor θ_r and the angular displacement of the *quadrature* magnetic axis θ are equal. Assuming that the initial conditions are zero, we have $\theta_r = \theta = \int_{t_0}^{t} \omega_r(\tau)d\tau = \int_{t_0}^{t} \omega(\tau)d\tau$.

The magnetizing reluctance $\Re_m$ is a function of the rotor angular displacement θ_r. Using the number of turns N_s, the magnetizing inductance is $L_m(\theta_r) = \dfrac{N_s^2}{\Re_m(\theta_r)}$. This magnetizing inductance varies twice per one revolution of the rotor and has minimum and maximum values as $L_{m\,\min} = \dfrac{N_s^2}{\Re_{m\,\max}(\theta_r)}\Big|_{\theta_r = 0,\pi,2\pi,\ldots}$ and $L_{m\,\max} = \dfrac{N_s^2}{\Re_{m\,\min}(\theta_r)}\Big|_{\theta_r = \frac{1}{2}\pi,\frac{3}{2}\pi,\frac{5}{2}\pi,\ldots}$. Hence, the magnetizing inductance is a function of the rotor angular displacement.

Assuming that the inductance variation is an ideally sinusoidal, we have

$$L_m(\theta_r) = \bar{L}_m - L_{\Delta m}\cos 2\theta_r,$$

where $\bar{L}_m$ is the average value of the magnetizing inductance and $L_{\Delta m}$ is the half of amplitude of the sinusoidal variation of the magnetizing inductance. The plot for $L_m(\theta_r)$ is documented in Figure 3.12b.

The electromagnetic torque, developed by single-phase reluctance motors, is found using the expression for the coenergy. For the studied example, $W_c(i_{as},\theta_r) = \frac{1}{2}\left(L_{ls} + \bar{L}_m - L_{\Delta m}\cos 2\theta_r\right)i_{as}^2$. Hence, one finds

$$T_e = \frac{\partial W_c(i_{as},\theta_r)}{\partial \theta_r} = \frac{\partial\left[\frac{1}{2}i_{as}^2\left(L_{ls} + \bar{L}_m - L_{\Delta m}\cos 2\theta_r\right)\right]}{\partial \theta_r} = L_{\Delta m}i_{as}^2\sin 2\theta_r.$$

For motors, one should ensure controlled *clockwise* and *counterclockwise* $360°$ rotation capabilities. The electromagnetic torque, to ensure $360°$ rotation, is not developed by synchronous reluctance motors if one feeds the DC current or voltage to the winding. We study control features that are based upon derived electromagnetics. The average value of T_e is not equal to zero if the current is a function of θ_r. As an illustration, let the phase current in the *as* winding be

$$i_{as} = i_M\mathrm{Re}\left(\sqrt{\sin 2\theta_r}\right) \quad \text{or} \quad i_{as} = \begin{cases} i_M\sqrt{\sin 2\theta_r} & \text{if } \sin 2\theta_r > 0 \\ 0 & \text{if } \sin 2\theta_r \leq 0. \end{cases}$$

For $i_{as} = i_M\mathrm{Re}\left(\sqrt{\sin 2\theta_r}\right)$, the electromagnetic torque is

$T_e = L_{\Delta m}i_{as}^2\sin 2\theta_r = L_{\Delta m}i_M^2\left(\mathrm{Re}\sqrt{\sin 2\theta_r}\right)^2\sin 2\theta_r \neq 0$. Feeding $i_{as} = \begin{cases} i_M\sqrt{\sin 2\theta_r} & \text{if } \sin 2\theta_r > 0 \\ 0 & \text{if } \sin 2\theta_r \leq 0' \end{cases}$

we have $T_{eav} = \dfrac{1}{\pi}\displaystyle\int_0^{\pi}\dfrac{1}{2}L_{\Delta m}i_{as}^2\sin 2\theta_r d\theta_r = \dfrac{1}{4}L_{\Delta m}i_M^2$ (one recalls that

$\int \sin^2 ax\,dx = \frac{1}{2}x - \frac{1}{2a}\cos ax\sin ax$, and, in general,

$\int \sin^{2m} ax\,dx = \dfrac{(2m)!}{2^{2m}(m!)^2}x - \dfrac{\cos ax}{a}\displaystyle\sum_{i=0}^{m-1}\dfrac{(2m)!(i!)^2}{2^{2(m-i)}(2i+1)!(m!)^2}\sin^{2i+1} ax$.

From $T_e = L_{\Delta m} i_{as}^2 \sin 2\theta_r$, one formally finds the phase current to maximize T_e and eliminate the torque ripple to be $i_{as} = i_M 1/\sqrt{\sin 2\theta_r}$. This phase current, mathematically, leads to $T_e = L_{\Delta m} i_M^2$. However, it is impossible to implement $i_{as} = i_M 1/\sqrt{\sin 2\theta_r}$ due to constraints on the maximum current, singularity, denominator is complex as $\sin 2\theta_r < 0$, etc. Therefore, the derived expression for the phase current $i_{as}(\theta_r)$ should be modified to ensure applicability and practicality.

The mathematical model of the single-phase reluctance motor is found by using the following Kirchhoff's and Newton's second laws

$$u_{as} = r_s i_{as} + \frac{d\psi_{as}}{dt} \ \text{(circuitry-electromagnetic equation of motion)},$$

$$J\frac{d^2\theta_r}{dt^2} = T_e - B_m \omega_r - T_L \ \text{(\textit{torsional-mechanical} equation of motion)}.$$

From $\psi_{as} = \left(L_{1s} + \bar{L}_m - L_{\Delta m}\cos 2\theta_r\right) i_{as}$, one obtains the following set of three first-order nonlinear differential equations:

$$\begin{aligned}
\frac{di_{as}}{dt} &= \frac{1}{L_{ls} + \bar{L}_m - L_{\Delta m}\cos 2\theta_r}\left(-r_s i_{as} - 2L_{\Delta m} i_{as}\omega_r \sin 2\theta_r + u_{as}\right), \\
\frac{d\omega_r}{dt} &= \frac{1}{J}\left(L_{\Delta m} i_{as}^2 \sin 2\theta_r - B_m \omega_r - T_L\right), \\
\frac{d\theta_r}{dt} &= \omega_r
\end{aligned} \tag{3.1}$$

which describe single-phase reluctance motor dynamic and steady-state behavior.

Example 3.5:

The parameters and other quantities of interest, for example $L_m(\theta_r)$ or $T_{\text{friction}}(\omega_r)$, of electromechanical motion devices can be estimated as machine is designed. Various coefficients and quantities can be measured as the machine is fabricated. For a single-phase synchronous reluctance motor, the parameters are $r_s = 1$ ohm, $L_{md} = 0.4$ H, $L_{mq} = 0.04$ H, $L_{ls} = 0.05$ H, $J = 0.00001$ kg-m^2, and $B_m = 0.00005$ N-m-s/rad. The voltage applied to the stator winding is $u_{as} = u_M \text{Re}\left[\sqrt{\sin 2(\theta_r - 0.2)}\right]$, where $u_M = 50$ V.

For no-load condition, the motor parameters are uploaded as

```
P=2; rs=1; Lmd=0.4; Lmq=0.04; Lls=0.05; J=0.00001; Bm=0.00005;
Lmb=(Lmq+Lmd)/3; Ldm=(Lmd-Lmq)/3; um=50;
TL=0.005; % Load torque TL is applied at 1.25 sec
```

The Simulink diagram, which corresponds to the differential equations (3.1), is documented in Figure 3.13a. The transient of the angular velocity $\omega_r(t)$ is plotted in Figure 3.13b using the statement

```
plot(wr(:,1),wr(:,2)); xlabel('Time [seconds]','FontSize',14);
ylabel('\omega_r','FontSize',14);
title('AngularVelocityDynamics,\omega_r [rad/sec]','FontSize',14);
```

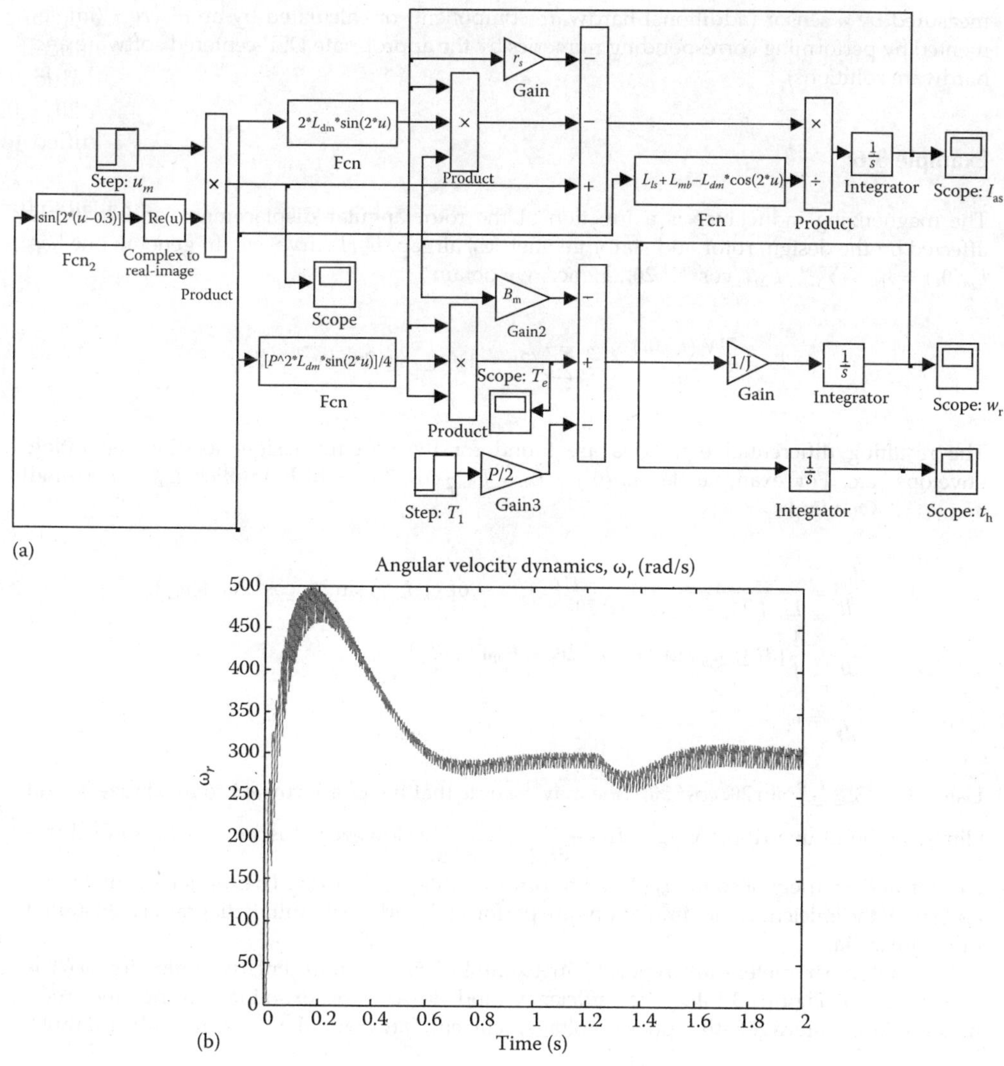

(a)

(b)

FIGURE 3.13

(a) Simulink diagram (ch3_01.mdl) to simulate a single-phase synchronous reluctance motor; (b) transient response for the angular velocity.

The motor reaches the steady-state 280 rad/s within 0.75 s. The load torque $T_L = 0.005$ N-m is applied at 1.25 s.

Examining transient dynamics, one observes the electromagnetic torque ripple and phase current chattering that were expected due to the reported torque production features. These effects lead to low efficiency, heating, vibration, noise, mechanical wearing, and other highly undesired features. Therefore, single-phase synchronous motors are not used due to the low performance and inadequate capabilities. This electromechanical motion device was studied as a *primer* as well as to introduce advanced topics. It is found that the phase voltage must be supplied as a function of the rotor angular displacement θ_r in order to develop the electromagnetic torque ensuring 360° rotation capabilities. The angular displacement θ_r must be measured by a sensor (additional hardware component) or calculated by an *observer* (implemented by performing corresponding numerics by the appropriate DSP-centered software and hardware solutions). ∎

Example 3.6:

The magnetizing inductance is a function of the rotor angular displacement, and $L_m(\theta_r)$ is affected by the design, rotor and stator geometries, airgap, $B\text{–}H$ curve, etc. In general, one has $L_m(\theta_r) = \bar{L}_m - \sum_{n=1}^{\infty} L_{\Delta mn} \cos^{2n-1} 2\theta_r$. Hence, we obtain

$$T_e = \frac{\partial W_c(i_{as}, \theta_r)}{\partial \theta_r} = i_{as}^2 \sum_{n=1}^{\infty} (2n - 1) L_{\Delta mn} \sin 2\theta_r \cos^{2n-2} 2\theta_r.$$

The resulting differential equations are found for the specific design, loading, operating envelope, etc. For example, let $L_m(\theta_r) = \bar{L}_m - L_{\Delta m2} \cos^3 2\theta_r$, which implies $L_{\Delta m2} \neq 0$ and $\forall L_{\Delta mn} = 0$. One finds

$$\frac{di_{as}}{dt} = \frac{1}{L_{ls} + \bar{L}_m - L_{\Delta m2} \cos^3 2\theta_r} \left(-r_s i_{as} - 6 L_{\Delta m2} i_{as} \omega_r \sin 2\theta_r \cos^2 2\theta_r + u_{as} \right),$$

$$\frac{d\omega_r}{dt} = \frac{1}{J} \left(3 L_{\Delta m2} i_{as}^2 \sin 2\theta_r \cos^2 2\theta_r - B_m \omega_r - T_L \right),$$

$$\frac{d\theta_r}{dt} = \omega_r.$$

Using $T_e = 3 L_{\Delta m2} i_{as}^2 \sin 2\theta_r \cos^2 2\theta_r$, one may assume that the phase current to maximize T_e and eliminate the torque ripple is $i_{as} = i_M \dfrac{1}{\sqrt{\sin 2\theta_r \cos^2 2\theta_r}}$. However, it is impossible to feed this current, and a variety of sound $i_{as}(\theta_r)$ with corresponding $u_{as}(\theta_r)$ exist. To examine the dynamics and study the efficiency, the simulations are performed, and the Simulink diagram is illustrated in Figure 3.14a.

The motor parameters are reported in Example 3.5. The transient dynamics for $\omega_r(t)$ is illustrated in Figure 3.14b. The efficiency and losses are examined using the root-mean-square values of the phase voltage, current, and angular velocity, for example,

$u_{as_{rms}} = \sqrt{\dfrac{1}{T} \displaystyle\int_0^T u_{as}^2 dt}$, $i_{as_{rms}} = \sqrt{\dfrac{1}{T} \displaystyle\int_0^T i_{as}^2 dt}$, and $\omega_{r_{rms}} = \sqrt{\dfrac{1}{T} \displaystyle\int_0^T \omega_r^2 dt}$. The detail analysis of effi-

ciency, which is a steady-state quantity, is reported in Chapter 4. The losses can be estimated as $P_l(t) = r_s i_{as}^2(t) + B_m \omega_r^2(t)$, while the efficiency is given as $\eta = P_{output}/P_{input} = T_L \omega_r / u_{as} i_{as}$, $P_{output} = P_{input} - P_l$. The load torque $T_L = 0.0025$ N-m is applied at 1.25 s. ∎

Example 3.7:

By applying a DC voltage, a single-phase synchronous reluctance motor can be used as a rotational relay. If $\theta_r \neq \pm\frac{1}{2}\pi$ and a DC voltage u_{as} is applied, to minimize the reluctance, the rotor will rotate *clockwise* or *counterclockwise* to take the angular position $\theta_r = \pm\frac{1}{2}\pi$ depending on its initial displacement θ_{r0}, see Figure 3.12a. This feature is in agreement with the expression for the electromagnetic torque $T_e = L_{\Delta m} i_{as}^2 \sin 2\theta_r$. Applying a DC voltage and using the returning spring, as studied in Section 3.3.1 for various translational devices, one may operate the studied motion device as a limited-angle rotational relay. ∎

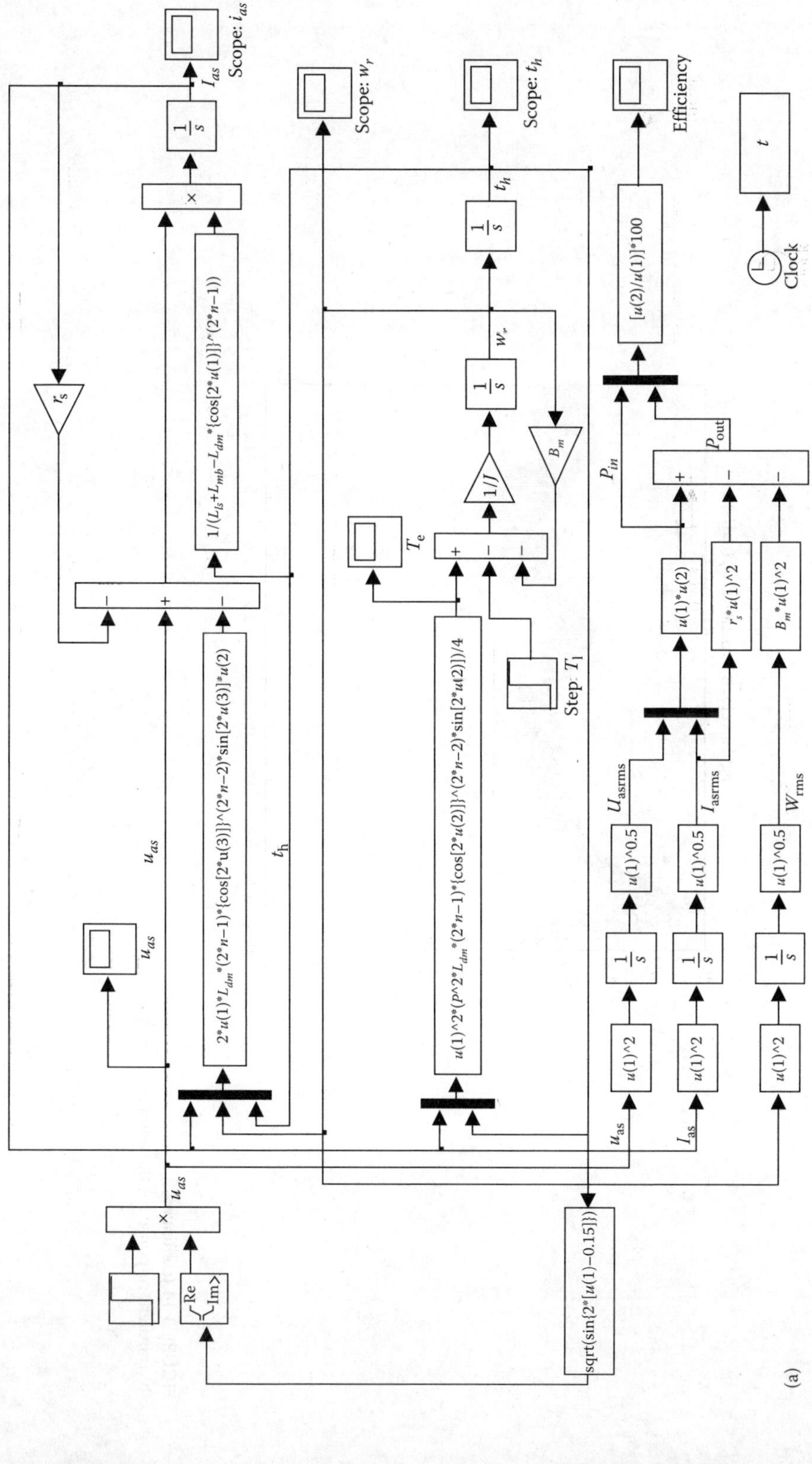

(a)

FIGURE 3.14

(a) Simulink diagram (ch3_02.mdl) to simulate single-phase synchronous reluctance motors and

(continued)

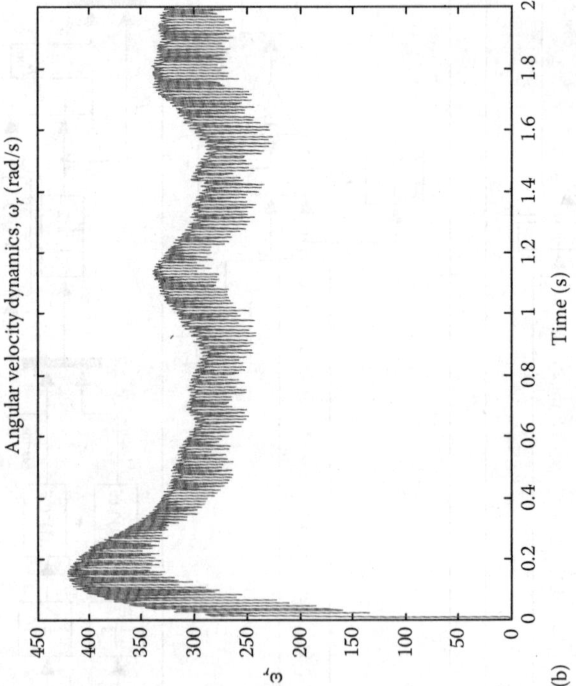

FIGURE 3.14 (continued)
(b) transient response for the angular velocity.

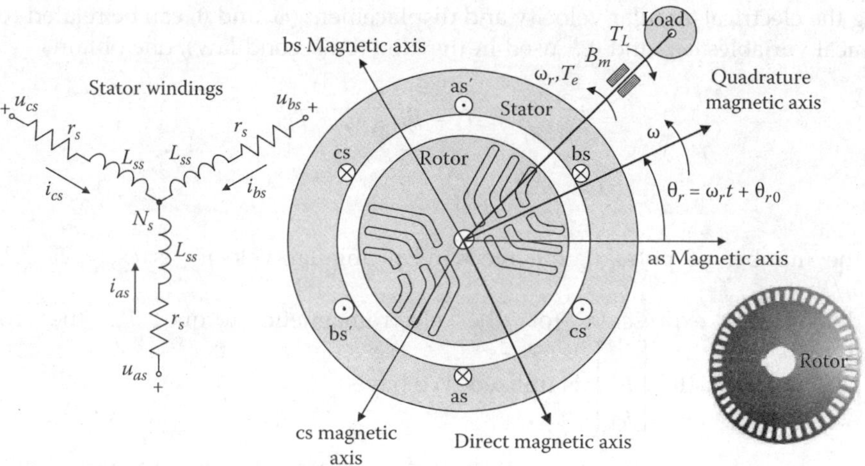

FIGURE 3.15
Three-phase synchronous reluctance motor and a rotor (cavities are filled with plastic or other nonmagnetic materials).

Consider three-phase synchronous reluctance motors, as illustrated in Figure 3.15. These motors ensure better performance as compared to single-phase synchronous reluctance motors. However, these synchronous reluctance motors cannot even closely compete with permanent-magnet and conventional synchronous, induction and DC motors in terms of performance, capabilities, etc.

The machine parameters are the stator resistance r_s (it is assumed that the phase resistances are equal), the magnetizing inductances in the *quadrature* and *direct* axes L_{mq} and L_{md}, the average magnetizing inductance $\bar{L}_m$, the leakage inductance L_{ls}, the moment of inertia J, and the viscous friction coefficient B_m. The circuitry-electromagnetic dynamics in the vector-matrix form is

$$\mathbf{u}_{abcs} = \mathbf{r}_s \mathbf{i}_{abcs} + \frac{d\boldsymbol{\Psi}_{abcs}}{dt}, \quad \boldsymbol{\Psi}_{abcs} = \mathbf{L}_s(\theta_r)\mathbf{i}_{abcs}, \tag{3.2}$$

where $\mathbf{u}_{abcs} = [u_{as}, u_{bs}, u_{cs}]^T$, u_{as}, u_{bs}, and u_{cs} are the phase voltages supplied to the as, bs, and cs stator windings, $\mathbf{i}_{abcs} = [i_{as}, i_{bs}, i_{cs}]^T$, i_{as}, i_{bs}, and i_{cs} are the phase currents, and $\boldsymbol{\psi}_{abcs} = [\psi_{as}, \psi_{bs}, \psi_{cs}]^T$, ψ_{as}, ψ_{bs}, and ψ_{cs} are the flux linkages.

In (3.2), the resistance matrix is $\mathbf{r}_s = \begin{bmatrix} r_s & 0 & 0 \\ 0 & r_s & 0 \\ 0 & 0 & r_s \end{bmatrix}$, while the inductance mapping $\mathbf{L}_s(\theta_r)$ is assumed to be

$$\mathbf{L}_s(\theta_r) = \begin{bmatrix} L_{ls} + \bar{L}_m - L_{\Delta m}\cos 2\theta_r & -\frac{1}{2}\bar{L}_m - L_{\Delta m}\cos 2\left(\theta_r - \frac{1}{3}\pi\right) & -\frac{1}{2}\bar{L}_m - L_{\Delta m}\cos 2\left(\theta_r + \frac{1}{3}\pi\right) \\ -\frac{1}{2}\bar{L}_m - L_{\Delta m}\cos 2\left(\theta_r - \frac{1}{3}\pi\right) & L_{ls} + \bar{L}_m - L_{\Delta m}\cos 2\left(\theta_r - \frac{2}{3}\pi\right) & -\frac{1}{2}\bar{L}_m - L_{\Delta m}\cos 2\left(\theta_r + \pi\right) \\ -\frac{1}{2}\bar{L}_m - L_{\Delta m}\cos 2\left(\theta_r + \frac{1}{3}\pi\right) & -\frac{1}{2}\bar{L}_m - L_{\Delta m}\cos 2\left(\theta_r + \pi\right) & L_{ls} + \bar{L}_m - L_{\Delta m}\cos 2\left(\theta_r + \frac{2}{3}\pi\right) \end{bmatrix},$$

$$\bar{L}_m = \frac{1}{3}\left(L_{mq} + L_{md}\right) \quad \text{and} \quad L_{\Delta m} = \frac{1}{3}\left(L_{md} - L_{mq}\right).$$

The expressions for the flux linkages and inductances are nonlinear functions of the electrical angular displacement θ_r.

Using the electrical angular velocity and displacement (ω_r and θ_r can be related to be the mechanical variables ω_{rm} and θ_{rm} used in the Newton second law), one obtains

$$J\frac{2}{P}\frac{d\omega_r}{dt} = T_e - B_m\frac{2}{P}\omega_r - T_L,$$

$$\frac{d\theta_r}{dt} = \omega_r. \tag{3.3}$$

Using the number of poles P, the mechanical angular velocity is $\omega_{rm} = 2\omega_r/P$, while $\theta_{rm} = 2\theta_r/P$.

To derive the expression for the electromagnetic torque T_e, the coenergy $W_c = \frac{1}{2}[\,i_{as}\ \ i_{bs}\ \ i_{cs}\,]\mathbf{L}_s(\theta_r)\begin{bmatrix} i_{as} \\ i_{bs} \\ i_{cs} \end{bmatrix}$ is utilized. We have

$$T_e = \frac{P}{2}\frac{\partial W_c}{\partial \theta_r} = \frac{P}{2}\frac{1}{2}[\,i_{as}\ \ i_{bs}\ \ i_{cs}\,]\begin{bmatrix} 2L_{\Delta m}\sin 2\theta_r & 2L_{\Delta m}\sin 2\left(\theta_r - \frac{1}{3}\pi\right) & 2L_{\Delta m}\sin 2\left(\theta_r + \frac{1}{3}\pi\right) \\ 2L_{\Delta m}\sin 2\left(\theta_r - \frac{1}{3}\pi\right) & 2L_{\Delta m}\sin 2\left(\theta_r - \frac{2}{3}\pi\right) & 2L_{\Delta m}\sin 2\theta_r \\ 2L_{\Delta m}\sin 2\left(\theta_r + \frac{1}{3}\pi\right) & 2L_{\Delta m}\sin 2\theta_r & 2L_{\Delta m}\sin 2\left(\theta_r + \frac{2}{3}\pi\right) \end{bmatrix}\begin{bmatrix} i_{as} \\ i_{bs} \\ i_{cs} \end{bmatrix}.$$

One obtains

$$T_e = \frac{P}{2}L_{\Delta m}\left[i_{as}^2\sin 2\theta_r + 2i_{as}i_{bs}\sin 2\left(\theta_r - \frac{1}{3}\pi\right) + 2i_{as}i_{cs}\sin 2\left(\theta_r + \frac{1}{3}\pi\right) + i_{bs}^2\sin 2\left(\theta_r - \frac{2}{3}\pi\right) \right.$$

$$\left. + 2i_{bs}i_{cs}\sin 2\theta_r + i_{cs}^2\sin 2\left(\theta_r + \frac{2}{3}\pi\right) \right]. \tag{3.4}$$

Hence, T_e is a nonlinear function of the motor variables (phase currents $\mathbf{i}_{abcs}$ and electrical angular displacement θ_r) and motor parameters (number of poles P and inductance $L_{\Delta m}$). For three-phase synchronous reluctance motors, $L_{mq} = \frac{3}{2}\left(\bar{L}_m - L_{\Delta m}\right)$ and $L_{md} = \frac{3}{2}\left(\bar{L}_m + L_{\Delta m}\right)$. Therefore, $\bar{L}_m = \frac{1}{3}\left(L_{mq} + L_{md}\right)$ and $L_{\Delta m} = \frac{1}{3}\left(L_{md} - L_{mq}\right)$. Using the trigonometric identities, one finds the following formula for the electromagnetic torque:

$$T_e = \frac{P\left(L_{md} - L_{mq}\right)}{6}\left[\left(i_{as}^2 - \frac{1}{2}i_{bs}^2 - \frac{1}{2}i_{cs}^2 - i_{as}i_{bs} - i_{as}i_{cs} + 2i_{bs}i_{cs} \right)\sin 2\theta_r \right.$$

$$\left. + \frac{\sqrt{3}}{2}\left(i_{bs}^2 - i_{cs}^2 - 2i_{as}i_{bs} + 2i_{as}i_{cs} \right)\cos 2\theta_r \right].$$

The resulting differential equations, which describe the dynamics of three-phase synchronous motors in non-Cauchy's form, are

$$\frac{di_{as}}{dt} = \frac{1}{L_{ls} + \bar{L}_m - L_{\Delta m}\cos 2\theta_r}\left\{ -r_s i_{as} + u_{as} + \left[\frac{1}{2}\bar{L}_m + L_{\Delta m}\cos 2\left(\theta_r - \frac{1}{3}\pi\right) \right]\frac{di_{bs}}{dt} \right.$$

$$+ \left[\frac{1}{2}\bar{L}_m + L_{\Delta m}\cos 2\left(\theta_r + \frac{1}{3}\pi\right) \right]\frac{di_{cs}}{dt} - 2L_{\Delta m}\omega_r\left[i_{as}\sin 2\theta_r \right.$$

$$\left.\left. + i_{bs}\sin 2\left(\theta_r - \frac{1}{3}\pi\right) + i_{cs}\sin 2\left(\theta_r + \frac{1}{3}\pi\right) \right] \right\},$$

$$\frac{di_{bs}}{dt} = \frac{1}{L_{ls} + \bar{L}_m - L_{\Delta m}\cos 2\left(\theta_r - \frac{2}{3}\pi\right)}\left\{-r_s i_{bs} + u_{bs} + \left[\frac{1}{2}\bar{L}_m + L_{\Delta m}\cos 2\left(\theta_r - \frac{1}{3}\pi\right)\right]\frac{di_{as}}{dt}\right.$$

$$+ \left(\frac{1}{2}\bar{L}_m + L_{\Delta m}\cos 2\theta_r\right)\frac{di_{cs}}{dt} - 2L_{\Delta m}\omega_r\left[\left(i_{as}\sin 2\left(\theta_r - \frac{1}{3}\pi\right)\right.\right.$$

$$\left.\left.\left. + i_{bs}\sin 2\left(\theta_r - \frac{2}{3}\pi\right) + i_{cs}\sin 2\theta_r\right)\right]\right\},$$

$$\frac{di_{cs}}{dt} = \frac{1}{L_{ls} + \bar{L}_m - L_{\Delta m}\cos 2\left(\theta_r + \frac{2}{3}\pi\right)}\left\{-r_s i_{cs} + u_{cs} + \left(\frac{1}{2}\bar{L}_m + L_{\Delta m}\cos 2\left(\theta_r + \frac{1}{3}\pi\right)\right)\frac{di_{as}}{dt}\right.$$

$$+ \left(\frac{1}{2}\bar{L}_m + L_{\Delta m}\cos 2\theta_r\right)\frac{di_{bs}}{dt} - 2L_{\Delta m}\omega_r\left[i_{as}\sin 2\left(\theta_r + \frac{1}{3}\pi\right)\right.$$

$$\left.\left. + i_{bs}\sin 2\theta_r + i_{cs}\sin 2\left(\theta_r + \frac{2}{3}\pi\right)\right]\right\},$$

$$\frac{d\omega_r}{dt} = \frac{P^2}{4J}L_{\Delta m}\left[i_{as}^2\sin 2\theta_r + 2i_{as}i_{bs}\sin 2\left(\theta_r - \frac{1}{3}\pi\right) + 2i_{as}i_{cs}\sin 2\left(\theta_r + \frac{1}{3}\pi\right)\right.$$

$$\left. + i_{bs}^2\sin 2\left(\theta_r - \frac{2}{3}\pi\right) + 2i_{bs}i_{cs}\sin 2\theta_r + i_{cs}^2\sin 2\left(\theta_r + \frac{2}{3}\pi\right)\right] - \frac{B_m}{J}\omega_r - \frac{P}{2J}T_L,$$

$$\frac{d\theta_r}{dt} = \omega_r.$$

The differential equations in Cauchy's form for three-phase synchronous motors are reported in Refs. [4,5]. To control the angular velocity, the electromagnetic torque must be regulated. To maximize the electromagnetic torque and reduce the torque ripple, one strives to feed the following phase currents as functions of the angular displacement θ_r measured or observed (sensorless control):

$$i_{as} = \sqrt{2}i_M\sin\left(\theta_r + \frac{1}{3}\varphi_i\pi\right), \quad i_{bs} = \sqrt{2}i_M\sin\left[\theta_r - \frac{1}{3}(2 - \varphi_i)\pi\right], \quad \text{and}$$

$$i_{cs} = \sqrt{2}i_M\sin\left[\theta_r + \frac{1}{3}(2 + \varphi_i)\pi\right].$$

For $\varphi_i = 0.3245$, one obtains $T_e = \sqrt{2}PL_{\Delta m}i_M^2$.

The derived expressions for the phase currents i_{as}, i_{bs}, and i_{cs}, which maximize T_e and eliminate (theoretically) torque ripple, are called the balanced current set. Power amplifiers usually control the phase voltages u_{as}, u_{bs}, and u_{cs}. The three-phase balanced voltage set is

$$u_{as} = \sqrt{2}u_M\sin\left(\theta_r + \frac{1}{3}\varphi_i\pi\right), \quad u_{bs} = \sqrt{2}u_M\sin\left[\theta_r - \frac{1}{3}(2 - \varphi_i)\pi\right], \quad \text{and}$$

$$u_{cs} = \sqrt{2}u_M\sin\left[\theta_r + \frac{1}{3}(2 + \varphi_i)\pi\right],$$

where u_M is the magnitude of the supplied voltages.

We conclude that T_e is theoretically maximized, and the magnitude of T_e is controlled by changing the magnitude of the phase currents i_M or voltages u_M. The angular rotor displacement θ_r must be measured to control synchronous motors, and balanced $\mathbf{i}_{abcs}$ and $\mathbf{u}_{abcs}$ are functions of θ_r. Assuming an ideal machine design and ideal implementation of sinusoidal $\mathbf{i}_{abcs}$ or $\mathbf{u}_{abcs}$ (by power electronics), one may not expect the torque ripple. In practice, experimental results show that there are current chattering, torque ripple, overheating, vibrations, noise, and other undesirable phenomena due to nonlinear magnetic system, cogging, fringing effects, electromagnetic field nonuniformity, nonuniformity of electromagnetic system and magnetic materials, PWM, nonsinusoidal applied voltage waveforms, and other phenomena. In general, even the high-fidelity modeling (three-dimensional Maxwell's equations and tensor calculus) and heterogeneous simulations do not allow one to ensure an absolute consistency. However, very important features were found. We establish a control concept to ensure a near-optimal operation of synchronous reluctance motors. Direct-current, conventional synchronous, induction, and permanent-magnet synchronous electromechanical motion devices ensure significant advantages as compared to synchronous reluctance motors.

In this chapter, we covered electrostatic and variable reluctance electromechanical motion devices. Our major objectives were to

1. Examine basic physics and functionality of electromechanical motion devices
2. Develop sound control concepts
3. Assess the baseline performance and capabilities
4. Introduce the reader to hardware design and integration, including power electronics

The specific objectives and related tasks were also covered. For example, modeling, simulation, and experimental results in evaluation of performance and capabilities were reported. It was illustrated that only soundly examining the overall objectives, multiple specific objectives and goals can be addressed, examined, and solved.

References

1. Fitzgerald, A. E., Kingsley, C., and Umans, S. D., *Electric Machinery*, McGraw-Hill, New York, 1990.
2. Krause, P. C. and Wasynczuk, O., *Electromechanical Motion Devices*, McGraw-Hill, New York, 1989.
3. Krause, P. C., Wasynczuk, O., and Sudhoff, S. D., *Analysis of Electric Machinery*, IEEE Press, New York, 1995.
4. Lyshevski, S. E., *Electromechanical Systems, Electric Machines, and Applied Mechatronics*, CRC Press, Boca Raton, FL, 1999.
5. Lyshevski, S. E., *Electromechanical Systems and Devices*, CRC Press, Boca Raton, FL, 2007.
6. White, D. C. and Woodson, H. H., *Electromechanical Energy Conversion*, Wiley, New York, 1959.

4

Permanent-Magnet Direct-Current Motion Devices and Actuators

4.1 Permanent-Magnet Motion Devices and Electric Machines: Introduction

The principle of energy conversion and electromagnetic electromechanical motion devices were first examined and demonstrated by Michael Faraday in 1821. One of the first commutator direct current (DC) electric motors was designed, tested, utilized, and commercialized by Anyos Jedlik in 1828 and William Sturgeon in 1832. Alternating current machines (synchronous and induction) were invented and demonstrated by Nicola Tesla in 1880s. By 1882, Nicola Tesla pioneered and developed the theory of the rotating magnetic field which is a cornerstone principle of electromechanical motion devices. He designed and demonstrated a two-phase induction motor in 1883. The first three-phase squirrel-cage induction motor was invented and demonstrated by Michail Osipovish Dolivo-Dobrovolski in 1890.

In this chapter, we study various high-performance translational and rotational DC electromechanical motion device and actuators which operate utilizing the electromagnetic interactions between windings and permanent magnets. As covered in Section 2.5, the torque tends to align the magnetic moment $\vec{m}$ with $\vec{B}$, and $\vec{T} = \vec{m} \times \vec{B}$. Various illustrative examples were reported and visualized in Figure 2.14. One also recalls that the torque is $\vec{T} = \vec{R} \times \vec{F}$, where for a filamentary closed loop the expression for the electromagnetic force is $\vec{F} = -i \oint_l \vec{B} \times d\vec{l}$. This equation is simplified to $\vec{F} = -i\vec{B} \times \oint d\vec{l}$ for a uniform magnetic flux density distribution. We examined electrostatic and variable-reluctance actuators in Chapter 3 emphasizing that other advanced actuation solutions exist. The device physics can be centered on utilization of the energy stored by permanent magnets which establish a strong stationary magnetic field. The electromagnetic force and torque production is evident from the equations for F and T reported. Various permanent-magnet motion devices were devised, designed, fabricated, and widely utilized.

The studied motion device and actuators have been called electric machines and electric motors. Usually, the windings are placed on the rotor (brushes and commutator are used to supply voltage to windings on the rotational rotor), while permanent magnets are on the stator (stationary member). From $\vec{F} = -i\vec{B} \times \oint d\vec{l}$ one concludes that the device physics is based on electromagnetic force or torque developed between windings on the moving (or stationary) member and permanent magnets on the stationary (or moving) member. As examples, permanent-magnet DC motors, limited-angle axial-topology actuators, and

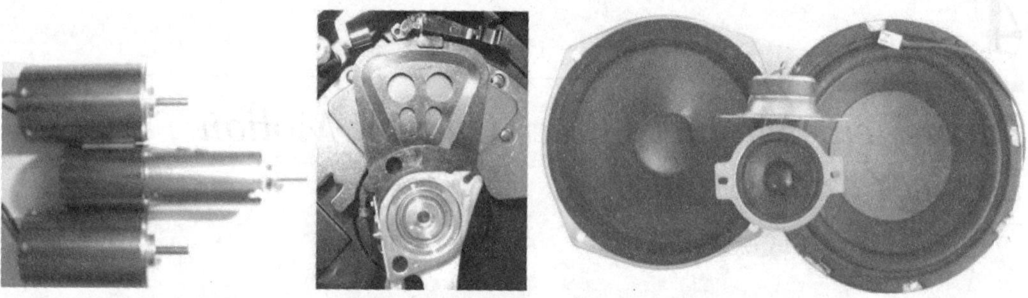

FIGURE 4.1
Permanent-magnet DC motors, limited-angle axial-topology actuator and speakers.

speakers are documented in Figure 4.1. In general, superior performance, excellent capabilities, and affordability are ensured by permanent-magnet DC and AC electromechanical motion devices. Correspondingly, these devices are the preferred choice and are widely applied in the majority of systems. The power range of permanent-magnet devices is up to ~100 kW with the overloading capability (for a relatively short time from seconds to minutes) reaching ~10. This chapter covers various permanent-magnet electromechanical motion devices.

4.2 Radial Topology Permanent-Magnet Direct-Current Electric Machines

4.2.1 Electric Machines and PWM Amplifiers

The basic electromagnetic principles and fundamental physical laws are used to devise, design, and examine various electromechanical motion devices [1–8]. The principle of energy conversion by electromechanical motion devices was first demonstrated by Michael Faraday in 1821 using the homopolar motor. The first practical commutator-type DC electric motor was invented and demonstrated by William Sturgeon in 1832. The DC generator-motor system was demonstrated in 1873 by Zenobe Gramme. In this book, we focus on high-performance electromechanical motion devices which utilize permanent magnets.

Permanent-magnet DC electric machines guarantee high power and torque densities, efficiency, affordability, reliability, ruggedness, overloading capabilities, and other advantages [1,3–6]. The power range of permanent-magnet DC electric machines (motors and generators) is from microwatts to ~100 kW, and the dimensions are from ~1 mm to ~1 m in diameter and length. The same permanent-magnet electric machine can be used as a motor or generator. Due to the above mentioned advantages and very high performance, permanent-magnet DC electric machines and motion devices are very widely used in aerospace, automotive, marine, power, robotics, and other applications. Only permanent-magnet synchronous machines, which do not have brushes and mechanical commutator, surpass permanent-magnet DC machines. Therefore, depending on applications, in high-performance drives and servos, predominantly permanent-magnet DC and synchronous electric machines are utilized.

To drive a computer/camera hard drive, household fan, small pump, and 60 ton tank (track), ~1 W, ~10 W, ~100 W, ~10 kW, and ~100 kW machines respectively, are needed.

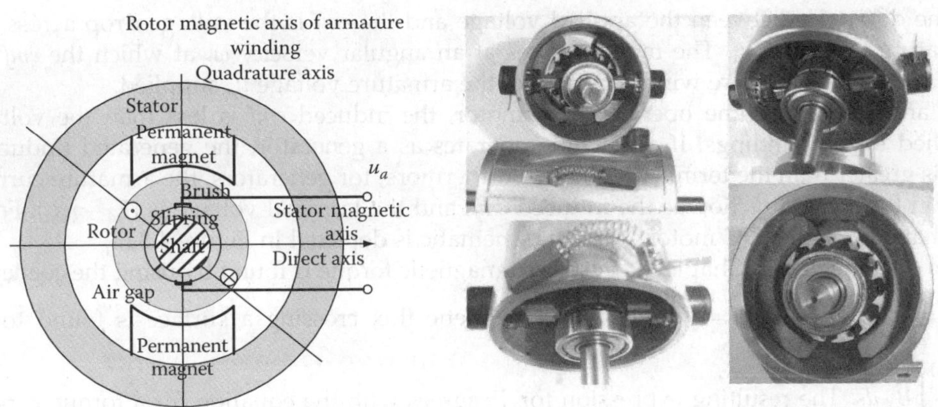

FIGURE 4.2
Permanent-magnet DC electric machine schematics and images.

Therefore, $\sim$1 µW to $\sim$100 kW power range covers the major consumer and industrial systems. In submegawatt and megawatt range applications (ships, locomotives, high-power energy systems, etc.), induction and synchronous machines are used.

Permanent-magnet DC electric machines are rotating energy-transforming electromechanical motion devices which convert energies. Motors (actuators) convert electrical energy to mechanical energy, while generators convert mechanical energy to electrical energy. As was emphasized, the same permanent-magnet electric machine can operate as the motor (if one applies the voltage) or as the generator (if the torque is applied to rotate the machine, the voltage is induced). Electric machines have stationary and rotating members separated by an air gap. The armature winding is placed in the rotor slots and connected to a rotating commutator, which rectifies the voltage, see Figure 4.2. One supplies the armature voltage u_a to the rotor windings. The rotor windings and permanent magnets on stator are magnetically coupled. The brushes ride on the commutator which is connected to the armature windings. The armature winding consists of identical uniformly distributed coils. The excitation stationary magnetic field is produced by permanent magnets. The images of a permanent-magnet DC electric machine, with the above mentioned components, are reported in Figure 4.2.

Due to the commutator (circular conducting copper segments on the rotor as depicted in Figure 4.2), the voltage is supplied to the armature windings on rotor. The armature windings and permanent magnets produce stationary *mmfs* which are displaced by 90 electrical degrees. The armature magnetic force is along the *quadrature* (rotor) magnetic axis, while the *direct* axis stands for a permanent-magnet magnetic axis. The electromagnetic torque is produced as a result of the interaction of the magnetic dipole moment and stationary magnetic field. For motors, using Kirchhoff's law, one obtains the following steady-state equation for the armature voltage u_a

$$u_a - E_a = r_a i_a$$

where r_a is the armature resistance, E_a is the *back emf* (which was also denoted as $\mathcal{E}$), and i_a is the currents in the armature winding.

The difference between the applied voltage and the *emf* is the voltage drop across the armature resistance r_a. The motor rotates at an angular velocity ω_r at which the *emf* E_a, induced in the armature winding, *balances* the armature voltage u_a supplied.

If an electric machine operates as a motor, the induced *emf* is less than the voltage applied to the windings. If a machine operates as a generator, the generated (induced) *emf* is greater than the terminal voltage. Furthermore, for generators, the armature current i_a is in the same direction as the induced *emf*, and the terminal voltage is $(E_a - r_a i_a)$. For a permanent-magnet DC motor, a circuit schematic is depicted in Figure 4.3a.

As documented in Chapter 2, the electromagnetic torque is found by using the coenergy $W_c = \oint_i \psi\, di$ as $T_e(i, \theta) = \dfrac{\partial W_c(i, \theta)}{\partial \theta}$. The magnetic flux crossing a surface is found to be $\Phi = \oint_s \vec{B} \cdot d\vec{s}$. The resulting expression for T_e agrees with the equation for a torque experienced by a current loop in the magnetic field, e.g., $\vec{T} = \vec{m} \times \vec{B} = \vec{a}_m m \times \vec{B} = iA\vec{a}_m \times \vec{B}$. In permanent-magnet electric machines, the stationary near-uniform magnetic field $\vec{B}$ is

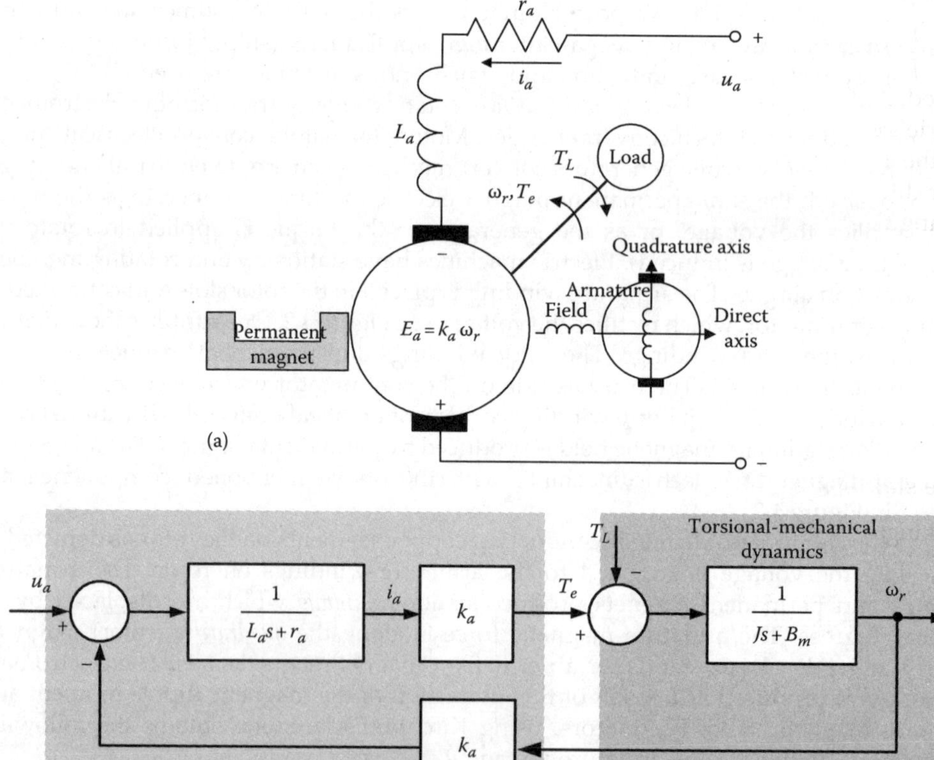

FIGURE 4.3
(a) Schematic diagram of a permanent-magnet DC electric machine (current direction corresponds to the motor operation); (b) s-domain diagram of permanent-magnet DC motors.

produced by permanent magnets, and B and A can be assumed to be constant. One also recalls that the *emf* and *mmf* are given as

$$emf = \oint_l \vec{E} \cdot \vec{dl} = \oint_l \left(\vec{v} \times \vec{B} \right) \cdot \vec{dl} - \oint_s \frac{\partial \vec{B}}{\partial t} \vec{ds} \quad \text{and} \quad mmf = \oint_l \vec{H} \cdot \vec{dl} = \oint_s \vec{J} \cdot \vec{ds} + \oint_s \frac{\partial \vec{D}}{\partial t} \vec{ds}.$$

Applying the above reported expressions for the *emf* and electromagnetic torque, under the common assumptions (magnetic field is stationary, magnetic system is linear, magnetic *susceptibility* is constant, etc.), we have the following expressions for the *back emf* and the electromagnetic torque

$$E_a = k_a \omega_r$$

and

$$T_e = k_a i_a$$

where k_a are the *back emf* and *torque* constants.

Using Kirchhoff's voltage law $u_a = r_a i_a + \dfrac{d\psi}{dt}$ and Newton's second law of motion $\dfrac{d\omega_r}{dt} = \dfrac{1}{J}(T_e - B_m \omega_r - T_L)$, the differential equations for permanent-magnet DC motors are derived. Assume that the *susceptibility* of permanent magnets is constant (in general, there is Curie's constant, and the *susceptibility* varies as a function of temperature), one concludes that the field, established by the permanent magnets, is constant. Thus, $k_a = \text{const}$. The linear differential equations which describe the transient behavior of the armature current and angular velocity are

$$\begin{aligned} \frac{di_a}{dt} &= -\frac{r_a}{L_a} i_a - \frac{k_a}{L_a} \omega_r + \frac{1}{L_a} u_a \\ \frac{d\omega_r}{dt} &= \frac{k_a}{J} i_a - \frac{B_m}{J} \omega_r - \frac{1}{J} T_L \end{aligned} \tag{4.1}$$

In the state-space matrix form, we have $\dfrac{d\mathbf{x}}{dt} = A\mathbf{x} + B\mathbf{u}$.

Taking note of $\mathbf{x} = [i_a \ \omega_r]^T$, $\mathbf{u} = u_a$, $A \in \mathbb{R}^{2 \times 2}$, and $B \in \mathbb{R}^{2 \times 1}$, from (4.1), we have

$$\begin{bmatrix} \dfrac{di_a}{dt} \\ \dfrac{d\omega_r}{dt} \end{bmatrix} = \begin{bmatrix} -\dfrac{r_a}{L_a} & -\dfrac{k_a}{L_a} \\ \dfrac{k_a}{J} & -\dfrac{B_m}{J} \end{bmatrix} \begin{bmatrix} i_a \\ \omega_r \end{bmatrix} + \begin{bmatrix} \dfrac{1}{L_a} \\ 0 \end{bmatrix} u_a - \begin{bmatrix} 0 \\ \dfrac{1}{J} \end{bmatrix} T_L. \tag{4.2}$$

An *s*-domain diagram of permanent-magnet DC motors is illustrated in Figure 4.3b.

From the differential equation $\dfrac{di_a}{dt} = -\dfrac{r_a}{L_a} i_a - \dfrac{k_a}{L_a} \omega_r + \dfrac{1}{L_a} u_a$, for the steady-state operation one has $0 = -r_a i_a - k_a \omega_r + u_a$. Hence, $\omega_r = \dfrac{u_a - r_a i_a}{k_a}$.

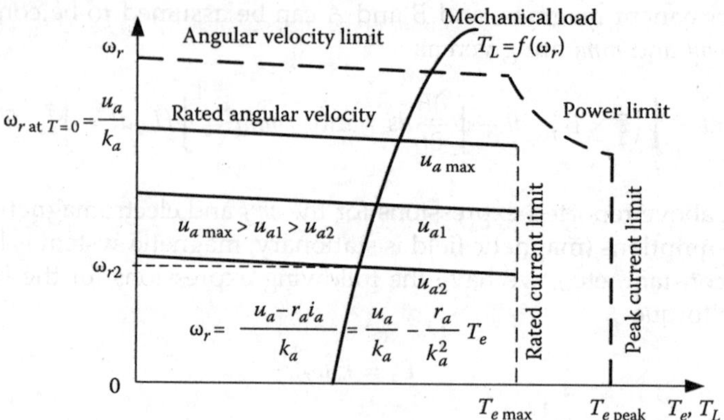

FIGURE 4.4

Torque–speed characteristics of permanent-magnet DC motors. Note that the overloading capabilities are specified by the manufacturer, and for a short time, $T_{e\,\text{peak}}/T_{e\,\text{max}}$ could reach ~10.

The electromagnetic torque is $T_e = k_a i_a$. In the steady-state operation $T_e = T_L$ or $T_e = T_{\text{friction}} + T_L$. Thus, the steady-state torque–speed characteristics are described by the following torque–speed equation

$$\omega_r = \frac{u_a - r_a i_a}{k_a} = \frac{u_a}{k_a} - \frac{r_a}{k_a^2} T_e. \tag{4.3}$$

Equation (4.3) illustrates that one changes the applied armature voltage u_a to vary the angular velocity. Furthermore, if the load is applied, the angular velocity reduces. The slope of the torque–speed characteristic is $-r_a/k_a^2$. The torque–speed characteristics are illustrated in Figure 4.4 for different u_a, and $|u_a| \le u_{a\,\text{max}}$, where $u_{a\,\text{max}}$ is the maximum (rated) voltage.

To reduce the angular velocity, one decreases u_a. The angular velocity at which motor rotates is found as the intersection of the torque–speed characteristic and the load characteristic. For example, for u_{a2} applied, the angular velocity is ω_{r2}. In fact, from Newton's second law, neglecting friction, one has $\dfrac{d\omega_r}{dt} = \dfrac{1}{J}(T_e - T_L)$. Thus, at $T_e = T_L$ motor rotates at the constant angular velocity. At no load, from (4.3) one finds that the angular velocity is $\omega_r = u_a/k_a$. The angular velocity can be reversed if the polarity of the applied voltage is changed (the direction of the field cannot be changed).

Example 4.1:

Calculate and plot the torque–speed characteristics for a 12 V (rated) permanent-magnet DC motor with the following parameters: $r_a = 2$ ohm and $k_a = 0.05$ V-s/rad or N-m/A. The load is a nonlinear function of the angular velocity and $T_L = f(\omega_r) = 0.02 + 0.000002\omega_r^2$ N m.

The torque–speed characteristics are governed by equation (4.3). Using different values for the armature voltage u_a in equation (4.3), the steady-state characteristics are calculated and plotted as depicted in Figure 4.5. The load curve is also illustrated. The following MATLAB® file to perform calculations and plotting is used.

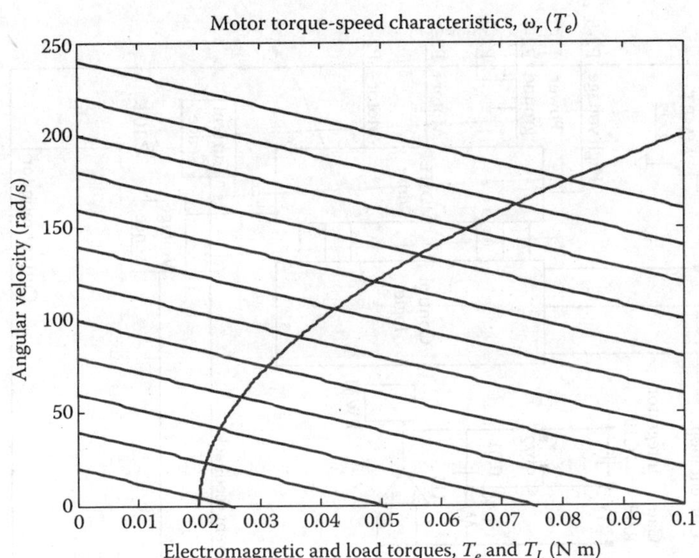

FIGURE 4.5
Torque–speed and load characteristics for a permanent-magnet DC motor.

```
% parameters of a permanent-magnet DC motor
ra = 2; ka = 0.05;
Te = 0:0.001:0.1; % torque in N-m
for ua = 1:1:12;   % applied voltage
wr = ua/ka-(ra/ka^2)*Te; % angular velocity for different voltages
wrl = 0:1:225; Tl = 0.02+2e-6*wrl.^2; % load torque at different velocities
plot(Te,wr,'-',Tl,wrl,'-','LineWidth',2);
title('Motor Torque-Speed Characteristics, \omega_r(T_e)','FontSize',14);
xlabel('Electromagnetic and Load Torques. T_e and T_L [N-m]','FontSize',14);
ylabel('Angular Velocity [rad/sec]','FontSize',14);
hold on; axis([0, 0.1, 0, 250]);
end;
```

To regulate the angular velocity, one changes u_a. The power electronics and PWM amplifiers are covered in Chapter 7. The four-quadrant H-configured power stages guarantee high performance, efficiency, and other desired capabilities. To rotate motor clockwise and counterclockwise, the bipolar voltage u_a should be applied to the armature winding. Permanent-magnet DC electric machines are made in different sizes. For ~500 W permanent-magnet DC motors, see an image on Figure 4.2, the schematic of a four-quadrant 25 A PWM servo amplifier (20–80 V, ±12.5 A continuous current, ±25 A peak current, 22 kHz, 129 × 76 × 25 mm dimensions) is documented in Figure 4.6. The motor armature winding is connected to P2-1 and P2-2. To control the angular velocity, one supplies the reference voltage to P1-4, which is compared with the voltage induced by the tachogenerator (tachometer voltage is proportional to the motor angular velocity) which is supplied to P1-6. This amplifier can be used in the servosystem applications, and the voltage proportional to the angular (or linear) displacement should be supplied to P1-6.

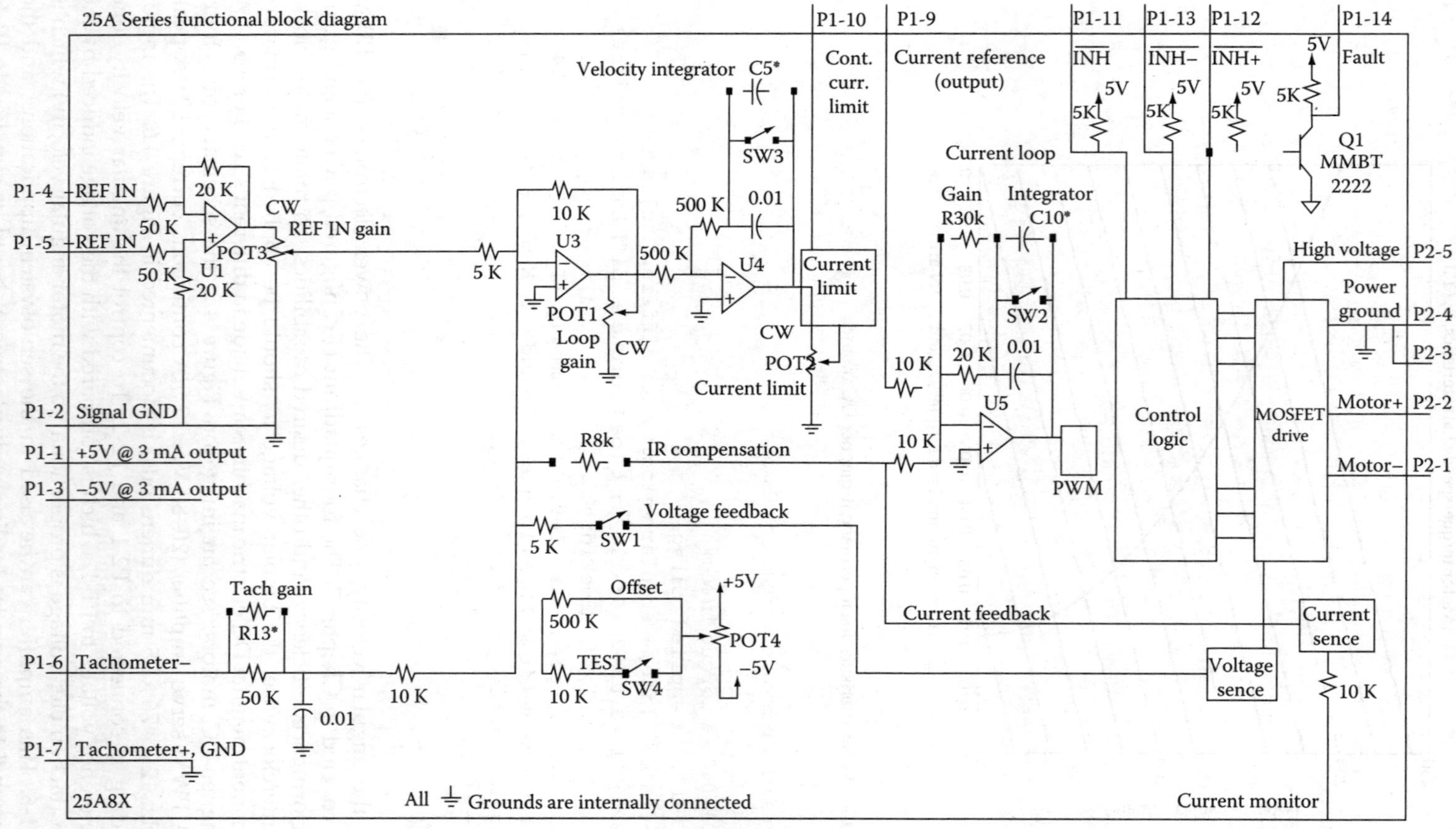

FIGURE 4.6

25 A PWM servo amplifier. (From Lyshevski, S.E., *Electromechanical Systems, Electric Machines, and Applied Mechatronics*, CRC Press, Boca Raton, FL, 1999. Courtesy of Advanced Motion Controls, www.a-m-c.com. With permission.)

The proportional-integral analog controller is integrated in the PWM amplifier. One can change the proportional and integral feedback gains adjusting the corresponding potentiometers (resistors). Various PWM amplifiers are available from Advanced Motion Controls and other companies. For example, the 12A8 amplifiers specifications are 20–80 V, $\pm$6 A continuous current, $\pm$12 A peak current, 36 kHz, and $129 \times 76 \times 25$ mm dimensions.

The size of permanent-magnet DC electric machine can be less that the size of operational amplifier. High-performance 2 and 4 mm in diameter permanent-magnet DC motors have been manufactured. One must estimate the load torque. To guarantee the rotation, the following condition $T_e > T_L$ must be met. Furthermore, the acceleration capability is found as $(T_e - T_L)/J$. Having derived the electromagnetic torque T_e and angular velocity ω_r required, the power is found as $P = T_e \omega_r$. The sizing and volumetric features can be estimated by taking note that the power density estimation is ~ 1 W/cm^3. The power density is significantly affected by the design, electromagnetic system, dimensionality, permanent-magnet used, magnet dimensions, angular velocity, and other features.

A small $\sim$1–10 W permanent-magnet DC motors can be driven by dual operational amplifiers, and the schematic is depicted in Figure 4.7a. The TCA0372 dual power operational amplifiers (40 V and 1 A) can be effectively used to ensure bidirectional rotation of motors, and the transient dynamics are reported in Figure 4.7a. The motor angular velocity is controlled by changing u_a. Monolithic PWM amplifiers with corresponding ICs are available. For example, the MC33030 DC servo motor controller/driver (36 V and 1 A) integrates on-chip operational amplifier and comparator, driving logics, PWM four-quadrant converter, and other circuitry. The built-in proportional controller changes the armature voltage u_a using the difference between the reference and actual angular velocity (drive application) or displacement (servo application). The schematic of the MC33030 is documented in Figure 4.7b. The VNH2SP30, VNH3SP30, and other fully integrated H-bridge permanent-magnet DC motor driver (STMicroelectronis www.st.com and other) can be used.

If the permanent-magnet DC electric machine is used as a generator with a resistive load R_L, we have

$$\frac{di_a}{dt} = -\frac{r_a + R_L}{L_a} i_a + \frac{k_a}{L_a} \omega_r.$$

The induced *emf* is $E_a = k_a \omega_r$. In the steady-state operation, the induced terminal voltage is proportional to the angular velocity. In order to generate the voltage, one rotates a permanent-magnet DC electric machine by using a prime mover, and the torque applied is denoted as T_{pm}. The applied torque T_{pm} could be the aerodynamic, hydrodynamic, thermal, or of any other origin. The resulting differential equations of motion for a permanent-magnet DC generator are

$$\frac{di_a}{dt} = -\frac{r_a + R_L}{L_a} i_a + \frac{k_a}{L_a} \omega_r,$$

$$\frac{d\omega_r}{dt} = -\frac{k_a}{J} i_a - \frac{B_m}{J} \omega_r + \frac{1}{J} T_{pm} \qquad (4.4)$$

4.2.2 Simulation and Experimental Studies of Permanent-Magnet Direct-Current Machines

The majority of electric machines are described by nonlinear differential equations. Permanent-magnet DC electric machines are among a very limited class of motion devices which can be described by linear differential equations. However, the linear theory may not always be applied to permanent-magnet DC machines because there are constraints on the applied voltage $|u_a| \leq u_{a\max}$. One recalls also that the friction is a nonlinear function.

The state-space model of permanent-magnet DC motors was found as given by (4.2). The simulation parameters (final time, initial conditions, and other) as well as electric machines parameters must be assigned. Let the initial conditions are zero and the final

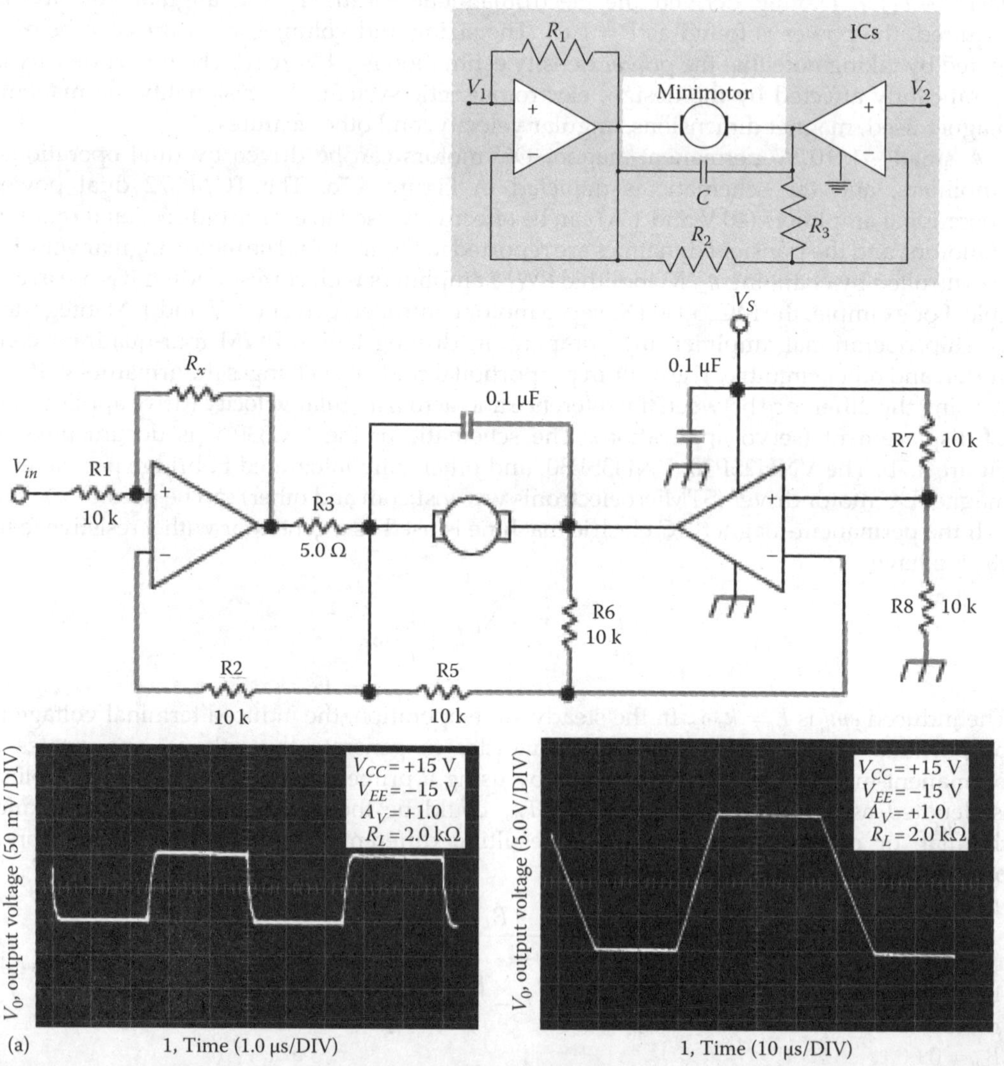

FIGURE 4.7
(a) Application of dual operational amplifier to control permanent-magnet DC minimotors, bidirectional speed control of permanent-magnet DC minimotors using TCA0372 dual power operational amplifier (40 V and 1 A), and the TCA0372 transient responses;

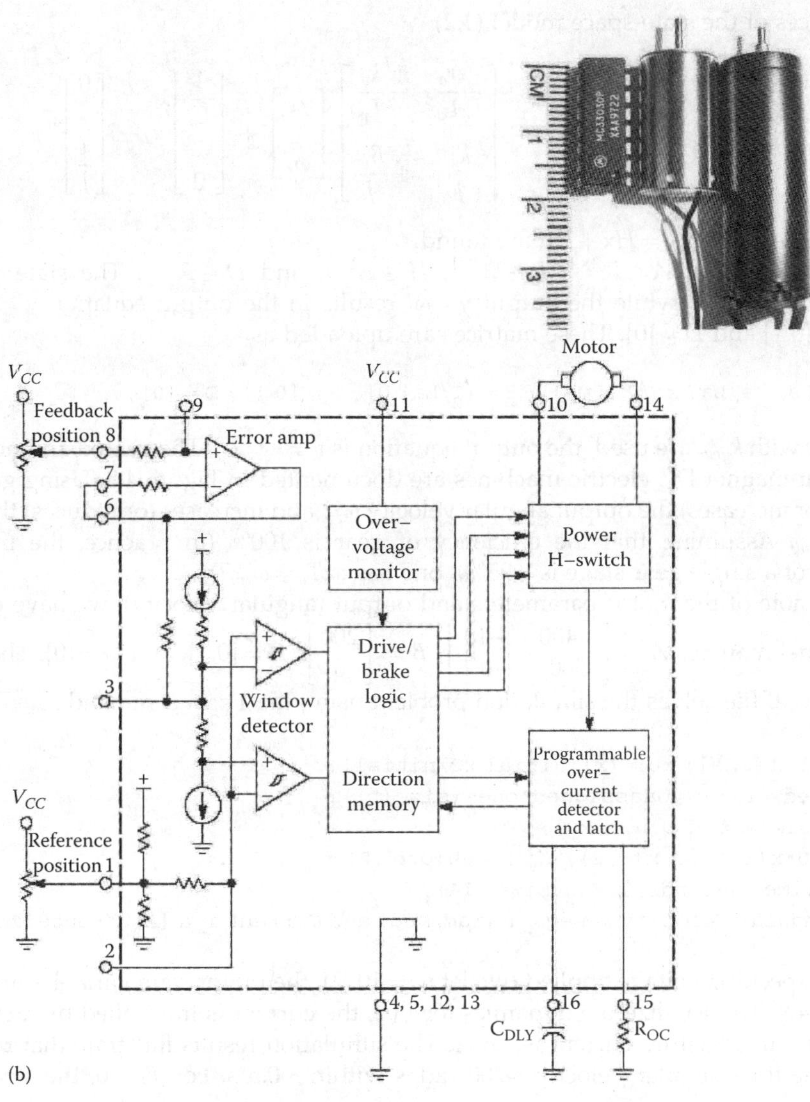

FIGURE 4.7 (continued)
(b) Image of the MC33030 DC servo motor controller/driver and ~1–3 W permanent-magnet DC minimotors, and the MC33030 DC servo motor controller/driver schematics. (From Lyshevski, S.E., *Electromechanical Systems, Electric Machines, and Applied Mechatronics*, CRC Press, Boca Raton, FL, 1999. Courtesy of Motorola.)

time is 0.5 s. The following motor coefficients for a 12 V (rated) permanent-magnet DC motor were experimentally found: $r_a = 2$ ohm, $k_a = 0.05$ V-s/rad (N-m/A), $L_a = 0.005$ H, $B_m = 0.0001$ N-m-s/rad, and $J = 0.0001$ kg-m^2. In the Command Window we enter these parameters as

```
tfinal = 0.5; x1initial = 0; x2initial = 0;
ra = 2; ka = 0.05; La = 0.005; Bm = 0.00001; J = 0.0001;
```

The matrices of the state-space model (4.2)

$$\frac{d\mathbf{x}}{dt} = A\mathbf{x} + B\mathbf{u}, \quad \begin{bmatrix} \dfrac{di_a}{dt} \\ \dfrac{d\omega_r}{dt} \end{bmatrix} = \begin{bmatrix} -\dfrac{r_a}{L_a} & -\dfrac{k_a}{L_a} \\ \dfrac{k_a}{J} & -\dfrac{B_m}{J} \end{bmatrix} \begin{bmatrix} i_a \\ \omega_r \end{bmatrix} + \begin{bmatrix} \dfrac{1}{L_a} \\ 0 \end{bmatrix} u_a - \begin{bmatrix} 0 \\ \dfrac{1}{J} \end{bmatrix} T_L$$

and output equation $\mathbf{y} = H\mathbf{x} + D\mathbf{u}$ are found.

It is obvious that $A \in \mathbb{R}^{2\times2}$, $B \in \mathbb{R}^{2\times1}$, $H \in \mathbb{R}^{1\times2}$, and $D \in \mathbb{R}^{1\times1}$. The state vector is $\mathbf{x} = [x_1\ x_2]^T = [i_a\ \omega_r]^T$, while the output $y = \omega_r$ results in the output equation $\mathbf{y} = H\mathbf{x} + D\mathbf{u}$ with $H = [0\ 1]$ and $D = [0]$. These matrices are uploaded as

```
A= [-ra/La -ka/La; ka/J -Bm/J]; B= [1/La; 0]; H= [0 1]; D= [0];
```

If the gear with k_{gear} are used, the output equation is $y = k_{gear}\omega_r$. The geared and not geared permanent-magnet DC electric machines are documented in Figure 4.8. Using gears, one reduces (or increases) the output angular velocity ω_{rm} and increases (or reduces) the output torque T_{em}. Assuming that the efficiency of gear is 100% (in practice, the maximum efficiency of a single gear stage is $\sim$95%), one has $\omega_r T_e = \omega_{rm} T_{em}$.

Taking note of the motor parameters and output (angular velocity), we have the state-space model matrices $A = \begin{bmatrix} -400 & -10 \\ 500 & -0.1 \end{bmatrix}$, $B = \begin{bmatrix} 200 \\ 0 \end{bmatrix}$, $H = [0\ 1]$, and $D = [0]$. The following MATLAB file solves the simulation problem using the `lsim` command

```
t=0:.001: tfinal; x0= [x1initial x2initial];
Uaassigned=10; u=Uaassigned*ones(size(t));
[y,x] =lsim(A,B,H,D,u,t,x0);
plot(t,10*x(:,1),t,x(:,2),':','LineWidth',2);
xlabel('Time (seconds)','FontSize',14);
title('Angular Velocity \omega_r [rad/sec] and Current i_a [A]','FontSize',14);
```

For the specified voltage applied (we let $u_a = 10$ V), the motor state variables are plotted in Figure 4.9. To visualize the dynamics for $i_a(t)$, the current is multiplied by factor of 10. The maximum armature current is 4.6 A. The simulation results illustrate that the motor reaches the final angular velocity $\sim$200 rad/s within $\sim$0.5 s. For $T_L = 0$, the steady-state

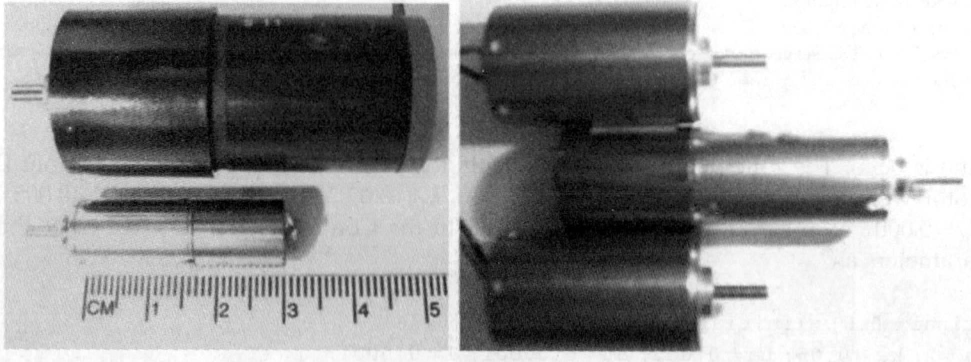

FIGURE 4.8
Geared and nongeared permanent-magnet DC motors with outputs $y = k_{gear}\omega_r$ and $y = \omega_r$.

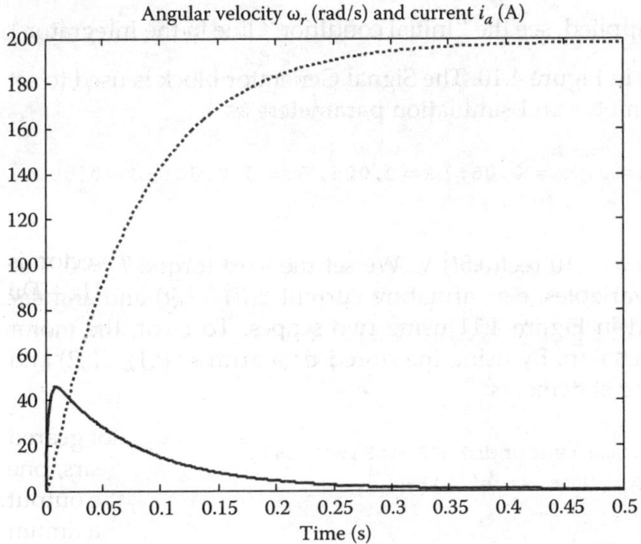

FIGURE 4.9
Dynamics of the state variables $i_a(t)$ (solid line) and $\omega_r(t)$ (dashed line).

value of the angular velocity should be $\omega_r = u_a/k_a = 200$ rad/s, see equation (4.3). However, as evident from Figure 4.9, the steady-state ω_r is slightly less than 200 rad/s. This is due to the friction torque which is $B_m\omega_r \approx 0.002$ N-m. The armature current i_a will not decrease to zero because $T_{e\,\text{steady-state}} = B_m\omega_r = k_a i_a$.

The simulation of permanent-magnet DC electric machines can also be performed using Simulink. One needs to develop a Simulink diagram. An s-domain diagram for permanent-magnet DC motors was reported in Figure 4.3b. Using this s-domain diagram, the corresponding Simulink model (`ch4_01.mdl`) is built and represented in Figure 4.10. The initial

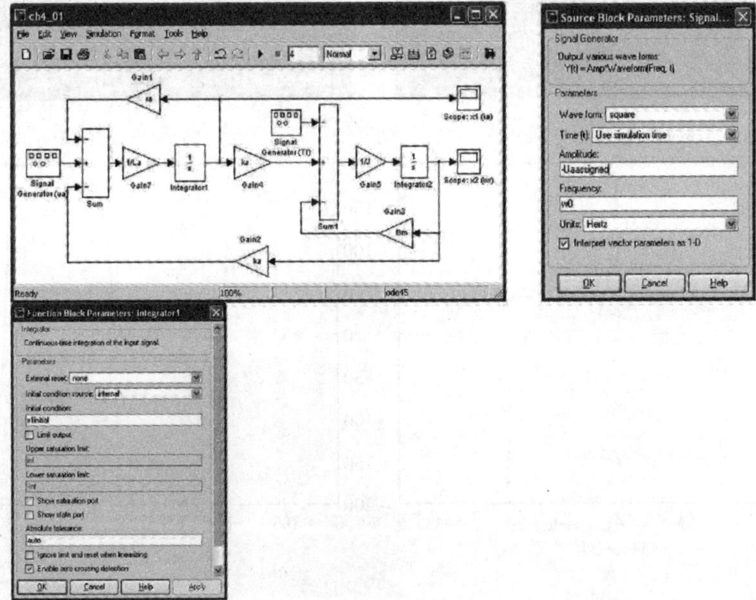

FIGURE 4.10
Simulink model to simulate permanent-magnet DC motors, Signal Generator, and Integrator blocks.

conditions $x_0 = \begin{bmatrix} x_{10} \\ x_{20} \end{bmatrix} = \begin{bmatrix} 0 \\ 0 \end{bmatrix}$ are applied, see the "Initial condition" line in the Integrator 1 (armature current) block as shown in Figure 4.10. The Signal Generator block is used to set the applied voltage. We enter the motor and simulation parameters as

```
x1initial = 0; x2initial = 0; ra = 2; ka = 0.05; La = 0.005; Bm = 0.00001; J = 0.0001;
Uaassigned = 10; w0 = 0.5;
```

The applied armature voltage is $u_a = 10 \, \text{rect}(0.5t)$ V. We set the load torque $T_L = 0$. The transient responses for two state variables, e.g., armature current $x_1(t) = i_a(t)$ and angular velocity $x_2(t) = \omega_r(t)$, are illustrated in Figure 4.11 using two scopes. To plot the motor dynamics, one may use the plot function. By using the stored data arrays x(:,1), x(:,2) and x1(:,1), x1(:,2) we type the following statements

```
plot(x(:,1),x(:,2)); xlabel('Time (seconds)','FontSize',14);
title('Armature Current i_a, [A]','FontSize',14);
```

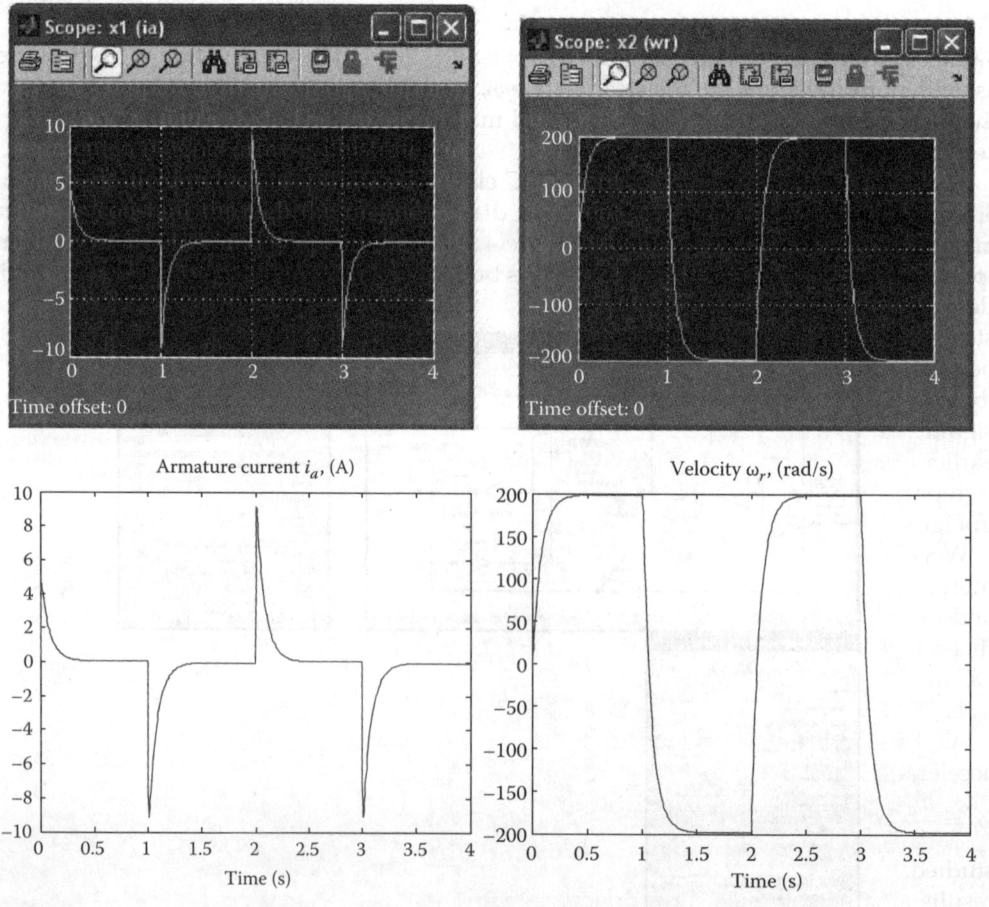

FIGURE 4.11
Permanent-magnet motor dynamics.

and

```
plot(x1(:,1),x1(:,2)); xlabel('Time (seconds)','FontSize',14);
title('Velocity \omega_r, [rad/sec]','FontSize',14);
```

The resulting plots are documented in Figure 4.11.

As simulations are performed, the analysis can be accomplished. One analyzes the steady-state and dynamics responses of the state variables, settling time, overshoot, stability, etc. For example, the efficiency and losses are of interest. The losses can be estimated as

$$P_1(t) = r_a i_a^2(t) + B_m \omega_r^2(t),$$

while the efficiency can be assessed at the steady-state operation using the input and output power as

$$\eta = P_{\text{output}}/P_{\text{input}} = T_L \omega_r / u_a i_a,$$
$$P_{\text{output}} = P_{\text{input}} - P_1 = u_a i_a - \left(r_a i_a^2 + B_m \omega_r^2\right),$$
$$P_{\text{input}} = P_1 + P_{\text{output}} = r_a i_a^2 + B_m \omega_r^2 + T_L \omega_r,$$

or

$$\eta = (P_{\text{output}}/P_{\text{input}}) \times 100\% = (T_L \omega_r / u_a i_a) \times 100\%.$$

Steady-state and dynamic analysis of efficiency, power, and losses can be accomplished studying the operation under varying inputs, loads, and disturbances. Only for motion devices which predominantly operate within the constant operating conditions, the steady-state analysis may be sufficient. For any systems, the steady-state analysis can be accomplished from the dynamic analysis, but, not vise versa. Therefore, the dynamic (behavioral) analysis is more general and should be prioritized. The losses increase significantly in the transient regimes. One may modify the Simulink model developed earlier to perform the analysis of power and losses. The resulting diagram (`ch4_011.mdl`) is reported in Figure 4.12a. The input power and losses are plotted and documented in Figure 4.12b.

We demonstrated the application of MATLAB to simulate permanent-magnet DC motors. These simulations must be carried out in order to estimate system performance and capabilities. However, the experimental studies are very important and validate any theoretical and applied results. The experimental results for JDH2250 permanent-magnet DC motors are conducted. Figure 4.13 documents the acceleration of the unloaded motor if u_a is 7.5 and 15 V.

As the load torque T_L is applied, the angular velocity decreases. Figure 4.14 reports the acceleration and loading of the motor. The equation for torque–speed characteristics $\omega_r = \dfrac{u_a - r_a i_a}{k_a} = \dfrac{u_a}{k_a} - \dfrac{r_a}{k_a^2} T_e$ illustrates the steady-state operation. The motor dynamics, as studied by solving the differential equations provide general results. The experimental results are in a complete correspondence with the transient analysis performed using the equations of motion derived. The motor acceleration, transient dynamics waveforms, settling time, and other performance characteristics and capabilities were examined.

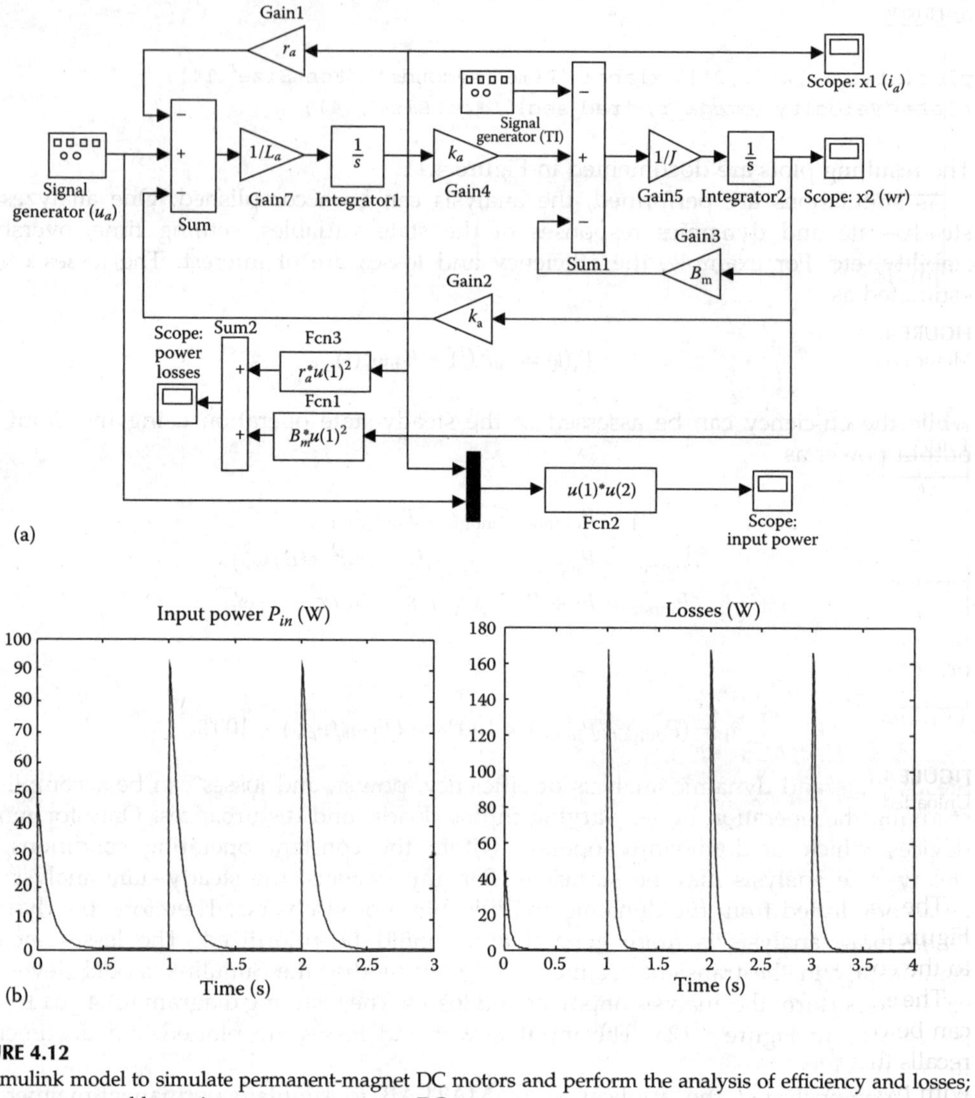

FIGURE 4.12

(a) Simulink model to simulate permanent-magnet DC motors and perform the analysis of efficiency and losses;
(b) input power and losses in a permanent-magnet DC motor.

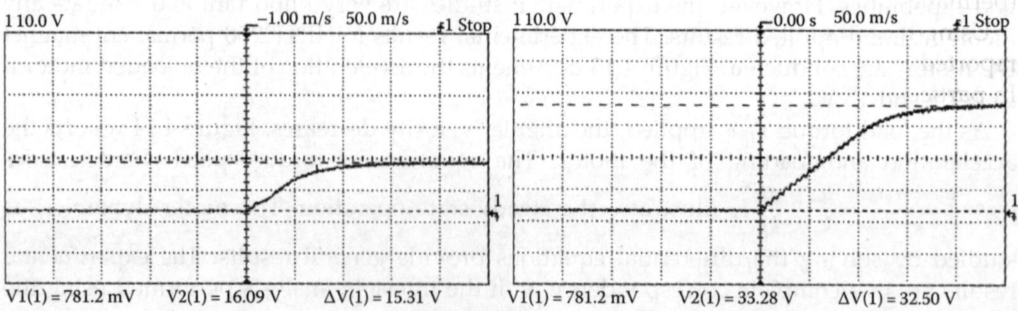

FIGURE 4.13

Motor acceleration to the angular velocity 150 and 300 rad/s.

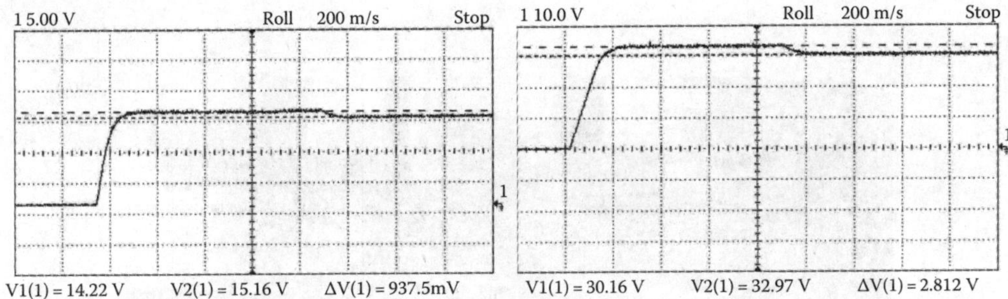

FIGURE 4.14
Motor acceleration and loading.

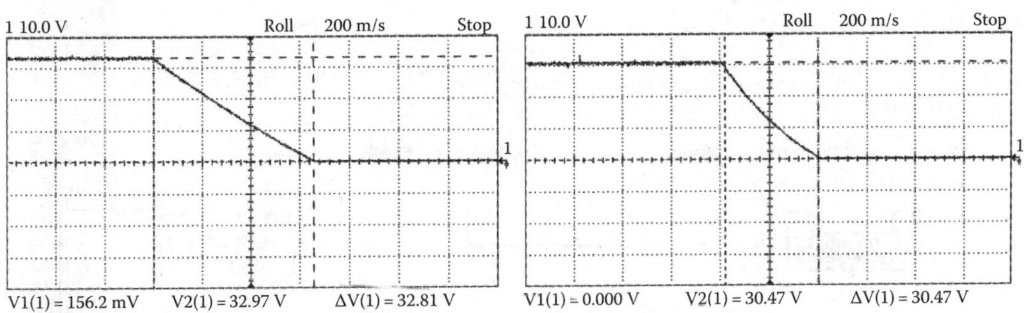

FIGURE 4.15
Unloaded and loaded motor disacceleration from 300 rad/s to stall.

The disacceleration dynamics of the unloaded and loaded motor are documented in Figure 4.15. The experimental results do not accurately match the simulated dynamics due to the complex friction phenomena emphasized in Example 2.7.

The experimental studies can be performed by utilizing the motor-generator system which can be composed from two identical (or different) permanent-magnet DC machines. One recalls that these machines can be utilized as motors or generators. The experimental setup with two identical 23 NEMA-size ($\sim$250 W rated, 70 V, 400 rad/s) permanent-magnet DC machines, coupled using a flexible coupler, is illustrated in Figure 4.16a. The prime mover (permanent-magnet DC motor) drives a generator, and the circuit is shown in Figure 4.16b.

Using the subscripts *pm* and g for a prime mover (motor) and generator, and notations reported in Figure 4.16b, one finds the equations of motion for a motor-generator system. In particular,

$$\frac{di_{apm}}{dt} = -\frac{r_{apm}}{L_{apm}}i_{apm} - \frac{k_{apm}}{L_{apm}}\omega_{rpm} + \frac{1}{L_{apm}}u_{apm},$$

$$\frac{di_{ag}}{dt} = -\frac{r_{ag} + R_L}{L_{ag}}i_{ag} + \frac{k_{ag}}{L_{ag}}\omega_{rpm},$$

$$\frac{d\omega_{rpm}}{dt} = \frac{k_{apm}}{J_{pm} + J_g}i_{apm} - \frac{B_{mpm} + B_{mg}}{J_{pm} + J_g}\omega_{rpm} - \frac{k_{ag}}{J_{pm} + J_g}i_{ag}.$$

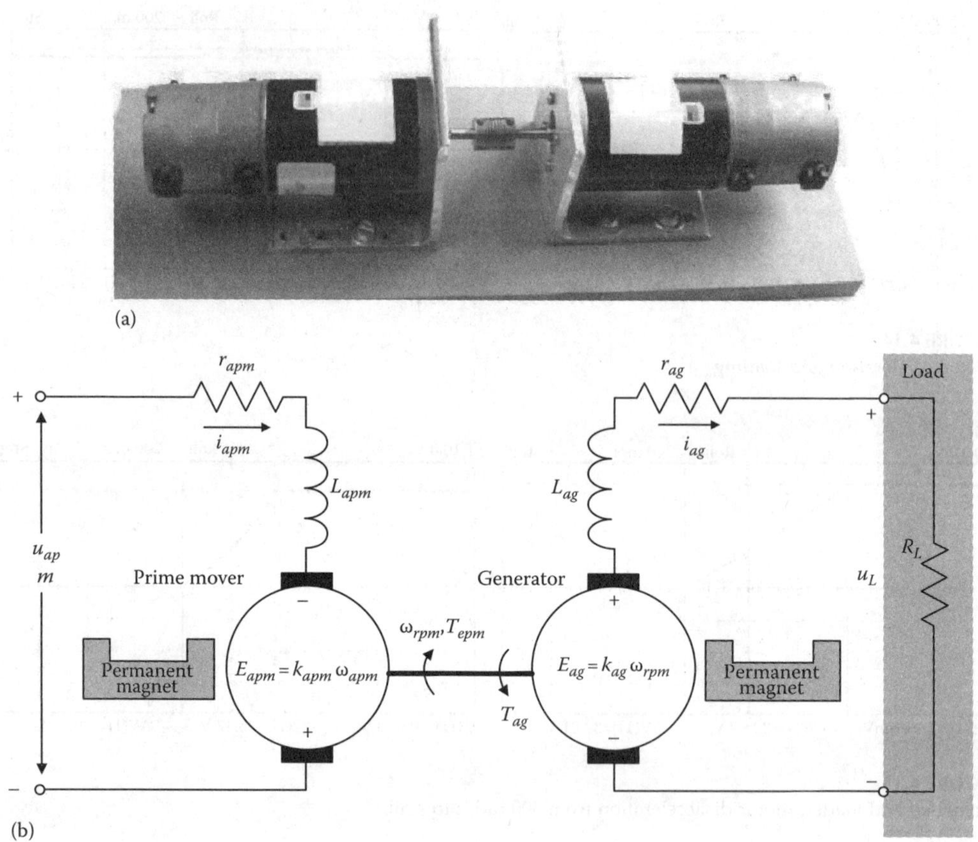

FIGURE 4.16
Motor-generator system. (a) two permanent-magnet DC machines, e.g., a prime mover (motor) and generator. The angular velocity is measured by a tachogenerator (tachometer) which is directly mounted on the shaft of each machine (voltage induced by a tachogenerator is proportional to the angular velocity); (b) permanent-magnet DC generator with a resistive load R_L is driven by a permanent magnet DC motor.

4.2.3 Electromechanical Devices with Power Electronics

The applied voltage to permanent-magnet DC motors is supplied by the high-frequency switching converters and PWM amplifiers. Therefore, motion devices are controlled utilizing the chosen power electronics solution. In this section, we briefly covered high-performance four-quadrant PWM amplifiers. Various one-quadrant converters are reported in Chapter 7. One can integrate and analyze electric machines controlled by PWM converters and amplifiers. For example, the schematics of a high-frequency *step-down* switching converter with a permanent-magnet DC motor is illustrated Figure 4.17.

The voltage applied to the motor winding u_a is regulated by controlling the switching *on* and *off* durations t_{on} and t_{off}. One changes the duty ratio d_D, and $d_D = \dfrac{t_{on}}{t_{on} + t_{off}}$. From (4.1), using the differential equations for the *buck* converter with a filter, as developed in Chapter 7 for the *RL* load

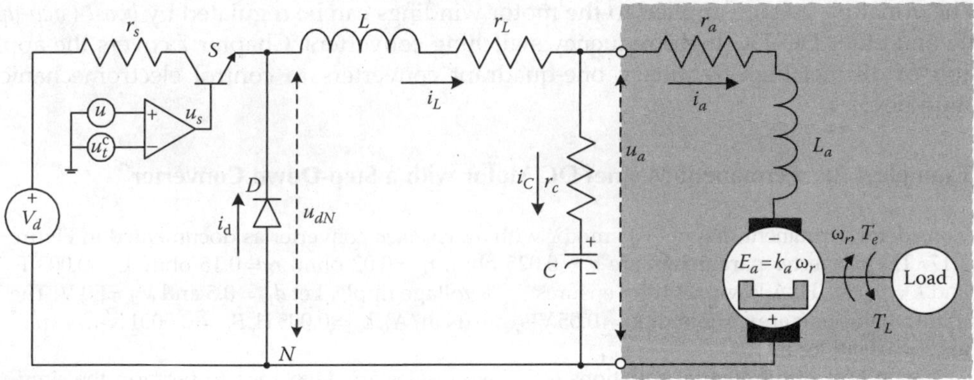

FIGURE 4.17
Permanent-magnet DC motor with *step-down* switching converter and filter.

$$\frac{du_C}{dt} = \frac{1}{C}(i_L - i_a),$$

$$\frac{di_L}{dt} = \frac{1}{L}(-u_C - (r_L + r_c)i_L + r_c i_a - r_s i_L d_D + V_d d_D),$$

$$\frac{di_a}{dt} = \frac{1}{L_a}(u_C + r_c i_L - (r_a + r_c)i_a - E_a),$$

we have a set of five first-order differential equations. In particular,

$$\frac{du_C}{dt} = \frac{1}{C}(i_L - i_a),$$

$$\frac{di_L}{dt} = \frac{1}{L}(-u_C - (r_L + r_c)i_L + r_c i_a - r_s i_L d_D + V_d d_D),$$

$$\frac{di_a}{dt} = \frac{1}{L_a}(u_C + r_c i_L - (r_a + r_c)i_a - k_a\omega_r), \tag{4.5}$$

$$\frac{d\omega_r}{dt} = \frac{1}{J}(k_a i_a - B_m \omega_r - T_L).$$

The duty ratio is regulated by changing the signal-level control voltage u_c which cannot be greater than $u_{t\,max}$ and less than $u_{t\,min}$. That is, $u_{t\,min} \leq u_c \leq u_{t\,max}$. Hence, u_c is bounded. For $u_{t\,min} = 0$, we have

$$d_D = \frac{u_c}{u_{t\,max}} \in [0\ 1], \quad u_c \in [0\ u_{c\,max}], \quad u_{c\,max} = u_{t\,max}.$$

The signal-level control voltage u_c is considered as a control input. The second equation in (4.5) is rewritten as $\dfrac{di_L}{dt} = \dfrac{1}{L}\left(-u_C - (r_L + r_c)i_L + r_c i_a - \dfrac{r_s}{u_{t\,max}}i_L u_c + \dfrac{V_d}{u_{t\,max}}u_c\right)$.

The mathematical model of permanent-magnet DC motors with the *buck* converter is nonlinear due to a nonlinear term $\dfrac{r_s}{L u_{t\,max}}i_L u_c$. Furthermore, the hard control bounds are imposed, e.g., $0 \leq u_c \leq u_{c\,max}$, $u_c \in [0\ u_{c\,max}]$ because $d_D \in [0\ 1]$.

The armature voltage applied to the motor windings can be regulated by *boost* (*step-up*), Cuk, and other DC–DC high-frequency switching converters. Chapter 7 covers the application of distinct high-frequency one-quadrant converters to control electromechanical motion devices.

Example 4.2:   Permanent-Magnet DC Motor with a Step-Down Converter

Consider a permanent-magnet DC motor with a *step-down* converter as documented in Figure 4.17. The converter parameters are $r_s = 0.025$ ohm, $r_L = 0.02$ ohm, $r_c = 0.15$ ohm, $C = 0.003$ F, and $L = 0.0007$ H. A low-pass filter ensures ∼5% voltage ripple. Let $d_D = 0.5$ and $V_d = 50$ V. The motor coefficients are $r_a = 2$ ohm, $k_a = 0.05$ V-s/rad (N m/A), $L_a = 0.005$ H, $B_m = 0.0001$ N-m-s/rad, and $J = 0.0001$ kg-m^2.

Taking note of differential equations (4.5), two m-files are developed to perform the simulation assuming that the unloaded ($T_L = 0$) motor accelerates from the stall. The MATLAB file ch4_1.m is

```
t0 = 0; tfinal = 0.4; tspan = [t0 tfinal]; y0 = [0 0 0 0]';
[t,y] = ode45('ch4_2',tspan,y0);
subplot(2,2,1); plot(t,y(:,1),'-');
xlabel('Time (seconds)','FontSize',12);
title('Voltage u_C, [V]','FontSize',12);
subplot(2,2,2); plot(t,y(:,2),'-');
xlabel('Time (seconds)','FontSize',12);
title('Current i_L, [A]','FontSize',12);
subplot(2,2,3); plot(t,y(:,3),'-');
xlabel('Time (seconds)','FontSize',12);
title('Current i_a, [A]','FontSize',12);
subplot(2,2,4); plot(t,y(:,4),'-');
xlabel('Time (seconds)','FontSize',12);
title('Angular Velocity \omega_r, [rad/sec]','FontSize',12);
```

while the second file ch4_2.m is

```
% Dynamics of the PM DC motor with buck converter
function yprime = difer(t,y);
% parameters
Vd = 50; D = 0.5; rs = 0.025; rl = 0.02; rc = 0.15; C = 0.003; L = 0.0007;
ra = 2; ka = 0.05; La = 0.005; Bm = 0.00001; J = 0.0001; Tl = 0;
% differential equations for PM DC Motor - Buck Converters
yprime = [(y(2,:)-y(3,:))/C; ...
(-y(1,:)-(rl+rc)*y(2,:)+rc*y(3,:)-rs*y(2,:)*D+Vd*D)/L; ...
(y(1,:)+rc*y(2,:)-(rc+ra)*y(3,:)-ka*y(4,:))/La; ...
(ka*y(3,:)-Bm*y(4,:)-Tl)/J];
```

The transient dynamics for the state variables $u_C(t)$, $i_L(t)$, $i_a(t)$, and $\omega_r(t)$ are illustrated in Figure 4.18. The settling time is ∼0.4 s, and the motor reaches the steady-state angular velocity 496 rad/s. To evaluate efficiency, thermodynamics, acceleration capabilities, and other important characteristics, the differential equations which describe the electromechanical system dynamics (and steady-state operation as well) must be solved, and experimental studies should be carried out.   ∎

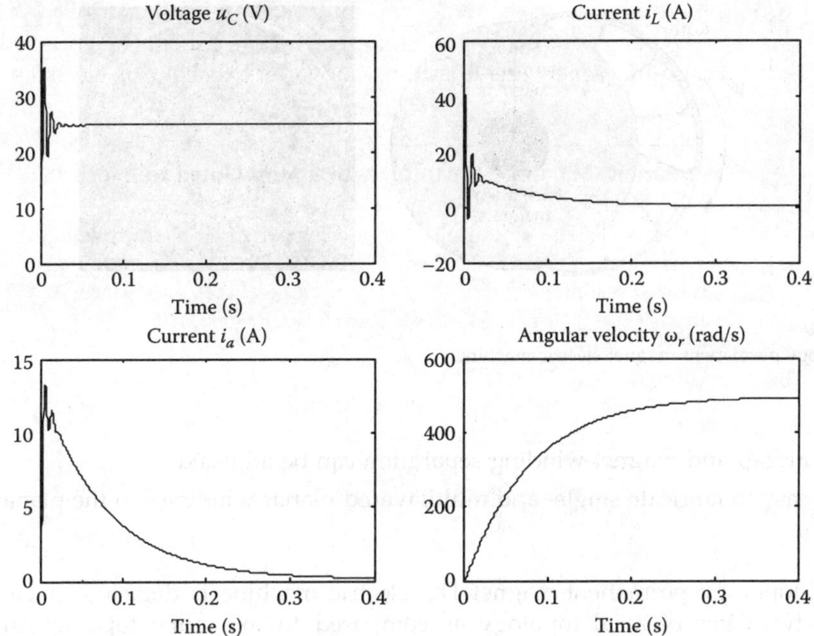

FIGURE 4.18
Transient dynamics of a permanent-magnet DC motor with a *buck* converter.

4.3 Axial Topology Permanent-Magnet Direct-Current Electric Machines

4.3.1 Fundamentals of Axial Topology Permanent-Magnet Machines

Having covered radial topology permanent-magnet DC electric machines, we will examine the axial topology electric machines commonly used as *disc* motors, hard drive actuators, and *limited angle* motors. The planar-segmented permanent magnets array is placed on the stator with the planar windings on the rotor. Brushes and commutator are used to supply the armature voltage to the windings on rotor. The planar windings on the rotor surface significantly simplify fabrication enabling one to fabricate affordable high-performance electromagnetic miniscale and conventional (high torque and power) electromechanical motion devices.

The advantages of the axial topology are

1. Affordability and simplicity to fabricate and assemble machines because permanent magnets and windings are planar

2. There are relaxed shape-geometry and sizing requirements imposed on the magnets and windings (however, device performance is significantly affected by the topology, geometry, spacing, etc.)

3. There is no rotor back ferromagnetic material required (silicon, polymers, or plastics can be applied to fabricate various electromechanical motion devices)

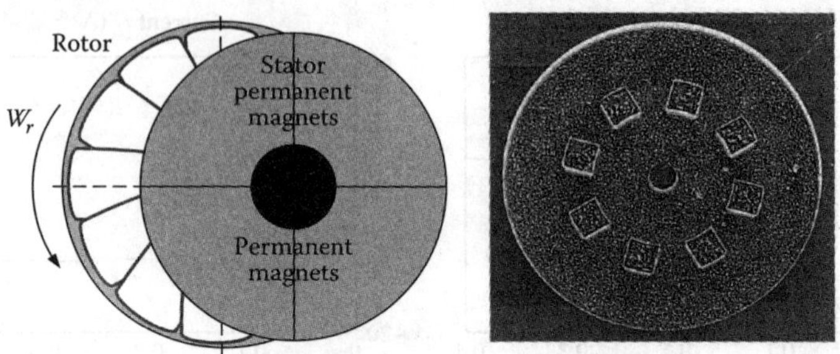

FIGURE 4.19
Axial topology permanent-magnet electric machine.

4. The airgap and magnet-winding separation can be adjusted

5. It is easy to fabricate single- and multilayered planar windings on the planar (flat) stator

The axial topology permanent-magnet DC electric machine is depicted in Figure 4.19. Due to advantages of axial topology as compared to the radial topology, mini- and microelectromechanical motion devices can be fabricated. The image of the fabricated microstructure (stator or rotor) with segmented magnet array is illustrated in Figure 4.19.

Consider a current loop in the magnetic field which is produced by a permanent magnet. Assume that the magnetic flux is constant through the magnetic plane (current loop). The torque on a planar current loop of any size and shape in the uniform magnetic field is

$$\vec{T} = i\vec{s} \times \vec{B} = \vec{m} \times \vec{B},$$

where i is the current in the loop (winding) and $\vec{m}$ is the magnetic dipole moment (A-m^2). The torque is given as $\vec{T} = \vec{R} \times \vec{F}$, where for a filamentary closed loop we have $\vec{F} = -i\oint_l \vec{B} \times d\vec{l}$ which is simplified to $\vec{F} = -i\vec{B} \times \oint d\vec{l}$ for a uniform magnetic flux density distribution.

The torque on the current loop tends to turn the loop to align the magnetic field produced by the loop with permanent-magnet magnetic field causing the resulting electromagnetic torque. For example, a 10×20 cm current loop in an uniform magnetic field with the flux density $\vec{B} = -0.6\vec{a}_y + 0.8\vec{a}_z$ is illustrated in Figure 4.20. The torque is $\vec{T} = i\vec{s} \times \vec{B} = \vec{m} \times \vec{B} = 10[(0.1)(0.2)\vec{a}_z] \times (-0.6\vec{a}_y + 0.8\vec{a}_z) = 0.12\vec{a}_x$ N-m. Hence, the loop tends to rotate around the axis parallel to the positive x-axis.

The electromagnetic force is $\vec{F} = \oint_l i\,d\vec{l} \times \vec{B} = -i\oint_l \vec{B} \times d\vec{l}$, or, in the differential form, $d\vec{F} = i\,d\vec{l} \times \vec{B}$. Furthermore $\vec{T} = \vec{R} \times \vec{F}$. For a straight filament (conductor) in a uniform magnetic field the expression $\vec{F} = i\vec{l} \times \vec{B}$ is found from $\vec{F} = -i\vec{B} \times \oint_l d\vec{l}$. The interaction of

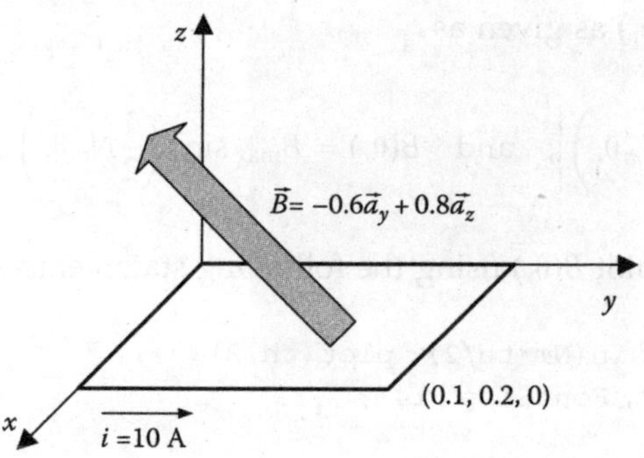

FIGURE 4.20
Rectangular planar current loop in an uniform magnetic field.

the current (in the windings) and magnets result in the electromagnetic force and torque. Permanent magnets are magnetized in the specified direction to ensure the torque production. Each magnet produces the stationary magnetic field, and one considers the interactions of winding filaments with a number of magnets N_m ($N_m = 2m$, m is the integer) which are magnetized in the specific axis.

Consider a one-dimensional problem assuming that the resulting electromagnetic force, acting on the filament, is perpendicular to filament producing the electromagnetic torque to rotate the rotor about a point O (center of the rotor). The expressions $\vec{F} = i\vec{l} \times \vec{B}$ and $\vec{T} = \vec{R} \times \vec{F}$ are simplified to a one-dimensional case assuming the ideal structural design and optimal topology.

We consider the rotational electromechanical motion devices. Under the sound simplifications made, assuming an optimal structural design, one applies the *effective* flux density $B(\theta_r)$ which varies as a function of the angular displacement θ_r due to the angular displacement of rotor with windings relative to the stator with magnets (which produce the stationary field). Depending on the magnet magnetization, geometry and shape, one applies distinct expressions for $B(\theta_r)$. For the permanent magnets illustrated in Figure 4.19, the flux density, as viewed from the windings, is a periodic function of θ_r. If there is no spacing between the magnets or strips of segmented magnets (which can be uniformly magnetized), or, if there is a spacing (or magnets geometry, shape, or magnetization vary), one may use

$$B(\theta_r) = B_{\max}\left|\sin\left(\frac{1}{2}N_m\theta_r\right)\right|$$

or

$$B(\theta_r) = B_{\max}\sin^n\left(\frac{1}{2}N_m\theta_r\right), \quad n = 1, 3, 5, \ldots,$$

where $B_{\max}$ is the maximum effective flux density produced by the magnets as viewed from the winding ($B_{\max}$ depends on the magnets used, magnet-winding separation, number of layers, temperature, etc.), N_m is the number of magnets (segments), and n is the integer which is a function of the magnet magnetization, geometry, shape, width, thickness, permanent-magnet segments separation, and other factors.

Consider an illustrative example. For three $B(\theta_r)$ as given as

$$B(\theta_r) = B_{\max} \sin\left(\frac{1}{2} N_m \theta_r\right), \quad B(\theta_r) = B_{\max}\left|\sin\left(\frac{1}{2} N_m \theta_r\right)\right|, \quad \text{and} \quad B(\theta_r) = B_{\max} \sin^5\left(\frac{1}{2} N_m \theta_r\right)$$

with $B_{\max} = 0.9$ T and $N_m = 4$, we calculate and plot $B(\theta_r)$ using the following statements

```
th = 0:.01:2*pi; Nm = 4; Bmax = 0.9; B = Bmax*sin(Nm*th/2); plot(th,B);
xlabel('Rotor Displacement, \theta_r [rad]','FontSize',14);
ylabel('B(\theta_r) [T]','FontSize',14);
title('Field as a Function on Displacement,B(\theta_r)','FontSize',14);
```

and

```
th = 0:.01:2*pi; Nm = 4; Bmax = 0.9; B = Bmax*sign(sin(Nm*th/2)); plot(th,B);
xlabel('Rotor Displacement, \theta_r [rad]','FontSize',14);
ylabel('B(\theta_r) [T]','FontSize',14);
title('Field as a Function on Displacement,B(\theta_r)','FontSize',14);
```

and

```
th = 0:.01:2*pi; Nm = 4; Bmax = 0.9; B = Bmax*sin(Nm*th/2).^5; plot(th,B);
xlabel('Rotor Displacement, \theta_r [rad]','FontSize',14);
ylabel('B(\theta_r) [T]','FontSize',14);
title('Field as a Function on Displacement,B(\theta_r)','FontSize',14);
```

The resulting plots for $B(\theta_r)$ are reported in Figure 4.21.

Hence, $B(\theta_r)$ can be described by

$$B(\theta_r) = B_{\max}\left|\sin\left(\frac{1}{2} N_m \theta_r\right)\right|, \quad B(\theta_r) = \sum_{n=1}^{\infty} B_{\max\,n} \sin^{2n-1}\left(\frac{1}{2} N_m \theta_r\right),$$

as well as other periodic continuous or discontinuous functions. The analytic (or numerical) expression for $B(\theta_r)$ is used to derive *emf* and T_e.

For a one-dimensional problem, the expressions $\vec{F} = i\vec{l} \times \vec{B}$ and $\vec{T} = \vec{R} \times \vec{F}$ may be simplified to

$$T_e = l_{eq} N i_a B(\theta_r),$$

where l_{eq} is the *effective* length which includes the equivalent winding filament length and lever arm and N is the number of turns.

One can also use the expression for the coenergy $W_c(i_a, \theta_r) = A_{eq}(\theta_r)B(\theta_r)i_a$ to derive the electromagnetic torque. Here, A_{eq} is the *effective* area which takes into account winding-magnet interaction, number of turns, magnetic field nonuniformity, and other features.

Due to the use of brushes and commutator, the electromagnetic torque (produced by multiple filaments) is maximized by properly commutating coils by supplying u_a.

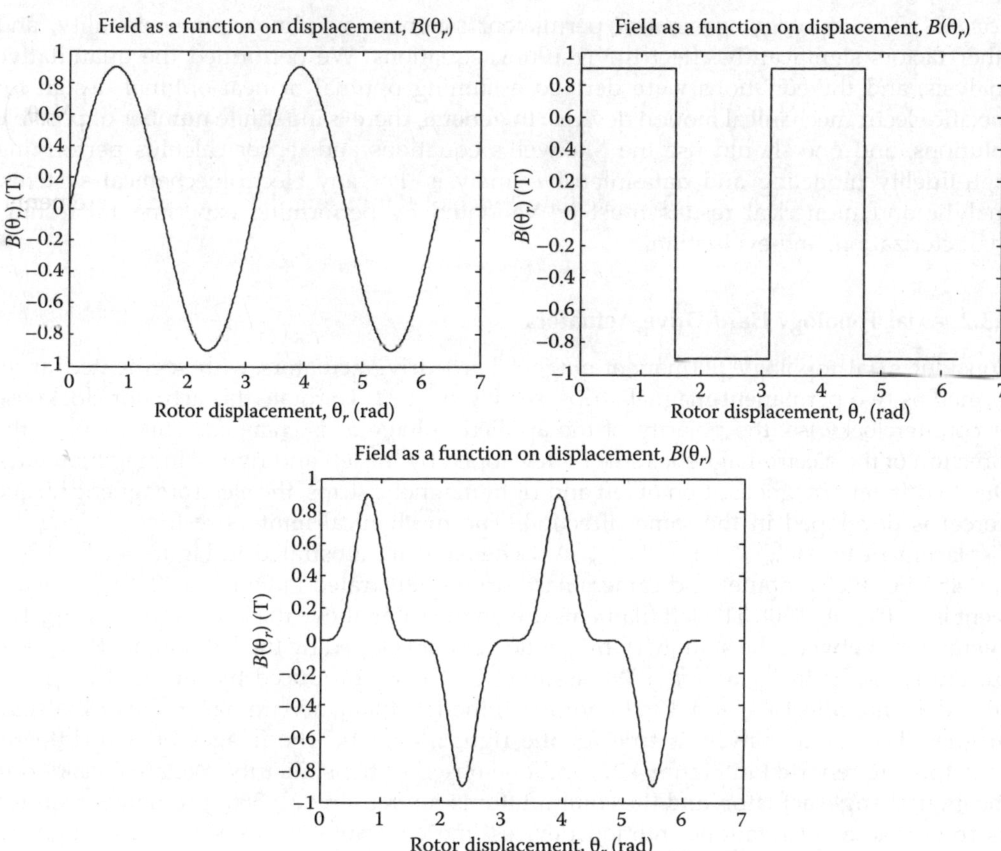

FIGURE 4.21
Plots of $B(\theta_r) = B_{\max}\sin\left(\frac{1}{2}N_m\theta_r\right)$, $B(\theta_r) = B_{\max}\left|\sin\left(\frac{1}{2}N_m\theta_r\right)\right|$, and $B(\theta_r) = B_{\max}\sin^5\left(\frac{1}{2}N_m\theta_r\right)$, $B_{\max} = 0.9$ T and $N_m = 4$.

Depending on design, for many axial and radial topology permanent-magnet DC motors, the expression for the electromagnetic torque is

$$T_e = k_a i_a,$$

where, for example, for $B(\theta_r) = B_{\max}\left|\sin\left(\frac{1}{2}N_m\theta_r\right)\right|$, one has $k_a = l_{eq}NB_{\max}$.

Using Kirchhoff's voltage law $u_a = r_a i_a + \dfrac{d\psi}{dt}$ and Newton's second law of motion $\dfrac{d\omega_r}{dt} = \dfrac{1}{J}(T_e - B_m\omega_r - T_L)$, the differential equations for the axial topology permanent-magnet DC motors (assuming conventional design) are

$$\frac{di_a}{dt} = -\frac{r_a}{L_a}i_a - \frac{k_a}{L_a}\omega_r + \frac{1}{L_a}u_a, \quad \frac{d\omega_r}{dt} = \frac{k_a}{J}i_a - \frac{B_m}{J}\omega_r - \frac{1}{J}T_L.$$

Remark. The topology, geometry, permanent magnets, windings, dimensionality, and other factors significantly affect the resulting equations. We performed the quantitative analysis, and the equations were derived assuming optimal or near-optimal design for specific electromechanical motion devices. In general, there is indefinite number of possible solutions, and one should use the Maxwell's equations and tensor calculus performing high-fidelity modeling and data-intensive analysis. For any electromechanical systems, analytic and numerical results must be validated by performing experimental testing, characterization, and evaluation. ∎

4.3.2 Axial Topology Hard Drive Actuators

Consider axial topology permanent-magnet hard drive actuators with segmented array formed as two permanent-magnet strips, see Figure 4.22. To rotate this actuator clockwise or counterclockwise, the polarity of the applied voltage u_a is changed. This changes the direction of the electromagnetic force F_e developed by the left and right winding filaments. Due to different magnetization of left and right magnetic strips, the electromagnetic torque (force) is developed in the same direction. The mechanical limiters restrict the angular displacement to $-\theta_{r\max} \leq \theta_r \leq \theta_{r\max}$. As schematically illustrated in Figure 4.22 $-45° \leq \theta_r \leq 45°$. For the computer and camera hard drives, illustrated in Figure 4.22, the displacement is $-10° \leq \theta_r \leq 10°$. The left filaments are on the rotor above the left magnetic strip. The interactions between the stationary magnetic field developed by the left magnet $B_L(\theta_r)$ and current i_a results in F_{eL} and T_{eL}. We assume that $B_R(\theta_r)$ produced by the right magnetic strip does not affect F_{eL} and T_{eL}. In addition, the left filaments are never above the right magnet. The same analysis is true for the right filaments. The images of two different actuators are reported in Figure 4.22, and one observes the similarity. We are considering the limited angle actuator, and the commutator is not required. If 360° rotation is required as rotor rotates in rotational motion devices, the commutator is needed to change the polarity of DC voltage supplied to the filaments in order to develop the electromagnetic torque.

The Kirchhoff's voltage law is $u_a = r_a i_a + \dfrac{d\psi}{dt}$. From $\psi = L_a i_a + A_{eq} B(\theta_r)$, we obtain

$$\frac{di_a}{dt} = \frac{1}{L_a}\left(-r_a i_a - A_{eq}\frac{dB(\theta_r)}{dt} + u_a\right).$$

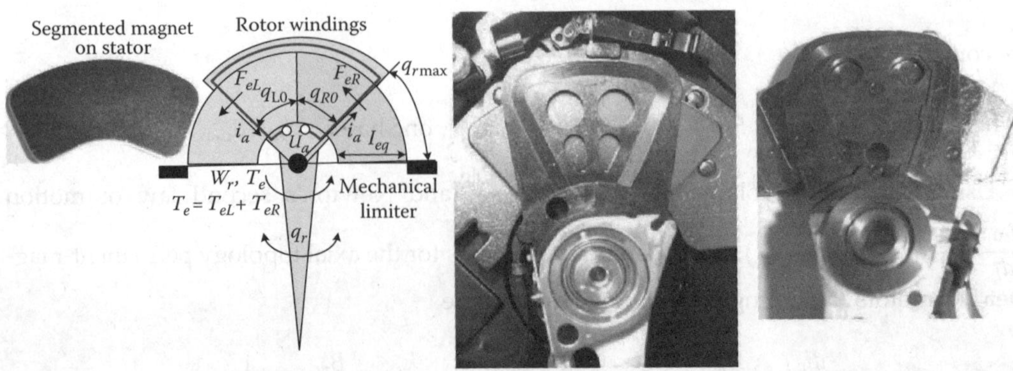

FIGURE 4.22
Limited-angle axial topology hard-drive actuator.

Remark. The magnetic flux is $\Phi = \oint_s \vec{B} \cdot d\vec{s}$, and equation $\Phi = AB$ can be applied with great caution. In general, $\oint_s \vec{B} \cdot d\vec{s} \neq BA$. $\blacksquare$

There are two (left and right) filaments. One may use two *emf* terms yielding

$$\frac{di_a}{dt} = \frac{1}{L_a}\left(-r_a i_a - A_{eq}\frac{dB_L(\theta_r)}{d\theta_r}\omega_r - A_{eq}\frac{dB_R(\theta_r)}{d\theta_r}\omega_r + u_a\right).$$

Newton's second law of motion results in the following equations

$$\frac{d\omega_r}{dt} = \frac{1}{J}(T_e - B_m\omega_r - T_L), \quad T_e = T_{eL} + T_{eR},$$

$$\frac{d\theta_r}{dt} = \omega_r.$$

The analysis is performed as one takes note of $B(\theta_r)$. The distribution of $B(\theta_r)$ significantly affects the overall performance and capabilities. This is reflected by model developed which describes the transient and steady-state behavior. We consider two cases when two magnetic strips are magnetized to ensure:

$$B(\theta_r) = k\theta_r, \quad k > 0; \quad \text{and} \quad (2)\ B(\theta_r) = B_{\max}\tanh(a\theta_r), \quad a > 0.$$

For $B(\theta_r) = k\theta_r$, let $k = 1$, while for $B(\theta_r) = B_{\max}\tanh(a\theta_r)$, we study $a = 10$ and $a = 100$ if $B_{\max} = 0.7$ T. Those $B(\theta_r)$ are plotted in Figure 4.23.

For $B(\theta_r) = k\theta_r$, we have $\dfrac{di_a}{dt} = \dfrac{1}{L_a}\left(-r_a i_a - 2A_{eq}k\omega_r + u_a\right)$.

The expression for the electromagnetic torque $T_e = T_{eL} + T_{eR}$ should be derived. We found $T_{ej} = l_{eq}Ni_a B_j(\theta_r)$. The angular displacements of two filaments are $\theta_L(t)$ and $\theta_R(t)$. As documented in Figure 4.22, there are mechanical limits, and $-\theta_{r\max} \leq \theta_r \leq \theta_{r\max}$. For $-10° \leq \theta_r \leq 10°$, one has $-0.175 \leq \theta_r \leq 0.175$ rad. Assuming $\theta_{r0} = 0$, $\theta_{L0} = 0.175$ rad and $\theta_{R0} = 0.175$ rad, for a symmetric kinematics studied, we have $\theta_L(t) = \theta_{L0} - \theta_r(t)$ and $\theta_R(t) = \theta_{R0} + \theta_r(t)$, $-0.175 \leq \theta_r \leq 0.175$ rad. The filament displacement angles $\theta_L(t)$ and $\theta_R(t)$ are constrained.

As illustrated in Figure 4.22, the current i_a flows in different directions in the left and right filaments. The direction of the developed electromagnetic forces F_{eL} and F_{eR} is the same. Thus, $T_e = T_{eL} + T_{eR} = l_{eq}Nk(\theta_L + \theta_R)i_a$, $\theta_L(t) = \theta_{L0} - \theta_r(t)$, $\theta_R(t) = \theta_{R0} + \theta_r(t)$, $-\theta_{r\max} \leq \theta_r \leq \theta_{r\max}$.

Using the nonideal Hook's law for the returning spring, we have

$$\frac{d\omega_r}{dt} = \frac{1}{J}\left[l_{eq}Nk(\theta_L + \theta_R)i_a - B_m\omega_r - k_s\theta_r - k_{s1}\theta_r^3 - T_{L\xi}\right],$$

$$\theta_L(t) = \theta_{L0} - \theta_r(t), \quad \theta_R(t) = \theta_{R0} + \theta_r(t),$$

$$\frac{d\theta_r}{dt} = \omega_r, \quad -\theta_{r\max} \leq \theta_r \leq \theta_{r\max},$$

where $T_{L\xi}$ denotes the stochastic load torque.

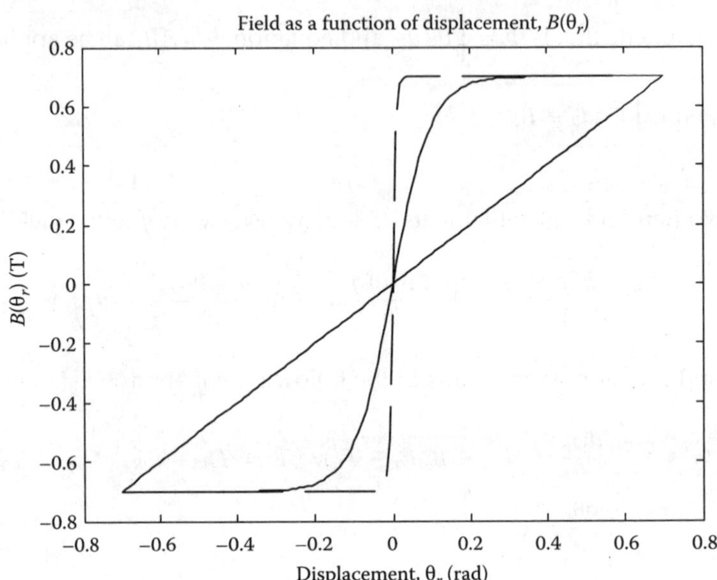

FIGURE 4.23
Plots of $B(\theta_r) = k\theta_r$, $k = 1$ and $B(\theta_r) = B_{\max}\tanh(a\theta_r)$, $B_{\max} = 0.7$ T, $a = 10$, and $a = 100$.

The parameters can be estimated and experimentally measured. We have $k = 1$, $r_a = 35$ ohm, $L_a = 0.0041$ H, $l_{eq} = 0.015$ m, $N = 300$, $A_{eq} = 0.000045$, $B_m = 0.0005$ N-m-s/rad, $k_s = 0.1$ N-m/rad, $k_{s1} = 0.05$ N-m/rad^3, and $J = 0.0000015$ kg-m^2.

These measured parameters can be easily justified. For example, for a copper winding with a radius of 0.05 mm and length $l_L + l_{\text{top}} + l_{\text{bottom}} + l_R$, we have

$$r_a = \frac{N(l_L + l_{\text{top}} + l_{\text{bottom}} + l_R)\sigma}{A} = \frac{300 \times 0.06 \times 1.72 \times 10^{-8}}{\pi(0.00005)^2} = 39 \text{ ohm.}$$

The circular loop self-inductance (loop radius is R_1 and the wire diameter is d), is estimated as

$$L_a = N^2 R_1 \mu_0 \mu_r \left(\ln\frac{16R_1}{d} - 2\right) = 300^2 \times 0.0075 \times 4\pi \times 10^{-7}\left(\ln\frac{8 \times 0.0075}{0.00005} - 2\right) = 0.0043 \text{ H.}$$

The equation for the moment of inertia of thin disk $J = mR_{\text{disk}}^2$ overestimates the J value. We have the disk-centered geometry with the goal to minimize J in order to increase the acceleration capabilities. Therefore, the rotating pointer is made from plastic and has cavities as illustrated in Figure 4.22. The estimated r_a, L_a, and J are in the correspondence with the measured parameter values.

One uploads the parameters and constants as

```
k=1; N=300; ra=35; La=4.1e-3; leq=1.5e-2; Aeq=4.5e-5; Bm=5e-4;
ks=0.1; ks1=0.05; J=1.5e-6;
```

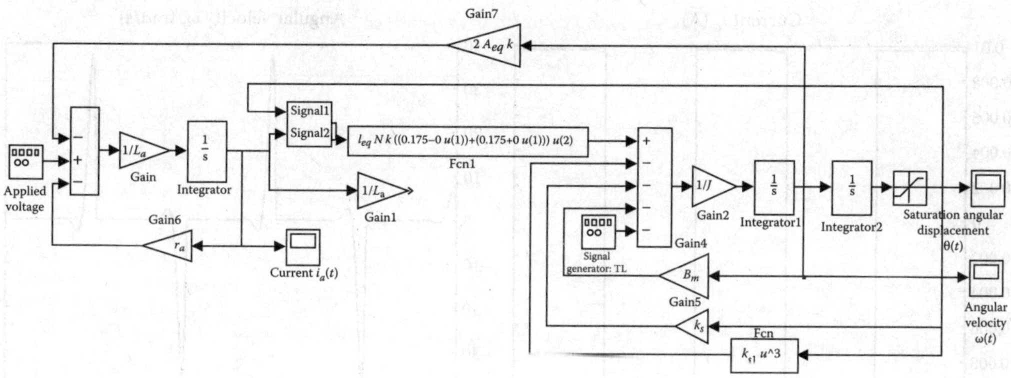

FIGURE 4.24
Simulink diagram to simulate the axial topology limited angle actuator when $B(\theta_r) = k\theta_r$.

For the resulting equations of motion

$$\frac{di_a}{dt} = \frac{1}{L_a}(-r_a i_a - 2A_{eq}k\omega_r + u_a),$$

$$\frac{d\omega_r}{dt} = \frac{1}{J}\left[l_{eq}Nk(\theta_L + \theta_R)i_a - B_m\omega_r - k_s\theta_r - k_{s1}\theta_r^3 - T_{L\xi}\right],$$

$$\theta_L(t) = \theta_{L0} - \theta_r(t), \quad \theta_R(t) = \theta_{R0} + \theta_r(t),$$

$$\frac{d\theta_r}{dt} = \omega_r, \quad -\theta_{r\,max} \leq \theta_r \leq \theta_{r\,max},$$

the corresponding Simulink diagram is depicted in Figure 4.24 (`ch4_04.mdl`).

The transient dynamics for $i_a(t)$, $\omega_r(t)$, and $\theta_r(t)$ are reported in Figure 4.25.

Having examined the case when $B(\theta_r) = k\theta_r$, we study the second case when the magnets are magnetized to ensure $B(\theta_r) = B_{max}\tanh(a\theta_r)$. From

$$\frac{di_a}{dt} = \frac{1}{L_a}\left(-r_a i_a - A_{eq}\frac{dB_L(\theta_r)}{d\theta_r}\omega_r - A_{eq}\frac{dB_R(\theta_r)}{d\theta_r}\omega_r + u_a\right),$$

we have

$$\frac{di_a}{dt} = \frac{1}{L_a}\left(-r_a i_a - A_{eq}aB_{max}\operatorname{sech}^2(a\theta_L)\omega_r - A_{eq}aB_{max}\operatorname{sech}^2(a\theta_R)\omega_r + u_a\right)$$

$$= \frac{1}{L_a}\left(-r_a i_a - A_{eq}aB_{max}\operatorname{sech}^2 a(\theta_{L0} - \theta_r)\omega_r - A_{eq}aB_{max}\operatorname{sech}^2 a(\theta_{R0} + \theta_r)\omega_r + u_a\right).$$

The electromagnetic torque is $T_e = T_{eL} + T_{eR}$.

Hence,

$$T_e = l_{eq}NB_{max}(\tanh a\theta_L + \tanh a\theta_R)i_a, \quad \theta_L(t) = \theta_{L0} - \theta_r(t), \quad \theta_R(t) = \theta_{R0} + \theta_r(t), \quad -\theta_{r\,max} \leq \theta_r \leq \theta_{r\,max}.$$

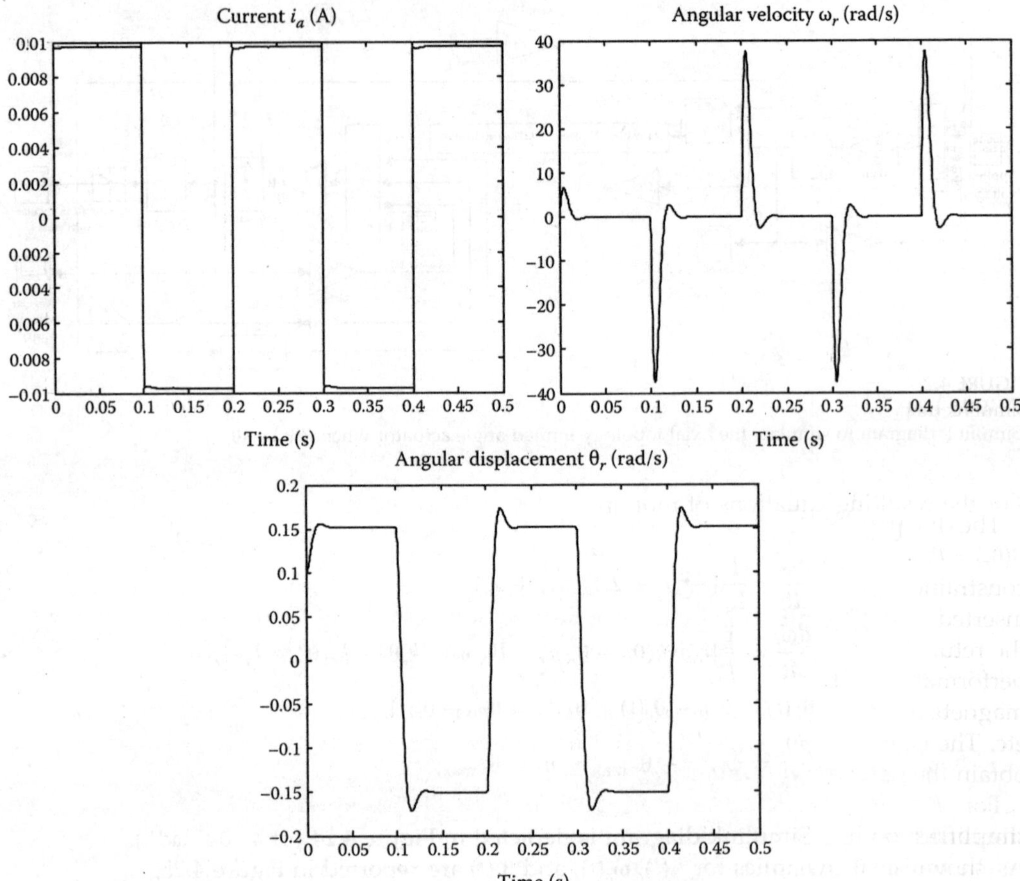

FIGURE 4.25
Transient dynamics of the state variables.

One obtains

$$\frac{di_a}{dt} = \frac{1}{L_a}\left(-r_a i_a - A_{eq}aB_{\max}\operatorname{sech}^2 a(\theta_{L0} - \theta_r)\omega_r - A_{eq}aB_{\max}\operatorname{sech}^2 a(\theta_{R0} + \theta_r)\omega_r + u_a\right),$$

$$\frac{d\omega_r}{dt} = \frac{1}{J}\left[l_{eq}NB_{\max}(\tanh a\theta_L + \tanh a\theta_R)i_a - B_m\omega_r - k_s\theta_r - k_{s1}\theta_r^3 - T_{L\xi}\right]$$

$$= \frac{1}{J}\left\{l_{eq}NB_{\max}[\tanh a(\theta_{L0} - \theta_r) + \tanh a(\theta_{R0} + \theta_r)]i_a - B_m\omega_r - k_s\theta_r - k_{s1}\theta_r^3 - T_{L\xi}\right\},$$

$$\frac{d\theta_r}{dt} = \omega_r, \quad -\theta_{r\,\max} \le \theta_r \le \theta_{r\,\max}.$$

The parameters and constants are uploaded as

```
Bmax = 0.7; a = 10; N = 300; ra = 35; La = 4.1e-3; leq = 1.5e-2;
Aeq = 4.5e-5; Bm = 5e-4; ks = 0.1; ks1 = 0.05; J = 1.5e-6;
```

The Simulink model (`ch4_05.mdl`) is reported in Figure 4.26.

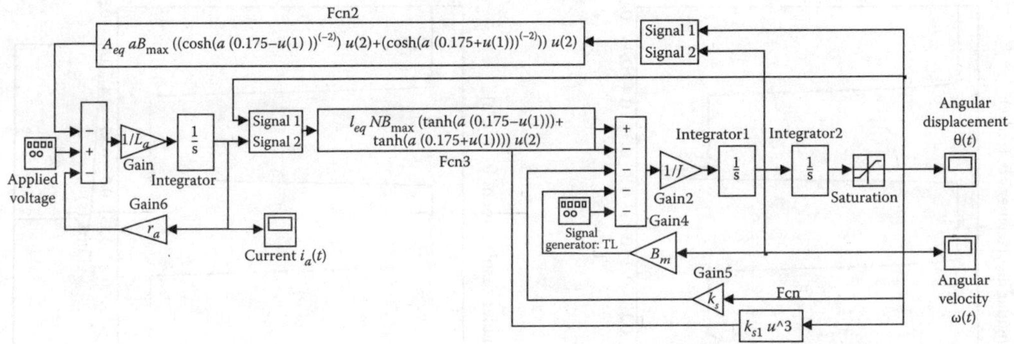

FIGURE 4.26
Simulink diagram to simulate the axial topology limited angle actuator when $B(\theta_r) = B_{max}\tanh(a\theta_r)$.

The dynamics of the state variables $i_a(t)$, $\omega_r(t)$, and $\theta_r(t)$ are represented in Figure 4.27 for $B(\theta_r) = B_{max}\tanh(a\theta_r)$ when $a = 10$ and $a = 100$, respectively. The angular displacement is constrained by the mechanical limits $-0.175 \leq \theta_r \leq 0.175$ rad, and the Saturation block is inserted in the Simulink diagram to account for these bounds. One observes the effect of the return spring on the actuator dynamics when the pointer moves to left and right. The performance of electromechanical motion devices is significantly affected by the magnet magnetization, parameters, and constants which depend on design, fabrication, materials, etc. The closed-loop systems are used to optimize the system dynamics with the goal to obtain the optimal performance.

For $B(\theta_r) = B_{max}\tanh(a\theta_r)$, if $a = 100$, the resulting differential equations can be simplified by simplifying the expressions for the back *emf* and electromagnetic torque. As shown in Figure 4.23, for even small displacement θ_r, $-\theta_{r\,max} \leq \theta_r \leq \theta_{r\,max}$, one has $B(\theta_r) = B_{max}\tanh(100\theta_r) \approx B_{max}\,\mathrm{sgn}(\theta_r)$, yielding $B_L(\theta_r) \approx -B_{max}$ and $B_R(\theta_r) \approx B_{max}$. From $B_L(\theta_r) \approx$ const and $B_R(\theta_r) \approx$ const, one yields *emf* ≈ 0. Using $T_e = T_{eL} + T_{eR} = l_{eq}NB_{max}(\tanh a\theta_L + \tanh a\theta_R)i_a \approx 2l_{eq}NB_{max}i_a$, we have the equations of motion as

$$\frac{di_a}{dt} = \frac{1}{L_a}(-r_a i_a + u_a),$$

$$\frac{d\omega_r}{dt} = \frac{1}{J}\left[2l_{eq}NB_{max}i_a - B_m\omega_r - k_s\theta_r - k_{s1}\theta_r^3 - T_{L\xi}\right],$$

$$\frac{d\theta_r}{dt} = \omega_r, \quad -\theta_{r\,max} \leq \theta_r \leq \theta_{r\,max}.$$

The Simulink model (`ch4_06.mdl`) is reported in Figure 4.28.

The dynamics of $i_a(t)$, $\omega_r(t)$, and $\theta_r(t)$ are illustrated in Figure 4.29. The comparison of the results reported in Figures 4.27b and 4.29 provides the evidence that though the overall dynamics is similar, in general complete mathematical models provides accurate results and must be used. For example, one uses $i_a(t)$ to evaluate the losses, thermodynamics, heating, etc. An accurate model provides higher values of $i_a(t)$ to be used ensuring coherency.

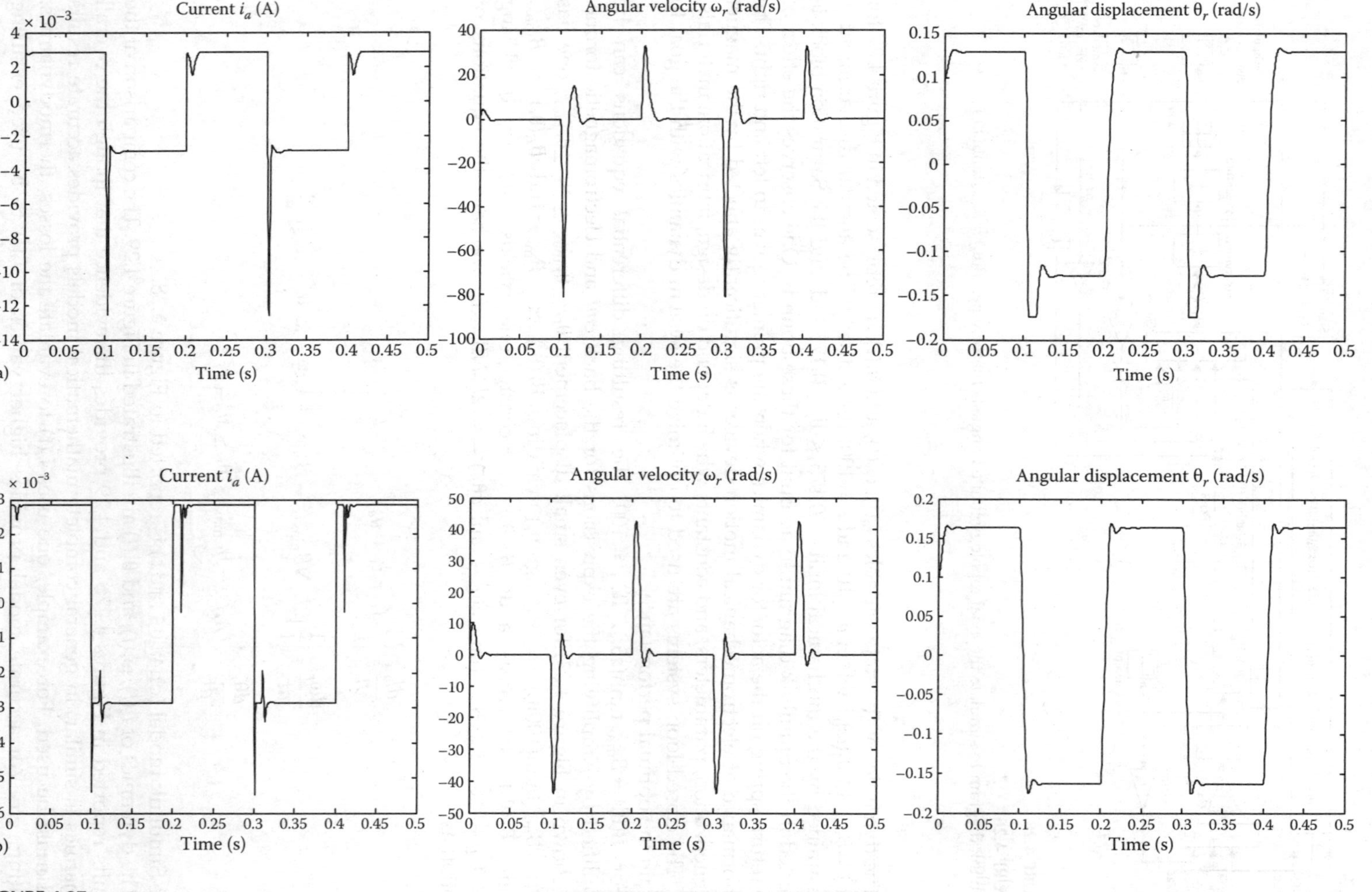

FIGURE 4.27
Transient dynamics of the state variables, $B(\theta_r) = B_{max}\tanh(a\theta_r)$: (a) $a = 10$; (b) $a = 100$.

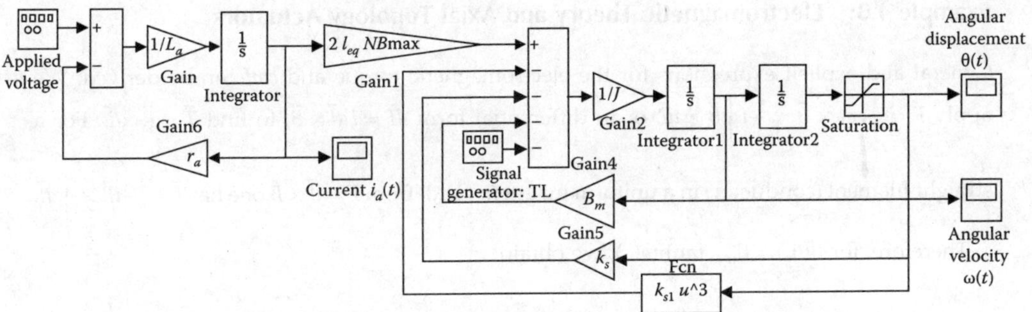

FIGURE 4.28
Simulink diagram to simulate the axial topology limited angle actuator described by a simplified set of differential equations.

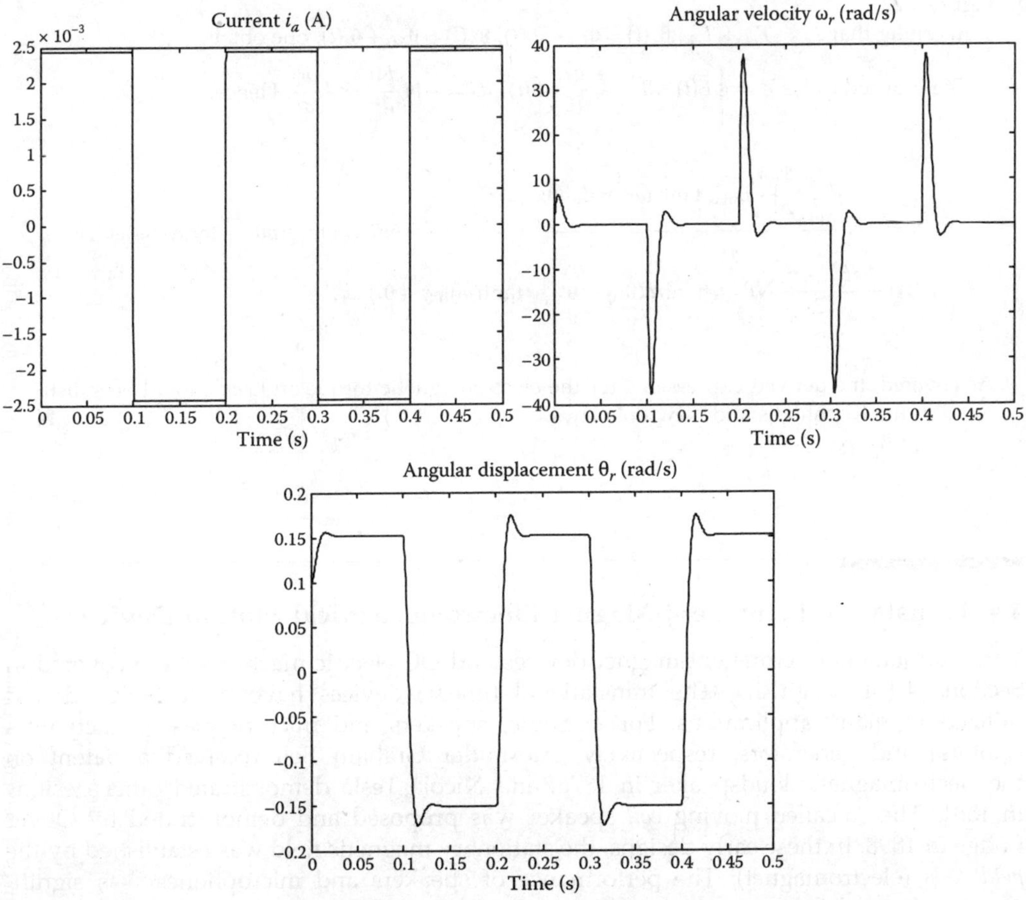

FIGURE 4.29
Transient dynamics of the state variables: Simplified linear model.

Example 4.3:   Electromagnetic Theory and Axial Topology Actuators

General and explicit expressions for the electromagnetic torque and *emf* can be derived. We apply $\vec{F} = \oint_l i\,d\vec{l} \times \vec{B} = -i\oint_l \vec{B} \times d\vec{l}$, or in differential form $d\vec{F} = i\,d\vec{l} \times \vec{B}$, to find $\vec{T} = \vec{R} \times \vec{F}$. For a straight filament (conductor) in a uniform magnetic field, from $\vec{F} = i\vec{l} \times \vec{B}$ one has $\vec{F} = -i\vec{B} \times \oint_l d\vec{l}$.

Therefore, for $B(\theta_r) = B_{max}\tanh(a\theta_r)$, we obtain

$$\vec{T} = \int_{r_{in}}^{r_{out}} i_a B_{max}\tanh(a\theta_r)\,dr\,\vec{a}_z$$

where r_{out} and r_{in} are the outer and inner radii of the magnets as evident from images on Figure 4.22.

Recalling that $T_e = T_{eL} + T_{eR}$, $\theta_L(t) = \theta_{L0} - \theta_r(t)$, $\theta_R(t) = \theta_{R0} + \theta_r(t)$, one obtains T_e

The induced *emf* is $\mathcal{E} = \oint_l \vec{E}(t) \cdot d\vec{l} = -\dfrac{d}{dt}\int_s \vec{B}(t) \cdot d\vec{s} = -N\dfrac{d\Phi}{dt} = -\dfrac{d\psi}{dt}$. Hence,

$$\mathcal{E} = -N\frac{d\displaystyle\int_{r_{in}}^{r_{out}}\int_{\theta_{i\,min}}^{\theta_{i\,max}} B_{max}\tanh(a\theta_r)r\,dr\,d\theta_i}{dt} = -\frac{r_{out}^2 - r_{in}^2}{2}NB_{max}(\tanh a\theta_L + \tanh a\theta_R)\omega_r$$

$$= -\frac{r_{out}^2 - r_{in}^2}{2}NB_{max}[\tanh a(\theta_{L0} - \theta_r) + \tanh a(\theta_{R0} + \theta_r)]\omega_r.$$

As covered, the derived expressions for the electromagnetic torque and *emf* should be substituted in the Kirchhoff's and Newton's laws.   ∎

4.4 Translation Permanent-Magnet Electromechanical Motion Devices

Various rotational permanent-magnet devices and DC electric machines were covered in Sections 4.1 through 4.3. The translational (linear) devices have been designed and utilized in many applications. For example, speakers and microphones are actuators (motors) and generators, respectively. Alexander Graham Bell received a patent on the electromagnetic loudspeaker in 1876, and Nicola Tesla demonstrated other designs in 1881. The so-called moving *coil* speaker was proposed and demonstrated by Oliver Lodge in 1898. In these early designs, the stationary magnetic field was established by the *field coils* (electromagnet). The performance of speakers and microphones was significantly enhanced by using permanent magnets to establish a stationary magnetic field. The images of speakers with radially magnetized permanent magnets are illustrated in Figure 4.30.

FIGURE 4.30
Speakers as limited-displacement permanent-magnet DC actuators.

A lightweight cone or dome shape diaphragm is connected to a rigid frame using a flexible suspension. A variety of different materials are used, and the most common are paper, plastic, and coated and composite materials. An *N*-turn winding (*voice coil*) is under the stationary magnetic field established by radially magnetized permanent magnet or magnets as illustrated in Figure 4.30. To displace a diaphragm, one applies the voltage to the winding, and the electromagnetic force is produced. The suspension system maintains the coil centered within the gap and provides a restoring force to make the speaker cone return to a neutral position (equilibrium) if voltage is not applied. Insulated copper and silver wire is used to fabricate a *voice coil* within a circular, rectangular, or hexagonal cross section. The coil is oriented coaxially inside the gap. Ceramic, ferrite, alnico, and rear-earth (samarium cobalt, neodymium iron boron, and other) permanent magnet are used.

The analysis, design, and optimization tasks can be performed by applying the results reported. In particular, the *emf* and electromagnetic force are

$$emf = \oint_l \vec{E} \cdot d\vec{l} = \oint_l \left(\vec{v} \times \vec{B} \right) \cdot d\vec{l} - \oint_s \frac{\partial \vec{B}}{\partial t} d\vec{s} \quad \text{and} \quad \vec{F} = \oint_l i\,d\vec{l} \times \vec{B} = -i \oint_l \vec{B} \times d\vec{l}. \quad \text{Applying the}$$

Kirchhoff's voltage law and Newtonian mechanics, one obtains the resulting equation of motions. These equations are significantly affected by the speaker and microphone design, magnetic system, geometry, kinematics, etc. Piezoelectric, electrostatic, thermal, and other speakers are also used in low-performance application particularly if the specifications on the low cost and small size are imposed.

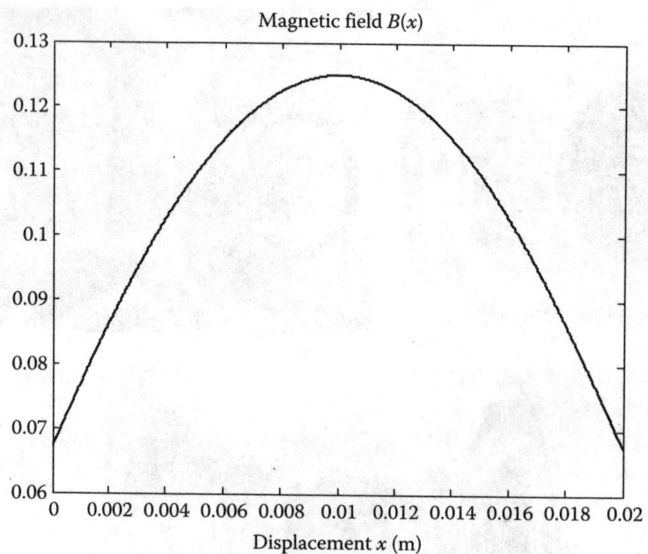

FIGURE 4.31
Plots of $B(x)$ with $x_0 = 0.01$ m, $l_{\min} \leq x \leq l_{\max}$, $l_{\min} = 0$, $l_{\max} = 0.02$ m, $B_{\max} = 0.25$ T, and $a = 10{,}000$.

Example 4.4:

Consider a speaker with a radially magnetized ring magnet which is reported in Figure 4.30. The distribution of $B(\mathbf{r})$, as viewed from the coil, significantly affects the overall performance and capabilities. Magnet is magnetized to ensure the desired $B(\mathbf{r})$, and one may approximate $B(\mathbf{r})$ using various continuous differentiable functions. For example, for one-dimensional field, trigonometric, exponential, sigmoid

$$\left(B = B_{\max} \frac{1}{1+e^{-ax}}, \; B = B_{\max} \frac{1}{1+e^{-a|x|}}, \quad B = B_{\max} \frac{1}{1+e^{-ax^2}}, \quad B = B_{\max} \frac{1}{1+e^{-ax^3}} \right), \; a > 0,$$ and other func-

tions can be used depending on the magnet magnetization, relative displacement of coils with respect to magnet, magnet and coil geometry and orientation, magnet–coil separation, etc. The plot for the axial field $B(x)$ is documented in Figure 4.31. At equilibrium when $\Sigma F = 0$, the displacement is denoted as x_0. Considering the motion in the x-direction, which is constrained within $l_{\min} \leq x \leq l_{\max}$, the distribution of $B(x)$, as viewed from the coils with a high (width) $l_h = (l_{\max} - l_{\min})$, is of a particular importance. The cone displacement x is usually less than l_h, e.g., $x < l_h$. The MATLAB statement used to perform calculations and plotting is

```
lmin=0; lmax=0.02; x=0:(lmax-lmin)/1000:lmax;
Bmax=0.25; a=10000; B=Bmax.*(1-1./(1+exp(-a*(x-(lmax-lmin)/2).^2)));
plot(x,B,'LineWidth',2);
xlabel('Displacement x [m]','FontSize',14);
title('Magnetic Field B(x)','FontSize',14);
```

Using Kirchhoff's voltage law and Newtonian dynamics, one obtains

$$\frac{di_a}{dt} = \frac{1}{L_a}\left(-r_a i_a - emf + u_a\right),$$

$$\frac{dv}{dt} = \frac{1}{m}\left(i_a \int B(x)dl - F_{air} - F_{elastic} - F_\xi\right),$$

$$\frac{dx}{dt} = v,$$

where F_{air} is the air friction force, $F_{elastic}$ is the elastic restoring force, and F_ξ is the disturbance force which is of a stochastic origin.

The electromagnetics of various electromechanical motion devices is covered. By using B ($\mathbf{r}$) and coil geometry, the *emf* and F_e are straightforwardly derived. The expressions for the air friction, elastic restoring forces, and stochastic forces of the mechanical origin can be obtained. Only for a preliminary design, one may apply the following approximations $F_{elastic} = k_{elastic}x$ and $F_{air} = k_{air}v$, where $k_{elastic}$ and k_{air} are the constants. Coherent expressions for $F_{elastic}$ and F_{air} should be used.

References

1. Chapman, S. J., *Electric Machinery Fundamentals*, McGraw-Hill, New York, 1999.
2. Fitzgerald, A. E., Kingsley, C., and Umans, S. D., *Electric Machinery*, McGraw-Hill, New York, 1990.
3. Krause, P. C. and Wasynczuk, O., *Electromechanical Motion Devices*, McGraw-Hill, New York, 1989.
4. Krause, P. C., Wasynczuk, O., and Sudhoff, S. D., *Analysis of Electric Machinery*, IEEE Press, New York, 1995.
5. Lyshevski, S. E., *Electromechanical Systems, Electric Machines, and Applied Mechatronics*, CRC Press, Boca Raton, FL, 1999.
6. Lyshevski, S. E., *Electromechanical Systems and Devices*, CRC Press, Boca Raton, FL, 2008.
7. Slemon, G. R., *Electric Machines and Drives*, Addison-Wesley Publishing Company, Reading, MA, 1992.
8. White, D. C. and Woodson, H. H., *Electromechanical Energy Conversion*, Wiley, New York, 1959.

5

Induction Machines

5.1 Introduction and Fundamentals

In high-performance drives and servos, permanent-magnet electric machines are used. For fractional-, medium-, and high-power (in some applications, ~10,000 kW) drives, three-phase induction machines are frequently utilized. Fractional horse-power industrial and consumer drives employ low-cost single- and three-phase induction machines [1–6]. Induction electric machines, compared with permanent-magnet motion devices, have much lower torque and power densities, may not be effectively used as generators, but require simple electronic and control solutions in the motor (drive) operation.

Induction motors were invented and demonstrated by a generous scientist and engineer Nicola Tesla in the 1880s. Nicola Tesla made indispensable contributions to science and engineering inventing many electric machines, electromagnetic devices, radars, etc. Nicola Tesla integrated his induction motors with the matching polyphase power electronics demonstrating the speed control capabilities in 1888. Having underlined extraordinary engineering and technological developments performed by Nicola Tesla, it must be emphasized that he made a significant contribution not only to device physics of electro-mechanical and electromagnetic motion devices, but also to theoretical electromagnetics. In particular, in 1882, Nicola Tesla pioneered and developed the theory of the rotating magnetic field which is a cornerstone principle of electromechanical motion devices. He designed and demonstrated a two-phase induction motor in 1883. The first three-phase squirrel-cage induction motor was invented and demonstrated by Michail Osipovish Dolivo-Dobrovolski in 1890. These induction motors have been used for almost 120 years.

In squirrel-cage induction electric motors, the phase voltages in the short-circuited rotor windings are induced due to time-varying stator magnetic field as well as motion of the rotor with respect to the stator. The phase voltages are supplied to the stator windings. The electromagnetic torque results due to the interaction of the time-varying electromagnetic fields. The images of 250 W induction motors (left and at the center machines) are illustrated in Figure 5.1. For the illustrative purposes, a 250 W permanent-magnet synchronous machine (which is much smaller as compared to induction machines of the same rated power, and allows much larger overloading capabilities) is provided as the third image.

To design electric machines, analyze them and evaluate their performance, one may perform sequential steps starting from machine synthesis and optimization to modeling, simulations, testing, characterization, etc. It is unlikely that one may significantly improve the technology-centered designs of electric machines which have been successfully per-formed by different manufacturers. Stand-alone books concentrate on induction machine

FIGURE 5.1
250 W induction and permanent-magnet synchronous (black) electric machines (NEMA 56 frame size induction motors and a NEMA 23 frame size permanent-magnet synchronous machine, NEMA 23 size means 2.3 in. or 57 mm diameter).

design, and various tasks involved (three-dimensional electromagnetic, mechanical, thermal, vibroacoustic, structural, and other designs) are supported by various electric machine design tools. In general, the machine design and optimization are of a great importance. The machine design tasks are beyond the scope of this book. For specific classes of electric machine, the designs are covered in highly specialized books. Though the need for a machine design is obvious, only a very small fraction of engineers are involved because an optimal machine design is a narrow highly specialized problem. Sound solutions are available and have been finalized within more than 100 years for existing broad lines and numerous series of DC, induction and synchronous machines. In contrast, the power electronics, microelectronics, and sensors have been rapidly developed providing tremendous opportunities to advance electromechanical systems. Furthermore, the application and market for electric drives and servos have being notably expanded (avionics, bioengineering, electronics, etc.) in addition to the conventional automotive, power, robotics, and transportation areas. Therefore, we focus on the electromechanical system design issues considering electric machines as a key ready-to-use component rightly assuming the existing *optimal-performance* design of stand-alone electric machines. We concentrate on the systems design in order to guarantee best performance and *achievable* capabilities. Other important point is that through optimization and synergetic control activities, one improves the machine performance and capabilities enabling better energy conversion, torque production, losses reduction, vibration, noise minimization, etc. That is, we enable and enhance machine capabilities refining stand-alone machine design deficiencies or imperfections.

This chapter reports sound analysis (modeling, simulation, performance/capabilities assessment, and other tasks) and control of induction motors in the *machine* (phase) variables. Though the *quadrature* and *direct* reference frame can be used [1–6], the related concept (such as *vector* control and other approaches) may not offer advantages and provide any benefits due to the need for the most advanced DSPs to perform the associated *direct* and *inverse* transformations in real time [5,6]. Furthermore, all electric machines are operated and controlled only by using *machine* variables. For example one varies the phase (*machine*) voltages u_{as}, u_{bs}, u_{cs} and u_{ar}, u_{br}, u_{cr} while the directly measured currents are the phase (*machine*) currents i_{as}, i_{bs}, i_{cs} and i_{ar}, i_{br}, i_{cr}. For the squirrel-cage induction motors, one uses u_{as}, u_{bs}, u_{cs} and i_{as}, i_{bs}, i_{cs} because u_{ar}, u_{br}, u_{cr} are not supplied and rotor windings are short-circuited. It will be documented that the highest acceleration capabilities and minimal settling time can be achieved utilizing the high torque pattern (the so-called frequency control), though to reduce the losses one may use the voltage–frequency control. These control concepts in the *machine* variable comply with the power electronic and hardware solutions.

5.2 Torque–Speed Characteristics and Control of Induction Motors

5.2.1 Torque–Speed Characteristics

The angular velocity of induction motors must be controlled, and the torque–speed curves $\omega_r = \Omega_T(T_e)$ are studied. The electromagnetic torque developed by induction motors is a function of the stator and rotor currents as well as rotor displacement. Induction motors are controlled by changing the frequency f and magnitude u_M of the voltages supplied to the phase windings. The magnitude of the voltages applied to the stator windings cannot exceed the rated voltage u_{Mmax}, and $u_{Mmin} \leq u_M \leq u_{Mmax}$. The angular frequency of the applied phase voltages ω_f is also bounded, and $\omega_f = 2\pi f$, $f_{min} \leq f \leq f_{max}$, $f_{min} > 0$ and $f_{max} > 0$.

The synchronous angular velocity ω_e of induction machines, as a function of f, is

$$\omega_e = \frac{4\pi f}{P}.$$

The electrical angular velocity of induction motors ω_r is less or equal (at no load and no friction) to ω_e. Hence, for induction motors $\omega_r \leq \omega_e$. In contrast, synchronous motors rotate at ω_e and $\omega_r = \omega_e$.

The steady-state curves $\omega_r = \Omega_T(T_e)$ are found by plotting the angular velocity versus the electromagnetic torque. The National Electric Manufacturers Association (NEMA) in the United States and the International Electromechanical Commission (IEC) in Europe have defined four basic classes of induction machines to be A, B, C, and D. For different classes, typical steady-state torque–speed characteristic curves in the motor region are depicted in Figure 5.2a, where the *slip* is given as

$$slip = \frac{\omega_e - \omega_r}{\omega_e} \quad \text{and} \quad \omega_r = (1 - slip)\omega_e.$$

These torque–speed characteristics can be found using the experimental data by measuring the torque and angular velocity which evolve in time, e.g., one uses $T_e(t)$ and $\omega_r(t)$. The measured experimental dynamic characteristics $\omega_r(t) = \Omega_T[T_e(t)]$ will be different as compared to the steady-state $\omega_r = \Omega_T(T_e)$. In fact, $\omega_r = \Omega_T(T_e)$ *averages* $\omega_r(t) = \Omega_T[T_e(t)]$. However, $\omega_r(t) = \Omega_T[T_e(t)]$ could be found to be more general and descriptive. To derive experimental dynamic characteristics $\omega_r(t) = \Omega_T[T_e(t)]$, if $T_e(t)$ is not directly measurable, one also may measure (or observe) the phase currents and angular displacement to obtain $T_e(t)$.

The steady-state angular velocity with which induction motor rotates is found as the intersection of the torque–speed $\omega_r = \Omega_T(T_e)$ and load $T_L(\omega_r)$ characteristics as illustrated in Figure 5.2b. From the Newton second law, neglecting the friction torque (which can be considered as a part of the load torque T_L), we have $\dfrac{d\omega_r}{dt} = \dfrac{1}{J}(T_e - T_L)$. One concludes that $\omega_r = $ const if $T_e = T_L$. Furthermore, assuming that $T_e > T_L$ and $T_{e\,start} > T_{L0}$, the motor

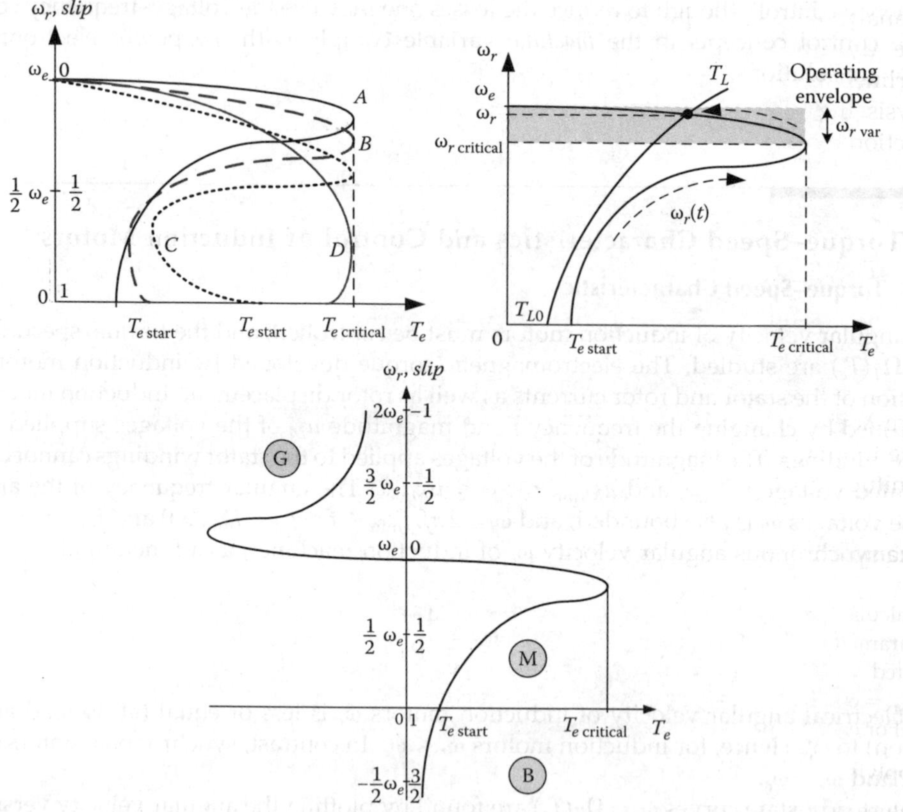

FIGURE 5.2
(a) Typical torque–speed characteristics of A, B, C, and D class induction motors; (b) Torque–speed and load curves: Motor accelerates to rotate at the angular velocity ω_r, $\omega_r \leq \omega_e$; (c) Torque–speed characteristic curves in the motor (M), generator (G), and breaking (B) regions.

accelerates until ω_r reaches an operation when $T_e = T_L$. The critical angular velocity $\omega_{r\,\text{critical}}$ is documented in Figure 5.2b. To ensure the acceleration and rotation, one must guarantee $T_e > T_L$ in the full operating envelope for all T_L, and the rated torque may be specified to be $T_{e\,\text{max}} = T_{e\,\text{critical}}$. The induction motor torque–speed operating envelope $[\omega_r\ T_e]$ is defined by $\omega_r \in [\omega_{r\,\text{critical}}\ \omega_e]$ and $T_e \in [0\ T_{e\,\text{critical}}]$, $T_e > T_L$, $\forall T_L$.

The industrial induction motors are usually designed to be either the A- or B-class machines. These motors have normal starting torque and low slip, which is usually ~0.05. In contrast, the C-class induction motors have higher starting torque due to double-rotor design, and the slip is greater than 0.05. The D-class induction motors have high rotor resistance, and the high starting torque results. Typically, the *slip* of the D-class induction motors is in the range from 0.5 to 0.9. The E- and F-class induction motors have very low starting torque, and the rotor bars are deeply buried resulting in high leakage inductances. Figure 5.2c depicts the torque–speed characteristic of A- and B-class induction machines in the motor, generator, and braking regions.

The steady-state torque–speed characteristics of induction motors can be obtained by using the equivalent circuits. It will be demonstrated that solving the differential equations and analyzing the dynamics for $\omega_r(t)$ and evolution of $T_e(t)$, one obtains $\omega_r(t) = \Omega_T[T_e(t)]$. These dynamic torque–speed characteristics $\omega_r(t) = \Omega_T[T_e(t)]$, which can be obtained experimentally, result in a coherent performance evaluations. To approach the preliminary analysis, the following formula can be applied to obtain the torque–speed characteristics of induction motors

$$T_e = \frac{3\left(u_M \dfrac{X_M}{X_s + X_M}\right)^2 \dfrac{r_r'}{slip}}{\omega_e\left[\left(r_s\left(\dfrac{X_M}{X_s + X_M}\right)^2 + \dfrac{r_r'}{slip}\right)^2 + (X_s + X_r')^2\right]}, \quad slip = \frac{\omega_e - \omega_r}{\omega_e}, \quad \omega_e = \frac{4\pi f}{P},$$

where X_s and X_r' are the stator and rotor reactances, and X_M is the magnetizing reactance.

The torque–speed characteristics $\omega_r = \Omega_T(T_e)$ are found assigning different values of the magnitude u_M and frequency f of the phase voltage supplied.

Example 5.1:

Calculate and plot the torque–speed characteristic for a four-pole induction motor. The motor parameters are $r_s = 24.5$ ohm, $r_r' = 23$ ohm, $X_s = 10$ ohm, $X_r' = 40$ ohm, and $X_M = 25$ ohm. The rated voltage is $u_{Mmax} = 110$ V. The maximum frequency of the supplied phase voltages is $f_{max} = 60$ Hz. Let, the phase voltages are supplied with frequencies 20, 40, and 60 Hz.

For each value of the assigned frequency, we calculate the synchronous angular velocity by applying the equation $\omega_e = \dfrac{4\pi f}{P}$. The torque–speed characteristics are found assigning different values for the angular velocity. The following MATLAB® file is developed to calculate and plot $\omega_r = \Omega_T(T_e)$

```
clear all
% parameters of an induction motor
rs=24.5; rr=23; Xs=10; Xr=40; Xm=25;
uM=110; f=60; P=4; we=4*pi*f/P;
% calculation of a torque-speed characteristic
for wr=[1:1:4*pi*f/P];   % angular velocity
slip=(we-wr)/we;          % slip
Te=3*(uM*Xm/(Xs+Xm))^2*(rr/slip)/(we*((rs*(Xm/(Xs+Xm))^2+rr/slip)^2+(Xs+Xr)^2));
plot(Te,wr,'o'); title('Torque-Speed Characteristics','FontSize',14);
xlabel('Electromagnetic Torque T_e, N-m','FontSize',14);
ylabel('Angular velocity \omega_r, rad/sec','FontSize',14);
hold on;
end;
```

The resulting torque–speed characteristics for 20, 40, and 60 Hz are documented in Figure 5.3. Using the plots $\omega_r = \Omega_T(T_e)$ reported, one assesses the acceleration capabilities, control features, etc. The frequency-centered control is a baseline principle in control of high-performance electric drives with induction motors. The starting $T_{e\,start}$ is maximum at the minimum frequency f_{min}, and $T_{e\,start}$ is ~1.8 N m.

The derived torque–speed characteristic introduces the reader to the frequency control of induction motors which should be viewed as a most important concept controlling the angular

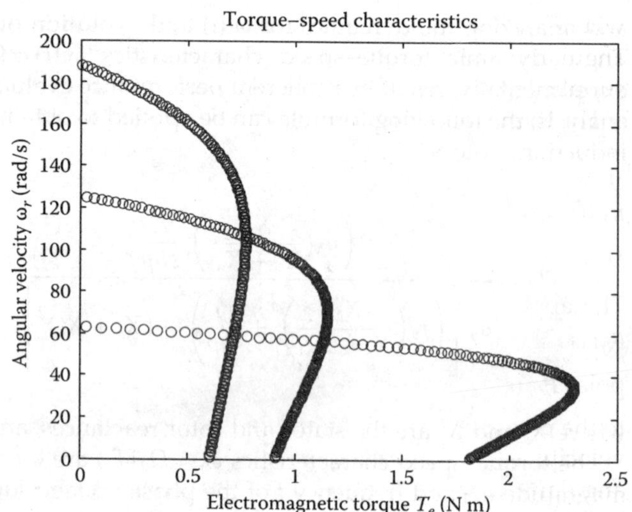

FIGURE 5.3
Torque–speed characteristics of a three-phase induction motor, f is 20, 40, and 60 Hz.

velocity in industrial induction motors and drives. It is evident that by changing the frequency f, one varies ω_r. The descriptive feature on this concept is depicted by equations used and results illustrated in Figure 5.3. ∎

5.2.2 Control of Induction Motors

To control induction motors, one must vary the electromagnetic torque T_e thereby changing the angular velocity. To vary T_e, one changes the frequency and magnitude of phase voltages applied to the windings. As will be derived, the electromagnetic torque T_e developed by two- and three-phase induction motors is given by (5.8) and (5.26), e.g.,

$$T_e = -\frac{P}{2}L_{ms}\left[\left(i_{as}i'_{ar} + i_{bs}i'_{br}\right)\sin\theta_r + \left(i_{as}i'_{br} - i_{bs}i'_{ar}\right)\cos\theta_r\right],$$

$$T_e = -\frac{P}{2}L_{ms}\left[\left(i_{as}i'_{ar} + i_{bs}i'_{br} + i_{cs}i'_{cr}\right)\sin\theta_r + \left(i_{as}i'_{cr} + i_{bs}i'_{ar} + i_{cs}i'_{br}\right)\sin\left(\theta_r - \frac{2}{3}\pi\right)\right.$$
$$\left. + \left(i_{as}i'_{br} + i_{bs}i'_{cr} + i_{cs}i'_{ar}\right)\sin\left(\theta_r + \frac{2}{3}\pi\right)\right].$$

To guarantee the balanced operation of two-phase induction motors, one supplies the following phase voltages to the stator windings

$$u_{as}(t) = \sqrt{2}u_M\cos(\omega_f t), \quad u_{bs}(t) = \sqrt{2}u_M\sin(\omega_f t),$$

and the sinusoidal steady-state phase currents are

$$i_{as}(t) = \sqrt{2}i_M\cos(\omega_f t - \varphi_i), \quad i_{bs}(t) = \sqrt{2}i_M\sin(\omega_f t - \varphi_i),$$

where u_M and i_M are the magnitude of the *as* and *bs* stator voltages and currents, ω_f is the angular frequency of the supplied phase voltages, $\omega_f = 2\pi f$, f is the frequency of the supplied voltage, and φ_i is the phase difference.

For three-phase induction motors, one supplies the following phase voltages

$$u_{as}(t) = \sqrt{2}u_M\cos(\omega_f t), \quad u_{bs}(t) = \sqrt{2}u_M\cos\left(\omega_f t - \frac{2}{3}\pi\right), \quad \text{and} \quad u_{cs}(t) = \sqrt{2}u_M\cos\left(\omega_f t + \frac{2}{3}\pi\right)$$

The applied voltage to the motor windings cannot exceed the rated voltage u_{Mmax}, e.g., $u_{Mmin} \leq u_M \leq u_{Mmax}$. The synchronous angular velocity ω_e is found by using the number of poles P and the frequency f as $\omega_e = \dfrac{4\pi f}{P}$. The frequency is constrained as $f_{min} \leq f \leq f_{max}$ due to the power electronics limits (define f_{min}) and mechanical limits on the maximum angular velocity ω_{rmax} which leads to f_{max}. The synchronous ω_e, electrical ω_r, and mechanical ω_{rm} angular velocities can be regulated by changing the frequency f. To vary ω_r, one can change both the magnitude of the applied voltages u_M and f. The torque–speed curves of induction motors can be studied using the equivalent circuits. Alternatively, from transient dynamics, as found from the experimental results or solving the derived differential equations, one can find the experimental and analytical evolution of $\omega_r = \Omega_T(T_e)$ by plotting the angular velocity ω_r versus the electromagnetic torque T_e.

The following principles are used to control the angular velocity of squirrel-cage induction motors. These principles are application specific, defined by induction motors used, depend on requirements imposed, and tailored to the hardware and software solutions.

5.2.2.1 Voltage Control

By changing the magnitude u_M of the supplied phase voltages to the stator windings, the angular velocity is regulated in the stable operating region, see Figure 5.4a. It was emphasized that $u_{Mmin} \leq u_M \leq u_{Mmax}$, where u_{Mmax} is the maximum allowed or rated voltage. By reducing u_M, one reduces $T_{e\,start}$ and $T_{e\,critical}$. The operating envelope for the angular velocity is $\omega_r \in [\omega_{r\,critical}\ \omega_e]$. For example, for A-, B-, and C-class induction motors, one is not able to effectively regulate the angular velocity. The voltage control is applicable only for low-efficiency D-class induction motors (frequently utilized in consumer electronics and household) in order to avoid the use of power converters.

5.2.2.2 Frequency Control

The magnitude of the supplied phase voltages is constant $u_M = \text{const}$, and the angular velocity is regulated by varying the frequency of the supplied voltages f, $f_{min} \leq f \leq f_{max}$ (usually f_{min} is $\sim$2 Hz). This concept is justified recalling that $\omega_e = \dfrac{4\pi f}{P}$, e.g., by varying f one changes the synchronous angular velocity ω_e. The angular frequency ω_f of the supplied voltages is related to the frequency f as $\omega_f = 2\pi f$, $f_{min} \leq f \leq f_{max}$. The torque–speed characteristics for different values of the frequency are shown in Figure 5.4b.

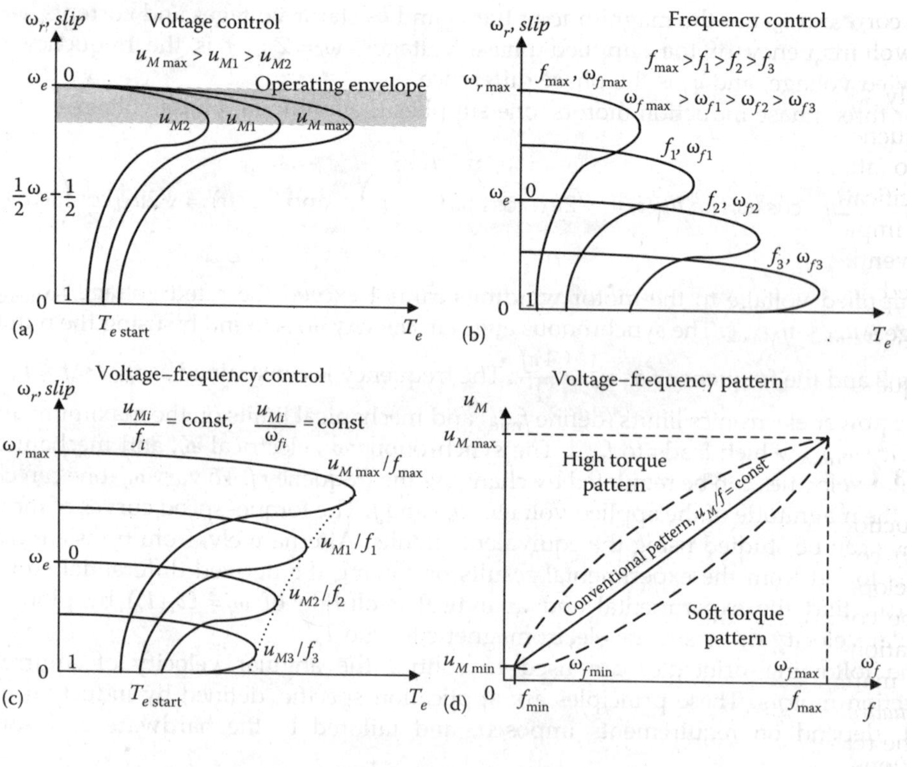

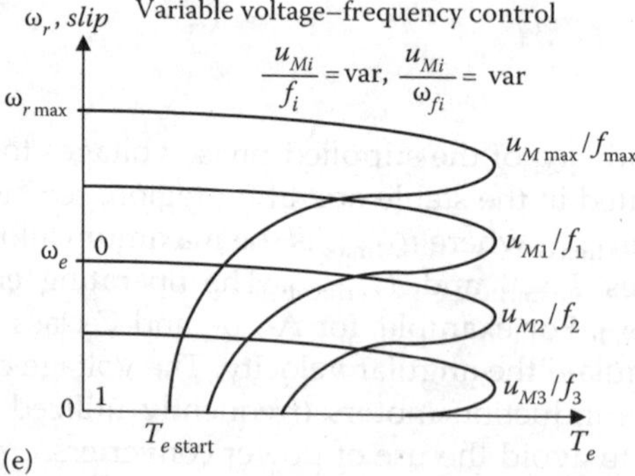

FIGURE 5.4

Torque–speed characteristics $\omega_r = \Omega_T(T_e)$: (a) Voltage control; (b) frequency control; (c) voltage–frequency control: *Constant-volts-per-hertz* control; (d) voltage–frequency patterns; (e) variable voltage–frequency control.

5.2.2.3 Voltage–Frequency Control

To minimize losses, the voltage magnitude u_M is regulated if the frequency f is changed. In particular, u_M can be decreased linearly as f is reduced. To guarantee the so-called *constant-volts-per-hertz* control, one maintains the following relationship

$$\frac{u_{Mi}}{f_i} = \text{const} \quad \text{or} \quad \frac{u_{Mi}}{\omega_{fi}} = \text{const}$$

The corresponding torque–speed characteristics are documented in Figure 5.4c. Regulating the voltage-frequency patterns, one varies the torque–speed curves. For example, one may apply $\sqrt{u_{Mi}/f_i} = \text{const}$ to correspondingly adjust the magnitude u_M with respect to frequency f.

To attain the specified acceleration, settling time, overshoot, rise time, and other specifications, the general purpose (conventional), soft- and high-starting torque patterns are implemented based upon the requirements imposed and operating envelope. The conventional (*constant-volts-per-hertz* control), soft- and high-torque patterns are illustrated in Figure 5.4d. By assigning $\omega_f = \varphi(u_M)$ with domain $u_{Mmin} \leq u_M \leq u_{Mmax}$ and range $\omega_{fmin} \leq \omega_f \leq \omega_{fmax}$ ($f_{min} \leq f \leq f_{max}$), one maintains $\dfrac{u_{Mi}}{f_i} = \text{var}$ or $\dfrac{u_{Mi}}{\omega_{fi}} = \text{var}$. The desired torque–speed characteristics, as documented in Figure 5.4e, can be guaranteed.

5.2.3 Control of Induction Motors in the *Arbitrary* Reference Frame and Vector Control

Induction motors can be analyzed and controlled using the *quadrature*, *direct*, and *zero* (*qd0*) components of voltages, currents, and flux linkages. This analysis concept was developed in 1930s to reduce the mathematical complexity and make the transient analysis to be computationally tractable allowing analytic solutions of the resulting differential equations [2,3,5,6]. The availability of high-performance software diminished the use of this mathematically elegant approach in analysis. From control standpoints, practicality of *quadrature*, *direct*, and *zero* variables is quite debatable. Example 6.1 could be of interest to the reader to comprehend the concept and underlying ideas.

Induction and synchronous machines can be examined in the *arbitrary* reference frame. In particular, the stationary, rotor, and synchronous frames are commonly specified [2–4,6]. In the synchronous reference frame (superscript e), the $qd0$-axis components of stator and rotor AC voltages, currents, and flux linkages have a DC form. Furthermore, $u^e_{ds}(t) = 0$ and $u^e_{os}(t) = 0$. To control induction motors, one may conclude that the DC *quadrature* voltage $u^e_{qs}(t)$ should be regulated because $u^e_{ds}(t) = 0$ and $u^e_{os}(t) = 0$. Though the deviations are mathematically accurate, the results may have a limited practicality and should not confuse the designer. We obtained the $qd0$ quantities which mathematically (formally) correspond to the physical *machine* variables as the Park transformations are applied. One measures $i_{as}(t)$, $i_{bs}(t)$, and $i_{cs}(t)$, while $i^e_{qs}(t)$, $i^e_{ds}(t)$, and $i^e_{os}(t)$ are not directly measured. In AC machines, voltages, currents, and flux linkages are AC *machine* quantities. These AC phase voltages u_{as}, u_{bs}, and u_{cs} are to be supplied to the stator windings. One does not directly measure or observe the *quadrature-*, *direct-*, and *zero*-axis components. One cannot supply $u^e_{qs}(t)$, $u^e_{ds}(t)$, and $u^e_{os}(t)$ to the phase windings. To rotate induction motors, the AC phase voltages are supplied to the phase windings.

Therefore, the $qd0$ voltages, thought can be viewed as controls, are not applied to the phase windings. If one decided to control induction machines in the $qd0$ frames deriving $\mathbf{u}^e_{qd0s}$, the designed must perform: Having somehow found $\mathbf{u}^e_{qd0s}$, calculate (in real-time using DSPs) $\mathbf{u}_{abcs}$ by using the Park transformation matrix as

$$\mathbf{u}_{abcs} = \left(\mathbf{K}^e_s\right)^{-1}\mathbf{u}^e_{qd0s}, \quad \mathbf{K}^e_s = \frac{2}{3}\begin{bmatrix} \cos\theta_e & \cos\left(\theta_e - \tfrac{2}{3}\pi\right) & \cos\left(\theta_e + \tfrac{2}{3}\pi\right) \\ \sin\theta_e & \sin\left(\theta_e - \tfrac{2}{3}\pi\right) & \sin\left(\theta_e + \tfrac{2}{3}\pi\right) \\ \tfrac{1}{2} & \tfrac{1}{2} & \tfrac{1}{2} \end{bmatrix}. \text{ The frequency } f \text{ (which}$$

affects ω_e and θ_e, $\theta_e = \omega_e t$) is varied by the power converter to control the angular velocity.

Thus, as one may derive $\mathbf{u}^e_{qd0s}$, the phase voltages u_{as}, u_{bs}, and u_{cs} should be calculated in real-time by applying an *inverse* Park transformation $\mathbf{u}_{abcs} = \left(\mathbf{K}^e_s\right)^{-1}\mathbf{u}^e_{qd0s}$. The most advanced DSPs are required if the designer decided to apply the synchronous (as well as stationary and rotor) reference frame. For example, the angular displacement θ_r must be measured (or observed) and utilized as one applies the rotor reference frame. Correspondingly additional hardware (sensor, conditioning ICs, etc.) is needed. The calculation of the phase voltages $\mathbf{u}_{abcs}$, which must be accomplished in real-time, significantly complicates software and hardware solutions. Though, the so-called *vector* control of induction motors can be applied, quite limited practical benefits (if any) may emerge because the stationary, rotor or synchronous reference frames imply the application of the *qd0* quantities [5,6].

The variable voltage-frequency control $\dfrac{u_{Mi}}{f_i} = \mathrm{var}$, $u_{M\min} \leq u_M \leq u_{M\max}$, $f_{\min} \leq f \leq f_{\max}$, $\omega_{f\min} \leq \omega_f \leq \omega_{f\max}$, as discussed in Section 5.2.2, guarantees the high-torque patterns which surpass (or at least matches) the capabilities of the *vector* control or other *qd0*-centered concepts [6]. The highest $T_{e\,\mathrm{start}}$ and $T_{e\,\mathrm{critical}}$ are developed using the well-established frequency control, and $f_{\min} \leq f \leq f_{\max}$. The $T_{e\,\mathrm{start\,max}}$ corresponds to $f_{\min}$. This $f_{\min}$ is defined mainly by the converter topologies, solid-state devices, driving ICs, efficiency, and other power electronics and hardware specifications. Therefore, the use of *machine* variables and sound concepts should be prioritized.

5.3 Two-Phase Induction Motors

5.3.1 Modeling of Two-Phase Induction Motors

We study a two-phase induction machine as illustrated in Figure 5.5. To rotate induction machines and control the angular velocity, for squirrel-cage motors, one varies the frequency and magnitude of the phase voltages u_{as} and u_{bs} supplied to the stator windings. For the wound-rotor induction machines, in addition to controlling the magnitude and frequency of u_{as} and u_{bs}, one may vary the voltages u_{ar} and u_{br} which are supplied to the rotor windings. One should examine the voltages induced in the rotor windings (as the induced *back emfs*) which are not the voltages supplied to the stator windings u_{ar} and u_{br}. The stator (*as* and *bs*) and rotor (*ar* and *br*) windings, as well as the stator–rotor magnetic coupling, are depicted in Figure 5.5. In squirrel-cage induction machines, the voltages are induced in the rotor windings due to the time-varying stator magnetic field and motion of the rotor with respect to the stator. The electromagnetic torque results due to the interaction of time-varying electromagnetic fields.

To derive the governing equations, we describe the stator and rotor circuitry-electromagnetic dynamics, model the *torsional-mechanical* behavior, and find the expression for electromagnetic torque. As the control and state variables, we use the voltages supplied to the *as*, *bs*, *ar*, and *br* windings, as well as the currents and flux linkages. Using Kirchhoff's voltage law, we have

$$
\begin{aligned}
u_{as} &= r_s i_{as} + \frac{d\psi_{as}}{dt}, \quad u_{bs} = r_s i_{bs} + \frac{d\psi_{bs}}{dt}, \\
u_{ar} &= r_r i_{ar} + \frac{d\psi_{ar}}{dt}, \quad u_{br} = r_r i_{br} + \frac{d\psi_{br}}{dt},
\end{aligned}
\tag{5.1}
$$

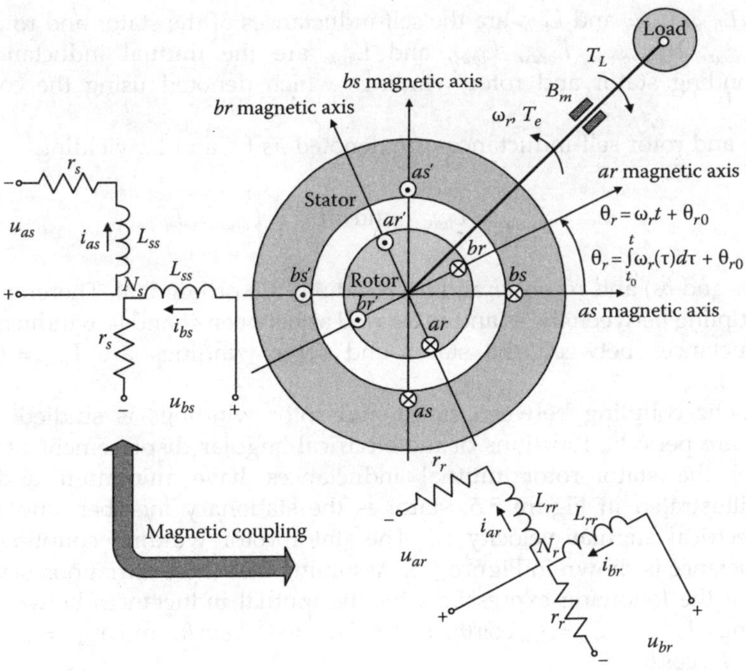

FIGURE 5.5
Two-phase symmetric induction motor.

where u_{as} and u_{bs} are the phase voltages supplied to the *as* and *bs* stator windings, u_{ar} and u_{br} are the phase voltages supplied to the *ar* and *br* rotor windings (for squirrel-cage motors, $u_{ar} = 0$ and $u_{br} = 0$), i_{as} and i_{bs} are the phase currents in the stator windings, i_{ar} and i_{br} are the phase currents in the rotor windings, ψ_{as} and ψ_{bs} are the stator flux linkages, ψ_{ar} and ψ_{br} are the rotor flux linkages, and r_s and r_r are the resistances of the stator and rotor winding.

From (5.1), using the vector notations, one obtains

$$\mathbf{u}_{abs} = \mathbf{r}_s \mathbf{i}_{abs} + \frac{d\boldsymbol{\psi}_{abs}}{dt}, \quad \mathbf{u}_{abr} = \mathbf{r}_r \mathbf{i}_{abr} + \frac{d\boldsymbol{\psi}_{abr}}{dt}, \tag{5.2}$$

where $\mathbf{u}_{abs} = \begin{bmatrix} u_{as} \\ u_{bs} \end{bmatrix}$, $\mathbf{u}_{abr} = \begin{bmatrix} u_{ar} \\ u_{br} \end{bmatrix}$, $\mathbf{i}_{abs} = \begin{bmatrix} i_{as} \\ i_{bs} \end{bmatrix}$, $\mathbf{i}_{abr} = \begin{bmatrix} i_{ar} \\ i_{br} \end{bmatrix}$, $\boldsymbol{\psi}_{abs} = \begin{bmatrix} \psi_{as} \\ \psi_{bs} \end{bmatrix}$, $\boldsymbol{\psi}_{abr} = \begin{bmatrix} \psi_{ar} \\ \psi_{br} \end{bmatrix}$ are the phase voltages, currents, and flux linkages; $\mathbf{r}_s = \begin{bmatrix} r_s & 0 \\ 0 & r_s \end{bmatrix}$ and $\mathbf{r}_r = \begin{bmatrix} r_r & 0 \\ 0 & r_r \end{bmatrix}$ are the matrices of the stator and rotor resistances.

Assuming that the magnetic system is linear, the flux linkages are expressed using the self- and mutual-inductances. We have

$$\psi_{as} = L_{asas}i_{as} + L_{asbs}i_{bs} + L_{asar}i_{ar} + L_{asbr}i_{br}, \quad \psi_{bs} = L_{bsas}i_{as} + L_{bsbs}i_{bs} + L_{bsar}i_{ar} + L_{bsbr}i_{br},$$
$$\psi_{ar} = L_{aras}i_{as} + L_{arbs}i_{bs} + L_{arar}i_{ar} + L_{arbr}i_{br}, \quad \psi_{br} = L_{bras}i_{as} + L_{brbs}i_{bs} + L_{brar}i_{ar} + L_{brbr}i_{br},$$

where, L_{asas}, L_{bsbs}, L_{arar}, and L_{brbr} are the self-inductances of the stator and rotor windings, and L_{asbs}, L_{asar}, L_{asbr}, ..., L_{bras}, L_{brbs}, and L_{brar} are the mutual inductances between the corresponding stator and rotor windings which denoted using the corresponding subscripts.

The stator and rotor self-inductances are denoted as L_{ss} and L_{rr} yielding

$$L_{ss} = L_{asas} = L_{bsbs} \quad \text{and} \quad L_{rr} = L_{arar} = L_{brbr}.$$

The stator (*as* and *bs*) and rotor (*ar* and *br*) windings are orthogonal. Therefore, there is no magnetic coupling between the *as* and *bs*, as well as between *ar* and *br* windings. Hence, the mutual inductances between the stator and rotor windings are $L_{asbs} = L_{bsas} = 0$ and $L_{arbr} = L_{brar} = 0$.

The magnetic coupling between stator and rotor windings is studied. The mutual inductances are periodic functions of the electrical angular displacement of the rotor θ_r. Furthermore, the stator–rotor mutual inductances have minimum and maximum values. As illustrated in Figure 5.5, stator is the stationary member, and rotor rotates with the electrical angular velocity ω_r. The stator–rotor winding coupling due to the mutual inductance is shown in Figure 5.5. Assuming that these variations obey the cosine law, we have the following expressions for the mutual inductances between stator and rotor windings $L_{asar} = L_{aras} = L_{sr} \cos \theta_r$, $L_{asbr} = L_{bras} = -L_{sr} \sin \theta_r$ and $L_{bsar} = L_{arbs} = L_{sr} \sin \theta_r$, $L_{bsbr} = L_{brbs} = L_{sr} \cos \theta_r$.

For the magnetically coupled windings we found the expressions for the flux linkages as

$$\psi_{as} = L_{ss} i_{as} + L_{sr} \cos \theta_r i_{ar} - L_{sr} \sin \theta_r i_{br}, \quad \psi_{bs} = L_{ss} i_{bs} + L_{sr} \sin \theta_r i_{ar} + L_{sr} \cos \theta_r i_{br},$$

$$\psi_{ar} = L_{sr} \cos \theta_r i_{as} + L_{sr} \sin \theta_r i_{bs} + L_{rr} i_{ar}, \quad \psi_{br} = -L_{sr} \sin \theta_r i_{as} + L_{sr} \cos \theta_r i_{bs} + L_{rr} i_{br}.$$

The following expression for the flux linkages results

$$\begin{bmatrix} \boldsymbol{\psi}_{abs} \\ \boldsymbol{\psi}_{abr} \end{bmatrix} = \begin{bmatrix} \mathbf{L}_s & \mathbf{L}_{sr}(\theta_r) \\ \mathbf{L}_{sr}^T(\theta_r) & \mathbf{L}_r \end{bmatrix} \begin{bmatrix} \mathbf{i}_{abs} \\ \mathbf{i}_{abr} \end{bmatrix},$$

where $\mathbf{L}_s$ is the matrix of the stator self-inductances, $\mathbf{L}_s = \begin{bmatrix} L_{ss} & 0 \\ 0 & L_{ss} \end{bmatrix}$, $L_{ss} = L_{ls} + L_{ms}$, $L_{ms} = \dfrac{N_s^2}{\mathfrak{R}_m}$;

$\mathbf{L}_r$ is the matrix of the rotor self-inductances, $\mathbf{L}_r = \begin{bmatrix} L_{rr} & 0 \\ 0 & L_{rr} \end{bmatrix}$, $L_{rr} = L_{lr} + L_{mr}$, $L_{mr} = \dfrac{N_r^2}{\mathfrak{R}_m}$; $\mathbf{L}_{sr}(\theta_r)$

is the stator–rotor mutual inductance mapping, $\mathbf{L}_{sr}(\theta_r) = \begin{bmatrix} L_{sr} \cos \theta_r & -L_{sr} \sin \theta_r \\ L_{sr} \sin \theta_r & L_{sr} \cos \theta_r \end{bmatrix}$, $L_{sr} = \dfrac{N_s N_r}{\mathfrak{R}_m}$;

L_{ms} and L_{mr} are the stator and rotor magnetizing inductances; L_{ls} and L_{lr} are the stator and rotor leakage inductances; N_s and N_r are the number of turns of the stator and rotor windings,

Using the number of turns in the stator and rotor windings, we have

$$\mathbf{i}'_{abr} = \frac{N_r}{N_s} \mathbf{i}_{abr}, \quad \mathbf{u}'_{abr} = \frac{N_s}{N_r} \mathbf{u}_{abr}, \quad \text{and} \quad \boldsymbol{\psi}'_{abr} = \frac{N_s}{N_r} \boldsymbol{\psi}_{abr}.$$

Applying the turn ratio, the flux linkages are

$$\begin{bmatrix} \boldsymbol{\psi}_{abs} \\ \boldsymbol{\psi}'_{abr} \end{bmatrix} = \begin{bmatrix} \mathbf{L}_s & \mathbf{L}'_{sr}(\theta_r) \\ \mathbf{L}'^{T}_{sr}(\theta_r) & \mathbf{L}'_r \end{bmatrix} \begin{bmatrix} \mathbf{i}_{abs} \\ \mathbf{i}'_{abr} \end{bmatrix},$$

$$\mathbf{L}'_r = \left(\frac{N_s}{N_r}\right)^2 \mathbf{L}_r = \begin{bmatrix} L'_{rr} & 0 \\ 0 & L'_{rr} \end{bmatrix}, \quad \mathbf{L}'_{sr}(\theta_r) = \left(\frac{N_s}{N_r}\right) \mathbf{L}_{sr}(\theta_r) = L_{ms} \begin{bmatrix} \cos\theta_r & -\sin\theta_r \\ \sin\theta_r & \cos\theta_r \end{bmatrix},$$

where $L'_{rr} = L'_{lr} + L'_{mr}$, $L_{ms} = \dfrac{N_s}{N_r} L_{sr}$, $L'_{mr} = \left(\dfrac{N_s}{N_r}\right)^2 L_{mr}$, $L'_{mr} = L_{ms} = \dfrac{N_s}{N_r} L_{sr}$, and $L'_{rr} = L'_{lr} + L_{ms}$.

From $\mathbf{L}_s$, $\mathbf{L}'_r$, and $\mathbf{L}'_{sr}(\theta_r)$, one obtains

$$\begin{bmatrix} \psi_{as} \\ \psi_{bs} \\ \psi'_{ar} \\ \psi'_{br} \end{bmatrix} = \begin{bmatrix} L_{ss} & 0 & L_{ms}\cos\theta_r & -L_{ms}\sin\theta_r \\ 0 & L_{ss} & L_{ms}\sin\theta_r & L_{ms}\cos\theta_r \\ L_{ms}\cos\theta_r & L_{ms}\sin\theta_r & L'_{rr} & 0 \\ -L_{ms}\sin\theta_r & L_{ms}\cos\theta_r & 0 & L'_{rr} \end{bmatrix} \begin{bmatrix} i_{as} \\ i_{bs} \\ i'_{ar} \\ i'_{br} \end{bmatrix}. \tag{5.3}$$

The differential equations (5.1) and (5.2) are written as

$$\mathbf{u}_{abs} = \mathbf{r}_s \mathbf{i}_{abs} + \frac{d\boldsymbol{\psi}_{abs}}{dt}, \quad \mathbf{u}'_{abr} = \mathbf{r}'_r \mathbf{i}'_{abr} + \frac{d\boldsymbol{\psi}'_{abr}}{dt}, \tag{5.4}$$

where $\mathbf{r}'_r = \dfrac{N_s^2}{N_r^2} \mathbf{r}_r = \dfrac{N_s^2}{N_r^2} \begin{bmatrix} r'_r & 0 \\ 0 & r'_r \end{bmatrix}$.

The self-inductances L_{ss} and L'_{rr} are time-invariant. Furthermore, L_{ms} is constant. From (5.4), using the expressions for the flux linkages (5.3), one obtains a set of four nonlinear differential equations

$$L_{ss}\frac{di_{as}}{dt} + L_{ms}\frac{d\left(i'_{ar}\cos\theta_r\right)}{dt} - L_{ms}\frac{d\left(i'_{br}\sin\theta_r\right)}{dt} = -r_s i_{as} + u_{as},$$

$$L_{ss}\frac{di_{bs}}{dt} + L_{ms}\frac{d\left(i'_{ar}\sin\theta_r\right)}{dt} + L_{ms}\frac{d\left(i'_{br}\cos\theta_r\right)}{dt} = -r_s i_{bs} + u_{bs},$$

$$L_{ms}\frac{d(i_{as}\cos\theta_r)}{dt} + L_{ms}\frac{d(i_{bs}\sin\theta_r)}{dt} + L'_{rr}\frac{di'_{ar}}{dt} = -r'_r i'_{ar} + u'_{ar},$$

$$-L_{ms}\frac{d(i_{as}\sin\theta_r)}{dt} + L_{ms}\frac{d(i_{bs}\cos\theta_r)}{dt} + L'_{rr}\frac{di'_{br}}{dt} = -r'_r i'_{br} + u'_{br}. \tag{5.5}$$

The *emf* is given as $emf = \displaystyle\oint_l \vec{E}\cdot d\vec{l} = \underbrace{\oint_l \left(\vec{v}\times\vec{B}\right)\cdot d\vec{l}}_{\text{motional induction (generation)}} - \underbrace{\oint_s \frac{\partial \vec{B}}{\partial t}d\vec{s}}_{\text{transformer induction}}$, and the

Faraday law of induction is $\mathcal{E} = \displaystyle\oint_l \vec{E}(t)\cdot d\vec{l} = -\frac{d}{dt}\int_s \vec{B}(t)\cdot d\vec{s} = -N\frac{d\Phi}{dt} = -\frac{d\psi}{dt}$. The unit for

the *emf* is volts. From (5.5), the induced *emf* terms for the rotor windings are

$$emf_{ar} = -L_{ms}\frac{d(i_{as}\cos\theta_r)}{dt} - L_{ms}\frac{d(i_{bs}\sin\theta_r)}{dt} - L'_{rr}\frac{di'_{ar}}{dt},$$

$$emf_{br} = L_{ms}\frac{d(i_{as}\sin\theta_r)}{dt} - L_{ms}\frac{d(i_{bs}\cos\theta_r)}{dt} - L'_{rr}\frac{di'_{br}}{dt}.$$

Hence, for the rotor windings, the motional *emf* terms in the steady-state operation are

$$emf_{ar\omega} = L_{ms}(i_{as}\sin\theta_r - i_{bs}\cos\theta_r)\omega_r \quad \text{and} \quad emf_{br\omega} = L_{ms}(i_{as}\cos\theta_r + i_{bs}\sin\theta_r)\omega_r.$$

These *emfs* justify the statement that the voltages are induced in the rotor windings.

From (5.5), Cauchy's form of differential equations are found, see the first four equations in (5.10). The electrical angular velocity ω_r and displacement θ_r are the state variables. Therefore, the *torsional-mechanical* equation of motion must be derived describing the evolution of ω_r and θ_r. From Newton's second law, we have

$$\frac{d\omega_{rm}}{dt} = \frac{1}{J}(T_e - B_m\omega_{rm} - T_L), \quad \frac{d\theta_{rm}}{dt} = \omega_{rm}. \tag{5.6}$$

The mechanical angular velocity of the rotor ω_{rm} is expressed by using the electrical angular velocity ω_r and the number of poles P. In particular, $\omega_{rm} = \frac{2}{P}\omega_r$.

For the mechanical angular displacement θ_{rm}, we have $\theta_{rm} = \frac{2}{P}\theta_r$.

It is convenient to derive the equations of motion using the electrical angular velocity ω_r and displacement θ_r in order to ease notations and ensure the generality of results. From the Newton second law of motion (5.6), one finds

$$\frac{d\omega_r}{dt} = \frac{P}{2J}T_e - \frac{B_m}{J}\omega_r - \frac{P}{2J}T_L,$$

$$\frac{d\theta_r}{dt} = \omega_r. \tag{5.7}$$

The electromagnetic torque, developed by induction motors, is $T_e = \frac{P}{2}\frac{\partial W_c\left(\mathbf{i}_{abs},\, \mathbf{i}'_{abr},\, \theta_r\right)}{\partial\theta_r}$.
Assuming that the magnetic system is linear, the coenergy is

$$W_c = W_f = \frac{1}{2}\mathbf{i}^T_{abs}(\mathbf{L}_s - L_{ls}\mathbf{I})\mathbf{i}_{abs} + \mathbf{i}^T_{abs}\mathbf{L}'_{sr}(\theta_r)\mathbf{i}'_{abr} + \frac{1}{2}\mathbf{i}'^T_{abr}(\mathbf{L}'_r - L'_{lr}\mathbf{I})\mathbf{i}'_{abr}.$$

The self-inductances L_{ss} and L'_{rr} as well as the leakage inductances L_{ls} and L'_{lr} are not functions of the angular displacement θ_r. From (5.3), we recall that the stator–rotor mutual inductances are assumed a pure sinusoidal variations,
$\mathbf{L}'_{sr}(\theta_r) = L_{ms}\begin{bmatrix} \cos\theta_r & -\sin\theta_r \\ \sin\theta_r & \cos\theta_r \end{bmatrix}$. Hence, for P-pole two-phase induction motors, the electromagnetic torque T_e is

$$T_e = \frac{P}{2}\frac{\partial W_c\left(\mathbf{i}_{abs},\, \mathbf{i}'_{abr},\, \theta_r\right)}{\partial\theta_r} = \frac{P}{2}\mathbf{i}^T_{abs}\frac{\partial\mathbf{L}'_{sr}(\theta_r)}{\partial\theta_r}\mathbf{i}'_{abr} = \frac{P}{2}L_{ms}[i_{as}\ i_{bs}]\begin{bmatrix} -\sin\theta_r & -\cos\theta_r \\ \cos\theta_r & -\sin\theta_r \end{bmatrix}\begin{bmatrix} i'_{ar} \\ i'_{br} \end{bmatrix}$$

$$= -\frac{P}{2}L_{ms}\left[(i_{as}i'_{ar} + i_{bs}i'_{br})\sin\theta_r + (i_{as}i'_{br} - i_{bs}i'_{ar})\cos\theta_r\right]. \tag{5.8}$$

Using (5.7) and (5.8), the *torsional-mechanical* equations are

$$\frac{d\omega_r}{dt} = -\frac{P^2}{4J}L_{ms}\left[(i_{as}i'_{ar} + i_{bs}i'_{br})\sin\theta_r + (i_{as}i'_{br} - i_{bs}i'_{ar})\cos\theta_r\right] - \frac{B_m}{J}\omega_r - \frac{P}{2J}T_L,$$

$$\frac{d\theta_r}{dt} = \omega_r.$$

(5.9)

Supplying the phase voltages u_{as} and u_{bs}, one rotates the motor in the desired direction, e.g., clockwise or counterclockwise. The electromagnetic torque T_e counteracts the load and friction torques. One recalls that the torques and forces are vectors. In actuators and motors, the friction torque acts against the electromagnetic torque, while the load torques may be bidirectional opposing or favoring T_e. Depending on the direction of rotation (clockwise and counterclockwise), the sign for T_e changes. The mutual inductances and expression $\mathbf{L}'_{sr}(\theta_r)$ can be refined based upon the direction of rotation and initial conditions (displacement of rotor with respect to stator).

The circuitry-electromagnetic and *torsional-mechanical* equations (5.5) and (5.9) are integrated obtaining the nonlinear differential equations which describe the dynamics of two-phase induction motors

$$\frac{di_{as}}{dt} = -\frac{L'_{rr}r_s}{L_\Sigma}i_{as} + \frac{L^2_{ms}}{L_\Sigma}i_{bs}\omega_r + \frac{L_{ms}L'_{rr}}{L_\Sigma}i'_{ar}\left(\omega_r\sin\theta_r + \frac{r'_r}{L'_{rr}}\cos\theta_r\right)$$

$$+ \frac{L_{ms}L'_{rr}}{L_\Sigma}i'_{br}\left(\omega_r\cos\theta_r - \frac{r'_r}{L'_{rr}}\sin\theta_r\right) + \frac{L'_{rr}}{L_\Sigma}u_{as} - \frac{L_{ms}}{L_\Sigma}\cos\theta_r u'_{ar} + \frac{L_{ms}}{L_\Sigma}\sin\theta_r u'_{br},$$

$$\frac{di_{bs}}{dt} = -\frac{L'_{rr}r_s}{L_\Sigma}i_{bs} - \frac{L^2_{ms}}{L_\Sigma}i_{as}\omega_r - \frac{L_{ms}L'_{rr}}{L_\Sigma}i'_{ar}\left(\omega_r\cos\theta_r - \frac{r'_r}{L'_{rr}}\sin\theta_r\right)$$

$$+ \frac{L_{ms}L'_{rr}}{L_\Sigma}i'_{br}\left(\omega_r\sin\theta_r + \frac{r'_r}{L'_{rr}}\cos\theta_r\right) + \frac{L'_{rr}}{L_\Sigma}u_{bs} - \frac{L_{ms}}{L_\Sigma}\sin\theta_r u'_{ar} - \frac{L_{ms}}{L_\Sigma}\cos\theta_r u'_{br},$$

$$\frac{di'_{ar}}{dt} = -\frac{L_{ss}r'_r}{L_\Sigma}i'_{ar} + \frac{L_{ms}L_{ss}}{L_\Sigma}i_{as}\left(\omega_r\sin\theta_r + \frac{r_s}{L_{ss}}\cos\theta_r\right) - \frac{L_{ms}L_{ss}}{L_\Sigma}i_{bs}\left(\omega_r\cos\theta_r - \frac{r_s}{L_{ss}}\sin\theta_r\right)$$

$$- \frac{L^2_{ms}}{L_\Sigma}i'_{br}\omega_r - \frac{L_{ms}}{L_\Sigma}\cos\theta_r u_{as} - \frac{L_{ms}}{L_\Sigma}\sin\theta_r u_{bs} + \frac{L_{ss}}{L_\Sigma}u'_{ar},$$

$$\frac{di'_{br}}{dt} = -\frac{L_{ss}r'_r}{L_\Sigma}i'_{br} + \frac{L_{ms}L_{ss}}{L_\Sigma}i_{as}\left(\omega_r\cos\theta_r - \frac{r_s}{L_{ss}}\sin\theta_r\right) + \frac{L_{ms}L_{ss}}{L_\Sigma}i_{bs}\left(\omega_r\sin\theta_r + \frac{r_s}{L_{ss}}\cos\theta_r\right)$$

$$+ \frac{L^2_{ms}}{L_\Sigma}i'_{ar}\omega_r + \frac{L_{ms}}{L_\Sigma}\sin\theta_r u_{as} - \frac{L_{ms}}{L_\Sigma}\cos\theta_r u_{bs} + \frac{L_{ss}}{L_\Sigma}u'_{br},$$

$$\frac{d\omega_r}{dt} = -\frac{P^2}{4J}L_{ms}\left[(i_{as}i'_{ar} + i_{bs}i'_{br})\sin\theta_r + (i_{as}i'_{br} - i_{bs}i'_{ar})\cos\theta_r\right] - \frac{B_m}{J}\omega_r - \frac{P}{2J}T_L,$$

$$\frac{d\theta_r}{dt} = \omega_r,$$

(5.10)

where $L_\Sigma = L_{ss}L'_{rr} - L^2_{ms}$.

The derived equations can be applied in various analysis, design, and optimization tasks.

5.3.2 Lagrange Equations of Motion for Induction Machines

The mathematical model can be derived using the Lagrange equations of motion. The generalized independent coordinates are four charges and the rotor angular displacement. Hence,

$$q_1 = \frac{i_{as}}{s}, \quad q_2 = \frac{i_{bs}}{s}, \quad q_3 = \frac{i'_{ar}}{s}, \quad q_4 = \frac{i'_{br}}{s}, \quad \text{and} \quad q_5 = \theta_r.$$

The generalized forces are the voltages and load torque, e.g.,

$$Q_1 = u_{as}, \quad Q_2 = u_{bs}, \quad Q_3 = u'_{ar}, \quad Q_4 = u'_{br}, \quad \text{and} \quad Q_5 = -T_L$$

The resulting five Lagrange equations $\dfrac{d}{dt}\left(\dfrac{\partial \Gamma}{\partial \dot{q}_i}\right) - \dfrac{\partial \Gamma}{\partial q_i} + \dfrac{\partial D}{\partial \dot{q}_i} + \dfrac{\partial \Pi}{\partial q_i} = Q_i$ are

$$\frac{d}{dt}\left(\frac{\partial \Gamma}{\partial \dot{q}_1}\right) - \frac{\partial \Gamma}{\partial q_1} + \frac{\partial D}{\partial \dot{q}_1} + \frac{\partial \Pi}{\partial q_1} = Q_1, \quad \frac{d}{dt}\left(\frac{\partial \Gamma}{\partial \dot{q}_2}\right) - \frac{\partial \Gamma}{\partial q_2} + \frac{\partial D}{\partial \dot{q}_2} + \frac{\partial \Pi}{\partial q_2} = Q_2,$$

$$\frac{d}{dt}\left(\frac{\partial \Gamma}{\partial \dot{q}_3}\right) - \frac{\partial \Gamma}{\partial q_3} + \frac{\partial D}{\partial \dot{q}_3} + \frac{\partial \Pi}{\partial q_3} = Q_3, \quad \frac{d}{dt}\left(\frac{\partial \Gamma}{\partial \dot{q}_4}\right) - \frac{\partial \Gamma}{\partial q_4} + \frac{\partial D}{\partial \dot{q}_4} + \frac{\partial \Pi}{\partial q_4} = Q_4, \qquad (5.11)$$

$$\frac{d}{dt}\left(\frac{\partial \Gamma}{\partial \dot{q}_5}\right) - \frac{\partial \Gamma}{\partial q_5} + \frac{\partial D}{\partial \dot{q}_5} + \frac{\partial \Pi}{\partial q_5} = Q_5.$$

Using the notations introduced, the total kinetic, potential, and dissipated energies, to be used in equations (5.11), are

$$\Gamma = \frac{1}{2} L_{ss}\dot{q}_1^2 + L_{ms}\dot{q}_1\dot{q}_3 \cos q_5 - L_{ms}\dot{q}_1\dot{q}_4 \sin q_5 + \frac{1}{2} L_{ss}\dot{q}_2^2 + L_{ms}\dot{q}_2\dot{q}_3 \sin q_5 + L_{ms}\dot{q}_2\dot{q}_4 \cos q_5$$

$$+ \frac{1}{2} L'_{rr}\dot{q}_3^2 + \frac{1}{2} L'_{rr}\dot{q}_4^2 + \frac{1}{2} J\dot{q}_5^2,$$

$$\Pi = 0,$$

and

$$D = \frac{1}{2}\left(r_s\dot{q}_1^2 + r_s\dot{q}_2^2 + r'_r\dot{q}_3^2 + r'_r\dot{q}_4^2 + B_m\dot{q}_5^2\right).$$

The derivative terms of (5.11) are

$$\frac{\partial \Gamma}{\partial q_1} = 0, \quad \frac{\partial \Gamma}{\partial \dot{q}_1} = L_{ss}\dot{q}_1 + L_{ms}\dot{q}_3 \cos q_5 - L_{ms}\dot{q}_4 \sin q_5,$$

$$\frac{\partial \Gamma}{\partial q_2} = 0, \quad \frac{\partial \Gamma}{\partial \dot{q}_2} = L_{ss}\dot{q}_2 + L_{ms}\dot{q}_3 \sin q_5 + L_{ms}\dot{q}_4 \cos q_5,$$

$$\frac{\partial \Gamma}{\partial q_3} = 0, \quad \frac{\partial \Gamma}{\partial \dot{q}_3} = L'_{rr}\dot{q}_3 + L_{ms}\dot{q}_1 \cos q_5 + L_{ms}\dot{q}_2 \sin q_5,$$

$$\frac{\partial \Gamma}{\partial q_4} = 0, \quad \frac{\partial \Gamma}{\partial \dot{q}_4} = L'_{rr}\dot{q}_4 - L_{ms}\dot{q}_1 \sin q_5 + L_{ms}\dot{q}_2 \cos q_5,$$

$$\frac{\partial \Gamma}{\partial q_5} = -L_{ms}\dot{q}_1\dot{q}_3 \sin q_5 - L_{ms}\dot{q}_1\dot{q}_4 \cos q_5 + L_{ms}\dot{q}_2\dot{q}_3 \cos q_5 - L_{ms}\dot{q}_2\dot{q}_4 \sin q_5$$

$$= -L_{ms}[(\dot{q}_1\dot{q}_3 + \dot{q}_2\dot{q}_4) \sin q_5 + (\dot{q}_1\dot{q}_4 - \dot{q}_2\dot{q}_3) \cos q_5],$$

$$\frac{\partial \Gamma}{\partial \dot{q}_5} = J\dot{q}_5,$$

$$\frac{\partial \Pi}{\partial q_1} = 0, \quad \frac{\partial \Pi}{\partial q_2} = 0, \quad \frac{\partial \Pi}{\partial q_3} = 0, \quad \frac{\partial \Pi}{\partial q_4} = 0, \quad \frac{\partial \Pi}{\partial q_5} = 0,$$

$$\frac{\partial D}{\partial \dot{q}_1} = r_s\dot{q}_1, \quad \frac{\partial D}{\partial \dot{q}_2} = r_s\dot{q}_2, \quad \frac{\partial D}{\partial \dot{q}_3} = r'_r\dot{q}_3, \quad \frac{\partial D}{\partial \dot{q}_4} = r'_r\dot{q}_4, \quad \frac{\partial D}{\partial \dot{q}_5} = B_m\dot{q}_5.$$

Having found the derivative terms of five Lagrange equations (5.11), expressing the generalized coordinates and forces using the machine variables ($\dot{q}_1 = i_{as}$, $\dot{q}_2 = i_{bs}$, $\dot{q}_3 = i'_{ar}$, $\dot{q}_4 = i'_{br}$, $\dot{q}_5 = \omega_r$ and $Q_1 = u_{as}$, $Q_2 = u_{bs}$, $Q_3 = u'_{ar}$, $Q_4 = u'_{br}$, $Q_5 = -T_L$), one obtains the following differential equations

$$L_{ss}\frac{di_{as}}{dt} + L_{ms}\frac{d(i'_{ar} \cos \theta_r)}{dt} - L_{ms}\frac{d(i'_{br} \sin \theta_r)}{dt} + r_s i_{as} = u_{as},$$

$$L_{ss}\frac{di_{bs}}{dt} + L_{ms}\frac{d(i'_{ar} \sin \theta_r)}{dt} + L_{ms}\frac{d(i'_{br} \cos \theta_r)}{dt} + r_s i_{bs} = u_{bs},$$

$$L_{ms}\frac{d(i_{as} \cos \theta_r)}{dt} + L_{ms}\frac{d(i_{bs} \sin \theta_r)}{dt} + L'_{rr}\frac{di'_{ar}}{dt} + r'_r i'_{ar} = u'_{ar}, \quad (5.12)$$

$$- L_{ms}\frac{d(i_{as} \sin \theta_r)}{dt} + L_{ms}\frac{d(i_{bs} \cos \theta_r)}{dt} + L'_{rr}\frac{di'_{br}}{dt} + r'_r i'_{br} = u'_{br},$$

$$J\frac{d^2\theta_r}{dt^2} + L_{ms}\left[(i_{as}i'_{ar} + i_{bs}i'_{br}) \sin \theta_r + (i_{as}i'_{br} - i_{bs}i'_{ar}) \cos \theta_r\right] + B_m\frac{d\theta_r}{dt} = -T_L.$$

From (5.12), for *P*-pole induction motors, by using $\frac{d\theta_r}{dt} = \omega_r$, six differential equations (5.5) and (5.9) result which were found applying other physical laws in Section 5.2.1. The advantage of the Lagrange concept is that Kirchhoff's, Newton's, Faraday's, Lorenz's, coenergy, or other laws are not used to derive the resulting models. The Lagrange equations of motion provide a general and coherent procedure. Furthermore, one may derive the *emf* terms, electromagnetic torque, etc. For example, from $\frac{d}{dt}\left(\frac{\partial \Gamma}{\partial \dot{q}_5}\right) - \frac{\partial \Gamma}{\partial q_5} + \frac{\partial D}{\partial \dot{q}_5} + \frac{\partial \Pi}{\partial q_5} = Q_5$, one concludes that the electromagnetic torque is

$$T_e = \frac{\partial \Gamma}{\partial q_5} = \frac{\partial \Gamma}{\partial \theta_r} = -L_{ms}[(\dot{q}_1\dot{q}_3 + \dot{q}_2\dot{q}_4) \sin q_5 + (\dot{q}_1\dot{q}_4 - \dot{q}_2\dot{q}_3) \cos q_5]. \quad (5.13)$$

Equation (5.13) results in $T_e = -L_{ms}\left[\left(i_{as}i'_{ar} + i_{bs}i'_{br}\right)\sin\theta_r + \left(i_{as}i'_{br} - i_{bs}i'_{ar}\right)\cos\theta_r\right]$. The same equation, as given by (5.8), was derived by applying the coenergy.

Example 5.2:

We simulate and analyze the performance of two different two-phase 115 V (*rms*), 60 Hz, four-pole ($P = 4$) induction motors. The dynamics is described by differential equations (5.5) through (5.9), (5.10), or (5.12). The objective is to examine the performance of induction motors analyzing the acceleration capabilities, settling time, efficiency, etc. The torque–speed characteristics will be studied by solving the differential equations to compare the dynamic and steady-state torque–speed characteristics $\omega_r(t) = \Omega_T[T_e(t)]$ and $\omega_r = \Omega_T(T_e)$. The parameters of the A-class motors are $r_s = 1.2$ ohm, $r'_r = 1.5$ ohm, $L_{ms} = 0.16$ H, $L_{ls} = 0.02$ H, $L_{ss} = L_{ls} + L_{ms}$, $L'_{lr} = 0.02$ H, $L'_{rr} = L'_{lr} + L_{ms}$, $B_m = 1 \times 10^{-6}$ N-m-s/rad, and $J = 0.005$ kg-m². For a D-class induction motor, we have $r_s = 24.5$ ohm, $r'_r = 23$ ohm, $L_{ms} = 0.27$ H, $L_{ls} = 0.027$ H, $L_{ss} = L_{ls} + L_{ms}$, $L'_{lr} = 0.027$ H, $L'_{rr} = L'_{lr} + L_{ms}$, $B_m = 1 \times 10^{-6}$ N-m-s/rad, and $J = 0.001$ kg-m².

To guarantee the balanced operation, the supplied phase voltages are

$$u_{as}(t) = \sqrt{2}u_M \cos(\omega_f t) \quad \text{and} \quad u_{bs}(t) = \sqrt{2}u_M \sin(\omega_f t).$$

For the rated voltage, we have $u_{as}(t) = \sqrt{2}115\cos(377t)$ and $u_{bs}(t) = \sqrt{2}115\sin(377t)$.

No load and loaded conditions are examined assigning the load torque to be 0 and 0.5 N m. The simulations are performed, and transient dynamics of the stator and rotor currents in the *as*, *bs*, *ar*, and *br* windings $i_{as}(t)$, $i_{bs}(t)$, $i'_{ar}(t)$, and $i'_{br}(t)$, as well as the mechanical angular velocity $\omega_{rm}(t)$, are potted in Figures 5.6 and 5.7. Figure 5.6 illustrates the transient dynamics of the A-class motor. The motor accelerates from stall, e.g., $\omega_{rm0} = 0$ rad/s. Figures 5.6a and b depict the motor dynamics if $T_L = 0$ and $T_L = 0.5$ N m (applied at $t = 0$ s), respectively. The dynamics and acceleration capabilities of the D-class motor are documented in Figure 5.7.

The A-class induction motor reaches the steady-state angular velocity within 0.8 s (with no load), while the settling time is 1.4 s if motor operates under $T_L = 0.5$ N m. Better acceleration capabilities are observed for the D-class motor with the settling time 0.5 and 0.8 s for no load and loaded conditions, respectively. Analyzing the torque–speed characteristics, it was emphasized that the D-class induction motors may develop higher starting electromagnetic torque as compared to the A- and B-class motors, see Figure 5.2a. However, the D-class motors may not necessarily possess higher $T_{e\,\text{start}}$ and $T_{e\,\text{critical}}$ as compared to the A-class motor. In fact, as illustrated in Figures 5.6 and 5.7, A-class induction motors possess higher $T_{e\,\text{start}}$ and $T_{e\,\text{critical}}$ as well as higher ratio $T_{e\,\text{critical}}/T_{e\,\text{start}}$. For the A- and D-class induction motors studied, the $T_{e\,\text{critical}}$ are $\sim$3 and 1.1 N m, respectively. Furthermore, the efficiency of D-class induction motors is low due to high r_r.

The moment of inertia significantly affects the acceleration capabilities. Using equation $\dfrac{d\omega_{rm}}{dt} = \dfrac{1}{J}(T_e - B_m\omega_{rm} - T_L)$ we conclude that A-class induction motor possesses better acceleration capabilities for the moments of inertia used (J are 0.005 and 0.001 kg-m²). One recalls that the equivalent J used integrates the moment of inertia of rotor and attached mechanisms. Figures 5.6 and 5.7 illustrate that the electromagnetic torque for A- and D-class motors reaches 4.1 and 1.8 N m, respectively. Hence, A-class induction motor ensures better performance. In the A-class motor, the magnitude of the stator phase currents are higher (the input power is higher, but the losses are lower due to low phase resistances), while L_{ms} is lower, as compared to the D-class motor.

This analysis is confirmed by assessing the torque–speed characteristics. The dynamics of $\omega_{rm}(t)$ and evolutions of $T_e(t)$ are documented in Figures 5.6 and 5.7. The resulting characteristics $\omega_{rm}(t) = \Omega_T[T_e(t)]$ are obtained by plotting the mechanical angular velocity versus the electromagnetic torque evolution. The torque–speed characteristics are derived by using the simulation results carried out studying the motor dynamics. Thus, the steady-state-centered analysis can be

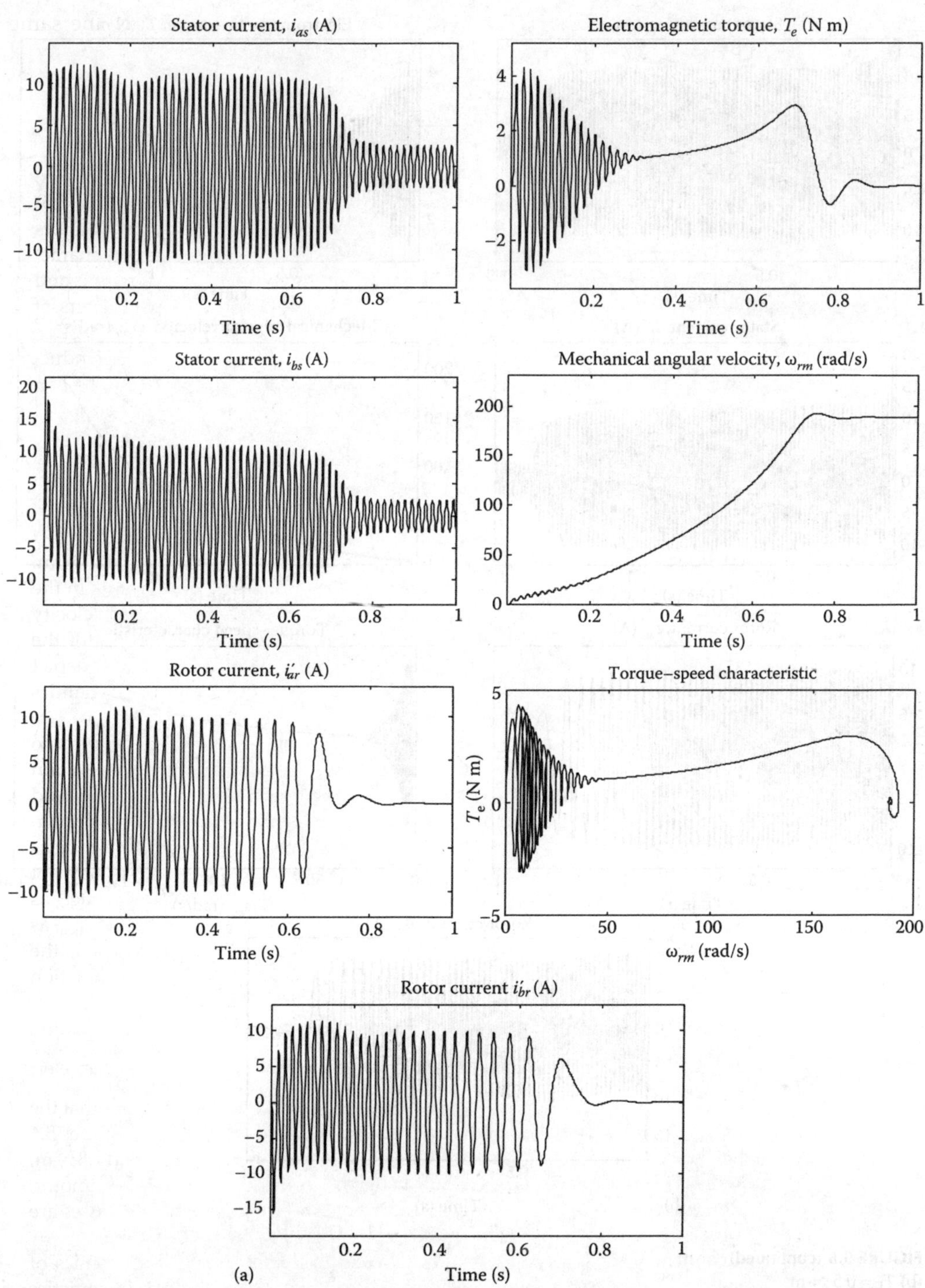

FIGURE 5.6

Dynamics and torque–speed characteristic of the A-class induction motor: (a) $T_L = 0$ N-m;

(*continued*)

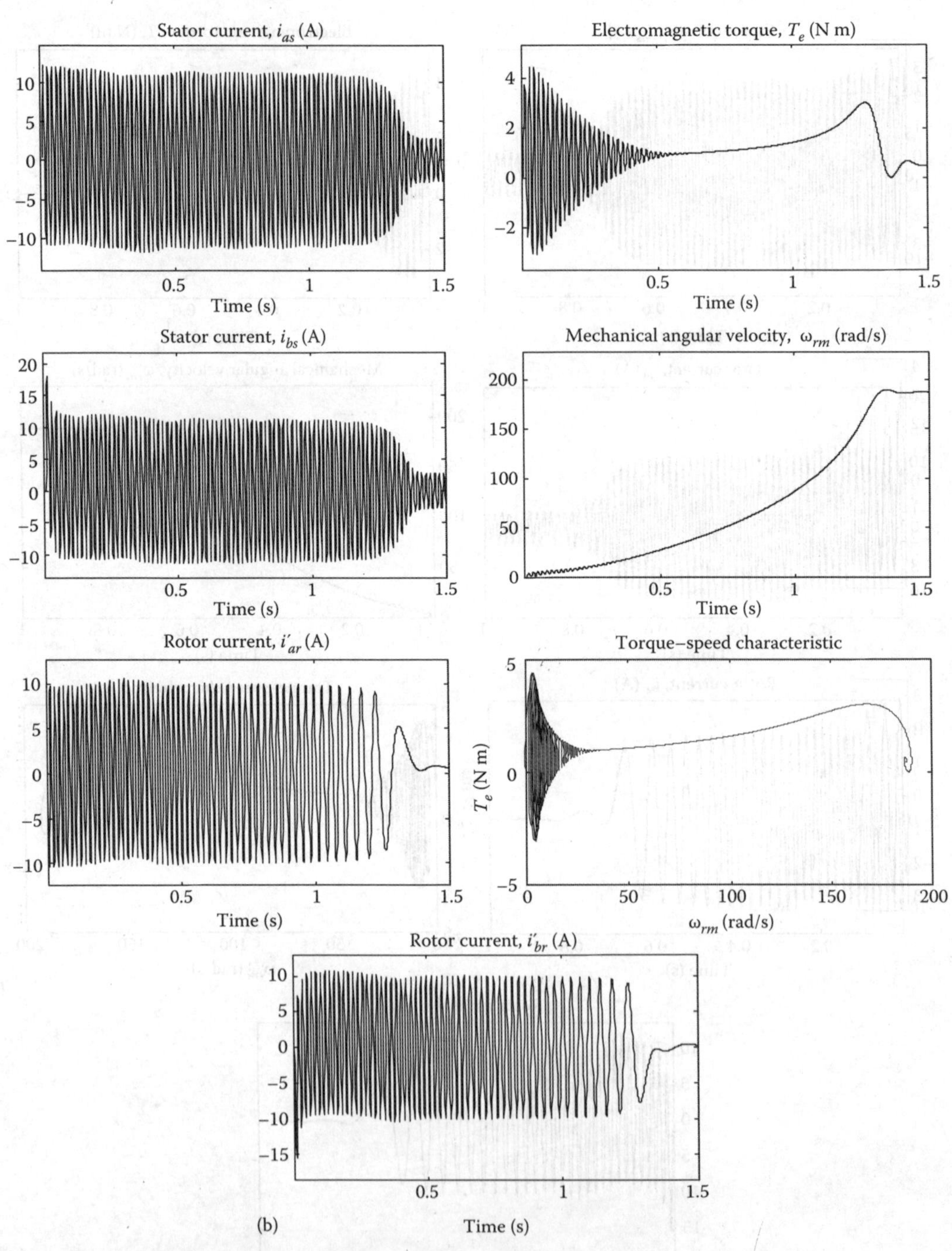

FIGURE 5.6 (continued)
(b) $T_L = 0.5$ N-m.

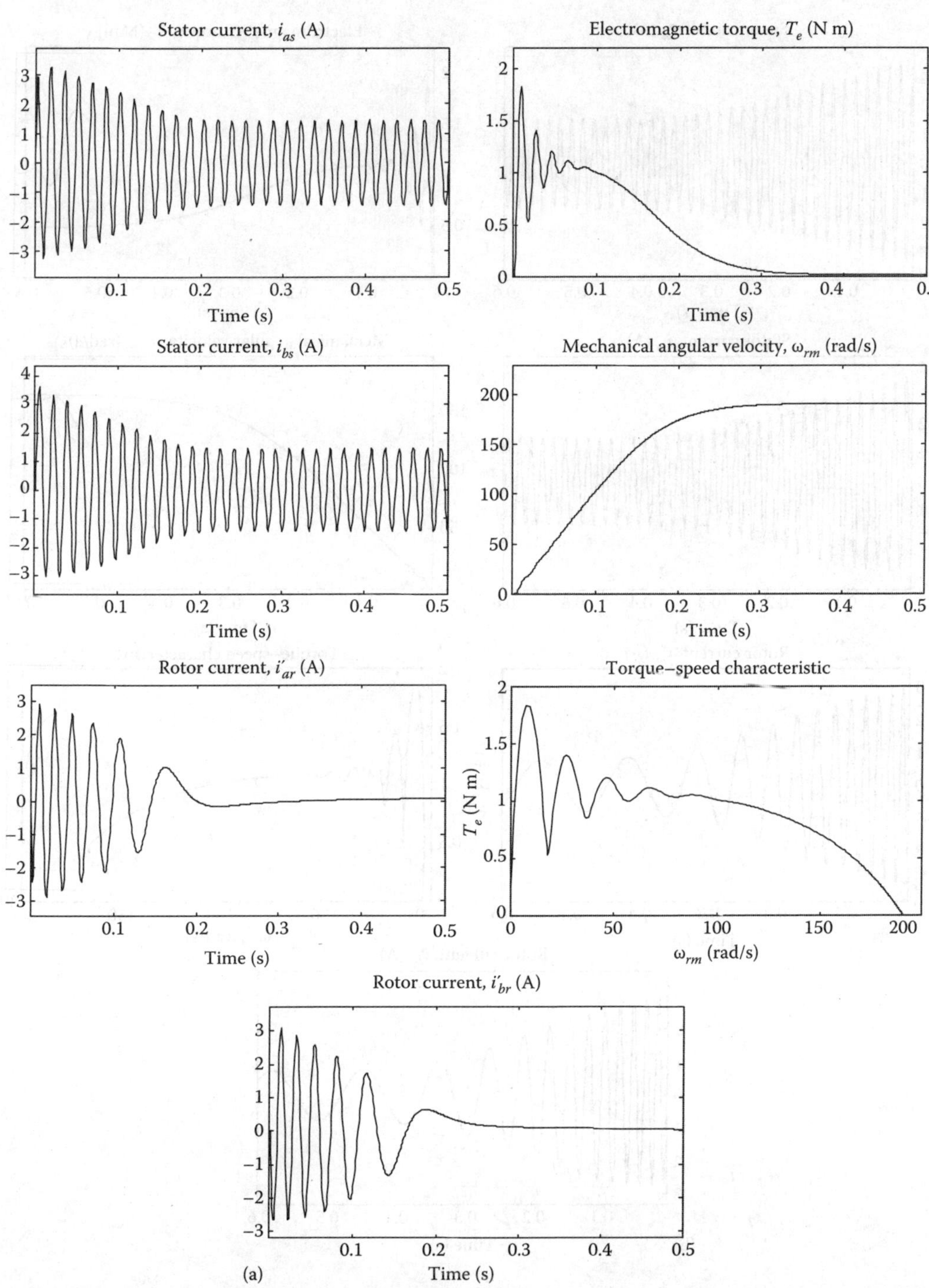

FIGURE 5.7
Dynamics and torque–speed characteristic of the D-class induction motor: (a) $T_L = 0$ N-m;

(continued)

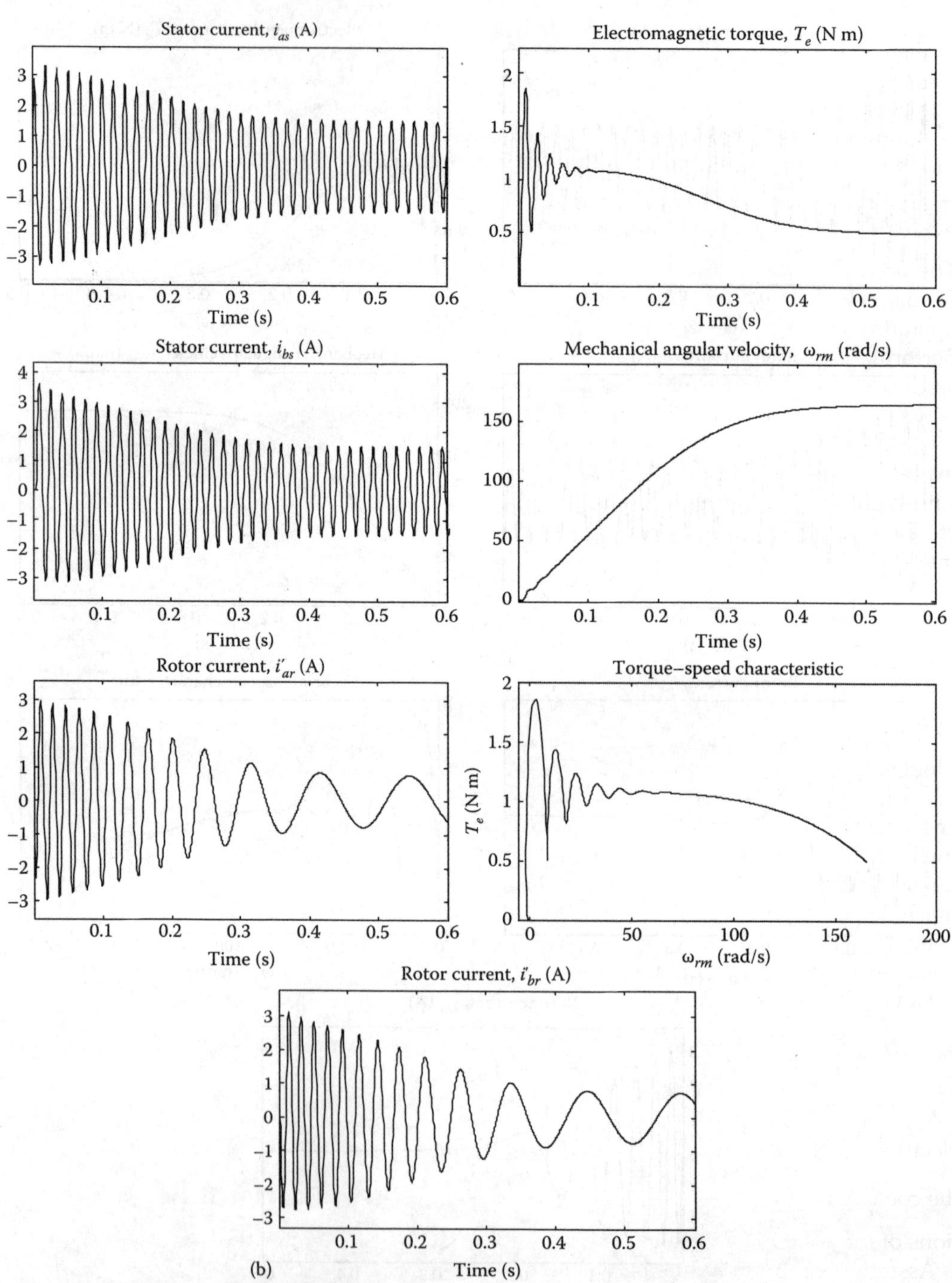

FIGURE 5.7 (continued)

(b) $T_L = 0.5$ N-m.

accomplished by examining the transient dynamics, but not vice versa. Figure 5.6 document the torque–speed characteristics of the A-class motor, while Figure 5.7 illustrate $\omega_{rm}(t) = \Omega_T[T_e(t)]$ for the D-class motor. One may conclude that the steady-state torque–speed characteristic curves, as plotted in Figures 5.3 and 5.4, may not result in coherent analysis of the evolution for $T_e(t)$ as induction motors are in transients (acceleration, disacceleration, loading, disturbances, etc.). Hence, the steady-sate analysis can be used mainly for preliminary studies. ∎

5.3.3 Advanced Topics in Analysis of Induction Machines

The analysis was performed assuming optimal design of induction machines, linearity of magnetic system, etc. The designer can achieve near-optimal design in the specified operating envelope. However, the undesired effects can degrade the machine and system performance and capabilities. In this section, we document how to perform the advanced studies examining near-optimal stator–rotor magnetic coupling. The magnetic coupling between stator and rotor windings may not obey the expressions $L_{asar} = L_{aras} = L_{sr} \cos \theta_r$, $L_{asbr} = L_{bras} = -L_{sr} \sin \theta_r$, $L_{bsar} = L_{arbs} = L_{sr} \sin \theta_r$, and $L_{bsbr} = L_{brbs} = L_{sr} \cos \theta_r$. One may not expect that in the full operating envelope, an optimal design will result in this ideal pure sinusoidal distribution which eliminates the torque ripple, current chattering, overheating, etc. Depending on the induction machine overall design, in the full operating envelope, one may have

$$L_{asar} = L_{aras} = \sum_{n=1}^{\infty} L_{srn} \cos^{2n-1} \theta_r, \quad L_{asbr} = L_{bras} = -\sum_{n=1}^{\infty} L_{srn} \sin^{2n-1} \theta_r,$$

$$L_{bsar} = L_{arbs} = \sum_{n=1}^{\infty} L_{srn} \sin^{2n-1} \theta_r, \quad L_{bsbr} = L_{brbs} - \sum_{n=1}^{\infty} L_{srn} \cos^{2n-1} \theta_r. \tag{5.14}$$

Modeling, simulations, and evaluation can be accomplished as the machine design is performed and experimental studies are conducted to finalize motor parameters, inductance mapping, and other descriptive features used in the analysis tasks. As reported, using the circuitry-electromagnetic equations of motion (5.4) and *torsional-mechanical* dynamics (5.7), the transient behavior can be described with the ultimate objective to examine machine performance and capabilities. The motor parameters, induced *emf*, torque, magnetic filed, and other qualitative and quantitative quantities can be experimentally obtained in the full operating envelope.

Having obtained (5.14), one finds the inductance mapping $\mathbf{L}'_{sr}(\theta_r)$. In (5.4) $\mathbf{u}_{abs} = \mathbf{r}_s \mathbf{i}_{abs} + \dfrac{d\boldsymbol{\psi}_{abs}}{dt}$, $\mathbf{u}'_{abr} = \mathbf{r}'_r \mathbf{i}'_{abr} + \dfrac{d\boldsymbol{\psi}'_{abr}}{dt}$, we apply the expression for the flux linkages $\begin{bmatrix} \boldsymbol{\psi}_{abs} \\ \boldsymbol{\psi}'_{abr} \end{bmatrix} = \begin{bmatrix} \mathbf{L}_s & \mathbf{L}'_{sr}(\theta_r) \\ \mathbf{L}'_{sr}\,T(\theta_r) & \mathbf{L}'_r \end{bmatrix} \begin{bmatrix} \mathbf{i}_{abs} \\ \mathbf{i}'_{abr} \end{bmatrix}$ to find the total derivative terms. The circuitry-electromagnetic equations of motion result. The electromagnetic torque is found using the coenergy $T_e = \dfrac{P}{2} \dfrac{\partial W_c\left(\mathbf{i}_{abs},\, \mathbf{i}'_{abr},\, \theta_r\right)}{\partial \theta_r} = \dfrac{P}{2} \mathbf{i}^T_{abs} \dfrac{\partial \mathbf{L}'_{sr}(\theta_r)}{\partial \theta_r} \mathbf{i}'_{abr}$, and the *torsional-mechanical* equations of motion (5.7) are used.

Assume that for an induction machine

$$L_{asar} = L_{aras} = L_{sr1} \cos \theta_r + L_{sr2} \cos^3 \theta_r, \quad L_{asbr} = L_{bras} = L_{sr1} \sin \theta_r - L_{sr2} \sin^3 \theta_r,$$

and

$$L_{bsar} = L_{arbs} = L_{sr1} \sin \theta_r + L_{sr2} \sin^3 \theta_r, \quad L_{bsbr} = L_{brbs} = L_{sr1} \cos \theta_r + L_{sr2} \cos^3 \theta_r$$

which correspond to (5.14). Taking note of the turn ratio, the flux linkages are

$$\begin{bmatrix} \boldsymbol{\psi}_{abs} \\ \boldsymbol{\psi}'_{abr} \end{bmatrix} = \begin{bmatrix} \mathbf{L}_s & \mathbf{L}'_{sr}(\theta_r) \\ \mathbf{L}'^{\,T}_{sr}(\theta_r) & \mathbf{L}'_r \end{bmatrix} \begin{bmatrix} \mathbf{i}_{abs} \\ \mathbf{i}'_{abr} \end{bmatrix}, \quad \mathbf{L}_s = \begin{bmatrix} L_{ss} & 0 \\ 0 & L_{ss} \end{bmatrix}, \quad \mathbf{L}'_r = \begin{bmatrix} L'_{rr} & 0 \\ 0 & L'_{rr} \end{bmatrix},$$

$$\mathbf{L}'_{sr}(\theta_r) = \left(\frac{N_s}{N_r}\right)\mathbf{L}_{sr}(\theta_r) = \begin{bmatrix} L_{ms1}\cos\theta_r + L_{ms2}\cos^3\theta_r & -L_{ms1}\sin\theta - L_{ms2}\sin^3\theta_r \\ L_{ms1}\sin\theta_r + L_{ms2}\sin^3\theta_r & L_{ms1}\cos\theta_r + L_{ms2}\cos^3\theta_r \end{bmatrix},$$

or

$$\begin{bmatrix} \psi_{as} \\ \psi_{bs} \\ \psi'_{ar} \\ \psi'_{br} \end{bmatrix} = \begin{bmatrix} L_{ss} & 0 & L_{ms1}\cos\theta_r + L_{ms2}\cos^3\theta_r & -L_{ms1}\sin\theta_r - L_{ms2}\sin^3\theta_r \\ 0 & L_{ss} & L_{ms1}\sin\theta_r + L_{ms2}\sin^3\theta_r & L_{ms1}\cos\theta_r + L_{ms2}\cos^3\theta_r \\ L_{ms1}\cos\theta_r + L_{ms2}\cos^3\theta_r & L_{ms1}\sin\theta_r + L_{ms2}\sin^3\theta_r & L'_{rr} & 0 \\ -L_{ms1}\sin\theta_r - L_{ms2}\sin^3\theta_r & L_{ms1}\cos\theta_r + L_{ms2}\cos^3\theta_r & 0 & L'_{rr} \end{bmatrix} \begin{bmatrix} i_{as} \\ i_{bs} \\ i'_{ar} \\ i'_{br} \end{bmatrix}$$

$$\tag{5.15}$$

From (5.4), $\mathbf{u}_{abs} = \mathbf{r}_s\mathbf{i}_{abs} + \dfrac{d\boldsymbol{\psi}_{abs}}{dt}$, $\mathbf{u}'_{abr} = \mathbf{r}'_r\mathbf{i}'_{abr} + \dfrac{d\boldsymbol{\psi}'_{abr}}{dt}$, and (5.15), one finds

$$L_{ss}\frac{di_{as}}{dt} + L_{ms1}\frac{d(i'_{ar}\cos\theta_r)}{dt} + L_{ms2}\frac{d(i'_{ar}\cos^3\theta_r)}{dt} - L_{ms1}\frac{d(i'_{br}\sin\theta_r)}{dt} - L_{ms2}\frac{d(i'_{br}\sin^3\theta_r)}{dt} = -r_s i_{as} + u_{as},$$

$$L_{ss}\frac{di_{bs}}{dt} + L_{ms1}\frac{d(i'_{ar}\sin\theta_r)}{dt} + L_{ms2}\frac{d(i'_{ar}\sin^3\theta_r)}{dt} + L_{ms1}\frac{d(i'_{br}\cos\theta_r)}{dt} + L_{ms2}\frac{d(i'_{br}\cos^3\theta_r)}{dt} = -r_s i_{bs} + u_{bs},$$

$$L_{ms1}\frac{d(i_{as}\cos\theta_r)}{dt} + L_{ms2}\frac{d(i_{as}\cos^3\theta_r)}{dt} + L_{ms1}\frac{d(i_{bs}\sin\theta_r)}{dt} + L_{ms2}\frac{d(i_{bs}\sin^3\theta_r)}{dt} + L'_{rr}\frac{di'_{ar}}{dt} = -r'_r i'_{ar} + u'_{ar},$$

$$-L_{ms1}\frac{d(i_{as}\sin\theta_r)}{dt} - L_{ms2}\frac{d(i_{as}\sin^3\theta_r)}{dt} + L_{ms1}\frac{d(i_{bs}\cos\theta_r)}{dt} + L_{ms2}\frac{d(i_{bs}\cos^3\theta_r)}{dt} + L'_{rr}\frac{di'_{br}}{dt} = -r'_r i'_{br} + u'_{br}.$$

$$\tag{5.16}$$

The rotor motional *emf* terms in the steady-state operation are

$$emf_{ar\omega} = \left(L_{ms1}i_{as}\sin\theta_r + 3L_{ms2}i_{as}\sin\theta_r\cos^2\theta_r - L_{ms1}i_{bs}\cos\theta_r - 3L_{ms2}i_{bs}\cos\theta_r\sin^2\theta_r\right)\omega_r$$

and

$$emf_{br\omega} = \left(L_{ms1}i_{as}\cos\theta_r + 3L_{ms2}i_{as}\cos\theta_r\sin^2\theta_r + L_{ms1}i_{bs}\sin\theta_r + 3L_{ms2}i_{bs}\sin\theta_r\cos^2\theta_r\right)\omega_r.$$

The *torsional-mechanical* dynamics is studied, and the expression for T_e is of a great importance. The electromagnetic torque developed is

$$T_e = \frac{P}{2}\frac{\partial W_c\left(\mathbf{i}_{abs},\,\mathbf{i}'_{abr},\,\theta_r\right)}{\partial\theta_r} = \frac{P}{2}\mathbf{i}^T_{abs}\frac{\partial\mathbf{L}'_{sr}(\theta_r)}{\partial\theta_r}\mathbf{i}'_{abr}$$

$$= \frac{P}{2}\begin{bmatrix} i_{as} & i_{bs} \end{bmatrix}\begin{bmatrix} -L_{ms1}\sin\theta_r - 3L_{ms2}\sin\theta_r\cos^2\theta_r & -L_{ms1}\cos\theta_r - 3L_{ms2}\cos\theta_r\sin^2\theta_r \\ L_{ms1}\cos\theta_r + 3L_{ms2}\cos\theta_r\sin^2\theta_r & -L_{ms1}\sin\theta_r - 3L_{ms2}\sin\theta_r\cos^2\theta_r \end{bmatrix}\begin{bmatrix} i'_{ar} \\ i'_{br} \end{bmatrix}$$

$$= -\frac{P}{2}\left\{L_{ms1}\left[\left(i_{as}i'_{ar} + i_{bs}i'_{br}\right)\sin\theta_r + \left(i_{as}i'_{br} - i_{bs}i'_{ar}\right)\cos\theta_r\right]\right.$$

$$\left. + 3L_{ms2}\left[\left(i_{as}i'_{ar} + i_{bs}i'_{br}\right)\sin\theta_r\cos^2\theta_r + \left(i_{as}i'_{br} - i_{bs}i'_{ar}\right)\cos\theta_r\sin^2\theta_r\right]\right\}.$$

$$\tag{5.17}$$

Using the Newtonian dynamics (5.7) and the derived expression for T_e (5.17), the *torsional-mechanical* equations are

$$\frac{d\omega_r}{dt} = -\frac{P^2}{4J}\left\{ L_{ms1}\left[\left(i_{as}i'_{ar} + i_{bs}i'_{br}\right)\sin\theta_r + \left(i_{as}i'_{br} - i_{bs}i'_{ar}\right)\cos\theta_r\right]\right.$$

$$\left. + 3L_{ms2}\left[\left(i_{as}i'_{ar} + i_{bs}i'_{br}\right)\sin\theta_r\cos^2\theta_r + \left(i_{as}i'_{br} - i_{bs}i'_{ar}\right)\cos\theta_r\sin^2\theta_r\right]\right\}$$

$$ - \frac{B_m}{J}\omega_r - \frac{P}{2J}T_L, \tag{5.18}$$

$$\frac{d\theta_r}{dt} = \omega_r.$$

The resulting equations of motion are found by using (5.16) and (5.18). Cauchy's form of differential equations can be obtained [6]. The analysis and simulation can be carried-out using the nonlinear differential equations which are in Cauchy's form or not in Cauchy's form.

Example 5.3:

We perform simulations for an A-class two-phase, 115 V (*rms*), 60 Hz, four-pole ($P=4$) induction motor [6]. To ensure the descriptive features, the motor parameters are in correspondence as used in Example 5.2, except $L_{ms} = L_{ms1} + 3L_{ms2}$. We have $r_s = 1.2$ ohm, $r'_r = 1.5$ ohm, $L_{ms1} = 0.145$ H, $L_{ms2} = 0.005$ H, $L_{ls} = 0.02$ H, $L_{ss} = L_{ls} + L_{ms1} + 3L_{ms2}$, $L'_{lr} = 0.02H$, $L'_{rr} = L'_{lr} + L_{ms1} + 3L_{ms2}$, $B_m = 1 \times 10^{-6}$ N-m-s/rad, and $J = 0.005$ kg-m^2. The supplied phase voltages are $u_{as}(t) = \sqrt{2}115\cos(377t)$ and $u_{bs}(t) = \sqrt{2}115\sin(377t)$.

Using the derived equations of motion in non-Cauchy's and Cauchy's forms, the Simulink models (diagrams) are developed. The transient behavior and dynamic torque–speed characteristics for a pure sinusoidal magnetic coupling are reported in Figure 5.8a. For the possible (near-optimal) coupling, the results are documented in Figure 5.8b and c for unloaded and loaded ($T_L = 0.5$ N m at $t = 1.5$ s) motor as it accelerates from the stall. One can assess (1) acceleration capabilities; (2) transient dynamics for all state variables $i_{as}(t)$, $i_{bs}(t)$, $i'_{ar}(t)$, $i'_{br}(t)$, $\omega_r(t)$ and $\theta_r(t)$; (3) evolution of the electromagnetic torque T_e; (4) dynamic torque–speed characteristics $\omega_{rm}(t) = \Omega_T[T_e(t)]$; (5) efficiency and losses; (6) thermodynamics; (7) motional $emf_{ar\omega}$ and $emf_{br\omega}$ induced in the rotor windings; etc. The results indicate that even a very small deviation from the ideal pure sinusoidal stator–rotor magnetic coupling results in a significant degradation of the motor performance and capabilities. We found a significant reduction of the efficiency, degradation of acceleration capabilities, torque ripple (which results in vibration, noise, mechanical wearing, etc.), and other undesirable effects. The analysis performed supports the need for a coherent structural design, accurate modeling, and nonlinear simulations with minimum level of simplifications and assumptions. The realistic motor design must be integrated with sound studies of motion devices and systems in the behavior domain. ∎

5.4 Three-Phase Induction Motors in the Machine Variables

We examined two-phase induction motors using the *machine* variables utilizing the *as*, *bs*, *ar*, and *br* physical quantities (voltage, current, and flux linkages). The large majority of industrial induction machines are three-phase motors. Our goal is to develop the equations of motion in order to accomplish various analysis and design tasks for three-phase induction motors shown in Figure 5.9.

Kirchhoff's voltage law gives the equations for the voltages, supplied to the *abc* stator and rotor windings, the *abc* stator and rotor currents, and flux linkages. We have

$$u_{as} = r_s i_{as} + \frac{d\psi_{as}}{dt}, \quad u_{bs} = r_s i_{bs} + \frac{d\psi_{bs}}{dt}, \quad u_{cs} = r_s i_{cs} + \frac{d\psi_{cs}}{dt},$$

$$u_{ar} = r_r i_{ar} + \frac{d\psi_{ar}}{dt}, \quad u_{br} = r_r i_{br} + \frac{d\psi_{br}}{dt}, \quad u_{cr} = r_r i_{cr} + \frac{d\psi_{cr}}{dt}. \tag{5.19}$$

where u_{as}, u_{bs}, and u_{cs} are the phase voltages supplied to the *as*, *bs*, and *cs* stator windings, u_{ar}, u_{br}, and u_{cr} are the phase voltages supplied to the *ar*, *br*, and *cr* rotor windings

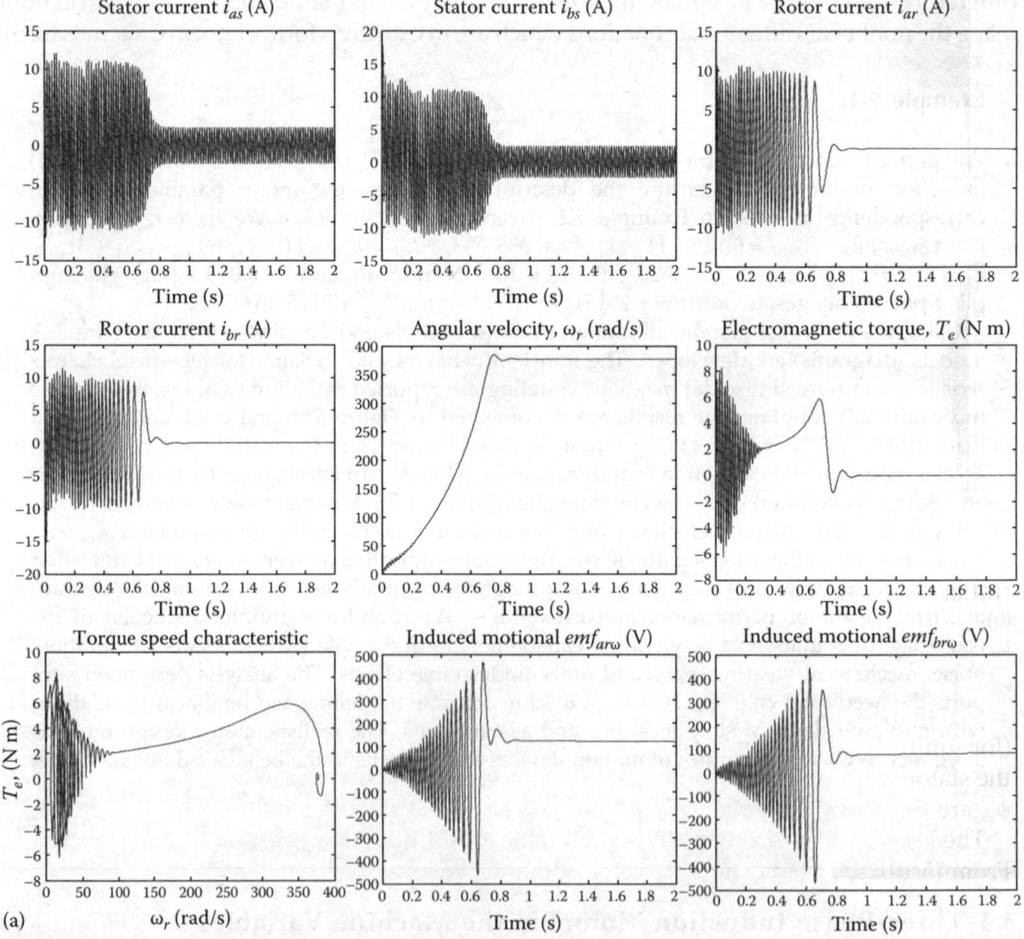

FIGURE 5.8

Dynamics of the A-class induction motor: (a) $T_L = 0$ N-m, $L_{asar} = L_{aras} = L_{sr}\cos\theta_r$, $L_{asbr} = L_{bras} = -L_{sr}\sin\theta_r$, $L_{bsar} = L_{arbs} = L_{sr}\sin\theta_r$ and $L_{bsbr} = L_{brbs} = L_{sr}\cos\theta_r$;

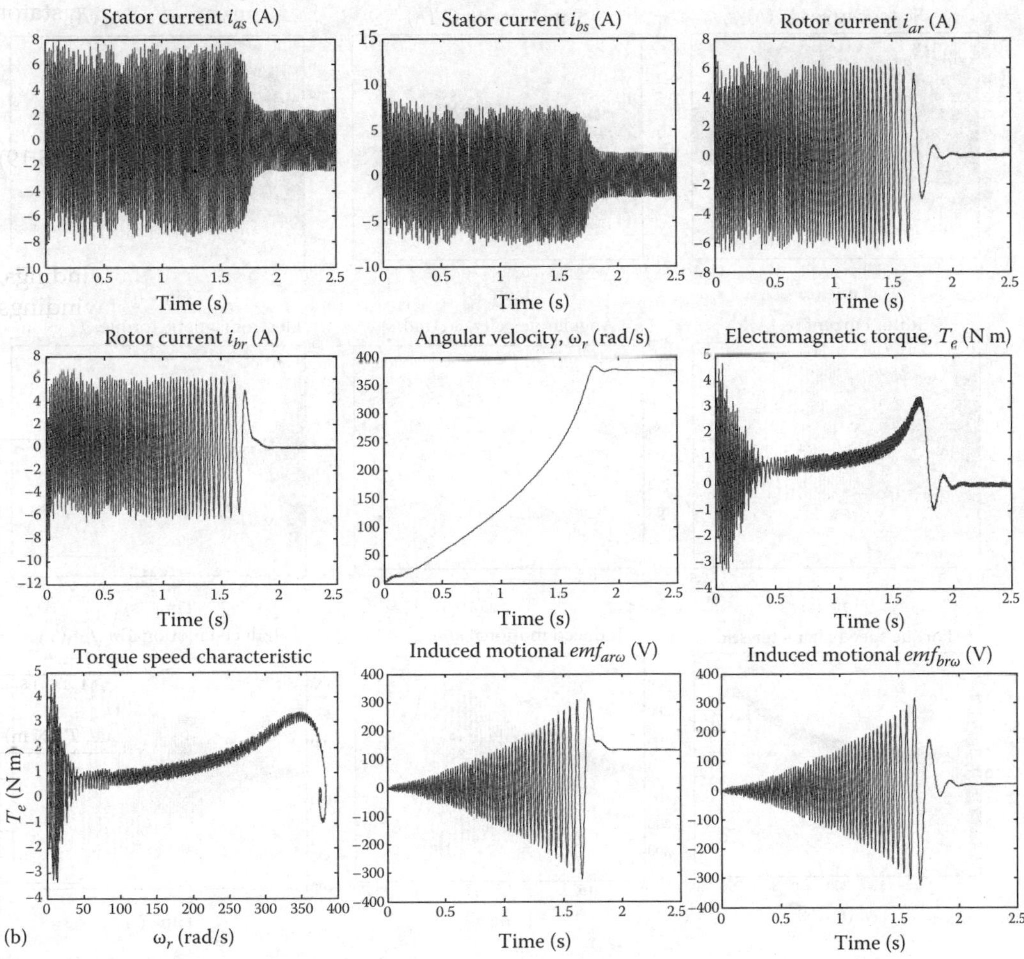

FIGURE 5.8 (continued)

(b) $T_L = 0$ N-m, $L_{asar} = L_{aras} = L_{sr1} \cos\theta_r + L_{sr2} \cos^3\theta_r$, $L_{asbr} = L_{bras} = -L_{sr1} \sin\theta_r - L_{sr2} \sin^3\theta_r$, $L_{bsar} = L_{arbs} = L_{sr1} \sin\theta_r + L_{sr2} \sin^3\theta_r$ and $L_{bsbr} = L_{brbs} = L_{sr1} \cos\theta_r + L_{sr2} \cos^3\theta_r$;

(for squirrel-cage motors $u_{ar} = 0$, $u_{br} = 0$, and $u_{cr} = 0$), i_{as}, i_{bs}, and i_{cs} are the phase currents in the stator windings, i_{ar}, i_{br}, and i_{cr} are the phase currents in the rotor windings, ψ_{as}, ψ_{bs}, and ψ_{cs} are the stator flux linkages, and ψ_{ar}, ψ_{br}, and ψ_{cr} are the rotor flux linkages.

The *abc* stator and rotor voltages, currents, and flux linkages are used as the variables. From (5.19), the application of vector notations yields

$$\mathbf{u}_{abcs} = \mathbf{r}_s \mathbf{i}_{abcs} + \frac{d\boldsymbol{\psi}_{abcs}}{dt},$$
$$\mathbf{u}_{abcr} = \mathbf{r}_r \mathbf{i}_{abcr} + \frac{d\boldsymbol{\psi}_{abcr}}{dt},$$

$$(5.20)$$

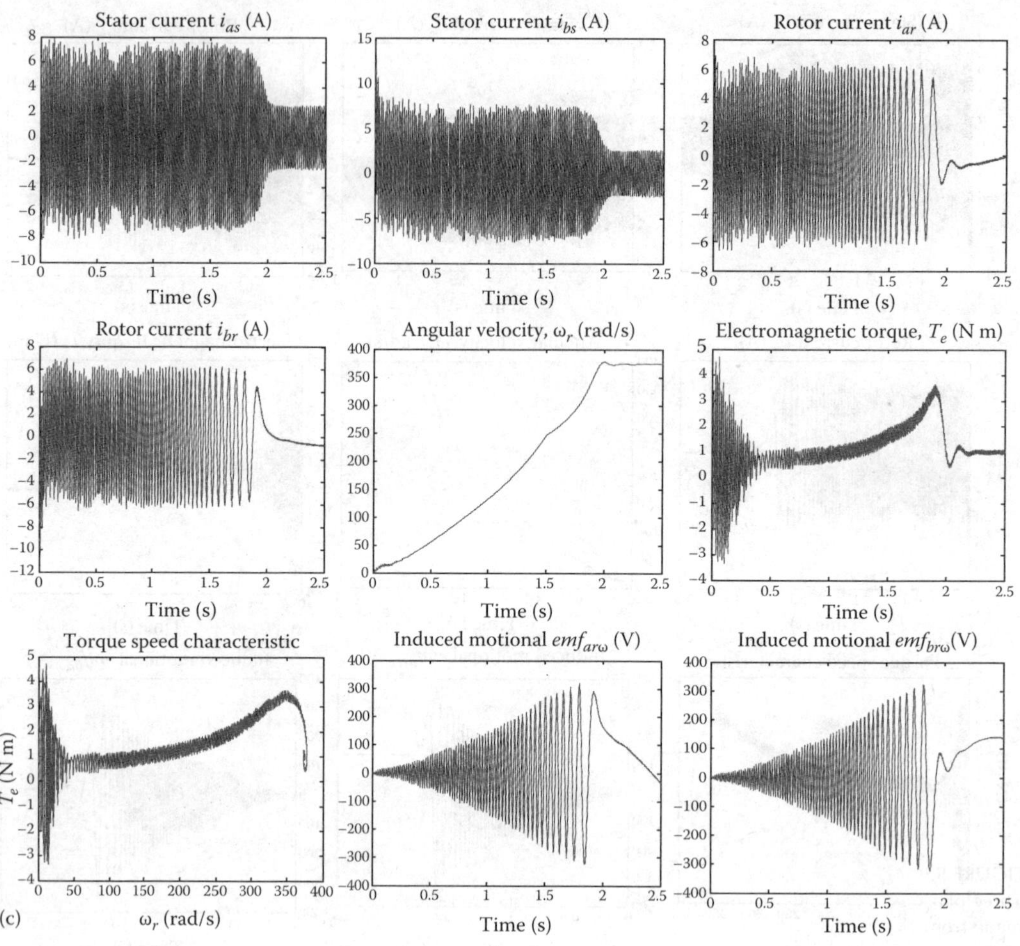

FIGURE 5.8 (continued)

(c) $T_L = 0.5$ N-m (at $t = 1.5$ s), $L_{asar} = L_{aras} = L_{sr1} \cos \theta_r + L_{sr2} \cos^3 \theta_r$, $L_{asbr} = L_{bras} = -L_{sr1} \sin \theta_r - L_{sr2} \sin^3 \theta_r$, $L_{bsar} = L_{arbs} = L_{sr1} \sin \theta_r + L_{sr2} \sin^3 \theta_r$ and $L_{bsbr} = L_{brbs} = L_{sr1} \cos \theta_r + L_{sr2} \cos^3 \theta_r$.

where the *abc* stator and rotor voltages, currents, and flux linkages vectors are

$$\mathbf{u}_{abcs} = \begin{bmatrix} u_{as} \\ u_{bs} \\ u_{cs} \end{bmatrix}, \mathbf{u}_{abcr} = \begin{bmatrix} u_{ar} \\ u_{br} \\ u_{cr} \end{bmatrix}, \mathbf{i}_{abcs} = \begin{bmatrix} i_{as} \\ i_{bs} \\ i_{cs} \end{bmatrix}, \mathbf{i}_{abcr} = \begin{bmatrix} i_{ar} \\ i_{br} \\ i_{cr} \end{bmatrix}, \boldsymbol{\psi}_{abcs} = \begin{bmatrix} \psi_{as} \\ \psi_{bs} \\ \psi_{cs} \end{bmatrix}, \text{and } \boldsymbol{\psi}_{abcr} = \begin{bmatrix} \psi_{ar} \\ \psi_{br} \\ \psi_{cr} \end{bmatrix}.$$

In (5.20), the diagonal stator and rotor resistances matrices are

$$\mathbf{r}_s = \begin{bmatrix} r_s & 0 & 0 \\ 0 & r_s & 0 \\ 0 & 0 & r_s \end{bmatrix} \quad \text{and} \quad \mathbf{r}_r = \begin{bmatrix} r_r & 0 & 0 \\ 0 & r_r & 0 \\ 0 & 0 & r_r \end{bmatrix}.$$

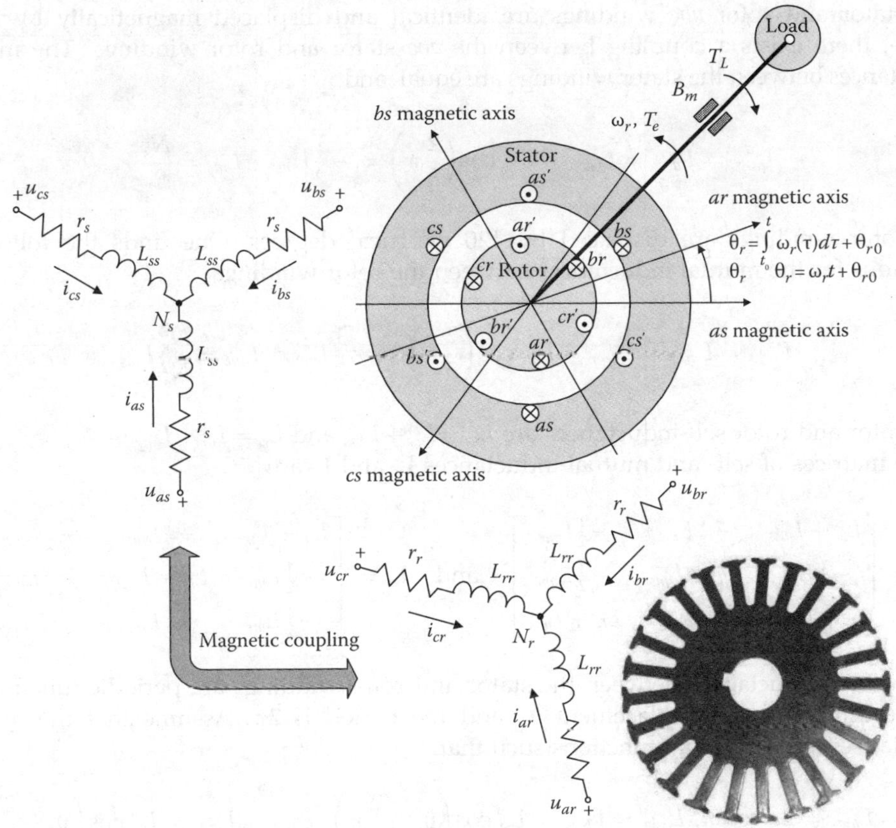

FIGURE 5.9
Three-phase symmetric induction motor: Rotor windings are placed in the slots in the laminated rotor usually made from electric steel.

The flux linkages are found as functions of the corresponding currents in the stator and rotor windings using the self- and mutual-inductances. The analysis of the stator and rotor magnetically coupled system, accomplished referencing to Figure 5.9, gives the following equations

$$\psi_{as} = L_{asas}i_{as} + L_{asbs}i_{bs} + L_{ascs}i_{cs} + L_{asar}i_{ar} + L_{asbr}i_{br} + L_{ascr}i_{cr},$$
$$\psi_{bs} = L_{bsas}i_{as} + L_{bsbs}i_{bs} + L_{bscs}i_{cs} + L_{bsar}i_{ar} + L_{bsbr}i_{br} + L_{bscr}i_{cr},$$
$$\psi_{cs} = L_{csas}i_{as} + L_{csbs}i_{bs} + L_{cscs}i_{cs} + L_{csar}i_{ar} + L_{csbr}i_{br} + L_{cscr}i_{cr},$$
$$\psi_{ar} = L_{aras}i_{as} + L_{arbs}i_{bs} + L_{arcs}i_{cs} + L_{arar}i_{ar} + L_{arbr}i_{br} + L_{arcr}i_{cr},$$
$$\psi_{br} = L_{bras}i_{as} + L_{brbs}i_{bs} + L_{brcs}i_{cs} + L_{brar}i_{ar} + L_{brbr}i_{br} + L_{brcr}i_{cr},$$
$$\psi_{cr} = L_{cras}i_{as} + L_{crbs}i_{bs} + L_{crcs}i_{cs} + L_{crar}i_{ar} + L_{crbr}i_{br} + L_{crcr}i_{cr},$$

where L_{asas}, L_{bsbs}, L_{cscs}, L_{arar}, L_{brbr}, and L_{crcr} are the stator and rotor self-inductances, and L_{asbs}, L_{ascs}, L_{asar}, L_{asbr}, L_{ascr}, ..., L_{cras}, L_{crbs}, L_{crcs}, L_{crar}, and L_{crbr} are the mutual inductances between stator–stator, stator–rotor, and rotor–rotor windings.

The stator and rotor *abc* windings are identical and displaced magnetically by $2\pi/3$. Hence, there exists a coupling between the *abc* stator and rotor windings. The mutual inductances between the stator windings are equal, and

$$L_{asbs} = L_{ascs} = L_{bscs} = L_{ms}\cos\left(\frac{2}{3}\pi\right) = -\frac{1}{2}L_{ms}, \quad L_{ms} = \frac{N_s^2}{\Re_m}.$$

The rotor windings are displaced by 120 electrical degrees. One finds the following equations for the mutual inductances between the rotor windings

$$L_{arbr} = L_{arcr} = L_{brcr} = L_{mr}\cos\left(\frac{2}{3}\pi\right) = -\frac{1}{2}L_{mr}, \quad L_{mr} = \frac{N_r^2}{N_s^2}L_{ms}.$$

The stator and rotor self-inductances are $L_{ss} = L_{ls} + L_{ms}$ and $L_{rr} = L_{lr} + L_{mr}$.
The matrices of self- and mutual-inductances $\mathbf{L}_s$ and $\mathbf{L}_r$ are

$$\mathbf{L}_s = \begin{bmatrix} L_{ls} + L_{ms} & -\frac{1}{2}L_{ms} & -\frac{1}{2}L_{ms} \\ -\frac{1}{2}L_{ms} & L_{ls} + L_{ms} & -\frac{1}{2}L_{ms} \\ -\frac{1}{2}L_{ms} & -\frac{1}{2}L_{ms} & L_{ls} + L_{ms} \end{bmatrix} \quad \text{and} \quad \mathbf{L}_r = \begin{bmatrix} L_{lr} + L_{mr} & -\frac{1}{2}L_{mr} & -\frac{1}{2}L_{mr} \\ -\frac{1}{2}L_{mr} & L_{lr} + L_{mr} & -\frac{1}{2}L_{mr} \\ -\frac{1}{2}L_{mr} & -\frac{1}{2}L_{mr} & L_{lr} + L_{mr} \end{bmatrix}.$$

The mutual inductances between the stator and rotor windings are periodic functions of the electrical angular displacement θ_r, and the period is 2π. Assume that the mutual inductances are sinusoidal functions such that

$$L_{asar} = L_{aras} = L_{sr}\cos\theta_r, \; L_{asbr} = L_{bras} = L_{sr}\cos\left(\theta_r + \frac{2}{3}\pi\right), \; L_{ascr} = L_{cras} = L_{sr}\cos\left(\theta_r - \frac{2}{3}\pi\right),$$

$$L_{bsar} = L_{arbs} = L_{sr}\cos\left(\theta_r - \frac{2}{3}\pi\right), \; L_{bsbr} = L_{brbs} = L_{sr}\cos\theta_r, \; L_{bscr} = L_{crbs} = L_{sr}\cos\left(\theta_r + \frac{2}{3}\pi\right),$$

$$L_{csar} = L_{arcs} = L_{sr}\cos\left(\theta_r + \frac{2}{3}\pi\right), \; L_{csbr} = L_{brcs} = L_{sr}\cos\left(\theta_r - \frac{2}{3}\pi\right), \; L_{cscr} = L_{crcs} = L_{sr}\cos\theta_r,$$

where $L_{sr} = \dfrac{N_s N_r}{\Re_m}$.

We have the following stator–rotor mutual inductance mapping

$$\mathbf{L}_{sr}(\theta_r) = L_{sr}\begin{bmatrix} \cos\theta_r & \cos\left(\theta_r + \frac{2}{3}\pi\right) & \cos\left(\theta_r - \frac{2}{3}\pi\right) \\ \cos\left(\theta_r - \frac{2}{3}\pi\right) & \cos\theta_r & \cos\left(\theta_r + \frac{2}{3}\pi\right) \\ \cos\left(\theta_r + \frac{2}{3}\pi\right) & \cos\left(\theta_r - \frac{2}{3}\pi\right) & \cos\theta_r \end{bmatrix}.$$

One obtains

$$\begin{bmatrix} \boldsymbol{\psi}_{abcs} \\ \boldsymbol{\psi}_{abcr} \end{bmatrix} = \begin{bmatrix} \mathbf{L}_s & \mathbf{L}_{sr}(\theta_r) \\ \mathbf{L}_{sr}^T(\theta_r) & \mathbf{L}_r \end{bmatrix}\begin{bmatrix} \mathbf{i}_{abcs} \\ \mathbf{i}_{abcr} \end{bmatrix}, \tag{5.21}$$

$$\boldsymbol{\psi}_{abcs} = \mathbf{L}_s\mathbf{i}_{abcs} + \mathbf{L}_{sr}(\theta_r)\mathbf{i}_{abcr}, \quad \boldsymbol{\psi}_{abcr} = \mathbf{L}_{sr}^T(\theta_r)\mathbf{i}_{abcs} + \mathbf{L}_r\mathbf{i}_{abcr}.$$

Using the number of turns N_s and N_r, we have

$$\mathbf{u}'_{abcr} = \frac{N_s}{N_r}\mathbf{u}_{abcr}, \quad \mathbf{i}'_{abcr} = \frac{N_r}{N_s}\mathbf{i}_{abcr} \quad \text{and} \quad \boldsymbol{\psi}'_{abcr} = \frac{N_s}{N_r}\boldsymbol{\psi}_{abcr}.$$

The inductances are $L_{ms} = \frac{N_s}{N_r}L_{sr}$, $L_{sr} = \frac{N_sN_r}{\mathfrak{R}_m}$, and $L_{ms} = \frac{N_s^2}{\mathfrak{R}_m}$. One finds

$$\mathbf{L}'_{sr}(\theta_r) = \frac{N_s}{N_r}\mathbf{L}_{sr}(\theta_r) = L_{ms}\begin{bmatrix} \cos\theta_r & \cos\left(\theta_r + \frac{2}{3}\pi\right) & \cos\left(\theta_r - \frac{2}{3}\pi\right) \\ \cos\left(\theta_r - \frac{2}{3}\pi\right) & \cos\theta_r & \cos\left(\theta_r + \frac{2}{3}\pi\right) \\ \cos\left(\theta_r + \frac{2}{3}\pi\right) & \cos\left(\theta_r - \frac{2}{3}\pi\right) & \cos\theta_r \end{bmatrix},$$

and

$$\mathbf{L}'_r = \frac{N_s^2}{N_r^2}\mathbf{L}_r = \begin{bmatrix} L'_{lr} + L_{ms} & -\frac{1}{2}L_{ms} & -\frac{1}{2}L_{ms} \\ -\frac{1}{2}L_{ms} & L'_{lr} + L_{ms} & -\frac{1}{2}L_{ms} \\ -\frac{1}{2}L_{ms} & -\frac{1}{2}L_{ms} & L'_{lr} + L_{ms} \end{bmatrix}$$

where $L'_{lr} = \frac{N_s^2}{N_r^2}L_{lr}$.

The matrix equation (5.21) for the flux linkages is rewritten as

$$\begin{bmatrix} \boldsymbol{\psi}_{abcs} \\ \boldsymbol{\psi}'_{abcr} \end{bmatrix} = \begin{bmatrix} \mathbf{L}_s & \mathbf{L}'_{sr}(\theta_r) \\ \mathbf{L}'^{\,T}_{sr}(\theta_r) & \mathbf{L}'_r \end{bmatrix}\begin{bmatrix} \mathbf{i}_{abcs} \\ \mathbf{i}'_{abcr} \end{bmatrix}. \tag{5.22}$$

Substituting the expressions for $\mathbf{L}_s$, $\mathbf{L}'_{sr}(\theta_r)$, and $\mathbf{L}'_r$ in (5.22), we have

$$\begin{bmatrix} \psi_{as} \\ \psi_{bs} \\ \psi_{cs} \\ \psi'_{ar} \\ \psi'_{br} \\ \psi'_{cr} \end{bmatrix} = \begin{bmatrix} L_{ls}+L_{ms} & -\frac{1}{2}L_{ms} & -\frac{1}{2}L_{ms} & L_{ms}\cos\theta_r & L_{ms}\cos\left(\theta_r+\frac{2}{3}\pi\right) & L_{ms}\cos\left(\theta_r-\frac{2}{3}\pi\right) \\ -\frac{1}{2}L_{ms} & L_{ls}+L_{ms} & -\frac{1}{2}L_{ms} & L_{ms}\cos\left(\theta_r-\frac{2}{3}\pi\right) & L_{ms}\cos\theta_r & L_{ms}\cos\left(\theta_r+\frac{2}{3}\pi\right) \\ -\frac{1}{2}L_{ms} & -\frac{1}{2}L_{ms} & L_{ls}+L_{ms} & L_{ms}\cos\left(\theta_r+\frac{2}{3}\pi\right) & L_{ms}\cos\left(\theta_r-\frac{2}{3}\pi\right) & L_{ms}\cos\theta_r \\ L_{ms}\cos\theta_r & L_{ms}\cos\left(\theta_r-\frac{2}{3}\pi\right) & L_{ms}\cos\left(\theta_r+\frac{2}{3}\pi\right) & L'_{lr}+L_{ms} & -\frac{1}{2}L_{ms} & -\frac{1}{2}L_{ms} \\ L_{ms}\cos\left(\theta_r+\frac{2}{3}\pi\right) & L_{ms}\cos\theta_r & L_{ms}\cos\left(\theta_r-\frac{2}{3}\pi\right) & -\frac{1}{2}L_{ms} & L'_{lr}+L_{ms} & -\frac{1}{2}L_{ms} \\ L_{ms}\cos\left(\theta_r-\frac{2}{3}\pi\right) & L_{ms}\cos\left(\theta_r+\frac{2}{3}\pi\right) & L_{ms}\cos\theta_r & -\frac{1}{2}L_{ms} & -\frac{1}{2}L_{ms} & L'_{lr}+L_{ms} \end{bmatrix}\begin{bmatrix} i_{as} \\ i_{bs} \\ i_{cs} \\ i'_{ar} \\ i'_{br} \\ i'_{cr} \end{bmatrix}.$$

Using (5.20) and (5.22), we obtain

$$\mathbf{u}_{abcs} = \mathbf{r}_s\mathbf{i}_{abcs} + \frac{d\boldsymbol{\psi}_{abcs}}{dt} = \mathbf{r}_s\mathbf{i}_{abcs} + \mathbf{L}_s\frac{d\mathbf{i}_{abcs}}{dt} + \frac{d\left(\mathbf{L}'_{sr}(\theta_r)\mathbf{i}'_{abcr}\right)}{dt},$$

$$\mathbf{u}'_{abcr} = \mathbf{r}'_r\mathbf{i}'_{abcr} + \frac{d\boldsymbol{\psi}'_{abcr}}{dt} = \mathbf{r}'_r\mathbf{i}'_{abcr} + \mathbf{L}'_r\frac{d\mathbf{i}'_{abcr}}{dt} + \frac{d\left(\mathbf{L}'_{sr}{}^T(\theta_r)\mathbf{i}_{abcs}\right)}{dt},$$

$$\tag{5.23}$$

where $\mathbf{r}'_r = \frac{N_s^2}{N_r^2}\mathbf{r}_r$.

The flux linkages total derivatives $\dfrac{d\boldsymbol{\psi}_{abcs}}{dt}$ and $\dfrac{d\boldsymbol{\psi}'_{abcr}}{dt}$ provide the expressions for *emf* terms. Equations (5.23) in expanded form are rewritten as

$$u_{as} = r_s i_{as} + (L_{ls} + L_{ms})\frac{di_{as}}{dt} - \frac{1}{2}L_{ms}\frac{di_{bs}}{dt} - \frac{1}{2}L_{ms}\frac{di_{cs}}{dt} + L_{ms}\frac{d(i'_{ar}\cos\theta_r)}{dt} + L_{ms}\frac{d(i'_{br}\cos(\theta_r + \frac{2\pi}{3}))}{dt} + L_{ms}\frac{d(i'_{cr}\cos(\theta_r - \frac{2\pi}{3}))}{dt},$$

$$u_{bs} = r_s i_{bs} - \frac{1}{2}L_{ms}\frac{di_{as}}{dt} + (L_{ls} + L_{ms})\frac{di_{bs}}{dt} - \frac{1}{2}L_{ms}\frac{di_{cs}}{dt} + L_{ms}\frac{d(i'_{ar}\cos(\theta_r - \frac{2\pi}{3}))}{dt} + L_{ms}\frac{d(i'_{br}\cos\theta_r)}{dt} + L_{ms}\frac{d(i'_{cr}\cos(\theta_r + \frac{2\pi}{3}))}{dt},$$

$$u_{cs} = r_s i_{cs} - \frac{1}{2}L_{ms}\frac{di_{as}}{dt} - \frac{1}{2}L_{ms}\frac{di_{bs}}{dt} + (L_{ls} + L_{ms})\frac{di_{cs}}{dt} + L_{ms}\frac{d(i'_{ar}\cos(\theta_r + \frac{2\pi}{3}))}{dt} + L_{ms}\frac{d(i'_{br}\cos(\theta_r - \frac{2\pi}{3}))}{dt} + L_{ms}\frac{d(i'_{cr}\cos\theta_r)}{dt},$$

$$u'_{ar} = r'_r i'_{ar} + L_{ms}\frac{d(i_{as}\cos\theta_r)}{dt} + L_{ms}\frac{d(i_{bs}\cos(\theta_r - \frac{2\pi}{3}))}{dt} + L_{ms}\frac{d(i_{cs}\cos(\theta_r + \frac{2\pi}{3}))}{dt} + (L'_{lr} + L_{ms})\frac{di'_{ar}}{dt} - \frac{1}{2}L_{ms}\frac{di'_{br}}{dt} - \frac{1}{2}L_{ms}\frac{di'_{cr}}{dt},$$

$$u'_{br} = r'_r i'_{br} + L_{ms}\frac{d(i_{as}\cos(\theta_r + \frac{2\pi}{3}))}{dt} + L_{ms}\frac{d(i_{bs}\cos\theta_r)}{dt} + L_{ms}\frac{d(i_{cs}\cos(\theta_r - \frac{2\pi}{3}))}{dt} - \frac{1}{2}L_{ms}\frac{di'_{ar}}{dt} + (L'_{lr} + L_{ms})\frac{di'_{br}}{dt} - \frac{1}{2}L_{ms}\frac{di'_{cr}}{dt},$$

$$u'_{cr} = r'_r i'_{cr} + L_{ms}\frac{d(i_{as}\cos(\theta_r - \frac{2\pi}{3}))}{dt} + L_{ms}\frac{d(i_{bs}\cos(\theta_r + \frac{2\pi}{3}))}{dt} + L_{ms}\frac{d(i_{cs}\cos\theta_r)}{dt} - \frac{1}{2}L_{ms}\frac{di'_{ar}}{dt} - \frac{1}{2}L_{ms}\frac{di'_{br}}{dt} + (L'_{lr} + L_{ms})\frac{di'_{cr}}{dt}.$$

We obtain the following set of equations which describe the circuitry-electromagnetic dynamics of three-phase induction motors

$$u_{as} = r_s i_{as} + (L_{ls} + L_{ms})\frac{di_{as}}{dt} - \frac{1}{2}L_{ms}\frac{di_{bs}}{dt} - \frac{1}{2}L_{ms}\frac{di_{cs}}{dt} + L_{ms}\cos\theta_r\frac{di'_{ar}}{dt} + L_{ms}\cos\left(\theta_r + \frac{2\pi}{3}\right)\frac{di'_{br}}{dt}$$
$$+ L_{ms}\cos\left(\theta_r - \frac{2\pi}{3}\right)\frac{di'_{cr}}{dt} - L_{ms}\left[i'_{ar}\sin\theta_r + i'_{br}\sin\left(\theta_r + \frac{2\pi}{3}\right) + i'_{cr}\sin\left(\theta_r - \frac{2\pi}{3}\right)\right]\omega_r,$$

$$u_{bs} = r_s i_{bs} - \frac{1}{2}L_{ms}\frac{di_{as}}{dt} + (L_{ls} + L_{ms})\frac{di_{bs}}{dt} - \frac{1}{2}L_{ms}\frac{di_{cs}}{dt} + L_{ms}\cos\left(\theta_r - \frac{2\pi}{3}\right)\frac{di'_{ar}}{dt} + L_{ms}\cos\theta_r\frac{di'_{br}}{dt}$$
$$+ L_{ms}\cos\left(\theta_r + \frac{2\pi}{3}\right)\frac{di'_{cr}}{dt} - L_{ms}\left[i'_{ar}\sin\left(\theta_r - \frac{2\pi}{3}\right) + i'_{br}\sin\theta_r + i'_{cr}\sin\left(\theta_r + \frac{2\pi}{3}\right)\right]\omega_r,$$

$$u_{cs} = r_s i_{cs} - \frac{1}{2}L_{ms}\frac{di_{as}}{dt} - \frac{1}{2}L_{ms}\frac{di_{bs}}{dt} + (L_{ls} + L_{ms})\frac{di_{cs}}{dt} + L_{ms}\cos\left(\theta_r + \frac{2\pi}{3}\right)\frac{di'_{ar}}{dt} + L_{ms}\cos\left(\theta_r - \frac{2\pi}{3}\right)\frac{di'_{br}}{dt}$$
$$+ L_{ms}\cos\theta_r\frac{di'_{cr}}{dt} - L_{ms}\left[i'_{ar}\sin\left(\theta_r + \frac{2\pi}{3}\right) + i'_{br}\sin\left(\theta_r - \frac{2\pi}{3}\right) + i'_{cr}\sin\theta_r\right]\omega_r,$$

$$u'_{ar} = r'_r i'_{ar} + L_{ms}\cos\theta_r\frac{di_{as}}{dt} + L_{ms}\cos\left(\theta_r - \frac{2\pi}{3}\right)\frac{di_{bs}}{dt} + L_{ms}\cos\left(\theta_r + \frac{2\pi}{3}\right)\frac{di_{cs}}{dt} + (L'_{lr} + L_{ms})\frac{di'_{ar}}{dt}$$
$$- \frac{1}{2}L_{ms}\frac{di'_{br}}{dt} - \frac{1}{2}L_{ms}\frac{di'_{cr}}{dt} - L_{ms}\left[i_{as}\sin\theta_r + i_{bs}\sin\left(\theta_r - \frac{2\pi}{3}\right) + i_{cs}\sin\left(\theta_r + \frac{2\pi}{3}\right)\right]\omega_r,$$

$$u'_{br} = r'_r i'_{br} + L_{ms}\cos\left(\theta_r + \frac{2\pi}{3}\right)\frac{di_{as}}{dt} + L_{ms}\cos\theta_r\frac{di_{bs}}{dt} + L_{ms}\cos\left(\theta_r - \frac{2\pi}{3}\right)\frac{di_{cs}}{dt} - \frac{1}{2}L_{ms}\frac{di'_{ar}}{dt}$$
$$+ (L'_{lr} + L_{ms})\frac{di'_{br}}{dt} - \frac{1}{2}L_{ms}\frac{di'_{cr}}{dt} - L_{ms}\left[i_{as}\sin\left(\theta_r + \frac{2\pi}{3}\right) + i_{bs}\sin\theta_r + i_{cs}\sin\left(\theta_r - \frac{2\pi}{3}\right)\right]\omega_r,$$

$$u'_{cr} = r'_r i'_{cr} + L_{ms}\cos\left(\theta_r - \frac{2\pi}{3}\right)\frac{di_{as}}{dt} + L_{ms}\cos\left(\theta_r + \frac{2\pi}{3}\right)\frac{di_{bs}}{dt} + L_{ms}\cos\theta_r\frac{di_{cs}}{dt} - \frac{1}{2}L_{ms}\frac{di'_{ar}}{dt}$$
$$- \frac{1}{2}L_{ms}\frac{di'_{br}}{dt} + (L'_{lr} + L_{ms})\frac{di'_{cr}}{dt} - L_{ms}\left[i_{as}\sin\left(\theta_r - \frac{2\pi}{3}\right) + i_{bs}\sin\left(\theta_r + \frac{2\pi}{3}\right) + i_{cs}\sin\theta_r\right]\omega_r. \quad (5.24)$$

Equations (5.24) yield the differential equations in Cauchy's form as

$$
\begin{bmatrix} \frac{di_{as}}{dt} \\ \frac{di_{bs}}{dt} \\ \frac{di_{cs}}{dt} \\ \frac{di'_{ar}}{dt} \\ \frac{di'_{br}}{dt} \\ \frac{di'_{cr}}{dt} \end{bmatrix}
= \frac{1}{L_{\Sigma L}}
\begin{bmatrix}
-r_s L_{\Sigma m} & -\frac{1}{2}r_s L_{ms} & -\frac{1}{2}r_s L_{ms} & 0 & 0 & 0 \\
-\frac{1}{2}r_s L_{ms} & -r_s L_{\Sigma m} & -\frac{1}{2}r_s L_{ms} & 0 & 0 & 0 \\
-\frac{1}{2}r_s L_{ms} & -\frac{1}{2}r_s L_{ms} & -r_s L_{\Sigma m} & 0 & 0 & 0 \\
0 & 0 & 0 & -r_r L_{\Sigma m} & -\frac{1}{2}r_r L_{ms} & -\frac{1}{2}r_r L_{ms} \\
0 & 0 & 0 & -\frac{1}{2}r_r L_{ms} & -r_r L_{\Sigma m} & -\frac{1}{2}r_r L_{ms} \\
0 & 0 & 0 & -\frac{1}{2}r_r L_{ms} & -\frac{1}{2}r_r L_{ms} & -r_r L_{\Sigma m}
\end{bmatrix}
\begin{bmatrix} i_{as} \\ i_{bs} \\ i_{cs} \\ i'_{ar} \\ i'_{br} \\ i'_{cr} \end{bmatrix}
$$

$$
+ \frac{1}{L_{\Sigma L}}
\begin{bmatrix}
0 & 0 & 0 & r_r L_{ms}\cos\theta_r & r_r L_{ms}\cos\left(\theta_r+\frac{2}{3}\pi\right) & r_r L_{ms}\cos\left(\theta_r-\frac{2}{3}\pi\right) \\
0 & 0 & 0 & r_r L_{ms}\cos\left(\theta_r-\frac{2}{3}\pi\right) & r_r L_{ms}\cos\theta_r & r_r L_{ms}\cos\left(\theta_r+\frac{2}{3}\pi\right) \\
0 & 0 & 0 & r_r L_{ms}\cos\left(\theta_r+\frac{2}{3}\pi\right) & r_r L_{ms}\cos\left(\theta_r-\frac{2}{3}\pi\right) & r_r L_{ms}\cos\theta_r \\
r_s L_{ms}\cos\theta_r & r_s L_{ms}\cos\left(\theta_r-\frac{2}{3}\pi\right) & r_s L_{ms}\cos\left(\theta_r+\frac{2}{3}\pi\right) & 0 & 0 & 0 \\
r_s L_{ms}\cos\left(\theta_r+\frac{2}{3}\pi\right) & r_s L_{ms}\cos\theta_r & r_s L_{ms}\cos\left(\theta_r-\frac{2}{3}\pi\right) & 0 & 0 & 0 \\
r_s L_{ms}\cos\left(\theta_r-\frac{2}{3}\pi\right) & r_s L_{ms}\cos\left(\theta_r+\frac{2}{3}\pi\right) & r_s L_{ms}\cos\theta_r & 0 & 0 & 0
\end{bmatrix}
\begin{bmatrix} i_{as} \\ i_{bs} \\ i_{cs} \\ i'_{ar} \\ i'_{br} \\ i'_{cr} \end{bmatrix}
$$

$$
+ \frac{1}{L_{\Sigma L}}
\begin{bmatrix}
0 & 1.299L_{ms}^2\omega_r & -1.299L_{ms}^2\omega_r & L_{\Sigma ms}\omega_r\sin\theta_r & L_{\Sigma ms}\omega_r\sin\left(\theta_r+\frac{2}{3}\pi\right) & L_{\Sigma ms}\omega_r\sin\left(\theta_r-\frac{2}{3}\pi\right) \\
-1.299L_{ms}^2\omega_r & 0 & 1.299L_{ms}^2\omega_r & L_{\Sigma ms}\omega_r\sin\left(\theta_r-\frac{2}{3}\pi\right) & L_{\Sigma ms}\omega_r\sin\theta_r & L_{\Sigma ms}\omega_r\sin\left(\theta_r+\frac{2}{3}\pi\right) \\
1.299L_{ms}^2\omega_r & -1.299L_{ms}^2\omega_r & 0 & L_{\Sigma ms}\omega_r\sin\left(\theta_r+\frac{2}{3}\pi\right) & L_{\Sigma ms}\omega_r\sin\left(\theta_r-\frac{2}{3}\pi\right) & L_{\Sigma ms}\omega_r\sin\theta_r \\
L_{\Sigma ms}\omega_r\sin\theta_r & L_{\Sigma ms}\omega_r\sin\left(\theta_r-\frac{2}{3}\pi\right) & L_{\Sigma ms}\omega_r\sin\left(\theta_r+\frac{2}{3}\pi\right) & 0 & -1.299L_{ms}^2\omega_r & 1.299L_{ms}^2\omega_r \\
L_{\Sigma ms}\omega_r\sin\left(\theta_r+\frac{2}{3}\pi\right) & L_{\Sigma ms}\omega_r\sin\theta_r & L_{\Sigma ms}\omega_r\sin\left(\theta_r-\frac{2}{3}\pi\right) & 1.299L_{ms}^2\omega_r & 0 & -1.299L_{ms}^2\omega_r \\
L_{\Sigma ms}\omega_r\sin\left(\theta_r-\frac{2}{3}\pi\right) & L_{\Sigma ms}\omega_r\sin\left(\theta_r+\frac{2}{3}\pi\right) & L_{\Sigma ms}\omega_r\sin\theta_r & -1.299L_{ms}^2\omega_r & 1.299L_{ms}^2\omega_r & 0
\end{bmatrix}
\begin{bmatrix} i_{as} \\ i_{bs} \\ i_{cs} \\ i'_{ar} \\ i'_{br} \\ i'_{cr} \end{bmatrix}
$$

$$
+ \frac{1}{L_{\Sigma L}}
\begin{bmatrix}
2L_{ms}+L'_{lr} & \frac{1}{2}L_{ms} & \frac{1}{2}L_{ms} & -L_{ms}\cos\theta_r & -L_{ms}\cos\left(\theta_r+\frac{2}{3}\pi\right) & -L_{ms}\cos\left(\theta_r-\frac{2}{3}\pi\right) \\
\frac{1}{2}L_{ms} & 2L_{ms}+L'_{lr} & \frac{1}{2}L_{ms} & -L_{ms}\cos\left(\theta_r-\frac{2}{3}\pi\right) & -L_{ms}\cos\theta_r & -L_{ms}\cos\left(\theta_r+\frac{2}{3}\pi\right) \\
\frac{1}{2}L_{ms} & \frac{1}{2}L_{ms} & 2L_{ms}+L'_{lr} & -L_{ms}\cos\left(\theta_r+\frac{2}{3}\pi\right) & -L_{ms}\cos\left(\theta_r-\frac{2}{3}\pi\right) & -L_{ms}\cos\theta_r \\
-L_{ms}\cos\theta_r & -L_{ms}\cos\left(\theta_r-\frac{2}{3}\pi\right) & -L_{ms}\cos\left(\theta_r+\frac{2}{3}\pi\right) & 2L_{ms}+L'_{lr} & \frac{1}{2}L_{ms} & \frac{1}{2}L_{ms} \\
-L_{ms}\cos\left(\theta_r+\frac{2}{3}\pi\right) & -L_{ms}\cos\theta_r & -L_{ms}\cos\left(\theta_r-\frac{2}{3}\pi\right) & \frac{1}{2}L_{ms} & 2L_{ms}+L'_{lr} & \frac{1}{2}L_{ms} \\
-L_{ms}\cos\left(\theta_r-\frac{2}{3}\pi\right) & -L_{ms}\cos\left(\theta_r+\frac{2}{3}\pi\right) & -L_{ms}\cos\theta_r & \frac{1}{2}L_{ms} & \frac{1}{2}L_{ms} & 2L_{ms}+L'_{lr}
\end{bmatrix}
\begin{bmatrix} u_{as} \\ u_{bs} \\ u_{cs} \\ u'_{ar} \\ u'_{br} \\ u'_{cr} \end{bmatrix}. \quad (5.25)
$$

Here, $L_{\Sigma L} = \left(3L_{ms}+L'_{lr}\right)L'_{lr}$, $L_{\Sigma m} = 2L_{ms}+L'_{lr}$, and $L_{\Sigma ms} = \frac{3}{2}L_{ms}^2 + L_{ms}L'_{lr}$.

The expression for the electromagnetic torque developed by induction motors is obtained using the coenergy $W_c\left(\mathbf{i}_{abcs},\, \mathbf{i}'_{abcr},\, \theta_r\right)$. For P-pole three-phase induction machines we have $T_e = \frac{P}{2}\dfrac{\partial W_c\left(\mathbf{i}_{abcs},\, \mathbf{i}'_{abcr},\, \theta_r\right)}{\partial \theta_r}$. The coenergy is given by the following expression

$$
W_c = W_f = \frac{1}{2}\mathbf{i}_{abcs}^T\left(\mathbf{L}_s - L_{ls}\mathbf{I}\right)\mathbf{i}_{abcs} + \mathbf{i}_{abcs}^T\mathbf{L}'_{sr}(\theta_r)\mathbf{i}'_{abcr} + \frac{1}{2}\mathbf{i}_{abcr}^{\prime\,T}\left(\mathbf{L}'_r - L'_{lr}\mathbf{I}\right)\mathbf{i}'_{abcr}.
$$

In W_c, matrices $\mathbf{L}_s$, $L_{ls}\mathbf{I}$, $\mathbf{L}'_r$, and $L'_{lr}\mathbf{I}$ are not functions of the electrical angular displacement θ_r. Using $\mathbf{L}'_{sr}(\theta_r)$, the expression for the electromagnetic torque is found to be

$$
T_e = \frac{P}{2}\mathbf{i}_{abcs}^T\frac{\partial \mathbf{L}'_{sr}(\theta_r)}{\partial \theta_r}\mathbf{i}'_{abcr}
$$

$$
= -\frac{P}{2}L_{ms}\begin{bmatrix} i_{as} & i_{bs} & i_{cs}\end{bmatrix}
\begin{bmatrix}
\sin\theta_r & \sin\left(\theta_r+\frac{2}{3}\pi\right) & \sin\left(\theta_r-\frac{2}{3}\pi\right) \\
\sin\left(\theta_r-\frac{2}{3}\pi\right) & \sin\theta_r & \sin\left(\theta_r+\frac{2}{3}\pi\right) \\
\sin\left(\theta_r+\frac{2}{3}\pi\right) & \sin\left(\theta_r-\frac{2}{3}\pi\right) & \sin\theta_r
\end{bmatrix}
\begin{bmatrix} i'_{ar} \\ i'_{br} \\ i'_{cr} \end{bmatrix}
$$

$$
= -\frac{P}{2}L_{ms}\left[\left(i_{as}i'_{ar} + i_{bs}i'_{br} + i_{cs}i'_{cr}\right)\sin\theta_r + \left(i_{as}i'_{cr} + i_{bs}i'_{ar} + i_{cs}i'_{br}\right)\sin\left(\theta_r - \frac{2}{3}\pi\right)\right.
$$

$$
\left. + \left(i_{as}i'_{br} + i_{bs}i'_{cr} + i_{cs}i'_{ar}\right)\sin\left(\theta_r + \frac{2}{3}\pi\right)\right], \quad (5.26)
$$

which can be also expressed as

$$T_e = -\frac{P}{2}L_{ms}\left\{\left[i_{as}\left(i'_{ar} - \frac{1}{2}i'_{br} - \frac{1}{2}i'_{cr}\right) + i_{bs}\left(i'_{br} - \frac{1}{2}i'_{ar} - \frac{1}{2}i'_{cr}\right) + i_{cs}\left(i'_{cr} - \frac{1}{2}i'_{br} - \frac{1}{2}i'_{ar}\right)\right]\sin\theta_r\right.$$

$$\left. + \frac{\sqrt{3}}{2}\left[i_{as}\left(i'_{br} - i'_{cr}\right) + i_{bs}\left(i'_{cr} - i'_{ar}\right) + i_{cs}\left(i'_{ar} - i'_{br}\right)\right]\cos\theta_r\right\}.$$

Having found T_e as given by (5.26), the *torsional-mechanical* equations are

$$\frac{d\omega_r}{dt} = \frac{P}{2J}T_e - \frac{B_m}{J}\omega_r - \frac{P}{2J}T_L$$

$$= -\frac{P^2}{4J}L_{ms}\left[\left(i_{as}i'_{ar} + i_{bs}i'_{br} + i_{cs}i'_{cr}\right)\sin\theta_r + \left(i_{as}i'_{cr} + i_{bs}i'_{ar} + i_{cs}i'_{br}\right)\sin\left(\theta_r - \frac{2}{3}\pi\right)\right.$$

$$\left. + \left(i_{as}i'_{br} + i_{bs}i'_{cr} + i_{cs}i'_{ar}\right)\sin\left(\theta_r + \frac{2}{3}\pi\right)\right] - \frac{B_m}{J}\omega_r - \frac{P}{2J}T_L,$$

$$\frac{d\theta_r}{dt} = \omega_r. \tag{5.27}$$

Combining differential equations (5.25) and (5.27), the resulting model for three-phase induction motors in the *machine* variables results. For induction motors, analysis tasks can be performed using Cauchy's and not Cauchy's forms of differential equations.

Example 5.4:   Simulation of Three-Phase Induction Motors Using Simulink

One may use Simulink to simulate three-phase induction motors modeled not in Cauchy's form [6,7]. We use the circuitry-electromagnetic equations (5.24) and the *torsional-mechanical* equations of motion (5.27). One finds the following differential equations

$$\frac{di_{as}}{dt} = \frac{1}{L_{ls} + L_{ms}}\left[-r_s i_{as} + \frac{1}{2}L_{ms}\frac{di_{bs}}{dt} + \frac{1}{2}L_{ms}\frac{di_{cs}}{dt} - L_{ms}\frac{d\left(i'_{ar}\cos\theta_r\right)}{dt} - L_{ms}\frac{d\left(i'_{br}\cos\left(\theta_r + \frac{2\pi}{3}\right)\right)}{dt} - L_{ms}\frac{d\left(i'_{cr}\cos\left(\theta_r - \frac{2\pi}{3}\right)\right)}{dt} + u_{as}\right],$$

$$\frac{di_{bs}}{dt} = \frac{1}{L_{ls} + L_{ms}}\left[-r_s i_{bs} + \frac{1}{2}L_{ms}\frac{di_{as}}{dt} + \frac{1}{2}L_{ms}\frac{di_{cs}}{dt} - L_{ms}\frac{d\left(i'_{ar}\cos\left(\theta_r - \frac{2\pi}{3}\right)\right)}{dt} - L_{ms}\frac{d\left(i'_{br}\cos\theta_r\right)}{dt} - L_{ms}\frac{d\left(i'_{cr}\cos\left(\theta_r + \frac{2\pi}{3}\right)\right)}{dt} + u_{bs}\right],$$

$$\frac{di_{cs}}{dt} = \frac{1}{L_{ls} + L_{ms}}\left[-r_s i_{cs} + \frac{1}{2}L_{ms}\frac{di_{as}}{dt} + \frac{1}{2}L_{ms}\frac{di_{bs}}{dt} - L_{ms}\frac{d\left(i'_{ar}\cos\left(\theta_r + \frac{2\pi}{3}\right)\right)}{dt} - L_{ms}\frac{d\left(i'_{br}\cos\left(\theta_r - \frac{2\pi}{3}\right)\right)}{dt} - L_{ms}\frac{d\left(i'_{cr}\cos\theta_r\right)}{dt} + u_{cs}\right],$$

$$\frac{di'_{ar}}{dt} = \frac{1}{L'_{lr} + L_{ms}}\left[-r'_r i'_{ar} - L_{ms}\frac{d\left(i_{as}\cos\theta_r\right)}{dt} - L_{ms}\frac{d\left(i_{bs}\cos\left(\theta_r - \frac{2\pi}{3}\right)\right)}{dt} - L_{ms}\frac{d\left(i_{cs}\cos\left(\theta_r + \frac{2\pi}{3}\right)\right)}{dt} + \frac{1}{2}L_{ms}\frac{di'_{br}}{dt} + \frac{1}{2}L_{ms}\frac{di'_{cr}}{dt} + u'_{ar}\right],$$

$$\frac{di'_{br}}{dt} = \frac{1}{L'_{lr} + L_{ms}}\left[-r'_r i'_{br} - L_{ms}\frac{d\left(i_{as}\cos\left(\theta_r + \frac{2\pi}{3}\right)\right)}{dt} - L_{ms}\frac{d\left(i_{bs}\cos\theta_r\right)}{dt} - L_{ms}\frac{d\left(i_{cs}\cos\left(\theta_r - \frac{2\pi}{3}\right)\right)}{dt} + \frac{1}{2}L_{ms}\frac{di'_{ar}}{dt} + \frac{1}{2}L_{ms}\frac{di'_{cr}}{dt} + u'_{br}\right],$$

$$\frac{di'_{cr}}{dt} = \frac{1}{L'_{lr} + L_{ms}}\left[-r'_r i'_{cr} - L_{ms}\frac{d\left(i_{as}\cos\left(\theta_r - \frac{2\pi}{3}\right)\right)}{dt} - L_{ms}\frac{d\left(i_{bs}\cos\left(\theta_r + \frac{2\pi}{3}\right)\right)}{dt} - L_{ms}\frac{d\left(i_{cs}\cos\theta_r\right)}{dt} + \frac{1}{2}L_{ms}\frac{di'_{ar}}{dt} + \frac{1}{2}L_{ms}\frac{di'_{br}}{dt} + u'_{cr}\right],$$

$$\frac{d\omega_r}{dt} = -\frac{P^2}{4J}L_{ms}\left[\left(i_{as}i'_{ar} + i_{bs}i'_{br} + i_{cs}i'_{cr}\right)\sin\theta_r + \left(i_{as}i'_{cr} + i_{bs}i'_{ar} + i_{cs}i'_{br}\right)\sin\left(\theta_r - \frac{2}{3}\pi\right) + \left(i_{as}i'_{br} + i_{bs}i'_{cr} + i_{cs}i'_{ar}\right)\sin\left(\theta_r + \frac{2}{3}\pi\right)\right]$$

$$- \frac{B_m}{J}\omega_r - \frac{P}{2J}T_L,$$

$$\frac{d\theta_r}{dt} = \omega_r. \tag{5.28}$$

The numerical simulations are performed for a 220 V, 60 Hz, two-pole induction motor. The parameters are

$r_s = 0.8$ ohm, $r_r = 1$ ohm, $L_{ms} = 0.1$ H, $L_{ls} = 0.01$ H, $L_{lr} = 0.01$ H, $B_m = 4 \times 10^{-4}$ N-m-s/rad, and $J = 0.002$ kg-m^2.

The balanced three-phase voltage set is

$$u_{as}(t) = \sqrt{2}u_M \cos(\omega_f t), \quad u_{bs}(t) = \sqrt{2}u_M \cos\left(\omega_f t - \frac{2}{3}\pi\right), \quad u_{cs}(t) = \sqrt{2}u_M \cos\left(\omega_f t + \frac{2}{3}\pi\right),$$

$$u_M = 220 \text{ V}.$$

The frequency of the supplied voltage is 60 Hz. Hence, $\omega_e = 4\pi f/P = 377$ rad/s. If $T_L = 0$ and the friction torque is negligible, the steady-state value of angular velocity should approach ω_e. Figure 5.10a illustrates the transient dynamics for the angular velocity. The dynamic torque–speed characteristic $\omega_r(t) = \Omega_T[T_e(t)]$ is documented in Figure 5.10b.

The dynamics of $\omega_r(t)$ and $\omega_r(t) = \Omega_T[T_e(t)]$ if the load torque $T_L = 4$ N-m is applied at 0.4 s (the motor rotates at $\sim$377 rad/s at no load) are plotted in Figure 5.11. One observes that the

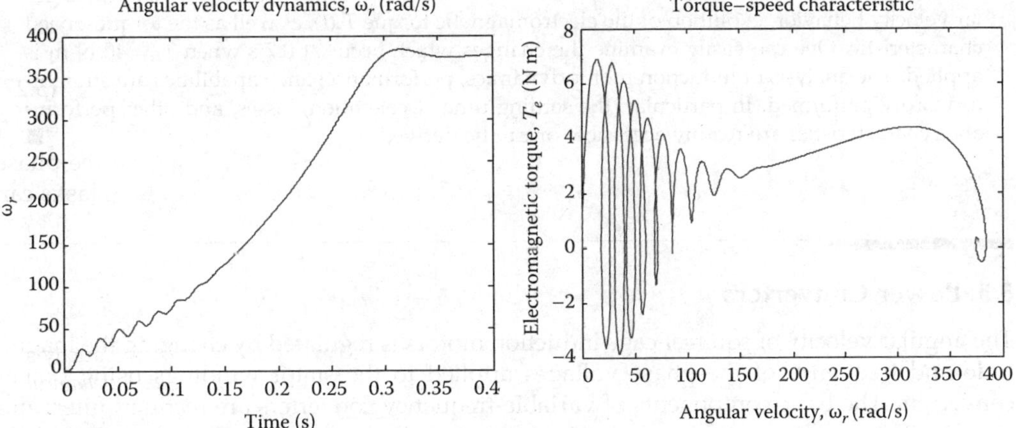

FIGURE 5.10
(a) Transient dynamic for the angular velocity $\omega_r(t)$; (b) dynamic torque–speed characteristic $\omega_r(t) = \Omega_T[T_e(t)]$.

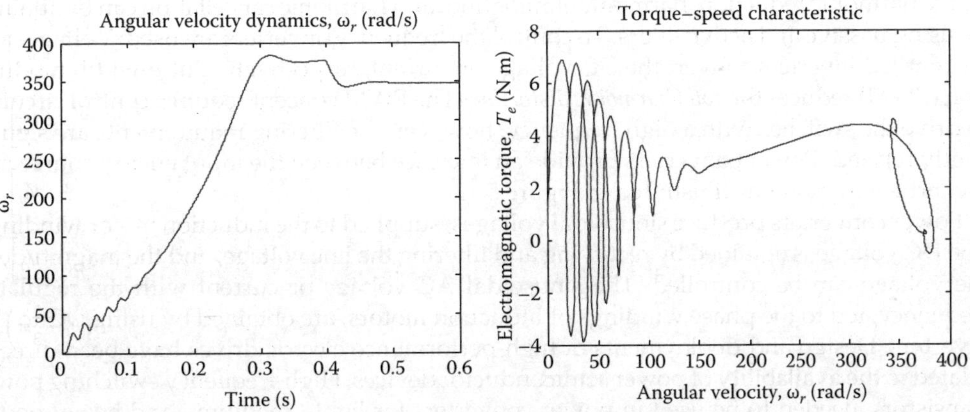

FIGURE 5.11
Transient dynamics for the angular velocity $\omega_r(t)$ and evolution of $\omega_r(t) = \Omega_T[T_e(t)]$.

angular velocity decreases as T_L is applied. One refers to Figure 5.2b to visualize that the steady-state value of ω_r at which rotor rotates decreases if T_L is applied. The critical angular velocity from the plot $\omega_r(t) = \Omega_T[T_e(t)]$ is ~315 rad/s. If ~4.5 N-m or higher load would be applied, $T_{e\max} < T_L$, and the motor will disaccelerate to the stall. The induction motor cannot start if $T_L > \sim 1.5$ N-m is applied when the motor at stall ($\omega_r = 0$ rad/s). Correspondingly, one uses the voltage–frequency or frequency control to start the motor at high load ($T_{L\max}$ is ~4.5 N-m) as well as operate the motor at the desired angular velocity.   ∎

Example 5.5:   Simulation of a Three-Phase Squirrel-Cage Induction Motor

Consider a 220 V, 60 Hz, two-pole induction motor. The parameters are $r_s = 0.3$ ohm, $r_r' = 0.2$ ohm, $L_{ms} = 0.035$ H, $L_{ls} = 0.003$ H, $L_{lr}' = 0.003$ H, $B_m = 1 \times 10^{-3}$ N-m-s/rad, and $J = 0.02$ kg-m^2. The three-phase balanced voltage set is applied with $u_M = 220$ V. We apply the load torque 40 N-m at 0.7 s.

Using the ode45 differential equations solver, we numerically solve a set of eight differential equations (5.25) and (5.27) [6]. The transients of the stator and rotor currents $i_{as}(t)$, $i_{bs}(t)$, $i_{cs}(t)$, $i_{ar}'(t)$, $i_{br}'(t)$, $i_{cr}'(t)$, as well as the angular velocity $\omega_r(t)$, are plotted using the corresponding statements. Figure 5.12 illustrates the phase voltages applied, phase currents dynamics, angular velocity behavior, evolution of the electromagnetic torque $T_e(t)$, as well as the torque–speed characteristic. One can easily examine the changes which occur at 0.7 s when $T_L = 40$ N-m is applied. The analysis of induction motor dynamics, performance, and capabilities are straightforwardly performed. In particular, the settling time, acceleration, losses, and other performance characteristics are readily accessible or can be derived.   ∎

5.5 Power Converters

The angular velocity of squirrel-cage induction motors is regulated by changing the magnitude and frequency of the phase voltages applied to the stator windings using power converters. The basic components of variable-frequency converters are rectifier, filter, and inverter. The simplest rectifiers are the single-phase half- and full-wave rectifiers. To control medium- and high-power induction motors, *polyphase* rectifiers are used. *Polyphase* rectifiers contain several AC sources, and the rectified voltage is summated at the output. The rectified voltage is filtered to reduce the harmonic content of the rectifier output voltage. Passive and active harmonic reduction, harmonic elimination, and harmonic cancellation can be attained by using passive and active filters. To control the frequency, inverters are used. Voltage- and current-fed inverters convert the DC voltage or current, respectively. Pulse-width modulation (PWM) reduces the *total harmonic distortion*. The PWM concept requires control circuitry to drive the switches with a high frequency, however, the filtering requirements are significantly relaxed. Power converters provide an interface between the input energy source and the induction motor as illustrated in Figure 5.13.

Power converters produce sinusoidal voltages, supplied to the induction motor windings. The DC voltage is obtained by rectifying and filtering the line voltage, and the magnitude of the voltage can be controlled. The sinusoidal AC voltage or current with the regulated frequency, fed to the phase windings of induction motors, are obtained by using DC-to-AC inverters. Design and deployments of high-performance electric drives have been directly related to the availability of power semiconductor devices. High-frequency switching power transistors, needed to be used in power converters for light-, medium-, and heavy-power applications, exist. The specialized design is available, and for example, 3000 V, 1000 A IGBTs are integrated with diodes in the same package. The ~200 kVA soft switching resonant-link

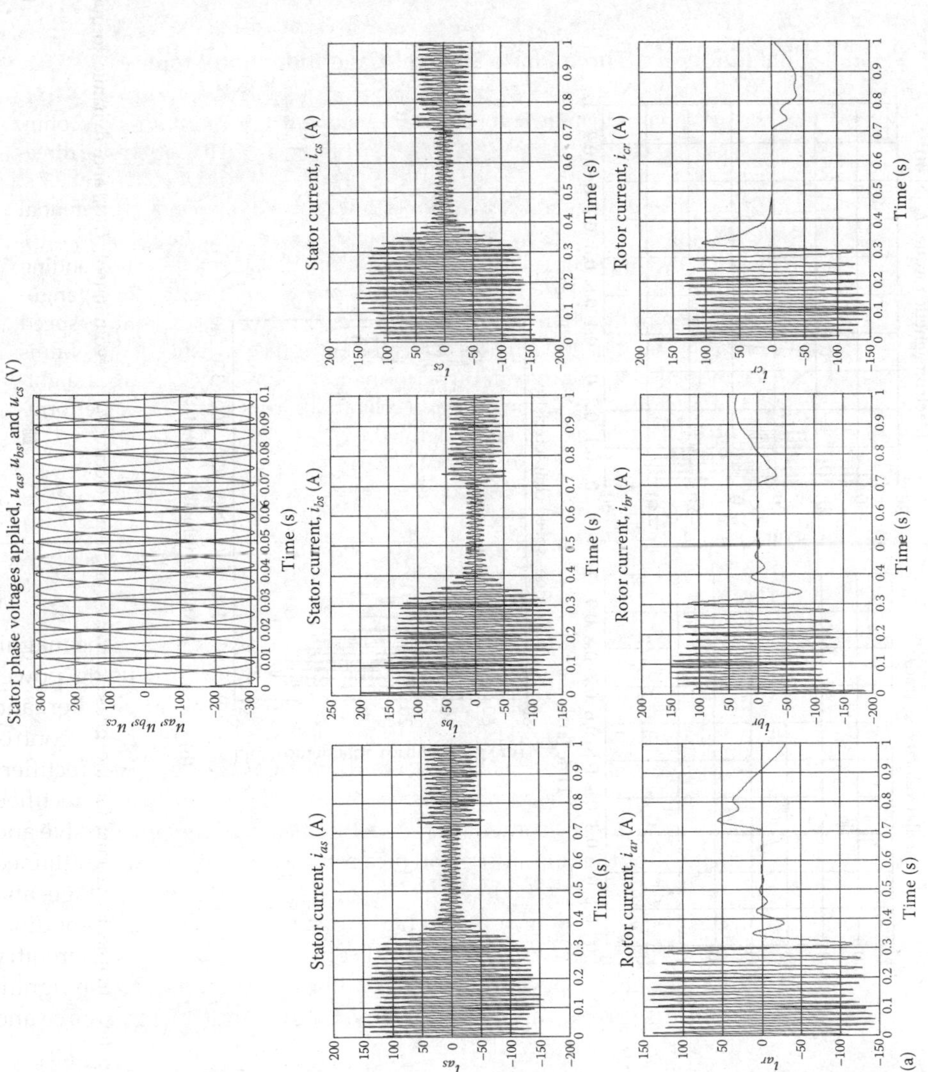

FIGURE 5.12

Balanced phase voltages applied, transient dynamic for the phase currents, angular velocity $\omega_r(t)$, evolution of $T_e(t)$, and dynamic torque–speed characteristic $\omega_r(t) = \Omega_T[T_e(t)]$.

(continued)

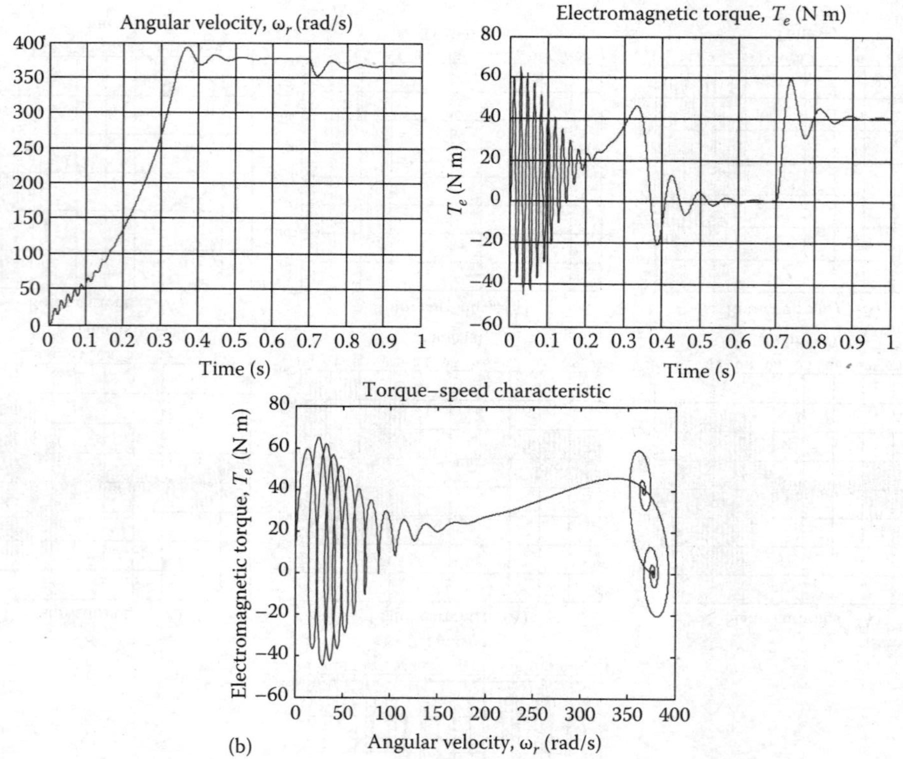

FIGURE 5.12 (continued)

Balanced phase voltages applied, transient dynamic for the phase currents, angular velocity $\omega_r(t)$, evolution of $T_e(t)$, and dynamic torque–speed characteristic $\omega_r(t) = \Omega_T[T_e(t)]$.

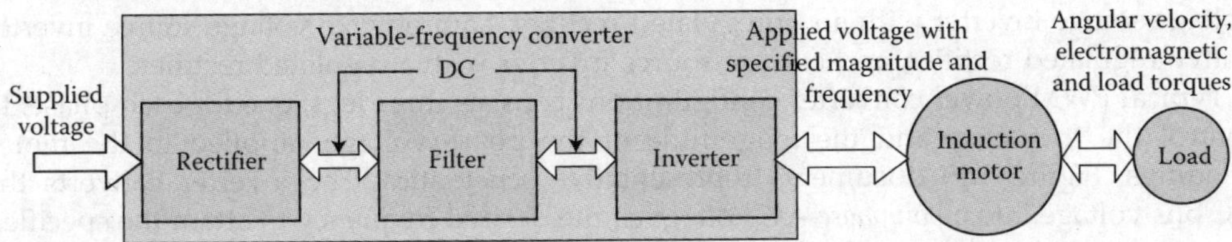

FIGURE 5.13
Variable-frequency power converter and loaded induction motor.

inverters have the switching frequency $\sim$70 kHz. The development of the gate turn-off (GTO) thyristor was the key to extend the power rating of electric drives with induction machines to the megawatt range. Power converters with GTO have found a widespread application in traction drives (electric drivetrains in ships and locomotives). Gate turn-off thyristors are current-controlled devices which require large gate current to enable turn-off the anode current. Hence, large snubbers are needed to ensure turn-off without failures.

There exist two basic types of inverters. The voltage source inverters supply induction motor windings with variable frequency phase voltages. In contrast, the variable frequency phase currents are fed to the induction motor windings by current source inverters. Figure 5.14 illustrates high-level diagrams of power converters which include PWM

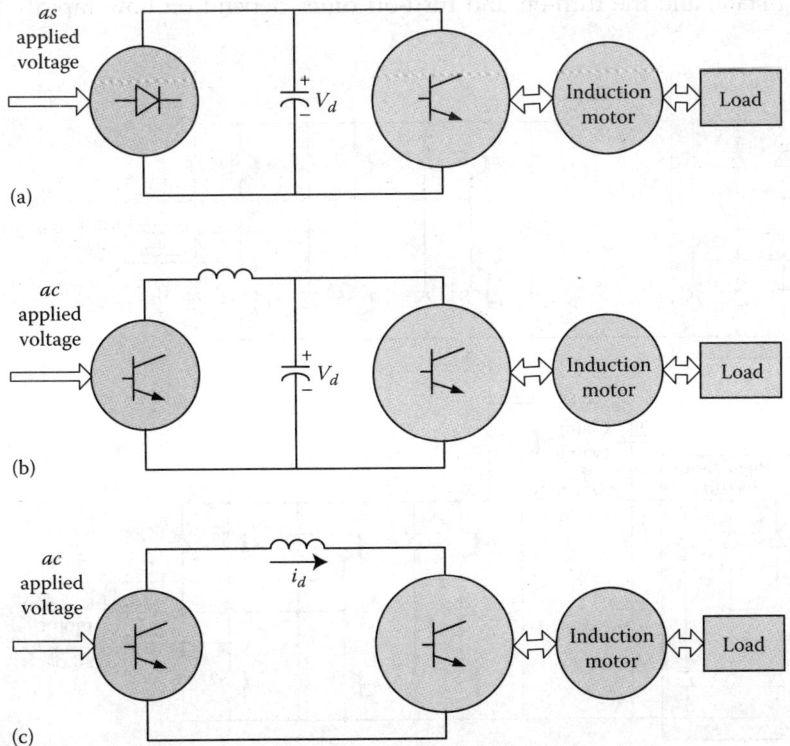

FIGURE 5.14
Variable-frequency power converters: (a) PWM voltage source inverter with an unregulated rectifier; (b) squire-wave voltage source inverter with a regulated rectifier; (c) current source inverter with a regulated rectifier.

voltage source inverter with an unregulated rectifier, squire-wave voltage source inverter with a regulated rectifier, and current source inverter with a regulated rectifier.

Typical PWM power converter configurations consists three legs, one for each phase, to control the frequency and the magnitude of the phase voltages applied to the motor windings. Figure 5.15 documents representative schematics. The inverter converts the DC bus voltage into a *polyphase* AC voltage at the desired frequency to attain the specified torque–speed characteristics, efficiency, starting capabilities, acceleration, etc. Switching stresses, losses, high-electromagnetic interference, extended operating areas, and some other drawbacks of hard-switching inverters lead to the implementation of soft-switching technology, as illustrated in Figure 5.15b. Soft-switching by using resonant-linked converters allows one to attain zero voltage across (current through) the switching device. That is, the semiconductor device is switched when the voltage across it, or the current through it, is zero. Hence, low losses and electromagnetic interference, high efficiency, and switching capabilities result compared with the hard-switching inverters shown in Figure 5.15a.

For hard- and soft-switching inverter configurations, shown in Figure 5.15, the phase voltage waveform, supplied to the motor winding, is illustrated in Figure 5.16a and b, respectively.

Transistors and thyristors are current-controlled solid-state devices. Transistors require continuous drive signals, while thyristors need a momentary gate current to turn-on and turn-off. For example, a base current must be regulated to maintain the BJT in the conducting state, and the turn-on and turn-off times depend on how rapidly the charge

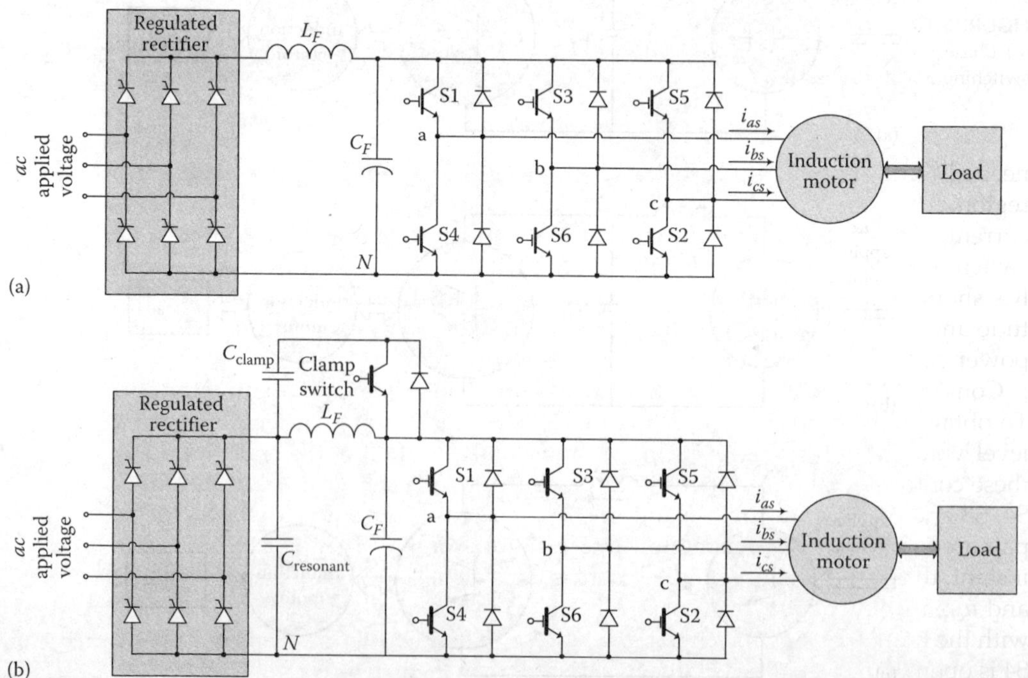

FIGURE 5.15
(a) Power converter with three-phase hard-switching inverter; (b) power converter with three-phase soft-switching inverter.

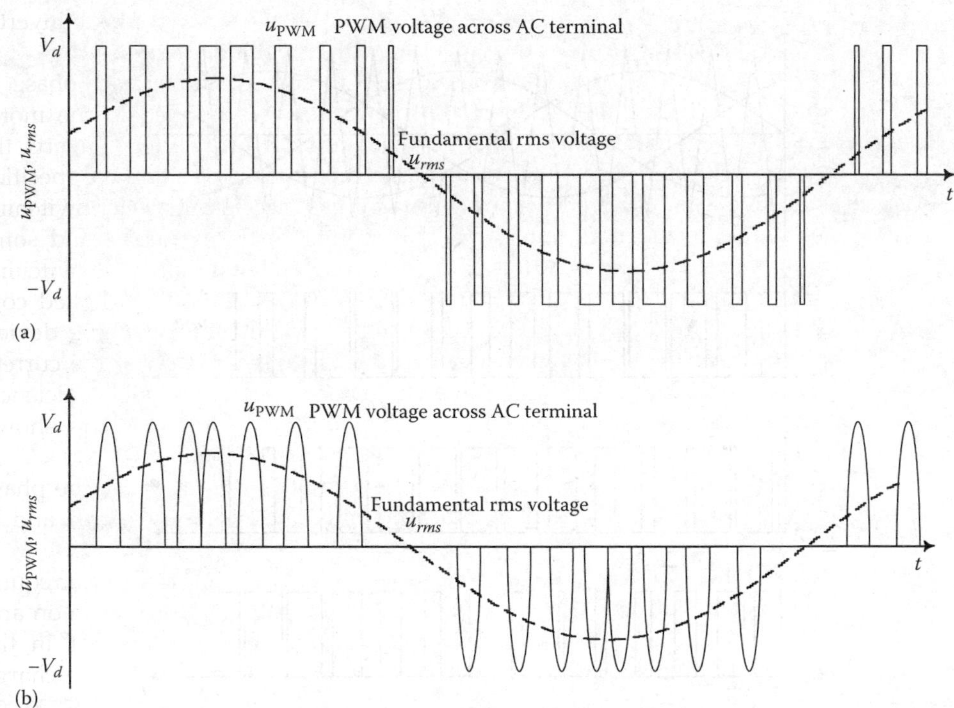

FIGURE 5.16
(a) Phase voltage waveforms in the hard-switching PWM inverter; (b) phase voltage waveforms in the soft-switching PWM inverter.

needed to be supplied (to turn-on) or removed (to turn-off) can be delivered to the base region. The turn-off switching speed is decreased by initially applying a spike of base current, and, then, reducing the current to the magnitude needed. In contrast, the turn-off switching speed is decreased by initially applying a spike of negative base current. Control ICs should drive high-frequency transistors with the overall objective to vary the magnitude and frequency of phase voltages. The closed-loop systems for induction motor with power converter are designed.

Consider the hard-switching inverter as shown in Figure 5.15a with three switch pairs. To obtain three-phase balanced output voltages using PWM concept, a triangular signal-level voltage is compared with three sinusoidal specified frequency control signals and these control signals are shifted by 120°, see Figure 5.17a. High-frequency switches S1 and S4, S3 and S6, S5 and S2 close and open opposite of each other. That is, switches in each pair are turned on and off simultaneously. If, for example, S1 and S4 are closed at the same instant, the circuit is short-circuited across the source. The instantaneous voltages u_{aN}, u_{bN}, and u_{cN} are either equal to V_d or 0. The signal level voltages u_{ac}, u_{bc}, and u_{cc} are compared with the triangular signal u_t. If, for example, u_{ac} is greater than u_t, then S1 is closed, whereas S4 is open. If the signal-level voltage u_{ac} is less than u_t, then S4 is closed whereas S1 is open. The resulting waveform for the phase voltage u_{aN} is shown in Figure 5.17b. In the similar manner, the phase voltages u_{bN} and u_{cN} are defined by comparing the signal-level voltages u_{bc} and u_{cc} with u_t to open or close switches S3–S6 and S5–S2. The resulting

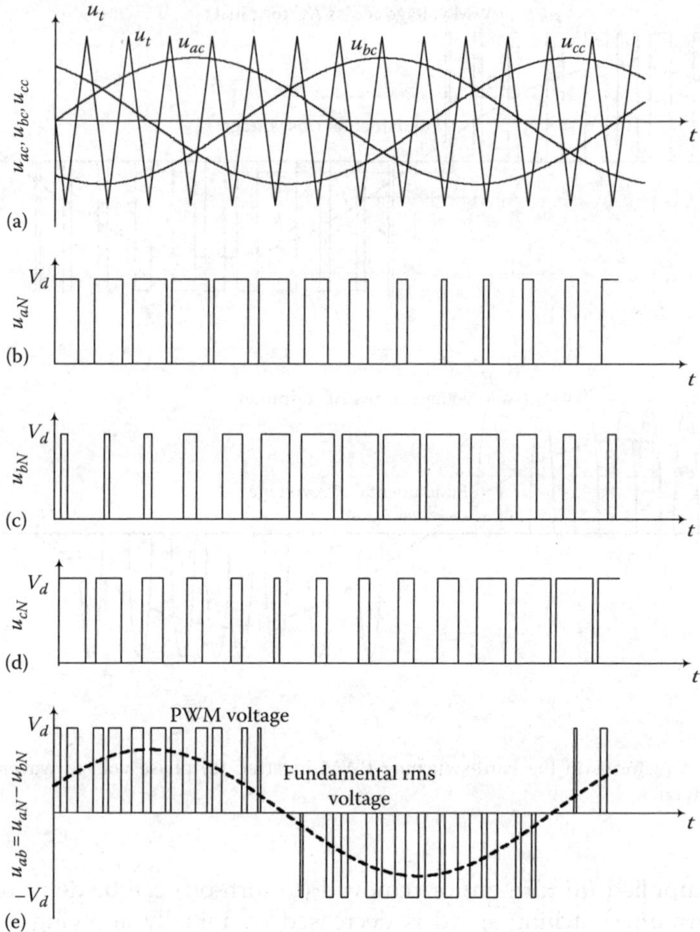

FIGURE 5.17
Voltage waveforms in three-phase hard-switching inverters.

voltages u_{bN} and u_{cN} possess the same pattern as the aN voltage, except that u_{bN} and u_{cN} are shifted by 120° and 240° as illustrated in Figure 5.17c and d. The voltages u_{aN}, u_{bN}, and u_{cN} are measured with respect to the negative DC bus. These DC components are canceled as one considers the line-to-line voltage, which is plotted in Figure 5.17e. The line-to-line voltage u_{ab} is found by subtracting voltage u_{bN} from u_{aN}. One can analyze the waveforms of the instantaneous and *rms* voltages, shown in Figures 5.16 and 5.17e.

The square-wave voltage source inverters, known as six-step inverters, are commonly used. The three-phase square-wave voltage source inverter bridge is shown in Figure 5.15. The rectifier rectifies the three-phase AC applied voltage, and a large electrolytic capacitor C_F maintains a near-constant DC voltage as well as provides a path for the rapidly changing currents drawn by the inverter. The inductor L_F attenuates current spikes. Assume that the inverter consists of six ideal switches. We consider the basic operation

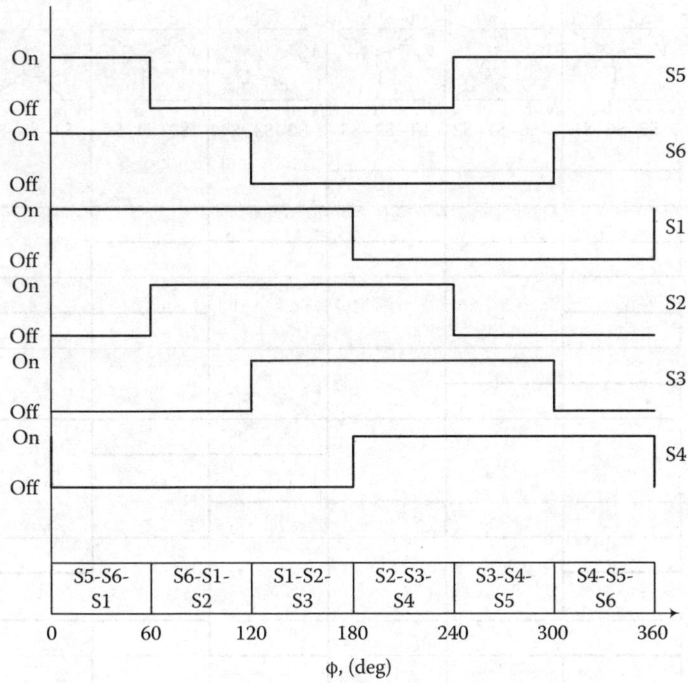

FIGURE 5.18
Switching pattern for three-phase six-step inverters.

of the square-wave voltage inverters. Each switch is closed for 180° and is opened for the remaining 180° in a cyclic pattern. Furthermore, S3 is closed 120° after S1, S5 is closed 120° after S3, S4 is closed 180° after S1, S6 is closed 180° after S3, and S2 is closed 180° after S5, see Figure 5.18. The result of this switching operation is that a combination of three switches is closed simultaneously for every 60° duration as shown in Figure 5.18. That is, in three-phase six-step inverters, the switching appears every 60° interval ($\frac{1}{6}T$ time interval).

To determine the voltage waveforms, applied to the *abc* windings, consider the six-step inverter and motor circuitry as illustrated in Figure 5.15. During the interval from 0° to 60°, where switches S5, S6, and S1 are closed, the phase *a* is in parallel with *c*, and they are connected to the phase *b* in series, which is connected to the source via S6. The voltage waveforms, as shown in Figure 5.19, result. In particular, $u_{aN} = u_{cN} = V_d$ and $u_{bN} = 0$. Hence, $u_{ab} = V_d$, $u_{bc} = -V_d$, and $u_{ca} = 0$. Because phases *a* and *c* are connected in parallel, the apparent impedance, seen from the neutral of the motor (depicted in Figure 5.19 as a point N'), is halved. Hence, the voltage drop across the phases *as* and *cs* is $\frac{2}{3}V_d$, whereas voltage drop across the phase *bs* is $\frac{1}{3}V_d$. That is, $u_{as} = \frac{1}{3}V_d$, $u_{bs} = -\frac{2}{3}V_d$, and $u_{cs} = \frac{1}{3}V_d$. Hence the voltage drop across the phase is always $\frac{1}{3}V_d$ or $\frac{2}{3}V_d$ depending on the

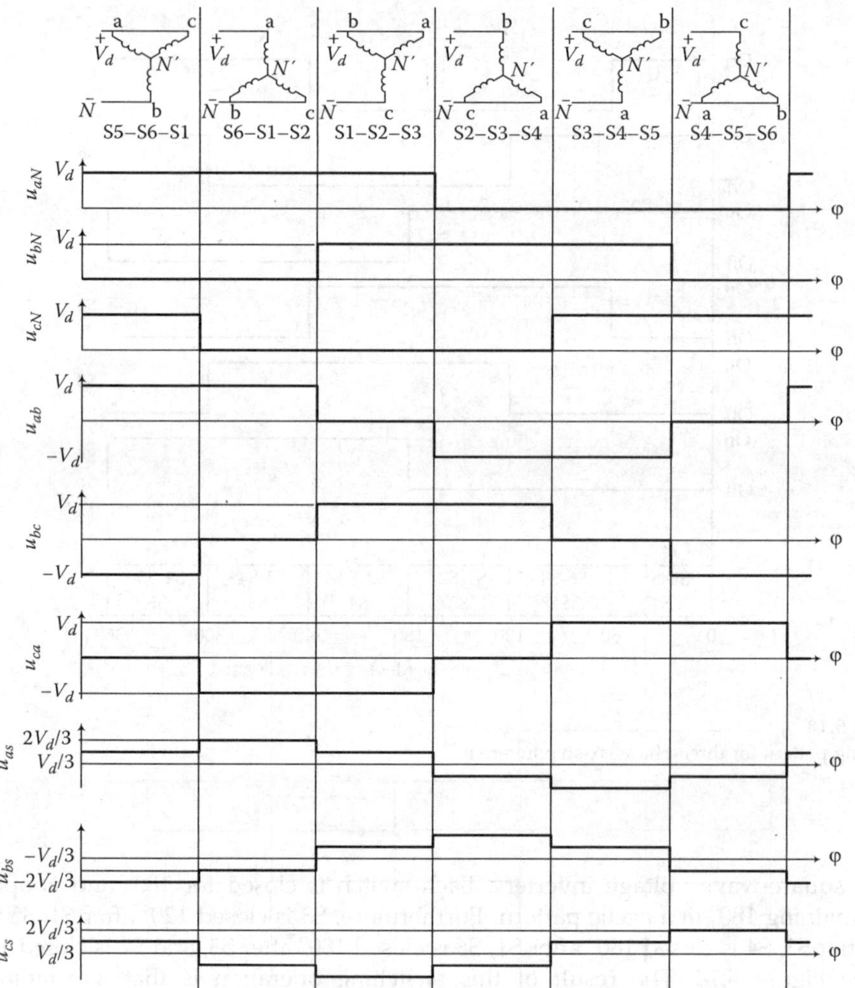

FIGURE 5.19

Voltage waveforms at the terminals aN, bN, and cN, line-to-line, and line-to-neutral voltages applied to the induction motor phase windings.

connection of the phases (series or parallel). The waveforms for u_{as}, u_{bs}, and u_{cs} are shown in Figures 5.19.

Compared with the voltage source inverters, the power converters, if the current source inverters are implemented, are different. In particular, the current source inverter is fed from a constant current source which is generated by a controlled rectifier with a large DC link inductor L to smooth the current. The representative schematics of a current source inverter is shown in Figure 5.20.

At any time, only two thyristors conduct. In particular, one of the thyristors is connected to the positive DC link and the other is connected to the negative DC link. The current is switched sequentially into one of the phases of a three-phase induction motor by the top half of the inverter and returns from another phase to the DC link by the bottom half of the

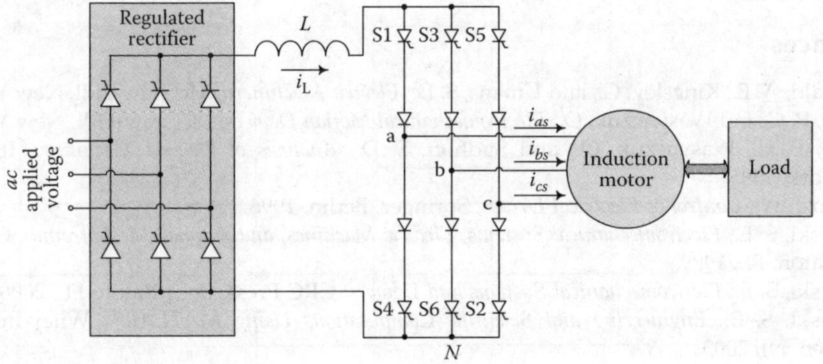

FIGURE 5.20
Power converter with a thyristor current source inverter.

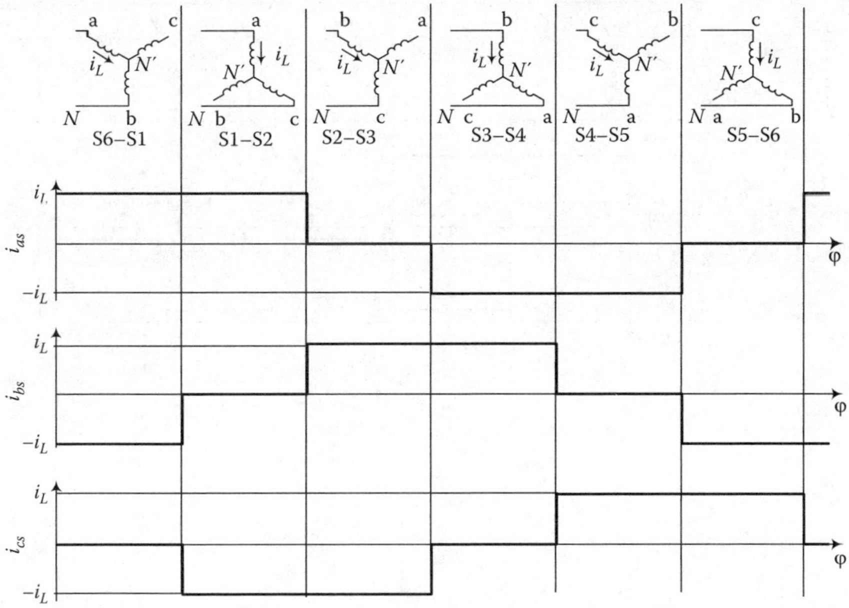

FIGURE 5.21
Phase currents for the induction motors fed by current source inverter.

inverter. Since the current is constant, there will be a constant voltage drop across the stator winding of the motor and zero voltage drop across the self-inductance of the winding. Hence, the motor terminal voltage is not set by the inverter but by the resistance of the stator winding of the motor. Since the motor is wound with sinusoidally distributed windings, the voltages that appear on the terminals of the motor are sinusoidal. The current waveforms of the motor connected to a current source inverter are shown in Figure 5.21.

References

1. Fitzgerald, A. E., Kingsley, C., and Umans, S. D., *Electric Machinery*, McGraw-Hill, New York, 1990.
2. Krause, P. C. and Wasynczuk, O., *Electromechanical Motion Devices*, McGraw-Hill, New York, 1989.
3. Krause, P. C., Wasynczuk, O., and Sudhoff, S. D., *Analysis of Electric Machinery*, IEEE Press, New York, 1995.
4. Leonhard, W., *Control of Electrical Drives*, Springer, Berlin, 1996.
5. Lyshevski, S. E., *Electromechanical Systems, Electric Machines, and Applied Mechatronics*, CRC Press, Boca Raton, FL, 1999.
6. Lyshevski, S. E., *Electromechanical Systems and Devices*, CRC Press, Boca Raton, FL, 2008.
7. Lyshevski, S. E., *Engineering and Scientific Computations Using MATLAB®*, Wiley-Interscience, Hoboken, NJ, 2003.

6

Permanent-Magnet Synchronous Machines and Their Applications

6.1 Introduction to Synchronous Machines

Synchronous machines can be classified as conventional, permanent-magnet, variable-reluctance, etc. The authors are focused on the high-performance electromechanical (mechatronic) systems. Therefore, this chapter concentrates on permanent-magnet synchronous machines which possess superior performance and capabilities surpassing any other electric machines such as permanent-magnet DC electric machines, induction motors, etc. Conventional synchronous machines are covered in Refs. [1–6].

In synchronous motors, the electromagnetic torque results due to the interaction of time-varying magnetic filed established by the stator windings and stationary magnetic field produced by the windings or magnets on the rotor [1–6]. There exist synchronous reluctance machines which exhibit, in general, low performance as documented in Chapter 3.

In high-performance drives, servos and power generation systems (up to $\sim$100 kW rated and $\sim$500 kW peak), three-phase permanent-magnet synchronous machines (motors and generators) is the preferable choice. In high-power (from hundreds of kW to MW range) generation systems, conventional three-phase synchronous generators are utilized [2–5].

There exist translational (linear) and rotational synchronous machines. Due to the fact that a great majority of mechatronic systems utilize rotational electric machines, we concentrate on the rotational motion devices. In this chapter, we consider radial and axial topology permanent-magnet synchronous machines. Two- and three-phase radial machines are illustrated in Figure 6.1a and b.

We examine the energy conversion, torque production, control, and other important issues. The angular velocity of synchronous motors is fixed with the frequency of the supplied phase voltages to the stator windings. The phase voltages must be applied as functions of the rotor angular displacement θ_r. The steady-state torque–speed characteristics can be represented as a family of horizontal lines as depicted in Figure 6.1c. The electrical angular velocity is equal to the synchronous angular velocity $\omega_e = 4\pi f/P$. The designer examines the maximum load torque $T_{L\,\mathrm{max}}$ to satisfy $T_{e\,\mathrm{rated}} > T_{L\,\mathrm{max}}$ or $T_{e\,\mathrm{peak}} > T_{L\,\mathrm{max}}$. For a short period of time ($\sim$1 min for many permanent-magnet synchronous motors), one may significantly overload motors, and the ratio $T_{e\,\mathrm{peak}}/T_{e\,\mathrm{rated}}$ could be from $\sim$2 to $\sim$10. If $T_{e\,\mathrm{critical}} < T_L$, the rotor magnetic field is no longer locked to the stator magnetic field (motor cannot produce the needed torque due to the constraints imposed including rated voltage and current of power converters). For $T_{e\,\mathrm{critical}} < T_L$, the rotor magnetic field slips behind the stator field. Due to the loss of synchronization, the electromagnetic torque developed by the motor reverses (surges). Therefore, the condition $T_{e\,\mathrm{critical}} > T_L$ must be

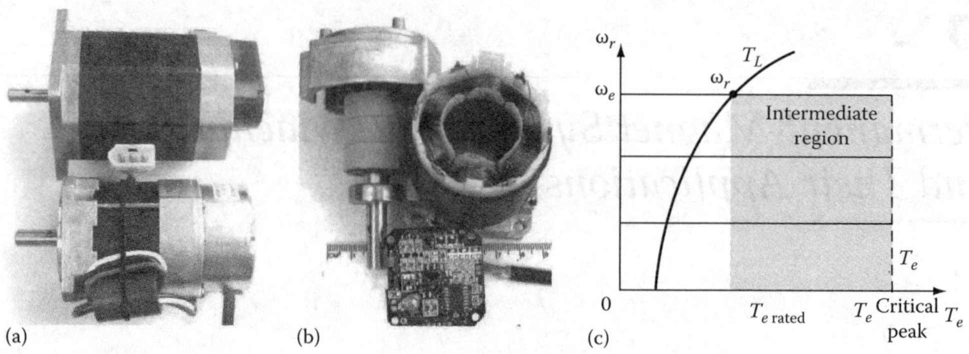

FIGURE 6.1

(a) Two- and three-phase permanent-magnet synchronous machines which can be used as motors and generators (23 NEMA size, ~100 W rated, and ~500 W peak); (b) three-phase permanent-magnet synchronous machine: rotor (with permanent magnets), stator (with windings), and encoder with circuitry; (c) torque–speed characteristic curves of synchronous motors.

always guaranteed, and the peak power converter capabilities (peak voltage and current) must be met. Though the permanent magnet synchronous machines can be overloaded by the factor of 10, the maximum current overloading capabilities of pulse-width modulation (PWM) amplifiers is usually up to ~2, while the peak voltage cannot exceed the bus voltage.

Radial and axial topology permanent-magnet synchronous machines are widely used as actuators (motors) and generators. The motors are controlled by power converters (PWM amplifiers). The implications of microelectronics to motion devices have received meticulous consideration as technologies to fabricate various advanced machines are becoming developed and widely deployed. Mini- and micromachines have been fabricated utilizing CMOS-centered and micromachining technologies and processes. The images of 2 and 4 mm in diameter permanent-magnet synchronous machines are reported in Figure 6.2a and b. These permanent-magnet synchronous machines are smaller than the operational amplifier or ICs-centered electronics to control them. However, to guarantee rotation and actuation, the condition $T_e > T_L$ and $F_e > F_L$ must be satisfied. Correspondingly, the operating envelope (torque/force, load, load profile, angular velocity, etc.) defines T_e thereby resulting in the motor dimensionality. The acceleration capability, which defines the settling time and repositioning, depends on the ratio $(T_e - T_L)/J$.

The torque and power densities, rated angular velocity, and other characteristics are defined by the machine design, dimensionality, materials, technologies, and other factors. For a preliminary power estimate, one may assume that the power density ~1 W/cm^3 can be achieved. Figure 6.2c documents the images of a permanent-magnet synchronous motor in the computer hard drive and VHS with the monolithic ICs driver. A two-phase permanent magnet synchronous motor (stepper motor) with the monolithic controller/driver is reported in Figure 6.2d.

We address and solve a spectrum of problems in analysis and control of various synchronous machines. A coherent analysis results in deriving sound control concepts ensuring best performance and *achievable* capabilities. For example, maximum efficiency, minimal losses, maximum torque and power densities, vibrations and noise minimization, as well as other critical improvements can be achieved for mechatronic systems.

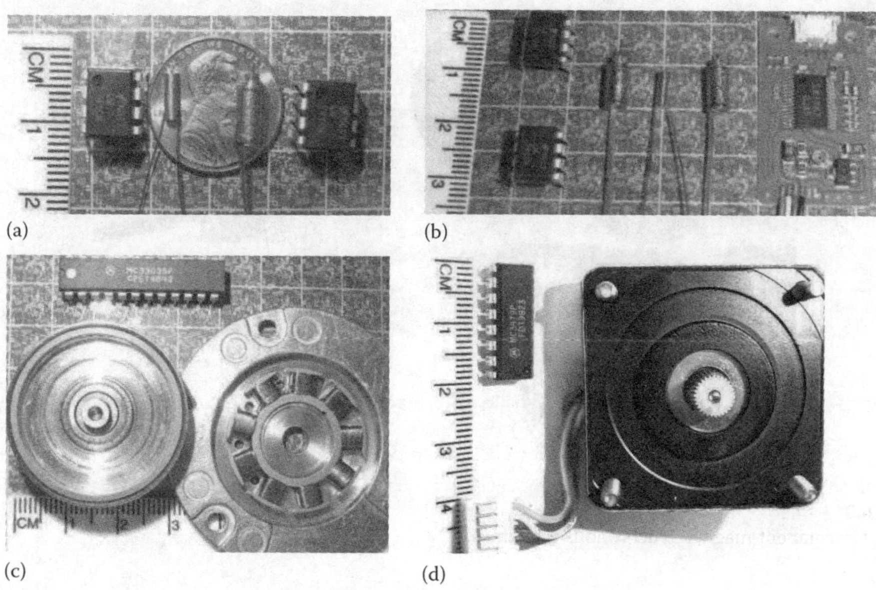

FIGURE 6.2
(a) Images of permanent-magnet synchronous machines and operational amplifiers on a silicon wafer with ICs on it; (b) operational amplifiers (on left), permanent-magnet synchronous motors, and power electronics board (on right) to control them; (c) permanent-magnet synchronous motor to drive a computer hard drive and a monolithic driver; (d) stepper motor (two-phase permanent-magnet synchronous machine) and a controller/driver.

6.2 Radial Topology Permanent-Magnet Synchronous Machines

Consider the radial topology permanent-magnet synchronous machines. These motion devices surpass other electric machines. High efficiency, high power and torque densities, superior overloading, robustness, expanded operating envelope, and other features contribute to excellent performance and superior capabilities. The studied permanent-magnet synchronous motion devices are brushless electric machines because the excitation field (flux) is produced by permanent magnets placed on the rotor. The images of permanent-magnet synchronous machines used in computer hard drives and VHS drives are illustrated in Figure 6.3. The permanent-magnet synchronous motors are frequently called the "brushless DC motors" perhaps due to the similarity of the torque–speed characteristics. However, the operating principles and design of permanent-magnet DC and synchronous machines are fundamentally distinct. Therefore, the terminology "brushless DC motor" could be deceptive, and, is not used in this book.

6.2.1 Two-Phase Permanent-Magnet Synchronous Machines and Stepper Motors

Stepper motors, illustrated in Figure 6.4, are two-phase permanent magnet synchronous machines. These electromechanical motion devices, though primarily used as motors, can be effectively utilized as two-phase AC generators if rotated by a prime mover, e.g., torque T_{pm} is applied to rotate a generator.

FIGURE 6.3
Images of permanent-magnet synchronous machines.

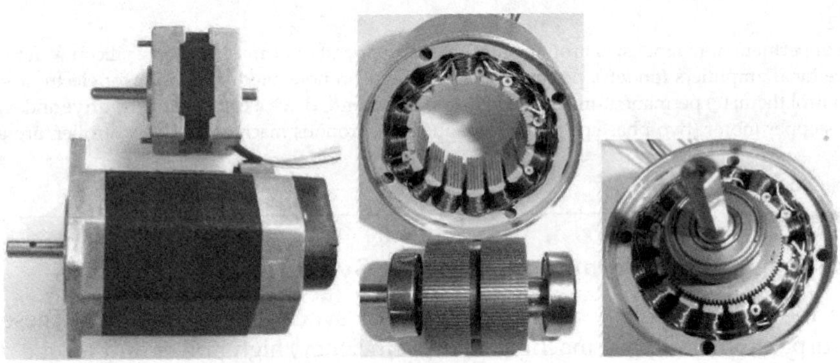

FIGURE 6.4
Stepper motors: There are various stepper motors of different designs, topologies, sizes, etc.

Consider radial topology two-phase permanent-magnet synchronous motors. Using Kirchhoff's voltage law, we have the following two equations:

$$u_{as} = r_s i_{as} + \frac{d\psi_{as}}{dt}, \quad u_{bs} = r_s i_{bs} + \frac{d\psi_{bs}}{dt}, \tag{6.1}$$

where the flux linkages are expressed as

$$\psi_{as} = L_{asas} i_{as} + L_{asbs} i_{bs} + \psi_{asm} \quad \text{and} \quad \psi_{bs} = L_{bsas} i_{as} + L_{bsbs} i_{bs} + \psi_{bsm}.$$

where u_{as} and u_{bs} are the phase voltages supplied to the stator windings as and bs, i_{as} and i_{bs} are the phase currents, ψ_{as} and ψ_{bs} are the stator flux linkages, r_s is the resistances of the stator windings, L_{asas} and L_{bsbs} are the self-inductances, and L_{asbs} and L_{bsas} are the mutual inductances.

The flux linkages are periodic functions of the angular displacement (rotor position), and let

$$\psi_{asm} = \psi_m \sin \theta_r \quad \text{and} \quad \psi_{bsm} = -\psi_m \cos \theta_r.$$

The self-inductances of the stator windings are $L_{ss} = L_{asas} = L_{bsbs} = L_{ls} + \bar{L}_m$.

The stator windings are displaced by 90 electrical degrees. Hence, the mutual inductances between the stator windings are $L_{asbs} = L_{bsas} = 0$. We have

$$\psi_{as} = L_{ss} i_{as} + \psi_m \sin \theta_r \quad \text{and} \quad \psi_{bs} = L_{ss} i_{bs} - \psi_m \cos \theta_r.$$

From (6.1), one finds

$$u_{as} = r_s i_{as} + \frac{d(L_{ss} i_{as} + \psi_m \sin \theta_r)}{dt} = r_s i_{as} + L_{ss} \frac{di_{as}}{dt} + \psi_m \omega_r \cos \theta_r$$

$$u_{bs} = r_s i_{bs} + \frac{d(L_{ss} i_{bs} - \psi_m \cos \theta_r)}{dt} = r_s i_{bs} + L_{ss} \frac{di_{bs}}{dt} - \psi_m \omega_r \sin \theta_r$$

(6.2)

Using Newton's second law, for the mechanical angular velocity and displacement, we have

$$\frac{d\omega_{rm}}{dt} = \frac{1}{J}(T_e - B_m \omega_{rm} - T_L),$$

$$\frac{d\theta_{rm}}{dt} = \omega_{rm}.$$

(6.3)

The expression for the electromagnetic torque developed by permanent-magnet motors is obtained by using the coenergy. Using

$$W_c = \frac{1}{2}\left(L_{ss} i_{as}^2 + L_{ss} i_{bs}^2\right) + \psi_m i_{as} \sin \theta_r - \psi_m i_{bs} \cos \theta_r + W_{PM}$$

and recalling that $T_e = \dfrac{\partial W_c}{\partial \theta_{rm}}$ one finds

$$T_e = \frac{P\psi_m}{2}(i_{as} \cos \theta_r + i_{bs} \sin \theta_r).$$

(6.4)

Integrating the circuitry-electromagnetic equations (6.2) with the *torsional–mechanical* dynamics (6.3), and using (6.4), we obtain

$$\frac{di_{as}}{dt} = -\frac{r_s}{L_{ss}} i_{as} - \frac{\psi_m}{L_{ss}} \omega_r \cos \theta_r + \frac{1}{L_{ss}} u_{as},$$

$$\frac{di_{bs}}{dt} = -\frac{r_s}{L_{ss}} i_{bs} + \frac{\psi_m}{L_{ss}} \omega_r \sin \theta_r + \frac{1}{L_{ss}} u_{bs},$$

$$\frac{d\omega_r}{dt} = \frac{P^2 \psi_m}{4J}(i_{as} \cos \theta_r + i_{bs} \sin \theta_r) - \frac{B_m}{J} \omega_r - \frac{P}{2J} T_L,$$

$$\frac{d\theta_r}{dt} = \omega_r.$$

(6.5)

For two-phase permanent-magnet synchronous motors (assuming the sinusoidal winding distributions and the sinusoidal *mmf* waveforms), the electromagnetic torque was found as given by (6.4). To guarantee the balanced operation, one strives to ensure

$$i_{as} = \sqrt{2}i_M \cos \theta_r \quad \text{and} \quad i_{bs} = \sqrt{2}i_M \sin \theta_r.$$

Hence, the phase voltages to be applied are $u_{as} = \sqrt{2}u_M \cos \theta_r$ and $u_{bs} = \sqrt{2}u_M \sin \theta_r$.

Using this balanced current set, we maximize the electromagnetic torque obtaining

$$T_e = \frac{P\psi_m}{2} \sqrt{2}i_M \left(\cos^2 \theta_r + \sin^2 \theta_r\right) = \frac{P\psi_m}{\sqrt{2}} i_M.$$

The mechanical angular displacement (rotor displacement) θ_{rm} is related to the electrical angular displacement as $\theta_{rm} = 2\theta_r/P$. The equations can be easily refined using θ_r or θ_{rm}. One prefer θ_{rm} because two-phase permanent-magnet synchronous motors are commonly utilized as stepper motors which usually operate in the open-loop configuration. The stepper motor is rotated step-by-step by properly *energizing* the windings by supplying u_{as} and u_{bs}. The full-, half-, and mini-step operations can be achieved. The stepper motors are designed with a high P. For example, if $P = 50$, one achieves 3.6° rotation using the full-step operation.

Permanent-magnet synchronous motors with high P develop high electromagnetic torque, while the mechanical angular velocity is relatively low because $\omega_{rm} = 2\omega_r/P$. These motors are effectively utilized as direct drives and servos. The direct connection of motors without matching mechanical coupling allows one to achieve a high level of efficiency, reliability, and performance.

Permanent-magnet synchronous motors are controlled to ensure the desired dynamics, steady-state operation, stability, precision, disturbance attenuation, etc. For stepper motors, which are illustrated in Figures 6.2d and 6.4, one *energizes* the stator windings supplying an adequate sequence of u_{as} and u_{bs}, and rotor rotates in the *counterclockwise* or *clockwise* direction due to the T_e developed. By supplying u_{as} and u_{bs}, one achieves the angular incremental rotor displacement equal to full or half step. The rotor repositioning (instead of using angular displacement or velocity, the number of steps per second are also frequently used and specified) is regulated by changing the frequency of the voltages supplied to the phase windings. Due to the possibility to operate stepper motors in the open-loop mode sequentially supplying u_{as} and u_{bs}, some warnings should be stated. The stepper motor can *miss* the step (or steps) if (1) the instantaneous torque $T_{e\,\text{inst}}$ is not sufficient; (2) $T_{e\,\text{inst}} < T_L$; (3) u_{as} and u_{bs} are supplied at high frequency with respect to the dynamic features defined by L_{ss}, J, and other parameters; etc. Furthermore, the stepper motor can *pass* the step (or steps) if (1) $T_e \gg T_L$; (2) high kinetic energy is stored in the moving rotor with the attached kinematics due to high J; etc. Other factors contribute to *missing* or *passing* steps such as varying J, varying and potentially bidirectional T_L, disturbances, parameter variations, etc. Therefore, open-loop electromechanical systems with stepper motors are utilized if $T_L \approx$ constant and $J \approx$ constant deriving the phase voltage switching frequency by coherently examining the system dynamics, disturbances, parameter variations, etc.

For stepper motors, the electrical angular velocity and displacement are found using the number of rotor tooth RT, e.g., $\omega_r = RT\omega_{rm}$ and $\theta_r = RT\theta_{rm}$. The flux linkages are the functions of the number of the rotor tooth and displacement. Let

$$\psi_{asm} = \psi_m \cos\left(RT\theta_{rm}\right) \quad \text{and} \quad \psi_{bsm} = \psi_m \sin\left(RT\theta_{rm}\right).$$

We have

$$\psi_{as} = L_{ss}i_{as} + \psi_m \cos\left(RT\theta_{rm}\right) \quad \text{and} \quad \psi_{bs} = L_{ss}i_{bs} + \psi_m \sin\left(RT\theta_{rm}\right)$$

Equation (6.1) yields

$$u_{as} = r_s i_{as} + \frac{d[L_{ss}i_{as} + \psi_m \cos\left(RT\theta_{rm}\right)]}{dt} = r_s i_{as} + L_{ss}\frac{di_{as}}{dt} - RT\psi_m \omega_{rm}\sin\left(RT\theta_{rm}\right),$$

$$u_{bs} = r_s i_{bs} + \frac{d[L_{ss}i_{bs} + \psi_m \sin\left(RT\theta_{rm}\right)]}{dt} = r_s i_{bs} + L_{ss}\frac{di_{bs}}{dt} + RT\psi_m \omega_{rm}\cos\left(RT\theta_{rm}\right).$$

Therefore,

$$\frac{di_{as}}{dt} = -\frac{r_s}{L_{ss}}i_{as} + \frac{RT\psi_m}{L_{ss}}\omega_{rm}\sin\left(RT\theta_{rm}\right) + \frac{1}{L_{ss}}u_{as},$$

$$\frac{di_{bs}}{dt} = -\frac{r_s}{L_{ss}}i_{bs} - \frac{RT\psi_m}{L_{ss}}\omega_{rm}\cos\left(RT\theta_{rm}\right) + \frac{1}{L_{ss}}u_{bs}. \tag{6.6}$$

Using coenergy $W_c = \dfrac{1}{2}\left(L_{ss}i_{as}^2 + L_{ss}i_{bs}^2\right) + \psi_m i_{as}\cos\left(RT\theta_{rm}\right) + \psi_m i_{bs}\sin\left(RT\theta_{rm}\right) + W_{PM}$, the expression for T_e is found to be

$$T_e = \frac{\partial W_c}{\partial \theta_{rm}} = RT\psi_m\left[-i_{as}\sin\left(RT\theta_{rm}\right) + i_{bs}\cos\left(RT\theta_{rm}\right)\right]. \tag{6.7}$$

From Newton's second law (6.3), applying (6.7), the behavior of the rotor angular velocity and displacement are described by

$$\frac{d\omega_{rm}}{dt} = \frac{RT\psi_m}{J}\left[-i_{as}\sin\left(RT\theta_{rm}\right) + i_{bs}\cos\left(RT\theta_{rm}\right)\right] - \frac{B_m}{J}\omega_{rm} - \frac{1}{J}T_L,$$

$$\frac{d\theta_{rm}}{dt} = \omega_{rm}. \tag{6.8}$$

From (6.6) and (6.8), one has

$$\frac{di_{as}}{dt} = -\frac{r_s}{L_{ss}}i_{as} + \frac{RT\psi_m}{L_{ss}}\omega_{rm}\sin\left(RT\theta_{rm}\right) + \frac{1}{L_{ss}}u_{as},$$

$$\frac{di_{bs}}{dt} = -\frac{r_s}{L_{ss}}i_{bs} - \frac{RT\psi_m}{L_{ss}}\omega_{rm}\cos\left(RT\theta_{rm}\right) + \frac{1}{L_{ss}}u_{bs},$$

$$\frac{d\omega_{rm}}{dt} = \frac{RT\psi_m}{J}\left[-i_{as}\sin\left(RT\theta_{rm}\right) + i_{bs}\cos\left(RT\theta_{rm}\right)\right] - \frac{B_m}{J}\omega_{rm} - \frac{1}{J}T_L,$$

$$\frac{d\theta_{rm}}{dt} = \omega_{rm}. \tag{6.9}$$

Using (6.9), an s-domain diagram is developed and illustrated in Figure 6.5.

The analysis of (6.7) leads one to the expressions for a balanced two-phase current sinusoidal set:

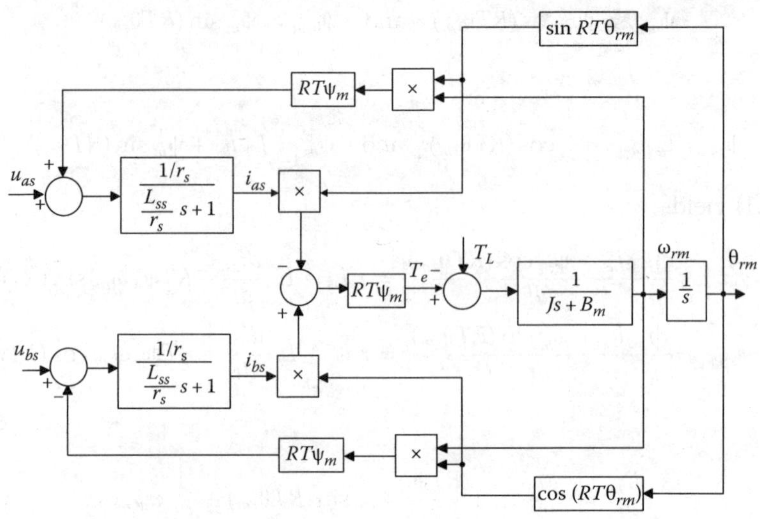

FIGURE 6.5
s-Domain diagram of permanent-magnet stepper motors.

$$i_{as} = -\sqrt{2}i_M \sin(RT\theta_{rm}) \quad \text{and} \quad i_{bs} = \sqrt{2}i_M \cos(RT\theta_{rm})$$

which results in $T_e = \sqrt{2}RT\psi_m i_M$.

The phase voltages u_{as} and u_{bs} should be supplied as functions of the rotor angular displacement

$$u_{as} = -\sqrt{2}u_M \sin(RT\theta_{rm}) \quad \text{and} \quad u_{bs} = \sqrt{2}u_M \cos(RT\theta_{rm})$$

to maximize the electromagnetic torque. However, to implement the displacement-dependent current or voltage sets, one needs to measure (or observe) the angular displacement using sensors or observers (observers can be designed and implemented using the induced *emf* or phase current measurements). The Hall sensors are commonly used to measure the rotor angular displacement in permanent-magnet synchronous motors. To eliminate the need for sensors or observers (sensorless control), stepper motors can operate in the open-loop configuration by supplying the phase voltages with the allowable frequency ω_a. For example, without measuring θ_{rm}, one, for example, may supply the pulses

$$u_{as} = -\sqrt{2}u_M \, \text{sgn}\left[\sin(\omega_a t)\right] \quad \text{and} \quad u_{as} = \sqrt{2}u_M \, \text{sgn}\left[\sin\left(\omega_a t - \frac{1}{2}\pi\right)\right].$$

Example 6.1: Stepper Motors in the Quadrature and Direct Variables

Having examined two-phase permanent-magnet synchronous motors, including stepper motors, in the *machine* variable, we study stepper motors in the quadrature and direct (qd) quantities. The rotor and synchronous reference frames are among the most frequently applied, and the resulting equations of motion are the same because $\omega_e = \omega_r$ [2–5]. In the *machine* variables, Kirchhoff's voltage law results in two nonlinear differential equations (6.6). We apply the *direct* Park formation [2–5]. In the synchronous reference frame one has

$$\begin{bmatrix} u_{qs}^e \\ u_{ds}^e \end{bmatrix} = \begin{bmatrix} -\sin(RT\theta_{rm}) & \cos(RT\theta_{rm}) \\ \cos(RT\theta_{rm}) & \sin(RT\theta_{rm}) \end{bmatrix} \begin{bmatrix} u_{as} \\ u_{bs} \end{bmatrix}, \quad \begin{bmatrix} i_{qs}^e \\ i_{ds}^e \end{bmatrix} = \begin{bmatrix} -\sin(RT\theta_{rm}) & \cos(RT\theta_{rm}) \\ \cos(RT\theta_{rm}) & \sin(RT\theta_{rm}) \end{bmatrix} \begin{bmatrix} i_{as} \\ i_{bs} \end{bmatrix}.$$

The following differential equations in the qd quantities result

$$u_{qs}^e = r_s i_{qs}^e + L_{ss}\frac{di_{qs}^e}{dt} + RT\psi_m\omega_{rm} + RTL_{ss}i_{ds}^e\omega_{rm}, \quad u_{ds}^e = r_s i_{ds}^e + L_{ss}\frac{di_{ds}^e}{dt} - RTL_{ss}i_{qs}^e\omega_{rm}.$$

Hence,

$$\frac{di_{qs}^e}{dt} = -\frac{r_s}{L_{ss}}i_{qs}^e - \frac{RT\psi_m}{L_{ss}}\omega_{rm} - RTi_{ds}^e\omega_{rm} + \frac{1}{L_{ss}}u_{qs}^e,$$

$$\frac{di_{ds}^e}{dt} = -\frac{r_s}{L_{ss}}i_{ds}^e + RTi_{qs}^e\omega_{rm} + \frac{1}{L_{ss}}u_{ds}^e. \tag{6.10}$$

From the derived T_e, as given by (6.10), using the *inverse* Park transformation $\begin{bmatrix} i_{as} \\ i_{bs} \end{bmatrix} = \begin{bmatrix} -\sin(RT\theta_{rm}) & \cos(RT\theta_{rm}) \\ \cos(RT\theta_{rm}) & \sin(RT\theta_{rm}) \end{bmatrix} \begin{bmatrix} i_{qs}^e \\ i_{ds}^e \end{bmatrix}$, we have

$$T_e = RT\psi_m i_{qs}^e.$$

From Newton's second law of motions (6.8), one finds

$$\frac{d\omega_{rm}}{dt} = \frac{RT\psi_m}{J}i_{qs}^e - \frac{B_m}{J}\omega_{rm} - \frac{1}{J}T_L,$$

$$\frac{d\theta_{rm}}{dt} = \omega_{rm}. \tag{6.11}$$

Differential equations (6.10) and (6.11) yield the resulting model. The phase voltages supplied to the *as* and *bs* windings must be found. Recalling that $i_{as} = -\sqrt{2}i_M\sin(RT\theta_{rm})$, $i_{bs} = \sqrt{2}i_M\cos(RT\theta_{rm})$, and $u_{as} = -\sqrt{2}u_M\sin(RT\theta_{rm})$, $u_{bs} = \sqrt{2}u_M\cos(RT\theta_{rm})$, we apply the Park transformations

$$\begin{bmatrix} i_{qs}^e \\ i_{ds}^e \end{bmatrix} = \begin{bmatrix} -\sin(RT\theta_{rm}) & \cos(RT\theta_{rm}) \\ \cos(RT\theta_{rm}) & \sin(RT\theta_{rm}) \end{bmatrix} \begin{bmatrix} i_{as} \\ i_{bs} \end{bmatrix}$$

$$\text{and} \quad \begin{bmatrix} u_{qs}^e \\ u_{ds}^e \end{bmatrix} = \begin{bmatrix} -\sin(RT\theta_{rm}) & \cos(RT\theta_{rm}) \\ \cos(RT\theta_{rm}) & \sin(RT\theta_{rm}) \end{bmatrix} \begin{bmatrix} u_{as} \\ u_{bs} \end{bmatrix}.$$

From $i_{qs}^e = -i_{as}\sin(RT\theta_{rm}) + i_{bs}\cos(RT\theta_{rm})$ and $i_{ds}^e = i_{as}\cos(RT\theta_{rm}) + i_{bs}\sin(RT\theta_{rm})$, one finds

$$i_{qs}^e = \sqrt{2}i_M\sin^2(RT\theta_{rm}) + \sqrt{2}i_M\cos^2(RT\theta_{rm}) = \sqrt{2}i_M$$

and $i_{ds}^e = -\sqrt{2}i_M\sin(RT\theta_{rm})\cos(RT\theta_{rm}) + \sqrt{2}i_M\sin(RT\theta_{rm})\cos(RT\theta_{rm}) = 0$.

Hence, the current components are $i_{qs}^e = \sqrt{2}i_M$ and $i_{ds}^e = 0$.

Similarly, the qd voltage components are $u_{qs}^e = \sqrt{2}u_M$ and $u_{ds}^e = 0$, which are DC voltage components.

To the best of author's knowledge, in industrial applications, stepper motors have not been controlled using the qd voltage or current components because the phase voltages u_{as} and u_{bs} must be supplied. The derived u_{qs}^e and u_{ds}^e illustrate a very limited practicality and obscurity of the application of the qd quantities because the most advanced DSPs must be utilized to perform the Park transformation in real time. In addition, the angular displacement must be measured, while stepper motors usually utilized as an open-loop system. ∎

To control the stepper motor output (angular velocity or displacement), the power amplifiers *energize* the *as* and *bs* windings ensuring the full step or other operations. As an example, the Motorola monolithic MC3479 Stepper Motor Driver (6 V and 0.35 A) can be used [4]. This driver is designed to drive a two-phase stepper motors bidirectionally. The pin connection for the plastic 648C case, representative block diagram, circuitry, and timing/output diagram are reported in Figure 6.6. Other solutions are available. For example, the L9942 integrated stepper motor driver (STMicroelectronics, www.st.com) may ensure the full-, half-, and mini-stepping modes. It should be emphasized again that not only mini- and microstepping, but even full- and half-stepping may be ensured only in a very limited operating envelope if there are no disturbances and loads.

This MC3479 driver is designed to drive a stepper minimotors in various applications such as positioning tables, disk drives, small robots, etc. As illustrated in Figure 6.6, the H-bridge topology power stage supplies the phase voltages u_{as} and u_{bs} to the motor windings (only one coil with terminals L1 and L2 is shown in Figure 6.6). The applied voltage polarity depends on which transistor (Q_H or Q_L) is *on*, and these transistors are driven by the signal-level voltages from the decoding circuitry. The maximum sink current is a function of the resistor between pin 6 and ground. When the outputs are in a high

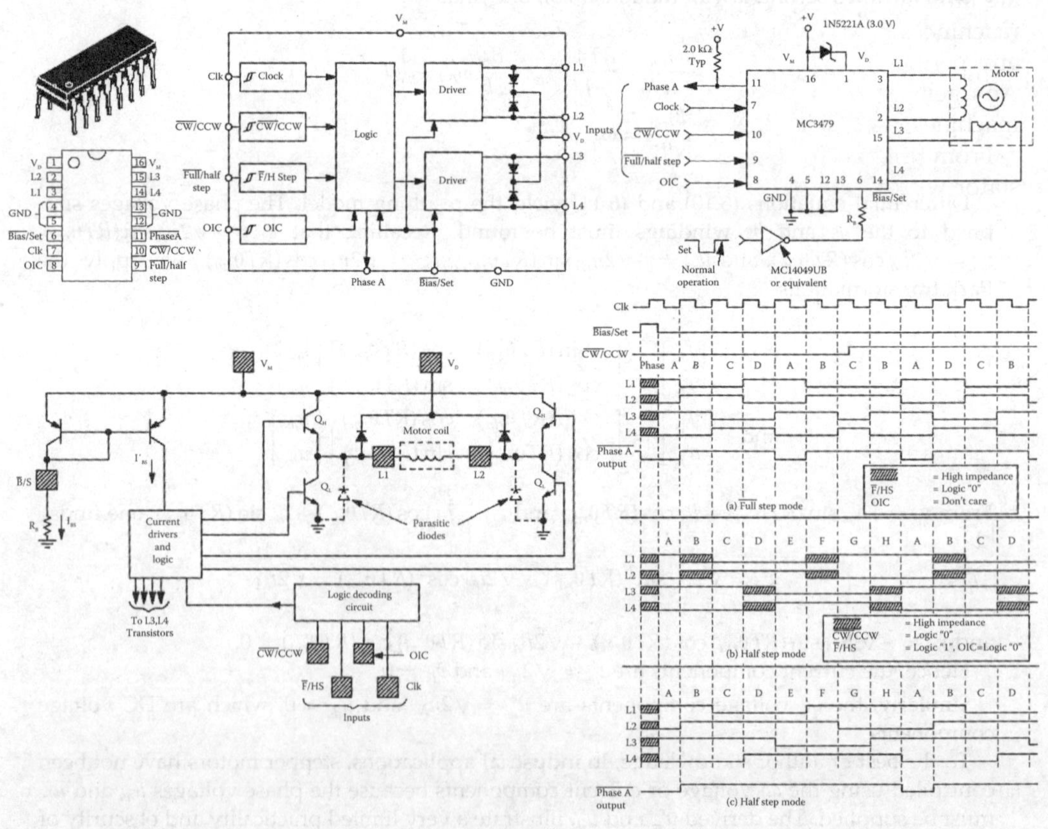

FIGURE 6.6

Motorola monolithic MC3479 Stepper Motor Driver. (From Lyshevski, S.E., *Electromechanical Systems, Electric Machines, and Applied Mechatronics*, CRC Press, Boca Raton, FL, 1999. Copyright of Motorola. With permission.)

impedance state, both transistors (Q_H or Q_L) are *off*. The pin V_D provides a current path for the winding (coil) current during transients (switchings) in order to attenuate the *back emf* (voltage) spikes. Pin V_D is normally connected to V_M (pin 16) through a diode, resistor, or directly. The peaks instantaneous voltage at the outputs must not exceed the value V_M which is 6 V. The parasitic diodes across Q_L of each output results in a circuit path for the switched current. When the input is at a Logic "0" (less than 0.8 V), the outputs correspond to a full step operation with each clock cycle. The direction depends on the CW/CCW input. There are four switching phases for each cycle of the sequencing logic. Phase voltages are supplied, and currents i_{as} and i_{bs} flow in the motor windings. For a Logic "1" (more than 2 V), the outputs change a half step during each clock cycle. Eight switching phases result for each complete cycle of the sequencing logic. The output sequences and timing diagrams are reported in Figure 6.6. A complete description, application notes, and important details are reported by the Motorola. We very briefly covered MC3479 Stepper Motor Driver, and other specialized ICs and motor controllers/drivers exist.

6.2.2 Radial Topology Three-Phase Permanent-Magnet Synchronous Machines

Three-phase two-pole permanent-magnet synchronous machines (motors and generators) are illustrated in Figures 6.7a and 6.8a. The images of the permanent-magnet synchronous machines (motor and generator) are also reported in Figures 6.7b and 6.8b. One recalls that any permanent-magnet synchronous machine can be used as a motor or as a generator. Any generator should be rotated by a prime mover, and the test-beds with two electric machines are documented in Figure 6.8b.

From Kirchhoff's second law, one obtains three differential equations for the *as*, *bs*, and *cs* stator windings as

$$u_{as} = r_s i_{as} + \frac{d\psi_{as}}{dt}, \quad u_{bs} = r_s i_{bs} + \frac{d\psi_{bs}}{dt}, \quad u_{cs} = r_s i_{cs} + \frac{d\psi_{cs}}{dt}. \tag{6.12}$$

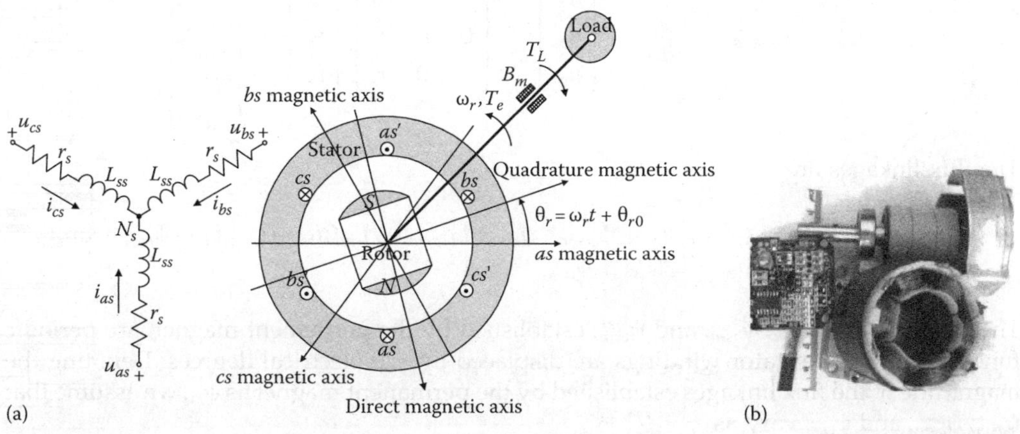

FIGURE 6.7
(a) Two-pole permanent-magnet synchronous motor; (b) image of the 23 NEMA size permanent-magnet synchronous machine: rotor, stator, and encoder with ICs.

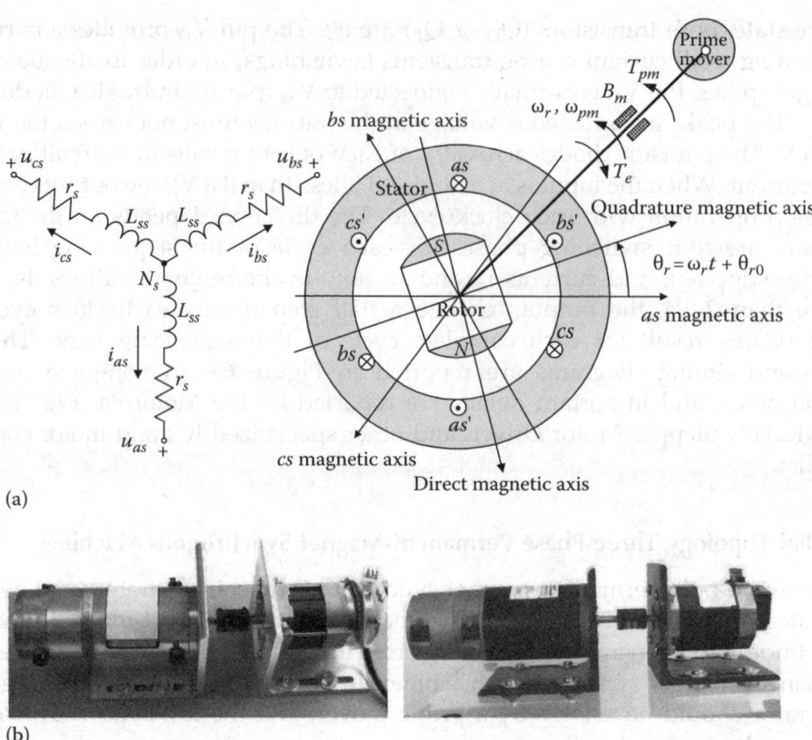

FIGURE 6.8
(a) Three-phase wye-connected synchronous generator; (b) images of the test-beds: permanent-magnet synchronous machine (generator on the right) driven by permanent-magnet DC motor (prime mover on the left).

Using vector notations, equations (6.12), yields

$$
\mathbf{u}_{abcs} = \mathbf{r}_s \mathbf{i}_{abcs} + \frac{d\psi_{abcs}}{dt}, \quad
\begin{bmatrix} u_{as} \\ u_{bs} \\ u_{cs} \end{bmatrix} =
\begin{bmatrix} r_s & 0 & 0 \\ 0 & r_s & 0 \\ 0 & 0 & r_s \end{bmatrix}
\begin{bmatrix} i_{as} \\ i_{bs} \\ i_{cs} \end{bmatrix} +
\begin{bmatrix} \dfrac{d\psi_{as}}{dt} \\[2ex] \dfrac{d\psi_{bs}}{dt} \\[2ex] \dfrac{d\psi_{cs}}{dt} \end{bmatrix}.
$$

The flux linkages are

$$
\psi_{as} = L_{asas} i_{as} + L_{asbs} i_{bs} + L_{ascs} i_{cs} + \psi_{asm}, \quad
\psi_{bs} = L_{bsas} i_{as} + L_{bsbs} i_{bs} + L_{bscs} i_{cs} + \psi_{bsm}, \quad \text{and}
$$
$$
\psi_{cs} = L_{csas} i_{as} + L_{csbs} i_{bs} + L_{cscs} i_{cs} + \psi_{csm}
$$

The flux linkages ψ_{asm}, ψ_{bsm}, and ψ_{csm}, established by the permanent magnet, are periodic functions of θ_r. The stator windings are displaced by 120 electrical degrees. Denoting the magnitude of the flux linkages established by the permanent magnet as ψ_m, we assume that ψ_{asm}, ψ_{bsm}, and ψ_{csm} vary as

$$
\psi_{asm} = \psi_m \sin \theta_r, \quad \psi_{bsm} = \psi_m \sin\left(\theta_r - \frac{2}{3}\pi\right), \quad \text{and} \quad \psi_{csm} = \psi_m \sin\left(\theta_r + \frac{2}{3}\pi\right).
$$

Self- and mutual-inductances for three-phase permanent-magnet synchronous machines can be derived. In particular, the equations for the magnetizing qd inductances are $L_{mq} = \dfrac{N_s^2}{\mathfrak{R}_{mq}}$ and $L_{md} = \dfrac{N_s^2}{\mathfrak{R}_{md}}$. The qd magnetizing reluctances can be different, and $\mathfrak{R}_{mq} > \mathfrak{R}_{md}$. Hence, we have $L_{mq} < L_{md}$.

The minimum value of L_{asas} occurs periodically at $\theta_r = 0, \pi, 2\pi, \ldots$, while the maximum value of L_{asas} occurs at $\theta_r = \dfrac{1}{2}\pi, \dfrac{3}{2}\pi, \dfrac{5}{2}\pi, \ldots$. One concludes that $L_{asas}(\theta_r)$, which is $L_{ls} + L_{mq} \le L_{asas} \le L_{ls} + L_{md}$, is a periodic function of θ_r. Hence, $L_{asas}(\theta_r)$ varies as a sine function and has a constant component. We let

$$L_{asas} = L_{ls} + \bar{L}_m - L_{\Delta m}\cos 2\theta_r,$$

where $\bar{L}_m$ is the average value of the magnetizing inductance, and $L_{\Delta m}$ is half the amplitude of the sinusoidal variation of the magnetizing inductance.

The relationships between L_{mq}, L_{md} and $\bar{L}_m$, $L_{\Delta m}$ are $L_{mq} = \dfrac{3}{2}(\bar{L}_m - L_{\Delta m})$ and $L_{md} = \dfrac{3}{2}(\bar{L}_m + L_{\Delta m})$. Therefore, $\bar{L}_m = \dfrac{1}{3}(L_{mq} + L_{md})$ and $L_{\Delta m} = \dfrac{1}{3}(L_{md} - L_{mq})$.

Using the expressions for L_{mq} and L_{md}, we have $\bar{L}_m = \dfrac{1}{3}\left(\dfrac{N_s^2}{\mathfrak{R}_{mq}} + \dfrac{N_s^2}{\mathfrak{R}_{md}}\right)$ and $L_{\Delta m} = \dfrac{1}{3}\left(\dfrac{N_s^2}{\mathfrak{R}_{md}} - \dfrac{N_s^2}{\mathfrak{R}_{mq}}\right)$. The following equations for the flux linkages result:

$$
\boldsymbol{\psi}_{abcs} = \mathbf{L}_s \mathbf{i}_{abcs} + \boldsymbol{\psi}_m
$$

$$
= \begin{bmatrix}
L_{ls} + \bar{L}_m - L_{\Delta m}\cos 2\theta_r & -\tfrac{1}{2}\bar{L}_m - L_{\Delta m}\cos 2\left(\theta_r - \tfrac{1}{3}\pi\right) & -\tfrac{1}{2}\bar{L}_m - L_{\Delta m}\cos 2\left(\theta_r + \tfrac{1}{3}\pi\right) \\
-\tfrac{1}{2}\bar{L}_m - L_{\Delta m}\cos 2\left(\theta_r - \tfrac{1}{3}\pi\right) & L_{ls} + \bar{L}_m - L_{\Delta m}\cos 2\left(\theta_r - \tfrac{2}{3}\pi\right) & -\tfrac{1}{2}\bar{L}_m - L_{\Delta m}\cos 2\theta_r \\
-\tfrac{1}{2}\bar{L}_m - L_{\Delta m}\cos 2\left(\theta_r + \tfrac{1}{3}\pi\right) & -\tfrac{1}{2}\bar{L}_m - L_{\Delta m}\cos 2\theta_r & L_{ls} + \bar{L}_m - L_{\Delta m}\cos 2\left(\theta_r + \tfrac{2}{3}\pi\right)
\end{bmatrix}
\begin{bmatrix} i_{as} \\ i_{bs} \\ i_{cs} \end{bmatrix}
+ \psi_m \begin{bmatrix} \sin\theta_r \\ \sin\left(\theta_r - \tfrac{2}{3}\pi\right) \\ \sin\left(\theta_r + \tfrac{2}{3}\pi\right) \end{bmatrix}
$$

Permanent-magnet synchronous motion devices are round-rotor electric machines. Hence, the magnetic paths in the qd magnetic axes are virtually identical yielding $\mathfrak{R}_{mq} = \mathfrak{R}_{md}$.

Thus, $\bar{L}_m = \dfrac{2N_s^2}{3\mathfrak{R}_{mq}} = \dfrac{2N_s^2}{3\mathfrak{R}_{md}}$, $L_{\Delta m} = 0$, and $\bar{L}_m = L_{ss} - L_{ls}$.

From the derived inductance mapping $\mathbf{L}_s(\theta_r)$ one obtains the inductance matrix as

$$
\mathbf{L}_s = \begin{bmatrix}
L_{ls} + \bar{L}_m & -\tfrac{1}{2}\bar{L}_m & -\tfrac{1}{2}\bar{L}_m \\
-\tfrac{1}{2}\bar{L}_m & L_{ls} + \bar{L}_m & -\tfrac{1}{2}\bar{L}_m \\
-\tfrac{1}{2}\bar{L}_m & -\tfrac{1}{2}\bar{L}_m & L_{ls} + \bar{L}_m
\end{bmatrix}.
$$

Hence, the expressions for the flux linkages are simplified to

$$\psi_{as} = (L_{ls} + \bar{L}_m)i_{as} - \dfrac{1}{2}\bar{L}_m i_{bs} - \dfrac{1}{2}\bar{L}_m i_{cs} + \psi_m \sin\theta_r,$$

$$\psi_{bs} = -\dfrac{1}{2}\bar{L}_m i_{as} + (L_{ls} + \bar{L}_m)i_{bs} - \dfrac{1}{2}\bar{L}_m i_{cs} + \psi_m \sin\left(\theta_r - \dfrac{2}{3}\pi\right), \tag{6.13}$$

$$\psi_{cs} = -\dfrac{1}{2}\bar{L}_m i_{as} - \dfrac{1}{2}\bar{L}_m i_{bs} + (L_{ls} + \bar{L}_m)i_{cs} + \psi_m \sin\left(\theta_r + \dfrac{2}{3}\pi\right),$$

or in matrix form

$$\boldsymbol{\psi}_{abcs} = \mathbf{L}_s \mathbf{i}_{abcs} + \boldsymbol{\psi}_m = \begin{bmatrix} L_{ls} + \bar{L}_m & -\frac{1}{2}\bar{L}_m & -\frac{1}{2}\bar{L}_m \\ -\frac{1}{2}\bar{L}_m & L_{ls} + \bar{L}_m & -\frac{1}{2}\bar{L}_m \\ -\frac{1}{2}\bar{L}_m & -\frac{1}{2}\bar{L}_m & L_{ls} + \bar{L}_m \end{bmatrix} \begin{bmatrix} i_{as} \\ i_{bs} \\ i_{cs} \end{bmatrix} + \psi_m \begin{bmatrix} \sin\theta_r \\ \sin\left(\theta_r - \frac{2}{3}\pi\right) \\ \sin\left(\theta_r + \frac{2}{3}\pi\right) \end{bmatrix}.$$

Using (6.12) and (6.13), we have

$$\mathbf{u}_{abcs} = \mathbf{r}_s \mathbf{i}_{abcs} + \frac{d\boldsymbol{\psi}_{abcs}}{dt} = \mathbf{r}_s \mathbf{i}_{abcs} + \mathbf{L}_s \frac{d\mathbf{i}_{abcs}}{dt} + \frac{d\boldsymbol{\psi}_m}{dt}, \quad \frac{d\boldsymbol{\psi}_m}{dt} = \psi_m \begin{bmatrix} \cos\theta_r \omega_r \\ \cos\left(\theta_r - \frac{2}{3}\pi\right)\omega_r \\ \cos\left(\theta_r + \frac{2}{3}\pi\right)\omega_r \end{bmatrix}.$$

Cauchy's form differential equations can be found by using $\mathbf{L}_s^{-1}$. In particular,

$$\frac{d\mathbf{i}_{abcs}}{dt} = -\mathbf{L}_s^{-1}\mathbf{r}_s\mathbf{i}_{abcs} - \mathbf{L}_s^{-1}\frac{d\boldsymbol{\psi}_m}{dt} + \mathbf{L}_s^{-1}\mathbf{u}_{abcs}.$$

The circuitry-electromagnetic dynamics is given as

$$\begin{aligned}
\frac{di_{as}}{dt} =& -\frac{r_s(2L_{ss} - \bar{L}_m)}{2L_{ss}^2 - L_{ss}\bar{L}_m - \bar{L}_m^2}i_{as} - \frac{r_s\bar{L}_m}{2L_{ss}^2 - L_{ss}\bar{L}_m - \bar{L}_m^2}i_{bs} - \frac{r_s\bar{L}_m}{2L_{ss}^2 - L_{ss}\bar{L}_m - \bar{L}_m^2}i_{cs} \\
& -\frac{\psi_m(2L_{ss} - \bar{L}_m)}{2L_{ss}^2 - L_{ss}\bar{L}_m - \bar{L}_m^2}\omega_r\cos\theta_r - \frac{\psi_m\bar{L}_m}{2L_{ss}^2 - L_{ss}\bar{L}_m - \bar{L}_m^2}\omega_r\cos\left(\theta_r - \frac{2}{3}\pi\right) - \frac{\psi_m\bar{L}_m}{2L_{ss}^2 - L_{ss}\bar{L}_m - \bar{L}_m^2}\omega_r\cos\left(\theta_r + \frac{2}{3}\pi\right) \\
& +\frac{2L_{ss} - \bar{L}_m}{2L_{ss}^2 - L_{ss}\bar{L}_m - \bar{L}_m^2}u_{as} + \frac{\bar{L}_m}{2L_{ss}^2 - L_{ss}\bar{L}_m - \bar{L}_m^2}u_{bs} + \frac{\bar{L}_m}{2L_{ss}^2 - L_{ss}\bar{L}_m - \bar{L}_m^2}u_{cs}, \\
\frac{di_{bs}}{dt} =& -\frac{r_s\bar{L}_m}{2L_{ss}^2 - L_{ss}\bar{L}_m - \bar{L}_m^2}i_{as} - \frac{r_s(2L_{ss} - \bar{L}_m)}{2L_{ss}^2 - L_{ss}\bar{L}_m - \bar{L}_m^2}i_{bs} - \frac{r_s\bar{L}_m}{2L_{ss}^2 - L_{ss}\bar{L}_m - \bar{L}_m^2}i_{cs} \\
& -\frac{\psi_m\bar{L}_m}{2L_{ss}^2 - L_{ss}\bar{L}_m - \bar{L}_m^2}\omega_r\cos\theta_r - \frac{\psi_m(2L_{ss} - \bar{L}_m)}{2L_{ss}^2 - L_{ss}\bar{L}_m - \bar{L}_m^2}\omega_r\cos\left(\theta_r - \frac{2}{3}\pi\right) - \frac{\psi_m\bar{L}_m}{2L_{ss}^2 - L_{ss}\bar{L}_m - \bar{L}_m^2}\omega_r\cos\left(\theta_r + \frac{2}{3}\pi\right) \\
& +\frac{\bar{L}_m}{2L_{ss}^2 - L_{ss}\bar{L}_m - \bar{L}_m^2}u_{as} + \frac{2L_{ss} - \bar{L}_m}{2L_{ss}^2 - L_{ss}\bar{L}_m - \bar{L}_m^2}u_{bs} + \frac{\bar{L}_m}{2L_{ss}^2 - L_{ss}\bar{L}_m - \bar{L}_m^2}u_{cs}, \\
\frac{di_{cs}}{dt} =& -\frac{r_s\bar{L}_m}{2L_{ss}^2 - L_{ss}\bar{L}_m - \bar{L}_m^2}i_{as} - \frac{r_s\bar{L}_m}{2L_{ss}^2 - L_{ss}\bar{L}_m - \bar{L}_m^2}i_{bs} - \frac{r_s(2L_{ss} - \bar{L}_m)}{2L_{ss}^2 - L_{ss}\bar{L}_m - \bar{L}_m^2}i_{cs} \\
& -\frac{\psi_m\bar{L}_m}{2L_{ss}^2 - L_{ss}\bar{L}_m - \bar{L}_m^2}\omega_r\cos\theta_r - \frac{\psi_m\bar{L}_m}{2L_{ss}^2 - L_{ss}\bar{L}_m - \bar{L}_m^2}\omega_r\cos\left(\theta_r - \frac{2}{3}\pi\right) - \frac{\psi_m(2L_{ss} - \bar{L}_m)}{2L_{ss}^2 - L_{ss}\bar{L}_m - \bar{L}_m^2}\omega_r\cos\left(\theta_r + \frac{2}{3}\pi\right) \\
& +\frac{\bar{L}_m}{2L_{ss}^2 - L_{ss}\bar{L}_m - \bar{L}_m^2}u_{as} + \frac{\bar{L}_m}{2L_{ss}^2 - L_{ss}\bar{L}_m - \bar{L}_m^2}u_{bs} + \frac{2L_{ss} - \bar{L}_m}{2L_{ss}^2 - L_{ss}\bar{L}_m - \bar{L}_m^2}u_{cs}.
\end{aligned} \tag{6.14}$$

The transient behavior of the mechanical system must be used. Newton's second law yields a set of two differential equations

$$\frac{d\omega_{rm}}{dt} = \frac{1}{J}(T_e - B_m\omega_{rm} - T_L), \quad \frac{d\theta_{rm}}{dt} = \omega_{rm}.$$

The expression for the electromagnetic torque is found using the coenergy

$$W_c = \frac{1}{2}[i_{as}\ i_{bs}\ i_{cs}]\mathbf{L}_s\begin{bmatrix} i_{as} \\ i_{bs} \\ i_{cs} \end{bmatrix} + [i_{as}\ i_{bs}\ i_{cs}]\begin{bmatrix} \psi_m \sin\theta_r \\ \psi_m \sin\left(\theta_r - \frac{2}{3}\pi\right) \\ \psi_m \sin\left(\theta_r + \frac{2}{3}\pi\right) \end{bmatrix} + W_{PM},$$

where W_{PM} is the energy stored in the permanent magnet.

For round-rotor synchronous machines, $\mathbf{L}_s$ and W_{PM} are not functions of θ_r. One obtains the following formula for the electromagnetic torque for P-pole three-phase permanent-magnet synchronous motors

$$T_e = \frac{P}{2}\frac{\partial W_c}{\partial\theta_r} = \frac{P\psi_m}{2}\left[i_{as}\cos\theta_r + i_{bs}\cos\left(\theta_r - \frac{2}{3}\pi\right) + i_{cs}\cos\left(\theta_r + \frac{2}{3}\pi\right)\right]. \tag{6.15}$$

Therefore,

$$\frac{d\omega_{rm}}{dt} = \frac{P\psi_m}{2J}\left[i_{as}\cos\theta_r + i_{bs}\cos\left(\theta_r - \frac{2}{3}\pi\right) + i_{cs}\cos\left(\theta_r + \frac{2}{3}\pi\right)\right] - \frac{B_m}{J}\omega_{rm} - \frac{1}{J}T_L,\ \frac{d\theta_{rm}}{dt} = \omega_{rm}.$$

Using the electrical angular velocity ω_r and displacement θ_r, related to the mechanical angular velocity and displacement as $\omega_{rm} = \frac{2}{P}\omega_r$ and $\theta_{rm} = \frac{2}{P}\theta_r$, the differential equations to describe the *torsional–mechanical* transient dynamics are

$$\frac{d\omega_r}{dt} = \frac{P^2\psi_m}{4J}\left[i_{as}\cos\theta_r + i_{bs}\cos\left(\theta_r - \frac{2}{3}\pi\right) + i_{cs}\cos\left(\theta_r + \frac{2}{3}\pi\right)\right] - \frac{B_m}{J}\omega_r - \frac{P}{2J}T_L,$$
$$\frac{d\theta_r}{dt} = \omega_r. \tag{6.16}$$

A nonlinear mathematical model of permanent-magnet synchronous motors in Cauchy's form is given by a system of five highly nonlinear differential equations (6.14) and (6.16).

To control the angular velocity, one regulates the currents fed or voltages supplied to the stator *abc* windings. Neglecting the viscous friction coefficient, the analysis of Newton's second law $J\frac{d\omega_{rm}}{dt} = T_e - T_L$ indicates that (1) the angular velocity ω_{rm} increases (motor accelerates) if $T_e > T_L$; (2) the angular velocity ω_{rm} decreases (motor disaccelerates) if $T_e < T_L$; (3) the angular velocity ω_{rm} is constant ($\omega_{rm} = $ constant, e.g., steady-state operation) if $T_e = T_L$. That is, to regulate motion devices, the electromagnetic torque (6.15) must be changed. A balanced three-phase current set is

$$i_{as} = \sqrt{2}i_M\cos(\omega_r t) = \sqrt{2}i_M\cos(\omega_e t) = \sqrt{2}i_M\cos\theta_r,$$
$$i_{bs} = \sqrt{2}i_M\cos\left(\omega_r t - \frac{2}{3}\pi\right) = \sqrt{2}i_M\cos\left(\omega_e t - \frac{2}{3}\pi\right) = \sqrt{2}i_M\cos\left(\theta_r - \frac{2}{3}\pi\right),$$
$$i_{cs} = \sqrt{2}i_M\cos\left(\omega_r t + \frac{2}{3}\pi\right) = \sqrt{2}i_M\cos\left(\omega_e t + \frac{2}{3}\pi\right) = \sqrt{2}i_M\cos\left(\theta_r + \frac{2}{3}\pi\right).$$

From the trigonometric identity $\cos^2\theta_r + \cos^2\left(\theta_r - \frac{2}{3}\pi\right) + \cos^2\left(\theta_r + \frac{2}{3}\pi\right) = \frac{3}{2}$, one yields

$$T_e = \frac{P\psi_m}{2}\sqrt{2}i_M\left[\cos^2\theta_r + \cos^2\left(\theta_r - \frac{2}{3}\pi\right) + \cos^2\left(\theta_r + \frac{2}{3}\pi\right)\right] = \frac{3P\psi_m}{2\sqrt{2}}i_M.$$

Hence, to ensure the controlled motion and regulate the output mechanical variables (angular velocity and displacement), i_M must be changed. Furthermore, the phase currents i_{as}, i_{bs}, and i_{cs}, should be functions of the electrical angular displacement θ_r measured by using the Hall-effect sensors. If the PWM amplifiers are used, one changes the magnitude u_M of the phase voltages u_{as}, u_{bs}, and u_{cs}. The angular displacement θ_r is needed to be measured (or estimated), and the *abc* voltages needed to be supplied are

$$u_{as} = \sqrt{2}u_M \cos\theta_r, \quad u_{bs} = \sqrt{2}u_M \cos\left(\theta_r - \frac{2}{3}\pi\right), \quad \text{and} \quad u_{cs} = \sqrt{2}u_M \cos\left(\theta_r + \frac{2}{3}\pi\right).$$

Utilizing the basic electromagnetics, we found that to control the angular velocity (in the drive application) or the displacement (in servosystem application), one should supply the phase voltages to the stator windings as a function of the angular displacement. The control concepts must be implemented by the hardware. For example, the balance voltage sets can be implemented by the "control logic" which regulates the output stage transistors in PWM amplifiers, while the angular displacement is measured by the Hall-effect sensors. Various PWM amplifiers have been designed and used. The motor windings *as*, *bs* (for two-phase motors) and *as*, *bs*, *cs* (for three-phase motors) are connected to the power stage outputs. Permanent-magnet synchronous electric machines are made in different sizes, and matching high-performance PWM power amplifiers are available. For ~250 W (80 V rated and ~15 A peak) permanent-magnet synchronous motors, the schematics of a B15A8 servo amplifier (20–80 V, ±7.5 A continuous current, ±15 A peak current, 2.5 kHz bandwidth, 129 × 76 × 25 mm dimensions) is documented in Figure 6.9. The motor phase windings are connected to P2-1, P2-2, and P2-3. One connects the Hall sensor outputs to P1-12, P1-13, and P1-14. The "control logic" utilizes the measured rotor angular displacement to develop appropriate phase voltages u_{as}, u_{bs}, and u_{cs} by driving the transistors.

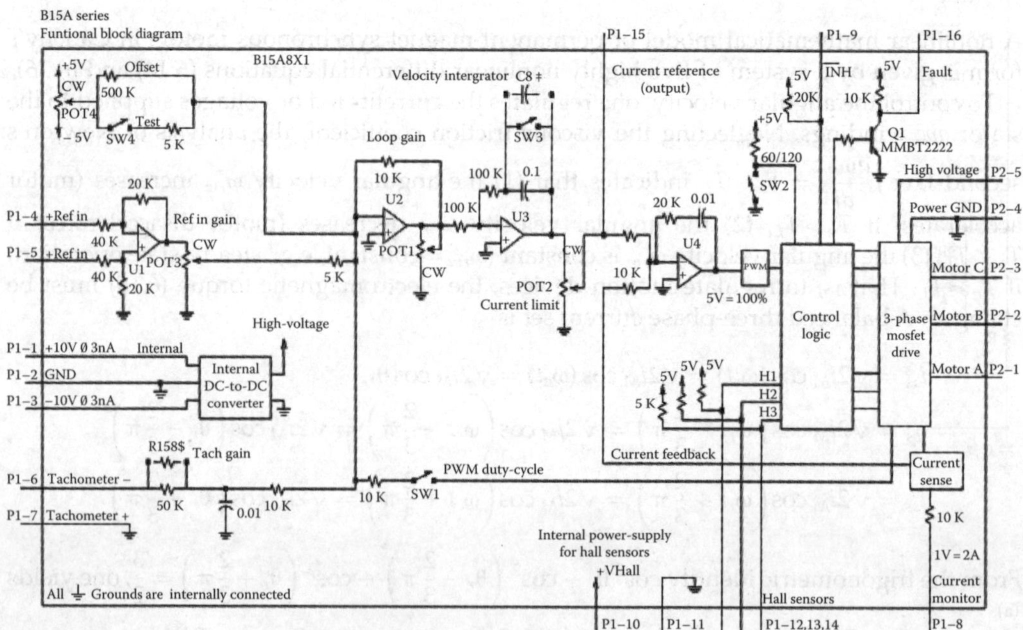

FIGURE 6.9

B15A8 PWM servo amplifier. (From Lyshevski, S.E., *Electromechanical Systems, Electric Machines, and Applied Mechatronics*, CRC Press, Boca Raton, FL, 1999. Courtesy of Advanced Motion Controls, www.a-m-c.com. With permission.)

The proportional-integral analog controller is integrated in the amplifier. The reference (command) voltage is supplied to P1-4. The voltage induced by the tachometer (proportional to the motor angular velocity) is supplied to P1-6. The reference and measured angular velocities are compared to obtain the tracking error $e(t)$. This $e(t)$ is utilized by the analog proportional-integral controller to develop the control signals which turn *on* and *off* transistors. One can change the proportional and integral feedback gains adjusting the potentiometers (resistors). The amplifier can be used in the servosystem applications, and the angular (or linear) displacement should be supplied to P1-6. Various PWM amplifiers are available from Advanced Motion Controls and other companies.

Small permanent-magnet synchronous motors (from mW to $\sim$1 W) are applied in various applications such as rotating and positioning stages, hard drives, robotics, appliances, etc. A small $\sim$1 W permanent-magnet synchronous motors, shown in Figure 6.2c, can be driven by monolithic PWM amplifiers such as MC33035 (40 V and 50 mA) [4], L6235 (SSTMicroelectronics, www.st.com), and others. The phase voltages are derived by using the rotor angular displacement decoder. The application-specific ICs drive power MOSFETs as reported in Figure 6.10. The PWM concept is implemented, and the Hall sensor signals are supplied to obtain u_{as}, u_{bs}, and u_{cs}. The representative block diagram, as documented in

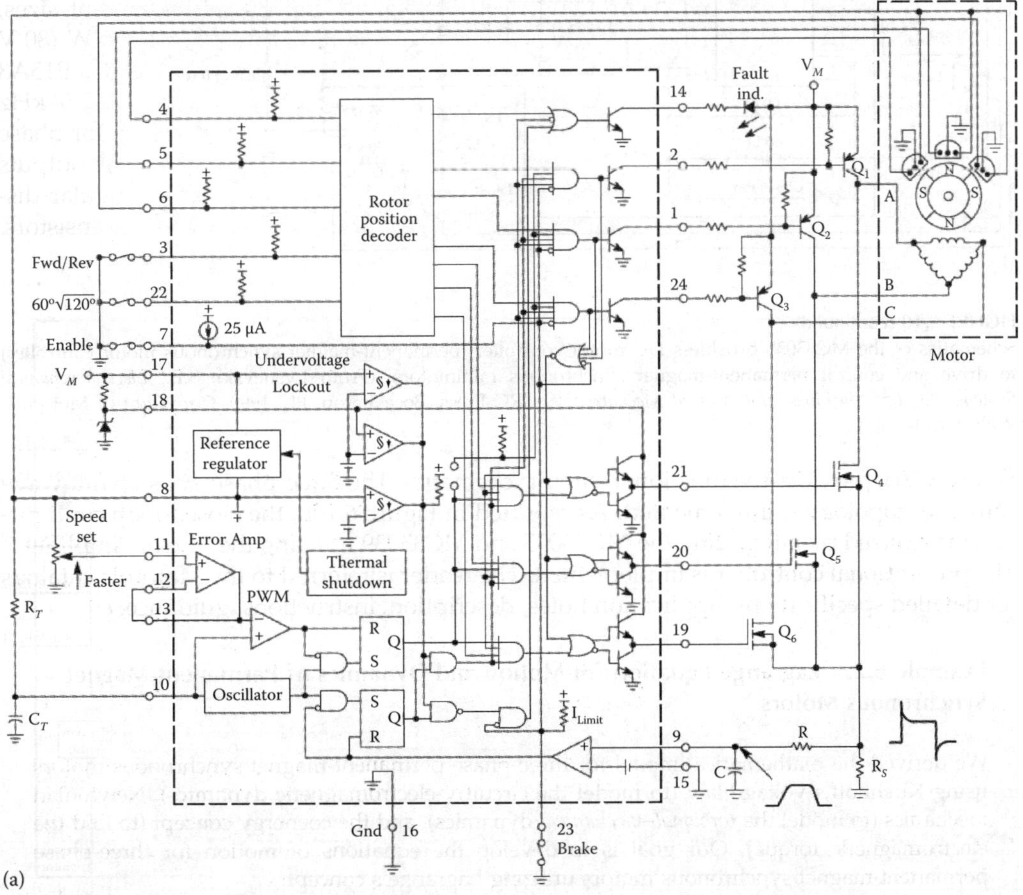

FIGURE 6.10

Schematics of the MC33035 brushless DC motor controller (permanent-magnet synchronous motor controller) to drive and control permanent-magnet synchronous minimotors.

(continued)

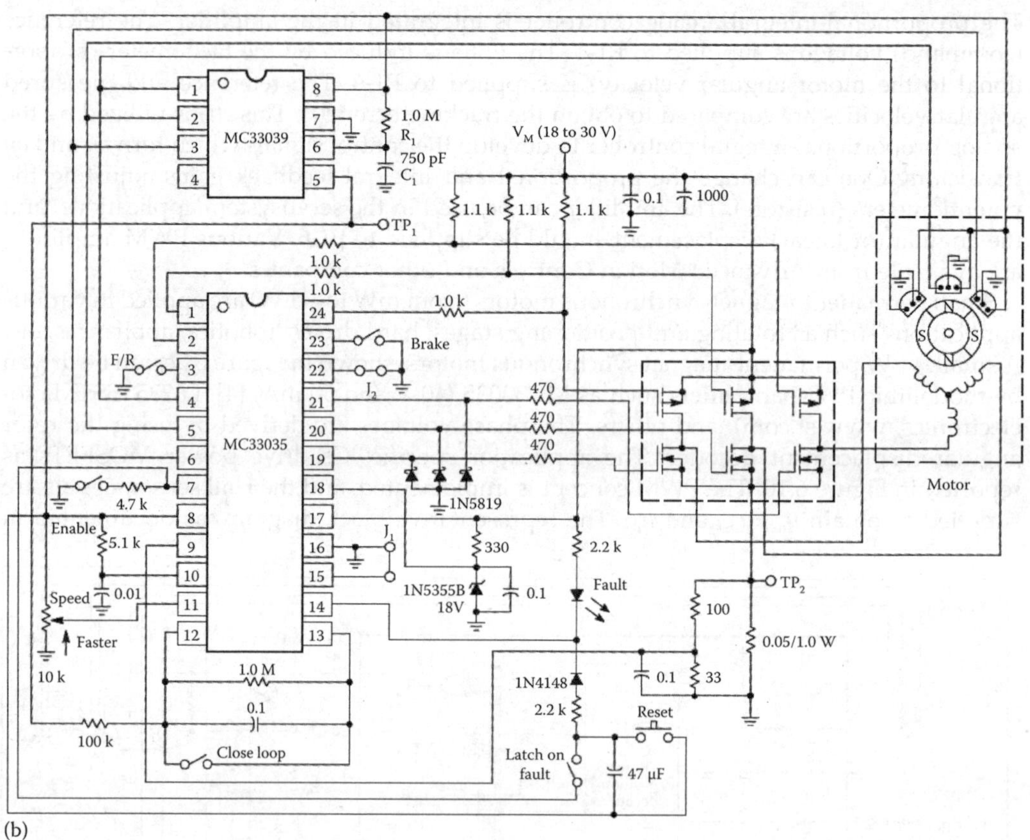

(b)

FIGURE 6.10 (continued)

Schematics of the MC33035 brushless DC motor controller (permanent-magnet synchronous motor controller) to drive and control permanent-magnet synchronous minimotors. (From Lyshevski, S.E., *Electromechanical Systems, Electric Machines, and Applied Mechatronics*, CRC Press, Boca Raton, FL, 1999. Copyright of Motorola. With permission.)

Figure 6.10a, provides the functional/circuit schematics. The three-phase, six-step full-wave converter topology is implemented. As reported in Figure 6.10b, the closed-loop configuration is realized by integrating the MC33035 and MC33039 utilizing the "Error Amplifier." The proportional controller is implemented. The reader is referred to the Motorola catalogs for detailed specifications, application notes, description, instructions, guidance, etc.

Example 6.2: Lagrange Equations of Motion and Dynamics of Permanent-Magnet Synchronous Motors

We derived the mathematical model for three-phase permanent-magnet synchronous motors using Kirchhoff's voltage law (to model the circuitry-electromagnetic dynamics), Newtonian mechanics (to model the *torsional–mechanical* dynamics), and the coenergy concept (to find the electromagnetic torque). Our goal is to develop the equations of motion for three-phase permanent-magnet synchronous motors utilizing Lagrange's concept.

The generalized coordinates are the electric charges in the *abc* stator windings $q_1 = \dfrac{i_{as}}{s}$, $\dot{q}_1 = i_{as}$, $q_2 = \dfrac{i_{bs}}{s}$, $\dot{q}_2 = i_{bs}$, $q_3 = \dfrac{i_{cs}}{s}$, $\dot{q}_3 = i_{cs}$, and the angular displacement $q_4 = \theta_r$, $\dot{q}_4 = \omega_r$.

The generalized forces are the applied voltages to the abc windings $Q_1 = u_{as}$, $Q_2 = u_{bs}$, $Q_3 = u_{cs}$, and the load torque $Q_4 = -T_L$.

The resulting Lagrange equations are

$$\frac{d}{dt}\left(\frac{\partial \Gamma}{\partial \dot{q}_1}\right) - \frac{\partial \Gamma}{\partial q_1} + \frac{\partial D}{\partial \dot{q}_1} + \frac{\partial \Pi}{\partial q_1} = Q_1, \quad \frac{d}{dt}\left(\frac{\partial \Gamma}{\partial \dot{q}_2}\right) - \frac{\partial \Gamma}{\partial q_2} + \frac{\partial D}{\partial \dot{q}_2} + \frac{\partial \Pi}{\partial q_2} = Q_2,$$

$$\frac{d}{dt}\left(\frac{\partial \Gamma}{\partial \dot{q}_3}\right) - \frac{\partial \Gamma}{\partial q_3} + \frac{\partial D}{\partial \dot{q}_3} + \frac{\partial \Pi}{\partial q_3} = Q_3, \quad \frac{d}{dt}\left(\frac{\partial \Gamma}{\partial \dot{q}_4}\right) - \frac{\partial \Gamma}{\partial q_4} + \frac{\partial D}{\partial \dot{q}_4} + \frac{\partial \Pi}{\partial q_4} = Q_4.$$

The total kinetic energy includes kinetic energies of electrical and mechanical systems. We have

$$\Gamma = \Gamma_E + \Gamma_M = \frac{1}{2}L_{asas}\dot{q}_1^2 + \frac{1}{2}(L_{asbs} + L_{bsas})\dot{q}_1\dot{q}_2 + \frac{1}{2}(L_{ascs} + L_{csas})\dot{q}_1\dot{q}_3 + \frac{1}{2}L_{bsbs}\dot{q}_2^2$$

$$+ \frac{1}{2}(L_{bscs} + L_{csbs})\dot{q}_2\dot{q}_3 + \frac{1}{2}L_{cscs}\dot{q}_3^2 + \psi_m\dot{q}_1 \sin q_4 + \psi_m\dot{q}_2 \sin\left(q_4 - \frac{2}{3}\pi\right)$$

$$+ \psi_m\dot{q}_3 \sin\left(q_4 + \frac{2}{3}\pi\right) + \frac{1}{2}J\dot{q}_4^2.$$

Therefore,

$$\frac{\partial \Gamma}{\partial q_1} = 0, \quad \frac{\partial \Gamma}{\partial \dot{q}_1} = L_{asas}\dot{q}_1 + \frac{1}{2}(L_{asbs} + L_{bsas})\dot{q}_2 + \frac{1}{2}(L_{ascs} + L_{csas})\dot{q}_3 + \psi_m \sin q_4,$$

$$\frac{\partial \Gamma}{\partial q_2} = 0, \quad \frac{\partial \Gamma}{\partial \dot{q}_2} = \frac{1}{2}(L_{asbs} + L_{bsas})\dot{q}_1 + L_{bsbs}\dot{q}_2 + \frac{1}{2}(L_{bscs} + L_{csbs})\dot{q}_3 + \psi_m \sin\left(q_4 - \frac{2}{3}\pi\right),$$

$$\frac{\partial \Gamma}{\partial q_3} = 0, \quad \frac{\partial \Gamma}{\partial \dot{q}_3} = \frac{1}{2}(L_{ascs} + L_{csas})\dot{q}_1 + \frac{1}{2}(L_{bscs} + L_{csbs})\dot{q}_2 + L_{cscs}\dot{q}_3 + \psi_m \sin\left(q_4 + \frac{2}{3}\pi\right),$$

$$\frac{\partial \Gamma}{\partial q_4} = \psi_m\dot{q}_1 \cos q_4 + \psi_m\dot{q}_2 \cos\left(q_4 - \frac{2}{3}\pi\right) + \psi_m\dot{q}_3 \cos\left(q_4 + \frac{2}{3}\pi\right), \quad \frac{\partial \Gamma}{\partial \dot{q}_4} = J\dot{q}_4.$$

The total potential energy is $\Pi = 0$.

The total dissipated energy is a sum of the heat energy dissipated by the electrical system and the heat energy dissipated by the mechanical system. That is,

$$D = \frac{1}{2}\left(r_s\dot{q}_1^2 + r_s\dot{q}_2^2 + r_s\dot{q}_3^2 + B_m\dot{q}_4^2\right).$$

The differentiation of D with respect to the generalized coordinates yields

$$\frac{\partial D}{\partial \dot{q}_1} = r_s\dot{q}_1, \quad \frac{\partial D}{\partial \dot{q}_2} = r_s\dot{q}_2, \quad \frac{\partial D}{\partial \dot{q}_3} = r_s\dot{q}_3, \quad \text{and} \quad \frac{\partial D}{\partial \dot{q}_4} = B_m\dot{q}_4.$$

The Lagrange equations result in differential equations

$$L_{asas}\frac{di_{as}}{dt} + \frac{1}{2}(L_{asbs} + L_{bsas})\frac{di_{bs}}{dt} + \frac{1}{2}(L_{ascs} + L_{csas})\frac{di_{cs}}{dt} + \psi_m\omega_r \cos\theta_r + r_s i_{as} = u_{as},$$

$$\frac{1}{2}(L_{asbs} + L_{bsas})\frac{di_{as}}{dt} + L_{bsbs}\frac{di_{bs}}{dt} + \frac{1}{2}(L_{bscs} + L_{csbs})\frac{di_{cs}}{dt} + \psi_m\omega_r \cos\left(\theta_r - \frac{2}{3}\pi\right) + r_s i_{bs} = u_{bs},$$

$$\frac{1}{2}(L_{ascs} + L_{csas})\frac{di_{as}}{dt} + \frac{1}{2}(L_{bscs} + L_{csbs})\frac{di_{bs}}{dt} + L_{cscs}\frac{di_{cs}}{dt} + \psi_m \omega_r \cos\left(\theta_r + \frac{2}{3}\pi\right) + r_s i_{cs} = u_{cs},$$

$$J\frac{d^2\theta_r}{dt^2} - \psi_m i_{as}\cos\theta_r - \psi_m i_{bs}\cos\left(\theta_r - \frac{2}{3}\pi\right) - \psi_m i_{cs}\cos\left(\theta_r + \frac{2}{3}\pi\right) + B_m\frac{d\theta_r}{dt} = -T_L.$$

For round-rotor permanent-magnet synchronous motors, one obtains

$$(L_{ls} + \bar{L}_m)\frac{di_{as}}{dt} - \frac{1}{2}\bar{L}_m\frac{di_{bs}}{dt} - \frac{1}{2}\bar{L}_m\frac{di_{cs}}{dt} + \psi_m \omega_r \cos\theta_r + r_s i_{as} = u_{as},$$

$$-\frac{1}{2}\bar{L}_m\frac{di_{as}}{dt} + (L_{ls} + \bar{L}_m)\frac{di_{bs}}{dt} - \frac{1}{2}\bar{L}_m\frac{di_{cs}}{dt} + \psi_m \omega_r \cos\left(\theta_r - \frac{2}{3}\pi\right) + r_s i_{bs} = u_{bs},$$

$$-\frac{1}{2}\bar{L}_m\frac{di_{as}}{dt} - \frac{1}{2}\bar{L}_m\frac{di_{bs}}{dt} + (L_{ls} + \bar{L}_m)\frac{di_{cs}}{dt} + \psi_m \omega_r \cos\left(\theta_r + \frac{2}{3}\pi\right) + r_s i_{cs} = u_{cs},$$

$$J\frac{d\omega_r}{dt} + B_m\omega_r - \psi_m\left[i_{as}\cos\theta_r + i_{bs}\cos\left(\theta_r - \frac{2}{3}\pi\right) + i_{cs}\cos\left(\theta_r + \frac{2}{3}\pi\right)\right] = -T_L,$$

$$\frac{d\theta_r}{dt} = \omega_r.$$

From $\dfrac{\partial\Gamma}{\partial q_4} = \psi_m\dot{q}_1\cos q_4 + \psi_m\dot{q}_2\cos\left(q_4 - \frac{2}{3}\pi\right) + \psi_m\dot{q}_3\cos\left(q_4 + \frac{2}{3}\pi\right)$, which is a term of the

fourth Lagrange equation $\dfrac{d}{dt}\left(\dfrac{\partial\Gamma}{\partial\dot{q}_4}\right) - \dfrac{\partial\Gamma}{\partial q_4} + \dfrac{\partial D}{\partial\dot{q}_4} + \dfrac{\partial\Pi}{\partial q_4} = Q_4$, one concludes that the electromag-

netic torque is given as $T_e = \psi_m\left[i_{as}\cos\theta_r + i_{bs}\cos\left(\theta_r - \frac{2}{3}\pi\right) + i_{cs}\cos\left(\theta_r + \frac{2}{3}\pi\right)\right]$.

For P-pole permanent-magnet synchronous motors, differential equations in Cauchy's form, as given by (6.14) and (6.16), results. ∎

Example 6.3: Experimental Studies, Simulation, and Analysis of a Permanent-Magnet Synchronous Motor in Simulink

The radial topology permanent-magnet synchronous motors are described by five nonlinear differential equations (6.14) and (6.16). The following phase voltages, as functions of θ_r, are applied to guarantee the balanced operation

$$u_{as} = \sqrt{2}u_M\cos\theta_r, \quad u_{bs} = \sqrt{2}u_M\cos\left(\theta_r - \frac{2}{3}\pi\right), \quad \text{and} \quad u_{cs} = \sqrt{2}u_M\cos\left(\theta_r + \frac{2}{3}\pi\right).$$

The Simulink diagram to simulate permanent-magnet synchronous motors is documented in Figure 6.11. To ensure a greatest degree of flexibility and effectiveness, the motor parameters are embedded by utilizing the corresponding symbols rather than numerical values.

For a 500 W four-pole permanent-magnet synchronous motor, the parameters are: $u_M = 50$, $r_s = 1$ ohm, $L_{ss} = 0.005$ H, $L_{ls} = 0.0005$ H, $\bar{L}_m = 0.0045$ H, $\psi_m = 0.15$ V-s/rad (N-m/A), $B_m = 0.0005$ N-m-s/rad, and $J = 0.0015$ kg-m^2. The motor parameters can be measured and derived by performing the experimental results. One can measure the stator resistance r_s. The constant ψ_m is calculated using the induced *emf* (terminal phase voltage induced operating machine as a generator) at the steady-state ω_r. The friction coefficient B_m is found by measuring the magnitude of the phase currents at no load (one applies the expression $T_e = B_m\omega_r$ at no load when $T_L = 0$). The moment of

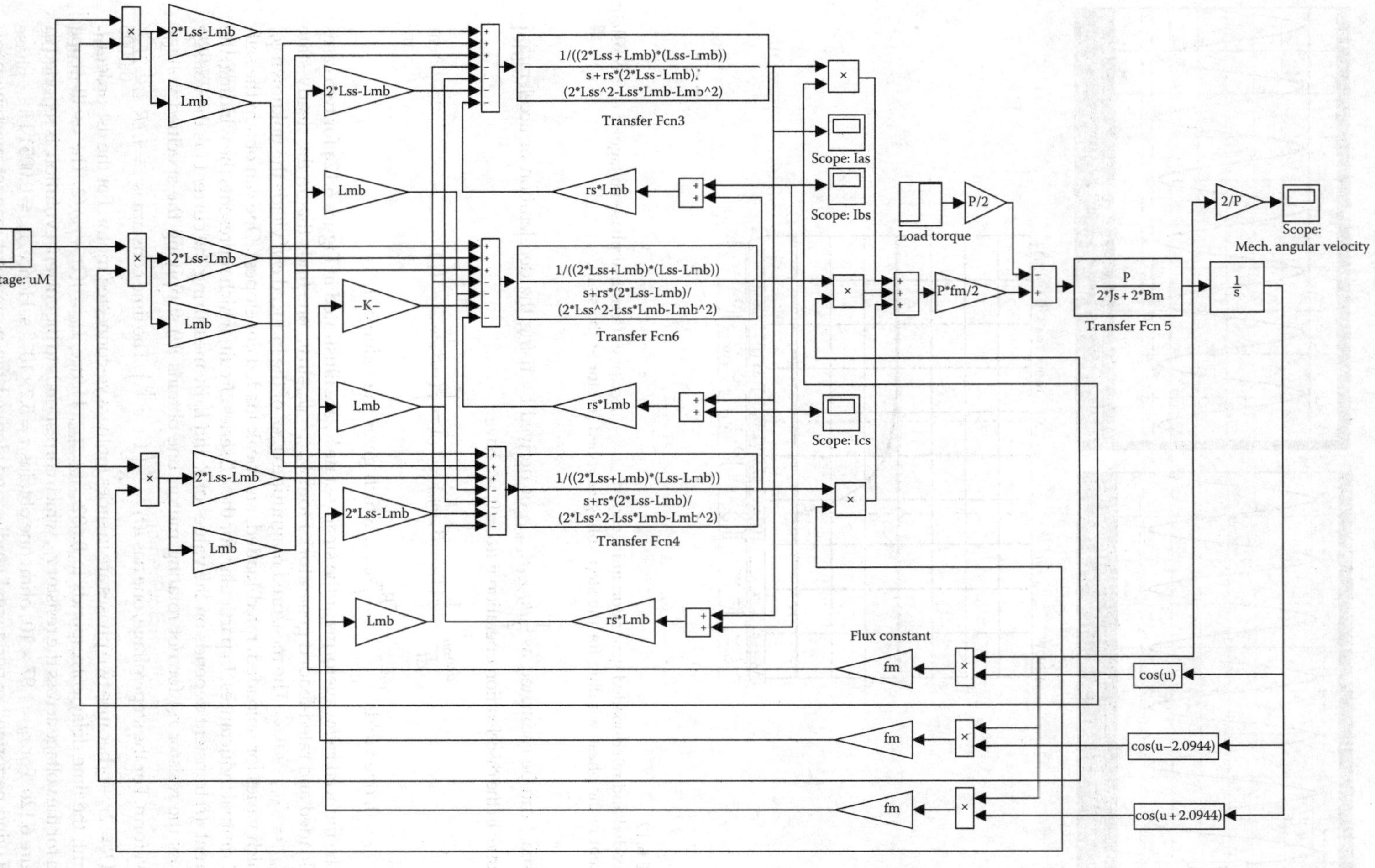

FIGURE 6.11
Simulink diagram to simulate permanent-magnet synchronous motors (ch6_02.mdl).

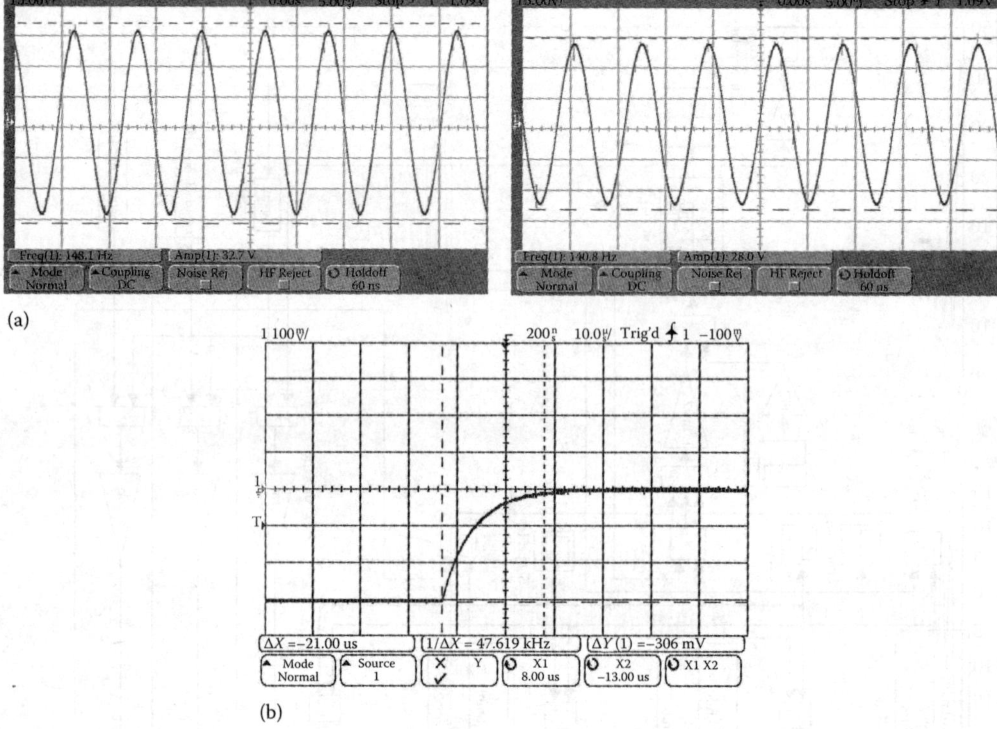

(a)

(b)

FIGURE 6.12

(a) Unloaded and rated-loaded generator: induced $emf_{as\omega} = -\psi_m \cos\theta_r\omega_r$ (terminal phase voltage) if $\omega_r = $ constant;
(b) current in the phase winding for the step voltage applied (motor at the stall).

inertia J can be estimated as $\frac{1}{2}m_{\text{rotor}}R_{\text{rotor}}^2$, or derived by using the deceleration or acceleration regime. In the disceleration operation if $u_M = 0$, we have

$$\frac{d\omega_{rm}}{dt} = \frac{1}{J}(T_e - B_m\omega_{rm} - T_L) = \frac{1}{J}(-B_m\omega_{rm} - T_L).$$

For $T_L = 0$, one yields $\frac{d\omega_{rm}}{dt} = -\frac{B_m}{J}\omega_{rm}$, which gives the value of J.

The induced *emf* waveforms at the steady-state ω_r are illustrated in Figure 6.12a for two cases (unloaded and rated-loaded generator). From $\psi_{asm} = \psi_m \sin\theta_r$, one finds that in the steady-state $emf_{as\omega} = -\psi_m\cos\theta_r\omega_r$. The measured magnitude $\psi_m\omega_r$ of the induced *emf* yields the unknown ψ_m (which varies for unloaded, rated-loaded, and peak-loaded envelopes). One can obtain the self- and mutual inductances. In particular, for the RL series circuit, which represents the winding, the current $i(t)$ transient depends on the values of R and L. By measuring the current (as the voltage across the resistor r_R) for not rotating motor, one obtains $i(t)$ supplying the specified voltage waveform. For the step voltage, one has $i(t) = \frac{u}{R}\left(1 - e^{-\frac{R}{L}t}\right)$. The time constant is $\tau = L/R$, $L = 2L_{ss}$, and $R = 2r_s + r_R$ because two phases are in series for the wye-connected motor. For the fist-order RL circuit, the time delay corresponds to $0.632i_{\text{steady-state}}$. Hence, $L_{ss} = \tau(2r_s + r_R)/2$. The oscilloscope data for the voltage across the resistor r_R, which corresponds to the current evolution, is reported in Figure 6.12b. For $r_R = 1.97 \times 10^3$ ohm, one obtains $\tau = 5.2 \times 10^{-6}$ s. Hence, $L_{ss} = 0.0051$ H.

Having performed experimental studies and derived the permanent-magnet machine parameters, the simulation is carried out. The motor dynamics is studied as the motor accelerates with the rated voltage applied. The motor parameters are uploaded using the following statement:

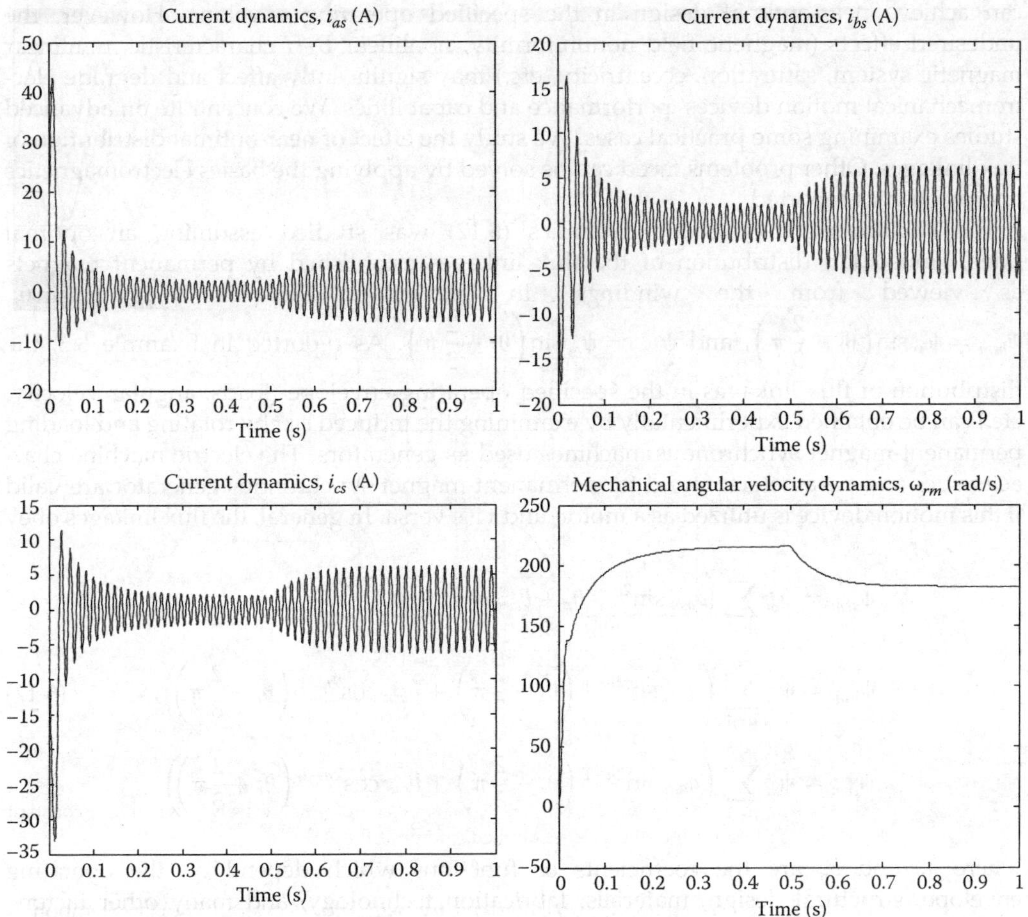

FIGURE 6.13
Transient dynamics of the permanent-magnet synchronous motor variables.

```
% Parameters of the permanent-magnet synchronous motor
P=4; uM=50; rs=1; Lss=0.005; Lls=0.0005; fm=0.15; Bm=0.0005; J=0.0015;
Lmb=Lss-Lls;
```

The motor accelerates from stall, $T_{L0} = 0.1$ N-m for $t \in [0\ 0.5)$ s, and $T_L = 0.5$ N-m is applied at $t = 0.5$ s. Figure 6.13 illustrates the evolution of the phase currents and mechanical angular velocity. The motor reaches the steady-state mechanical angular velocity (215.3 rad/s with the load $T_{L0} = 0.1$ N-m) at 0.3 s. The angular velocity reduces and phase currents magnitude increases as the higher load is applied. The current and angular velocity dynamics, reported in Figure 6.13, allow one to assess the current magnitude, efficiency, acceleration rate, starting capabilities, etc. ∎

6.2.3 Advanced Topics in Analysis of Permanent-Magnet Synchronous Machines

We documented various analysis tasks by assuming an optimal design of permanent-magnet synchronous machines, linearity of magnetic system, etc. To the best, the designer can achieve near-optimal design in the specified operating envelope. However, the undesired effects (magnetic field nonuniformity, nonlinear B–H characteristic, nonlinear magnetic system, saturation, eccentricity, etc.) may significantly affect and degrade electromechanical motion devices' performance and capabilities. We concentrate on advanced studies examining some practical cases. We study the effect of near-optimal distribution of flux linkages. Other problems faced can be solved by applying the basics electromagnetics and mechanics reported.

The circuitry-electromagnetic dynamics (6.12) was studied assuming an optimal (ideal sinusoidal) distribution of the flux linkages established by permanent magnets as viewed from the windings. In particular, we let $\psi_{asm} = \psi_m \sin \theta_r$, $\psi_{bsm} = \psi_m \sin\left(\theta_r - \frac{2}{3}\pi\right)$, and $\psi_{csm} = \psi_m \sin\left(\theta_r + \frac{2}{3}\pi\right)$. As reported in Example 6.3, the distribution of flux linkages in the specified operating envelope (loads, angular velocity, etc.) can be obtained experimentally by examining the induced *emf* by rotating and loading permanent-magnet synchronous machines used as generators. The electric machine characteristics and parameters, found for permanent-magnet synchronous generator are valid if this motion device is utilized as a motor, and vise versa. In general, the flux linkages obey

$$\psi_{asm} = \psi_m \sum_{n=1}^{\infty} \left(a_{asn} \sin^{2n-1}\theta_r + b_{asn} \cos^{2n-1}\theta_r\right),$$

$$\psi_{bsm} = \psi_m \sum_{n=1}^{\infty} \left(a_{bsn} \sin^{2n-1}\left(\theta_r - \frac{2}{3}\pi\right) + b_{bsn} \cos^{2n-1}\left(\theta_r - \frac{2}{3}\pi\right)\right), \qquad (6.17)$$

$$\psi_{csm} = \psi_m \sum_{n=1}^{\infty} \left(a_{csn} \sin^{2n-1}\left(\theta_r + \frac{2}{3}\pi\right) + b_{csn} \cos^{2n-1}\left(\theta_r + \frac{2}{3}\pi\right)\right)$$

where a_n and b_n are the coefficients or functions which depend on the operating envelope, structural design, materials, fabrication technology, and many other factors, e.g., $a_n(\mathbf{E}, \mathbf{D}, \mathbf{B}, \mathbf{H}, \mathbf{i}_{abcs}, \omega_r, \mathbf{T}_L, \varepsilon, \mu, \Sigma)$ and $b_n(\mathbf{E}, \mathbf{D}, \mathbf{B}, \mathbf{H}, \mathbf{i}_{abcs}, \omega_r, \mathbf{T}_L, \varepsilon, \mu, \Sigma)$, Σ denotes the machine structural design, topology, sizing, materials, and other factors.

In practice, (6.17) usually results in

$$\psi_{asm} = \psi_m \sum_{n=1}^{\infty} a_n \sin^{2n-1}\theta_r, \quad \psi_{bsm} = \psi_m \sum_{n=1}^{\infty} a_n \sin^{2n-1}\left(\theta_r - \frac{2}{3}\pi\right),$$

$$\psi_{csm} = \psi_m \sum_{n=1}^{\infty} a_n \sin^{2n-1}\left(\theta_r + \frac{2}{3}\pi\right). \qquad (6.18)$$

From (6.18), one yields the expressions for the flux linkages as

$$\begin{bmatrix} \psi_{as} \\ \psi_{bs} \\ \psi_{cs} \end{bmatrix} = \begin{bmatrix} L_{ls} + \bar{L}_m & -\frac{1}{2}\bar{L}_m & -\frac{1}{2}\bar{L}_m \\ -\frac{1}{2}\bar{L}_m & L_{ls} + \bar{L}_m & -\frac{1}{2}\bar{L}_m \\ -\frac{1}{2}\bar{L}_m & -\frac{1}{2}\bar{L}_m & L_{ls} + \bar{L}_m \end{bmatrix} \begin{bmatrix} i_{as} \\ i_{bs} \\ i_{cs} \end{bmatrix} + \psi_m \begin{bmatrix} \sum_{n=1}^{\infty} a_n \sin^{2n-1}\theta_r \\ \sum_{n=1}^{\infty} a_n \sin^{2n-1}\left(\theta_r - \frac{2}{3}\pi\right) \\ \sum_{n=1}^{\infty} a_n \sin^{2n-1}\left(\theta_r + \frac{2}{3}\pi\right) \end{bmatrix}. \qquad (6.19)$$

Using (6.19), the total derivatives $d\psi_{as}/dt$, $d\psi_{bs}/dt$, and $d\psi_{cs}/dt$ can be found and used in (6.12)

$$u_{as} = r_s i_{as} + \frac{d\psi_{as}}{dt}, \quad u_{bs} = r_s i_{bs} + \frac{d\psi_{bs}}{dt}, \quad u_{cs} = r_s i_{cs} + \frac{d\psi_{cs}}{dt},$$

yielding the circuitry-electromagnetic equations of motion in the *machine* variables.

One needs to derive the expression for the electromagnetic torque. From $T_e = \dfrac{P}{2}\dfrac{\partial W_c}{\partial \theta_r}$, we obtain

$$T_e = \frac{P\psi_m}{2}\left[i_{as}\sum_{n=1}^{\infty}(2n-1)a_n \cos\theta_r \sin^{2n-2}\theta_r + i_{bs}\sum_{n=1}^{\infty}(2n-1)a_n \cos\left(\theta_r - \frac{2}{3}\pi\right) \right.$$
$$\left. \sin^{2n-2}\left(\theta_r - \frac{2}{3}\pi\right) + i_{cs}\sum_{n=1}^{\infty}(2n-1)a_n \cos\left(\theta_r + \frac{2}{3}\pi\right)\sin^{2n-2}\left(\theta_r + \frac{2}{3}\pi\right)\right]. \quad (6.20)$$

Newton's second law (6.3) and (6.20) result in the *torsional–mechanical* dynamics as

$$\frac{d\omega_r}{dt} = \frac{P^2\psi_m}{4J}\left[i_{as}\sum_{n=1}^{\infty}(2n-1)a_n \cos\theta_r \sin^{2n-2}\theta_r + i_{bs}\sum_{n=1}^{\infty}(2n-1)a_n \cos\left(\theta_r - \frac{2}{3}\pi\right) \right.$$
$$\left. \sin^{2n-2}\left(\theta_r - \frac{2}{3}\pi\right) + i_{cs}\sum_{n=1}^{\infty}(2n-1)a_n \cos\left(\theta_r + \frac{2}{3}\pi\right)\sin^{2n-2}\left(\theta_r + \frac{2}{3}\pi\right)\right]$$
$$- \frac{B_m}{J}\omega_r - \frac{P}{2J}T_L, \quad (6.21)$$
$$\frac{d\theta_r}{dt} = \omega_r.$$

The current and voltage sets to be applied are derived using the expression for T_e (6.20). The trigonometric identities are applied. From (6.20) one finds the following phase currents and voltages which theoretically guarantee balanced operations:

$$i_{as} = \sqrt{2}i_M \cos\theta_r \left[\sum_{n=1}^{\infty}(2n-1)a_n \sin^{2n-2}\theta_r\right]^{-1},$$

$$i_{bs} = \sqrt{2}i_M \cos\left(\theta_r - \frac{2}{3}\pi\right)\left[\sum_{n=1}^{\infty}(2n-1)a_n \sin^{2n-2}\left(\theta_r - \frac{2}{3}\pi\right)\right]^{-1}, \quad (6.22)$$

$$i_{cs} = \sqrt{2}i_M \cos\left(\theta_r + \frac{2}{3}\pi\right)\left[\sum_{n=1}^{\infty}(2n-1)a_n \sin^{2n-2}\left(\theta_r + \frac{2}{3}\pi\right)\right]^{-1},$$

and

$$u_{as} = \sqrt{2}u_M \cos\theta_r \left[\sum_{n=1}^{\infty}(2n-1)a_n \sin^{2n-2}\theta_r\right]^{-1},$$

$$u_{bs} = \sqrt{2}u_M \cos\left(\theta_r - \frac{2}{3}\pi\right)\left[\sum_{n=1}^{\infty}(2n-1)a_n \sin^{2n-2}\left(\theta_r - \frac{2}{3}\pi\right)\right]^{-1}, \quad (6.23)$$

$$u_{cs} = \sqrt{2}u_M \cos\left(\theta_r + \frac{2}{3}\pi\right)\left[\sum_{n=1}^{\infty}(2n-1)a_n \sin^{2n-2}\left(\theta_r + \frac{2}{3}\pi\right)\right]^{-1}.$$

The phase currents and voltages are constrained. Therefore, the saturation effect (bounds) should be integrated using the conditional statements. The singularity problem can be resolved by normalizing sets (6.22) and (6.23). It should be emphasized that the current and voltage sets (6.22) and (6.23) are the mathematical expressions to implement, if possible, by the hardware (power electronics, DSP, etc.) and software. Utilizing specific output stages (usually 6- or 12-step) and converter topologies (hard- or soft-switching, etc.), though one strives to ensure the appropriate solutions, hardware largely defines the voltage waveforms. Advanced DSPs are required to integrate $a_n(\mathbf{E}, \mathbf{D}, \mathbf{B}, \mathbf{H}, \mathbf{i}_{abcs}, \omega_r, \mathbf{T}_L, \boldsymbol{\varepsilon}, \boldsymbol{\mu}, \boldsymbol{\Sigma})$ as conditional logics and look-up table to realize (6.22) or (6.23).

Within the existing converter topologies one cannot ensure sinusoidal-like voltage waveforms. Furthermore, the PWM concept implies the voltage *averaging*, and the rated solid-state device voltage, current, switching frequency, and other characteristics affect voltage waveforms. The hardware-centered (implementable) phase voltages must be used in the performance analysis and capability assessments. Even the ideal sinusoidal voltage set as $u_{as} = \sqrt{2}u_M \cos\theta_r$, $u_{bs} = \sqrt{2}u_M \cos\left(\theta_r + \dfrac{2}{3}\pi\right)$, and $u_{cs} = \sqrt{2}u_M \cos\left(\theta_r - \dfrac{2}{3}\pi\right)$ is not implementable. However, we advanced the analysis tasks providing the justification and foundation for advanced studies. One may integrate the derived basic electromagnetics, control concepts, and hardware solutions ensuring near-optimal performance.

Example 6.4:

For two-phase permanent-magnet synchronous machines it is desired to ensure the machine design which leads to

$$\psi_{asm} = \psi_m \sin\theta_r \quad \text{and} \quad \psi_{bsm} = \psi_m \cos\theta_r.$$

The derived electromagnetic torque $T_e = \dfrac{P\psi_m}{2}(\cos\theta_r i_{as} - \sin\theta_r i_{bs})$ yields the balanced current set as

$$i_{as} = i_M \cos\theta_r \quad \text{and} \quad i_{bs} = i_M \sin\theta_r.$$

Let the flux linkages, established by the permanent magnets as viewed from the windings, are

$$\psi_{asm} = \psi_m \sum_{n=1}^{\infty} a_n \sin^{2n-1}\theta_r \quad \text{and} \quad \psi_{bsm} = \psi_m \sum_{n=1}^{\infty} a_n \cos^{2n-1}\theta_r.$$

The electromagnetic torque is given as

$$T_e = \frac{P\psi_m}{2}\left[i_{as} \sum_{n=1}^{\infty} (2n-1)a_n \cos\theta_r \sin^{2n-2}\theta_r - i_{bs} \sum_{n=1}^{\infty} (2n-1)a_n \sin\theta_r \cos^{2n-2}\theta_r \right].$$

Let $a_1 \neq 1$, $a_2 \neq 0$, and $\forall a_n = 0$, $n > 2$. Hence,

$$T_e = \frac{P\psi_m}{2}\left[i_{as} \cos\theta_r \left(a_1 + 3a_2 \sin^2\theta_r\right) - i_{bs} \sin\theta_r \left(a_1 + 3a_2 \cos^2\theta_r\right) \right].$$

The phase voltages u_{as} and u_{bs}, as functions of θ_r, which ensure the near-balanced operating conditions are

$$u_{as} = u_M \cos \theta_r / \left(a_1 + 3a_2 \sin^2 \theta_r + \varepsilon \right) \quad \text{if } \left| u_M \cos \theta_r / \left(a_1 + 3a_2 \sin^2 \theta_r + \varepsilon \right) \right| \leq u_{max},$$

$$u_{as} = u_{max} \quad \text{or} \quad u_{as} = -u_{max} \text{ otherwise,}$$

$$u_{bs} = -u_M \sin \theta_r / \left(a_1 + 3a_2 \cos^2 \theta_r + \varepsilon \right) \quad \text{if } \left| -u_M \sin \theta_r / \left(a_1 + 3a_2 \cos^2 \theta_r + \varepsilon \right) \right| \leq u_{max},$$

$$u_{bs} = -u_{max} \quad \text{or} \quad u_{bs} = u_{max} \text{ otherwise.}$$

If $a_1 \gg a_2$, one may utilize $u_{as} = u_M \cos \theta_r$ and $u_{bs} = -u_M \sin \theta_r$. ∎

Example 6.5: Simulation and Analysis of a Permanent-Magnet Synchronous Motor

We simulate and study the performance of a radial topology three-phase synchronous motor. The motor parameters are: $P = 4$, $u_M = 50$, $r_s = 1$ ohm, $L_{ss} = 0.002$ H, $L_{ls} = 0.0002$ H, $\bar{L}_m = 0.0018$ H, $\psi_m = 0.1$ V-s/rad (N-m/A), $B_m = 0.00008$ N-m-s/rad, and $J = 0.00004$ kg-m^2. For no load and light load conditions (T_L is ~ 0.1 N-m), the constants are $a_1 = 1$ and all other a_n are zeros ($\forall a_n = 0$, $n > 1$). For the loaded motor, we have $a_1 = 1$, $a_2 = 0.05$, $a_3 = 0.02$, and $\forall a_n = 0$, $n > 3$.

The studied permanent-magnet synchronous motor is described by five nonlinear differential equations (6.12) and (6.21). For $a_1 \neq 0$, $a_2 \neq 0$, $a_3 \neq 0$, and $\forall a_n = 0$, $n > 3$, the resulting expressions for flux linkages

$$\psi_{as} = L_{ss} i_{as} - \frac{1}{2} \bar{L}_m i_{bs} - \frac{1}{2} \bar{L}_m i_{cs} + \psi_m \left(a_1 \sin \theta_r + a_2 \sin^3 \theta_r + a_3 \sin^5 \theta_r \right),$$

$$\psi_{bs} = -\frac{1}{2} \bar{L}_m i_{as} + L_{ss} i_{bs} - \frac{1}{2} \bar{L}_m i_{cs}$$
$$+ \psi_m \left[a_1 \sin \left(\theta_r + \frac{2}{3}\pi \right) + a_2 \sin^3 \left(\theta_r + \frac{2}{3}\pi \right) + a_3 \sin^5 \left(\theta_r + \frac{2}{3}\pi \right) \right],$$

$$\psi_{cs} = -\frac{1}{2} \bar{L}_m i_{as} - \frac{1}{2} \bar{L}_m i_{bs}$$
$$+ L_{ss} i_{cs} + \psi_m \left[a_1 \sin \left(\theta_r - \frac{2}{3}\pi \right) + a_2 \sin^3 \left(\theta_r - \frac{2}{3}\pi \right) + a_3 \sin^5 \left(\theta_r - \frac{2}{3}\pi \right) \right],$$

are substituted in the Kirchhoff's second law (6.12) yielding

$$u_{as} = r_s i_{as} + L_{ss} \frac{di_{as}}{dt} - \frac{1}{2} \bar{L}_m \frac{di_{bs}}{dt} - \frac{1}{2} \bar{L}_m \frac{di_{cs}}{dt} + \psi_m \cos \theta_r \left(a_1 + 3a_2 \sin^2 \theta_r + 5a_3 \sin^4 \theta_r \right) \omega_r,$$

$$u_{bs} = r_s i_{bs} - \frac{1}{2} \bar{L}_m \frac{di_{as}}{dt} + L_{ss} \frac{di_{bs}}{dt} - \frac{1}{2} \bar{L}_m \frac{di_{cs}}{dt} + \psi_m \cos \left(\theta_r + \frac{2}{3}\pi \right) \left[a_1 + 3a_2 \sin^2 \left(\theta_r + \frac{2}{3}\pi \right) + 5a_3 \sin^4 \left(\theta_r + \frac{2}{3}\pi \right) \right] \omega_r$$

$$u_{cs} = r_s i_{cs} - \frac{1}{2} \bar{L}_m \frac{di_{as}}{dt} - \frac{1}{2} \bar{L}_m \frac{di_{bs}}{dt} + L_{ss} \frac{di_{cs}}{dt} + \psi_m \cos \left(\theta_r - \frac{2}{3}\pi \right) \left[a_1 + 3a_2 \sin^2 \left(\theta_r - \frac{2}{3}\pi \right) + 5a_3 \sin^4 \left(\theta_r - \frac{2}{3}\pi \right) \right] \omega_r$$

The circuitry-electromagnetic dynamics is

$$\frac{di_{as}}{dt} = \frac{1}{L_{ss}} \left[-r_s i_{as} + \frac{1}{2} \bar{L}_m \frac{di_{bs}}{dt} + \frac{1}{2} \bar{L}_m \frac{di_{cs}}{dt} - \psi_m \cos \theta_r \left(a_1 + 3a_2 \sin^2 \theta_r + 5a_3 \sin^4 \theta_r \right) \omega_r + u_{as} \right], \tag{6.24}$$

$$\frac{di_{bs}}{dt} = \frac{1}{L_{ss}} \left[-r_s i_{bs} + \frac{1}{2} \bar{L}_m \frac{di_{as}}{dt} + \frac{1}{2} \bar{L}_m \frac{di_{cs}}{dt} - \psi_m \cos \left(\theta_r + \frac{2}{3}\pi \right) \left(a_1 + 3a_2 \sin^2 \left(\theta_r + \frac{2}{3}\pi \right) + 5a_3 \sin^4 \left(\theta_r + \frac{2}{3}\pi \right) \right) \omega_r + u_{bs} \right]$$

$$\frac{di_{cs}}{dt} = \frac{1}{L_{ss}} \left[-r_s i_{cs} + \frac{1}{2} \bar{L}_m \frac{di_{as}}{dt} + \frac{1}{2} \bar{L}_m \frac{di_{bs}}{dt} - \psi_m \cos \left(\theta_r - \frac{2}{3}\pi \right) \left(a_1 + 3a_2 \sin^2 \left(\theta_r - \frac{2}{3}\pi \right) + 5a_3 \sin^4 \left(\theta_r - \frac{2}{3}\pi \right) \right) \omega_r + u_{cs} \right]$$

From (6.20), the expression for the electromagnetic torque is

$$T_e = \frac{P\psi_m}{2}\left[i_{as}\cos\theta_r(a_1 + 3a_2\sin^2\theta_r + 5a_3\sin^4\theta_r) + i_{bs}\cos(\theta_r + \tfrac{2}{3}\pi)(a_1 + 3a_2\sin^2(\theta_r + \tfrac{2}{3}\pi)\right.$$
$$\left. +5a_3\sin^4(\theta_r + \tfrac{2}{3}\pi)) + i_{cs}\cos(\theta_r - \tfrac{2}{3}\pi)(a_1 + 3a_2\sin^2(\theta_r - \tfrac{2}{3}\pi) + 5a_3\sin^4(\theta_r - \tfrac{2}{3}\pi))\right]$$

The *torsional–mechanical* equations of motion (6.21) are

$$\frac{d\omega_r}{dt} = \frac{P^2\psi_m}{4J}\left[i_{as}\cos\theta_r(a_1 + 3a_2\sin^2\theta_r + 5a_3\sin^4\theta_r).\right.$$
$$+i_{bs}\cos\left(\theta_r + \frac{2}{3}\pi\right)\left(a_1 + 3a_2\sin^2\left(\theta_r + \frac{2}{3}\pi\right) + 5a_3\sin^4\left(\theta_r + \frac{2}{3}\pi\right)\right)$$
$$\left. +i_{cs}\cos\left(\theta_r - \frac{2}{3}\pi\right)\left(a_1 + 3a_2\sin^2\left(\theta_r - \frac{2}{3}\pi\right) + 5a_3\sin^4\left(\theta_r - \frac{2}{3}\pi\right)\right)\right] - \frac{B_m}{J}\omega_r - \frac{P}{2J}T_L,$$
$$\frac{d\theta_r}{dt} = \omega_r. \tag{6.25}$$

Using (6.24) and (6.25), the Simulink diagram to simulate permanent-magnet synchronous motors ($a_1 \neq 0$, $a_2 \neq 0$, $a_3 \neq 0$, and $\forall a_n = 0$, $n > 3$) is developed as reported in Figure 6.14. The following phase voltages, as functions of θ_r, are supplied

$$u_{as} = \sqrt{2}u_M\cos\theta_r, \quad u_{bs} = \sqrt{2}u_M\cos\left(\theta_r + \frac{2}{3}\pi\right), \quad \text{and} \quad u_{cs} = \sqrt{2}u_M\cos\left(\theta_r - \frac{2}{3}\pi\right).$$

This voltage set can be used because $a_1 \gg (2n-1)a_n$, $\forall n > 1$.

The motor dynamics is studied as the motor accelerates from stall with the rated voltage applied ($u_M = 50$ V) at no load. The load torque $T_L = 0.1$ N-m is applied at $t = 0.025$ s. Figure 6.15 illustrates the evolution of the phase currents and electrical angular velocity. The motor reaches the steady-state ω_r (500 rad/s with no load, and 490 rad/s if $T_L = 0.1$ N-m, respectively) within 0.02 s. The angular velocity reduces and phase currents magnitude increases as T_L is applied. The current dynamics, reported in Figure 6.15, allows one to assess the motor performance and capabilities when $a_1 = 1$ and $\forall a_n = 0$, $n > 1$.

For the loaded motor, a_n coefficients (found by examining the induced *emf* when the motor is operated as a generator) are $a_1 = 1$, $a_2 = 0.05$, $a_3 = 0.02$, and $\forall a_n = 0$, $n > 3$. The motor accelerates with the rated voltage applied, e.g., $u_M = 50$ V. At $t = 0$ s, the load torque is $T_{L0} = 0.2$ N-m. At $t = 0.025$ s, the load increases to $T_L = 0.5$ N-m. The evolution of the phase currents and the electrical angular velocity is documented in Figure 6.16. The motor reaches the steady-state operation within 0.02 s. One observes the phase current chattering and the electromagnetic torque ripple. This leads to the reduction of efficiency, losses, vibration, noise, heating, etc. There are many other secondary effects which degrade the motor performance, for example, for the overheated motor, ψ_m reduces leading to the reduction of T_e as well as torque density. Using the conditional statements and look-up table, the near-balanced voltage sets can be implemented in a full operating envelope. The closed-loop electromechanical systems must be designed to ensure optimal performance guarantying *achievable* capabilities. ∎

Example 6.6:　Modeling, Analysis, and Simulation of a Permanent-Magnet Synchronous Generator

Experimental results may indicate that in the full operating envelope $a_1 \neq 0$, $a_2 \neq 0$, $a_3 \neq 0$, and $\forall a_n = 0$, $n > 3$. The equations of motion for permanent-magnet synchronous generators is developed by using the circuitry-electromagnetic equations (6.12) with (6.19), expression for

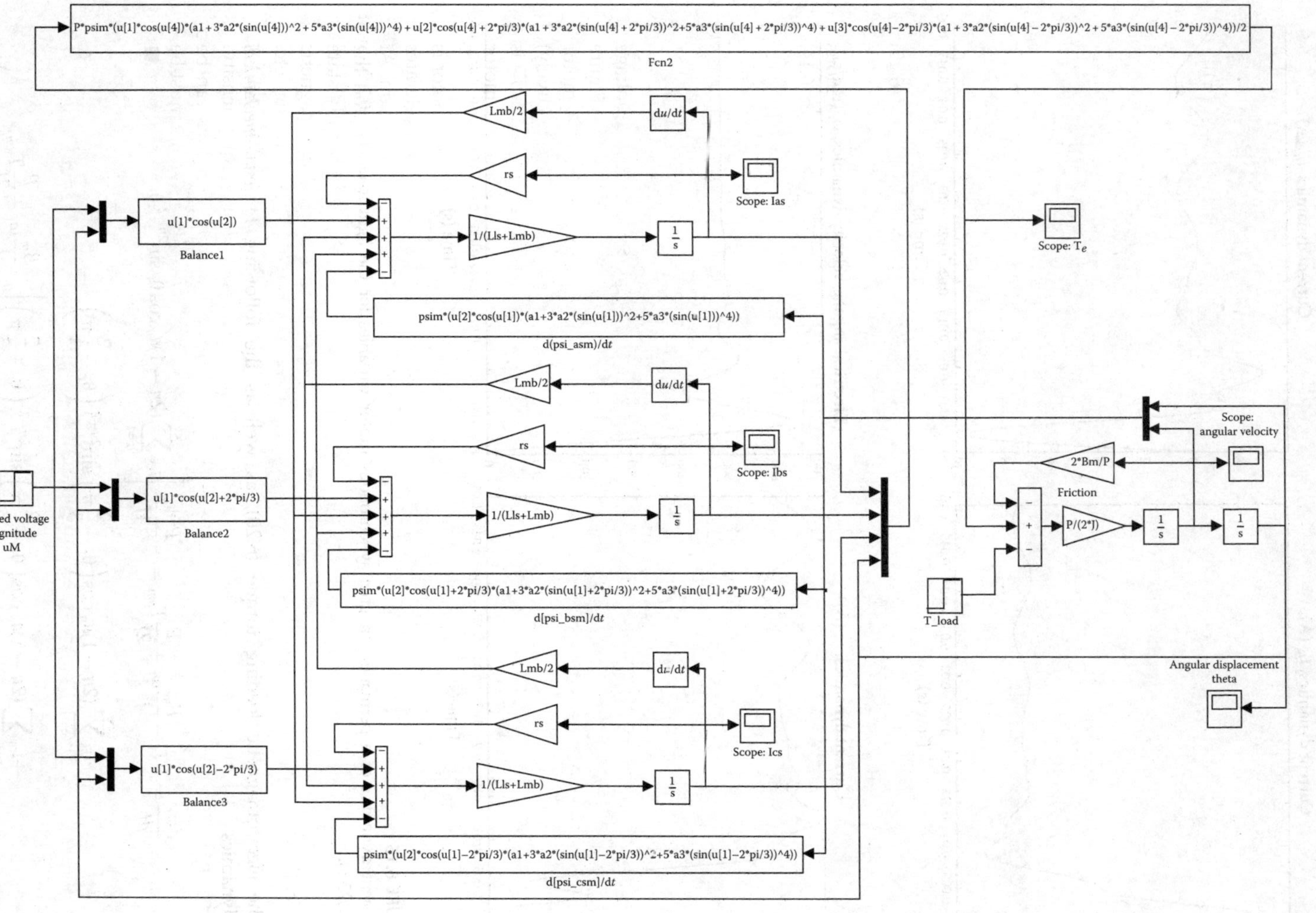

FIGURE 6.14

Simulink diagram to simulate permanent-magnet synchronous motors when $a_1 \neq 0$, $a_2 \neq 0$, $a_3 \neq 0$, and $\forall a_n = 0$, $n > 3$ (ch6_03.mdl).

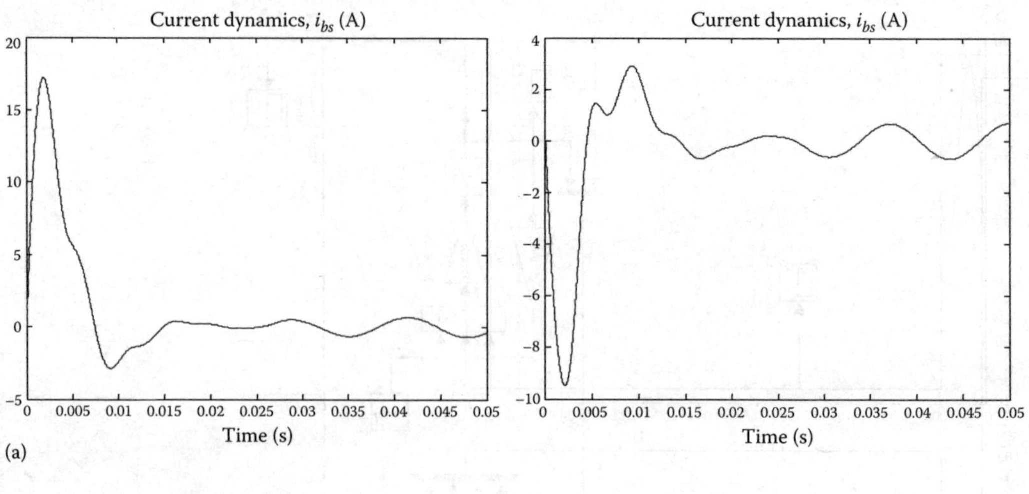

(a)

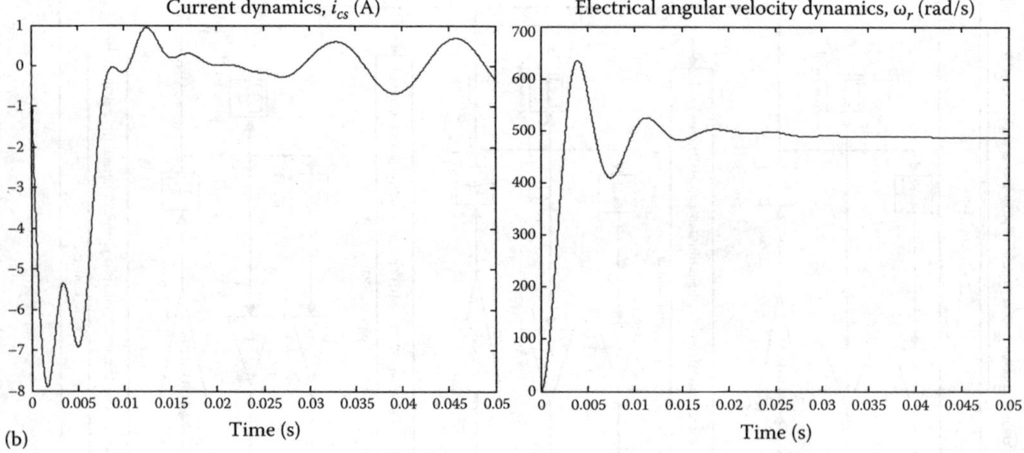

(b)

FIGURE 6.15

Transient dynamics of a permanent-magnet synchronous motor variables for the light load ($T_L = 0.1$ N-m at $t = 0.025$ s).

the electromagnetic loading torque (6.20), as well as the following *torsional–mechanical* dynamics

$$\frac{d\omega_r}{dt} = -T_e - \frac{B_m}{J}\omega_r + \frac{P}{2J}T_{pm} = -\frac{P^2\psi_m}{4J}\left[i_{as}\sum_{n=1}^{\infty}(2n-1)a_n\cos\theta_r\sin^{2n-2}\theta_r \right.$$

$$+ i_{bs}\sum_{n=1}^{\infty}(2n-1)a_n\cos\left(\theta_r - \frac{2}{3}\pi\right)\sin^{2n-2}\left(\theta_r - \frac{2}{3}\pi\right)$$

$$\left. + i_{cs}\sum_{n=1}^{\infty}(2n-1)a_n\cos\left(\theta_r + \frac{2}{3}\pi\right)\sin^{2n-2}\left(\theta_r + \frac{2}{3}\pi\right)\right] - \frac{B_m}{J}\omega_r + \frac{P}{2J}T_{pm},$$

$$\frac{d\theta_r}{dt} = \omega_r.$$

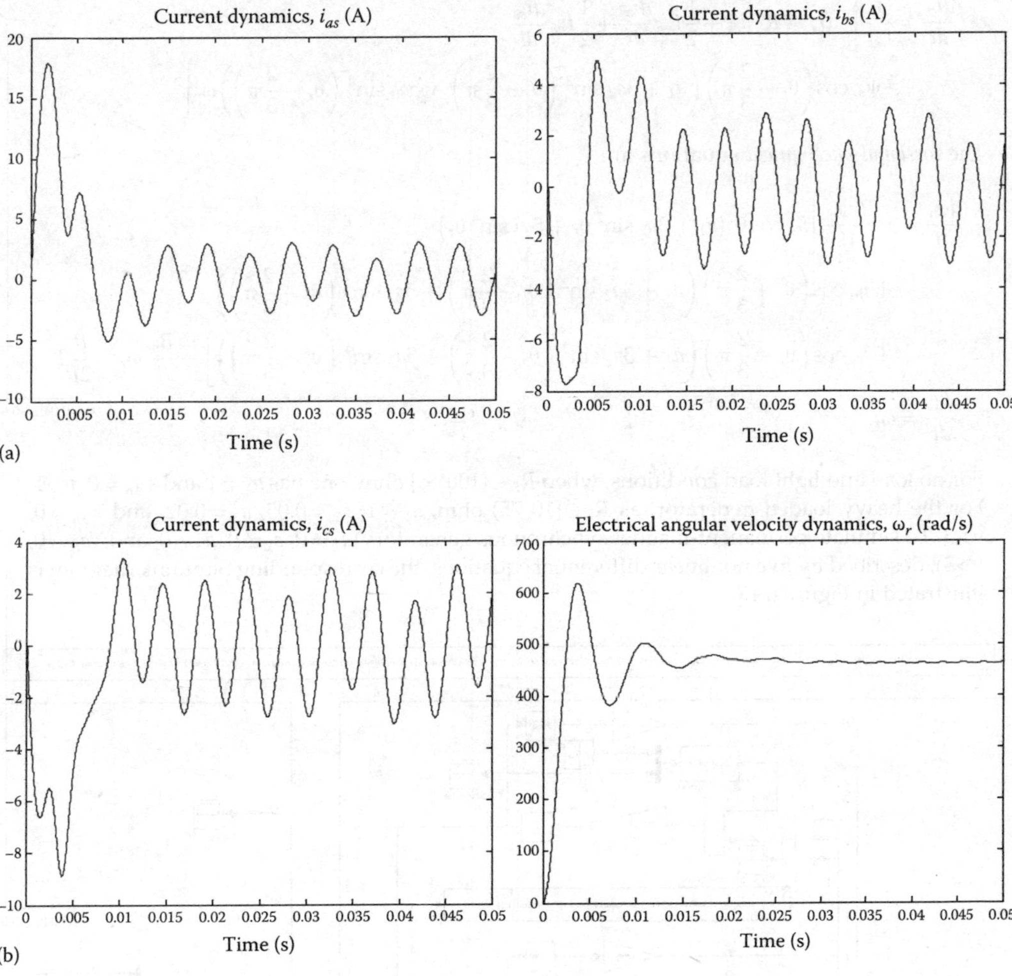

FIGURE 6.16
Transient dynamics of a permanent-magnet synchronous motor variables for the rated load ($T_{L0} = 0.2$ N-m and $T_L = 0.5$ N-m at $t = 0.025$ s).

The torque developed by the prime mover T_{pm} rotates the generator, and the terminal phase voltages are the induced *emfs*. We simulate and analyze a radial topology three-phase permanent-magnet synchronous generator driven by a prime mover which develops T_{pm}. The generator parameters are the same as used in Example 6.5. Using the expressions for the flux linkages, one finds the circuitry-electromagnetic dynamics as

$$\frac{di_{as}}{dt} = \frac{1}{L_{ss}}\left[-(r_s + R_L)i_{as} + \frac{1}{2}\bar{L}_m\frac{di_{bs}}{dt} + \frac{1}{2}\bar{L}_m\frac{di_{cs}}{dt} - \psi_m\cos\theta_r\left(a_1 + 3a_2\sin^2\theta_r + 5a_3\sin^4\theta_r\right)\omega_r\right],$$

$$\frac{di_{bs}}{dt} = \frac{1}{L_{ss}}\left[-(r_s + R_L)i_{bs} + \frac{1}{2}\bar{L}_m\frac{di_{as}}{dt} + \frac{1}{2}\bar{L}_m\frac{di_{cs}}{dt}\right.$$
$$\left. -\psi_m\cos\left(\theta_r + \frac{2}{3}\pi\right)\left(a_1 + 3a_2\sin^2\left(\theta_r + \frac{2}{3}\pi\right) + 5a_3\sin^4\left(\theta_r + \frac{2}{3}\pi\right)\right)\omega_r + u_{bs}\right]$$

$$\frac{di_{cs}}{dt} = \frac{1}{L_{ss}}\left[-(r_s + R_L)i_{cs} + \frac{1}{2}\bar{L}_m\frac{di_{as}}{dt} + \frac{1}{2}\bar{L}_m\frac{di_{bs}}{dt} \right.$$

$$\left. -\psi_m\cos\left(\theta_r - \frac{2}{3}\pi\right)\left(a_1 + 3a_2\sin^2\left(\theta_r - \frac{2}{3}\pi\right) + 5a_3\sin^4\left(\theta_r - \frac{2}{3}\pi\right)\right)\omega_r \right]$$

The *torsional–mechanical* equations are

$$\frac{d\omega_r}{dt} = -\frac{P^2\psi_m}{4J}\left[i_{as}\cos\theta_r\left(a_1 + 3a_2\sin^2\theta_r + 5a_3\sin^4\theta_r\right) \right.$$

$$+ i_{bs}\cos\left(\theta_r + \frac{2}{3}\pi\right)\left(a_1 + 3a_2\sin^2\left(\theta_r + \frac{2}{3}\pi\right) + 5a_3\sin^4\left(\theta_r + \frac{2}{3}\pi\right)\right)$$

$$\left. + i_{cs}\cos\left(\theta_r - \frac{2}{3}\pi\right)\left(a_1 + 3a_2\sin^2\left(\theta_r - \frac{2}{3}\pi\right) + 5a_3\sin^4\left(\theta_r - \frac{2}{3}\pi\right)\right)\right] - \frac{B_m}{J}\omega_r + \frac{P}{2J}T$$

$$\frac{d\theta_r}{dt} = \omega_r.$$

For no load and light load conditions, when $R_L \in [100\ \infty]$ ohm, one has $a_1 = 1$ and $\forall a_n = 0$, $n > 1$. For the heavy loaded generator, as $R_L \in [10\ 75)$ ohm, $a_1 = 1$, $a_2 = 0.05$, $a_3 = 0.02$, and $\forall a_n = 0$, $n > 3$. To simulate permanent-magnet synchronous generators ($a_1 \neq 0$, $a_2 \neq 0$, $a_3 \neq 0$, and $\forall a_n = 0$, $n > 3$), described by five nonlinear differential equations, the corresponding Simulink diagram is illustrated in Figure 6.17.

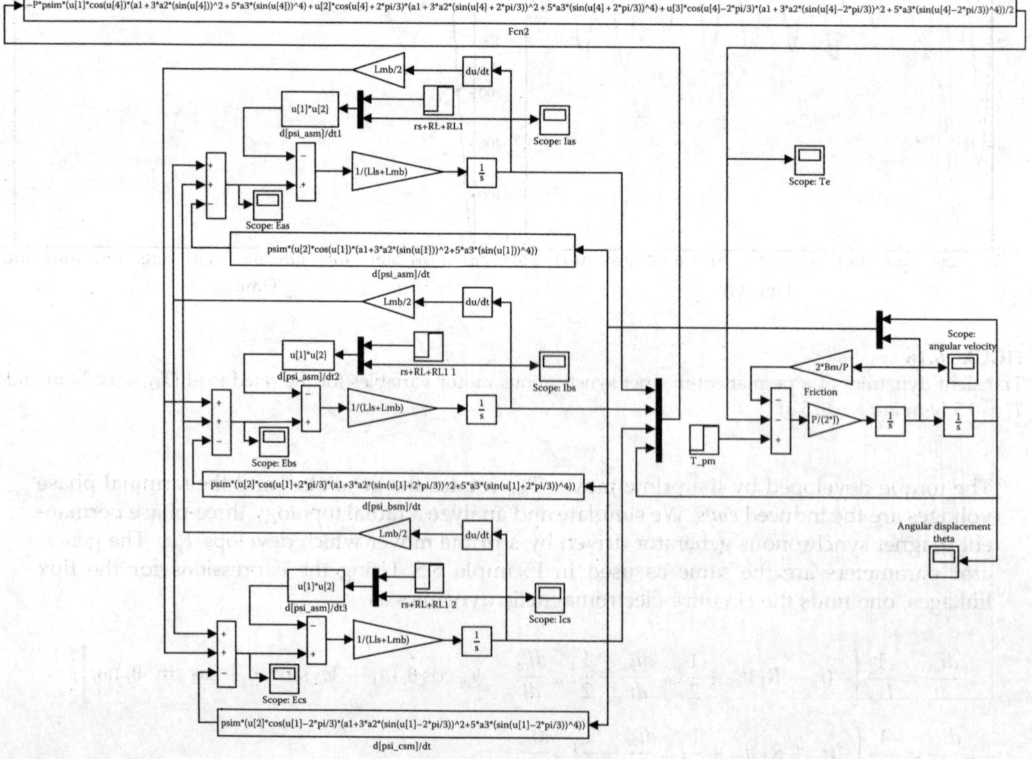

FIGURE 6.17
Simulink diagram to simulate permanent-magnet synchronous generators if $a_1 \neq 0$, $a_2 \neq 0$, $a_3 \neq 0$, and $\forall a_n = 0$, $n > 3$ (ch6_04.mdl).

The generator dynamics and voltage generation are studied as the generator accelerates from stall with $T_{pm} = 0.05$ N-m. The symmetric three-phase load resistors $R_L = 150$ ohm and $R_L = 100$ ohm are inserted at $t = 0$ s and $t = 0.5$ s, respectively. Figure 6.18 illustrates the evolution of the induced *emfs*, which define the terminal phase voltages, and the electrical angular velocity. The generator reaches the steady-state ω_r when $T_{pm} = T_e + T_{\text{friction}}$. The generator is loaded by the symmetric wye-connected three-phase resistors R_L, which are in series with r_s of the *abc* phases. As the load increases (R_L are reduced), the angular velocity and induced *emfs* decrease. The generator dynamics, reported in Figure 6.18, allows one to assess the lightly loaded generator performance and capabilities when $a_1 = 1$ and $\forall a_n = 0$, $n > 1$.

For the loaded generator, when $R_L \in [10\ 75)$ ohm (peak and rated loads, respectively), we have $a_1 = 1$, $a_2 = 0.05$, $a_3 = 0.02$, and $\forall a_n = 0$, $n > 3$. The generator performance is studied if $R_L = 25$ ohm at $t = 0$ s, and $R_L = 75$ ohm at $t = 0.5$ s. The evolution of the induced *emfs* and the electrical angular velocity are documented in Figure 6.19. As R_L increases at $t = 0.5$ s, the load reduces, and the angular velocity increases. This results in the increase of terminal phase voltages. The induced *emfs* can be significantly distorted if the coefficients a_n, $n > 1$ are relatively high as compared to a_1. This results in the efficiency reduction, losses, and other undesirable

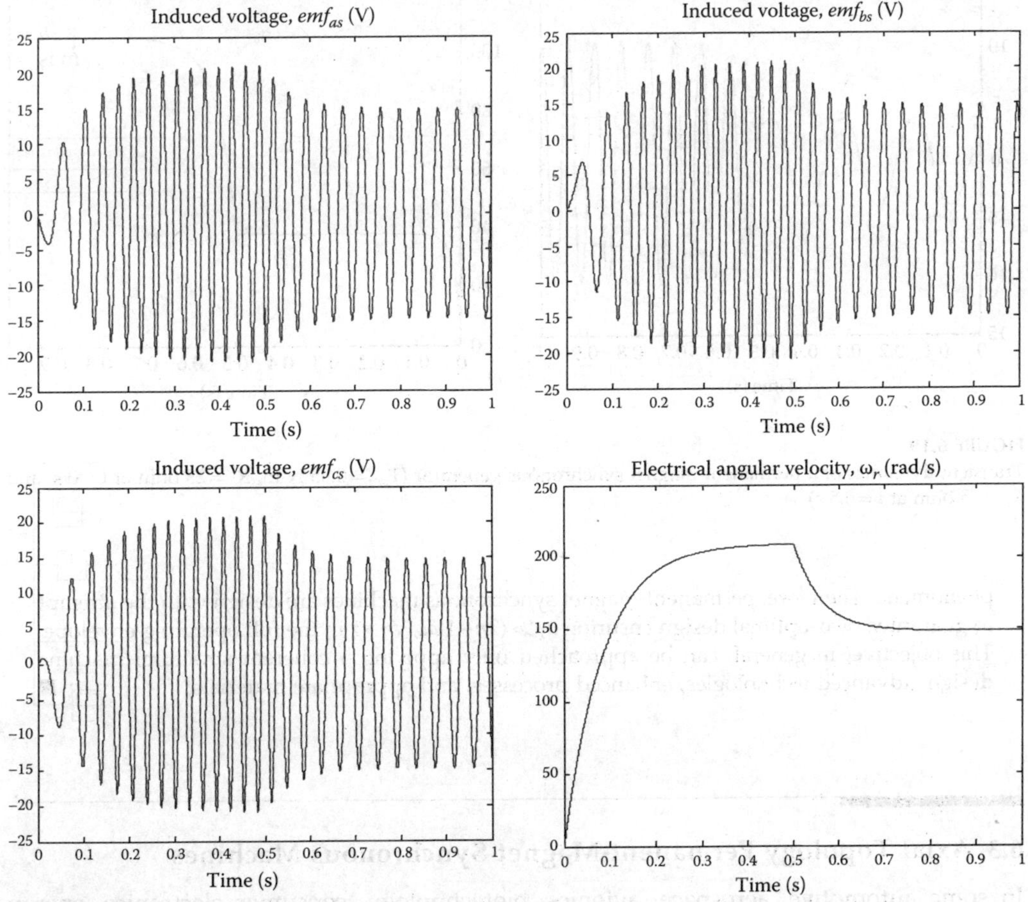

FIGURE 6.18
Transient dynamics of a permanent-magnet synchronous generator ($T_{pm} = 0.05$ N-m, $R_L = 150$ ohm at $t = 0$ s, and $R_L = 100$ ohm at $t = 0.5$ s).

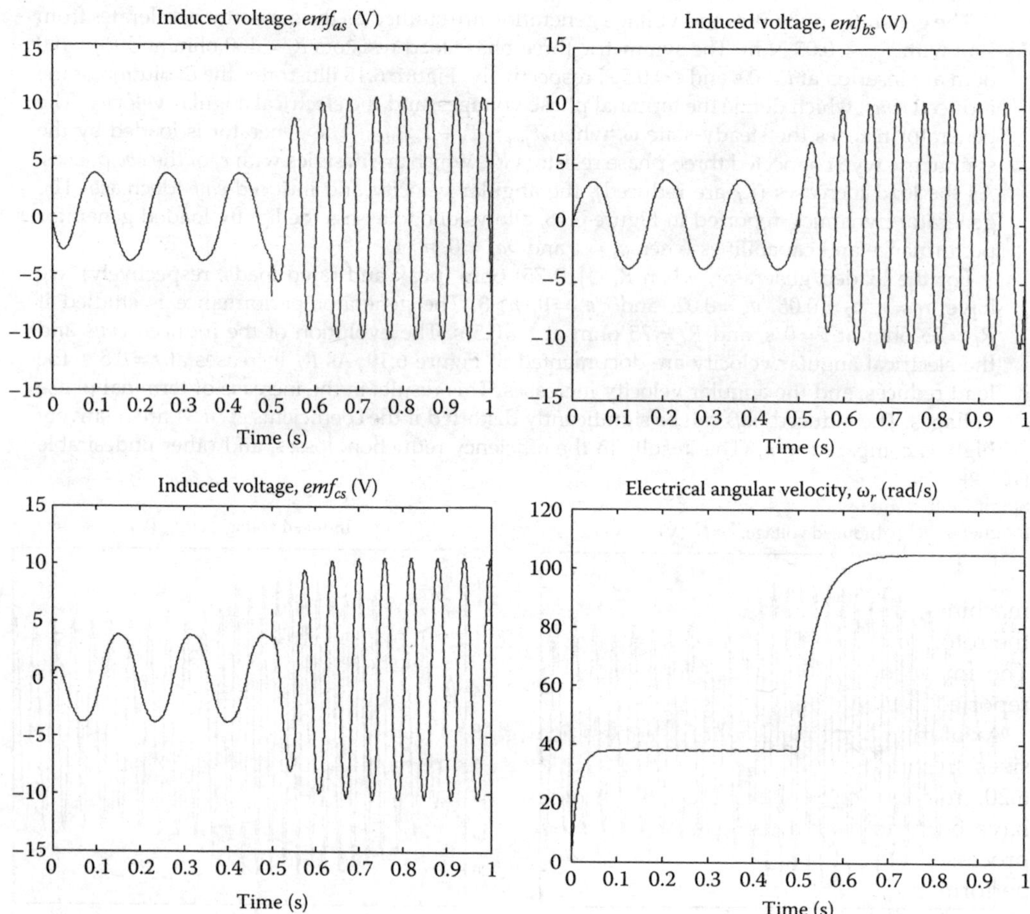

FIGURE 6.19

Transient dynamics of a permanent-magnet synchronous generator ($T_{pm} = 0.05$ N-m, $R_L = 25$ ohm at $t = 0$ s, and $R_L = 75$ ohm at $t = 0.5$ s).

phenomena. Therefore, permanent-magnet synchronous machines are design with the attempt to guarantee near-optimal design ensuring $a_1 \gg (2n-1)a_n$, $\forall n > 1$ in the full operating envelope. This objective, in general, can be approached only applying a coherent structural machine design, advanced technologies, enhanced processes, and appropriate materials. ∎

6.3 Axial Topology Permanent-Magnet Synchronous Machines

In some automotive, aerospace, avionics, biotechnology, consumer electronics, energy systems, marine, medical, robotics, and other applications, axial topology permanent-magnet synchronous machines could be a preferable solution. The axial topology permanent-magnet DC electromechanical motion devices were examined in Chapter 4. In synchronous

FIGURE 6.20
Stator (with planar windings) and rotor (with magnet segments array) of a three-phase axial topology permanent-magnet synchronous machine which can be used as a motor and generator.

machines, the stationary magnetic field is established by permanent magnets placed on the rotor, and the AC phase voltages are applied to the stator windings as functions of θ_r. The image of a three-phase axial topology permanent-magnet synchronous machine is reported in Figure 6.20.

Axial topology permanent-magnet synchronous machines can be fabricated in different sizes. In contrast to the conventional technology as applied to a device illustrated in Figure 6.20, micro- and miniscale axial topology permanent-magnet synchronous machines have been fabricated using *surface* and *bulk* micromachining applying enhanced CMOS processes. The advantages of axial topology electromechanical motion devices are excellent performance, enabling capabilities, affordable high-yield fabrication, assembly and packaging simplicity, etc. This simplicity results because (1) segmented arrays are formed by planar magnets; (2) there are no very strict three-dimensional geometry and magnetization requirements imposed on magnets (though inadequate geometry and magnetization will significantly degrade the overall performance); (3) rotor back ferromagnetic material is not required; and (4) it is easy to make planar windings on the planar stator. The axial topology permanent-magnet synchronous machines are reported in Figure 6.21.

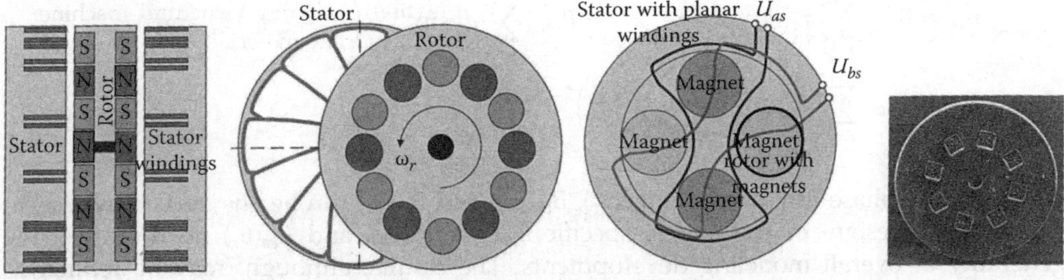

FIGURE 6.21
Axial topology permanent-magnet synchronous machines and fabricated ministructure (rotor or stator) with planar segmented magnets.

Two- and three-phase axial permanent-magnet synchronous machines are illustrated in Figures 6.20 and 6.21.

Assuming that the magnetic flux is constant through the magnetic plane (current loop), the torque on a planar current loop of any size and shape in the uniform magnetic field is $\vec{T} = i\vec{s} \times \vec{B} = \vec{m} \times \vec{B}$. In permanent-magnet DC motion devices, brushes are needed to supply the voltage to the armature windings on the rotor, and, a commutator is utilized. In permanent-magnet synchronous motors, phase windings are on the stator (stationary part). Correspondingly, brushes are not needed. However, to develop the electromagnetic torque, the phase voltages are supplied as functions of the angular displacement θ_r. Hence, there is a need for the angular displacement sensors and advanced power electronics. For three-phase synchronous machines, one supplies phase voltages u_{as}, u_{bs}, and u_{cs}. For two-phase machines, phase voltages u_{as} and u_{bs} are applied.

Consider axial topology permanent-magnet synchronous machines. The *effective* phase flux density varies as a function of θ_r due to the angular displacement of the rotor with magnets relative to the stator with windings. For three-phase axial topology permanent-magnet synchronous machines, depending on the topology, magnet magnetization, geometry, and shape, one finds distinct expressions for the *effective* phase flux densities $B_{as}(\theta_r)$, $B_{bs}(\theta_r)$, and $B_{cs}(\theta_r)$, which are periodic functions of θ_{rm}. For AC machines, it is convenient to use the electric angular displacement θ_r which is related to the mechanical angular displacement θ_{rm}, $\theta_{rm} = 2\theta_r/N_m$. Here, N_m is the number of magnets (segments). If an optimal structural design is accomplished and magnets are ideally magnetized, one may have

$$B_{as}(\theta_r) = B_{\max} \sin(\theta_r), \quad B_{bs}(\theta_r) = B_{\max} \sin\left(\theta_r - \frac{2}{3}\pi\right), \quad \text{and} \quad B_{cs}(\theta_r) = B_{\max} \sin\left(\theta_r + \frac{2}{3}\pi\right),$$

where $B_{\max}$ is the maximum *effective* flux density produced by the magnets as viewed from the winding ($B_{\max}$ depends on the magnets used, magnet-winding separation, number of layers, temperature, etc.).

For the specific magnet (magnet array) topologies and configurations, particular *effective* phase flux densities result. For example, one may strive to ensure $B_{as}(\theta_r) = B_{\max}|\sin \theta_r|$, $B_{bs}(\theta_r) = B_{\max}\left|\sin\left(\theta_r - \frac{2}{3}\pi\right)\right|$, and $B_{cs}(\theta_r) = B_{\max}\left|\sin\left(\theta_r + \frac{2}{3}\pi\right)\right|$.

In the full operating envelope, one experimentally and analytically derives $a_n(\mathbf{E}, \mathbf{D}, \mathbf{B}, \mathbf{H}, i_{abcs}, \omega_r, \mathbf{T}_L, \varepsilon, \mu, \Sigma)$. Therefore, we have

$$B_{as} = B_{\max} \sum_{n=1}^{\infty} a_n \sin^{2n-1} \theta_r, \quad B_{bs} = B_{\max} \sum_{n=1}^{\infty} a_n \sin^{2n-1}\left(\theta_r - \frac{2}{3}\pi\right), \quad \text{and}$$

$$B_{cs} = B_{\max} \sum_{n=1}^{\infty} a_n \sin^{2n-1}\left(\theta_r + \frac{2}{3}\pi\right).$$

$$(6.26)$$

Other *effective* phase flux densities $B_{as}(\theta_r)$, $B_{bs}(\theta_r)$, and $B_{cs}(\theta_r)$ can be ensured. However, in general, any design- or technology-specific $B_{as}(\theta_r)$, $B_{bs}(\theta_r)$, and $B_{cs}(\theta_r)$ do not affect the generality of overall modeling developments. The results, although, may be refined as the expressions for *effective* phase flux densities may vary.

From (6.26), using the number of turns and the *effective* area, one obtains the expression for the flux linkages

$$
\begin{bmatrix} \psi_{as} \\ \psi_{bs} \\ \psi_{cs} \end{bmatrix} = \begin{bmatrix} L_{ss} & 0 & 0 \\ 0 & L_{ss} & 0 \\ 0 & 0 & L_{ss} \end{bmatrix} \begin{bmatrix} i_{as} \\ i_{bs} \\ i_{cs} \end{bmatrix} + \psi_m \begin{bmatrix} \sum_{n=1}^{\infty} a_n \sin^{2n-1} \theta_r \\ \sum_{n=1}^{\infty} a_n \sin^{2n-1} \left(\theta_r - \frac{2}{3}\pi \right) \\ \sum_{n=1}^{\infty} a_n \sin^{2n-1} \left(\theta_r + \frac{2}{3}\pi \right) \end{bmatrix}. \tag{6.27}
$$

In (6.27), the mutual inductances between the planar windings are zero (or negligibly small), while there are mutual inductances $-\frac{1}{2}\bar{L}_m$ in radial topology permanent-magnet synchronous machines due to the ferromagnetic core. Substituting (6.27) in the Kirchhoff's second law (6.12), the circuitry-electromagnetic equations of motion in the *machine* variables result.

The expression $T_e = \dfrac{N_m}{2} \dfrac{\partial W_c}{\partial \theta_r}$ yields

$$
\begin{aligned}
T_e = \frac{N_m \psi_m}{2} \Bigg[& i_{as} \sum_{n=1}^{\infty} (2n-1)a_n \cos \theta_r \sin^{2n-2} \theta_r \\
& + i_{bs} \sum_{n=1}^{\infty} (2n-1)a_n \cos\left(\theta_r - \frac{2}{3}\pi \right) \sin^{2n-2} \left(\theta_r - \frac{2}{3}\pi \right) \\
& + i_{cs} \sum_{n=1}^{\infty} (2n-1)a_n \cos\left(\theta_r + \frac{2}{3}\pi \right) \sin^{2n-2} \left(\theta_r + \frac{2}{3}\pi \right) \Bigg].
\end{aligned} \tag{6.28}
$$

Hence, the *torsional–mechanical* dynamics is

$$
\begin{aligned}
\frac{d\omega_r}{dt} = \frac{N_m^2 \psi_m}{4J} \Bigg[& i_{as} \sum_{n=1}^{\infty} (2n-1)a_n \cos \theta_r \sin^{2n-2} \theta_r \\
& + i_{bs} \sum_{n=1}^{\infty} (2n-1)a_n \cos\left(\theta_r - \frac{2}{3}\pi \right) \sin^{2n-2} \left(\theta_r - \frac{2}{3}\pi \right) \\
& + i_{cs} \sum_{n=1}^{\infty} (2n-1)a_n \cos\left(\theta_r + \frac{2}{3}\pi \right) \sin^{2n-2} \left(\theta_r + \frac{2}{3}\pi \right) \Bigg] - \frac{B_m}{J}\omega_r - \frac{N_m}{2J}T_L,
\end{aligned}
$$

$$
\frac{d\theta_r}{dt} = \omega_r. \tag{6.29}
$$

The near-balanced current and voltage sets are given by (6.22) and (6.23).

Example 6.7: Axial Topology Single-Phase Permanent-Magnet Synchronous Machine

Consider a single-phase permanent-magnet synchronous motor with the segmented array of the permanent magnets as illustrated in Figure 6.22.

Let the *effective* flux density be $B_{as}(\theta_{rm}) = B_{\max} \sin\left(\frac{1}{2}N_m \theta_{rm} \right)$, or $B_{as}(\theta_r) = B_{\max} \sin \theta_r$. The electromagnetic torque, developed by single-phase axial topology permanent-magnet synchronous motors, is found by using $\vec{T} = i\vec{s} \times \vec{B} = \vec{m} \times \vec{B}$ or $\vec{T} = \vec{R} \times \vec{F}$. Alternatively, the expression for

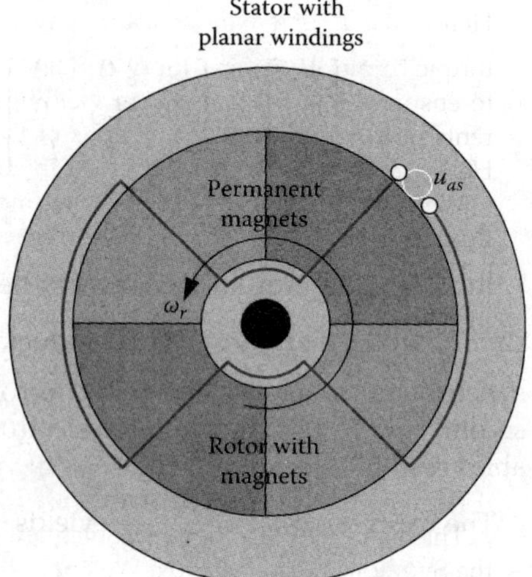

FIGURE 6.22
Axial topology single-phase permanent-magnet
synchronous machine.

coenergy $W_c(i_{as}, \theta_r) = N A_{eq} B_{as}(\theta_r) i_{as}$ can be used to derive T_e. Here, N is the number of turns; A_{eq} is the *effective* area which takes into account magnetic field nonuniformity, magnets nonuniformity, etc.

We have $T_e = \dfrac{\partial W_c(i_{as}, \theta_r)}{\partial \theta_r} = \dfrac{N_m \psi_m}{2} i_{as} \cos \theta_r$.

The electromagnetic torque is not developed by the synchronous motors if one feeds the DC current or voltage to the winding. The average value of T_e is not equal to zero if the current is a function of the rotor displacement θ_r. As an illustration, we fed the phase current $i_{as} = i_M \cos \theta_r$. The electromagnetic torque is $T_e = \dfrac{N_m \psi_m}{2} i_M \cos^2 \theta_r$ and $T_{e\,av} \neq 0$. However, there will be the torque ripple.

The equations of motion are found by using the Kirchhoff's law $u_{as} = r_s i_{as} + \dfrac{d\psi_{as}}{dt}$ and Newtonian mechanics. From the flux linkage equation $\psi_{as} = L_{ss} i_{as} + \psi_m \sin \theta_r$, we have

$$\frac{di_{as}}{dt} = \frac{1}{L_{ss}}(-r_s i_{as} - \psi_m \cos \theta_r \omega_r + u_{as})$$

$$\frac{d\omega_r}{dt} = \frac{N_m^2 \psi_m}{4J} i_{as} \cos \theta_r - \frac{B_m}{J} \omega_r - \frac{N_m}{2J} T_L,$$

$$\frac{d\theta_r}{dt} = \omega_r.$$

∎

Example 6.8:   Axial Topology Two-Phase Permanent-Magnet Synchronous Motors

Consider two-phase permanent-magnet synchronous machines as depicted in Figure 6.21. Let the *effective* phase flux densities be

$$B_{as} = B_{\max} \sin \theta_r \quad \text{and} \quad B_{bs} = B_{\max} \cos \theta_r.$$

The electromagnetic torque is found to be $T_e = \dfrac{N_m \psi_m}{2}(i_{as} \cos \theta_r - i_{bs} \sin \theta_r)$.
   The balanced current set is

$$i_{as} = i_M \cos \theta_r, \quad i_{bs} = i_M \sin \theta_r,$$

Hence, the T_e is maximized. From $T_e = \dfrac{N_m \psi_m}{2} i_M$ one concludes that theoretically there is no torque ripple. Therefore, for two-phase permanent-magnet synchronous machines, it is desired to ensure the structural design yielding $B_{as} = B_{max} \sin \theta_r$ and $B_{bs} = B_{max} \cos \theta_r$. The phase currents or voltages are the functions of the electrical angular displacement θ_r measured by the Hall-effect sensors or calculated by the observers.

We refine our analysis considering more practical case. Let $a_3 = 1$ and all other $\forall a_n = 0$. This flux distribution may correspond to the rotor with planar segmented magnets shown in Figure 6.21. One obtains

$$\psi_{asm} = \psi_m \sin^5 \theta_r \quad \text{and} \quad \psi_{bsm} = \psi_m \cos^5 \theta_r.$$

The electromagnetic torque is $T_e = \dfrac{5 N_m \psi_m}{2} \left(i_{as} \cos \theta_r \sin^4 \theta_r - i_{bs} \sin \theta_r \cos^4 \theta_r \right)$.

Let the motor parameters be $N = 20$, $A_{eq} = 0.001$, $B_{max} = 1$, and $N_m = 8$.

Assume that the phase currents are $i_{as} = i_M \cos \theta_r$ and $i_{bs} = -i_M \sin \theta_r$, $i_M = 2$ A.

The expression for the electromagnetic torque, deviations, and plotting are performed using the Symbolic Math Toolbox. We apply the equation for the coenergy to obtain the electromagnetic torque performing the differentiation $T_e = \dfrac{\partial W_c(i_{as}, \theta_r)}{\partial \theta_r}$. The following MATLAB® file with comments is used

```
% To use a symbolic variable, create an object of type SYM
x = sym ('x');
N = 20; Aeq = 0.001; Bmax = 1; psim = N*Aeq*Bmax; Nm = 8; iM = 2;
y1 = N*Aeq*Bmax*sin (x)^5; y2 = N*Aeq*Bmax*cos (x)^5;
% Differentiate y1 and y2 using the DIFF command
d1 = diff (y1); d2 = diff (y2);
% Phase currents
ias = iM*cos (x); ibs = -iM*sin (x);
% Derive and plot the electromagnetic torque
Te = Nm* (d1*ias+d2*ibs)/2, Te = simplify (Te), ezplot (Te)
```

The results of the calculations are reported in the Command Window as

```
Te = 4/5*sin (x)^4*cos (x)^2+4/5*cos (x)^4*sin (x)^2
Te = -4/5* (-1+cos (x)^2) *cos (x)^2
```

One concludes that the electromagnetic torque is $T_e = -\dfrac{4}{5} \left(-1 + \cos^2 \theta_r \right) \cos^2 \theta_r$ N-m.

The plot for the electromagnetic torque is illustrated in Figure 6.23a. The electromagnetic torque varies as a sinusoidal-like function of the rotor angular displacement. The torque ripple is an undesirable phenomenon due to losses, noise, vibration, etc. To minimize the torque ripple, one can perform structural redesign attempting to ensure sinusoidal flux distribution. Alternatively, we may derive and fed (if implementable) the proper phase currents within the obtained balanced set.

Having derived $T_e = \dfrac{5 N_m \psi_m}{2} \left(i_{as} \cos \theta_r \sin^4 \theta_r - i_{bs} \sin \theta_r \cos^4 \theta_r \right)$, the current set

$$i_{as} = i_M \cos \theta_r / \sin^4 \theta_r \quad \text{and} \quad i_{bs} = -i_M \sin \theta_r / \cos^4 \theta_r$$

leads to the maximization of the electromagnetic torque and elimination of the torque ripple. Using

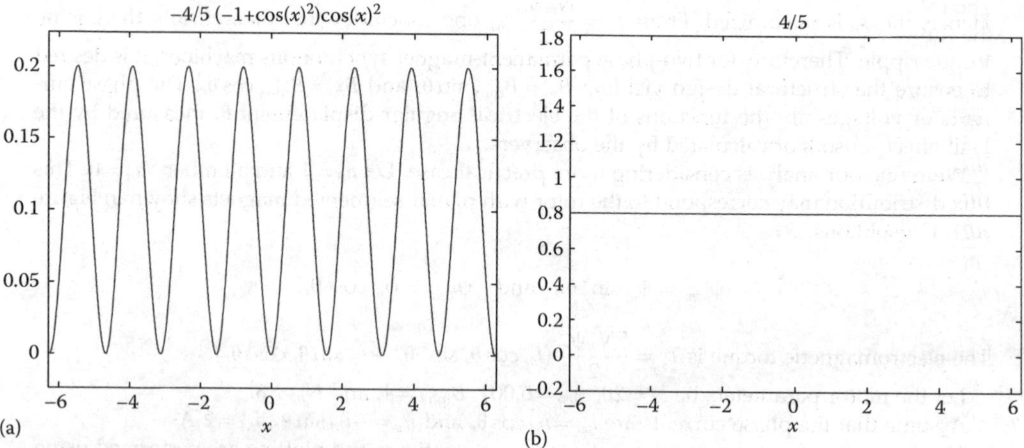

FIGURE 6.23

Electromagnetic torque: (a) $T_e = -\dfrac{4}{5}\left(-1 + \cos^2 \theta_r\right)\cos^2 \theta_r$ N-m; (b) $T_e = 0.8$ N-m.

```
% Phase currents
ias = iM*cos(x)/sin(x)^4; ibs = -iM*sin(x)/cos(x)^4;
% Derive and plot the electromagnetic torque
Te = Nm*(d1*ias+d2*ibs)/2, Te = simplify(Te), ezplot(Te)
```

we obtain the following results

```
Te = 4/5*cos(x)^2+4/5*sin(x)^2
Te = 4/5
```

That is, $T_e = 0.8$ N-m. This T_e is plotted in Figure 6.23b. Due to singularity and saturation, it is impossible to implement the current set $i_{as} = i_M \cos \theta_r / \sin^4 \theta_r$ and $i_{bs} = -i_M \sin \theta_r / \cos^4 \theta_r$. The following near-balanced current set can be applied:

$$i_{as} = i_M \cos \theta_r / \left(\sin^4 \theta_r + \varepsilon\right) \quad \text{if } \left|i_M \cos \theta_r / \left(\sin^4 \theta_r + \varepsilon\right)\right| \leq i_{\max}, \quad i_{as} = i_{\max} \text{ or } i_{as} = i_{\max} \text{ otherwise,}$$

$$i_{bs} = i_M \sin \theta_r / \left(\cos^4 \theta_r + \varepsilon\right) \quad \text{if } \left|i_M \sin \theta_r / \left(\cos^4 \theta_r + \varepsilon\right)\right| \leq i_{\max}, \quad i_{bs} = i_{\max} \text{ or } i_{bs} = i_{\max} \text{ otherwise.}$$

Example 6.9   Axial Topology Three-Phase Permanent-Magnet Synchronous Motors: Analytic and Experimental Analysis

Our goal is to analyze three-phase synchronous motors. For an axial topology motor, depicted in Figure 6.20, in the full operating envelope, we have $a_1 \neq 0$, $a_2 \neq 0$, $a_3 \neq 0$ while $\forall a_n = 0$, $n > 3$. These a_n are found experimentally as reported below. From (6.27) one finds

$$\psi_{as} = L_{ss}i_{as} + L_{asbs}i_{bs} + L_{ascs}i_{cs} + \psi_m\left(a_1 \sin \theta_r + a_2 \sin^3 \theta_r + a_3 \sin^5 \theta_r\right),$$

$$\psi_{bs} = L_{bsas}i_{as} + L_{ss}i_{bs} + L_{bscs}i_{cs} + \psi_m\left[a_1 \sin\left(\theta_r + \frac{2}{3}\pi\right) + a_2 \sin^3\left(\theta_r + \frac{2}{3}\pi\right) + a_3 \sin^5\left(\theta_r + \frac{2}{3}\pi\right)\right],$$

$$\psi_{cs} = L_{csas}i_{as} + L_{csbs}i_{bs} + L_{ss}i_{cs} + \psi_m\left[a_1 \sin\left(\theta_r - \frac{2}{3}\pi\right) + a_2 \sin^3\left(\theta_r - \frac{2}{3}\pi\right) + a_3 \sin^5\left(\theta_r - \frac{2}{3}\pi\right)\right].$$

The mutual inductances between the phase windings are zero. Using the total derivatives for the flux linkages in the Kirchhoff's second law (6.12), we have

$$\frac{di_{as}}{dt} = \frac{1}{L_{ss}}\left[-r_s i_{as} - \psi_m \cos\theta_r \left(a_1 + 3a_2 \sin^2\theta_r + 5a_3 \sin^4\theta_r\right)\omega_r + u_{as}\right],$$

$$\frac{di_{bs}}{dt} = \frac{1}{L_{ss}}\left[-r_s i_{bs} - \psi_m \cos\left(\theta_r + \frac{2}{3}\pi\right)\left(a_1 + 3a_2 \sin^2\left(\theta_r + \frac{2}{3}\pi\right) + 5a_3 \sin^4\left(\theta_r + \frac{2}{3}\pi\right)\right)\omega_r + u_{bs}\right]$$

$$\frac{di_{cs}}{dt} = \frac{1}{L_{ss}}\left[-r_s i_{cs} - \psi_m \cos\left(\theta_r - \frac{2}{3}\pi\right)\left(a_1 + 3a_2 \sin^2\left(\theta_r - \frac{2}{3}\pi\right) + 5a_3 \sin^4\left(\theta_r - \frac{2}{3}\pi\right)\right)\omega_r + u_{cs}\right].$$

$$(6.30)$$

Using the expression for the electromagnetic torque (6.28), one yields

$$T_e = \frac{N_m \psi_m}{2}\left[i_{as}\cos\theta_r\left(a_1 + 3a_2 \sin^2\theta_r + 5a_3 \sin^4\theta_r\right) + i_{bs}\cos\left(\theta_r + \frac{2}{3}\pi\right)\left(a_1 + 3a_2 \sin^2\left(\theta_r + \frac{2}{3}\pi\right) + 5a_3 \sin^4\left(\theta_r + \frac{2}{3}\pi\right)\right)\right.$$

$$\left. + i_{cs}\cos\left(\theta_r - \frac{2}{3}\pi\right)\left[a_1 + 3a_2 \sin^2\left(\theta_r - \frac{2}{3}\pi\right) + 5a_3 \sin^4\left(\theta_r - \frac{2}{3}\pi\right)\right]\right]$$

The *torsional–mechanical* equations of motion (6.29) are

$$\frac{d\omega_r}{dt} = \frac{N_m^2 \psi_m}{4J}\left[i_{as}\cos\theta_r\left(a_1 + 3a_2 \sin^2\theta_r + 5a_3 \sin^4\theta_r\right) + i_{bs}\cos\left(\theta_r + \frac{2}{3}\pi\right)\left(a_1 + 3a_2 \sin^2\left(\theta_r + \frac{2}{3}\pi\right) + 5a_3 \sin^4\left(\theta_r + \frac{2}{3}\pi\right)\right)\right.$$

$$\left. + i_{cs}\cos\left(\theta_r - \frac{2}{3}\pi\right)\left[a_1 + 3a_2 \sin^2\left(\theta_r - \frac{2}{3}\pi\right) + 5a_3 \sin^4\left(\theta_r - \frac{2}{3}\pi\right)\right]\right] - \frac{B_m}{J}\omega_r - \frac{P}{2J}T_L,$$

$$\frac{d\theta_r}{dt} = \omega_r.$$

$$(6.31)$$

Using (6.30) and (6.31), the Simulink diagram to simulate axial topology permanent-magnet synchronous motors ($a_1 \neq 0$, $a_2 \neq 0$, $a_3 \neq 0$, $\forall a_n = 0$, $n>3$) is built as depicted in Figure 6.24. The phase voltages supplied are

$$u_{as} = \sqrt{2}u_M \cos\theta_r, \quad u_{bs} = \sqrt{2}u_M \cos\left(\theta_r + \frac{2}{3}\pi\right), \quad \text{and} \quad u_{cs} = \sqrt{2}u_M \cos\left(\theta_r - \frac{2}{3}\pi\right).$$

The motor parameters are experimentally found. We have $N_m = 8$, $r_s = 13.5$ ohm, $L_{ss} = 0.035$ H, $B_m = 0.0000005$ N-m-s/rad, and $J = 0.00001$ kg-m^2. The values of ψ_m and a_n are found by measuring the induced *emf* when the synchronous machine was rotated by the prime mover. The experimental results are illustrated in Figure 6.25. At no load, for two different steady-state angular velocities ω_r (956 and 1382 rad/s) the terminal phase voltage is 29.9 and 39.7 V, respectively. Even at the same load, ψ_m varies due to different operating envelope, and $\psi_m \in [0.029\ 0.031]$ V-s/rad assuming that $a_1 \neq 0$, $\forall a_n = 0$, $n>1$.

For no load and light load conditions (T_L is ~ 0.01 N-m), $a_1 = 1$ and all other a_n are zeros ($\forall a_n = 0$, $n>1$). For the loaded motor, $a_1 = 0.85$, $a_2 = 0.06$, $a_3 = 0.04$, and $\forall a_n = 0$, $n>3$. The parameters are uploaded as

```
% Optimal distribution: Light loads
psim = 0.03; a1 = 1; a2 = 0; a3 = 0;
% Near-optimal distribution: Heavy loads
psim = 0.03; a1 = 0.85; a2 = 0.06; a3 = 0.04;
% Motor parameters
Nm = 8; uM = 50; rs = 13.5; Lss = 0.035; Bm = 0.0000005; J = 0.00001;
```

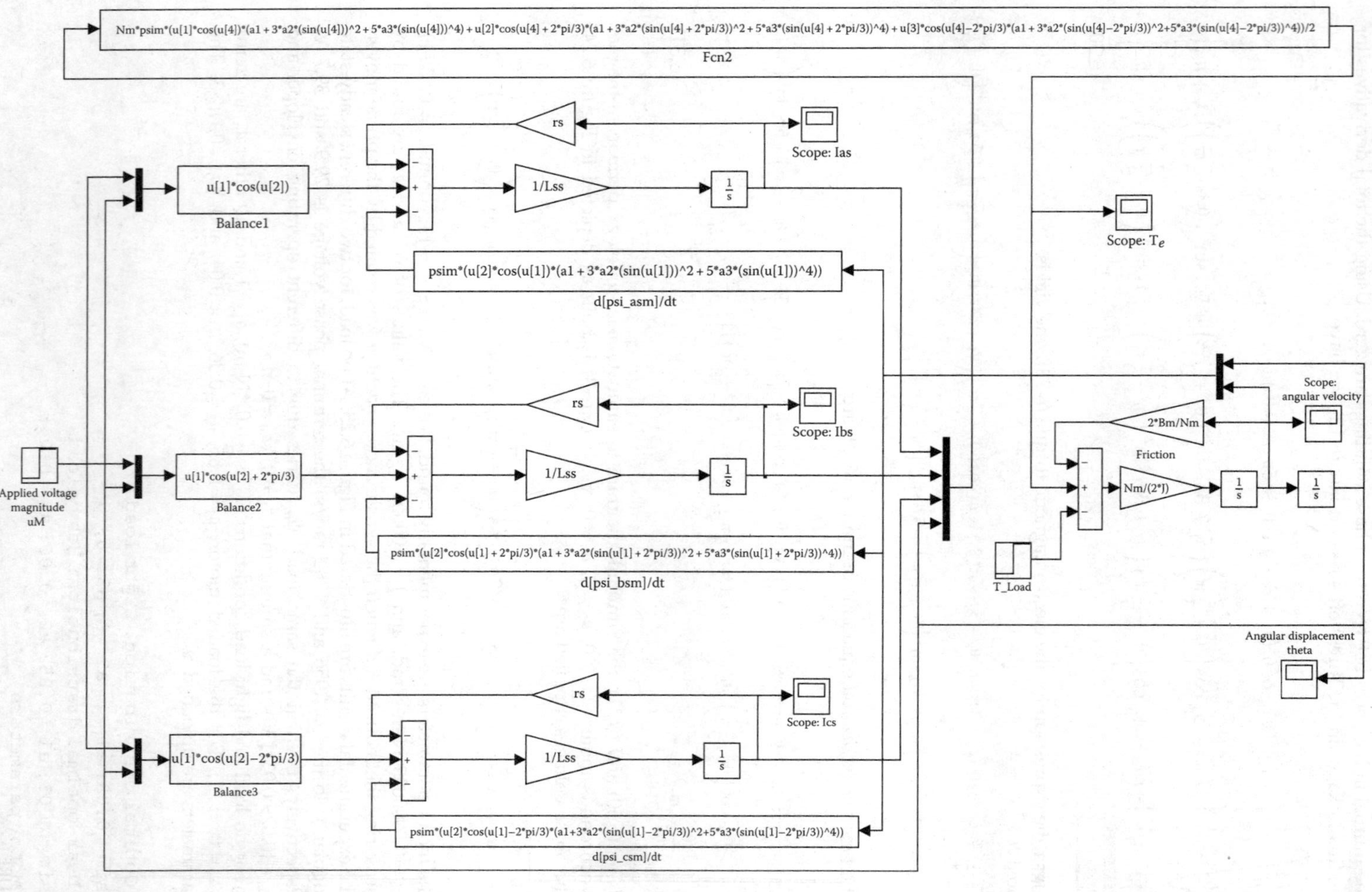

FIGURE 6.24

Simulink diagram for axial topology permanent-magnet synchronous motors ($a_1 \neq 0$, $a_2 \neq 0$, $a_3 \neq 0$ while $\forall a_n = 0$, $n > 3$), ch6_05.mdl.

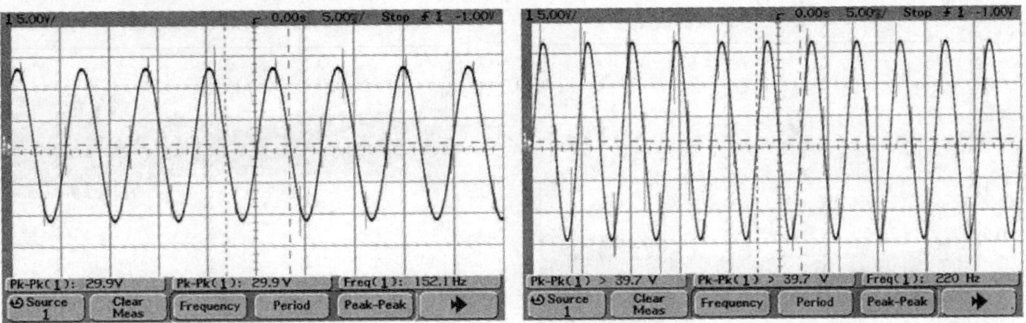

FIGURE 6.25
Induced $emf_{as\omega} = \psi_m \cos\theta_r \omega_r$ if the generator is rotated at 956 and 1382 rad/s.

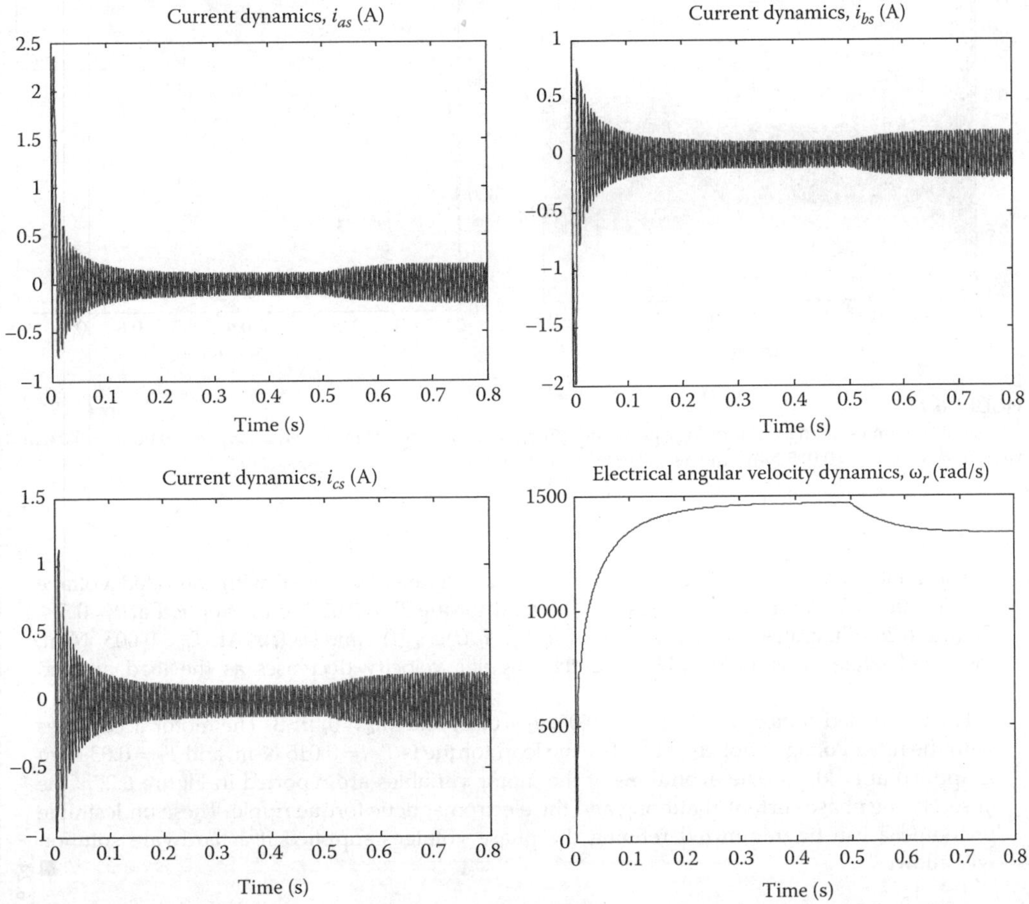

Current dynamics, i_{as} (A)

Current dynamics, i_{bs} (A)

Current dynamics, i_{cs} (A)

Electrical angular velocity dynamics, ω_r (rad/s)

FIGURE 6.26
Transient dynamics of an axial topology permanent-magnet synchronous motor if $a_1 = 1$ and $\forall a_n = 0$, $n > 1$.

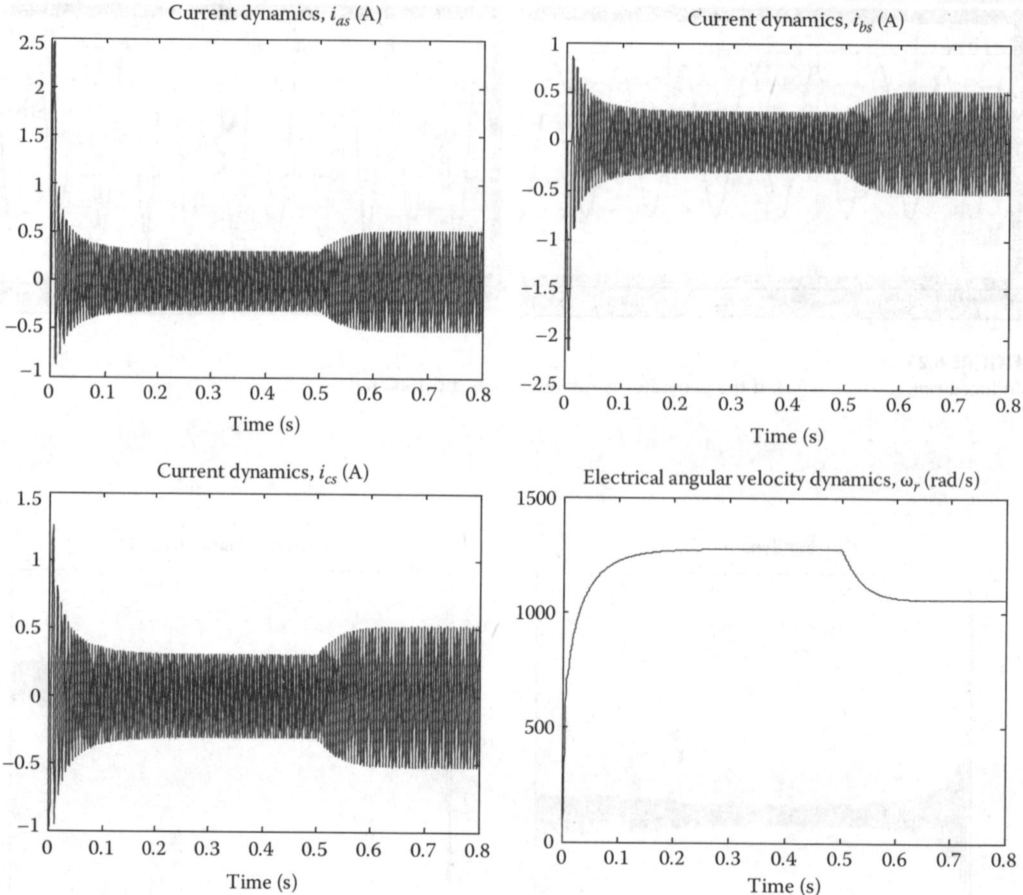

FIGURE 6.27
Transient dynamics of an axial topology permanent-magnet synchronous motor ($a_1 = 0.85$, $a_2 = 0.06$, $a_3 = 0.04$, and $\forall a_n = 0$, $n > 3$.), $T_{L0} = 0.015$ N-m and $T_L = 0.03$ N-m at $t = 0.5$ s.

The motor dynamics is studied as the motor accelerates from stall with the rated voltage applied ($u_M = 50$ V) if $T_{L0} = 0.005$ N-m. The load torque $T_L = 0.01$ N-m is applied at $t = 0.5$ s. Figure 6.26 illustrates the dynamics of $i_{as}(t)$, $i_{bs}(t)$, $i_{cs}(t)$, and $\omega_r(t)$. At $T_L = 0.005$ N-m, the steady-state ω_r is 1470 rad/s, and the angular velocity decreases as the load applied at $t = 0.5$ s.

For the loaded motor, $a_1 = 0.85$, $a_2 = 0.06$, $a_3 = 0.04$, and $\forall a_n = 0$, $n > 3$. The motor accelerates with the rated voltage applied. At $t = 0$ s, the load torque is $T_{L0} = 0.015$ N-m, and $T_L = 0.03$ N-m is applied at $t = 0.5$ s. The evolutions of the motor variables are reported in Figure 6.27. One observes the phase current chattering and the electromagnetic torque ripple. These undesirable phenomena can be minimized refining the phase voltages supplied if a hardware solution will allow. ∎

References

1. Chapman, S. J., *Electric Machinery Fundamentals*, McGraw-Hill, New York, 1999.
2. Krause, P. C. and Wasynczuk, O., *Electromechanical Motion Devices*, McGraw-Hill, New York, 1989.
3. Krause, P. C., Wasynczuk, O., and Sudhoff, S. D., *Analysis of Electric Machinery*, IEEE Press, New York, 1995.
4. Lyshevski, S. E., *Electromechanical Systems, Electric Machines, and Applied Mechatronics*, CRC Press, Boca Raton, FL, 1999.
5. Lyshevski, S. E., *Electromechanical Systems and Devices*, CRC Press, Boca Raton, FL, 2008.
6. Slemon, G. R., *Electric Machines and Drives*, Addison-Wesley Publishing Company, Reading, MA, 1992.

7

Electronics and Power Electronics Solutions in Mechatronic Systems

7.1 Operational Amplifiers

In electromechanical systems, signal processing, signal conditioning, and other tasks are accomplished by ICs. The digital controllers and filters are implemented using microcontrollers and DSPs. Electromechanical systems are predominantly continuous, and analog controllers and filters can be implemented utilizing operational amplifiers and specialized ICs. These controllers and filters are integrated within power amplifiers. The electromechanical motion devices and solid-state devices are continuous. The use of digital controllers and sensors results in the so-called *hybrid* closed-loop electromechanical systems as introduced in Chapter 1, see system configurations documented in Figures 1.2 and 1.4.

As reported in Chapters 3 and 4, in electromechanical systems, analog proportional-integral-derivative controllers are implemented using operational amplifiers. This section examines and documents the use of operational amplifiers to implement various analog controllers and filters with specified transfer functions. Various operations on signals and variables may be required. For example, sensors convert time-varying physical quantities (displacement, velocity, acceleration, force, torque, pressure, temperature, and others) in electric signals (voltage or current). The signal-level sensor output must be amplified and filtered. The single operational amplifier is a two-port network which has *noninverting* and *inverting* input terminals (3 and 2) as well as one output terminal (6) as depicted in Figure 7.1. Two (or one) DC voltages are needed, and terminal 7 is connected to a positive voltage u_+, while a negative voltage (or ground) u_- is supplied to the terminal 4. The pin connections of the single, dual, and quad low-power operational amplifiers MC33171, MC33172, and MC33174 are reported in Figure 7.1, which also illustrates 8- and 14-pin plastic packages (cases 626 and 646). These operational amplifiers are also available in the surface mount packages (cases 751 and 948). Operational amplifiers, which consist of dozens of transistors, are fabricated using the complementary metal oxide semiconductor (CMOS) or biCMOS fabrication technologies [1]. Figure 7.1 depicts the representative schematic diagram.

The amplifier output is the difference between two input voltages, $u_1(t)$ and $u_2(t)$, applied to the *inverting* input terminal and the *noninverting* input terminal, multiplied by the differential open-loop coefficient k_{og}. That is, the resulting output voltage is

$$u_0(t) = k_{og}[u_2(t) - u_1(t)]$$

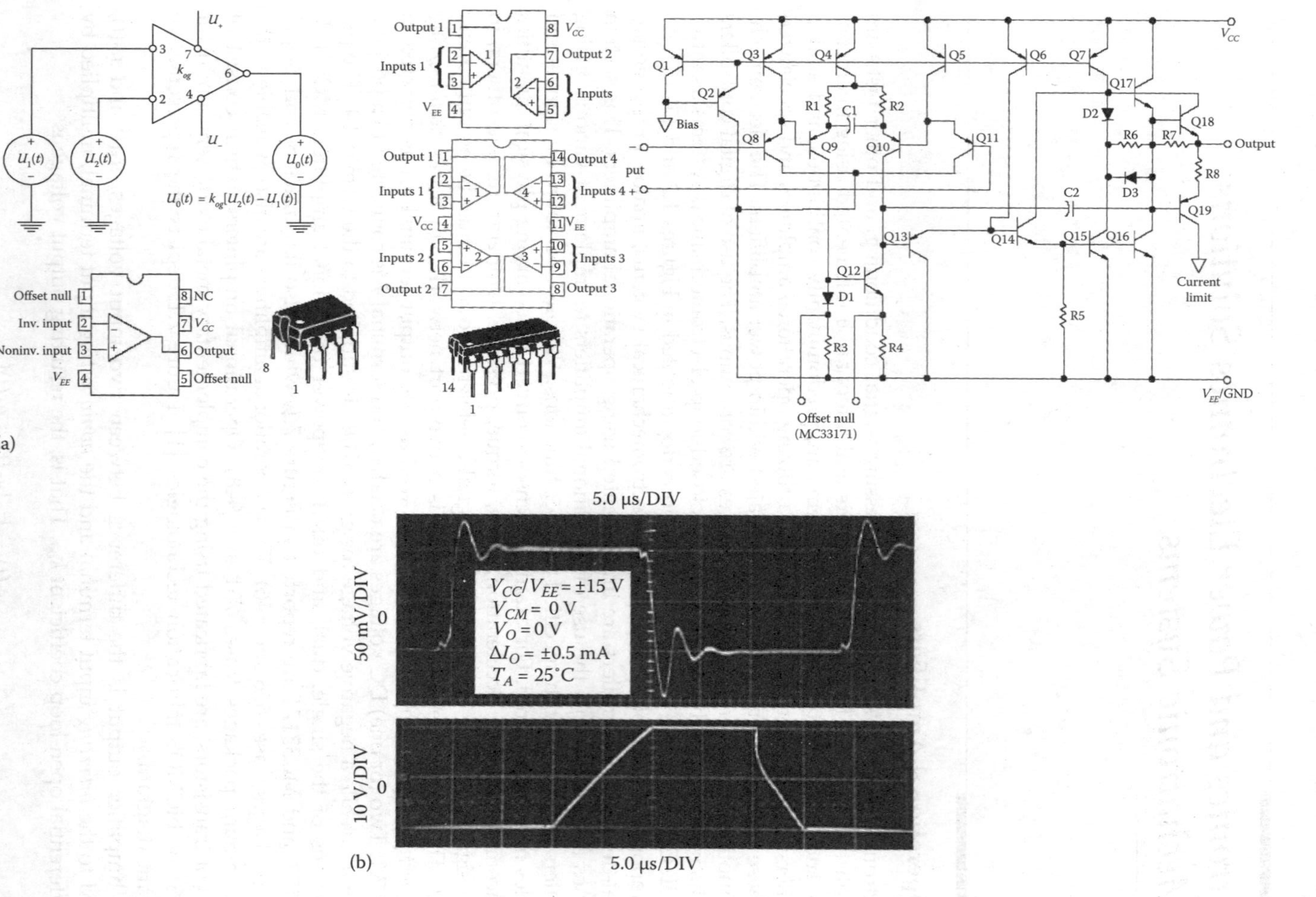

FIGURE 7.1

Operational amplifiers, pin connections, packages, representative schematics, and transient responses. (From Lyshevski, S.E., *Electromechanical Systems, Electric Machines, and Applied Mechatronics*, CRC Press, Boca Raton, FL, 1999. Courtesy of Motorola.)

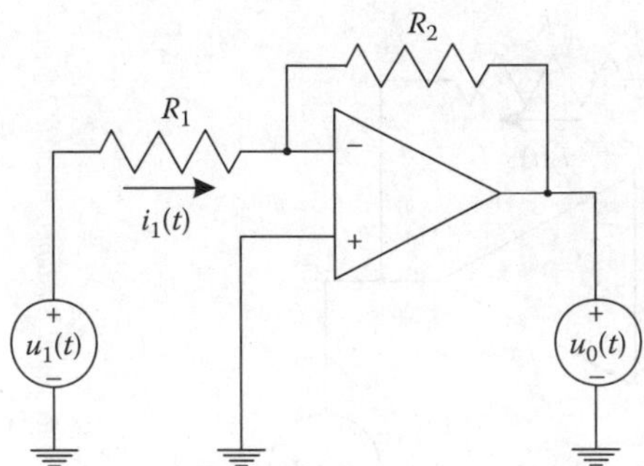

FIGURE 7.2
Inverting configuration of the operational amplifier.

The differential open-loop coefficient is positive. The value of k_{og} is very large, and k_{og} varies as $\sim[1 \times 10^5 \text{ to } 1 \times 10^7]$. The general purpose operational amplifiers have input and output resistances $\sim[1 \times 10^5 \text{ to } 1 \times 10^{12}]$ and $\sim[10, 1000]$ ohm, respectively. The *inverting* and *noninverting* input terminals are distinguished using "−" and "+" signs. Supplying the signal-level input voltage $u_1(t)$ to the *inverting* input terminal using external resistor R_1, and grounding the *noninverting* input terminal, one can find the differential closed-loop coefficient k_{cg} if a negative feedback is used. The output terminal is connected to the inverting input terminal, and the resistor R_2 is inserted as depicted in Figure 7.2.

To find the differential closed-loop coefficient k_{cg}, one has to obtain the ratio between the output and input voltages $u_0(t)$ and $u_1(t)$. The voltage between two input terminals is $u_0(t)/k_{og}$, and the voltage at the *inverting* input terminal is $-u_0(t)/k_{og}$ because the *noninverting* input terminal is grounded. Therefore, we have $i_1(t) = \dfrac{u_1(t) + \dfrac{u_0(t)}{k_{og}}}{R_1}$, and the

output voltage is $u_0(t) = -\dfrac{u_0(t)}{k_{og}} - \dfrac{u_1(t) + \dfrac{u_0(t)}{k_{og}}}{R_1} R_2$. Hence, the differential closed-loop coef-

ficient is found to be $k_{cg} = \dfrac{u_0(t)}{u_1(t)} = -\dfrac{\dfrac{R_2}{R_1}}{1 + \dfrac{1}{k_{og}} + \dfrac{R_2}{k_{og}R_1}}$.

The configuration studied allows one to invert the signal-level input signal. Taking note that the differential open-loop coefficient k_{og} is very large (hundred thousands), we have

$$k_{cg} = \frac{u_0(t)}{u_1(t)} \approx -\frac{R_2}{R_1}.$$

The inverting summing amplifier (the so-called weighted summer) is shown in Figure 7.3. We have an m input signals. In particular, the applied voltages are $u_{1,1}(t), \ldots, u_{1,m}(t)$. The currents are found to be $i_{1,1}(t) = u_{1,1}(t)/R_{1,1}, \ldots, i_{1,m}(t) = u_{1,m}(t)/R_{1,m}$. The current in the feedback path is given by $i_2(t) = i_{1,1}(t) + \cdots + i_{1,m}(t)$. Hence, the amplifier output is

$$u_0(t) = -\left(\frac{R_2}{R_{1,1}} u_{1,1}(t) + \frac{R_2}{R_{1,2}} u_{1,2}(t) + \cdots + \frac{R_2}{R_{1,m-1}} u_{1,m-1}(t) + \frac{R_2}{R_{1,m}} u_{1,m}(t) \right).$$

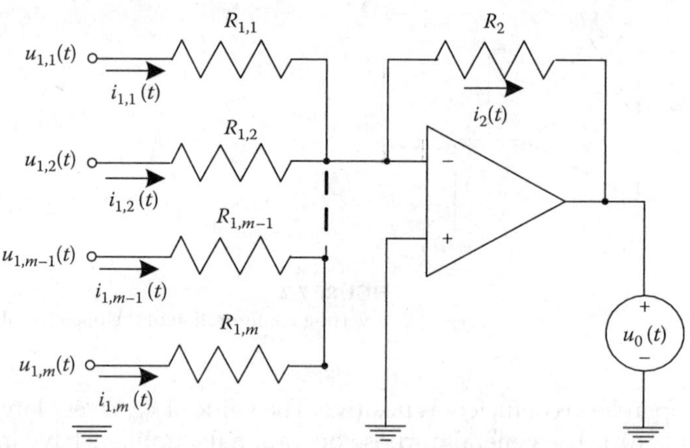

FIGURE 7.3
Summing amplifier.

To generalize the results derived from studying the inverting configuration, we use the input impedance $Z_1(s)$ and the feedback path impedance $Z_2(s)$ as illustrated in Figure 7.4. The impedance is the ratio of the phasor voltage to the phasor current at the two terminals. In particular, the impedances of the resistor, capacitor, and inductor are

$$Z_R(s) = R, \quad Z_R(j\omega) = R, \quad Z_C(s) = \frac{1}{sC}, \quad Z_C(j\omega) = \frac{1}{j\omega C}, \quad \text{and}$$

$$Z_L(s) = sL, \quad Z_L(j\omega) = j\omega L.$$

The transfer function of the closed-loop amplifier configuration is $G(s) = \dfrac{U_0(s)}{U_1(s)} = -\dfrac{Z_2(s)}{Z_1(s)}.$

As an example, consider the operational amplifier with $Z_1(s)$ and $Z_2(s)$ as shown in Figure 7.5a.

The transfer function is found by using the expressions for impedances $Z_1(s) = R_1$ and $Z_2(s) = \dfrac{R_2}{R_2C_2s + 1}.$ We have $G(s) = \dfrac{U_0(s)}{U_1(s)} = -\dfrac{Z_2(s)}{Z_1(s)} = -\dfrac{\dfrac{R_2}{R_1}}{R_2C_2s + 1}.$

The closed-loop gain coefficient for the inverting operational amplifier studied is $k_{cg} = -\dfrac{R_2}{R_1},$ while the time constant is R_2C_2. In the frequency domain, by substituting

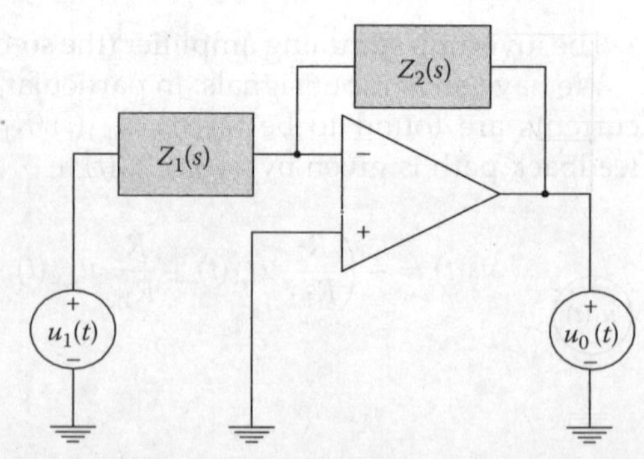

FIGURE 7.4
Inverting configuration of the operational amplifier
with impedances $Z_1(s)$ and $Z_2(s)$.

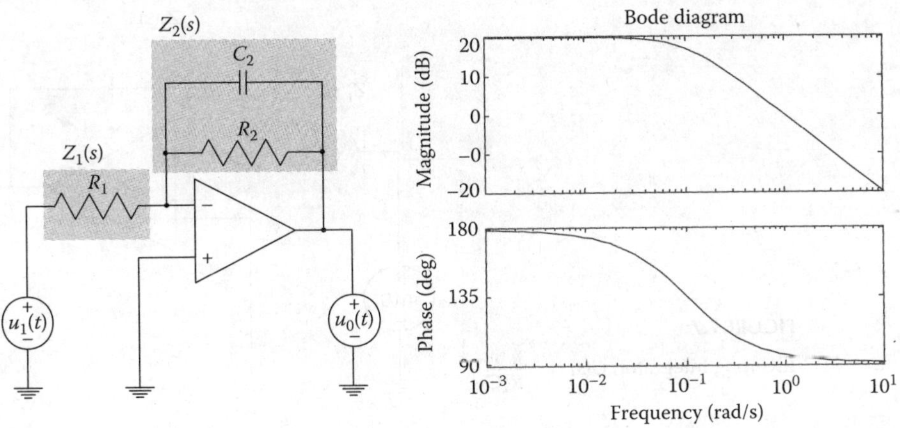

FIGURE 7.5
(a) Inverting operational amplifier with $Z_1(s)$ and $Z_2(s)$; (b) Bode plots.

$s = j\omega$, we have $G(j\omega) = \dfrac{U_0(j\omega)}{U_1(j\omega)} = -\dfrac{Z_2(j\omega)}{Z_1(j\omega)} = -\dfrac{\dfrac{R_2}{R_1}}{R_2 C_2 j\omega + 1}$. The Bode plots of the resulting low-pass filter can be easily found and plotted. For example, let $R_1 = 1{,}000$, $R_2 = 10{,}000$, and $C_2 = 0.001$. Using a MATLAB® statement

```
R1 = 1000; R2 = 10000; C2 = 0.001; num = [-R2/R1]; den = [R2*C2 1]; bode(num,den)
```

one finds two Bode plots as documented in Figure 7.5b.

Filters with specific transfer functions are used to ensure the noise immunity, performing the signal conditioning of signals which are utilized to implement control algorithms or any other functions. In fact, sensor signals contain the noise which is due to different origins. The low-, medium-, and high-frequency noise can be attenuated by filters. The filters must be properly designed and implemented. The possible filter configuration is demonstrated in Figure 7.6.

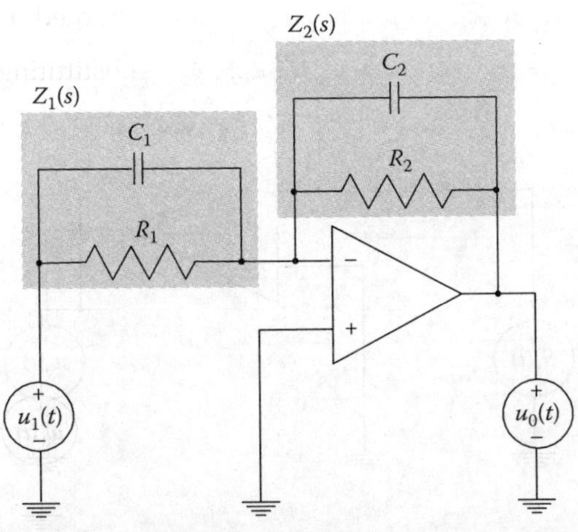

FIGURE 7.6
Analog filter implemented utilizing an inverting operational amplifier with $Z_1(s) = \dfrac{R_1}{R_1 C_1 s + 1}$ and $Z_2(s) = \dfrac{R_2}{R_2 C_2 s + 1}$.

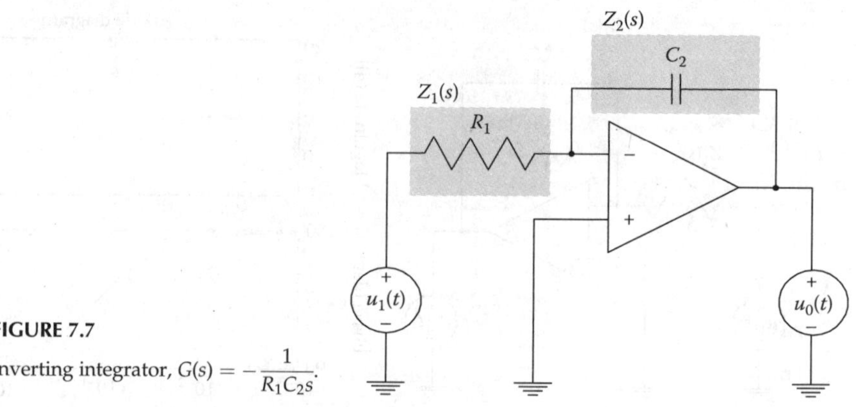

FIGURE 7.7

Inverting integrator, $G(s) = -\dfrac{1}{R_1 C_2 s}$.

The transfer function is found using the input and feedback impedances. In particular,

$$G(s) = \frac{U_0(s)}{U_1(s)} = -\frac{Z_2(s)}{Z_1(s)} = -\frac{\dfrac{R_2}{R_2 C_2 s + 1}}{\dfrac{R_1}{R_1 C_1 s + 1}} = -\frac{\dfrac{R_2}{R_1}(R_1 C_1 s + 1)}{R_2 C_2 s + 1}.$$

Different transfer functions can be implemented by operational amplifiers utilizing passive elements in the input and feedback paths. Operational amplifiers are widely used to implement analog control laws. An inverting integrator is obtained by placing a capacitor C_2 in the feedback path, see Figure 7.7. The resulting transfer function is $G(s) = \dfrac{U_0(s)}{U_1(s)} = -\dfrac{Z_2(s)}{Z_1(s)} = -\dfrac{1}{R_1 C_2 s}$. Denoting the initial value of the capacitor voltage as $u_C(t_0)$, the amplifier output voltage is $u_0(t) = -u_C(t_0) - \dfrac{1}{R_1 C_2} \displaystyle\int_{t_0}^{t_f} u_1(\tau) d\tau$.

The operational differentiator performs the differentiation of the input signal. The current through the input capacitor is $C_1 \dfrac{du_1(t)}{dt}$, see Figure 7.8; that is, the output voltage

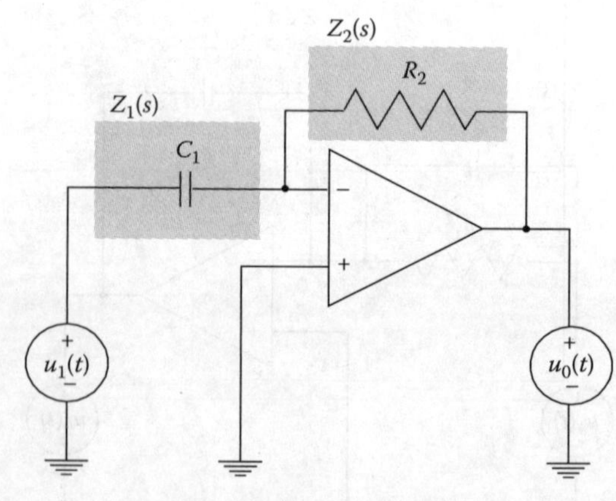

FIGURE 7.8

Inverting differentiator, $G(s) = -R_2 C_1 s$.

is proportional to the derivative of the input voltage with respect to time. Hence, $u_0(t) = -R_2 C_1 \dfrac{du_1(t)}{dt}$. The transfer function is found to be $G(s) = \dfrac{U_0(s)}{U_1(s)} = -\dfrac{Z_2(s)}{Z_1(s)} = -R_2 C_1 s$.

For various input and feedback path impedances $Z_1(s)$ and $Z_2(s)$, Table 7.1 provides the transfer functions of the inverting operational amplifier. The passive elements are utilized in the input and feedback paths.

TABLE 7.1

Transfer Functions of the Inverting Amplifier

Input circuit with impedance $Z_1(s)$	Feed circuit with impedance $Z_2(s)$	Transfer function
$Z_1(s)$ — R_1; $Z_1(s) = R_1$	$Z_2(s)$ — R_2; $Z_2(s) = R_2$	$G(s) = \dfrac{U_0(s)}{U_1(s)} = -\dfrac{R_2}{R_1}$
$Z_1(s)$ — R_1; $Z_1(s) = R_1$	$Z_2(s)$ — C_2; $Z_2(s) = \dfrac{1}{C_2 s}$	$G(s) = \dfrac{U_0(s)}{U_1(s)} = -\dfrac{1}{R_1 C_2 s}$
$Z_1(s)$ — R_1; $Z_1(s) = R_1$	$Z_2(s)$ — C_2 parallel R_2; $Z_2(s) = \dfrac{R_2}{R_2 C_2 s + 1}$	$G(s) = \dfrac{U_0(s)}{U_1(s)} = -\dfrac{\dfrac{R_2}{R_1}}{R_2 C_2 s + 1}$
$Z_1(s)$ — R_1; $Z_1(s) = R_1$	$Z_2(s)$ — R_2, C_2; $Z_2(s) = \dfrac{R_2 C_2 s + 1}{C_2 s}$	$G(s) = \dfrac{U_0(s)}{U_1(s)} = -\dfrac{R_2 C_2 s + 1}{R_1 C_2 s}$
$Z_1(s)$ — C_1; $Z_1(s) = \dfrac{1}{C_1 s}$	$Z_2(s)$ — R_2; $Z_2(s) = R_2$	$G(s) = \dfrac{U_0(s)}{U_1(s)} = -R_2 C_1 s$

(continued)

TABLE 7.1 (continued)

Transfer Functions of the Inverting Amplifier

Input circuit with impedance $Z_1(s)$	Feed circuit with impedance $Z_2(s)$	Transfer function
$Z_1(s)$ — C_1, R_1 $Z_1(s) = \dfrac{R_1}{R_1 C_1 s + 1}$	$Z_2(s)$ — R_2 $Z_2(s) = R_2$	$G(s) = \dfrac{U_0(s)}{U_1(s)} = -\dfrac{R_1 R_2 C_1 s + R_2}{R_1}$
$Z_1(s)$ — C_1, R_1 $Z_1(s) = \dfrac{R_1}{R_1 C_1 s + 1}$	$Z_2(s)$ — R_2, C_2 $Z_2(s) = \dfrac{R_2 C_2 s + 1}{C_2 s}$	$G(s) = \dfrac{U_0(s)}{U_1(s)}$ $= -\dfrac{(R_1 C_1 s + 1)(R_2 C_2 s + 1)}{R_1 C_2 s}$
$Z_1(s)$ — C_1, R_1 $Z_1(s) = \dfrac{R_1}{R_1 C_1 s + 1}$	$Z_2(s)$ — C_1, R_2 $Z_2(s) = \dfrac{R_2}{R_2 C_2 s + 1}$	$G(s) = \dfrac{U_0(s)}{U_1(s)} = -\dfrac{\dfrac{R_2}{R_1}(R_1 C_1 s + 1)}{R_2 C_2 s + 1}$

The transfer function of the proportional-integral-derivative (PID) control law $G_{PID}(s)$ is given as

$$G_{PID}(s) = k_p + \frac{k_i}{s} + k_d s = \frac{k_d s^2 + k_p s + k_i}{s}.$$

The PID controller can be implemented using the configuration as depicted in Figure 7.9. The transfer function realized by an inverting operational amplifier is

$$G(s) = \frac{U_0(s)}{U_1(s)} = -\frac{(R_1 C_1 s + 1)(R_2 C_2 s + 1)}{R_1 C_2 s} = -\frac{R_2 C_1 s^2 + \dfrac{R_1 C_1 + R_2 C_2}{R_1 C_2}s + \dfrac{1}{R_1 C_2}}{s}.$$

For the derived transfer function $G(s)$, k_p, k_i, and k_d are defined by the values of resistors and capacitors of the input and feedback paths. In particular,

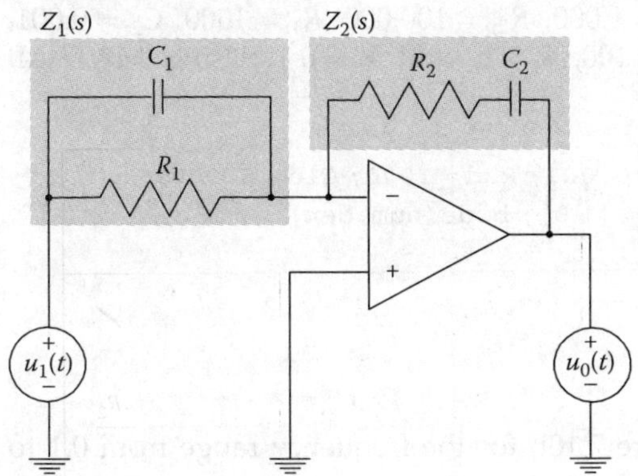

FIGURE 7.9
Inverting operational amplifier circuitry.

$$k_p = -\frac{R_1 C_1 + R_2 C_2}{R_1 C_2}, \quad k_i = -\frac{1}{R_1 C_2}, \text{ and } k_d = -R_2 C_1.$$ These k_p, k_i, and k_d must be positive and this can be accomplished by utilizing an additional inverting operational amplifier.

To guarantee the stability, the proportional, integral, and derivative feedback gains (k_p, k_i, and k_d) must be positive. In engineering practice, the configuration, illustrated in Figure 7.10a, is commonly used to implement the PID controller. One obtains the expression for the overall transfer function as

$$G(s) - \frac{U_0(s)}{U_1(s)} = \frac{R_{2p}}{R_{1p}} + \frac{1}{R_{1i}C_{2i}s} + R_{2d}C_{1d}s.$$

The proportional, integral, and derivative feedback coefficients are $k_p = \dfrac{R_{2p}}{R_{1p}}$, $k_i = \dfrac{1}{R_{1i}C_{2i}}$, and $k_d = R_{2d}C_{1d}$.

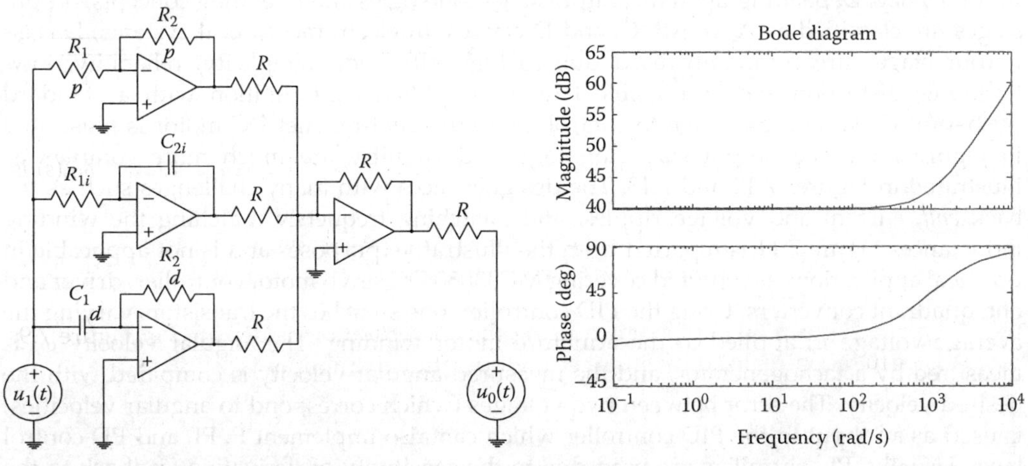

FIGURE 7.10
(a) Implementation of an analog PID control law; (b) Bode plots.

The Bode plot can be found. Letting $R_{1p} = 1{,}000$, $R_{2p} = 100{,}000$, $R_{1i} = 1000$, $C_{2i} = 0.001$, $R_{2d} = 1{,}000$, and $C_{1d} = 0.0001$, one has $k_p = 100$, $k_i = 1$, and $k_d = 0.1$. Using MATLAB statements

```
R1p = 1000; R2p = 100000; R1i = 1000; C2i = 0.001; R2d = 1000; C1d = 0.0001;
num = [R2d*C1d R2p/R1p 1/(R1i*C2i)]; den = [1 0]; bode(num,den)
```

or

```
G = tf([num],[den]); bode(G,{0.1, 1e4})
```

one finds the Bode plots as reported in Figure 7.10b for the frequency range from 0.1 to 10,000 rad/s.

The catalog data, schematics, characteristics, frequency response, and transient dynamics for low power, single supply, dual and quad operational amplifiers manufactured by a great number of companies are available. The possible configurations (*noninverting, inverting*, notch, bandpass, and other) are easily examined.

7.2 Power Amplifiers and Power Converters

7.2.1 Power Amplifiers and Analog Controllers

Power amplifiers allow one to perform the amplification [2–5]. Various classes of amplifiers are available. The most important part of power amplifiers is the output stage which deals with relatively large voltages and currents. For example, for 100 W (rated), 50 V permanent-magnet motors, the rated current is ~2 A, while the pear current could be ~20 A. The power dissipated in the output stage power transistors should be minimized to guarantee the efficiency. Depending upon the output stage topologies and operating concepts, output stages are classified as A, B, AB, C, and D classes. In electromechanical systems, D-class output stages are commonly used due to high efficiency, simplicity, reliability, low-harmonic distortion, etc. A possible largely simplified configuration with a standard push–pull D-class output stage to control the permanent-magnet DC motor is illustrated in Figure 7.11. The output stage topology and circuitry are much more complex as illustrated in Figures 7.12 and 7.13. The designer faces with many challenges such as the back *emf*, current and voltage ripples, and switching frequency matching the winding inductances Figure 7.11 is reported from the illustrative purposes and is not applicable in practical applications as reported covering MC33030 DC servo motor controller/driver and one-quadrant converters. Using the PID controller, one switches the transistors varying the average voltage u_a applied to the armature motor winding. The angular velocity ω_r is measured by a tachogenerator, and the measured angular velocity is compared with the desired velocity. The error between two voltages (which correspond to angular velocities) is used as an input to the PID controller which can also implement P, PI, and PD control laws. Usually, PI controllers are used due to the sensitivity of derivative feedback to the noise which in practice may not be attenuated utilizing filters.

If the signal-level voltage u_{PID} of the output inverting operational amplifier of the PID controller is positive, transistor S_1 is *off* because the voltage between the base and emitter is

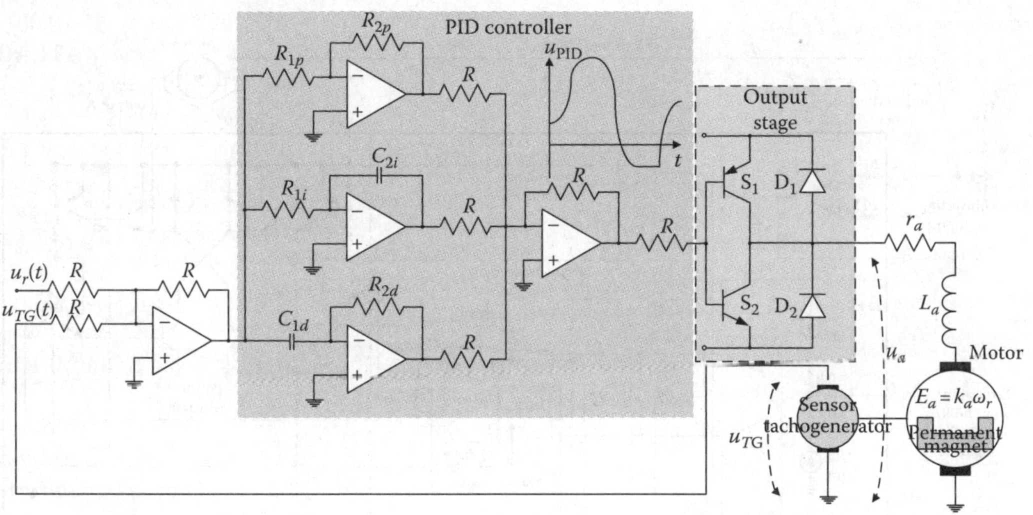

FIGURE 7.11
Application of the D-class power amplifier to control an electric drive with a permanent-magnet DC motor: Closed-loop configuration with a PID controller and sensor (tachogenerator) to measure the angular velocity ω_r.

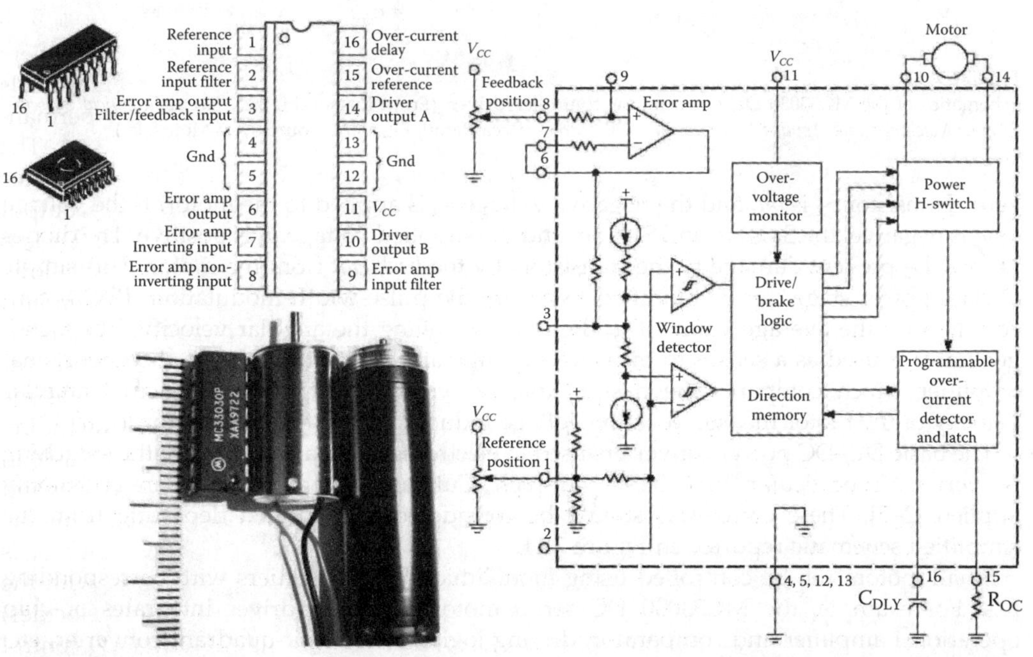

FIGURE 7.12
Pin connection and block diagram of the MC33030 DC servo motor controller/driver for permanent-magnet DC motors. (From Lyshevski, S.E., *Electromechnical Systems, Electric Machines, and Applied Mechatronics*, CRC Press, Boca Raton, FL, 1999. Courtesy of Motorola.)

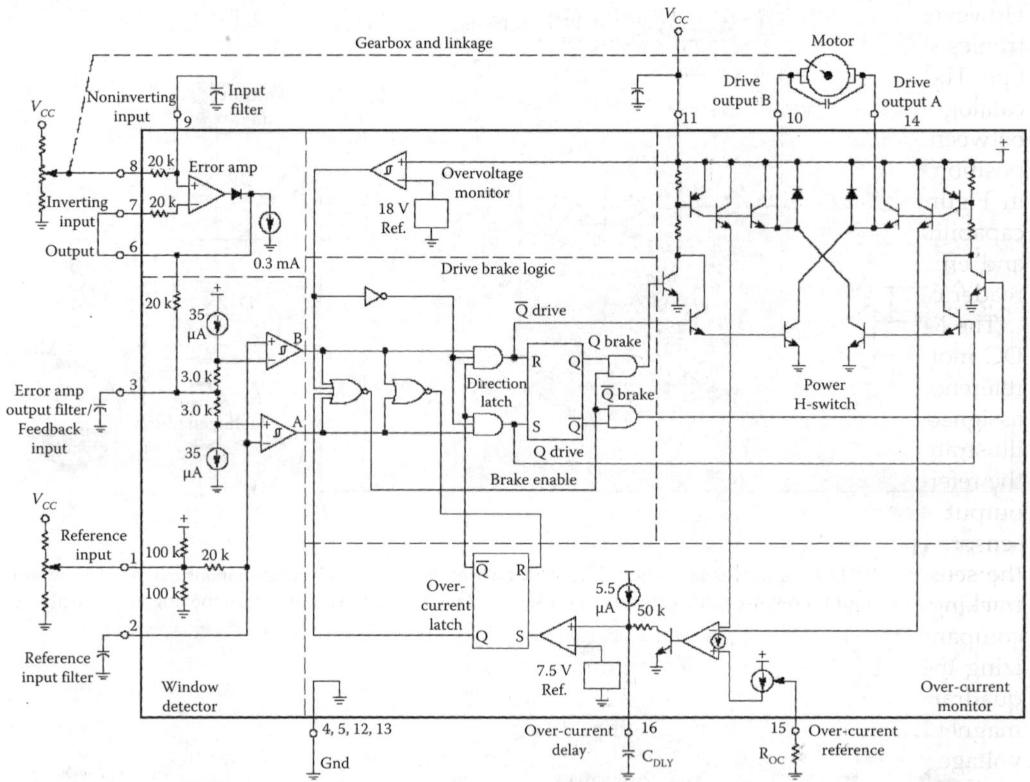

FIGURE 7.13
Schematics of the MC33030 DC servo motor controller/driver. (From Lyshevski, S.E., *Electromechnical Systems, Electric Machines, and Applied Mechatronics*, CRC Press, Boca Raton, FL, 1999. Courtesy of Motorola.)

zero. Transistor S_2 is *on*, and the negative voltage u_a is applied to the motor. If the voltage u_{PID} is negative, the S_1 is *on*, and S_2 is *off*, and the supplied voltage u_a is positive. The diodes D_1 and D_2 prevent damage to the transistors by the *back emf* from the motor. This simple D-class power stage can be modified to ensure the pulse-width-modulation (PWM) concept to vary the average value of u_a thereby controlling the angular velocity. The tachogenerator is used as a sensor to measure the angular velocity to be fed to the operational amplifier which compares the desired angular velocity $u_r(t)$ with the actual angular velocity of the motor measured as the voltage induced by the tachogenerator $u_{TG}(t)$.

The basic DC–DC power converters used in electromechanical systems are the switching converters. In particular, *buck, boost, buck-boost*, Cuk, and other converters are commonly applied [2–5]. These converters should be considered and studied departing from the simplified schematic reported in Figure 7.11.

Small motors can be controlled using monolithic PWM amplifiers with corresponding ICs. For example, the MC33030 DC servo motor controller/driver integrates on-chip operational amplifier and comparator, driving logics, PWM four-quadrant converter, etc. The rated (peak) output voltage and current are 36 V and 1 A. Hence, one can use MC33030 for small ∼10 W DC motors. It should be emphasized that for a short period permanent-magnet DC and synchronous electric machines can operate at higher voltage and current. For motors, T_{epeak}/T_{erated} and P_{peak}/P_{rated} could be ∼10, i.e., the ratio i_{peak}/i_{rated} is ∼10.

However, for power electronics, the ratio I_{peak}/I_{rated} is up to ~ 2. Therefore, power electronics should accommodate the peak motor current within the specific operating envelope. The monolithic MC33030 servo-motor driver contains 119 active transistors, and the catalog data provides a great deal of details with explicit description. The difference between the reference and actual angular velocity or displacement, linear velocity or position, is compared by the "error amplifier," and two comparators are used as shown in Figure 7.12 [2]. A *pnp* differential output power stage ensures driving and braking capabilities. The four-quadrant H-configured power stage guarantees high performance and efficiency. For a complete description of the MC33030 motor controller/driver, the reader can refer to www.motorola.com.

The MC33030, as a PWM amplifier, can be utilized to drive ~ 10 W permanent-magnet DC motors in electric drives and servos application. For electric drives and servos, the reference (command) and output are the angular velocity and displacement, which can be assigned and measured as voltages. Schematic of a drive/servo system with MC33030 is illustrated by a representative block diagram reported in Figure 7.13. One sets a voltage on the reference input (pin 1). The velocity or displacement sensor is installed to measure the output velocity or displacement. A tachogenerator or potentiometer can be used as a sensor. The sensor voltage is supplied to pin 3. The reference voltage is compared with the sensor voltage. By using this difference, the system is controlled, e.g., one uses the tracking error as given by $e(t) = r(t) - y(t)$. The "window detector" is composed of A and B comparators with hysteresis. The proportional controller $u(t) = k_p e(t)$ is implemented utilizing the "error amplifier." A *pnp* differential input stage ensures grounding. The four-quadrant power stage provides the armature voltage to rotate the motor. The permanent-magnet DC motor is connected to pins 10 and 14. The current limit is set on pin 15 and the voltage protection is ensured by the "over-voltage monitor" which is important to use due to the induced *back emf*. This schematic can be modified and enhanced by adding additional filters, control, motor, and other circuitry.

The dual power operational amplifiers can be utilized in various output stage topologies. Figure 7.14 documents the image of a 7-pin, 40 V, 1.5 A, heatsink-mount dual power operational amplifier which can be used in half and full bridge motor drivers. In general, the application-specific ICs (control, filtering, signal conditioning, and other tasks) and output stages need to be designed. For miniscale electric machines, though a

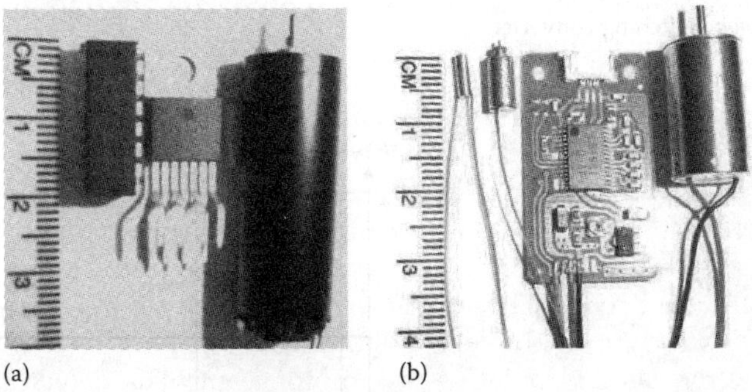

(a)　　　　　　　　　　(b)

FIGURE 7.14

Power electronics and permanent-magnet electric machines: (a) MC30330 servo motor controller/driver and dual operational amplifier (~ 30 W peak) to control the armature voltage applied to permanent-magnet DC electric machines (3 W rated and ~ 30 W peak); (b) application specific ICs-PWM amplifier and 2, 4, and 10 mm diameter permanent-magnet synchronous (2 and 4 mm) and DC (10 mm) electric motors.

state-of-the-art design is performed and enabling CMOS technology nodes are used to fabricate those integrated ICs-PWM amplifiers, the size of minimachines can be less than electronics. For example, 2 and 4 mm diameter electric machines, illustrated in Figure 7.14, are fabricated using CMOS and surface micromachining processes used in micro-electronics.

7.2.2 Switching Converter: Buck Converter

By using PWM switching concept the voltage at the load terminal can be effectively regulated, ensuring high performance. A high-frequency *buck* (*step-down*) switching converter is shown in Figure 7.15.

In the *step-down* switching converter, as documented in Figure 7.15, the switch S is open and closed. The switching frequency is given as $f = \dfrac{1}{t_{on} + t_{off}}$, where t_{on} and t_{off} are the switching *on* and *off* durations. Assuming that the switch is lossless, the voltage u_{dN} is equal to the supplied voltage V_d when the switch is closed, and the output voltage is zero if the switch is open, see Figure 7.16.

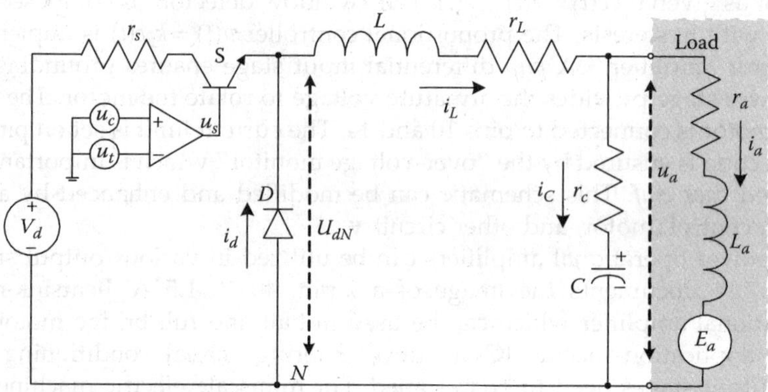

FIGURE 7.15

High-frequency *step-down* switching converter.

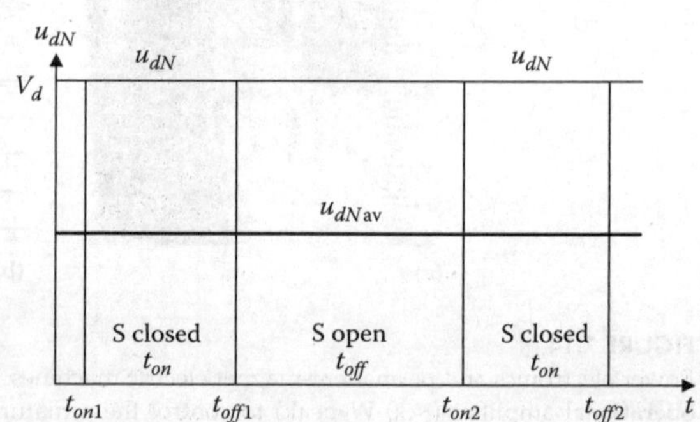

FIGURE 7.16

Voltage waveforms.

One concludes that the voltage u_{dN}, as well as the voltage applied to the load u_a, can be regulated by controlling the switching *on* and *off* durations (t_{on} and t_{off}). The average voltage applied to the load depends on t_{on} and t_{off}. In the steady state, we have

$$u_{dN\ av} = \frac{t_{on}}{t_{on} + t_{off}} V_d = d_D V_d,$$

where d_D is the duty ratio (duty cycle), which is a function of the switching frequency and the time during which the switch is *on*, $d_D = \dfrac{t_{on}}{t_{on} + t_{off}}$.

One has $d_D \in [0\ 1]$, and $d_D = 0$ if $t_{on} = 0$, while $d_D = 1$ if $t_{off} = 0$. By changing the duty ratio d_D (switching activity), which is bounded as $d_D \in [0\ 1]$, the average voltage supplied to the load u_a, is regulated. To establish the PWM switching, one can use a so-called control-triangle concept. The switching signal u_s, which drives the switch, is generated by comparing a signal-level control voltage u_c with a repetitive triangular signal u_t. A comparator is shown in Figures 7.15 and 7.17. The duration of the output pulses u_s represents the *weighted* value between the triangular voltage u_t with the assigned switching frequency and control signal u_c. The output voltage of the comparator u_s drives the switch S. Hence, the *on* and *off* durations result by comparing u_c and u_t. Figure 7.17 illustrates the voltage waveforms.

The Motorola dual operational amplifier and dual comparator MC3405 is reported illustrated in Figure 7.18. The comparator schematics, operation, and waveforms explained above are reported shown.

For the high-frequency *step-down* switching converter, illustrated in Figure 7.15, a low-pass first-order filter with inductance L and capacitance C ensures the specified voltage ripple. We have

$$\frac{\Delta u_a}{u_a} = \frac{1 - d_D}{8LCf^2} \quad \text{and} \quad L_{\min} = \frac{(1 - d_D)r_a}{2f}.$$

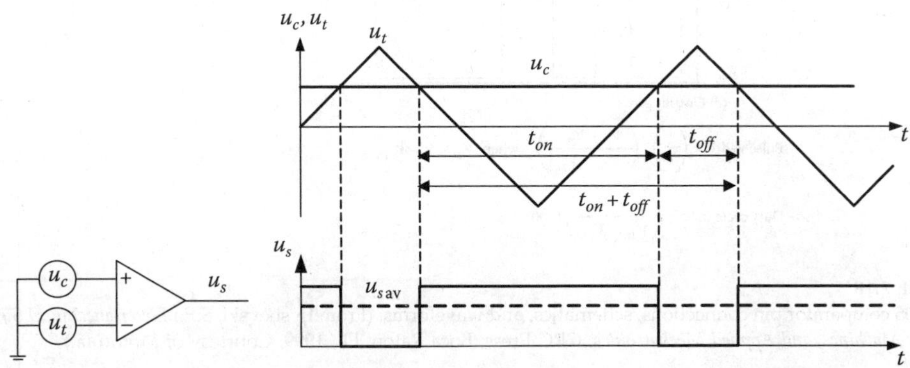

FIGURE 7.17
Comparator in the PWM concept and the voltage waveforms.

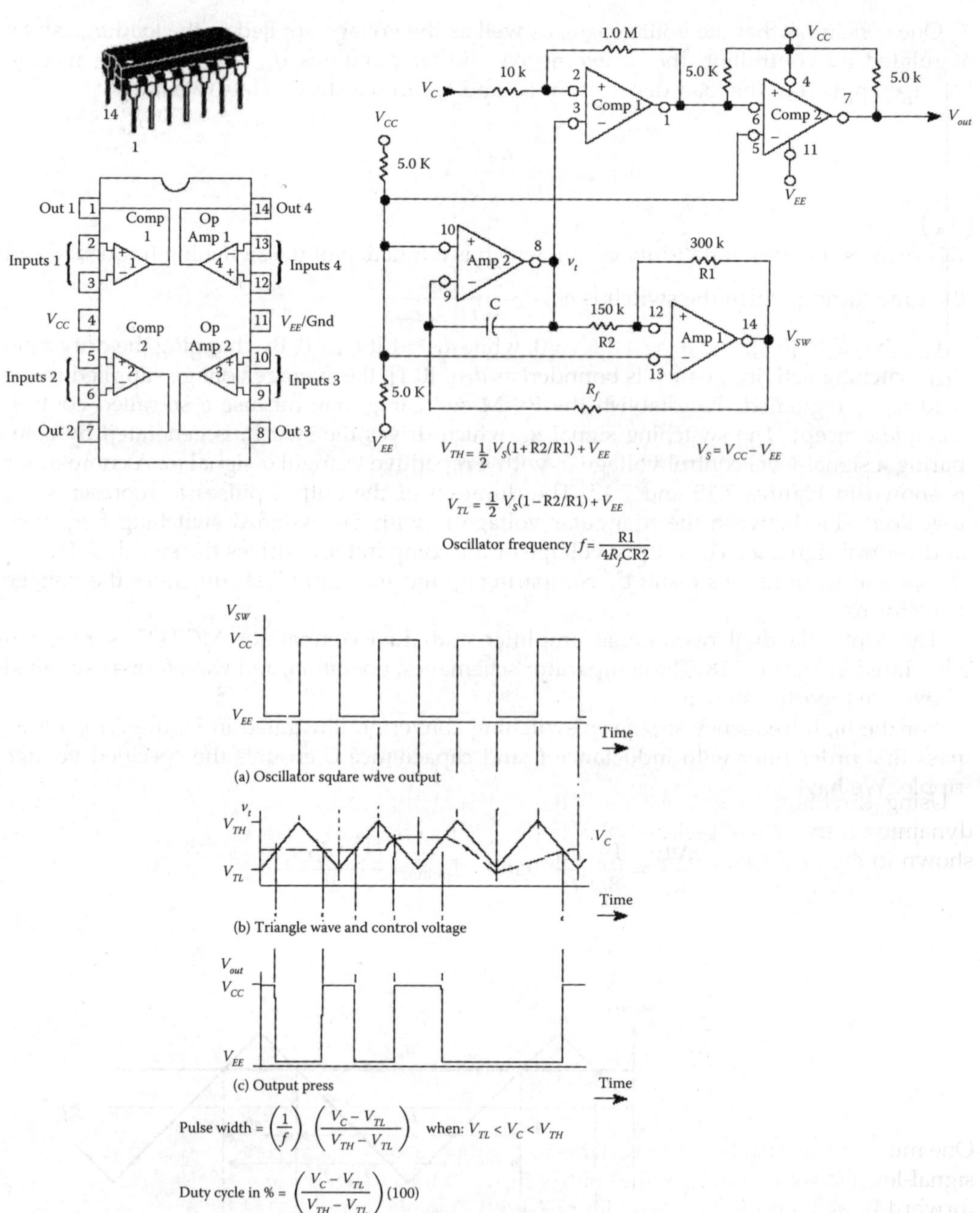

$$V_{TH} = \tfrac{1}{2} V_S(1 + R2/R1) + V_{EE} \qquad V_S = V_{CC} - V_{EE}$$

$$V_{TL} = \tfrac{1}{2} V_S(1 - R2/R1) + V_{EE}$$

$$\text{Oscillator frequency } f = \frac{R1}{4R_f CR2}$$

$$\text{Pulse width} = \left(\frac{1}{f}\right)\left(\frac{V_C - V_{TL}}{V_{TH} - V_{TL}}\right) \quad \text{when: } V_{TL} < V_C < V_{TH}$$

$$\text{Duty cycle in \%} = \left(\frac{V_C - V_{TL}}{V_{TH} - V_{TL}}\right)(100)$$

FIGURE 7.18

MC3405 comparator pin connections, schematics, and waveforms. (From Lyshevski, S.E., *Electromechanical Systems, Electric Machines, and Applied Mechatronics*, CRC Press, Boca Raton, FL, 1999. Courtesy of Motorola.)

Switches, inductors, and capacitors have resistances, which are denoted as r_s, r_L, and r_c. Two circuits, one when the switch is closed and one when it is open, are illustrated in Figure 7.19.

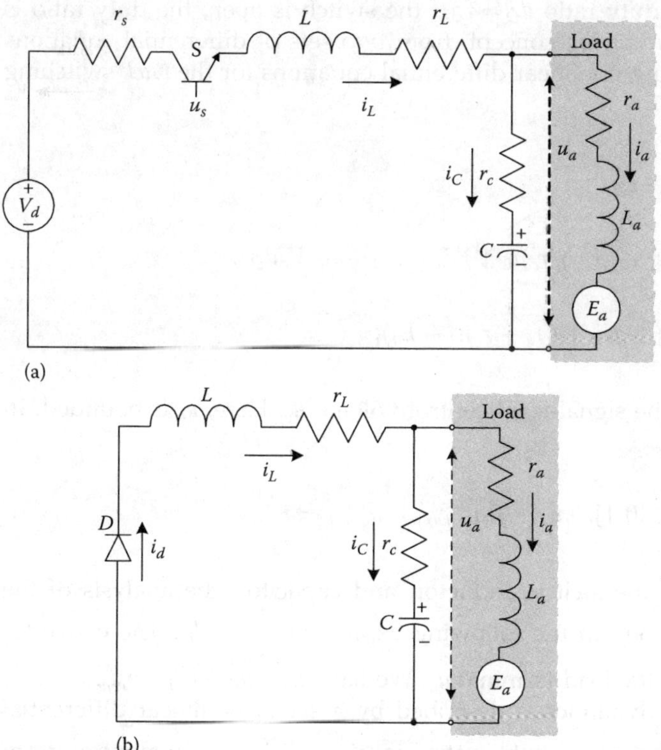

FIGURE 7.19
Circuits of the *buck* DC–DC converter:
(a) Switch is closed; (b) switch is open.

Using Kirchhoff's laws, one finds the differential equations to describe the converter dynamics. If the switch is closed, the diode D is reverse biased. For the resulting circuit, as shown in Figure 7.19a, we have the following set of differential equations

$$\frac{du_C}{dt} = \frac{1}{C}(i_L - i_a),$$

$$\frac{di_L}{dt} = \frac{1}{L}(-u_C - (r_L + r_c)i_L + r_c i_a - r_s i_L + V_d),$$

$$\frac{di_a}{dt} = \frac{1}{L_a}(u_C + r_c i_L - (r_a + r_c)i_a - E_a).$$

One must distinguish the state variable u_C (voltage across the capacitor C) and comparator signal-level input u_c, which is the control input. If the switch is open, the diode D becomes forward biased, and $i_d = i_L$. (See Figure 7.19b.) One obtains

$$\frac{du_C}{dt} = \frac{1}{C}(i_L - i_a),$$

$$\frac{di_L}{dt} = \frac{1}{L}(-u_C - (r_L + r_c)i_L + r_c i_a),$$

$$\frac{di_a}{dt} = \frac{1}{L_a}(u_C + r_c i_L - (r_a + r_c)i_a - E_a).$$

When the switch is closed, the duty ratio $d_D = 1$. If the switch is open, the duty ratio is zero, i.e., $d_D = 0$. By using the *averaging* concept, from two sets of differential equations derived, one obtains the following nonlinear differential equations for the *buck* switching converter

$$\frac{du_C}{dt} = \frac{1}{C}(i_L - i_a),$$

$$\frac{di_L}{dt} = \frac{1}{L}(-u_C - (r_L + r_c)i_L + r_c i_a - r_s i_L d_D + V_d d_D),$$

$$\frac{di_a}{dt} = \frac{1}{L_a}(u_C + r_c i_L - (r_a + r_c)i_a - E_a).$$

The duty ratio is regulated by the signal-level control voltage u_c. Here, u_c is bounded. In particular,

$$d_D = \frac{u_c}{u_{t\,max}} \in [0\ 1], \quad u_c \in [0\ u_{c\,max}], \quad u_{c\,max} = u_{t\,max}.$$

Neglecting small resistances of the switch, inductor, and capacitor, the analysis of the steady-state performance leads one to the following expression $\frac{u_a}{V_d} = d_D$. The converter output is the voltage applied to the load terminal u_a. We have $u_a = u_C + r_c i_L - r_a i_a$.

The resulting *buck* converter dynamics is described by a set of nonlinear differential equations. Having derived that the duty ratio is $d_D = \frac{u_c}{u_{t\,max}}$, a nonlinear term $\frac{r_s}{L}i_L d_D = \frac{r_s}{L u_{t\,max}}i_L u_c$ is the multiplication of the state variable i_L and control u_c. The control limit is imposed because $0 \leq u_c \leq u_{c\,max}$, $u_c \in [0\ u_{c\,max}]$.

Example 7.1:

Simulate and examine the *step-down* converter steady-state and dynamic behavior applying MATLAB. The converter parameters are $r_s = 0.025$ ohm, $r_L = 0.02$ ohm, $r_c = 0.15$ ohm, $r_a = 3$ ohm, $C = 0.003$ F, $L = 0.0007$ H, and $L_a = 0.005$ H. The duty ratio is 0.5. The supplied DC voltage is $V_d = 50$ V, and $E_a = 10$ V.

Using the differential equations derived, the following m-files are developed to perform the simulations using the ode45 differential equation solver (command). Two MATLAB files which solve the differential equations and perform plotting are reported below.

MATLAB file (ch7_01.m)

```
t0 = 0; tfinal = 0.03; tspan = [t0 tfinal]; y0 = [0 0 0]';
[t,y] = ode45('ch7_02',tspan,y0);
subplot(2,2,1); plot(t,y);
xlabel('Time (seconds)','FontSize',10);
title('Transient Dynamics of State Variables','FontSize',10);
subplot(2,2,2); plot(t,y(:,1),'-');
xlabel('Time (seconds)','FontSize',10);
title('Voltage u_C, [V]','FontSize',10);
```

```
subplot(2,2,3);plot(t,y(:,2),'-');
xlabel('Time (seconds)','FontSize',10);
title('Current i_L, [A]','FontSize',10);
subplot(2,2,4);plot(t,y(:,3),'-');
xlabel('Time (seconds)','FontSize',10);
title('Current i_a, [A]','FontSize',10);
```

MATLAB file (ch7_02.m)

```
% Dynamics of the buck converter
function yprime=difer(t,y);
% parameters
Vd=50; Ea=10; rs=0.025; rl=0.02; rc=0.15; ra=3;
C=0.003; L=0.0007; La=0.005; D=0.5;
% differential equations for buck converters
yprime=[(y(2,:)-y(3,:))/C; ...
(-y(1,:)-(rl+rc)*y(2,:)+rc*y(3,:)-rs*y(2,:)*D+Vd*D)/L; ...
(y(1,:)+rc*y(2,:)-(rc+ra)*y(3,:)-Ea)/La];
```

The transient dynamics for the state variables $u_C(t)$, $i_L(t)$, and $i_a(t)$ are illustrated in Figure 7.20, and the settling time is 0.025 s. The steady-state value of the output voltage is 25 V because the applied voltage is 50 V and the duty ratio was assigned $d_D = 0.5$.

Using $u_a = u_C + r_c i_L - r_a i_a$, the voltage at the load terminal u_a can be calculated and plotted. In particular, one types in the Command Window

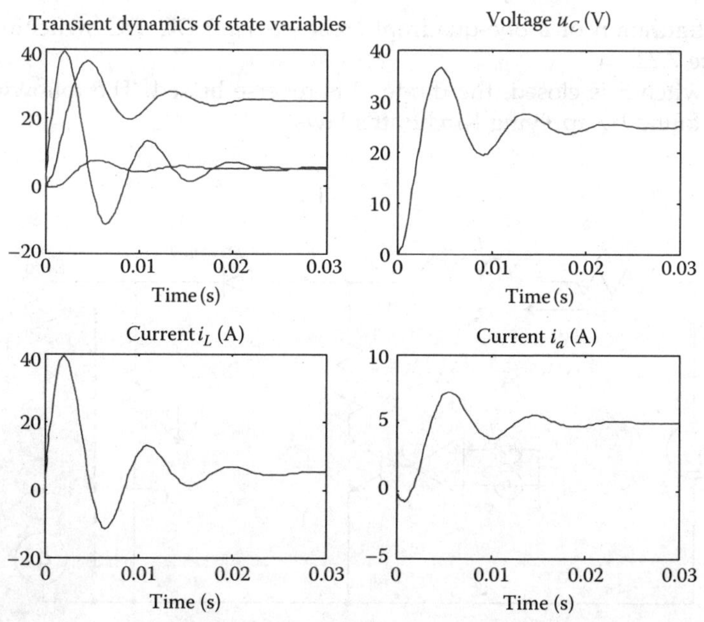

FIGURE 7.20

Transient dynamics of the *buck* converter, $V_d = 50$ V and $d_D = 0.5$.

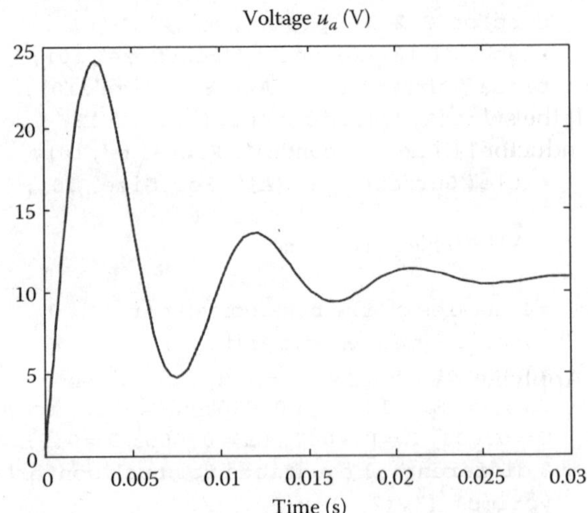

FIGURE 7.21
Voltage on the load terminal.

```
rc=0.15; ra=3; plot(t,y(:,1)+rc*y(:,2)-ra*y(:,3),'-');
xlabel('Time (seconds)', 'FontSize',14);
title('Voltage u_a, [V]','FontSize',14);
```

The plot for $u_a(t)$ results as illustrated in Figure 7.21.

7.2.3 Boost Converter

A typical configuration of a one-quadrant *boost* (*step-up*) DC–DC switching converter is given in Figure 7.22.

When the switch S is closed, the diode D is reverse biased. The following differential equations are found by applying Kirchhoff's laws

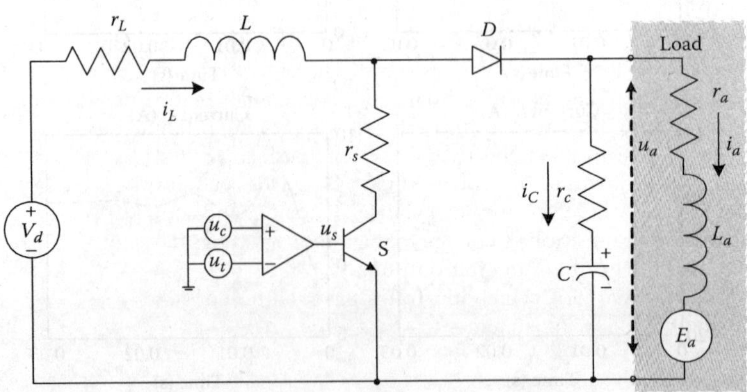

FIGURE 7.22
High-frequency *boost* converter.

$$\frac{du_C}{dt} = -\frac{1}{C}i_a, \quad \frac{di_L}{dt} = \frac{1}{L}(-(r_L + r_s)i_L + V_d), \quad \frac{di_a}{dt} = \frac{1}{L_a}(u_C - (r_a + r_c)i_a - E_a).$$

If the switch is open, the diode is forward biased because the direction of the current in the inductor i_L does not change instantly. Hence,

$$\frac{du_C}{dt} = \frac{1}{C}(i_L - i_a), \quad \frac{di_L}{dt} = \frac{1}{L}(-u_C - (r_L + r_c)i_L + r_c i_a + V_d),$$

$$\frac{di_a}{dt} = \frac{1}{L_a}(u_C + r_c i_L - (r_a + r_c)i_a - E_a).$$

Applying the *averaging* concept, using d_D, one finds

$$\frac{du_C}{dt} = \frac{1}{C}(i_L - i_a - i_L d_D),$$

$$\frac{di_L}{dt} = \frac{1}{L}(-u_C - (r_L + r_c)i_L + r_c i_a + u_C d_D + (r_c - r_s)i_L d_D - r_c i_a d_D + V_d),$$

$$\frac{di_a}{dt} = \frac{1}{L_a}(u_C + r_c i_L - (r_a + r_c)i_a - r_c i_L d_D - E_a).$$

The steady-state analysis results in the following relationship $\dfrac{u_a}{V_d} = \dfrac{1}{1 - d_D}$. The expression for the voltage ripple is $\dfrac{\Delta u_a}{u_a} = \dfrac{d_D}{r_a C f^2}$. The minimum value of the inductance depends on the switching frequency and load resistance. One has $L_{\min} = \dfrac{d_D(1 - d_D)^2 r_a}{2f}$.

Example 7.2:

We perform simulations of the *boost* converter if the parameters are $r_s = 0.025$ ohm, $r_L = 0.02$ ohm, $r_c = 0.15$ ohm, $r_a = 3$ ohm, $C = 0.003$ F, $L = 0.0007$ H, and $L_a = 0.005$ H. Assume $d_D = 0.5$, $V_d = 50$ V and $E_a = 10$ V.

Using the differential equations found, two m-files are developed. The transients for $u_C(t)$, $i_L(t)$, and $i_a(t)$ are plotted in Figure 7.23. The settling time is 0.038 s. One concludes that the transient dynamics of the *boost* converters is fast. Therefore, the converter equations of motion are not usually integrated in the differential equations to analyze the transients of conventional electromechanical systems. The analysis performed is very important to design and analyze the power electronics which affect the overall system performance such as efficiency, loading capabilities, etc. We illustrate the correspondence and dependence between the motor and converter parameters. Therefore, the compliance, matching, and compatibility are emphasized. We demonstrated that the applied voltage to the motor windings may not be considered as the control input, and $u_a(t)$ is the converter output.

The three-dimensional plot, which illustrates the evolution of states $u_C(t)$, $i_L(t)$, and $i_a(t)$ can be plotted using the following statement

```
% 3-D plot using x1, x2 and x3 as the variables
plot3(y(:,1),y(:,2),y(:,3));
xlabel('Voltage \rmu_C','FontSize',14);
```

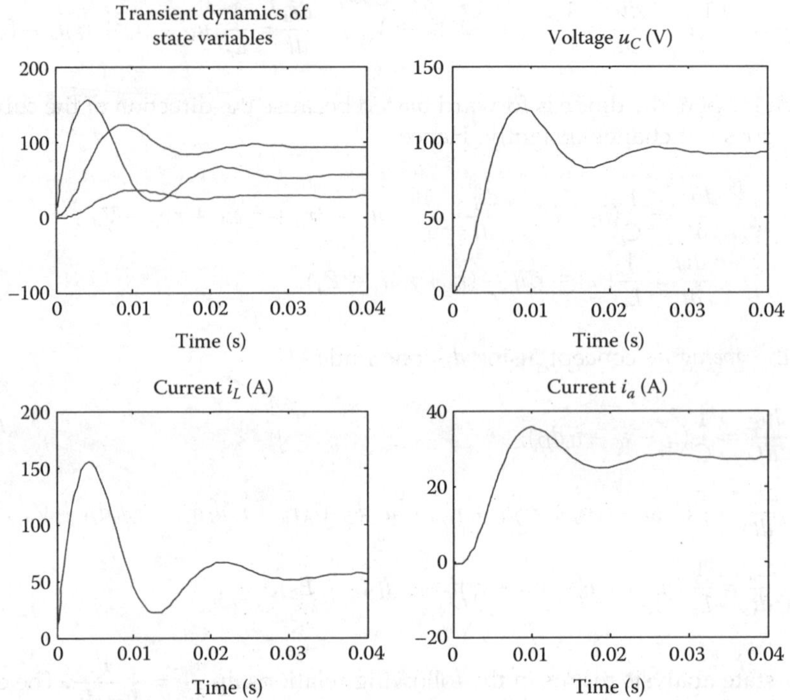

FIGURE 7.23
Transient dynamics of the *boost* converter.

```
ylabel('Current \rmi_L','FontSize',14);
zlabel('Current \rmi_a','FontSize',14);
text(-50,5,8,'Initial Conditions, \itx_0','FontSize',14);
```

The resulting plot is depicted in Figure 7.24.

We applied Kirchhoff's laws to derive the mathematical model in the form of non-linear differential equations for a one-quadrant *boost* DC–DC converter as illustrated in

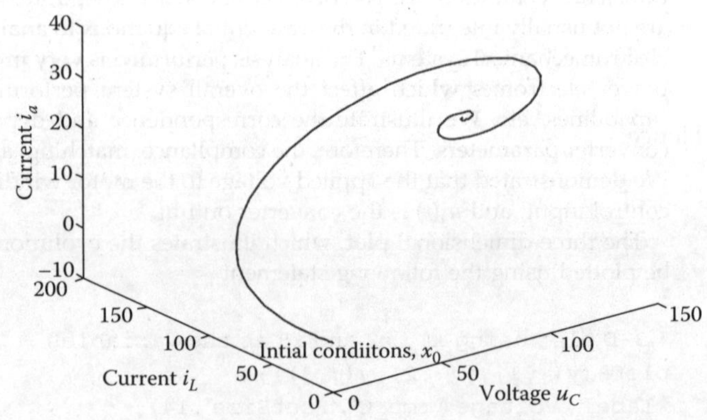

FIGURE 7.24
Evolution of the state variables.

Figure 7.22. Alternatively, the Lagrange concept can be applied. The Lagrange equations of motion are

$$\frac{d}{dt}\left(\frac{\partial \Gamma}{\partial \dot{q}_1}\right) - \frac{\partial \Gamma}{\partial q_1} + \frac{\partial D}{\partial \dot{q}_1} + \frac{\partial \Pi}{\partial q_1} = Q_1 \quad \text{and} \quad \frac{d}{dt}\left(\frac{\partial \Gamma}{\partial \dot{q}_2}\right) - \frac{\partial \Gamma}{\partial q_2} + \frac{\partial D}{\partial \dot{q}_2} + \frac{\partial \Pi}{\partial q_2} = Q_2.$$

In these equations, the electric charges in the first and the second loops are denoted as q_1 and q_2 respectively. That is, we have $i_L = \dot{q}_1$ and $i_a = \dot{q}_2$. The generalized forces are denoted as Q_1 and Q_2, e.g., $Q_1 = V_d$ and $Q_2 = -E_a$.

When the switch is closed, the total kinetic Γ, potential Π, and dissipated D energies are

$$\Gamma = \frac{1}{2}\left(L\dot{q}_1^2 + L_a\dot{q}_2^2\right), \quad \Pi = \frac{1}{2}\frac{q_2^2}{C}, \quad \text{and} \quad D = \frac{1}{2}\left[(r_L + r_s)\dot{q}_1^2 + (r_c + r_a)\dot{q}_2^2\right].$$

Assuming that the resistances, inductances, and capacitances are not varying (time-constant), we have

$$\frac{\partial \Gamma}{\partial q_1} = 0, \quad \frac{\partial \Gamma}{\partial q_2} = 0, \quad \frac{\partial \Gamma}{\partial \dot{q}_1} = L\dot{q}_1, \quad \frac{\partial \Gamma}{\partial \dot{q}_2} = L_a\dot{q}_2, \quad \frac{d}{dt}\left(\frac{\partial \Gamma}{\partial \dot{q}_1}\right) = L\ddot{q}_1, \quad \frac{d}{dt}\left(\frac{\partial \Gamma}{\partial \dot{q}_2}\right) = L_a\ddot{q}_2,$$

$$\frac{\partial \Pi}{\partial q_1} = 0, \quad \frac{\partial \Pi}{\partial q_2} = \frac{q_2}{C}, \quad \text{and} \quad \frac{\partial D}{\partial \dot{q}_1} = (r_L + r_s)\dot{q}_1, \quad \frac{\partial D}{\partial \dot{q}_2} = (r_c + r_a)\dot{q}_2.$$

The application of the Lagrange equations of motion yields

$$L\ddot{q}_1 + (r_L + r_s)\dot{q}_1 = Q_1,$$

$$L_a\ddot{q}_2 + (r_c + r_a)\dot{q}_2 + \frac{1}{C}q_2 = Q_2.$$

One obtains

$$\ddot{q}_1 = \frac{1}{L}\left(-(r_L + r_s)\dot{q}_1 + Q_1\right),$$

$$\ddot{q}_2 = \frac{1}{L_a}\left(-(r_c + r_a)\dot{q}_2 - \frac{1}{C}q_2 + Q_2\right),$$

If the switch is open, we have

$$\Gamma = \frac{1}{2}\left(L\dot{q}_1^2 + L_a\dot{q}_2^2\right), \quad \Pi = \frac{1}{2}\frac{(q_1 - q_2)^2}{C}, \quad \text{and} \quad D = \frac{1}{2}\left(r_L\dot{q}_1^2 + r_c(\dot{q}_1 - \dot{q}_2)^2 + r_a\dot{q}_2^2\right).$$

Hence,

$$\frac{\partial \Gamma}{\partial q_1} = 0, \quad \frac{\partial \Gamma}{\partial q_2} = 0, \quad \frac{\partial \Gamma}{\partial \dot{q}_1} = L\dot{q}_1, \quad \frac{\partial \Gamma}{\partial \dot{q}_2} = L_a\dot{q}_2, \quad \frac{d}{dt}\left(\frac{\partial \Gamma}{\partial \dot{q}_1}\right) = L\ddot{q}_1, \quad \frac{d}{dt}\left(\frac{\partial \Gamma}{\partial \dot{q}_2}\right) = L_a\ddot{q}_2,$$

$$\frac{\partial \Pi}{\partial q_1} = \frac{q_1 - q_2}{C}, \quad \frac{\partial \Pi}{\partial q_2} = -\frac{q_1 - q_2}{C},$$

$$\text{and} \quad \frac{\partial D}{\partial \dot{q}_1} = (r_L + r_c)\dot{q}_1 - r_c\dot{q}_2, \quad \frac{\partial D}{\partial \dot{q}_2} = -r_c\dot{q}_1 + (r_c + r_a)\dot{q}_2.$$

The resulting equations are

$$L\ddot{q}_1 + (r_L + r_c)\dot{q}_1 - r_c\dot{q}_2 + \frac{q_1 - q_2}{C} = Q_1,$$

$$L_a\ddot{q}_2 - r_c\dot{q}_1 + (r_c + r_a)\dot{q}_2 - \frac{q_1 - q_2}{C} = Q_2.$$

Therefore,

$$\ddot{q}_1 = \frac{1}{L}\left(-(r_L + r_c)\dot{q}_1 + r_c\dot{q}_2 - \frac{q_1 - q_2}{C} + Q_1\right),$$

$$\ddot{q}_2 = \frac{1}{L_a}\left(r_c\dot{q}_1 - (r_c + r_a)\dot{q}_2 + \frac{q_1 - q_2}{C} + Q_2\right).$$

From the differential equations derived when the switch is closed and open, Cauchy's form of differential equations are found by using $i_L = \dot{q}_1$ and $i_a = \dot{q}_2$. That is, $\frac{dq_1}{dt} = i_L$ and $\frac{dq_2}{dt} = i_a$. The voltage across the capacitor u_C can be expressed using the charges. That is, when the switch is closed, $u_C = -\frac{q_2}{C}$, and if the switch is open, one obtains $u_C = \frac{q_1 - q_2}{C}$. The analysis of the differential equations derived using Kirchhoff's voltage law and the Lagrange equations of motion leads one to the conclusion that the mathematical model of the *boost* converter can be found using different state variables. In particular, u_C, i_L, i_a and q_1, i_L, q_2, i_a are used. However, the resulting differential equations, which describe the converter dynamics, are related, and the solutions are the same.

7.2.4 Buck-Boost Converters

The *buck-boost* switching converter is illustrated in Figure 7.25.

If the switch is closed, the diode is reverse biased, and when the switch is open, the diode is forward biased. One easily derives a set of differential equations using Kirchhoff's law or Lagrange equations of motion. The steady-state ratio between the supplied and terminal voltage is $\frac{u_a}{V_d} = \frac{-d_D}{1 - d_D}$. That is, depending on the duty ratio, the output voltage u_a is less or

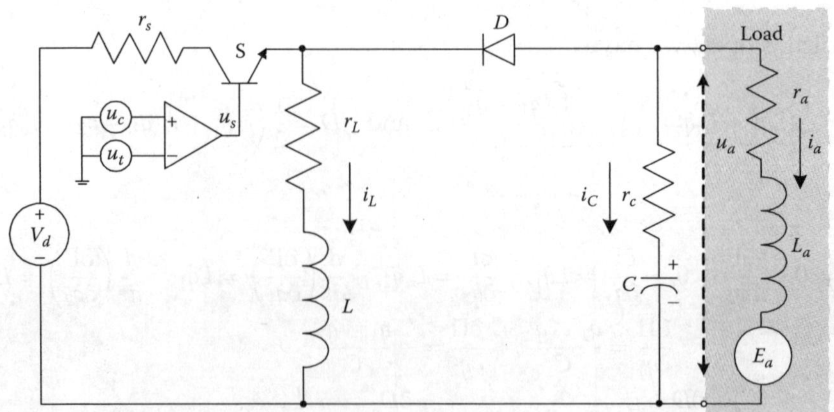

FIGURE 7.25
High-frequency *buck-boost* switching converter.

greater than V_d. The expressions for the voltage ripple and minimum inductance are $\dfrac{\Delta u_a}{u_a} = \dfrac{d_D}{r_a C f}$ and $L_{min} = \dfrac{(1 - d_D)^2 r_a}{2f}$.

7.2.5 Cuk Converters

The Cuk converter is based on a capacitive energy transfer while the *buck, buck-boost, boost,* and *flyback* converter topologies are based on the inductive energy transfer. If the switch is *on* or *off*, the currents in the input and output inductors L_1 and L are continuous. The output voltage, applied to the load can be either smaller or greater than the supplied voltage V_d. When the switch is turned *off*, the diode is forward biased. The voltage V_d is supplied, and the capacitor C_1 is charged through the inductor L_1. (See Figure 7.26.) To study how the converter operates, assume that the switch is turned *on*. The current through the inductor L_1 rises, and at the same time, the voltage of capacitor C_1 reverses bias the diode and turns the diode *off*. The capacitor C_1 discharges the stored energy through the circuit formed by capacitors C_1, C, the load r_a–L_a–E_a, and the inductor L. Consider the situation if the switch is turned *off*. The voltage V_d is applied, and the capacitor C_1 charges. The energy stored in the inductor L transfers to the load. The diode and switch provide a synchronous switching action, and the capacitor C_1 is an element for transferring energy from the energy source to the load.

From Kirchhoff's laws, using the differential equations which describe the transient dynamics when the switch is open and closed, one finds the following set of differential equations

$$\frac{du_{C1}}{dt} = \frac{1}{C_1}(i_{L1} - i_{L1}d_D + i_L d_D),$$

$$\frac{du_C}{dt} = \frac{1}{C}(i_L - i_a),$$

$$\frac{di_{L1}}{dt} = \frac{1}{L_1}(-u_{C1} - (r_{L1} + r_{c1})i_{L1} + u_{C1}d_D + (r_{c1} - r_s)i_{L1}d_D + r_s i_L d_D + V_d),$$

$$\frac{di_L}{dt} = \frac{1}{L}(-u_C - (r_L + r_c)i_L + r_c i_a - u_{C1}d_D + r_s i_{L1}d_D - (r_{c1} + r_s)i_L d_D),$$

$$\frac{di_a}{dt} = \frac{1}{L_a}(u_C + r_c i_L - (r_c + r_a)i_a - E_a).$$

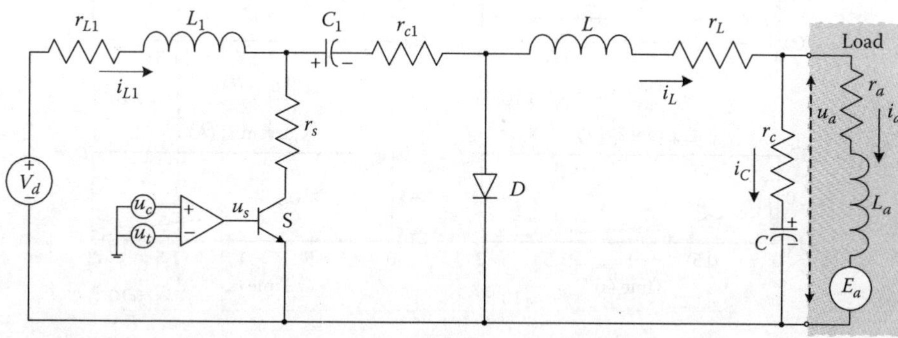

FIGURE 7.26
Cuk high-frequency switching converter.

From the differential equations derived, neglecting the resistances of the switch, inductors, and capacitors, one finds the following steady-state equations important in the converter design and converter-load matching: $\dfrac{u_a}{V_d} = \dfrac{-d_D}{1-d_D}$, $\dfrac{\Delta u_a}{u_a} = \dfrac{1-d_D}{8LCf^2}$, $L_{1\,min} = \dfrac{(1-d_D)^2 r_a}{2d_D f}$, and $L_{min} = \dfrac{(1-d_D)r_a}{2f}$.

Example 7.3:

Perform simulations of the Cuk converter if $V_d = 50$ V, $E_a = 10$ V, $r_{L1} = 0.035$ ohm, $r_L = 0.02$ ohm, $r_c = 0.15$ ohm, $r_s = 0.03$ ohm, $r_{c1} = 0.018$ ohm, $r_a = 3$ ohm, $C_1 = 2 \times 10^{-5}$ F, $C = 3.5 \times 10^{-6}$ F, $L_1 = 5 \times 10^{-6}$ H, $L = 7 \times 10^{-6}$ H, and $L_a = 0.005$ H. The duty ratio is $d_D = 0.5$.

Using the differential equations found, the simulation is performed. The transient dynamics for voltages and currents $u_{C1}(t)$, $u_C(t)$, $i_{L1}(t)$, $i_L(t)$, and $i_a(t)$, which are considered as the converter state variables are illustrated in Figure 7.27. The settling time is 0.0005 s. The steady-state value of the converter variables are easily assessed and examined. ◼

7.2.6 Flyback and Forward Converters

To avoid the interference between the input and output, which is a serious disadvantage, *flyback* and *forward* converters are used to magnetically isolate the input and output using transformers in the switching scheme. The application of transformers increases the size and cost. However, *flyback* and *forward* converters are commonly used ensuring isolation between input and output. The energy is stored in the inductor when the switch is closed, and the stored energy is transformed to the load when the switch is open. The *flyback* and *forward* magnetically coupled DC–DC converters are illustrated in Figures 7.28 and 7.29.

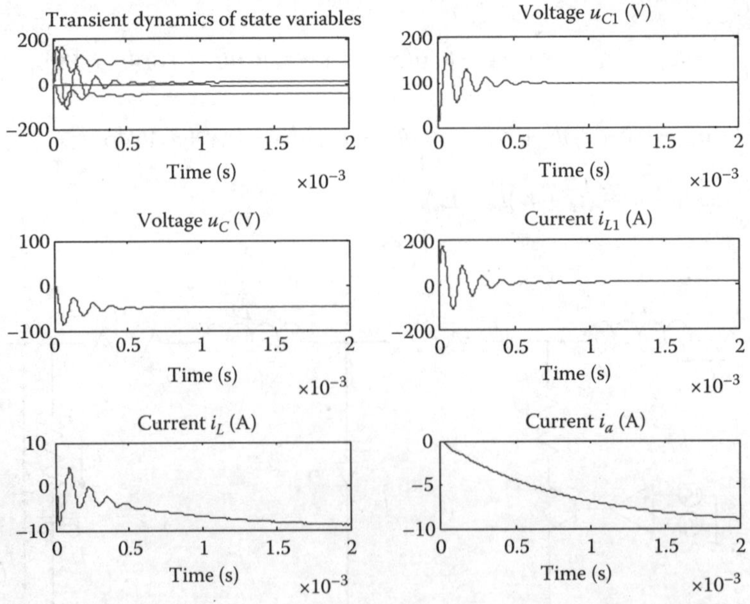

FIGURE 7.27
Transient dynamics of the Cuk converter.

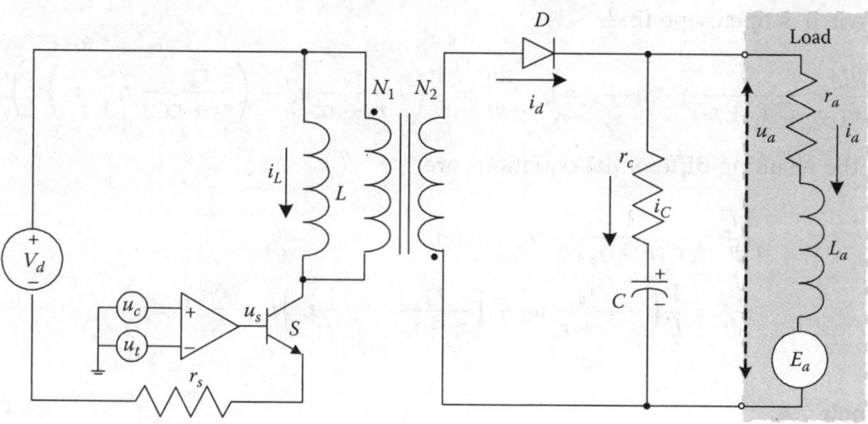

FIGURE 7.28
Flyback DC–DC converter.

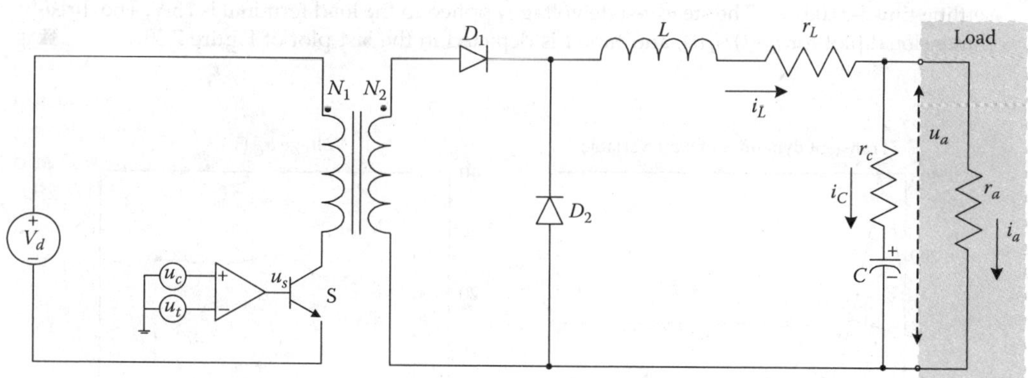

FIGURE 7.29
Forward DC–DC converter.

When the switch is closed, the diode is reverse biased. If the switch is open, the diode is forward biased. The switch is closed for time $\dfrac{d_D}{f}$ and open for $\dfrac{1-d_D}{f}$. When the switch is open $i_d = \dfrac{N_1}{N_2} i_L$, and using the duty ratio d_D, the differential equations can be found.

The differential equations when the switch is closed are

$$\frac{du_C}{dt} = \frac{1}{C(r_c + r_a)}(-u_C + r_a i_L),$$

$$\frac{di_L}{dt} = \frac{1}{L}\left(-\frac{r_a}{r_c + r_a} u_C + \left(\frac{r_a^2}{r_c + r_a} - r_L - r_a\right) i_L + \frac{N_2}{N_1} V_d\right).$$

If the switch is open, one finds

$$\frac{du_C}{dt} = \frac{1}{C(r_c + r_a)}(-u_C + r_a i_L), \quad \frac{di_L}{dt} = \frac{1}{L}\left(-\frac{r_a}{r_c + r_a}u_C + \left(\frac{r_a^2}{r_c + r_a} - r_L - r_a\right)i_L\right).$$

Hence, the resulting differential equations are

$$\frac{du_C}{dt} = \frac{1}{C(r_c + r_a)}(-u_C + r_a i_L),$$

$$\frac{di_L}{dt} = \frac{1}{L}\left(-\frac{r_a}{r_c + r_a}u_C + \left(\frac{r_a^2}{r_c + r_a} - r_L - r_a\right)i_L + \frac{N_2}{N_1}V_d d_D\right).$$

Example 7.4:

Using a set of differential equations derived, we simulate the *forward* converter if the applied voltage is $V_d = 50$ V and the duty ratio is $d_D = 0.5$. The parameters are $r_L = 0.02$ ohm, $r_C = 0.01$ ohm, $r_a = 3$ ohm, $L = 0.000005$ H, $C = 0.003$ F, and $N_2/N_1 = 1$.

Solving the derived differential equations for the *forward* converter, the dynamics is studied. The transient dynamics for the state variables $u_C(t)$ and $i_L(t)$ are documented in Figure 7.30. The settling time is 0.0015 s. The steady-state voltage applied to the load terminal is 25 V. The three-dimensional plot for $u_C(t)$, $i_L(t)$, and time t is depicted in the last plot of Figure 7.30. ∎

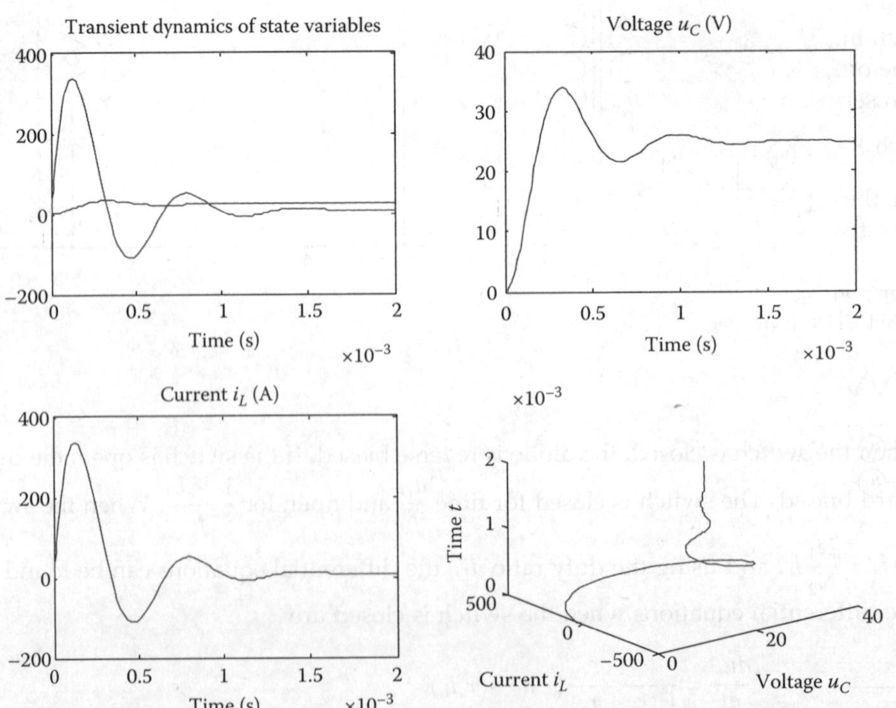

FIGURE 7.30
Transient dynamics of the *forward* converter and three-dimensional states evolution.

7.2.7 Resonant and Switching Converters

The current trends in the development of advanced switching converters have facilitated unified activities in topology design, nonlinear analysis, optimization, and control. To attain high efficiency and power density, new topologies have been developed. Nonlinear analysis and design must be performed to guarantee a spectrum of specifications and requirements imposed on the converter dynamics. It was documented that the output voltage of converters is regulated by changing the duty ratio, which is constrained by lower and upper limits. To approach the design trade-offs and enhance converter performance (settling time, overshoot, stability, robustness, power losses, and other quantities), advanced concepts, topologies, and nonlinearities should be examined. The resonant converters have been widely used in high-performance electromechanical systems. For example, the zero-voltage zero-current switching has become the enabling technology for the majority of converters to improve power density, efficiency, reliability, and other performance characteristics. Recent innovations include development of advanced converter topologies and control algorithms to maximize efficiency, minimize losses, increase power density, etc. The problem of controlling high-frequency switching converters is a very important one in many applications. Nonlinear analysis and control are central issues to be solved to improve the steady state and dynamic characteristics. A great variety of resonant converter topologies and filters have been developed. The resulting nonlinear dynamics cannot be linearized, and the hard bounds imposed cannot be neglected. Specific requirements are assigned, and the *absolute* limit of converter performance must be placed in the scope of practical design. The steady-state and dynamic characteristics of converters can be improved through coherent topology synthesis. There exist a great number of converter topologies and filter configurations. Consider the resonant converter which is shown in Figure 7.31.

The output voltage at the load terminal (formed by a resistor r_a and inductor L_a which are in series) is regulated by controlling the switching *on* and *off* durations t_{on} and t_{off}. The switch S is open and closed, and the switching frequency is $\dfrac{1}{t_{on} + t_{off}}$. When the switch is open, the diode D is forward biased to carry the output inductor current i_L, and the voltage across the capacitor C_1 is zero. When the switch is closed, the diode remains forward

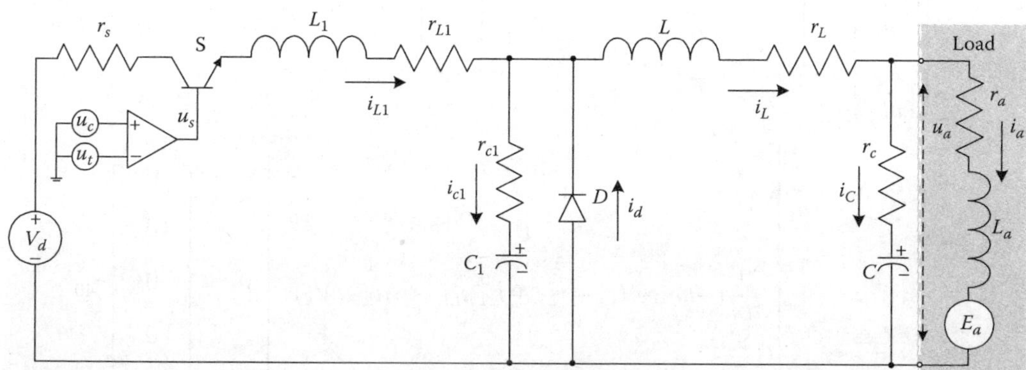

FIGURE 7.31
Resonant converter with zero-current switching.

biased while $i_{L1} < i_L$. As i_{L1} reaches i_L, the diode turns *off*. Hence, the switch turns *off* and *on* at zero-current (i_{L1} is zero). A control-triangle concept is used to establish the PWM switching. The switching signal u_s, which drives the switch, is generated by comparing a signal-level control voltage u_c with a repetitive triangular signal u_t. In resonant converters, the frequency is usually controlled to regulate the output voltage. A set of differential equations to describe the resonant converter dynamics is

$$\frac{du_{C1}}{dt} = \frac{1}{C_1}(i_{L1} - i_L)d_D,$$

$$\frac{du_C}{dt} = \frac{1}{C}(i_L - i_a),$$

$$\frac{di_{L1}}{dt} = \frac{1}{L_1}[-u_{C1} - (r_s + r_{L1} + r_{c1})i_{L1} + r_{c1}i_L + V_d]d_D,$$

$$\frac{di_L}{dt} = \frac{1}{L}[u_{C1} - u_C + r_{c1}i_{L1} - (r_{c1} + r_L + r_c)i_L + r_c i_a]d_D,$$

$$\frac{di_a}{dt} = \frac{1}{L_a}(u_C + r_c i_L - (r_c + r_a)i_a - E_a).$$

A nonlinear mathematical model results due to the multiplication of the state variables $u_{C1}(t)$, $u_C(t)$, $i_{L1}(t)$, $i_L(t)$, $i_a(t)$, and duty ratio d_D. From the differential equations obtained, we have the following nonlinear state-space model

$$
\begin{bmatrix} \dfrac{du_{C1}}{dt} \\[2mm] \dfrac{du_C}{dt} \\[2mm] \dfrac{di_{L1}}{dt} \\[2mm] \dfrac{di_L}{dt} \\[2mm] \dfrac{di_a}{dt} \end{bmatrix}
=
\begin{bmatrix}
0 & 0 & 0 & 0 & 0 \\[2mm]
0 & 0 & 0 & \dfrac{1}{C} & -\dfrac{1}{C} \\[2mm]
0 & 0 & 0 & 0 & 0 \\[2mm]
0 & 0 & 0 & 0 & 0 \\[2mm]
0 & \dfrac{1}{L_a} & 0 & \dfrac{r_c}{L_a} & -\dfrac{r_c + r_a}{L_a}
\end{bmatrix}
\begin{bmatrix} u_{C1} \\[2mm] u_C \\[2mm] i_{L1} \\[2mm] i_L \\[2mm] i_a \end{bmatrix}
$$

$$
+
\begin{bmatrix}
\dfrac{1}{C_1}(i_{L1} - i_L) \\[3mm]
0 \\[3mm]
\dfrac{1}{L_1}(-u_{C1} - (r_s + r_{L1} + r_{c1})i_{L1} + r_{c1}i_L + V_d) \\[3mm]
\dfrac{1}{L}(u_{C1} - u_C + r_{c1}i_{L1} - (r_{c1} + r_L + r_c)i_L + r_c i_a) \\[3mm]
0
\end{bmatrix} d_D
-
\begin{bmatrix} 0 \\[3mm] 0 \\[3mm] 0 \\[3mm] 0 \\[3mm] \dfrac{1}{L_a}E_a \end{bmatrix}.
$$

Example 7.5:

Our goal is to perform simulations and examine the resonant converter which is illustrated in Figure 7.31. Let $V_d = 50$ V and $E_a = 10$ V. The parameters are $r_s = 0.025$ ohm, $r_{L1} = 0.01$ ohm, $r_{c1} = 0.04$ ohm, $r_L = 0.02$ ohm, $r_c = 0.02$ ohm, $L_1 = 0.000005$ H, $L = 0.0007$ H, $C_1 = 0.000003$ F, and $C = 0.003$ F. The duty ratio is $d_D = 0.5$. The resistive-inductive load is formed by r_a and L_a, $r_a = 3$ ohm and $L_a = 0.005$ H.

Using the derived set of differential equations, the MATLAB files are developed to simulate the converter. Transient dynamics of the resonant converter with zero-current switching is studied. The transient responses for the state variables $u_{C1}(t)$, $u_C(t)$, $i_{L1}(t)$, $i_L(t)$, and $i_a(t)$ are plotted in Figure 7.32. ∎

There is a great variety of high-performance PWM resonant and switching converters. We examined one-quadrant converters. In electromechanical systems, to ensure the high performance, two- and four-quadrant converters are used. These converters were covered, see Figures 7.12 and 7.13. As the converter topology, switching and energy storage mechanisms, filter circuitry, and other key components are developed, the corresponding data-intensive analysis is performed by deriving, solving, and examining nonlinear differential equations. Kirchhoff's laws and Lagrange equations of motion are used to describe the converter and filter dynamics. For various converters, nonlinear analysis and design are performed.

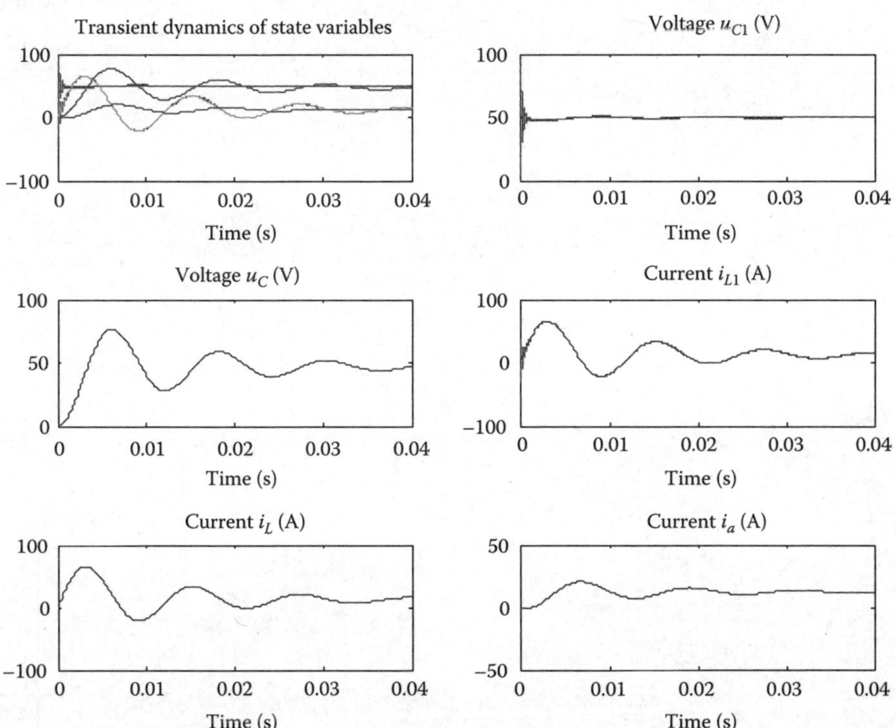

FIGURE 7.32
Transient dynamics of the resonant converter.

References

1. Sedra, A. S. and Smith, K. C., *Microelectronic Circuits*, Oxford University Press, New York, 1997.
2. Lyshevski, S. E., *Electromechanical Systems, Electric Machines, and Applied Mechatronics*, CRC Press, Boca Raton, FL, 1999.
3. Hart, D. W., *Introduction to Power Electronics*, Prentice Hall, Upper Saddle River, NJ, 1997.
4. Kassakian, J. G., Schlecht, M. F., and Verghese, G. C., *Principles of Power Electronics*, Addison-Wesley Publishing Company, Reading, MA, 1991.
5. Mohan, N., Undeland, T. M., and Robbins, W. P., *Power Electronics: Converters, Applications, and Design*, John Wiley and Sons, New York, 1995.

8

Control and Optimization of Mechatronic Systems

8.1 Basics and Introduction to Control and Optimization

Control of mechatronic systems is aimed to ensure best performance and *achievable* capabilities (functionality, controllability, stability, robustness, and immunity) by designing sound control laws and implementing them by analog or digital controllers. One needs to design closed-loop electromechanical systems optimizing their dynamic behavior, thereby enabling the steady-state performance as well. As was emphasized in Chapter 1, optimization may have a broader meaning and implies various aspects leading to distinct tasks. The optimization in the behavioral domain by means of controlling is considered with the ultimate objective to ensure best performance and *achievable* capabilities assuming optimal design in the structural and functional domains. In this chapter, the dynamic (behavioral) optimization problems, under their narrow meaning, will be solved by synthesizing optimal control laws which should be sound from device physics and implementable from hardware standpoints. We focus on the solution of various control problems to improve dynamics, efficiency, acceleration, and accuracy. Control and optimization in the behavioral domain are closely related problems, and frequently, they are used interchangeably. In fact, in order to design optimal control algorithms (called control laws and sometime controllers), one solves the optimization problem by minimizing the performance functionals and uses various stability criteria.

This chapter documents various methods in the design of closed-loop system applying sound methods. Some methods of control theory result in control laws which assume that all state variables are directly or indirectly measurable or observable. Many variables cannot be measured or observed. Therefore, *minimal-complexity* optimal control laws are synthesized to (1) guarantee near-optimal performance (capabilities), (2) minimize system complexity, and (3) ensure hardware–software soundness.

Mechatronic systems integrate a variety of components. Due to the fast dynamics of ICs and sensors, the overall behavior of electromechanical systems is usually predefined by the dynamics of electromechanical motion devices with the attached kinematics. The basics of electromagnetics, energy conversion, torque production, and other important descriptive features were studied. The derived mathematical models, in the form of nonlinear differential equations, allow one to accomplish analysis and control tasks.

Closed-loop systems are designed to ensure the best performance as measured against a spectrum of specifications and requirements. Analog and digital controllers can be derived and implemented for a large class of dynamic systems. The hardware solutions predefine system performance and capabilities, while control laws affect the system performance and capabilities. For example, for a high-performance ~30 kW electric drive (automotive application), through the structural design one defines that a permanent-magnet synchronous motor with a pulse-width modulation (PWM) amplifier should ensure the best performance and capabilities. This electric drive is open-loop stable and operational. However, the tracking control should be ensured by synthesizing a tracking controller, thereby guaranteeing cruise control features. Unsound design may lead to control laws which may destabilize the stable system (leading to unstable closed-loop system) or to control laws which may degrade the overall performance. Efficiency, stability, robustness, accuracy, and disturbance attenuation are obvious criteria among other requirements. Minimization of tracking error and settling time may result in an attempt to apply some methods yielding high feedback gains or discontinuous (relay-type) control laws. The chattering phenomena, oscillatory dynamics, losses, low efficiency, and other undesirable phenomena are observed in relay-type and high-gain control laws. Therefore, control laws which could be sound from mathematical prospective may not be applicable or may be inadequate to various electromechanical systems. The designer must apply sound methods including *minimal-complexity* control which result in *minimal-complexity* hardware solutions.

The specifications imposed on closed-loop systems are given in the performance (behavioral) and capability domains, which are related. The features and criteria under the consideration can be

- Electromagnetics-centered control soundness and efficiency: Control laws should be designed utilizing device physics integrating energy conversion, torque production, etc.
- Stability with the desired stability margins in the full operating envelope
- Robustness (low sensitivity) to parameter variations in the full operating envelope
- Robustness to structural, kinematical, and environmental changes
- Tracking accuracy, minimal dynamic, and steady-state errors
- Disturbance, load, and noise attenuation
- Transient response specifications (settling times, overshoot, etc.)

The systems performance is measured and assessed against multiple criteria (stability, robustness, transient behavior, accuracy, disturbance attenuation, and steady-state responses). The requirements and specifications are defined in the full operating envelope. Some performance characteristics are assigned and assessed using criteria and metrics as given by the performance functionals (for continuous-time systems) and performance indexes (for discrete-time systems). For example, in the behavioral domain, one can examine and optimize the input–output transient dynamics. Denoting the reference (command) and output variables as $r(t)$ and $y(t)$, the tracking error $e(t) = r(t) - y(t)$ can be minimized and the system can be controlled by using different control laws $u(t)$. These control laws can be found by applying various performance and stability criteria. The electromechanical system with the reference $r(t)$ and output $y(t)$ is illustrated in Figure 8.1a emphasizing the input (reference)–output dynamics. Systems are studied in the behavioral domain examining and optimizing the dynamic transients implying adequate steady-state responses. It is assumed that the optimal structural design was performed, for example, sound system organization, advanced hardware components (motion devices, power electronics, DSPs, etc.), and other features are guaranteed. By using the synthesis taxonomy, as reported in Figure 8.1b, the behavioral optimization and near-optimal capabilities assurance are to be performed.

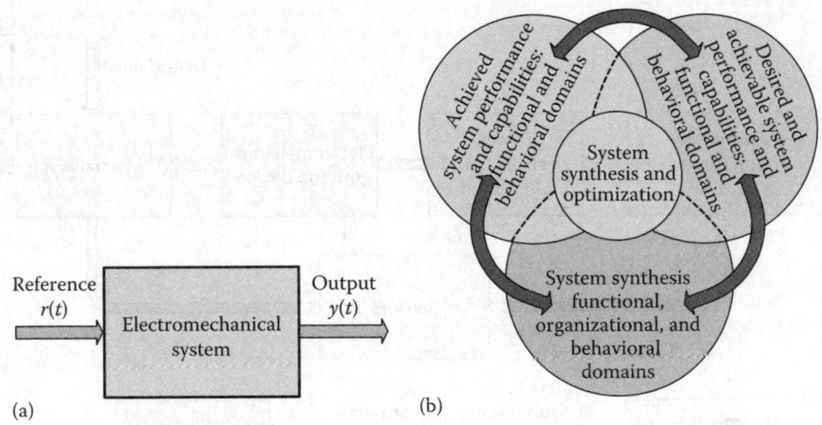

FIGURE 8.1

(a) Electromechanical system with input $r(t)$ and output $y(t)$; (b) synthesis taxonomy to ensure behavioral optimization and capabilities assurance.

Various components of electromechanical system were reported and covered. The high-level system diagram is shown in Figure 8.2a. The electromechanical systems performance and capabilities are measured using many criteria, for example, functionality, controllability, efficiency, stability, robustness, sensitivity, transient behavior, accuracy, disturbance attenuation, noise immunity, thermodynamics, and overloading. The typical output dynamics $y(t)$ is depicted assuming $r(t) = $ constant. The *generic* functional diagram of controlled (closed-loop) systems and a possible closed-loop system organization are reported in Figure 8.2b. Control laws must be designed, examined, verified, and implemented. We will apply and demonstrate different design methods. To implement analog and digital control laws, different ICs, microcontrollers, and DSPs are used as controller hardware solutions.

The designer examines system performance and capabilities applying the input–output, state, and other mappings. For example, to optimize system dynamics and attain other performance specifications, one can minimize the tracking error and settling time imposing requirements on efficiency. Steady-state and dynamic behavior can be optimized using different methods. The state-space methods imply the use of the state, output, control, reference, disturbance, and other variables utilizing the nonlinear equations of motion in the time domain. In contrast, the frequency- and s-domain methods (Laplace and Fourier transforms, transfer function, characteristic equation, pole placement, etc.) are applicable mainly to linear and near-linear systems. The majority of electromechanical systems are nonlinear, and these systems cannot be linearized. Therefore, nonlinear methods are prioritized.

Linear, nonlinear, and bounded proportional–integral–derivative (PID) control laws will be examined. The desired requirements and specifications may not be met for the specific solutions and hardware. Redesign should be performed to attain better system performance, if needed. Some solutions (e.g., the attempt to measure or observe all system variables) may significantly complicate the hardware and software, thereby increasing the overall system complexity which leads to the obvious drawbacks.

To implement PID control laws, only the tracking error $e(t)$ is be measured or estimated. The $e(t)$ usually is directly assessable or can be derived using $y(t)$ and $r(t)$. The advanced

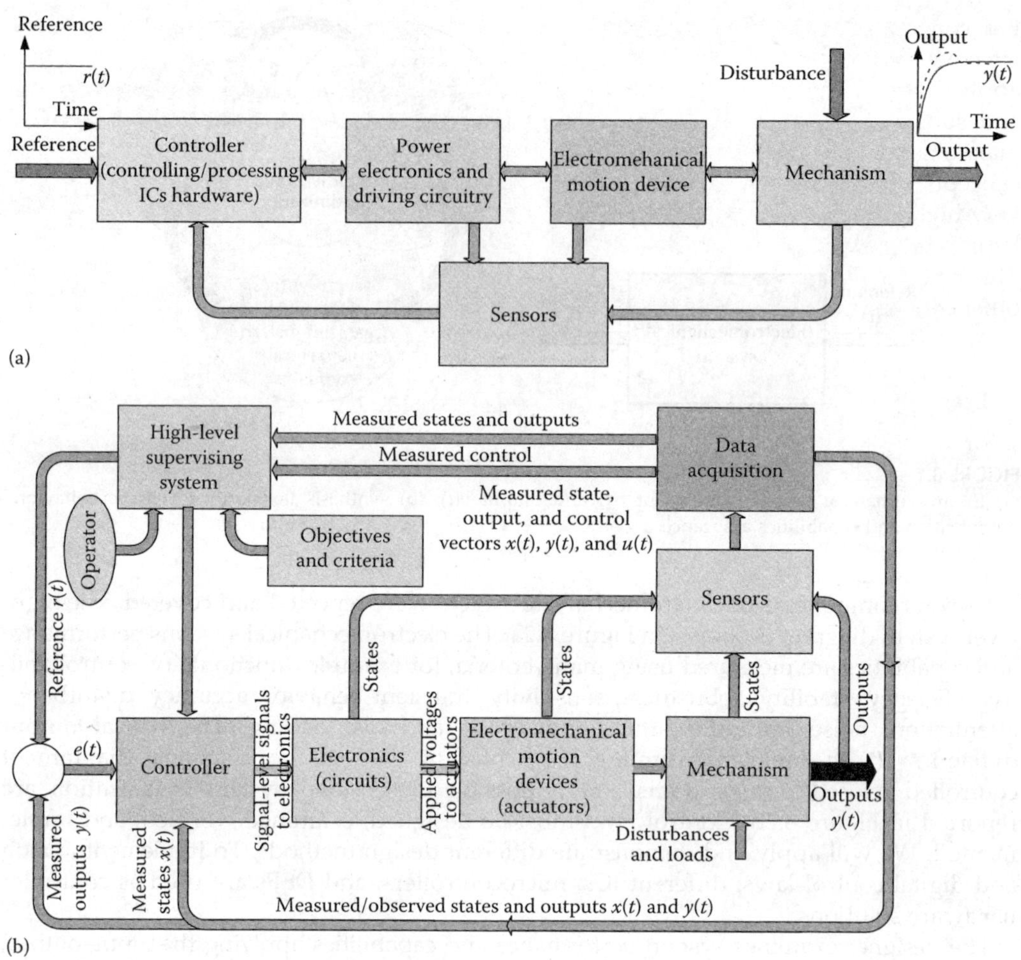

FIGURE 8.2
High-level functional diagrams of closed-loop electromechanical systems.

control systems design is an important task for high-performance systems for which multiobjective optimization is studied and strengthened specifications are imposed. System performance and capabilities, which are largely defined by the hardware, can be improved by the control laws. The departure from PID control laws may result in the substantial requirements on the advanced hardware such as sensors, ICs, DSPs, etc. If the state variables $x(t)$ are used to implement controllers, these $x(t)$ must be measured and utilized to realize control $u(t)$ in real time. For some systems, one may achieve a substantial performance improvement as advanced control laws are applied. For other systems, a simplest proportional–integral (PI) control law may guarantee the desired, acceptable, or *near-optimal* performance ensuring minimal hardware and software complexities. The physical limits, constraints, and bounds must be coherently examined. For many open-loop stable and unstable systems, PID controllers have been successfully utilized for centuries. Due to inherent physical limits and increase in the overall system complexity, some presumably advanced control laws may ensure a quite moderate improvement.

For conventional electromechanical systems, PID-centered controllers are found to be effective, valuable, and sufficient. The advanced methods in the design of control laws are applied if the strengthened specifications are imposed, and these specifications are supported and substantiated by hardware. For multiobjective problems (high accuracy, efficiency maximization, disturbance attenuation, and vibration and noise minimization), advanced concepts are applied. Advanced control laws, which should be *implementable*, are deployed in very-high-accuracy pointing systems, advanced multi-degree-of-freedom robots (manipulators), high-precision positioning systems, audio systems, high-performance drives, etc. This chapter reports the design of optimal, nonlinear, bounded, *minimal-complexity*, and other control laws.

Example 8.1:   System Performance and Its Evaluation Using Performance Functionals

Control laws can be synthesized (designed) or evaluated using the performance functionals and indexes that represent and assess specifications and requirements. Various methods in the analytic design of control laws are covered. Many analytic methods are based on the minimization of performance functionals which predefine control laws u, for example, their feedback maps and feedback gains.

By using the settling time and the tracking error $e(t) = r(t) - y(t)$, the performance criteria can be expressed as $J = \min\limits_{e} \int\limits_{0}^{\infty} |e|\,dt$, $J = \min\limits_{e} \int\limits_{0}^{\infty} e^2\,dt$, $J = \min\limits_{e} \int\limits_{0}^{\infty} (e^2 + e^4 + e^6)\,dt$, and $J = \min\limits_{t,e} \int\limits_{0}^{\infty} t|e|\,dt$.

These functionals utilize only e and t. Other functionals should be used if other performance requirements are imposed. Various performance and capability features (efficiency, losses, control efforts, and state transient evolutions) should be integrated and utilized. For example,

- The control efforts can be assessed using the positive-definite integrands u^{2n} ($n = 1, 2, 3, \dots$) or $|u|$.
- The control rate is evaluated as $(du/dt)^{2n}$ ($n = 1, 2, 3, \dots$) or $|du/dt|$.
- The torque ripple is assessed and integrated as ΔT_e^{2n} ($n = 1, 2, 3, \dots$) or $|\Delta T_e|$, $\Delta T_e = (T_e - T_{e\,\text{average}})$. The analysis of the torque ripple leads to the qualitative and quantitative assessment of efficiency, heating, vibration, noise, and wearing.

The state variables x are used in the performance functionals to integrate the dissipated energy, torque ripple, and other criteria. One recalls that T_e is a nonlinear function of the state variables (current, and angular displacement). One may use $J = \min\limits_{e,T_e,u} \int\limits_{0}^{\infty} (e^6 + T_e^4 + u^2)\,dt$. The synthesis of performance functionals and their integrands are covered in this chapter. The device physics was introduced to demonstrate the necessity to integrate and unify the physics, mathematics and engineering.   ∎

A great number of specifications are imposed. Stability, efficiency, robustness, and output dynamics (accuracy, settling time, and overshoot) are usually prioritized examining the constraints imposed on the system variables. We consider the output transient dynamics and the two evolution envelopes. The output transient response is illustrated in Figure 8.3 for the step reference command $r(t) = \text{constant}$. The system dynamics is stable because the output is bounded and converges to the steady-state value $y_{\text{steady-state}}$, for example,

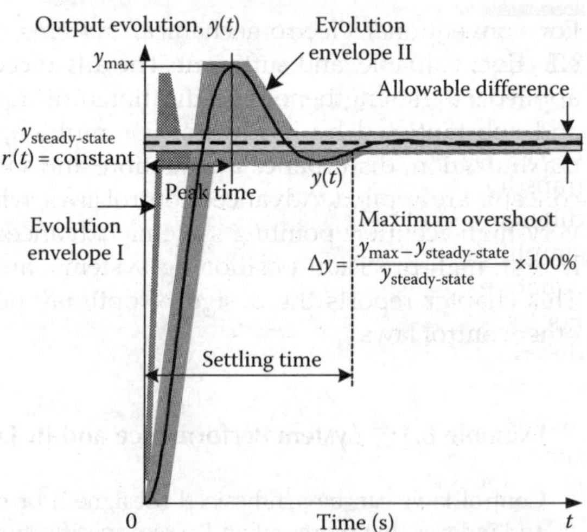

FIGURE 8.3
Output transient response and evolution envelopes.

$\lim_{t \to \infty} y(t) = y_{steady-state}$ and $e(t) \to 0$. The output dynamics $y(t)$ is studied within the evolution envelopes. Two evolution envelopes (I and II) are assigned to specify the desired accuracy, settling time, overshoot, etc. Let an optimal design be performed, and the best performance for $y(t)$ shown in Figure 8.3. The systems dynamics is within the evolution envelope II, which is the *achievable* envelope. The desired dynamics $y(t)$, specified by the evolution envelope I, is not achieved. The *desired* evolution envelope may or may not be achieved, and the requirements and specifications imposed may or may not be met. Due to the limits (peak, rated and maximum torque, force, power, voltage, current, acceleration, and other variables), the designer may not be able to achieve the *desired* performance while guaranteeing the best performance. The system structural redesign (organization and hardware) may be needed if necessary, and options may or may not exist. The hardware physical limits may not be surpassed or overcome by software or control solutions. The designer may refine the system organization and utilize advanced hardware components (if they exist) which may ensure the performance and capabilities improvements. The designer must be aware of the hardware physical limits, device specificity, technological constraints, affordability, and effect of control laws.

The system performance and capabilities in the behavioral domain are examined using the well-defined criteria, for example, settling time, overshoot, accuracy, and acceleration capabilities as represented in Figure 8.3 [1–4]. The settling time is the time needed for the system output $y(t)$ to reach and stay within the steady-state value $y_{steady-state}$. The allowable difference between $y(t)$ and $y_{steady-state}$ is used to find the settling time. This difference may vary from ~5% to ~0.001% or less. For example, in high-accuracy pointing systems, the required accuracy can be µrad for the overall displacements (repositioning) in the rad range. The settling time is the minimum time after which the system response remains within the desired accuracy, taking into account the steady-state value $y_{steady-state}$ and the command $r(t)$. The maximum overshoot is the difference between the maximum peak value of the systems output $y(t)$ and the steady-state value $y_{steady-state}$ divided by $y_{steady-state}$, for example, $\Delta y = \dfrac{y_{max} - y_{steady-state}}{y_{steady-state}} \times 100\%$. The peak time is the time required for the system output $y(t)$ to reach the first peak of the overshoot.

8.2 Equations of Motion: Electromechanical Systems Dynamics in the State-Space Form and Transfer Functions

Device physics and dynamics of electromechanical motion devices (actuators, sensors, and transducers), amplifiers, converters, and other components were described by nonlinear differential equations. For a very limited class of devices, under some assumptions and simplifications, linear differential equations were obtained. Specifically, linear differential equations were found for some radial and axial topologies of permanent-magnet DC motors. For these linear differential equations, transfer functions can be found. Linear systems can be described in the s and z domains using transfer functions $G_{sys}(s)$ and $G_{sys}(z)$. The Laplace operator $s = d/dt$ is used for continuous-time systems. The transfer functions can be used to synthesize linear PID-type control laws. For nonlinear systems, one cannot apply transfer functions and other methods of linear theory. However, many nonlinear electromechanical systems can be controlled using PID control laws ensuring adequate performance.

For linear systems, described in the state-space form, we use n states $x \in \mathbb{R}^n$ and m controls $u \in \mathbb{R}^m$. The transient dynamics of linear systems is described by a set of n linear first-order differential equations

$$\frac{dx_1}{dt} = a_{11}x_1 + a_{12}x_2 + \cdots + a_{1n-1}x_{n-1} + a_{1n}x_n + b_{11}u_1$$
$$+ b_{12}u_2 + \cdots + b_{1m-1}u_{m-1} + b_{1m}u_m, \quad x_1(t_0) = x_{10},$$

$$\vdots$$

$$\frac{dx_n}{dt} = a_{n1}x_1 + a_{n2}x_2 + \cdots + a_{nn-1}x_{n-1} + a_{nn}x_n + b_{n1}u_1$$
$$+ b_{n2}u_2 + \cdots + b_{nm-1}u_{m-1} + b_{nm}u_m, \quad x_n(t_0) = x_{n0}.$$

In the matrix form, we have

$$\frac{dx}{dt} = \begin{bmatrix} \dfrac{dx_1}{dt} \\[2mm] \dfrac{dx_2}{dt} \\[2mm] \vdots \\[2mm] \dfrac{dx_{n-1}}{dt} \\[2mm] \dfrac{dx_n}{dt} \end{bmatrix} = \begin{bmatrix} a_{11} & a_{12} & \cdots & a_{1n-1} & a_{1n} \\ a_{21} & a_{22} & \cdots & a_{2n-1} & a_{2n} \\ \vdots & \vdots & \ddots & \vdots & \vdots \\ a_{n-11} & a_{n-12} & \cdots & a_{n-1n-1} & a_{n-1n} \\ a_{n1} & a_{n2} & \cdots & a_{nn-1} & a_{nn} \end{bmatrix} \begin{bmatrix} x_1 \\ x_2 \\ \vdots \\ x_{n-1} \\ x_n \end{bmatrix}$$

$$+ \begin{bmatrix} b_{11} & b_{12} & \cdots & b_{1m-1} & b_{1m} \\ b_{21} & b_{22} & \cdots & b_{2m-1} & b_{2m} \\ \vdots & \vdots & \ddots & \vdots & \vdots \\ b_{n-11} & b_{n-12} & \cdots & b_{n-1m-1} & b_{n-1m} \\ b_{n1} & b_{n2} & \cdots & b_{nm-1} & b_{nm} \end{bmatrix} \begin{bmatrix} u_1 \\ u_2 \\ \vdots \\ u_{m-1} \\ u_m \end{bmatrix} = Ax + Bu,$$

$$x(t_0) = x_0.$$

Assuming that matrices $A \in \mathbb{R}^{n \times n}$ and $B \in \mathbb{R}^{n \times m}$ are constant-coefficients (system parameters are constant), one finds the characteristic equation as

$$|sI - A| = 0 \quad \text{or} \quad |a_n s^n + a_{n-1} s^{n-1} + \ldots + a_1 s + a_0| = 0.$$

Here, $I \in \mathbb{R}^{n \times n}$ is the identity matrix.

Solving the characteristic equation, we obtain the eigenvalues which are also called the characteristic roots and poles. The system is stable if the real parts of all eigenvalues are negative. The stability analysis using the eigenvalues is valid only for linear dynamic systems.

The transfer function can be found using the state-space equations. Consider the linear time-invariant system as described by

$$\frac{dx}{dt} = Ax + Bu, \quad y = Hx.$$

For the output vector $y \in Y \subset \mathbb{R}^b$, the output equation is $y = Hx$, where $H \in \mathbb{R}^{b \times n}$ is the matrix of the constant coefficients. The Laplace transform for the state-space $\dot{x} = Ax + Bu$ and output $y = Hx$ equations yields

$$sX(s) - x(t_0) = AX(s) + BU(s), \quad Y(s) = HX(s).$$

Assuming that the initial conditions are zero, we have $X(s) = (sI - A)^{-1} BU(s)$.

Using the system output $y(t)$, we have $Y(s) = HX(s) = H(sI - A)^{-1} BU(s)$. The transfer function is

$$G(s) = \frac{Y(s)}{U(s)} = H(sI - A)^{-1}B.$$

Assuming that the initial conditions are zero, we apply the Laplace transform to both sides of the n-order differential equation

$$\sum_{i=0}^{n} a_i \frac{d^i y(t)}{dt^i} = \sum_{i=0}^{m} b_i \frac{d^i u(t)}{dt^i}. \quad \text{From} \left(\sum_{i=0}^{n} a_i s^i \right) Y(s) = \left(\sum_{i=0}^{m} b_i s^i \right) U(s)$$

one finds

$$G(s) = \frac{Y(s)}{U(s)} = \frac{b_m s^m + b_{m-1} s^{m-1} + \cdots + b_1 s + b_0}{a_n s^n + a_{n-1} s^{n-1} + \cdots + a_1 s + a_0}.$$

By equating the denominator polynomial of the transfer function to zero, one obtains the characteristic equation. The stability of linear time-invariant systems is guaranteed if all characteristic eigenvalues, obtained by solving the characteristic equation,

$$|a_n s^n + a_{n-1} s^{n-1} + \cdots + a_1 s + a_0| = 0,$$

have negative real parts.

In general, electromechanical systems are described by nonlinear differential equations with control bounds. In the state-space form, we have

$$\dot{x}(t) = F(x, r, d) + B(x)u, \quad y = H(x), \quad u_{\min} \leq u \leq u_{\max}, \quad x(t_0) = x_0,$$

where $x \in X \subset \mathbb{R}^n$ is the state vector (displacement, position, velocity, current, and voltage) which evolves in X; $u \in U \subset \mathbb{R}^m$ is the bounded control vector (voltage, duty cycle, signal-level voltage to the comparator); $r \in R \subset \mathbb{R}^b$, and $y \in Y \subset \mathbb{R}^b$ are the measured reference and output vectors; $d \in D \subset \mathbb{R}^v$ is the disturbance vector (load, noise, etc.); $F(\cdot): \mathbb{R}^n \times \mathbb{R}^b \times \mathbb{R}^v \to \mathbb{R}^n$ and $B(\cdot): \mathbb{R}^n \to \mathbb{R}^{n \times m}$ are the nonlinear maps; $H(\cdot): \mathbb{R}^n \to \mathbb{R}^b$ is the nonlinear map defined in the neighborhood of the origin, $H(0) = 0$.

The output equation $y = H(x)$ illustrates that the system output $y(t)$ is a nonlinear function of the state variables $x(t)$. The control bounds are represented as $u_{\min} \leq u \leq u_{\max}$.

The majority of electromechanical motion devices are continuous and described by differential equations. For discrete motion devices, or, if digital control laws are to be designed, one studies the discrete systems which are described by difference equations. The differential equations can be discretized. For n-dimensional state, m-dimensional control, and b-dimensional output vectors, the electromechanical system states, controls,

and outputs variables are $x_k = \begin{bmatrix} x_{k1} \\ x_{k2} \\ \vdots \\ x_{kn-1} \\ x_{kn} \end{bmatrix}$, $u_k = \begin{bmatrix} u_{k1} \\ u_{k2} \\ \vdots \\ u_{km-1} \\ u_{km} \end{bmatrix}$, and $y_k = \begin{bmatrix} y_{k1} \\ y_{k2} \\ \vdots \\ y_{kb-1} \\ y_{kb} \end{bmatrix}$. In matrix form,

the state-space equations are

$$
x_{k+1} = \begin{bmatrix} x_{k+1,1} \\ x_{k+1,2} \\ \vdots \\ x_{k+1,n-1} \\ x_{k+1,n} \end{bmatrix} = \begin{bmatrix} a_{k11} & a_{k12} & \cdots & a_{k1n-1} & a_{k1n} \\ a_{k21} & a_{k22} & \cdots & a_{k2n-1} & a_{k2n} \\ \vdots & \vdots & \ddots & \vdots & \vdots \\ a_{kn-11} & a_{kn-12} & \cdots & a_{kn-1n-1} & a_{kn-1n} \\ a_{kn1} & a_{kn2} & \cdots & a_{knn-1} & a_{knn} \end{bmatrix} \begin{bmatrix} x_{k1} \\ x_{k2} \\ \vdots \\ x_{kn-1} \\ x_{kn} \end{bmatrix}
$$

$$
+ \begin{bmatrix} b_{k11} & b_{k12} & \cdots & b_{k1m-1} & b_{k1m} \\ b_{k21} & b_{k22} & \cdots & b_{k2m-1} & b_{k2m} \\ \vdots & \vdots & \ddots & \vdots & \vdots \\ b_{kn-11} & b_{kn-12} & \cdots & b_{kn-1m-1} & b_{kn-1m} \\ b_{kn1} & b_{kn2} & \cdots & b_{knm-1} & b_{knm} \end{bmatrix} \begin{bmatrix} u_{k1} \\ u_{k2} \\ \vdots \\ u_{km-1} \\ u_{km} \end{bmatrix}
$$

$$
= A_k x_k + B_k u_k,
$$

$$
x_{k=k_0} = x_{k0}.
$$

Here, $A_k \in \mathbb{R}^{n \times n}$ and $B_k \in \mathbb{R}^{n \times m}$ are the matrices of coefficients.

The output equation which integrates the system outputs and states variables is

$$
y_k = H_k x_k,
$$

where $H_k \in \mathbb{R}^{b \times n}$ is the matrix of the constant coefficients.

The n-order linear difference equation is $\displaystyle\sum_{i=0}^{n} a_i y_{n-i} = \sum_{i=0}^{m} b_i u_{n-i}$, $n \geq m$.

Assuming that the coefficients are time-invariant (constant), using the z-transform and letting the initial conditions to be zero, one has $\left(\sum_{i=0}^{n} a_i z^i \right) Y(z) = \left(\sum_{i=0}^{m} b_i z^i \right) U(z)$. Therefore, the transfer function is

$$G(z) = \frac{Y(z)}{U(z)} = \frac{b_m z^m + b_{m-1} z^{m-1} + \cdots + b_1 z + b_0}{a_n z^n + a_{n-1} z^{n-1} + \cdots + a_1 z + a_0}.$$

Nonlinear discrete electromechanical systems are described using nonlinear difference equations

$$x_{k+1} = F(x_k, r_k, d_k) + B(x_k) u_k, \quad y_k = H(x_k), \quad u_{k\,\min} \le u \le u_{k\,\max}.$$

8.3 Analog and Digital Proportional–Integral–Derivative Control

8.3.1 Analog Proportional–Integral–Derivative Control Laws

Electromechanical systems can be controlled using PID controllers. The majority of electromechanical motion devices and amplifiers are analog, and they evolve in continuous-time domain. The simple and effective control laws, utilized in electromechanical systems for decades, are the PID-type controllers. The linear analog PID control law is

$$u(t) = \underbrace{k_p e(t)}_{\text{proportional feedback}} + \underbrace{k_i \int e(t)dt}_{\text{integral feedback}} + \underbrace{k_d - \frac{de(t)}{dt}}_{\text{derivative feedback}}, \tag{8.1}$$

where $e(t)$ is the error between the reference signal and the system output, $e(t) = r(t) - y(t)$ and k_p, k_i, and k_d are the proportional, integral, and derivative feedback gains.

The diagram of the analog PID control law (8.1) is shown in Figure 8.4. The Laplace operator $s = d/dt$ is used to obtain the corresponding equations in the s-domain. For simplicity and illustrative purposes, occasionally, we will *mix* the time- and s-domain notations meaning that s represents the differentiation and $1/s$ means integration. For example, the PID control law (8.1) in s-domain is $U(s) = \left(k_p + \frac{k_i}{s} + k_d s \right) E(s)$, and the transfer function is $G_{PID}(s) = \frac{U(s)}{E(s)} = \frac{k_d s^2 + k_p s + k_i}{s}$. However, for notations simplicity, for example $u(t) = k_p e(t) + k_i \iint \int e(t)dt + k_d \frac{de(t)}{dt}$, with some excuses, can be written as $u(t) = k_p e(t) + k_i \frac{e}{s^3} + k_d \frac{de(t)}{dt}$. The reader can easily refine this time- and s-domain notation inconsistency when they appear.

Various control laws can be obtained utilizing (8.1). Setting k_d equal to zero, the PI control law is $u(t) = k_p e(t) + k_i \int e(t)dt$.

Assigning the integral feedback coefficient k_i to be zero, we have the proportional–derivative (PD) control law as $u(t) = k_p e(t) + k_d \frac{de(t)}{dt}$.

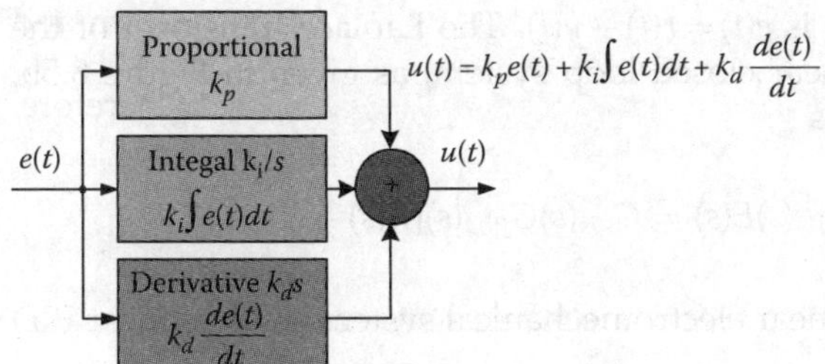

FIGURE 8.4
Linear analog PID control law.

If $k_i = 0$ and $k_d = 0$, the proportional (P) control law is $u(t) = k_p e(t)$.

Using the Laplace transform, from (8.1), we have $U(s) = \left(k_p + \dfrac{k_i}{s} + k_d s\right) E(s)$. One finds the transfer function of the analog PID control law to be

$$G_{PID}(s) = \frac{U(s)}{E(s)} = \frac{k_d s^2 + k_p s + k_i}{s}.$$

Different linear and nonlinear analog PID-type control laws can be designed and implemented. The closed-loop electromechanical system with a PID-type control law in the time- and s-domains are represented in Figure 8.5a and b. In the time-domain, we use a nonlinear system with control bounds. Using transfer functions for the open-loop system and control law, we assume that the system is linear or can be linearized, and a transfer function $G_{sys}(s)$ is used. The closed-loop electromechanical system which consists of open-loop electromechanical system (devices, components, and modules) with $G_{sys}(s)$, controller with $G_{PID}(s)$, and feedback (implying that the measured reference is compared with the measured or observed output) as depicted in Figure 8.5b.

If the system output $y(t)$ converges to the bounded reference signal $r(t)$ as time approaches infinity, the tracking of the reference input is accomplished. Ideally, the error vector $e(t)$ approaches zero. Tracking is achieved if $e(t) = [r(t) - y(t)] \to 0$ as $t \to \infty$. Ideally, $\lim\limits_{t\to\infty} e(t) = 0$, while, in practice, $\lim\limits_{t\to\infty} |e(t)| \leq \varepsilon$.

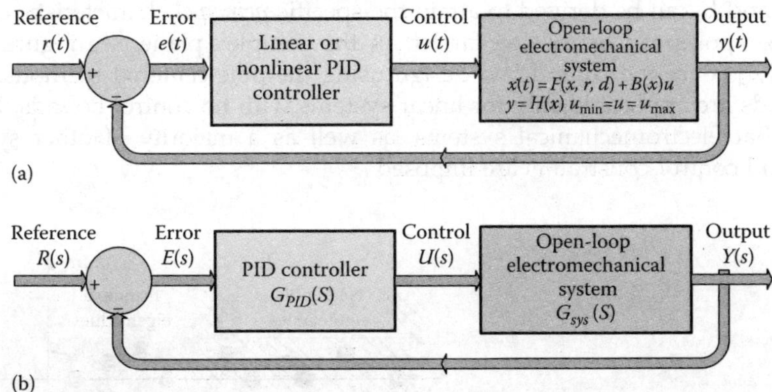

FIGURE 8.5
(a) Time-domain diagrams for a nonlinear closed-loop system with a PID control law; (b) s-domain diagrams of the linear closed-loop system with an analog PID control law.

In the time domain, the tracking error is $e(t) = r(t) - y(t)$. The Laplace transform of the error signal is $E(s) = R(s) - Y(s)$. For a linear closed-loop system, as given in Figure 8.5b, the Laplace transform of the output $y(t)$ is

$$Y(s) = G_{sys}(s)U(s) = G_{sys}(s)G_{PID}(s)E(s) = G_{sys}(s)G_{PID}(s)[R(s) - Y(s)].$$

The transfer function of the closed-loop linear electromechanical systems with a linear PID control law is

$$G(s) = \frac{Y(s)}{R(s)} = \frac{G_{sys}(s)G_{PID}(s)}{1 + G_{sys}(s)G_{PID}(s)}.$$

In the frequency domain, using $s = j\omega$, one obtains $G(j\omega) = \dfrac{Y(j\omega)}{R(j\omega)} = \dfrac{G_{sys}(j\omega)G_{PID}(j\omega)}{1 + G_{sys}(j\omega)G_{PID}(j\omega)}.$

The characteristic equation of the linear closed-loop system can be found. The stability can be ensured by the controller $G_{PID}(s)$, while the performance, which depends on $G_{PID}(s)$, can be refined by adjusting the proportional, integral, and derivative feedback gains. In particular, k_p, k_i, and k_d coefficients affect the characteristic equation and closed-loop system performance.

Using the constant factor k, poles at the origin, as well as real and complex-conjugate poles and zeros, one has

$$G(s) = \frac{k(T_{n1}s + 1)(T_{n2}s + 1)\cdots\left(T_{n,l-1}^2 s^2 + 2\xi_{n,l-1}T_{n,l-1}s + 1\right)\left(T_{n,l}^2 s^2 + 2\xi_{n,l}T_{n,l}s + 1\right)}{s^M(T_{d1}s + 1)(T_{d2}s + 1)\cdots\left(T_{d,p-1}^2 s^2 + 2\xi_{d,p-1}T_{d,p-1}s + 1\right)\left(T_{d,p}^2 s^2 + 2\xi_{d,p}T_{d,p}s + 1\right)},$$

where T_i and ξ_i are the time constants and damping coefficients and M is the order of the poles at the origin.

The controller transfer function $G_{PID}(s)$ can be found assigning the *desired* $G(s)$ which specifies the location of the poles and zeros affecting the settling time, stability margins, accuracy, and overshoot. For example, if the PID control law (8.1) is used, the feedback gains k_p, k_i, and k_d can be derived to attain the specific *principal* characteristic eigenvalues because other poles can be located far left in the complex plane, see Figure 8.6. Many textbooks [1,4] cover the control law design using the pole-centered methods. However, these methods are applicable only for linear systems with no control bounds. In contrast, we found that electromechanical systems, as well as a majority of other systems, are nonlinear and control constraints are imposed.

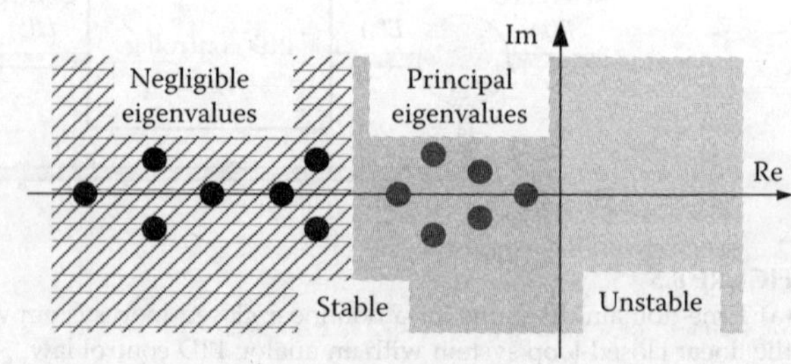

FIGURE 8.6
Eigenvalues in the complex plane.

Example 8.2:

We study an example of the so-called *force* or *torque control* when electromagnetic system is neglected. That is, only translational or rotational equations of motion are applied. Consider one-dimensional motion of a rigid-body mechanical system described by a set of two first-order differential equations

$$\frac{dx_1}{dt} = x_2, \quad \frac{dx_2}{dt} = u.$$

Here, the state variables $x_1(t)$ and $x_2(t)$ denote the displacement and velocity, while u is the force or torque to be applied to control the system motion.

The transfer function of the open-loop system is $G_{sys}(s) = \dfrac{1}{s^2}$.

Let the PD tracking control law be used. In particular, $u(t) = k_p e(t) + k_d \dfrac{de(t)}{dt}$. Hence,

$$G_{PD}(s) = \frac{U(s)}{E(s)} = k_p + k_d s.$$

The transfer function of the closed-loop system is

$$G(s) = \frac{Y(s)}{R(s)} = = \frac{G_{sys}(s)G_{PD}(s)}{1 + G_{sys}(s)G_{PD}(s)} = \frac{\frac{1}{s^2}(k_p + k_d s)}{1 + \frac{1}{s^2}(k_p + k_d s)} = \frac{k_p + k_d s}{s^2 + k_d s + k_p}.$$

Hence, the characteristic equation is $|s^2 + k_d s + k_p| = 0$.

One can specify the settling time which results in the desired characteristic eigenvalues. Let the desired poles be -1 and -1. Therefore, the desired characteristic equation is

$$|(s+1)(s+1)| = |s^2 + 2s + 1| = 0.$$

From the obtained characteristic equations

$$||s^2 + k_d s + k_p| = 0 \quad \text{and} \quad |s^2 + 2s + 1| = 0,$$

one finds the corresponding feedback gain coefficients to be $k_p = 1$ and $k_d = 2$.

The system is stable, and the analytic expressions for $x_1(t)$ and $x_2(t)$ can be found by using the Laplace transform, as $r(t)$ and initial conditions are specified. However, the use of the derivative feedback results in the sensitivity of the system to noise and dependence on the waveform of $r(t)$. Furthermore, the control bounds must be integrated. ∎

Example 8.3:

Consider an electric drive with a permanent-magnet motor.

The torque–speed characteristics are given by equation (4.3) $\omega_r = \dfrac{u_a - r_a i_a}{k_a} = \dfrac{u_a}{k_a} - \dfrac{r_a}{k_a^2} T_e$. As the armature voltage u_a is applied, the motor rotates at the particular angular velocity ω_r defined by k_a, r_a, B_m, and T_L. The motor is open-loop stable, and the experimental results were reported in Figures 4.13 through 4.15. We study the stability using the linear control system theory, examine how the characteristic equation changes as the derivative control law is used, and discuss the implication of the derivative feedback gain emphasizing the fact that positive k_d leads to unstable closed-loop system (feedback must be negative).

Using Kirchhoff's voltage law and Newton's second law of motion, the differential equations are

$$\frac{di_a}{dt} = -\frac{r_a}{L_a}i_a - \frac{k_a}{L_a}\omega_r + \frac{1}{L_a}u_a, \quad \frac{d\omega_r}{dt} = \frac{k_a}{J}i_a - \frac{B_m}{J}\omega_r - \frac{1}{J}T_L.$$

The transfer function for an open-loop electric drive (the output is ω_r and $T_L = 0$) is

$$G_{sys}(s) = \frac{Y(s)}{U(s)} = \frac{k_a}{L_aJs^2 + (r_aJ + L_aB_m)s + r_aB_m + k_a^2}.$$

The characteristic equation is $\left| L_aJs^2 + (r_aJ + L_aB_m)s + r_aB_m + k_a^2 \right| = 0$.

We have the second-order quadratic equation $as^2 + bs + c = 0$, where $a = L_aJ$, $b = r_aJ + L_aB_m$, and $c = r_aB_m + k_a^2$. The solution of the algebraic quadratic equation is $s_{1,2} = \dfrac{-b \pm \sqrt{b^2 - 4ac}}{2a}$.

The stability of the open-loop system (electric drive) is guaranteed because the real parts of all the characteristic roots are negative. All motor parameters are positive, for example, $a > 0$, $b > 0$, and $c > 0$. One concludes that for any possible values of a, b, and c, the real parts of the characteristic roots are negative.

Any PID control laws with negative feedback will guarantee stability of the closed-loop systems. As an abstract exercise, we study the derivative tracking control law $u_a(t) = k_d\dfrac{de(t)}{dt}$ with $G_D(s) = k_ds$ (derivative control laws are not used in drives and servo applications). The transfer function for the closed-loop system is

$$G(s) = \frac{Y(s)}{R(s)} = = \frac{G_{sys}(s)G_{PID}(s)}{1 + G_{sys}(s)G_{PID}(s)} = \frac{k_ak_ds}{L_aJs^2 + (r_aJ + L_aB_m + k_ak_d)s + r_aB_m + k_a^2}.$$

The characteristic equation is $\left| L_aJs^2 + (r_aJ + L_aB_m + k_ak_d)s + r_aB_m + k_a^2 \right| = 0$.

The stability is guaranteed only if the real parts of all the characteristic roots $s_{1,2} = \dfrac{-b \pm \sqrt{b^2 - 4ac}}{2a}$ are negative. Here, $b = r_aJ + L_aB_m + k_ak_d$. Thus, systems become unstable if $b \leq 0$. This can occur only if $r_aJ + L_aB_m + k_ak_d \leq 0$. All motor parameters are positive. Hence, the negative value of k_d (positive destabilizing feedback) at which a system becomes unstable is $k_d \leq -\dfrac{r_aJ + L_aB_m}{k_a}$. One can obtain the characteristics roots using the `roots` command, for example, `Eigenvalues = roots(den_s)`. ∎

The linear PID control law can be written as

$$u(t) = \underbrace{k_pe(t)}_{\text{proportional}} + \underbrace{\sum_{j=1}^{N_i} k_{ij}\frac{e}{s^j}}_{\text{integral}} + \underbrace{\sum_{j=1}^{N_d} k_{dj}\frac{d^je(t)}{dt^j}}_{\text{derivative}}, \tag{8.2}$$

where N_i and N_d are the positive integers and k_{ij} and k_{dj} are the integral and derivative feedback coefficients, respectively.

From (8.2), we obtain $U(s) = k_pE(s) + \sum_{j=1}^{N_i} k_{ij}\dfrac{E(s)}{s^j} + \sum_{j=1}^{N_d} k_{dj}s^jE(s)$ which results in the transfer function $G_{PID}(s)$. Taking note of $G_{sys}(s)$ and $G_{PID}(s)$, one finds $G(s)$.

Nonlinear PID control laws can be designed and implemented. For example, one may define the nonlinear tracking feedback mappings, and

$$u(t) = \underbrace{\sum_{k=1}^{K_p} k_{p(2k-1)} e^{2k-1}(t)}_{\text{proportional}} + \underbrace{\sum_{j=1}^{N_i} \sum_{k=1}^{K_i} k_{ij(2k-1)} \frac{e^{2k-1}}{s^j}}_{\text{integral}} + \underbrace{\sum_{j=1}^{N_d} \sum_{k=1}^{K_d} k_{dj(2k-1)} \frac{d^j e^{2k-1}(t)}{dt^j}}_{\text{derivative}}, \tag{8.3}$$

where K_p, K_i, and K_d are the positive integers and $k_{p(2k-1)}$, $k_{ij(2k-1)}$, and $k_{dj(2k-1)}$ are the proportional, integral, and derivative feedback coefficients, respectively.

In (8.3), integers K_p, K_i, and K_d are assigned by the designer defining the power for the tracking error mappings. Setting $N_i = 1$, $N_d = 1$, $K_p = 1$, $K_i = 1$, and $K_d = 1$, we have the PID control law as given by (8.1). Letting $N_i = 2$, $N_d = 1$, $K_p = 3$, $K_i = 2$, and $K_d = 1$, from (8.3), one obtains the nonlinear PID control law as given by

$$u(t) = k_{p1} e(t) + k_{p3} e^3(t) + k_{p5} e^5(t) + k_{i1,1} \frac{e}{s} + k_{i2,1} \frac{e}{s^2} + k_{i1,2} \frac{e^3}{s} + k_{i2,2} \frac{e^3}{s^2} + k_{d1,1} \frac{de(t)}{dt},$$

or, in the preferable mathematically consistent time-domain form

$$u(t) = k_{p1} e(t) + k_{p3} e^3(t) + k_{p5} e^5(t) + k_{i1,1} \int e(t)dt + k_{i2,1} \iint e(t)dt + k_{i1,2} \int e^3(t)dt$$

$$+ k_{i2,2} \iint e^3(t)dt + k_{d1,1} \frac{de(t)}{dt}.$$

Control $u(t)$ is a nonlinear function of $e(t)$. Nonlinear control laws can be applied to improve the system dynamics, enhance stability, ensure robustness, guarantee disturbance attenuation, etc. The power-series nonlinear PID-type control law is

$$u(t) = \underbrace{\sum_{k=1}^{K_p} k_{p(2k-1)} e^{\frac{2k-1}{2a_p+1}}(t)}_{\text{proportional}} + \underbrace{\sum_{j=1}^{N_i} \sum_{k=1}^{K_i} k_{ij(2k-1)} \frac{e^{\frac{2k-1}{2a_i+1}}}{s^j}}_{\text{integral}} + \underbrace{\sum_{j=1}^{N_d} \sum_{k=1}^{K_d} k_{dj(2k-1)} \frac{d^j e^{\frac{2k-1}{2a_d+1}}(t)}{dt^j}}_{\text{derivative}}, \tag{8.4}$$

where a_p, a_i, and a_d are nonnegative integers.

The linear PID control law (8.1) results if $N_i = 1$, $N_d = 1$, $K_p = 1$, $K_i = 1$, $K_d = 1$, $a_p = 0$, $a_i = 0$, and $a_d = 0$. Letting $a_p = 2$ and $a_i = 1$, one obtains nonlinear feedback maps with $e^{1/5}(t)$ or $e^{1/3}(t)$ which ensure the large control signal $u(t)$ for small values of the tracking error, while reducing $u(t)$ for large $e(t)$. As $e(t) < 0$, the conditional statement and look-up table (implemented by analog ICs, microcontrollers, or DSPs) are used to avoid the complex values. In general, PID-type control law (8.4) may provide one with the optimal performance and high accuracy relaxing the effect of control constraints.

The control bounds $u_{\min} \leq u \leq u_{\max}$ are defined by the physical and hardware limits, while $u(t)$, as given by (8.1) through (8.4), can exceed the hardware capabilities. One recalls that the duty ratio in power amplifiers or voltage applied to the motor windings are constrained as

$$d_{D\min} \leq d_D \leq d_{D\max}, \quad d_D \in [0 \ 1] \quad \text{or} \quad d_D \in [-1 \ 1],$$

$$u_{a\min} \leq u_a \leq u_{a\max} \quad \text{or} \quad u_{M\min} \leq u_M \leq u_{M\max}.$$

$$u(t) = \mathrm{sat}_{u_{\min}}^{u_{\max}}\left(\sum_{k=1}^{k_p} k_{p(2k-1)} e^{2k-1}(t) + \sum_{j=1}^{N_i}\sum_{k=1}^{K_i} k_{i\,j(2k-1)}\frac{e^{2k-1}}{s^j} + \sum_{j=1}^{N_d}\sum_{k=1}^{K_d} k_{d\,j(2k-1)}\frac{d^j\,e^{2k-1}(t)}{dt^j}\right), \quad u_{\min} \le u \le u_{\max}$$

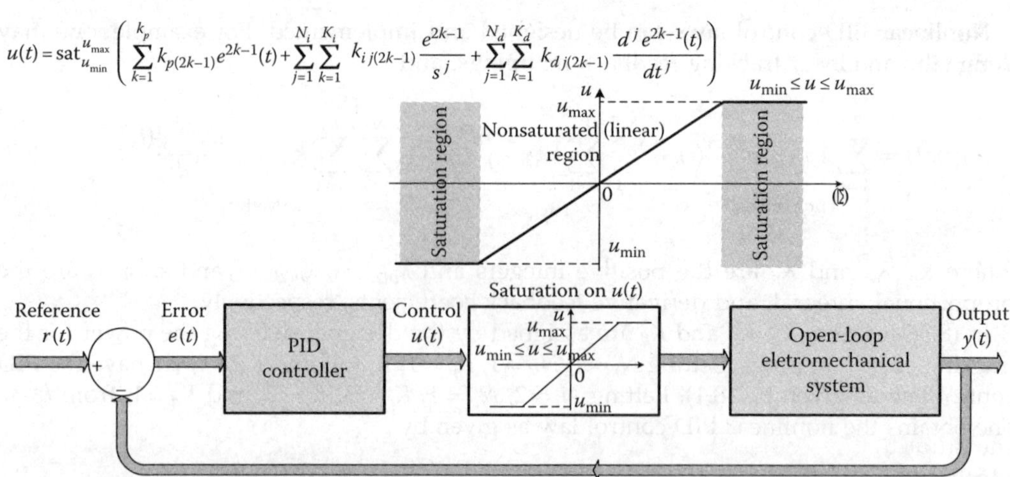

FIGURE 8.7
Closed-loop electromechanical system based on the bounded PID control law, $u_{\min} \le u \le u_{\max}$.

If nonlinear PID control laws (8.3) or (8.4) are utilized, linear methods (transfer functions, eigenvalues, and pole-placement) cannot be applied because the closed-loop system is nonlinear. Due to the control bounds $u_{\min} \le u \le u_{\max}$, even linear PID control law (8.1) usually leads to saturation in the rated operating envelope. Correspondingly, linear analysis must be applied with a great care.

In electromechanical systems, the majority of physical variables (controls, states, and outputs) are bounded. For any actuators and electromechanical motion devices, the voltages, applied to the windings are bounded. The duty ratio of PWM power amplifiers is constrained. The bounds on the current, charge, force, torque, power, acceleration, and other physical quantities are imposed. Mechanical limits are imposed on the maximum angular and linear velocities. These rated and peak (maximum allowed) voltages, currents, velocities, and displacements are specified. Due to the limits imposed, the allowed control is bounded, and the system variables must be within the maximum admissible (rated) set. The closed-loop electromechanical system with a saturated control is shown in Figure 8.7.

The bounded control is expressed as

$$u(t) = \mathrm{sat}_{u_{\min}}^{u_{\max}}\left(k_p e(t) + k_i \int e(t)dt + k_d \frac{de(t)}{dt}\right), \quad u_{\min} \le u \le u_{\max}. \qquad (8.5)$$

Thus, the control signal $u(t)$ varies between the minimum and maximum values, and $u_{\min} \le u \le u_{\max}$, $u_{\min} \le 0$ and $u_{\max} > 0$. In the linear region, the control varies between the maximum $u_{\max}$ and minimum $u_{\min}$ values, and $u(t) = k_p e(t) + k_i \int e(t)dt + k_d \frac{de(t)}{dt}$. If $k_p e(t) + k_i \int e(t)dt + k_d \frac{de(t)}{dt} > u_{\max}$, the control is bounded as $u(t) = u_{\max}$. For $k_p e(t) + k_i \int e(t)dt + k_d \frac{de(t)}{dt} < u_{\min}$, we have $u(t) = u_{\min}$. Due to control bounds, one must coherently use the control theory even if the electromechanical system itself is described or approximated by linear differential equations or transfer functions.

The constrained PID-type control laws with nonlinear feedback mappings are found using (8.3) and (8.4) as

$$u(t) = \mathrm{sat}_{u_{\min}}^{u_{\max}}\left(\sum_{k=1}^{K_p} k_{p(2k-1)}e^{2k-1}(t) + \sum_{j=1}^{N_i}\sum_{k=1}^{K_i} k_{ij(2k-1)}\frac{e^{2k-1}}{s^j} + \sum_{j=1}^{N_d}\sum_{k=1}^{K_d} k_{dj(2k-1)}\frac{d^j e^{2k-1}(t)}{dt^j}\right),$$

$$u_{\min} \leq u \leq u_{\max}$$

$$u(t) = \mathrm{sat}_{u_{\min}}^{u_{\max}}\left(\sum_{k=1}^{K_p} k_{p(2k-1)}e^{\frac{2k-1}{2a_p+1}}(t) + \sum_{j=1}^{N_i}\sum_{k=1}^{K_i} k_{ij(2k-1)}\frac{e^{\frac{2k-1}{2a_i+1}}}{s^j} + \sum_{j=1}^{N_d}\sum_{k=1}^{K_d} k_{dj(2k-1)}\frac{d^j e^{\frac{2k-1}{2a_d+1}}(t)}{dt^j}\right),$$

$$u_{\min} \leq u \leq u_{\max}$$

The studied PID control laws can be viewed as *minimal-complexity* control algorithms which can be implemented by analog and digital controllers. The reference and output signals are usually directly measured, and $e(t)$ is readily available to be utilized. Various circuits and hardware solutions exist and are available. Some implementation issues were covered in Chapter 7.

Example 8.4: Control of an Electromechanical System with a Permanent-Magnet DC Motor Using PID Control Law

A permanent-magnet DC machine under some assumptions can be described by linear differential equations. Consider a servosystem with a permanent-magnet DC motor which actuates a rotating stage as documented in Figure 8.8. This geared motor (with planetary gearhead) is attached to the rotating stage. Our goal is to design the control law to attain the desired performance (stability, fast displacement, and repositioning of the stage, disturbance attenuation, minimal steady-state tracking error, etc.).

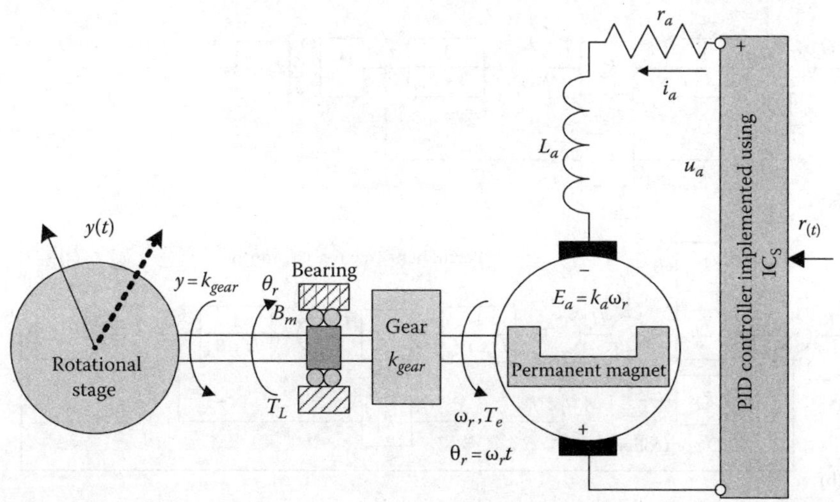

FIGURE 8.8

Schematic diagram of a servosystem with a permanent-magnet DC motor.

The stage angular displacement is a function of the rotor displacement. Using the gear ratio k_{gear}, one obtains the output equation $y = Hx$ as $y(t) = k_{gear}\theta_r(t)$. To change the angular velocity and displacement, one regulates the voltage applied to the armature winding u_a. The analog PID control law should be designed, and the feedback coefficients must be found. The rated armature voltage for the motor is $\pm u_{max}$. The rated current is i_{amax}, and the maximum angular velocity is ω_{rmax}. For a DC motor, from the experiments and catalog data we have: $u_{max} = 30$ V $(-30 \leq u_a \leq 30$ V$)$, $i_{amax} = 0.15$ A, $\omega_{rmax} = 150$ rad/s, $r_a = 200\ \Omega$, $L_a = 0.002$ H, $k_a = 0.2$ V-s/rad, (N-m/A), $J = 0.00000002$ kg-m^2, and $B_m = 0.00000005$ N-m-s/rad. The reduction gear ratio is 100:1.

For permanent-magnet DC motors, in Chapter 4, we derived the following differential equations

$$\frac{di_a}{dt} = \frac{1}{L_a}(-r_a i_a - k_a \omega_r + u_a), \quad \frac{d\omega_r}{dt} = \frac{1}{J}(T_e - T_{viscous} - T_L) = \frac{1}{J}(k_a i_a - B_m \omega_r - T_L), \quad \frac{d\theta_r}{dt} = \omega_r,$$

Using the Laplace operator $s = d/dt$, we have the resulting equations in the s-domain:

$$\left(s + \frac{r_a}{L_a}\right)I_a(s) = -\frac{k_a}{L_a}\Omega_r(s) + \frac{1}{L_a}U_a(s), \quad \left(s + \frac{B_m}{J}\right)\Omega_r(s) = \frac{1}{J}k_a I_a(s) - \frac{1}{J}T_L(s), \quad s\Theta_r(s) = \Omega_r(s).$$

From the output equation $y(t) = k_{gear}\theta_r(t)$, $Y(s) = k_{gear}\Theta_r(s)$, one obtains the s-domain diagram of the open-loop servosystem as documented in Figure 8.9a. The transfer function of a open-loop system is

$$G_{sys}(s) = \frac{Y(s)}{U_a(s)} = \frac{k_{gear}k_a}{s(L_a J s^2 + (r_a J + L_a B_m)s + r_a B_m + k_a^2)}.$$

Using the linear analog PID control law $u_a(t) = k_p e(t) + k_i \int e(t)dt + k_d \dfrac{de(t)}{dt}$, we have

$G_{PID}(s) = \dfrac{U_a(s)}{E(s)} = \dfrac{k_d s^2 + k_p s + k_i}{s}$. The closed-loop s-domain diagram is documented in Figure 8.9b.

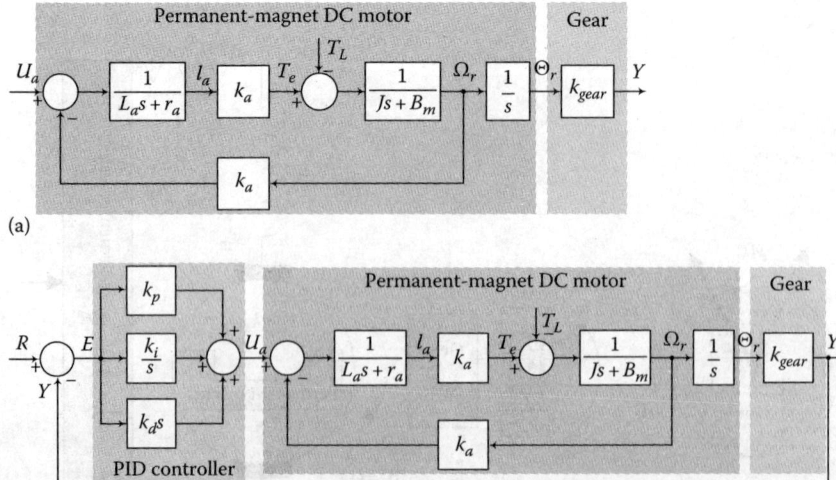

FIGURE 8.9

s-Domain diagram (a) of the open-loop system; (b) of the closed-loop system with an analog PID control law.

The closed-loop transfer function is

$$G(s) = \frac{Y(s)}{R(s)} = \frac{G_{sys}(s)G_{PID}(s)}{1 + G_{sys}(s)G_{PID}(s)} = \frac{k_{gear}k_a\left(k_d s^2 + k_p s + k_i\right)}{s^2\left(L_a J s^2 + (r_a J + L_a B_m)s + r_a B_m + k_a^2\right) + k_{gear}k_a\left(k_d s^2 + k_p s + k_i\right)}$$

$$= \frac{\dfrac{k_d}{k_i}s^2 + \dfrac{k_p}{k_i}s + 1}{\dfrac{L_a J}{k_{gear}k_a k_i}s^4 + \dfrac{(r_a J + L_a B_m)}{k_{gear}k_a k_i}s^3 + \dfrac{\left(r_a B_m + k_a^2 + k_{gear}k_a k_d\right)}{k_{gear}k_a k_i}s^2 + \dfrac{k_p}{k_i}s + 1}.$$

The numerical values of the numerator and denominator coefficients in the transfer function $G_{sys}(s)$ are found running the following MATLAB® statements

```
%        Motor parameters
ra = 200; La = 0.002; ka = 0.2; J = 0.00000002; Bm = 0.00000005; kgear = 0.01;
%        Numerator and denominator of the open-loop transfer function
format short e
num_s = [ka*kgear]; den_s = [La*J ra*J + La*Bm ra*Bm + ka^2 0];
num_s, den_s
```

The transfer function of the open-loop system is

$$G_{sys}(s) = \frac{Y(s)}{U(s)} = \frac{2 \times 10^{-3}}{s(4 \times 10^{-11}s^2 + 4 \times 10^{-6}s + 4 \times 10^{-2})}.$$

The open-loop system is unstable because one of the eigenvalues is at origin. In particular, using the `roots` command, we have

```
>> Eigenvalues = roots(den_s)
Eigenvalues =
                0
-8.8729e + 004
-1.1273e + 004
```

To stabilize the servo and attain the desired dynamic performance, control laws must be designed. The characteristic equation of the closed-loop transfer function $G(s)$ with an analog PID control law (8.1) is

$$\frac{L_a J}{k_{gear}k_a k_i}s^4 + \frac{(r_a J + L_a B_m)}{k_{gear}k_a k_i}s^3 + \frac{\left(r_a B_m + k_a^2 + k_{gear}k_a k_d\right)}{k_{gear}k_a k_i}s^2 + \frac{k_p}{k_i}s + 1 = 0.$$

The proportional k_p, integral k_i, and derivative k_d feedback coefficients of the control law $u_a(t) = k_p e(t) + k_i \int e(t)dt + k_d \dfrac{de(t)}{dt}$ affect the location of the eigenvalues. Let $k_p = 25{,}000$, $k_i = 250$, and $k_d = 25$. Hence, the PID control law is

$$u_a(t) = 25000e(t) + 250 \int e(t)dt + 25\frac{de(t)}{dt}.$$

The characteristic eigenvalues of the closed-loop system are of our interest. To derive the eigenvalues, the following MATLAB file is used

```
%       Motor parameters
ra=200; La=0.002; ka=0.2; J=0.00000002; Bm=0.00000005; kgear=0.01;
%    Feedback coefficients
kp=25000; ki=250; kd=25;
%       Denominator of the closed-loop transfer function
den_c=[(La*J)/(kgear*ka*ki) (ra*J+La*Bm)/(kgear*ka*ki) ...
(ra*Bm+ka^2+kgear*ka*kd)/(kgear*ka*ki) kp/ki 1];
%       Eigenvalues of the closed-loop system
Eigenvalues_Closed_Loop=roots(den_c)
```

The eigenvalues of the closed-loop system are

```
Eigenvalues_Closed_Loop=
-6.6393e+004
-3.3039e+004
-5.6983e+002
-1.0000e-002
```

All four eigenvalues are real. The closed-loop system is stable because the real parts of the poles are negative. Since all eigenvalues are real, there should be no overshoot (overshoot is usually an undesirable phenomenon in high-performance positioning systems). The transient dynamics is studied to assess the closed-loop system performance. The following MATLAB file allows the user to simulate the closed-loop electromechanical system

```
ra=200; La=0.002; ka=0.2; J=0.00000002; Bm=0.00000005; kgear=0.01;
kp=25000; ki=250; kd=25;
ref=1; % reference (command) displacement is 1 rad
%       Numerator and denominator of the closed-loop transfer function
num_c=[kd/ki kp/ki 1];
den_c=[(La*J)/(kgear*ka*ki) (ra*J+La*Bm)/(kgear*ka*ki) ...
(ra*Bm+ka^2+kgear*ka*kd)/(kgear*ka*ki) kp/ki 1];
t=0:0.0001:0.02;
u=ref*ones(size(t));
y=lsim(num_c,den_c,u,t);
plot(t,y,'-',y,u,':');
title('Angular Displacement, y(t)=0.01\theta_r, r(t)=1 [rad]','FontSize',14);
xlabel('Time (seconds)','FontSize',14);
ylabel('Output y(t) and Reference r(t)','FontSize',14);
axis([0 0.02,0 1.2]) % axis
```

The closed-loop servo output (angular displacement) and reference $r(t)$ are illustrated in Figure 8.10a, for $r(t)=1$ rad.

The feedback gains significantly affect the stability and dynamics. We reduce the proportional gain k_p and increase the integral feedback k_i. Let $k_p=2500$, $k_i=250{,}000$, and $k_d=25$. We calculate the eigenvalues and simulate the system. The characteristic eigenvalues are

```
Eigenvalues_Closed_Loop=
-6.5869e+004
-3.4079e+004
-2.7719e+001 +6.9284e+001i
-2.7719e+001 -6.9284e+001i
```

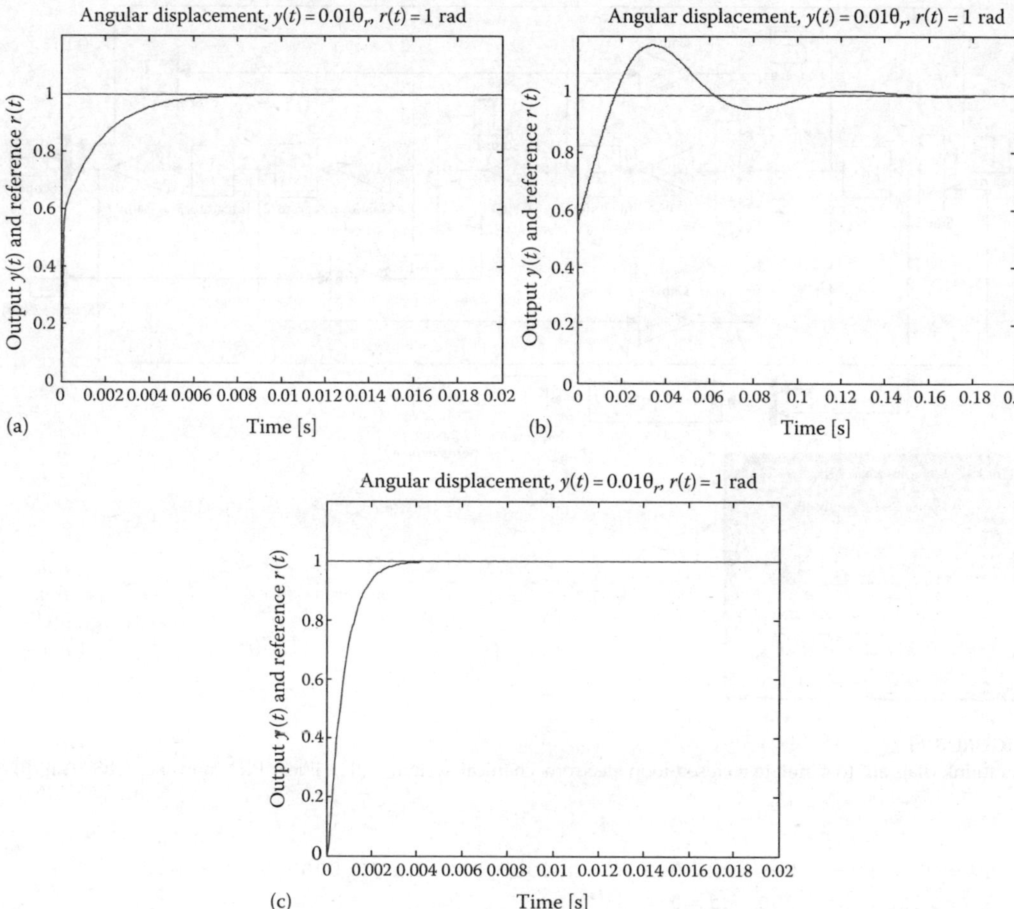

FIGURE 8.10
Dynamics of the closed-loop system with an analog PID control law (a) $k_p = 25{,}000$, $k_i = 250$, and $k_d = 25$; (b) $k_p = 2500$, $k_i = 250{,}000$, and $k_d = 25$; (c) $k_p = 25{,}000$, $k_i = 250$, and $k_d = 0$.

The *principal* eigenvalues are complex resulting in the overshoot and longer settling time. The system dynamics for $r(t) = 1$ rad is illustrated in Figure 8.10b. One must soundly design the PID control law and coherently derive the feedback gains. The derivative feedback is not usually used. Let $k_p = 25{,}000$, $k_i = 250$, and $k_d = 0$. The system dynamics is reported in Figure 8.10c, and the eigenvalues are real and found to be

```
Eigenvalues_Closed_Loop =
-8.8911e+004
-9.6324e+003
-1.4596e+003
-1.0000e-002
```

Different approaches can be used to simulate the electromechanical systems in MATLAB. The Simulink® diagram to perform simulations is documented in Figure 8.11. We use the PID controller block. The motor parameters and feedback gain coefficients are downloaded in the Command Window. In particular,

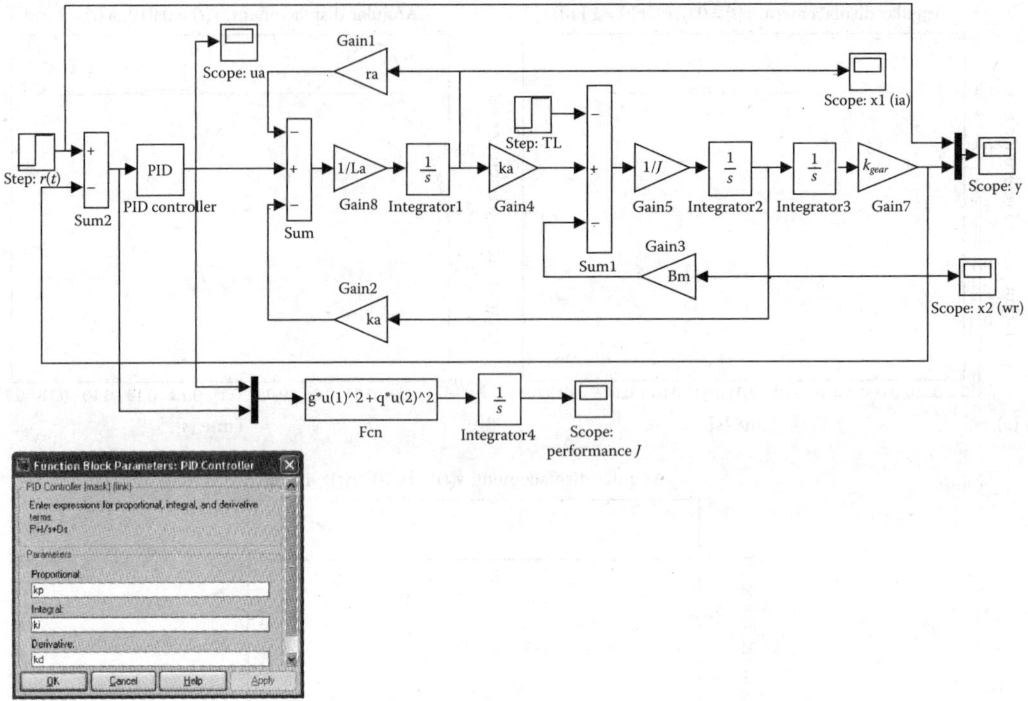

FIGURE 8.11
Simulink diagram to simulate a closed-loop electromechanical system with a linear PID controller (ch8_01.mdl).

```
ra = 200; La = 0.002; ka = 0.2; J = 0.00000002; Bm = 0.00000005; kgear = 0.01;
kp = 25000; ki = 250; kd = 0;
```

The transient dynamics of the system variables, as well as the output and voltage evolutions, are documented in Figure 8.12a. The quadratic performance functional, which can be used to assess the system dynamics, can be given as $J = \int_0^\infty \left(qe^2 + gu_a^2 \right) dt$. Let $q = 1$ and $g = 0$. The value of J is calculated to be $J = 0.00045$, and the evolution of $J(t)$ is depicted on Figure 8.12b. The Simulink diagram, as given in Figure 8.11, performs the integration and ensures visualization.

The analysis of the transients indicates that the settling time is 0.0042 s and there is no overshoot. The closed-loop system is stable. However, the armature current and voltage reach ~102 A and 24,750 V, while for the motor the peak current is ~0.2 A and the rated voltage is ±30 V. The armature current $i_a(t)$, angular velocity $\omega_r(t)$, and applied voltage $u_a(t)$ significantly exceed the rated and peak values. The saturation $u_{min} \leq u \leq u_{max}$, $-30 \leq u_a \leq 30$ V must be integrated. The bounded PI control law is

$$u_a(t) = \mathrm{sat}_{-30}^{+30}\left[25{,}000 e(t) + 250 \int e(t) dt \right], \quad -30 \leq u_a \leq 30 \text{ V}.$$

With these control bounds, the simulation is performed. The Simulink model is built utilizing the `Saturation` block as illustrated in Figure 8.13.

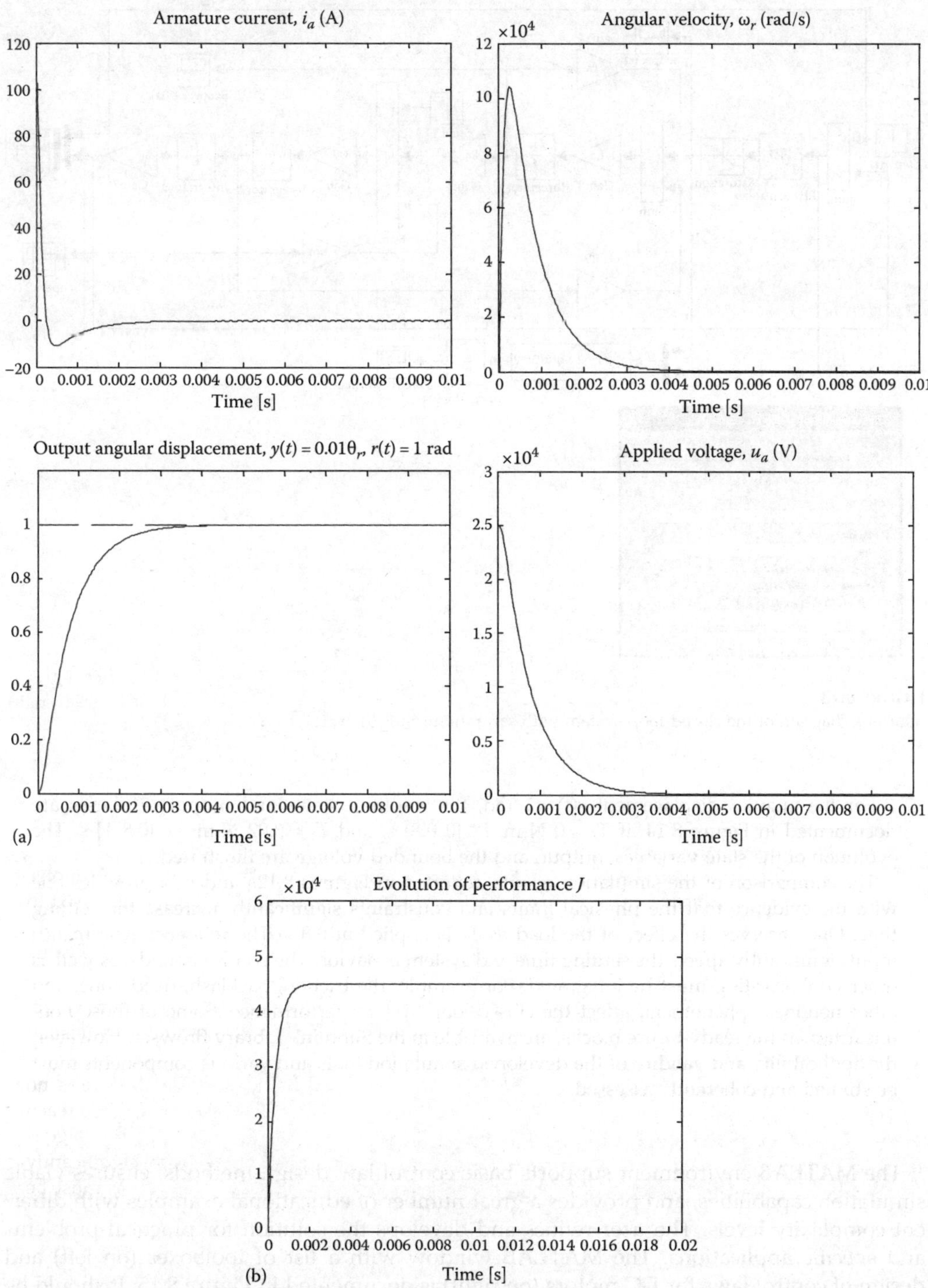

FIGURE 8.12

(a) Dynamics of the closed-loop system with a PID control law, $k_p = 25{,}000$, $k_i = 250$, and $k_d = 0$; (b) Evolution of the performance functional $J = \int\limits_0^\infty \left(qe^2 + gu_a^2\right)dt$, $q = 1$ and $g = 0$.

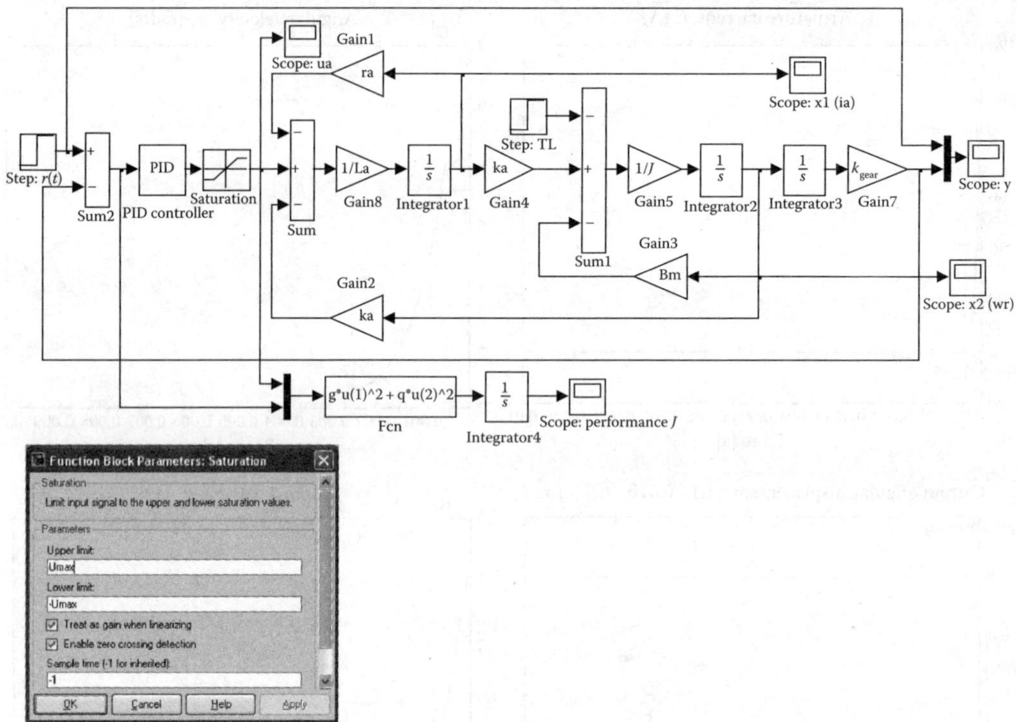

FIGURE 8.13

Simulink diagram of the closed-loop system with saturation (ch8_02.mdl).

For the angular displacement $r(t) = 1$ rad, the resulting states and output responses are documented in Figure 8.14, if $T_L = 0$ N-m, $t \in [0\ 0.8]$ s, and $T_L = 0.02$ N-m, $t \in [0.8\ 1]$ s. The evolution of the state variables, output, and the bounded voltage are illustrated.

The comparison of the simulation results, reported in Figures 8.12a and 8.14 provides one with the evidence that the physical limits and constraints significantly increase the settling time. One observes the effect of the load as T_L is applied at 0.8 s. The reference (command) input significantly affects the settling time and system behavior. The control bounds, as well as other nonlinearities, must be integrated. For example, the friction, backlash, dead zone, and other nonlinear phenomena affect the closed-loop system performance. Some of those nonlinearities, as the ready-to-use blocks, are available in the Simulink Library Browser. However, the applicability and validity of the developed simulation tools and various components must be studied and coherently assessed. ∎

The MATLAB environment supports basic control law design methods, ensures viable simulation capabilities, and provides a great number of educational examples with different complexity levels. The user refines and develops the solution for practical problems and specific applications. The MATLAB window with a list of toolboxes (on left) and design of control laws for DC motors (on right) is documented in Figure 8.15. It should be emphasized that engineering design and educational exercises are different, and one may not relay on postulates and assumptions in engineering practice.

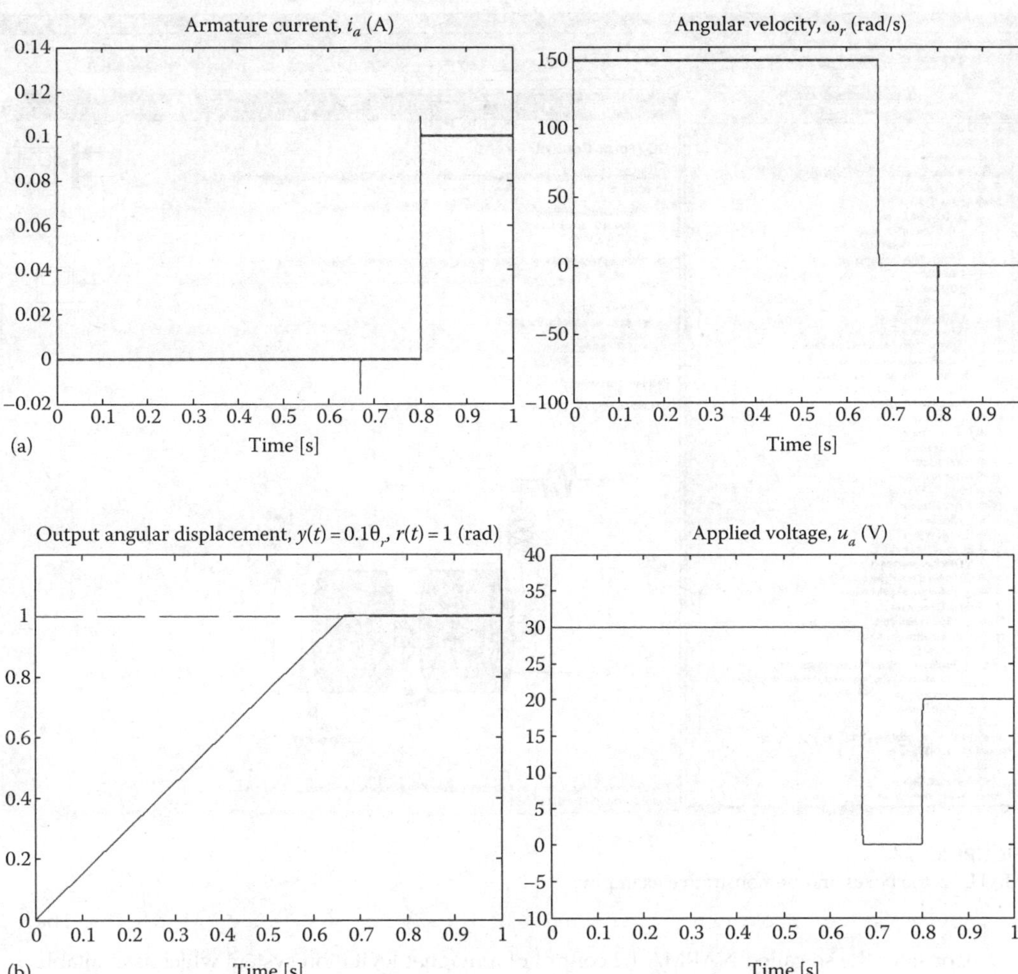

FIGURE 8.14

Dynamics of the closed-loop system with the bounded PID control law ($k_p = 25{,}000$, $k_i = 250$, and $k_d = 0$) for $r(t) = 1$ rad.

Example 8.5: Control of a Magnetic Levitation System

We studied various reluctance electromechanical motion devices such as solenoids, relays, and synchronous reluctance motors in Chapters 2 and 3. Nonlinear differential equations were derived and utilized to perform design and analysis tasks. Many magnetic levitation systems utilize the reluctance electromagnetics. To design optimal systems, various control and optimization methods can be used. In addition to sound Hamilton–Jacobi, Lyapunov, and other methods covered in this chapter, neural network, fuzzy logic, and some other concepts were developed. In general, control methods, control laws, and controllers have a very different degree of soundness, *implementability*, and practicality. The MATLAB environment offers various examples.

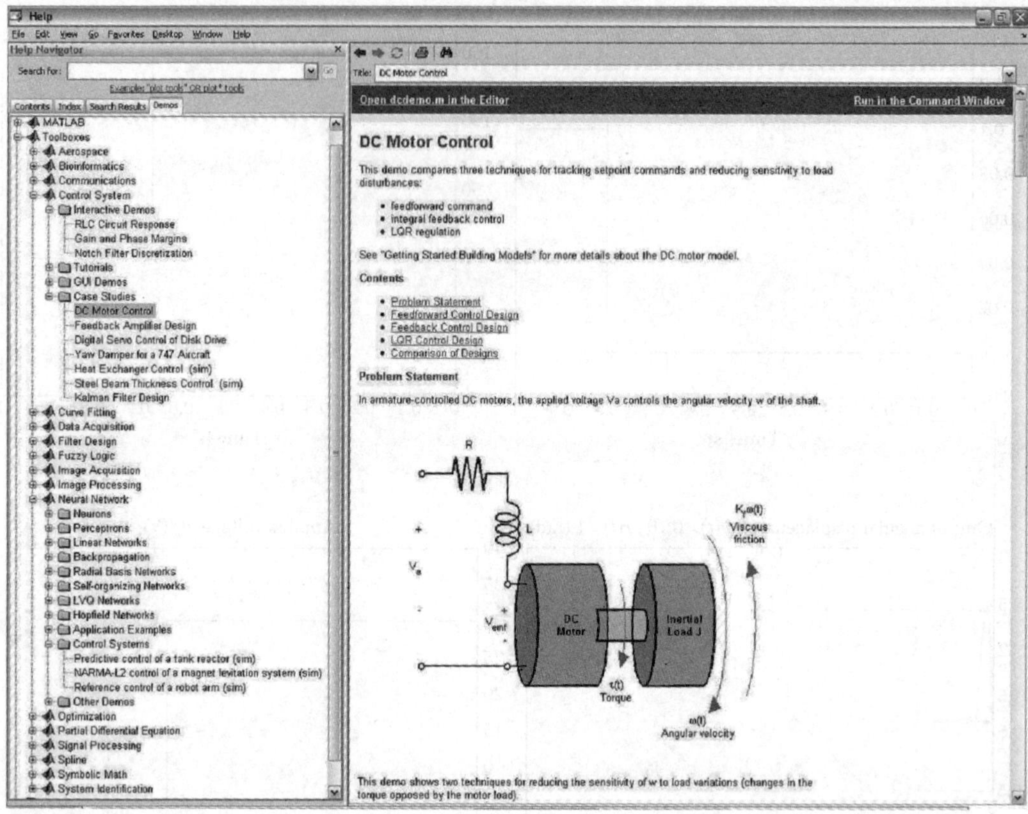

FIGURE 8.15

MATLAB toolboxes and demonstrative examples.

Consider the so-called NARMA-L2 control of a magnet levitation system which is available as an example in the Neural Network Toolbox. One may refine a "plant" model, derive alternative equations of motion for practical levitation systems, and use the voltage applied as a control variable. However, we use the provided "plant" as a ready-to-use example developed by the Neural Networks Toolbox developers. No changes are made to the original model and simulation settings. Our goal is to compare a neural network centered controller (which is extremely difficult or virtually impossible to implement) with a simplest proportional controller which can be implemented by using a single operational amplifier. The Simulink diagram is reported in Figure 8.16a. The simulation results are documented for trained neural network control laws and a proportional control algorithm with $k_p = 1000$. The behavior is illustrated in Figure 8.16b and c, respectively. One concludes that a simplest proportional control law guarantees a very good performance and may likely be considered as a primary control law for implementation as compared to a neural network alternative solution. In general, PID-centered control laws ensure *implementability* and practicality.   ■

8.3.2 Proportional–Integral–Derivative Digital Control Laws and Transfer Functions

Microcontrollers and DSPs can be utilized to implement control algorithms using continuous- or discrete-time variables or physical quantities observed, or measured by the sensors.

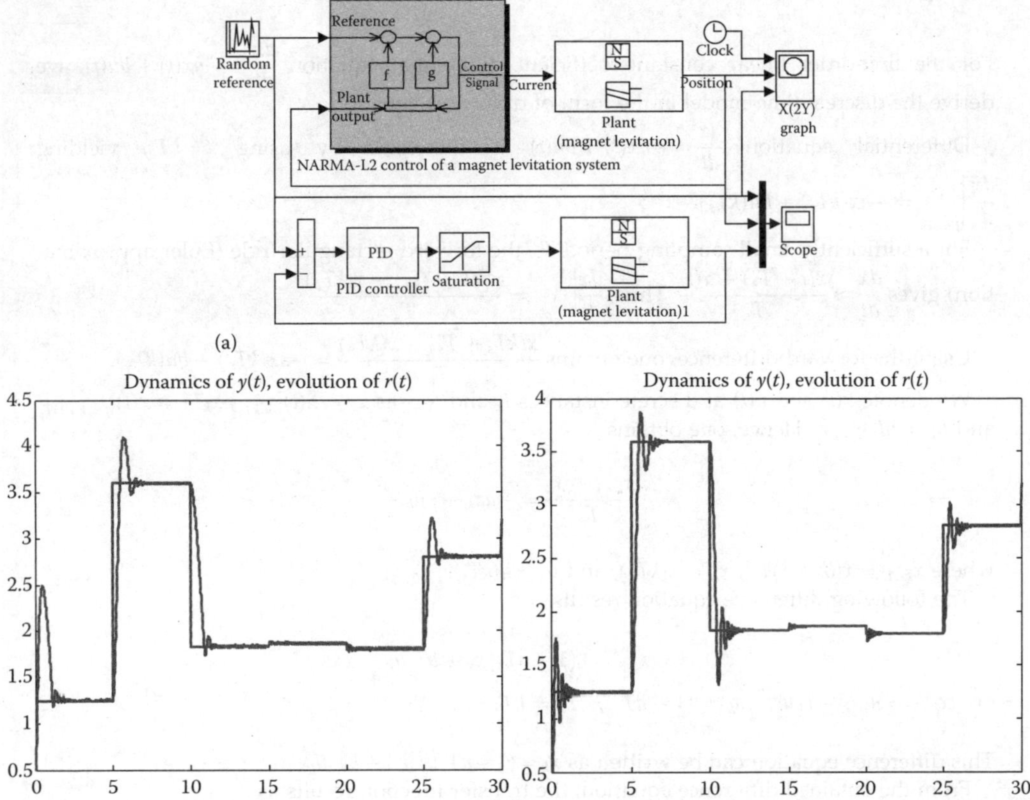

FIGURE 8.16
(a) Simulink diagram (`ch801.mdl`); (b) "Trained" neural network control: dynamics of displacement $y(t)$ and evolution of reference $r(t)$; (c) P control law: dynamics of displacement $y(t)$ and evolution of reference $r(t)$.

Diagnostics, filtering, data acquisition, and other tasks can be performed using discrete mathematics and digital processing offered by DSPs. Digital control algorithms can be designed, and discrete-time systems are studied.

Continuous-time signals $x(t)$, $y(t)$, and $e(t)$ can be sampled with the sampling period T_s, and the continuous- and discrete-time domains are related as $t = kT_s$, where k is the integer. Majority of electromechanical system components (actuators, sensors, etc.) are analog and described by differential equations. Optical encoders and digital ICs utilize digital quantities, and the evolution of these systems is analyzed in the discrete-time domain. Discrete-time systems are described by difference equations. Many systems can be studied as hybrid systems, for example, systems integrating analog and digital components, devices, and modules. To design digital control laws, the differential equations can be discretized, difference equations can be used, or hybrid models are applied.

Example 8.6:

For the first-order linear constant-coefficient differential equation $\dfrac{dx}{dt} = -ax(t) + bu(t)$, we derive the discrete-time model in the form of difference equation.

Differential equation $\dfrac{dx}{dt} = -ax(t) + bu(t)$ is dicretized by using $t = kT_s$, yielding $\left.\dfrac{dx}{dt}\right|_{t=kT_s} = -ax(kT_s) + bu(kT_s)$.

For a sufficiently small sampling period T_s, the forward rectangular rule (Euler approximation) gives $\dfrac{dx}{dt} \approx \dfrac{x(t + T_s) - x(t)}{T_s}$. Thus, $\left.\dfrac{dx}{dt}\right|_{t=kT_s} = \dfrac{x(kT_s + T_s) - x(kT_s)}{T_s}$.

Using the forward difference, one obtains $\dfrac{x(kT_s + T_s) - x(kT_s)}{T_s} = -ax(kT_s) + bu(kT_s)$.

We denote $x(t)$ and $u(t)$ at discrete instances t_k and t_{k+1} as $x_k = x(t)|_{t=kT_s}$, $x_{k+1} = x(t)|_{t=(k+1)T_s}$ and $u_k = u(t)|_{t=kT_s}$. Hence, one obtains

$$\frac{x_{k+1} - x_k}{T_s} = -ax_k + bu_k,$$

where $x_{k+1} = x[(k + 1)T_s]$, $x_k = x(kT_s)$, and $u_k = u(kT_s)$.

The following difference equation results

$$x_{k+1} = (1 - aT_s)x_k + bT_s u_k$$

or $x_{k+1} = a_k x_k + b_k u_k$, $a_k = (1 - aT_s)$, $b_k = bT$.

This difference equation can be written as $x_k = (1 - aT_s)\,x_{k-1} + bT_s u_{k-1}$.

From the obtained difference equation, the transfer function results as

$$G(z) = \frac{X(z)}{U(z)} = \frac{bT_s z^{-1}}{1 - (1 - aT_s)z^{-1}} = \frac{bT_s}{z - (1 - aT_s)}.$$

Thus, the continuous-time system was represented in the discrete-time domain by the difference equation. The z-domain transfer function was derived.   ∎

Hybrid systems integrate analog and digital components as shown in Figure 8.17. Nonlinear and linear continuous-time systems with digital controllers, hybrid circuits (including A/D and D/A converters, data hold circuits, etc.), power electronics, and analog electromechanical motion devices are represented by Figures 8.17a and b, respectively.

Assume that the electromechanical motion device dynamics is described by linear constant-coefficient (time-invariant) differential equations. The closed-loop system is documented in Figure 8.17b using the transfer function for the electronics-actuator-mechanism system $G_{sys}(s)$, data hold circuit $G_H(s)$, and digital controller $G_C(z)$. To convert the discrete-time signals from microcontrollers or DSPs to piecewise continuous signals to drive transistors in PWM amplifiers, distinct data hold circuits are used. Zero- and first-order data hold circuits are usually implemented to avoid the complexity and time delay associated with the application of high-order data hold circuits. The N-order data hold circuit with a zero-order data hold is documented in Figure 8.18.

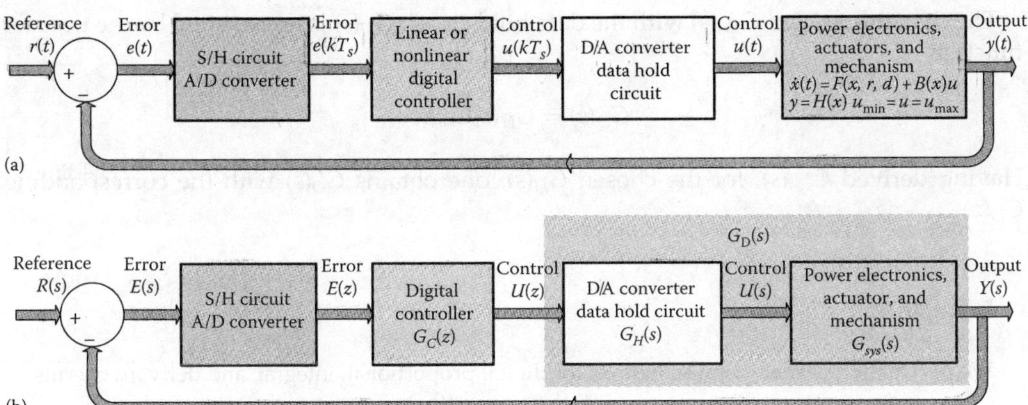

FIGURE 8.17
Block diagrams of nonlinear and linear hybrid systems with digital controllers.

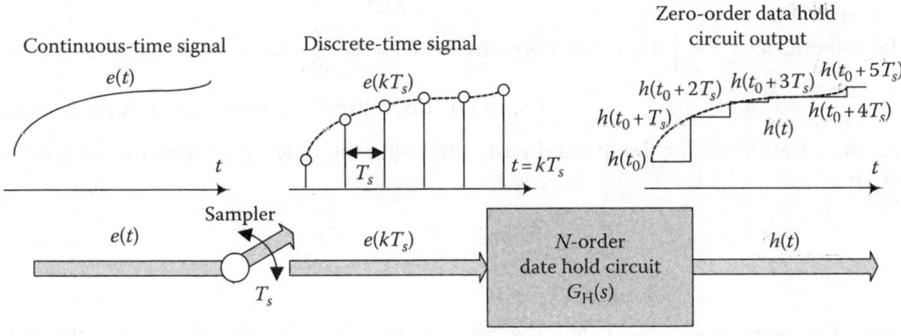

FIGURE 8.18
Sampler and N-order data hold circuit with zero-order data hold.

For the zero-order data hold circuit, the piecewise continuous data hold output is

$$h(t) = \sum_{k=0}^{\infty} e(kT_s)[1(t - kT_s) - 1(t - (k+1)T_s)].$$

The transfer function of the zero-order data hold is $G_H(s) = \dfrac{1 - e^{-T_s s}}{s}$

The first-order data hold, which can be used to perform the direct linear extrapolation, is expressed in the time domain as $h(t) = 1(t) + \dfrac{t}{T_s}1(t) - \dfrac{t - T_s}{T_s}1(t - T_s) - 1(t - T_s)$. Hence, the transfer function is $G_H(s) = \dfrac{1}{s} + \dfrac{1}{T_s s^2} - \dfrac{1}{T_s s^2}e^{-T_s s} - \dfrac{1}{s}e^{-T_s s} = \left(1 - e^{-T_s s}\right)\dfrac{T_s s + 1}{T_s s^2}.$

The dynamic system $G_{sys}(s)$ with the data hold circuit $G_H(s)$ is represented by the transfer function

$$G_D(s) = G_H(s)G_{sys}(s).$$

Having derived $G_{sys}(s)$, for the chosen $G_H(s)$, one obtains $G_D(s)$ with the corresponding $G_D(z)$.

Example 8.7:

We derive the z-domain representations for digital proportional, integral, and derivative terms of the PID control law $u(t) = k_p e(t) + k_i \int e(t)dt + k_d \dfrac{de(t)}{dt}$.

The transfer function of an analog PID control law is $G_{PID}(s) = \dfrac{U(s)}{E(s)} = \dfrac{k_d s^2 + k_p s + k_i}{s}$.

For the P control law, one has $u_p(t) = k_p e(t)$ and $G_P(s) = \dfrac{U_p(s)}{E(s)} = k_p$. Thus, the proportional digital control law is $u_p(kT_s) = k_p e(kT_s)$ and $G_p(z) = \dfrac{U_p(z)}{E(z)} = k_p$.

The integral $u_i(t) = k_i \int e(t)dt$ and derivative $u_d(t) = k_d \dfrac{de(t)}{dt}$ terms, with transfer functions $G_i(s) = \dfrac{U_i(s)}{E(s)} = \dfrac{k_i}{s}$ and $G_d(s) = \dfrac{U_d(s)}{E(s)} = k_d s$, can be discretized and represented in the z-domain. Using the z-transform, for the integral part, and using the Euler approximation, the transfer function is

$$G_i(z) = \frac{U_i(z)}{E(z)} = \frac{T_s}{1 - z^{-1}} = \frac{T_s z}{z - 1}.$$

To find the derivative term, using the trapezoidal approximation the first difference is obtained, and

$$G_d(z) = \frac{U_d(z)}{E(z)} = \frac{1 - z^{-1}}{T_s} = \frac{z - 1}{T_s z}.$$

Performing the summation of the derived terms, PI, PD, or PID control laws are found. ∎

There exist a great variety of analog PID-type control laws with the corresponding transfer functions $G_{PID}(s)$. For a PID control law $u(t) = k_p e(t) + k_i \int e(t)dt + k_d \dfrac{de(t)}{dt}$ with $G_{PID}(s) = \dfrac{U(s)}{E(s)} = \dfrac{k_d s^2 + k_p s + k_i}{s}$, one finds the z-domain representation of the control signal $U(z)$ and the transfer functions $G_{PID}(z)$. In the error form (the error e is commonly used to calculate the control), the following expressions result:

$$U(z) = \left[k_{dp} + \frac{k_{di}}{1 - z^{-1}} + k_{dd}\left(1 - z^{-1}\right) \right] E(z).$$

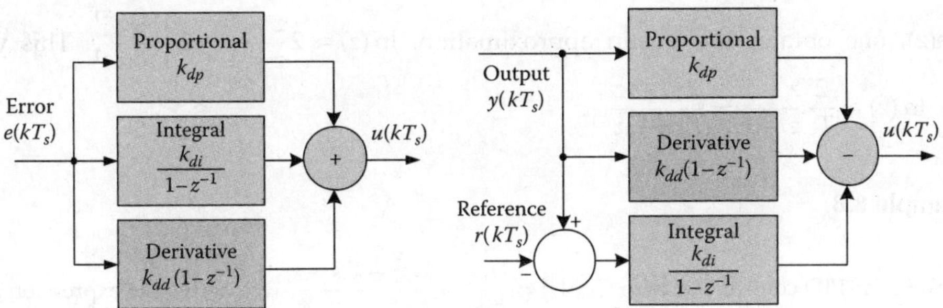

FIGURE 8.19
Error and reference-output forms of the digital PID control law.

From $G_{PID}(z) = \dfrac{U(z)}{E(z)} = k_{dp} + \dfrac{k_{di}}{1 - z^{-1}} + k_{dd}(1 - z^{-1})$, we have

$$G_{PID}(z) = \frac{(k_{dp} + k_{di} + k_{dd})z^2 - (k_{dp} + 2k_{dd})z + k_{dd}}{z^2 - z}.$$

The *reference-output* form of the digital PID control law is

$$U(z) = -k_{dp}Y(z) - k_{di}\frac{Y(z) - R(z)}{1 - z^{-1}} - k_{dd}(1 - z^{-1})Y(z).$$

For the *error* and *reference-output* forms, the z-domain diagrams of the digital PID control laws are illustrated in Figure 8.19.

The feedback gains k_{dp}, k_{di}, and k_{dd} of the digital control laws are related to the proportional, integral, and derivative coefficients of the analog PID control law (k_p, k_i, and k_d) as well as the sampling period T_s. The relationships between k_{dp}, k_{di}, k_{dd} and k_p, k_i, k_d can be obtained utilizing various analytical and numerical approaches, for example, applying specific formulas which relate z and s. Approximating the integral term by the trapezoidal summation and derivative term by a two-point difference, yields $k_{dp} = k_p - \dfrac{1}{2}k_{di}$, $k_{di} = k_i T_s$ and $k_{dd} = k_d/T_s$. Approximating the integration (rectangular, trapezoidal Tustin, bilinear, etc.) and differentiation (Euler, Taylor, backward difference, etc.), one may obtain other $G_{PID}(z)$ and expressions for feedback gains.

Using microcontrollers and DSPs, the PID control law can be implemented as

$$u(kT_s) = \underbrace{k_{dp}e(kT_s)}_{\text{proportional}} + \underbrace{\frac{1}{2}k_i T_s \sum_{j=1}^{k}\left\{e[(j-1)T_s] + e(iT_s)\right\}}_{\text{integral}} + \underbrace{\frac{k_d}{T_s}\left\{e(kT_s) - e[(k-1)T_s]\right\}}_{\text{derivative}}.$$

To find the transfer function for systems and controllers in the z-domain, the Tustin approximation is commonly applied to the transfer functions in the s-domain. In particular, from $z = e^{sT_s}$, we have $s = \dfrac{1}{T_s}\ln(z)$. The series expansion of $\ln(z)$ is

$$\ln(z) = 2\left[\frac{z-1}{z+1} + \frac{1}{3}\left(\frac{z-1}{z+1}\right)^3 + \frac{1}{5}\left(\frac{z-1}{z+1}\right)^5 + \cdots\right], \; z > 0. \text{ By truncating this series expansion}$$

for $\ln(z)$, one obtains the Tustin approximation: $\ln(z) \approx 2\dfrac{z-1}{z+1} = 2\dfrac{1-z^{-1}}{1+z^{-1}}$. This yields

$$s = \frac{1}{T_s}\ln(z) \approx \frac{2}{T_s}\frac{z-1}{z+1} = \frac{2}{T_s}\frac{1-z^{-1}}{1+z^{-1}}.$$

Example 8.8:

For a linear PID control law with $G_{PID}(s) = \dfrac{U(s)}{E(s)} = \dfrac{k_d s^2 + k_p s + k_i}{s}$, we derive the expression for

$G_{PID}(z)$ by applying the Tustin approximation. From $s \approx \dfrac{2}{T_s}\dfrac{1-z^{-1}}{1+z^{-1}}$, we have

$$
\begin{aligned}
G_{PID}(z) &= \frac{U(z)}{E(z)} \\[2mm]
&= \frac{k_d\left(\dfrac{2}{T_s}\dfrac{1-z^{-1}}{1+z^{-1}}\right)^2 + k_p\dfrac{2}{T_s}\dfrac{1-z^{-1}}{1+z^{-1}} + k_i}{\dfrac{2}{T_s}\dfrac{1-z^{-1}}{1+z^{-1}}} \\[2mm]
&= \frac{\left(2k_p T_s + k_i T_s^2 + 4k_d\right) + \left(2k_i T_s^2 - 8k_d\right)z^{-1} + \left(-2k_p T_s + k_i T_s^2 + 4k_d\right)z^{-2}}{2T_s(1 - z^{-2})}
\end{aligned}
$$

Thus, $U(z) - U(z)z^{-2} = k_{e0}E(z) + k_{e1}E(z)z^{-1} + k_{e2}E(z)z^{-2}$,

where, $k_{e0} = k_p + \dfrac{1}{2}k_i T_s + 2\dfrac{k_d}{T_s}$, $k_{e1} = k_i T_s - 4\dfrac{k_d}{T_s}$, and $k_{e2} = -k_p + \dfrac{1}{2}k_i T_s + 2\dfrac{k_d}{T_s}$.

The expression to implement the digital control law is

$$u(k) = u(k-2) + k_{e0}e(k) + k_{e1}e(k-1) + k_{e2}e(k-2).$$

Hence, to implement a digital control law, $e(k)$, $e(k-1)$, $e(k-2)$, and $u(k-2)$ are used.　■

The closed-loop system with a digital control law $G_C(z)$ is illustrated in Figure 8.17b. The transfer function of the closed-loop systems is

$$G(z) = \frac{Y(z)}{R(z)} = \frac{G_C(z)G_D(z)}{1 + G_C(z)G_D(z)}.$$

For a digital PID control law, one has $G(z) = \dfrac{Y(z)}{R(z)} = \dfrac{G_{PID}(z)G_D(z)}{1 + G_{PID}(z)G_D(z)}$. The analysis of linear discrete-time systems is by applying the methods of linear control theory.

Example 8.9:　Digital Electromechanical Servosystem with a Permanent-Magnet DC Motor

Consider a pointing system actuated by a permanent-magnet DC motor. For this system, analog control was examined in Example 8.4. Our goal is to study the digital PID control laws and analyze the system behavior. The objectives are to guarantee stability, attain the fast displacement (rapid repositioning), minimize tracking error, etc. The block diagram of the

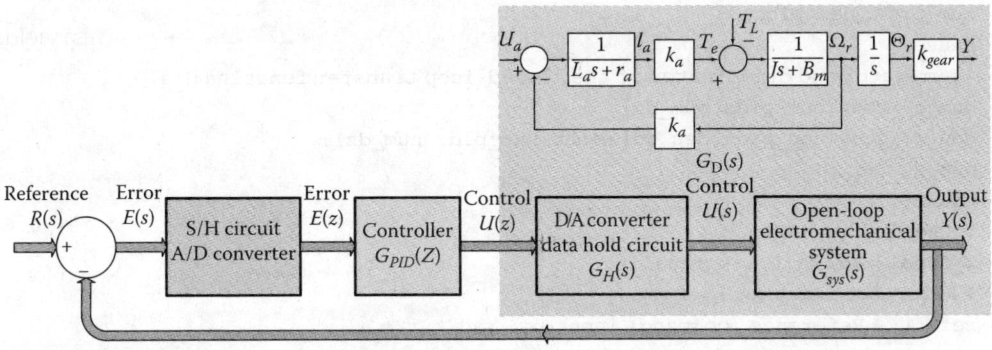

FIGURE 8.20
Block diagrams of the closed-loop systems with the digital PID controller.

closed-loop system with a digital PID controller, A/D and D/A converters, and data hold circuit is documented in Figure 8.20.

The transfer function of the open-loop system (permanent-magnet DC motor with a gearhead) is $G_{sys}(s) = \dfrac{Y(s)}{U(s)} = \dfrac{k_{gear}k_a}{s\left(L_aJs^2 + (r_aJ + L_aB_m)s + r_aB_m + k_a^2\right)}$. The transfer function of the zero-order data hold is $G_H(s) = \dfrac{1 - e^{-T_s s}}{s}$. Hence, one obtains

$$G_D(s) = G_H(s)G_{sys}(s) = \frac{1 - e^{-T_s s}}{s} \frac{k_{gear}k_a}{s\left(L_aJs^2 + (r_aJ + L_aB_m)s + r_aB_m + k_a^2\right)}.$$

The DC motor parameters are: $r_a = 200$ ohm, $L_a = 0.002$ H, $k_a = 0.2$ V-s/rad (N-m/A), $J = 0.00000002$ kg-m^2, and $B_m = 0.00000005$ N-m-s/rad.

The transfer function in the z-domain $G_D(z)$ is found from $G_D(s)$ by using the `c2dm` command. The `filter` command is used to simulate the dynamics. The following MATLAB file was developed to discretize the system, perform simulations, and plot the system evolution.

```
% Motor parameters
ra=200; La=0.002; ka=0.2; J=0.00000002; Bm=0.00000005; kgear=0.01;
% Numerator and denominator of the open-loop transfer function
format short e
num_s = [ka*kgear]; den_s = [La*J ra*J+La*Bm ra*Bm+ka^2 0];
num_s, den_s
pause;
% Numerator and denominator of GD(z) with zero-order data hold
Ts=0.0002;    % Sampling time (sampling period) Ts
[num_dz,den_dz] =c2dm(num_s,den_s,Ts,'zoh');
num_dz, den_dz
pause;
% Feedback coefficient gains of the analog PID control law
kp=25000; ki=250; kd=0.25;
% Feedback coefficient gains of the digital PID control law
kdi=ki*Ts; kdp=kp-kdi/2; kdd=kd/Ts;
% Numerator and denominator of the transfer function of the PID controller
num_pidz=[(kdp+kdi+kdd) -(kdp+2*kdd) kdd]; den_pidz=[1 -1 0];
```

```
num_pidz, den_pidz
pause;
% Numerator and denominator of the closed-loop transfer function G(z)
num_z = conv(num_pidz,num_dz);
den_z = conv(den_pidz,den_dz)+conv(num_pidz,num_dz);
num_z, den_z
pause;
% Samples, t = k*Ts
k_final = 20; k = 0:1:k_final;
% Reference input r(t) = 1 rad
ref = 1; % Reference (command) input is 1 rad
r = ref*ones(1,k_final+1);
% Modeling of the servo-system output y(k)
y = filter(num_z,den_z,r);
% Plotting statement
plot(k,y,'o',k,y,'-',k,r,':');
title('Angular Displacement, y(t) = 0.01\theta_r, r(t) = 1 [rad]','FontSize',14);
xlabel('Discrete Time k, Continuous Time t = kT_s [seconds]', 'FontSize',14);
ylabel('Output y(k) and Reference r(k)', 'FontSize',14);
axis([0 20,0 1.2])    % Axis limits
```

We found the transfer function for the open-loop system

$$G_{sys}(s) = \frac{Y(s)}{U(s)} = \frac{2 \times 10^{-3}}{s(4 \times 10^{-11}s^2 + 4 \times 10^{-6}s + 4 \times 10^{-2})}.$$ The sampling time is assigned to be

0.0002 s, for example, $T_s = 0.0002$ s. The transfer function $G_D(z)$ in the z-domain is

$$G_D(z) = \frac{5.53 \times 10^{-6}z^2 + 3.41 \times 10^{-6}z + 8.6 \times 10^{-9}}{z^3 - 1.1z^2 + 0.105z - 2.06 \times 10^{-9}}.$$

The transfer function of the digital PID controller is

$$G_{PID}(z) = \frac{(k_{dp} + k_{di} + k_{dd})z^2 - (k_{dp} + 2k_{dd})z + k_{dp}}{z^2 - z}, \quad k_{dp} = k_p - \frac{1}{2}k_{di}, \quad k_{di} = k_i T_s, \quad k_{dd} = k_d/T_s.$$

Let the feedback gains of the analog PID control law be $k_p = 25{,}000$, $k_i = 250$, and $k_d = 0.25$. The feedback coefficients of the digital control law are found using $k_{dp} = k_p - \frac{1}{2}k_{di}$, $k_{di} = k_i T_s$, and

$k_{dd} = k_d/T_s$. Hence, $G_{PID}(z) = \dfrac{2.63 \times 10^4 z^2 - 2.75 \times 10^4 z + 1.25 \times 10^3}{z^2 - z}.$

The transfer function of the closed-loop system is $G(z) = \dfrac{Y(z)}{R(z)} = \dfrac{G_{PID}(z)G_D(z)}{1 + G_{PID}(z)G_D(z)}.$ The following numerical results are found:

```
num_z =
0   1.4524e-001   -6.2713e-002   -8.6556e-002   4.0225e-003   1.0753e-005
den_z =
1.0000e+000    -1.9597e+000    1.1471e+000    -1.9147e-001
4.0225e-003    1.0753e-005
```

Thus, $G(z) = \dfrac{0.145z^4 - 0.063z^3 - 0.087z^2 + 0.004z + 1.07 \times 10^{-5}}{z^5 - 1.96z^4 + 1.15z^3 - 0.19z^2 + 0.004z + 1.07 \times 10^{-5}}.$

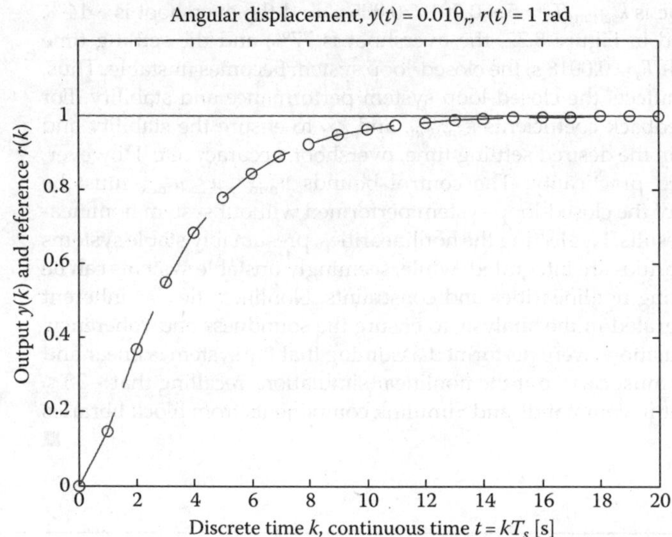

FIGURE 8.21
Output dynamics of the system with a digital PID control law, $T_s = 0.0002$ s.

The output dynamics for the reference input $r(kT_s) = 1$ rad, $k \geq 0$ is shown in Figure 8.21. The settling time is $k_{settling}T_s = 15 \times 0.0002 = 0.003$ s, and there is no overshoot.

The sampling time, defined by the microcontroller or DSP processing abilities, significantly affects the system dynamics. We increase the sampling time, and $T_s = 0.001$ s. Using the MATLAB file reported, for the sampling time $T_s = 0.001$ s, the closed-loop transfer function is

$$G(z) = \frac{1.14z^4 - 1.02z^3 - 0.12z^2 + 0.0012z + 2.61 \times 10^{-10}}{z^5 - 0.86z^4 - 0.021z^3 - 0.12z^2 + 0.0012z + 2.61 \times 10^{-10}}.$$

The output of the servosystem $y(kT_s)$ is plotted in Figure 8.22 for $T_s = 0.001$ s and $T_s = 0.0015$ s.

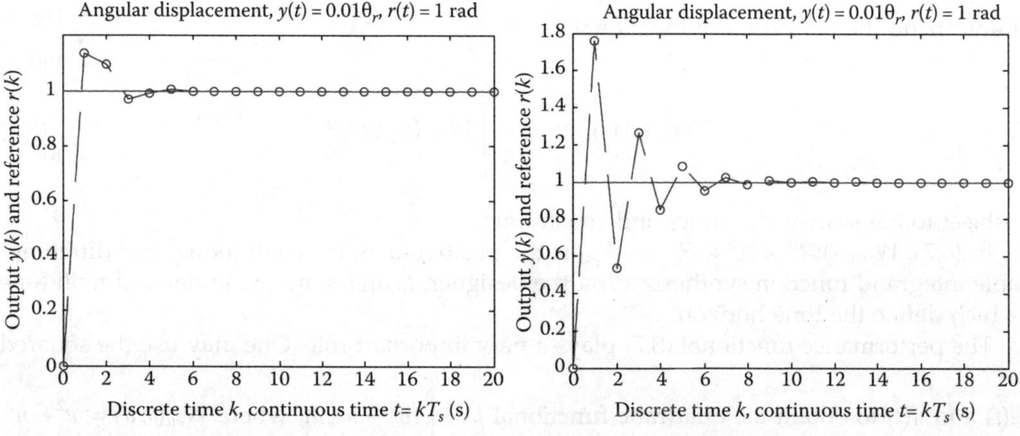

FIGURE 8.22
Output of a servo with a digital PID control law, $T_s = 0.001$ s and $T_s = 0.0015$ s.

If, $T_s = 0.001$ s, the settling time is $k_{settling}T_s = 5 \times 0.001 = 0.005$ s, and the overshoot is $\sim$14%. For $T_s = 0.0015$ s, as documented in Figure 8.22, the overshoot is 77%, and the settling time $k_{settling}T_s = 10 \times 0.0015 = 0.015$ s. If $T_s = 0.0018$ s, the closed-loop system becomes unstable. Thus, the sampling time significantly affects the closed-loop system performance and stability. For large T_s, one must refine the feedback coefficients k_{dp}, k_{di}, and k_{dd} to ensure the stability and desired dynamic responses within the desired settling time, overshoot, accuracy, etc. However, the linear analysis has a limited practicality. The control bounds $u_{min} \leq u \leq u_{max}$ must be examined. The stability analysis of the closed-loop system performed without system nonlinearities does not provide accurate results. Neglecting the nonlinearities, presumably stable systems may become unstable as nonlinearities are integrated, while, seemingly unstable systems can be stable as one integrates the existing nonlinearities and constraints. Nonlinearities, as inherent hardware features, must be integrated in the analysis to ensure the soundness and coherency. Using the `filter` command, simulations were performed assuming that the system is linear and no constraints are imposed. One must carry out the nonlinear simulations recalling that $-30 \leq u_a \leq 30$ V. Various MATLAB built-in commands and Simulink components from block libraries can be effectively utilized. ∎

8.4 Hamilton–Jacobi Theory and Optimal Control

We design control algorithms by minimizing performance functionals which quantitatively and qualitatively describe the system performance and capabilities. The following system variables are commonly utilized: states x, outputs y, tracking errors e, and control u. Considering multi-input/multi-output systems, one has $x \in \mathbb{R}^n$, $y \in \mathbb{R}^b$, $e \in \mathbb{R}^b$, and $u \in \mathbb{R}^m$. The variables tuple (x, e, u) or quadruple (x, y, e, u) describe, specify, quantify, and qualify the system performance and capabilities.

A general problem formulation in designing control laws using the Hamilton–Jacobi concept is

Analytically, the *admissible* (bounded or unbounded) time-invariant or time-varying control law as a nonlinear function of error $e(t)$ and state $x(t)$ vectors is desired

$$u = \phi(t, e, x), \tag{8.6}$$

minimizing the performance functional

$$J(x(\cdot),\, e(\cdot),\, u(\cdot)) = \int_{t_0}^{t_f} W_{xeu}(x, e, u)dt, \tag{8.7}$$

subject to the system dynamics and constraints.

In (8.7), $W_{xeu}(\cdot):\mathbb{R}^n \times \mathbb{R}^b \times \mathbb{R}^m \to \mathbb{R}_{\geq 0}$ is the positive-definite, continuous, and differentiable integrand function synthesized by the designer, t_0 and t_f are the initial and final time which define the time horizon.

The performance functional (8.7) plays a very important role. One may use the squared $e(t)$ and $u(t)$ to obtain the quadratic functional $J = \int_{t_0}^{t_f} (e^2 + u^2)dt$, where $W_{eu}(e, u) = e^2 + u^2$.

This performance functional is positive-definite, and the performance integrands are differentiable ensuring the analytic solutions. A positive-definite $W_{eu}(e, u) = |e| + |u|$ may not be used with an ease to analytically synthesize the control laws applying the Hamilton–Jacobi and many other concepts, though discontinuous integrands in $W_{xeu}(\cdot)$ can be applied in search, parametric, and numeric optimization methods. For the aforementioned methods, one may formulate the problem as

For given *admissible* control law $u = \phi(e, x)$ or *minimal-complexity* control law $u = \phi$ (e, x_m), minimize the performance functional $J(x(\cdot), e(\cdot), u(\cdot)) = \displaystyle\int_{t_0}^{t_f} W_{xeu}(x, e, u)dt$ subject to the system dynamics and constraints deriving the feedback coefficients. Here, x_m are the directly measured states.

For example, feedback coefficients of linear and nonlinear PID control laws can be derived.

One can design and minimize a great variety of functionals with an attempt to specify-and-achieve better performance. For example, nonquadratic functionals $J = \displaystyle\int_{t_0}^{t_f} (e^8 + u^6)dt$, $J = \displaystyle\int_{t_0}^{t_f} (|e|e^2 + |u|u^2)dt$, or $J = \displaystyle\int_{t_0}^{t_f} (|e|e^4 + |u|u^6)dt$ may be used with differ-

entiable $W_{eu}(\cdot)$. The quadratic functionals, for example, $J = \displaystyle\int_{t_0}^{t_f} (x^2 + e^2 + u^2)dt$, are com-

monly used. These quadratic functionals ease the analytic design ensuring that the problem is solvable. In general, the application of nonquadratic integrands results in mathematical complexity. However, the system performance may be improved if sound nonlinear functionals are used. The analytic design of constrained control laws is centered on the use of nonquadratic functionals.

As a performance functional (8.7) is chosen by synthesizing the integrand functions, the design centers on the minimization or maximization problem using the Hamilton–Jacobi concept, dynamic programming, maximum principle, calculus of variations, nonlinear programming, or other concepts. The ultimate objective is to design control laws (8.6) to attain optimal system performance, for example, stability, robustness, accuracy, etc.

Let the electromechanical system dynamics be described by nonlinear differential equations

$$\dot{x}(t) = F(x) + B(x)u, \quad x(t_0) = x_0. \tag{8.8}$$

To find an optimal control, the necessary conditions for optimality are applied. For $J = \displaystyle\int_{t_0}^{t_f} W_{xu}(x, u)dt$, the Hamiltonian function is

$$H\left(x, u, \frac{\partial V}{\partial x}\right) = W_{xu}(x, u) + \left(\frac{\partial V}{\partial x}\right)^T (F(x) + B(x)u), \tag{8.9}$$

where $V(\cdot){:}\mathbb{R}^n \to \mathbb{R}_{\geq 0}$ is the continuous and differentiable return function, $V(0) = 0$.

Control law, as given by equation (8.6) can be formed using the following first-order necessary condition for optimality

$$\frac{\partial H\left(x, u, \dfrac{\partial V}{\partial x}\right)}{\partial u} = 0. \tag{8.10}$$

The control function $u(\cdot):[t_0, t_f) \to \mathbb{R}^m$ is obtained from (8.10) as the performance functional (8.7) is defined. Using the quadratic performance functional

$$J = \frac{1}{2} \int_{t_0}^{t_f} (x^T Q x + u^T G u)\, dt, \quad Q \in \mathbb{R}^{n \times n}, \quad Q \geq 0, \quad G \in \mathbb{R}^{n \times m}, \quad G > 0, \tag{8.11}$$

from (8.9) and (8.10) one finds $\dfrac{\partial H}{\partial u} = u^T G + \left(\dfrac{\partial V}{\partial x}\right)^T B(x)$. Hence, the control law is

$$u = -G^{-1} B^T(x) \frac{\partial V}{\partial x}. \tag{8.12}$$

The second-order necessary conditions for optimality

$$\frac{\partial^2 H\left(x, u, \dfrac{\partial V}{\partial x}\right)}{\partial u \times \partial u^T} > 0 \tag{8.13}$$

is satisfied because $\dfrac{\partial^2 H}{\partial u \times \partial u^T} = G > 0$.

Example 8.10:

Consider a moving object assuming that the applied force F_a is a control variable u. Hence, we consider the *force control* commonly applied by mechanical engineers. Let the velocity v be the state variable x. From $\dot{x}(t) = \dfrac{1}{m} \sum F$, taking note of the viscous friction, the dynamics is

$$\dot{x}(t) = \frac{1}{m} \sum F = \frac{1}{m}(F_a - B_m x).$$

Using the state and control variables, one has the first-order differential equation which describes the input–output dynamics. In particular,

$$\dot{x}(t) = ax + bu,$$

where $a = -B_m/m$ and $b = 1/m$.

For a first-order system, we minimize the quadratic functional (8.11) with the weighting coefficients q and g. The functional

$$J = \frac{1}{2} \int_{t_0}^{t_f} (q x^2 + g u^2)\, dt, \quad q \geq 0, \quad g > 0,$$

yields the Hamiltonian function (8.9) as

$$H\left(x,u,\frac{\partial V}{\partial x}\right) = \underbrace{\frac{1}{2}\left(qx^2 + gu^2\right)}_{\substack{\text{performance functional}\\ J(x,u)=\frac{1}{2}\int_{t_0}^{t_f}(qx^2+gu^2)dt}} + \frac{\partial V}{\partial x}\underbrace{\frac{dx}{dt}}_{\substack{\text{system dynamics}\\ \dot{x}(t)=ax+bu}} = \frac{1}{2}\left(qx^2 + gu^2\right) + \frac{\partial V}{\partial x}(ax + bu).$$

The Hamiltonian function $H(\cdot)$ is minimized using the first-order necessary condition for optimality (8.10). From (8.10), one obtains $gu + \dfrac{\partial V}{\partial x}b = 0$. Hence,

$$u = -\frac{b}{g}\frac{\partial V}{\partial x} = -g^{-1}b\frac{\partial V}{\partial x}.$$

Let the continuous and differentiable return function $V(x)$ be given in the quadratic form, for example, $V(x) = \dfrac{1}{2}kx^2$, where k is the positive-definite unknown coefficient.

From $u = -g^{-1}b\dfrac{\partial V}{\partial x}$, the control law is

$$u = -g^{-1}bkx, \quad k > 0.$$

The unknown coefficient k is found by solving the Riccati differential equation $-dk/dt = q + 2ak - g^{-1}b^2k^2$ or the quadratic algebraic equation $-q - 2ak + g^{-1}b^2k^2 = 0$, $k > 0$, as will be illustrated later.

Making use of the system dynamics $\dot{x}(t) = ax + bu$ with control law $u = -g^{-1}bkx$, the closed-loop system is

$$\dot{x}(t) = \left(a - g^{-1}b^2k\right)x.$$

The closed-loop system is stable if $\left(a - g^{-1}b^2k\right) < 0$. This condition for stability is guaranteed because $a < 0$, $g > 0$, and $k > 0$.

The second-order necessary conditions for optimality (8.13) are guaranteed because $\dfrac{\partial^2 H}{\partial u^2} = g > 0$.

The electromechanical systems are nonlinear. For example, if the object is attached to a nonlinear spring with the restoring force $k_s x^3$. We have

$$\dot{x}(t) = \frac{1}{m}\sum F = \frac{1}{m}\left(F_a - B_m x - k_s x^3\right).$$

Minimizing the quadratic functional, the control law is found to be $u = g^{-1}bkx$, $k > 0$. The closed-loop system dynamics is $\dot{x}(t) = ax - \dfrac{1}{m}k_s x^3 - g^{-1}b^2kx$. This system is stable which can be easily proven by applying the Lyapunov stability theory covered in Section 8.12. In general, one may not be able to utilize only quadratic return functions $V(x)$. The nonquadratic $V(x)$ results in nonlinear control laws. ∎

8.5 Stabilization Problem for Linear Systems Using Hamilton–Jacobi Concept

A linear control law was synthesized in Example 8.10 for the system described by the first-order linear and nonlinear differential equations. Consider a linear time-invariant electromechanical system described by the linear differential equations

$$\dot{x}(t) = Ax + Bu, \quad x(t_0) = x_0, \tag{8.14}$$

where $A \in \mathbb{R}^{n \times n}$ and $B \in \mathbb{R}^{n \times m}$ are the constant-coefficient matrices.

Using the quadratic integrands, the quadratic performance functional is expressed as

$$J = \frac{1}{2} \int_{t_0}^{t_f} (x^T Q x + u^T G u)\, dt, \quad Q \geq 0, \quad G > 0, \tag{8.15}$$

where, $Q \in \mathbb{R}^{n \times n}$ is the positive semi-definite constant-coefficient weighting matrix and $G \in \mathbb{R}^{m \times m}$ is the positive-definite constant-coefficient weighting matrix.

From (8.14) and (8.15), the Hamiltonian function is

$$H\left(x, u, \frac{\partial V}{\partial x}\right) = \frac{1}{2}(x^T Q x + u^T G u) + \left(\frac{\partial V}{\partial x}\right)^T (Ax + Bu). \tag{8.16}$$

The Hamilton–Jacobi functional equation is

$$-\frac{\partial V}{\partial t} = \min_u \left[\frac{1}{2}(x^T Q x + u^T G u) + \left(\frac{\partial V}{\partial x}\right)^T (Ax + Bu)\right]. \tag{8.17}$$

The derivative of the Hamiltonian H exists, and the control function $u(\cdot):[t_0, t_f) \to \mathbb{R}^m$ is found by using the first-order necessary condition for optimality (8.10). From $\frac{\partial H}{\partial u} = u^T G + \left(\frac{\partial V}{\partial x}\right)^T B$, one finds

$$u = -G^{-1} B^T \frac{\partial V}{\partial x}. \tag{8.18}$$

The second-order necessary condition for optimality (8.13) is guaranteed. In fact, the weighting matrix G is positive-definite, yielding $\dfrac{\partial^2 H}{\partial u \times \partial u^T} = G > 0$.

Substituting the control law, (8.18) in (8.17), we obtain the partial differential equation

$$-\frac{\partial V}{\partial t} = \frac{1}{2}\left(x^T Q x + \left(\frac{\partial V}{\partial x}\right)^T BG^{-1}B^T \frac{\partial V}{\partial x}\right) + \left(\frac{\partial V}{\partial x}\right)^T Ax - \left(\frac{\partial V}{\partial x}\right)^T BG^{-1}B^T \frac{\partial V}{\partial x}$$

$$= \frac{1}{2}x^T Q x + \left(\frac{\partial V}{\partial x}\right)^T Ax - \frac{1}{2}\left(\frac{\partial V}{\partial x}\right)^T BG^{-1}B^T \frac{\partial V}{\partial x}. \tag{8.19}$$

The solution of (8.19) is satisfied by the quadratic return function

$$V(x) = \frac{1}{2}x^T K(t)x,\tag{8.20}$$

where $K \in \mathbb{R}^{n \times n}$ is the symmetric matrix, $K = K^T$.

The matrix $K = \begin{bmatrix} k_{11} & k_{12} & \cdots & k_{1\,n-1} & k_{1\,n} \\ k_{21} & k_{22} & \cdots & k_{2\,n-1} & k_{2\,n} \\ \vdots & \vdots & \ddots & \vdots & \vdots \\ k_{n-1\,1} & k_{n-1\,2} & \cdots & k_{n-1\,n-1} & k_{n-1\,n} \\ k_{n1} & k_{n2} & \cdots & k_{n\,n-1} & k_{n\,n} \end{bmatrix}$, $k_{ij} = k_{ji}$ must be positive-definite

because positive semi-definite and positive-definite constant-coefficient weighting matrices Q and G are used in the quadratic performance functional (8.15) yielding $J > 0$. The positive definiteness of the quadratic return function $V(x)$ can be verified using the Sylvester criterion.

From (8.20), applying the matrix identity $x^T KAx = \frac{1}{2}x^T\left(A^T K + KA\right)x$ in (8.19), one obtains

$$-\frac{\partial\left(\frac{1}{2}x^T Kx\right)}{\partial t} = \frac{1}{2}x^T Qx + \frac{1}{2}x^T A^T Kx + \frac{1}{2}x^T KAx - \frac{1}{2}x^T KBG^{-1}B^T Kx.\tag{8.21}$$

Using the boundary conditions

$$V\left(t_f, x\right) = \frac{1}{2}x^T K\left(t_f\right)x = \frac{1}{2}x^T K_f x,\tag{8.22}$$

the following nonlinear differential equation, called the Riccati equation, must be solved to find the unknown symmetric matrix K

$$-\dot{K} = Q + A^T K + K^T A - K^T BG^{-1}B^T K, \quad K\left(t_f\right) = K_f.\tag{8.23}$$

From (8.18) and (8.20), the control law is

$$u = -G^{-1}B^T Kx.\tag{8.24}$$

The feedback gain matrix K_F results, and $K_F = G^{-1}B^T K$. Using (8.14) and (8.24), we have the closed-loop system as

$$\dot{x}(t) = Ax + Bu = Ax - BG^{-1}B^T Kx = \left(A - BG^{-1}B^T K\right)x = \left(A - BK_F\right)x.\tag{8.25}$$

The eigenvalues of the matrix $\left(A - BG^{-1}B^T K\right) = \left(A - BK_F\right) \in \mathbb{R}^{n \times n}$ have negative real parts ensuring stability.

If in functional (8.15) $t_f = \infty$, the matrix K can be found by solving the nonlinear algebraic equation

$$0 = -Q - A^T K - K^T A + K^T BG^{-1}B^T K.\tag{8.26}$$

To solve (8.26), the MATLAB Riccati equation solver lqr can be applied. Hence, using the lqr command, one finds the feedback gain matrix K_F, matrix K, and eigenvalues of the closed-loop system.

Example 8.11:

Consider the system studied in Example 8.10, assuming $m=1$ and $B_m=0$ (the viscous friction is neglected). The differential equation is

$$\frac{dx}{dt} = u.$$

In the state-space model $\dot{x}(t) = Ax + Bu$, the matrices A and B are $A=[0]$ and $B=[1]$.

Let in the functional $J = \frac{1}{2} \int_{t_0}^{t_f} (qx^2 + gu^2)dt$ the weighting coefficients be $q=1$ and $g=1$. Hence, $Q=[1]$ and $G=[1]$.

The unknown k of the quadratic return function (8.20) $V(x) = \frac{1}{2}kx^2$ is found by solving (8.23) which is expressed as

$$-\dot{k}(t) = 1 - k^2(t), \quad k(t_f) = 0.$$

The solution of this nonlinear differential equation is $k(t) = \dfrac{1 - e^{-2(t_f - t)}}{1 + e^{-2(t_f - t)}}$.

A control law, which guarantees the minimum of the quadratic functional $J = \dfrac{1}{2}\int_0^{t_f}(x^2 + u^2)dt$

subject to the system dynamics $\dfrac{dx}{dt} = u$, is obtained using (8.24) as

$$u = -k(t)x = -\frac{1 - e^{-2(t_f - t)}}{1 + e^{-2(t_f - t)}}x.$$

By applying the lqr MATLAB command, for $t_f = \infty$, one finds the feedback gain, return function coefficient k, as well as the eigenvalue. Minimizing the functional $J = \dfrac{1}{2}\int_0^{\infty}(x^2 + u^2)dt$, using the statement

```
[K_feedback,K,Eigenvalues] = lqr(0,1,1,1,0)
```

we have the following numerical results

```
K_feedback =      1
K =               1
Eigenvalues =    -1
```

One concludes that the control law is $u = -x$, and the characteristic eigenvalue (pole) is -1. The pole has a negative real part, and the closed-loop system is stable. The numerical results correspond to the analytic solution found. ■

Example 8.12:

Consider one-dimensional motion of a rigid-body mechanical system described by a set of two first-order differential equations. The state variables are the displacement $x_1(t)$ and velocity $x_2(t)$. Neglecting the viscous friction, one has

$$\frac{dx_1}{dt} = x_2, \quad \frac{dx_2}{dt} = u,$$

where u represents the force or torque to be applied to control the system.

Using the state-space notations (8.14), we have the matrix differential equation

$$\dot{x}(t) = Ax + Bu, \quad \begin{bmatrix} \dot{x}_1 \\ \dot{x}_2 \end{bmatrix} = \begin{bmatrix} 0 & 1 \\ 0 & 0 \end{bmatrix} \begin{bmatrix} x_1 \\ x_2 \end{bmatrix} + \begin{bmatrix} 0 \\ 1 \end{bmatrix} u, \quad A = \begin{bmatrix} 0 & 1 \\ 0 & 0 \end{bmatrix}, \quad B = \begin{bmatrix} 0 \\ 1 \end{bmatrix}.$$

The quadratic functional (8.15) is

$$J = \frac{1}{2} \int_0^\infty (q_{11}x_1^2 + q_{22}x_2^2 + gu^2)dt, \quad q_{11} \geq 0, \quad q_{22} \geq 0 \quad \text{and} \quad g > 0$$

$$\text{or} \quad J = \frac{1}{2} \int_0^\infty \left([x_1 \ \ x_2] \begin{bmatrix} q_{11} & 0 \\ 0 & q_{22} \end{bmatrix} \begin{bmatrix} x_1 \\ x_2 \end{bmatrix} + uGu \right)dt, \quad Q = \begin{bmatrix} q_{11} & 0 \\ 0 & q_{22} \end{bmatrix}, \quad G = g.$$

Using the quadratic return function (8.20)

$$V(x) = \frac{1}{2}k_{11}x_1^2 + k_{12}x_1x_2 + \frac{1}{2}k_{22}x_2^2 = \frac{1}{2}[x_1 \ \ x_2] \begin{bmatrix} k_{11} & k_{12} \\ k_{21} & k_{22} \end{bmatrix} \begin{bmatrix} x_1 \\ x_2 \end{bmatrix}, \quad k_{12} = k_{21},$$

the control law (8.24) is

$$u = -G^{-1}B^TKx = -g^{-1}[0 \ \ 1] \begin{bmatrix} k_{11} & k_{12} \\ k_{21} & k_{22} \end{bmatrix} \begin{bmatrix} x_1 \\ x_2 \end{bmatrix} = -\frac{1}{g}(k_{21}x_1 + k_{22}x_2).$$

The unknown matrix $K = \begin{bmatrix} k_{11} & k_{12} \\ k_{21} & k_{22} \end{bmatrix}$, $k_{12} = k_{21}$, is found by solving the matrix Riccati equation (8.26) which yields

$$-Q - A^TK - K^TA + K^TBG^{-1}B^TK$$

$$= -\begin{bmatrix} q_{11} & 0 \\ 0 & q_{22} \end{bmatrix} - \begin{bmatrix} 0 & 0 \\ 1 & 0 \end{bmatrix} \begin{bmatrix} k_{11} & k_{12} \\ k_{21} & k_{22} \end{bmatrix} - \begin{bmatrix} k_{11} & k_{21} \\ k_{12} & k_{22} \end{bmatrix} \begin{bmatrix} 0 & 1 \\ 0 & 0 \end{bmatrix}$$

$$+ \begin{bmatrix} k_{11} & k_{21} \\ k_{12} & k_{22} \end{bmatrix} \begin{bmatrix} 0 \\ 1 \end{bmatrix} g^{-1}[0 \ \ 1] \begin{bmatrix} k_{11} & k_{12} \\ k_{21} & k_{22} \end{bmatrix} = \begin{bmatrix} 0 & 0 \\ 0 & 0 \end{bmatrix}.$$

Three algebraic equations to be solved are

$$\frac{k_{12}^2}{g} - q_{11} = 0, \quad -k_{11} + \frac{k_{12}k_{22}}{g} = 0, \quad \text{and} \quad -2k_{12} + \frac{k_{22}^2}{g} - q_{22} = 0.$$

The solution is $k_{12} = k_{21} = \pm\sqrt{q_{11}g}$, $k_{22} = \pm\sqrt{g(q_{22} + 2k_{12})}$ and $k_{11} = \dfrac{k_{12}k_{22}}{g}$.

The performance functional $J = \dfrac{1}{2}\displaystyle\int_0^\infty (q_{11}x_1^2 + q_{22}x_2^2 + gu^2)\,dt$ is positive-definite because the quadratic terms are used, and $q_{11} \geq 0$, $q_{22} \geq 0$, $g > 0$. Hence,

$$k_{11} = \sqrt{q_{11}\left(q_{22} + 2\sqrt{q_{11}g}\right)}, \quad k_{12} = k_{21} = \sqrt{q_{11}g}, \quad \text{and} \quad k_{22} = \sqrt{g\left(q_{22} + 2\sqrt{q_{11}g}\right)}.$$

The control law is

$$u = -\frac{1}{g}\left(\sqrt{q_{11}g}\,x_1 + \sqrt{g\left(q_{22} + 2\sqrt{q_{11}g}\right)}\,x_2\right) = -\sqrt{\frac{q_{11}}{g}}\,x_1 - \sqrt{\frac{q_{22} + 2\sqrt{q_{11}g}}{g}}\,x_2.$$

Having obtained the analytic solution in the symbolic form, we calculate the feedback gains and eigenvalues by applying the lqr command. Let $q_{11} = 100$, $q_{22} = 10$, and $g = 1$. We have

```
A= [0 1;0 0]; B= [0;1]; Q= [100 0;0 10]; G= [1];
[K_feedback,K,Eigenvalues] =lqr(A,B,Q,G)
```

with resulting

```
K_feedback=
    10.0000      5.4772
K=
    54.7723     10.0000
    10.0000      5.4772
Eigenvalues=
  -2.7386+1.5811i
  -2.7386-1.5811i
```

Hence, $K = \begin{bmatrix} k_{11} & k_{12} \\ k_{21} & k_{22} \end{bmatrix} = \begin{bmatrix} 54.77 & 10 \\ 10 & 5.48 \end{bmatrix}$, $k_{11} = 54.77$, $k_{12} = k_{21} = 10$, $k_{22} = 5.48$.

The control law is given as $u = -10x_1 - 5.48x_2$.

The stability of the closed-loop system

$$\frac{dx_1}{dt} = x_2, \quad \frac{dx_2}{dt} = -10x_1 - 5.48x_2,$$

is guaranteed. The eigenvalues have negative real parts, and the complex pole found is $-2.74 \pm 1.58i$. ∎

Example 8.13:

Consider the system described by the following state-space (8.14)

$$\dot{x} = Ax + Bu = \begin{bmatrix} -10 & 0 & -20 & 0 \\ 0 & -10 & -10 & 0 \\ 10 & 5 & -1 & 0 \\ 0 & 0 & 1 & 0 \end{bmatrix}\begin{bmatrix} x_1 \\ x_2 \\ x_3 \\ x_4 \end{bmatrix} + \begin{bmatrix} 10 & 0 \\ 0 & 10 \\ 0 & 0 \\ 0 & 0 \end{bmatrix}\begin{bmatrix} u_1 \\ u_2 \end{bmatrix}.$$

The output is x_4, for example, $y = x_4$. Hence, the output equation is

$$y = [0 \quad 0 \quad 0 \quad 1] \begin{bmatrix} x_1 \\ x_2 \\ x_3 \\ x_4 \end{bmatrix} + [0 \quad 0] \begin{bmatrix} u_1 \\ u_2 \end{bmatrix}.$$

The quadratic performance functional (8.15) is

$$J = \frac{1}{2} \int_0^\infty (x^T Q x + u^T G u) dt = \frac{1}{2} \int_0^\infty \left([x_1 \quad x_2 \quad x_3 \quad x_4] \begin{bmatrix} 0.05 & 0 & 0 & 0 \\ 0 & 0.1 & 0 & 0 \\ 0 & 0 & 0.01 & 0 \\ 0 & 0 & 0 & 1 \end{bmatrix} \begin{bmatrix} x_1 \\ x_2 \\ x_3 \\ x_4 \end{bmatrix} \right.$$

$$\left. + [u_1 \quad u_2] \begin{bmatrix} 0.001 & 0 \\ 0 & 0.001 \end{bmatrix} \begin{bmatrix} u_1 \\ u_2 \end{bmatrix} \right) dt$$

$$= \frac{1}{2} \int_0^\infty (0.05x_1^2 + 0.1x_2^2 + 0.01x_3^2 + x_4^2 + 0.001u_1^2 + 0.001u_2^2) dt.$$

The MATLAB m-file to design the control law and simulate the system is

```
% Constant-coefficient matrices A, B, C and D of system
A= [−10 0 −20 0;0 −10 −10 0;10 5 −1 0;0 0 1 0];
disp('eigenvalues_A'); disp(eig(A)); % Eigenvalues of matrix A
B= [10 0;0 10;0 0;0 0];
C= [0 0 0 1]; D= [0 0 0 0];
% Weighting matrices Q and G
Q= [0.05 0 0 0;0 0.1 0 0;0 0 0.01 0;0 0 0 1];
G= [0.001 0;0 0.001];
% Feedback and return function coefficients, eigenvalues
[K_feedback,K,Eigenvalues] = lqr(A,B,Q,G);
disp('K_feedback'); disp(K_feedback);
disp('K'); disp(K);
disp('eigenvalues A-BK_feedback'); disp(Eigenvalues);
% Closed-loop system
A_closed_loop=A-B*K_feedback;
% Dynamics
t=0:0.002:1;
uu= [0*ones(max(size(t)),4)]; % Applied inputs
x0= [20 10 -10 -20]; % Initial conditions
[y,x] =lsim(A_closed_loop,B*K_feedback,C,D,uu,t,x0);
plot(t,x,'LineWidth',2);
title('System Dynamics, x_1, x_2, x_3, x_4','FontSize',16);
xlabel('Time [seconds]', 'FontSize',16); disp('End')
```

The feedback gain matrix K_F, the return function matrix K, and the eigenvalues of the closed-loop system $(A - BG^{-1}B^T K) = (A - BK_F)$ are found. The control law is

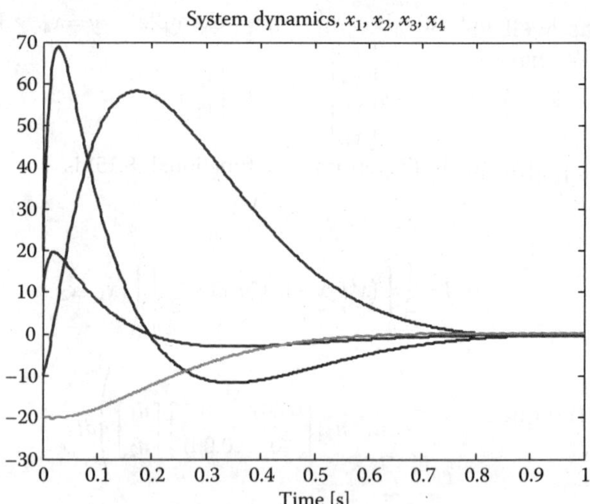

FIGURE 8.23
Dynamics of the state variables.

$$u = \begin{bmatrix} u_1 \\ u_2 \end{bmatrix} = -K_F x = - \begin{bmatrix} 6.78 & 0.21 & 4.77 & 29.7 \\ 0.21 & 9.12 & 1.46 & 10.9 \end{bmatrix} \begin{bmatrix} x_1 \\ x_2 \\ x_3 \\ x_4 \end{bmatrix}.$$

The dynamics of the closed-loop system states are plotted in Figure 8.23, for initial conditions $x_{10}=20$, $x_{20}=10$, $x_{30}=-10$, and $x_{40}=-20$. The state variables and output $(y=x_4)$ converge to the steady-state values. The closed-loop system is stable. In this section, the stabilization problem is considered. The reference $r(t)$ and tracking error $e(t)$ are not used.   ■

8.6 Tracking Control of Linear Systems

The stabilization problem and design of stabilizing control laws using the Hamilton–Jacobi theory are covered. The tracking control laws are synthesized using the tracking error $e(t)=r(t)-y(t)$.

The tracking control problem can be formulated as follows: For electromechanical systems, modeled as (8.14) $\dot{x}(t)=Ax+Bu$ with the output equation $y(t)=Hx(t)$, the tracking optimal control law $u=\phi(e,x)$ is synthesized by minimizing the performance functional.

Using the output equation $y(t)=Hx(t)$, for multivariable systems we have

$$e(t) = Nr(t) - y(t) = Nr(t) - Hx(t),$$

where $N \in \mathbb{R}^{b \times b}$ and $H \in \mathbb{R}^{b \times n}$ are the constant-coefficient matrices.

Denoting $e(t)=\dot{x}^{ref}(t)$, consider the dynamics of the *exogeneous* system

$$\dot{x}^{ref}(t) = Nr - y = Nr - Hx, \tag{8.27}$$

as well as the system equations of motion (8.14) $\dot{x}(t) = Ax + Bu$, $y = Hx$, $x_0(t_0) = x_0$. We have

$$\dot{x}_\Sigma(t) = A_\Sigma x_\Sigma + B_\Sigma u + N_\Sigma r, \quad y = Hx, \quad x_{\Sigma 0}(t_0) = x_{\Sigma 0}, \tag{8.28}$$

where

$$x_\Sigma = \begin{bmatrix} x \\ x^{ref} \end{bmatrix} \in \mathbb{R}^c \ (c = n + b); \quad A_\Sigma = \begin{bmatrix} A & 0 \\ -H & 0 \end{bmatrix} \in \mathbb{R}^{c \times c},$$

$$B_\Sigma = \begin{bmatrix} B \\ 0 \end{bmatrix} \in \mathbb{R}^{c \times m}, \quad \text{and} \quad N_\Sigma = \begin{bmatrix} 0 \\ N \end{bmatrix} \in \mathbb{R}^{c \times b}.$$

The quadratic performance functional is

$$J = \frac{1}{2} \int_{t_0}^{t_f} \left(\begin{bmatrix} x \\ x^{ref} \end{bmatrix}^T Q \begin{bmatrix} x \\ x^{ref} \end{bmatrix} + u^T G u \right) dt. \tag{8.29}$$

From (8.28) and (8.29), the Hamiltonian function is

$$H\left(x_\Sigma, u, r, \frac{\partial V}{\partial x_\Sigma}\right) = \frac{1}{2}\left(x_\Sigma^T Q x_\Sigma + u^T G u\right) + \left(\frac{\partial V}{\partial x_\Sigma}\right)^T (A_\Sigma x_\Sigma + B_\Sigma u + N_\Sigma r). \tag{8.30}$$

Using the first-order necessary condition for optimality (8.10), from (8.30), one finds $\frac{\partial H}{\partial u} = u^T G + \left(\frac{\partial V}{\partial x_\Sigma}\right)^T B_\Sigma$. Thus, the control law is

$$u = -G^{-1} B_\Sigma^T \frac{\partial V(x_\Sigma)}{\partial x_\Sigma} = -G^{-1} \begin{bmatrix} B \\ 0 \end{bmatrix}^T \frac{\partial V\left(\begin{bmatrix} x \\ x^{ref} \end{bmatrix}\right)}{\partial \begin{bmatrix} x \\ x^{ref} \end{bmatrix}}. \tag{8.31}$$

The solution of the Hamilton–Jacobi partial differential equation

$$-\frac{\partial V}{\partial t} = \frac{1}{2} x_\Sigma^T Q x_\Sigma + \left(\frac{\partial V}{\partial x_\Sigma}\right)^T A x_\Sigma - \frac{1}{2}\left(\frac{\partial V}{\partial x_\Sigma}\right)^T B_\Sigma G^{-1} B_\Sigma^T \frac{\partial V}{\partial x_\Sigma} \tag{8.32}$$

is satisfied by the quadratic return function

$$V(x_\Sigma) = \frac{1}{2} x_\Sigma^T K(t) x_\Sigma. \tag{8.33}$$

From (8.32) and (8.33), the Riccati equation, the solution of which provides the unknown symmetric matrix K, is

$$-\dot{K} = Q + A_\Sigma^T K + K^T A_\Sigma - K^T B_\Sigma G^{-1} B_\Sigma^T K, \quad K(t_f) = K_f. \tag{8.34}$$

The control law is found from (8.31) and (8.33) as

$$u = -G^{-1}B_\Sigma^T K x_\Sigma = -G^{-1}\begin{bmatrix} B \\ 0 \end{bmatrix}^T K \begin{bmatrix} x \\ x^{ref} \end{bmatrix}. \tag{8.35}$$

Recalling that $\dot{x}^{ref}(t) = e(t)$, one has $x^{ref}(t) = \int e(t)dt$. Hence, we have a control law with the state feedback and integral term for $e(t)$. In particular,

$$u(t) = -G^{-1}B_\Sigma^T K x_\Sigma(t) = -G^{-1}\begin{bmatrix} B \\ 0 \end{bmatrix}^T K \begin{bmatrix} x(t) \\ \int e(t)dt \end{bmatrix}. \tag{8.36}$$

This *integral* tracking control law usually does not ensure suitable performance because there is no proportional term for the tracking error $e(t)$. Other design concepts are of our particular interest as reported below.

8.7 State Transformation Method and Tracking Control

The tracking control problem is solved by designing the PI control laws using the *state transformation* method. We define the tracking error vector as

$$e(t) = Nr(t) - y(t) = Nr(t) - Hx^{sys}(t).$$

For linear systems

$$\dot{x}^{sys} = A^{sys}x^{sys} + B^{sys}u, \tag{8.37}$$

with the output equation $y(t) = Hx^{sys}(t)$, we have

$$\dot{e}(t) = N\dot{r}(t) - \dot{y}(t) = N\dot{r}(t) - H\dot{x}^{sys}(t) = N\dot{r}(t) - HA^{sys}x^{sys} - HB^{sys}u. \tag{8.38}$$

From (8.37) and (8.38), applying the expanded state vector $x(t) = \begin{bmatrix} x^{sys}(t) \\ e(t) \end{bmatrix}$, one finds

$$\dot{x}(t) = \begin{bmatrix} \dot{x}^{sys}(t) \\ \dot{e}(t) \end{bmatrix} = \begin{bmatrix} A^{sys} & 0 \\ -HA^{sys} & 0 \end{bmatrix}\begin{bmatrix} x^{sys} \\ e \end{bmatrix} + \begin{bmatrix} B^{sys} \\ -HB^{sys} \end{bmatrix}u + \begin{bmatrix} 0 \\ N \end{bmatrix}\dot{r} = Ax + Bu + \begin{bmatrix} 0 \\ N \end{bmatrix}\dot{r}. \tag{8.39}$$

The *space transformation* method utilizes the z and v vectors as defined by

$$z = \begin{bmatrix} x \\ u \end{bmatrix} \quad \text{and} \quad v = \dot{u}. \tag{8.40}$$

Here, $z \in \mathbb{R}^{c+m}$ and $v \in \mathbb{R}^m$ are the *state transformation* variables.

Using z and v, one obtains the system model as

$$\dot{z}(t) = \begin{bmatrix} A & B \\ 0 & 0 \end{bmatrix} z + \begin{bmatrix} 0 \\ I \end{bmatrix} v = A_z z + B_z v, \quad y = Hx^{sys}, \quad z(t_0) = z_0 \tag{8.41}$$

where $A_z \in \mathbb{R}^{(c+m)\times(c+m)}$ and $B_z \in \mathbb{R}^{(c+m)\times m}$ are the constant-coefficient matrices. Minimizing the quadratic functional

$$J = \frac{1}{2} \int_{t_0}^{t_f} (z^T Q_z z + v^T G_z v)\,dt, \quad Q_z \in \mathbb{R}^{(c+m)\times(c+m)}, \quad Q_z \geq 0, \quad G_z \in \mathbb{R}^{m\times m}, \quad G > 0 \tag{8.42}$$

the application of the first-order necessary condition for optimality (8.10) yields

$$v = -G_z^{-1} B_z^T \frac{\partial V}{\partial z}. \tag{8.43}$$

The solution of the Hamilton–Jacobi partial differential equation

$$-\frac{\partial V}{\partial t} = \frac{1}{2} z^T Q_z z + \left(\frac{\partial V}{\partial z}\right)^T A_z z - \frac{1}{2}\left(\frac{\partial V}{\partial z}\right)^T B_z G_z^{-1} B_z^T \frac{\partial V}{\partial z} \tag{8.44}$$

is satisfied by the continuous and differentiable quadratic return function

$$V(z) = \frac{1}{2} z^T K(t) z. \tag{8.45}$$

Using (8.43) and (8.45), one obtains the control function as

$$v = -G_z^{-1} B_z^T K z. \tag{8.46}$$

From (8.44), the Riccati equation to find the unknown matrix $K \in \mathbb{R}^{(c+m)\times(c+m)}$ is

$$-\dot{K} = KA_z + A_z^T K - KB_z G_z^{-1} B_z^T K + Q_z, \quad K(t_f) = K_f.$$

From (8.46) and (8.40), one has

$$\dot{u}(t) = -G_z^{-1} B_z^T K z = -G_z^{-1} \begin{bmatrix} 0 \\ I \end{bmatrix}^T \begin{bmatrix} K_{11} & K_{21}^T \\ K_{21} & K_{22} \end{bmatrix} \begin{bmatrix} x \\ u \end{bmatrix} = -G_z^{-1} K_{21} x - G_z^{-1} K_{22} u = K_{f1} x + K_{f2} u. \tag{8.47}$$

Using (8.39) $\dot{x}(t) = Ax + Bu$, we have $u = B^{-1}[\dot{x}(t) - Ax]$. Thus,

$$u = B^{-1}[\dot{x}(t) - Ax] = (B^T B)^{-1} B^T [\dot{x}(t) - Ax]. \tag{8.48}$$

Applying (8.48) in (8.47), one obtains

$$\dot{u}(t) = K_{f1}x + K_{f2}u = K_{f1}x + K_{f2}(B^T B)^{-1}B^T(\dot{x}(t) - Ax)$$

$$= \lfloor K_{f1} - K_{f2}(B^T B)^{-1}B^T A \rfloor x(t) + K_{f2}(B^T B)^{-1}B^T \dot{x}(t) = (K_{f1} - K_{F1}A)x(t) + K_{F1}\dot{x}(t)$$

$$= K_{F2}x(t) + K_{F1}\dot{x}(t). \tag{8.49}$$

Hence, the control law is derived as

$$u(t) = K_{F1}x(t) - K_{F1}x_0 + \int K_{F2}x(\tau)d\tau + u_0. \tag{8.50}$$

We designed a PI tracking control law with state feedback, as given by (8.50), because $x(t) = \begin{bmatrix} x^{sys}(t) \\ e(t) \end{bmatrix}$. The states $x^{sys}(t)$ and tracking error $e(t)$ are utilized. Hence, to implement control law (8.50), these $x^{sys}(t)$ and $e(t)$ directly measured or observed must be available.

For nonlinear electromechanical systems, the proposed procedure can be used to derive control laws. In particular, by using

$$v = -G_z^{-1}B_z^T(z)\frac{\partial V}{\partial z}, \tag{8.51}$$

we obtain the PI control law $u(t)$.

Example 8.14:

Design the tracking control law for a system with a PZT actuator controlled by changing the applied voltage u. The second-order equation of motion for a PZT actuator is

$$m\frac{d^2y}{dt^2} + b\frac{dy}{dt} + k_e y = k_e d_e u,$$

where y is the actuator displacement (output).

Hence, we have a set of two first-order differential equations

$$\frac{dy}{dt} = v,$$

$$\frac{dv}{dt} = -\frac{k_e}{m}y - \frac{b}{m}v + \frac{k_e d_e}{m}u.$$

The reference input is $r(t)$, and the tracking error is $e(t) = Nr(t) - y(t)$, $N = 1$. We have the system equations (8.37) and (8.38).

$$\dot{x}^{sys} = A^{sys}x^{sys} + B^{sys}u, \quad \dot{e} = N\dot{r} - HA^{sys}x^{sys} - HB^{sys}u,$$

where

$$x^{sys} = \begin{bmatrix} y \\ v \end{bmatrix}, \quad A^{sys} = \begin{bmatrix} 0 & 1 \\ -\frac{k_e}{m} & \frac{-b}{m} \end{bmatrix}, \quad B^{sys} = \begin{bmatrix} 0 \\ \frac{k_e d_e}{m} \end{bmatrix} \quad \text{and} \quad H = \begin{bmatrix} 1 & 0 \end{bmatrix}$$

One obtains (8.39) as

$$\dot{x} = \begin{bmatrix} \dot{x}^{sys} \\ \dot{e} \end{bmatrix} = \begin{bmatrix} A^{sys} & 0 \\ -HA^{sys} & 0 \end{bmatrix} \begin{bmatrix} x^{sys} \\ e \end{bmatrix} + \begin{bmatrix} B^{sys} \\ -HB^{sys} \end{bmatrix} u + \begin{bmatrix} 0 \\ N \end{bmatrix} \dot{r}, \quad y = [H \quad 0] \begin{bmatrix} x^{sys} \\ e \end{bmatrix}.$$

Applying the *state transformation* method, from (8.40), one has $z = \begin{bmatrix} x^{sys} \\ e \\ u \end{bmatrix}$. Therefore, the

control function (8.46) is $\dot{u}(t) = -K_f z(t) = -K_f \begin{bmatrix} y(t) \\ v(t) \\ e(t) \\ u(t) \end{bmatrix}$. The PI tracking control law (8.50) is

$$u(t) = K_{F1} \begin{bmatrix} y(t) \\ v(t) \\ e(t) \end{bmatrix} + \int K_{F2} \begin{bmatrix} y(\tau) \\ v(\tau) \\ e(\tau) \end{bmatrix} d\tau.$$

Using the PZT actuator parameters $k_e = 3000$, $b = 1$, $d_e = 0.000001$, and $m = 0.02$, the feedback gains are found by using the lqr MATLAB command. The weighting matrices of the quadratic

functional (8.42) are $Q_z = \begin{bmatrix} 1 & 0 & 0 & 0 \\ 0 & 1 & 0 & 0 \\ 0 & 0 & 1 \times 10^{10} & 0 \\ 0 & 0 & 0 & 1 \end{bmatrix}$ and $G_z = 10$. The closed-loop actuator dynam-

ics is documented in Figure 8.24. The settling time is 0.022 s, and the tracking error converges to zero.

The differential equation, which models the PZT actuators

$$m\frac{d^2y}{dt^2} + b\frac{dy}{dt} + k_e y = k_e(d_e u - z_h)$$

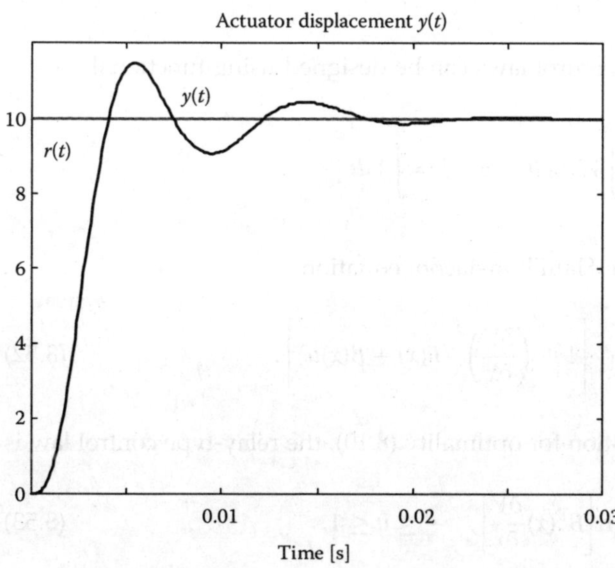

FIGURE 8.24

Closed-loop system dynamics if $r(t) = 10$ μm.

may be integrated with the hysteresis model

$$\dot{z}_h = \alpha d_e \dot{u} - \beta |\dot{u}| z_h - \gamma \dot{u} |z_h|,$$

where α and β are the constants.

We have the state-space nonlinear model of PZT actuators as

$$\frac{dy}{dt} = v, \quad \frac{dv}{dt} = -\frac{k_e}{m} y - \frac{b}{m} v - \frac{k_e}{m} z_h + \frac{kd_e}{m} u, \quad \frac{dz_h}{dt} = -\beta |\dot{u}| z_h + \alpha d_e \dot{u} - \gamma |z_h| \dot{u},$$

for which a tracking control law can be designed. However, z_h cannot be directly measured. For PZT actuators, for which one can derive high-fidelity mathematical models, the tracking control law $u(t) = K_{F1} \begin{bmatrix} y(t) \\ v(t) \\ e(t) \end{bmatrix} + \int K_{F2} \begin{bmatrix} y(\tau) \\ v(\tau) \\ e(\tau) \end{bmatrix} d\tau$ will ensure tracking, stability, and near-optimal performance utilizing the measured and physically well-defined displacement y, velocity v, and tracking error e. One may derive high-fidelity models and may design corresponding control algorithms. The soundness, needs, *implementability*, and practicality of any synthesized control laws must be studied. There is no end for enhancing the model accuracy, which, in turn, may result in considerable system complexity. The utilized model is (i) sound and sufficiently accurate in design of control laws (ii) results in the implementable control law. The *minimal-complexity* control law can be a PI algorithm which can be synthesized using the Lyapunov's theory, search, or nonlinear optimization methods. The PI *minimal-complexity* control may likely not ensure optimal performance. The trade-off between hardware complexity (affected by control laws to be implemented), system performance, and capabilities must be studied. It should be emphasized again that the electromechanical systems usually are open-loop stable systems for which a simple P or PI controls may ensure near-optimal performance. ∎

8.8 Time-Optimal Control

For dynamic systems, time-optimal control laws can be designed using functional

$$J = \frac{1}{2} \int_{t_0}^{t_f} W(x) dt \quad \text{or} \quad J = \int_{t_0}^{t_f} 1\, dt.$$

For nonlinear systems (8.8), from the Hamilton–Jacobi equation

$$-\frac{\partial V}{\partial t} = \min_{-1 \le u \le 1} \left[1 + \left(\frac{\partial V}{\partial x} \right)^T (F(x) + B(x)u) \right], \tag{8.52}$$

using the first-order necessary condition for optimality (8.10), the relay-type control law is

$$u = -\text{sgn} \left[B^T(x) \frac{\partial V}{\partial x} \right], \quad - \le u \le 1. \tag{8.53}$$

Control law (8.53) cannot be applied to systems due to the chattering phenomena, switching, losses, and other undesirable effects. Therefore, relay-type control laws with dead zone

$$u = -\text{sgn}\left[B(x)^T \frac{\partial V}{\partial x}\right]\Bigg|_{\text{dead zone}}, \quad -1 \leq u \leq 1, \tag{8.54}$$

may be considered as possible candidates. However, the application of relay-type control laws for mechatronic systems is very limited.

Example 8.15:

Synthesize the time-optimal control law for the system described by the following differential equations:

$$\dot{x}_1(t) = x_1^5 u_1 + x_2^7, \quad -1 \leq u_1 \leq 1,$$
$$\dot{x}_2(t) = x_1^3 x_2^5 u_2 - x_2^3, \quad -1 \leq u_2 \leq 1.$$

The performance functional is $J = \int_{t_0}^{t_f} 1 \, dt$. The Hamilton–Jacobi equation is

$$-\frac{\partial V}{\partial t} = \min_{u \in U} \left[1 + \left(\frac{\partial V}{\partial x}\right)^T (F(x) + B(x)u)\right]$$
$$= \min_{\substack{-1 \leq u_1 \leq 1 \\ -1 \leq u_2 \leq 1}} \left[1 + \frac{\partial V}{\partial x_1}\left(x_1^5 u_1 + x_2^7\right) + \frac{\partial V}{\partial x_2}\left(x_1^3 x_2^5 u_2 - x_2^3\right)\right].$$

From the first-order necessary condition for optimality (8.10), an optimal control law (8.53) is

$$u_1 = -\text{sgn}\left(x_1^5 \frac{\partial V}{\partial x_1}\right), \quad u_2 = -\text{sgn}\left(x_1^3 x_2^5 \frac{\partial V}{\partial x_2}\right).$$

Synthesis of $V(x)$ is a very challenging problem. Assume that the Hamilton–Jacobi equation is approximated by the quadratic function $V(x) = k_{11}x_1^2 + 2k_{12}x_1 x_2 + k_{22}x_2^2$. One finds $k_{11} = 0.25$, $k_{12} = 0.5$, and $k_{22} = 0.25$. For the derived

$$u_1 = -\text{sgn}\left[x_1^5(0.5x_1 + 0.5x_2)\right], \quad u_2 = -\text{sgn}\left[x_1^3 x_2^5(0.5x_1 + 0.5x_2)\right]$$

the simulations are performed. The Simulink diagram is reported in Figure 8.25a. We numerically solve the nonlinear differential equations using the chosen numerical methods and accuracy. The Configuration Parameters icon is displayed in Figure 8.26a. The transient dynamics is documented in Figure 8.25b for the initial conditions $x_{10} = 5$ and $x_{20} = -5$. One observes a significant switching activity of $u_1(t)$ and $u_2(t)$, which is even affected by the numerical method used and accuracy assigned (relative tolerance is 0.0000001). In practice, systems cannot exhibit the mathematically-derived control function (force, torque, voltage, current, etc.) switching. Relay-type (*bang-bang*) switching, if exhibited, is a highly undesirable phenomenon for majority of systems. Figure 8.25c reports the system evolution for a continuous control

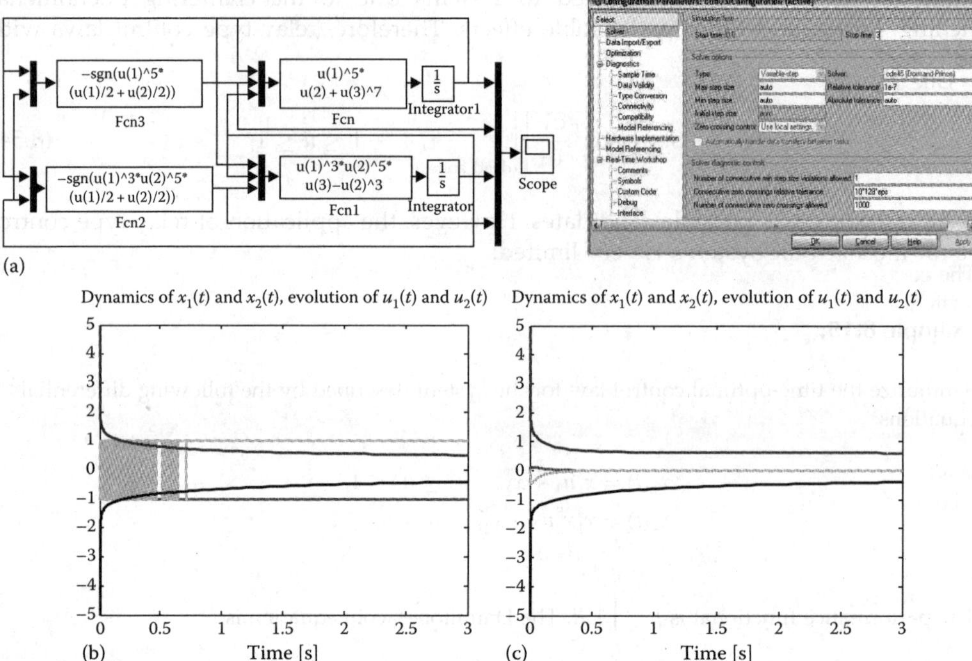

FIGURE 8.25
(a) Simulink diagram (`ch8_03.mdl`) and Configuration Parameters icon; (b) time-optimal control: dynamics of system variables $x_1(t)$ and $x_2(t)$, and control $u_1(t)$ and $u_2(t)$ activity if $x_{10}=5$ and $x_{20}=-5$; and (c) Continuous control: behavior of $x_1(t)$, $x_2(t)$, $u_1(t)$, and $u_2(t)$.

$$u_1 = -x_1^5(0.5x_1 + 0.5x_2), \quad u_2 = -x_1^3 x_2^5(0.5x_1 + 0.5x_2).$$

The closed-loop system is stable, and the system behavior with a continuous control law, as compared to the time-optimal solution with $u_1=\pm 1$ and $u_2=\pm 1$, is preferable.

The plotting statements are

```
plot(xu(:,1),xu(:,3),'m',xu(:,1),xu(:,5),'c',xu(:,1),xu(:,2),'b',xu(:,1),xu(:,4),'k','LineWidth',2);
title('Dynamics of x_1(t) and x_2(t), Evolution of u_1(t) and u_2(t)','FontSize',14);
xlabel('Time [seconds]', 'FontSize',14);
```

Example 8.16:

We design an optimal relay-type control law for a system studied in Example 8.12. The equations of motion are

$$\dot{x}_1(t) = x_2, \quad \dot{x}_2(t) = u, \quad -1 \leq u \leq 1.$$

The control takes the values $u=1$ and $u=-1$. If $u=1$, from $\dot{x}_1(t)=x_2$, $\dot{x}_2(t)=1$, one has $\dfrac{dx_2}{dx_1} = \dfrac{1}{x_2}$. The integration gives $x_2^2 = 2x_1 + c_1$.

If $u = -1$, from $\dot{x}_1(t) = x_2$, $\dot{x}_2(t) = -1$, we obtain $\dfrac{dx_2}{dx_1} = -\dfrac{1}{x_2}$. The integration yields $x_2^2 = -2x_1 + c_2$.

Due to the switching ($u = 1$ or $u = -1$), the switching curve is derived as a function of the state variables. The comparison of $x_2^2 = 2x_1 + c_1$ and $x_2^2 = -2x_1 + c_2$ gives the switching curve as

$$-x_2^2 - 2x_1 \mathrm{sgn}(x_2) = 0 \quad \text{or} \quad -x_1 - \frac{1}{2}x_2^2|x_2| = 0.$$

The control takes the values $u = 1$ and $u = -1$. Making use of the derived expression for the switching curve, one finds the time-optimal (relay) control law as

$$u = -\mathrm{sgn}\left(x_1 + \frac{1}{2}x_2^2|x_2|\right), \quad -1 \le u \le 1.$$

We found an optimal control law using the calculus of variations performing analysis of the solutions of the differential equations with the relay control law.

The Hamilton–Jacobi theory is applied. We minimize the functional $J = \displaystyle\int_{t_0}^{t_f} 1 dt$.

From the Hamilton–Jacobi equation $-\dfrac{\partial V}{\partial t} = \displaystyle\min_{-1 \le u \le 1}\left[1 + \frac{\partial V}{\partial x_1}x_2 + \frac{\partial V}{\partial x_2}u\right]$, an optimal control law derived using (8.10) is

$$u = -\mathrm{sgn}\left(\frac{\partial V}{\partial x_2}\right).$$

The solution of the Hamilton–Jacobi partial differential equation is given by the nonquadratic, continuous, and differentiable return function

$$V(x_1, x_2) = k_{11}x_1^2 + k_{12}x_1x_2 + k_{22}x_2^3|x_2|.$$

The control law is

$$u = -\mathrm{sgn}\left(x_1 + \frac{1}{2}x_2^2|x_2|\right).$$

The application of the return function $V(x_1, x_2) = \dfrac{1}{2}k_{11}x_1^2 + k_{12}x_1x_2 + \dfrac{1}{2}k_{22}x_2^2$ results in the control law $u = -\mathrm{sgn}(x_1 + x_2)$ which also stabilizes the closed-loop system. The switching surface $v(x)$ will be reported in Section 8.9, and for $u = -\mathrm{sgn}(x_1 + x_2)$, $v(x) = x_1 + x_2$. One has

$$u = \begin{cases} 1 & \text{if } v(x_1, x_2) < 0 \\ -1 & \text{if } v(x_1, x_2) > 0 \end{cases}.$$

The time-optimal design using the Hamilton–Jacobi theory matches with the results obtained using the calculus of variations. The transient dynamics is analyzed. The switching curve, the phase-plane evolution of the variables, and the transient behavior for different initial conditions are documented in Figure 8.26a. Figure 8.26b also illustrates the Simulink model which provides different options to implement the control law designed. The evolution of $x_1(t)$, $x_2(t)$, and $u(t)$ are also reported. Saving the data using the variable $x12u$ in the array format, the plotting statement is

```
plot(x12u(:,1),x12u(:,2),x12u(:,1),x12u(:,3), x12u(:,1),x12u(:,4),'LineWidth',2);
title('Dynamics of x_1(t) and x_2(t), Evolution of u(t)','FontSize',14);
xlabel('Time [seconds]','FontSize',14);
```

The hard switching of control $u(t)$ is undesirable or unacceptable. Section 8.9 reports the design of soft-switching control laws using the continuous real-analytic functions. For the second-order system $\dot{x}_1(t) = x_2, \dot{x}_2(t) = u, -1 \leq u \leq 1$, the soft-switching control law is

$$u = -\tanh\left[1000\left(x_1 + \frac{1}{2}x_2^2|x_2|\right)\right].$$

The evolution of $x_1(t)$, $x_2(t)$, and $u(t)$, illustrated in Figure 8.26c, provides an evidence of significant advantages of soft switching. The settling time remains the same guaranteeing near-minimal-time dynamics, but the *bang-bang* switching is eliminated.   ■

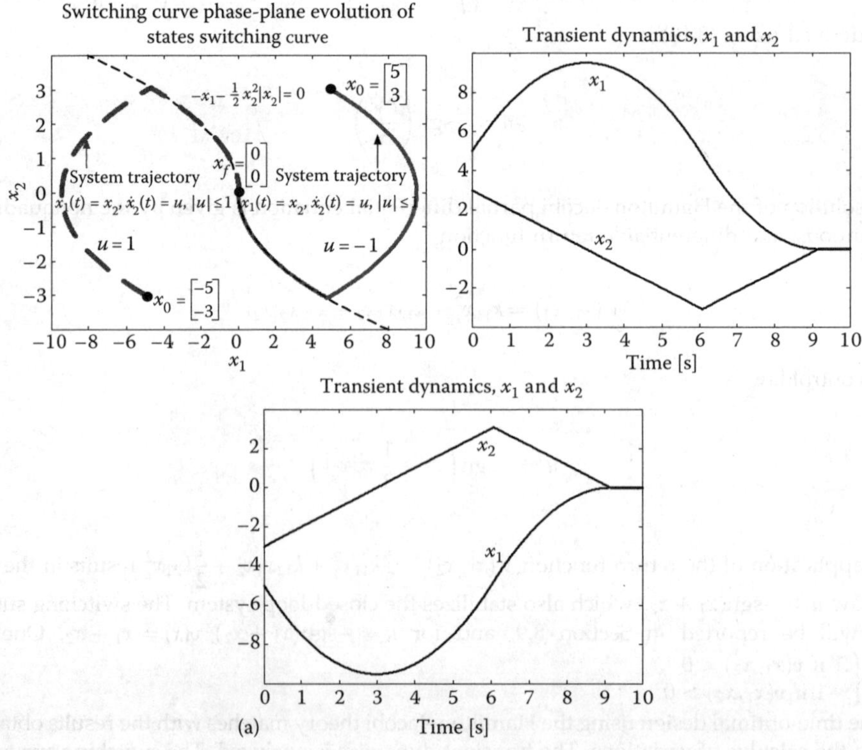

FIGURE 8.26
(a) Phase-plane evolution and system dynamics with time-optimal (switching) control;

(*continued*)

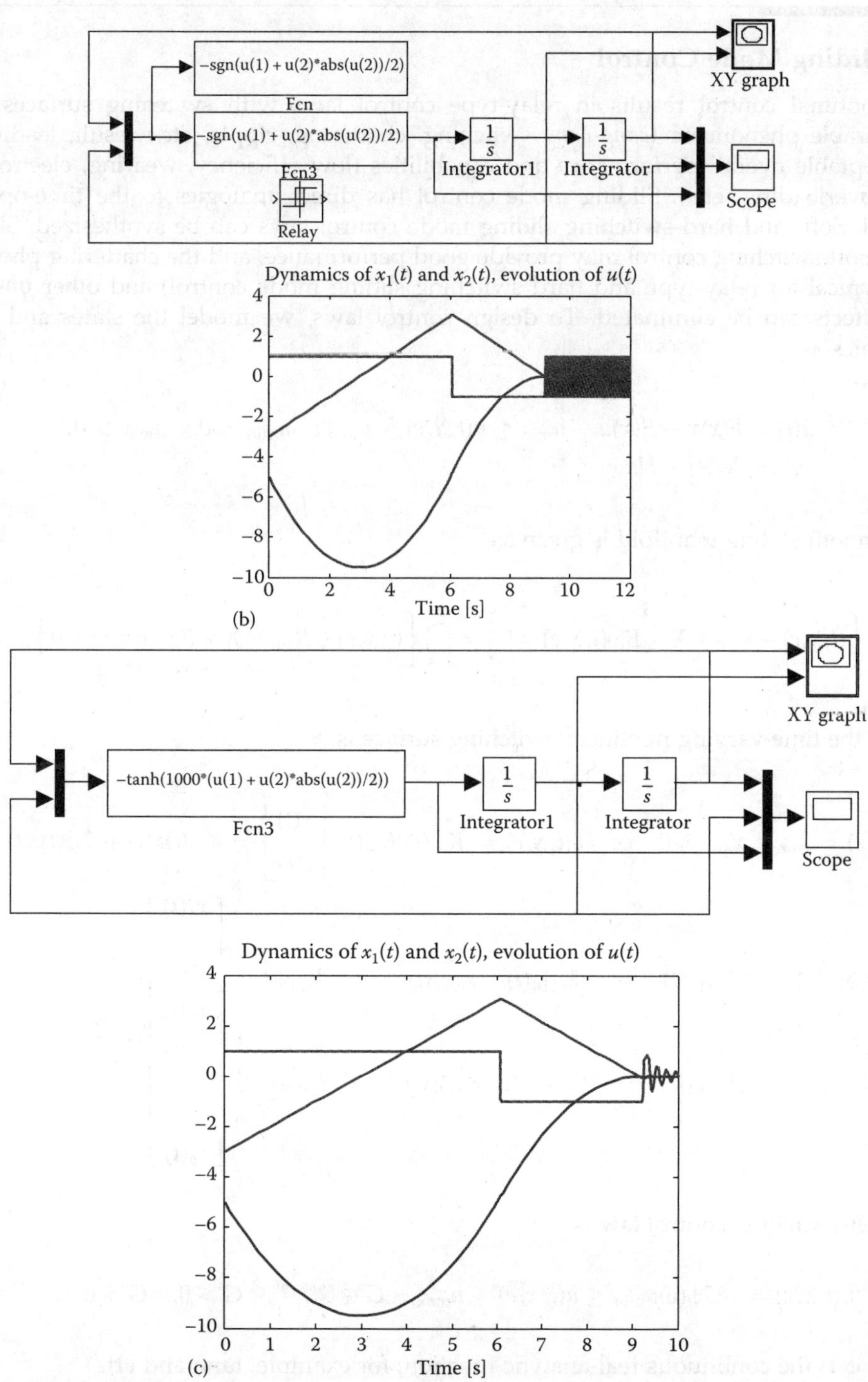

FIGURE 8.26 (continued)

(b) Simulink diagram (ch8_03.mdl), and evolutions of $x_1(t)$, $x_2(t)$, and $u(t)$ if $u = -\mathrm{sgn}\left(x_1 + \frac{1}{2}x_2^2|x_2|\right)$; (c) Simulink

diagram (ch8_04.mdl) for the soft-switching control law $u = -\tanh\left[1000\left(x_1 + \frac{1}{2}x_2^2|x_2|\right)\right]$, and evolutions of $x_1(t)$,

$x_2(t)$, and $u(t)$.

8.9 Sliding Mode Control

Time-optimal control results in relay-type control laws with switching surfaces. The undesirable phenomena (*bang-bang* switching, chattering, ripple, etc.) result, leading to unacceptable overall performance and capabilities (low efficiency, wearing, electromagnetic overloading, etc.). Sliding mode control has direct analogies to the time-optimal control. Soft- and hard-switching sliding mode control laws can be synthesized. Sliding mode soft-switching control may provide good performance, and the chattering phenomena (typical for relay-type and hard switching sliding mode control) and other undesirable effects can be eliminated. To design control laws, we model the states and error dynamics as

$$\dot{x}(t) = F(x)x + B(x)u, \quad u_{\min} \le u(t,x,e) \le u_{\max}, \quad u_{\min} < 0, \quad u_{\max} > 0,$$
$$\dot{e}(t) = N\dot{r}(t) - H\dot{x}. \tag{8.55}$$

The smooth sliding manifold is given as

$$M = \left\{ (t,x,e) \in R_{\ge 0} \times X \times E | v(t,x,e) = 0 \right\} = \bigcap_{j=1}^{m} \left\{ (t,x,e) \in R_{\ge 0} \times X \times E | v_j(t,x,e) = 0 \right\}, \tag{8.56}$$

where the time-varying nonlinear switching surface is

$$v(t,x,e) = K_{vxe}(t,x,e) = 0 \quad \text{or} \quad v(t,x,e) = [K_{vx}(t) \; K_{ve}(t)] \begin{bmatrix} x(t) \\ e(t) \end{bmatrix} = K_{vx}(t)x(t) + K_{ve}(t)e(t) = 0$$

$$\begin{bmatrix} v_1(t,x,e) \\ \vdots \\ v_m(t,x,e) \end{bmatrix} = \begin{bmatrix} k_{vx11}(t) & \cdots & k_{vx1n}(t) & k_{ve11}(t) & \cdots & k_{ve1b}(t) \\ \vdots & \vdots & \vdots & \vdots & \vdots & \vdots \\ k_{vxm1}(t) & \cdots & k_{vxmn}(t) & k_{vem1}(t) & \cdots & k_{vemb}(t) \end{bmatrix} \begin{bmatrix} x_1(t) \\ \vdots \\ x_n(t) \\ e_1(t) \\ \vdots \\ e_b(t) \end{bmatrix} = 0$$

The soft-switching control law is

$$u(t,x,e) = -G\phi(v)u_{\min} \le u(t,x,e) \le u_{\max}, \quad G \in \mathbb{R}^{m \times m}, \quad G > 0, \quad G > 0 \tag{8.57}$$

where ϕ is the continuous real-analytic function, for example, tanh and erf.

In contrast, the discontinuous (hard-switching) tracking control laws with constant and varying gains are expressed as

$$u(t, x, e) = -G\text{sgn}(v), \quad G \in \mathbb{R}^{m \times m}, \quad G > 0, \tag{8.58}$$

$$\text{or} \quad u(t, x, e) = -G(t, x, e)\text{sgn}(v), \quad G(\,\cdot\,):\mathbb{R}_{\geq 0} \times \mathbb{R}^n \times \mathbb{R}^b \to \mathbb{R}^{m \times m}.$$

The simplest hard-switching tracking control law (8.58) is

$$u(t, x, e) = \begin{cases} u_{max}, & \forall v(t, x, e) > 0 \\ 0, & \forall v(t, x, e) = 0, \quad u_{min} \leq u(t, x, e) \leq u_{max}, \\ u_{min}, & \forall v(t, x, e) < 0 \end{cases} \tag{8.59}$$

and a polyhedron with 2^m vertexes results in the control space.

The design of soft-switching control laws can be performed by utilizing the Hamilton–Jacobi or Lyapunov methods as reported in the following sections. It will be illustrated that the *admissible* control laws can be designed, and the nonlinear switching surface can be derived using $V(x)$ and $V(x, e)$.

In Example 8.16, we reported the use of soft-switching control. In particular, for

$$\dot{x}_1(t) = x_2, \dot{x}_2(t) = u, \quad -1 \leq u \leq 1,$$

the control law is $u = -\tanh\left[1000\left(x_1 + \dfrac{1}{2}x_2^2|x_2|\right)\right]$.

The analysis performed and comparison reported in Figure 8.26 demonstrated the advantages of soft switching. The examples of application of a soft-switching sliding mode control to electromechanical motion devices are reported in Ref. [5].

Example 8.17:

Rigid-body mechanical systems are described as

$$\frac{dx_1}{dt} = ax_1 + bu, \quad \frac{dx_2}{dt} = x_1, \quad -1 \leq u \leq 1,$$

where x_1 is the velocity and x_2 is the displacement.

For translational and rotational motions, the coefficients a and b can be expressed as $a = -B_v/m$ or $a = -B_m/J$, and $b = 1/m$ or $b = 1/J$, respectively.

In general, system parameters can vary. This leads to a need to examine system robustness and *sensitivity* to parameter variations. One may apply the so-called *robust control* techniques. Sound control techniques, emphasized and covered in this chapter, lead to *robust* control laws which will ensure near-optimal performance and capabilities in the full operating envelope. The designer should avoid the attempt to utilize questionable control laws (high gain, relay type, etc.) as reported. Furthermore, for practical systems, the design and analysis are straight-forward as reported below.

One recalls that the parameters (friction coefficient, mass, and moment of inertia) vary. For varying $a(\cdot)$ and $b(\cdot)$, we have

$$a_{min} \leq a(\,\cdot\,) \leq a_{max}, \quad a_{min} > 0, \quad a_{max} > 0, \quad a \in [a_{min}a_{max}] \quad \text{and} \quad b_{min} \leq b(\,\cdot\,) \leq b_{max},$$
$$b_{min} > 0, \quad b_{max} > 0, \quad b \in [b_{min}b_{max}].$$

One may design the hard and soft-switching control laws. Utilizing the linear switching surface, $v(x) = 100(x_1 + x_2)$, from (8.58) and (8.57), one derives the hard- and soft-switching control laws as

$$u = -\mathrm{sgn}[100(x_1 + x_2)]$$

and

$$u = -\tanh[100(x_1 + x_2)].$$

The variations of $a(\cdot)$ and $b(\cdot)$ can be continuous, piecewise continuous, deterministic, stochastic, etc. To illustrate the basic features, we let $a = a_0 - a_v \sin(\pi t)$ and $b = b_0 + b_v \cos(\pi t)$, where $a_0 = -1$, $a_v \in [0\ 0.5]$ and $b_0 = 1$, $b_v \in [0\ 0.5]$. The Simulink models to perform simulations for the closed-loop systems with the hard- and soft-switching control laws are reported in Figure 8.27a and b.

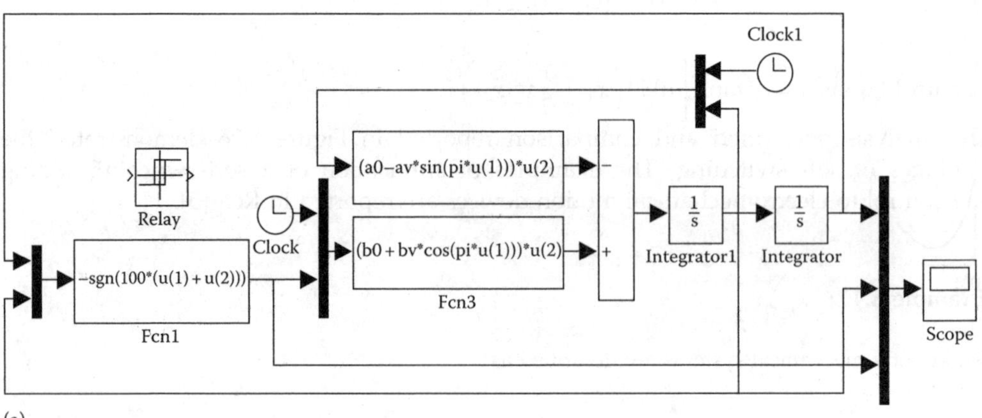

(a)

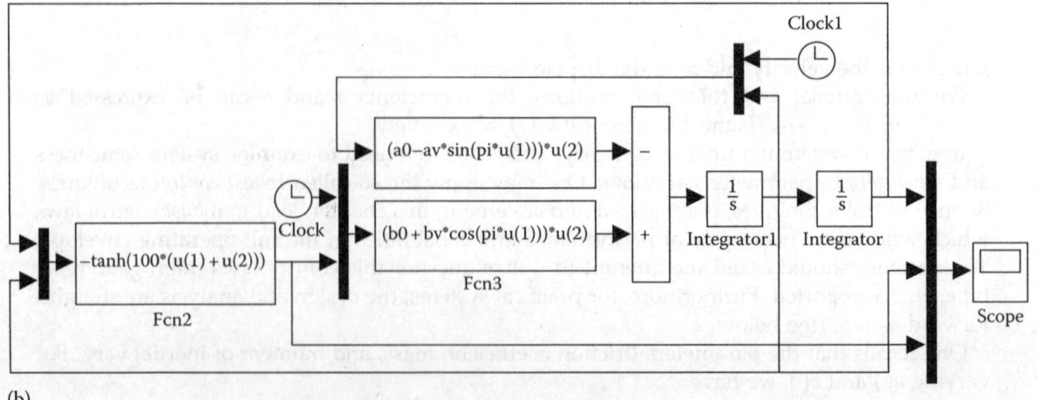

(b)

FIGURE 8.27
(a) Simulink model (`ch8_05.mdl`): closed-loop system with a hard-switching control law $u = -\mathrm{sgn}[100(x_1 + x_2)]$;
(b) Simulink model (`ch8_06.mdl`): closed-loop system with a soft-switching control law $u = -\tanh[100(x_1 + x_2)]$.

The simulation results for the initial conditions $x_{10} = -3$ and $x_{20} = -5$ are documented in Figure 8.28 for three cases: (1) $a_v = 0$ and $b_v = 0$; (2) $a_v = 0.25$ and $b_v = 0.25$; and (3) $a_v = 0.5$ and $b_v = 0.5$. The plotting statement is

```
plot(x12u(:,1),x12u(:,4),'m',x12u(:,1),x12u(:,2),'k',x12u(:,1),x12u(:,3),'b','LineWidth',2);
title('Dynamics of x_1(t) and x_2(t), Evolution of u(t)','FontSize',14);
xlabel('Time [seconds]','FontSize',14);
```

The hard-switching control law $u = -\text{sgn}(x_1 + x_2)$ results in continual switching activity with $u = \pm 1$, while soft switching leads to a preferable continuous-time control activity guaranteeing superior performance and capabilities. The soft-switching control law $u = -\tanh[100(x_1 + x_2)]$ is

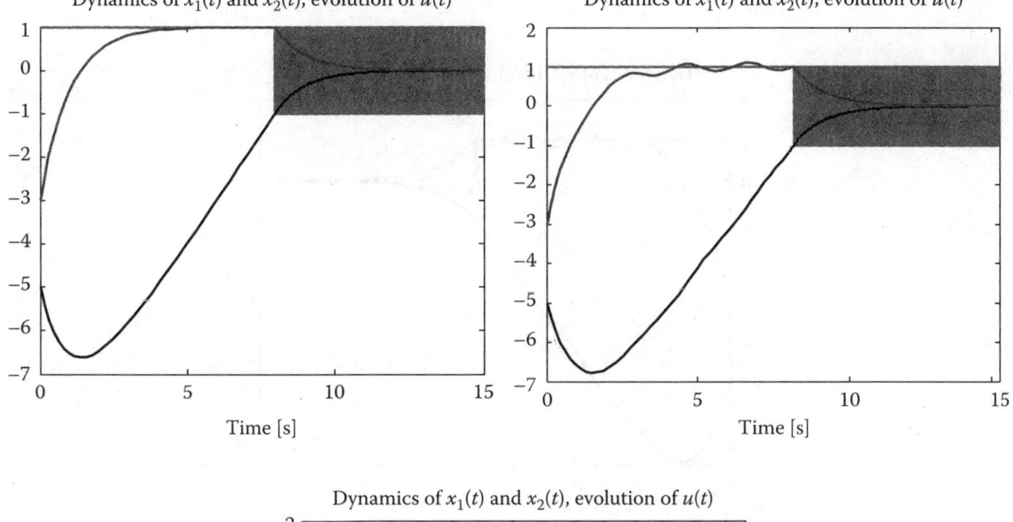

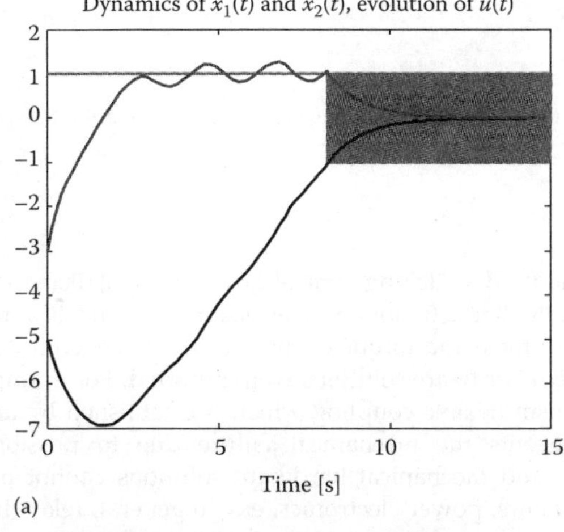

(a)

FIGURE 8.28

(a) Closed-loop system dynamics with a hard-switching control law $u = -\text{sgn}[100(x_1 + x_2)]$;

(continued)

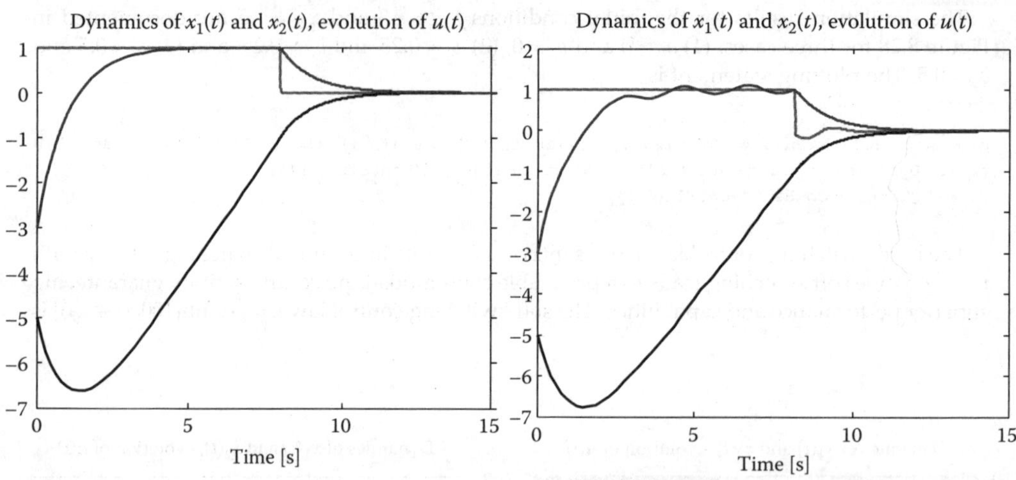

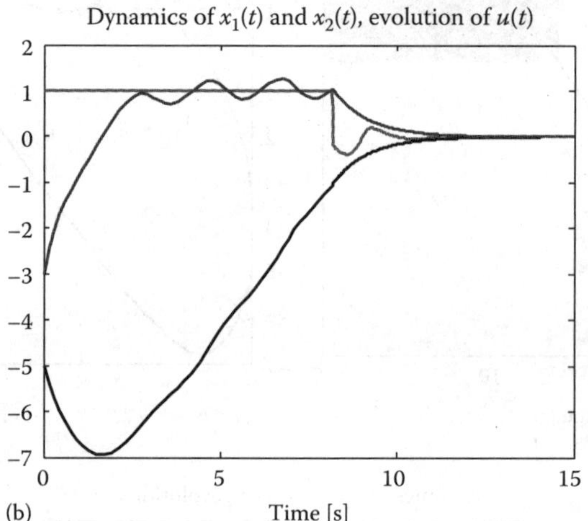

FIGURE 8.28 (continued)

(b) closed-loop system dynamics with a soft-switching control law $u = -\tanh[100(x_1 + x_2)]$, (1) $a_v = 0$ and $b_v = 0$; (2) $a_v = 0.25$ and $b_v = 0.25$; (3) $a_v = 0.5$ and $b_v = 0.5$.

implementable, while the hard-switching control law $u = -\mathrm{sgn}[100(x_1 + x_2)]$ may not be easily implemented. One recalls that actuators, power electronics, and ICs must be integrated, and the *bang-bang* changes of force and torque cannot be ensured even if desired. Fully-integrated electromechanical system hardware solutions were reported. For example, in Figure 1.4, there is the actuator–mechanism flexible coupling which is established by utilizing a *torque limiter*. This *torque limiter* prevents the mechanical failure due to possibly high instantaneous torques. The electronic and mechanical hardware solutions cannot provide hard-switching control activity for actuators, power electronics, etc. In general, relay (hard switching) control laws, as compared to continuous-time (soft switching) control laws, may not improve the system performance and capabilities. ∎

8.10 Constrained Control of Nonlinear Electromechanical Systems

In general, electromechanical systems are modeled by nonlinear differential equations, and saturation must be examined. Our goal is to minimize the functional

$$J = \int_{t_0}^{t_f} W_{xu}(x, u)dt, \tag{8.60}$$

subject to the system dynamics described by nonlinear differential equations

$$\dot{x}(t) = F(x) + B(x)u, \quad u_{\min} \leq u \leq u_{\max}, \quad u_{\min} < 0, \quad u_{\max} > 0, \quad x(t_0) = x_0. \tag{8.61}$$

The positive-definite, continuous, and differentiable integrand function $W_{xu}(\cdot): \mathbb{R}^n \times \mathbb{R}^m \to \mathbb{R}_{\geq 0}$ is used. In (8.61), $F(\cdot): \mathbb{R}_{\geq 0} \times \mathbb{R}^n \to \mathbb{R}^n$ and $B(\cdot): \mathbb{R}_{\geq 0} \times \mathbb{R}^n \ \mathbb{R}^{n \times m}$ are continuous and Lipschitz. The Hamiltonian is

$$H\left(x, u, \frac{\partial V}{\partial x}\right) = W_{xu}(x, u) + \left(\frac{\partial V}{\partial x}\right)^T [F(x) + B(x)u]. \tag{8.62}$$

Using the first-order necessary condition for optimality (8.10), one derives the control function $u(\cdot): [t_0, t_f) \to \mathbb{R}^m$ which minimizes the functional (8.60). Constrained optimization of electromechanical systems is a topic of great practical interest. We consider the systems modeled by nonlinear differential equations:

$$\dot{x}^{sys}(t) = F_s(x^{sys}) + B_s(x^{sys})u^{2w+1}, \quad y = H(x^{sys}), \quad u_{\min} \leq u \leq u_{\max}, \quad x^{sys}(t_0) = x_0^{sys}, \tag{8.63}$$

where, $x^{sys} \in X_s$ is the state vector, $u \in U$ is the vector of control inputs, $y \in Y$ is the measured output, and w is the nonnegative integer, $w = 0, 1, 2, \ldots$ For majority of the electromechanical devices and systems $w = 1$.

Using the Hamilton–Jacobi theory, the bounded control laws can be synthesized for continuous-time systems (8.63). To design the tracking control law, we integrate the system and *exogenous* dynamics, for example, we have

$$\dot{x}^{sys}(t) = F_s(x^{sys}) + B_s(x^{sys})u^{2w+1}, \quad y = H(x^{sys}), \quad u_{\min} \leq u \leq u_{\max}, \quad x^{sys}(t_0) = x_0^{sys}, \tag{8.64}$$

$$\dot{x}^{ref}(t) = Nr - y = Nr - H(x^{sys}).$$

Using the state vector $x = \begin{bmatrix} x^{sys} \\ x^{ref} \end{bmatrix} \in X$, from (8.64), one obtains

$$\dot{x}(t) = F(x, r) + B(x)u^{2w+1}, \quad u_{\min} \leq u \leq u_{\max}, \quad x(t_0) = x_0,$$

$$F(x, r) = \begin{bmatrix} F_s(x^{sys}) \\ -H(x^{sys}) \end{bmatrix} + \begin{bmatrix} 0 \\ N \end{bmatrix} r, \quad B(x) = \begin{bmatrix} B_s(x^{sys}) \\ 0 \end{bmatrix}. \tag{8.65}$$

We describe the control bounds by a bounded, integrable, one-to-one globally Lipschitz, vector-valued continuous function ϕ. Our goal is to analytically design the bounded admissible state-feedback control law in the form $u = \phi(x)$. As ϕ, one may apply the algebraic and transcendental (exponential, hyperbolic, logarithmic, and trigonometric) continuously differentiable, integrable, one-to-one functions. For example, the odd one-to-one integrable function tanh with domain $(-\infty, +\infty)$ describes the control bounds. This function has the corresponding inverse function $\tanh^{-1}$ with range $(-\infty, +\infty)$.

The performance functional to be minimized is

$$J = \int_{t_0}^{\infty} [W_x(x) + W_u(u)]dt = \int_{t_0}^{\infty} \left[W_x(x) + (2w + 1)\int (\phi^{-1}(u))^T G^{-1} \mathrm{diag}(u^{2w})du \right] dt, \qquad (8.66)$$

where $G^{-1} \in \mathbb{R}^{m \times m}$ is the positive-definite diagonal matrix, $G^{-1} > 0$.

Performance integrands W_x and W_u are real-valued, positive-definite, and continuously differentiable integrand functions. Using the properties of ϕ, one concludes that inverse function ϕ^{-1} is integrable. Hence, the integral $\int (\phi^{-1}(u))^T G^{-1} \mathrm{diag}(u^{2w})du$ exists.

Example 8.18:

Synthesize the performance functional to design a bounded control law for the following system:

$$\frac{dx}{dt} = ax + bu^3, \quad u_{\min} \le u \le u_{\max}.$$

Using the performance integrand

$$W_u(u) = (2w + 1)\int (\phi^{-1}(u))^T G^{-1} \mathrm{diag}(u^{2w})du,$$

and applying the integrable tanh function, one has the following positive-definite integrand

$$W_u(u) = 3\int \tanh^{-1} u G^{-1} u^2 du.$$

For $G^{-1} = 1/3$, we have $W_u(u) = 3\int \tanh^{-1} u G^{-1} u^2 du = \frac{1}{3} u^3 \tanh^{-1} u + \frac{1}{6} u^2 + \frac{1}{6} \ln(1 - u^2)$.

The hyperbolic tangent can used to describe the saturation effect. In general, one has

$$W_u(u) = (2w + 1)\int u^{2w} \tanh^{-1} \frac{u}{k} du = u^{2w+1} \tanh^{-1} \frac{u}{k} - k\int \frac{u^{2w+1}}{k^2 - u^2} du. \qquad \blacksquare$$

First- and second-order necessary conditions for optimality (8.10) and (8.13) that the control guarantees a minimum to the Hamiltonian

$$H = W_x(x) + (2w + 1)\int (\phi^{-1}(u))^T G^{-1} \mathrm{diag}(u^{2w})du + \frac{\partial V(x)^T}{\partial x}\left[F(x, r) + B(x)u^{2w+1}\right] \qquad (8.67)$$

are $\dfrac{\partial H}{\partial u} = 0$ and $\dfrac{\partial^2 H}{\partial u \times \partial u^T} > 0$. The positive-definite return function is $V(x_0) = \inf_{u \in U} J(x_0, u) = \inf J(x_0, \phi(\cdot)) \geq 0$. The function $V(x)$, $V(\cdot):\mathbb{R}^c \to \mathbb{R}_{\geq 0}$ is found using the Hamilton–Jacobi equation

$$-\frac{\partial V}{\partial t} = \min_{u \in U} \left\{ W_x(x) + (2w + 1) \int \left(\phi^{-1}(u)\right)^T G^{-1} \mathrm{diag}\left(u^{2w}\right) du + \frac{\partial V(x)^T}{\partial x} \left[F(x, r) + B(x)u^{2w+1}\right] \right\}.$$

$$(8.68)$$

The control law is derived by minimizing the nonquadratic functional (8.66). The first-order necessary condition (8.10) leads us to a bounded control law:

$$u = -\phi\left(GB^T(x) \frac{\partial V(x)}{\partial x} \right), \quad u \in U. \tag{8.69}$$

The second-order necessary condition for optimality (8.13) is met because the matrix G^{-1} is positive-definite. Hence, a unique, bounded, real-analytic, and continuous control law is designed. The solution of the functional equation should be found using nonquadratic return functions. To obtain $V(x)$, the performance functional is evaluated at the allowed values of the states and control. Linear and nonlinear functionals admit the final values. The minimum value of the nonquadratic functional (8.66) is given in a power-series form $J_{\min} = \sum_{i=1}^{\eta} v(x_0)^{2i}$, where η is the integer, $\eta = 1, 2, 3, \dots$. The solution of the partial differential equation (8.68) with (8.69) is satisfied by a continuously differentiable positive-definite return function

$$V(x) = \sum_{i=1}^{\eta} \frac{1}{2i} \left(x^i\right)^T K_i x^i, \quad K_i \in \mathbb{R}^{c \times c}, \tag{8.70}$$

where matrices K_i are found by solving the Hamilton–Jacobi equation.

From (8.69) and (8.70), the nonlinear bounded control law is given as

$$u = -\phi\left(GB^T(x) \sum_{i=1}^{\eta} \mathrm{diag}\left[x^{i-1}\right] K_i x^i \right), \quad \mathrm{diag}\left[x^{i-1}\right] = \begin{bmatrix} x_1^{i-1} & 0 & \cdots & 0 & 0 \\ 0 & x_2^{i-1} & \cdots & 0 & 0 \\ \vdots & \vdots & \ddots & \vdots & \vdots \\ 0 & 0 & \cdots & x_{c-1}^{i-1} & 0 \\ 0 & 0 & \vdots & 0 & x_c^{i-1} \end{bmatrix}. \tag{8.71}$$

The constraints on u are due to the hardware (solid-state devices, ICs, electromechanical motion devices, etc.) limits. Those constraints are integrated in the control law design. One does not need to *implement* those bounds by means of additional software or hardware. The control law to be implemented by the analog or digital controller is $GB^T(x)\dfrac{\partial V(x)}{\partial x}$ or $GB^T(x) \sum_{i=1}^{\eta} \mathrm{diag}\left[x^{i-1}\right] K_i x^i$. For example, the nonlinearity ϕ is the existing inherent constraint. However, the control law design complies with the system nonlinearities, and u is found as a nonlinear function of x and e.

Example 8.19:

Consider a dynamic system with constraints on control as described by

$$\dot{x}(t) = ax + bu, \quad 1 \leq u \leq 1, \quad u \in U,$$

Using the hyperbolic tangent function to describe the saturation at u, the performance functional (8.66) is

$$J = \int\limits_0^\infty \left(qx^2 + g \int \tanh^{-1} u \, du \right) dt, \quad q \geq 0, \quad g > 0.$$

The Hamiltonian function (8.67) is $H = qx^2 + g \int \tanh^{-1} u \, du + \dfrac{\partial V}{\partial x}(ax + bu)$.

The first-order necessary condition for optimality (8.10) results in $g \tanh^{-1} u + \dfrac{\partial V}{\partial x} b = 0$. Hence, the bounded control law is

$$u = -\tanh\left(g^{-1} b \frac{\partial V}{\partial x} \right), \quad u \in U.$$

The solution of the Hamilton–Jacobi equation (8.68) is found using (8.70). Approximating the solution by the quadratic return function $V(x) = \dfrac{1}{2} k x^2$, the bounded control law is found to be

$$u = -\tanh(g^{-1} b k x), \quad k > 0.$$

However, the solution of the Hamilton–Jacobi partial differential equation (8.68) should be approximated in $x \in X$ and $u \in U$. From (8.70), using $V(x) = \dfrac{1}{2} k_1 x^2 + \dfrac{1}{4} k_2 x^4 + \dfrac{1}{6} k_3 x^6$, one obtains

$$u = -\tanh\left[g^{-1} b (k_1 x + k_2 x^3 + k_3 x^5) \right]. \qquad\blacksquare$$

8.11 Optimization of Systems Using Nonquadratic Performance Functionals

The Hamilton–Jacobi theory, maximum principle, dynamic programming, and Lyapunov concept provide the designer with a general setup to solve linear and nonlinear optimal control problems for electromechanical systems. The general results can be derived using different performance functionals. In particular, quadratic and nonquadratic integrands have been applied. These functionals lead to a solution of optimization problems. The importance of synthesis of performance functionals lies on the matter that the control laws are predefined by the functionals used. Furthermore, the closed-loop system performance is defined by the performance integrands applied.

The closed-loop system performance is optimal (with respect to the minimizing functional), and stability margins are assigned in the specific sense as implied by the performance functionals. The performance integrands which allow one to measure system performance as well as to design bounded control laws were reported in Section 8.10. The system optimality and performance depend to a large extent on the specifications imposed (desired steady-state and dynamic performance, for example, settling time, accuracy, steady-state error, overshoot, etc.) and the inherent system capabilities which include state bounds, control constraints, etc.

In Section 8.10, to design the *admissible* (bounded) control laws, we minimized

$$J = \int_{t_0}^{t_f} \left[\frac{1}{2} x^T Q x + \int \left(\phi^{-1}(u) \right)^T G^{-1} du \right] dt \text{ or similar functionals using the bounded, integrable,}$$

one-to-one, real-analytic, globally Lipschitz continuous function $\phi(\cdot):\mathbb{R}^n \to \mathbb{R}^m$, $\phi \in U \subset \mathbb{R}^m$.

For linear (8.14) and nonlinear (8.8) systems with $u_{\min} \leq u \leq u_{\max}$, $u \in U$, the minimization of the nonquadratic functional

$$J = \int_{t_0}^{t_f} \left[\frac{1}{2} x^T Q x + \int \left(\phi^{-1}(u) \right)^T G \, du \right] dt \tag{8.72}$$

gives

$$-\frac{\partial V}{\partial t} = \min_{u \in U} \left\{ \frac{1}{2} x^T Q x + \int \left(\phi^{-1}(u) \right)^T G \, du + \frac{\partial V^T}{\partial x} (Ax + Bu) \right\} \quad \text{for } \dot{x}(t) = Ax + Bu,$$

and $\quad -\dfrac{\partial V}{\partial t} = \min\limits_{u \in U} \left\{ \dfrac{1}{2} x^T Q x + \int \left(\phi^{-1}(u) \right)^T G \, du + \dfrac{\partial V^T}{\partial x} [F(x) + B(x)u] \right\} \quad \text{for } \dot{x}(t) = F(x) + B(x)u.$

Using the first-order necessary condition for optimality (8.10), the *admissible* control laws are

$$u = -\phi \left(G^{-1} B^T \frac{\partial V(x)}{\partial x} \right), \quad u \in U \tag{8.73}$$

and

$$u = -\phi \left(G^{-1} B^T(x) \frac{\partial V(x)}{\partial x} \right), \quad u \in U. \tag{8.74}$$

As documented, for linear and nonlinear systems, bounded control laws (8.73) and (8.74) can be designed. We further concentrate on the synthesis of performance functionals and design of control laws. Consider electromechanical systems modeled by linear or nonlinear differential equations. We apply the following performance functional:

$$J = \int_{t_0}^{t_f} \frac{1}{2} \left[\omega(x)^T Q \omega(x) + \dot{\omega}(x)^T P \dot{\omega}(x) \right] dt, \quad Q \geq 0, \quad P > 0, \tag{8.75}$$

where $\omega(\cdot): \mathbb{R}^n \to \mathbb{R}_{\geq 0}$ is the differentiable real-analytic continuous function and $Q \in \mathbb{R}^{n \times n}$ and $P \in \mathbb{R}^{n \times n}$ are the positive-definite diagonal weighting matrices.

Using (8.75), the system transient performance and stability are specified by two integrands, for example, $\omega(x)^T Q \omega(x)$ and $\dot{\omega}(x)^T P \dot{\omega}(x)$. These integrands are given as the nonlinear functions of the states and the rate of change of the variables. The performance functional (8.75) depends on the system dynamics (states and control variables), control efforts, energy, etc. For linear systems, described by (8.14), we have

$$\dot{\omega}(x) = \frac{\partial \omega}{\partial x}\dot{x} = \frac{\partial \omega}{\partial x}(Ax + Bu).$$

Using (8.75), one has the following functional:

$$J = \int_{t_0}^{t_f} \frac{1}{2}\left[\omega(x)^T Q \omega(x) + \dot{x}^T \frac{\partial \omega^T}{\partial x} P \frac{\partial \omega}{\partial x}\dot{x}\right]dt = \int_{t_0}^{t_f} \frac{1}{2}\left[\omega(x)^T Q \omega(x) + (Ax + Bu)^T \frac{\partial \omega^T}{\partial x} P \frac{\partial \omega}{\partial x}(Ax + Bu)\right]dt$$

$$(8.76)$$

In (8.76), $\omega(x)$ is the differentiable and integrable real-valued continuous function. For example, one may use

- $\omega(x) = x$, which leads to the quadratic integrand functions
- $\omega(x) = x^3$, $\omega(x) = \tanh(x)$, or $\omega(x) = e^{-x}$, which result in nonquadratic functionals

For linear systems (8.14) and performance functional (8.75), the Hamiltonian function is

$$H\left(x, u, \frac{\partial V}{\partial x}\right) = \frac{1}{2}\omega(x)^T Q \omega(x) + \frac{1}{2}\dot{\omega}(x)^T P \dot{\omega}(x) + \frac{\partial V^T}{\partial x}(Ax + Bu). \qquad (8.77)$$

The application of the first-order condition for optimality (8.10) gives the following *optimal* control law:

$$u = -\left(B^T \frac{\partial \omega^T}{\partial x} P \frac{\partial \omega}{\partial x} B\right)^{-1} B^T \left(\frac{\partial \omega^T}{\partial x} P \frac{\partial \omega}{\partial x} Ax + \frac{\partial V}{\partial x}\right) \qquad (8.78)$$

The second-order condition for optimality (8.13) is guaranteed. In particular, from (8.77), one has $\dfrac{\partial^2 H}{\partial u \times \partial u^T} = B^T \dfrac{\partial \omega^T}{\partial x} P \dfrac{\partial \omega}{\partial x} B > 0$ because $\omega(x)$ is chosen such that $\dfrac{\partial \omega}{\partial x} B$ has full rank, and $P > 0$.

Synthesis of the performance integrands $\omega(x)^T Q \omega(x)$ and $\dot{\omega}(x)^T P \dot{\omega}(x)$ results in devising and utilizing of integrable and differentiable function $\omega(x)$. For example, applying $\omega(x) = x$, from (8.75) we have

$$J = \int_{t_0}^{t_f} \frac{1}{2}\left[x^T Q x + (Ax + Bu)^T P(Ax + Bu)\right]dt. \qquad (8.79)$$

Using (8.79), one obtains the functional equation

$$-\frac{\partial V}{\partial t} = \min_{u} \left\{ \frac{1}{2} \left[x^T Q x + (Ax + Bu)^T P(Ax + Bu) \right] + \frac{\partial V^T}{\partial x} (Ax + Bu) \right\}.$$ (8.80)

From

$$H\left(x, u, \frac{\partial V}{\partial x}\right) = \frac{1}{2} \left[x^T Q x + (Ax + Bu)^T P(Ax + Bu) \right] + \frac{\partial V^T}{\partial x} (Ax + Bu),$$ (8.81)

the control law is found by using the first-order necessary condition for optimality (8.10). One finds

$$u = -\left(B^T P B\right)^{-1} B^T \left(PAx + \frac{\partial V}{\partial x} \right).$$ (8.82)

The solution of the functional equation (8.80) is given by the quadratic return function (8.20) $V = \frac{1}{2} x^T K x$. From (8.82), we have the following linear control law:

$$u = -\left(B^T P B\right)^{-1} B^T (PA + K)x.$$ (8.83)

Using (8.80) and (8.83), the unknown symmetric matrix $K \in \mathbb{R}^{n \times n}$ is obtained by solving the following nonlinear differential equation:

$$-\dot{K} = Q - KB\left(B^T P B\right)^{-1} B^T K, \quad K(t_f) = K_f.$$ (8.84)

The control law designed (8.82) is different when compared with the conventional linear control law (8.24). Furthermore, the equations to compute matrix K are different, see (8.23) and (8.84).

We study nonlinear system dynamics as given by (8.8). The performance functional is

$$J = \int_{t_0}^{t_f} \frac{1}{2} \left[\omega(x)^T Q \omega(x) + [F(x) + B(x)u]^T \frac{\partial \omega^T}{\partial x} P \frac{\partial \omega}{\partial x} [F(x) + B(x)u] \right] dt$$ (8.85)

For a dynamic system (8.8) and performance functional (8.85), we have the following Hamiltonian function

$$H\left(x, u, \frac{\partial V}{\partial x}\right) = \frac{1}{2} \omega(x)^T Q \omega(x) + \frac{1}{2} [F(x) + B(x)u]^T \frac{\partial \omega^T}{\partial x} P \frac{\partial \omega}{\partial x} [F(x) + B(x)u] + \frac{\partial V^T}{\partial x} [F(x) + B(x)u]$$

(8.86)

The positive-definite return function $V(\cong):\mathbb{R}^n \to \mathbb{R}_{\geq 0}$ satisfies the following differential equation:

$$-\frac{\partial V}{\partial t} = \min_u \left\{ \frac{1}{2}\omega(x)^T Q\omega(x) + \frac{1}{2}[F(x) + B(x)u]^T \frac{\partial \omega^T}{\partial x} P \frac{\partial \omega}{\partial x}[F(x) + B(x)u] + \frac{\partial V^T}{\partial x}[F(x) + B(x)u] \right\}$$

$$(8.87)$$

From (8.87), using (8.10), one finds the control law as

$$u = -\left(B^T(x)\frac{\partial \omega^T}{\partial x} P \frac{\partial \omega}{\partial x} B(x) \right)^{-1} B^T(x)\left(\frac{\partial \omega^T}{\partial x} P \frac{\partial \omega}{\partial x} F(x) + \frac{\partial V}{\partial x} \right) \qquad (8.88)$$

The control law (8.88) is an *optimal* control, and the second-order necessary condition for optimality is met because $\dfrac{\partial^2 H}{\partial u \times \partial u^T} = B^T(x)\dfrac{\partial \omega^T}{\partial x} P \dfrac{\partial \omega}{\partial x} B(x) > 0$. The solution of the partial differential equation (8.87) is approximated by the nonquadratic return function:

$$V(x) = \sum_{i=1}^{\eta} \frac{1}{2i}\left(x^i\right)^T K_i x^i, \quad K_i \mathbb{R}^{n \times n}.$$

Example 8.20:

For the first-order system studied in Example 8.10

$$\frac{dx}{dt} = ax + bu,$$

the generalized quadratic performance functional (8.75) is synthesized assigning $\omega(x) = x$. From (8.76), we have

$$J = \int_{t_0}^{\infty} \frac{1}{2}\left[Q\omega(x)^2 + P\left(\frac{\partial \omega}{\partial x}\right)^2 (a^2 x^2 + 2abxu + b^2 u^2) \right] dt = \int_{t_0}^{\infty} \frac{1}{2}\left(x^2 + a^2 x^2 + 2abxu + b^2 u^2 \right) dt,$$

where we let $Q = 1$ and $P = 1$.

Using the quadratic return function $V = \frac{1}{2}kx^2$, the linear control law (8.83) is

$$u = -\frac{1}{b}(a + k)x.$$

Solving the differential equation (8.84) $-\dot{k} = 1 - k^2$, we obtain $k = 1$.

Thus, the control law is $u = -\frac{1}{b}(a + 1)x$.

The closed-loop system is stable and evolves as $\dfrac{dx}{dt} = -x$.   ∎

Example 8.21:

For the system $\dfrac{dx}{dt} = ax + bu$, we minimize the performance functional (8.76)

$$J = \int_{t_0}^{\infty} \frac{1}{2}\left[Q\omega(x)^2 + P\left(\frac{\partial \omega}{\partial x}\right)^2 (a^2x^2 + 2abxu + b^2u^2) \right] dt.$$

The nonquadratic integrands are designed using $\omega(x) = \tanh(x)$. Let $Q = 1$ and $P = 1$. Hence,

$$J = \int_{t_0}^{\infty} \frac{1}{2}\left[\tanh^2 x + \operatorname{sech}^4 x \,(a^2x^2 + 2abxu + b^2u^2) \right] dt.$$

This performance functional with the synthesized integrands is examined. For $x \ll 1$, $\tanh^2 x \approx x^2$ and $\operatorname{sech}^4 x \approx 1$. That is, if $x \ll 1$, the performance functional can be expressed as $J \approx \int_{t_0}^{\infty} \frac{1}{2}[x^2 + a^2x^2 + 2abxu + b^2u^2]\,dt$. This generalized quadratic-like functional was used in Example 8.20 when we let $\omega(x) = x$. However, this conclusion is accurate for $x \ll 1$.

If $x \gg 1$, we have $\tanh^2 x \approx 1$ and $\operatorname{sech}^4 x \approx 0$. Hence, for $x \gg 1$, the performance functional is $J \approx \frac{1}{2}\int_{t_0}^{\infty} dt$ which is commonly used to solve the time-optimal (minimum time) problem.

The first-order necessary condition for optimality (8.10) is applied. One finds a control law (8.78) as

$$u = -\frac{a}{b}x - \frac{1}{b\operatorname{sech}^4 x}\frac{\partial V}{\partial x}.$$

The functional equation to be solved is

$$-\frac{\partial V}{\partial t} = \frac{1}{2}\tanh^2 x - \frac{1}{2\operatorname{sech}^4 x}\frac{\partial^2 V}{\partial x^2}.$$

For $x \in X$, one approximates continuous functions $\tanh^2 x$ and $\operatorname{sech}^4 x$. The quadratic and nonquadratic return functions are used to solve the Hamilton–Jacobi differential equation. Letting $V = \frac{1}{2}kx^2$, we have

$$u = -\frac{a}{b}x - \frac{1}{b\operatorname{sech}^4 x}kx.$$

The closed-loop system evolves as

$$\frac{dx}{dt} = -\frac{k}{\operatorname{sech}^4 x}x.$$

If $x \ll 1$, $\operatorname{sech}^4 x \approx 1$, and thus, $u \approx -\dfrac{a+k}{b}x$. The system dynamics is $\dfrac{dx}{dt} = -kx$.

For $x \gg 1$, we have the nonlinear control law which can be typified as a high feedback gain control.

Solving $-\dfrac{\partial V}{\partial t} = \dfrac{1}{2}\tanh^2 x - \dfrac{1}{2\operatorname{sech}^4 x}\dfrac{\partial^2 V}{\partial x^2}$, $V = \dfrac{1}{2}kx^2$, we found $k = 1$.

Though the quadratic return function $V = \dfrac{1}{2}kx^2$ may approximate the solution of nonlinear functional partial differential equation in the specified $x \in X$, in general, nonquadratic return functions must be used. Letting $V = \dfrac{1}{2}k_1 x^2 + \dfrac{1}{4}k_2 x^4$, one has

$$u = -\frac{a}{b}x - \frac{1}{b\operatorname{sech}^4 x}\left(k_1 x + k_2 x^3\right).$$

The solution of $-\dfrac{\partial V}{\partial t} = \dfrac{1}{2}\tanh^2 x - \dfrac{1}{2\operatorname{sech}^4 x}\dfrac{\partial^2 V}{\partial x^2}$ gives $k_1 = 1$ and $k_2 = 0.5$. The transient dynamics of the closed-loop system for four different initial conditions x_0 ($x_0 = 0.1$, $x_0 = 1$, $x_0 = 5$, and $x_0 = 10$) are reported in Figure 8.29. The analysis indicates that the settling time is almost the same for any initial conditions. This is due to the fact that a novel nonlinear optimal control law

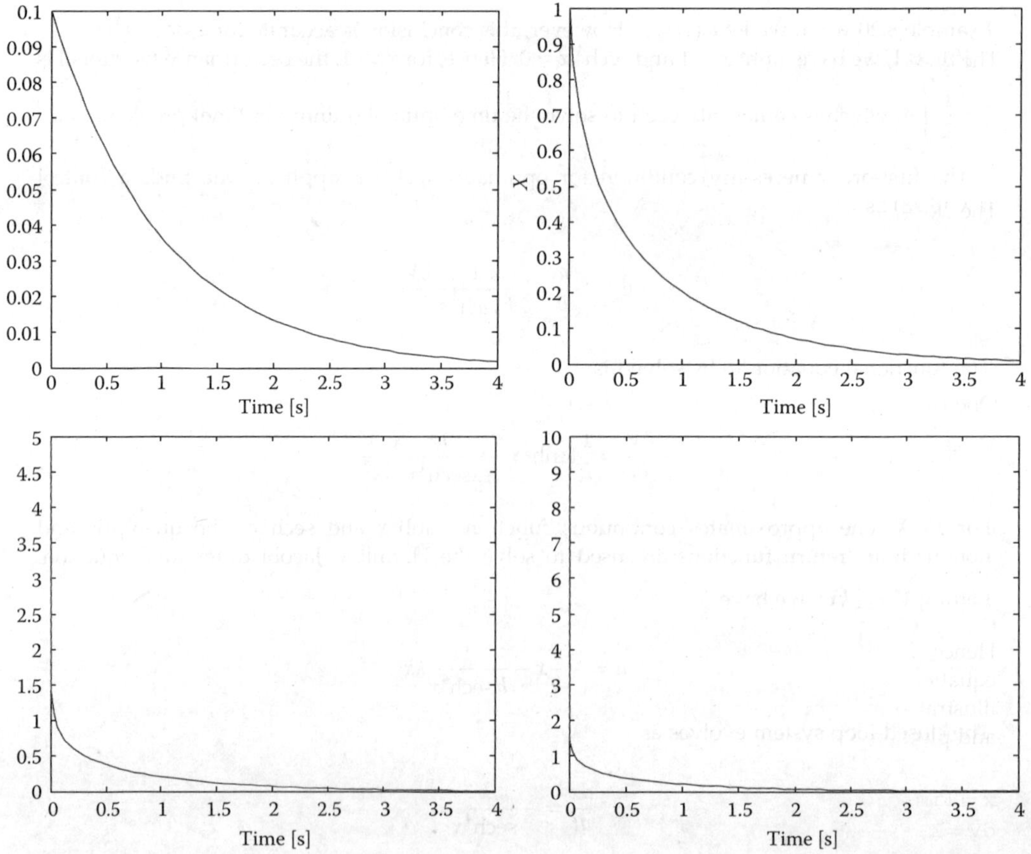

FIGURE 8.29

Transient dynamics of the closed-loop system if $x_0 = 0.1$, $x_0 = 1$, $x_0 = 5$ and $x_0 = 10$.

that possesses unique features, as discussed above, is designed. However, the control bounds which should be integrated will increase the settling time. ∎

Example 8.22:

The differential equations which describe a rigid-body mechanical system are

$$\frac{dx_1}{dt} = ax_1 + bu, \quad \frac{dx_2}{dt} = x_1,$$

where x_1 and x_2 are the velocity and displacement.

The performance integrands are designed utilizing $\omega(x) = \tanh(x)$. Using the identity matrices $Q = I$ and $P = I$, the functional (8.75) is expressed as

$$J = \int_{t_0}^{t_f} \frac{1}{2} \left\{ [\tanh x_1 \, \tanh x_2] \begin{bmatrix} 1 & 0 \\ 0 & 1 \end{bmatrix} \begin{bmatrix} \tanh x_1 \\ \tanh x_2 \end{bmatrix} + [\dot{x}_1 \text{sech}^2 x_1 \, \dot{x}_2 \text{sech}^2 x_2] \begin{bmatrix} 1 & 0 \\ 0 & 1 \end{bmatrix} \begin{bmatrix} \dot{x}_1 \text{sech}^2 x_1 \\ \dot{x}_2 \text{sech}^2 x_2 \end{bmatrix} \right\} dt$$

$$= \int_{t_0}^{t_f} \frac{1}{2} \left[\tanh^2 x_1 + \tanh^2 x_2 + \text{sech}^4 x_1 \left(a^2 x_1^2 + 2abx_1 u + b^2 u^2 \right) + x_1^2 \text{sech}^4 x_2 \right] dt.$$

The first-order necessary condition for optimality (8.10) results in the following control law:

$$u = -\frac{a}{b} x_1 - \frac{1}{b \, \text{sech}^4 x_1} \frac{\partial V}{\partial x_1}.$$

The closed-loop system is described as

$$\frac{dx_1}{dt} = -\frac{1}{\text{sech}^4 x_1} \frac{\partial V}{\partial x_1},$$

$$\frac{dx_2}{dt} = x_1.$$

One can examine the stability for this system. We use a positive-definite Lyapunov function $V_L = \frac{1}{2} x_1^2 + \frac{1}{2} x_2^2$. To derive an explicit expression for u, the quadratic return function $V = \frac{1}{2} k_{11} x_1^2 + k_{12} x_1 x_2 + \frac{1}{2} k_{22} x_2^2$ is used. For the closed-loop system, the total derivative is

$$\frac{dV_L}{dt} = -\frac{1}{\text{sech}^4 x_1} \left(\frac{\partial V}{\partial x_1} \right)^2 + \frac{\partial V}{\partial x_2} x_1 = -\frac{1}{\text{sech}^4 x_1} (k_{11} x_1 + k_{12} x_2)^2 + x_1 x_2.$$

Hence, dV_L/dt is negative-definite. It will be reported that the solution of the Hamilton–Jacobi equation results in $k_{11} = 1$ and $k_{12} = 0.25$. The total derivative of the Lyapunov function is illustrated in Figure 8.30a. The following MATLAB statement is used to perform calculations and plotting

```
x=linspace(-1,1,25); y=x; [X,Y]=meshgrid(x,y); k11=1 ; k12=0.25 ;
dV=X.*Y-((k11*X+k12*Y).^2)./sech(X).^4; surf(x,y,dV);
xlabel('x_1','FontSize',14); ylabel('x_2','FontSize',14);
zlabel('dV_L/dt','FontSize',14);
title('Lyapunov Function Total Derivative, dV_L/ dt','FontSize',14);
```

We solve the differential equation

$$-\frac{\partial V}{\partial t} = \frac{1}{2}\tanh^2 x_1 + \frac{1}{2}\tanh^2 x_2 - \frac{1}{2\mathrm{sech}^4 x_1}\left(\frac{\partial V}{\partial x_1}\right)^2 + \frac{1}{2}x_1^2\mathrm{sech}^4 x_2 + \frac{\partial V}{\partial x_2}x_1$$

by approximating its solution by the nonquadratic return function

$$V = \frac{1}{2}k_{11}x_1^2 + k_{12}x_1 x_2 + \frac{1}{2}k_{22}x_2^2 + \frac{1}{4}k_{41}x_1^4 + \frac{1}{4}k_{42}x_2^4.$$

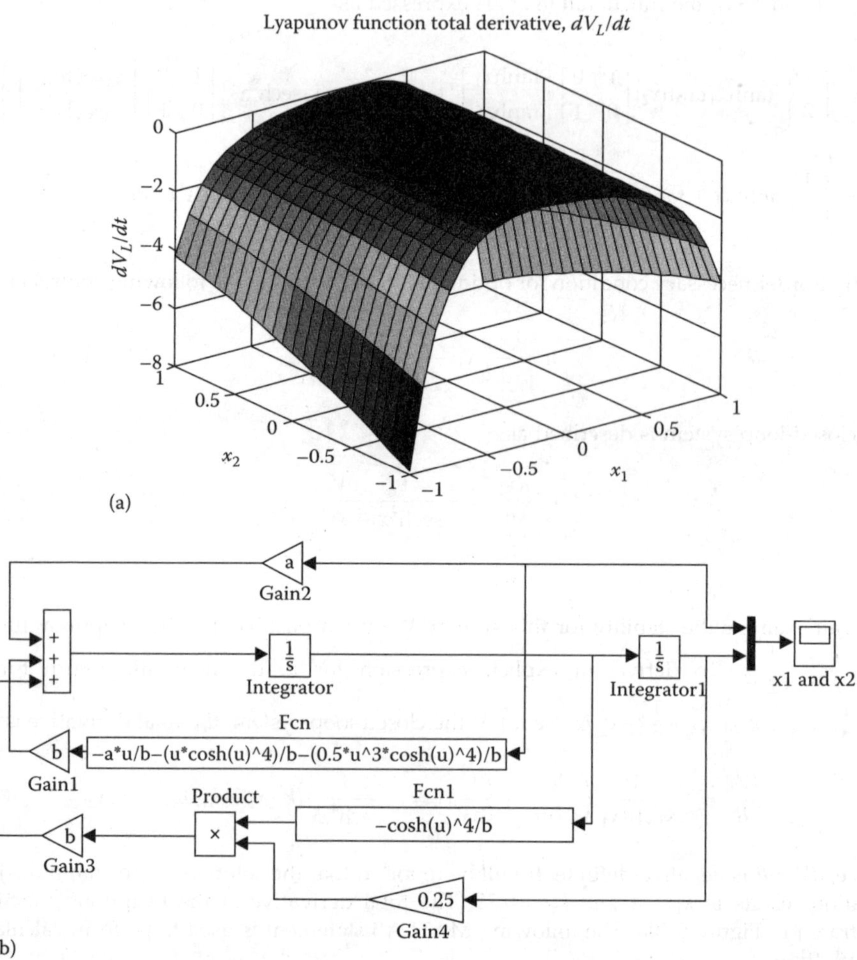

FIGURE 8.30

(a) Total derivative of the Lyapunov function; (b) Simulink diagram to simulate the closed-loop system (ch8_01.mdl); and

(continued)

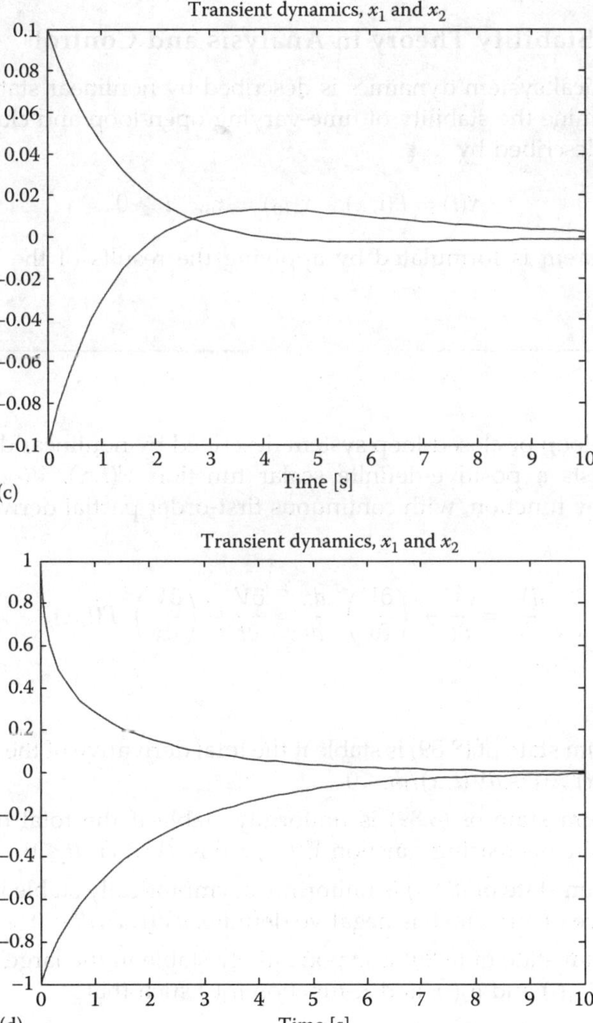

FIGURE 8.30 (continued)
(c–d) transient dynamics of the closed-loop system.

Having found the coefficients of $V(x)$, we have

$$u = -\frac{a}{b}x_1 - \frac{1}{b\,\mathrm{sech}^4 x_1}\left(x_1 + 0.25x_2 + 0.5x_1^3\right).$$

The Simulink model is reported in Figure 8.30b. Though there are no a and b coefficients are $a = 1$ and $b = 1$. The transient dynamics of the closed-loop system for the initial conditions $\begin{bmatrix} x_{10} \\ x_{20} \end{bmatrix} = \begin{bmatrix} 0.1 \\ -0.1 \end{bmatrix}$ and $\begin{bmatrix} x_{10} \\ x_{20} \end{bmatrix} = \begin{bmatrix} 1 \\ -1 \end{bmatrix}$ are reported in Figures 8.30c and d.

We conclude that the control law designed ensures optimal system evolution with respect to the functional minimized. ∎

8.12 Lyapunov Stability Theory in Analysis and Control

The electromechanical system dynamics is described by nonlinear state-space differential equations. We examine the stability of time-varying open-loop and closed-loop nonlinear dynamic systems described by

$$\dot{x}(t) = F(t, x), \quad x(t_0) = x_0, \quad t \geq 0. \tag{8.89}$$

The following theorem is formulated by applying the results of the Lyapunov stability theory.

THEOREM 8.1

Consider the open-loop or closed-loop system described by nonlinear differential equation (8.89). If there exists a positive-definite scalar function $V(t, x)$, $V(\cdot): \mathbb{R}_{\geq 0} \times \mathbb{R}^n \to \mathbb{R}_{\geq 0}$, called the Lyapunov function, with continuous first-order partial derivatives with respect to t and x

$$\frac{dV}{dt} = \frac{\partial V}{\partial t} + \left(\frac{\partial V}{\partial x}\right)^T \frac{dx}{dt} = \frac{\partial V}{\partial t} + \left(\frac{\partial V}{\partial x}\right)^T F(t, x),$$

then

- The equilibrium state of (8.89) is stable if the total derivative of the positive-definite function $V(t, x) > 0$ is $dV(t, x)/dt \leq 0$
- The equilibrium state of (8.89) is uniformly stable if the total derivative of the positive-definite decreasing function $V(t, x) > 0$ is $dV(t, x)/dt \leq 0$
- The equilibrium state of (8.89) is uniformly asymptotically stable in the large if the total derivative of $V(t, x) > 0$ is negative definite, $dV(t, x)/dt < 0$
- The equilibrium state of (8.89) is exponentially stable in the large if there exist the K_∞-functions $\rho_1(\cdot)$ and $\rho_2(\cdot)$, and K-function $\rho_3(\cdot)$ such that

$$\rho_1(\| x \|) \leq V(t, x) \leq \rho_2(\| x \|) \quad \text{and} \quad \frac{dV(x)}{dt} \leq -\rho_3(\| x \|). \qquad \blacksquare$$

Example 8.23:

Consider a system which is described by two nonlinear time-invariant differential equations:

$$\dot{x}_1(t) = x_1 - x_1^5 - x_1^3 x_2^4, \quad \dot{x}_2(t) = x_2 - x_2^9, \quad t \geq 0.$$

These differential equations describe the uncontrolled or controlled (closed-loop) dynamics. For example, let

$$\dot{x}_1(t) = x_1 + u_1, \quad \dot{x}_2(t) = x_2 + u_2,$$

where $u_1 = -x_1^5 - x_1^3 x_2^4$ and $u_2 = -x_2^9$.

A scalar positive-definite function is expressed in the quadratic form as $V(x_1, x_2) = \frac{1}{2}\left(x_1^2 + x_2^2\right)$.

The total derivative is found to be

$$\frac{dV(x_1, x_2)}{dt} = \left(\frac{\partial V}{\partial x}\right)^{\mathrm{T}}\frac{dx}{dt} = \left(\frac{\partial V}{\partial x}\right)^{\mathrm{T}}F(x) = \frac{\partial V}{\partial x_1}\left(x_1 - x_1^5 - x_1^3 x_2^4\right) + \frac{\partial V}{\partial x_2}\left(x_2 - x_2^9\right)$$

$$= x_1^2 - x_1^6 - x_1^4 x_2^4 + x_2^2 - x_2^{10}.$$

The total derivative of a positive-definite $V(x_1, x_2) > 0$ is $\dfrac{dV(x_1, x_2)}{dt} < 0$. For example, dV/dt is negative definite. Therefore, the equilibrium state of the system is uniformly asymptotically stable. ∎

Example 8.24:

Consider time-varying nonlinear differential equations

$$\dot{x}_1(t) = -x_1 + x_2^3, \quad \dot{x}_2(t) = -e^{-10t}x_1 x_2^2 - 5x_2 - x_2^3, \quad t \geq 0.$$

A scalar positive-definite function $V(t, x_1, x_2) > 0$ is chosen in the quadratic form as

$$V(t, x_1, x_2) = \frac{1}{2}\left(x_1^2 + e^{10t}x_2^2\right).$$

The total derivative is

$$\frac{dV(t, x_1, x_2)}{dt} = \frac{\partial V}{\partial t} + \frac{\partial V}{\partial x_1}\left(-x_1 + x_2^3\right) + \frac{\partial V}{\partial x_2}\left(-e^{-10t}x_1 x_2^2 - 5x_2 - x_2^3\right) = -x_1^2 - e^{10t}x_2^4.$$

The total derivative is negative definite, $\dfrac{dV(x_1, x_2)}{dt} < 0$. Using Theorem 8.1, one concludes that the equilibrium state is uniformly asymptotically stable. ∎

Example 8.25:

The system dynamics is described by the differential equations

$$\dot{x}_1(t) = -x_1 + x_2,$$
$$\dot{x}_2(t) = -x_1 - x_2 - x_2|x_2|, \quad t \geq 0.$$

The positive-definite scalar Lyapunov candidate is chosen in the following form:

$$V(x_1, x_2) = \frac{1}{2}\left(x_1^2 + x_2^2\right).$$

Thus, $V(x_1, x_2) > 0$. The total derivative is $\dfrac{dV(x_1, x_2)}{dt} = x_1 \dot{x}_1 + x_2 \dot{x}_2 = -x_1^2 - x_2^2(1 + |x_2|)$. Therefore, $\dfrac{dV(x_1, x_2)}{dt} < 0$.

Hence, the equilibrium state of the system is uniformly asymptotically stable, and the quadratic function $V(x_1, x_2) = \dfrac{1}{2}(x_1^2 + x_2^2)$ is the Lyapunov function. ∎

Example 8.26:   Stability of Permanent-Magnet Synchronous Motors

Consider a permanent-magnet synchronous motor in the rotor reference frame [5]. The mathematical model, assuming that $T_L = 0$, is

$$\frac{di_{qs}^r}{dt} = -\frac{r_s}{L_{ls} + \frac{3}{2}\bar{L}_m} i_{qs}^r - \frac{\psi_m}{L_{ls} + \frac{3}{2}\bar{L}_m} \omega_r - i_{ds}^r \omega_r + \frac{1}{L_{ls} + \frac{3}{2}\bar{L}_m} u_{qs}^r,$$

$$\frac{di_{ds}^r}{dt} = -\frac{r_s}{L_{ls} + \frac{3}{2}\bar{L}_m} i_{ds}^r + i_{qs}^r \omega_r + \frac{1}{L_{ls} + \frac{3}{2}\bar{L}_m} u_{ds}^r,$$

$$\frac{d\omega_r}{dt} = \frac{3P^2\psi_m}{8J} i_{qs}^r - \frac{B_m}{J} \omega_r.$$

For an open-loop system, formally $u_{qs}^r = 0$ and $u_{ds}^r = 0$. For a closed-loop system, one has $u_{qs}^r \neq 0$ and $u_{ds}^r = 0$. For example, $u_{qs}^r = -k_\omega \omega_r$. For $u_{qs}^r = 0$ and $u_{ds}^r = 0$, we have

$$\frac{di_{qs}^r}{dt} = -\frac{r_s}{L_{ls} + \frac{3}{2}\bar{L}_m} i_{qs}^r - \frac{\psi_m}{L_{ls} + \frac{3}{2}\bar{L}_m} \omega_r - i_{ds}^r \omega_r,$$

$$\frac{di_{ds}^r}{dt} = -\frac{r_s}{L_{ls} + \frac{3}{2}\bar{L}_m} i_{ds}^r + i_{qs}^r \omega_r,$$

$$\frac{d\omega_r}{dt} = \frac{3P^2\psi_m}{8J} i_{qs}^r - \frac{B_m}{J} \omega_r.$$

Using the quadratic positive-definite Lyapunov function

$$V\left(i_{qs}^r, i_{ds}^r, \omega_r\right) = \frac{1}{2}\left(i_{qs}^{r\,2} + i_{ds}^{r\,2} + \omega_r^2\right),$$

the expression for the total derivative is

$$\frac{dV\left(i_{qs}^r, i_{ds}^r, \omega_r\right)}{dt} = -\frac{r_s}{L_{ss}}\left(i_{qs}^{r\,2} + i_{ds}^{r\,2}\right) - \frac{B_m}{J}\omega_r^2 - \frac{8J\psi_m - 3P^2 L_{ss}\psi_m}{8JL_{ss}} i_{qs}^r \omega_r.$$

Thus, $\dfrac{dV\left(i_{qs}^r, i_{ds}^r, \omega_r\right)}{dt} < 0$. One concludes that the equilibrium state of an open-loop drive is uniformly asymptotically stable.

Consider the closed-loop system. To guarantee a balanced operation, we define the control law assigning the qd voltage components to be $u_{qs}^r = -k_\omega \omega_r$ and $u_{ds}^r = 0$. The following differential equations result

$$\frac{di_{qs}^r}{dt} = -\frac{r_s}{L_{ls} + \frac{3}{2}\bar{L}_m} i_{qs}^r - \frac{\psi_m}{L_{ls} + \frac{3}{2}\bar{L}_m} \omega_r - i_{ds}^r \omega_r - \frac{1}{L_{ls} + \frac{3}{2}\bar{L}_m} k_\omega \omega_r,$$

$$\frac{di_{ds}^r}{dt} = -\frac{r_s}{L_{ls} + \frac{3}{2}\bar{L}_m} i_{ds}^r + i_{qs}^r \omega_r,$$

$$\frac{d\omega_r}{dt} = \frac{3P^2 \psi_m}{8J} i_{qs}^r - \frac{B_m}{J} \omega_r.$$

Using the quadratic positive-definite Lyapunov function $V\left(i_{qs}^r, i_{ds}^r, \omega_r\right) = \frac{1}{2}\left(i_{qs}^{r\,2} + i_{ds}^{r\,2} + \omega_r^2\right)$, we

obtain $\dfrac{dV\left(i_{qs}^r, i_{ds}^r, \omega_r\right)}{dt} = -\dfrac{r_s}{L_{ss}}\left(i_{qs}^{r\,2} + i_{ds}^{r\,2}\right) - \dfrac{B_m}{J}\omega_r^2 - \dfrac{8J(\psi_m + k_\omega) - 3P^2 L_{ss}\psi_m}{8JL_{ss}} i_{qs}^r \omega_r.$

Hence, $V\left(i_{qs}^r, i_{ds}^r, \omega_r\right) > 0$ and $\dfrac{dV\left(i_{qs}^r, i_{ds}^r, \omega_r\right)}{dt} < 0$. Therefore, the conditions for asymptotic

stability are guaranteed. The rate of decrease of $V\left(i_{qs}^r, i_{ds}^r, \omega_r\right)$, for example, $\dfrac{dV\left(i_{qs}^r, i_{ds}^r, \omega_r\right)}{dt}$,

affects the drive dynamics. The derived expression for dV/dt illustrates the role of the proportional feedback gain k_ω. ∎

We will study the relationship between the Lyapunov and Hamilton–Jacobi concepts. This analysis is very important because it integrates optimization, stability, system dynamics, system complexity, and other features.

Example 8.27: Hamilton–Jacobi and Lyapunov Theories in Control Design and Introduction to Minimal-Complexity Control

As the foundations of the Lyapunov stability theory were covered, we apply the Lyapunov concept to design the control algorithms for linear systems (8.14). The minimization of the positive-definite quadratic performance functional (8.15), using the first-order necessary condition for optimality (8.10), yields a linear control law (8.24) $u = -G^{-1}B^T Kx$. Using the control law derived and the feedback gain matrix $K_F = G^{-1}B^T K$, we rewrite the functional (8.15) as

$$J = \frac{1}{2}\int_{t_0}^{\infty} \left(x^T Qx + u^T Gu\right)dt = \frac{1}{2}\int_{t_0}^{\infty} \left(x^T Qx + x^T K_F^T GK_F x\right)dt = \frac{1}{2}\int_{t_0}^{\infty} x^T \left(Q + K_F^T GK_F\right)x\,dt.$$

Using the quadratic, positive-definite function (8.20) $V(x) = \frac{1}{2}x^T Kx$ for the system (8.14), one finds

$$\frac{dV(x)}{dt} = \frac{1}{2}\left(\dot{x}^T Kx + x^T K\dot{x}\right) = \frac{1}{2}\left[(Ax - BK_F x)^T Kx + x^T K(Ax - BK_F x)\right].$$

We express the total derivative of a positive-definite $V(x)$ as $\dfrac{dV(x)}{dt} = -x^T \left(Q + K_F^T GK_F\right)x$. One specifies the rate of change of the Lyapunov function $V(x)$. The positive semi-definite and positive-definite constant-coefficient weighting matrices Q and G are used. Hence, $\dfrac{dV(x)}{dt} < 0$, for example, the total derivative of $V(x)$ is negative definite. One has

$$\frac{1}{2}\left[(Ax - BK_Fx)^T Kx + x^T K(Ax - BK_Fx)\right] = -\frac{1}{2}x^T(Q + K_F^T GK_F)x.$$

Using $K_F = G^{-1}B^T K$, we conclude that the algebraic Riccati equation, which we need to solve to obtain the unknown matrix K, is $-Q - A^T K - K^T A + K^T BG^{-1}B^T K = 0$. This equation is in correspondence with (8.23), and the unknown positive-definite matrix K is symmetric, $K = K^T$.

From $V(x) = \frac{1}{2}x^T Kx$, one concludes that $V(x) > 0$, and the total derivative $dV(x)/dt < 0$. That is, the closed-loop system (8.14) through (8.24)

$$\dot{x}(t) = Ax + Bu = Ax - BG^{-1}B^T Kx = (A - BG^{-1}B^T K)x = (A - BK_F)x$$

is stable.

The comparison of the results, obtained using the Hamilton–Jacobi and Lyapunov theories, indicates that they are related and complement each other. The Lyapunov concept may be integrated with other analytic methods to analytically design the ontrol laws. Alternatively, control laws (8.6) $u = \phi(t, e, x)$ may be defined by the designer, and the feedback gains can be derived using the requirements imposed on the Lyapunov pair, for example, $V(x) > 0$ and $dV/dt < 0$.

The use of the Hamilton–Jacobi theory and many other optimization methods results in optimal control laws $u = \phi(t, e, x)$ designed minimizing the performance functionals. To implement these control laws, as analog or digital controllers, one needs all variables (e and x) to be measured or observed.

In many practical applications, the use of the Lyapunov theory could be a preferable choice. In particular, the designer may apply the Lyapunov concept if not all variables are measured, or one synthesizes the *minimal-complexity* control laws in order to accomplish one or all of the following features:

1. Use only the directly measured variables

2. Reduce the system complexity and simplify hardware solutions (minimizing the number of sensors, ICs, etc.)

3. Reduce software complexity relaxing the requirements on microcontrollers, DSPs, etc ■

8.13 Minimal-Complexity Control Laws Design

The need to design *minimal-complexity* control laws was emphasized. We study the motion control problem with the ultimate goal to synthesize the tracking control laws by applying Lyapunov's stability theory and utilizing analytic design methods. The control laws affect the system dynamics and change the total derivative of the Lyapunov function $V(t, x, e)$. For $V(t, x, e) > 0$, one can derive a control function u to guarantee $dV(t, x, e)/dt < 0$. The Lyapunov theory can be applied to derive the *admissible* control laws within the constrained control set $U = \{u \in \mathbb{R}^m : u_{\min} \leq u \leq u_{\max}, \, u_{\min} < 0, \, u_{\max} > 0\} \subset \mathbb{R}^m$. The feedback gains are found by solving nonlinear matrix inequality assigning (specifying) the negative value or rate of $dV(t, x, e)/dt$.

The *minimal-complexity* control implies the synthesis of control laws with a minimal or practical number of physical variables to implement u. The designer may utilize the tracking error $e(t)$, directly measured states $x_m(t)$, or sensed performance/capabilities quantities $q_s(t)$. This performance/capabilities vector $q_s(t)$ is usually a function of $e(t)$ and $x_m(t)$. The designer may also wish to simplify the hardware and software complexity by minimizing the number of sensors, ICs, etc. The problem is formulated as

Using system stability and optimality measures (system performance and capabilities are quantified by a Lyapunov pair which also defines stability), find the *admissible minimal-complexity* control law as a nonlinear function of error $e(t)$ and measured states $x_m(t)$

$$u = \phi(e, x_m) \tag{8.90}$$

subject to the system dynamics and constraints.

Control equation (8.90) can be rewritten as $u = \phi(e, x_m, q_s)$.

THEOREM 8.2

Consider a closed-loop system (8.8) with control laws (8.90) or (8.6) under the references $r \in R$ and disturbances $d \in D$. For the closed-loop system (8.8)–(8.90) or (8.8)–(8.6)

1. Solutions are uniformly ultimately bounded

2. Equilibrium is exponentially stable in the convex and compact state evolution set $X(X_0, U, R, D) \subset \mathbb{R}^n$

3. Tracking is ensured and disturbance attenuation is guaranteed in the state-error evolution set $XE(X_0, E_0, U, R, D) \subset \mathbb{R}^n \times \mathbb{R}^b$

if there exists a continuous differentiable function $V(t, x, e)$, $V(\cdot): \mathbb{R}_{\geq 0} \times \mathbb{R}^n \times \mathbb{R}^b \to \mathbb{R}_{\geq 0}$ in XE such that for all $x \in X$, $e \in E$, $u \in U$, $r \in R$, and $d \in D$ on $[t_0, \infty)$

a. $V(t, x, e) > 0$,

b. $\dfrac{dV(t, x, e)}{dt} \leq -\rho_1 \parallel x \parallel -\rho_2 \parallel e \parallel$ hold.

Here, $\rho_1(\cdot): \mathbb{R}_{\geq 0} \to \mathbb{R}_{\geq 0}$ and $\rho_2(\cdot): \mathbb{R}_{\geq 0} \to \mathbb{R}_{\geq 0}$ are the K-functions. ∎

The quadratic and nonquadratic Lyapunov candidates are applied. The criteria (a) and (b), imposed on the Lyapunov pair, are examined. Using the system dynamics, the total derivative of the Lyapunov candidate $V(t, x, e)$ is obtained. The inequality $\dfrac{dV(t, x, e)}{dt} \leq -\rho_1 \parallel x \parallel - \rho_2 \parallel e \parallel$ may be solved to find the feedback coefficients. The Lyapunov candidate functions should be designed. For example, the nonquadratic scalar Lyapunov function $V(x, e)$, $V(\cdot): \mathbb{R}^n \times \mathbb{R}^b \to \mathbb{R}_{\geq 0}$ is

$$V(x, e) = \sum_{i=1}^{\eta} \frac{1}{2i} \left(x^i\right)^T K_{xi} x^i + \sum_{i=1}^{\varsigma} \frac{1}{2i} \left(e^i\right)^T K_{ei} e^i, \quad K_{xi} \in \mathbb{R}^{n \times n}, \quad K_{ei} \in \mathbb{R}^{b \times b}, \quad \eta = 1, 2, 3, \ldots$$

$$\text{and} \quad \zeta = 1, 2, 3, \ldots .$$

Using the matrix-functions $K_{xi}(\cdot)$: $\mathbb{R}_{\geq 0} \to \mathbb{R}^{n \times n}$ and $K_{ei}(\cdot)$: $\mathbb{R}_{\geq 0} \to \mathbb{R}^{b \times b}$, the time-varying Lyapunov function $V(t, x, e)$, $V(\cdot)$: $\mathbb{R}_{\geq 0} \times \mathbb{R}^n \times \mathbb{R}^b \to \mathbb{R}_{\geq 0}$ can be given as

$$V(t, x, e) = \sum_{i=1}^{\eta} \frac{1}{2i} \left(x^i\right)^T K_{xi}(t) x^i + \sum_{i=1}^{\varsigma} \frac{1}{2i} \left(e^i\right)^T K_{ei}(t) e^i, \quad K_{xi}(\cdot):\mathbb{R}_{\geq 0} \to \mathbb{R}^{n \times n}$$

and $\quad K_{ei}(\cdot):\mathbb{R}_{\geq 0} \to \mathbb{R}^{b \times b}$.

The scalar Lyapunov function $V(t, x, e)$, $V(\cdot)$: $\mathbb{R}_{\geq 0} \times \mathbb{R}^n \times \mathbb{R}^b \to \mathbb{R}_{\geq 0}$ can be expressed as

$$V(t, x, e) = \sum_{i=1}^{\eta} \frac{1}{2i} \left(x^i\right)^T K_{xi}(t) x^i + \sum_{i=1}^{\lambda} \frac{1}{2i} \left(x^i\right)^T K_{xei}(t) e^i + \sum_{i=1}^{\varsigma} \frac{1}{2i} \left(e^i\right)^T K_{ei}(t) e^i,$$

$$\lambda = 1, 2, 3, \ldots, K_{xei}(\cdot):\mathbb{R}_{\geq 0} \to \mathbb{R}^{n \times b}.$$

The results of analytic design of control laws can be applied as reported in illustrative Examples. From (8.51) or (8.69), one may derive unconstrained control laws $u = f(t, e, x_m)$ or constrained control laws $u = \phi(t, e, x_m)$ using the directly measurable (or observable) x_m and e. For example, using (8.51), we have $u = -G_z^{-1} B_z^T(z) \dfrac{\partial V_m(x, e)}{\partial z}$ where $V_m(x, e)$ is designed to ensure $u = f(e, x_m)$. Thus, departing from (8.50), $u(t) = K_{F1} \begin{bmatrix} x(t) \\ e(t) \end{bmatrix} + \displaystyle\int K_{F2} \begin{bmatrix} x(\tau) \\ e(\tau) \end{bmatrix} d\tau$, using the concept reported, one has a linear PI tracking control law with nonlinear feedback mappings using $x_m(t)$ and $e(t)$

$$u(t) = K_{mF1} \begin{bmatrix} x_m(t) \\ e(t) \end{bmatrix} + \int K_{mF2} \begin{bmatrix} x_m(\tau) \\ e(\tau) \end{bmatrix} d\tau \quad \text{or} \quad u(t) = \phi\left(K_{mF1} \begin{bmatrix} x_m(t) \\ e(t) \end{bmatrix} + \int K_{mF2} \begin{bmatrix} x_m(\tau) \\ e(\tau) \end{bmatrix} d\tau \right).$$

These control laws utilize measurable $x_m(t)$ and $e(t)$. To study the closed-loop system stability, the closed-loop dynamics is examined using the Lyapunov function $V_m(\cdot)$: $\mathbb{R}^n \times \mathbb{R}^b \to \mathbb{R}_{\geq 0}$. The system performance and capabilities in the operating envelope $XE(X_0, E_0, U, R, D)$ are quantitatively and qualitatively quantified by $V(x, e)$ and $dV(x, e)/dt$ which correspond to the performance functionals.

Example 8.28:

Consider an electric drive with a permanent-magnet DC motor (30 V, 300 rad/s, $r_a = 2$ ohm, $k_a = 0.1$ V-s/rad, $L_a = 0.005$ H, $B_m = 0.0001$ N-m-s/rad, and $J = 0.0001$ kg-m^2) and a *step-down* converter as covered in Sections 4.1.3 and 7.2.2. Using the Kirchhoff law and the *averaging* concept, we have the following nonlinear state-space model:

$$\begin{bmatrix} \dfrac{du_a}{dt} \\[2mm] \dfrac{di_L}{dt} \\[2mm] \dfrac{di_a}{dt} \\[2mm] \dfrac{d\omega_r}{dt} \end{bmatrix} = \begin{bmatrix} 0 & \dfrac{1}{C} & -\dfrac{1}{C} & 0 \\[2mm] -\dfrac{1}{L} & -\dfrac{r_L + r_c}{L} & \dfrac{r_c}{L} & 0 \\[2mm] \dfrac{1}{L_a} & \dfrac{r_c}{L_a} & -\dfrac{r_a + r_c}{L_a} & -\dfrac{k_a}{L_a} \\[2mm] 0 & 0 & \dfrac{k_a}{J} & -\dfrac{B_m}{J} \end{bmatrix} \begin{bmatrix} u_a \\[2mm] i_L \\[2mm] i_a \\[2mm] \omega_r \end{bmatrix} + \begin{bmatrix} 0 \\[2mm] \dfrac{V_d}{L u_{t\,max}} - \dfrac{r_s}{L u_{t\,max}} i_L \\[2mm] 0 \\[2mm] 0 \end{bmatrix} u_c - \begin{bmatrix} 0 \\[2mm] 0 \\[2mm] 0 \\[2mm] \dfrac{1}{J} \end{bmatrix} T_L,$$

$u_c \in [0\ 10]$ V.

The positive-definite Lyapunov function is

$$V(x,e) = \frac{1}{2}[u_a \quad i_L \quad i_a \quad \omega_r] K_{x1} \begin{bmatrix} u_a \\ i_L \\ i_a \\ \omega_r \end{bmatrix} + \frac{1}{2} k_{e1} e^2 + \frac{1}{4} k_{e2} e^4, \quad \text{where } K_{x1} = I \in \mathbb{R}^{4 \times 4}.$$

Let the measured variables to be utilized by a *minimal-complexity* control law is the tracking error $e(t)$ and angular velocity $\omega_r(t)$. A bounded *minimal-complexity* control law is synthesized as

$$u = \text{sat}_0^{+10}\left(k_{p1} e + k_{p2} e^3 + k_{i1} \int e \, dt + k_{i2} \int e^3 \, dt - k_{14} \omega_r \right).$$

The criteria (a) and (b) of Theorem 8.2, imposed on the Lyapunov pair, are guaranteed to satisfy the stability conditions. The positive-definite nonquadratic Lyapunov function was used. The feedback gains are found by solving the inequality $\dfrac{dV(e,x)}{dt} \leq -\dfrac{1}{2} \| x \|^2 - \dfrac{1}{2} \| e \|^2 - \dfrac{1}{4} \| e \|^4$ yielding $k_{p1} = 1.8$, $k_{p2} = 0.25$, $k_{i1} = 7.3$, $k_{i2} = 0.92$, and $k_{14} = 0.085$. The $dV(x,e)/dt$ quantitatively and qualitatively imposes the specifications on stability, efficiency (minimizing the current thereby reducing the losses), tracking error (e should converge to zero for $r = $ constant), etc.

The designed *minimal-complexity* control law is verified and experiments are performed. Different operating conditions are studied to analyze the dynamic performance. The system performance degrades at high temperature. For $T = 140°C$ (the maximum operating temperature), the resistance of the armature winding reaches the maximum value, and the *torque constant* k_a is minimum. The disturbance attenuation features are also of our interest because the angular velocity is required to remain equal to the reference value if the load torque T_L is applied. Figure 8.31 depicts the measured transient dynamics for control $u_c(t)$ and states $u_C(t)$, $i_a(t)$, $\omega_r(t)$ for the closed-loop system when $r = \omega_{reference} = 255$ rad/s. Here, though $u_C(t)$ and $i_a(t)$ are measured, in practical applications, one may simplify the system complexity avoiding to use these variables. A motor reaches the desired (reference) angular velocity within 0.05 s with overshoot of 4.5%, and the steady-state error $e(t)$ is zero. The disturbance attenuation is studied. The load torque 0.075 N-m is applied at $t = 0.073$ s. Tracking accuracy and disturbance attenuation are achieved, see the evolution of the angular velocity illustrated in Figure 8.31. By analyzing the motor angular velocity as the load applied, one concludes that the settling time is 0.01 s with 2.5% deflection from $\omega_{reference}$, and the steady-state error is zero. From the experimental data reported when motor temperature is 140°C, one concludes that the desired performance and capabilities have been achieved.

We compare the designed *minimal-complexity* control law with the conventional linear PI control law. A high frequency (36 kHZ with 2.5 kHz bandwidth) PWM *servo-amplifier* 25A8 (Advanced Motion Controls) is used to implement a linear PI control $u = 1.8e + 7.3 \int e \, dt$. The experimental results are illustrated in Figure 8.32 (motor starts from stall, and the load torque is applied at 0.08 s). The settling time is 0.075 s with overshoot, 5.2%. As the load torque of 0.075 N-m is applied at 0.08 s, the steady-state error of 1.9% results, and the settling time is 0.01 s with 3.4% maximum deflection from the assigned $\omega_{reference}$. The analysis of two experimentally examined control algorithms indicates that the acceleration rate remains the same. In particular, the motor reaches the maximum angular velocity at $\sim$0.04 s when the nonlinear and linear PI control laws are used. This can be easily justified because the maximum voltage is applied to the armature winding as the motor starts, for example, as the motor accelerates, u_a is 30 V and control u_c is 10 V. It is virtually impossible to refine the dynamics as the system is saturated to the maximum control. However, as the system reaches operating conditions where saturation is not a factor, the documented experimental results illustrate that *admissible minimal-complexity* control law improves the closed-loop dynamic and steady-state performance due to the use of nonlinear proportional and integral error feedback. We conclude that system dynamics,

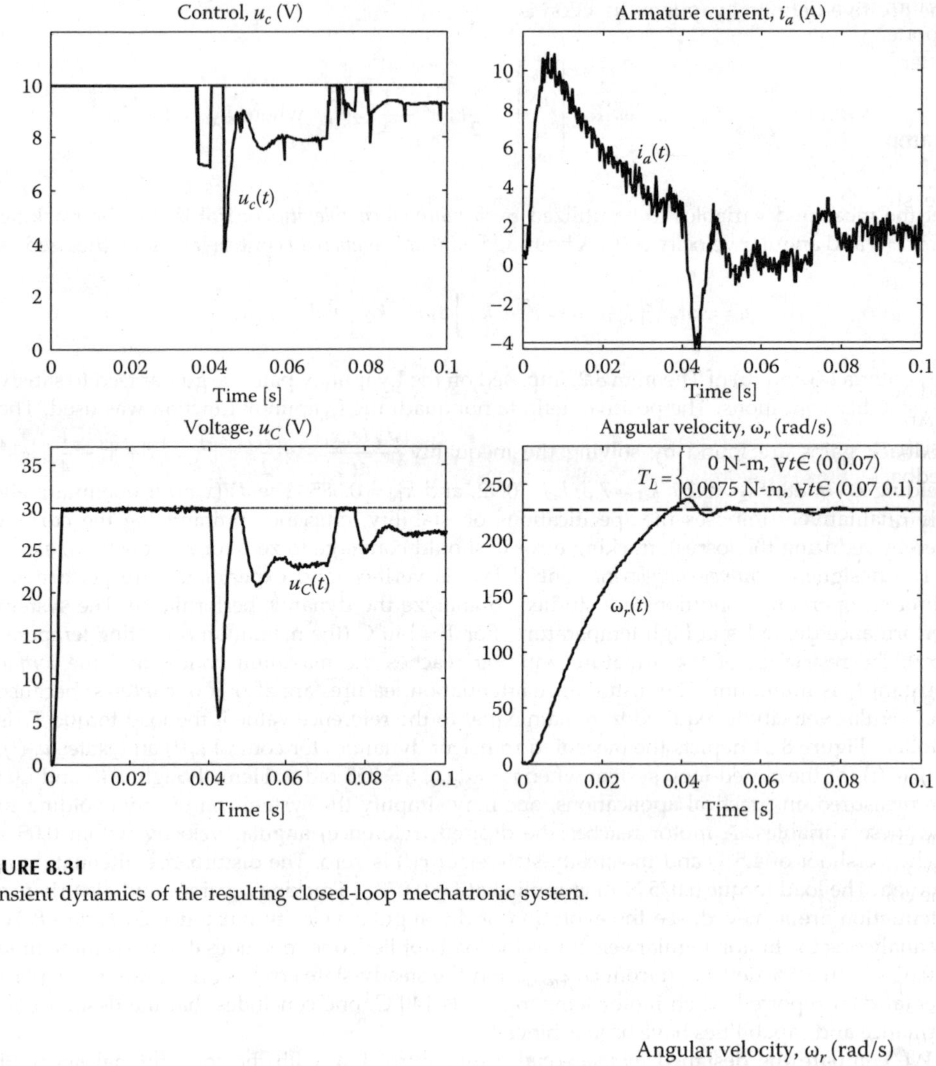

FIGURE 8.31
Transient dynamics of the resulting closed-loop mechatronic system.

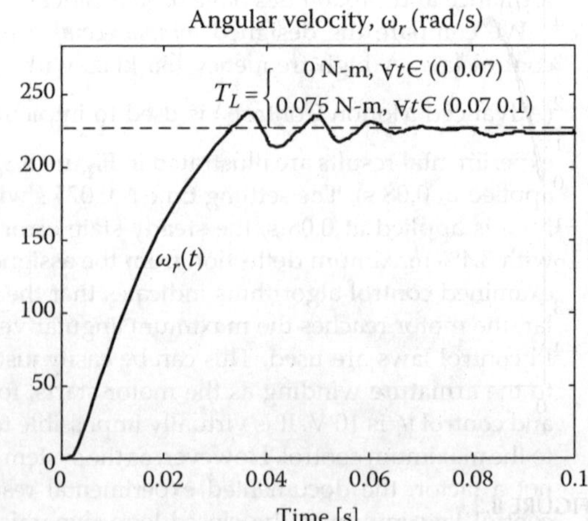

FIGURE 8.32
Dynamics of the closed-loop mechatronic system with
the 25A8 servo-amplifier.

performance, and capabilities can be improved due to the use of nonlinear feedback. The reported analytical and experimental results illustrate that nonlinear control laws guarantee better dynamic responses, precise tracking, disturbance attenuation, robustness, stability, etc. ∎

Example 8.29:

We study the eight-layered lead magnesium niobate actuator (3 mm diameter, 0.25 mm thickness). A set of differential equations to model the actuator dynamics is

$$\frac{dF}{dt} = -8500F + 14Fu + 450u, \qquad \frac{dv}{dt} = 1{,}000F - 100{,}000v - 2{,}500v^3 - 2{,}750x, \qquad \frac{dx}{dt} = v.$$

The control is bounded, and $-100 \leq u \leq 100$ V. The error is the difference between the reference and actuator linear displacements. That is, $e(t) = r(t) - y(t)$, where $y(t) = x(t)$.

An *admissible minimal-complexity* control law is synthesized using nonlinear error feedback. Using the criteria imposed on the Lyapunov pair $V(x,e) > 0$ and $dV/dt < 0$, the normalized feedback gains (using the displacement and error in micrometers, not in meters) are found by solving the inequality $\dfrac{dV(e,x)}{dt} \leq - \| e \|^2 - \| e \|^4 - \| x \|^2$. Here,

$$V(x,e) = \frac{1}{2}[F \ v \ x]K_{x1}\begin{bmatrix} F \\ v \\ x \end{bmatrix} + \frac{1}{2}k_{e1}e^2 + \frac{1}{4}k_{e2}e^4, \quad K_{x1} = I \in \mathbb{R}^{3\times3}.$$

We have

$$u = \mathrm{sat}_{-100}^{+100}\left(0.95e + 0.26e^3 + 0.045\int e\,dt + 0.0084\int e^3\,dt\right).$$

The criteria (a) and (b) imposed on the Lyapunov pair are satisfied. Hence, the bounded control law guarantees stability, ensures tracking, and guarantees the desired negativeness of dV/dt. The control law is experimentally tested. Figure 8.33 illustrates the transient dynamics for $x(t)$ if

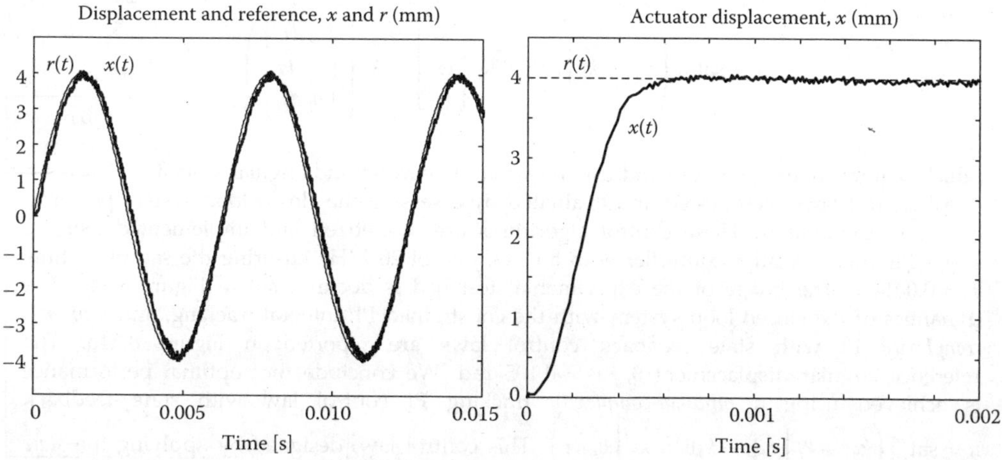

FIGURE 8.33
Output dynamics if $r(t) = 4 \times 10^{-6}\sin 1000t$ and $r(t) = \mathrm{const} = 4 \times 10^{-6}$ m.

the reference signal (desired position) is $r(t) = 4 \times 10^{-6}\sin 1000t$ and $r(t) = \text{constant} = 4 \times 10^{-6}$ m. We conclude that the stability is guaranteed, desired performance is achieved, and the output precisely follows the reference $r(t)$. ∎

Example 8.30:

We design and implement a closed-loop system for an axial-topology hard drive. The requirements are to guarantee accuracy, ensure fast repositioning, eliminate steady-state positional error (tracking error), attenuate disturbances, and minimize the settling time. The angular displacement θ_r is measured by the high-accuracy sensor, while the angular velocity ω_r can be estimated using an *observer*. The actuator parameters are $N = 108$, $r_a = 6.4$ ohm, $L_a = 3.32 \times 10^{-5}$ H, $B_{max} = 0.4$ T, $B_m = 5 \times 10^{-8}$ N-m-s/rad, and $J = 1.4 \times 10^{-6}$ kg-m^2 (the equivalent moment of inertia integrates the inertia of the sensor).

Using the tracking error $e(t)$, constrained tracking control laws are designed, examined, and tested. One has the following PI control law

$$u = \text{sat}_{-1}^{+1}\left(k_p e + k_i \int e\, d\tau\right) \text{ with } k_p = 20 \text{ and } k_i = 21.$$

The integral tracking control law (8.36) with control bounds is designed. We have

$$u = \text{sat}_{-1}^{+1}\left(-k_1 i_a - k_2 \omega_r - k_3 \theta_r + k_i \int e\, d\tau\right),\ k_1 = 0.47,\ k_2 = 0.02,\ k_3 = 2.8 \text{ and } k_i = 21.$$

The constrained *minimal-complexity* tracking PI control law with state feedback is designed using the *state transformation* method applying the Lyapunov theory. In particular, we have

$$u = \text{sat}_{-1}^{+1}\left(k_p e + k_i \int e\, d\tau - k_4 \theta_r - k_5 \int \theta_r\, d\tau\right),\ k_4 = 2.8,\ k_5 = 1.5,\ k_p = 20, \text{ and } k_{i,} = 21.$$

It should be emphasized that the concept reported in Section 8.7 results in control

$$u = \text{sat}_{-1}^{+1}\left(k_p e + k_i \int e\, d\tau + K_p \begin{bmatrix} i_a \\ \omega_r \\ \theta_r \end{bmatrix} + K_i \begin{bmatrix} \int i_a\, d\tau \\ \int \omega_r\, d\tau \\ \int \theta_r\, d\tau \end{bmatrix}\right)$$

which requires one to measure and use the armature current and angular velocity.

All control laws were tested and evaluated by assessing the closed-loop system performance and capabilities. These control algorithms are discretized and implemented using a 16-bit PIC16F877A microcontroller with a clock rate of 20 MHz ensuring the sampling time $T_s = 0.0014$ s. The image of the experimental test bed is documented in Figure 8.34a. The dynamics of the closed-loop system with the constrained PI, integral tracking, and *minimal-complexity* PI with state feedback control laws are reported in Figure 8.34b. The reference angular displacement θ_{ref} is ~ 0.105 rad. We conclude that optimal performance is achieved using a *minimal-complexity* tracking PI control law with state feedback $u = \text{sat}_{-1}^{+1}\left(k_p e + k_i \int e\, d\tau - k_4 \theta_r - k_5 \int \theta_r\, d\tau\right)$. This control law, designed by applying the *state transformation* and Lyapunov concepts, ensures highest accuracy, fastest repositioning and settling time, disturbance attenuation, and minimal overshoot. ∎

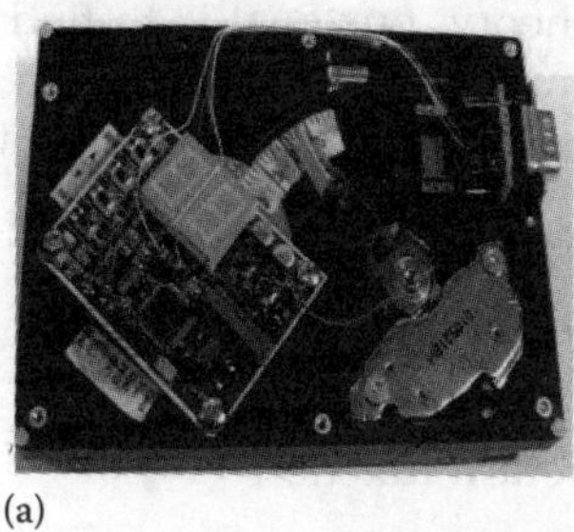

(a)

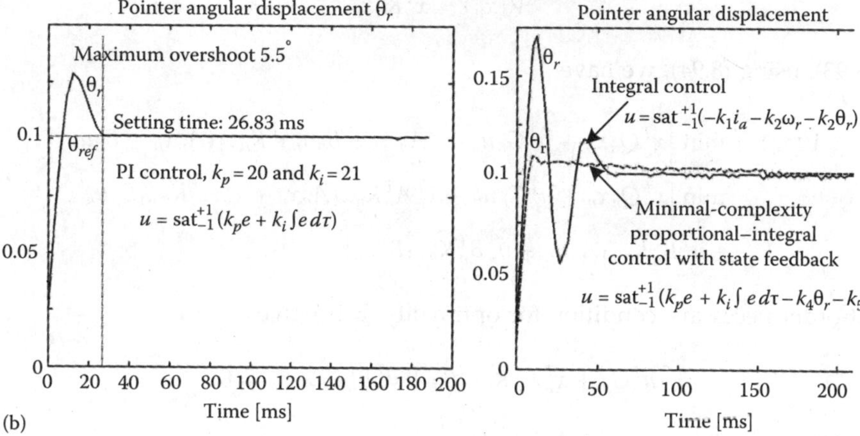

FIGURE 8.34
(a) Closed-loop servo with axial topology actuator, sensors, ICs, and microcontroller; (b) Transient dynamics of the closed-loop servo with the bounded PI $u = \mathrm{sat}_{-1}^{+1}\left(k_p e + k_i \int e\,d\tau\right)$, integral $u = \mathrm{sat}_{-1}^{+1}\left(-k_1 i_a - k_2 \omega_r - k_3 \theta_r + k_i \int e\,d\tau\right)$, and *minimal-complexity* tracking PI control law with state feedback $u = \mathrm{sat}_{-1}^{+1}\left(k_p e + k_i \int e\,d\tau - k_4 \theta_r - k_5 \int \theta_r\,d\tau\right)$.

8.14 Control of Linear Discrete-Time Systems Using the Hamilton–Jacobi Theory

8.14.1 Linear Discrete-Time Systems

Consider a discrete-time system described by the state-space difference equation

$$x_{n+1} = A_n x_n + B_n u_n, \quad n \geq 0. \tag{8.91}$$

The notations for discrete-time case are similar for those that were used for continuous-time systems. Different performance indexes are applied to optimize the closed-loop system dynamics. For example, the quadratic performance index to be minimized is

$$J = \sum_{n=0}^{N-1} \left[x_n^T Q_n x_n + u_n^T G_n u_n \right], \quad Q_n \geq 0, \quad G_n > 0. \tag{8.92}$$

Using the Hamilton–Jacobi theory, our goal is to find the control law which guarantees that the value of the performance index is minimum or maximum. For linear dynamic systems (8.91) and quadratic performance index (8.92), the solution of the Hamilton–Jacobi equation

$$V(x_n) = \min_{u_n} \left[x_n^T Q_n x_n + u_n^T G_n u_n + V(x_{n+1}) \right] \tag{8.93}$$

is satisfied by the quadratic return function

$$V(x_n) = x_n^T K_n x_n. \tag{8.94}$$

From (8.93), using (8.94), we have

$$
\begin{aligned}
V(x_n) &= \min_{u_n} \left[x_n^T Q_n x_n + u_n^T G_n u_n + (A_n x_n + B_n u_n)^T K_{n+1}(A_n x_n + B_n u_n) \right] \\
&= \min_{u_n} \left[x_n^T Q_n x_n + u_n^T G_n u_n + x_n^T A_n^T K_{n+1} A_n x_n + x_n^T A_n^T K_{n+1} B_n u_n \right. \\
&\quad \left. + u_n^T B_n^T K_{n+1} A_n x_n + u_n^T B_n^T K_{n+1} B_n u_n \right].
\end{aligned}
\tag{8.95}
$$

The first-order necessary condition for optimality (8.10) gives

$$u_n^T G_n + x_n^T A_n^T K_{n+1} B_n + u_n^T B_n^T K_{n+1} B_n = 0. \tag{8.96}$$

From (8.96), the digital control law is

$$u_n = -\left(G_n + B_n^T K_{n+1} B_n \right)^{-1} B_n^T K_{n+1} A_n x_n. \tag{8.97}$$

The second-order necessary condition for optimality (8.13) is guaranteed because

$$\frac{\partial^2 H(x_n, u_n, V(x_{n+1}))}{\partial u_n \times \partial u_n^T} = \frac{\partial^2 \left(u_n^T G_n u_n + u_n^T B_n^T K_{n+1} B_n u_n \right)}{\partial u_n \times \partial u_n^T} = 2G_n + 2B_n^T K_{n+1} B_n > 0, \quad K_{n+1} > 0.$$

Using the control law derived as (8.97), from (8.95), one finds

$$x_n^T K_n x_n = x_n^T Q_n x_n + x_n^T A_n^T K_{n+1} A_n x_n - x_n^T A_n^T K_{n+1} B_n \left(G_n + B_n^T K_{n+1} B_n \right)^{-1} B_n K_{n+1} A_n x. \tag{8.98}$$

From (8.98), the difference equation to obtain the unknown symmetric matrix of the quadratic return function is found to be

$$K_n = Q_n + A_n^T K_{n+1} A_n - A_n^T K_{n+1} B_n \left(G_n + B_n^T K_{n+1} B_n \right)^{-1} B_n K_{n+1} A_n. \tag{8.99}$$

If in performance index (8.92) $N = \infty$, we have

$$J = \sum_{n=0}^{\infty} \left[x_n^T Q_n x_n + u_n^T G_n u_n \right], \quad Q_n \geq 0, \quad G_n > 0. \tag{8.100}$$

The optimal control law is

$$u_n = -\left(G_n + B_n^T K_n B_n\right)^{-1} B_n^T K_n A_n x_n, \tag{8.101}$$

where the unknown symmetric matrix K_n is found by solving the nonlinear equation

$$-K_n + Q_n + A_n^T K_n A_n - A_n^T K_n B_n \left(G_n + B_n^T K_n B_n\right)^{-1} B_n K_n A_n = 0, \quad K_n = K_n^T. \tag{8.102}$$

Matrix K_n is positive-definite, and the MATLAB $\texttt{dlqr}$ command can be used to derive the matrix K_n, feedback matrix $\left(G_n + B_n^T K_n B_n\right)^{-1} B_n^T K_n A_n$ and eigenvalues. The closed-loop systems (8.91) with (8.101) is expressed as

$$x_{n+1} = \left[A_n - B_n\left(G_n + B_n^T K_{n+1} B_n\right)^{-1} B_n^T K_{n+1} A_n\right] x_n. \tag{8.103}$$

Example 8.31:

For the second-order discrete-time system

$$x_{n+1} = \begin{bmatrix} x_{1n+1} \\ x_{2n+1} \end{bmatrix} = A_n x_n + B_n u_n = \begin{bmatrix} 1 & 2 \\ 0 & 3 \end{bmatrix} \begin{bmatrix} x_{1n} \\ x_{2n} \end{bmatrix} + \begin{bmatrix} 4 & 5 \\ 6 & 7 \end{bmatrix} \begin{bmatrix} u_{1n} \\ u_{2n} \end{bmatrix},$$

we find the digital control law by minimizing the performance index

$$J = \sum_{n=0}^{\infty} \left[x_n^T Q_n x_n + u_n^T G_n u_n\right] = \sum_{n=0}^{\infty} \left[\begin{bmatrix} x_{1n} & x_{2n} \end{bmatrix} \begin{bmatrix} 10 & 0 \\ 0 & 10 \end{bmatrix} \begin{bmatrix} x_{1n} \\ x_{2n} \end{bmatrix} + \begin{bmatrix} u_{1n} & u_{2n} \end{bmatrix} \begin{bmatrix} 5 & 0 \\ 0 & 5 \end{bmatrix} \begin{bmatrix} u_{1n} \\ u_{2n} \end{bmatrix}\right]$$

$$= \sum_{n=0}^{\infty} \left(10x_{1n}^2 + 10x_{2n}^2 + 5u_{1n}^2 + 5u_{2n}^2\right).$$

Using the $\texttt{dlqr}$ command, we have

```
A= [1 2;0 3];B= [4 5;6 7]; Q=10*eye(size(A)); G=5*eye(size(B));
[Kfeedback,Kn,Eigenvalues] =dlqr(A,B,Q,G)
```

One obtains $K_n = \begin{bmatrix} 20 & -0.71 \\ -0.71 & 10.6 \end{bmatrix}$, and the control law is

$$u_n = -\left(G_n + B_n^T K_{n+1} B_n\right)^{-1} B_n^T K_n A_n x_n = -\begin{bmatrix} -0.285 & 0.231 \\ 0.332 & 0.222 \end{bmatrix} \begin{bmatrix} x_{1n} \\ x_{2n} \end{bmatrix},$$

$$u_{1n} = 0.285x_{1n} - 0.231x_{2n}, \quad u_{2n} = -0.332x_{1n} - 0.222x_{2n}.$$

The system is stable because the eigenvalues are within the unit circle. In particular, the eigenvalues are 0.522 and 0.0159. ∎

Example 8.32:

For the third-order system $x_{n+1} = \begin{bmatrix} x_{1n+1} \\ x_{2n+1} \\ x_{3n+1} \end{bmatrix} = A_n x_n + B_n u_n = \begin{bmatrix} 1 & 1 & 2 \\ 3 & 3 & 4 \\ 5 & 5 & 6 \end{bmatrix} \begin{bmatrix} x_{1n} \\ x_{2n} \\ x_{2n} \end{bmatrix} + \begin{bmatrix} 10 \\ 20 \\ 30 \end{bmatrix} u_n,$

we will find a control law by minimizing the quadratic performance index.

$$J = \sum_{n=0}^{\infty} \left[x_n^T Q_n x_n + u_n^T G_n u_n \right] = \sum_{n=0}^{\infty} \left[\begin{bmatrix} x_{1n} & x_{2n} & x_{3n} \end{bmatrix} \begin{bmatrix} 1 & 0 & 0 \\ 0 & 10 & 0 \\ 0 & 0 & 100 \end{bmatrix} \begin{bmatrix} x_{1n} \\ x_{2n} \\ x_{3n} \end{bmatrix} + 1000 u_n^2 \right]$$

$$= \sum_{n=0}^{\infty} \left(x_{1n}^2 + 10 x_{2n}^2 + 100 x_{3n}^2 + 1000 u_n^2 \right).$$

Our goal is also to simulate the closed-loop system. The matrices of the output equation

$$y_n = C_n x_n + H_n u_n,$$

are $C_n = [1\ 1\ 1]$ and $H_n = [0]$.
    We have

```
A=[1 1 2; 3 3 4; 5 5 6]; B=[10; 20; 30];
Q=eye(size(A)); Q(2,2)=10; Q(3,3)=100; G=1000;
[Kfeedback,Kn,Eigenvalies]=dlqr(A,B,Q,G)
```

One finds $K_n = \begin{bmatrix} 48.4 & 47.4 & 31.1 \\ 47.4 & 57.4 & 31.1 \\ 31.1 & 31.1 & 140 \end{bmatrix}$, and the control law is

$$u_n = -0.155 x_{1n} - 0.155 x_{2n} - 0.2 x_{3n}.$$

The dynamics of the closed-loop system, which is stable because the eigenvalues are within the unit circle, is simulated using the filter command. Having derived the closed-loop system dynamics as $x_{n+1} = \left[A_n - B_n \left(G_n + B_n^T K_{n+1} B_n \right)^{-1} B_n^T K_n A_n \right] x_n,$ one finds the numerator and denominator of the transfer function in the z-domain. The MATLAB statement is

```
A_closed=A-B*Kfeedback;C=[1 1 1]; H=[0];
[num,den]=ss2tf(A_closed,B,C,H);
x0=[10 0 -10];
k=0:1:20; u=1*[ones(1,21)];
x=filter(num,den,u,x0);
plot(k,x,'-',k,x,'o','LineWidth',2);
title('System Dynamics, x(k)','FontSize',16);
xlabel('Discrete Time, k','FontSize',16);
```

The simulation results are illustrated in Figure 8.35.
    The stabilization problem was solved, and the closed-loop system is stable. The tracking control problem must be approached and solved to guarantee that the system output follows the reference. ∎

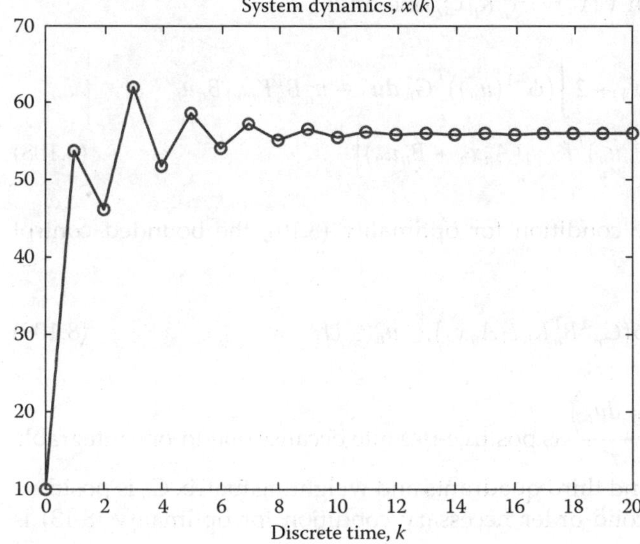

FIGURE 8.35
Output dynamics of the closed-loop system.

8.14.2 Constrained Optimization of Discrete-Time Systems

Due to the constraints imposed on controls $u_{nmin} \leq u_n \leq u_{nmax}$, the designer must synthesize bounded control laws. Our goal is to analytically design the constrained digital control laws using the Hamilton–Jacobi theory. The system is described by the difference equation

$$x_{n+1} = A_n x_n + B_n u_n, \quad x_{n0} \in X_0, \quad u_n \in U, \quad n \geq 0, \tag{8.104}$$

and the control constraints are described by continuous integrable one-to-one bounded functions ϕ, $\phi \in U$, for which the inverse function ϕ^{-1} exist.

The nonquadratic performance index to be minimized is

$$J = \sum_{n=0}^{N-1} \left[x_n^T Q_n x_n + 2 \int \left(\phi^{-1}(u_n) \right)^T G_n du_n - u_n^T B_n^T K_{n+1} B_n u_n \right]. \tag{8.105}$$

Performance indexes must be positive-definite. Hence, to attain the positive-definiteness,

$$\left[x_n^T Q_n x_n + 2 \int \left(\phi^{-1}(u_n) \right)^T G_n \, du_n \right] > u_n^T B_n^T K_{n+1} B_n u_n, \quad \text{for all } x_n \in X \quad \text{and} \quad u_n \in U. \tag{8.106}$$

The Hamilton–Jacobi recursive equation is

$$V(x_n) = \min_{u_n \in U} \left[x_n^T Q_n x_n + 2 \int \left(\phi^{-1}(u_n) \right)^T G_n \, du_n - u_n^T B_n^T K_{n+1} B_n u_n + V(x_{n+1}) \right]. \tag{8.107}$$

Using the quadratic return function $V(x_n) = x_n^T K_n x_n$, one finds

$$x_n^T K_n x_n = \min_{u_n \in U} \left[x_n^T Q_n x_n + 2 \int \left(\phi^{-1}(u_n) \right)^T G_n \, du_n - u_n^T B_n^T K_{n+1} B_n u_n \right.$$
$$\left. + (A_n x_n + B_n u_n)^T K_{n+1} (A_n x_n + B_n u_n) \right]. \tag{8.108}$$

Applying the first-order necessary condition for optimality (8.10), the bounded control law is

$$u_n = -\phi \left(G_n^{-1} B_n^T K_{n+1} A_n x_n \right), \quad u_n \in U. \tag{8.109}$$

It is evident that $\dfrac{\partial^2 \left(2 \int \left(\phi^{-1}(u_n) \right)^T G_n \, du_n \right)}{\partial u_n \times \partial u_n^T}$ is positive-definite because one-to-one integrable functions ϕ and ϕ^{-1} lie in the first and third quadrants and weighting matrix G_n is positive-definite. We conclude that the second-order necessary condition for optimality (8.13) is satisfied.

From (8.108) and (8.109), we have

$$x_n^T K_n x_n = x_n^T Q_n x_n + 2 \int x_n^T A_n^T K_{n+1} B_n d \left(\phi \left(G_n^{-1} B_n^T K_{n+1} A_n x_n \right) \right) + x_n^T A_n^T K_{n+1} A_n x_n$$
$$- 2 x_n^T A_n^T K_{n+1} B_n \phi \left(G_n^{-1} B_n^T K_{n+1} A_n x_n \right). \tag{8.110}$$

The second term of the right-hand side of (8.110) is

$$2 \int x_n^T A_n^T K_{n+1} B_n d \left(\phi \left(G_n^{-1} B_n^T K_{n+1} A_n x_n \right) \right)$$
$$= 2 x_n^T A_n^T K_{n+1} B_n \phi \left(G_n^{-1} B_n^T K_{n+1} A_n x_n \right) - 2 \int \left(\phi \left(G_n^{-1} B_n^T K_{n+1} A_n x_n \right) \right)^T d \left(B_n^T K_{n+1} A_n x_n \right). \tag{8.111}$$

From (8.110), using (8.111), one concludes that the unknown matrix K_{n+1} is found by solving

$$x_n^T K_n x_n = x_n^T Q_n x_n + x_n^T A_n^T K_{n+1} A_n x_n - 2 \int \left(\phi \left(G_n^{-1} B_n^T K_{n+1} A_n x_n \right) \right)^T d \left(B_n^T K_{n+1} A_n x_n \right). \tag{8.112}$$

Describing the control bounds imposed by using the continuous integrable one-to-one bounded functions $\phi \in U$, one finds the expression for $2 \int \left(\phi \left(G_n^{-1} B_n^T K_{n+1} A_n x_n \right) \right)^T d \left(B_n^T K_{n+1} A_n x_n \right)$. For example, using the tanh function to describe the saturation-type constraints, one obtains

$$\int \tanh z \, dz = \log \cosh z \quad \text{and} \quad \int \tanh^g z \, dz = -\frac{\tanh^{g-1} z}{g-1} + \int \tanh^{g-2} z \, dz, \quad g \neq 1.$$

Matrix K_{n+1} is found by solving the recursive (8.112), and the feedback gains result.

Minimizing the nonquadratic performance index (8.105) for $N = \infty$, we obtain the bounded control law as

$$u_n = -\phi\left(G_n^{-1}B_n^T K_n A_n x_n\right), \quad u_n \in U. \tag{8.113}$$

The *admissibility* concept, which is based on the Lyapunov stability theory, is applied to verify the stability of the resulting closed-loop system in the operating envelope $x_n \in X$ and $u_n \in U$. This problem is of a particular importance for open-loop unstable systems (8.104). The resulting closed-loop system (8.104) through (8.113) evolves in X. A subset of the admissible domain of stability $S \subset \mathbb{R}^n$ is found by using the Lyapunov stability theory as

$$S = \{x_n \in R^n : x_{n0} \in X_0,\ u_n \in U | V(0) = 0,\ V(x_n) > 0,\ \Delta V(x_n) < 0,\ \forall x_n \in X(X_0, U)\}.$$

The region of attraction can be studied, and S is an *invariant* domain. The quadratic Lyapunov function $V(x_n) = x_n^T K_n x_n$ is positive-definite if $K_n > 0$. Hence, the first difference, as given by

$$\Delta V(x_n) = V(x_{n+1}) - V(x_n) = x_n^T A_n^T K_{n+1} A_n x_n - 2x_n^T A_n^T K_{n+1} B_n \phi\left(G_n^{-1}B_n^T K_{n+1} A_n x_n\right)$$
$$+ \phi\left(G_n^{-1}B_n^T K_{n+1} A_n x_n\right)^T B_n^T K_{n+1} B_n \phi\left(G_n^{-1}B_n^T K_{n+1} A_n x_n\right) - x_n^T K_n x_n, \tag{8.114}$$

must be negative-definite for all $x_n \in X$, to ensure the stability. The evolution of the closed-loop systems depends on the initial conditions, constraints, references, disturbances, parameter variations, etc. For the initial conditions within the operating envelope and control, for a closed-loop system we have the evolution set $X(X_0, U)$. The *sufficiency* analysis of stability is performed studying the sets $S \subset \mathbb{R}^n$ and $X(X_0, U) \subset \mathbb{R}^n$. Stability is guaranteed if $X \subseteq S$.

The constrained optimization problem must be solved for nonlinear systems. Nonlinear discrete-time electromechanical systems are described as

$$x_{n+1} = F(x_n) + B(x_n)u_n, \quad x_{n0} \in X_0, \quad u_n \in U, \quad u_{n\min} \leq u_n \leq u_{n\max}, \quad n \geq 0 \tag{8.115}$$

The Hamilton–Jacobi theory is applied to design bounded control laws using nonquadratic performance indexes. To design a nonlinear *admissible* control law $u_n \in U$, we describe the imposed control bounds by a continuous integrable one-to-one bounded function $\phi \in U$. The nonquadratic performance index is

$$J = \sum_{n=0}^{N-1} \left[x_n^T Q_n x_n - u_n^T B^T(x_n) K_{n+1} B(x_n) u_n + 2 \int \left(\phi^{-1}(u_n)\right)^T G_n\, du_n \right]. \tag{8.116}$$

The integrand $2 \int \left(\phi^{-1}(u_n)\right)^T G_n\, du_n$ is positive-definite because the integrable one-to-one function ϕ lies in the first and third quadrants, integrable function ϕ^{-1} exists, and $G_n > 0$. The positive definiteness of the performance index is guaranteed if

$$x_n^T Q_n x_n + 2 \int \left(\phi^{-1}(u_n)\right)^T G_n\, du_n > u_n^T B^T(x_n) K_{n+1} B(x_n) u_n \quad \text{for all } x_n \in X \quad \text{and} \quad u_n \in U$$

$$\tag{8.117}$$

The positive definiteness of the performance index in $x_n \in X$ and $u_n \in U$ can be studied as the positive-definite symmetric matrix K_{n+1} is found. The inequality (8.117) is ensured by using Q_n and G_n. The first- and second-order necessary conditions for optimality are applied. For the quadratic return function (8.94), we have

$$V(x_{n+1}) = x_{n+1}^T K_{n+1} x_{n+1} = (F(x_n) + B(x_n)u_n)^T K_{n+1}(F(x_n) + B(x_n)u_n). \tag{8.118}$$

Therefore,

$$x_n^T K_n x_n = \min_{u_n \in U} \left[x_n^T Q_n x_n - u_n^T B^T(x_n) K_{n+1} B(x_n) u_n + 2 \int \left(\phi^{-1}(u_n) \right)^T G_n \, du_n \right.$$
$$\left. + (F(x_n) + B(x_n)u_n)^T K_{n+1}(F(x_n) + B(x_n)u_n) \right]. \tag{8.119}$$

Using the first-order necessary condition for optimality (8.10), a bounded control law is

$$u_n = -\phi\left(G_n^{-1} B^T(x_n) K_{n+1} F(x_n) \right), \quad u_n \in U. \tag{8.120}$$

The second-order necessary condition for optimality (8.13) is satisfied because $\dfrac{\partial^2 \left(2 \int \left(\phi^{-1}(u_n) \right)^T G_n \, du_n \right)}{\partial u_n \times \partial u_n^T} > 0$. From (8.119), using the bounded control law (8.120), we have the following recursive equation:

$$x_n^T K_n x_n = x_n^T Q_n x_n + 2 \int F^T(x_n) K_{n+1} B(x_n) d\left(\phi\left(G_n^{-1} B^T(x_n) K_{n+1} F(x_n) \right) \right) + F^T(x_n) K_{n+1} F(x_n)$$
$$- 2F^T(x_n) K_{n+1} B(x_n) \phi\left(G_n^{-1} B^T(x_n) K_{n+1} F(x_n) \right). \tag{8.121}$$

The integration by parts gives

$$2 \int F^T(x_n) K_{n+1} B(x_n) d\left(\phi\left(G_n^{-1} B^T(x_n) K_{n+1} F(x_n) \right) \right)$$
$$= 2F^T(x_n) K_{n+1} B(x_n) \phi\left(G_n^{-1} B^T(x_n) K_{n+1} F(x_n) \right)$$
$$- 2 \int \left(\phi\left(G_n^{-1} B^T(x_n) K_{n+1} F(x_n) \right) \right)^T$$
$$d\left(B^T(x_n) K_{n+1} F(x_n) \right).$$

The equation to find the unknown symmetric matrix K_{n+1} is

$$x_n^T K_n x_n = x_n^T Q_n x_n + F^T(x_n) K_{n+1} F(x_n) - 2 \int \left(\phi\left(G_n^{-1} B^T(x_n) K_{n+1} F(x_n) \right) \right)^T d\left(B^T(x_n) K_{n+1} F(x_n) \right)$$
$$\tag{8.122}$$

By describing the control bounds imposed by the continuous integrable one-to-one bounded functions $\phi \in U$, one finds $2 \int \left(\phi\left(G_n^{-1} B^T(x_n) K_{n+1} F(x_n) \right) \right)^T d\left(B^T(x_n) K_{n+1} F(x_n) \right)$. Hence, (8.122) can be solved.

The *admissibility* concept is applied to verify the stability of the resulting closed-loop system. The closed-loop system evolves in $X \subset \mathbb{R}^c$, and

$$\left\{ x_{n+1} = F(x_n) - B(x_n)\phi\left(G_n^{-1}B^T(x_n)K_{n+1}F(x_n)\right), \ x_{n0} \in X_0 \right\} \in X(X_0, U).$$

Using the Lyapunov stability theory, the domain of stability $S \subset \mathbb{R}^n$ is found by applying the sufficient conditions under which the discrete-time closed-loop system (8.115) through (8.120) is stable. The positive-definite quadratic function (8.94) is used. To guarantee the stability, the first difference

$$\Delta V(x_n) = V(x_{n+1}) - V(x_n) = F^T(x_n)K_{n+1}F(x_n) - 2F^T(x_n)K_{n+1}B(x_n)\phi\left(G_n^{-1}B^T(x_n)K_{n+1}F(x_n)\right)$$
$$+ \phi\left(G_n^{-1}B^T(x_n)K_{n+1}F(x_n)\right)^T B^T(x_n)K_{n+1}B(x_n)\phi\left(G_n^{-1}B^T(x_n)K_{n+1}F(x_n)\right) - x_n^T K_n x_n$$

$$(8.123)$$

must be negative-definite for all $x_n \in X$ and $u_n \in U$. One defines the set $S = \{x_n \in R^n : x_{n0} \in X_0, u \in U | V(0) = 0, V(x_n) > 0, \Delta V(x_n) < 0, \forall x \in X(X_0, U)\}$ The stability analysis is performed by studying S and $X(X_0, U)$. The constrained optimization problem is solved via the bounded admissible control law (8.120) and the stability is guaranteed if $X \subseteq S$.

8.14.3 Tracking Control of Discrete-Time Systems

We study systems modeled by the following difference equation in the state-space form:

$$x_{n+1}^{system} = A_n x_n^{system} + B_n u_n, \quad x_{n0}^{system} \in X_0, \quad u_n \in U, \quad n \geq 0. \tag{8.124}$$

The output equation is $y_n = H_n x_n^{system}$. The *exogeneous* system is given as

$$x_n^{ref} = x_{n-1}^{ref} + r_n - y_n. \tag{8.125}$$

Using (8.124) and (8.125), one finds

$$x_{n+1}^{ref} = x_n^{ref} + r_{n+1} - y_{n+1} = x_n^{ref} + r_{n+1} - H_n\left(A_n x_n^{system} + B_n u_n\right). \tag{8.126}$$

Hence,

$$x_{n+1} = \begin{bmatrix} x_{n+1}^{system} \\ x_{n+1}^{ref} \end{bmatrix} = \begin{bmatrix} A_n & 0 \\ -H_n A_n & I_n \end{bmatrix} x_n + \begin{bmatrix} B_n \\ -H_n B_n \end{bmatrix} u_n + \begin{bmatrix} 0 \\ I_n \end{bmatrix} r_{n+1}, \quad x_n = \begin{bmatrix} x_n^{system} \\ x_n^{ref} \end{bmatrix}. \tag{8.127}$$

To synthesize the bounded control law, we minimize the nonquadratic performance index

$$J = \sum_{n=0}^{N-1} \left[x_n^T Q_n x_n + 2\int \left(\phi^{-1}(u_n)\right)^T G_n \, du_n - u_n^T \begin{bmatrix} B_n \\ -H_n B_n \end{bmatrix}^T K_{n+1} \begin{bmatrix} B_n \\ -H_n B_n \end{bmatrix} u_n \right]. \tag{8.128}$$

For the quadratic return function (8.94), from the Hamilton–Jacobi equation

$$x_n^T K_n x_n = \min_{u_n \in U} \left[x_n^T Q_n x_n + 2 \int \left(\phi^{-1}(u_n) \right)^T G_n \, du_n - u_n^T \begin{bmatrix} B_n \\ -H_n B_n \end{bmatrix}^T K_{n+1} \begin{bmatrix} B_n \\ -H_n B_n \end{bmatrix} u_n \right.$$

$$\left. + \left(\begin{bmatrix} A_n & 0 \\ -H_n A_n & I_n \end{bmatrix} x_n + \begin{bmatrix} B_n \\ -H_n B_n \end{bmatrix} u_n \right)^T K_{n+1} \left(\begin{bmatrix} A_n & 0 \\ -H_n A_n & I_n \end{bmatrix} x_n + \begin{bmatrix} B_n \\ -H_n B_n \end{bmatrix} u_n \right) \right]$$

$$(8.129)$$

we apply the first-order necessary condition for optimality. The following bounded tracking control law yields

$$u_n = -\phi \left(G_n^{-1} \begin{bmatrix} B_n \\ -H_n B_n \end{bmatrix}^T K_{n+1} \begin{bmatrix} A_n & 0 \\ -H_n A_n & I_n \end{bmatrix} x_n \right), \quad u_n \in U. \tag{8.130}$$

The unknown matrix K_{n+1} is found by solving the equation

$$x_n^T K_n x_n = x_n^T Q_n x_n + x_n^T \begin{bmatrix} A_n & 0 \\ -H_n A_n & I_n \end{bmatrix}^T K_{n+1} \begin{bmatrix} A_n & 0 \\ -H_n A_n & I_n \end{bmatrix} x_n$$

$$- 2 \int \left(\phi \left(G_n^{-1} \begin{bmatrix} B_n \\ -H_n B_n \end{bmatrix}^T K_{n+1} \begin{bmatrix} A_n & 0 \\ -H_n A_n & I_n \end{bmatrix} x_n \right) \right)^T d \left(\begin{bmatrix} B_n \\ -H_n B_n \end{bmatrix}^T K_{n+1} \begin{bmatrix} A_n & 0 \\ -H_n A_n & I_n \end{bmatrix} x_n \right)$$

$$(8.131)$$

The tracking control problem can be solved for linear and nonlinear discrete-time systems applying the *state transformation* concept reported in Section 8.7. One recalls that the PI control law with the state feedback results. The problem, however, is to implement the control law designed using analog and digital controller hardware. Not all state variables can be directly measured or observed. Therefore, the *minimal-complexity* control laws are considered as a viable solution. Among various practical solutions, the PID-centered controllers usually are prioritized. For nonlinear systems the corresponding control laws can be designed by using the Lyapunov theory. The Hamilton–Jacobi method can also be used to derive the control laws. The designer attempts to minimize the sampling period which affects the dynamics of electromechanical systems. The design of the control laws must be integrated with the hardware ensuring codesign coherency and consistency.

8.15 Discussions on Physics and Essence of Control

Basics of electromechanical motion devices, power electronics hardware, system optimization, and other cornerstone fundamentals were covered. There is a need to discuss the essence of control laws *u* designed as analog or digital controllers from the hardware and software standpoints. Control laws are designed using the device physics which is modeled by system models in the form of differential and difference equations. Distinct mathematical models with different level of hierarchy, coherency, accuracy, details, and other descriptive features can be used. In these mathematical models, u means the control variable (or vector) which is a physical quantity to be varied in order to control a system. From classical mechanics, u may imply the applied force F or torque T (using the Newtonian mechanics, one can approach the so-called *force* and *torque control*). To develop these F and T, actuators are used. For electromagnetic, electrostatic, piezoelectric, and other actuators, the voltage V and current i could be considered as the control variables u. Actuators are controlled by using power electronics, and the output stage transistors are switched (controlled) by using the signal-level continuous-time voltage u_c developed by ICs.

To control the electromechanical system components, continuous-time u is used (digital control laws result in u_k which is converted to analog u utilizing the digital-to-analog converters, and the microcontrollers and DSPs outputs are analog signals). Sound PID, *admissible*, soft switching, *minimal-complexity*, and other control laws are reported and covered in detail. Relay and hard-switching control laws have a limited practicality due to abstract and largely hypothetical problem formulation under a great number of simplifications and assumptions. From the hardware standpoints, it is impossible to develop force F and torque T which would evolve as high-frequency relay-like (*bang-bang*) switching activity. Even if intended, one is not able to generate high-frequency switching of F and T of any origin. The use of the force and torque *limiters* was emphasized to reduce the instantaneous applied force/torque, load, and disturbances. Having emphasized mechanical features, we briefly recall electromagnetic actuators. Electromechanical motion devices are not controlled by applying the voltage constantly reversing its polarity. Furthermore, filters are always integrated to smooth (*average*) the PWM-like voltage of the output stages. One also recalls that the switching frequency of transistors is limited, and there is a transient dynamics for switching, for example, due to device physics, the output PWM voltage waveform is not an ideal train *pulses*, and the settling time of output stage transistors is within nano- or microseconds. For example, the transient dynamics of the TCA0372 dual power operational amplifier was reported in Figure 4.7a, while Figure 7.1 provides the transient behavior of the MC33171 operational amplifier.

Correspondingly, our major focus was concentrated on sound control concepts which ensure the best performance and *achievable* capabilities. Sound and practical PID, soft switching, and *minimal-complexity* control laws were designed, and these control algorithms can be implemented by analog and digital controllers ensuring near-optimal overall system perfromance.

References

1. Kuo, B. C., *Automatic Control Systems*, Prentice Hall, Englewood Cliffs, NJ, 1995.
2. Lyshevski, S. E., *Control Systems Theory with Engineering Applications*, Birkhauser, Boston, MA, 2000.
3. Ogata, K., *Discrete-Time Control Systems*, Prentice-Hall, Upper Saddle River, NJ, 1995.
4. Ogata, K., *Modern Control Engineering*, Prentice-Hall, Upper Saddle River, NJ, 1997.
5. Lyshevski, S. E., *Electromechanical Systems and Devices*, CRC Press, Boca Raton, FL, 2008.

9

Electroactive and Magnetoactive Materials

9.1 Introduction

Electroactive and magnetoactive materials are materials that modify their shape in response to electric or magnetic stimuli. Such materials permit induced-strain actuation and strain sensing which are of considerable importance in micromechatronics. On one hand, induced-strain actuation allows us to create motion at the microscale without pistons, gears, or other mechanisms. Induced-strain actuation relies on the direct conversion of electric or magnetic energy into mechanical energy. Induced-strain actuation is a solid-state actuation, has much fewer parts than conventional actuation, and is much more reliable. Induced-strain actuation offers the opportunity for creating micromechatronics systems that are miniaturized, effective, and efficient. On the other hand, strain sensing with electroactive and magnetoactive materials creates direct conversion of mechanical energy into electric and magnetic energy. With piezoelectric strain sensors, strong and clear voltage signals are obtained directly from the sensor without the need for intermediate gauge bridges, signal conditioners, and signal amplifiers. These direct sensing properties are especially significant in dynamics, vibration, and audio applications in which alternating effects occur in rapid succession, thus preventing charge leakage. Other applications of active materials are in sonic and ultrasonic transduction, in which the transducer acts both as a sensor and an actuator, first transmitting a sonic or ultrasonic pulse, and then listening for the echoes received from the defect or target.

In this chapter, we will discuss several types of active materials: piezoelectric ceramics, electrostrictive ceramics, piezoelectric polymers, and magnetostrictive compounds. Various formulations of these materials are currently available commercially. The names PZT (a piezoelectric ceramic), PMN (an electrostrictive ceramic), Terfenol-D (a magnetostrictive compound), and PVDF (a piezoelectric polymer) have become widely used. In this chapter, we attempt a review of the principal active material types. We will treat each material type separately, will present their salient features, and introduce the modeling equations. In our discussion, we will start with a general perspective on the overall subject of piezoelectricity and ferroelectric ceramics, explaining some of the physical behavior underpinning their salient features, especially in relation to perovskite crystalline structures. Then, we will discuss manufacturing and quality control issues related to ferroelectric ceramics. We will continue by considering separately the piezoceramics and electrostrictive ceramics commonly used in current applications and commercially available to the interested user. The focus of the discussion is then switched toward piezoelectric polymers, such as PVDF, with their interesting properties, such as flexibility, resilience, and durability, which make them preferable to ferroelectric ceramics in certain applications. The discussion of magnetostrictive materials, such as Terfenol-D, concludes our review of the active materials spectrum. This chapter paves way toward the next chapters, in which the use of active materials in the construction of induced-strain actuators and active sensors for micromechatronics applications will be discussed.

9.2 Piezoelectricity

Piezoelectricity (discovered in 1880 by Jacques and Pierre Curie) describes the phenomenon of generating an electric field when the material is subjected to a mechanical stress (direct effect) or, conversely, generating a mechanical strain in response to an applied electric field. The *direct piezoelectric effect* predicts how much electric field is generated by a given mechanical stress. This *sensing effect* is utilized in the development of piezoelectric sensors. The *converse piezoelectric effect* predicts how much mechanical strain is generated by a given electric field. This *actuation effect* is utilized in the development of piezoelectric induced-strain actuators.

Piezoelectric properties occur naturally in some crystalline materials, e.g., quartz crystals (SiO_2) and Rochelle salt. The latter is a natural ferroelectric material, possessing an orientable domain structure that aligns under an external electric field and thus enhances its piezoelectric response. Piezoelectric response can also be induced by electrical poling of certain polycrystalline materials, such as piezoceramics. In recent years, many piezoceramic formulations have become commercially available at an affordable price.

9.2.1 Actuation Equations

For linear piezoelectric materials, the interaction between the electrical and mechanical variables can be described by linear equations (ANSI/IEEE Standard 176-1987). A linear constitutive relation between mechanical and electrical variables may take the following tensorial form

$$S_{ij} = s^E_{ijkl}T_{kl} + d_{kij}E_k + \delta_{ij}\alpha^E_i\theta \tag{9.1}$$

$$D_i = d_{ikl}T_{kl} + \varepsilon^T_{ik}E_k + \tilde{D}_i\theta \tag{9.2}$$

where S_{ij} and T_{ij} are the strain and stress, E_k and D_i are the electric field and electric displacement, and θ is the temperature. The stress and strain variables are second-order tensors, while the electric field and the electric displacement are first-order tensors. The coefficient s_{ijkl} is the compliance, i.e., the strain per unit stress. The coefficients d_{kij} and d_{ikl} signify coupling between the electrical and the mechanical variables, i.e., the charge per unit stress and the strain per unit electric field. The coefficient α_i is the coefficient of thermal expansion. The coefficient $\tilde{D}_i$ is the electric displacement temperature coefficient. Since thermal effects only influence the diagonal terms, the respective coefficients, α_i and $\tilde{D}_i$, have single subscripts. The term δ_{ij} is the Kroneker delta ($\delta_{ij} = 1$ if $i = j$; zero otherwise). The Einstein summation convention for repeated tensor indices (Knowles, 1997) is employed. The superscripts T, D, E signify that the quantities are measured at zero stress ($T = 0$), zero electric displacement ($D = 0$), or zero electric field ($E = 0$), respectively. In practice, the zero electric displacement condition corresponds to open circuit (zero current across the electrodes), while the zero electric field corresponds to closed circuit (zero voltage across the electrodes). The strain is defined as

$$S_{ij} = \frac{1}{2}(u_{i,j} + u_{j,i}) \tag{9.3}$$

where u_i is the displacement, and comma followed by an index signifies partial differentiation with respect to the coordinate associated with that index.

Equation (9.1) is the *actuation equation*; it is used to predict how much strain will result for given stress, electric field, and temperature. The terms proportional with stress and temperature are common with the formulations of classical thermoelasticity. The term proportional with the electric field is specific to piezoelectricity and represents the induced-strain actuation effect, i.e.,

$$S_{ij}^{ISA} = d_{kij}E_k \tag{9.4}$$

For this reason, the coefficient d_{kij} can be interpreted as the *piezoelectric strain coefficient*.

Equation (9.2) is used to predict how much electric displacement, i.e., charge per unit area, is required to accommodate the simultaneous state of stress, electric field, and temperature. In particular, the term $d_{ikl}T_{kl}$ indicates how much charge is being produced by the application of the mechanical stress T_{kl}. For this reason, the coefficient d_{ikl} can be interpreted as the *piezoelectric charge coefficient*. Note that d_{kij} and d_{ikl} represent the same third-order tensor only that the indices have been named appropriately to the respective equations in which they are used.

9.2.2 Sensing Equations

So far, we have expressed the strain and electric displacement in terms of applied stress, electric field, and temperature using the constitutive tensorial Equations (9.1) and (9.2). However, these equations can be replaced by an equivalent set of equations that highlight the sensing effect, i.e., predict how much electric field will be generated by a given state of stress, electric displacement, and temperature. (Since the electric voltage is directly related to the electric field, such equations would be preferred for sensing applications.) The sensing equivalent of Equations (9.1) and (9.2) is

$$S_{ij} = s_{ijkl}^D T_{kl} + g_{kij}D_k + \delta_{ij}\alpha_i^D\theta \tag{9.5}$$

$$E_i = g_{ikl}T_{kl} + \beta_{ik}^T D_k + \tilde{E}_i\theta \tag{9.6}$$

Equation (9.6) predicts how much electric field, i.e., voltage per unit thickness, is generated by "squeezing" the piezoelectric material, i.e., it represents the direct piezoelectric effect. This formulation is useful in piezoelectric sensor design. Equation (9.6) is called the *sensor equation*. The coefficient g_{ikl} is the *piezoelectric voltage constant*, and represent how much electric field is induced per unit stress. The coefficient $\tilde{E}_i$ is the *pyroelectric voltage coefficient* and represents the amount of electric field is induced per unit temperature change.

9.2.3 Stress Equations

The piezoelectric constitutive equations can also be rearranged in such a way as to express stress and electric displacement in terms of strain, electric field, and temperature. This formulation is especially useful for including the piezoelectric effects in stress and strength analyses. The stress formulation of the piezoelectric constitutive equations is

$$T_{ij} = c^E_{ijkl}S_{kl} - e_{kij}E_k - c^E_{ijkl}\delta_{kl}\alpha^E_k\theta \tag{9.7}$$

$$D_i = e_{ikl}S_{kl} + \varepsilon^T_{ik}E_k + \tilde{D}_i\theta \tag{9.8}$$

where c^E_{ijkl} is the stiffness tensor, and e_{kij} is the piezoelectric stress constant. The term $c^E_{ijkl}\delta_{kl}\alpha^E_k\theta$ represents the stress induced in a piezoelectric material by temperature changes when the strain is forced to be zero. This corresponds to the material being fully constrained against deformation. Such stresses, which are induced by temperature effects, are also known as *residual thermal stresses*. They are very important in calculating the strength of piezoelectric materials, especially when they are processed at elevated temperatures.

9.2.4 Actuator Equations in Terms of Polarization

The use of electric field, E_i, and electric displacement, D_i, is convenient in practical piezoelectric sensor and actuator design, since these variables relate directly to the voltage and current that can be measured experimentally. However, a theoretical explanation of the observed phenomena through solid-state physics is more direct when using the polarization, P_i, instead of the electric displacement, D_i. The polarization P_i, electric displacement D_i, and electric field E_i are connected by the relation

$$D_i = \varepsilon_0 E_i + P_i \tag{9.9}$$

where ε_0 is the free space dielectric permittivity. The electric field and electric displacement are also connected by the relation

$$D_i = \varepsilon_{ik}E_k \tag{9.10}$$

where ε_{ik} is the effective dielectric permittivity of the material. Combining Equations (9.9) and (9.10) yields the polarization in terms of the electric field only, i.e.,

$$P_i = (\varepsilon_{ik} - \delta_{ik}\varepsilon_0)E_k = \kappa_{ik}E_k \tag{9.11}$$

where $\kappa_{ik} = \varepsilon_{ik} - \delta_{ik}\varepsilon_0$. Using Equation (9.11), we write Equations (9.1) and (9.2) in the form

$$S_{ij} = s^E_{ijkl}T_{kl} + d_{kij}E_k + \delta_{ij}\alpha_i\theta \tag{9.12}$$

$$P_i = d_{ikl}T_{kl} + \kappa^T_{ik}E_k + \tilde{P}_i\theta \tag{9.13}$$

where $\tilde{P}_i$ is the pyroelectric polarization coefficient. One notes that, in Equation (9.13), the coefficient d_{ikl} signifies the induced polarization per unit stress, hence it can be viewed as *polarization coefficient*.

9.2.5 Compressed Matrix Notations

The compressed matrix notations (Voigt notations) allow us to write the linear equations of piezoelectricity in matrix form instead of tensor form. The compressed matrix notations consist of replacing the double indices ij or kl by the single indices p or q, where $i, j, k, l = 1, 2, 3$ and $p, q = 1, 2, 3, 4, 5, 6$ (Table 9.1). Thus, the 3×3 stress and strain tensors, T_{ij} and S_{ij}, are replaced by six-element long column matrices T_p and S_p. The $3 \times 3 \times 3 \times 3$ fourth-order stiffness and compliance tensors c_{ijkl}^E and s_{ijkl}^E are replaced by the 6×6 stiffness and compliance matrices c_{pq}^E and s_{pq}^E. Similarly, c_{ijkl}^D and s_{ijkl}^D are replaced by c_{pq}^D and s_{pq}^D. The $3 \times 3 \times 3$ piezoelectric tensors, d_{ikl}, e_{ikl}, g_{ikl}, and h_{ikl}, are replaced by the 3×6 piezoelectric matrices d_{ip}, e_{ip}, g_{ip}, h_{ip}. The following correspondence rules apply

$$T_p = T_{ij}, \quad p = 1, 2, 3, 4, 5, 6 \quad \text{and} \quad i, j = 1, 2, 3 \ \text{(stress)} \tag{9.14}$$

and

$$\begin{aligned} S_p &= S_{ij}, \quad i = j, \quad p = 1, 2, 3 \\ S_p &= 2S_{ij}, \quad i \neq j, \quad p = 4, 5, 6 \end{aligned} \ \text{(strain)} \tag{9.15}$$

The factor of 2 in Equation (9.15) is related to that in the definition of shear strains in the tensor and matrix formulation. We also have

$$c_{pq}^E = c_{ijkl}^E, \quad c_{pq}^D = c_{ijkl}^D, \quad p = 1, 2, 3, 4, 5, 6 \quad \text{and} \quad i, j = 1, 2, 3 \ \text{(stiffness coefficients)} \tag{9.16}$$

and

$$\begin{cases} s_{pq}^E = s_{ijkl}^E, & i = j \ \text{and} \ k = l, \ p, q = 1, 2, 3 \\ s_{pq}^E = 2s_{ijkl}^E, & i = j \ \text{and} \ k \neq l, \ p = 1, 2, 3, \ q = 4, 5, 6 \\ s_{pq}^E = 4s_{ijkl}^E, & i \neq j \ \text{and} \ k \neq l, \ p, q = 4, 5, 6 \end{cases} \ \text{(compliance coefficients)} \tag{9.17}$$

TABLE 9.1

Conversion from Tensor to Matrix Indices for the Voigt Notations

ij or kl	p or q
11	1
22	2
33	3
23 or 32	4
31 or 13	5
12 or 21	6

where $i, j, k, l = 1, 2, 3$. Similar expressions can be derived for s_{pq}^D. The factors of 2 and 4 are associated with the factor of 2 from Equation (9.15). Similarly,

$$e_{ip} = e_{ikl}, \quad h_{ip} = h_{ikl} \text{ (piezoelectric stress constants)} \tag{9.18}$$

$$\begin{cases} d_{iq} = d_{ikl}, & k = l, \quad q = 1, 2, 3 \\ d_{iq} = 2d_{ikl}, & k \neq l, \quad q = 4, 5, 6 \end{cases} \text{ (piezoelectric strain constants)} \tag{9.19}$$

$$\begin{cases} g_{iq} = g_{ikl}, & k = l, \quad q = 1, 2, 3 \\ g_{iq} = 2g_{ikl}, & k \neq l, \quad q = 4, 5, 6 \end{cases} \text{ (piezoelectric voltage constants)} \tag{9.20}$$

where $i, k, l = 1, 2, 3$. The compressed matrix notations of Equations (9.14) through (9.20) have the advantage of brevity. They are commonly used in engineering applications. The values of the elastic and piezoelectric constants provided by the active material manufacturers in their product specifications, are given in compressed matrix notations.

9.2.6 Piezoelectric Equations in Compressed Matrix Notations

In engineering practice, Equations (9.1) and (9.2), which were written in tensorial form, can be rearranged in matrix form using the compressed matrix (Voigt) notations discussed in Section 2.5; the stress and strain tensors are arranged as six-component vectors, with the first three components representing *direct* stress and strain, while the last three components representing *shear* stress and strain. Thus,

$$\begin{Bmatrix} S_{11} \\ S_{22} \\ S_{33} \\ S_{23} \\ S_{31} \\ S_{12} \end{Bmatrix} \Longrightarrow \begin{Bmatrix} S_1 \\ S_2 \\ S_3 \\ S_4 \\ S_5 \\ \frac{1}{2}S_6 \end{Bmatrix}, \quad \begin{Bmatrix} T_{11} \\ T_{22} \\ T_{33} \\ T_{23} \\ T_{31} \\ T_{12} \end{Bmatrix} \Longrightarrow \begin{Bmatrix} T_1 \\ T_2 \\ T_3 \\ T_4 \\ T_5 \\ T_6 \end{Bmatrix} \tag{9.21}$$

Hence, the constitutive equations (9.1) and (9.2) take the matrix form

$$\begin{Bmatrix} S_1 \\ S_2 \\ S_3 \\ S_4 \\ S_5 \\ S_6 \end{Bmatrix} = \begin{bmatrix} s_{11} & s_{12} & s_{13} & 0 & 0 & 0 \\ s_{21} & s_{22} & s_{23} & 0 & 0 & 0 \\ s_{31} & s_{32} & s_{33} & 0 & 0 & 0 \\ 0 & 0 & 0 & s_{44} & 0 & 0 \\ 0 & 0 & 0 & 0 & s_{55} & 0 \\ 0 & 0 & 0 & 0 & 0 & s_{66} \end{bmatrix} \begin{Bmatrix} T_1 \\ T_2 \\ T_3 \\ T_4 \\ T_5 \\ T_6 \end{Bmatrix} + \begin{bmatrix} d_{11} & d_{21} & d_{31} \\ d_{12} & d_{22} & d_{32} \\ d_{13} & d_{23} & d_{33} \\ d_{14} & d_{24} & d_{34} \\ d_{15} & d_{25} & d_{35} \\ d_{16} & d_{26} & d_{36} \end{bmatrix} \begin{Bmatrix} E_1 \\ E_2 \\ E_3 \end{Bmatrix} + \begin{Bmatrix} \alpha_1 \\ \alpha_2 \\ \alpha_3 \\ 0 \\ 0 \\ 0 \end{Bmatrix} \theta \tag{9.22}$$

$$\begin{Bmatrix} D_1 \\ D_2 \\ D_3 \end{Bmatrix} = \begin{bmatrix} d_{11} & d_{12} & d_{13} & d_{14} & d_{15} & d_{16} \\ d_{21} & d_{22} & d_{23} & d_{24} & d_{25} & d_{26} \\ d_{31} & d_{32} & d_{33} & d_{34} & d_{35} & d_{36} \end{bmatrix} \begin{Bmatrix} T_1 \\ T_2 \\ T_3 \\ T_4 \\ T_5 \\ T_6 \end{Bmatrix} + \begin{bmatrix} \varepsilon_{11} & \varepsilon_{12} & \varepsilon_{13} \\ \varepsilon_{21} & \varepsilon_{22} & \varepsilon_{23} \\ \varepsilon_{31} & \varepsilon_{32} & \varepsilon_{33} \end{bmatrix} \begin{Bmatrix} E_1 \\ E_2 \\ E_3 \end{Bmatrix} + \begin{Bmatrix} \tilde{D}_1 \\ \tilde{D}_2 \\ \tilde{D}_3 \end{Bmatrix} \theta \tag{9.23}$$

Equations (9.22) and (9.23) can be written in matrix format, i.e.,

$$\{S\} = [s]\{T\} + [d]^t\{E\} + \{\alpha\}\theta \tag{9.24}$$

$$\{D\} = [d]\{T\} + [\varepsilon]\{E\} + \{\tilde{D}\}\theta \tag{9.25}$$

Note that the piezoelectric matrix $[d]^t$ used in Equation (9.24) is the transpose of the piezoelectric matrix $[d]$ in Equation (9.25). Equations (9.22) and (9.23) can be also written in compact form using index notations, i.e.,

$$S_p = s_{pq}^E T_q + d_{kp}E_k + \delta_{pq}\alpha_q^E\theta, \qquad p, q = 1,\ldots,6; \quad k = 1,2,3 \tag{9.26}$$

$$D_i = d_{iq}T_q + \varepsilon_{ik}^T E_k + \tilde{D}_i\theta, \qquad q = 1,\ldots, 6; \quad i, k = 1,2,3 \tag{9.27}$$

Compressed matrix expressions similar to Equations (9.22) through (9.25) can be derived for the other constitutive equations such as Equations (9.5) through (9.8), (9.12) and (9.13).

The values of the piezoelectric coupling coefficients differ from material to material. Most piezoelectric materials of interest are crystalline solids. These can be single crystals (either natural or synthetic) or polycrystalline materials like ferroelectric ceramics. In certain crystalline piezoelectric materials, the piezoelectric coefficients d_{ji} ($i=1,\ldots,6$; $j=1,2,3$) may be enhanced or diminished through preferred crystal cuts. The various possible crystal cuts for piezoelectric quartz are shown in Figure 9.1. The quartz crystal microbalance (QCM) sensor uses the AT cut which enhances the shear strain response

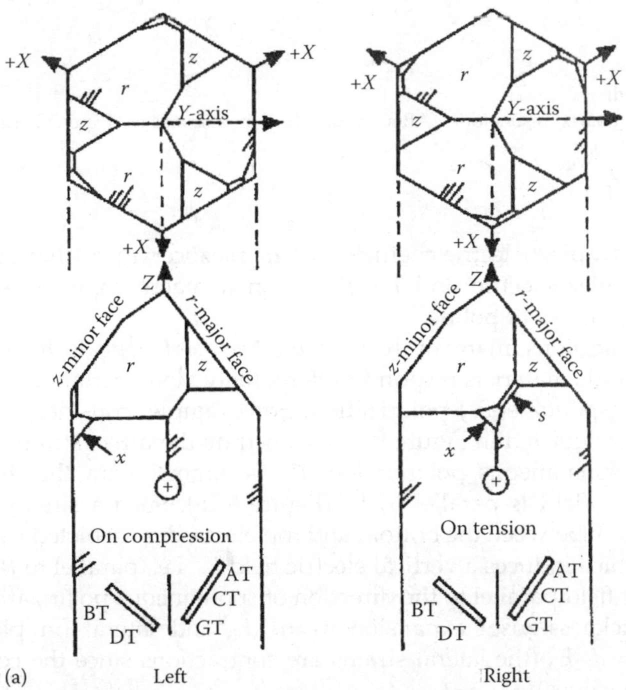

FIGURE 9.1
Directional dependence of quart crystal piezoelectricity: (a) crystallographic cuts of quartz crystal;

(continued)

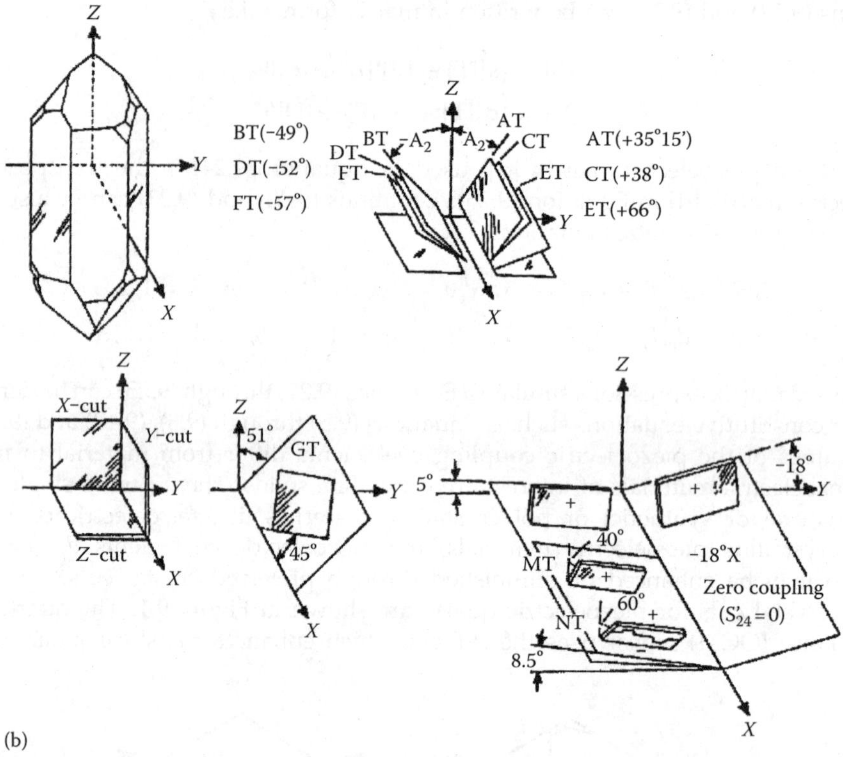

(b)

FIGURE 9.1 (continued)
(b) practical cuts of quartz resonators. (After Ikeda, T., *Fundamentals of Piezoelectricity*, Oxford University Press, Oxford, U.K., 1996.)

(i.e., increase the d_{15} piezoelectric coefficient). The piezoceramics, thought polycrystalline, can be rendered piezoelectric and be given, on a macroscopic scale, a single-crystal symmetry by the process of poling.

In practical applications, many of the piezoelectric coefficients d_{ji} have negligible values since the piezoelectric materials respond preferentially along certain directions depending on their intrinsic (spontaneous) polarization. For example, consider the situation of the piezoelectric wafer depicted in Figure 9.2. To illustrate the direct strain effects d_{33} and d_{31}, assume that the spontaneous polarization, P_S, is aligned with the three-axis and that the applied electric field is parallel to P_S (Figure 9.2a). Such a situation is obtained by applying a voltage V between the bottom and top electrodes (depicted by the grey shading in Figure 9.2a), which induces a vertical electric field E_3, i.e., parallel to P_S. The application of such an electric field parallel to the direction of spontaneous polarization ($E_3 \| P_S$) results in a vertical (thickness-wise) expansion $\varepsilon_3 = d_{33}E_3$ and lateral (in plane) contractions $\varepsilon_1 = d_{31}E_3$ and $\varepsilon_2 = d_{32}E_3$ (the lateral strains are contractions since the coefficients d_{31} and d_{32} have negative values).

To illustrate the shear strain piezoelectric effect d_{15}, assume that the electric field is applied perpendicular to the direction of spontaneous polarization. This can be obtained by electroding the lateral faces of the piezoelectric wafer. The application of a voltage to the

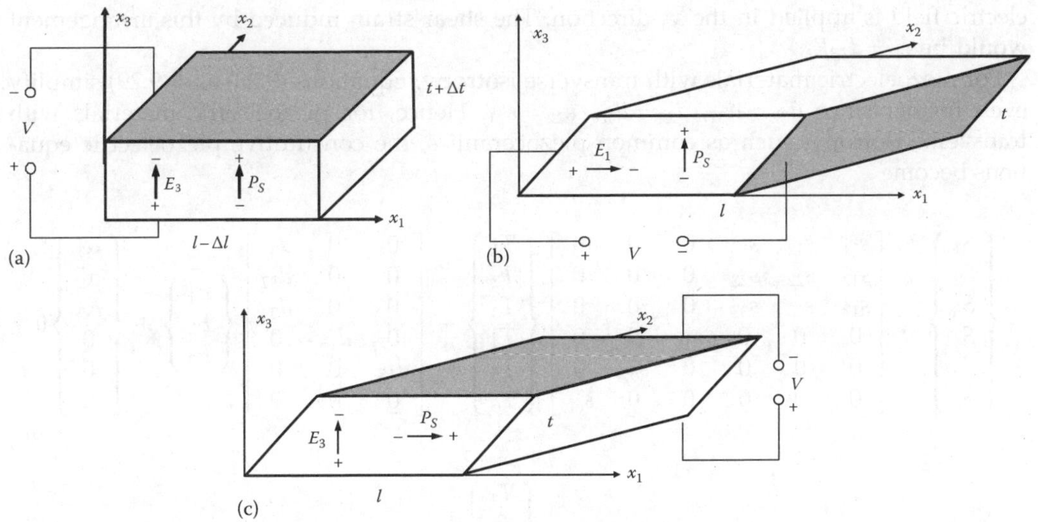

FIGURE 9.2
Basic induced-strain responses of piezoelectric materials: (a) direct strains $\varepsilon_3 = d_{33}E_3$ (thickness), $\varepsilon_1 = d_{31}E_3$, $\varepsilon_2 = d_{32}E_3$ (in plane); (b) shear strain $\varepsilon_5 = d_{15}E_1$; (c) shear strain $\varepsilon_5 = d_{35}E_3$ (Note: grey shading depicts the electrodes).

lateral electrodes shown in Figure 9.2b results in an in-plane electric field, E_1 that is perpendicular to the spontaneous polarization, $(E_1 \perp P_S)$. This produces a shear strain $\varepsilon_5 = d_{15}E_1$. Similarly, if the electrodes were applied to the front and back faces, the resulting electric field would be E_2 and the resulting strain would be $\varepsilon_4 = d_{24}E_2$. The constitutive equations associated with the direct and shear piezoelectric effects are

$$
\begin{Bmatrix} S_1 \\ S_2 \\ S_3 \\ S_4 \\ S_5 \\ S_6 \end{Bmatrix} =
\begin{bmatrix}
s_{11} & s_{12} & s_{13} & 0 & 0 & 0 \\
s_{12} & s_{22} & s_{23} & 0 & 0 & 0 \\
s_{13} & s_{23} & s_{33} & 0 & 0 & 0 \\
0 & 0 & 0 & s_{44} & 0 & 0 \\
0 & 0 & 0 & 0 & s_{55} & 0 \\
0 & 0 & 0 & 0 & 0 & s_{66}
\end{bmatrix}
\begin{Bmatrix} T_1 \\ T_2 \\ T_3 \\ T_4 \\ T_5 \\ T_6 \end{Bmatrix} +
\begin{bmatrix}
0 & 0 & d_{31} \\
0 & 0 & d_{32} \\
0 & 0 & d_{33} \\
0 & d_{24} & 0 \\
d_{15} & 0 & 0 \\
0 & 0 & 0
\end{bmatrix}
\begin{Bmatrix} E_1 \\ E_2 \\ E_3 \end{Bmatrix} +
\begin{Bmatrix} \alpha_1 \\ \alpha_2 \\ \alpha_3 \\ 0 \\ 0 \\ 0 \end{Bmatrix} \theta
\tag{9.28}
$$

$$
\begin{Bmatrix} D_1 \\ D_2 \\ D_3 \end{Bmatrix} =
\begin{bmatrix}
0 & 0 & 0 & 0 & d_{15} & 0 \\
0 & 0 & 0 & d_{24} & 0 & 0 \\
d_{31} & d_{32} & d_{33} & 0 & 0 & 0
\end{bmatrix}
\begin{Bmatrix} T_1 \\ T_2 \\ T_3 \\ T_4 \\ T_5 \\ T_6 \end{Bmatrix} +
\begin{bmatrix}
\varepsilon_{11} & 0 & 0 \\
0 & \varepsilon_{22} & 0 \\
0 & 0 & \varepsilon_{33}
\end{bmatrix}
\begin{Bmatrix} E_1 \\ E_2 \\ E_3 \end{Bmatrix} +
\begin{Bmatrix} \tilde{D}_1 \\ \tilde{D}_2 \\ \tilde{D}_3 \end{Bmatrix} \theta
\tag{9.29}
$$

The use of lateral electrodes may not be feasible in the case of a thin wafer. In this case, top and bottom electrodes can be used again, but the spontaneous polarization of the wafer must be aligned in an in-plane direction. This latter situation is depicted in Figure 9.2c, where the spontaneous polarization is shown in the x_1 direction, while the electric field is applied in the x_3 direction. The shear strain induced by this arrangement would be $\varepsilon_5 = d_{35}E_3$.

For piezoelectric materials with transverse isotropy, Equations (9.28) and (9.29) simplify even further since $d_{32} = d_{31}$, $d_{24} = d_{15}$, $\varepsilon_{22} = \varepsilon_{11}$. Hence, for piezoelectric materials with transverse isotropy, such as common piezoceramics, the constitutive piezoelectric equations become

$$
\begin{Bmatrix} S_1 \\ S_2 \\ S_3 \\ S_4 \\ S_5 \\ S_6 \end{Bmatrix} = \begin{bmatrix} s_{11} & s_{12} & s_{13} & 0 & 0 & 0 \\ s_{12} & s_{22} & s_{23} & 0 & 0 & 0 \\ s_{13} & s_{23} & s_{33} & 0 & 0 & 0 \\ 0 & 0 & 0 & s_{44} & 0 & 0 \\ 0 & 0 & 0 & 0 & s_{55} & 0 \\ 0 & 0 & 0 & 0 & 0 & s_{66} \end{bmatrix} \begin{Bmatrix} T_1 \\ T_2 \\ T_3 \\ T_4 \\ T_5 \\ T_6 \end{Bmatrix} + \begin{bmatrix} 0 & 0 & d_{31} \\ 0 & 0 & d_{32} \\ 0 & 0 & d_{33} \\ 0 & d_{15} & 0 \\ d_{15} & 0 & 0 \\ 0 & 0 & 0 \end{bmatrix} \begin{Bmatrix} E_1 \\ E_2 \\ E_3 \end{Bmatrix} + \begin{Bmatrix} \alpha_1 \\ \alpha_2 \\ \alpha_3 \\ 0 \\ 0 \\ 0 \end{Bmatrix} \theta
$$

$$(9.30)$$

$$
\begin{Bmatrix} D_1 \\ D_2 \\ D_3 \end{Bmatrix} = \begin{bmatrix} 0 & 0 & 0 & 0 & d_{15} & 0 \\ 0 & 0 & 0 & d_{15} & 0 & 0 \\ d_{31} & d_{32} & d_{33} & 0 & 0 & 0 \end{bmatrix} \begin{Bmatrix} T_1 \\ T_2 \\ T_3 \\ T_4 \\ T_5 \\ T_6 \end{Bmatrix} + \begin{bmatrix} \varepsilon_{11} & 0 & 0 \\ 0 & \varepsilon_{11} & 0 \\ 0 & 0 & \varepsilon_{33} \end{bmatrix} \begin{Bmatrix} E_1 \\ E_2 \\ E_3 \end{Bmatrix} + \begin{Bmatrix} \tilde{D}_1 \\ \tilde{D}_2 \\ \tilde{D}_3 \end{Bmatrix} \theta
$$

$$(9.31)$$

Compressed matrix (Voigt) expressions similar to Equations (9.30) and (9.31) can be derived for the other constitutive equations such as Equations (9.5) through (9.8), (9.12), and (9.13).

9.2.7 Relations between the Constants

The constants that appear in the equations described in Sections 9.2.1 through 9.2.6 can be related to each other. For example, the stiffness tensor, c_{ijkl}, is the inverse of the strain tensor, s_{ijkl}. Similar relations can be established for the other constants and coefficients. In writing these relations, we use the compressed matrix notation with i, j, k, $l = 1$, 2, 3 and p, q, $r = 1, \ldots, 6$. We will also use the 3×3 unit matrix δ_{ij} and the 6×6 unit matrix δ_{pq}. As before, Einstein convention of implied summation over the repeated indices applies. Hence,

$$c_{pr}^E s_{qr}^E = \delta_{pq}, \quad c_{pr}^D s_{qr}^D = \delta_{pq} \text{ (stiffness–compliance relations)} \tag{9.32}$$

$$\varepsilon_{ik}^S \beta_{jk}^S = \delta_{ij}, \quad \beta_{ik}^T \varepsilon_{jk}^T = \delta_{ij} \text{ (permittivity–impermittivity relations)} \tag{9.33}$$

$$c_{pq}^D = c_{pq}^E + e_{kp}h_{kq}, \quad s_{pq}^D = s_{pq}^E - d_{lp}g_{kq}$$

$$\text{(close circuit–open circuit effects on elastic constants)} \tag{9.34}$$

$$\varepsilon_{ij}^T = \varepsilon_{ij}^S + d_{iq}e_{jq}, \quad \beta_{ij}^T = \beta_{ij}^S - g_{iq}h_{jq} \text{ (stress–strain effects on dielectric constants)} \tag{9.35}$$

$$\begin{cases} e_{ip} = d_{iq}c_{qp}^E, & d_{ij} = \varepsilon_{ik}^T g_{kp} \\ g_{ip} = \beta_{ik}^T d_{kq}, & h_{ip} = g_{iq}\varepsilon_{qp}^D \end{cases}, \text{ (relations between piezoelectric constants)} \tag{9.36}$$

9.2.8 Electromechanical Coupling Coefficient

Electromechanical coupling coefficient is defined as the square root of the ratio between the mechanical energy stored and the electrical energy applied to a piezoelectric material

$$k = \sqrt{\frac{\text{Mechanical energy stored}}{\text{Electrical energy applied}}} \tag{9.37}$$

For direct actuation, we have $k_{33} = \sqrt{\dfrac{d_{33}^2}{s_{33}\varepsilon_{33}}}$, for transverse actuation, $k_{31} = \sqrt{\dfrac{d_{31}^2}{s_{11}\varepsilon_{33}}}$, and for

shear actuation, $k_{15} = \sqrt{\dfrac{d_{15}^2}{s_{55}\varepsilon_{33}}}$. For uniform in-plane deformation, we have the planar

coupling coefficient, $\kappa_p = \kappa_{13}\sqrt{\dfrac{2}{1-\nu}}$, where ν is the Poisson ratio.

9.2.9 Higher Order Models of the Electroactive Response

Higher order models of the electroactive response contain both linear and quadratic terms. The linear terms are associated with the conventional *piezoelectric response*. The quadratic terms are associated with the *electrostrictive response*, for which the application of an electric field in one direction induces constriction (squeezing) of the material. The electrostrictive effect is not limited to piezoelectric materials; it is present in all materials, though with different amplitudes. The electrostrictive response is quadratic in the electric field; hence, the direction of the electrostriction does not switch as the polarity of the electric field is switched. The constitutive equations that incorporate both piezoelectric and electrostrictive responses have the form

$$S_{ij} = s_{klij}^E T_{kl} + d_{kij}E_k + M_{klij}E_kE_l. \tag{9.38}$$

Note that the first two terms are the same as for piezoelectric materials. The third term is due to electrostriction. The coefficients M_{klij} are the electrostrictive coefficients.

9.3 Piezoelectric Phenomena

9.3.1 Polarization

Polarization is a phenomenon observed in dielectrics and it consists in the separation of positive and negative electric charges at different ends of the dielectric material, upon the application of an external electric field. A typical example is the polarization of the dielectric material inside a capacitor upon the application of an electric voltage across the capacitor plates. Polarization is the explanation for the fact that the dielectric capacitor can hold more charge than the vacuum capacitor, since

$$D = \varepsilon_0 E + P \tag{9.39}$$

In Equation (9.39), D represents the electric displacement, i.e., charge per unit area; E represents the electric field, i.e., voltage divided by the distance between the capacitor plates; ε_0 represents the electric permittivity of vacuum. It is apparent from Equation (9.39) that the polarization P represents the additional charge stored in a dielectric capacitor as compared to a vacuum capacitor.

9.3.2 Spontaneous Polarization

Spontaneous polarization is the phenomenon by which polarization occurs without the application of an external electric field. Spontaneous polarization has been observed in certain crystals in which the centers of positive and negative charges do not coincide. Crystals are classified into 32 point groups according to their crystallographic symmetry (international and Schonflies crystallographic symbols). These 32 point groups can be divided into two large classes, one containing point groups that have a center of symmetry, the other containing point groups that do not have a center of symmetry, and hence display some spontaneous polarization. Of the 21 point groups that do not display a center of symmetry, 20 groups contain crystals that may display spontaneous polarization. Spontaneous polarization can occur more easily in perovskite crystal structures.

9.3.3 Permanent Polarization

Permanent polarization is the phenomenon by which the polarization is retained even in the absence of an external electric field. The process through which permanent polarization is induced in a material is known as poling.

9.3.4 Paraelectric Materials

Paraelectric materials do not display permanent polarization, i.e., they have zero polarization in the absence of an external electric field. When an external field is applied, their polarization is roughly proportional with the applied electric field. It increases when the electric field is increased, and decreases back to zero when the field is reduced. If the field is reversed, the polarization also reverses (Figure 9.3a). Paraelectric behavior represents the behavior of common dielectrics.

9.3.5 Ferroelectric Materials

Ferroelectric materials have permanent polarization that can be altered by the application of an external electric field. The term "ferroelectric materials" was derived by analogy with the term "ferromagnetic materials," in which the permanent magnetization is altered by the application of an external magnetic field. Figure 9.3b describes graphically the ferroelectric behavior during the cyclic application of an electric field. As the electric field is increased beyond the critical value, called coercive field, E_c, the polarization suddenly increases to a high value. This value is roughly maintained when the electric field is decreased, such that at zero electric field the ferroelectric material retains a permanent spontaneous polarization P_S. When the electric field is further reduced beyond the negative value $-E_c$, the polarization suddenly switches to a large negative value, which is roughly maintained as the electric field is decreased. At zero electric field, the permanent spontaneous polarization is $-P_S$. As the electric field is again increased in the positive range, the polarization is again switched to a positive value, as the field is increased beyond E_c.

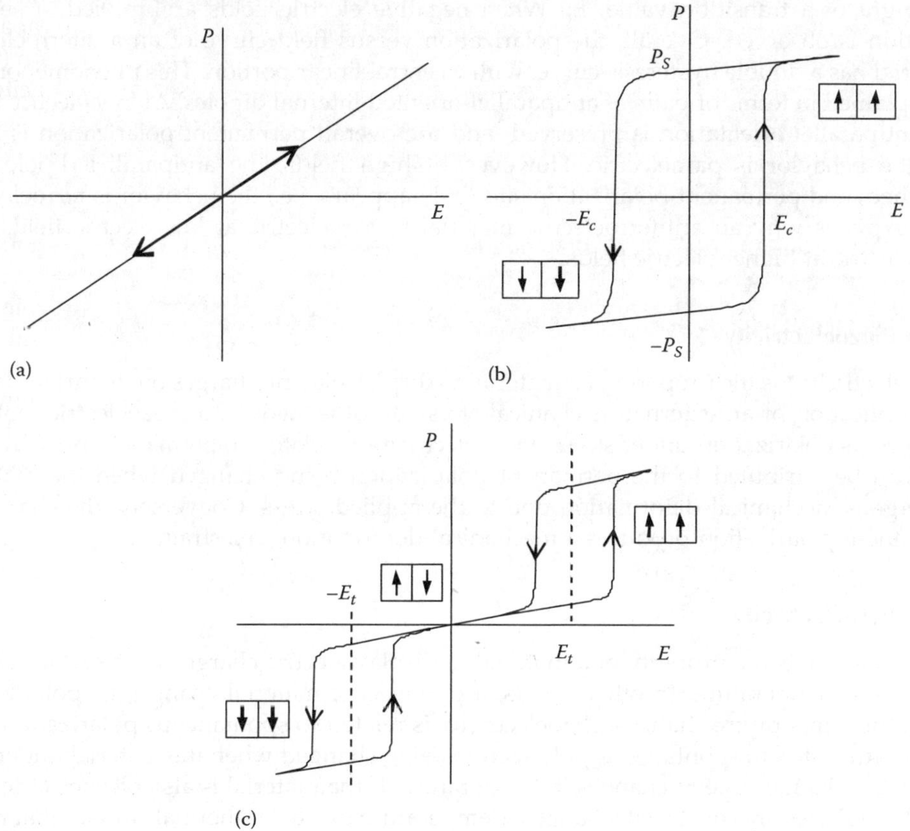

FIGURE 9.3
Polarization versus applied electric field for three types of materials: (a) paraelectric behavior; (b) ferroelectric behavior; (c) antiferroelectric behavior.

Characteristic of this behavior is the high hysteresis loop traveled during a cycle. The ferroelectric behavior can be explained through the existence of aligned internal dipoles that have their direction switched when the electric field is sufficiently strong. The slight horizontal slopes observed in Figure 9.3b are attributable to the paraelectric component of the total polarization.

9.3.6 Antiferroelectric Materials

Antiferroelectric materials have paraelectric behavior at low electric fields, and ferroelectric behavior at high electric fields. Figure 9.3c shows that, at low fields, an antiferroelectric material displays polarization that increases and decreases proportionally with the applied electric field. However, at higher fields, the polarization displays a ferroelectric behavior, characterized by a sudden increase in the polarization value and the presence of a hysteresis loop. As the field is decreased back to low values, the large polarization is suddenly reduced, the paraelectric behavior is recovered, and no permanent polarization is observed at zero field. This shows that the ferroelectric behavior is confined to a high field, to the left and right of a transition value, E_t. When negative electric fields are applied, a similar situation is observed. Overall, the polarization versus field curve of an antiferroelectric material has a double hysteresis curve, with a central linear portion. This phenomenon can be explained in terms of pairs of antiparallel-oriented internal dipoles. At low electric field, the antiparallel orientation is preserved, and the overall permanent polarization is zero, i.e., the behavior is paraelectric. However, at high fields, the antiparallel dipole gets switched, and permanent polarization suddenly appears, i.e., the behavior is ferroelectric. This explains why an antiferroelectric material is paraelectric at low electric fields, but ferroelectric at higher electric fields.

9.3.7 Piezoelectricity

Piezoelectricity* is the property of a material to display electric charges on its surface under the application of an external mechanical stress. In other words, a piezoelectric material changes its polarization under stress. Piezoelectricity is related to permanent polarization, and can be attributed to the permanent polarization being changed when the material undergoes mechanical deformation due to the applied stress. Conversely, the change in permanent polarization produces a mechanical deformation, i.e., strain.

9.3.8 Pyroelectricity

Pyroelectricity is the property of a material to display electric charge on its surface due to changes in temperature. In other words, a pyroelectric material changes its polarization when the temperature changes. Pyroelectricity is related to spontaneous polarization, and can be attributed to spontaneous polarization being changed when the material undergoes geometric changes due to changes in temperature. If the material is also piezoelectric, and if its boundaries are constraint, change in temperature produces thermal stresses that result in high polarization through the piezoelectric effect. Rochelle salt was one of the first observed ferroelectric materials. Most ferroelectric materials are both piezoelectric and pyroelectric. Remarkable about the Rochelle salt was that its piezoelectric coefficient was much larger than that of quartz. However, quartz is much more stable and rugged.

9.4 Ferroelectric Perovskites

Perovskites are a large family of crystalline oxides with the metal to oxygen ratio 2/3. Perovskites derive their name from a specific mineral known as perovskite. The parent material, perovskite, was first described in the 1839 by the geologist Gustav Rose, who named it after the famous Russian mineralogist Count Lev Aleksevich von Perovski (1792–1856). The simplest perovskite lattice has the expression, $X_m Y_n$, in which the X atoms are rectangular close-packed and the Y atoms occupy the octahedral interstices. The rectangular close-packed X atoms may be a combination of various species, X^1, X^2, X^3, etc. For example, in the barium titanate (BT) perovskite, $BaTiO_3$, we have $X^1 = Ba^{2+}$ and $X^2 = Ti^{4+}$, while $Y = O^{2-}$ (Figure 9.4). In the lattice structure, the Ba^{2+} divalent cations are at the corners, the Ti^{4+} tetravalent cation is in the center, while the O^{2-} anions are on the faces.

---

* The prefixes "piezo" and "pyro" are derived from the Greek words for "force" and "fire," respectively.

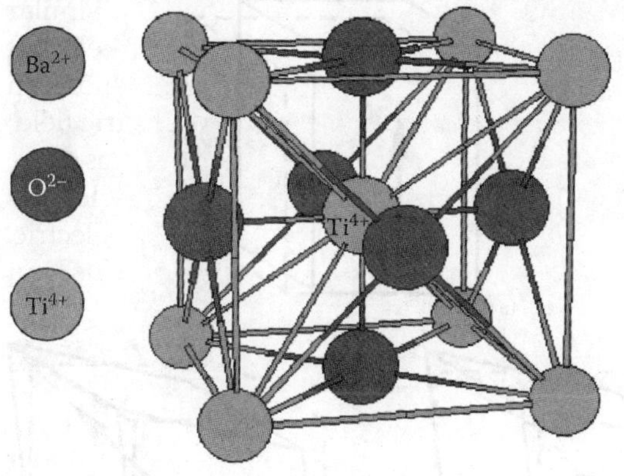

FIGURE 9.4
Crystal structure of a typical perovskite, BaTiO$_3$: the Ba^{2+} cations are at the cube corners, the Ti^{4+} cation is in the cube center, and the O^{2-} anions on the cube faces.

The Ba^{2+} cations are larger, while the Ti^{4+} cations are smaller. The size of the Ba^{2+} cation affects the overall size of the lattice structure. Perovskite arrangements like in BaTiO$_3$ are generically designated ABO$_3$. Their main commonality is that they have a small, tetravalent metal ion, e.g., titanium or zirconium, in a lattice of larger, divalent metal ions, e.g., lead or barium, and oxygen ions, e.g., lead or barium, and oxygen ions. Under conditions that confer tetragonal or rhombohedral symmetry, each crystal has a dipole moment.

9.4.1 Spontaneous Strain and Spontaneous Polarization of the Perovskite Structure

At elevated temperatures, the primitive perovskite arrangement is symmetric facedcentered cubic (FCC) and does not display electric polarity (Figure 9.5a). This symmetric lattice arrangement forms the *paraelectric phase* of the perovskite, which exists at elevated temperatures. As the temperature decreases, the lattice shrinks and the symmetric arrangement is no longer stable. In BT, i.e., BaTiO$_3$, the Ti^{4+} cation snaps from the cube center to other minimum-energy locations situated off center. This is accompanied by corresponding motion of the O^{2-} anions. Shifting of the Ti^{4+} and O^{2-} ions causes the structure to be

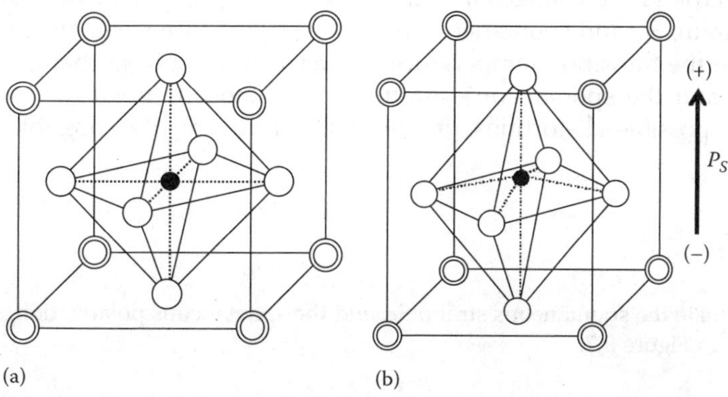

FIGURE 9.5
Spontaneous strain and polarization in a perovskite structure: (a) above the Curie point, the crystal has cubic lattice, displaying a symmetric arrangement of positive and negative charges and no polarization (paraelectric phase); (b) below the Curie point, the crystal has tetragonal lattice, with asymmetrically placed central atom, thus displaying polarization (ferroelectric phase).

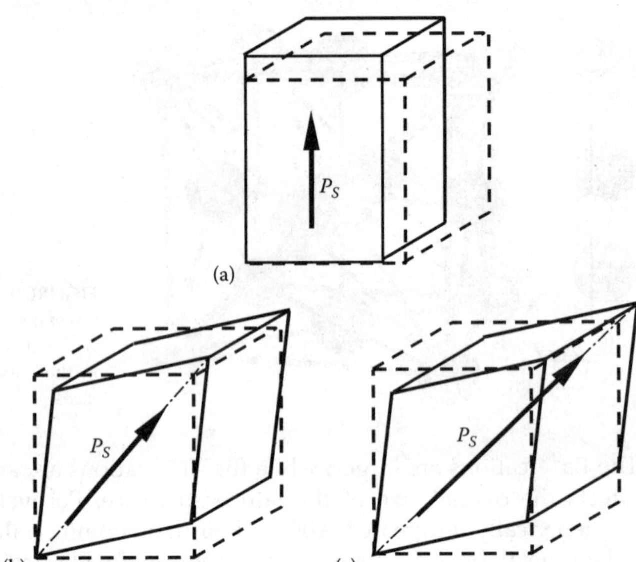

FIGURE 9.6
Possible distortions of the basic FCC lattice in ferroelectric perovskites: (a) tetragonal distortion; (b) orthorhombic distortion; (c) rhombohedral distortion.

altered, creating strain, and electric dipoles. The crystal lattice distorts and becomes slightly elongated in one direction, i.e., tetragonal (Figure 9.5b). In BT, the distortion ratio is $c/a = 1.01$, corresponding to 1% strain in the c-direction with respect to the a-direction.

This change in dimensions along the c-axis is called *spontaneous strain*, S_S. The orthorhombic tetragonal structure has polarity because the centers of the positive and negative charges no longer coincide, yielding a net electric dipole and spontaneous polarization, P_S. This polar lattice arrangement forms the *ferroelectric phase* of the perovskite, which exists at lower temperatures. The transition from one phase into the other takes place at the phase transition temperature, also called the *Curie temperature*, T_c. In BT, the phase transition temperature is around 130°C. As the perovskite is cooled below the transition temperature, T_C, the paraelectric phase changes into the ferroelectric phase, and the material displays spontaneous strain, S_S, and spontaneous polarization, P_S. Vice versa, when the perovskite is heated above the transition temperature, the ferroelectric phase changes into the paraelectric phase, and the spontaneous strain and spontaneous polarization are no longer present. Other possible distortions of the FCC lattice are also possible, as shown in Figure 9.6.

Example 9.1:

Problem: Calculate the spontaneous strain, S_S, and the spontaneous polarization, P_S for the BT lattice shown in Figure 9.7.

SOLUTION:

    (1) Spontaneous strain is calculated by assuming that the undistorted cell height was a, while the distorted cell height is c. Hence,

$$S_S = \frac{c - a}{a} = \frac{403.6 - 399.2}{399.2} \simeq 0.01 = 1\%$$

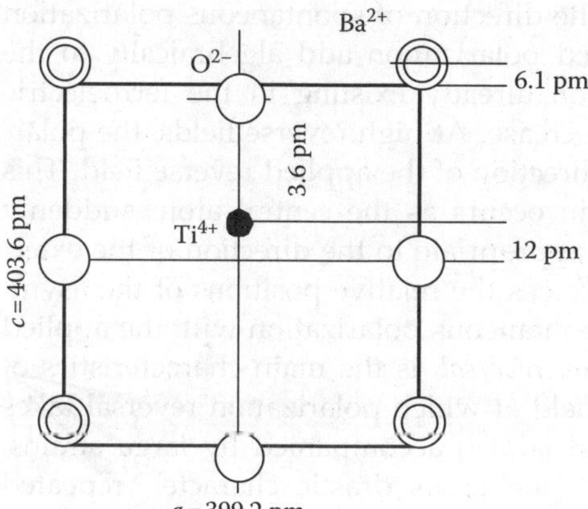

FIGURE 9.7
Ionic shifts inducing spontaneous strain and spontaneous polarization in BT.

(2) Spontaneous polarization is defined as the polarization per unit volume. The polarization is calculated as the sum of the dipole moments of each ion, weighted according to their contribution to the unit cell. Thus,

$$\text{For the Ba}^{2+} \text{ cations,} \quad p_{\text{Ba}^{2+}} = 8\frac{2e}{8}6.1 = 12.2\ e\text{pm}$$

$$\text{For the Ti}^{4+} \text{ cations,} \quad p_{\text{Ti}^{4+}} = 4e \times 12.0 = 48.0\ e\text{pm}$$

$$\text{For the O}^{2-} \text{ anions,} \quad p_{\text{O}^{2-}} = 2\frac{-2e}{2} \times (-3.6) = 7.2\ e\text{pm}$$

where $e = 1.602 \times 10^{-19}$ C is the value of the unit charge and 1 pm $= 10^{-12}$ m. The total polarization is

$$p = p_{\text{Ba}^{2+}} + p_{\text{Ti}^{4+}} + p_{\text{O}^{2-}} = 67.4\ e\text{pm} = 1.080 \times 10^{-29}\ \text{Cm}$$

Thus, the spontaneous polarization is

$$P_S = \frac{p}{a^2c} = \frac{1.080 \times 10^{-29}}{(399.2 \times 10^{-12})^2(403.6 \times 10^{-12})} = 0.170\ \text{C/m}^2$$

This predicted value is not too far off from the experimental value of 0.250 C/m². ∎

9.4.2 Induced Strain and Induced Polarization

When the perovskite is in the ferroelectric phase, strain and polarization can be induced by the application of an electric field. When applied in the direction of spontaneous polarization, the electric field increases the net polarization and the lattice distortion. The additional strain and polarization are called *induced strain* and *induced polarization*, respectively. At first, the induced strain and induced polarization increase linearly with the applied electric field. However, as the field increases to higher values, saturation-induced nonlinear effects set in. The maximum displacement allowed by the crystal structure is termed as the *polarization saturation*.

When the applied electric field is contrary to the direction of spontaneous polarization (i.e., *reverse field*), the induced strain and induced polarization add algebraically to the spontaneous strain and spontaneous polarization already existing in the ferroelectric perovskite, and the net strain and polarization decrease. At high reverse fields, the polarization and strain may suddenly increase in the direction of the applied reverse field. This spontaneous switching of polarization and strain occurs as the central atom suddenly jumps into an opposite off-center location, more appropriate to the direction of the externally applied electric field. The sudden jump reverses the relative positions of the asymmetric ions in the crystal lattice and aligns the spontaneous polarization with the applied electric field. This phenomenon, called *polarization reversal*, is the main characteristics of ferroelectric materials. The value of the electric field at which polarization reversal takes place is called *coercive field*. Polarization reversal is also accompanied by large strains. However, it results in a large hysteresis loop. Due to its drastic character, repeated application of polarization reversal subjects the crystal lattice to considerable internal stresses, increases the lattice fatigue, and shortens the life of the ferroelectric materials.

When the perovskite is in the paraelectric phase, linear piezoelectricity is absent. However, strain can be still induced through the *electrostrictive effect*, which is quadratic in the applied field.

9.4.3 Perovskite Compounds

Perovskite arrangements like in $BaTiO_3$ are generically designated $A^{2+}B^{4+}O_3^{2-}$. In $BaTiO_3$, the sites A^{2+} are occupied by Ba^{2+} cations, while the sites B^{4+} are occupied by Ti^{4+} cations. However, the A^{2+} and B^{4+} sites can also be occupied by other similar cations of similar sizes. For example the A^{2+} site can be occupied by other large-size divalent metallic cations, i.e., $A^{2+} = Ba^{2+}$, Sr^{2+}, Pb^{2+}, Sn^{2+}.... The B^{4+} site can be occupied by other small-size tetravalent metallic cations, i.e., $B^{4+} = Ti^{4+}$, Zr^{4+}, etc. Through such replacement, other perovskite compounds are obtained.

9.4.3.1 Solid-Solution Perovskites

Besides the basic replacement of the cations in the formulation, $A^{2+}B^{4+}O_3^{2-}$, it is also possible to have mixtures of similar cations forming solid solution alloys of various proportions. For example, the solid solution alloying of the B site can yield a combination of the form $A^{2+}\left(B_{1-x}^{4+}B_x'4+\right)O_3^{2-}$. In solid solutions, the distribution of the B and B′ cations can be *ordered* or *disordered*. Solid-solution perovskites may have several ferroelectric phases. The alloying proportions of the B and B′ components influence the type of phases present in the solid solution.

9.4.3.2 Complex Perovskites

Complex perovskites, of the form $A^{2+}\left(B_{1/2}^{3+}B_{1/2}'^{5+}\right)O_3^{2-}$, $A^{2+}\left(B_{1/3}^{2+}B_{2/3}'^{5+}\right)O_3^{2-}$ and $\left(A_{3/4}^{1+}A_{1/4}'^{3+}\right)\left(B_{1/6}^{2+}Nb_{5/6}^{5+}\right)O_3^{2-}$, are possible, for example, $A^{1+} = K^{1+}$; $A'^{3+} = Bi^{3+}$; $B^{2+} = Zn^{2+}$; Mg^{2+}, $B^{3+} = Mg^{3+}$; $B'^{5+} = Nb^{5+}$, etc. The complex perovskites can be *ordered* or *disordered*. In a disordered complex perovskite, the larger B atoms locally prop open the lattice, and the smaller B′ atoms have "rattling" space around them. In the ordered complex perovskite, the large B atoms and the smaller B′ atoms are placed in an ordered arrangement that eliminates any extra space; the structure becomes close-packed; and the smaller B′ atoms cannot rattle.

When an electric field is applied to a disordered complex perovskite, the B′ atoms with a large "rattling space" can shift easily without distorting the oxygen octahedron. The shift of the B′ atoms will produce polarization. Since this shift can be easily achieved, large polarization per unit electric field, i.e., large dielectric permittivity, is to be expected. In addition, the strain per unit polarization, i.e., the electrostrictive coefficient Q, is also expected to be small. The disordered complex perovskite have another interesting property: they exhibit a diffuse phase transition over a wide temperature range.

In ordered complex perovskite, the application of an electric field cannot make either the B or the B′ atoms shift without distorting the oxygen octahedron. In this case, small polarization coefficients and large electrostrictive coefficients are to be expected.

9.4.3.3 Other Perovskites

Mixed valence solid solutions on the A sublattice, or B ions in their less stable oxidation states (e.g., Ti^{3+} in $LaTiO_3$), can exhibit oxygen nonstoichiometry and metallic, semiconducting, or superconducting properties. Other ferroelectrics based on the perovskite structure exhibit rhombohedral distortion below the Curie temperature. The perovskite structure is conductive to solid solution alloying (e.g., metallic $La_{1-x}Sr_xMnO_3$) and polytypoid formation (e.g., the $Ba_2YCu_3O_{7-x}$ superconductor). In this way, a large variety of perovskite compounds has been synthesized with various crystal lattices.

9.4.3.4 Mechatronics Applications of Perovskite Materials

Most perovskites have high electrical resistivities, which makes them useful as dielectrics. In the paraelectric phase just above the Curie temperature, the very high dielectric constant finds useful applications in compact high-power capacitors. Perovskite materials (e.g., $Sr_{1-x}Ba_xTiO_3$) are being developed as the high-permittivity dielectrics for Gbit dynamic random access memory (DRAM). The ferroelectric perovskites have piezoelectric and electrostrictive properties that confer them many applications as resonators and induced-strain actuation. Ferroelectric perovskites such as $Pb(Zr_{1-x}Ti_x)O_3$ are candidates for non-volatile memories based on the storage of binary information in the orientation of the spontaneous electric dipole. The same materials are used for their piezoelectric properties in microphones, speakers, pressure and acceleration sensors, and high precision positioners. Some of the perovskites may be good conductors and semiconductors. For example, good perovskite conductors are the cubic sodium tungsten bronze. Mixed conductors (ionic and electronic) in this family are being developed as chemical and gas sensors. Other perovskites are superconductors. The high-temperature superconductors show promise in fast, low-power information processing and in power and signal transmission. Some common mechatronics applications of perovskite materials are

- Multilayer capacitor, $BaTiO_3$
- Piezoelectric transducer, $(Zr,Yi)O_3$
- Positive Temperature Coefficient (PTC) Thermistor, $BaTiO_3$
- Electro-optical modulator, $(Pb,La)(Zr,Ti)O_3$
- Switch, $LiNbO_3$
- Dielectric resonator, $BaZrO_3$

- Thick film resistor, $BaRuO_3$
- Electrostrictive actuator, $Pb(Mg,Nb)O_3$
- Superconductor, $Ba(Pb,Bi)O_3$ layered cuprates
- Magnetic bubble memory, $GdFeO_3$
- Laser host, $YAlO_3$
- Ferromagnet, $(Ca,La)MnO_3$
- Refractory electrode, $LaCoO_3$
- Second harmonic generator, $KNbO_3$

9.4.4 Phenomenological Representation of Piezoelectric and Electrostrictive Behavior

For small values of the external electric field, E, the following phenomenological model can be used

$$P = P_S + \varepsilon E \tag{9.40}$$

$$S = QP^2 \tag{9.41}$$

where Q is the electrostrictive constant with respect to polarization. Two situations have to be considered: (1) the ferroelectric phase and (2) the paraelectric phase.

9.4.4.1 Ferroelectric Phase

In the ferroelectric phase, i.e., below the Curie temperature, the material displays nonzero spontaneous polarization, $P_S \neq 0$. Substitution of Equation (9.40) into Equation (9.41) and expansion of the squared binomial yield

$$S = Q(P_S + \varepsilon E)^2$$
$$= QP_S^2 + (2\varepsilon QP_S)E + (Q\varepsilon^2)E^2 \tag{9.42}$$

This can be expressed as

$$S = S_S + dE + ME^2 \tag{9.43}$$

where

$$d = 2\varepsilon QP_S \text{ (piezoelectric strain constant)} \tag{9.44}$$
$$M = Q\varepsilon^2 \text{ (electrostrictive strain constant)} \tag{9.45}$$
$$S_S = QP_S^2 \text{ (spontaneous strain)} \tag{9.46}$$

9.4.4.2 Paraelectric Phase

In the paraelectric phase, i.e., above the Curie temperature, the spontaneous polarization is zero, and the net polarization is entirely due to the applied field

$$P = \varepsilon E \tag{9.47}$$

Substituting Equation (9.47) into Equation (9.46) yields $S_S = 0$, i.e., no spontaneous strain, as expected. However, some induced strain is still obtained through the electrostrictive effect. Substituting $P_S = 0$ into Equation (9.42) yields

$$S = \left(Q\varepsilon^2\right)E^2 = ME \tag{9.48}$$

A similar result could have been obtained by substituting Equation (9.47) directly into Equation (9.48) and then identifying the terms. This proves the common root of both piezoelectric and electrostrictive behaviors. In piezoelectric materials, the piezoelectric strain, both spontaneous and induced, is much larger than the electrostrictive strain, and hence the latter is normally ignored. However, the electrostrictive strain becomes important when the piezoelectric strain is absent. In certain formulations, the electrostrictive strain can be very large, as for example in electrostrictive ceramics.

9.4.5 Temperature Dependence of Spontaneous Polarization, Spontaneous Strain, and Dielectric Permittivity

In the ferroelectric phase, the perovskite displays spontaneous polarization, which decreases with temperature (Figure 9.8a). At the same time, the dielectric permittivity increases with temperature. At the transition temperature, i.e., at the Curie point, T_C, the spontaneous polarization vanishes, and the permittivity suddenly jumps to a very large value. These phenomena are associated with the transition from the ferroelectric phase into

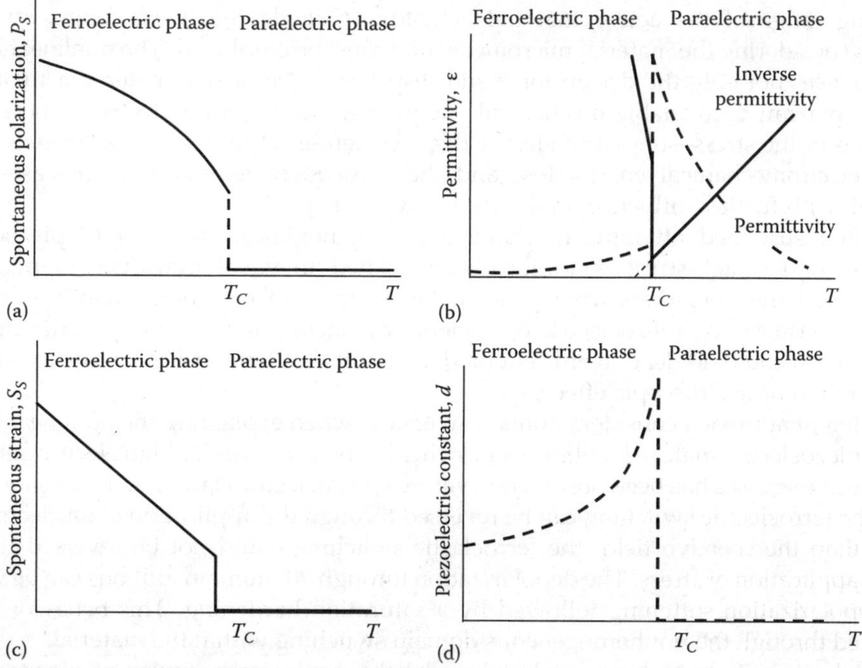

FIGURE 9.8
Temperature dependence of various properties: (a) spontaneous polarization; (b) dielectric permittivity; (c) spontaneous strain; (d) piezoelectric constant.

the paraelectric phase, i.e., from the distorted orthotropic lattice to the symmetric FCC lattice. As the temperature further increases, the permittivity decreases drastically with temperature following a $1/T$ rule. Indeed, a plot of the inverse permittivity, $1/\varepsilon$, versus temperature yields a straight line, which hits zero at a value T_0 below the Curie temperature. T_0 is the *Curie–Weiss temperature*. The corresponding equation is

$$\varepsilon = C \frac{1}{T - T_0} \tag{9.49}$$

where C is the Curie–Weiss constant. It is interesting to note that the value of the product between the electrostrictive constant with respect to polarization, Q, given in Equation (9.41), and the Curie–Weiss constant, C, given in Equation (9.49), is about the same for all perovskite crystals, i.e., $QC = 3.1(0.4) \times 10^3$ m^{4}C^{-2}K. In addition, the electrostrictive coefficient, Q, is nearly proportional to the square of the thermal expansion coefficient, α.

The spontaneous strain, S_S, which only exists in the ferroelectric phase, decreases with temperature (Figure 9.8c). As the Curie temperature, T_C, is crossed, transition from ferroelectric phase into paraelectric phase takes place and the spontaneous strain vanishes. The piezoelectric strain coefficient, d, increases with temperature up to T_C, and then vanishes (Figure 9.8d).

9.4.6 Nonlinear Behavior of Polycrystalline Ferroelectric Ceramics

The linear electromechanical models can accurately describe only a fraction of the full operating range of the active material actuators. Outside the linear range, advanced theories considering the material micromechanics must be employed. The nonlinear behavior influences not only the design force and displacement of active material actuator, but also the maximum available mechanical energy and the required electrical energy. The hysteresis of the stress–strain and electric displacement-electric field characteristics relates to the electromechanical energy loss and the subsequent temperature increase of the material, with further influence on the strain capabilities.

The first sustained attempts to characterize the nonlinear behavior of piezoelectric materials under high-stress conditions were related to naval transducer applications. They related the deviations from a linear behavior to macroscopic quantities such as dielectric permittivity, piezoelectric coefficient, loss factor, and coupling coefficients. But this approach could not lead to efficient modeling of the material behavior due to exclusive consideration of macroscopic effects.

A different approach considers domain switching when explaining the nonlinear behavior of a piezoelectric material. Polarization switching behavior under high electric fields and compressive stresses has been observed in various piezoelectric ceramics. It was shown that while the ferroelectric switching can be reversed through the application of an electric field bigger than the coercive field, the ferroelastic switching could not be reversed through further application of stress. The depolarization through 90° domain motions can be viewed as a depolarization softening followed by a saturation hardening. This behavior can be explained through the nonhomogeneous domain switching within the material.

Several attempts have been made to model the nonlinear behavior of piezoceramics under high electromechanical driving conditions. Whether these models involve finite element methods or just algebraic manipulations, they are usually tuned to experiments rather than predict the outcome of such experiments using first-principles models. This is mainly due to the scarcity of experimental data from the active materials and actuators manufacturers, regarding the nonlinear material behavior.

Regarding the electrical energy and electrical power needed to drive the actuator, early studies pointed out a strong dependence of dielectric permittivity on stress and AC field amplitude. The capacitance values of piezoelectric wafers driven at different frequency and driving fields seem to display an almost linear increase with voltage. The aspects strongly affect the power requirements of a piezoelectric actuator.

9.4.7 Induced-Strain Activation in Ferroelectric Ceramics

A poled ferroelectric ceramic responds with typical piezoelectric behavior to the application of an electric field or a mechanical stress. When a mechanical strain is applied, the polarization is changed due to the *direct piezoelectric effect*. When an electric field is applied, the mechanical strain is changed due to the *converse piezoelectric effect*, i.e., induced-strain actuation.

In a poled ferroelectric ceramic, electric domains exist mainly in two varieties:

1. Electric domains that are more or less aligned in the direction of the electric field during the poling operation, and are more or less parallel to the direction of spontaneous polarization, P_S

2. 90° electric domains that did not orient themselves during poling and are left perpendicular to the direction of spontaneous polarization, P_S

If an external electric field is applied in the direction of spontaneous polarization, the induced-strain actuation process takes place in three major steps (Figure 9.9). First, through the *intrinsic effect*, the strain of the piezoelectric domains increases under the influence of the applied electric field (Curves *a* and *b* in Figure 9.10). This effect is rather linear, and relates to conventional piezoelectricity. The induced strain adds to the already existing spontaneous strain, $S_S \simeq 0.275\%$, already created during the poling process (Figure 9.10).

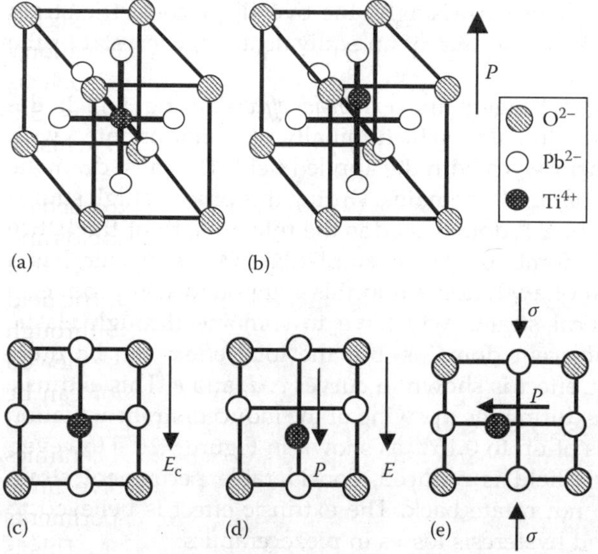

FIGURE 9.9
Behavior of lead titanate (PT) crystal: (a) cubic (paraelectric) phase above the Curie transition temperature, T_c; (b) tetragonal (ferroelectric) phase below T_c showing the Ti atom squeezed upward, which results in a spontaneous polarization, P pointing upward; (c) depolarization under coercive field E_c; (d) 180° domain switch for electric field higher than the coercive field, $E_c < E$; (e) 90° domain switch for stresses higher than the coercive stress, $\sigma_c < \sigma$.

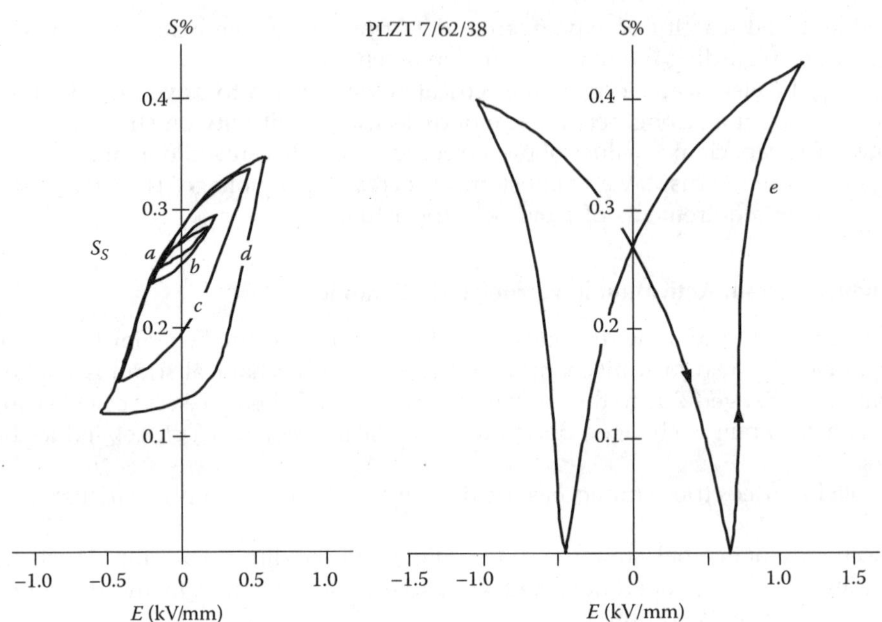

FIGURE 9.10
Induced-strain curves in PLZT 7/62/38 for various levels of electric field.

During the intrinsic response, the electric domains that are better aligned with the electric field deform more, while those that are less aligned deform less, because the field strength is projected through the individual orientation angle of each domain. In addition, the resulting strain in each domain has to be projected back onto the main strain direction, which in tetragonal lattices coincides with the electric field direction. Thus, the overall effect is strongly affected by the percentage orientation of the electric domains with respect to the direction of overall polarization. Not surprisingly, the overall piezoelectric strain coefficient of a ferroelectric ceramic is less than that of an equivalent single crystal of the same formulation.

Further increase in the electric field triggers the *extrinsic effect* during which the domains undergo rotation. Ferroelectric domains, which initially were not oriented with the applied field, now tend to orient themselves with the applied field. The most dramatic reorientation that can take place is that of the 90° domains. During this process, high strains can be produced since the rotation of the 90° domains adds the full strength of the lattice spontaneous strain existing in the 90° domains. (For example, $BaTiO_3$ perovskite has a spontaneous strain of 1%. The rotation of a 90° domain in this compound will produce a local strain of 1%.) Of course, the local strains will have to combine through elastic interactions with the strains of the adjacent domains, but the total effect can be quite significant. In Figure 9.10, the extrinsic effect is shown in curves *c*, *d*, and *e*. This extrinsic effect produces very spectacular results during the upswing of the induced-strain actuation process, and induces additional strains of up to 0.15% as shown in Figure 9.10. However, during the downswing, as the electric field is reduced, considerable permanent strain remains, since the ex-90° domains do not rotate back. The extrinsic effect is believed to be one of the causes of nonlinearity and hysteresis losses in piezoceramics.

In addition, the material undergoes electrostriction. This is a volumetric effect, which is proportional to the square of the electric field. In conventional piezoceramics, the electrostrictive effect is negligible. However, it becomes quite significant in electrostrictive ceramics, as discussed in a later section of this chapter.

If, after being decreased to zero, the electric field is increased in the opposite direction, the strain decreases at first, until all the polarization-induced spontaneous strain is canceled, and the overall strain is zero. Further increase of the reverse electric field beyond this point produces the phenomenon of domain switching, whereby the crystal lattice snaps into a new position, which is now aligned with the applied reversed field. At this point, the reverse coercive field, $-E_c$, was attained. If the reverse field is further increased, small additional strains may still be obtained through the intrinsic effect and through electrostriction, but these are diminishing returns. If the electric field is brought back to zero, the strain decreases, but significant permanent strain remains. Increasing the field in the positive direction will, at first, induce further reduction of the strain because the electric domains are almost all of reverse polarity. Then, as the field is further increased, domain switching will again happen as the value of forward coercive field, $+E_c$, is attained. The result is the "butterfly curve" *e* in Figure 9.10.

The high-field nonlinear behavior is frequency dependent. The typical butterfly curve (Figure 9.10, curve *e*) can only be attained under quasi static application of the electric field, in which case sufficient time is given to the electric domain switching to propagate through the full body of the piezoceramic. If the frequency is increased, the domain reorientation and domain switching cannot fully develop before the electric field is reversed. As a result, as the frequency increases, the maximum attainable strains under maximum field tend to diminish. However, hysteresis also decreases.

9.5 Fabrication of Electroactive Ceramics

Ceramics are hard, brittle, heat-resistant, and corrosion-resistant materials made by "green-state" shaping and high-temperature firing of nonmetallic minerals, such as clay. Ceramics are polycrystalline materials. Common ceramics are electrical and thermal insulators.

Electroactive ceramics are a class of ceramics that display strong piezoelectric and/or electrostrictive response. Electroactive ceramics consist of polycrystalline structures of ferroelectric perovskites with strong piezoelectric and/or electrostrictive properties. Most piezoelectric materials are crystalline solids. Some piezoelectric materials are single crystals, either natural or synthetic. Others are polycrystalline materials that are given macroscale single-crystal like symmetry through poling. The electroactive ceramics are synthetic compounds that can be fabricated with engineered propertied tailored to meet specific requirements. At a macroscale, the ferroelectric ceramics are given single-crystal symmetry by *poling*. Poled ferroelectric ceramics are commonly called piezoelectric ceramics (ANSI/IEEE Std. 176).

Electroactive ceramics display significant mechanical response under applied electric field, and electrical response under applied mechanical action. Typical strain response of commercially available electroactive ceramics is around 0.1% in the quasi linear range. Higher strain response can be obtained by taking the electroactive ceramics in the strongly nonlinear range at high electric fields. Nonlinear strains of up to 0.2% and higher have been reported in certain electroactive ceramics. The operation in the high nonlinear range is usually associated with a marked increase in hysteresis. This results in significant internal heating under high frequency operation. Operation in the high nonlinear domain also results in a marked decrease in fatigue life.

The ceramic perovskites display both piezoelectric and electrostrictive behavior. One or the other of these two properties is usually enhanced through chemical formulation and processing. Some of these electroactive ceramics can display, according to detailed formulation and processing, either a predominantly piezoelectric or a predominantly electrostrictive behavior.

9.5.1 Conventional Fabrication of Ferroelectric Ceramics

Fabrication of ferroelectric ceramics is done in several stages:

1. Synthesis of the ferroelectric perovskite powders
2. Sintering and compaction of the perovskite powders into ferroelectric ceramics
3. Electric poling of the ferroelectric ceramics

9.5.1.1 Synthesis of the Ferroelectric Perovskite Powders

For example, let us consider the conventional fabrication of the PZT compound, which has the chemical formula $Pb(Zr_xTi_{1-x})O_3$. The synthesis of the ferroelectric perovskite compound uses the oxide mixing technique. A stoichiometric mixture of lead oxide, PbO, zirconium dioxide, ZrO_2, and titanium dioxide, TiO_2 powders is calcined in an oven at 800°C–1000°C for 1–2 h. The calcination process causes a solid-state reaction that results in aggregates of the PZT perovskite. Note that calcination happens at high temperature but below the melting point of the constituents. The resulting PZT perovskite aggregates are crushed into smaller size and then milled into a fine powder.

9.5.1.2 Sintering and Compaction of the Perovskite Powders into Ferroelectric Ceramics

PZT perovskite powder is mixed with a binder and laid into shapes called green preforms. To obtain thin sheets, the mixture is cast into a film using the doctor blade method, in which the distance of a blade above the carrier determines the film thickness. The preforms are subjected to sintering and compaction/densification at elevated temperatures. The sintering is a diffusion process that takes place at elevated temperatures, but below the melting point of the material. Thus, a ferroelectric ceramic is obtained through sintering of the fine particles of ferroelectric PZT perovskite. The sintering process provides sufficient mechanical strength to the final product without distorting significantly from the initial shape. During sintering, accelerated diffusion of the constituent atoms takes place at the interface between particles. This diffusion reconstructs the crystal bonding at the interface and provides binding between the particles. The diffusion rate is accelerated by temperature. Another important factor is the grain size. This is related to the speed of diffusion at the interface between adjacent particles. Since the surface to volume ratio increases significantly when the grain size is reduced, the use of finer particles provides higher specific area and promotes faster sintering and better internal structure. Faster sintering is beneficial because it prevents excessive grain growth. This results in a higher density ceramic with more effective ferroelectric properties.

The sintering of PZT perovskite powder is done in a furnace at 1200°C. During the ramping up of the furnace temperature, the binder is volatized at around 500°C ("bisque firing"). The careful control of this intermediate stage is important for the quality of the final product. After the binder has volatized, the temperature is ramped up again ("high firing") and then maintained at 1200°C during the sintering process. Hot isostatic pressing (HIP process) could also be applied to increase the ceramic density by compacting out the intergranular pores and voids. HIP also increases the ceramic performance as a ferroelectric material.

9.5.1.3 Electric Poling of the Ferroelectric Ceramics

During cooling, the PZT ceramic undergoes phase transformation from paraelectric state to ferroelectric state. This transformation takes place as the material cools below the Curie temperature, T_C. The resulting ferroelectric ceramic has a polycrystalline structure (grains) with randomly oriented ferroelectric domains (Figure 9.11a). If the grains are large, ferroelectric domains can exist even inside each grain. Due to the random orientation of the electric domains, individual polarizations cancel each other, and the net polarization of the virgin ferroelectric ceramic is zero.

This random orientation can be transformed into a preferred orientation through poling. Poling aligns the dipole domains, and gives permanent polarization to the piezoceramic material. A poled ferroelectric ceramic behaves more or less like a single crystal. Poling of piezoceramics is attained at elevated temperatures in the presence of a high electric field, e.g., 1–3 kV/mm DC in a silicon oil bath. The application of a high electric field at elevated temperatures results in the alignment of the crystalline domains. This alignment is locked in place when the piezoceramic is cooled with the high electric field still applied; thus, we obtain *permanent polarization*. During poling, the orientation of the piezoelectric domains

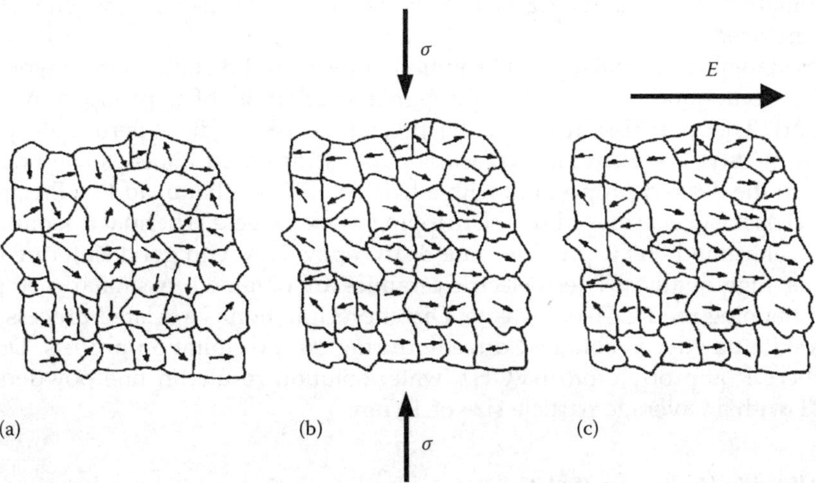

FIGURE 9.11
Piezoelectric effect in polycrystalline perovskite ceramics: (a) in the absence of stress and electric field, the electric domains are randomly oriented; (b) application of stress produces orientation of the electric domains perpendicular to the loading direction. The oriented electric domains yield a net polarization; (c) application of an electric field orients the electric domains along the field lines and produces induced strain.

also produces a mechanical deformation. When the piezoceramic is cooled, this deformation is locked in place (permanent strain).

9.5.2 Novel Methods for the Fabrication of Ferroelectric Ceramics

A number of novel fabrication methods have been recently developed for ferroelectric ceramics. Some of these methods aim at generating better ferroelectric powders. Others aim at improving the sintering process. Another group of methods aim at producing the final product, i.e., the ferroelectric ceramic, directly, without intermediate steps. Other methods aim at growing ferroelectric single crystals, which have much better properties than the ferroelectric ceramics. Some of these methods are briefly discussed next.

9.5.2.1 Doping

Doping is used to improve the solid-reaction process. Magnesium oxide, MgO, when in a few mol% excess during the PMN–PT fabrication is particularly effective in obtaining a perfect perovskite structure. Lead oxide, PbO, when in a few mol% excess during the PMN sintering, suppresses the unwanted pyrochlore structure and promotes the predominant perovskite structure.

9.5.2.2 Coprecipitation

Coprecipitation and *alkoxide hydrolysis* are wet chemical methods that can produce better perovskite powder precursors to the sintering process. The conventional high-temperature solid-state reaction process yields perovskite conglomerates that have to be crushed and milled into fine powders before being used in the sintering process. However, this process cannot ensure uniform microscopic composition, cannot go below 1 μm size, and is prone to contamination from the milling tools. Such problems are alleviated when wet chemical methods are used.

Coprecipitation starts with a liquid solution of the mixed metallic salts in which precipitation of a compound is promoted through the addition of a precipitation promoter (precipitant). Thermal dissolution of the precipitate results in a perovskite powder of very good quality. For example, $BaTiO_3$ is obtained from an aqueous solution of $BaCl_2$ and $TiCl_4$ in the desired proportion, with a half volume of ethanol added. Precipitation of $BaTiO(C_2H_4)g4H_2O$ is achieved by the dripping of oxalic acid. Thermal dissolution of this precipitate at around 800°C provides fine $BaTiO_3$ powders with good sintering characteristics. Thus, high-quality BT ferroelectric ceramics are obtained. A similar method can be used for obtaining the PLZT perovskite. Direct precipitation is a related process, in which the perovskite powder is obtained directly during the precipitation process. Dripping Ti$(OR)_4$, where R is propyl, into $Ba(OH)_4$ water solution results in fine powders of high-quality BT with an average particle size of 10 nm.

9.5.2.3 Alkoxide Hydrolysis (Sol–Gel)

Alkoxide hydrolysis, also called sol–gel method, starts with metal alkoxides $M(OR)_n$, where M is a metal and R is an alkyl, mixed in alcohol in appropriate proportions. When water is added, the hydrolytic reaction produces alcohol and metal oxide or metal hydrate. The method produces very fine high-purity powders. Since metal alkoxides are evaporative, purification is easy through distillation. Purity of 99.98% has been reported, with remarkable increase in the permittivity of the sintered ceramic, in comparison with oxide-mixing products. To obtain BT powders, $Ba(OC_3H_7)_2$ and $Ti(OC_5H_{11})_4$ compounds are diluted with isopropyl alcohol (or benzene) and hydrolyzed at the appropriate pH. Very fine stoichiometric BT powders with good crystallinity and small particle size (1–10 nm particles in 1 μm agglomerates) can be obtained.

To produce the PZT ferroelectric ceramic, a partial sol–gel method is used, since Pb alkoxide is more difficult to obtain than Ti and Zr alkoxides. A two-stage process is applied. In the first stage, the $(Zr,Ti)O_2$ compound is obtained in powder form through the sol–gel method. In the second stage, the $(Zr,Ti)O_2$ powder is mixed with PbO powders and subjected to solid-state reaction through calcination.

9.5.2.4 Thin-Film Fabrication

Thin-film fabrication methods are applied to obtain very thin films of ferroelectric perovskites. Thin-film fabrication methods fall into two large classes:

1. Liquid-phase methods
2. Gas-phase methods

The liquid-phase methods are epitaxial, i.e., promote the growth of the perovskite layer on the face of a substrate, such that the perovskite layer has the same crystal orientation as the substrate. Several methods have been used, such as the count liquid phase epitaxial method, the melting epitaxial method, and the capillarity epitaxial method. Sol–gel method has also been used. The gas phase methods include electron beam sputtering, RF sputtering, pulse-laser deposition, and chemical vapor deposition.

9.5.2.5 Single-Crystal Growth

Single-crystal fabrication has traditionally been used for the growth of quartz, SiO_2, crystals through hydrothermal synthesis, and of $LiNbO_3$ and $LiTaO_3$ perovskites, by the Czochralski method. More recently, the growth of PMN–PT and PZT–PT single crystals grown through the flux method has been reported. Very large electromechanical coupling factors (~95%) and remarkable piezoelectric constants (1570 μC/MN) can be obtained when poled in a certain crystal direction.

9.6 Piezoelectric Ceramics

Piezoelectric ceramics are ferroelectric perovskites in which the piezoelectric response is dominant. Commercial piezoelectric ceramics are typically made of simple perovskites and solid-solution perovskite alloys. Typical examples of simple perovskites are

- BT with chemical formula $BaTiO_3$
- PT with chemical formula $PbTiO_3$

Typical examples of solid-solution perovskites are

- Lead zirconate titanate (PZT) with chemical formula $Pb(Zr,Ti)O_3$
- Lead lanthanum zirconate titanate (PLZT) with chemical formula $(Pb,La)(Zr,Ti)O_3$
- Ternary ceramics, e.g., $BaO\text{-}TiO_2\text{-}R_2O_3$, where R is a rare earth

9.6.1 Soft and Hard Piezoelectric Ceramics

When the piezoceramics are classified according to their coercive field during induced-strain actuation, two main categories emerge. If the coercive field is large, say greater than 1 kV/mm, than the piezoceramic is *"hard."* A hard piezoceramic shows an extensive linear drive region, but a relatively small strain magnitude. If the coercive field is moderate, say between 0.1 and 1 kV/mm, then the piezoceramic is classified as *"soft."* A soft piezoceramic shows a large field-induced strain, but relatively large hysteresis. Finally, if the coercive field is less than 0.1 kV/mm, the material is rather an electro-strictor, which displays an approximately quadratic dependence of strain on the electric field. The hard and soft behavior is also related to the Curie temperature, T_C. Hard piezoceramics tend to have a higher Curie point, $T_C > 250°C$, while soft piezoceramics have a moderate Curie point, $150°C < T_C < 250°C$ (Table 9.2). The hard PZT compositions are acceptor doped, whereas the soft PZT compositions are donor doped. The stress-induced 90° domains switching for hard PZT compositions occurs at higher stresses than in the case of soft PZT compositions

9.6.2 Navy Types Designation of Piezoelectric Ceramics

The standard DOD-STD-1376B(SH) defines six piezoelectric ceramic types also known as Navy Type I–VI. This standard is used by piezoelectric ceramics manufacturers and suppliers as a minimum quality requirement for their products (Table 9.3). The following definitions apply:

TYPE I   A modified PZT composition is generally recommended for medium- to high-power acoustic applications. Its "resistance" to depoling at high electric drive and mechanical stress makes it suitable for deep-submersion acoustic applications.

TYPE II   A PZT composition modified to yield higher charge sensitivity, but one that is not suitable for high electric drive due to dielectric heating. This material is more suitable for passive devices such as hydrophones. Advantages also include better time stability.

TYPE III   Similar to Type I, but greatly improved for use at high electric drive because of lower losses. Its field dependency on dielectric and mechanical losses is substantially reduced. However, at low to moderate electric-drive levels, Type I material may actually be a better choice because of greater electromechanical activity

TYPE IV   A modified BT composition for use in moderate electric-drive applications. It is characterized by lower piezoelectric activity and lower Curie temperature than any of the PZT compositions.

TYPE V   A composition intermediate to Types II and VI and thus to be used accordingly.

TYPE VI   Similar to Type II with higher charge sensitivity and dielectric constant, at the expense of a reduced Curie temperature.

TABLE 9.2

Typical Piezoelectric, Dielectric, and Mechanical Properties for Piezoelectric Ceramics PZT-5 and PZT-8; Transverse Isotropy Is Assumed, $(\ldots)_{23} = (\ldots)_{13}$

Property	Soft PZT-5 Navy Type VI	Hard PZT-8 Navy Type III
ρ (kg/m^3)	7600	7600
$d_{33}(\times 10^{-12}$ m/V)	550	220
$d_{31}(\times 10^{-12}$ m/V)	−270	−100
$d_{15}(\times 10^{-12}$ m/V)	720	320
$g_{33}(\times 10^{-3}$ C/N)	18	25
$g_{31}(\times 10^{-3}$ C/N)	−9	−11
$g_{15}(\times 10^{-3}$ C/N)	24	36
$s_{11}^{E}(\times 10^{-12}$ m^2/N)	16	11
$s_{33}^{E}(\times 10^{-12}$ m^2/N)	20	13
$\varepsilon_{33}/\varepsilon_0$	3400	1000
k_{33}	0.71	0.61
k_{31}	0.36	0.31
k_{15}	0.67	0.54
Young's modulus Y_{11}^{E} (GPa)	62	91
Young's modulus Y_{33}^{E} (GPa)	50	77
Poisson ratio ν	0.31	0.31
Tensile strength X_t (MPa)	75	75
Compressive strength X_c (MPa)	>520	>520
Depoling compression stress T_C (MPa)	30	150
Poling field, DC, E_c^{DC} (kV/mm)	1.2	0.5
Depoling field, AC, E_c^{AC} (kV/mm)	0.7	1.5
Curie temperature T_C (°C)	200	300
Mechanical quality factor Q_M	75	900

Note: $\varepsilon_0 = 8.85 \times 10^{-12}$ F/m.

TABLE 9.3

Navy Type Designation of Hard and Soft Piezoelectric Material

Navy Type Designation	Commercial Designation	Material Type	Notes
Navy Type I	PZT-4	PZT	High-power hard piezoelectric
Navy Type II	PZT-5A/PC5	PZT	High-sensitivity "soft" material
Navy Type III	PZT-8	PZT	High-power "hard" piezoelectric
Navy Type IV	PC-3	BT	High-power "hard" piezoelectric
Navy Type V	PZT-5J	PZT	High-sensitivity "soft" material
Navy Type VI	PZT-5H	PZT	High-sensitivity "soft" material

Source: http://www.morganelectroceramics.com/piezomaterials/index.html

9.6.3 PZT Piezoceramics

PZT is a solid-solution ferroelectric perovskite with wide applications in induced-strain actuation. The acronym PZT stands for *lead zirconate titanate*, i.e., $Pb(Zr_{1-x}Ti_x)O_3$. In a PZT perovskite unit cell, the lead ion Pb^{2+} occupies the corners, the oxygen ion O^{2-} occupies the faces, and the titanium/zirconium ions Zr^{4+}/Ti^{4+} occupy the octahedral voids. To date, many PZT formulations exist, the main differentiation being between soft (e.g., PZT 5-H) and hard (e.g., PZT 8). PZT attains the highest piezoelectric coupling and the maximum electric permittivity near the morphotropic phase boundary (MPB), which happens when the Zr/Ti ratio is approximately 53/47 (Figure 9.12). This corresponds to the change in the crystal structure from the tetragonal phase to the rhombohedral phase (Figure 9.6c). The explanation for this phenomenon is as follows.

Above the Curie temperature, PZT has a cubic lattice and is paraelectric. The Curie temperature varies with the alloying proportion, from ~250°C for pure $PbZrO_3$ to ~500°C for pure $PbTiO_3$ (Figure 9.12). Below the Curie temperature, PZT is ferroelectric but its lattice can be either tetragonal or rhombohedral, according to the alloying proportion. In the phase diagram, the line separating the two phases is called the MPB. Figure 9.12 shows that the tetragonal lattice has six distortion variants, i.e., the central cation can be displaced in any one of the six possible positions parallel to the three lattice axes. The rhombohedral

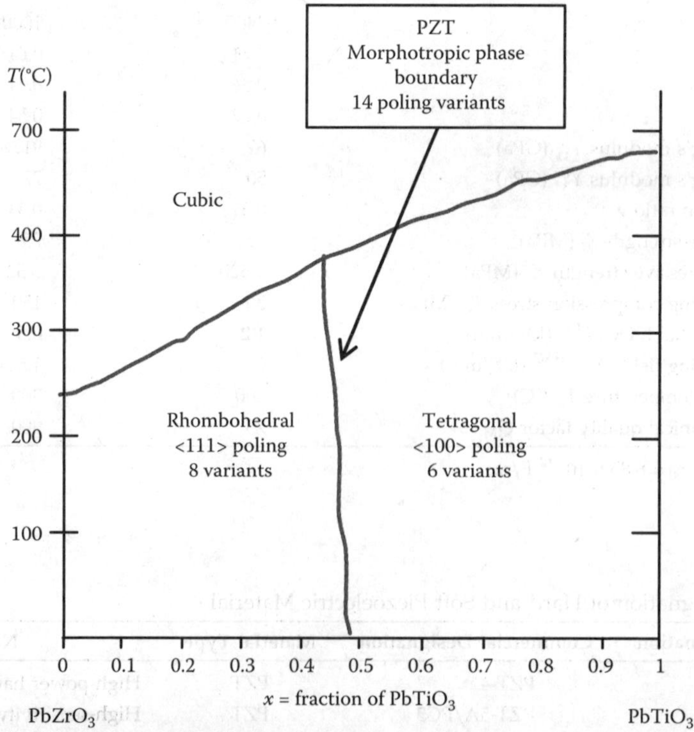

FIGURE 9.12
Phase diagram of $Pb(Zr_{1-x}Ti_x)O_3$ perovskite, showing the number of distortion variants in each of the ferroelectric phases and the PZT formulation at the MPB corresponding to $x = 0.53$ at room temperature.

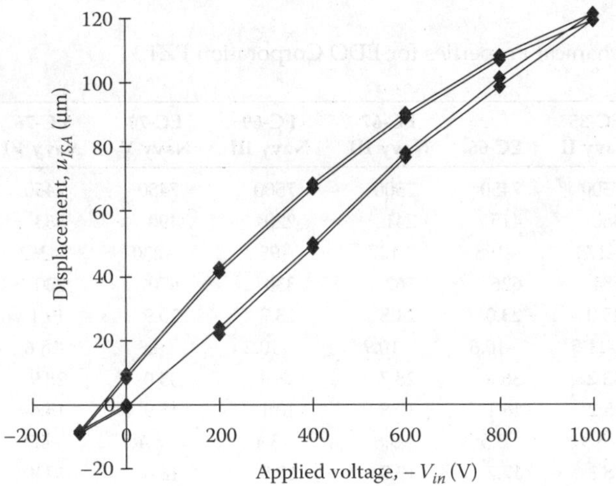

FIGURE 9.13
The induced-strain displacement versus applied voltage for a Polytec PI P-245.70 long PZT stack (stack length is 99 mm, 100 μm corresponds to 0.1% strain).

lattice has eight distortion variants, i.e., the central cation can be displaced in any one of the eight possible positions parallel to the four diagonals. On the line separating the two phases in the phase diagram, i.e., on the MPB, both the tetragonal phase and the rhombo-hedral phase may exist. Hence the total number of distortion variants on MPB is 14, which is the cumulative effect of both phases. Having more distortion variants increases the material responsiveness, since the material has more options to deform under the action of external factors, e.g., electric field or mechanical pressure. At room temperature, the MPB is placed around the 47/53 alloying ratio.

Within the linear range, PZT-like piezoelectric ceramics produce strains that are more or less proportional to the applied electric field or voltage (Figure 9.13). Induced strains of more than 1000 μ strain (0.1%) are encountered. These features make piezoelectric ceramics very attractive for actuator applications.

9.6.4 Commercially Available Piezoceramic Materials

Many commercially available piezoceramic formulations exist. For example, Table 9.4 presents the material description and typical applications for a number of PZT piezoelectric ceramics produced by EDO Corporation:

- EC-63 is a PZT formulation developed for moderate power applications. It features low dissipation, high d constants, and high Curie point. These features make this material desirable for ultrasonic cleaning equipment and other cavitation products, as well as acoustic projectors.

- EC-64 is a Navy Type I composition of PZT material that was developed for high-power acoustic projectors. It has high coercive field, high coupling coefficient, and low dielectric loss under high driving fields. It is well suited for high-power, low-frequency broadband projectors and other high-power electroacoustic devices.

- EC-65 is a Navy Type II composition of PZT material that was specially developed for its high dielectric permittivity, high sensitivity, and low aging rate. It is well suited for high-sensitivity hydrophones and other receiving devices.

TABLE 9.4

Typical Piezoelectric, Dielectric, and Mechanical Properties for EDO Corporation PZT Piezoelectric Ceramics

Property	EC-63	EC-64 Navy I	EC-65 Navy II	EC-66	EC-67 Navy III	EC-69 Navy III	EC-70 Navy V	EC-76 Navy VI
ρ (kg/m^3)	7500	7500	7500	7450	7500	7500	7450	7450
$d_{33}(\times 10^{-12}$ m/V)	270	295	380	415	241	220	490	583
$d_{31}(\times 10^{-12}$ m/V)	−120	−127	−173	−198	−107	−95	−230	−262
$d_{15}(\times 10^{-12}$ m/V)	475	506	584	626	362	330	670	730
$g_{33}(\times 10^{-3}$ C/N)	24.1	25.0	25.0	23.0	24.8	23.7	20.9	19.1
$g_{31}(\times 10^{-3}$ C/N)	−10.3	−10.7	−11.5	−10.6	−10.9	−10.2	−9.8	−8.6
$g_{15}(\times 10^{-3}$ C/N)	37.0	39.8	38.2	36.6	28.7	28.9	35.0	28.9
$s_{11}^{E}(\times 10^{-12}$ m^2/N)	11.3	12.8	15.2	16.1	10.8	10.1	15.9	15.6
$s_{12}^{E}(\times 10^{-12}$ m^2/N)	−3.7	−4.2	−5.3	−5.5	−3.6	−3.4	−5.4	−4.7
$s_{33}^{E}(\times 10^{-12}$ m^2/N)	14.3	15.0	18.3	17.7	13.7	13.5	18.0	19.8
$\varepsilon_{33}^{T}/\varepsilon_0$	1250	1300	1725	2125	1100	1050	2750	3450
k_{33}	0.68	0.71	0.72	0.72	0.66	0.62	0.74	0.75
k_{31}	0.34	0.35	0.36	0.36	0.33	0.31	0.37	0.38
k_{15}	0.69	0.72	0.69	0.68	0.59	0.55	0.67	0.68
k_p	0.58	0.60	0.62	0.62	0.56	0.52	0.63	0.64
Young's modulus Y_{11}^{E} (GPa)	89	78	66	62	93	99	63	64
Poisson ratio σ	0.31	0.32	0.33	0.33	0.31	0.29	0.31	0.29
Curie temperature T_C (°C)	320	320	350	270	300	300	220	190
Mechanical Q_M	500	400	100	80	900	960	75	65

Note: $\varepsilon_0 = 8.85 \times 10^{-12}$ F/m ; Poisson ratio calculated from the formula, $k_p^2 = \dfrac{2}{1-\sigma}k_{31}^2$.

- EC-66 is a PZT composition modified to provide a compromise between EC-65 and EC-70, by having high dielectric permittivity and relatively high g constant. This material is intended for hydrophones, low-power projectors, and other receiving devices.

- EC-67 is a Navy Type III composition of PZT material that meets high-drive requirements yet with high coupling properties. It was developed through the modification of the EC-69 composition. It is well suited for low-frequency high-power applications in flextensional designed actuators.

- EC-69 is a Navy Type III composition of PZT material that provides maximum stability for high driving fields and high pressure operations. It is intended for high-power acoustic projectors or deep water applications.

- EC-70 is a Navy Type I composition of PZT material that provides high dielectric permittivity. It is well suited for hydrophones or low-power projectors requiring low impedance and high sensitivity.

- EC-76 is Navy Type VI composition of PZT material that provides an extremely high dielectric permittivity and high coupling coefficient. This formulation is preferred for applications requiring high d constants or large displacements.

TABLE 9.5

Typical Piezoelectric, Dielectric, and Mechanical Properties for EDO Corporation BT Piezoelectric Ceramics

Property	EC-21	EC-31	EC-55(Navy IV)	EC-57
ρ (kg/m^3)	5700	5500	5500	5300
$d_{33}(\times10^{-12}$ m/V)	117	152	150	87
$d_{31}(\times10^{-12}$ m/V)	−49	−59	−58	−32
$d_{15}(\times10^{-12}$ m/V)	191	248	245	154
$g_{33}(\times10^{-3}$ C/N)	12.4	14.8	14.3	16.2
$g_{31}(\times10^{-3}$ C/N)	−5.2	−5.8	−5.6	−5.5
$g_{15}(\times10^{-3}$ C/N)	15.7	20.4	20.1	28.0
$s_{11}^{E}(\times10^{-12}$ m^2/N)	8.8	9.3	8.6	8.0
$s_{12}^{E}(\times10^{-12}$ m^2/N)	−2.7	−2.9	−2.6	−2.2
$s_{33}^{E}(\times10^{-12}$ m^2/N)	10.0	9.7	9.8	8.3
$\varepsilon_{33}/\varepsilon_0$	1070	1170	1220	640
k_{33}	0.38	0.48	0.46	0.38
k_{31}	0.17	0.19	0.19	0.15
k_{15}	0.37	0.49	0.48	0.23
k_p	0.26	0.32	0.31	0.25
Young's modulus Y_{11}^{E} (GPa)	114	107	116	125
Poisson ratio ν	0.14	0.29	0.25	0.28
Curie temperature T_C (°C)	130	115	115	140
Mechanical Q_M	1400	400	550	600

Note: Poisson ratio calculated from the formula, $k_p^2 = \dfrac{2}{1-\sigma}k_{31}^2$.

Table 9.5 presents the typical piezoelectric, dielectric, and mechanical properties for EDO Corporation BT piezoelectric ceramics:

- EC-21 is a modified BT composition developed for high-power operations. It is characterized by its low impedance and low dielectric loss under high driving fields. This formulation is well suited to sonar and specialized transducers.

- EC-31 is a modified BT composition, which combines high voltage sensitivity with good temperature characteristics in the range −10°C to 60°C. This formulation is a proven material for hydrophones and other receiving devices and low-power projectors, e.g., for fish-finding applications.

- EC-55 is a modification of the EC-31 composition, which was developed to permit operation at higher power levels. It is characterized by its low dielectric losses under high driving fields. This material is well suited for high-power projectors in the −10°C to 60°C temperature range.

- EC-57 is a modified BT composition developed to provide temperature stability of parameters in the −40°C to 100°C temperature range. This material is well suited for high-power ultrasonic cleaning applications and low-loss high-stability sonar projectors.

Table 9.6 presents typical piezoelectric, dielectric, and mechanical properties for American Piezoceramic (APC) Inc. piezoelectric ceramics.

TABLE 9.6

Properties of American Piezoceramics, Inc., Piezoelectric Ceramics

Property	APC 840	APC 841	APC 850	APC 855	APC 856	APC 880
ρ (kg/m^3)	7600	7600	7700	7500	7500	7600
d_{33} ($\times 10^{-12}$ m/V)	290	275	400	580	620	215
d_{31} ($\times 10^{-12}$ m/V)	−125	109	−175	270	260	−95
d_{15} ($\times 10^{-12}$ m/V)	480	450	590	720	710	330
g_{33} ($\times 10^{-3}$ C/N)	26.5	25.5	26	19.5	18.5	25
g_{31} ($\times 10^{-3}$ C/N)	−11	10.5	−12.4	8.8	8.1	−10
g_{15} ($\times 10^{-3}$ C/N)	38	35	36	27	25	28
s_{11}^E ($\times 10^{-12}$ m^2/N)	11.8	11.7	15.3	14.8	15.0	10.8
s_{33}^E ($\times 10^{-12}$ m^2/N)	17.4	17.3	17.3	16.7	17.0	15.0
$\varepsilon_{33}^T/\varepsilon_0$	1250	1350	1750	3250	4100	1000
k_{33}	0.72	0.68	0.72	0.74	0.73	0.62
k_{31}	0.35	0.33	0.36	0.38	0.36	0.30
k_{15}	0.70	0.67	0.68	0.66	0.65	0.55
k_p	0.59	0.60	0.63	0.65	0.65	0.50
Poisson ratio σ	0.30	0.40	0.35	0.32	0.39	0.28
Young's modulus Y_{11}^E (GPa)	80	76	63	61	58	90
Young's modulus Y_{33}^E (GPa)	68	63	54	48	45	72
Curie temperature T_C (°C)	325	320	360	195	150	310
Dissipation factor, tan δ (%)	0.4	0.35	1.4	2	2.7	0.35
Mechanical Q_M	500	1400	80	75	72	1000

Source: www.americanpiezo.com

Note: $\varepsilon_0 = 8.85 \times 10^{-12}$ F/m ; Poisson ratio calculated from the formula, $k_p^2 = \dfrac{2}{1-\sigma}k_{31}^2$.

9.7 Electrostrictive Ceramics

9.7.1 Electrostrictive Ceramics Relaxor Ferroelectrics

Electrostrictive ceramics are perovskite materials in which the electrostrictive response is dominant. Perovskites that display a large electrostrictive response are the disordered complex perovskites, which have high electrostrictive coefficient with respect to electric field, M, and a diffuse transition temperature. Such ferroelectric ceramics are also called *relaxor ferroelectrics*, because they display large dielectric relaxation, i.e., frequency dependence of the dielectric permittivity. In a relaxor ferroelectric, the permittivity decreases as the test frequency increases. In addition, the value of temperature at which the permittivity reaches a peak is noticed to shift upward. This behavior is in contrast with that of conventional ferroelectrics, for which the temperature at which the permittivity peaks hardly changes with frequency. The dielectric relaxation phenomenon can be attributed to the presence of microdomains in the crystal structure. Mulvihill et al. (1995) subjected a <111> single crystal of lead zirconate niobate, $Pb(Zr_{1/3},Nb_{2/3})O_3$, to dielectric constant measurements in two states: (1) unpoled and (2) poled. The unpoled state, which has only microdomains, exhibited the dielectric relaxation phenomenon. The poled state, which has macrodomains induced by the applied electric field, did not exhibit dielectric relaxation, and behaves more like a conventional dielectric.

In relaxor materials, the transition between piezoelectric behavior and nonpiezoelectric behavior does not occur at a specific temperature (Curie point), but instead occurs over a temperature range (Curie range). Thus, electrostrictive ceramics have a rather diffused phase transition that spans a temperature range around the transition temperature. Hence, the temperature dependence of electrostrictive ceramics around the transition temperature is markedly less than that of piezoelectric ceramics. Sometimes, their transition temperature range is lower than the room temperature, which is beneficial for stable operation at elevated temperatures.

Lead magnesium niobate, lead magnesium niobate/lanthanum formulations, and lead nickel niobate currently are among the most studied relaxor electrostrictive ferroelectrics. They have very high dielectric permittivity and polarization. The coercive field of electrostrictive ceramics is much smaller than that of piezoelectric ceramics.

A common electrostrictive ceramic is lead-magnesium-niobate, $Pb(Mg_{1/3}Nb_{2/3})O_3$, also known as PMN. Another commonly used electrostrictive ceramic is $PbTiO_3$, also known as PT. Combination of these two formulations are also common, under the designation PMN–PT. Another electrostrictive ceramic is $(Pb,La)(Zr,Ti)O_3$, also known as PLZT. Other ferroelectric ceramic systems that have been found to display strong electrostrictive behavior include lead barium zirconate titanate, $(Pb,Ba)(Zr,Ti)O_3$, and barium stannate titanate, $Ba(Sn,Ti)O_3$. In order to obtain large electrostriction, it is essential that ferroelectric microdomains in the ceramic structure are generated. Various methods are used to attain this effect, such as the doping with ions of a different valence or ionic radius, or the creation of vacancies, which introduce microscopic spatial inhomogeneity.

9.7.2 Properties and Constitutive Equations of Electrostrictive Ceramics

Properties of typical electrostrictive ceramics are given in Table 9.7. The strain-field curves of electrostrictive ceramics display the typical quadratic behavior (Figure 9.14a). On such curves, a positive mechanical strain is obtained for both positive and negative electric fields. However, the strain field curve is strongly nonlinear, as appropriate to quadratic behavior. What is remarkable about electrostrictive ceramics is their very low hysteresis. Figure 9.14a shows that the increasing and decreasing curves superpose almost everywhere, with only a small exception in a limited region at relatively low fields.

The constitutive equations of electrostrictive ceramics are similar to those for piezoelectric ceramics, only that significant second-order terms are also included, i.e.,

$$S_{ij} = s^E_{ijkl}T_{kl} + d_{kij}E_k + m_{klij}E_kE_l \qquad (9.50)$$

$$D_m = d_{mkl}T_{kl} + 2m_{mnij}E_nT_{ij} + \varepsilon^T_{mn}E_n \qquad (9.51)$$

In Equation (9.50), the first two terms are the same as those used to describe the piezoelectric constitutive law, i.e., Hook's law and the converse piezoelectric effect. The third term represents the electrostriction effect. The constants m_{klij} are the electrostrictive coefficients. Equation (9.50) indicates that electrostriction appears as a quadratic addition to the linear piezoelectric effect. In fact, the two effects are separable because the piezoelectric effect is possible only in noncentrosymmetric materials, whereas the electrostrictive effects are not limited by symmetry and are present in all materials. In addition to the direct electrostrictive effect, the converse electrostrictive effect also exists, as shown by Equation (9.51).

Commercially available PMN formulations are internally biased and optimized to give a quasi-linear behavior. In this situation, they display much less nonlinearity than the

TABLE 9.7

Typical Piezoelectric, Dielectric, and Mechanical Properties for Electrostrictive Ceramics (EDO Corporation PMN EC-97 and EC-98); Transverse Isotropy Is Assumed, $(\ldots)_{23} = (\ldots)_{13}$

Property	PT EC-97	PMN EC-98
ρ (kg/m^3)	6700	7850
$d_{33}(\times 10^{-12}$ m/V)	68	730
$d_{31}(\times 10^{-12}$ m/V)	−3	−312
$d_{15}(\times 10^{-12}$ m/V)	67	825
$g_{33}(\times 10^{-3}$ C/N)	32.0	15.6
$g_{31}(\times 10^{-3}$ C/N)	−1.7	−6.4
$g_{15}(\times 10^{-3}$ C/N)	33.5	17.0
$s_{11}^{E}(\times 10^{-12}$ m^2/N)	—	16.3
$s_{12}^{E}(\times 10^{-12}$ m^2/N)	—	−5.6
$s_{33}^{E}(\times 10^{-12}$ m^2/N)	7.7	21.1
$\varepsilon_{33}^{T}/\varepsilon_0$	270	5500
k_{33}	0.53	0.72
k_{31}	0.01	0.35
k_{15}	0.35	0.67
k_p	0.01	0.61
Young's modulus Y_{11}^{E} (GPa)	128	61
Poisson ratio σ		0.34
Curie temperature T_C (°C)	240	170
Mechanical Q_M	950	70

Note: $\varepsilon_0 = 8.85 \times 10^{-12}$ F/m; Poisson ratio calculated from the formula,

$$k_p^2 = \frac{2}{1-\sigma}k_{31}^2.$$

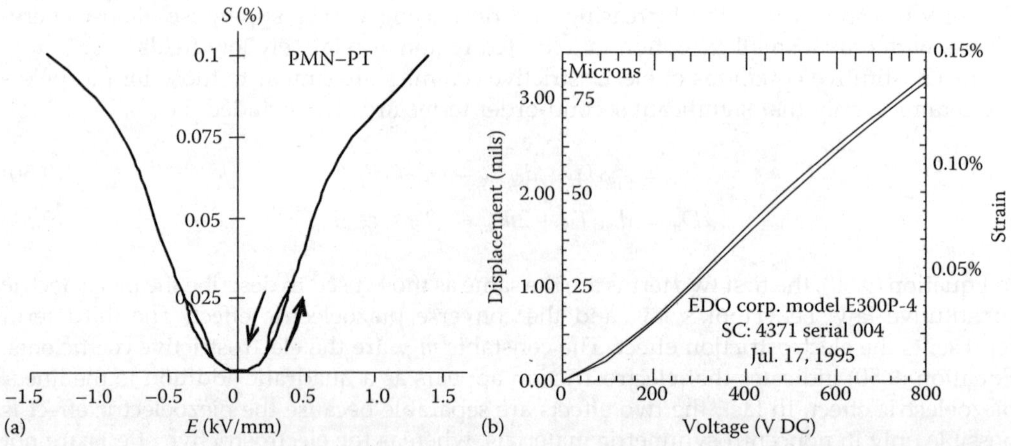

FIGURE 9.14
Electrostrictive ceramics behavior: (a) field induced strain in 90-10 PMN–PT electrostrictive ceramic; (b) induced-strain displacement versus applied voltage for a 54 mm long actuator made of EDO Corporation EC-98 electrostrictive PMN ceramic.

conventional quadratic electrostriction, and resemble more the conventional linear piezo-electricity (Figure 9.14b). Linearized electrostrictive ceramics retain the very low hysteresis of quadratic electrostrictive ceramics. From this standpoint, they are superior to conventional piezoelectric ceramics. However, linearized electrostrictive ceramics do not accept field reversal. After linearization, the constitutive equations of electrostrictive ceramics resemble those of conventional piezoceramics, i.e.,

$$S_{ij} = s^E_{ijkl}T_{kl} + \tilde{d}_{ijk}E_k \tag{9.52}$$

$$D_m = \tilde{d}_{mkl}T_{kl} + \varepsilon^T_{mn}E_n \tag{9.53}$$

The tilde symbol indicates that the piezoelectric constants $\tilde{d}_{ijk}$ of Equations (9.52) and (9.53) are different from the corresponding constants d_{ijk} in the original Equations (9.50) and (9.51). This difference is due to the linearization process. In Equations (9.50) and (9.51), the d_{ijk} constants were quite small, since the main effect was due to the quadratic effects represented by the m_{klij} constants. In Equations (9.52) and (9.53), the $\tilde{d}_{ijk}$ constants are quite significant, since they represent the effect of the linearization of Equations (9.50) and (9.51).

Figure 9.15 compares induced-strain response of some commercially available piezoelectric, electrostrictive, and magnetostrictive actuation materials. It can be seen that the electrostrictive materials have less hysteresis, but more nonlinearity. The little hysteresis of electrostrictive ceramics can be an important plus in certain applications, especially at high frequencies. However, one should be aware that this low hysteresis is strongly temperature dependent. As the temperature decreases, the hysteresis of electrostrictive ceramics increases, such that, below a certain temperature, the hysteresis of electrostrictive ceramics may exceed that of piezoelectric ceramics. In general, since the beneficial behavior of the

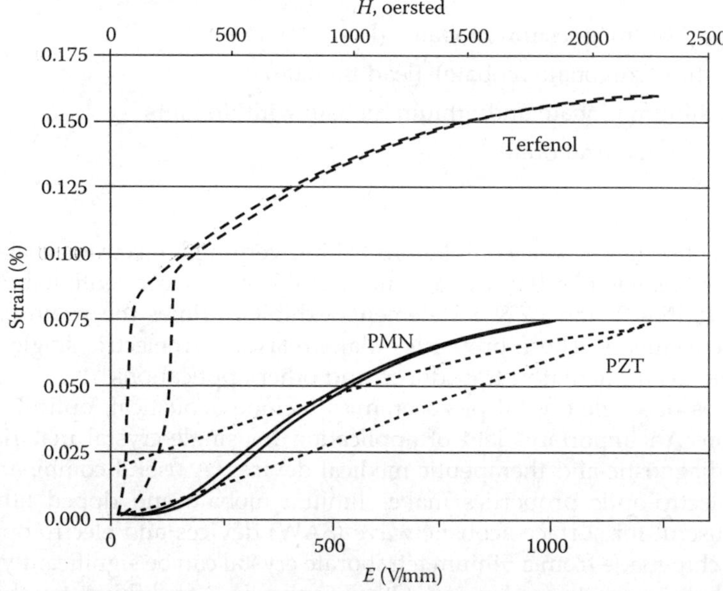

FIGURE 9.15
Strain versus electric field behavior of currently available induced strain materials.

electrostrictive ceramics is related to the diffuse phase transition in the relaxor range, their properties degrade as the operation temperature gets outside the relaxor phase transition range.

9.7.3 Typical Applications of Electrostrictive Ceramics

The use of electrostrictive ceramics is closely related to the use of piezoelectric ceramics. In biased quasi linear formulations, they can be used as sensors and actuators, either as stacks, when the d_{33} effect is being utilized, or as bonded wafer, when the d_{31} effect is utilized. The nonlinear electromechanical coupling of the electrostrictive ceramic material can be utilized to develop transducers with tunable sensitivity. By adjusting the bias electric field, the slope of the transduction relation between stress/strain and electric field can be adjusted at will. This tunable transduction property can be used for both sensor and actuation applications. Other applications of electrostrictive ceramics include shape-control of deformable mirrors, tunable ultrasonic probe for medical applications, impact dot-matrix and ink-jet printing.

9.8 Single-Crystal Piezoceramics

New electroactive materials with much larger induced-strain capabilities are currently developed. Very promising are the single-crystal piezoceramics. The properties of a single-crystal piezoelectric approach those that might be offered by a polycrystalline ceramic element if all of its domains were perfectly aligned. Single-crystal piezoceramic research includes the following compounds

- PMN–PT, (lead magnesium niobate)–(lead titanate)
- PZN–PT, (lead zirconate niobate)–(lead titanate)
- $LiNbO_3$, lithium niobate and lithium niobate with dopants
- $Li_2B_4O_7$, lithium tetraborate
- Quartz

Single crystals of some relaxor formulations exhibit very high electromechanical coupling factors—values greater than 0.9, versus values of 0.7–0.8 for conventional PZT ceramics. Single-crystal PMN–PT and PZN–PT elements exhibit 10 times the strain of comparable polycrystalline elements. These properties make relaxor ferroelectric single crystals very attractive materials for actuator, transducer, and other applications.

Potential uses of single-crystal piezoceramics include acoustical, optical, and wireless communication. An important field of application for single-crystal materials is that of actuators for diagnostic and therapeutic medical devices. A useful combination of piezoelectric and electro-optic properties makes lithium niobate and doped lithium niobate crystals very useful for surface acoustic wave (SAW) devices and electro-optical applications. A SAW chip made from a lithium tetraborate crystal can be significantly smaller than its lithium niobate or quartz counterpart. Other applications for lithium tetraborate crystals include bulk acoustic wave (BAW) devices, pagers, cordless and cellular telephones, and

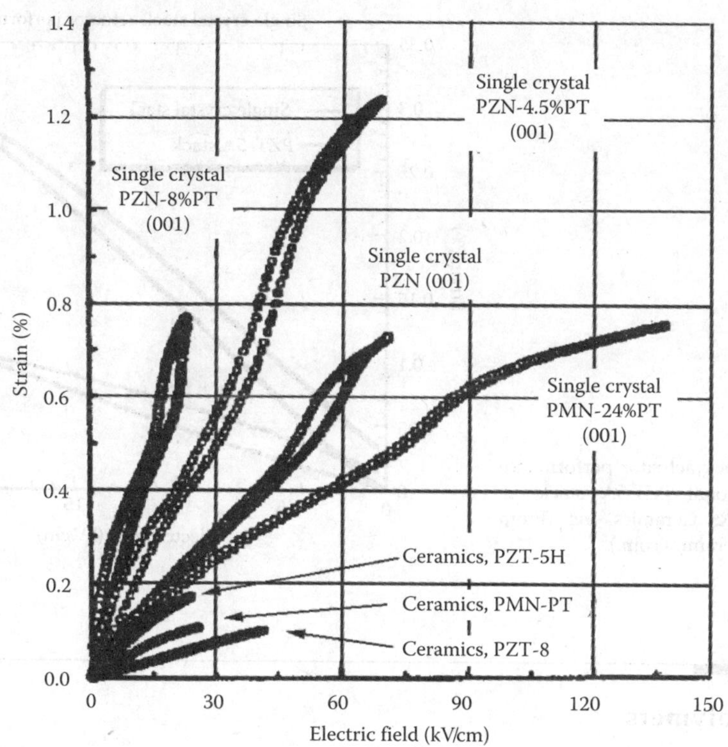

FIGURE 9.16
Strain versus electric field behavior of the new <001> oriented rhombohedral crystals of PZN–PT and PMN–PT compared to current piezoelectric ceramics. (From Park, S-.E., and Shrout, T.R., *J. Appl. Phys.*, 82, 1804, 1997. With permission.)

data communication devices. Applications for quartz crystals include timing mechanisms for watches and clocks and delay lines for electrical circuits. The performance of a single-crystal element depends on the direction in which the raw crystal is cut. A cut normal to the x axis will produce maximal potential for expanding in thickness; a crystal cut normal to the y axis will have maximal potential for shear distortion.

Park and Shrout (1997) studied single crystals of PZN–PT, a relaxor perovskite Pb $(Zn_{1/3}Nb_{2/3})O_3$–$PbTiO_3$. Strain levels of up to 1.5%, and reduced hysteresis have been reported. The response of these new materials can be an order of magnitude larger than that of conventional PZT materials (see curves in Figure 9.16). When building actuators from PZN–PT material, one must take into account the strong dependence of the piezoelectric properties on the crystal orientation. This imposes certain design restrictions in comparison with conventional piezoceramics. Commercial production of induced-strain actuators based on these new materials has been undertaken by TRS Ceramics, Inc., and a prototype PZN–PT actuator with a maximum strain of around 0.3% has been reported (Figure 9.17). However, the reliability of PZN–PT single crystals is still being studied.

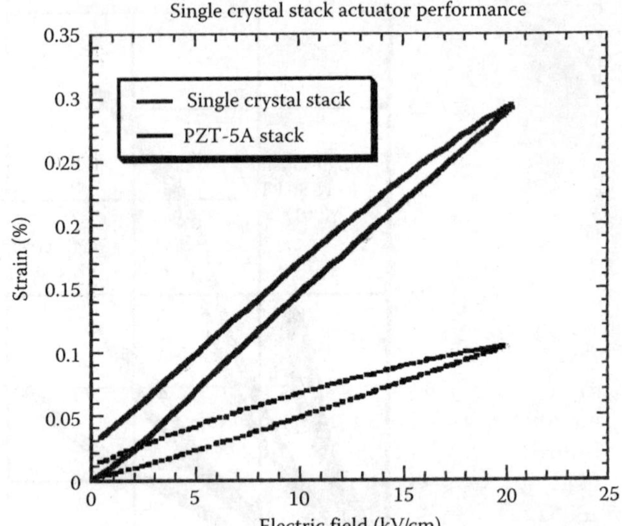

FIGURE 9.17

Single-crystal stack actuator performance versus conventional PZT-5A stack, as measured by TRS Ceramics, Inc. (From http://www.trsceramics.com.)

9.9 Piezopolymers

Piezoelectric polymers are polymers that display piezoelectric properties similar to those of quartz and piezo ceramics. Piezoelectric polymers are supplied in the form of thin films. They are flexible and show large compliance. Piezoelectric polymers are cheaper and easier to fabricate than piezoceramics. The flexibility of the piezopolymers overcomes some of the drawbacks associated with the piezoelectric ceramics brittleness. Piezopolymers have proven useful in headphones, speakers, and high-frequency ultrasonic transducers. However, the elastic modulus of piezopolymers is low and hence they are inappropriate for actuator applications.

9.9.1 Fabrication of Piezopolymers

A typical piezoelectric polymer is the polyvinylidene fluoride, abbreviated PVDF or PVF_2. This polymer has strong piezoelectric and pyroelectric properties. Its chemical formulation is $(-CH_2-CF_2-)_n$. This polymer displays a crystallinity of 40%–50%. The PVDF crystal is dimorphic, the two types designated as I (or β) and II (or α). In the β phase (i.e., type I), PVDF is polar and piezoelectric. In the α phase, PVDF is not polar and is commonly used as an electrical insulator. To impart piezoelectric properties, the α phase is converted into the β phase and then polarized. Stretching the α-phase material produces the β-phase. The symmetry of PVDF is $mm2$. Remarkable progress has recently been made in developing piezoelectric polymeric materials using copolymers. The copolymer VDF/TrFE consists mostly of piezoelectric β-phase with high crystallinity (~90%). The film is then cut into various sizes to form piezopolymer sensors (Figure 9.18).

The piezopolymer surface is metallized to produce the surface electrodes. Silver ink can be screen-printed in patterns onto a clear PVDF film. Leads are attached according

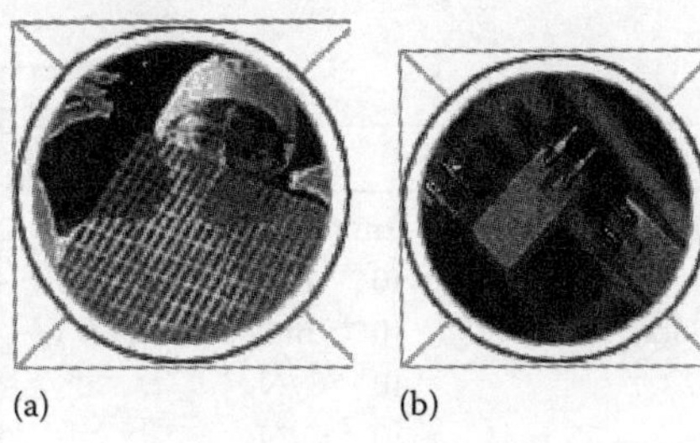

(a)   (b)

FIGURE 9.18
(a) Piezopolymer film screen printed with silver ink, ready to be cut out into individual sensors; (b) piezopolymer sensors. (From http://www.msiusa.com.)

to customer's specifications. Crimp or eyelet lead attachments are used. After surface metallization, polarization is obtained through the application of a strong electric field. Improved polarization is obtained through *field cooling*, i.e., cooling with the electric field applied to the sample. To achieve this, the PVDF specimen is to be held at 90°C–130°C for 15–120 min under a field of 500–1000 kV/cm. It is then cooled to room temperature while maintaining the same electric field level.

9.9.2 Piezopolymer Properties and Constitutive Equations

The piezoelectric properties of piezopolymers are comparable to those of piezoceramics (Tables 9.8 and 9.9). However, its modulus of elasticity is much lower. Remarkable about piezopolymers is their large pyroelectric constant, which makes them good candidates for infrared sensor applications. Another beneficial property is that, unlike piezoceramics, they can be operated at high structural strains; applied strain of up to 0.2% has been reported.

The constitutive relations for PVDF can be described as

$$S_{ij} = s^E_{ijkl} T_{kl} + d_{kij} E_k + \alpha^E_i \theta$$
$$D_j = d_{jkl} T_{kl} + \varepsilon^T_{jk} E_k + \tilde{D}_m \theta \tag{9.54}$$

where S_{ij} is the mechanical strain, T_{ij} is the mechanical stress, E_k is the electrical field, and D_i is the electrical displacement (charge per unit area). The coefficient s^E_{ijkl} is the mechanical compliance of the material, measured at zero electric field ($E = 0$). The coefficient ε^E_{jk} is the dielectric constant measured at zero mechanical stress ($T = 0$). The coefficient d_{ijk} is the *piezoelectric strain constant* (also known as the *piezoelectric charge constant*), which couples the electrical and mechanical variables and expresses how much strain is obtained per unit applied electric field, θ is the absolute temperature, α^E_i is the coefficient of thermal expansion under constant electric field, $\tilde{D}_i$ is the temperature coefficient of the electric displacement.

One important advantage of piezopolymer films is their low acoustic impedance ($Z = \rho c$). The acoustic impedance of piezopolymers is closer to that of water, bio-tissue, and other organic materials. For example, the acoustic impedance of PVDF film is only 2.6 times larger than that of water, whereas the acoustic impedance of piezoceramics is typically 11 times

TABLE 9.8

Typical Properties of Polymeric Piezo Films

Symbol	Parameter	PVDF	Copolymer	Units
t	Thickness	9, 28, 52, 110	<1 to 1200	μm (micron, 10^{-6} m)
d_{31}	Piezo strain constant	23	11	10^{-12} m/V
d_{33}	Piezo strain constant	−33	−38	10^{-12} m/V
g_{31}	Piezo stress constant	216	162	10^{-3} C/N
g_{33}	Piezo stress constant	−330	−542	10^{-3} C/N
k_{31}	Electromechanical coupling factor	0.12	0.20	—
k_{33}	Electromechanical coupling factor	0.14	0.25–0.29	—
C	Electric capacitance	380 for 28 μm	68 for 100 μm	pF/cm^2 at 1 kHz
E	Young's modulus	2–4	3–5	GPa
c	Speed of sound	1.5	2.3	km/s (stretch)
		2.2	2.4	km/s (thickness)
ρ	Pyroelectric coefficient	30	40	10^{-6} C/m^2/°K
ε	Permittivity	106–113	65–75	10^{-12} F/m
$\varepsilon/\varepsilon_0$	Relative permittivity	12–13	7–8	—
ρ_m	Mass density	1780	1820	kg/m
ρ_e	Volume resistivity	>10^{13}	>10^{14}	Ω/m
R_{NiAl}	Surface metallization resistivity	<3.0	<3.0	Ω/m^2 for NiAl metallization
R_{Ag}	Surface metallization resistivity	0.1	0.1	Ω/m^2 for Ag Ink
$\tan \delta_e$	Loss tangent	0.02	0.015	@ 1 kHz
	Yield strength	45–55	20–30	MPa (stretch axis)
	Temperature range	−40 to 80…100	−40 to 115…145	°C
	Water absorption	<0.02	<0.02	% H$_2$O
	Maximum operating voltage	30 (750)	30 (750)	V/μm (V/mil) DC @ 25°C
	Breakdown voltage	80 (2000)	80 (2000)	V/μm (V/mil) DC @ 25°C

TABLE 9.9

Comparison of PVDF Properties with Those of Piezoelectric Ceramics

Property	Units	PVDF Film	PZT (PbZrTiO$_3$)	BaTiO$_3$
ρ_m	kg/m^3	1780	7500	5700
$\varepsilon/\varepsilon_0$	—	12	1200	1700
d_{31}	10^{-12} m/V	23	110	78
g_{31}	10^{-3} Vm/N	216	10	5
k_{31}	@ 1 kHz	0.12	0.30	0.21
Young's modulus	GPa	~3	~60	~110
Acoustic impedance	10^6 kg/m^2 s	2.7	30	30

that of water. When the acoustic impedance of two media have similar values, transmission between the two media is enhanced, and reflection at the interface is reduced.

9.9.3 Typical Piezopolymer Applications

Unlike piezoelectric ceramics, piezopolymers are flexible, and not brittle. This property is especially important for applications involving complicated shapes and significant structural strains. As a sensor, PVDF provides higher voltage/electric field in response to mechanical stress than piezoceramics. The piezoelectric g-constant (i.e., the voltage generated per unit mechanical stress) is typically 10–20 times larger than for piezoceramics. The PVDF film also produces an electric voltage in response to infrared light, due to its strong pyroelectric coefficient. Hence, piezopolymers have found wide application as sensors (Figure 9.18).

When used as an actuator, the piezopolymer gives a much lower force than piezoceramics, due to its much lower modulus (Table 9.9). Hence, it is best used in the actuation of compliant microstructures, with a low inherent stiffness. It is inappropriate for structural control applications involving conventional structural materials.

Piezopolymers are often used for sensing. PVDF can be formed in thin films and bonded to many surfaces. Uniaxial films, which are electrically poled in one direction, can measure stresses along one axis, while biaxial films can measure stresses in a plane. A PVDF sensor is used as a strain gauge but does not require a conditioning power supply. The output signal is also comparable to that of an amplified strain gauge signal. This high sensitivity is due to the low thickness of the typical PVDF film (25 μm). Because of its good sensor properties (i.e., the high g constant), lightweight, flexibility, and toughness, PVDF is used in numerous sensor applications.

PVDF is commonly used in switching applications, where it develops a voltage in response to a pressure (e.g., ~10 V under finger pressure). Keyboards make use of this property. One surface of the film is completely metallized, whereas the other surface is metallized only in areas corresponding to the keys and their associated signal-detection wires. The film is stretched under the metallic keyboard with recesses at various key locations; the finger pressure on the film produces a signal corresponding to a specific key, which is then used to perform appropriate keyboard tasks.

PVDF film is also used for acoustic applications as compact microphones, earphones, and loudspeakers. Specially shaped PVDF sensors have also been used as modal sensors for sensing the natural modes of vibration of structural elements such as beams, plates, and cylinders. They can be used to monitor machine vibrations, as contact microphones, gas and fluid sensors, and in antitheft system. PVDF ultrasonic transducers can transmit ultrasonic waves better into the living tissue than do piezoceramic ultrasonic transducers.

Many surveillance systems make use of the pyroelectric properties of the PVDF sensor. This application typically uses two closely coupled sensors. Much as the ambient-temperature compensation scheme used with strain gauge measurements, only one sensor is exposed to the infrared emissions, whereas both sensors respond equally to other stimuli, such as noise, vibration, and global temperature changes. Signals from both sensors are combined to detect the infrared signal alone.

The sensitivity of PVDF films to pressure changes has been utilized in tactile sensors that can read the Braille alphabet and distinguish different grades of sandpaper. Tactile sensors with ultra thin (200–300 μɛ) PVDF films have been proposed for use in robotics. A skin-like sensor that replicates the temperature and pressure sensing capabilities of human skin may be used in different modes to detect edges, corners, and geometric features or to distinguish between different grades of fabric. The pyroelectric effect, which allows piezoelectric polymers to sense temperature, also limits their use to lower temperature ranges.

9.10 Magnetostrictive Materials

In simple terms, magnetostriction is a material property that causes certain ferromagnetic materials to change shape when an external magnetic field is applied. Magnetostrictive materials expand in the presence of a magnetic field, as their magnetic domains align with the field lines. Magnetostriction was initially observed in nickel, cobalt, iron, and their alloys, but the values were small (<50 $\mu\varepsilon$). Large strains ($\sim$10,000 $\mu\varepsilon$) were observed in the rare-earth elements terbium (Tb) and dysprosium (Dy) at cryogenic temperatures (i.e., below 180°K). Large room-temperature magnetostriction exists in the terbium–iron alloy $TbFe_2$. The binary alloy Terfenol-D ($Tb_{1-x}Dy_xFe_{1.9}$), developed at Ames Laboratory and the Naval Ordnance Laboratory, displays magnetostriction of up to 2000 $\mu\varepsilon$ at up to 80°C and higher. Terfenol-D binary alloy formulations are of the form $Tb_{1-x}Dy_xFe_{1.9-2}$, where x is the relative proportion of dysprosium, while the proportion of iron can vary between 1.9 and 2. In the foregoing discussion, we will use the generic value 2, while understanding that the actual value may be between 1.9 and 2, according to the detailed formulation of the particular Terfenol-D alloy. The magnetostrictive constitutive equations contain both linear and quadratic terms

$$S_{ij} = s^E_{ijkl}T_{kl} + d_{kij}H_k + m_{klij}H_kH_l \tag{9.55}$$

$$B_j = d_{jkl}T_{kl} + \mu^T_{jk}H_k \tag{9.56}$$

where, in addition to the already defined variables, H_k is the magnetic field intensity, B_j is the magnetic flux density, and μ^T_{jk} is the magnetic permeability under constant stress. The coefficients d_{kij} and m_{klij} are defined in terms of magnetic units. For a rod surrounded by a coil with n turns per unit length, the magnetic field intensity H is related to the current I by the relation

$$H = nI \tag{9.57}$$

9.10.1 Physical Explanation of the Magnetostriction Phenomenon

Spontaneous magnetism occurs because of an unusual imbalance in the magnetic moments of the electron shells in an atomic structure. The electrons can get ordered in such a way that the net magnetic moment points in a certain direction, thus lowering the crystal symmetry. New properties result. One of these new properties is magnetostriction.

Magnetism at an atomic level results from adding up of the magnetic dipoles generated by the rotation of the negatively charged orbital electrons around the atomic nucleus. In addition, each electron has its own magnetic dipole, due to spinning around its own axis.

Thus, the atomic magnetism results from the addition of the orbital and spin magnetic moments of all the electrons making up the atom.

The magnetism of solid materials results from a similar process; however, it is not the straight addition of the atomic magnetism of each individual atom considered separately. The reason for this is that, when atoms combine to form a solid, the electrons become shared in molecular orbitals and band states. The outer atomic orbitals which have the smallest energy binding them to the nucleus, overlap in space with similar orbital of neighboring atoms, and produce extended electron states ("electron cloud"). Even if the original atomic orbitals had atomic magnetism due to unbalanced electron population in the outer shell, the expanded electron states do not usually result in magnetism, since the individual contributions of the individual electrons balance out due to their large numbers. However, in the case of transition metals (rare earths and actinides), magnetic moments also exist in the inner shell. This is a result of orbital shells filled "out of order," due to the rules of quantum mechanics. Thus, one of the inner orbitals is left partially filled with electrons, resulting in a net magnetic moment due to the unbalanced electron spin population in this shell. These elements and their compounds may exhibit magnetic properties even when the outer shell of electrons has been shared into the extended electron states. This type of magnetism is called *paramagnetism*.

Another result of the unbalanced inner atomic orbital shells is the geometric distortion of the atomic shape. Thus, after the outer electronic shell has been removed and the atoms have become ions, the resulting ionic shapes in the crystal lattice are not spherical, but distorted. This geometric distortion of the inner orbital contributes to a net magnetic moment of the lattice ions. The net magnetic moment of the ion (*ionic moment*) is made up of two contributions:

1. Magnetic moment due to the net spin of the electrons in the unbalanced orbital shell; this is the net *spin moment*.
2. Magnetic moment due to the distortion of the inner orbital shell; this is the *orbital moment*.

The ionic magnetic moments add up to form the material magnetism. When the ionic magnetic moments align themselves and produce a net magnetic moment in the material, the behavior is called *ferromagnetic*. Ferromagnetic behavior is displayed by some transition metals such as iron (Fe), cobalt (Co), and nickel (Ni). It is also displayed by some rare earth elements at low temperatures. As the temperature increases, the ferromagnetic behavior diminishes, and suddenly disappears at the Curie temperature, T_C.

Above the Curie temperature, T_C, the materials are not spontaneously magnetic. Although spontaneous magnetization does not exist, magnetization is induced when an external magnetic field is applied. This indicates that, although the ionic magnetic moments do not spontaneously align, they can be induced to align under the influence of an external magnetic field. When the external magnetic field is removed, the material magnetic moment also disappears. This type of behavior is called *paramagnetic*. When a paramagnetic material is cooled below the Curie temperature, while subjected to a strong magnetic field, the magnetism becomes "frozen" in the material and the material becomes *ferromagnetic*. The material has been *magnetized*. Magnetization can also be induced by applying a strong magnetic field to the material while it is below the Curie temperature. The stronger the field, the more the ionic moments gets aligned. When all the ionic moments are aligned, the saturation magnetization, M_S, is reached. The saturation magnetization may depend on temperature.

The strength of the required field to induce saturation magnetization varies from material to material. In iron, cobalt, and nickel, which are well below the Curie temperature and at room temperature, the magnetization field required to induce ferromagnetism is not very strong.

When the magnetic moments align antiparallel to each other, the net magnetic moment is zero. This type of behavior is called *antiferromagnetic*. It is observed in chromium (Cr) and manganese (Mn). However, if the applied magnetic field is very strong, the ionic magnetic moments can be induced to align, and the material snaps from antiferromagnetic into ferromagnetic behavior.

Antiparallel ordering of ionic magnetic moments in a compound such as $TbFe_2$ results in a state in which the Tb moments are aligned antiparallel to the Fe moments. Since the Tb and Fe moments are unequal, and since there are two Fe contributions to each Tb contribution, the result is a net magnetic moment in the $TbFe_2$ compound. This type of behavior is called *ferrimagnetic*. Ferrimagnetic behavior is less strong than the ferromagnetic behavior.

The individual ionic moments add up to create the net magnetism of the material. How much of the individual ionic moments end up in a certain compound depends on the crystal structure and chemistry. This aspect is quite evident in the behavior of the common transition metals, iron (Fe), cobalt (Co), and nickel (Ni), chrome (Cr), and manganese (Mn). For example, Fe in its 2+ ionic state has six electrons filling the five $3d$ states. Their occupancy is $n_+ = 5$ and $n_- = 1$. The resulting spin moment of the ion is 4 μ_B. However, in bulk body-centered cubic (bcc) lattice structure, iron has a measured spin moment of 2.2 μ_B/atom. The reduction from 4 μ_B to 2.2 μ_B is due to the bonding chemistry. Similar reductions of the spin magnetic moments are observed in Co and Ni. This indicates that the electron orbitals are extended and overlapped during the bonding process. The degree of extension and overlap of electron orbitals also determines how much the moments on neighboring atoms interact with each other through *magnetic exchange*. In Fe, Co, and Ni, the magnetic exchange is strong. In rare earth elements, the magnetic exchange is weaker.

Partially filled orbitals lead to both magnetic moments and nonspherical ion shapes. In certain elements, such as the rare earths, the *spin-orbit coupling* provides a strong coupling between the magnetic spin direction and the orientation of the anisotropic (oblate) shape. Thus, ions of these elements display *magnetic anisotropy*, which can be defined as the tendency of the magnetic moment to point in a certain crystalline direction due to the electrical interaction between its electron charge cloud and the neighboring charged ions in the crystal lattice (Figure 9.19a). The *magnetoelastic coupling* is defined as the tendency of neighboring ions to shift their position in response to the rotation of the magnetic moment and the attached anisotropic charge cloud. The magnetoelastic coupling produces the *Joule magnetostriction*, which is an anisotropic change in length due to the application of a magnetic field.

The design of large-magnetostriction materials consist in combining the two groups of magnetic elements (rare earths and transition metals) to get large magnetostrictive response. The following guidelines apply:

1. Intrinsic magnetoelastic coupling must be large.

2. Curie temperature (magnetic ordering temperature) must be well above the desired operating temperature.

3. Magnetization induced in the material should be reasonably strong.

4. Magnetization direction should be easily changed, i.e., the magnetic coercive fields should be reasonably small. In practice, this requires a small magnetic anisotropy.

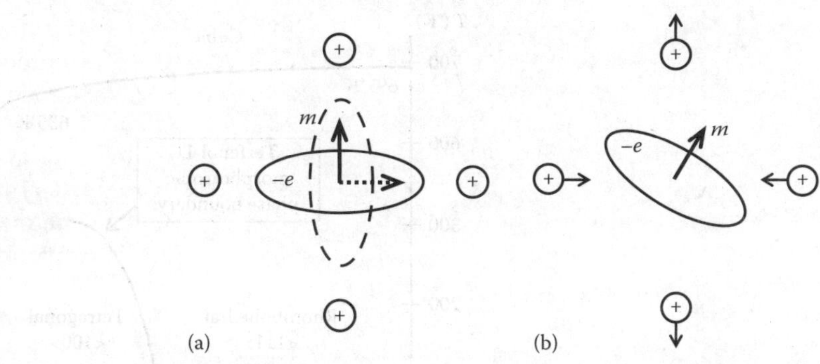

FIGURE 9.19
Schematic of a oblate-shaped charge density rotating in a magnetic field: (a) magnetic field vector m is oriented vertically, while the negative charge, $-e$, is attracted toward the $+$ ions in the lattice corners; (b) as the magnetic field is rotated, the lattice is deformed, resulting in magnetostriction. (The oblate shape of the charge density can be due to partially filled inner shells, such as the $4f$ shell in Tb ions).

The physical explanation given above applies primordially to magnetostrictive materials in homogenous crystalline phase. However, most commercially available magnetostrictive materials are polycrystalline, separated by grain boundaries. In addition, domains of magnetic polarization may exist in the crystalline grains, thus creating a more complex situation which needs further attention. Magnetic domains are volumes in the crystalline grain that are uniformly magnetized along particular crystal directions known as *easy axes*. The crystal within each domain volume is subjected to strain along the easy axis, resulting in lattice distortion. *Domain walls* are the regions separating the magnetic domains. Since adjacent domains have drastically different magnetic orientations, a rapid change in magnetization occurs across the domain wall. The application of either stress/strain or magnetic field to magnetostrictive material produces a change in the anisotropic energy. The magnetization in the material is redistributed in order to minimize the overall energy of the material under the new external conditions. This redistribution is achieved through

1. Movements of the magnetic domain walls, thus extending the domains with magnetization favorable to the new external conditions

2. Rotation of the magnetization inside the domains to align to the new external conditions

The changes in the domain configurations are impeded by various factors such as domain wall thickness, crystallographic defects and dislocations. Under cyclic application of external stress/strain or magnetic field, significant hysteresis is observed.

9.10.2 Commercially Available Magnetostrictive Materials

In practice, remarkable results have been obtained with the $Tb_{1-x}Dy_x\,Fe_2$ binary alloy. The $TbFe_2$ alloy was found to have the largest magnetostrains at room temperature. However, the magnetization anisotropy of $TbFe_2$ is also very large. Thus, $TbFe_2$ requires very large magnetic fields in order to obtain the desired large strains. The magnetic anisotropy can be compensated by partial substitution of Tb with other rare earth elements that have a

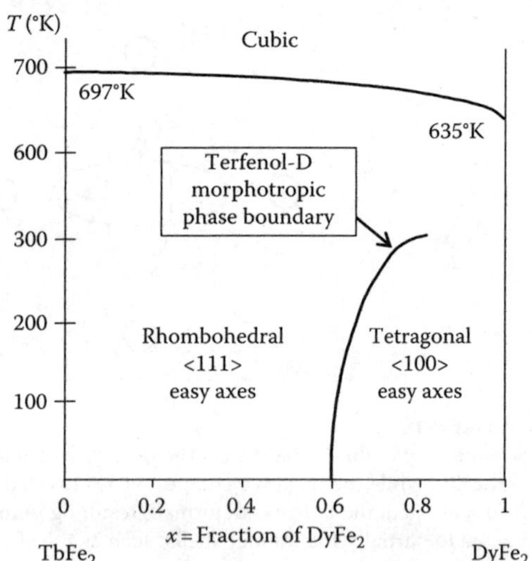

FIGURE 9.20

Phase diagram of $Tb_{1-x}Dy_xFe_2$ compound, showing the MPB for the Terfenol-D formulation corresponding to $x = 0.7$ at room temperature.

magnetic anisotropy of opposite sign. Among the substitution candidates, dysprosium (Dy) was found most beneficial. $Tb_{1-x}Dy_xFe_2$ binary alloy attains the highest magnetostrictive coupling and the maximum magnetic permeability near the MPB (Figure 9.20). This corresponds to the change in the crystal structure from the tetragonal phase to the rhombohedral phase, which happens when a fraction of $DyFe_2$ is approximately $x = 0.7$. The explanation for this phenomenon is as follows. Above the Curie temperature, $Tb_{1-x}Dy_xFe_2$ has a cubic lattice and is paramagnetic. The Curie temperature varies slightly with the alloying proportion, from ~697°K for pure $TbFe_2$ to 635°K for pure $DyFe_2$ (Figure 9.20). Below the Curie temperature, $Tb_{1-x}Dy_xFe_2$ is ferromagnetic, but its lattice can be either tetragonal or rhombohedral, according to the alloying proportion. In the phase diagram, the line separating the two phases is called the MPB.

Figure 9.20 shows that the tetragonal lattice has < 100 > easy axes, i.e., six possible distortion variants, while the rhombohedral lattice has < 111 > easy axes, i.e., eight distortion variants. On the line separating the two phases in the phase diagram, i.e., on the MPB, both the tetragonal phase and the rhombohedral phase may exist. Hence, the total number of distortion variants on MPB is 14, which is the cumulative effect of both phases. Having more distortion variants increases the material responsiveness, since the material has more options to deform under the action of external factors, e.g., magnetic field or mechanical stress. At room temperature, the MPB for $Tb_{1-x}Dy_xFe_2$ is placed around the 30/70 alloying ratio ($x = 0.7$). This formulation is commonly known as Terfenol-D. Terfenol-D displays "giant" magnetostriction, with up to 0.2% strain. Terfenol-D formulation types offered by Etrema Corp. are given in Table 9.10.

Terfenol-D materials are fabricated in crystal growth machines that produce solid metallic rods, of round or square cross section. When activated by a magnetic field, the Terfenol-D rods from the crystal machine expand along the growth axis and contract in the plane perpendicular to this axis. Plates and discs can be cut from the as-grown rod (Figure 9.21). The cut can be made along different directions and planes, such that the response corresponds to the desired application. Sputtering targets can be also produced. Terfenol-D powders are obtained through milling in inert atmosphere. Various mesh sizes are

TABLE 9.10

Terfenol-D Formulation Types and the Associated
Manufacturing Process

Type Numbers	Material Grades
Terfenol-D Type I	Bridgman (BRDG)/Bottom Pour
Terfenol-D Type II	Free Stand Zone melt (FSZM)
Terfenol-D Type III	ETREMA Crystal Grower (ECG)
Terfenol-D Type IV	Powdered Sintered (not available at this time)
Terfenol-D Type V	"Pin" less than 28 sq. mm. cross-sectional area

Source: http://etrema-usa.com/terfenol

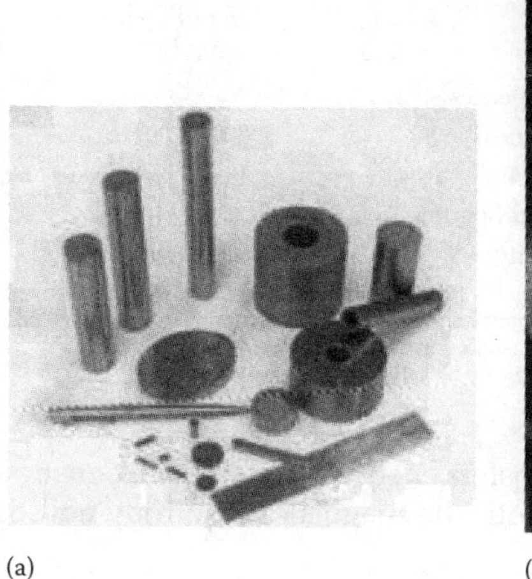

(a)   (b)

FIGURE 9.21

Presentation of Terfenol-D products: (a) various geometric shapes; (b) powder form. (From http://etrema-usa.com.)

available. For eddy current control, lamination of the rods can be used. Holes may be produced along the rod axis to enhance cooling.

The physical properties of a typical Terfenol-D material are given in Table 9.11.

9.10.3 Use of Magnetostrictive Materials

The fundamental properties of magnetostrictive materials impact their application. A summary of these properties is given next:

1. The fundamental effect utilized in commercially available magnetostrictive materials is the Joule magnetostriction, which is orders of magnitude larger than bulk magnetostriction.

2. Joule magnetostriction is due to the magnetoelastic anisotropy associated with unfilled inner shells in the electronic structure of rare earth elements which creates nonspherical electron clouds with associated electric polarity and magnetic moment.

TABLE 9.11

Physical Properties of Terfenol-D

Nominal Composition	$Tb_{0.3}Dy_{0.7}Fe_{1.92}$
Mechanical properties	
Young's modulus	25–35 MPa
Sound speed	1640–1940 m/s
Tensile strength	28 MPa
Compressive strength	700 MPa
Thermal properties	
Coefficient of thermal expansion	$12 \times 10^{-6}/°C$
Specific heat	0.35 kJ/kg/°C
Thermal conductivity	13.5 W/m/°C
Electrical properties	
Resistivity	58×10^{-8} Ωm
Curie temperature	380°C
Magnetostrictive properties	
Strain (estimated linear)	800–1200 $\mu\varepsilon$
Energy density	14–25 kJ/m³
Magnetomechanical properties	
Relative permeability	3–10
Coupling factor	0.75

Source: http://etrema-usa.com/terfenol.

3. The magnetoelastic coupling transfers energy between the magnetic and the elastic fields and thus contributes to both the magnetic anisotropy and the elastic constants.

4. The application of an external magnetic field to a magnetostrictive material results in alignment of the magnetic domains through rotation of the oblate ionic electron clouds. This produces magnetostriction. The phenomenon is self-saturating, and the maximum is attained at some large field when all the domains have aligned.

5. Conversely, the application of stress or strain to a magnetostrictive material produces a magnetic field through the deformation-induced rotation of the magnetic domains.

6. The application of an initial compressive stress significantly increases the first magnetostrictive response.

7. Magnetostriction is independent of the sign of the applied magnetic field.

8. By applying a bias field, a *piezomagnetic response,* i.e., a response that follows the sign of the applied field can be obtained. This is a local linearization of the overall magnetostrictive response curve. The bias field must be at least as strong as the magnitude of the maximum reverse field expected to appear in the desired application.

9. Magnetostrictive materials cannot be intrinsically biased, as with piezoelectric materials. The very large coercive fields that would be required to achieve intrinsic bias of magnetostrictive materials would quench the magnetostrictive response.

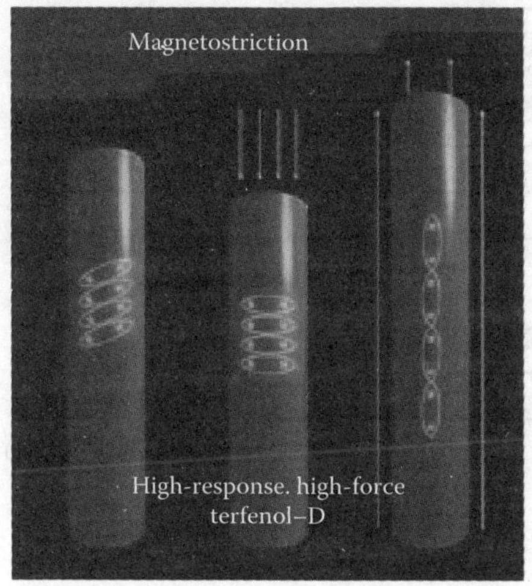

FIGURE 9.22
Schematic representation of Terfenol-D material response indicating magnetic domain rotation under an applied magnetic field: (a) in the absence of an external field, the magnetic domains are oriented according to internal magnetic bias; (b) external magnetic field opposite to internal bias reduces dipole rotation resulting in a negative strain that shortens the Terfenol-D bar; (c) magnetic field in the direction of the internal bias increases the dipole rotation resulting in a positive strain that lengthens the Terfenol-D bar.

Hence, magnetostrictive actuators require an external magnetic bias achievable with the use of a strong permanent magnet.

10. Magnetostriction response per unit field is highly dependent on temperature. Rare earth compositions have been constructed to optimize the magnetostrictive response per unit field, based on compensatory ionic interplay in the crystalline structure. However, this effect is temperature dependent. Operating a magnetostrictive material over a wide temperature range would require much larger fields than at the compensating temperature.

Figure 9.22 presents a schematic representation of the magnetostrictive response of magnetically biased Terfenol-D material, showing the magnetic dipole domains rotation under an applied magnetic field. Figure 9.22a shows a magnetically biased Terfenol-D rod with the magnetic domains in a midpoint rotational state. Figure 9.22b shows that the application of an external magnetic field in a direction opposite to the bias field reduces the dipoles rotations and results in shortening of the bar. Figure 9.22c shows that the application of an external magnetic field in the direction of the magnetic bias increases the dipole rotation and results in extension of the bar.

The field displacement curves for Terfenol-D magnetostrictive material is strongly dependent on compressive prestress. At zero prestress, the magnetostrictive response does not exceed $\sim$800 $\mu\varepsilon$. Addition of compressive prestress, can almost double the magnetostrictive response. For example, addition of 1 ksi ($\sim$6.9 MPa) of compressive prestress can produce almost 1600 $\mu\varepsilon$ strain (Figure 9.23). As the prestress is increased, the maximum strain increases slightly. For example, a prestress of 4 ksi can result in a maximum strain of 1700 $\mu\varepsilon$, which is slightly higher than the 1600 ppm obtained with 1 ksi ($\sim$6.9 MPa) prestress. As the prestress is increased even further, an opposite effect starts to develop, and the maximum strain starts to decrease, e.g., curve 8 ksi ($\sim$55 MPa) in Figure 9.23. At high prestress, the induced-magnetostrictive strain is mostly suppressed by the prestress, and the overall strain is quite small. For example, Figure 9.23 shows that, for 16 ksi ($\sim$110 MPa) prestress, the magnetostrictive strain that does not exceed $\sim$400 $\mu\varepsilon$. Another aspect that needs to be mentioned is the shape of the magnetostrictive strain curve

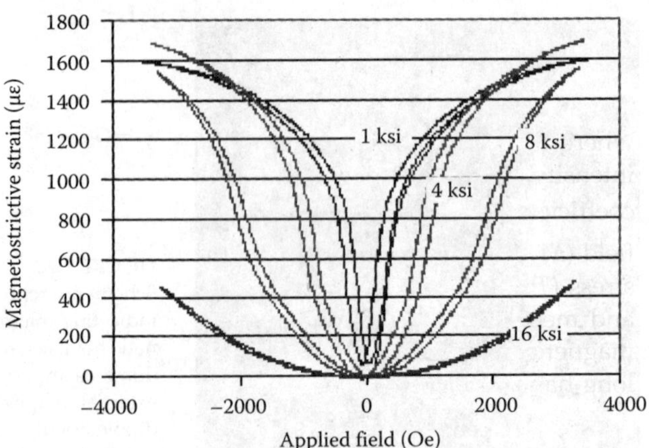

FIGURE 9.23
Typical strain–magnetic field response for Terfenol-D at various levels of prestress (1000 ppm = 0.1%). (From http://etrema-usa.com.)

changes as the amount of prestress is increased. At low prestress, e.g., the curve 1 ksi (~6.9 MPa) in Figure 9.23, the curve is convex and it levels off at high magnetic field values. In this case, the strain response per unit field diminishes as the field increases. At high prestress levels, e.g., curve 16 ksi (~110 MPa) in Figure 9.23, the curve is concave, and the strain response per unit field increases as the field increases. At intermediate values, the curves take shapes in between these two extremes. Thus, values of prestress exist for which the curve takes a shallow S shape, which is closer to a quasi-linear curve than the other cases, e.g., curve 8 ksi (~55 MPa) in Figure 9.23.

9.10.4 Linearized Equations of Piezomagnetism

Magnetostrictive material response is essentially quadratic in magnetic field, i.e., the magnetostrictive response does not change sign when the magnetic field is reversed. However, the nonlinear magnetostrictive behavior can be linearized about an operating point through the application of a bias magnetic field. In this case, piezomagnetic behavior, in which response reversal accompanies field reversal, can be obtained. The equations of linear piezomagnetism can be obtained from Equations (9.55) and (9.56), by removing the quadratic terms and updating the linear terms to represent the values obtained by linearization about a biased position, i.e.,

$$S_{ij} = s^E_{ijkl}T_{kl} + \tilde{d}_{kij}H_k \tag{9.58}$$

$$B_j = \tilde{d}_{jkl}T_{kl} + \mu^T_{jk}H_k \tag{9.59}$$

The tilde symbol indicates that the piezoelectric constants $\tilde{d}_{ijk}$ of Equations (9.58) and (9.59) are different from the corresponding constants d_{ijk} in the original Equations (9.55) and (9.56). This difference is due to the linearization process. In Equations (9.55) and (9.56), the d_{ijk} constants were quite small, since the main effect was due to the quadratic effects represented by the m_{klij} constants. In Equations (9.58) and (9.59), the $\tilde{d}_{ijk}$ constants are quite significant, since they represent the effect of the linearization of Equations (9.55) and (9.56). In a compact matrix the (Voigt) notations are

$$S_p = s_{pq}^H T_q + \tilde{d}_{kp} H_k, \quad p, q = 1, \ldots, 6; \quad k = 1, 2, 3 \tag{9.60}$$

$$B_i = \tilde{d}_{iq} T_q + \mu_{ik}^T H_k, \quad q = 1, \ldots, 6; \quad i, k, = 1, 2, 3 \tag{9.61}$$

where S_p is the mechanical strain, T_q is the mechanical stress, H_k is the magnetic field intensity, B_i is the magnetic flux density, and μ_{ik}^T is the magnetic permeability. The coefficient s_{pq}^H is the mechanical compliance of the material, measured at zero magnetic field ($M = 0$). The coefficient μ_{ik}^T is the magnetic permeability measured at zero mechanical stress ($T = 0$). The coefficient d_{ik} is the *piezomagnetic constant*, which couples the magnetic and mechanical variables, and expresses the amount of strain obtained per unit applied magnetic field. For common magnetoactive materials, Equations (9.60) and (9.61) take the long-hand form

$$\begin{Bmatrix} S_1 \\ S_2 \\ S_3 \\ S_4 \\ S_5 \\ S_6 \end{Bmatrix} = \begin{bmatrix} s_{11} & s_{12} & s_{13} & 0 & 0 & 0 \\ s_{12} & s_{11} & s_{13} & 0 & 0 & 0 \\ s_{13} & s_{13} & s_{33} & 0 & 0 & 0 \\ 0 & 0 & 0 & s_{44} & 0 & 0 \\ 0 & 0 & 0 & 0 & s_{55} & 0 \\ 0 & 0 & 0 & 0 & 0 & s_{66} \end{bmatrix} \begin{Bmatrix} T_1 \\ T_2 \\ T_3 \\ T_4 \\ T_5 \\ T_6 \end{Bmatrix} + \begin{bmatrix} 0 & 0 & d_{31} \\ 0 & 0 & d_{31} \\ 0 & 0 & d_{33} \\ 0 & d_{15} & 0 \\ d_{15} & 0 & 0 \\ 0 & 0 & 0 \end{bmatrix} \begin{Bmatrix} H_1 \\ H_2 \\ H_3 \end{Bmatrix} \tag{9.62}$$

$$\begin{Bmatrix} B_1 \\ B_2 \\ B_3 \end{Bmatrix} = \begin{bmatrix} 0 & 0 & 0 & 0 & d_{15} & 0 \\ 0 & 0 & 0 & d_{15} & 0 & 0 \\ d_{31} & d_{31} & d_{33} & 0 & 0 & 0 \end{bmatrix} \begin{Bmatrix} T_1 \\ T_2 \\ T_3 \\ T_4 \\ T_5 \\ T_6 \end{Bmatrix} + \begin{bmatrix} \mu_{11}^T & 0 & 0 \\ 0 & \mu_{11}^T & 0 \\ 0 & 0 & \mu_{33}^T \end{bmatrix} \begin{Bmatrix} H_1 \\ H_2 \\ H_3 \end{Bmatrix} \tag{9.63}$$

The magnetomechanical coupling coefficient, k, is defined as a ratio of the magnetoelastic energy to the geometric mean of the elastic and magnetic energies, i.e.,

$$k = \frac{U_{me}}{\sqrt{U_e U_m}} \tag{9.64}$$

where U_e is the elastic energy, U_m is the magnetic energy, and U_{me} is the magnetoelastic energy in the material. In common applications of magnetostrictive materials, the magnetic field and external stress are applied in the z direction (three-axis). In this case, the only nonzero components are $H_3 \neq 0$, $T_3 \neq 0$, $B_3 \neq 0$, $S_3 \neq 0$, $S_1 = S_2 \neq 0$. Under these assumptions, the constitutive equations can be simplified to the form

$$S = s^H T + d_{33} H \tag{9.65}$$

$$B = d_{33} T + \mu^T H \tag{9.66}$$

where the subscripts 3 and 33 have been omitted, except for d_{33} where it was retained in order to avoid confusion with the differential operator d. Also omitted are the S_1 and S_2 equations. The magnetomechanical coupling coefficient of Equation (9.64) takes the form

$$k_{33} = \frac{d_{33}}{\sqrt{s_{33}^H \mu_{33}^T}} \tag{9.67}$$

9.10.5 Structural Magnetostrictive Materials

The conventional magnetostrictive materials, such as TERFENOL, show a very large ("giant") magnetostrictive response, but are brittle and can be easily broken under tension. For this reason, such magnetostrictive material can be used only under compressive stress, which is practically achieved with the use of prestress compressive springs. It is apparent that such brittle materials cannot be directly incorporated into structural design, since they cannot act as tension loaded members. By being brittle, conventional magnetostrictive materials cannot survive mechanical shocks and hence, need to be protected from dynamic loading. In addition, conventional magnetostrictive materials ignite in air and cannot be easily processed with usual machine tools, which poses additional barrier for their use in practical applications.

In recent years, a new class of magnetostrictive materials that could be directly used in structural applications is being developed. Of particular interest is the gallium–iron alloy known under the acronym GALFENOL. Though the magnetostrictive response of GALFENOL is not as high as that of TERFENOL ($\sim$400 $\mu\varepsilon$ vs. $\sim$2000 $\mu\varepsilon$), its structural and processing properties are much better. In addition, due to the low cost of iron, the cost of GALFENOL could potentially be an order of magnitude lower than that of TERFENOL. Samples of GALFENOL rods in single and polycrystalline form have already been produced. The use of stress annealing has been shown to produce enhanced magnetostrictive response through uniaxial anisotropy in rods. GALFENOL can be rolled into sheets, which can be then annealed under optimized combinations of temperature, time, and controlled atmosphere to simultaneously achieve surface texture control for magnetostrictive properties enhancement, and grain size control for mechanical properties improvement. More details about the current efforts in GALFENOL development are given by Flatau (2006).

9.11 Summary and Conclusions

This chapter has reviewed the equations of piezoelectricity and piezomagnetism, and has briefly discussed the basic types of electroactive and magnetoactive materials. Electroactive and magnetoactive materials are materials that modify their shape in response to electric or magnetic stimuli. Such materials permit induced-strain actuation and strain sensing which are of considerable importance in minimechatronics.

On one hand, induced-strain actuators are based on active materials that display dimensional changes when energized by electric, magnetic, or thermal fields. Piezoelectric, electrostrictive, and magnetostrictive materials have been presented and analyzed. Of these, piezoelectric (PZT), electrostrictive (PMN), and magnetostrictive (Terfenol-D) materials have been shown to have excellent frequency response (1 to 10 kHz, depending on actuator length), high force (up to 50 kN on current models), but small induced-strain stroke capabilities (typically, 0.1 mm for 0.1% strain on a 100 mm actuator). With this class of induced-strain actuators, displacement amplification devices need always to be incorporated into the application design. Induced-strain actuators based on such active materials (PZT, PMN, and Terfenol-D) are essentially small-stroke large-force solid-state actuators that have wide frequency bandwidth. However, they also display certain limitations. The most obvious one is that, in actuation applications, some form of mechanical amplification is required.

On the other hand, strain sensing with electroactive and magnetoactive materials creates direct conversion of mechanical energy into electric and magnetic energy. Strong and clear voltage signals are obtained under dynamic conditions directly from the sensor without the need for intermediate gauge bridges, signal conditioners, and signal amplifiers. These direct sensing properties are especially significant in dynamics, vibration, and audio applications in which high-frequency alternating effects happen in rapid succession thus preventing charge leaking. Other applications of active materials are in sonic and ultrasonic transduction, in which the transducer acts as both sensor and actuator, first transmitting a sonic or ultrasonic pulse, and then listening for the echoes received from the defect or target.

In summary, one can conclude that the potential of active materials for sensing and actuation applications has been demonstrated in several successful applications. However, this field is still in its infancy and further research and development is being undertaken to establish active materials as reliable, durable, and cost-effective options for large-scale engineering applications.

9.11.1 Advantages and Limitations of Piezoelectric and Electrostrictive Active Materials

Piezoelectric ceramics, e.g., PZT, are essentially small-stroke large-force solid-state actuators with very good high-frequency performance. However, they also display certain limitations. The most obvious limitation is that, in many engineering applications, some form of mechanical amplification is required. Other limitations are associated with electrical breakdown, depoling, Curie temperature, nonlinearity, and hysteresis.

Electrical breakdown may happen when an electric field applied in the poling direction exceeds the dielectric strength of the material, resulting in electrical arcing through the material and short circuit. Electrical breakdown also destroys the piezoelectric properties of the material.

Depoling may happen when an electric field is applied opposite to the poling direction, resulting in degradation of the piezoelectric properties or even polarization in the opposite direction. The depoling field (*aka* coercive field) may be as low as half of the electrical breakdown field.

Curie temperature: At temperatures close to the Curie temperature, depoling is facilitated, aging and creep are accelerated, and the maximum safe mechanical stress is reduced. For typical PZT materials, the Curie temperature is about 350°C. The operating temperature should generally be at least 50°C lower than the Curie temperature.

Nonlinearity and hysteresis: Actual piezoceramics are nonlinear and hysteretic (Figure 9.10). Hysteresis is due to several mechanisms. Upon removal of the electric field, remnant mechanical strain is observed. Hysteresis of common piezoelectric may range from 1% to 10%. Under high-frequency operation, hysteresis may generate excessive heat, and loss of performance may occur if the Curie temperature is exceeded.

The main advantage of electrostrictive materials over piezoelectric materials is their very low hysteresis. This could be especially beneficial in high-frequency dynamic applications, which could involve considerable hysteresis-associated heat dissipation. The main disadvantage of electrostrictive materials is the temperature dependence of their properties.

9.11.2 Advantages and Limitations of Magnetostrictive Active Materials

The main advantage of magnetoactive actuation materials over electroactive materials may be found in the fact, that it is sometimes easier to create a high intensity magnetic field than a high intensity electric field. High electric fields require high voltages, which raise important insulation and electric safety issues. High magnetic fields could be realized with lower voltages, since the magnetic field does not depend on the applied voltage but only on the current passing through the coil. Thus, using coils with a large number of turns per unit length, or using a high-amperage current, may yield considerable magnetic field, while still under low voltage. In this respect, the use of superconductive materials could provide additional benefits.

An important limitation of the magnetoactive materials is that they cannot be easily energized in the two dimensional topology, as when applied to a structural surface. This limitation stems from the difficulty of creating high-density magnetic fields without a closed magnetic circuit armature. For the same reason, the magnetoactive induced-strain actuators will always require additional construction elements besides the magnetoactive materials. While a bare-bones electroactive actuator need not contain anything more than just the active material, the bare-bones magnetoactive actuator always needs the energizing coil and the magnetic circuit armature. For this reason, the power density (either per unit volume or per unit mass) of magnetoactive induced-strain actuators is always below that of their electroactive counterparts.

9.12 Problems and Exercises

PROBLEM 9.1

Explain the difference between tensor notations and Voigt matrix notations in the writing of the compliance and stiffness matrices

PROBLEM 9.2

Explain the following difference in subscripts usage: the (1, 3) term in the compliance matrix is denoted s_{13}, whereas the (1, 3) term in the piezoelectric coefficient matrix is denoted d_{31}

PROBLEM 9.3

Calculate the spontaneous strain, S_S, and the spontaneous polarization, P_S, for the hypothetical lattice shown in Figure 9.24.

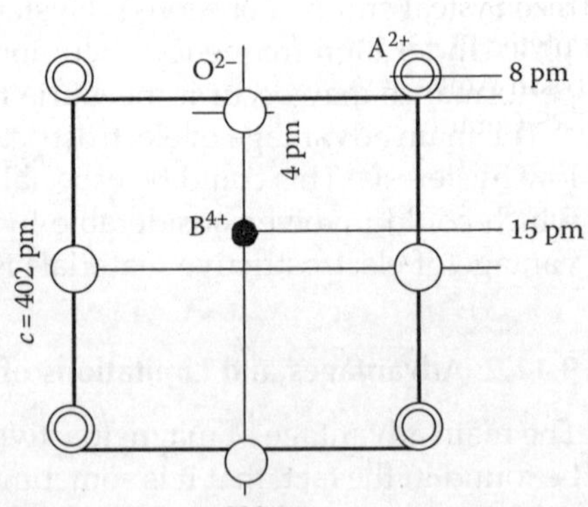

FIGURE 9.24
Ionic shifts inducing spontaneous strain and spontaneous polarization in a hypothetical material.

References

ANSI/IEEE Std. 176, *IEEE Standard on Piezoelectricity*, The Institute of Electrical and Electronics Engineers, Inc., New York, 1987.

Engdahl, G., *Handbook of Giant Magnetostrictive Materials*, Academic Press, San Diego, CA, 2000.

Flatau, A. B., Structural Magnetostrictive Alloys–ONR MURI, 2006, http://www.aerosmart.umd.edu/MURI/index.html

Ikeda, T., *Fundamentals of Piezoelectricity*, Oxford University Press, Oxford, U.K, 2006.

Janocha, H., *Adaptronics and Smart Structures*, Springer Verlag, Berlin, 1999.

Knowles, J. K., *Linear Vector Spaces and Cartesian Tensors*, Oxford University Press, Inc., New York, September 1997.

Lachisserie, Etienne du Tremolet de, *Magnetostriction—Theory and Applications*, CRC Press, London, 1993.

Lines, M. E. and Glass, A. M., *Principles and Applications of Ferroelectrics and Related Materials*, Clarendon Press, Oxford, 2001.

MIL-STD-1376B (SH) *Military Standard Piezoelectric Ceramic Materials and Measurements Guidelines for Sonar Transducers*, Defense Quality and Standardization Office, Falls Church, VA, 24 February 1995.

Mitsui, T. and Nomura, S., *Ferroelectrics and Related Substances*, Landolt-Bornstein Series, Group III, Vol. 16, Springer-Verlag, Berlin, 1981.

Park, S-.E. and Shrout, T. R., Ultra high strain and piezoelectric behavior in relaxor ferroelectric single crystals, *Journal of Applied Physics*, 82, 4, 1804–1811, 1997.

Taylor, G. W., Gagnepain, J. J., Meeker, T. R., Nakamura, T., and Shuvalov, L. A., *Piezoelectricity*, Gordon and Breach Science Publishers, New York, 1985.

Uchino, K., *Piezoelectric Actuators and Ultrasonic Motors*, Kluwer Academic Publishers, Morwell, MA, 1997.

List of Commercial Suppliers of Electroactive and Magnetoactive Materials

APC International Ltd and American Piezo Ceramics, Inc., Mackeyville, PA, email APCSales@aol.com., Web www.americanpiezo.com

Burleigh Instruments, Inc., Burleigh Park, Fishers, New York.

EDO Corporation, Salt Lake City, UT, http://www.nycedo.com/edocorp/pageba5_edoecp.htm

Etrema Products, Inc., Ames, IA, http://etrema-usa.com

Kinetic Ceramics, Inc., Hayward, CA, http://www.kineticceramics.com/contact_us/contact_us.htm

Measurement Specialties, Inc., Wayne, PA, sensors@msiusa.com, http://www.msiusa.com

Micromega Dynamics, SA, Angleur, Belgium, e-mail info@micromega-dynamics.com, http://www.micromega-dynamics.com/profile.htm

Piezo Kinetics, Inc., Bellefonte, PA, http://www.piezo-kinetics.com/

Piezo Systems, Inc., Cambridge, MA, http://www.piezo.com/

Polytec PI, Inc., Costa Mesa, CA, http://www.polytechpi.com/

Tokin America, Inc., Union City, CA, http://www.nec-tokinamerica.com/products.html

TRS Ceramics Inc., State College, PA, http://www.trsceramics.com

References

ANSI/IEEE Std 176, IEEE Standard on Piezoelectricity, The Institute of Electrical and Electronics Engineers, Inc., New York, 1987.

[illegible], Computer Online Encyclopedia, Academic Press, San Diego, CA, [illegible]; [illegible] Magnetoelectric Type Upset-CNS-NIST, 2006, [illegible].html

[illegible], PZT [illegible] Ceramic and the [illegible] Fiber, [illegible], 2006.

[illegible]

[illegible]

[illegible]

[illegible]

[illegible]

[illegible]

List of Commercial Suppliers of Electroactive and Magnetoactive Materials

[illegible], International Ltd. and [illegible] Piezo Ceramic Inc., [illegible], the PZT and [illegible], Sensor Web, www.[illegible].com

[illegible]

[illegible]

[illegible]

[illegible]

[illegible]

[illegible]

[illegible]

10

Induced-Strain Actuators

10.1 Introduction

The induced-strain active-material actuators are the enabling technology for a number of micromechatronics applications involving micropositioning, vibration control, etc. In this chapter, we will describe the principles and applications of induced-strain active-material actuators. The chapter will start with some generic concepts about induced-strain active-material actuators in comparison with the more conventional electromechanical and hydraulic actuators. Next, the construction details of piezoelectric, electrostrictive, and magnetostrictive induced-strain actuators will be examined. The modeling of induced-strain actuators will be developed from first principles. The typical performance of commercially available induced-strain actuators will be examined. A linearized model for the electromechanical behavior will be derived. The principles of induced-strain actuation (ISA) of a compliant structure will be analyzed. The displacement analysis and the electric response will be derived for both static and dynamic applications. Displacement-amplified induced-strain actuators will be introduced, and their analysis will be discussed. The electrical power and energy flow in ISA applications will be studied. The analysis of mechanical power and energy extraction will be followed by the analysis of the electrical power and energy demands. The power and energy conversion efficiency will be examined. Criteria for optimal energy conversion will be derived. A comparative study of commercially available induced-strain actuators will be presented. The efficient design of ISA applications will be studied. Guidelines for the effective design and construction of ISA solutions will be provided. Electric power supplies for energizing induced-strain actuators will be briefly discussed. The last part of the chapter will deal with shape memory alloy (SMA) actuators. The presentation of the shape memory effect and the superplasticity behavior of certain materials (e.g., nickel–titanium alloy Nitinol) will be followed by a development of the basic analytical principles for estimation of the SMA actuator behavior under load when heat activated. The chapter will finish with summary and conclusions, followed by a bibliography and a list of commercial suppliers of induced-strain actuators. The chapter contains a number of worked out examples inserted in the text, and a number of problems and exercises grouped at the end of the chapter.

10.2 Active-Material Induced-Strain Actuators

Actuators are devices that produce mechanical action. Actuators convert input energy (electric, hydraulic, pneumatic, thermal, etc.) into mechanical output (force and displacement). Actuators rely on a "prime mover" (e.g., the displacement of high-pressure fluid in a hydraulic cylinder or the electromagnetic force in an electric motor) and on mechanisms to convert the prime mover into the desired action. If the prime mover can impart a large stroke, direct actuation can be effected (e.g., a hydraulic ram). When the prime mover has a small stroke but high frequency bandwidth, a switching principle is employed to produce continuous motion through the addition of switched incremental steps (e.g., stepper motors). Like everywhere in engineering, the quest for simpler, more reliable, more powerful, easier to maintain, and cheaper actuators is continuously on. In this respect, the use of active-materials solid-state induced-strain actuators has recently seen a significant increase. Initially developed for high-frequency, low-displacement acoustic applications, these revolutionary actuators are currently expanding their field of application into other areas of mechanical and aerospace design. Compact and reliable, induced-strain actuators directly transform input electrical energy into output mechanical work. One application area in which solid-state induced-strain devices have a very promising future is that of translational actuation for vibration and aeroelastic control.

At present, the translational actuation market is dominated by hydraulic and pneumatic pressure cylinders, and by electromagnetic solenoids and shakers. Hydraulic and pneumatic cylinders offer reliable performance, with high force and large displacement capabilities. When equipped with servovalves, hydraulic cylinders can deliver variable stroke output. Servovalve-controlled hydraulic devices are the actuator of choice for most aerospace applications (Figure 10.1) as well as for many automotive and robotic applications. However, a major drawback of conventional hydraulic actuators is the need for a separate hydraulic power supply equipped with large electric motors and hydraulic pumps that

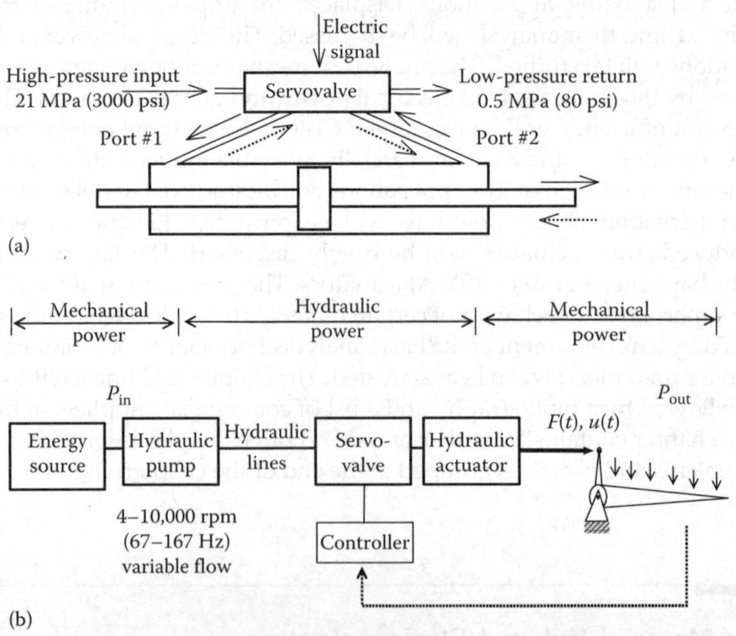

FIGURE 10.1
Conventional hydraulic actuation system: (a) flight controls of modern aircraft; (b) details of the servovalve controlled hydraulic cylinder.

send the high-pressure hydraulic fluid to the actuators through hydraulic lines. These features can be a major drawback in certain applications. For example, a 300-passenger airplane has over a kilometer of hydraulic lines spanning its body, from the engines to the most remote wing tip. In such a network, a vulnerable hydraulic piping can present a major safety liability under both civilian and military operations. In ground transportation, similar considerations have spurred automobile designers to promote the "brake-by-wire" concept that is scheduled to enter the commercial market in the next few years. In some other applications, the use of conventional actuation is simply not an option. For example, the actuation of an aerodynamic servo-tab at the tip of a rotating blade, such as in helicopter applications, cannot be achieved through conventional hydraulic or electric methods due to the prohibitive high-g centrifugal forces generated during the blade rotation.

Electromechanical actuation is increasingly preferred in several industrial applications because it can directly convert electrical energy into mechanical energy. The most widely used high-power electromechanical actuators are the electric motors. However, they can deliver only rotary motion and need gearboxes and rotary-to-translational conversion mechanisms to achieve translational motion. This route is unyielding, leads to additional weight, and has low frequency bandwidth. Direct conversion of electrical energy into translational force and motion is possible, but its practical implementation in the form of solenoids and electrodynamic shakers is marred by low-force performance. The use of solenoids or electrodynamic shakers to perform the duty cycle of hydraulic cylinders does not seem conceivable.

Solid-state induced-strain actuators offer a viable alternative (Figure 10.2). Though their output displacement is relatively small, they can produce remarkably high force. With well-architectured displacement amplification, induced-strain actuators can achieve output strokes similar to those of conventional hydraulic actuators, but over much wider bandwidth. In addition, solid-state induced-strain actuators do not require separate hydraulic power units and long hydraulic lines used by conventional hydraulic actuators because they only need access to electric supply at the actuator site.

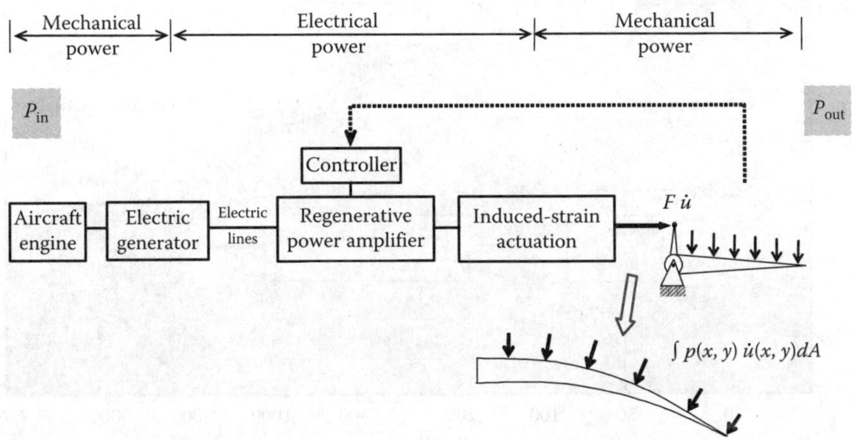

FIGURE 10.2
Schematic representation of a solid-state induced-strain actuated flight control system using electroactive materials.

The development of solid-state induced-strain actuators has entered the production stage; actual actuation devices based on these concepts are likely to reach the applications market in the next few years. An increasing number of vendors are producing and marketing solid-state actuation devices based on induced-strain principles.

An important characteristic and possible drawback of induced-strain actuator is the fact that they operate with a finite stroke. This is different from conventional actuators, where usually the force, and not the stroke, is the limiting factor. The finite-stroke principle of induced-strain actuators requires a different design approach. The effect of actuator compressibility becomes important. Under large loads, much of the induced-strain stroke may be absorbed internally, and only a small portion remains available to do useful work. These aspects are not usually considered in conventional actuators. However, they become very important when designing applications that include induced-strain actuators (Figure 10.3).

Another important characteristic of induced-strain active-material actuators is their relatively small stroke. The active material induced strain is around 0.12%. For a 100 mm long actuator, this usually results in 0.12 mm induced-strain displacement, which is quite small. However, the resulting force can be very large, since it is proportional with the cross-sectional area and the actuator modulus; values of 10–30 kN are quite common for many induced-strain actuators. For practical applications, displacement amplification devices are used; these devices trade some of the force for an increase in effective displacement. Figure 10.4 presents the "family tree" of displacement-amplified induced-strain actuators. It is apparent that a variety of options based on solid mechanics and fluid mechanics principles

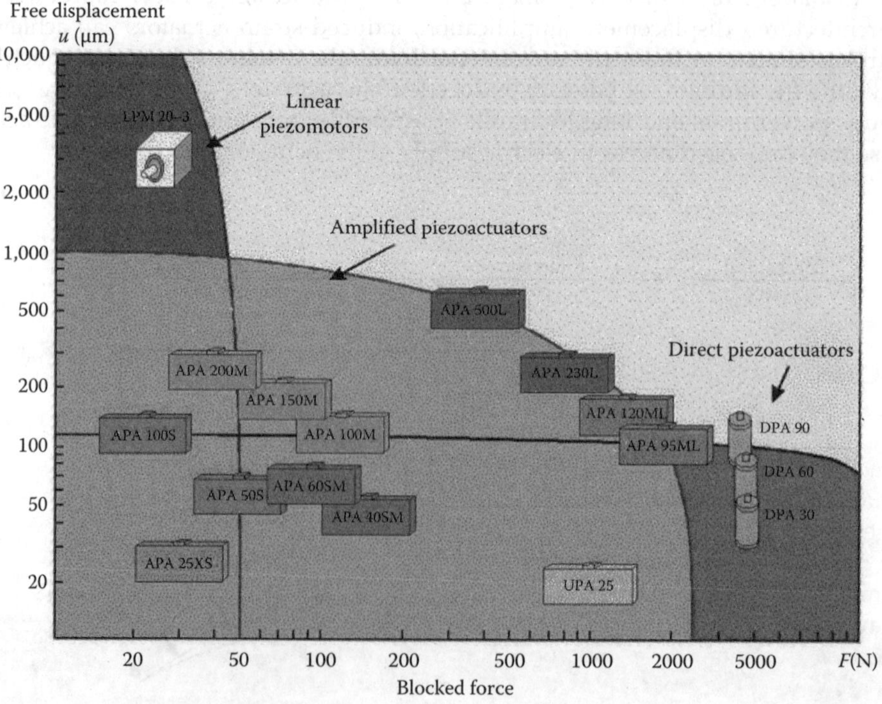

FIGURE 10.3
Induced-strain actuators selection chart. (From http://www.micromega-dynamics.com/profile.htm.)

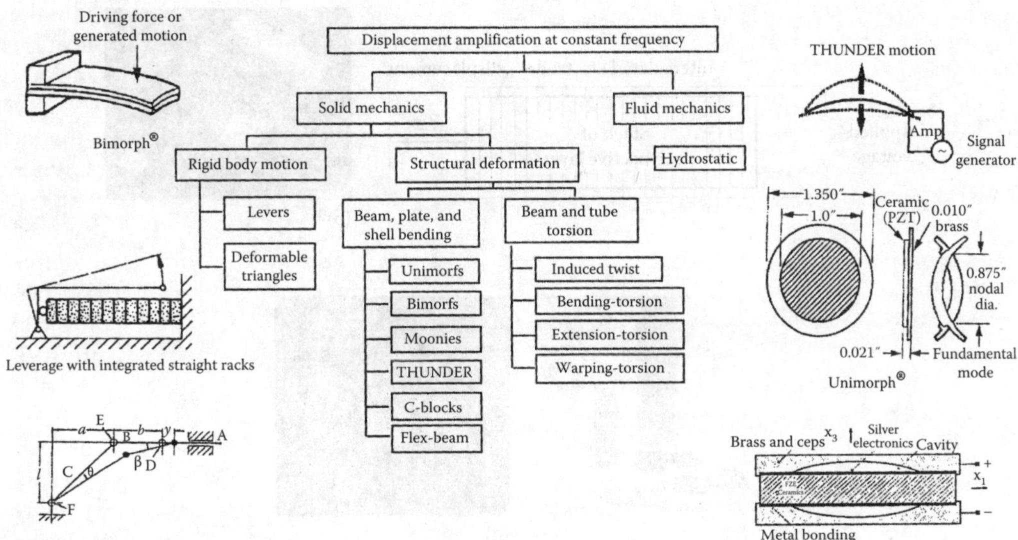

FIGURE 10.4
Family tree of displacement amplified induced-strain actuators.

exist. Some trade names, such as "bimorph," "unimorph," "moonie," "THUNDER," etc., have entered the technical parlance. A common characteristic of all these devices is the inevitable retention of energy in the elastic deformation of the device itself. Hence, the displacement amplification is always accompanied by a reduction in the available energy that can be used to drive the external load.

Due to the large variety of displacement amplification methods, it is unfeasible to treat each of them in detail. Rather, it is more useful to derive general analysis methods that can be subsequently particularized to specific situations. For this reason, this chapter will present the analysis of induced-strain actuators for generic situations.

10.3 Construction of Induced-Strain Actuators

10.3.1 Electroactive Stacks

An electroactive solid-state actuator consists of a stack of many layers of electroactive material (e.g., PZT or PMN) connected to the positive and negative terminals of a high voltage source (Figure 10.5a). The layers in an electroactive stack are energized and loaded in the 3-direction. The 3-direction corresponds to the direction of initial polarization and gives maximum ISA. The layers are very thin such that a high electric field is applied across each layer. Such an electroactive stack behaves like an electrical capacitor. When activated, the electroactive material expands and produces output displacement. The electroactive stacks are constructed by two methods. In the first method, the layers of active material and the electrodes are mechanically assembled and glued together using a structural adhesive. The adhesive modulus (typically, 2–5 GPa) is at least an order of magnitude lower than the modulus of the ceramic (typically, 70–90 GPa). This aspect may lead to the

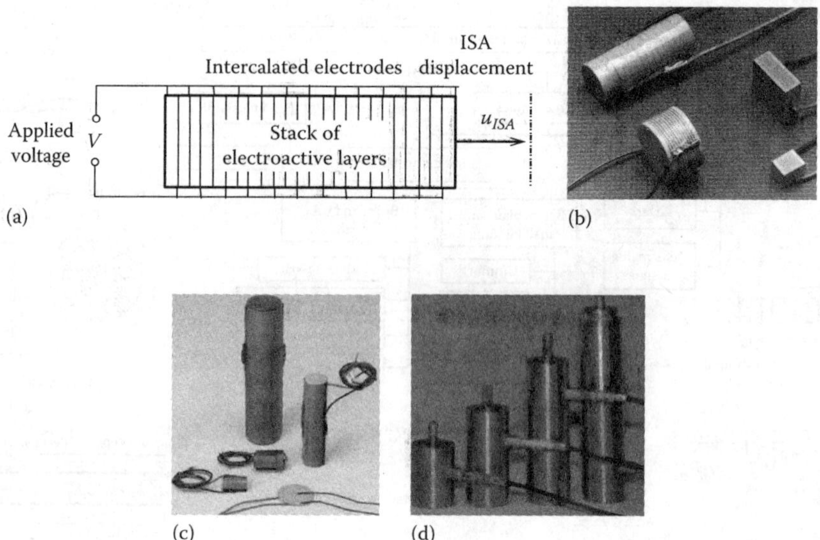

FIGURE 10.5

Induced-strain actuator using a PZT or PMN electroactive stack: (a) stack schematic; (b) typical commercially available cofired stack from EDO Corporation (http://www.nycedo.com/edocorp/pageba5_edoecp.htm); (c) various-size stacks from Kinetic Ceramics, Inc., and (d) heavy-duty actuators from Kinetic Ceramics. (From http://www.kineticceramics.com/contact_us/contact_us.htm.)

stack stiffness being lower than the stiffness of the basic ceramic material due to displacement lost in the adhesive compression. In the second method, the ceramic layers and the electrodes are assembled in the "green" state. Then, they are fired together (cofired) under a high isostatic pressure (HIP process). This process ensures a much stiffer final product and, hence, a better actuator performance. However, the processing limitations, such as oven and hot-press sizes. Allow the application of this process only to small stacks (Figure 10.5b).

The electroactive stacks may be surrounded by a protective polymeric or elastomeric wrapping. Electrical connection lead wires protrude from the wrapping. Steel washers, one at each end, are also provided to distribute the load into the brittle ceramic material (Figure 10.5c). When mounted in the application structure, these stacks must be handled with specialized knowledge. Protection from accidental impact damage must be assured. Adequate structural support and alignment are needed. Mechanical connection to the application structure must be such that neither tension stresses nor bending are induced in the stack since the active ceramic material has low tension strength. Hence, the load applied to the stack must always be compressive and perfectly centered. If tension loading is expected, adequate prestress must be provided through springs or other means. For generic applications, the stack may be encapsulated into a steel casing which provides a prestress mechanism and the electrical and mechanical connections (Figure 10.5d).

Figure 10.6 compares the response of two commercially available actuators, one piezoelectric, the other electrostrictive. It can be seen that both have about the same maximum induced-strain value, as it can be readily verified by dividing the maximum displacement by the actuator length. Both material types display quasi linear behavior. The piezoelectric actuator allows some field reversal (up to 25%, according to manufacturer), which makes the total stroke larger. On the other hand, the electrostrictive actuator has much less hysteresis, that is, less energy losses and less heating in the high-frequency regime.

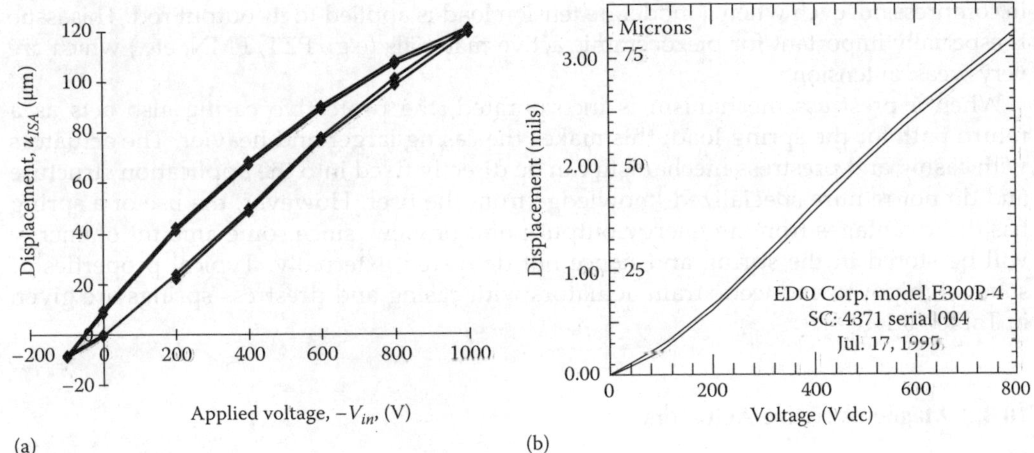

FIGURE 10.6

The induced-strain displacement versus applied voltage for typical electroactive actuators: (a) Polytec PI P-245.70 PZT stack, 99-mm long; (b) EDO Corporation model E300-P4 PMN stack, 57 mm long.

10.3.2 Electroactive Actuators with Casing and Prestress Mechanism

Some commercially available solid-state actuators are just the drive units described above and shown in Figure 10.5a through c. Other commercially available solid-state actuators also include a protective casing and a prestress mechanism (Figure 10.5d). The casing provides protection for the active material and its electrical connections. It also facilitates the mechanical connection between the actuator and the application structure. The prestress mechanism (Figure 10.7b) is necessary to ensure that the active material is loaded only

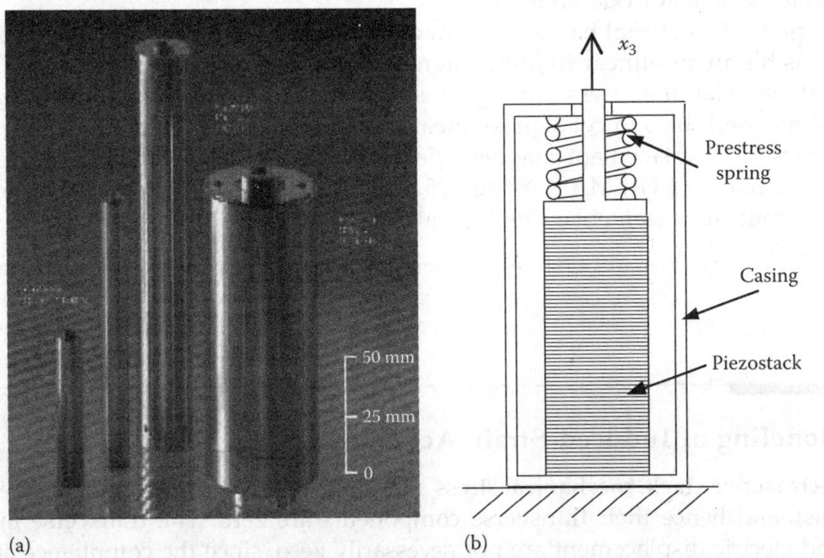

FIGURE 10.7

Piezoelectric actuators with casing and prestress spring: (a) Commercially available actuators (From Politec PI, Inc., http://www.polytecpi.com); (b) schematic of the actuator construction.

in compression, even when a moderate tension load is applied to its output rod. This issue is especially important for piezoceramic active materials (e.g., PZT, PMN, etc.) which are very weak in tension.

When a prestress mechanism is incorporated, the protective casing also acts as a return path for the spring load; this makes the casing larger and heavier. The actuators with casing and prestress mechanism can be directly fixed into the application structure and do not require specialized knowledge from the user. However, the use of a spring has disadvantages from an energy output point of view, since some amount of energy will be stored in the spring and hence not delivered externally. Typical properties of some high-power induced-strain actuators with casing and prestress springs are given in Table 10.1.

10.3.3 Magnetostrictive Actuators

Magnetostrictive materials can also be used for induced-strain actuation. Figure 10.8a shows the typical layout of a magnetostrictive actuator. It consists of a Terfenol-D bar surrounded by an electric coil and enclosed into an annular magnetic armature. The magnetic circuit is closed through the end caps. In this arrangement, the magnetic field is strongest in the cylindrical inner region filled by the Terfenol-D bar. When the coil is activated, the Terfenol-D expands and produces output displacement. The Terfenol-D bar, the coil, and the magnetic armature are assembled with prestress between two steel-washers and put inside a protective wrapping to form the basic magnetoactive induced-strain actuator. An important design consideration for magnetostrictive actuators is the engineering of the *magnetic circuit*. The magnetic circuit consists of the solenoid coil (which provides the actuation magnetic field), permanent magnets (which provide magnetic bias), and other parts through which the magnetic loop is closed. A good magnetic circuit ensures the highest magnetic flux density in the Terfenol-D and uniform magnetic flux throughout the actuator operating cycle.

The Terfenol-D material has been shown to be capable of up to 2000 $\mu\varepsilon$; however, its behavior is highly nonlinear in both magnetic field response and the effect of compressive prestress. Manufacturers of magnetostrictive actuators optimize the internal prestress and magnetic bias to get a quasi linear behavior in the range of 750–1000 $\mu\varepsilon$. Figure 10.8b shows the displacement-magnetic field response for a typical large-power magnetostrictive actuator (ETREMA AA-140J025, 200 mm long, ~1 kg weight, 0.140 mm peak-to-peak output displacement). The typical response curve of this actuator is shown in Figure 10.9b.

10.4 Modeling of Induced-Strain Actuators

In an electroactive stack, mechanical stress and electric field act only in the 3-direction (the stack axis), and hence their transverse components are zero. The transverse mechanical strain and electric displacement are not necessarily zero, since the compliance and piezoelectric coupling matrices are fully populated. However, since the focus of the analysis is

TABLE 10.1

Technical Data for Polytec PI Extra-High-Load HVPZT Induced-Strain Actuators

Models	P-243.10	P-243.20	P-243.30	P-243.40	P-247.20	P-247.30	P-247.50	P-247.70	Units
Free stroke for 0 to −1000 V	10	20	40	60	20	40	80	120	μm ±20%
Static large-signal stiffness	1130	560	290	210	1280	680	360	240	N/μm ±20%
Dynamic small-signal stiffness[a]	1700	840	435	315	1920	1020	540	370	
Push/pull force capacity	30/2	30/2	30/2	30/2	30/3.5	30/3.5	30/3.5	30/3.5	kN
Max. operating voltage	−1500	−1500	−1500	−1500	−1000	−1000	−1000	−1000	V
Electrical capacitance	360	720	1330	1800	1220	2300	4400	6560	nF ±20%
Dynamic operating current coefficient (DOCC)	45	45	45	45	76	76	76	76	μA/(Hz × μm)
Unloaded resonant frequency ($f0$)	11	8	5	3.5	8	6.5	4.5	3	kHz ±20%
Standard operating temperature range (°C)	−40 to +80	−40 to +80	−40 to +80	−40 to +80	−40 to +80	−40 to +80	−40 to +80	−40 to +80	°C
Weight	460	630	850	960	580	660	830	960	g ±5%
Length	41.5	63	116.5	144	56	73	116.5	144	mm ±0.5

Source: http://www.polytechpi.com/

[a] Dynamic small-signal stiffness ~50% higher than static large-signal stiffness.

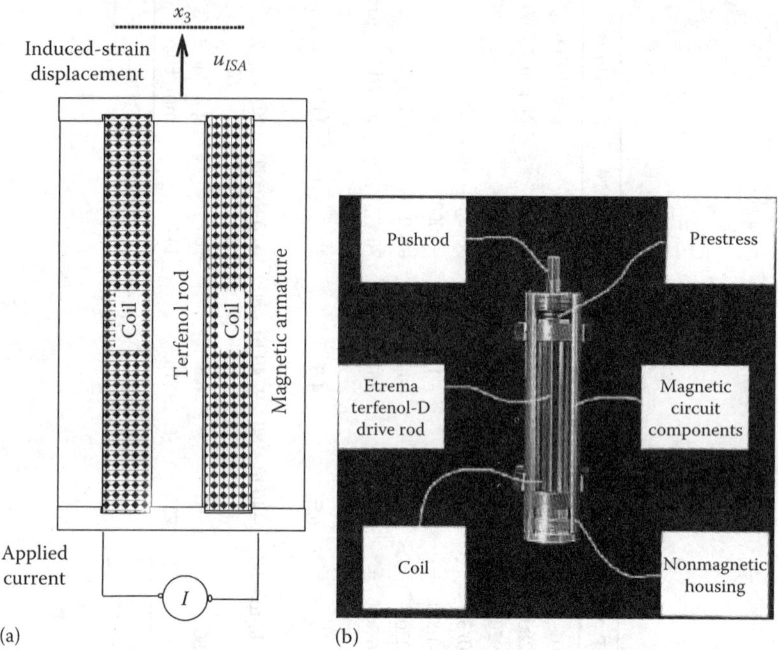

FIGURE 10.8
Internal construction of a magnetostrictive (Terfenol-D) solid-state actuator: (a) line-art schematic; (b) functionality of various components. (From http://etrema-usa.com.)

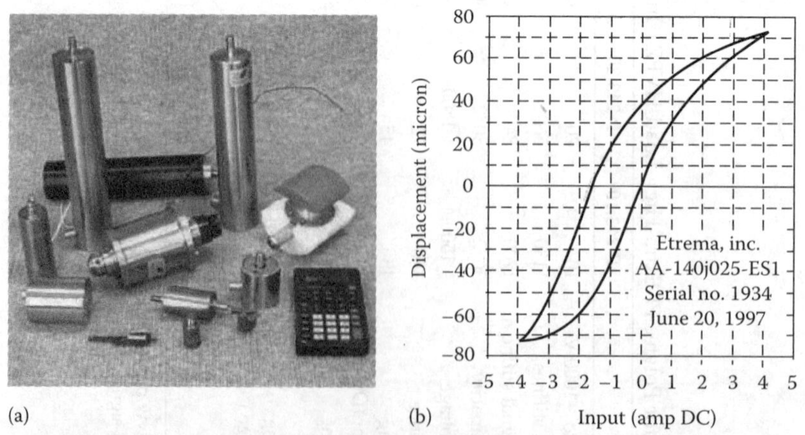

FIGURE 10.9
(a) Magnetostrictive Terfenol-D solid-state actuators; (b) and typical response curve (actuator model number AA140J025-ES1, ETREMA, Inc.). (From http://etrema-usa.com.)

directed toward the axial behavior, the transverse strains and electric displacements are not relevant. Hence, simpler, 1-D constitutive equations can be used, that is,

$$S = sT + d_{33}E$$
$$D = d_{33}T + \varepsilon E \tag{10.1}$$

where S is the axial strain, T is the axial stress, E is the electric field (voltage per unit distance between electrodes), D is the electric displacement (charge per unit area), ε is the dielectric permittivity, and d_{33} is the piezoelectric induced-strain coefficient. To simplify notations, the subscripts 3 and 33 were removed from the S, s, T, E variables. However, the subscript 33 was retained on d_{33} in order to avoid confusion with the differential operator d. The compliance s is measured at zero electric field, whereas the permittivity ε is measured at zero mechanical stress.

For magnetoactive ISA, the 1-D constitutive equations take the form

$$S = sT + d_{33}H$$
$$B = d_{33}T + \mu H \tag{10.2}$$

where H is the magnetic field intensity, B is the magnetic flux density, μ is the magnetic permeability under constant stress, and d_{33} is the piezomagnetic induced-strain coefficient.

10.4.1 Typical Performance of Solid-State Induced-Strain Actuators

Figures 10.10 through 10.12 illustrate typical performance of commercially available solid-state induced-strain actuators under no-load condition. Figure 10.10 presents the

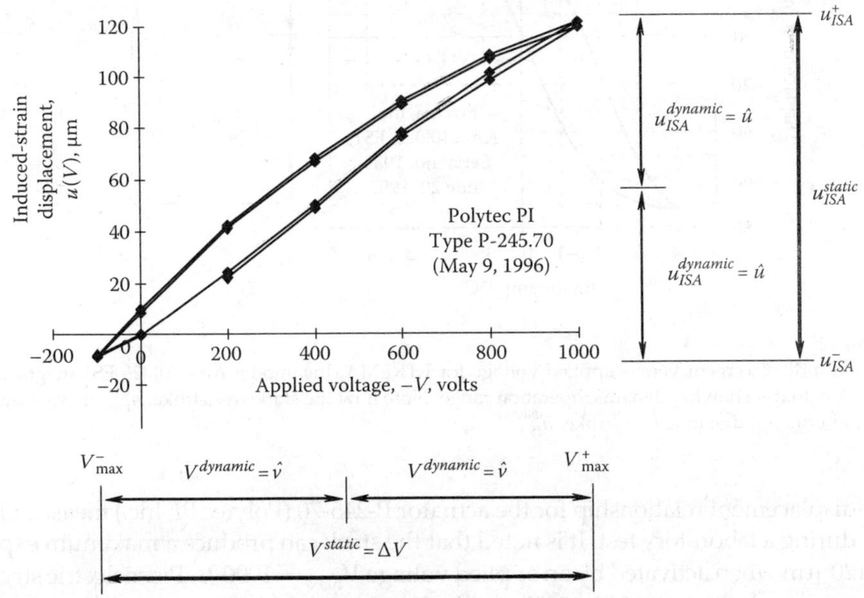

FIGURE 10.10
Induced-strain displacement versus applied voltage for Polytec PI model P-245.70 piezoelectric (PZT) actuator. Note how the static free stroke, u_{ISA}^{static}, is split into two to obtain the alternating dynamic free stroke, $u_{ISA}^{dynamic}$.

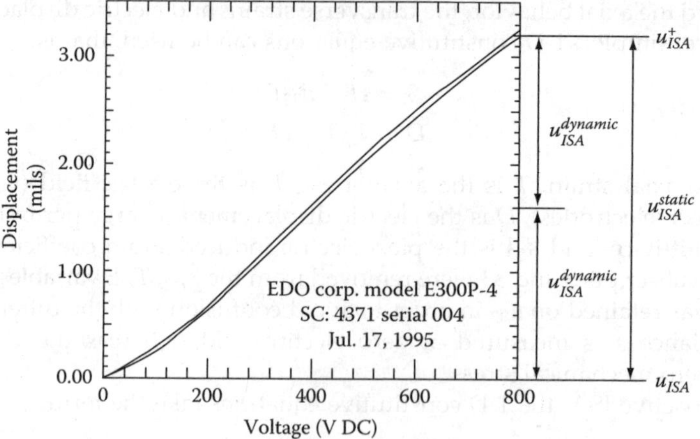

FIGURE 10.11

Induced-strain displacement versus applied voltage for EDO Corp. model E300P-4 electrostrictive (PMN) actuator showing dynamic operation range. Note how the static free stroke, u^{static}_{ISA}, is split into two to obtain the alternating dynamic free stroke, $u^{dynamic}_{ISA}$.

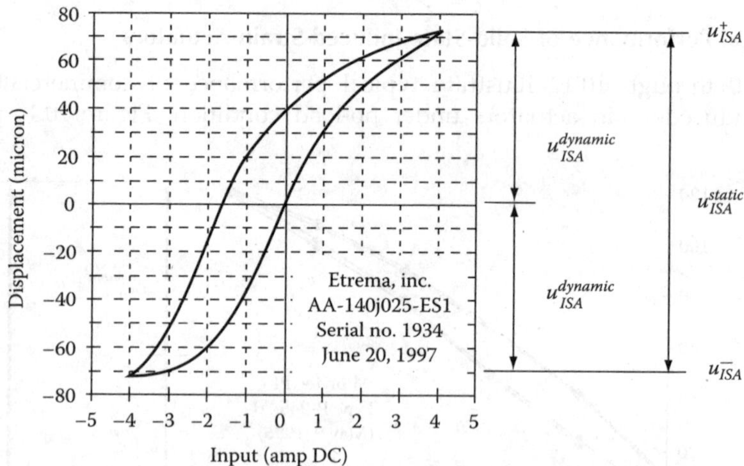

FIGURE 10.12

Induced-strain displacement versus applied voltage for ETREMA, Inc. model AA-140J025-ES1 magnetostrictive (Terfenol-D) actuator showing dynamic operation range. Note how the static free stroke, u^{static}_{ISA}, is split into two to obtain the alternating dynamic free stroke, $u^{dynamic}_{ISA}$.

voltage-displacement relationship for the actuator P-245-70 (Polytec PI, Inc.) measured by the authors during a laboratory test. It is noted that the stack can produce a maximum expansion $u^+_{ISA} = 120$ μm when activated by an applied voltage $V^+_{max} = 1000$ V. Piezoelectric stacks also accept reverse polarity up to 25%–30%, that is, $V^-_{max} = -250$ V; the corresponding maximum contraction would be $u^-_{ISA} = -30$ μm. (For stack-safety reasons, our experiments shown in Figure 10.10 were not conducted to the maximum contraction level.) The total static travel of this actuator is $u^{static}_{ISA} = 150$ μm. Since the stack length was $l = 100$ mm, the static free strain of this stack would be $\varepsilon_{ISA}^{static} = 0.150\%$. For dynamic applications, the actuator is electrically biased about a mid-range position and then an alternating voltage is superimposed. The bias position would be $u_0 = \frac{1}{2}\left(u_{ISA}^+ + u_{ISA}^-\right) = 45$ μm, whereas the alternating component would be $u_{ISA}^{dynamic} = \pm 75$ μm. The corresponding dynamic free strain is $\varepsilon_{ISA}^{dynamic} = \pm 0.075\%$. Figure 10.10 also shows that this induced-strain actuator has significant hysteresis, which is typical of PZT stacks.

Figure 10.11 presents the voltage–displacement curve for the PMN actuator E300P-4 (EDO Corporation) as measured by the manufacturer. The maximum expansion is $u_{ISA}^+ = 3.2$ mil $= 81$ μm. PMN material do not accept reversed polarity, hence $u_{ISA}^- = 0$. Thus, the total free stroke of this induced-strain actuator is merely $u_{ISA}^{static} = 81$ μm. Since the stack length was $l = 2.4$-in $= 61$ mm, the corresponding free strain of this stack is $\varepsilon_{ISA}^{static} = 0.133\%$. If the actuator were to be used for dynamic applications, the dynamic free stroke would be $u_{ISA}^{dynamic} = \pm 40$ μm, with the corresponding dynamic free strain $\varepsilon_{ISA}^{dynamic} = \pm 0.067\%$. Figure 10.11 also shows that this induced-strain actuator has very little hysteresis, typical of PMN stacks.

Figure 10.12 presents the displacement versus electric current curve for the Terfenol actuator AA140J013-ES1 (ETREMA, Inc.) as measured by the manufacturer. This actuator was specially biased for dynamic operation. Thus, the maximum expansion and the maximum contraction have equal magnitudes, that is, $u_{ISA}^+ = -u_{ISA}^- = u_{ISA}^{dynamic} = 70$ μm. The total free stroke of this induced-strain actuator is $u_{ISA}^{static} = 140$ μm. Since the length of the Terfenol rod was $l = 140$ mm, the corresponding free strain of this induced-strain actuator is $\varepsilon_{ISA}^{static} = 0.100\%$. For dynamic operation, the dynamic free strain is $\varepsilon_{ISA}^{dynamic} = \pm 0.050\%$. Figure 10.12 also shows that this induced-strain actuator has significant hysteresis.

10.4.2 Linearized Electromechanical Behavior of Solid-State Induced-Strain Actuators

As illustrated in Figures 10.10 through 10.12, the displacement of induced-strain actuators is not perfectly linear. However, quasi linear behavior maybe assumed using the secant approximation to the full-stroke values. As a result, apparent electromechanical constants are obtained. Thus, the overall performance of the induced-strain actuators can be defined in terms of readily available vendor data, such as free-stroke, u_{ISA}, the corresponding voltage, V (or current, I for magnetoactive actuators), the internal stiffness, k_i, and the capacitance, C_0 (or inductance, L_0 for magnetoactive actuators). In doing so, one is effectively using the secant method to linearize the full-stroke behavior of the actuator. This approach predicts the power and energy ratings of induced-strain actuators under full-stroke operation in a straightforward easy-to-understand manner. Consider a piezo-electric stack of N wafers of thickness h and area A. The overall stack length is

$$l = Nh \tag{10.3}$$

Assume an effective compliance s; hence, the effective overall stiffness of the actuator is

$$k_i = \frac{A}{ls} \tag{10.4}$$

Since each wafer is covered with electrodes, which are connected in parallel, the overall electrode area is NA. Assume an effective dielectric permittivity ε; hence, the overall capacitance of the piezoelectric stack is

$$C_0 = N\frac{\varepsilon A}{h} \quad \text{(stress-free capacitance)} \tag{10.5}$$

Since the applied voltage V acts across each wafer of thickness h, the effective electric field strength is

$$E = \frac{V}{h} \tag{10.6}$$

Assume the free-stroke is due to an effective induced-strain piezoelectric coefficient, d_{33}. Hence,

$$u_{ISA} = Nd_{33}Eh = d_{33}El \quad \text{(free stroke)} \tag{10.7}$$

If we have to take into account the presence of the adhesive glue joining the piezoelectric wafers present in the stack, then the internal stiffness and the effective induced strain will be modified accordingly.

Usual vendor data contains the free-stroke, u_{ISA}, the corresponding maximum voltage, V, the internal stiffness, k_i, and the capacitance, C_0. Based on this vendor data, we can calculate apparent full-stroke elastic compliance s, electrical permittivity ε, and piezoelectric strain coefficient d_{33}, as

$$s = \frac{A}{lk_i} \tag{10.8}$$

$$\varepsilon = \frac{C_0 h^2}{Al} \tag{10.9}$$

$$d_{33} = \frac{u_{ISA}h}{Vl} \tag{10.10}$$

The electromechanical coupling coefficient k_{33} relevant to expansion in the 3-direction under electric field in the same 3-direction is given by

$$k_{33}^2 = \frac{d_{33}^2}{s_{33}^E \varepsilon_{33}^T} \tag{10.11}$$

Consistent with the linearization scheme, the constants d_{33}, ε_{33}^T, and s_{33}^E are given by expressions of Equation (10.8) through Equation (10.10) as d_{33}, ε, and s. Denoting k_{33} by κ (to avoid confusion with the notation, k, already used for the mechanical stiffness) and substituting from the above Equations (10.3) through (10.10), we obtain, after some manipulations, the following simple expression for the *effective full-stroke electromechanical coupling coefficient* of the piezoelectric stack actuator:

$$\kappa^2 = \frac{d_{33}^2}{s\varepsilon} = \frac{\frac{1}{2}k_i u_{ISA}^2}{\frac{1}{2}C_0 V^2} \quad \text{(effective electromechanical coupling coefficient)} \tag{10.12}$$

Since u_{ISA} and k_i are reference parameters of the induced-strain actuator, we define the *reference mechanical energy* of the induced-strain actuator, that is,

$$E_{mech}^{ref} = \tfrac{1}{2} k_i u_{ISA}^2 \tag{10.13}$$

Similarly, the capacitance C_0 and the maximum applied voltage V are reference electrical parameters of an induced-strain actuator; hence, we define the *reference electrical energy* of an induced-strain actuator, that is,

$$E_{elec}^{ref} = \tfrac{1}{2} C_0 V^2 \tag{10.14}$$

Note that E_{elec}^{ref} is the input electrical energy required to make the free induced-strain actuator reach the free stroke u_{ISA}. In the more general case of operation with bias voltage between limits V_{max}^{+} and V_{max}^{-}, the voltage to be used in Equation (10.14) should be V_{max}^{+}.

Substitution of Equations (10.13) and (10.14) into Equation (10.12) yields

$$\kappa^2 = \frac{E_{mech}^{ref}}{E_{elec}^{ref}} \tag{10.15}$$

Equation (10.15) states that the effective electromechanical coupling coefficient of an induced-strain actuator, κ, can be calculated as the ratio between the reference mechanical energy, E_{mech}^{ref}, and the reference electrical energy, E_{elec}^{ref}.

For *magnetoactive actuators*, a similar process is employed using electric current I and inductance L_0 instead of voltage V and capacitance C_0. As a result, the effective full-stroke electromechanical coupling coefficient of the magnetoactive actuator is obtained as

$$\kappa^2 = \frac{d_{33}^2}{s\mu} = \frac{\tfrac{1}{2} k_i u_{ISA}^2}{\tfrac{1}{2} L_0 I^2} \tag{10.16}$$

Example 10.1: Actuator Basic Properties Estimation

Consider a piezoelectric actuator made up of a stack APC 850 square wafers with the properties given in Tables 10.2 and 10.3. Find

TABLE 10.2

Properties of Piezoelectric Material APC-850

Property	Symbol	Value
Compliance, in plane	s_{11}^E	15.30×10^{-12} Pa^{-1}
Compliance, thickness wise	s_{33}^E	17.30×10^{-12} Pa^{-1}
Dielectric constant	ε_{33}^T	$\varepsilon_{33}^T = 1750\varepsilon_0$
Thickness-wise induced-strain coefficient	d_{33}	400×10^{-12} m/V
In-plane induced-strain coefficient	d_{31}	-175×10^{-12} m/V
Coupling factor, parallel to electric field	k_{33}	0.77
Coupling factor, transverse to electric field	k_{31}	0.36
Coupling factor, transverse to electric field, polar motion	k_p	0.63
Poisson ratio	ν	0.35
Density	ρ	7700 kg/m^3

Note: $\varepsilon_0 = 8.85 \times 10^{-12}$ F/m.

TABLE 10.3

Definition of Example Piezoelectric Stack Actuator

Property	Value
Number of wafer layers in the stack	$N = 100$
Square-wafer size	$a_{wafer} = 10$ mm
Wafer thickness	$h_{wafer} = 250$ μm
Allowable electric field	$E_{max}^+ = 1.0$ kV/mm
Allowable reverse electric field	$E_{max}^- = 0.0$ kV/mm
Thickness of bonding adhesive layer	$h_{glue} = 4$ μm
Elastic modulus of bonding adhesive layer	$E_{glue} = 5$ GPa
Thickness of electrode	$h_{electrode} = 1.0$ μm
Elastic modulus of electrode	$E_{electrode} = 30$ GPa
Mechanical hysteretic damping in the actuator	$\eta = 2.0\%$
Electrical hysteretic damping in the actuator	$\delta = 1.0\%$

Note: The wafers are plated on both sides with electrodes, which are wired in parallel such that the voltage applied to the stack terminals is also applied to each wafer.

1. Stack length, l, and internal stiffness, k_i (neglect the effect of electrodes and glue)

SOLUTION
If we neglect the effect of electrodes and glue, the length of the actuator is simply the summation of the wafer thicknesses, that is, $l = 100 \times 0.250 = 25$ mm. The internal stiffness is calculated as follows:

   a. Calculate the wafer compliance $s_{wafer} = s \times h_{wafer}/A$, where $A = 10 \times 10 = 100$ mm^2 is the wafer area; upon calculations, we get $s_{wafer} = 43.250$ nm/kN

   b. Add the compliances of all wafers and take the inverse to get the stack stiffness, that is, $s_{stack} = N \times s_{wafer} = 4.325$ μm/kN, and $k_i = 1/s_{stack} = 231.2$ kN/mm.

2. Stress-free capacitance, C_0

SOLUTION
Use Equation (10.5) to calculate $C_0 = 619.5$ nF.

3. Maximum allowable voltage, V_{max}

SOLUTION
Use Equation (10.6) to calculate $V_{max} = 250$ V.

4. Free stroke, u_{ISA}, and free strain, S_{ISA}, in percentage induced by the application of the voltage V_{max}

SOLUTION
Use Equation (10.7) to calculate the free stroke $u_{ISA} = 10.0$ μm. Use Equation (10.21) to calculate $S_{ISA} = 0.040\%$ (Note that the same strain value is obtained if we divide the free stroke by the stack length.).

5. Reference mechanical energy, E_{mech}^{ref}, and reference electrical energy, E_{elec}^{ref}

SOLUTION

Use Equations (10.13) and (10.14) to calculate $E_{mech}^{ref} = 11.561$ mJ and $E_{elec}^{ref} = 19.259$ mJ. ∎

Example 10.2: Actuator Equivalent Properties from Vendor Data

Consider the piezo stack actuator E400P-3 with properties given in Tables 10.3 and 10.4. Use full stroke values. Find

1. Electroactive material used in the actuator construction

SOLUTION

From Table 10.3, we read the active material to be PMN-EC9 8.

2. Internal stiffness, k_i, electric capacitance, C_0, maximum voltage, V_{max}, free stroke, u_{ISA}

SOLUTION

From Table 10.3, we read the full-stroke stiffness, $k_i = 166.7$ kN/mm; electric capacitance, $C_0 = 0.580$ μF; maximum voltage, $V_{max} = V^+ - V^- = 800$ V; free stroke, $u_{ISA} = 40$ μm.

TABLE 10.4

Induced-Strain Actuators Selected in the Study

Manufacturer	Model
Polytec PI	P-245.70
	P-246.70
	P-247.70
	P-844.60
EDO Corp.	E100P-2
	E200P-3
	E300P-4
	E400P-3
Kinetic Ceramics	D125160
	D125200
	A125160
	A125200
TOKIN	AE1010D16
	ASB171C801
PiezoSystems Jena	PAHL 120/20
Kinetic Ceramics (Nonlinear range)[a]	D125160 (NL)
	D125200 (NL)
	A125160 (NL)
	A125200 (NL)

[a] Nonlinear range is used to get enhanced performance from the actuator. In the nonlinear range, the piezoceramics undergo large domain rotations, have larger hysteresis, but may have shorter fatigue life. For the models quoted in the table, endurance of 10^7 cycles without failure was proven experimentally on a few specimens.

3. Layer thickness, h, number of layers, N, active material area, A, active material length, l

   Solution From Table 10.4, we read layer thickness $h = 0.500$ mm; number of layers $N = 56$; active material area $A = 247.3$ mm^2; active material length $l = 28$ mm.

4. Maximum electric field, E_{max}

   SOLUTION
   Use Equation (10.6) to calculate $E_{max} = V_{max}/h = 1.6$ kV/mm.

5. Apparent zero-field compliance, s

   SOLUTION
   Use Equation (10.8) to calculate $s = A/(lk_i) = 52.98$/TPa.

6. Apparent zero-stress electric permittivity, ε

   SOLUTION
   Use Equation (10.9) to calculate $\varepsilon = C_0 h^2/(Al) = 20.94$ nF/m.

7. Apparent piezoelectric strain coefficient, d_{33}

   SOLUTION
   Use Equation (10.10) to calculate $d_{33} = u_{ISA}h/(V_{max}l) = 0.893$ $\mu\varepsilon$/(kV/m).

8. Reference mechanical energy, E_{mech}^{ref}

   SOLUTION
   Use Equation (10.13) to calculate $E_{mech}^{ref} = \frac{1}{2}k_i u_{ISA}^2 = 133.4$ mJ.

9. Reference electrical energy, E_{elec}^{ref}

   SOLUTION
   Use Equation (10.14) to calculate $E_{elec}^{ref} = \frac{1}{2}C_0 k_i V_{max}^2 = 185.6$ mJ.

10. Apparent piezoelectric coupling coefficient, κ

    SOLUTION
    Use Equation (10.15) to calculate $\kappa^2 = E_{mech}^{ref}/E_{elec}^{ref} = 0.719$; hence, $\kappa = 0.848$.   ■

Example 10.3:   Effect of Glue and Electrodes on a Piezo Stack Induced-Strain Actuator

Consider the piezoelectric actuator made up of a stack APC 850 square wafers with the properties given in Tables 10.2 and 10.3. This is the same actuator considered in Example 10.1, only that we will now include the effect of glue and electrodes. Find

1. Internal stiffness, k_i, taking into account the effect of electrodes and interlayer glue

   SOLUTION
   The internal stiffness is obtained by first calculating the additive compliances of each component and then inverting the compliance to get the stiffness, that is,

a. Compliance and number of the piezo wafers:

$$s_{wafer} = \frac{sh_{wafer}}{A} = 43.250 \text{ mm/GN}, \quad N_{wafer} = N = 100$$

b. Compliance of the glue:

$$s_{glue} = \frac{h_{glue}}{AE_{glue}} = 8.0 \text{ mm/GN}, \quad N_{glue} = N - 1 = 99$$

c. Compliance of the piezo wafers:

$$s_{elec} = \frac{h_{elec}}{AE_{elec}} = 0.333 \text{ mm/GN}, \quad N_{elec} = 2N = 200$$

d. Additive compliance of the assembly:

$$s_{stack} = N_{wafer}s_{wafer} + N_{glue}s_{glue} + N_{elec}s_{elec} = 5.184 \text{ } \mu m/kN$$

e. Stack stiffness:

$$k_i = 1/s_{stack} = 192.9 \text{ kN/mm}.$$

2. Stress-free capacitance of the stack, C_0

SOLUTION
Use Equation (10.5) to calculate $C_0 = 619.5$ nF.

3. The total displacement (full stroke) u_{ISA} that is achieved when the actuator is energized with a voltage inducing the maximum allowable electric field given in Table 10.3

SOLUTION
Use Equation (10.7) to calculate the free stroke $u_{ISA} = 10.0$ μm.

4. Total length of the stack, L, in millimeters, and the effective induced strain in the stack, S_{ISA}, in percentage.

SOLUTION
Calculate the additive length of the assembly as $l = N_{wafer}h_{wafer} + N_{glue}h_{glue} + N_{elec}h_{elec} = 25.6$ mm. Then, divide the free stroke $u_{ISA} = 10.0$ μm by this stack length to get $S_{ISA} = 0.0469\%$.

5. Reference mechanical energy, E_{mech}^{ref}, and reference electrical energy, E_{elec}^{ref}

SOLUTION
Use Equations (10.13) and (10.14), to calculate $E_{mech}^{ref} = 9.646$ mJ, $E_{elec}^{ref} = 19.359$ mJ.

6. Compare the results obtained here with the corresponding results from the previous problem in which the effect of glue and electrodes was ignored and comment on this comparison

SOLUTION
Due to the effect of glue and electrodes, the internal stiffness, k_i, has reduced by 16.6% and the length, l, has increased by 2.4%. Consequently, the effective strain, S_{ISA}, has reduced by ~2.3% and the reference mechanical energy, E_{mech}^{ref}, has reduced by ~16.6%. The capacitance, C_0, and the reference electrical energy, E_{elec}^{ref}, have not changed. ∎

10.5 Principles of Induced-Strain Structural Actuation

In structural applications, the induced-strain actuators must work together with the actuated structure, and special attention must be given to their interaction. The main differentiating feature between active materials actuators and conventional actuators lies in the amount of available displacement. The induced-strain effect present in active materials results in output displacements that seldom exceed 100 microns (0.1 mm). In a conventional actuator, for example, a hydraulic cylinder, displacement of the order of several millimeters can be easily achieved; if more displacement is needed, additional hydraulic fluid could be pumped in. In contrast, an induced-strain actuator has at its disposal only the very limited amount of displacement generated by the induced-strain effect. This limited displacement needs to be carefully managed if the desired action is to be achieved. Under reactive service loads, the internal compressibility of the active materials actuator "eats-up" part of the induced-strain displacement, and leads to reduced output displacement (Figure 10.13a). If the external stiffness, k_e, is reduced, the force in the actuator is also reduced, and more displacement is seen at the actuator output

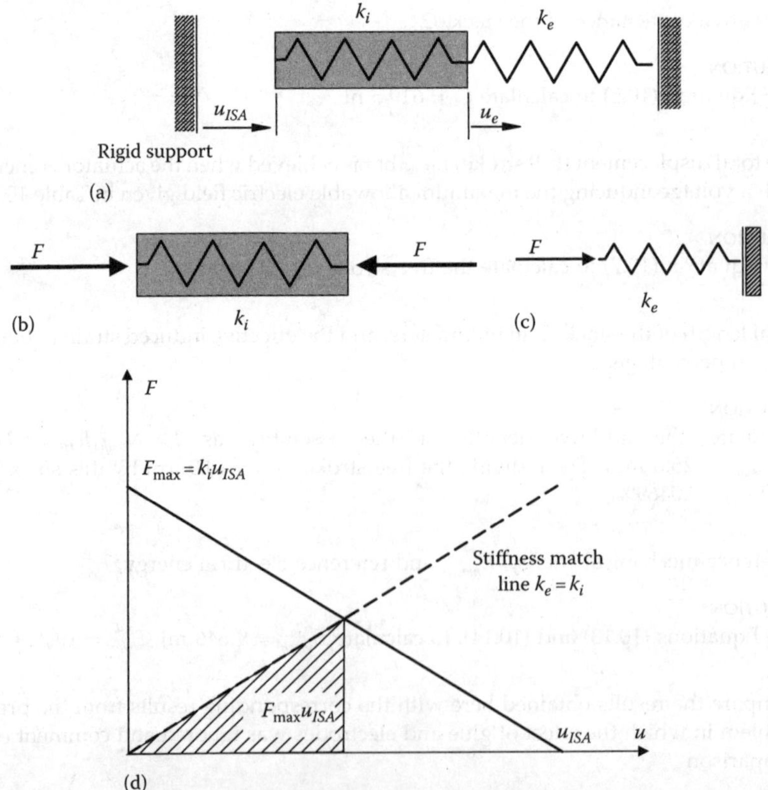

FIGURE 10.13
Schematic representation of an induced-strain actuator with rigid support: (a) the assembly of actuator and external spring; (b) the forces acting on the actuator; (c) the forces acting on the external spring; (d) force-displacement matching diagram.

end. For a free actuator, that is, under no external reaction, the output displacement reaches a maximum (*free stroke*). However, no active work is being done in this case since the force is zero. At the other extreme, when the actuator is fully constrained, the force is maximum (*blocked force*), but no work is again performed since the displacement is zero. An optimum is attained between these two extremes, and this optimum can be best described in terms of the *stiffness match* principle. Under static conditions, the stiffness match principle implies that the external stiffness (i.e., the stiffness of the application) and the internal stiffness of the actuator are equal, which gives a stiffness ratio with value $r = 1$. Under dynamic conditions, this principle can be expressed as *dynamic stiffness match* or *impedance match*.

10.5.1 Output Displacement under Load

Recall the 1-D constitutive Equation (10.1)

$$S = sT + d_{33}E \tag{10.17}$$

$$D = d_{33}T + \varepsilon E \tag{10.18}$$

Figure 10.13 presents an induced-strain actuator (ISA device) of internal stiffness k_i acting against an external spring load of stiffness k_e. The compressive force acting in the actuator is

$$F = k_e u_e \tag{10.19}$$

where u_e is the compression of the external spring k_e. Inside the induced-strain actuator, the force, F, produces the compressive stress.

$$T = -\frac{F}{A} \tag{10.20}$$

where A is the cross-sectional area of the electroactive induced-strain actuator. The negative sign indicates that the induced-strain actuator is in compression. The strain, S, is given by

$$S = \frac{du}{dx} \tag{10.21}$$

Introduce the symbol S_{ISA} to denote the strain induced in the electroactive material by the applied electric field E, that is,

$$S_{ISA} = d_{33}E \tag{10.22}$$

Substitute Equations (10.21) and (10.22) into Equation (10.17) to get

$$\frac{du}{dx} = s\left(-\frac{F}{A}\right) + S_{ISA} \tag{10.23}$$

On integration,

$$u(x) = -s\frac{F}{A}x + S_{ISA}x \tag{10.24}$$

The displacement at the end of the stack, which we have defined as the "external displacement," is

$$u_e = u(l) = -s\frac{F}{A}l + S_{ISA}l \tag{10.25}$$

Introduce the symbol u_{ISA} to denote the induced-strain displacement in the stack (free-stroke), that is,

$$u_{ISA} = S_{ISA}l = d_{33}El \tag{10.26}$$

Also introduce the symbol u_i to denote the compression of the induced-strain actuator, that is,

$$u_i = \frac{F}{k_i} \tag{10.27}$$

where k_i is the internal stiffness of the induced-strain actuator, that is,

$$k_i = \frac{A}{sl} \tag{10.28}$$

Hence, Equation (10.25) becomes

$$u_e = -u_i + u_{ISA} \tag{10.29}$$

This equation expresses the fact that the sum of the internal and external displacements equals the total induced-strain displacement, u_{ISA}, that is,

$$u_i + u_e = u_{ISA} \tag{10.30}$$

This simple analysis reveals that the total induced-strain displacement, u_{ISA}, is partly consumed as an internal displacement, u_i, due to the compressibility of the ISA device, and partly delivered as useful output displacement, u_e. Eliminate the force F between Equations (10.19) and (10.27) to get

$$u_i = \frac{k_e}{k_i}u_e = ru_e \tag{10.31}$$

where r is the *stiffness ratio* given by

$$r = \frac{k_e}{k_i} \tag{10.32}$$

Substitution of Equation (10.31) into Equation (10.30) yields

$$u_e = \frac{1}{1+r} u_{ISA} \tag{10.33}$$

Equation (10.33) indicates that the output displacement depends not only on the induced-strain displacement, u_{ISA}, but also on the relative stiffness, r, between the ISA device and the external spring. Define the *output displacement coefficient* $\eta(r)$ as

$$\eta(r) = \frac{u_e}{u_{ISA}} \tag{10.34}$$

Substitution of Equation (10.34) into Equation (10.33) yields

$$u_e = \frac{1}{1+r} u_{ISA} = \eta(r) u_{ISA} \tag{10.35}$$

where

$$\eta(r) = \frac{1}{1+r} \tag{10.36}$$

Variation of the output displacement coefficient $\eta(r)$ with stiffness ratio, r, is shown as the curve u_e/u_{ISA} in Figure 10.14. As the external stiffness increases, the fraction of the induced-strain displacement, u_{ISA}, which reaches the output, diminishes. For very large external stiffness, one gets the *blocked* condition, that is, $u_e \rightarrow 0$ as $r \rightarrow \infty$.

Substitution of Equation (10.35) into Equation (10.31) and use of Equation (10.32) yields the internal displacement u_i, that is,

$$u_i = \frac{r}{1+r} u_{ISA} \tag{10.37}$$

Substitute Equation (10.35) into Equation (10.19) to calculate the force in actuator, that is,

$$F = k_e \eta(r) u_{ISA} \tag{10.38}$$

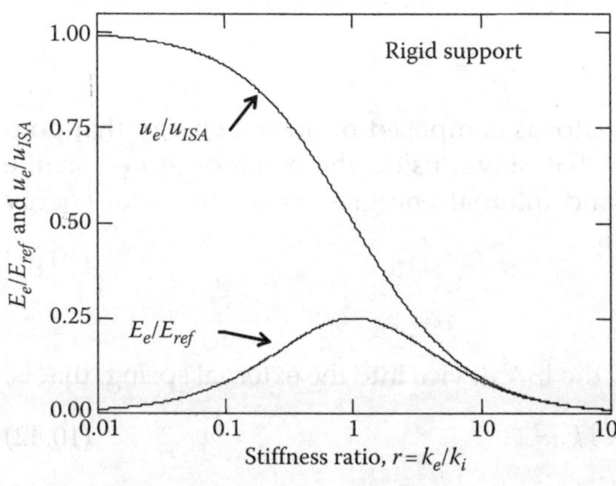

FIGURE 10.14
Variation of output displacement coefficient, $\eta = u_e/u_{ISA}$, and output energy coefficient, $E'_e = E_e/E_{ref}$, with stiffness ratio, $r = k_e/k_i$ for an induced-strain actuator on a rigid support.

Substitute Equations (10.22), (10.26), (10.36), and (10.38) into Equation (10.24) to get

$$u(x) = -s\frac{F}{A}x + S_{ISA}x = -\left(\frac{sl}{A}k_e\right)\frac{1}{1+r}d_{33}Ex + d_{33}Ex \tag{10.39}$$

Using Equations (10.28) and (10.32), Equation (10.39) gives the displacement $u(x)$ along the actuator as

$$u(x) = \frac{1}{1+r}d_{33}Ex \tag{10.40}$$

Differentiation of Equation (10.40) with respect to x yields the strain $S(x)$ along the actuator, that is,

$$S(x) = u'(x) = \frac{1}{1+r}d_{33}E \tag{10.41}$$

Note that the strain along the stack is constant; however, it depends on the stiffness ratio r. The displacement increases linearly with the position along the stack, starting from zero at actuator root and reaching u_e at the actuator tip. The displacement also depends on the stiffness ratio r.

Example 10.4:

Consider again the piezoelectric actuator discussed in Example 10.1; the actuator is made up of a stack APC 850 square wafers with the properties given in Tables 10.2 and 10.3. Find

1. Stiffness ratio, r, when the actuator works against an external elastic spring of stiffness $k_e = 60$ kN/mm

   SOLUTION
   Recall the internal stiffness $k_i = 231.2$ kN/mm calculated for this actuator in Example 10.1. Then, use Equation (10.33) to calculate $r = 0.260$.

2. Output displacement u_e when the actuator works at full stroke against the elastic spring of stiffness $k_e = 60$ kN/mm. Express result in physical units and as percentage of u_{ISA}

   SOLUTION
   Use Equations (10.34) and (10.35) to calculate $u_e = 7.94$ μm, that is, $u_e = 79.4\%$ of u_{ISA}. ∎

10.5.2 Mechanical Energy

Mechanical energy of an induced-strain actuator is composed of output energy that goes into the external load and internal energy that stays inside the actuator stored in the actuator compliance. The sum of external and internal energies makes the total energy developed during the induced-strain process.

10.5.2.1 Output Energy under Load

The output energy is the energy delivered by the ISA device into the external spring, that is,

$$E_e = \tfrac{1}{2}k_e u_e^2 \tag{10.42}$$

On substitution,

$$E_e = \frac{r}{(1+r)^2}\left(\tfrac{1}{2}k_i u_{ISA}^2\right) \tag{10.43}$$

Note that the output energy expression given in Equation (10.43) consists of a variable coefficient that depends on the stiffness ratio, r, and a constant energy term $\left(\tfrac{1}{2}k_i u_{ISA}^2\right)$ that depends only on the free stroke, u_{ISA}, and internal stiffness, k_i, of the induced-strain device; this constant term is the reference mechanical energy, $E_{mech}^{ref} = \tfrac{1}{2}k_i u_{ISA}^2$, defined by Equation (10.13). Hence, Equation (10.43) can be written as

$$E_e(r) = \frac{r}{(1+r)^2} E_{mech}^{ref} \tag{10.44}$$

The variable term in Equation (10.43) is the output energy coefficient, $E_e' = E_e/E_{mech}^{ref}$, that is,

$$E_e'(r) = \frac{r}{(1+r)^2} \tag{10.45}$$

The governing design factor in the construction of a conventional ISA device is the stiffness ratio, r. Variation of the energy coefficient with the stiffness ratio, r, is shown as the curve E_e/E_{ref} in Figure 10.14. The maximum displacement output is obtained as r approaches zero, that is, when the external stiffness is so weak that practically no resistance is presented to the ISA device and hence the full induced-strain displacement, u_{ISA}, is delivered externally. This case is of limited interest for actuator applications since the actuation force is zero and no output energy is delivered. In the other extreme, when the external stiffness is very much greater than the internal stiffness, the output force reaches a maximum, but the output displacement is zero. Hence, the output energy is again zero. Between these two extremes, an optimum solution may be reached. Differentiation of the energy coefficient with respect to the stiffness ratio, r, yields

$$\frac{dE_e'}{dr} = 0 \rightarrow 1 - \frac{2r}{1+r} = 0 \rightarrow r_{opt} = 1 \tag{10.46}$$

This is the *stiffness match principle*. When the external stiffness matches the internal stiffness, the energy output reaches a maximum and its value is

$$E_e^{max} = \tfrac{1}{4} E_{mech}^{ref} \tag{10.47}$$

Example 10.5:

Consider again the piezoelectric actuator discussed in Example 10.1; the actuator is made up of a stack of APC 850 square wafers with the properties given in Tables 10.2 and 10.3. Find

1. Output mechanical energy, E_e, expression as a function of stiffness ratio, r, when working at full stroke against an external elastic spring; maximum mechanical energy, E_e^{max}, that could be extracted from the actuator by tuning the stiffness ratio r; the value of stiffness ratio r to achieve this maximum mechanical energy output

SOLUTION

Output mechanical energy, E_e, expression as a function of stiffness ratio, r, when working at full stroke against an external elastic spring is given by Equations (10.42) and (10.43). Maximum mechanical energy, E_e^{max}, that could be extracted from the actuator by tuning the stiffness ratio r is given by Equation (10.47); recall the reference mechanical energy $E_{mech}^{ref} = 11.561$ mJ calculated in Example 10.1; on substitution into Equation (10.47) we get $E_e^{max} = 2.890$ mJ. The value of stiffness ratio r to achieve this maximum mechanical energy output is $r_{opt} = 1$, as indicated by Equation (10.46).

2. Output mechanical energy, E_e, when the actuator works at full stroke against the elastic spring of stiffness $k_e = 60$ kN/mm. Express result in physical units and in percentage of E_e^{max}

SOLUTION

Use Equations (10.42) through (10.44) to calculate $E_e = 1.89$ mJ, that is, $E_e = 65.4\%$ of E_e^{max}.

3. The value of the elastic spring stiffness, k_e, that will ensure stiffness match; the corresponding value of the stiffness ratio, r_{match}. The output mechanical energy, E_e^{match}, under the stiffness-match condition

SOLUTION

Stiffness match corresponds to the situation when the external stiffness equals the internal stiffness; the corresponding value of the stiffness ratio, $r_{match} = 1$. Substitution into Equation (10.44) gives the output mechanical energy under stiffness-match conditions as $E_e^{match} = 2.890$ mJ; note that this is the maximum mechanical energy E_e^{max} that could be extracted from the actuator as defined by Equations (10.46) and (10.47). This indicates that the stiffness match configuration is the optimum configuration for energy transfer from the induced-strain actuator into the external load. ∎

10.5.2.2 Internal Energy Under Load

The energy stored internally in the actuator under load is given by

$$E_i = \tfrac{1}{2} k_i u_i^2 \tag{10.48}$$

Using Equations (10.13) and (10.37), we write Equation (10.48) as

$$E_i = \frac{r^2}{(1+r)^2} \tfrac{1}{2} k_i u_{ISA}^2 = \frac{r^2}{(1+r)^2} E_{mech}^{ref} \tag{10.49}$$

One notes that, for a free actuator ($r = 0$), the internal energy E_i is zero, since no force is present to do work, whereas for a blocked actuator ($r \to 0$), the internal energy equals the reference energy E_{mech}^{ref}.

10.5.2.3 Total Induced-Strain Energy under Load

The total energy generated by the induced-strain effect in an actuator under load is the sum of the energy transferred to the external load, E_e, and the energy stored internally, E_i, that is,

$$E_{ISA} = E_e + E_i = \frac{r}{(1+r)^2} E_{mech}^{ref} + \frac{r^2}{(1+r)^2} E_{mech}^{ref} = \frac{r}{1+r} E_{mech}^{ref} \tag{10.50}$$

One notes that for a free actuator ($r=0$), the total ISA energy is zero since no force is present to do work, whereas for a blocked actuator ($r \to 0$), the ISA energy equals the reference energy, E_{mech}^{ref}. For stiffness match ($r=1$), the external and internal energies are equal to $\frac{1}{4} E_{mech}^{ref}$, and the total ISA energy is $E_{ISA} = \frac{1}{2} E_{mech}^{ref}$.

10.5.3 Electrical Response under Load

The electric displacement (charge per unit area) is calculated using Equation (10.1). First, recall Equation (10.1), that is,

$$S = sT + d_{33}E \tag{10.51}$$

$$D = d_{33}T + \varepsilon E \tag{10.52}$$

Equation (10.52) indicates that the electric displacement D comprises two terms: (a) εE, which is of purely electrical nature, and (b) $d_{33}T$, which represents the effect of mechanical stress upon the electrical displacement. In our case, the stress is negative (compression) as indicated by Equation (10.20); hence, the external load effect on the induced-strain actuator is to reduce the electrical displacement (i.e., charge per unit area) stored in the actuator. For zero external stiffness ($r=0$), this reduction is zero. Under these stress-free conditions, the electric charge accepted by the piezoelectric material reaches a maximum. As the load increases ($r \neq 0$), the electric charge stored in the actuator decreases. As the "blocked" conditions are approached ($r \to \infty$), we obtain the maximum reduction in electric charge that can be stored in the actuator.

Solve Equation (10.51) for stress as a function of strain and electric field, that is,

$$T = \frac{1}{s}(S - d_{33}E) \tag{10.53}$$

Substitute Equation (10.53) into Equation (10.52) to get

$$D(x) = d_{33}\frac{1}{s}[u'(x) - d_{33}E] + \varepsilon E \tag{10.54}$$

Recalling Equation (10.12), Equation (10.54) can be rearranged as

$$D(x) = \varepsilon E \left[1 - \kappa^2 + \kappa^2 \frac{u'(x)}{d_{33}E} \right] \tag{10.55}$$

We note that the electric displacement varies with position x along the induced-strain actuator. Recall that the induced-strain actuator is made up of a stack of N piezoelectric wafers with thickness h and electroded faces of area A (Figure 10.15).

The electrodes are wired in such a way that all the wafers are connected in parallel. Since the wafers are very thin, it is appropriate to approximate the strain inside each wafer with its average value; for the nth wafer, we have

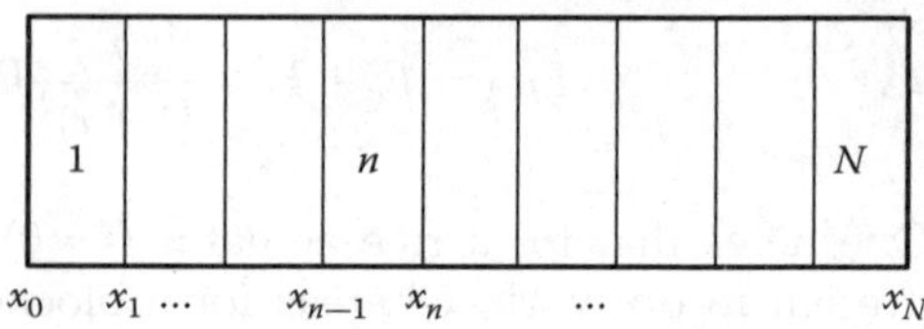

FIGURE 10.15
Schematic representation of electrode numbering in a stack actuator.

$$u'_n = \int_{x_{n-1}}^{x_n} u'(x)dx = \frac{1}{h}(u_n - u_{n-1}) \tag{10.56}$$

where x_{n-1} and x_n are the coordinates at the bottom and top of the nth wafer in the stack and u_{n-1} and u_n are the corresponding displacements. Substitution of Equation (10.56) into Equation (10.55) yields the electric displacement in the nth wafer as

$$D_n = \varepsilon E\left(1 - \kappa^2 + \kappa^2 \frac{u_n - u_{n-1}}{hd_{33}E}\right) \tag{10.57}$$

Multiplication of Equation (10.57) by the electrode area A yields the charge associated with the nth wafer, that is,

$$Q_n = A\varepsilon E\left(1 - \kappa^2 + \kappa^2 \frac{u_n - u_{n-1}}{hd_{33}E}\right) \tag{10.58}$$

The total charge is obtained by summation over all the wafers, that is,

$$Q = \sum_{n=1}^{N} Q_n = A\varepsilon E\left[N(1 - \kappa^2) + \kappa^2 \frac{1}{hd_{33}E} \sum_{n=1}^{N}(u_n - u_{n-1})\right] = A\varepsilon E\left[N(1 - \kappa^2) + \kappa^2 \frac{u_N - u_0}{hd_{33}E}\right] \tag{10.59}$$

where u_0 and u_N are the displacements at the bottom and top of the stack. In our case, $u_0 = 0$ and $u_N = u_e = ld_{33}E/(1 + r)$; hence, Equation (10.59) yields

$$Q = A\varepsilon E\left[N(1 - \kappa^2) + \kappa^2 \frac{1}{1 + r} \frac{l\,d_{33}E}{h\,d_{33}E}\right] \tag{10.60}$$

Recall Equation (10.3), that is, $l = Nh$, Equation (10.6), that is, $E = V/h$, and Equation (10.5); then, Equation (10.60) becomes

$$Q = N\varepsilon \frac{A}{h}V\left(1 - \kappa^2 + \kappa^2 \frac{1}{1 + r}\right) = C_0 V\left(1 - \kappa^2 + \kappa^2 \frac{1}{1 + r}\right) \tag{10.61}$$

10.5.3.1 Effective Capacitance

Equation (10.61) reflects the *effective capacitance* of the induced-strain actuator under load, which can be defined as

$$C = C_0\left(1 - \kappa^2 + \kappa^2\frac{1}{1+r}\right) = C_0\left(1 - \kappa^2\frac{r}{1+r}\right) \tag{10.62}$$

It is apparent that the effect of the external load is to reduce the effective capacitance of the piezoelectric stack. As the stiffness of the external load increases, the effective capacitance decreases. Let us consider two extreme cases: *blocked actuator* and *free actuator*. For an infinitely stiff external constraint ($r \to \infty$), we approach the *blocked capacitance*, that is,

$$C_{block} = C_0\left(1 - \kappa^2\right) \tag{10.63}$$

For a zero external stiffness ($r = 0$), we approach the free capacitance, C_0, of Equation (10.5).

10.5.3.2 Capacitance Reduction due to Load

It is apparent from Equation (10.62) that a reduction of the effective capacitance will take place under load. We can express Equation (10.62) using the *capacitance reduction*, ΔC_{ISA}, defined as

$$\Delta C_{ISA} = \kappa^2\frac{r}{1+r}C_0, \quad C = C_0 - \Delta C_{ISA} \tag{10.64}$$

It is apparent that the capacitance reduction, ΔC_{ISA}, is zero for the free actuator ($r = 0$) and reaches a maximum, $\Delta^{max}C_{ISA} = \kappa^2 C_0$, for the blocked actuator ($r \to \infty$).

Example 10.6:

Consider again the piezoelectric actuator discussed in Example 10.1; the actuator is made up of a stack APC 850 square wafers with the properties given in Tables 10.2 and 10.3. Find the effective capacitance under load, C, in physical units and as a percentage of C_0

SOLUTION
Use Equation (10.62) to calculate $C = 543.3$ nF, that is, $C = 87.7\%$ of C_0. ∎

10.5.3.3 Electric Energy

In Section 10.5.2, we examined the mechanical energy that can be delivered by the induced-strain actuator. However, this output mechanical energy results from electromechanical conversion of the input electrical energy. (This conversion is done through the piezoelectric or piezomagnetic effects, as appropriate.) For free electroactive induced-strain actuators, the reference electrical energy defined by Equation (10.14) is $E_{elec}^{ref} = \frac{1}{2}C_0V^2$. When an external load is applied, the effective capacitance is modulated by the stiffness ratio as indicated by Equation (10.62), that is,

$$C = C_0\left(1 - \kappa^2\frac{r}{1+r}\right) \tag{10.65}$$

The electrical energy input to the actuator is

$$E_{elec}(r) = \tfrac{1}{2}CV^2 = \left(1 - \kappa^2\frac{r}{1+r}\right)\left(\tfrac{1}{2}C_0V^2\right) = \left(1 - \kappa^2\frac{r}{1+r}\right)E_{elec}^{ref} \tag{10.66}$$

Similarly, for magnetoactive solid-state actuators,

$$E_{elec}(r) = \tfrac{1}{2}LI^2 = \left(1 - \kappa^2\frac{r}{1+r}\right)\left(\tfrac{1}{2}LI^2\right) = \left(1 - \kappa^2\frac{r}{1+r}\right)E_{elec}^{ref} \tag{10.67}$$

We have seen in Section 10.5.2 that the maximum output of mechanical energy is attained when the stiffness ratio is unity, $(r = 1)$. This condition is also known as the *stiffness match point*. At the stiffness match point, the electrical energy input takes the value

$$E_{elec}^{r=1} = E_{elec}(r)|_{r=1} = \left(1 - \tfrac{1}{2}\kappa^2\right)E_{elec}^{ref}. \tag{10.68}$$

10.5.4 Energy Conversion Efficiency

The energy conversion efficiency is simply defined as the ratio between the output and input energies, that is,

$$\eta = \frac{E_{out}}{E_{in}} \tag{10.69}$$

In our case, the output energy is the mechanical energy in the external spring, E_e, and the input energy is the electrical energy, E_{elec}. Recall the mechanical energy Equation (10.44) of Section 10.5.2, that is,

$$E_e = \frac{r}{(1+r)^2}\left(\frac{1}{2}k_i u_{ISA}^2\right) \tag{10.70}$$

Also, recall the electrical energy Equation (10.66), that is,

$$E_{elec}(r) = \left(1 - \kappa^2\frac{r}{1+r}\right)E_{elec}^{ref} \tag{10.71}$$

Substituting Equations (10.70) and (10.71) into Equation (10.69) yields

$$\eta = \frac{\dfrac{r}{(1+r)^2}\left(\dfrac{1}{2}k_i u_{ISA}^2\right)}{\left(1 - \kappa^2\dfrac{r}{1+r}\right)\left(\dfrac{1}{2}CV^2\right)} = \frac{\dfrac{r}{(1+r)^2}}{\left(1 - \kappa^2\dfrac{r}{1+r}\right)}\frac{E_{mech}^{ref}}{E_{elec}^{ref}} \tag{10.72}$$

Recall Equation (10.15) of Section 10.4.2, that is,

$$\kappa^2 = \frac{E_{mech}^{ref}}{E_{elec}^{ref}} \tag{10.73}$$

Substitution of Equation (10.73) into Equation (10.72) yields, after simplification, the expression of the conversion efficiency as a function of stiffness ratio and electromechanical coupling coefficient, that is,

$$\eta(r, \kappa) = \frac{r\kappa^2}{(1 + r)[1 + r(1 - \kappa^2)]} \tag{10.74}$$

For a given electromechanical coupling coefficient, κ, we can seek the optimum stiffness ratio, r, that would yield the best energy conversion efficiency. Differentiation of Equation (10.74) with respect to r yields

$$\frac{\partial \eta}{\partial r} = \frac{\left[1 - r^2(1 - \kappa^2)\right]\kappa^2}{(1 + r)^2[1 + r(1 - \kappa^2)]^2} \tag{10.75}$$

Equation (10.75) becomes zero for

$$r_\eta(\kappa) = \frac{1}{\sqrt{1 - \kappa^2}} \tag{10.76}$$

This indicates that Equation (10.74) has a maximum at $r_\eta(\kappa)$. Substitution of Equation (10.76) into Equation (10.74) yields the *maximum attainable energy conversion efficiency* of an induced-strain actuator, that is,

$$\eta_{\max}(\kappa) = \eta(r, \kappa)\big|_{r=r_\eta} = \frac{\kappa^2}{\left(1 + \sqrt{1 - \kappa^2}\right)^2} \tag{10.77}$$

The maximum conversion efficiency given in Equation (10.77) can be compared with the *conversion efficiency at the stiffness match point*, which is obtained by putting $r = 1$ into Equation (10.74), that is,

$$\eta_{r=1} = \frac{1}{2}\frac{\kappa^2}{2 - \kappa^2}. \tag{10.78}$$

Note that the maximum energy conversion efficiency, $\eta_{\max}$, and the conversion efficiency at the stiffness match point, $\eta_{r=1}$, are different. In design applications, either one of the two efficiencies can be optimized, but not both. In other words, one can either design for maximum energy output by imposing the stiffness match condition ($r = 1$), or can design for maximum conversion efficiency by choosing $r_\eta = 1/\sqrt{1 - \kappa^2}$. However, in some practical applications, the numerical difference between the two results may be small.

Example 10.7:

Consider again the piezoelectric actuator discussed in Example 10.1; the actuator is made up of a stack APC 850 square wafers with the properties given in Tables 10.2 and 10.3. Find

1. Input electrical energy, E_{elec}, when the actuator works at full stroke against the elastic spring of stiffness $k_e = 60$ kN/mm. Express result in physical units and in percentage of E_{elec}^{ref}. Discuss results.

SOLUTION

Use Equation (10.66) to calculate $E_{elec} = 16.977$ mJ, that is, $E_{elec} = 87.7\%$ of E_{elec}^{ref}. It is apparent that the input electrical energy required when operating under load is less than that required for stress-free operation.

2. Input electrical energy at the stiffness-match point, $E_{elec}^{r=1}$. Express result in physical units and in percentage of E_{elec}^{ref}. Discuss results.

SOLUTION

Use Equation (10.68) to calculate $E_{elec}^{r=1} = 13.579$ mJ, that is, $E_{elec}^{r=1} = 70.1\%$ of E_{elec}^{ref}. It is apparent that the input electrical energy required at the stiffness match is much less than that required for operation under the other load condition.

3. Energy conversion efficiency, η, maximum attainable energy conversion efficiency, η_{max}, and conversion efficiency at the stiffness match point, $\eta_{r=1}$. Discuss results.

SOLUTION

Use Equation (10.70) to calculate $\eta = 11.1\%$. Use Equation (10.73) to calculate $\eta_{max} = 22.3\%$. Then, use Equation (10.74) to calculate $\eta_{r=1} = 21.3\%$. It is apparent that the maximum attainable conversion efficiency is much larger than the conversion efficiency under any other loading conditions. It is also apparent that the stiffness match conversion efficiency is less than the maximum attainable conversion efficiency, but the difference is only 1%. ∎

10.5.5 Compliant Support Effects

Figure 10.16 shows an induced-strain actuator in which the support structure is not infinitely rigid. A stiffness value, k_s, is assumed to be applied at the support end of the ISA device. The presence of finite support stiffness, k_s, can significantly modify the output displacement and output energy of the induced-strain actuator, as seen in the following analysis.

10.5.5.1 Displacement Analysis

The induced-strain displacement, u_{ISA}, is consumed in the support stiffness, k_s, in the internal stiffness of the actuator, k_i, and in the external stiffness, k_e, that is,

$$u_{ISA} = \frac{F}{k_s} + \frac{F}{k_i} + u_e \qquad (10.79)$$

The external displacement is given by

$$F = k_e u_e \qquad (10.80)$$

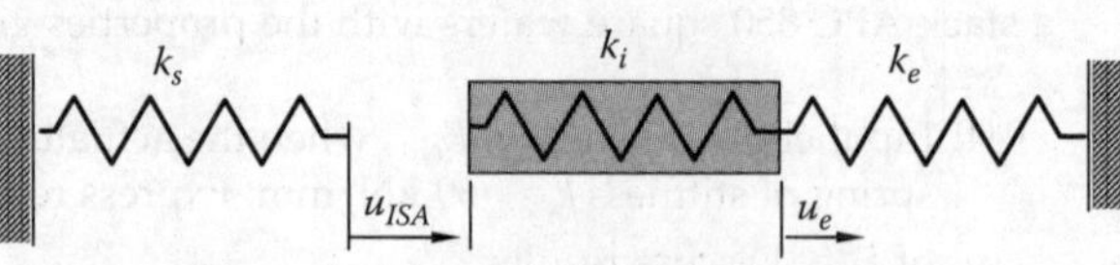

FIGURE 10.16
Schematic representation of an induced-strain actuator with compliant support.

Upon elimination,

$$u_e = \left(1 + \frac{k_e}{k_s} + \frac{k_e}{k_i}\right)^{-1} u_{ISA} \tag{10.81}$$

Introducing the *structural stiffness ratio*,

$$r_s = \frac{k_s}{k_i} \tag{10.82}$$

Hence, write the output displacement coefficient, η, as a function of both r and r_s, that is,

$$\eta(r, r_s) = \frac{1}{1 + r(1 + 1/r_s)} \tag{10.83}$$

The output displacement, u_e, is expressed in terms of the reference (free-stroke) displacement, u_{ISA}, as

$$u_e = \eta(r, r_s)u_{ISA} \tag{10.84}$$

A plot of the output displacement coefficient, η, versus the stiffness ratio, r, for two values of the support stiffness ratio, r_s, is given as the curves u_e/u_{ISA} in Figure 10.17. It can be seen that the output displacement coefficient, η, increases as r_s increases, that is, a stiffer support will give a better displacement output. As $r_s \to \infty$, the expression of η approaches the simpler expression $\eta(r) = 1/(1 + r)$, which was derived as Equation (10.36) in Section 10.5.1.

10.5.5.2 Output Energy Analysis

The output energy delivered by the induced-strain device is, as before, $E_e = \frac{1}{2}k_e u_e^2$. Upon substitution,

$$E_e = \frac{r}{[1 + r(1 + 1/r_s)]^2} \left(\tfrac{1}{2} k_i u_{ISA}^2\right) \tag{10.85}$$

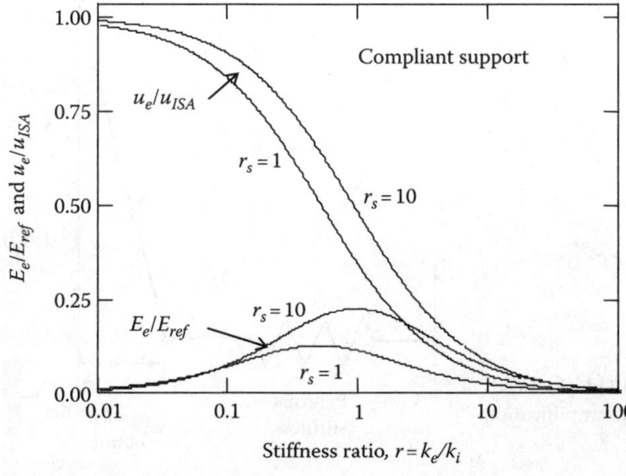

FIGURE 10.17
Variation of output displacement coefficient, $\eta = u_e/u_{ISA}$, and output energy coefficient, $E_e' = E_e/E_{ref}$, with stiffness ratio, $r = k_e/k_i$: for an induced-strain actuator on a compliant support with various values of the support stiffness ratio, r_s.

The last parenthesis in Equation (10.85) is the reference energy, E_{mech}^{ref}, defined by Equation (10.13). Dividing Equation (10.85) by E_{mech}^{ref} yields the output energy coefficient,

$$E'_e(r, r_s) = \frac{r}{[1 + r(1 + 1/r_s)]^2} \qquad (10.86)$$

Note that the output energy coefficient depends on the stiffness ratio, r, and the structural stiffness coefficient, r_s. The curves E_e/E_{ref} in Figure 10.17 presents the variations of the displacement and energy coefficients with the stiffness ratio, r, for two values of the structural stiffness coefficient, $r_s = 10$ and $r_s = 1$. For the case of "stiff" support ($r_s = 10$), the displacement and energy curves in Figure 10.17 are almost identical with those given in the previous section for an induced-strain actuator with rigid support (Figure 10.14). For the case of "elastic" support ($r_s = 1$), a significant shift is observed in the output displacement and output energy curves. More importantly, the peak of the output energy curve is reduced by about a factor of 2. This 50% reduction in energy output is explainable, since half of the available energy ends up as elastically stored in the support.

10.5.6 Displacement–Amplification Effects

10.5.6.1 Displacement Analysis

Displacement amplifiers are currently used to increase the displacement output of ISA devices. The simplest representation of a displacement amplification concept is that of a lever with unequal arms (Figure 10.18).

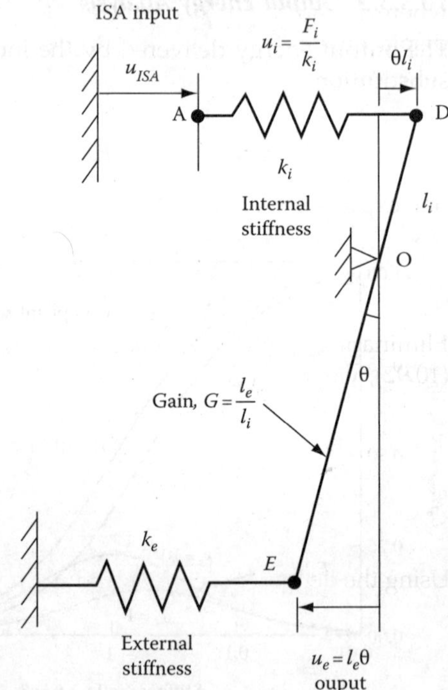

FIGURE 10.18
Conceptual models for displacement-amplified induced-strain actuator on rigid support and compliant amplification mechanism.

This simple representation can be used to analyze even more complicated displacement amplifiers (flextensional, deformable triangles, hydrostatic, etc.). At the input end, the lever is actuated by an ISA device, which has an internal stiffness, k_i, and produces an induced-strain displacement, u_{ISA}. The displacement of the input lever arm is $u_{ISA} - F_i/k_i$, that is, the induced-strain displacement minus the internal compressibility. Note that F_i, is the input force in the amplification mechanism and the force in the ISA device. This displacement produces a rotation θ of the lever and hence,

$$l_i\theta = u_{ISA} - \frac{F_i}{k_i} \tag{10.87}$$

If the amplifying lever is considered rigid, then the displacement at the output arm, u_e, is given by simple rigid body rotation, that is,

$$u_e = l_e\theta \tag{10.88}$$

where u_e is the external displacement. For an equivalent external stiffness, k_i, the external reaction, F_e, is

$$F_e = k_e u_e \tag{10.89}$$

The input force, F_i, and the output reaction, F_e, satisfy the lever equilibrium relation

$$F_i l_i = F_e l_e \tag{10.90}$$

where l_i and l_e are the effective lever arms associated with the internal and external displacements, u_i and u_e, respectively. Introducing the kinematic gain

$$G = \frac{l_e}{l_i} \tag{10.91}$$

one expresses the internal force in terms of the external reaction, and *vice versa*, that is,

$$F_i = F_e G \quad \text{and} \quad F_e = F_i/G \tag{10.92}$$

Eliminating θ between Equations (10.87) and (10.88), and using Equations (10.90) through (10.92), yields

$$u_e = \frac{G}{1 + G^2 \dfrac{k_e}{k_i}} u_{ISA} \tag{10.93}$$

Using the definitions, $r = k_e/k_i$ and $\eta = u_e/u_{ISA}$, one writes the output displacement coefficient

$$\eta(G, r) = \frac{G}{1 + rG^2} \tag{10.94}$$

The output displacement coefficient, $\eta(G, r)$, is always less than the kinematic gain, G, since part of the amplification effect is always lost due to the internal compliance of the induced-strain device.

10.5.6.2 Output Energy Analysis

The energy output from a displacement-amplified induced-strain actuator is simply the energy stored in the external spring, that is,

$$E_e = \tfrac{1}{2} k_e u_e^2 \tag{10.95}$$

Substituting $k_e = r \cdot k_i$ and $u_e = \eta \cdot u_{ISA}$ into Equation (10.95), and then using Equation (10.94), yields

$$E_e = \frac{rG^2}{(1 + rG^2)^2} \left(\frac{1}{2} k_i u_{ISA}^2 \right) \tag{10.96}$$

Dividing the output energy, E_e, by the reference energy, $E_{mech}^{ref} = \tfrac{1}{2} k_i \cdot u_{ISA}^2$, one gets the expression of the output energy coefficient, E_e', for a displacement-amplified induced-strain actuator

$$E_e' = \frac{rG^2}{(1 + rG^2)^2} \tag{10.97}$$

The energy coefficient is a function of stiffness ratio, r, and kinematic gain, G.

Figure 10.19 shows plots of the energy coefficient versus stiffness ratio, r, for various values of the kinematic gain, G. For $G = 1$, that is, for an ISA device without amplification, the peak value of the energy coefficient is reached for $r_{opt} = 1$, that is, the *stiffness match concept*. For values of G greater than 1, that is, for devices with displacement amplification, the value of the optimal stiffness ratio, r_{opt}, changes and the peak energy output is reached at lower values of r_{opt}. For a 10 times kinematic gain ($G = 10$), the optimal stiffness ratio is about 1/100. Exact expressions of the optimal stiffness ratio in terms of the kinematic

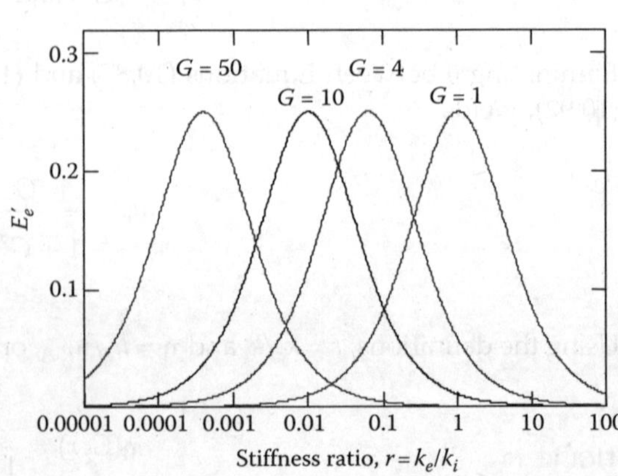

FIGURE 10.19
Output energy coefficient, E_e', versus stiffness ratio, $r = k_e/k_i$, for various values of the kinematic gain, G.

gain, G, are obtained by differentiating the energy coefficient with respect to r and setting the derivative equal to zero, that is,

$$\frac{d}{dr}E'_e = \left[\frac{G^2}{(1+rG^2)^2}\right]\left(1 - 2r\frac{G^2}{1+rG^2}\right) = 0 \tag{10.98}$$

hence,

$$r_{opt} = \frac{1}{G^2} \tag{10.99}$$

Conversely, for fixed external and internal stiffness values, the optimum gain is

$$G_{opt} = \frac{1}{r^2} \tag{10.100}$$

The output displacement coefficient (overall amplification ratio) corresponding to the optimum kinematic gain is obtained by substituting the expression of G_{opt} into Equation (10.94). On simplification, we obtain

$$\eta_{opt} = \frac{1}{2\sqrt{r}} \tag{10.101}$$

It should be noted that the value of the optimum amplification ratio, η_{opt}, is half the value of the optimum gain, G_{opt}. To obtain a certain overall amplification ratio at the optimum operating point, one has to provide twice as much kinematic gain in the internal construction of the displacement amplifier. Plots of the optimal kinematic gain, G_{opt}, and of the amplification ratio, η_{opt}, *versus* the inverse stiffness ratio, $1/r$, are given in Figure 10.20.

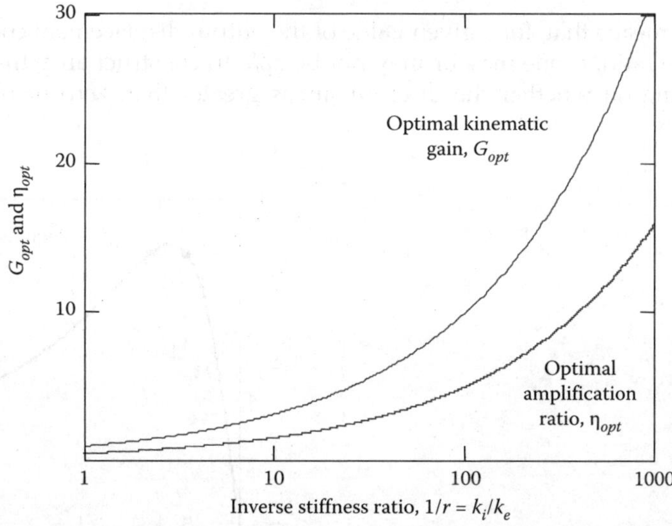

FIGURE 10.20
Optimal kinematic gain and amplification ratio versus inverse stiffness ratio.

10.5.6.3 Optimal Kinematic Gain, G, for a Given Value of η

When the output displacement coefficient, $\eta = u_e/u_{ISA}$, and the stiffness ratio, $r = k_e/k_i$, are specified, the value of the kinematic gain, G, must be selected. The expression of the output displacement coefficient given in Equation (10.94) can be solved for a given η to obtain the desired value of G. A plot of η versus G for a fixed value of r is given in Figure 10.21.

The kinematic gain, G, that will produce a required output displacement coefficient, η, is obtained by intersecting the η–G curve of Figure 10.21 with a $\eta =$ constant line. Two solutions are generally possible, G_1 and G_2. The same result can be obtained analytically by solving Equation (10.94) for η in terms of G with r as a parameter. One gets a quadratic equation in G, which accepts two solutions

$$G_1(\eta, r) = \frac{1}{2\eta}\frac{1}{r}\left(1 + \sqrt{1 - 4\eta^2 r}\right)$$
$$G_2(\eta, r) = \frac{1}{2\eta}\frac{1}{r}\left(1 - \sqrt{1 - 4\eta^2 r}\right)$$

$$(10.102)$$

Of the two possible solutions given by Equation (10.102), the one containing the minus sign, that is, G_2, is desired because it achieves the same output coefficient, η, but with a lower kinematic gain. Hence, we conclude that the kinematic gain, G, required to obtain a certain output coefficient, η, can be calculated with the formula

$$G(\eta, r) = \frac{1}{2\eta}\frac{1}{r}\left(1 - \sqrt{1 - 4\eta^2 r}\right)$$

$$(10.103)$$

The existence of the solution is determined by the discriminant being greater than zero, that is,

$$\Delta = 1 - 4\eta^2 r \geq 0$$

$$(10.104)$$

In practice, this means that, for a given value of the output displacement coefficient, η, and a given stiffness ratio, r, one may or may not be able to construct an actual amplification device depending on whether the discriminant is greater than zero or not. For a given

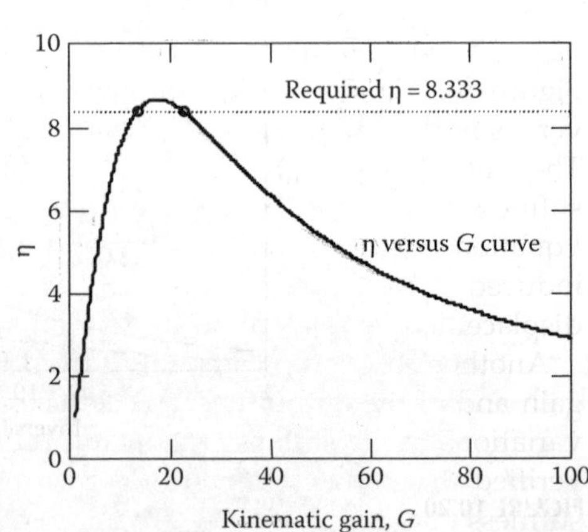

FIGURE 10.21
Variation of output displacement coefficient, η, with kinematic gain, G, for a fixed stiffness ratio, $r = 1/300$.

displacement coefficient, real values of the kinematic gain, G, are only obtained if the stiffness ratio, r, is less than a critical value, r_{cr}, that is,

$$r \leq r_{cr} \tag{10.105}$$

where

$$r_{cr} = \frac{1}{4\eta^2} \tag{10.106}$$

Equation (10.106) means that real values of the kinematic gain, G, only exist if the internal stiffness of the ISA device, k_i, is greater than a critical value, that is,

$$k_i \geq (k_i)_{cr} \tag{10.107}$$

where

$$(k_i)_{cr} = \frac{k_e}{r_{cr}} = \frac{k_e}{1/\eta^2} = \eta^2 k_e \tag{10.108}$$

Thus, the condition of Equation (10.107) can be written as

$$k_i \geq \eta^2 k_e \tag{10.109}$$

At the critical stiffness ratio, r_{cr}, the discriminant Δ is zero, and hence, the kinematic gain, G, is given by the expression

$$G_{cr} = \frac{1}{2\eta} \frac{1}{r_{cr}} \tag{10.110}$$

The output energy coefficient, E'_e, given by Equation (10.97), can also be expressed in terms of the critical stiffness ratio, r_{cr}, in the form

$$E'_e = \frac{1}{4} \frac{r}{r_{cr}} \tag{10.111}$$

Figure 10.22 gives plots of the normalized kinematic gain and output energy coefficient versus normalized stiffness ratio for a fixed value of the output displacement coefficient. The output energy reaches a maximum when the stiffness ratio matches the critical stiffness ratio given by Equation (10.106). Hence, the critical stiffness ratio defined by Equation (10.106) is actually the optimum stiffness ratio for the displacement-amplified induced-strain actuator. This observation is of utmost importance in the design of a displacement-amplified induced-strain actuator.

Another observation stemming from Figure 10.22 is that the behavior of the kinematic gain and of the output energy coefficient around the optimum point is not robust. Small variations in the stiffness ratio can produce large variations in G and E_e. It can be easily verified that the derivative of G with respect to r has an infinite value at the optimum stiffness ratio point, $r/r_{cr} = 1$. The use of optimum stiffness ratio in displacement-amplified

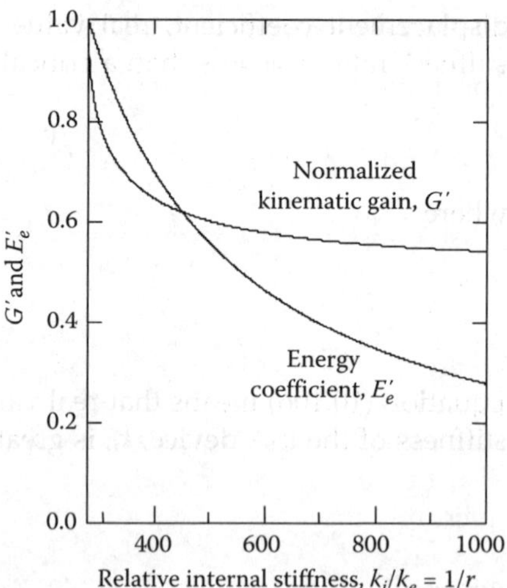

FIGURE 10.22
Variation of normalized kinematic gain, G', and of output energy coefficient, E', with relative internal stiffness, $k_i/k_e = 1/r$, for a fixed value of η ($\eta = 8.333$).

induced-strain actuators, though attractive, may not be desirable since small manufacturing variations easily lead to "de-tuning" and loss of performance. A design point slightly away from the optimum may offer a better choice from this point of view and may lead to a more robust behavior.

10.5.7 Displacement-Amplified Induced-Strain Actuator with Support and Transmission Flexibility

Real-life displacement-amplified induced-strain actuators incorporate a host of parasitic flexibilities that diminish their performance. The support structure, the lever arms, and the lever fulcrum may present significant compliance. Figure 10.23 shows these additional effects modeled by the finite stiffness values, k_{si}, k_{li}, k_{le}, and k_{s0}. The effect of these additional compliances upon the output displacement and output energy of the displacement-amplified induced-strain actuator is studied next.

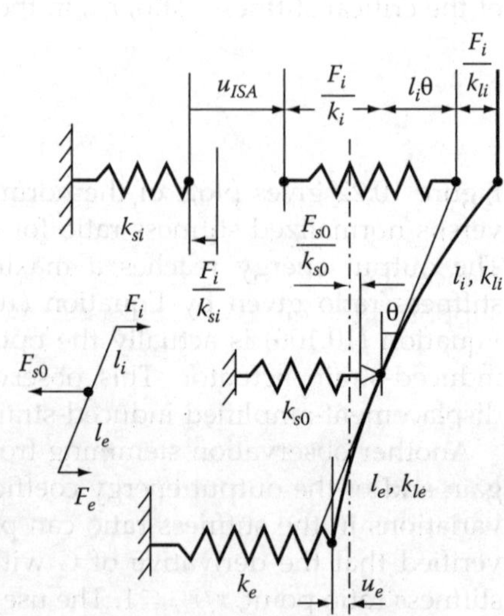

FIGURE 10.23
Conceptual model of displacement-amplified induced-strain actuator having compliant support and compliant amplification mechanism.

10.5.7.1 Displacement Analysis

The induced-strain displacement, u_{ISA}, is consumed into the support elasticity, k_{si}, the internal elasticity of the ISA stack, k_i, the elasticity of the input lever arm, k_{li}, the elasticity of the fulcrum support, k_{s0}, and the lever rotation, θ. Hence,

$$u_{ISA} = \frac{F_i}{k_{si}} + \frac{F_i}{k_i} + \frac{F_i}{k_{li}} + \frac{F_{s0}}{k_{s0}} + l_i\theta \qquad (10.112)$$

where F_i is the force in the ISA device and F_{s0} is the force in the fulcrum support. Note that, at the fulcrum support, one encounters the combined action of the input and output forces, that is, $F_{s0} = F_i + F_e$. The rotation, θ, produces an amplified displacement, $l_e\theta$, which is distributed into the displacement of the output lever arm, F_e/k_{le}, the displacement of the fulcrum support, F_{s0}/k_{s0}, and in the output displacement, u_e. Thus,

$$l_e\theta = \frac{F_e}{k_{le}} + \frac{F_{s0}}{k_{s0}} + u_e \qquad (10.113)$$

Recalling that $F_e = k_e u_e$, and $F_e l_e = F_i l_i$, one eliminates θ between Equations (10.112) and (10.113), and gets the constitutive relation of the displacement-amplified induced-strain actuator with support and amplification elasticity. Introducing notations for the equivalent input and output stiffness

$$k_i^* = \left[\frac{1}{k_i} + \frac{1}{k_{li}} + \frac{1}{k_{si}} + \frac{(1+G)}{G}\frac{1}{k_{s0}}\right]^{-1}, \quad k_e^* = \left[\frac{1}{k_e} + \frac{1}{k_{le}} + (1+G)\frac{1}{k_{s0}}\right]^{-1} \qquad (10.114)$$

yields

$$u_e = G\left(\frac{k_e}{k_e^*} + G^2\frac{k_e}{k_i^*}\right)^{-1} u_{ISA} \qquad (10.115)$$

Equation (10.115) can be simplified by defining the input and output stiffness ratios, r_i^* and r_e^*, as

$$r_i^* = \frac{k_e}{k_i^*} \quad \text{and} \quad r_e^* = \frac{k_e}{k_e^*} \qquad (10.116)$$

Hence, the constitutive relation of the displacement-amplified induced-strain actuator with support and amplification elasticity becomes

$$u_e = G\left(r_e^* + r_i^* G^2\right)^{-1} u_{ISA} \qquad (10.117)$$

for a displacement-amplified induced-strain actuator with support and amplification elasticity, the output displacement coefficient, $\eta = u_e/u_{ISA}$, is given by

$$\eta = G\left(r_e^* + r_i^* G^2\right)^{-1} \qquad (10.118)$$

Note that the output displacement coefficient, η, is only a fraction of the kinematic gain, G, since part of the amplification effect is lost in the elasticity of the ISA material, the structural support, and the lever arms. Also, note the similarity of Equations (10.118) and (10.114) giving the displacement amplification coefficient for a displacement-amplified ISA device with and without support and amplification elasticity, respectively. The improved equation Equation (10.118), which takes into account support and amplification compliances, uses the modified stiffness ratios, r_i^* and r_e^*, defined by Equations (10.113), (10.114), and (10.116). When the support and amplification compliances are ignored, $r_i^* \rightarrow 1$ and $r_e^* \rightarrow r$ and Equation (10.118) reduces to the simpler equation Equation (10.93).

10.5.7.2 Output Energy Analysis

Substituting Equation (10.117) into Equation (10.42), and using $k_e = r k_i$ and $u_e = \eta u_{ISA}$, and the expression for η given by Equation (10.118) yields the output energy expression for a displacement-amplified induced-strain actuator with support and amplification elasticity

$$E_e = \left[r G^2 \left(r_e^* + r_i^* G^2 \right)^{-2} \right] \left(\tfrac{1}{2} k_i u_{ISA}^2 \right) \tag{10.119}$$

Dividing the output energy, E_e, by the reference energy, $E_{mech}^{ref} = \dfrac{1}{2} k_i u_{ISA}^2$, yields the output energy coefficient, E_e', in the form

$$E_e'(r, r_i^*, r_e^*) = r G^2 \left(r_e^* + r_i^* G^2 \right)^{-2} \tag{10.120}$$

In maximizing the output energy coefficient of a displacement-amplified induced-strain actuator with support and amplification elasticity, one needs to take into account not only the conventional stiffness ratio, $r = k_e/k_i$, but also the additional stiffness ratios, $r_i^* = k_e/k_i^*$ and $r_e^* = k_e/k_e^*$, that account for the elasticity of the support structure and the amplification arms.

10.6 Induced-Strain Actuation under Dynamic Operation

In this section, we discuss the analysis of induced-strain actuators for dynamic actuation applications. The types of applications under consideration are assumed to operate at frequencies well below the resonance frequency of the free induced-strain actuator. However, resonance of the complete system, consisting of the induced-strain actuator and the dynamic external load, is possible.

During dynamic operation, alternating electric voltage is applied and the electrical charge flows as electric current in and out of the induced-strain actuator. The voltage and current are not necessarily in phase, hence,

$$V(t) = \hat{V} \cos \omega t$$
$$I(t) = \hat{I} \cos (\omega t + \phi) \tag{10.121}$$

Using complex notations with the implied convention $\cos \omega t = \mathrm{Re}(e^{i\omega t})$, we write

$$V(t) = \hat{V}e^{i\omega t}$$
$$I(t) = \hat{I}e^{i(\omega t + \phi)}$$

(10.122)

Recalling Euler's identity, $e^{i\alpha} = \cos\alpha + i\sin\alpha$, we understand that Equation (10.122) implies the convention of using only the real part of the complex exponential since

$$V(t) = \hat{V}\cos\omega t = \mathrm{Re}\left(\hat{V}e^{i\omega t}\right)$$
$$I(t) = \hat{I}\cos(\omega t + \phi) = \mathrm{Re}\left(\hat{I}e^{i(\omega t + \phi)}\right)$$

(10.123)

We can write Equation (10.121) using the complex amplitudes, $\bar{V}$ and $\bar{I}$, that incorporate the phase difference between the two variables, $V(t)$ and $I(t)$, that is,

$$V(t) = \bar{V}e^{i\omega t} \qquad \bar{V} = \hat{V}$$
$$\text{where}$$
$$I(t) = \bar{I}e^{i\omega t} \qquad \bar{I} = \hat{I}e^{i\phi}$$

(10.124)

where $\bar{V}$ and $\bar{I}$ are complex numbers. Associated with the harmonic voltage of Equation (10.111), we have a harmonic electric field

$$E(t) = \hat{E}e^{i\omega t}$$

(10.125)

where

$$\hat{E} = \hat{V}/h$$

(10.126)

The electric field of Equation (10.125) induces a harmonic expansion in the piezoelectric actuator, that is,

$$u(x,t) = \hat{u}(x)e^{i\omega t}$$

(10.127)

When the actuator acts against an external load, a reaction force is generated. Figure 10.24 shows a solid-state induced-strain actuator operating against a dynamic load of parameters, $k_e(\omega)$, $m_e(\omega)$, and $c_e(\omega)$, which are frequency-dependent effective stiffness, mass, and damping, respectively. Denoted, by $\bar{k}_e(\omega)$, the effective structural stiffness under dynamic conditions is

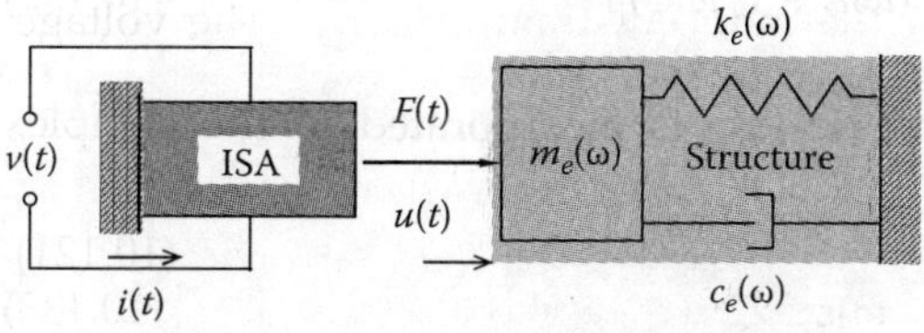

FIGURE 10.24
Schematic of the interaction between an induced-strain actuator (ISA device) and an elastic structure under dynamic conditions showing the frequency dependent dynamic stiffness.

$$\bar{k}_e(\omega) = \left[k_e(\omega) - \omega^2 m_e(\omega)\right] + i\omega c_e(\omega) \quad \text{(external dynamic stiffness)} \tag{10.128}$$

A material under dynamic operation displays internal heating due to several loss mechanisms. A simplified representation of this behavior assumes the complex compliance

$$\bar{s} = (1 - i\eta)s \tag{10.129}$$

Associated with the complex compliance, $\bar{s}$, we define the complex internal stiffness, $\bar{k}_i$, given by

$$\bar{k}_i = \frac{A}{\bar{s}l} = (1 + i\eta)\frac{A}{sl} = (1 + i\eta)k_i \quad \text{(complex internal stiffness)} \tag{10.130}$$

where k_i is the static internal stiffness given by Equation (10.28). Hence, we define the complex dynamic stiffness ratio as

$$\bar{r}(\omega) = \frac{\bar{k}_e(\omega)}{\bar{k}_i} \quad \text{(dynamic stiffness ratio)} \tag{10.131}$$

The dynamic stiffness ratio given by Equation (10.131) resembles the static stiffness ratio, $r = k_e/k_i$, given by Equation (10.32) that uses the complex dynamic stiffness except, $\bar{k}_e(\omega)$, and complex internal stiffness, $\bar{k}_i$.

The actuator is energized by an AC power supply that applies a time varying voltage, $V(t)$, at the actuator terminals and supplies a time-varying current, $I(t)$. As the charge builds up, the voltage and the electric field increase. Under the action of the electric field, the electroactive material expands and produces an output displacement, $u(t)$, which generates a reaction force from the mechanical system, $F(t)$. The reaction force, $F(t)$, acts on the induced-strain actuator. This produces loss of output displacement through the actuator compressibility and the counter electric motive force (counter emf) due to the piezoelectric effect. Phase lag may also happen due to damping effects. Hence, an actuator under load always has a lower output displacement than a load-free actuator energized by the same voltage. In complex notations, in-phase (conservative) mechanical response is manifested as a *real* quantity, whereas the out-of-phase (dissipative) response is manifested as an *imaginary* quantity.

Similar real and imaginary concepts exist for the electrical quantities, but with a different meaning. A real electrical quantity signifies dissipation, whereas an imaginary electrical quantity signifies conservation. In an LRC circuit, the complex voltage amplitude $\bar{V}$ and complex current amplitude $\bar{I}$ are related by the complex impedance $Z(\omega) = R + (i\omega L + 1/i\omega C)$ in the form

$$\bar{V} = Z(\omega)\bar{I} = [R + (i\omega L + 1/i\omega C)]\bar{I} \tag{10.132}$$

The electric dissipation associated with dielectric loss is incorporated in the complex permittivity

$$\bar{\varepsilon} = (1 - i\delta)\varepsilon \tag{10.133}$$

In view of the above, we write the *piezoelectric constitutive equations* in *complex form*, that is,

$$S = \bar{s}T + d_{33}E$$
$$D = d_{33}T + \bar{\varepsilon}E \tag{10.134}$$

Complex compliance and permittivity will be assumed henceforth (even if, sometimes, the bar sign over the quantity will be omitted for simplicity). Subsequently, we consider the complex stress-free capacitance, $\bar{C}_0$, that is,

$$\bar{C}_0 = (1 - i\delta)C_0 \tag{10.135}$$

Since $\hat{u}_{ISA}$ and k_i are reference parameters of the induced-strain actuator under dynamic operation, we define the *reference mechanical energy* of the induced-strain actuator, that is,

$$E_{mech}^{ref} = \tfrac{1}{2}k_i\hat{u}_{ISA}^2 \tag{10.136}$$

The quantity, E_{mech}^{max}, is a frequency-independent metric that can be effectively used in comparing the dynamic performance of various active-material induced-strain actuators. Similarly, the capacitance C_0 and the maximum dynamic voltage amplitude, $\hat{V}$, are reference electrical parameters of the induced-strain actuator under dynamic operation; hence, we define the reference electrical energy of an induced-strain actuator, that is,

$$E_{elec}^{ref} = \tfrac{1}{2}C_0\hat{V}^2 \tag{10.137}$$

Note that E_{elec}^{ref} is the input electrical energy required to make the free induced-strain actuator reach the free stroke $\hat{u}_{ISA}$. We also consider the complex coupling factor

$$\bar{\kappa}^2 = \frac{d_{33}^2}{\bar{s}\bar{\varepsilon}} \tag{10.138}$$

The analysis for dynamic operation has many similarities with the analysis for static operation. Of course, the mechanical applications considered in this analysis are well below the resonant frequency of the actuator itself, such that elastic wave effects in the actuator are negligible and can be ignored. The behavior of the external load, on the other hand, is truly dynamic, such that the effective structural stiffness, $\bar{k}_e(\omega)$, is strongly dependent on frequency, and the system may go through resonance within the operational frequency range.

In considering the dynamic operation of practical induced-strain actuators, two other aspects are important:

1. Dynamic stroke
2. Resonance of the complete system

The dynamic stroke is an important aspect since most induced-strain materials do not display symmetrical behavior with respect to polarity reversal. For example, electroactive stacks accept some reverse polarity (Figure 10.10), whereas PMN stacks do not accept any polarity reversal (Figure 10.11). To deal with this situation and create a symmetric stroke, the induced-strain actuator is usually biased about a midpoint position. For example, the Terfenol actuator shown in Figure 10.12 was internally biased to accept complete polarity reversal. In other cases, the bias is done through the external power supply. In a generic case, dynamic operation of a solid-state actuator can be assumed to take place about a mid-range position by superposing dynamic voltage amplitude onto a bias voltage component

$$V(t) = V_0 + \hat{V} \cos \omega t \tag{10.139}$$

where V_0 is the bias voltage and $\hat{V}$ is the dynamic voltage amplitude. The corresponding induced-strain displacement will be

$$u(t) = u_0 + \hat{u}_{ISA} \cos \omega t \tag{10.140}$$

where u_0 is the bias position and $\hat{u}_{ISA}$ is the dynamic displacement (Figure 10.25).

For dynamic operation, the total static displacement is equally divided on the two sides of the bias point such that equal positive and negative excursions are achieved. Thus, the effective maximum displacement for dynamic operation (*dynamic stroke*) is usually half the effective maximum displacement for static operation (*static stroke*).

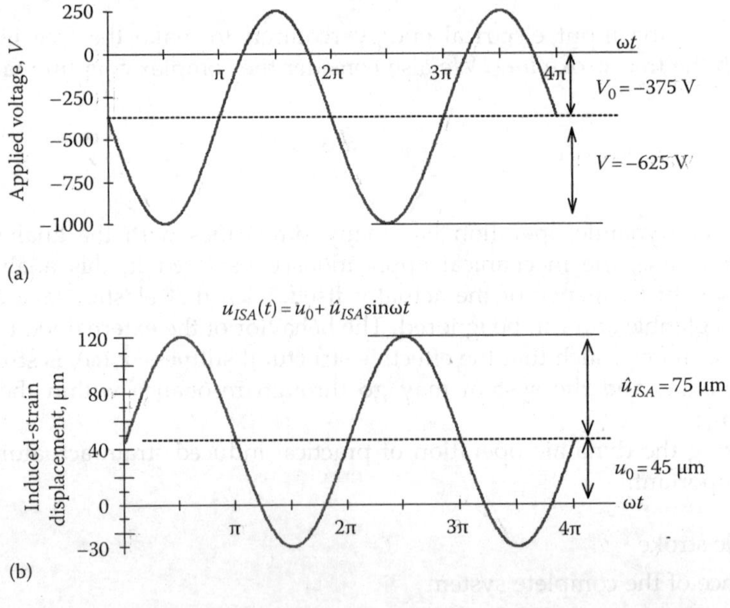

FIGURE 10.25
Dynamic operation of Polytec PI P247.70 induced-strain actuator: (a) applied voltage, $v(t)$, has bias and dynamic components, V_0 and $\hat{V}$; (b) the corresponding induced-strain displacement, $u_{ISA}(t)$.

Example 10.8:

Consider again the piezoelectric actuator discussed in Example 10.1 made up of a stack APC 850 square wafers with the properties given in Tables 10.2 and 10.3. Recall the stack length, l; static internal stiffness, k_i; stress-free capacitance, C_0; electromechanical coupling coefficient, κ; maximum voltage, V_{max}; calculated in Example 10.1. Find

1. Complex internal stiffness, $\bar{k}_i$; complex stress-free capacitance, $\bar{C}_0$; complex electro-mechanical coupling coefficient, $\bar{\kappa}^2$

   #### SOLUTION
   Recall the values of static internal stiffness, $k_i = 231.2$ kN/mm; stress-free capacitance, $C_0 = 619.5$ nF; electromechanical coupling coefficient, $\kappa = 0.773$ calculated in the previous problem. Use Equation (10.130) and Table 10.3 to get $\bar{k}_i = 231.2 + i4.62$ kN/mm. Use Equation (10.135) to get $\bar{C}_0 = 619.5 - i6.195$ nF. Use Equation (10.138) to get $\bar{\kappa} = 0.773 + i0.0116$.

2. Bias voltage, V_0, and the alternating voltage amplitude, $\hat{V}$, to be applied to the actuator for dynamic operation

   #### SOLUTION
   Recall the maximum voltage, $V_{max} = 250$ V, calculated previously in Example 10.1. Use Equation (10.139) to get $V_0 = 125$ V and $\hat{V} = 125$ V.

3. Bias position, u_0, and the amplitude of free dynamic stroke, $\hat{u}_{ISA}$, under free dynamic operation (When the actuator is sitting at the bias position, the spring of the external load is relaxed.)

   #### SOLUTION
   Recall the static free stroke, $u_{ISA} = 10.0$ μm, calculated previously in Example 10.1. Use Equation (10.140) to get $u_0 = 5.0$ μm and $\hat{u}_{ISA} = 5.0$ μm.

4. Reference mechanical energy, E_{mech}^{ref}, and reference electrical energy, E_{elec}^{ref}, that apply to this dynamic operation

   #### SOLUTION
   Use Equations (10.136) and (10.137) to calculate $E_{mech}^{ref} = 2.890$ mJ, $E_{elec}^{ref} = 4.840$ mJ. ∎

10.6.1 Mechanical Response

Recall the 1-D constitutive Equation (10.1), that is,

$$S = sT + d_{33}E$$
$$D = d_{33}T + \varepsilon E \tag{10.141}$$

Newton's law of motion and strain-displacement compatibility are applied to an infinitesimal element of thickness dx, to get

$$\frac{dT}{dx}A = \rho A \frac{d^2u}{d^2t} \tag{10.142}$$

$$S = \frac{du}{dx} \tag{10.143}$$

where A is the cross-sectional area. Substituting Equations (10.142) and (10.143) into the constitutive relations of Equation (10.141) yields

$$\frac{d^2u}{dx^2} = s\rho\frac{d^2u}{dt^2} + d\frac{dE}{dx} \tag{10.144}$$

Assuming uniform electric field, that is, $dE/dx = 0$, Equation (10.144) becomes

$$\frac{d^2u}{dx^2} = s\rho\frac{d^2u}{dt^2} \tag{10.145}$$

Equation (10.145) is the wave equation and has the general solution

$$u(x,t) = (C_1\sin\gamma x + C_2\cos\gamma x)e^{i\omega t} = \hat{u}(x)e^{i\omega t} \tag{10.146}$$

where γ is the *wave number* given by

$$\gamma = \omega\sqrt{\rho s} \text{ (wave number)} \tag{10.147}$$

If internal damping is taken into account according to Equation (10.129), then we encounter the *complex wave number*, $\bar{\gamma}$ defined by

$$\bar{\gamma} = \omega\sqrt{\rho\bar{s}} \text{ (complex wave number)} \tag{10.148}$$

The constants C_1 and C_2 of Equation (10.146) are found from boundary conditions. Taking $u(0,t) = 0$ yields $C_2 = 0$, and Equation (10.146) gives

$$\hat{u}(x) = C_1\sin\gamma x \tag{10.149}$$

At $x = l$, the actuator interacts with the external load represented by the external dynamic stiffness $\bar{k}_e(\omega)$ given by Equation (10.128). For a 1-dof external system, the parameters, k_e, m_e, and c_e are constant and the external system has a resonance frequency given by

$$\omega_0 = \sqrt{\frac{k_e}{m_e}} \tag{10.150}$$

The dynamic stiffness $\bar{k}_e(\omega)$ is a complex function of the operation frequency, ω. For structures with multiple degrees of freedom, one can approximate the dynamic behavior by a series expansion of the dynamic stiffness, and a 1-dof behavior can sometimes be identified by taking the dominant term corresponding to the resonance frequency closest to the operating frequency. Using the resonance frequency ω_0 given by Equation (10.150) and the associated damping factor, $\zeta = c_e/2m\omega_0$, we express the equivalent dynamic stiffness of Equation (10.128) in the form

$$\bar{k}_e(\omega) = \left[1 - (\omega/\omega_0)^2 + i2\zeta(\omega/\omega_0)\right]k_e = \left[(1-p^2) + i2\zeta p\right]k_e \tag{10.151}$$

where p is the frequency ratio $p = \omega/\omega_0$. The boundary condition at $x = l$ can be expressed as the equilibrium between external forces and internal stress resultants, that is,

$$\bar{T}(l)A = -\bar{k}_e(\omega)\hat{u}(l) \tag{10.152}$$

where the minus sign signifies that the external reaction produces a compression in the induced-strain actuator. Evaluation of Equation (10.141) at $x = l$ with the use of Equations (10.143), (10.149), and (10.152) gives

$$C_1 \gamma \cos \gamma l = s \frac{-\bar{k}_e(\omega)C_1 \sin \gamma l}{A} + d_{33}\hat{E} \tag{10.153}$$

Upon solution, Equation (10.153) yields

$$C_1 = \frac{d_{33}\hat{E}}{\gamma \cos \gamma l + \dfrac{s}{A}\bar{k}_e(\omega)\sin \gamma l} = \frac{d_{33}\hat{E}l}{\gamma l \cos \gamma l + \dfrac{sl}{A}\bar{k}_e(\omega)\sin \gamma l} = \frac{d_{33}\hat{E}l}{\gamma l \cos \gamma l + \bar{r}(\omega)\sin \gamma l} \tag{10.154}$$

where Equations (10.130) and (10.131) were invoked.

10.6.1.1 Output Mechanical Displacement

Substitution of Equation (10.154) into Equation (10.149) yields

$$\hat{u}(x;\omega) = \frac{d_{33}\hat{E}l}{\gamma l \cos \gamma l + \bar{r}(\omega)\sin \gamma l}\sin \gamma x \tag{10.155}$$

We are going to denote by $\hat{S}_{ISA}$ the strain induced in the electroactive material by the applied electric field $\hat{E}$, and by $\hat{u}_{ISA}$ the corresponding displacement induced at the end of the stack (free displacement), that is,

$$\hat{S}_{ISA} = d_{33}\hat{E}, \quad \hat{u}_{ISA} = \hat{S}_{ISA}l = d_{33}\hat{E}l \tag{10.156}$$

Using Equation (10.156), we can express Equation (10.155) as

$$\hat{u}(x;\omega) = \frac{\hat{u}_{ISA}}{\cos \gamma l + \bar{r}(\omega)\dfrac{\sin \gamma l}{\gamma l}}\frac{\sin \gamma x}{\gamma l} \tag{10.157}$$

The output displacement of the actuator is obtained by putting $x = l$ in Equation (10.157), that is,

$$\hat{u}_e(\omega) = \hat{u}(l;\omega) = \frac{\hat{u}_{ISA}}{\cos \gamma l + \bar{r}(\omega)\dfrac{\sin \gamma l}{\gamma l}}\frac{\sin \gamma l}{\gamma l} = \frac{\hat{u}_{ISA}\sin \gamma l}{\gamma l \cos \gamma l + \bar{r}(\omega)\sin \gamma l} = \frac{1}{\gamma l \cot \gamma l + \bar{r}(\omega)}\hat{u}_{ISA} \tag{10.158}$$

Note that the displacement variation along the stack, $\hat{u}(x; \omega)$, is a complex function containing ω as parameter. Equation (10.157) is complete in as much as it takes into account the wave propagation inside the induced-strain actuator and dynamic effects of the external load. However, in some mechanical engineering applications, the operating frequency is below 100 Hz. At this frequency the wave propagation effects are negligible. For example, an external resonance frequency $f_0 = 25$ Hz and an actuator length $l = 288$ mm yield $\gamma l = 0.018 = 1.8\%$. In such a typical case, the imaginary part of $\hat{u}(x; \omega)$ is around 1.6% of the real part and the wave propagation effects may be ignored.

If wave propagation effects are ignored ($\gamma l \ll 1$), then $\sin \gamma x \approx \gamma x$, $(\sin \gamma l)/\gamma l \approx 1$ and Equation (10.157) yield the *quasi-static* solution

$$\hat{u}_{qs}(x; \omega) = \frac{1}{1 + \bar{r}(\omega)} \frac{x}{l} \hat{u}_{ISA} \text{ (quasi-static solution)} \qquad (10.159)$$

For the quasi-static solution of Equation (10.159), the strain is constant and the displacement varies linearly along the actuator. For a typical actuator driven at low frequencies below 100 Hz, the difference between the exact solution of Equation (10.157) and the quasi steady solution of Equation (10.159) is insignificant. For example, at $x = l$, $|\hat{u}_{qs}(l; \omega)|/|\hat{u}(l; \omega)| = 99.991\%$. This shows that the use of the quasi steady solution is justified for some low-frequency mechanical engineering applications. Correspondingly, the quasi-static actuator displacement amplitude under load is obtained by substituting $x = l$ into Equation (10.159), that is,

$$\hat{u}_e(\omega) = \frac{1}{1 + \bar{r}(\omega)} \hat{u}_{ISA} \text{ (output dynamic displacement under load)} \qquad (10.160)$$

Example 10.9:

Consider again the piezoelectric actuator discussed in Example 10.1 made up of a stack APC 850 square wafers with the properties given in Tables 10.2 and 10.3. Recall the stack length, l; static internal stiffness, k_i; stress-free capacitance, C_0; electromechanical coupling coefficient, κ; maximum voltage, V_{max}; calculated in Example 10.1, as well as other parameters calculated in previous examples. Find

1. Expression of the complex dynamic stiffness, $\bar{k}_e(\omega)$, of an external load of spring of stiffness $k_e = 60$ kN/mm, resonance frequency $f_0 = 20$ Hz, and damping ratio $\zeta = 3\%$

   SOLUTION
   The desired expression is Equation (10.151), that is, $\bar{k}_e(\omega) = \left[1 - (\omega/\omega_0)^2 + i2\zeta(\omega/\omega_0)\right]k_e$, where $\omega_0 = 2\pi f_0 = 125.7$ rad/s.

2. Values of the complex stiffness, $\bar{k}_e$, and complex stiffness ratio, $\bar{r}$, when the operating frequency is half the resonance frequency of the external load, ω_0

   SOLUTION
   Use Equation (10.151) with $p = \omega/\omega_0 = 0.5$ to get $\bar{k}_e|_{\omega=0.5\omega_0} = 45 + i1.8$ kN/mm. Then use this result in Equation (10.131) to get $\bar{r}_e|_{\omega=0.5\omega_0} = 0.1947 + i0.0039$.

3. Complex wavenumber, $\bar{\gamma}$, and the value $|\bar{\gamma}l|$ associated with operation at half the resonance frequency of the external system, ω_0. Verify whether or not the quasi-static assumption can be applied.

SOLUTION

The desired expression is Equation (10.148). Use this equation to calculate $\bar{\gamma}|_{\omega=0.5\omega_0} = 22.93 + i0.229/\text{km}$. Recall the actuator length calculated in Example 10.1, that is, $l = 25$ mm; calculate $|\bar{\gamma}l| = 0.0006$. Since $|\bar{\gamma}l| \ll 1$, the quasi-static assumption can be applied.

4. Complex expression of the output dynamic displacement amplitude, $\hat{u}_e(\omega)$, during full-stroke dynamic operation against the external load $\bar{k}_e(\omega)$. Value of $\hat{u}$ when the operating frequency is half the resonance frequency of the external load, ω_0. Express the result in physical units, first. Then, express the absolute value of the result as a percentage of the free dynamic stroke, $\hat{u}_{ISA}$.

SOLUTION

Since the quasi-static assumption applies, the desired expression is Equation (10.160); use this equation to calculate the numerical value $\hat{u}_e(\omega)|_{\omega=0.5\omega_0} = 4.185 - i0.0136$ μm. In terms of $\hat{u}_{ISA}$, we find $\left|\hat{u}_e(\omega)|_{\omega=0.5\omega_0}\right| = 83.7\%$ of $\hat{u}_{ISA}$. ∎

10.6.1.2 Output Mechanical Power

The power produced by the actuator is used to move the external load. For a mass–spring–damper external load (Figure 10.24), the output power will have two components: (a) the active power, representing energy dissipated into the damper, and (b) the reactive power, representing energy stored into the spring and mass during some parts of the cycle and returned to the actuator during other parts of the cycle. By definition, *mechanical power = force × velocity*. For harmonic motion, $u(t) = \hat{u}\cos\omega t$, velocity and force have the expressions

$$\dot{u}(t) = \hat{u}\frac{d}{dt}\cos\omega t = \hat{u}(-\omega)\sin\omega t = \omega\hat{u}\cos(\omega t + \pi/2)$$

$$F(t) = \hat{F}\cos(\omega t + \psi) \tag{10.161}$$

Note that the displacement, $u(t)$, is taken as the reference variable, whereas the force $F(t)$ is assumed to have a phase difference ϕ with respect to $u(t)$. Using complex notations with the implied convention $\cos\omega t = \text{Re}(e^{i\omega t})$, we write

$$\dot{u}(t) = i\omega\hat{u}e^{i\omega t}$$

$$F(t) = \hat{F}e^{i(\omega t + \psi)} \tag{10.162}$$

We can write Equation (10.161) using the complex amplitudes $\bar{u}$ and $\bar{F}$ that incorporate the phase difference between the two variables, $u(t)$ and $F(t)$, that is,

$$\dot{u}(t) = i\omega\bar{u}\,e^{i\omega t} \qquad\qquad \bar{u} = \hat{u}$$

$$\text{where} \tag{10.163}$$

$$F(t) = \bar{F}e^{i\omega t} \qquad\qquad \bar{F} = \hat{F}e^{i\psi}$$

where $\bar{u}$ and $\bar{F}$ are complex numbers.

$$P(t) = F(t)\dot{u}(t) = \hat{F}\cos(\omega t + \psi)\hat{u}(-\omega)\sin\omega t \tag{10.164}$$

Using standard trigonometric identities, we express Equation (10.164) in the form

$$P(t) = \tfrac{1}{2}\hat{F}\hat{u}(-\omega)(-\sin\psi + \sin(2\omega t + \psi)) = \tfrac{1}{2}\hat{F}\hat{u}\omega\sin\psi - \tfrac{1}{2}\hat{F}\hat{u}\sin(2\omega t + \psi) \qquad (10.165)$$

The function $P(t)$ defined by Equation (10.165) is a harmonic function that contains a constant part, $\dfrac{1}{2}\hat{F}\hat{u}\sin\psi$, and a harmonic part, $-\dfrac{1}{2}\hat{F}\hat{u}\sin(2\omega t + \psi)$, oscillating at twice the frequency, ω, of $u(t)$ and $F(t)$ variables. When integrated over a period T, the harmonic part cancels out and only the constant part remains; this represents the active (real) power, that is,

$$P_{mech}^{active} = P_{mech}^{avg} = \frac{1}{T}\int_0^T P(t)dt = \tfrac{1}{2}\hat{F}\hat{u}\omega\sin\psi \qquad (10.166)$$

The same derivation can be worked out in terms of velocity $v(t)$, that is,

$$\begin{aligned} v(t) &= \hat{v}\cos\omega t \\ F(t) &= \hat{F}\cos(\omega t + \phi) \end{aligned} \qquad (10.167)$$

Using complex notations with the implied convention, $\cos\omega t = \mathrm{Re}(e^{i\omega t})$, we write

$$\begin{aligned} v(t) &= \hat{v}e^{i\omega t} \\ F(t) &= \hat{F}e^{i(\omega t + \phi)} \end{aligned} \qquad (10.168)$$

We can write Equation (10.161) using the complex amplitudes $\bar{v}$ and $\bar{F}$ that incorporate the phase difference between the two variables, $v(t)$ and $F(t)$, that is,

$$\begin{aligned} v(t) &= \bar{v}e^{i\omega t} \\ F(t) &= \bar{F}e^{i\omega t} \end{aligned} \quad \text{where} \quad \begin{aligned} \bar{v} &= \hat{v} \\ \bar{F} &= \hat{F}e^{i\phi}, \end{aligned} \quad \text{and} \quad \bar{v} = i\omega\bar{u} \qquad (10.169)$$

where $\bar{v}$ and $\bar{F}$ are complex numbers. Recalling *mechanical power = force × velocity*, we write

$$P(t) = F(t) \cdot v(t) = \hat{F}\cos(\omega t + \phi) \cdot \hat{v}\cos\omega t \qquad (10.170)$$

Using standard trigonometric identities, we express Equation (10.164) in the form

$$P(t) = \tfrac{1}{2}\hat{F}\hat{v}(\cos\phi + \cos(2\omega t + \phi)) = \tfrac{1}{2}\hat{F}\hat{v}\cos\phi + \tfrac{1}{2}\hat{F}\hat{v}\cos(2\omega t + \phi) \qquad (10.171)$$

The function $P(t)$ defined by Equation (10.171) is a harmonic function that contains a constant part, $\tfrac{1}{2}\hat{F}\hat{v}\cos\phi$, and a varying part, $\tfrac{1}{2}\hat{F}\hat{v}\cos(2\omega t + \phi)$, that oscillates at twice the frequency of $v(t)$ and $F(t)$. When integrated over a period T, the harmonic part cancels out and only the constant part remains, which represents the active (real) power, that is,

$$P_{mech}^{active} = P_{mech}^{avg} = <P> = \frac{1}{T} \int_0^T P(t)dt = \tfrac{1}{2}\hat{F}\hat{v}\cos\phi \tag{10.172}$$

Equations (10.172) and (10.166) are equivalent since $\hat{v} = \omega\hat{u}$, $\phi = \psi - \pi/2$, and $\cos\phi = \cos(\psi - \pi/2) = \sin\psi$. Note that Equation (10.172) expresses the general fact that the time-averaged product of the two harmonic variables, $v(t)$ and $F(t)$, is given by

$$<P> = \tfrac{1}{2}\hat{F}\hat{v}\cos\phi \tag{10.173}$$

The result of Equation (10.173) can be also expressed in terms of the complex notations (10.168). Note that

$$\mathrm{Re}\left(e^{i(\omega t + \phi)}e^{-i\omega t}\right) = \mathrm{Re}\left(e^{i\phi}\right) = \cos\phi \tag{10.174}$$

Recall the definition of the conjugate $\tilde{z}$ of a complex number z, that is,

$$\tilde{z} = x - iy, \quad \text{where} \quad z = x + iy \tag{10.175}$$

we can write

$$e^{i\phi} = \hat{F}e^{i(\omega t + \phi)}\hat{v}e^{-i\omega t} = F\tilde{v} \tag{10.176}$$

Substitution of Equations (10.174) and (10.176) into Equation (10.173) yields

$$<P> = P_{mech}^{avg} = \mathrm{Re}\left(\tfrac{1}{2}F\tilde{v}\right) = \mathrm{Re}\left(\tfrac{1}{2}\tilde{F}v\right) \tag{10.177}$$

where $\mathrm{Re}(F\tilde{v}) = \mathrm{Re}(\tilde{F}v)$ since $\cos(-\alpha) = \cos(\alpha)$. Equation (10.177) states the general fact that the *time-averaged product of the two harmonic variables is one half the product of one variable times the conjugate of the other*. Since the harmonic exponential $e^{i\omega t}$ is common in both F and v, the products $F\tilde{v}$ and $\tilde{F}v$ do not contain it because $e^{i\omega t}$ and $e^{-i\omega t}$ cancel out. Hence, Equation (10.177) can be written in terms of just the complex amplitudes to get

$$P_{mech}^{avg} = <Fv> = \mathrm{Re}\left(\tfrac{1}{2}\bar{F}\tilde{\bar{v}}\right) = \mathrm{Re}\left(\tfrac{1}{2}\tilde{\bar{F}}\bar{v}\right) \tag{10.178}$$

According to Equation (10.160), the complex displacement amplitude at the actuator tip acting against an external load is expressed in terms of the free displacement amplitude, $\hat{u}_{ISA}$, and the frequency-dependent complex stiffness ratio, $\bar{r}(\omega)$, that is,

$$u_e(t) = \bar{u}_e(\omega)e^{i\omega t} = \frac{1}{1 + \bar{r}(\omega)}\hat{u}_{ISA}e^{i\omega t} \tag{10.179}$$

Differentiation of Equation (10.179) with respect to time t yields the velocity, that is,

$$v(t) = \dot{u}_e(t) = i\omega\bar{u}_e(\omega)e^{i\omega t} = \frac{i\omega}{1 + \bar{r}(\omega)}\hat{u}_{ISA}e^{i\omega t} \tag{10.180}$$

The output force applied to the external load is calculated as the product between external stiffness and output displacement, that is,

$$\bar{F} = \bar{k}_e \bar{u}_e = \bar{r}(\omega) k_i \bar{u}_e \tag{10.181}$$

Substituting Equations (10.179) and (10.181) into Equation (10.178) yields the output mechanical power in the form

$$P_{mech}^{avg}(\omega) = \mathrm{Re}\left[\tfrac{1}{2} k_i \bar{\tilde{r}}(\omega) \bar{\tilde{u}}_e (i\omega \bar{u}_e)\right] = \mathrm{Re}\left(\frac{\bar{\tilde{r}}(\omega)}{1+\bar{\tilde{r}}(\omega)} \frac{i\omega}{1+\bar{r}(\omega)}\right)\left(\tfrac{1}{2} k_i \hat{u}_{ISA}^2\right) \tag{10.182}$$

The value $\cos\phi$ can be determined by comparing Equations (10.173) and (10.182), that is,

$$\cos\phi = \mathrm{Re}\left(\frac{\bar{\tilde{r}}(\omega)}{1+\bar{\tilde{r}}(\omega)} \frac{i\omega}{1+\bar{r}(\omega)}\right) \bigg/ \left|\frac{\bar{\tilde{r}}(\omega)}{1+\bar{\tilde{r}}(\omega)} \frac{i\omega}{1+\bar{r}(\omega)}\right| \tag{10.183}$$

Equation (10.182) shows that the power output varies with the complex stiffness, $\bar{r}(\omega)$, which depends strongly on frequency. The last factor in Equation (10.182) is the reference mechanical energy of an active-material actuator operating under dynamic conditions, as given by Equation (10.136), that is,

$$E_{mech}^{ref} = \tfrac{1}{2} k_i \hat{u}_{ISA}^2 \tag{10.184}$$

Consistent with the definition of reference mechanical energy, E_{mech}^{ref}, we define the reference mechanical power as

$$P_{mech}^{ref} = \omega_0 E_{mech}^{ref} \tag{10.185}$$

where ω_0 is the resonance frequency of the external load described in Equation (10.151). Substitution of Equation (10.184) into Equation (10.182) yields

$$P_{mech}^{avg} = \mathrm{Re}\left(\frac{\bar{\tilde{r}}(\omega)}{1+\bar{\tilde{r}}(\omega)} \frac{i\omega}{1+\bar{r}(\omega)}\right) E_{mech}^{ref} \tag{10.186}$$

For low-damping mechanical systems driven well below the mechanical resonance frequency, the complex stiffness ratio is predominantly real, that is, $\bar{r} \approx r$. In this case, Equation (10.186) gives $\bar{P}_{mech} \approx 0$, which means that the average output power is practically zero since the mechanical power is predominantly reactive. Under the action of the induced-strain actuator, energy flows into the external spring during actuation expansion and returns into the actuator during actuator contraction, such that the net balance is practically zero.

Although the average mechanical power may be quite small or even zero, the instantaneous power that flows in and out of the actuator can be quite large. Of interest to the actuator designer is the *peak power* during the cyclic operation. To calculate the peak power, seek the maximum of the instantaneous power over a cycle of operation; examination of Equation (10.171) reveals that the maximum is attained as

$$P_{mech}^{peak} = \tfrac{1}{2} \hat{F} \hat{v} (1 + \cos\phi) = \tfrac{1}{2} \left|\bar{F}\bar{v}\right| (1 + \cos\phi) \tag{10.187}$$

Using Equations (10.169), (10.179), and (10.181) into Equation (10.187) yields

$$P_{mech}^{peak} = \omega \left| \frac{\tilde{\tilde{r}}(\omega)}{[1 + \bar{r}(\omega)][1 + \tilde{\tilde{r}}(\omega)]} \right| (1 + \cos \phi) E_{mech}^{ref} \tag{10.188}$$

where $\cos \phi$ is given by Equation (10.183). For low-damping mechanical systems driven well below the mechanical resonance frequency, the complex stiffness ratio is predominantly real, that is, $\bar{r} \approx r$ and $\cos \phi = 0$. In this case, Equation (10.188) takes the simpler form

$$P_{mech}^{peak} = \omega \frac{r}{(1 + r)^2} E_{mech}^{ref} = \omega E_e \tag{10.189}$$

where E_e is given by Equation (10.44). The mechanical output power expressions given by Equations (10.186), (10.188), and (10.189) also depend on the external loading conditions through the stiffness ratio, r. For example, the plot of Equation (10.189) shown in Figure 10.26 indicates that the output power first increases with the stiffness ratio, r, reaches a maximum, and then starts to decrease.

The maximum peak output power is obtained at $r = 1$ and it has the value

$$P_{mech}^{maxpeak} = \omega \tfrac{1}{4} E_{mech}^{ref} \tag{10.190}$$

It can be seen that the maximum peak output power is the product of the angular frequency, ω, and $1/4$ of the reference mechanical output energy, E_{mech}^{ref}, given by Equation (10.136). On the same account, the maximum output energy per cycle is given by

$$\hat{E}_{mech}^{max} = \tfrac{1}{4} E_{mech}^{ref} \tag{10.191}$$

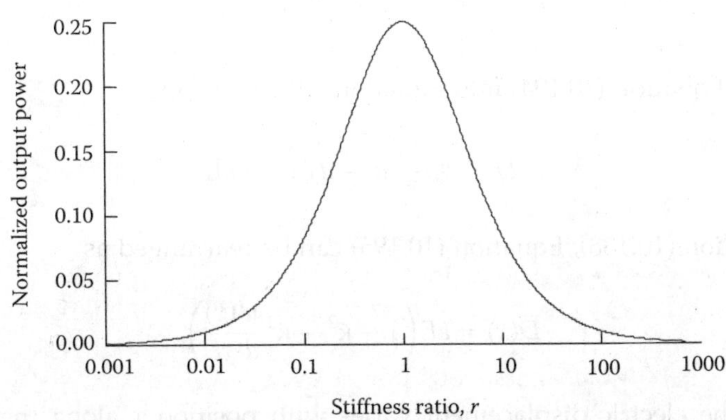

FIGURE 10.26
Variation of output power with stiffness ratio, r.

Example 10.10:

Consider again the piezoelectric actuator discussed in Example 10.1 made up of a stack of APC 850 square wafers with the properties given in Tables 10.2 and 10.3. Recall the stack length, l; static internal stiffness, k_i; stress-free capacitance, C_0; electromechanical coupling coefficient, κ; maximum voltage, $V_{\max}$; calculated in Example 10.1, as well as other parameters calculated in previous examples. Find

1. Expressions of the average mechanical power output, P^{avg}_{mech}, and $\cos\phi$. The values of P^{avg}_{mech}, $\cos\phi$, and ϕ when the actuator is operating at half the resonance of the external system, ω_0.

   SOLUTION
   The desired expressions are Equations (10.182) and (10.183). The numerical values are $P^{avg}_{mech}|_{\omega=0.5\omega_0} = 0.495$ mW and $\cos\phi|_{\omega=0.5\omega_0} = 0.020$, that is, $\phi = 88.855°$.

2. Expression of the peak mechanical power output per cycle, P^{peak}_{mech} and its value when the actuator is operating at half the resonance of the external system, ω_0. Comment on the relative numerical values of P^{avg}_{mech} and P^{peak}_{mech}.

   SOLUTION
   The desired expression is Equation (10.188). The numerical value is $P^{peak}_{mech}|_{\omega=0.5\omega_0} = 25.271$ mW. It is apparent that the value P^{peak}_{mech} is orders of magnitude larger than P^{avg}_{mech}. ∎

10.6.2 Electrical Response under Load

To calculate the electric displacement (charge per unit area), recall Equation (10.134), that is,

$$S = \bar{s}T + d_{33}E \tag{10.192}$$

$$D = d_{33}T + \bar{\varepsilon}E \tag{10.193}$$

Equation (10.192) yields the stress as a function of strain and electric field, that is,

$$T = \frac{1}{\bar{s}}(S - d_{33}E) = \frac{1}{\bar{s}}(u' - d_{33}E) \tag{10.194}$$

Substitution of Equation (10.194) into Equation (10.193) yields

$$D = d_{33}\frac{1}{\bar{s}}(u' - d_{33}E) + \bar{\varepsilon}E \tag{10.195}$$

Recalling Equation (10.138), Equation (10.195) can be rearranged as

$$D(x) = \bar{\varepsilon}E\left(1 - \bar{\kappa}^2 + \bar{\kappa}^2\frac{u'(x)}{d_{33}E}\right) \tag{10.196}$$

We note that the electric displacement varies with position x along the induced-strain actuator. Recall that the induced-strain actuator is made up of a stack of N piezoelectric wafers of thickness h and electroded faces of area A. The electrodes are wired in such a way that all the wafers are connected in parallel. Since the wafers are very thin, it is appropriate to approximate the strain inside each wafer with its average value; for the nth wafer, we have

$$u'_n = \int_{x_{n-1}}^{x_n} u'(x)dx = \frac{1}{h}(u_n - u_{n-1}) \tag{10.197}$$

where x_{n-1} and x_n are the coordinates at the bottom and top of the nth wafer in the stack, and u_{n-1} and u_n are the corresponding displacements. Substitution of Equation (10.197) into Equation (10.196) yields the electric displacement in the nth wafer as

$$D_n = \bar{\varepsilon}E\left(1 - \bar{\kappa}^2 + \bar{\kappa}^2 \frac{u_n - u_{n-1}}{hd_{33}E}\right) \tag{10.198}$$

Multiplication of Equation (10.198) by the electrode area A yields the charge associated with the nth wafer, that is,

$$Q_n = A\bar{\varepsilon}E\left(1 - \bar{\kappa}^2 + \bar{\kappa}^2 \frac{u_n - u_{n-1}}{hd_{33}E}\right) \tag{10.199}$$

The total charge is obtain by summation over all the wafers, that is, in terms of amplitudes,

$$\begin{aligned}
\hat{Q} = \sum_{n-1}^{N} \hat{Q}_n &= A\bar{\varepsilon}E\left[N(1 - \bar{\kappa}^2) + \bar{\kappa}^2 \frac{1}{hd_{33}E} \sum_{n=1}^{N}(\hat{u}_n - \hat{u}_{n-1})\right] \\
&= A\bar{\varepsilon}E\left[N(1 - \bar{\kappa}^2) + \bar{\kappa}^2 \frac{\hat{u}_N - \hat{u}_0}{hd_{33}E}\right]
\end{aligned} \tag{10.200}$$

where $\hat{u}_0$ and $\hat{u}_N$ are the displacement amplitudes at the bottom and top of the stack actuator. In our case, $u_0 = 0$, whereas $\hat{u}_N = \hat{u}_e(\omega) = ld_{33}\hat{E}/[\gamma l \cot \gamma l + \bar{r}(\omega)]$ according to Equation (10.158); hence, Equation (10.200) yields

$$\bar{Q} = A\bar{\varepsilon}\hat{E}\left[N(1 - \bar{\kappa}^2) + \bar{\kappa}^2 \frac{1}{\gamma l \cot \gamma l + \bar{r}(\omega)} \frac{l}{h} \frac{d_{33}\hat{E}}{d_{33}\hat{E}}\right] \tag{10.201}$$

where $\bar{Q} = \hat{Q}e^{i\phi}$ is the complex amplitude that incorporates the modulus $\hat{Q}$ and generic phase ϕ. Recall Equation (10.3), that is, $h = l/N$, and Equation (10.126), that is, $\hat{E} = \hat{V}/h$; hence, Equation (10.201) becomes

$$\bar{Q} = N\bar{\varepsilon}\frac{A}{h}\hat{V}\left(1 - \bar{\kappa}^2 + \bar{\kappa}^2 \frac{1}{\gamma l \cot \gamma l + \bar{r}(\omega)}\right) = \bar{C}_0\hat{V}\left(1 - \bar{\kappa}^2 + \bar{\kappa}^2 \frac{1}{\gamma l \cot \gamma l + \bar{r}(\omega)}\right) \tag{10.202}$$

where $\bar{C}_0$ is the stress-free complex capacitance given by

$$\bar{C}_0 = N\frac{\bar{\varepsilon}A}{h} \tag{10.203}$$

10.6.2.1 Effective Capacitance

Equation (10.202) reflects the *effective complex capacitance*, $\bar{C}$, of the induced-strain actuator, defined as

$$\bar{C}(\omega) = \bar{C}_0\left[1 - \bar{\kappa}^2 + \bar{\kappa}^2\frac{1}{\bar{\gamma}l\cot\bar{\gamma}l + \bar{r}(\omega)}\right] \tag{10.204}$$

where, for completeness, the complex wave number $\bar{\gamma}$ defined in Equation (10.147) was used. It is apparent that the effect of the external load is to reduce the effective capacitance of the induced-strain actuator. As the stiffness of the external load increases, the effective capacitance decreases. Let us consider two extreme cases: *blocked actuator* and *free actuator*. For an infinitely stiff external constrain ($r \to \infty$), we approach the *blocked capacitance*, that is,

$$\bar{C}_{block} = C_0\left(1 - \bar{\kappa}^2\right) \tag{10.205}$$

Note that in determining Equation (10.205), we assumed that the driving frequency is well away from the resonant frequency of the blocked stack.

For a zero external stiffness ($r = 0$), we approach the free capacitance C_0 of Equation (10.5).

10.6.2.2 Quasi-Static Capacitance

If wave propagation effects can be ignored ($\gamma l \ll 1$), then $\gamma l \cot \gamma l \approx 1$ and Equation (10.204) yield the *quasi-static capacitance*, that is,

$$\bar{C}_{qs}(\omega) = \bar{C}_0\left(1 - \kappa^2\frac{\bar{r}(\omega)}{1 + \bar{r}(\omega)}\right) \tag{10.206}$$

10.6.2.3 Capacitance Reduction due to Load

It is apparent from Equations (10.204) and (10.206) that a reduction of the effective capacitance will take place under load. Denoting this *capacitance reduction* by ΔC_{ISA}, we can write Equation (10.206) as

$$\bar{C}(\omega) = \bar{C}_0 - \Delta C_{ISA}(\omega) \quad \text{where} \quad \Delta C_{ISA}(\omega) = \bar{\kappa}^2\frac{\bar{r}(\omega)}{1 + \bar{r}(\omega)}\bar{C}_0 \tag{10.207}$$

This effect can be expressed in nondimensional form through the introduction of a capacitance reduction coefficient, C'_{ISA}, defined as

$$C'_{ISA}(\omega) = \frac{\Delta C_{ISA}(\omega)}{\bar{C}_0} = \bar{\kappa}^2\frac{\bar{r}(\omega)}{1 + \bar{r}(\omega)} \tag{10.208}$$

Variation of the capacitance reduction coefficient, C'_{ISA}, with the frequency ratio and static stiffness ratio are given in Figure 10.27. It can be seen that a resonance peak is noticed at $p = \omega/\omega_0 = 1.33$. This phenomenon strongly influences the behavior of the electroactive stack during dynamic operation.

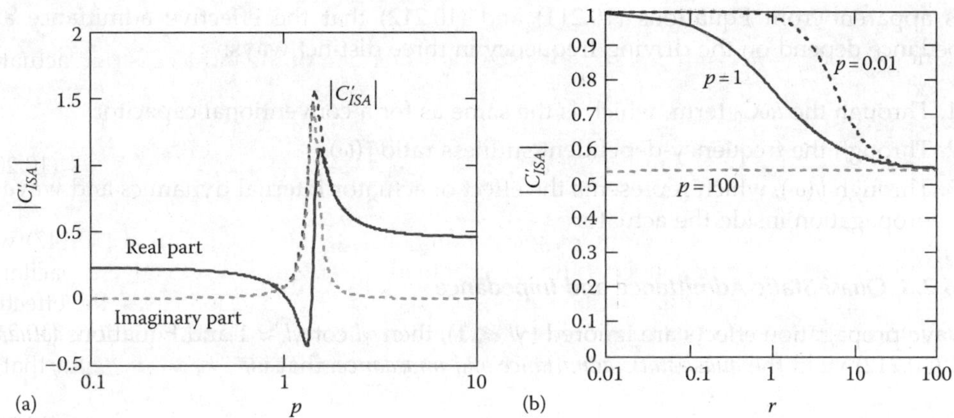

FIGURE 10.27
Variation of capacitance reduction coefficient, C'_{ISA}, with frequency ratio, p, and static stiffness ratio, r: (a) variation with p for $r = 1$; (b) variation with r, for three values of p.

10.6.2.4 Effective Admittance and Impedance

Recall Equation (10.202), that is,

$$\bar{Q} = C_0 \ddot{V}\left[1 - \bar{\kappa}^2 + \bar{\kappa}^2 \frac{1}{\gamma l \cot \gamma l + \bar{r}(\omega)}\right] \tag{10.209}$$

The current, I, is the time derivative of charge, Q; for harmonic motion, $I = \dot{Q} = i\omega Q$. Hence, Equation (10.209) yields

$$\bar{I} = i\omega \bar{C}_0 \hat{V}\left[1 - \bar{\kappa}^2 + \bar{\kappa}^2 \frac{1}{\gamma l \cot \gamma l + \bar{r}(\omega)}\right] \tag{10.210}$$

The admittance, Y, is defined as the ratio between current and voltage, that is,

$$Y(\omega) = \frac{I}{V} = i\omega \bar{C}_0\left[1 - \bar{\kappa}^2 + \bar{\kappa}^2 \frac{1}{\gamma l \cot \gamma l + \bar{r}(\omega)}\right] \tag{10.211}$$

Note that the wave number in Equation (10.211) also varies with frequency ω as indicated by Equation (10.148), that is, $\bar{\gamma}(\omega) = \omega\sqrt{\rho \bar{s}}$.

The impedance, Z, is defined as the ratio between voltage and current, that is, the reciprocal of the admittance, $Y(\omega)\,Z(\omega) = 1$. Inversion of Equation (10.211) yields

$$Z(\omega) = \frac{V}{I} = \frac{1}{Y(\omega)} = \frac{1}{i\omega \bar{C}_0}\left[1 - \bar{\kappa}^2 + \bar{\kappa}^2 \frac{1}{\gamma l \cot \gamma l + \bar{r}(\omega)}\right]^{-1} \tag{10.212}$$

It is apparent from Equations (10.211) and (10.212) that the effective admittance and impedance depend on the driving frequency in three distinct ways:

1. Through the $i\omega C_0$ term, which is the same as for a conventional capacitor
2. Through the frequency-dependent stiffness ratio $\bar{r}(\omega)$
3. Through $\bar{y}(\omega)$, which represents the effect of actuator internal dynamics and wave propagation inside the actuator

10.6.2.5 Quasi-Static Admittance and Impedance

If wave propagation effects are ignored ($\gamma l \ll 1$), then $\gamma l \cot \gamma l \approx 1$ and Equations (10.211) and (10.212) yield the *quasi-static admittance and impedance*, that is,

$$\bar{Y}_{qs}(\omega) = i\omega\bar{C}_0\left[1 - \bar{\kappa}^2\frac{\bar{r}(\omega)}{1+\bar{r}(\omega)}\right] \tag{10.213}$$

$$\bar{Z}_{qs}(\omega) = \frac{1}{i\omega\bar{C}_0}\left[1 - \bar{\kappa}^2\frac{\bar{r}(\omega)}{1+\bar{r}(\omega)}\right]^{-1} \tag{10.214}$$

It is apparent from Equations (10.213) and (10.214) that the quasi-static admittance and impedance depend on the driving frequency in only two distinct ways:

1. Through the $i\omega C_0$ term, which is the same as for a conventional capacitor
2. Through the frequency-dependent stiffness ratio $\bar{r}(\omega)$

For magnetoactive actuators, a similar derivation leads to the expression

$$Z_L(\omega) = i\omega\bar{L}_0\left[1 - \bar{\kappa}_L^2\frac{\bar{r}(\omega)}{1+\bar{r}(\omega)}\right]^{-1} \tag{10.215}$$

$$Y_L(\omega) = \frac{1}{i\omega\bar{L}_0}\left[1 - \bar{\kappa}_L^2\frac{\bar{r}(\omega)}{1+\bar{r}(\omega)}\right]^{-1} \tag{10.216}$$

where $\bar{L}_0$ is the magnetic inductance and $\bar{\kappa}_L^2$ is the effective piezomagnetic coupling coefficient defined as

$$\bar{\kappa}_L^2 = \frac{\frac{1}{2}k_i u_{ISA}^2}{\frac{1}{2}\bar{L}_0 I^2} \tag{10.217}$$

Example 10.11:

Consider again the piezoelectric actuator discussed in Example 10.1 made up of a stack APC 850 square wafers with the properties given in Tables 10.2 and 10.3. Recall the stack length, l; static internal stiffness, k_i; stress-free capacitance, C_0; electromechanical coupling coefficient, κ; maximum voltage, V_{max}; calculated in Example 10.1, as well as other parameters calculated in previous examples. Find:

1. Effective capacitance, $\bar{C}(\omega)$, expression and its value when the actuator is operating at half the resonance of the external system, ω_0

SOLUTION

Since quasi-static assumption applies, the desired expression is Equation (10.206). On calculation, we get $\bar{C}(\omega)|_{\omega=0.5\omega_0} = 559.3 - i8.41$ nF.

2. Expressions for effective admittance and impedance; their values when the actuator is operating at half the resonance of the external system, ω_0

SOLUTION

Since quasi-static assumption applies, the desired expressions are Equations (10.213) and (10.214). On calculation, we get $Y(\omega)|_{\omega=0.5\omega_0} = 0.528 - 35.139$ µS and $Z(\omega)|_{\omega=0.5\omega_0} = 0.428 - i35.139$ kΩ. $\blacksquare$

10.6.2.6 Electrical Input Power

Under dynamic operation, the flow of electric energy, that is, the *electric power*, must be analyzed. The electrical power to be discussed in connection with induced-strain actuators can have several descriptors: power rating, active power, reactive power, and maximum instantaneous power.

Consider the expression of instantaneous electrical power defined by the product of voltage and current. Recalling Equation (10.121), we write

$$P(t) = V(t)I(t) = (\hat{V}\cos\omega t)(\hat{I}\cos(\omega t + \phi)) \tag{10.218}$$

Using trigonometric identities, we express Equation (10.218) in the form

$$P(t) = \hat{V}\cos\omega t \cdot \hat{I}\cos(\omega t + \phi) = \tfrac{1}{2}\hat{V}\hat{I}(\cos\phi + \cos(2\omega t + \phi))$$
$$= \tfrac{1}{2}\hat{V}\hat{I}\cos\phi + \tfrac{1}{2}\hat{V}\hat{I}\cos(2\omega t + \phi) \tag{10.219}$$

Using the complex notations defined by Equations (10.122) through (10.124), we can also define the *complex power*, $\bar{P}$, that is,

$$\bar{P} = \tfrac{1}{2}\bar{V}\bar{I} = \tfrac{1}{2}\hat{V}\hat{I}e^{i\phi} \tag{10.220}$$

If the electrical system is *voltage controlled*, then the voltage is the reference signal and Equations (10.121) through (10.124) apply. The electric current is expressed in terms of voltage and admittance, that is,

$$I(t) = Y(\omega)V(t) \quad \text{and} \quad \phi = \arg(Y), \quad \cos\phi = Y_R/|Y| \tag{10.221}$$

In this case, the power expression of Equation (10.219) takes the *admittance-voltage power* form, that is,

$$P(t) = \tfrac{1}{2}Y_R\hat{V}^2 - \tfrac{1}{2}|Y|\hat{V}^2\cos(2\omega t - \phi) \tag{10.222}$$

$$\bar{P} = \tfrac{1}{2}Y\hat{V}^2 \tag{10.223}$$

If the electrical system is *current controlled*, then the current is the reference signal and a modification of Equations (10.121) through (10.124) apply, that is,

$$I(t) = \hat{I}e^{i\omega t}$$
$$V(t) = \hat{V}e^{i(\omega t + \psi)} \tag{10.224}$$

The electric voltage is expressed in terms of current and impedance, that is,

$$V(t) = Z(\omega)I(t) \quad \text{and} \quad \psi = \arg(Z), \quad \cos\psi = Z_R/|Z| \tag{10.225}$$

Since admittance and impedance are in reciprocal relation, $Y(\omega)\,Z(\omega) = 1$, the phase angles ψ and ϕ are related by the relations

$$\psi = -\phi, \quad \cos\psi = \cos\phi \tag{10.226}$$

In this case, the power expression of Equation (10.219) takes the *impedance-current* form, that is,

$$P(t) = \tfrac{1}{2}Z_R\hat{I}^2 - \tfrac{1}{2}|Z|\hat{I}^2\cos(2\omega t + \psi) \tag{10.227}$$

$$\bar{P} = \tfrac{1}{2}Z\hat{I}^2 \tag{10.228}$$

10.6.2.6.1 Active Power

Equation (10.219) defines a harmonic function $P(t)$ that contains a constant part, $\tfrac{1}{2}\hat{V}\hat{I}\cos\phi$, and a harmonic part, $\tfrac{1}{2}\hat{V}\hat{I}\cos(2\omega t + \phi)$, oscillating at twice the frequency ω of $V(\omega)$ and $I(\omega)$. The constant part of Equation (10.219), $\tfrac{1}{2}\hat{V}\hat{I}\cos\phi$, is called *active power* or *real* power; it can be obtained by taking the time average of power, that is,

$$P_{active} = P_{average} = \frac{1}{T}\int_0^T P(t)\mathrm{d}t = \tfrac{1}{2}\hat{V}\hat{I}\cos\phi = \tfrac{1}{2}Z_R\hat{I}^2 = \tfrac{1}{2}|Y|\hat{V}^2\cos\phi = \tfrac{1}{2}|Z|\hat{I}^2\cos\phi \tag{10.229}$$

In view of Equations (10.219) through (10.227), it is apparent that the active power is related to the real part of the admittance and impedance, that is,

$$P_{active} = \tfrac{1}{2}\mathrm{Re}(\bar{V}\bar{I}) = \tfrac{1}{2}\hat{V}\hat{I}\cos\phi = \tfrac{1}{2}Y_R\hat{V}^2 = \tfrac{1}{2}Z_R\hat{I}^2 \tag{10.230}$$

The fact that the active power is related to the real part of admittance and impedance explains why the active power is also called real power.

10.6.2.6.2 Reactive Power

The previous section has described the active power a.k.a. real power. A related concept is that of *reactive power* or *imaginary power*. This concept is related to the imaginary parts of admittance and impedance, that is,

$$P_{reactive} = \tfrac{1}{2}\mathrm{Im}(\bar{V}\bar{I}) = \tfrac{1}{2}Y_I\hat{V}^2 = \tfrac{1}{2}Z_I\hat{I}^2 \tag{10.231}$$

The concept of reactive power is associated with the reactive elements present in an electric circuit. The reactive power is not consumed, but is recirculated between the power supply and the actuator. In conventional electrical engineering applications, the users try to keep the reactive power as small as possible because the recirculation of large reactive power is associated with high losses in the transmission lines. In induced-strain actuator applications, the reactive power is actually the dominant factor, since the actuator impedance is dominated by its capacitive behavior, $i\omega C$. Managing high reactive power requirements is one of the challenges of using induced-strain actuators.

10.6.2.6.3 Power Rating

The power rating is usually defined as the product

$$P_{rating} = \tfrac{1}{2}\hat{V}\hat{I} \tag{10.232}$$

In view of Equation (10.220), the power rating is equal to the *complex power magnitude*, that is,

$$P_{rating} = |\bar{P}| = \tfrac{1}{2}\hat{V}\hat{I} = \tfrac{1}{2}|Y|\hat{V}^2 = \tfrac{1}{2}|Z|\hat{I}^2 \tag{10.233}$$

Comparing Equations (10.230), (10.231), and (10.233) yields the following relation between active power, reactive power, and power rating

$$P_{rating}^2 = P_{active}^2 + P_{reactive}^2 \tag{10.234}$$

10.6.2.6.4 Peak Power

Of great importance in sizing an ISA system is to know the maximum instantaneous power required by the system. Examination of Equation (10.219) reveals that it has a constant part, $\tfrac{1}{2}\hat{V}\hat{I}\cos\phi$, and oscillatory part, $\tfrac{1}{2}\hat{V}\hat{I}\cos(2\omega t + \phi)$. The constant part is associated with the power uniformly dissipated by the actuator, that is, the active power. This active power is due to internal mechanical and electrical losses inside the actuator as well as external damping losses in the mechanical load driven by the actuator.

The oscillatory part of Equation (10.219) represents power that flows in and out of the induced-strain actuator during its cyclic operation. Although its average value is zero, its instantaneous component can be quite high and hence it has to be taken into consideration when designing power supplies for induced-strain actuators. Overall, one is interested in finding the maximum instantaneous power, a.k.a., the *peak power*, P_{elec}^{peak}. Examination of Equation (10.219) reveals that

$$P_{elec}^{peak} = \tfrac{1}{2}\hat{V}\hat{I}(1 + \cos\phi) \tag{10.235}$$

Comparison of Equations (10.230), (10.231), and (10.235) yields

$$P_{max} = P_{elec}^{peak} = P_{rating} + P_{active} \tag{10.236}$$

10.6.2.6.5 Variation of Power with Frequency in the Presence of a Resonant External Load

Define the reference electrical power, P_{elec}^{ref}, as

$$P_{elec}^{ref} = \tfrac{1}{2}\omega_0 C_0 \hat{V}^2 \tag{10.237}$$

Using the reference power defined by Equation (10.237) as a scaling factor, we obtain nondimensional expressions for the active power and rating power, that is,

$$P'_{rating}(p,r) = \frac{P_{rating}(p,r)}{P_{elec}^{ref}} = \left| ip\left(1 - \kappa^2 \frac{\bar{r}(p,r)}{1 + \bar{r}(p,r)}\right) \right| \tag{10.238}$$

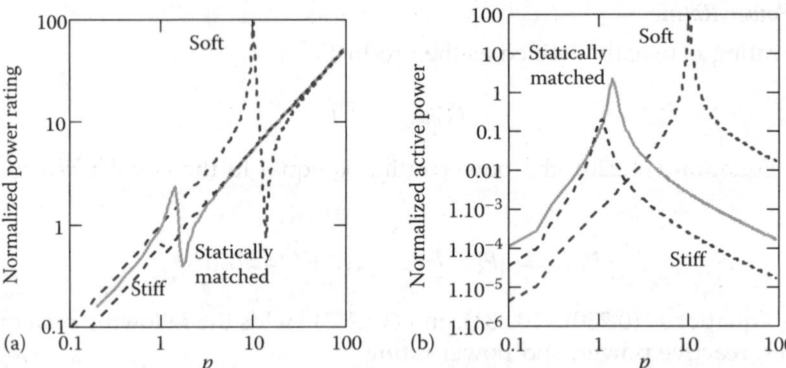

FIGURE 10.28
Normalized (a) power rating and (b) active power for three stiffness ratios: soft ($r = 0.01$), statically matched ($r = 1$), and stiff ($r = 10$).

$$P'_{active}(p, r) = \frac{P_{active}(p, r)}{P^{ref}_{elec}} = \mathrm{Re}\left[ip\left(1 - \bar{\kappa}^2 \frac{\bar{r}(p, r)}{1 + \bar{r}(p, r)}\right)\right] \tag{10.239}$$

Note that the independent variables in Equations (10.238) and (10.239) are the two non-dimensional parameters, (1) frequency ratio, $p = \omega/\omega_0$; and (2) the static stiffness ratio, r.

Figure 10.28 shows a plot of power rating and active power versus frequency for three stiffness ratio assumptions: (a) soft external system, (b) statically matched system; and (c) stiff external system. One notices the resonant-like behavior at certain frequency ratios that strongly depends on the static stiffness ratio. This resonance behavior is due to the electromechanical coupling between the induced-strain actuator and the external structure. The peak in active power consumption around this resonance is due to the large amount of energy transmitted externally in this narrow frequency band.

10.6.2.7 Electrical Power with Bias Voltage

The harmonic operation of the induced-strain actuator considered in Section 10.6.2.6 was assumed symmetrical about the zero-voltage axis. Actual induced-strain actuators do not display a symmetrical behavior when the polarity of the applied voltage is reversed under full-stroke operation. This is due to mechanical and electrical factors such as low tensional strength and depolarization under opposing field. A preferred polarity exists which generates the maximum expansion. Under reversed polarity, some limited contraction may be achieved. Hence, one needs to analyze an induced-strain actuator subjected to DC bias and, if required, a prestress spring. Dynamic operation takes place about a mid-range position, which is achieved by superposing a bias component onto the dynamic component. For biased operation, the applied voltage is given by Equation (10.139), that is,

$$V(t) = V_0 + \hat{V} \cos \omega t \tag{10.240}$$

where V_0 is the bias voltage and $\hat{V}$ is the dynamic voltage amplitude. The corresponding induced-strain displacement is given by Equation (10.140), that is,

$$u_{ISA}(t) = u_0 + \hat{u}_{ISA} \cos \omega t \tag{10.241}$$

where u_0 is the bias position and $\hat{u}_{ISA}$ is the dynamic displacement amplitude. The values V_0, $\hat{V}$, u_0, and $\hat{u}_{ISA}$ are calculated from manufacturers' specifications. Correspondingly, one has the electric current

$$I(t) = \hat{I} \cos(\omega t - \phi) \tag{10.242}$$

By definition, *electric power* = *voltage* $\times$ *current*, and hence,

$$P(t) = V(t) \cdot I(t) = (V_0 + \hat{V} \cos \omega t)\, \hat{I} \cos(\omega t - \phi) \tag{10.243}$$

Expansion of Equation (10.243) gives

$$\begin{aligned}
P(t) &= \tfrac{1}{2}\hat{V}\hat{I}\cos\phi + \tfrac{1}{2}\hat{V}\hat{I}\cos(2\omega t - \phi) + V_0\hat{I}\cos(\omega t - \phi) \\
&= P_{active} + P_{reactive}(t)
\end{aligned} \tag{10.244}$$

Comparing Equation (10.244) with Equation (10.219) we note that the active component of power is not influenced by the bias voltage and has the expression of Equation (10.230), that is,

$$P_{active} = \tfrac{1}{2}\hat{V}\hat{I}\cos\phi = \tfrac{1}{2}\operatorname{Re}\bar{P} = \tfrac{1}{2}Y_R\hat{V}^2 = \tfrac{1}{2}Z_R\hat{I}^2 \tag{10.245}$$

Substitution of Equation (10.213) into Equation (10.245) yields

$$P_{active} = \operatorname{Re}\left[i\omega\bar{C}_0\left(1 - \kappa^2\frac{\bar{r}(\omega)}{1 + \bar{r}(\omega)}\right)\right]\tfrac{1}{2}\hat{V}^2 \tag{10.246}$$

The reactive component of power has the expression

$$P_{reactive}(t) = \tfrac{1}{2}\hat{V}\hat{I}\cos(2\omega t - \phi) + V_0\hat{I}\cos(\omega t - \phi) \tag{10.247}$$

The influence of the bias voltage, V_0, is to significantly increase the reactive power component. Recall the complex power magnitude of Equation (10.220), that is,

$$|\bar{P}| = \tfrac{1}{2}\hat{V}\hat{I} = \tfrac{1}{2}|Y|\hat{V}^2 = \tfrac{1}{2}|Z|\hat{I}^2 \tag{10.248}$$

Using Equation (10.248) into Equation (10.247), one can factor out $|\bar{P}|$ and get

$$P_{reactive}(t) = |\bar{P}|[\cos(2\omega t - \phi) + 2v_0\cos(\omega t - \phi)] \tag{10.249}$$

where v_0 is the *bias voltage coefficient* defined as

$$v_0 = \frac{V_0}{\hat{V}} \tag{10.250}$$

Electroactive induced-strain actuators are mainly capacitive, and their phase angle is close to $-90°$. For $\phi = -90°$, Equation (10.249) becomes

$$P_{reactive}(t) = |\bar{P}|[\sin(2\omega t) + 2v_0\sin(\omega t)] \tag{10.251}$$

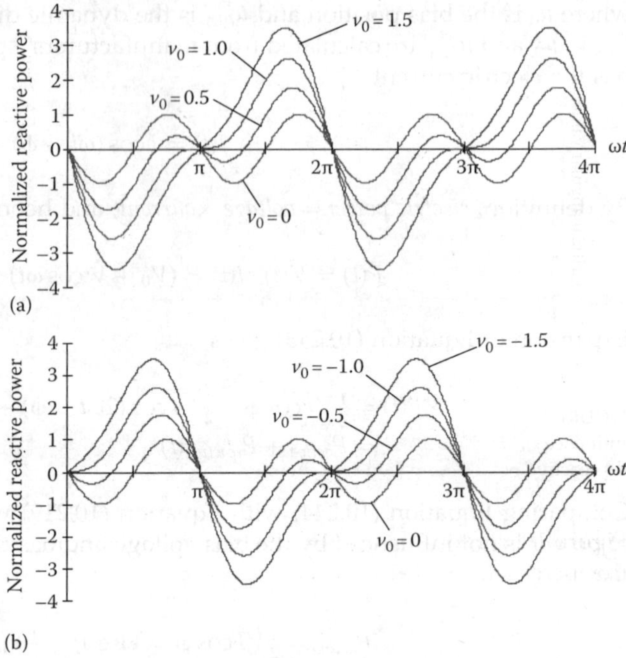

FIGURE 10.29
Variation of normalized reactive power for $\phi = -90°$: (a) positive bias coefficient; (b) negative bias coefficient.

Figure 10.29 presents the time-variation of the normalized reactive power for various values of the bias voltage coefficient, v_0. Note that, for $-1 < v_0 < 1$, the reactive power curve shows a pair of local maxima, of which only one is also global maximum, that is, the peak value of the reactive power per cycle. For $v_0 = 1$, the curve has a horizontal-tangent inflection point at $3\pi/2$. For $v_0 > 1$ and for $v_0 < -1$, the reactive power curve presents only one maximum since its behavior is dominated by the bias-voltage component. For ϕ slightly different from 90°, curves similar to those given in Figure 10.29 can be obtained.

To calculate the peak value per cycle of the reactive power, one can differentiate Equation (10.251) with respect to ωt and set the derivative to zero. The resulting quadratic equation has a pair of solutions of which only one corresponds to the peak reactive power. Thus, the peak reactive power takes place at the following critical angular values:

$$(\omega t)_{crt} = \cos^{-1}\left(-\frac{v_0}{4} + \frac{1}{4}\sqrt{v_0^2 + 8}\right), \quad \text{for } v_0 > 0 \tag{10.252}$$

$$(\omega t)_{crt} = \pi - \cos^{-1}\left(-\frac{v_0}{4} - \frac{1}{4}\sqrt{v_0^2 + 8}\right), \quad \text{for } v_0 < 0 \tag{10.253}$$

The variation of the critical angular values, ωt_{cr}, with the bias voltage coefficient is shown in Figure 10.30a. For positive bias $v_0 > 0$, ωt_{cr} starts at 45° and increases with the bias voltage coefficient, v_0; for negative bias $v_0 < 0$, ωt_{cr} starts at 135° and decreases with the bias voltage coefficient, v_0. This behavior is consistent with the shift of the power peaks in Figure 10.29. (For zero bias, both 45° and 135° are possible valid solutions; we choose to use 45°.) The normalized peak reactive power varies with v_0, as shown in Figure 10.30b. Using

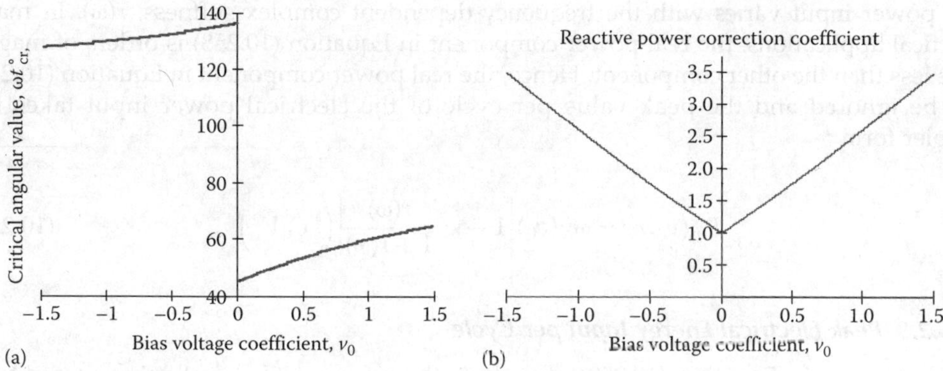

FIGURE 10.30
Influence of the bias voltage coefficient on the reactive power: (a) variation of the critical angular values, ωt_{cr}; (b) reactive power correction coefficient.

Figure 10.30b, one can define a reactive power correction factor, $\chi(v_0)$, which accounts for the increase in the peak reactive power due to the bias voltage operation. Thus,

$$P_{reactive}^{peak} = \chi(v_0)|\bar{P}| \tag{10.254}$$

Substitution of Equations (10.213) and (10.233) into Equation (10.254) yields

$$P_{reactive}^{peak} = \chi(v_0)\left|i\omega\bar{C}_0\left(1 - \bar{\kappa}^2\frac{\bar{r}(\omega)}{1 + \bar{r}(\omega)}\right)\right|\tfrac{1}{2}\hat{V}^2 \tag{10.255}$$

For a limited range of v_0, the curve shown in Figure 10.30b is almost linear, though its algebraic expression (not given here for brevity) is quite elaborate. A reasonable approximation for the reactive power correction factor in the range $-1.5 < v_0 < 1.5$ is given by the formula

$$\chi(v_0) \approx 1 + 1.62|v_0| \tag{10.256}$$

10.6.2.8 Peak Electrical Power with Bias Voltage

Combining Equations (10.245) and (10.254), we identify the peak electrical power with bias voltage as

$$P_{elec}^{peak} = P_{active} + P_{reactive}^{peak} = \mathrm{Re}\bar{P} + \chi(v_0)|\bar{P}| \tag{10.257}$$

Upon substitution of Equations (10.246) and (10.255) into Equation (10.257), we get the complete expression of the peak input power per cycle of an electroactive induced-strain actuator

$$P_{elec}^{peak}(v_0, \bar{r}(\omega)) = \mathrm{Re}\bar{P}(\omega) + \chi(v_0)|\bar{P}(\omega)|\tfrac{1}{2}\hat{V}^2$$
$$= \left\{\mathrm{Re}\left[i\omega\bar{C}_0\left(1 - \bar{\kappa}^2\frac{\bar{r}(\omega)}{1 + \bar{r}(\omega)}\right)\right] + \chi(v_0)\left|i\omega\bar{C}_0\left(1 - \bar{\kappa}^2\frac{\bar{r}(\omega)}{1 + \bar{r}(\omega)}\right)\right|\right\} \tag{10.258}$$

The power input varies with the frequency-dependent complex stiffness, $\bar{r}(\omega)$. In many practical applications, the real power component in Equation (10.258) is orders of magnitude less than the other component. Hence, the real power component in Equation (10.258) can be ignored and the peak value per cycle of the electrical power input takes the simpler form

$$P_{in}(v_0, r) = \omega \chi(v_0)\left|1 - \bar{\kappa}^2 \frac{\bar{r}(\omega)}{1 + \bar{r}(\omega)}\right|\left(\tfrac{1}{2}\bar{C}_0\hat{V}^2\right) \tag{10.259}$$

10.6.2.9 Peak Electrical Energy Input per Cycle

The last factor in Equation (10.259) represents the reference electrical energy amplitude given by

$$E_{elec}^{ref} = \tfrac{1}{2}\left|\bar{C}_0\right|\hat{V}^2 \tag{10.260}$$

Equation (10.260) can be easily calculated, for a given active-material stack, from manufacturers' specifications. Let us express Equation (10.259) in the form

$$P_{in}(v_0, \bar{r}) = \omega \cdot E_{elec}(v_0, \bar{r}) \tag{10.261}$$

where $E_{elec}(v_0, r)$ is the peak electrical energy per cycle given by

$$E_{elec}^{peak}(v_0, \bar{r}(\omega)) = \chi(v_0)\left|1 - \bar{\kappa}^2 \frac{\bar{r}(\omega)}{1 + \bar{r}(\omega)}\right|E_{elec}^{ref} \tag{10.262}$$

Equation (10.261) shows that the input power is directly dependent on the peak electrical energy per cycle, $E_{elec}(v_0, r)$. We have seen in the previous sections that the maximum output of mechanical energy is attained when the stiffness ratio is unity ($r = 1$). This condition is also known as the stiffness match point. At the stiffness match point, the peak electrical energy per cycle takes the value

$$E_{elec}^{r=1} = E_{elec}^{peak}(v_0, r)\big|_{r=1} = \chi(v_0)\left|1 - \frac{1}{2}\bar{\kappa}^2\right|E_{elec}^{ref} \tag{10.263}$$

The first factor in Equation (10.259) shows that the input power increases linearly with frequency, as expected for a predominantly reactive electrical load. The second and third factors in Equation (10.259) are modifiers that take into account the bias voltage effects and the external loading conditions, respectively. Equation (10.259) shows that, at a given frequency, the peak input power per cycle varies strongly with the stiffness ratio, r. Figure 10.31 gives a plot of the normalized peak input power versus stiffness ratio, r. It is apparent that the peak input power per cycle decreases as the stiffness ratio increases. For a fully blocked actuator ($r \to \infty$), the relative reduction in peak input power is maximum, and equal to κ^2. For practical values of κ, the power reduction can be as much as 50% for a fully blocked actuator. However, this reduction may not be of great practical significance since a fully blocked actuator has zero output displacement and hence does not deliver any mechanical power. Further discussion of these effects can be found in the subsequent section dealing with electromechanical power conversion efficiency.

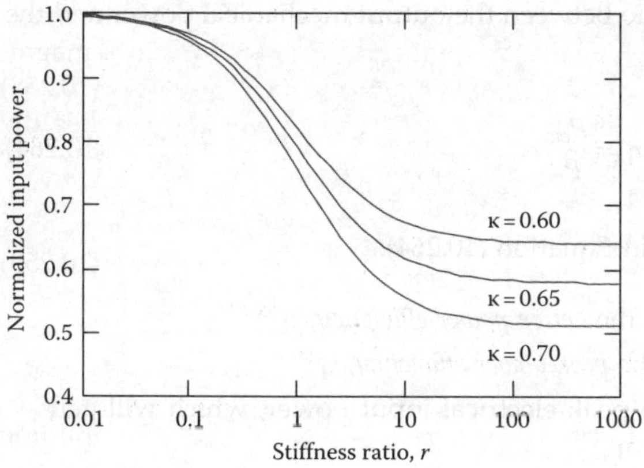

FIGURE 10.31
Variation of input power with stiffness ratio for three values of the electromechanical coupling coefficient, κ.

Example 10.12:

Consider again the piezoelectric actuator discussed in Example 10.1 made up of a stack of APC 850 square wafers with the properties given in Tables 10.2 and 10.3. Recall the stack length, l; static internal stiffness, k_i; stress-free capacitance, C_0; electromechanical coupling coefficient, κ; maximum voltage, V_{max}; calculated in Example 10.1, as well as other parameters calculated in previous examples. Find

1. Expressions for electrical complex power, $\bar{P}$, active power, P_{active}, bias-voltage power correction factor, $\chi(v_0)$ and peak power, P_{elec}^{peak}; their values when the actuator is operating at half the resonance of the external system, ω_0

   SOLUTION
   Since quasi-static assumption applies, the desired expressions are Equations (10.223), (10.230), (10.256), and (10.258). On calculation, we get $\bar{P}(\omega)|_{\omega=0.5\omega_0} = 4.127 + 274.5$ mW, $P_{active}(\omega)|_{\omega=0.5\omega_0} = 4.13$ mW, $\chi(v_0)|_{v_0=1} = 2.62$, $P_{elec}^{peak}(\omega)|_{\omega=0.5\omega_0} = 723.5$ mW.

2. Expression of peak electrical energy per cycle, E_{elec}^{peak}, and its values when the actuator is operating at half the resonance of the external system, ω_0

   SOLUTION
   Since quasi-static assumption applies, the desired expression is Equation (10.262). Upon calculation, we get $E_{elec}^{peak}(\omega)|_{\omega=0.5\omega_0} = 11.45$ mJ.

3. Expression and value of electrical energy at the stiffness match point, $E_{elec}^{r=1}$

   SOLUTION
   The desired expression is Equation (10.263). Upon calculation, we get $E_{elec}^{r=1} = 8.90$ mJ. ∎

10.6.3 Power Conversion Efficiency

Power conversion efficiency describes how efficient the conversion of electrical input power into mechanical output power is during dynamic operation of induced-strain actuators. The power conversion efficiency is the correspondent of the energy conversion efficiency discussed in Section 10.5.4 for static operation. The electromechanical conversion efficiency of the system consisting of the electroactive induced-strain actuator and the mechanical load can be defined as the ratio between the output mechanical power and the input electrical power, that is,

$$\eta = \frac{P_{out}}{P_{in}} \tag{10.264}$$

Several power expressions may be used in Equation (10.264):

1. Active powers, which will generate the *active power efficiency*, η^{active}
2. Peak powers, which will generate the *peak power efficiency*, η^{peak}
3. Active output mechanical power to peak electrical input power, which will generate the *active/peak power efficiency*, η^{active}_{peak}

The choice of appropriate efficiency expression depends to a large extent on the power supply used to energize the actuator. Specialized power supplies, purposely-built for energizing highly reactive loads, may be able to handle reactive powers efficiently and recirculate the reactive energy that flows in and out of the induced-strain actuator; in this case, the power conversion efficiency may be evaluated based on the ratio between the active power transmitted to the mechanical load divided by the active electrical power injected in the actuator. Alternatively, one can consider the ratio between the peak mechanical power applied to the mechanical load and the peak electrical power. Conventional linear power supplies are unable to handle reactive loads efficiently, that is, they cannot recirculate the high reactive power sent into the actuator at the upstroke. In this case, the relevant efficiency may be the ratio between the active mechanical power transmitted to the mechanical load and the peak power that has to be supplied to the actuator during each cycle. In this case, a ratio between the active mechanical power and peak electrical power may be more relevant. Let us consider each of these cases in turn.

10.6.3.1 Active Power Conversion Efficiency

Consider the active (average) mechanical and electrical power expressions given by Equations (10.186) and (10.246). Substitution into Equation (10.264) yields

$$\eta^{active} = \frac{P^{avg}_{mech}}{P^{avg}_{elec}} = \frac{\mathrm{Re}\left(\dfrac{\tilde{\bar{r}}(\omega)}{1 + \tilde{\bar{r}}(\omega)} \dfrac{i\omega}{1 + \bar{r}(\omega)}\right)\left(\frac{1}{2} k_i \hat{u}^2_{ISA}\right)}{\mathrm{Re}\left[i\omega\bar{C}_0\left(1 - \bar{\kappa}^2 \dfrac{\bar{r}(\omega)}{1 + \bar{r}(\omega)}\right)\right]\frac{1}{2}\hat{V}^2} \tag{10.265}$$

Equation (10.265) shows that the power conversion efficiency varies with the frequency-dependent complex stiffness ratio, $\bar{r}(\omega)$. For low-damping mechanical systems driven well below the mechanical resonance frequency, the complex stiffness ratio is predominantly real, that is, $\bar{r} \approx r$ and the instantaneous mechanical power becomes dominantly reactive. In this case, active (average) mechanical power vanishes and the conversion efficiency defined by Equation (10.265) also vanishes.

In a more general sense, Equation (10.265) represents the ratio of the mechanical energy dissipated due to mechanical damping divided by the electrical energy dissipated due to electrical losses.

10.6.3.2 Peak Power Conversion Efficiency

Consider the peak mechanical and electrical power expressions given by Equations (10.188) and (10.255); substitution into Equation (10.264) yields

$$\eta^{peak} = \frac{\left|P_{mech}^{peak}\right|}{\left|P_{elec}^{peak}\right|} = \frac{\omega \left|\dfrac{\tilde{\tilde{r}}(\omega)}{[1+\bar{r}(\omega)][1+\tilde{\tilde{r}}(\omega)]}\right|(1+\cos\phi_{mech})E_{mech}^{ref}}{\chi(v_0)\left|i\omega\bar{C}_0\left[1-\bar{\kappa}^2\dfrac{\bar{r}(\omega)}{1+\bar{r}(\omega)}\right]\right|\frac{1}{2}\hat{V}^2} \tag{10.266}$$

where $\cos\phi_{mech}$ is given by Equation (10.183), that is,

$$\cos\phi = \mathrm{Re}\left(\frac{\tilde{\tilde{r}}(\omega)}{1+\tilde{\tilde{r}}(\omega)}\frac{i\omega}{1+\bar{r}(\omega)}\right)\bigg/\left|\frac{\tilde{\tilde{r}}(\omega)}{1+\tilde{\tilde{r}}(\omega)}\frac{i\omega}{1+\bar{r}(\omega)}\right| \tag{10.267}$$

Upon simplification, Equation (10.266) yields

$$\eta^{peak} = \left|\frac{\bar{\kappa}^2\tilde{\tilde{r}}(\omega)}{[1+\bar{r}(\omega)(1-\bar{\kappa}^2)][1+\tilde{\tilde{r}}(\omega)]}\right|\frac{(1+\cos\phi_{mech})}{\chi(v_0)} \tag{10.268}$$

The power conversion efficiency varies with the frequency-dependent complex stiffness ratio, $\bar{r}(\omega)$. For low-damping mechanical systems driven well below the mechanical resonance frequency, the complex stiffness ratio is predominantly real, that is, $\bar{r} \approx r$, $\cos\phi_{mech} \approx 0$. Thus, the peak power conversion efficiency takes the simpler form

$$\eta^{peak} = \frac{\kappa^2 r}{\chi(v_0)[1+r(1-\kappa^2)](1+r)} \tag{10.269}$$

A plot of Equation (10.269) is given in Figure 10.32. It is apparent that, as the stiffness ratio increases, the conversion efficiency also increases at first up to a maximum, and then decreases.

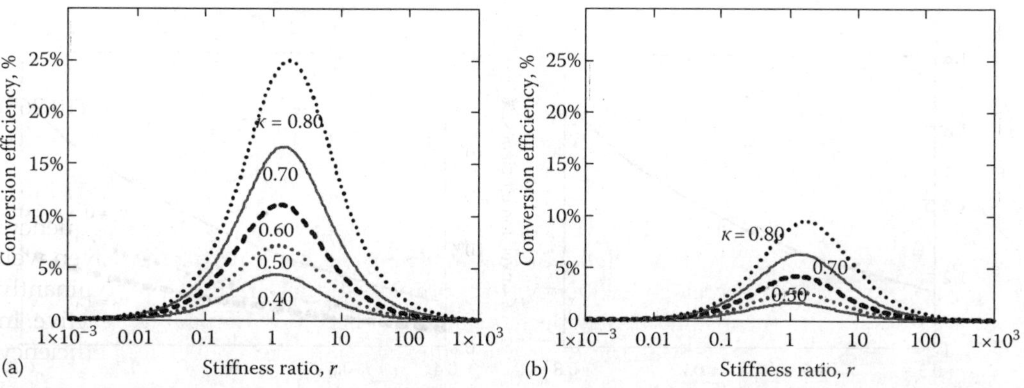

FIGURE 10.32
Variation of electromechanical power conversion efficiency with r and κ: (a) no bias voltage, $v_0 = 0$; (b) with bias voltage, $v_0 = 1$.

The peak-power conversion efficiency has a maximum at the stiffness ratio value r_η, which is close to $r = 1$ but not exactly it. This stiffness ratio, r_η, is an optimal stiffness ratio that maximizes the power conversion efficiency; its value can be found by setting to zero the derivative of the conversion efficiency with respect to r. Upon calculation, we obtain

$$r_\eta(\kappa) = \frac{1}{\sqrt{1 - \kappa^2}} \tag{10.270}$$

Note that the optimal stiffness ratio, r_η, varies with the electromechanical coupling coefficient, κ. For $\kappa = 0$, the optimal stiffness ratio is $r_\eta = 1$, that is, it coincides with the optimal stiffness ratio for maximum power output. However, for $\kappa = 0$, Equation (10.269) indicates that the conversion efficiency is exactly zero. Figure 10.33a shows the variation of optimal stiffness ratio, r_η, with electromechanical coupling coefficient, κ. It can be seen that practical values of the optimal stiffness ratio, r_η, are found in the range 1.25–1.4. Substitution of Equation (10.270) into Equation (10.269) gives the best conversion efficiency, η_{max}, in terms of the electromechanical coupling coefficient, κ, and the bias voltage coefficient, v_0, that is,

$$\eta_{max}(\kappa, v_0) = \frac{1}{\chi(v_0)} \frac{\kappa^2}{\left(1 + \sqrt{1 - \kappa^2}\right)^2} \tag{10.271}$$

Plots of Equation (10.271) are given in Figure 10.33b. For $v_0 = 0$ (i.e., operation without bias voltage), the best conversion efficiency may vary from 4% to 26%. As bias voltage is applied, the conversion efficiency decreases; for $v_0 = 1$ (i.e., when bias voltage equals dynamic voltage amplitude), the electromechanical conversion efficiency varies between 2% and 10%.

The maximum conversion efficiency given by Equation (10.271) can be compared with the *conversion efficiency at the stiffness match point*, which is obtained by putting $r = 1$ into Equation (10.269), that is,

$$\eta_{r=1}(\kappa, v_0) = \frac{\kappa^2}{2(2 - \kappa^2)\chi(v_0)} \tag{10.272}$$

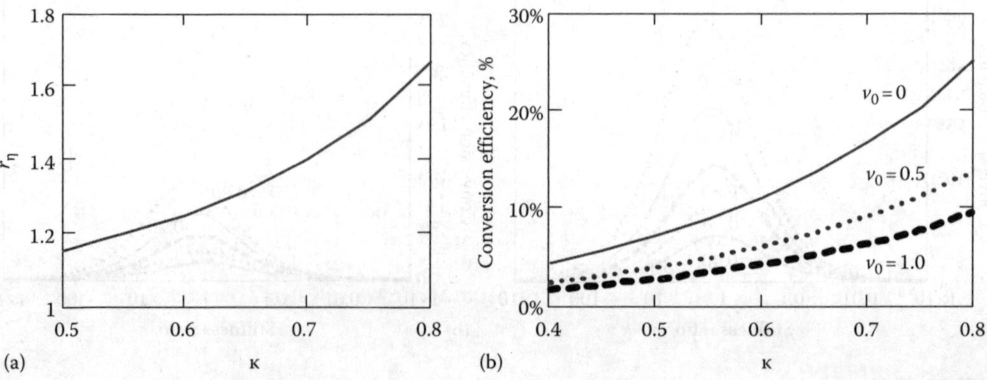

FIGURE 10.33
(a) Variation of r_η with κ; (b) Variation of electromechanical power conversion efficiency with κ and v_0.

Note that the maximum conversion efficiency, η_{max}, and the conversion efficiency at the stiffness match point, $\eta_{r=1}$, are different. In design applications, either one of the two efficiencies can be optimized, but not both. In other words, one can either design for maximum power output by imposing the stiffness match condition ($r = 1$), or can design for maximum conversion efficiency by choosing $r_\eta = 1/\sqrt{1 - \kappa^2}$. However, in some practical applications, the numerical difference between the two results may be small.

10.6.3.3 Peak Electrical Power to Active Mechanical Power Conversion Efficiency

Consider the active (average) mechanical power given by Equation (10.186) and the peak electrical power given by Equation (10.255). Substitution into Equation (10.264) yields

$$\eta_{peak}^{active} = \frac{P_{mech}^{avg}}{P_{elec}^{peak}} = \frac{\text{Re}\left(\dfrac{\tilde{\tilde{r}}(\omega)}{1 + \tilde{\tilde{r}}(\omega)} \dfrac{i\omega}{1 + \bar{r}(\omega)}\right)\left(\frac{1}{2} k_i \hat{u}_{ISA}^2\right)}{\chi(v_0)\left|i\omega\bar{C}_0\left(1 - \bar{\kappa}^2 \dfrac{\bar{r}(\omega)}{1 + \bar{r}(\omega)}\right)\right|\frac{1}{2}\hat{V}^2} \tag{10.273}$$

Upon simplification, we obtain

$$\eta_{peak}^{active} = \bar{\kappa}^2 \frac{\text{Re}\left(\dfrac{\tilde{\tilde{r}}(\omega)}{1 + \tilde{\tilde{r}}(\omega)} \dfrac{i}{1 + \bar{r}(\omega)}\right)}{\chi(v_0)\left|1 - \bar{\kappa}^2 \dfrac{\bar{r}(\omega)}{1 + \bar{r}(\omega)}\right|} \tag{10.274}$$

For low-damping mechanical systems driven well below the mechanical resonance frequency, the complex stiffness ratio is predominantly real, that is, $\bar{r} \approx r$ and the instantaneous mechanical power becomes dominantly reactive. In this case, active mechanical power vanishes and the conversion efficiency defined by Equation (10.274) vanishes too.

Example 10.13:

Consider again the piezoelectric actuator discussed in Example 10.1 made up of a stack APC 850 square wafers with the properties given in Tables 10.2 and 10.3. Recall the stack length, l; static internal stiffness, k_i; stress-free capacitance, C_0; electromechanical coupling coefficient, κ; maximum voltage, V_{max}; calculated in Example 10.1, as well as other parameters calculated in previous examples. Find

1. Expression of peak power conversion efficiency, η^{peak}. Values of peak power conversion efficiency when the actuator is operating at half the resonance of the external system, ω_0, and in two bias voltage conditions: (a) no bias voltage; (b) with bias voltage $V_0 = \hat{V}$. Discuss the effect of bias voltage on peak power conversion efficiency. Repeat using the simplified expression for low-damping systems. Discuss the effect of using the simplified formula.

SOLUTION
The desired expression is Equation (10.268). The two values of bias voltage are (a) $v_0 = 0$ and (b) $v_0 = 1$. Using Equation (10.256), we get the following values for the bias-voltage

power correction factor: (a) $\chi(0) = 1$, and (b) $\chi(1) \approx 2.62$. Upon substitution in Equation (10.268) we get (a) $\eta^{peak}|_{\omega=0.5\omega_0;v_0=0} = 9.203\%$, and (b) $\eta^{peak}|_{\omega=0.5\omega_0;v_0=1} = 3.512\%$. One notices that the consideration of bias voltage results in a reduction in the power conversion efficiency due to increase in the peak electric power demand.

The process is repeated with the simplified Equation (10.269). In this case, we get (a) $\eta^{peak}|_{\omega=0.5\omega_0;v_0=0} = 9.026\%$, and (b) $\eta^{peak}|_{\omega=0.5\omega_0;v_0=1} = 3.445\%$. We note a small but rather insignificant change in values in comparison with using the full complex Equation (10.268). This is consistent with the fact that the external load is a load-damping system

2. Expression of the optimal stiffness ratio for optimal peak power conversion r_η and of the corresponding best conversion efficiency, $\eta_{\max}(v_0)$. Values of these variables in two bias voltage conditions: (a) no bias voltage and (b) with bias voltage $V_0 = \hat{V}$. Discuss results.

SOLUTION

The desired expressions are Equations (10.270) and (10.271). Using Equation (10.270), we get $r_\eta = 1.576$; this value does not depend on the bias voltage situation. Using Equation (10.271), we get (a) $\eta_{\max}|_{v_0=0} = 22.347\%$ and (b) $\eta_{\max}|_{v_0=1} = 8.529\%$. When these values are compared with the results obtained above, it becomes apparent that the use of the optimum stiffness ratio would give much better conversion efficiency than the stiffness ratio of this particular application.

3. Expression of the peak power conversion efficiency at stiffness match, $\eta_{r=1}(v_0)$. Values of these variables in two bias voltage conditions: (a) no bias voltage and (b) with bias voltage $V_0 = \hat{V}$. Discuss results.

SOLUTION

The desired expression is Equation (10.272). Using Equation (10.272), we get (a) $\eta_{r=1}|_{v_0=0} = 21.284\%$, and (b) $\eta_{r=1}|_{v_0=1} = 8.124\%$. When these values are compared with the results obtained above, it becomes apparent that the difference between using the optimum stiffness ratio r_η and using the stiffness match, $r = 1$, is only slight. ■

10.7 Energy-Based Comparison of Induced-Strain Actuators

10.7.1 Data Collection

A large variety of induced-strain actuators (ISA devices) is presently available in the commercial market. A study was conducted to collect performance data from the actuator vendors and manufacturers. A template of relevant input data was drafted. Data regarding both, the induced-strain actuator and the active material inside the actuator, was sought. Such data falls broadly into two categories: general data of the ISA device and data about the active material (PZT, PMN, etc.). Data in each of these categories is as follows:

1. General data of the ISA device

   a. Manufacturer (name, address, fax/phone, contact point)

   b. Device identification number

   c. Description

   d. Maximum (free-stroke) displacement, mm

   e. Maximum force, N

   f. Stiffness, kN/mm

   g. Length, mm

   h. Outside diameter, mm (or width and thickness, for rectangular cross-section, mm × mm)

   i. Mass, kg

   j. Volume, cm^3

   k. Voltage, V, or current, A, as appropriate

   l. Capacitance, μF, or inductance, mH, as appropriate

   m. Price, $

2. Data about the active material (PZT, PMN, etc.)

   a. Active material diameter, mm (or width and thickness, for rectangular cross-section, mm × mm)

   b. Active material length, mm

   c. For stacked actuators, the layer thickness, mm, and the number of layers

   d. Nonlinearity index, or a representative curve of the output displacement against electrical input

This input data was used to calculate certain representative characteristics of the ISA device:

- Apparent free strain, %
- Apparent volume, cm^3
- Apparent density, $10^3 \times kg/m^3$
- Maximum deliverable energy per unit mass J/kg
- Maximum deliverable energy per unit volume, J/cm^3
- Maximum deliverable energy per unit cost, J/$1000

Out of a large number of entries, a few representative actuators were selected. Both top performers and lower performers were selected. However, it should be noted that the data is as obtained from the manufacturers. We ran some consistency checks, but otherwise we had no way to independently verify the data. Table 10.4 presents the model numbers of the induced-strain actuators selected for analysis, sorted by manufacturer.

Nonlinear range is used by some manufacturers to get enhanced performance from the induced-strain actuators. These situations were dealt with by making additional entries in Table 10.4 with the model number suffixed with (NL). In the nonlinear range, the piezo-ceramics undergo large domain rotations, experience larger hysteresis, but may have shorter fatigue life. For the models quoted in the table, endurance of 10^7 cycles without failure was proven experimentally on a few specimens.

The basic data for these selected induced-strain actuators is given in Table 10.5. Model number, active material type, expansion and contraction voltages, maximum expansion and contraction, electric capacitance, stiffness, maximum force, and price are given. The maximum force entered in this table may represent either the blocked force or the maximum

Micromechatronics: Modeling, Analysis, and Design with MATLAB®

TABLE 10.5

Basic Data for Selected Induced-Strain Actuators

Model	Active Material Type V^+ (V)	Expansion Voltage V^- (V)	Contraction Voltage u_{ISA+} (μm)	Max. Expansion u_{ISA-} (μm)	Max. Contraction (μF)	Electric Capacitance k_i^* (kN/mm)	Small-Amplitude Stiffness k_i (kN/mm)	Full-Stroke Stiffness F (N)	Max. Force[a] ($)	Price
P-245.70	HVPZT	−1,000	250	120	−30.0	0.45	8.0	5.3	2,000	2,010
P-246.70	HVPZT	−1,000	250	120	−30.0	3.28	190.0	126.7	12,500	4,975
P-247.70	HVPZT	−1,000	250	120	−30.0	6.56	370.0	246.7	30,000	7,900
P-844.60	LVPZT	−100	25	90	−22.5	43.00	33.0	22.0	3,000	4,665
E100P-2	PMN-EC98	800	0	18	0	0.060	100.0	66.7	1,500	280
E200P-3	PMN-EC98	800	0	40	0	0.240	100.0	66.7	3,500	385
E300P-4	PMN-EC98	800	0	66	0	1.600	160.0	106.7	9,000	675
E400P-3	PMN-EC98	800	0	40	0	0.580	250.0	166.7	8,600	430
D125160	PZWT100	1,000	0	210	0.0	9.82	153.2	102.2	28,000	4,400
D125200	PZWT101	1,000	0	262	0.0	12.78	122.6	81.7	28,000	5,280
A125160	PZWT100	1,000	0	210	0.0	9.82	153.2	102.2	28,000	4,980
A125200	PZWT101	1,000	0	262	0.0	12.78	122.6	81.7	28,000	5,880
AE1010D16	LVPZT	150	0	18.4	0	5.4	190.2	126.8	3,500	366
ASB171C801	LVPZT	150	0	170	0	15	5.5	3.7	800	588
PAHL 120/20	LVPZT	150	−10	120	−8	42	30	20.0	3500	2344

[a] "Max. force" may represent either the "blocked force" or the "max. allowable compression force," depending on vendor.

allowable compression force, depending on vendor. The stiffness values entered in this table fall into two categories: (a) small-amplitude stiffness and (b) full-stroke stiffness. The small-amplitude stiffness values apply to high-frequency small-amplitude utilization of induced-strain actuators, such as in ultrasonic devices and sonars. The active material expands and contracts with small amplitudes about its unbiased zero position. In this case, the material has a high gradient of the force-displacement, that is, a high stiffness. Full-stroke stiffness values apply to low-frequency quasi-static operation of the induced-strain actuators, such as in nano positioning, and mechanical actuation applications. In this case, large amplitude displacements are encountered. Operation can take place from zero, or about a biased non-zero neutral position. Because of the large amplitudes involved in this type of operation, the force–displacement curve is nonlinear. For a nonlinear force–displacement curve, the stiffness varies considerably, and can become considerably smaller at the maximum force–displacement points. The full-stroke stiffness values were considered in this study. However, both small-amplitude and full-stroke stiffness values were entered in Table 10.5. When only the small-amplitude stiffness was provided by the manufacturer, the full-stroke stiffness had to be evaluated. In this case, the high-amplitude stiffness was evaluated using a 50% reduction factor with respect to the small-amplitude stiffness.

Table 10.6 presents the basic data for the active-material stacks of the selected induced-strain actuators. Model number, active material type, number of layers in the stack, layer thickness, outside and inside diameters, length, cross-sectional area, volume, and mass are given.

A number of the actuators included in our study could be delivered with casing and prestress mechanism. The prestress spring ensures that the net load seen by the active material is only compressive. This precaution is important for active materials that resist better in compression than in tension, such as the piezoelectric/electrostrictive ceramics and the magnetostrictive metallic compounds. If an induced-strain actuator does not include a prestressing spring, care must be taken in the application to ensure that only compressive loads are applied. This can be achieved through an external prestressing of the actuators. For these actuators, the basic data must also include the overall length, diameter, volume, and mass of the actuator. The data for these actuators is given in Table 10.7. Note that the overall dimensions, volume, and mass of the complete actuators are considerably larger than those for the active material alone. It is expected that this aspect will make the energy density of the actuator with casing and prestress mechanism lower than that of the active material alone.

10.7.2 Results

10.7.2.1 Data Reduction

The collected data was processed using the theory developed in this chapter. The following results were calculated:

- Effective electromechanical coupling coefficient, κ
- Maximum output energy (mechanical energy), E_e^{max}
- Required input energy (electrical energy), E_{elec}
- Energy conversion efficiency, η
- Volume-based energy density, defined as reference energy per unit volume
- Mass-based energy density defined as reference energy per unit mass
- Cost-based energy density defined as reference energy per unit cost

TABLE 10.6

Basic Data for the Active-Material Stacks of the Selected Induced-Strain Actuators

Model	Active Material Type	Number of Layers N	Layer Thickness t (mm)	Active Material Outside Diameter D_o (mm)	Active Material Inside Diameter D_i (mm)	Active Material Length (mm)	Active Material Density ρ_{ISA} (kg/m³)	Active Material Area A (mm²)	Active Material Volume V_{ISA} (mm³)	Active Material Mass m_{ISA} (g)
P-245.70	HVPZT	183	0.500	10	0	100	7,800	78.5	7,850	61
P-246.70	HVPZT	178	0.500	25	0	100	7,800	490.6	49,063	383
P-247.70	HVPZT	178	0.500	35	0	100	7,800	961.6	96,163	750
P-844.60	LVPZT	999	0.110	9.9 × 10.9	0	109.9	7,800	100	10,990	86
E100P-2	PMN	28	0.500	8	0	14	7,850	50.2	703	9
E200P-3	PMN	56	0.500	12	5	28	7,850	93.4	2,616	30
E300P-4	PMN	92	0.500	19	0	46	7,850	283.4	13,036	140
E400P-3	PMN	56	0.500	22	13	28	7,850	247.3	6,924	113
D125160	PZWT100	280	0.500	31.75	0	144.78	7,000	791.3	114,569	802
D125200	PZWT100	350	0.500	31.75	0	180.34	7,000	791.3	142,708	999
A125160	PZWT100	280	0.500	31.75	0	144.78	7,000	791.3	114,569	802
A125200	PZWT100	350	0.500	31.75	0	180.34	7,000	791.3	142,708	999
AE1010D16	LVPZT	N/A	N/A	11.5×11.5	N/A	20	8,000	132.3	2,645	21
ASB171C801	LVPZT	N/A	N/A	5×5	N/A	200	8,000	25.0	5,000	40
PAHL 120/20	LVPZT	1180	0.1	9	0	118	7,800	64	2,950	59

TABLE 10.7

Overall Dimensions and Mass for Selected Induced-Strain
Actuators with Casing and Prestress Springs

Model S	Overall Diameter (mm)	Overall Length (mm)	Overall Mass m (g)
P-245.70	18	125	154
P-246.70	39.8	140	830
P-247.70	50	142	980
P-844.60	20	137	215
A125160	50.8	196	1163
A125200	38.1	233	1450
ASB171C801	19.6	213	500
PAHL 120/20	20	126	228

- Energy-based price defined as the cost of a unit of reference energy
- Reactive power correction factor for bias voltage operation, $\chi(v_0)$
- Best stiffness ratio for maximum energy conversion with bias voltage, r_η
- Best energy conversion efficiency for operation with bias voltage, $\eta_{\max}$

These results were calculated for both static and dynamic operations. The reactive power correction factor for bias voltage operation, $\chi(v_0)$, the best stiffness ratio for maximum energy conversion with bias voltage, r_η, and the best energy conversion efficiency for operation with bias voltage, $\eta_{\max}$, apply only to dynamic operation.

10.7.2.2 Effective Full-Stroke Electromechanical Coupling Coefficient

The effective full-stroke electromechanical coupling coefficient was calculated with the simple formula developed in Section 10.4.2, that is,

$$\kappa^2 = \frac{d_{33}^2}{s\varepsilon} = \frac{k_i u_{ISA}^2}{C_0 V^2} \tag{10.275}$$

where u_{ISA} is the actuator free stroke (maximum excursion), k_i is the actuator internal stiffness, C is the actuator capacitance, and V is the voltage that must be applied to the actuator to obtain the free stroke u_{ISA}. It is apparent that the effective full-stroke electromechanical coupling coefficient depends on the electrical parameters, C and V, and the mechanical parameters, u_{ISA} and k_i. As already discussed in previous sections, the full-stroke behavior of the present-day active materials is significantly nonlinear. For example, the proportionality between induced displacement and applied voltage diminishes as the voltage increases, resulting in a leveling off of the response at high voltage levels. Similarly, the proportionality between the applied force and elastic compression increases as the force increases, resulting in an apparent "softening" of the actuator at high stress levels. As much as 50% stiffness reduction has been reported. Similar effects are observed in the electrical capacitance. In our study, we are applying a *secant linearization* scheme, whereby the full-stroke values are used. This allows us to compute the effective full-stroke electromechanical coupling coefficient using Equation (10.275). The results of these calculations are given in Table 10.8.

TABLE 10.8

Electromechanical Parameters for Selected Induced-Strain Actuators

		Static Operation		Dynamic Operation		
Model	Electromechanical Coupling Coefficient κ	Energy Conversion Efficiency at Stiffness Match η (%)	Reactive Power Correction Factor $\chi(V_0)$	Power Conversion Efficiency with Bias Voltage at Stiffness Match η' (%)	Best Stiffness Ratio for Maximum Power Conversion with Bias Voltage r_η	Best Power Conversion Efficiency with Bias Voltage η_{max} (%)
P-245.70	0.413	4.66	1.97	2.37	1.10	2.37
P-246.70	0.746	19.26	1.97	9.77	1.50	10.16
P-247.70	0.736	18.56	1.97	9.41	1.48	9.76
P-844.60	0.644	13.07	1.97	6.63	1.31	6.74
E100P-2	0.750	19.57	2.62	7.47	1.51	7.78
E200P-3	0.833	26.60	2.62	10.15	1.81	10.99
E300P-4	0.674	14.67	2.62	5.60	1.35	5.73
E400P-3	0.848	28.03	2.62	10.70	1.88	11.70
D125160	0.677	14.88	2.62	5.68	1.36	5.81
D125200	0.663	14.06	2.62	5.37	1.34	5.48
A125160	0.677	14.88	2.62	5.68	1.36	5.81
A125200	0.663	14.06	2.62	5.37	1.34	5.48
AE1010D16	0.594	10.73	2.62	4.09	1.24	4.14
ASB171C801	0.560	9.31	2.62	3.55	1.21	3.59
PAHL 120/20	0.552	8.99	2.42	3.72	1.20	3.75

10.7.2.3 Maximum Energy Output Capabilities

A focus of our study was to determine how induced-strain actuators can be best used in mechanical application. The output of the induced-strain actuator is mechanical energy. The maximum output energy, E_e^{max}, was calculated from the data using the formula

$$E_e^{max} = \tfrac{1}{4}\left(\tfrac{1}{2}k_i u_{ISA}^2\right) \tag{10.276}$$

where k_i is the internal stiffness of the actuator and u_{ISA} is the maximum displacement (free stroke). When the induced-strain actuator is operated in dynamic regime, the energy of Equation (10.276) represents the peak energy per cycle. In this case, u_{ISA} represents the dynamic displacement amplitude. Results of these calculations for static and dynamic operation are given in Tables 10.9 and 10.10, respectively.

10.7.2.4 Active Material Energy Density

Tables 10.9 and 10.10 present the mechanical performance of selected induced-strain actuators under static and dynamic operations, respectively. The input electrical energy, output mechanical energy, energy conversion efficiency, and energy densities per unit

TABLE 10.9

Static Performance of Selected Induced-Strain Actuators

Model	Voltage[a] V (V)	Stroke u_{ISA} (μm)	Induced Strain[a] %	Input Electrical Energy[a] E_{elec} (J)	Output Mechanical Energy[b] E_e (J)	Energy Conversion Efficiency η (%)	Output Energy per Active Material Volume E_e/V_{ISA} (J/dm^3)	Output Energy per Active Material Mass E_e/m_{ISA} (J/kg)	Output Energy per Unit Cost E_e/Price (mJ/\$1000)	Cost of Energy Unit Price/E_e (\$/mJ)
P-245.70	−1250	150	0.150	0.3216	0.0150	4.7	1.911	0.245	7.5	\$134.00
P-246.70	−1250	150	0.150	1.8500	0.3563	19.3	7.261	0.931	71.6	\$13.96
P-247.70	−1250	150	0.150	3.7375	0.6938	18.6	7.214	0.925	87.8	\$11.39
P-844.60	−125	112.5	0.102	0.2663	0.0348	13.1	3.167	0.406	7.5	\$134.03
E100P-2	800	18	0.129	0.0138	0.0027	19.6	3.839	0.300	9.6	\$103.70
E200P-3	800	40	0.143	0.0501	0.0133	26.6	5.098	0.444	34.6	\$28.88
E300P-4	800	66	0.143	0.3958	0.0581	14.7	4.455	0.415	86.0	\$11.62
E400P-3	800	40	0.143	0.1189	0.0333	28.0	4.814	0.295	77.5	\$12.90
D125160	1000	210	0.145	3.7838	0.5631	14.9	4.915	0.702	128.0	\$7.81
D125200	1000	262	0.145	4.9876	0.7012	14.1	4.914	0.701	132.8	\$7.53
A125160	1000	210	0.145	3.7838	0.5631	14.9	4.915	0.702	113.1	\$8.84
A125200	1000	262	0.145	4.9876	0.7012	14.1	4.914	0.701	119.3	\$8.39
AE1010D16	150	18.4	0.092	0.0500	0.0054	10.7	2.029	0.254	14.7	\$68.20
ASB171C801	150	170	0.085	0.1423	0.0132	9.3	2.649	0.331	22.5	\$44.39
PAHL 120/20	160	128	0.102	0.4557	0.0360	7.9	3.886	0.498	15.4	\$65.11

[a] Total excursion.

[b] Calculated at stiffness match condition.

TABLE 10.10

Dynamic Performance of Selected Induced-Strain Actuators

Model	Bias Voltage V_0 (V)	AC Voltage V (V)	Midpoint Position u_0 (μm)	Dynamic Displacement u_{ISA} (μm)	Dynamic Induced Strain S_{ISA} (%)	Input Electrical Energy E_{elec} (J)	Output Mechanical Energy[a] E_e (J)	Power Conversion Efficiency η(%)	Output Energy per Active Material Volume E_e/V_{ISA} (J/dm^3)	Output Energy per Active Material Mass E_e/m_{ISA} (J/kg)	Output Energy per Unit Cost E_e/Price (mJ/\$1000)	Cost of Energy Unit Price/E_e (\$/mJ)
P-245.70	−375	−625	45	75	0.075	0.1585	0.0038	2.4	0.478	0.061	1.9	\$536.00
P-246.70	−375	−625	45	75	0.075	0.9121	0.0891	9.8	1.815	0.233	17.9	\$55.86
P-247.70	−375	−625	45	75	0.075	1.8426	0.1734	9.4	1.804	0.231	22.0	\$45.55
P-844.60	−37.5	−62.5	33.75	56.25	0.051	0.1313	0.0087	6.6	0.792	0.102	1.9	\$536.13
E100P-2	400	400	9	9	0.064	0.0090	0.0007	7.5	0.960	0.075	2.4	\$414.81
E200P-3	400	400	20	20	0.071	0.0328	0.0033	10.2	1.274	0.111	8.7	\$115.50
E300P-4	400	400	33	33	0.072	0.2593	0.0145	5.6	1.114	0.104	21.5	\$46.49
E400P-3	400	400	20	20	0.071	0.0779	0.0083	10.7	1.204	0.074	19.4	\$51.60
D125160	500	500	105	105	0.073	2.4784	0.1408	5.7	1.229	0.176	32.0	\$31.25
D125200	500	500	131	131	0.073	3.2669	0.1753	5.4	1.228	0.175	33.2	\$30.12
A125160	500	500	105	105	0.073	2.4784	0.1408	5.7	1.229	0.176	28.3	\$35.37
A125200	500	500	131	131	0.073	3.2669	0.1753	5.4	1.228	0.175	29.8	\$33.54
AE1010D16	75	75	9.2	9.2	0.046	0.0328	0.0013	4.1	0.507	0.063	3.7	\$272.82
ASB171C801	75	75	85	85	0.043	0.0932	0.0033	3.6	0.662	0.083	5.6	\$177.57
PAHL 120/20	70	80	56	64	0.054	0.2754	0.0102	3.7	1.105	0.142	4.4	\$228.91

[a] Calculated at stiffness match condition.

volume, mass, and cost are given. The energy densities mass were obtained by dividing the output mechanical energy by the active material volume and mass. Thus, the output energy densities based on active material volume and active material mass are given. The energy density per unit cost was obtained by dividing the output energy by the cost of the actuator. Also given is the cost of a unit of output energy, which is the inverse of the energy density per unit cost.

10.7.2.5 Energy Conversion Efficiency

Energy conversion efficiency is an important component of the actuator study since it presents an overall evaluation of the whole transduction process. Tables 10.9 and 10.10 have listed the maximum output mechanical energy, and the electrical energy necessary to produce this output under static and dynamic operations, respectively. Division of the output mechanical energy by the input electrical energy yields the energy conversion efficiency of the induced-strain actuator. The energy conversion efficiency is given also in Tables 10.9 and 10.10.

10.7.2.6 Power and Power Density

We have shown in Section 6.1.2 that the output power of an induced-strain actuator can be directly related to the peak energy per cycle under dynamic operation and the angular frequency, $\omega = 2\pi f$, that is,

$$P_e = \omega E_e \tag{10.277}$$

Hence, the maximum output power is directly related to the maximum peak power per cycle, that is,

$$P_e^{\max} = \omega E_e^{\max} \tag{10.278}$$

where $E_e^{\max}$ is given by Equation (10.276). It is apparent that the output power increases linearly with frequency. For comparative studies, the same frequency must be used for all the actuators under consideration. In our study, we used the frequency of 1 kHz as a reference frequency, and calculated the maximum output power and the corresponding power densities per active material volume and mass. The results are presented in Table 10.11. However, it must be mentioned that full stroke operation at 1 kHz may result in considerable heat dissipation that needs to be removed through adequate cooling.

10.7.2.7 Energy Density of the Complete Actuator: Effect of Casing and Prestress Spring

For the actuators that are provided with casing and prestress mechanism, the actuator energy density must be calculated by dividing by the volume and mass of the entire actuator, which are larger than the volume and mass of the active material inside the actuator. Results for energy density per unit volume and energy density per unit mass are given in Table 10.12. Examination of Table 10.12 reveals that the addition of casing and

TABLE 10.11

Output Power of Selected Induced-Strain Actuators during Full-Stroke Operation at 1 kHz

Model	Output Power P_e (W)	Output Power per Active Material Volume P_e/V_{ISA} (W/dm^3)	Output Power per Active Material Mass P_e/m_{ISA} (kW/kg)
P-245.70	23.6	149	3.0
P-246.70	559.6	614	11.4
P-247.70	1089.7	591	11.3
P-844.60	54.7	416	5.0
E100P-2	4.2	469	6.0
E200P-3	20.9	638	8.0
E300P-4	91.2	352	7.0
E400P-3	52.4	672	7.6
D125160	884.5	357	7.7
D125200	1101.4	337	7.7
A125160	884.5	357	7.7
A125200	1101.4	337	7.7
AE1010D16	8.4	257	3.2
ASB171C801	20.8	223	4.2
PAHL 120/20	64.3	234	6.9

Note: Full stroke operation at 1 kHz may result in considerable heat dissipation that needs to be removed effectively through adequate cooling.

TABLE 10.12

Effect of Casing and Prestress Mechanism on the Energy Density of Selected Induced-Strain Actuators

Model	Output Energy per Volume E_e/V_{ISA} (J/dm^3)		Output Energy per Mass E_e/m_{ISA} (J/kg)	
	per Active Material Volume	per Overall Volume	per Active Material Mass	per Overall Mass
P-245.70	0.717	0.177	0.092	0.037
P-246.70	2.723	0.767	0.349	0.161
P-247.70	2.705	0.934	0.347	0.265
P-844.60	1.188	0.303	0.152	0.061
A125160	1.843	0.532	0.263	0.182
A125200	1.843	0.989	0.263	0.181
ASB171C801	0.993	0.077	0.124	0.010
PAHL 120/20	5.207	0.388	0.261	0.154

prestress mechanism significantly lowers the overall energy density values. The effect of actuator casing and prestress mechanisms on the energy density can be quite significant. When the energy density is calculated with respect to the complete actuator (and not just the active material inside the actuator), the energy density values diminish considerably.

10.7.2.8 Consistency Checks

To give credibility to the numerical results and to filter out any inadvertent discrepancies, a number of consistency checks were performed. First, we examined the credibility of the effective full-stroke electromechanical coupling coefficient, κ. The values of this coefficient are given in the second column of Table 10.8. It is noticed that none of these values exceed 1, and most of them are below 0.75, which is an accepted upper limit of conventional piezoelectric material formulations. Those few values above 0.75 may be due to better than usual piezoelectric material formulations. On total, we conclude that, from this point of view, data consistency is acceptable.

Another consistency test that can be applied is associated with the energy density. In our study, we computed the energy density of active material by simply dividing the maximum output energy by the active material volume and mass provided by the manufacturer. However, the energy density could also be correlated with other basic material data, such as

- Free strain, ε_{ISA}, defined as the ratio between the free displacement, u_{ISA}, and the length, L
- Apparent Young's modulus, E, defined from the stiffness formula $k_i = EA/L$, where A is the cross-sectional area of the stack

The free strain was calculated by dividing the free displacement by the active material length. The apparent Young's modulus was calculated from the formula $k_i = EA/L$ when the active material stiffness was available. This is especially the case with glued stacks, where the compliance of the adhesive layer lowers significantly the stiffness of the stack. In certain situations, measured values of the active material stiffness were not available from the manufacturer. However, the manufacturer could provide common values for the active material Young's modulus. This Young's modulus was used to evaluate the active material stiffness.

For consistency checks, we also note that the energy density per unit volume can also be calculated from the strain energy formula, that is,

$$\frac{E_e^{max}}{Vol} = \frac{1}{4}\left(\tfrac{1}{2}E\varepsilon_{ISA}^2\right) \tag{10.279}$$

Dividing the volume density by the equivalent mass density, ρ, yields the energy density per unit mass, that is,

$$\frac{E_e^{max}}{mass} = \frac{1}{\rho}\frac{1}{4}\left(\tfrac{1}{2}E\varepsilon_{ISA}^2\right) \tag{10.280}$$

These observations were used to perform further consistency checks on our results.

10.7.2.9 Nonlinear Operation

Enhanced stroke performance of induced-strain actuators can be achieved by driving them into the nonlinear range. In the nonlinear range, the piezoceramics undergo large domain rotations, have larger hysteresis, and have shorter fatigue life. Not all the manufacturers allow their actuators to be driven into the nonlinear range, due to the detrimental effects than such operation can have on the actuator (heating, large internal stresses, cracks, etc.). However, some manufacturers have provided data for the operation of their actuators in

TABLE 10.13

Enhanced Stroke Performance of Induced-Strain Actuators Achieved by Driving into the Nonlinear Range

Model	Active Material Type	Expansion Voltage V^+ (V)	Contraction Voltage V^- (V)	Max. Expansion u_{ISA}^+ (μm)	Max. Contraction u_{ISA}^+ (μm)	Full Stroke (μm)
D125160	PZWT100	1000	0	210	0.0	210
D125160 (NL)	PZWT100	1200	−250	324	−68	392
D125200	PZWT100	1000	0	262	0.0	262
D125200 (NL)	PZWT100	1000	−250	406	−84	490

Notes: Nonlinear range is used to get enhanced performance from the actuator. In the nonlinear range, the piezoceramics undergo large domain rotations, have larger hysteresis, and have shorter fatigue life. For the models quoted in the table, endurance of 10^7 cycles without failure was proven experimentally on a few specimens

the nonlinear range and this data has been included in this study. For the models included in this study, endurance of 10^7 cycles without failure was proven experimentally on a few specimens. The results are presented in Table 10.13.

10.7.3　Discussion of Results

The results presented in Tables 10.7 through 10.10 were used to construct the comparative charts presented in Figures 10.34 through 10.41. These charts give a quick visual presentation of the relative performance of the actuators in terms of effective electromechanical coupling coefficient, maximum output energy capabilities, output energy densities, and energy conversion efficiency.

Figure 10.34 presents a comparative bar chart of the effective full-stroke electromechanical coupling coefficient, κ. It is apparent that the plotted results present κ values varying from ∼0.4 through ∼0.9. Most of the plotted values are around κ = 0.7, which is consistent with commonly accepted values for piezoelectric materials. The lower values presented in Figure 10.34 may be due to piezoelectric material formulation that have a lower electromechanical coupling coefficient, but may excel in other properties, such as, say, endurance

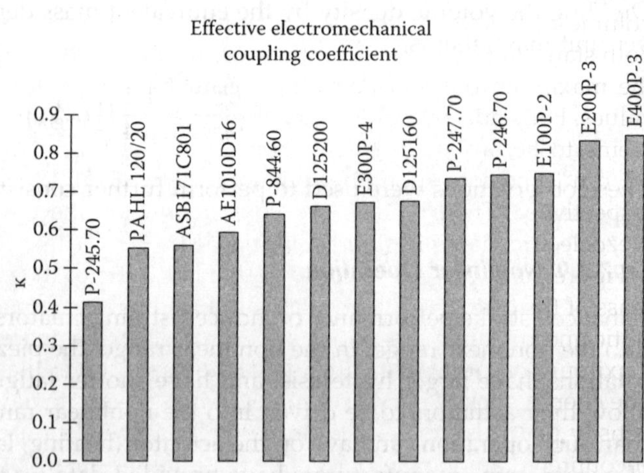

FIGURE 10.34

Effective electromechanical coupling coefficient of commercially available induced-strain actuators.

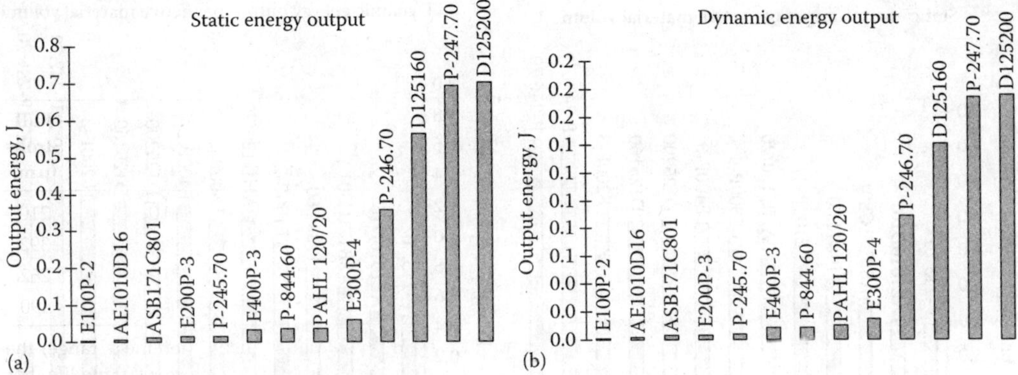

FIGURE 10.35

Maximum output energy capability of a selection of commercially available induced-strain actuators: (a) static operation; (b) dynamic operation.

and fatigue. The higher values presented in Figure 10.34 may be due to piezoelectric material formulation that have a higher than average electromechanical response.

Of major significance in the design of an effective induced-strain actuator is the amount of energy available for performing external mechanical work. Figure 10.35 presents a comparative bar chart of the maximum energy output capabilities of the commercially available induced-strain actuators considered in this study. Both static and dynamic operations are examined. It is apparent that the dynamic values of the maximum output energy are about four times lower than the corresponding static values. This is mainly due to the fact that the maximum dynamic displacement amplitude is about half of the maximum static displacement amplitude (free stroke). As detailed in previous sections, the dynamic displacement amplitude is half of the peak-to-peak total excursion during a dynamic cycle, whereas the static displacement amplitude is the peak-to-peak total excursion between a maximum retraction and a maximum expansion point. Hence, the maximum output energies, which are roughly proportional to the square of the maximum displacement, are in the ratio 4:1, as apparent from Figure 10.35.

Figure 10.36 presents a comparative bar chart of the energy density per active material volume of the commercially available induced-strain actuators considered in this study. Both static and dynamic operations are examined. It is apparent that the dynamic values of the maximum output energy are about four times lower than the corresponding static values. For static operations, a representative value for the energy density per unit volume seems to be around 5 J/dm^3. Top performers seem to be able to produce as much as 7 J/dm^3. The corresponding dynamic values would be around 1.2 J/dm^3 and 1.8 J/dm^3, respectively. These values are consistent with the output energy density of the bulk piezoelectric materials at 0.1% induced strain (6 J/dm^3 static, 1.5 J/dm^3 dynamic).

Figure 10.37 presents a comparative bar chart of the energy density per active material mass of the commercially available induced-strain actuators considered in this study. Both static and dynamic operations are examined. It is apparent that the dynamic values of the maximum output energy are about four times lower than the corresponding static values. The energy density per active material mass seems to have a wider spread than the energy density per active material volume. This can be attributed to the spread of mass densities encountered among various active material formulations. For static operations, the values

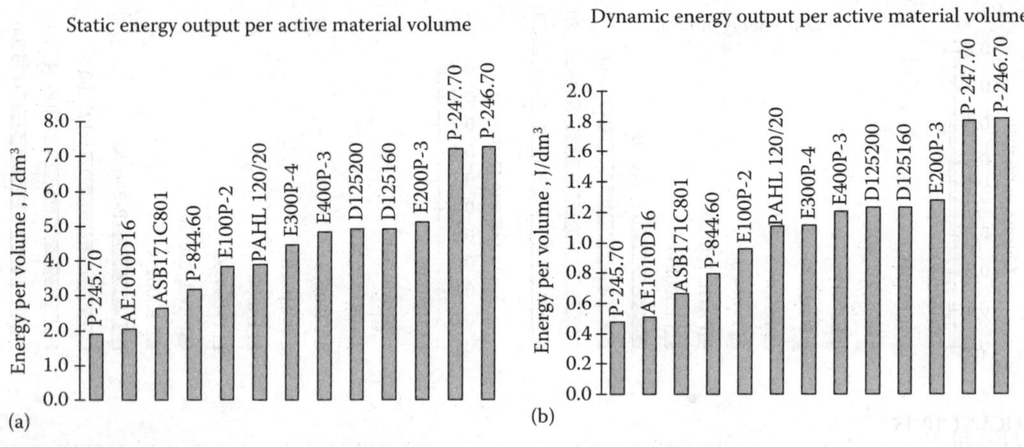

FIGURE 10.36

Maximum output energy density per active material volume of commercially available induced-strain actuators: (a) static operation; (b) dynamic operation.

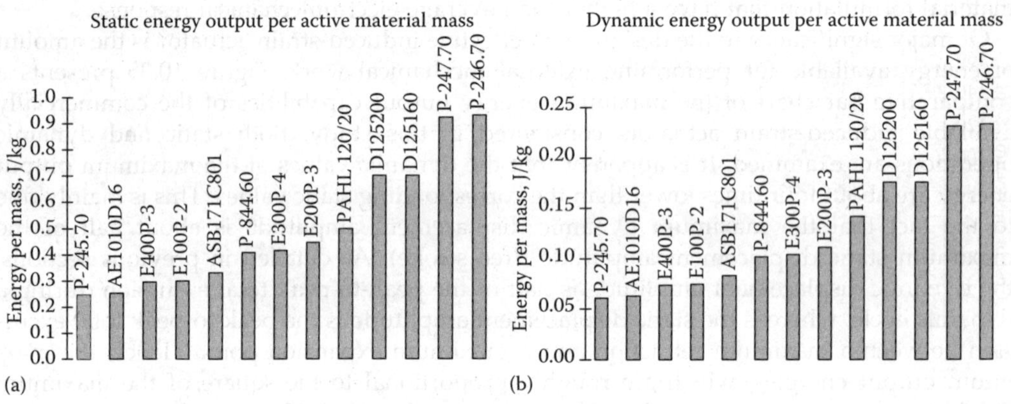

FIGURE 10.37

Maximum output energy density per active material mass of commercially available induced-strain actuators: (a) static operation; (b) dynamic operation.

seem to vary between 0.25 and 0.9 J/kg. Top performers seem to be able to produce energy density values around 0.93 J/kg. The corresponding dynamic values would be between 0.06 and 0.23 J/kg for the range, with 0.233 J/kg for the top performer.

Figure 10.38 presents a comparative bar chart of the energy output per unit cost of the commercially available induced-strain actuators considered in this study. Both static and dynamic operations are examined. It is apparent that the dynamic values of the maximum output energy are about four times lower than the corresponding static values. For static operation, the energy density per unit cost varies between 7.5 and 120 mJ/$1000. For dynamic operation, the range is between ~2 and 30 mJ/$1000. This wide variation in energy density per unit cost could be attributed to the cost being made up of a fixed component and a variable component. The variable component is expected to vary somehow proportionally with the quantity of active material inside the induced-strain actuator,

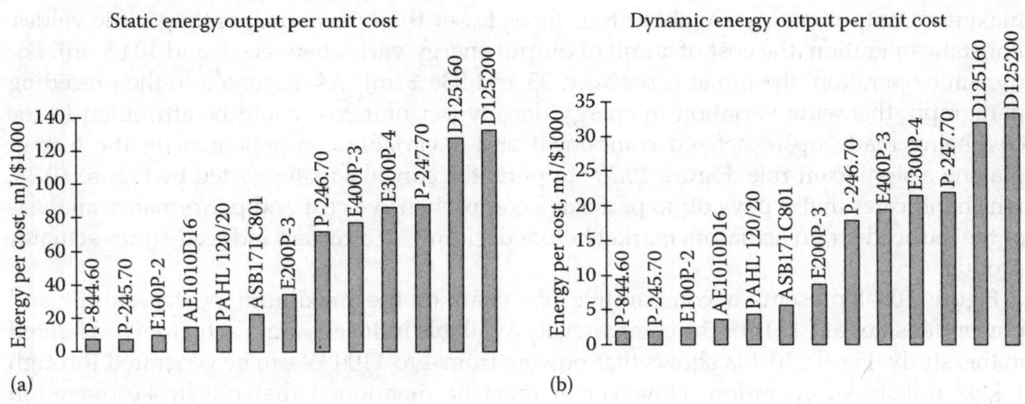

FIGURE 10.38

Maximum output energy density per unit cost of commercially available induced-strain actuators: (a) static operation; (b) dynamic operation.

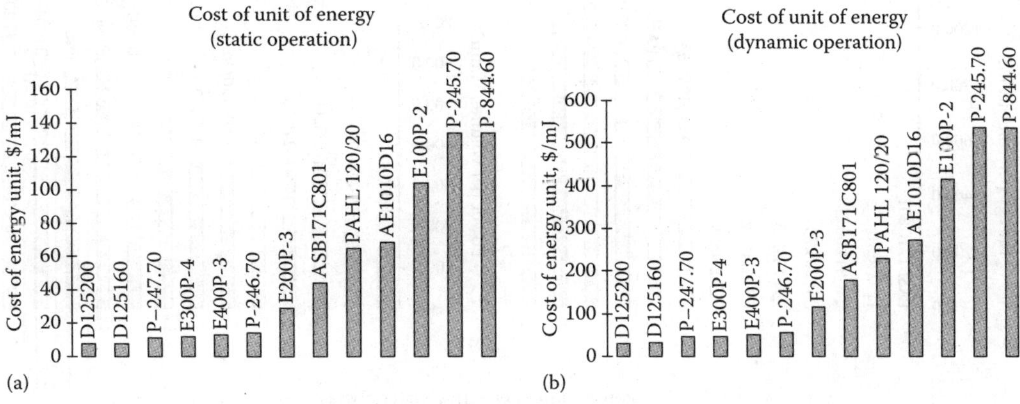

FIGURE 10.39

Cost of unit of output energy for commercially available induced-strain actuators: (a) static operation; (b) dynamic operation.

whereas the fixed component depends on other manufacturing factors. Since the induced-strain actuators are still specialty products that are not yet produced in large quantities, the fixed cost component seems to represent a very significant part of the total price. Therefore, the large size actuators have a better energy density per unit cost (i.e., offer a better deal!) than the small size actuators. In addition to the variable cost versus fixed cost, another factor that might influence this wide variation in energy density per unit cost may be related to the manufacturer's or vendor's way of doing business. In conclusion, Figure 10.38 indicates that it is advantageous to perform a comprehensive cost and performance investigation of the induced-strain actuators market before deciding on a certain induced-strain actuator product.

Figure 10.39 presents a comparative bar chart of the cost of a unit of output energy of the commercially available induced-strain actuators considered in this study. Both static and dynamic operations are examined. As previously observed, the dynamic values of the maximum output energy are about four times lesser than the corresponding static values. For static operation, the cost of a unit of output energy varies between 8 and 134 \$/mJ. For dynamic operation, the range is between 33 and 536 \$/mJ. As discussed in the preceding paragraph, this wide variation in energy density per unit cost could be attributed to the cost being made up of a fixed component and a variable component, with the former playing a significant role. Figure 10.39 supports the conclusion suggested by Figure 10.38, which indicates that it pays off to perform a comprehensive cost and performance analysis of the induced-strain actuators market before deciding on a certain induced-strain actuator product.

Figure 10.40 presents a comparative bar chart of the maximum power output and power densities at 1 kHz of the commercially available induced-strain actuators considered in this study. Figure 10.40a shows that powers from 4 to 1100 W can be generated through 1 kHz full-stroke operation. However, it must be mentioned that full-stroke operation

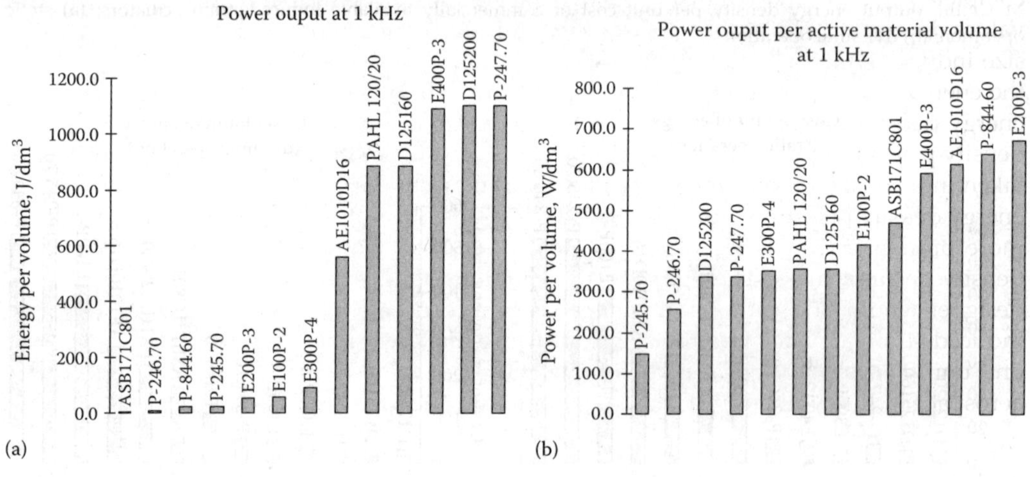

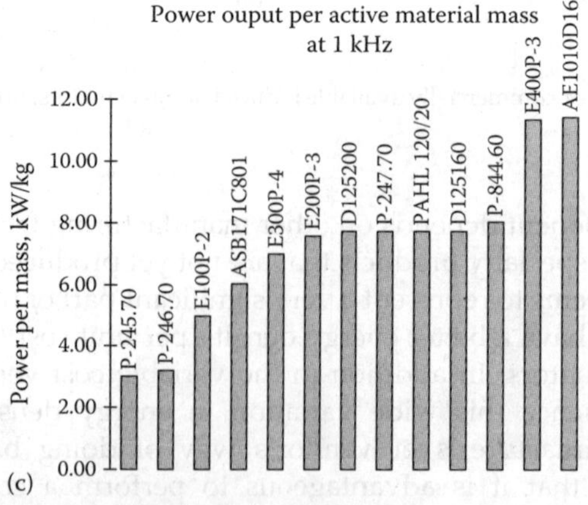

FIGURE 10.40

Maximum output power and power density and mass of commercially available induced-strain actuators operating at 1 kHz: (a) output power; (b) power density per active material volume; (c) power density per active material mass (Note: full stroke operation at 1 kHz may result in considerable heat dissipation that needs to be removed effectively through adequate cooling).

at 1 kHz may result in considerable heat dissipation that needs to be removed effectively through adequate cooling. The corresponding power density per active material volume is presented in Figure 10.40b. Typical values are of the order of 350 W/dm^3 for many of the actuators. High-performance induced-strain actuators seem to be able to produce 1 kHz power density as high as 670 W/dm^3. Figure 10.40c presents a plot of the power density per active material mass. We notice that a large number of the actuators have a 1 kHz power density around 7.7 kW/kg. The high-performance induced-strain actuators seem to be able to produce 1 kHz power output as high as 11.4 kW/kg.

Figure 10.41 illustrates the effect of actuator casing and prestress mechanisms on the energy density values. It presents a comparative bar chart of the energy density with respect to the active material and the energy densities with respect to the complete actuator. Figure 10.41a presents the results pertaining to energy density per unit volume, whereas Figure 10.41b presents the results pertaining to energy density per unit mass. It is apparent from Figure 10.41 that when the energy density is calculated with respect to the complete actuator (and not just the active material inside the actuator), the energy density values diminish considerably. As expected, this decrease is more pronounced in the small size induced-strain actuators than in the large size induced-strain actuators. In addition, the energy density per unit volume is affected much more by this phenomenon than the energy density per unit mass. In fact, for large size actuators (e.g., P-247.70), the energy density per unit mass diminishes by only about 30% when the total mass of the actuator is taken into consideration. In contrast, for the same actuator, the corresponding loss in the energy density per unit volume is around 64%. This illustrates the fact that the reduction is more pronounced in terms of energy density per unit volume than in terms of energy density per unit mass. These observations indicate that, when the volume is critical, the designer would find it more advantageous to incorporate the active-material core of the induced-strain actuator into the active structure directly, rather than trying to use a "bolt on" unit. For applications where volume and mass are essential, as, for example, in the aerospace industry, the direct incorporation of the induced-strain actuator without casing

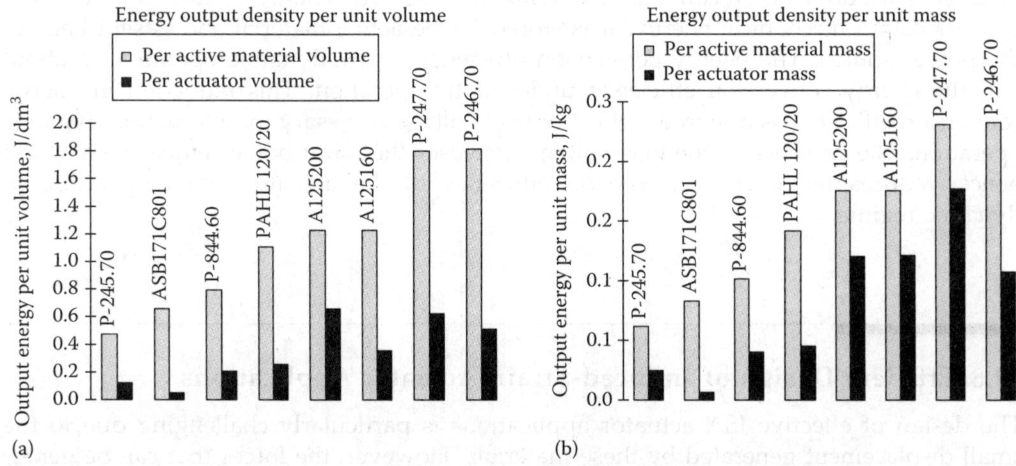

FIGURE 10.41
Comparison of output energy density per unit mass for seven induced-strain actuators with casing and prestress mechanism, showing that the addition of casing reduces the mass-based energy density due to inactive mass of the casing and prestress mechanism: (a) per unit volume; (b) per unit mass.

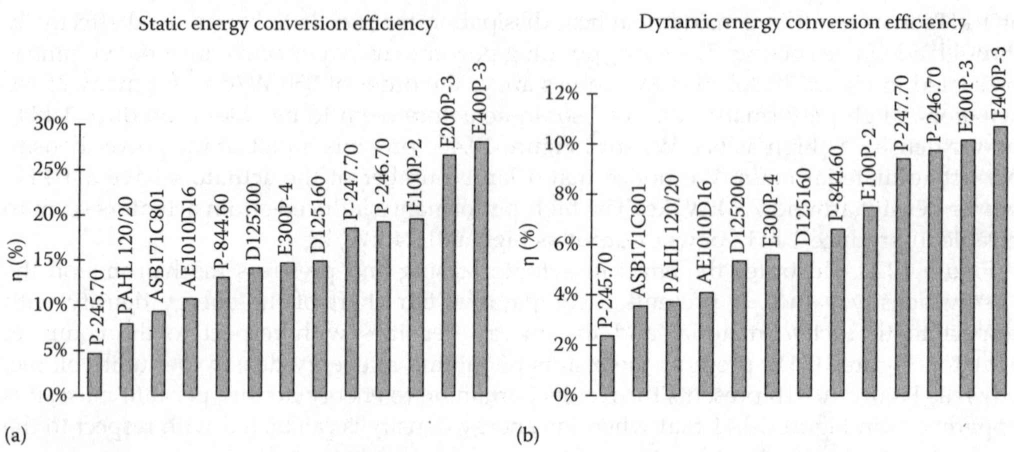

FIGURE 10.42
Energy conversion efficiency of commercially available induced-strain actuators: (a) static operation; (b) dynamic operation.

and prestress mechanism in the host structure is highly desirable since it leads to important volume and mass savings.

Figure 10.42 presents a comparison of the energy conversion efficiency from input electrical energy into output mechanical energy. It should be mentioned that the formulae used to estimate the required input electrical energy are only first-order approximations since they ignored the variation of capacitance and inductance in the presence of an applied external load. In addition, the energy dissipation in internal electric resistance and in hysteresis is, for the moment, ignored. Figure 10.42 shows that the energy conversion efficiency under static operation can be between 15% and 20%. High-performance induced-strain actuators can reach energy conversion efficiencies as high as 28%. This does not mean that the remaining 80% of energy is lost, but it is not converted into mechanical energy. It is stored in the active material and is sent back to the power source. The energy conversion efficiency under dynamic operation is about half the energy conversion efficiency under static operation. This reduction in energy conversion efficiency is due to the effect of bias voltage necessary for alternating dynamic operation. The presence of the bias voltage increases the input power requirements, and hence, reduces the overall conversion efficiency of the actuator when operating in dynamic regime.

10.8 Efficient Design of Induced-Strain Actuator Applications

The design of effective ISA actuator applications is particularly challenging due to the small displacement generated by these materials. However, the forces that can be generated by an ISA actuator can be very large. These forces are limited only by the inherent stiffness and the compressive strength of the ISA material. For a stack of a given length, the stiffness increases proportionally with the effective area. For most practical applications, a displacement amplification device (displacement amplifier) is employed to increase the small induced-strain displacement of the basic actuator. The displacement amplification is basically a lever mechanism, though various constructive variants may be employed. In a generic formulation, the displacement amplifier can be viewed as a compliant mechanism (CM). Hence, the effective amplification, η, will depend on both the mechanism geometric ratio, G, and its internal stiffness. The larger the internal stiffness of the compliant mechanism, the closer the effective amplification, η, will be to the geometric ratio, G.

The effective design of the displacement amplifier can "make or break" the practical effectiveness of an ISA actuator. The most important parameter that needs to be optimized during such a design is the energy extraction coefficient defined as the ratio between the effective mechanical energy delivered by the actuator and the maximum possible energy that can be delivered by the actuator in the stiffness-match condition.

(The overall efficiency of active-material actuation depends, to a great extent, on the efficiency of the entire system, which includes the active-material transducer, the displacement amplification mechanisms, and the power supply.)

10.8.1 Efficient Static Design

Consider an active material actuator of nominal (free-stroke) displacement u_{ISA}, and internal stiffness k_i. During static operation, the total induced energy, E_{ISA}, gets divided between the internal and external subsystems: part of it, E_e, gets transmitted to the external application, whereas the rest, E_i, remains stored in the internal compressibility of the stack. The value of E_{ISA}, and its distribution between E_e and E_i, depends on the stiffness ratio $r = k_e/k_i$. It has been shown in Section 10.5.2 that

$$E_e(r) = \frac{r}{(1+r)^2} E_{mech}^{ref}, \quad E_i(r) = \frac{r^2}{(1+r)^2} E_{mech}^{ref}$$

$$E_{ISA}(r) = E_e(r) + E(r) = \frac{r}{1+r} E_{mech}^{ref} \tag{10.281}$$

where $E_{mech}^{ref} = \frac{1}{2} k_i u_{ISA}^2$ is the reference mechanical energy defined by Equation (10.13) in terms of actuator characteristics, u_{ISA} and k_i. Dividing throughout by E_{mech}^{ref} yields the nondimensional energy coefficients, E'_e, E'_i, E'_{ISA}, that is,

$$E'_e(r) = \frac{r}{(1+r)^2}, \quad E'_i(r) = \frac{r^2}{(1+r)^2}, \quad E'_{ISA}(r) = \frac{r}{1+r} \tag{10.282}$$

Figure 10.43a shows the variation of the total induced energy coefficient, E'_{ISA}, and of the internal and external energy coefficients, E'_i and E'_e, with the stiffness ratio, r. It can be seen that the external energy coefficient, E'_e, reaches a maximum at $r = 1$. The $r = 1$ point corresponds to the *stiffness match* condition. Beyond the $r = 1$ point, the energy delivery starts to decrease in spite of continued increase in the overall induced-strain energy, E'_{ISA}. The explanation is that beyond $r = 1$ the induced-strain energy is retained in the internal compressibility of the ISA stack, that is, it remains stored internally as E_i. Hence, under static conditions, an ISA actuator cannot be efficiently used beyond $r = 1$. Figure 10.43b shows the variation of the energy transmission efficiency, $\eta(r)$, with the stiffness ratio, r. Below the stiffness match condition, the energy delivered externally, E_e, is always greater than the energy stored internally, E_i. Hence, for $r < 1$, the energy transmission efficiency is greater than 50%. The energy transmission efficiency approaches 100% as $r \to 0$, but this

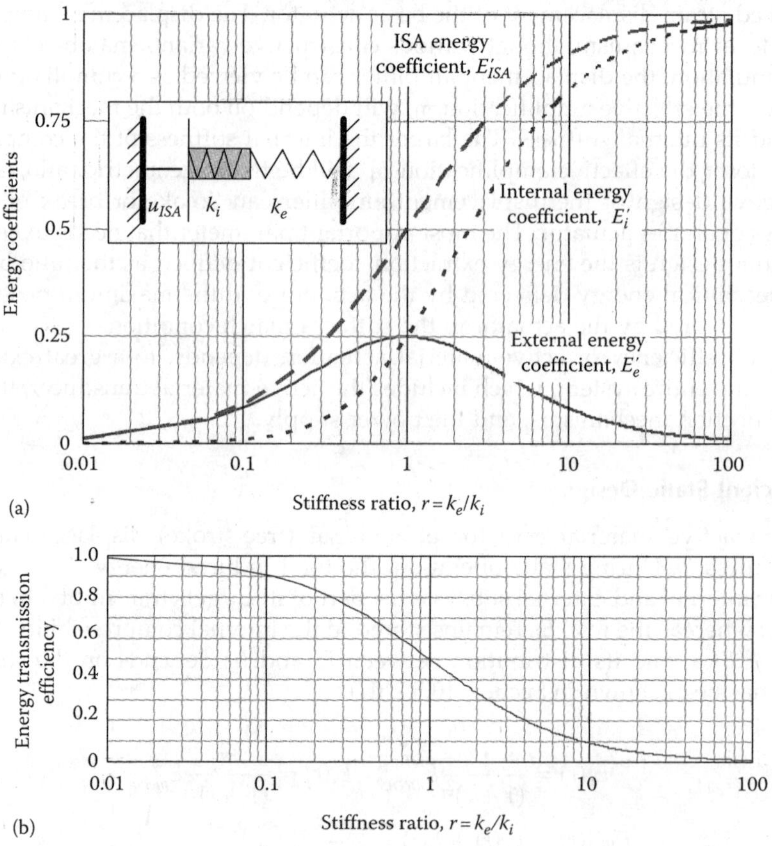

FIGURE 10.43
Variation of energy coefficients and energy transmission efficiency with stiffness ratio, r, under static operation.

corresponds to the *free actuator* operation, that is, with no actuation force, and hence no energy being actually transmitted. Thus, the 100% efficiency is not practically possible. In the extreme, $r \to \infty$, the actuator is blocked and hence has no output stroke. In this condition, the actuator produces maximum ISA energy, but none of it gets delivered externally either.

As the optimum (stiffness match) condition is approached, the delivered energy reaches its peak, but the transmission efficiency continues to decrease. Hence, at the stiffness-match point, the energy transmission efficiency is only 50%. This is consistent with the fact that, since the stiffness is matched, the internal and external energies are equal, and hence half the ISA energy remains stored internally. Beyond $r = 1$, the energy transmission efficiency, η, continues to decrease steadily. Figure 10.43b also illustrates another important concept, *viz.* that of *conjugate stiffness ratios*. Consider $r_1 = 1/4$ and $r_2 = 4$. The corresponding external energy delivery is the same in both cases: $E(r_1) = E(r_2) = 0.16 E_{mech}^{ref}$. But the energy transmission efficiency is widely different, *viz.* $\eta(r_1) = 80\%$ whereas $\eta(r_2) = 20\%$. The same external effect, $E_e = 0.16 E_{mech}^{ref}$, can be obtained with two widely different values of energy efficiency. The *soft* design ($r_1 = 1/4$) is four times more efficient than the *stiff* design ($r_2 = 4$). This example only illustrates a more general principle that, for static operation away from the stiffness match condition, there are always two conjugate solutions that produce the same output energy, but one is more efficient than the other. The more efficient solution is usually the softer design.

In conclusion, under static conditions, the best use of an actuator is made when the internal and external stiffness are matched. In this case, maximum attainable energy delivery is attained, and its value is $(E_e)_{max} = \frac{1}{2}E_{mech}^{ref}$. If stiffness match cannot be obtained, a soft design $(r < 1)$ will always be more efficient than a stiff design $(r > 1)$. These observations are essential for the successful design of efficient ISA devices for static applications.

10.8.2 Efficient Dynamic Design

Consider next a typical configuration for ISA operation under dynamic condition at frequencies typical to mechanical applications. Figure 10.44 recalls Figure 10.24 and presents an ISA stack coupled with an external dynamic system consisting of mass, spring, and damper. In aero-servo-elastic control, the mass, spring, and damper values vary with the operation frequency, and also with airspeed and length scale. For the present study, constant mass, spring, and damper values are considered. Hence, the equivalent dynamic stiffness can be written as $\bar{k}_e(\omega) = (k_e - \omega^2 m_e) + i\omega c$. Assuming the natural frequency of the external system as ω_0, one recalls Equation (10.151) and expresses the complex dynamic stiffness $\bar{k}_e(\omega)$ as

$$\bar{k}_e(\omega) = \left[1 - (\omega/\omega_0)^2 + i2\zeta(\omega/\omega_0)\right]k_e = \left[(1 - p^2) + i2\zeta p\right]k_e \qquad (10.283)$$

where $p = \omega/\omega_0$ the frequency ratio and ζ is the internal damping of the external system. The *real* and *imaginary* parts of the complex dynamic stiffness signify in-phase and out-of-phase force reaction components. At low frequency $(p \to 0)$, the inertial and damping terms in the dynamic stiffness expression vanish, and the static stiffness k_e is predominant. This is the *quasi-static dynamic* operation. At higher frequencies, but still below external resonance $(0 < p < 1)$, the effective stiffness is reduced by $1 - p^2$, whereas the dissipative (out-of-phase) component grows as $2\zeta p$.

As we approach the mechanical resonance $(p \to 1)$, the reactive inertial forces balance the reactive elastic forces, and hence the real part of the dynamic stiffness vanishes. At resonance $(p = 1)$, only the imaginary part of the dynamic stiffness is nonzero, that is, dissipative forces predominate. Above resonance $(p > 1)$, the inertial forces dominate, and stiffness magnitude increases parabolically. A phase shift of 90° is recorded at resonance, and an overall phase shift of 180° takes place as the operating point passes from the sub-resonance into the post-resonance regimes. Using the frequency-dependent expression of the complex dynamic stiffness, we define a frequency-dependent stiffness ratio, that is, the *dynamic stiffness ratio*, defined by Equation (10.131) as $\bar{r}(\omega) = \bar{k}_e(\omega)/\bar{k}_i$. The dynamic stiffness ratio is a complex number, just as the dynamic stiffness.

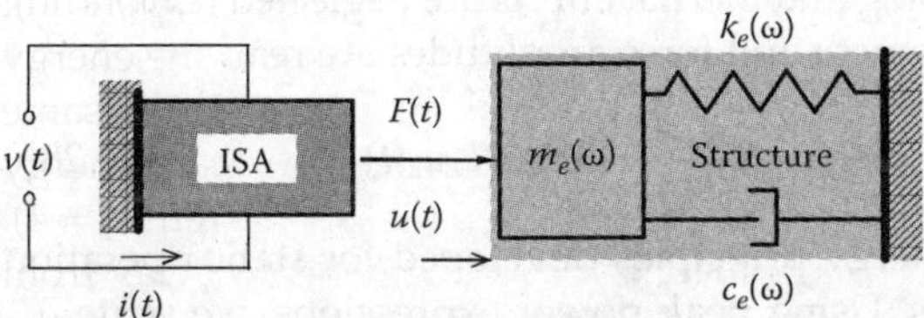

FIGURE 10.44
A dynamic ISA system consisting of an ISA device and a dynamic structure.

The ISA is driven by alternating voltage, $v(t)$, and current, $i(t)$, which induce an alternating strain. The resulting dynamic displacement, $u(x, t; \omega)$, varies with time and position. Neglecting wave propagation effects inside the induced-strain actuator yields a linear variation of the displacement along the actuator length. Displacement compatibility and force balance between the actuator and the external mechanical impedance are imposed at $x = l$. At frequencies well below the internal actuator resonance, the dynamic displacement expression takes the form

$$u(x, t; \omega) = \frac{1}{1 + \bar{r}(\omega)} \frac{x}{l} u_{ISA} e^{i\omega t} = \hat{u}(x; \omega) e^{i\omega t} \tag{10.284}$$

where $\hat{u}$ signifies motion amplitude and u_{ISA} is the free stroke. At the interface between the actuator and the external load we have the external displacement given by

$$u_e(t; \omega) = \bar{u}_e e^{i\omega t} = \frac{1}{1 + \bar{r}(\omega)} u_{ISA} e^{i\omega t}, \quad \bar{u}_e = \frac{1}{1 + \bar{r}(\omega)} u_{ISA}. \tag{10.285}$$

where $\bar{u}$ signifies complex number. This remarkably simple expression is ideally suited for studying the power and energy transmitted to the external system during dynamic operation.

The *stiffness match* concept from static analysis is extended to dynamic analysis, using the *dynamic stiffness concept*. The dynamic stiffness concept allows direct analytical continuation between the static and dynamic regimes. Three questions will be addressed:

1. How is *optimum stiffness ratio* defined under dynamic conditions?
2. Does a statically matched system maintain its superiority when operated dynamically?
3. How should one design a system to be optimal and robust over a frequency range?

In addressing the first question, we first define the *optimum condition* under dynamic operation. To obtain a step-by-step understanding, three situations with increasing degree of complexity will be considered:

1. Quasi-static dynamic operation
2. Undamped dynamic operation
3. Damped dynamic operation

Details are given in the following sections.

10.8.3 Quasi-Static Dynamic Operation

Under quasi-static dynamic operation, damping and inertial effects are neglected ($\bar{k}_e(\omega) \approx k_e$). Hence, no phase shift occurs, and all displacement and force amplitudes are real:

$$u_{ISA}(t) = \hat{u}_{ISA} e^{i\omega t}, \quad u_e(t) = \hat{u}_e e^{i\omega t}, \quad u_i(t) = \hat{u}_i e^{i\omega t}, \quad F(t) = k_e u_e(t) \tag{10.286}$$

Note that for quasi steady operation, the energy principles developed for static operation are directly translated into power principles. Using peak power expressions, we write

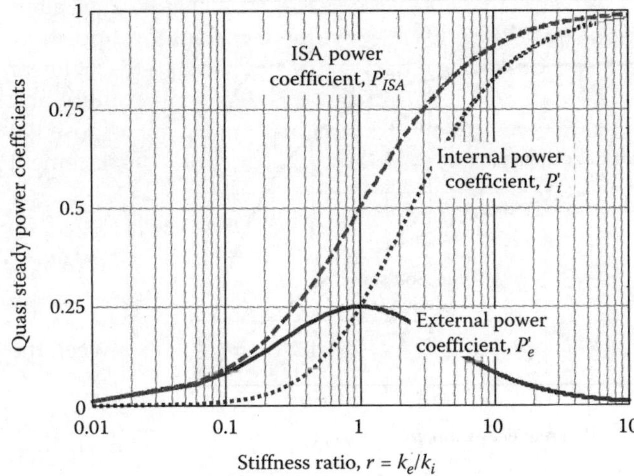

FIGURE 10.45
Variation of quasi steady power with static stiffness ratio.

$$P_e(r) = \frac{r}{(1+r)^2} P_{mech}^{ref}, \quad P_i(r) = \frac{r^2}{(1+r)^2} P_{mech}^{ref}$$

$$\hat{P}_{ISA}(r) = P_e(r) + P_i(r) = \frac{r}{1+r} P_{mech}^{ref} \tag{10.287}$$

where $P_{mech}^{ref} = \omega E_{mech}^{ref}$ as given by Equation (10.186). A plot of these expressions versus stiffness ratio, r, is given in Figure 10.45. The plot closely resembles that of Figure 10.43, only that energy coefficients were replaced by power coefficients. For quasi-static dynamic operation, the maximum power delivered by an ISA device is achieved when the internal and external stiffnesses are matched. Under quasi-static dynamic operation, the *static stiffness match principle* still applies.

10.8.4 Undamped Dynamic Operation

Many dynamic utilizations of ISA technology do not take place under quasi-static conditions. For mechanical resonances in the 10–50 Hz frequency range, inertial forces cannot be neglected. As the mechanical resonance of the external system is approached, the reactive inertial forces get subtracted from the reactive elastic forces, and hence the static stiffness match is upset. A system with statically matched stiffness is not expected to retain its optimal condition under fully dynamic operation. This is illustrated in Figure 10.46.

For a statically matched system, the external power coefficient at $p = 0$ (quasi-static operation) is optimum, that is, it has the maximum possible value, 0.5. As the frequency increases, the power coefficient of a statically matched system decreases. At $p = 1$ (i.e., at external resonance, $\omega = \omega_0$), the power coefficient becomes zero. This behavior is expected, since the equivalent dynamic stiffness of the external system decreases as resonance conditions approach. At resonance, the equivalent dynamic stiffness is virtually zero; hence no force is transmitted, and thus no power. For a soft external system ($r = 1/4$), the same behavior is encountered. The starting point at $p = 0$ (quasi-static conditions) is lower since the unmatched static stiffness gives less than optimal quasi-static behavior. For a stiff external system ($r = 4$), the behavior is different. Under quasi-static conditions, the power coefficient of this stiff system is the same as that of the conjugate soft system. As frequency

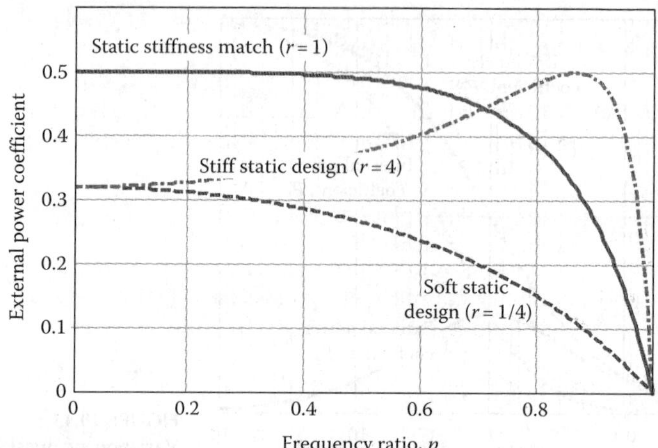

FIGURE 10.46
External power coefficient versus frequency ratio for an undamped dynamic system.

increases, the behavior of the stiff system is drastically different from that of the soft system. Its power coefficient increases, whereas that of the conjugate soft system decreases. This behavior is expected due to the *dynamic softening* of the stiff system. As frequency increases, the statically stiff system progressively approaches the optimum (stiffness-match) condition. When the dynamic external and internal stiffnesses are equal, *dynamic stiffness* match is attained. The frequency value at which dynamic stiffness match occurs is obtained by setting the value of the dynamic stiffness ratio to 1, that is,

$$p_{match}(r) = \sqrt{1 - 1/r} \tag{10.288}$$

Figure 10.47 shows the variation of the dynamic stiffness match frequency ratio with static stiffness ratio. For large static stiffness ratios, the frequency ratio for dynamic stiffness match approaches asymptotically the value 1. For moderate static stiffness ratios, values between 0 and 1 can be obtained. For example, for a static stiffness ratio $r = 4$ (that is, stiff static design), the dynamic stiffness match takes place at $p_{match}(4) = \sqrt{3/2} = 0.866$. This value corresponds to the local maximum of the power coefficient for a stiff static design, as shown in Figure 10.46.

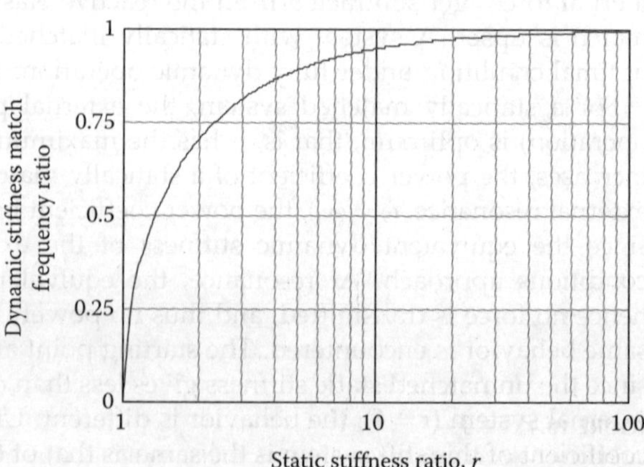

FIGURE 10.47
Variation of dynamic stiffness match frequency with static stiffness ratio.

The dynamic stiffness match principle is a powerful design tool. Depending on the design degrees of freedom under consideration, either the static stiffness ratio or the operating frequency ratio can be selected in such a manner that the operating point is as close as possible to the optimum condition. Adequate use of the dynamic match principle in ISA systems design can produce significant weight savings and increased performance.

10.8.5 Damped Dynamic System

For a system that presents internal and external losses, the dynamic stiffness is a complex quantity with real and imaginary components. In such a case, the dynamic stiffness match principle has to be extended to take into account the additional aspects of complex power analysis. We define

$$\bar{P} = \bar{F}\dot{\bar{u}}, \quad \bar{F} = -\bar{k}_e \bar{u}_e, \quad \dot{\bar{u}} = i\omega\bar{u}, \quad P_{rating} = |\bar{P}|, \quad \cos\varphi = \mathrm{Re}\bar{P}/|\bar{P}|, \quad P_{av} = |\bar{P}|\cos\varphi \qquad (10.289)$$

The complex power $\bar{P}$ varies with frequency and static stiffness ratio. Its expression is

$$\bar{P}(r,p) = -i\frac{\bar{r}(r,p)}{[1 + \bar{r}(r,p)]^2}P_{ref} \qquad (10.290)$$

Figure 10.48 presents the variation of power rating with frequency for a statically matched and two statically unmatched conjugate systems. It can be seen that the statically matched and the soft systems display a decrease in power rating with frequency, as expected. The stiff system displays a power rating increase up to the dynamic frequency match point, followed by a decrease. These trends are consistent with the behavior displayed by the undamped system (Figure 10.47). An aspect particular to the damped systems is the variation of the power dissipation factor, $\cos\phi$, given in Figure 10.49. For a statically matched design, the power dissipation factor is positive ($\cos\phi > 0$) and increases with

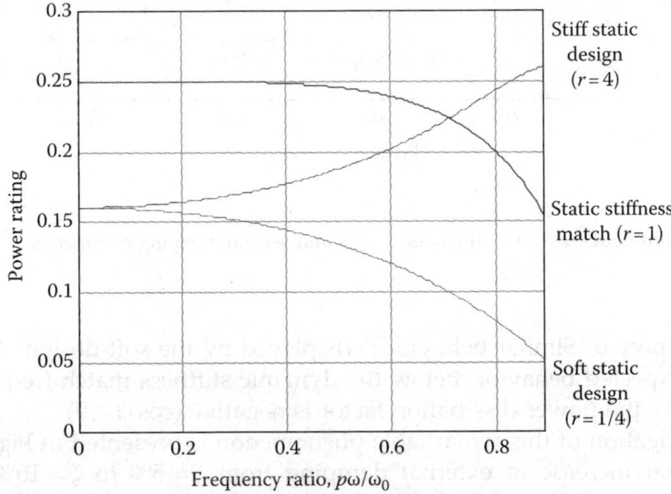

FIGURE 10.48
Variation of power rating with frequency ratio for three stiffness ratios ($\zeta = 5\%$, $\eta = 0$).

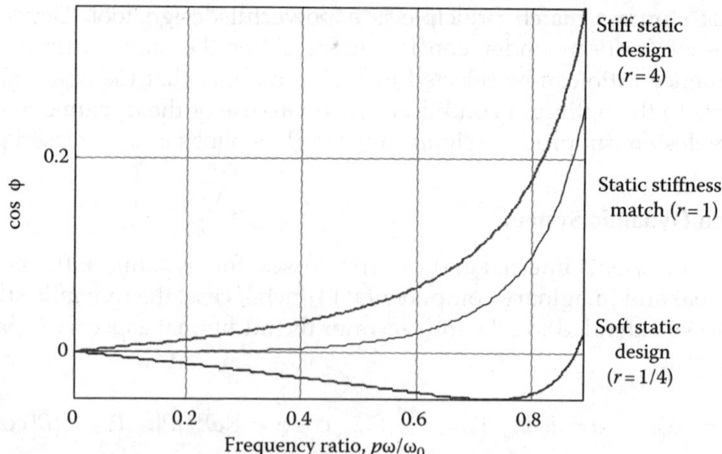

FIGURE 10.49

Variation of cos φ with frequency ratio for three stiffness ratios ($\zeta = 5\%$, $\eta = 0$).

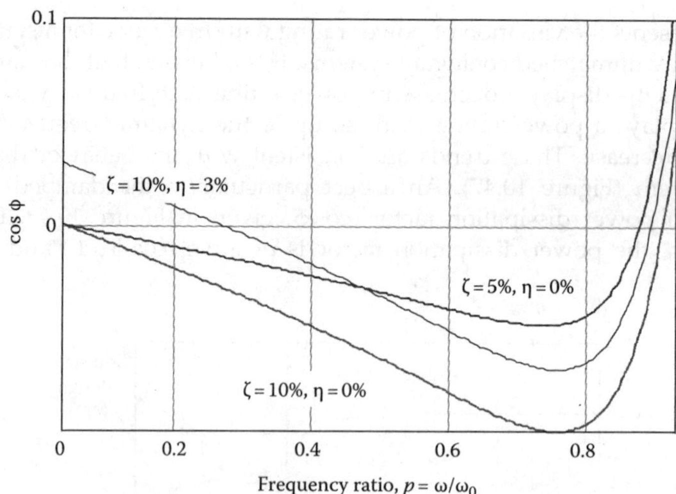

FIGURE 10.50

Variation of cos φ with frequency for a stiff static design and several damping parameters.

frequency, as expected. Similar behavior is displayed by the soft design. The stiff design presents an unexpected behavior: below the dynamic stiffness match frequency given by Equation (10.288), the power dissipation factor is negative ($\cos \phi < 0$).

Further investigation of this remarkable phenomenon is presented in Figure 10.50. This figure shows that increase in external damping from $\zeta = 5\%$ to $\zeta = 10\%$ increases the amplitude of the negative power dissipation factor. Increase of internal damping from $\eta = 0\%$ to $\eta = 3\%$ has the opposite effect, and decreases the negative power dissipation factor.

10.8.6 Design Example of Induced-Strain Actuation Application

We consider a design example in which an airfoil vane has to be actuated with the parameters given in Table 10.14. Figure 10.51 presents the design flowchart used to meet the requirements of the actuation application.

The actuation can be done using harmonic, step, or ramp excitations (Figure 10.52). Alternatively, special amplitude-modulated signals can be used to meet the power requirements. The variations of the required energy and power within a cycle for the above-mentioned excitations are used to determine whether existing induced-strain actuators are capable of meeting the actuation requirements with a given input signal.

Table 10.15 compares the peak values of required energy and power for each actuation signal shape. These values are then compared with the energy/power capabilities of existing induced-strain actuators.

We will assume an induced-strain actuator that acts through a displacement amplification mechanism in order to produce the displacements required to meet the design specification. The metric for selection is the maximum energy available from the smart material device. The limitation on the available power comes mostly from the maximum current capabilities of the power amplifier.

In the description of the displacement amplification mechanism, we assumed a simple model (Figure 10.53) that transforms the linear motion of the active material device into the rotary output motion required in the design specification. In this model, the displacement amplification mechanism was taken as a "black box" with two design variables: the gain, G, and the work efficiency, η_m, defined as

$$G = \frac{u_e}{u_e'}, \quad \eta_m = \frac{F_e u_e}{F_e' u_e'} \tag{10.291}$$

The output of the displacement amplifier was fed into a rotor arm of radius $r_0 = 5$ mm.

TABLE 10.14

Preliminary Requirements for an
Induced-Strain Actuation Application

Moment	$M_{peak} = 2.5$ Nm
Deflection	$\delta = +/-3°$
Rate	$\dot{\delta} = 3.5$ rad/s

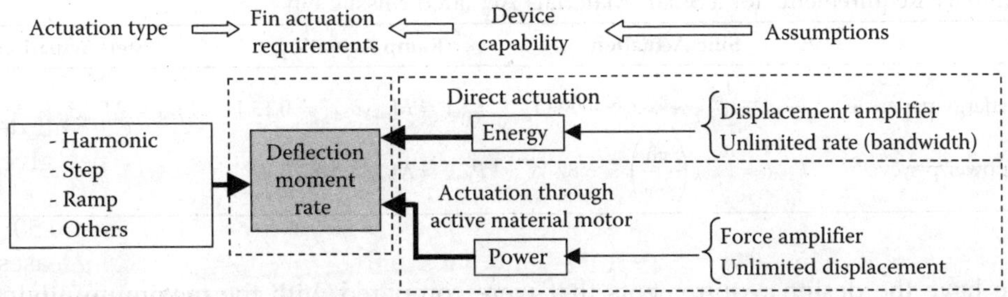

FIGURE 10.51
Meeting the requirements involves input signal design and smart-materials device capabilities.

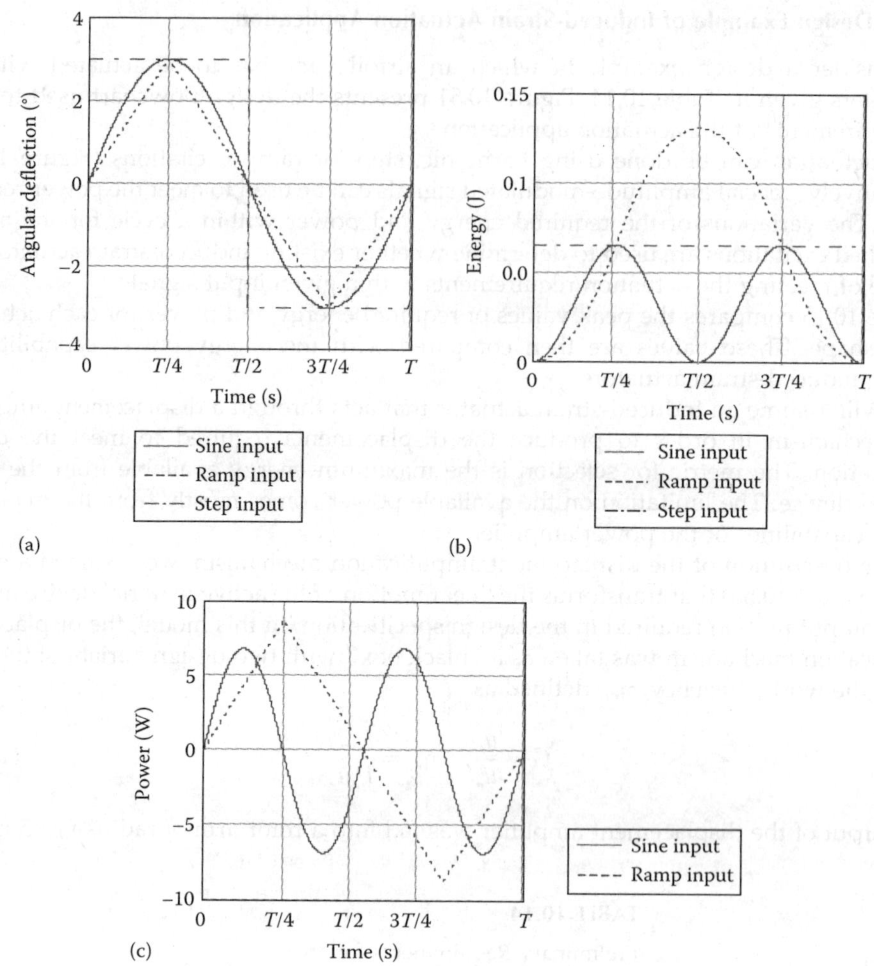

FIGURE 10.52
Variation of the required actuation energy and power within a cycle: (a) input signal; (b) required energy within a cycle; (c) required power within a cycle.

TABLE 10.15

Preliminary Requirements for a Smart-Materials Actuated Missile Fin

	Sine Actuation	Ramp Actuation	Step Actuation
Peak energy per cycle	$E_{peak} = \dfrac{1}{2} M_{peak}\delta_{peak} = 0.065$ J	$E_{peak} = M_{peak}\delta_{peak} = 0.13$ J	$E_{peak} = M_{peak}\delta_{peak} = 0.13$ J
Peak power per cycle	$P_{peak} = E_{peak}\left(\dfrac{\pi\dot{\delta}}{2\delta_p}\right) = 6.87$ W	$P_{peak} = M_{peak}\dot{\delta} = 8.75$ W	—

We have the design requirements that were compared with the maximum attainable performances of the active material devices, using the review described in Section 10.8. It was found that the required energy values fall within the capabilities of commercially

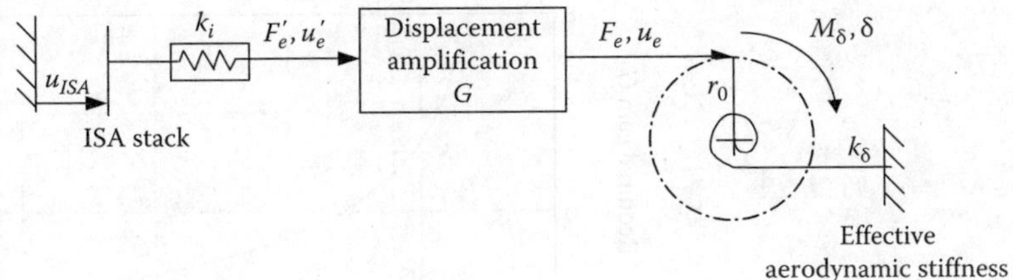

FIGURE 10.53
Schematic drawing of the displacement amplification process.

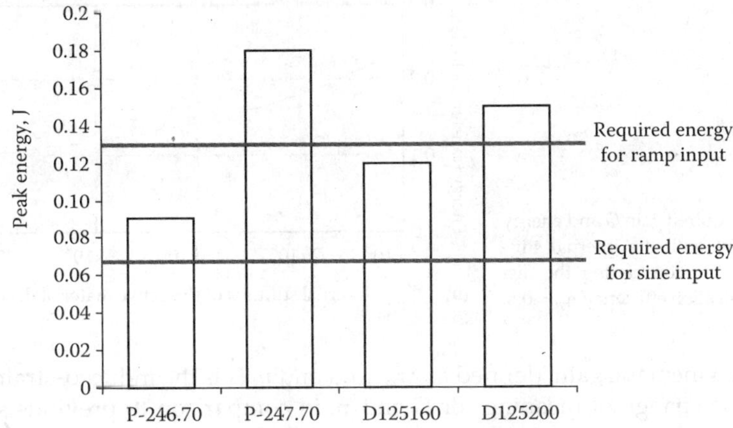

FIGURE 10.54
Commercially available actuators meeting the energy requirements for sine and ramp signals.

available induced-strain actuators for both sine and ramp actuation (Figure 10.54). Hence, two candidates were identified: P-247.70 and D125200.

The design of a displacement amplifier for the case of direct actuation must address several other parameters besides the required energy output. This is mainly because displacement amplifiers have a smaller-than-unity energy transmission efficiency, η_m. For example, if we use the P-245.70 actuator and model the load response solely as a spring system, the displacement amplification should have the energy transfer efficiency, η_m, in excess of 0.36. For the D125200 actuator, the same parameter η_m should be greater than 0.42.

These aspects outline the need to choose another design metric specific to the design amplification mechanism besides the energy transfer coefficient which is a smart material device characteristic. If the induced-strain actuator can be designed at will, the internal stiffness of the actuator can be considered as a design parameter. We define the energy transfer coefficient, E'_e, as the ratio of the available energy to the reference energy

$$E'_e = \frac{\frac{1}{2}k_\delta\delta^2}{\frac{1}{2}k_iu_{ISA}^2} = \frac{rG^2}{\left(1+r\dfrac{G^2}{\eta_m^2}\right)^2} \tag{10.292}$$

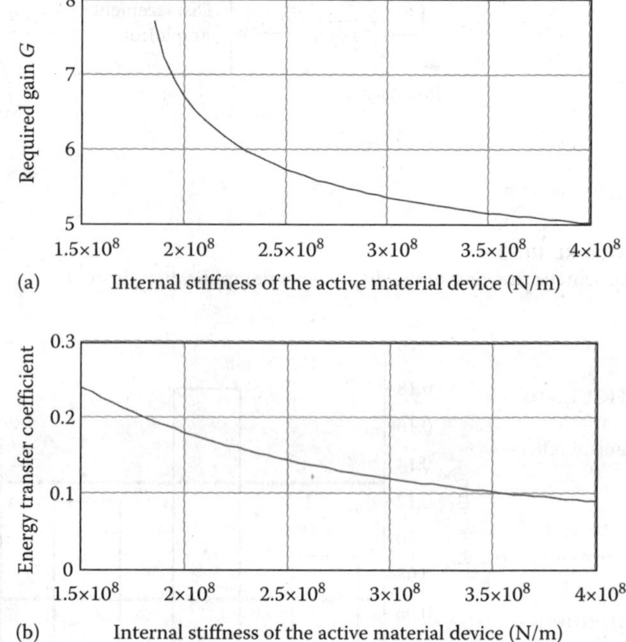

FIGURE 10.55
Variation of the required gain G and energy transfer coefficient with the internal stiffness of the actuator, considering the displacement amplification efficiency $\eta_m = 0.8$.

where η is the kinematic gain, defined as $\delta r_o / u_{ISA}$ and u_{ISA} is the induced-strain free stroke. Please note the change of meaning for G and η, in comparison to previous sections. The overall gain is defined as

$$G = \eta_m \left(\frac{1 - \sqrt{1 - 4\frac{r\eta^2}{\eta_m}}}{2r\eta} \right) \tag{10.293}$$

Figure 10.55 illustrates how the gain and the energy transfer coefficient vary with the internal stiffness of the actuator. Since an optimal design should tend toward minimum gain and maximum transfer, it is apparent from Figure 10.55 that an optimum point can be achieved. If we define the metric to characterize the optimum point as the ratio of the energy transfer coefficient to the gain of the displacement amplification mechanism, an optimal internal stiffness can be found for any allowed energy transfer efficiency (Figure 10.56).

10.9 Power Supply Issues in Induced-Strain Actuation

The power delivery capability of induced-strain actuators depends on internal and external factors. For the best utilization of a given induced-strain actuator, several aspects should be considered. Under static operation, the maximum energy output from a given actuator is obtained when the internal and external stiffness are matched. Under dynamic operation,

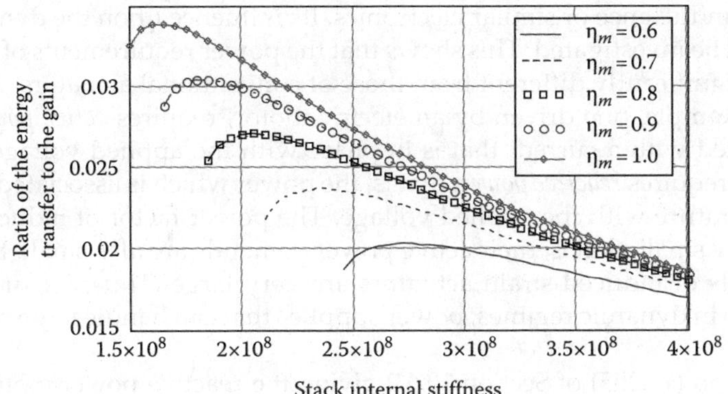

FIGURE 10.56
Variation of the optimum metric $R = E'_e/G$ with the internal stiffness and the displacement amplification mechanism efficiency.

this simple rule no longer applies, and the complex ratio between the internal and external complex stiffness has to be considered. Since the external stiffness also contains inertial (reactive) terms, significant variation of the apparent external stiffness takes place over the frequency domain. Hence, the concept of dynamic stiffness match should be considered in any dynamic application of induced-strain actuators. This match is frequency dependent, and hence the concept of stiffness tuning for optimal performance at the nominal application frequency should be considered.

Another important aspect is that of the power supply. For static operations, the power supply needs only to provide the input voltage necessary to obtain the required expansion of the electroactive material inside the induced-strain actuator. Since the operation is static (or quasi-static), the power supply current is of little consequence; adequate results can be obtained with power supplies of relatively low power ratings. Once the desired voltage has been attained, the power supply needs only provide sufficient residual current to compensate for the charge leakage.

For dynamic operations, the power supply must be able to supply not only the voltage but also a sizable current. During a harmonic operation, the electroactive material inside the induced-strain actuator is cyclically charged and discharged. This means that electric charge has to be carried into the induced-strain actuator during the expansion part of the cycle, and then carried out of the induced-strain actuator during the contraction part of the cycle. This cyclic charging and discharging of the induced-strain actuator may require considerable current since the induced-strain actuator has a sizable capacitance. The shorter the cycle, that is, the higher the frequency, the larger is the current required to achieve adequate dynamic operation.

The operating characteristics of the electrical power supply are of major importance in the dynamic applications of induced-strain actuators. At low frequencies (quasi-static operation), the voltage of the power supply is the controlling factor. Since the induced-strain actuator admittance increases linearly with frequency, dynamic operation may lead to significant electric current demands that have to be met by the power supply. The high values of peak current required taking the induced-strain actuator through the harmonic cycle present a challenge to the power supply. Additionally, the back-and-forth circulation of high currents through the supply leads may produce significant power loss through electrical heating. The consumption of reactive power can be reduced through the use of a compensating inductance or similar electronics. Its influence upon the dynamic matching concept should be investigated. This shows that the power requirements of induced-strain actuators are significantly different from those of conventional actuators. A conventional actuator, for example, one driven by an electric motor, requires *active power*, that is, the power associated with a current that is in phase with the applied voltage. An induced-strain actuator requires *reactive power*, that is, the power which is associated with a current that is in quadrature with the applied voltage. The power factor of induced-strain actuators is typically small, that is, their active power demands are also small. But the reactive power demands of induced-strain actuators are very large. Thus, for driving induced-strain actuators in dynamic regimes, power supplies that can handle large reactive powers are needed.

Recall Equation (10.255) of Section 10.6.2 giving the reactive power input requirements for an induced-strain actuator, that is,

$$P_{in} = \omega\chi(v_0)\left(1 - \kappa^2\frac{r}{1+r}\right)\left(\tfrac{1}{2}C_0V^2\right) \tag{10.294}$$

where C is the capacitance, V is the voltage amplitude, ω is the angular frequency, r is the dynamic stiffness ratio, and $\chi(v_0)$ is the reactive power correction factor that depends on the bias voltage coefficient, v_0. The reactive power correction factor takes values between 1 and 3.5 for v_0 between 0 and 1.5. Since induced-strain actuators can operate in the kilohertz range, Equation (10.294) indicates that very large power densities may be necessary. Thus, a number of practical barriers need to be overcome in the dynamic operation of induced-strain actuators: (a) the capability of power supply to deliver reactive power in the kilovolt-ampere range; (b) the dissipation of the heat generated as this reactive power is transported in and out of the induced-strain actuator; and (c) the electromechanical system resonance that may set its own limits on the frequency that can be achieved during the dynamic operation.

The power supply aspects of ISA are currently being addressed by using specialized power supplies (switching amplifiers) that are able to handle reactive loads much better than conventional linear amplifiers. The switching amplifiers (Figure 10.57) utilize high frequency pulse width modulation (PWM) principles and solid-state switching technology to drive current in and out of the capacitive load presented by the induced-strain piezo-electric actuator. In this process, a high-value inductance ballast is utilized (Figure 10.57a). To simultaneously achieve optimal actuator performance and minimum weight, the switching amplifier design has to be performed using a complete model that includes adequate representation of the actuator and external load dynamics (Figure 10.57b).

The difficulty connected with the electric-supply may seem less severe when using magnetoactive devices, such as Terfenol-D actuators, because they are current-driven and require lower voltages at low frequencies. At present, the low-voltage power supply technology is predominant and cheaper than the high-voltage technology, which somewhat facilitates the use of magnetostrictive actuators.

The effect of adaptive excitation on the current and power requirements and on the displacement output from a heavy-duty piezoelectric actuator driving a smart structure near the electromechanical resonance is shown in Figure 10.58. The actuator was assumed to have $k_i = 370$ kN/mm internal stiffness, $C = 5.6$ μF internal capacitance, 3.5 kHz resonance, and output displacement $u_m \pm u_a = 22.5 \pm 37.5$ μm when driven by a voltage $V_m \pm V_a = -375 \mp 625$ V. The external mechanical load was assumed to have a matched

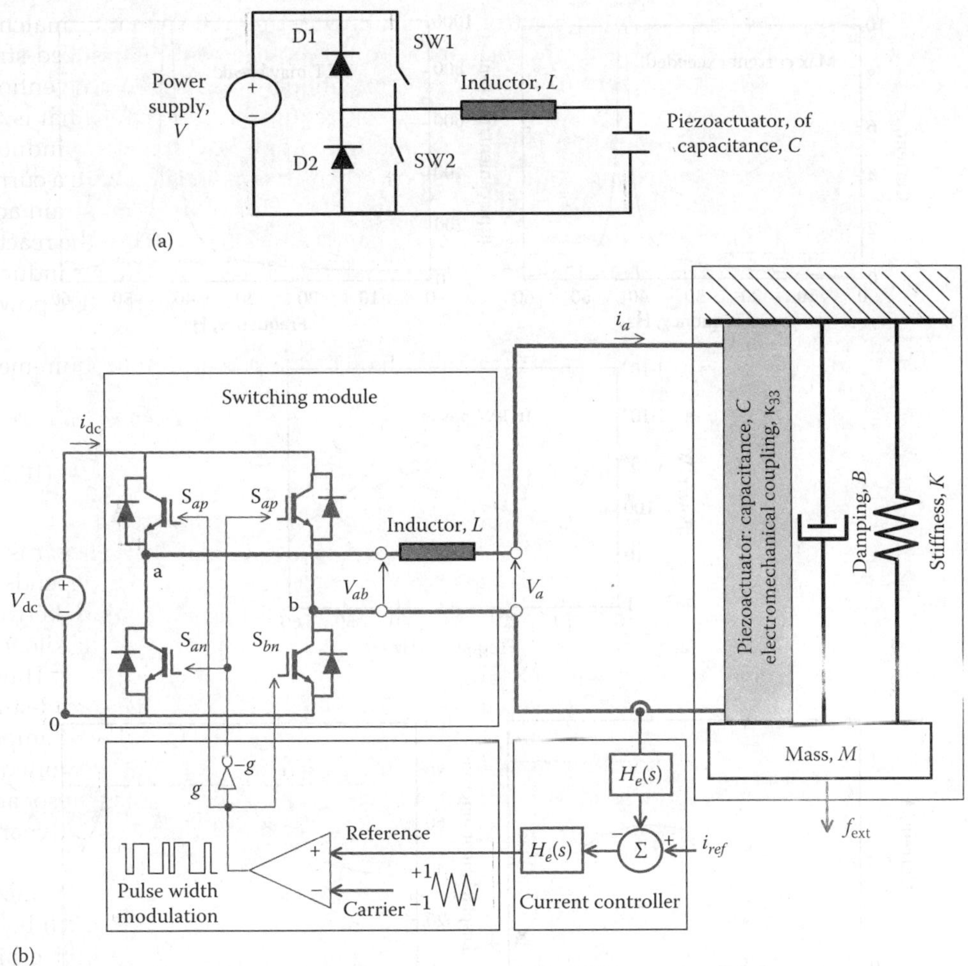

FIGURE 10.57
Power supplies for active material actuators: (a) principle of switching power supplies for high reactive load; (b) schematic of the supply system incorporating the switching module, current controller, pulse width modulator, and the piezo actuator-external load assembly.

static stiffness, a mechanical resonance of 30 Hz, and a 1% internal damping. The system was assumed to be driven by a 1 kVA amplifier, with up to -1000 V voltage and 1 A current. As shown in Section 10.10.2, the electromechanical system incorporating the structure and the embedded actuator displays an electromechanical resonance at 42.42 Hz. Without adaptive excitation, that is, under constant voltage supply, the current demands are very large, the actuator experiences a displacement peak that may lead to its destruction, and the required power is excessive. When adaptive excitation was simulated, both the displacement and the current could be kept within bounds; the power requirements became more manageable and stayed within the 1 kVA capability of the power amplifier.

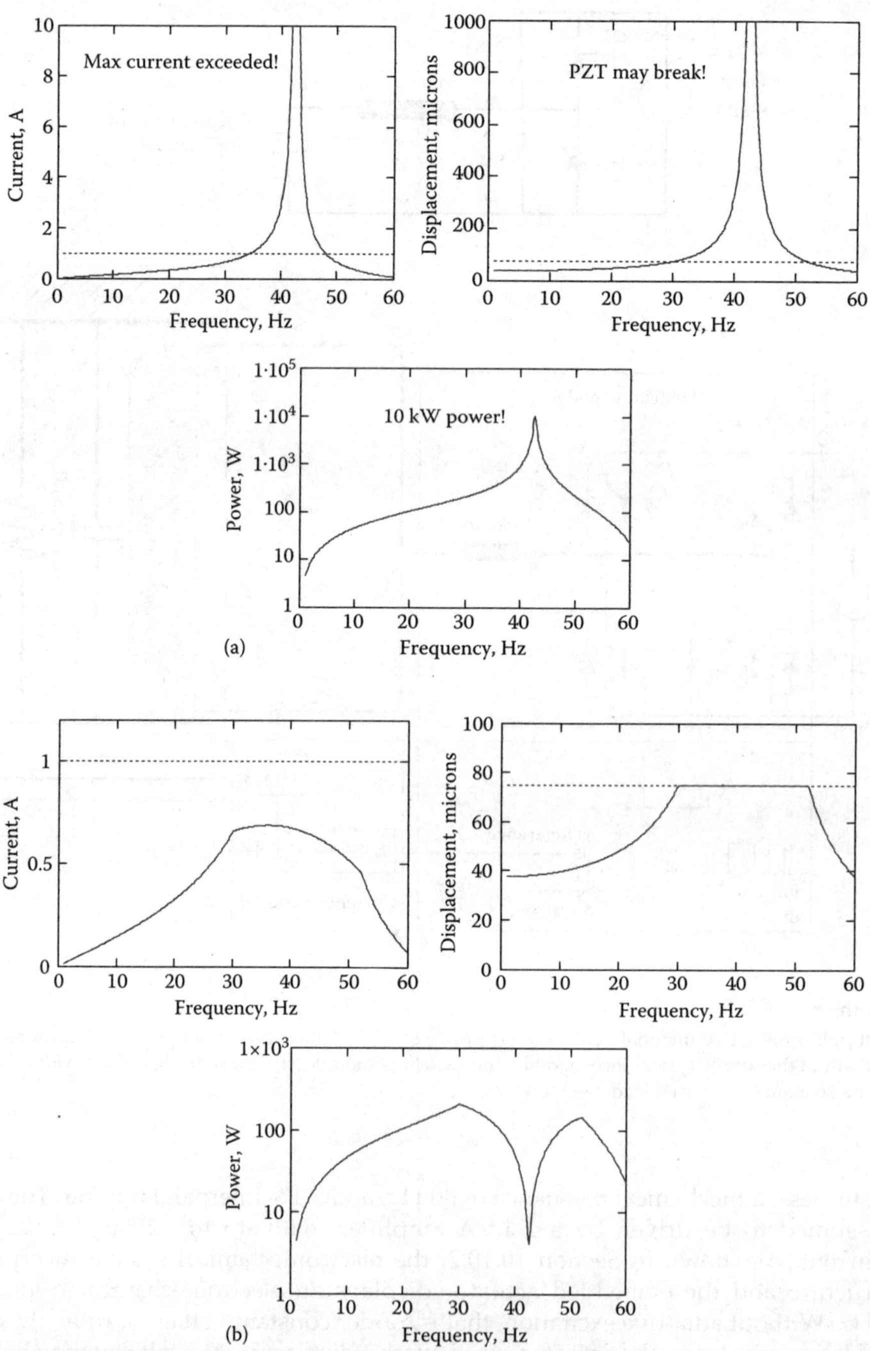

FIGURE 10.58
Adaptive excitation of piezoelectric actuators near embedded electromechanical resonance: (a) under constant voltage excitation, the current demands are very large, the actuator may break, and the required power is excessive; (b) with adaptive excitation, both the displacement and the current are kept within bounds, and the power becomes manageable.

10.10 Shape Memory Alloy Actuators

A class of induced-strain actuating materials with much larger strain response but low frequency bandwidth is that shape memory alloys (SMA). The SMA effect is a highly nonlinear phenomenon; hence, the analysis of these actuators cannot be easily presented in terms of conventional linear analysis. This section will first introduce the SMA phenomenon, then discuss some common SMA materials, and end by presenting the principles of SMA induced-strain actuators. A simplified analysis will be presented at first; then, a more elaborate nonlinear analysis will be briefly discussed. Examples will be used to drive home the main points of SMA actuators design and utilization.

10.10.1 Shape Memory Phenomenon

SMA materials are thermally activated ISA materials that undergo phase transformation when the temperature passes certain values. The metallurgical phases involved in this process are the low-temperature martensite and the high-temperature austenite. When phase transformation takes place, the SMA material modifies its shape, that is, it has "memory." The SMA process starts with the material being annealed at high-temperature in the austenite phase (Figure 10.59). In this way, a certain shape is "locked" into the material. Upon cooling, the material transforms into the martensite phase and adopts a twinned crystallographic structure. When mechanical deformation is applied, the twinned crystallographic structure switches to a skew crystallographic structure. Strains as high as 8% can be achieved through this de-twinning process. This process gives the appearance of permanent plastic deformation though no actual plastic flow took place. For this reason, it is called quasiplasticity. In typical actuator applications, this process is used to store mechanical energy by stretching the SMA wires.

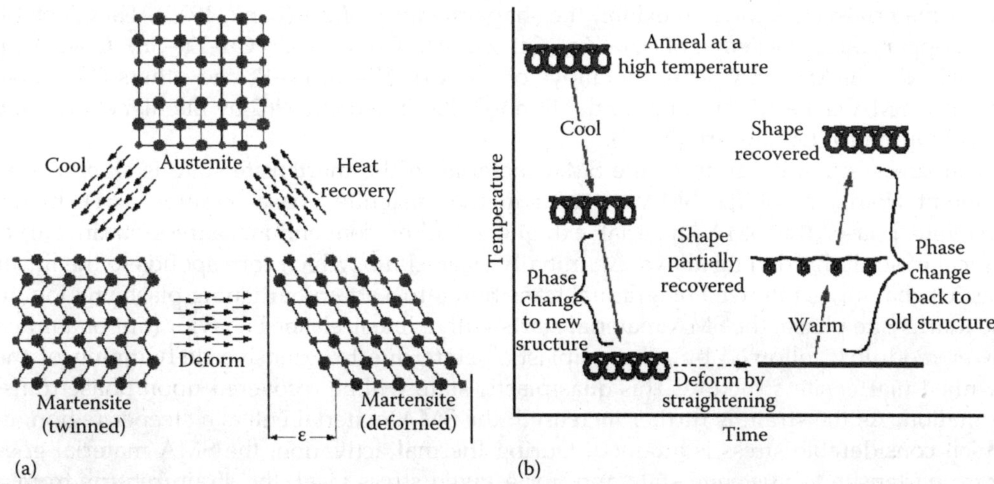

FIGURE 10.59
Principles of SMA materials: (a) change in crystallographic structure during cooling and heating; (b) associated component-shape changes, using a coil spring as an example.

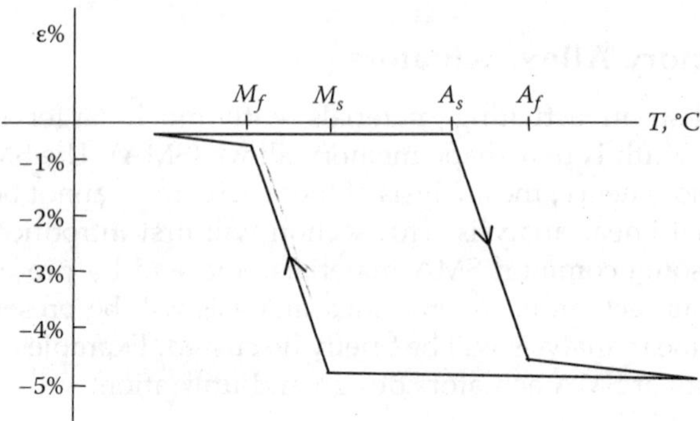

FIGURE 10.60
Schematic representation of strain recovery through the SMA effect in a two-way shape memory alloy.

Upon heating, the martensite phase changes into austenite phase and the shape initially imposed by annealing is recovered. In this way, the permanent deformation created through de-twinning of the martensite phase is removed, and the material returns to its initial state memorized during annealing at austenite temperatures. If recovery is mechanically prevented, restraining stresses of up to 700 MPa can be developed. This is the one-way shape memory effect, which is a one-time only deployment at the end of which full recovery of the initial state is obtained. For dynamic operations, a two-way shape memory effect is preferred, in order that cyclic loading and unloading is attained. Two-way shape memory effect can be achieved through special thermomechanical training of the material, in which one part of the material is stretched while another part is compressed, acting as recovery spring. In such cases, repeated activation of SMA strains is possible in synch with heating–cooling cycles (Figure 10.60).

10.10.2 Common SMA Materials

Many materials are known to exhibit the shape-memory effect (Bank, 1975). They include the copper alloy systems of Cu–Zn, Cu–Zn–Al, Cu–Zn–Ga, Cu–Zn–Sn, Cu–Zn–Si, Cu–Al–Ni, Cu–Au–Zn, Cu–Sn; the alloys of Au–Cd, Ni–Al, Fe–Pt and others. The most common SMA is the Ni–Ti compound, Nitinol™, discovered in 1962 by Buehler et al. at the Naval Ordnance Laboratory (NOL).

The stress–strain diagram of the SMA material in the martensite state is significantly different from that of the SMA material in the austenite phase (Figure 10.61). In the austenite phase, the SMA material exhibits a rather conventional stress–strain curve, typical of strain-hardening alloys. An initially steep climb, which corresponds to the linear elastic behavior, is followed by gradual transition into a strain-hardening plastic region. In the martensite phase, the SMA material starts with a linear-elastic behavior (but of a much lower modulus) followed by a "quasiplastic" state due to extensive de-twinning of the twinned martensite structure. This quasiplastic state will be recovered upon phase transformation. As the strain is further increased, the SMA material enters a steeper region, in which considerable stress is attained. During thermal activation, the SMA material goes from martensite to austenite state and, for a given stress level, the strain returns from a high value in the martensite state to a low value in the austenite state. This is the ISA strain, S_{ISA}, used in applications.

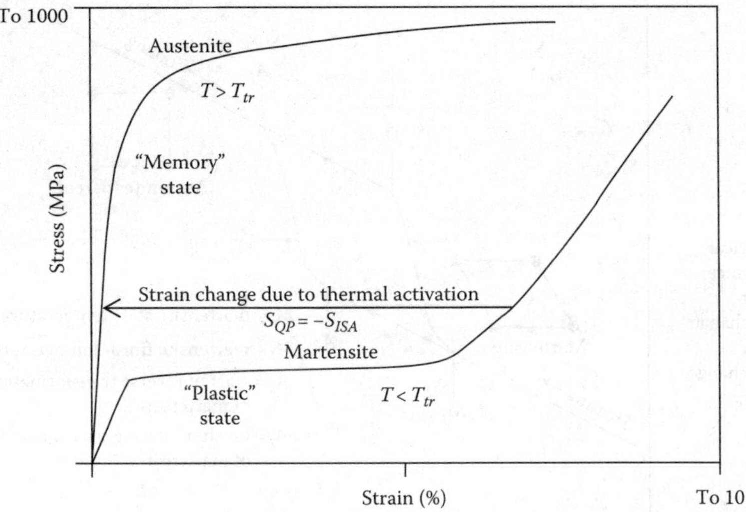

FIGURE 10.61
Stress–strain curves for a shape memory alloy in austenite and martensite phases.

During heating, the change from martensite to austenite starts at temperature A_s and finishes at temperature A_f. During cooling, the reverse change from austenite to martensite starts at temperature M_s and finishes at temperature M_f. Note that $A_s < A_f$, whereas $M_s > M_f$. In the interval (A_s, A_f) on heating, and (M_s, M_f) on cooling, two phases coexist in temperature-dependent volume fractions. The values of the phase-transition temperatures can be varied with alloy composition and stress level. In fact, in SMA materials, stress and temperature act in opposition such that an SMA material brought to austenite by heating can be isothermally forced back to martensite by stressing (Figure 10.62b). SMA composition can be tuned to start the austenite transformation at almost any predetermined temperature in the range $-50°C$ to $62°C$ (Duerig et al., 1990) and even up to $100°C$ (http://www.nitinol.com/). Even higher transition temperatures ($200°C–300°C$) have recently been reported by NASA Glenn Research Center in Ohio using the NiTiPt family of SMA materials (www.techbriefs.com, March 2008). An example of Nitinol SMA material properties is given in Table 10.16.

10.10.3 New Shape Memory Materials

In recent years, new classes of shape memory materials have been developed. One of these classes is the *magnetostrictive SMAs*. Magnetostrictive SMAs are being developed that respond to magnetic excitation. Because magnetic excitation can be rapidly switched on and off, the magnetic SMA materials should have a much fast frequency response than the thermal SMA materials. Furuya et al. (1998) proposed a rapidly solidified ferromagnetic shape memory Fe-29.6at%Pd alloy that shows magnetostriction of up to 1800 microstrain at an applied magnetic field of 8×10^5 A/m. The material also holds its dependency on temperature (see Figure 10.63). Furuya et al. (1998) also investigated the dynamic response under the alternative magnetic field and have confirmed that the magnitude of striction of

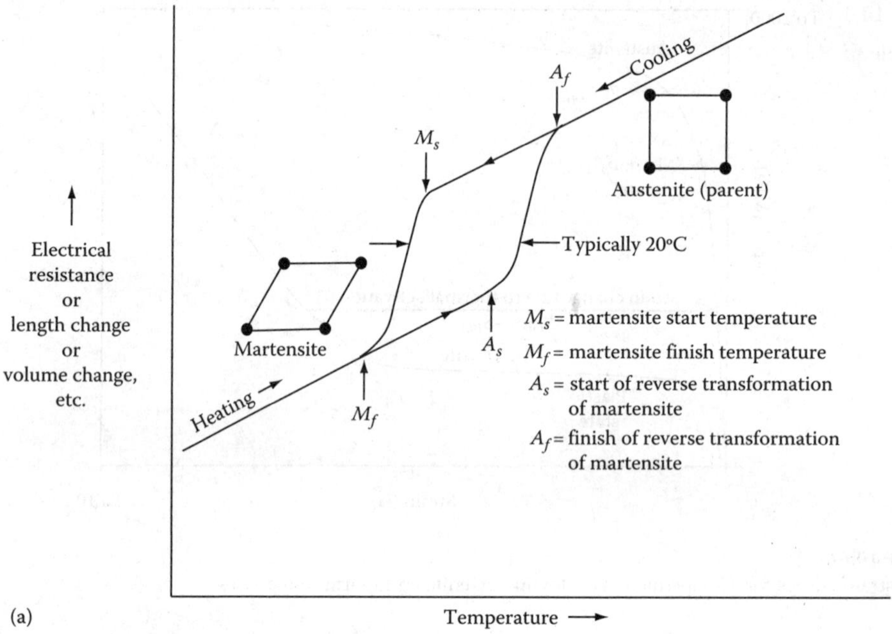

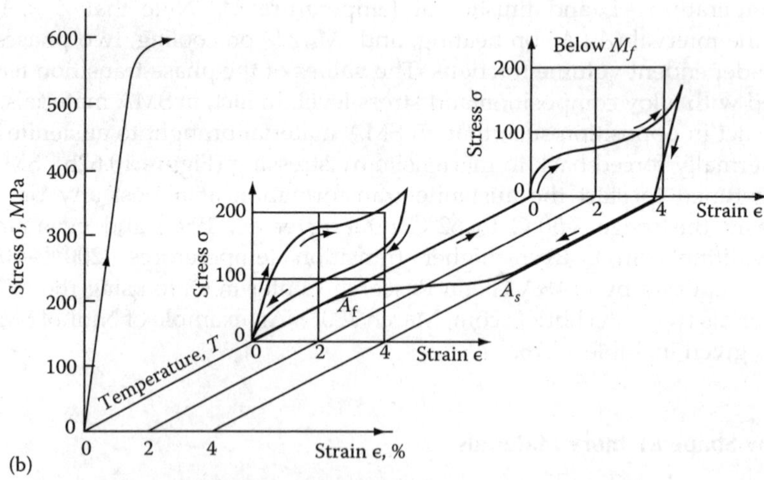

FIGURE 10.62
Schematic of phase transition effects in SMA materials: (a) temperature-induced phase transition between the cubic austenite phase (parent phase) and rhombic martensite phase (product phase); (b) effect of temperature on the stress–strain curve.

the Fe–Pd foil is at least 10 times greater than that of the conventional Fe-based magnetostrictive materials at 10 Hz (Figure 10.64). The response signal can be confirmed up to 100 Hz, which is at least 20 times faster than that of thermal SMA materials.

TABLE 10.16

Example of Nitinol SMA Material Properties

Elastic Modulus (GPa)	Phase Transformation Temperatures (°C)	Other Material Properties
Austenite	Martensite finish $M_f = 10$	Electric resistivity $\rho_\Omega = 70\ \mu\Omega$ cm
$E^A = 65$	Martensite start $M_s = 18$	Relative mass density $\rho = 6.5$
Martensite	Austenite start $A_s = 34$	Specific heat $c_p = 0.25$ kcal/kg/°C
$E^M = 20$	Austenite finish $A_f = 45$	

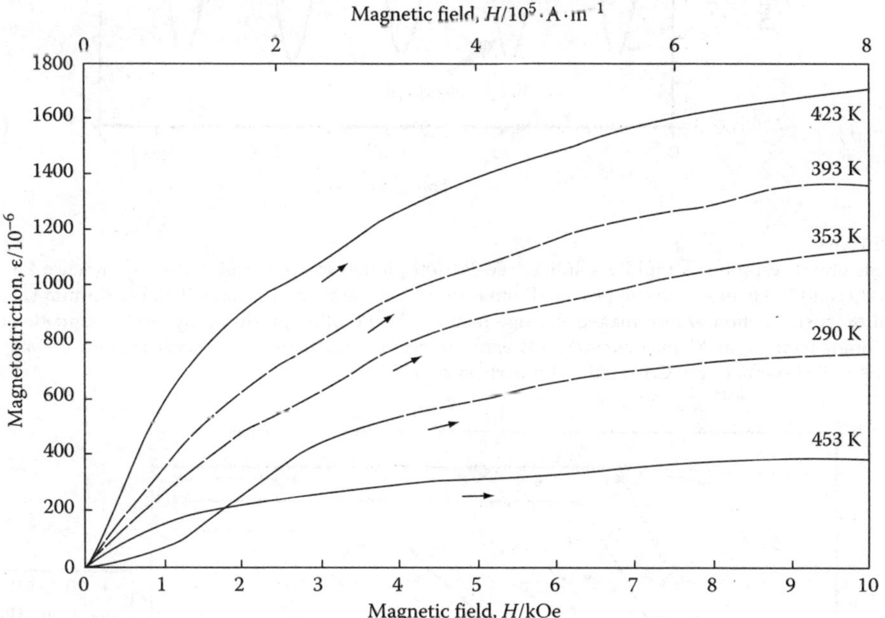

FIGURE 10.63

Changes of magnetostriction versus magnetic field curves with increasing temperature in melt-spun Fe–29.6at%Pd alloy thin plate (roll speed 28.3 m/s, 900°C, 1 h annealed, phase transformation temperature 163°C). (From Furuya, Y., Watanabe T., Hagood N. W., Kimura H., and Tani J., Giant magnetostriction of ferromagnetic shape memory Fe–Pd alloy produced by electromagnetic nozzleless melt-spinning method, in *9th International Conference on Adaptive Structures and Technologies*, Boston, MA, October 14–16, 1998, Technomics Pub. Co., 1998. With permission.)

Another SMA displaying ferromagnetic properties is Ni–Mn–Ga (Murray et al., 2000). The mechanism of field-induced deformation is the rearrangement of the variant structure of a twinned martensite through the motion of twin boundaries to accommodate the applied field and/or load. Field-induced twin boundary motion can result in deformation of several percent (Figure 10.65).

Another new class of shape memory materials is the *shape memory polymers* (SMP), which are obtained by using organic chemistry rather than metallurgy to obtain the shape memory effect. The shape memory behavior of SMP is explained in terms of entanglement and disentanglement of long polymeric molecules rather than phase transformation and twinning/de-twinning as in SMAs.

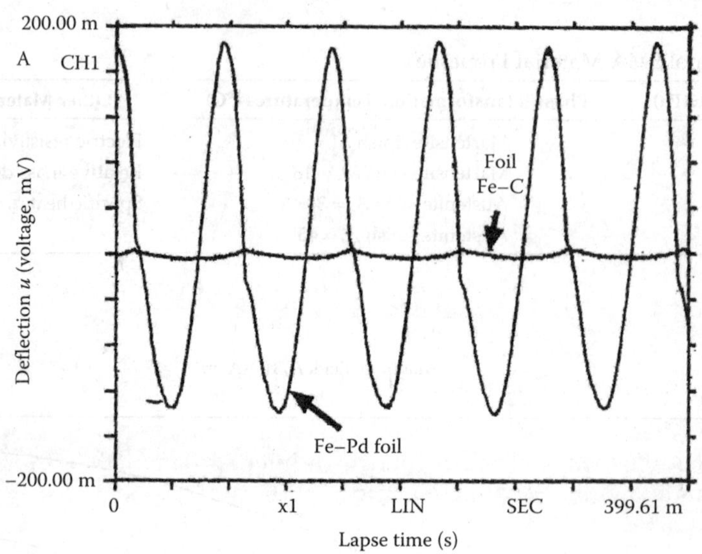

FIGURE 10.64

The cyclic strain response of rapidly solidified Fe–Pd foil plotted in comparison with a reference Fe–C foil at $H = 0.3$ kOe and 10 Hz in solenoid-type coil. (From Furuya, Y., Watanabe T., Hagood N. W., Kimura H., and Tani J., Giant magnetostriction of ferromagnetic shape memory Fe–Pd alloy produced by electromagnetic nozzleless melt-spinning method, in 9*th International Conference on Adaptive Structures and Technologies*, Boston, MA, October 14–16, 1998, Technomics Pub. Co., 1998. With permission.)

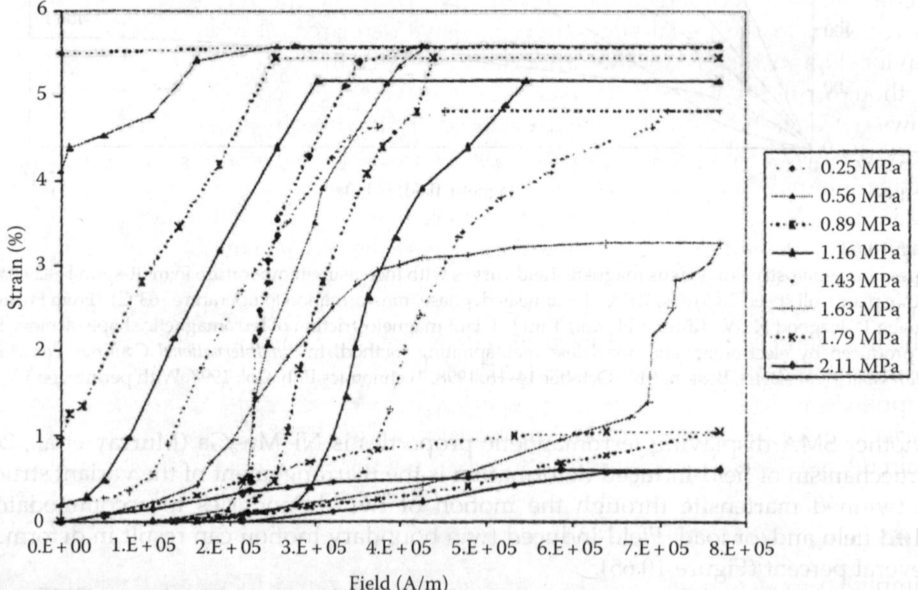

FIGURE 10.65

Magneto-mechanical response of Ni–Mn–Ga at various constant stress levels. (From Murray, S. J., Allen, S. M., O'Handley, R. C., and Lograsso, T. A., Magnetomechanical performance and mechanical properties of Ni–Mn–Ga ferromagnetic shape memory alloy, in *Smart Structures and Materials 2000, Proc. SPIE*, vol. 3992, pp. 387–395. With permission.)

TABLE 10.17

Comparative Properties of Three Metallic Wires

Wire	Initial Young's Modulus, GPa	Tensile Strength, MPa	Strain to Failure, %	Energy Dissipation at Failure, J/cm^3	Relative Density
Superelastic	66–80 (A)	1600	14.5	125	6.5
Nitinol	20–30 (M)				
304 Stainless steel, CR 50%	193	1030	4.5	41	8
5036 Aluminum	70	340	10	31	2.7

10.10.4 Superelasticity of SMA Materials

In addition to "memory," SMA materials have other remarkable properties: phase-dependent elastic modulus, superelasticity, and high internal damping. The phase-dependent elastic modulus allows the SMA materials to be stiffer at higher temperatures than at lower temperatures (Table 10.17).

In most metals, the elastic modulus decreases with temperature. In SMA materials, it actually increases, since the modulus of the high-temperature austenite phase can be up to 3 times larger than that of the low-temperature martensite phase. Superelasticity is associated with the fact that strains of up to 12% can be accommodated before actual plastic deformation begins. Superelasticity is displayed by low-A_f SMA materials, which are austenite at room temperature. As stress is applied, the superelastic SMA material undergoes phase transformation and austenite changes into martensite. Upon unloading, the austenite phase is recovered, and the material returns to the zero-stress zero-strain state. This recovery to the initial state after extensive deformation represents the superelastic behavior (Figure 10.66). Another important aspect of the superelastic behavior is the fact that though full strain recovery has been achieved the recovery path is nonlinear and follows a hysteretic path. The area enclosed during the hysteresis loop represents the energy dissipated during this process. The dissipated energy can be associated with an effective internal damping coefficient. As seen in Figure 10.66, the area of the hysteresis loop is quite large. This means that the energy dissipation and the associated internal damping are also large. The internal damping of the SMA materials associated with the hysteresis curve enclosed by the loading–unloading cycle is orders of magnitude larger than that of conventional elastic materials.

The interplay between superelasticity effect and shape memory effect depends on the transition temperatures of the material (M_f, M_s, A_s, A_f), on the ambient temperature, and on the applied stress values. *In general terms, an increase in stress is equivalent to a reduction in the ambient temperature* (Figure 10.62). The transition temperatures can be manipulated to a certain degree through metallurgical processes (alloying and heat treatment).

10.10.5 Simplified Modeling of SMA Actuators

A simplified modeling of the SMA material behavior can be developed assuming that it consists of the superposition of two separate effects:

- Conventional elastic deformation ruled by Hooke's law which takes place with the appropriate moduli (or compliances) corresponding to the martensite or the austenite phase, as appropriate.

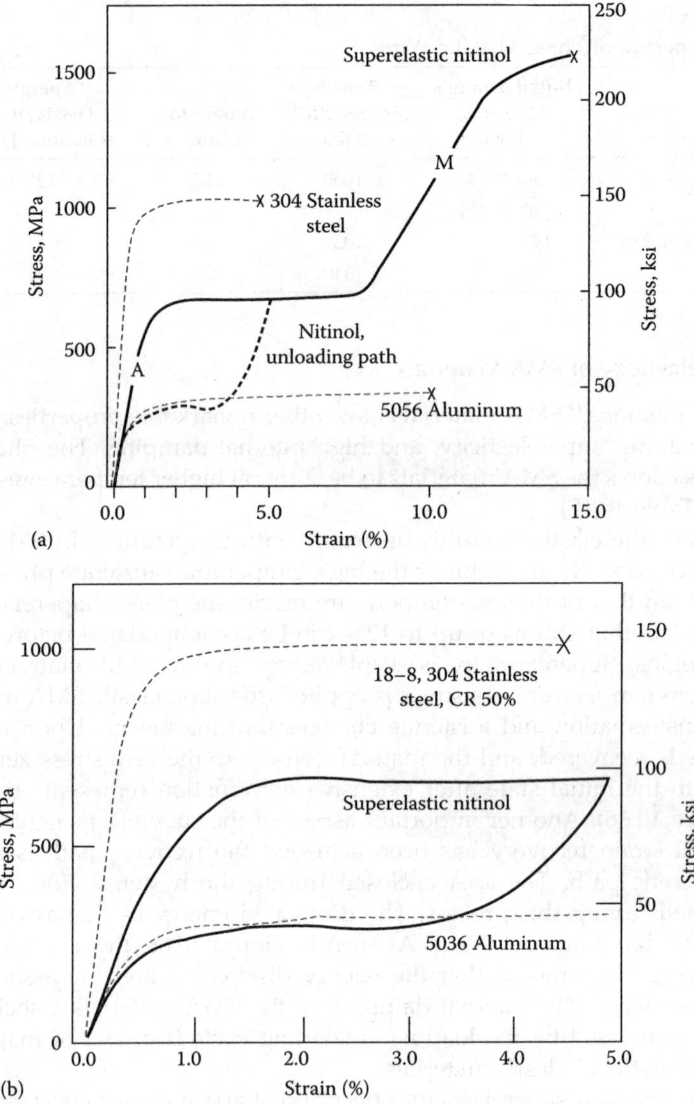

FIGURE 10.66
Stress–strain curves of superelastic Nitinol and other metallic wires: (a) overall behavior; (b) zoom-in on the low-strain range.

- Quasiplastic deformation due to de-twinning while the material is in the martensite phase. The quasiplastic deformation produces quasiplastic strain S_{QP}. When the phase transition is thermally activated, the quasiplastic strain is recovered producing the induced strain, $S_{ISA} = -S_{QP}$, which generates the ISA effect.

Under these assumptions, the strain equation can be written as:

$$S_1^M = s_{11}^M T_1 + S_{QP}$$
$$S_1^A = s_{11}^A T_1$$

(10.295)

where S_1 is the strain, T_1 is the stress, s_{11} is the elastic compliance, and S_{QP} is the quasiplastic strain generated through the de-twining of the SMA lattice under stress while in the martensite phase. Superscripts M and A signify that the material is in the martensite and austenite phase, respectively.

Upon thermal activation, the material recovers the austenite state, thus producing a change of shape, that is, shrinking by the apparent strain

$$\Delta S_1 = S_1^A - S_1^M$$
$$= s_{11}^A T_1 - s_{11}^M T_1 - S_{QP}$$
$$= \left(s_{11}^A - s_{11}^M\right)T_1 - S_{QP} \tag{10.296}$$

It is apparent that the shrinking strain has two components, one due to the recovery of the quasiplastic strain, the other due to the difference in compliance between the two metallurgical phases. When this effect is used to produce actuation, the apparent strain change is expressed in terms of the induced strain, S_{ISA}, and the elastic deformation of the material in each phase, that is,

$$\Delta S_1 = \left(s_{11}^A - s_{11}^M\right)T_1 + S_{ISA} \tag{10.297}$$

where

$$S_{ISA} = -S_{QP} \tag{10.298}$$

It is apparent from Equation (10.298) that, in the case of SMA materials, the induced strain is negative, that is, the induced-strain effect is a contraction. This is fundamentally different from the induced-strain effect observed in electroactive and magnetoactive materials, where the basic effect was an expansion. This aspect underlines a basic difference in how the two material types are used in actuation applications: the SMA materials are used to pull onto something, whereas the electro/magneto active materials are used to push onto something.

The simplified modeling equations (10.295) through (10.298) apply when the SMA activation is complete, that is, all the martensite phase has been transformed into austenite phase upon adequate heating. If the shrinking strain ΔS_1 is known, then Equation (10.297) can be used to calculate the induced strain, S_{ISA}, that is,

$$S_{ISA} = \Delta S_1 - \left(s_{11}^A - s_{11}^M\right)T_1 \tag{10.299}$$

Subsequently, Equation (10.298) can be used to get the quasiplastic strain, S_{QP}.

The blocked stress, T_{block}, and the blocked force, F_{block}, of the SMA actuating wire are determined as follows. First, assume that the wire in martensite phase is stretched to attain S_{QP}; then, the load is released and the end of the wire is fixed solidly and not allowed to move at all. When the SMA wire is thermally activated, the martensite to austenite phase transition takes place and the wire will want to shrink, but will not be allowed because it is blocked. The blocked stress, T_{block}, is the stress developed in the blocked wire after the austenite phase has been fully established. Its expression is obtained by making the second expression in Equation (10.295) equal to S_{QP}, that is,

$$S_{QP} = s_{11}^A T_{block} \rightarrow T_{block} = \frac{S_{QP}}{s_{11}^A} \tag{10.300}$$

Multiplication of Equation (10.300) by the wire cross-sectional area yields the blocked force, F_{block}, that is,

$$F_{block} = T_{block}A \tag{10.301}$$

10.10.6 SMA Actuator Experiment

As an example, Figure 10.67 presents a simple SMA actuation experiment in which a Nitinol SMA wire is used to lift a weight. The experimental setup shown in Figure 10.67a consists of a Nitinol SMA wire, which is attached to an adjustable support and stretched by a weight of mass $m = 3.175$ kg. A DC power supply is attached with insulated cables to the end of the wire. The wire attached to the adjustable support is electrically insulated from the ground. The dimensions of the wire are length, $L = 740$ mm and diameter, $d = 0.4064$ mm.

Typical Nitinol material properties are given in Table 10.16. When the weight is attached, the wire stretches; the support is adjusted such that the weight rests at $h_M = 75$ mm above the ground (Figure 10.67b). In this situation, the power supply is set at zero such that no current flows through the SMA wire and the wire is cool. When the power supply is activated, the wire starts to heat and the SMA effect comes into play. As a result, the wire shrinks and pulls the weight upward. Maximum upward displacement is attained when the current through the wire reaches $I_{max} = 1.5$ A. Under these conditions, the weight has

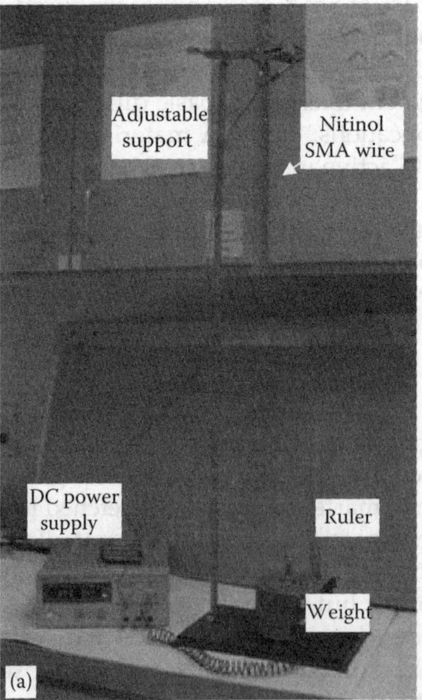

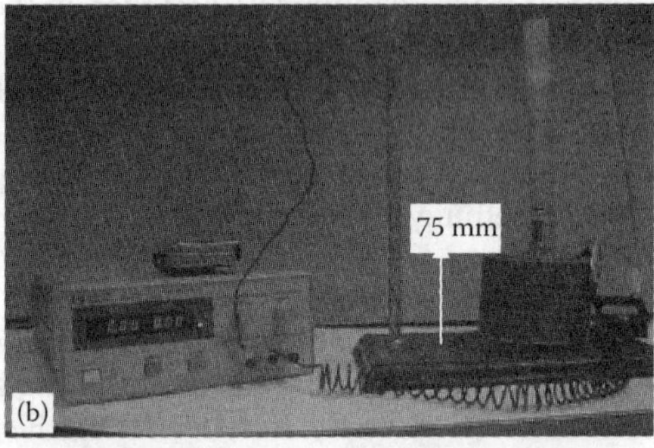

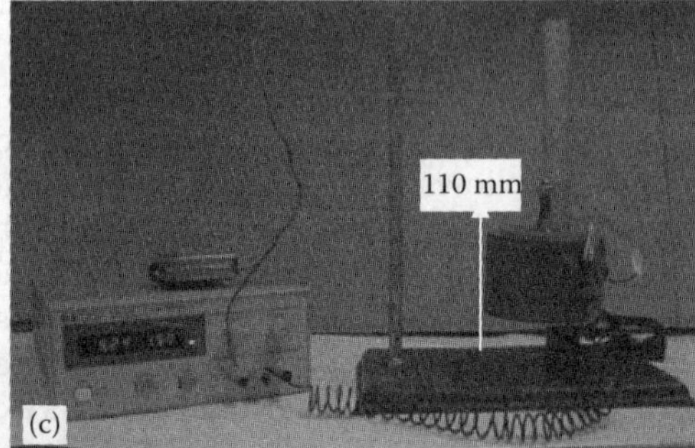

FIGURE 10.67
SMA actuator experiment: (a) experimental setup; (b) initial weight position at 75 mm above ground; (c) higher weight position at 110 mm above ground when the SMA wire was activated by a current of 1.5 A.

been lifted to the position $h_A = 110$ mm above the ground (Figure 10.67c). The wire length change due to thermal activation is calculated as

$$\Delta L = -h_A + h_M = -110 \text{ mm} + 75 \text{ mm} = -35 \text{ mm} \tag{10.302}$$

Subsequently, the apparent strain change is given by

$$\Delta S_1 = \frac{\Delta L}{L} = -4.7\% \tag{10.303}$$

The strain calculated in Equation (10.303) is negative, that is, it represents shrinking of the wire.

When the electric current is brought back to zero, the wire cools, and the weight pulls it back down. When the process is repeated, the same behavior is observed. However, small variations in the actual values are observed (Table 10.18). If this cycle is repeated several more times, an asymptotic behavior is observed where predictable values are obtained in each cycle (that is, the *SMA wire has been trained*). The data recorded in this experiment can be used to determine some important facts about this SMA wire during its training process. First, we will calculate the quasiplastic strain S_{QP}; second, we will calculate the blocked stress, T_{block}, and blocked force, F_{block}.

To calculate the quasiplastic strain, S_{QP}, we use the weight of the attached mass and the wire diameter to calculate the stress in the SMA wire, that is,

$$A = \frac{\pi}{4} d^2 = 1.30 \text{ mm}^2$$
$$F = mg = 31.148 \text{ N} \tag{10.304}$$
$$T_1 = F/A = 240 \text{ MPa}$$

Then, we calculate the compliance values in martensitic and austenitic states, that is,

$$s_1^M = 1/E_M = 0.0500 \text{ 1/GPa}$$
$$s_{11}^A = 1/E_A = 0.0154 \text{ 1/GPa} \tag{10.305}$$

Using Equation (10.299), we calculate the induced strain, S_{ISA}, that is,

$$S_{ISA} = \Delta S_1 - \left(s_{11}^A - s_1^M\right) T_1 = -3.9\% \tag{10.306}$$

TABLE 10.18

Effect of SMA Training by Repeated Shrinking under Heat Activation and Stretching under Cooling of a Nitinol SMA Wire of $L = 740$ mm Original Length

Cool ($I = 0.0$ A)	Hot ($I = 1.5$ A)	ΔL
75 mm	110 mm	35 mm
75 mm	110 mm	35 mm
82 mm	118 mm	36 mm
81 mm	118 mm	37 mm

Finally, Equation (10.298) gives the quasiplastic strain, S_{QP}, as the negative of the induced strain, S_{ISA}, that is,

$$S_{QP} = -S_{ISA} = 3.9\% \tag{10.307}$$

The blocked stress is calculated with Equation (10.300), that is,

$$T_{block} = \frac{S_{QP}}{s_{11}^{A}} = 2534 \text{ MPa} \tag{10.308}$$

Comparing this value with the value $UTS = 1600$ MPa given in Table 10.16 indicates that the Nitinol wire will break before the full value of the blocked stress can be achieved. For the argument's sake, we can use Equation (10.301) to calculate the blocked force, that is,

$$F_{block} = T_{block}A = 329 \text{ N} \tag{10.309}$$

10.10.7 Extensive Modeling of SMA Actuators

Extensive modeling of the SMA materials aims at taking into account the effects of partial transformation when the operating temperature is somewhere between the M_f and A_f temperatures. In this case, partial transformation occurs, and the martensite fraction, ξ, has to be taken into consideration. In addition, the effect of stress on the phase transition temperatures must be also considered. Since the mechanical behavior of SMA materials is closely related to the microscopic martensite phase transformation, the constitutive relations developed for conventional materials such as Hooke's law and plastic flow theory are not directly applicable. Hence, specific constitutive relations, which take into consideration the phase transformation behavior of SMA, have been developed (Cory, 1978; Muller, 1979; Tanaka and Nagaki, 1982, 1985). Two approaches are generally used: (a) the phenomenological (macroscopic) approach, based on extensive experimental work; and (b) the physical (microscopic) approach using fundamental physical concepts. Hybrid approaches that combine both approaches to obtain a more accurate description and prediction of the SMA material behavior have also been used. Tanaka's model, based on the concept of the free-energy driving force, considers a one-dimensional metallic material undergoing phase transformation. The state variables for the material are strain, ε, temperature, T, and martensite fraction, ξ. Then, a general state variable, Λ, is defined as

$$\Lambda = (\bar{\varepsilon}, T, \xi) \tag{10.310}$$

The Helmholtz free energy is a function of the state variable Λ. The general constitutive relations are then derived from the first and second laws of thermodynamics as

$$\bar{\sigma} = \rho_0 \frac{\partial \Phi}{\partial \bar{\varepsilon}} = \sigma(\bar{\varepsilon}, T, \xi) \tag{10.311}$$

The stress is a function of the martensite fraction, an internal variable. Differentiation of Equation (10.311) yields the stress rate equation:

$$\bar{\sigma} = \frac{\partial \sigma}{\partial \bar{\varepsilon}}\dot{\bar{\varepsilon}} + \frac{\partial \sigma}{\partial T}\dot{T} + \frac{\partial \sigma}{\partial \xi}\dot{\xi} = D\dot{\bar{\varepsilon}} + \Theta\dot{T} + \Omega\dot{\xi} \tag{10.312}$$

where D is Young's modulus, Θ is the thermoelastic tensor, and Ω is the transformation tensor, a metallurgical quantity that represents the change of strain during phase transformation. The martensite fraction, ξ, can be assumed as an exponential function of stress and temperature:

$$\xi_{M \to A} = \exp[A_a(T - A_s) + B_a\sigma], \quad \xi_{A \to M} = 1 - \exp[A_m(T - M_s) + B_m\sigma] \tag{10.313}$$

where A_a, A_m, B_a; B_m are material constants in terms of the transition temperatures, A_s, A_f, M_s; M_f.

Liang and Rogers (1990) derived a constitutive equation by the time-integration of Equation (10.312)

$$\sigma - \sigma_0 = D(\varepsilon - \varepsilon_0) + \theta(T - T_0) + \Omega(\xi - \xi_0). \tag{10.314}$$

For this, a cosine model for the phase transformation was adopted, that is,

$$\xi_{A \to M} = \frac{\xi_M}{2}\{\cos[a_A(T - A_s) + b_A\sigma] + 1\},$$

$$\xi_{M \gets A} = \frac{1 - \xi_A}{2}\{\cos[a_M(T - M_s) + b_M\sigma] + 1\} + \frac{1 + \xi_A}{2} \tag{10.315}$$

where $a_A = \dfrac{\pi}{A_f - A_s}$, $a_M = \dfrac{\pi}{M_s - M_f}$, $b_A = -\dfrac{a_A}{C_A}$, $b_M = -\dfrac{a_M}{C_M}$ while C_M and C_A are slopes.

10.10.8 Ballistic Protection Using the Superelastic Behavior of Shape Memory Alloys

The superelastic behavior of Nitinol alloys, shown in Figure 10.66, presents two distinct characteristics. First, it has an ultimate strain of around 15% concurrent with an ultimate stress above 1500 MPa. As shown in Figure 10.66, these values are well above the best values for steel and aluminum alloys. Second, upon unloading from a stress value below the ultimate stress, the strain is recovered elastically, and the unloading path stays below the loading path. Thus, a hysteresis loop is closed, and energy dissipation is recorded.

At temperatures where superelasticity appears, the stress–strain curve due to loading and unloading forms a hysteresis curve (Figure 10.66). The area surrounded by the hysteresis loop and the area below the unloading curve represent dissipated strain energy and recoverable strain energy per unit volume, respectively. Compared with the strain energy of normal metals within the range of recoverable deformation, the dissipative and recoverable strain energies of SMA are very large (Lin et al., 1994).

The superelastic behavior of Nitinol can be used to arrest and prevent penetration from an impacting object (Figure 10.68). The kinetic energy of the impacting object is partially

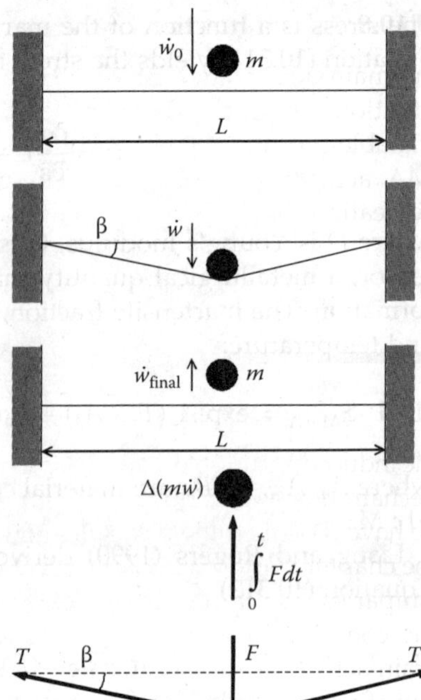

FIGURE 10.68
Dynamics of a superelastic Nitinol wire hit by an impacting mass: (a) capture, arresting, and repelling phases; (b) analysis of the interaction between the impacting object and the superelastic Nitinol wire.

dissipated through the hysteresis phenomenon, and partially recovered as spring-back. The analysis is performed using the following equations:

$$T(\beta) = \frac{F}{2\sin\beta} \qquad \text{(tension force as function of impact force)} \qquad (10.316)$$

$$u(\beta) = L\frac{1-\cos\beta}{\cos\beta} \qquad \text{(axial displacement)} \qquad (10.317)$$

$$\beta(t) = \sin^{-1}\left(\frac{w(t)}{L/2}\right) \qquad \text{(deflection angle)} \qquad (10.318)$$

$$\Delta E = \int_0^w F\,dw - \int_0^u T\,du + Q \qquad \text{(impact energy balance)} \qquad (10.319)$$

$$Q = AL\,c\,\Delta T + Q_{loss} \qquad \text{(heat balance)} \qquad (10.320)$$

$$\int_0^u T(u)\,du = \int_0^\varepsilon AL\sigma(\varepsilon)\,d\varepsilon \qquad \text{(strain energy balance)} \qquad (10.321)$$

$$\Delta(m\dot{w}) = \int_0^t F(t)\,dt \qquad \text{(impulse-momentum)} \qquad (10.322)$$

where Q_{loss} is the energy dissipated in the hysteresis loop of the SMA wire.

10.10.9 Advantages and Limitations of SMA Actuation

The main advantage of SMA materials is their capability to produce sizable (up to 8%) actuation strains. In addition, they have inherent simplicity since only heating (readily available through the electric Joule effect) is needed for actuation. Main limitations of the SMA actuators are the poor energy conversion efficiency and the low bandwidth of the heating/cooling process, which can only achieve a few hertz, at the very best.

10.11 Summary and Conclusions

The induced-strain active-material actuators are the enabling technology for a number of micro-mechatronics applications involving micropositioning, vibration control, etc. In this chapter, we have described the principles and applications of induced-strain active-material actuators. The chapter started with generic concepts about induced-strain active-material actuators in comparison with the more conventional electromechanical and hydraulic actuators. The construction details of piezoelectric, electrostrictive, and magnetostrictive induced-strain actuators were examined. The modeling of induced-strain actuators was developed from first principles. The typical performance of commercially available induced-strain actuators was examined. A linearized model for their basic electromechanical behavior was derived. The principles of ISA of a compliant structure were analyzed. The displacement analysis and the electric response were studied for both static and dynamic applications. Displacement-amplified induced-strain actuators were introduced and their analysis was discussed. The electrical power and energy flow in ISA applications was studied. The analysis of mechanical power and energy extraction was followed by the analysis of the electrical power and energy demands. The power and energy conversion efficiency were examined. Criteria for optimal energy conversion were derived. A comparative study of commercially available induced-strain actuators was presented. The efficient design of ISA applications was studied. Guidelines for the effective design and construction of ISA solutions were discussed. Electric power supplies for energizing induced-strain actuators were briefly discussed. A list of commercial suppliers of induced-strain actuators has been provided.

It was found that piezoelectric (PZT), electrostrictive (PMN), and magnetostrictive (Terfenol-D) materials have excellent frequency response (1–10 kHz, depending on actuator length), high force (up to 50 kN on current models), but small induced-strain stroke capabilities (typically, 0.1 mm for 0.1% strain on a 100 mm actuator). With this class of induced-strain actuators, displacement amplification devices must always be incorporated into the application design. The effective implementation of ISA depends on achieving optimal energy extraction. Induced-strain actuators are fundamentally different from conventional actuators. Induced-strain actuators have a limited amount of active stroke that has to be carefully managed. We have shown that the stiffness and impedance matching principles can produce the maximum energy extraction from the induced-strain actuator and ensure its transmission into the external application. Details of these principles, together with typical energy density values for various induced-strain actuators have been given. It was found that, for static applications, as much as 7 J/dm^3 (0.9 J/kg) can be obtained with commercially available induced-strain actuators. For dynamic applications, the corresponding values are 1.8 J/dm^3 (0.233 J/kg). The power delivery capabilities of induced-strain actuators were also examined. It was found that high-performance induced-strain actuators could be able to produce as much as 670 W/dm³ at 1 kHz full-stroke operation, provided adequate heat-dissipation of the hysteresis heat output is provided. In addition, such a high performance will also depend on the availability of an adequate electric power supply able to handle high reactive loads at high frequencies. One can conclude that the potential of active materials for micromechatronics actuation applications has been clearly demonstrated. However, this field is still in its infancy and further research and development is required to establish active materials actuators as reliable, durable, and cost-effective replacements for conventional actuators in large-scale engineering applications.

The last part of the chapter dealt with SMA actuators. The introduction of the shape memory effect and the superplasticity behavior of certain materials (e.g., nickel–titanium alloy Nitinol) was followed by a development of the basic analytical principles for estimation of the SMA actuator behavior under load and heat activation.

The chapter contains a number of worked out examples inserted in the text and a number of problems and exercises grouped at the chapter end.

10.12 Problems and Exercises

Quiz: Short Questions

1. What is the main characteristic that differentiates an induced-strain actuator from a conventional actuator (e.g., a hydraulic cylinder) and requires specialized design principles to be applied?

2. What is the principle that ensures that maximum energy is extracted from an induced-strain actuator and transferred into the external load? What is the corresponding principle for dynamic applications?

3. True or false: the induced-strain actuator displacement reaches a maximum under the stiffness match condition. Explain your answer

4. Do maximum energy extraction and the maximum energy conversion efficiency take place under the same conditions? Explain.

5. What are the main differences between a piezo induced-strain actuator and an SMA induced-strain actuator?

6. Which of the following induced-strain actuator types would be preferred for a high-frequency application: (a) piezo; (b) SMA?

7. Which of the following induced-strain actuator types would be preferred for a long-stroke application: (a) piezo; (b) SMA?

8. What are the main limitations of piezo induced-strain actuators?

9. What are the main limitations of SMA induced-strain actuators?

PROBLEM 10.1
Actuator Equivalent Properties

Consider the piezo stack actuator E100P-2 with properties given in Tables 10.3 and 10.4. Using the full stroke values, find

1. Electroactive material used in the actuator construction
2. Internal stiffness, k_i, electric capacitance, C_0, maximum voltage, V_{max}, free stroke, u_{ISA}
3. Layer thickness, h, number of layers, N, active material area, A, active material length, l
4. Maximum electric field, E_{max}
5. Apparent zero-field compliance, s
6. Apparent zero-stress electric permittivity, ε
7. Apparent piezoelectric strain coefficient, d_{33}
8. Reference mechanical energy, E_{mech}^{ref}
9. Reference electrical energy, E_{elec}^{ref}
10. Apparent piezoelectric coupling coefficient, κ

PROBLEM 10.2
Static Analysis of an Induced-Strain Actuator

Consider a piezoelectric actuator made up of a stack APC 850 square wafers with the properties given in Tables 10.2 and 10.19. Find:

1. Stack length, l, and internal stiffness, k_i neglecting the effect of electrodes and glue
2. Stress-free capacitance, C_0
3. Maximum allowable voltage, V_{max}
4. Free stroke, u_{ISA}, and free strain, S_{ISA}, in percentage induced by the application of the voltage, V_{max}
5. Reference mechanical energy, E_{mech}^{ref}, and reference electrical energy, E_{elec}^{ref}

TABLE 10.19

Piezoelectric Stack Actuator Definition for Problems 10.2 and 10.4

Property	Value
Number of wafer layers in the stack	$N = 200$
Square-wafer size	$a_{wafer} = 7$ mm
Wafer thickness	$h_{wafer} = 200$ μm
Allowable electric field	$E_{max}^{+} = 1.2$ kV/mm
Allowable reverse electric field	$E_{max}^{-} = 0.0$ kV/mm
Thickness of bonding adhesive layer	$h_{glue} = 3$ μm
Elastic modulus of bonding adhesive layer	$E_{glue} = 6$ GPa
Thickness of electrode	$h_{electrode} = 0.5$ μm
Elastic modulus of electrode	$E_{electrode} = 40$ GPa
Mechanical hysteretic damping in the actuator	$\eta = 1\%$
Electrical hysteretic damping in the actuator	$\delta = 0.5\%$

Note: The wafers are plated on both sides with electrodes, which are wired in parallel such that the voltage applied to the stack terminals is also applied to each wafer.

6. Stiffness ratio, r, when the actuator works against an external elastic spring of stiffness $k_e = 20$ kN/mm

7. Output displacement, u_e, when the actuator works at full stroke against the elastic spring of stiffness $k_e = 20$ kN/mm. Express result in physical units and as percentage of u_{ISA}

8. Output mechanical energy, E_e, expression as a function of stiffness ratio, r, when working at full stroke against an external elastic spring. Maximum mechanical energy, E_e^{max}, that could be extracted from the actuator by tuning the stiffness ratio, r. The value of stiffness ratio, r, to achieve this maximum mechanical energy output.

9. Output mechanical energy, E_e, when the actuator works at full stroke against the elastic spring of stiffness $k_e = 20$ kN/mm. Express result in physical units and as percentage of E_e^{max}.

10. The value of the elastic spring stiffness, k_e, that will ensure stiffness match; the corresponding value of the stiffness ratio, r_{match}. The output mechanical energy, E_e^{match}, under the stiffness-match condition

11. Effective capacitance, C, in physical units and as a percentage of C_0

12. Input electrical energy, E_{elec}, when the actuator works at full stroke against the elastic spring of stiffness $k_e = 20$ kN/mm. Express result in physical units and as percentage of E_{elec}^{ref}. Discuss results.

13. Input electrical energy at the stiffness-match point, $E_{elec}^{r=1}$. Express result in physical units and as percentage of E_{elec}^{ref}. Discuss results.

14. Energy conversion efficiency, η, maximum attainable energy conversion efficiency, η_{max}, and conversion efficiency at the stiffness match point, $\eta_{r=1}$. Discuss results.

PROBLEM 10.3
Effect of Glue and Electrodes on a Piezo Stack Induced-Strain Actuator

Consider the piezoelectric actuator made up of a stack APC 850 square wafers with the properties given in Tables 10.2 and 10.19. This is the same actuator considered in Problem 10.2, only that we will now analyze the effect of glue and electrodes. Find

1. Internal stiffness, k_i, taking into account the effect of electrodes and interlayer glue

2. Stress-free capacitance of the stack, C_0

3. The total displacement (full stroke), u_{ISA}, that is achieved when the actuator is energized with a voltage inducing the maximum allowable electric field given in Table 10.19

4. Total length of the stack, L, in millimeter, and the effective induced strain in the stack, S_{ISA}, in percentage

5. Reference mechanical energy, E_{mech}^{ref}, reference electrical energy, E_{elec}^{ref}, and maximum possible energy output, E_e^{max}

6. Compare the results obtained here with the corresponding results from Problem 10.2 in which the effect of glue and electrodes was ignored and comment on this comparison

PROBLEM 10.4
Dynamic Analysis of Induced-Strain Actuator

Consider the piezoelectric actuator made up of a stack APC 850 square wafers with the properties given in Tables 10.2 and 10.19. This is the same actuator considered in the previous problem, only that we will now analyze its dynamic behavior. Recall the stack length, l; static internal stiffness, k_i; stress-free capacitance, C_0; electromechanical coupling coefficient, κ; maximum voltage, $V_{\max}$; calculated in previous problem. Find

1. Complex internal stiffness, $\bar{k}_i$; complex stress-free capacitance, $\bar{C}_0$; complex electromechanical coupling coefficient, $\bar{\kappa}^2$

2. Bias voltage, V_0, and the alternating voltage amplitude, $\hat{V}$, to be applied to the actuator for dynamic operation

3. Bias position, u_0, and the amplitude of free dynamic stroke, $\hat{u}_{ISA}$, under free dynamic operation (When the actuator is sitting at the bias position, the spring of the external load is relaxed.)

4. Reference mechanical energy, E_{mech}^{ref}, and reference electrical energy, E_{elec}^{ref}, that apply to this dynamic operation

5. Expression of the complex dynamic stiffness, $\bar{k}_e(\omega)$, of an external load of spring of stiffness $k_e = 20$ kN/mm, resonance frequency $f_0 = 15$ Hz, and damping ratio $\zeta = 2\%$

6. Complex stiffness ratio when the operating frequency is half the resonance frequency of the external load, ω_0

7. Complex wavenumber, $\bar{\gamma}$, and the value $|\gamma l|$ associated with operation at half the resonance frequency of the external system, ω_0. Verify whether or not the quasi-static assumption can be applied

8. Complex expression of the output dynamic displacement amplitude, $\hat{u}(\omega)$, during full-stroke dynamic operation against the external load, $\bar{k}_e(\omega)$. Value of $\hat{u}$ when the operating frequency is half the resonance frequency of the external load, ω_0. Express the result in physical units, first. Then, express the absolute value of the result as a percentage of the free dynamic stroke, $\hat{u}_{ISA}$.

9. Expressions of the average mechanical power output, P_{mech}^{avg}, and $\cos\phi$. The values of P_{mech}^{avg}, $\cos\phi$, and ϕ when the actuator is operating at half the resonance of the external system, ω_0.

10. Expression of the peak mechanical power output per cycle, P_{mech}^{peak}, and its value when the actuator is operating at half the resonance of the external system, ω_0. Comment on the relative numerical values of P_{mech}^{avg} and P_{mech}^{peak}.

11. Effective capacitance, $\bar{C}(\omega)$, expression and its value when the actuator is operating at half the resonance of the external system, ω_0.

12. Expressions for effective admittance and impedance; their values when the actuator is operating at half the resonance of the external system, ω_0.

13. Expressions for electrical complex power, $\bar{P}$, active power, P_{active}, bias-voltage power correction factor, $\chi(v_0)$, and peak power, P_{elec}^{peak}; their values when the actuator is operating at half the resonance of the external system, ω_0.

14. Expression of peak electrical energy per cycle, E_{elec}^{peak}, and its values when the actuator is operating at half the resonance of the external system, ω_0.

15. Expression and value of electrical energy at the stiffness match point, $E_{elec}^{r=1}$.

16. Expression of peak power conversion efficiency, η^{peak}. Values of peak power conversion efficiency when the actuator is operating at half the resonance of the external system, ω_0 and in two bias voltage conditions: (a) no bias voltage; (b) with bias voltage $V_0 = \hat{V}$. Discuss the effect of bias voltage on peak power conversion efficiency. Repeat using the simplified expression for low-damping systems. Discuss the effect of using the simplified formula.

17. Expression of the optimal stiffness ratio for optimal peak power conversion, r_η and of the corresponding best conversion efficiency, $\eta_{max}(v_0)$. Values of these variables in two bias voltage conditions: (a) no bias voltage; (b) with bias voltage $V_0 = \hat{V}$. Discuss results.

18. Expression of the peak power conversion efficiency at stiffness match, $\eta_{r=1}(v_0)$. Values of these variables in two bias voltage conditions: (a) no bias voltage and (b) with bias voltage $V_0 = \hat{V}$. Discuss results.

PROBLEM 10.5
SMA Actuator Number 1

Consider an SMA wire with the following characteristics: $L = 250$ mm, $d = 0.40$ mm, $E^M = 20$ GPa, $E^A = 80$ GPa, $UTS = 1600$ MPa. The SMA wire was strained by stretching while in the martensite state. The quasiplastic strain introduced during straining is $S_{QP} = 4\%$. Find

1. The compliance constants in the martensite and austenite states, s_{11}^M, s_{11}^A. Express result in 1/GPa.

2. The strain corresponding to an applied stress, $T_1 = 100$ MPa, in the martensite and austenite conditions. Express result in percentage.

3. The apparent ISA strain, S_{ISA}, produced during the thermal activation which induces phase transformation from the martensitic into the austenitic state under the constant applied stress $T_1 = 100$ MPa. Express result in percentage.

4. The lift distance (total shrinkage), Δu_1, upon thermal activation of the phase transition from the martensite into austenite phases under the constant applied stress $T_1 = 100$ MPa. Express result in millimeters.

5. The mass load, m, which will produce the effect at number 4. Express result in kilograms.

6. The "blocked stress," T_{block}, and the "blocked force," F_{block}, of the SMA actuating wire. To determine the blocked force, assume that the wire in martensite phase is stretched to attain $S_{QP} = 4\%$, and then the end of the wire is fixed solidly and not allowed to move at all. When the SMA wire is thermally activated, the martensite to austenite phase transition takes place and the wire will want to shrink, but will not be allowed because it is blocked. The blocked force, F_{block}, is the force developed in the blocked wire after the austenite phase has been fully established.

7. The mass, m_{block}, that would produce the weight F_{block}.

8. Extra work for graduate students:

   a. Write expressions and give values for the partial derivatives of the total strain with respect to stress, T_1, and the induced strain, S_{ISA}.

   b. Explain why the results of number 6 are unrealistic.

PROBLEM 10.6
SMA Actuator Number 2

Consider an SMA wire with the following characteristics: $L = 600$ mm, $d = 0.5$ mm, $E^M = 30$ GPa, $E^A = 70$ GPa, $UTS = 1600$ MPa. The SMA wire is initially in the martensite state. One end of the SMA wire is attached to a fixed point whereas the other end is attached to a weight of mass $m = 5$ kg. When the weight is attached, the SMA wire stretches through quasiplasticity and rests with the weight at position $h_M = 70$ mm above the ground. Then, the SMA wire is heated by an electric current that passes through it. Due to heating, the SMA wire undergoes phase transformation into the austenitic state and shrinks. At the end of the phase-transformation process, the weight has been lifted to the new position $h_A = 100$ mm above the ground. Find

1. The compliance constants in the martensite and austenite states, s_{11}^M, s_{11}^A. Express result in 1/GPa.

2. The load, F, acting on the SMA wire and the corresponding stress, T_1. Round off stress result to the nearest megapascal.

3. The lift distance (total shrinkage), Δu_1, upon thermal activation of the phase transition from the martensite into austenite phases. Express result in millimeters.

4. The apparent ISA strain, S_{ISA}, produced during the thermal activation which induces phase transformation from the martensitic into austenitic the state. Express result in percentage.

5. The quasiplastic strain, S_{QP}, introduced during the loading of the wire in the cool martensitic state.

6. The "blocked stress," T_{block}, and the "blocked force," F_{block}, of the SMA actuating wire. To determine the blocked force, assume that after the wire is stretched in the cool martensitic phase to attain S_{QP}, the weight is removed and the end of the wire is fixed solidly and not allowed to move at all. When the SMA wire is thermally activated into the austenitic phase, the wire will want to shrink, but will not be allowed because it is blocked. The blocked force, F_{block}, is the force developed in the blocked wire after the austenitic phase has been fully established.

7. The mass, m_{block}, that would produce the weight F_{block}.

8. Extra work for graduate students:

    a. Write expressions and give values for the partial derivatives of the total strain with respect to stress, T_1, and the induced strain, S_{ISA}.

    b. Explain why the results of 6 are unrealistic.

Disclaimer

The data presented in some of the tables and used to calculate the results presented in some of the figures was obtained from the manufacturers. Some simple consistency checks were run, but otherwise there was no way to independently verify the data in depth. Any derived results are consequently affected by the reliability of the input data.

References

Anon, IEEE Standard on Piezoelectricity, ANSI/IEEE Std. 176-1987, IEEE, NY, 1987.

Bank, R., *Shape Memory Effects in Alloys*, Plenum, New York, 1975, p. 537, 1975.

Cory, J. S., Nitinol thermodynamic state surfaces *J. Energy* 2, No. 5, 1978.

Duerig, T. W., Melton, K. N., Stockel, D., and Wayman, C. M., *Engineering Aspects of Shape Memory Alloys*, Butterworth-Heinemann, London, 1990.

Furuya, Y., Watanabe, T., Hagood, N. W., Kimura, H., and Tani, J., Giant magnetostriction of ferromagnetic shape memory Fe-Pd Alloy produced by electromagnetic nozzleless melt-spinning method, *9th International Conference on Adaptive Structures and Technologies*, Boston, MA, October 14–16, 1998, Technomics Publishing Company, Lancaster, PA, 1998.

Giurgiutiu, V., Active-materials induced-strain actuation for aeroelastic vibration control, *Shock and Vibration Dig.*, 32(5), 355–368, September 2000.

Giurgiutiu, V. and Pomirleanu, R., Energy-Based Comparison of Solid-State Actuators, University of South Carolina, Department of Mechanical Engineering, Laboratory for Active Materials and Smart Structures, Report # USC-ME-LAMSS-2001-101, 2001.

Liang, C. and Rogers, C. A., A one-dimensional thermomechanical constitutive relation of shape memory materials, *J. Intelligent Mater. Syst. Struct.*, 1(2), 207–234, 1990.

Lin, P. H., Tobushi, H., Tanaka, K., Hattori, T., and Makita, M., Pseudoelastic behavior of TiNi shape memory alloy subjected to strain variations, *J. Intelligent Mater. Syst. Struct.*, 5, 694–701, September 1994.

Muller, I., A model for a body with shape memory, *Arch. Ration. Mech. Anal.* 70, 61–77, 1979.

Murray, S. J., Allen, S. M., O'Handley, R. C., and Lograsso, T. A., Magnetomechanical performance and mechanical properties of Ni-Mn-Ga ferromagnetic shape memory alloys, *Smart Structures and Materials 2000*, SPIE Proc., vol. 3992, pp. 387–395, 2000.

Rosen, C. Z., Hiremath, B. V., and Newnham, R., (Eds.) *Key Papers in Physics: Piezoelectricity*, American Institute of Physics, College Park, MD, 1992.

Tanaka, K. and Nagaki, S., A thermomechanical description of materials with internal variables in the process of phase transformation, *Ing. Arch.*, 51, 287–299, 1982.

Tanaka, K. and Iwasaki, R., A phenomenological theory of transformation superplasticity, *Eng. Fracture Mech.*, 21(4), 709–720, 1985.

List of Commercial Suppliers of Electroactive and Magnetoactive Induced-Strain Actuators

APC International Ltd. and American Piezo Ceramics, Inc., Duck Run, Mackeyville, PA, APCSales@aol.com., www.americanpiezo.com

Burleigh Instruments, Inc., Burleigh Park, Fishers, New York 14453

EDO Corporation, Salt Lake City, Utah, http://www.nycedo.com/edocorp/pageba5_edoecp.htm

Etrema Products, Inc., Ames, Iowa, http://etrema-usa.com

Kinetic Ceramics, Inc., Hayward, CA, http://www.kineticceramics.com/contact_us/contact_us.htm

Measurement Specialties, Inc., *Sensor Products Division*, Wayne, PA, http://www.msiusa.com

Micromega Dynamics, SA, Angleur, Belgium, http://www.micromega-dynamics.com/profile.htm

Piezo Kinetics, Inc., Bellefonte, PA, http://www.piezo-kinetics.com

Piezo Systems, Inc., Cambridge, MA, http://www.piezo.com

Polytec PI, Inc., Costa Mesa, CA, http://www.polytechpi.com

Tokin America, Inc., Union City, CA, http://www.nec-tokinamerica.com/products.html

TRS Ceramics Inc., State College, PA, http://www.trsceramics.com

11

Piezoelectric Wafer Active Sensors

11.1 Introduction

Piezoelectric wafer active sensors (PWAS) are inexpensive transducers that operate on the piezoelectric principle. Initially, PWAS were used for vibration control as pioneered by Crawley and deLuis (1987) and Fuller et al. (1990). Tzou and Tseng (1990) and Lester and Lefebvre (1993) modeled the piezoelectric sensor/actuator design for dynamic measurement/control. For damage detection, Banks et al. (1996) used PZT wafers to excite a structure and then sense the free decay response. The use of PWAS for structural health monitoring has followed three main paths: (1) modal analysis and transfer function, (2) electromechanical impedance, and (3) wave propagation. The use of PWAS for damage detection with Lamb-wave propagation was pioneered by Chang and his coworkers (Chang, 1995, 1998, 2001; Wang and Chang, 2000; Ihn and Chang, 2002). They have studied the use of PWAS for generation and reception of elastic waves in composite materials. Passive reception of elastic waves was used for impact detection. Pitch-catch transmission–reception of low-frequency Lamb waves was used for damage detection. PWAS wave propagation was also studied by Culshaw et al. (1998), Lin and Yuan (2001), Osmont et al. (2000), and Diamanti et al. (2002). The use of PWAS for high-frequency local modal sensing with the electromechanical impedance method was pursed by Liang et al. (1994), Sun et al. (1994), Chaudhry et al. (1995), Park and Inman (2001), Giurgiutiu et al. (1998–2002), and others.

PWAS couple the electrical and mechanical effects (mechanical strain, S_{ij}, mechanical stress, T_{kl}, electrical field, E_k, and electrical displacement D_j) through the tensorial piezoelectric constitutive equations

$$
\begin{aligned}
S_{ij} &= s_{ijkl}^{E} T_{kl} + d_{kij} E_k \\
D_j &= d_{jkl} T_{kl} + \varepsilon_{jk}^{T} E_k
\end{aligned}
\tag{11.1}
$$

where s_{ijkl}^{E} is the mechanical compliance of the material measured at zero electric field ($E=0$), ε_{jk}^{T} is the dielectric permittivity measured at zero mechanical stress ($T=0$), and d_{kij} represents the piezoelectric coupling effect. As apparent in Figure 11.1, PWAS are small and unobtrusive. PWAS utilize the d_{31} coupling between in-plane strain and transverse electric field. A 7 mm diameter PWAS, 0.2 mm thin, weighs a bare 78 mg. At less than \$10 each, PWAS are no more expensive than conventional high-quality resistance strain gauges. However, the PWAS performance exceeds by far that of conventional resistance strain gauges. This is especially apparent in high-frequency applications at hundreds of kilohertz and beyond. There are several ways in which PWAS can be used, as shown next.

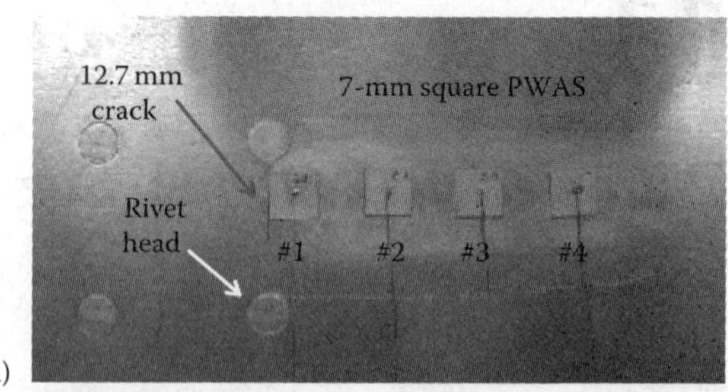

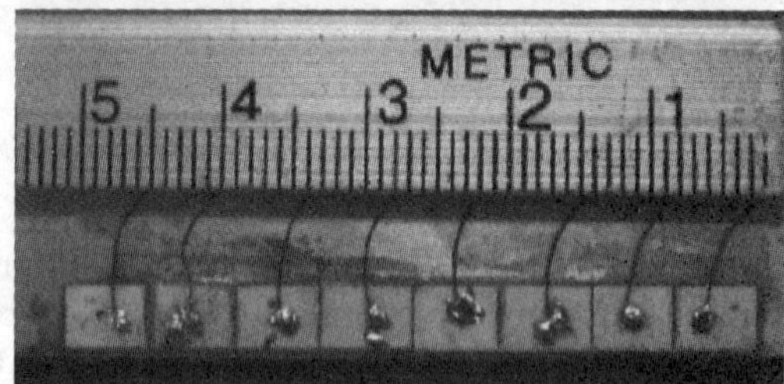

FIGURE 11.1
Piezoelectric wafer active sensors (PWAS) options:
(a) network of independant 7-mm square PWAS
on an aircraft panel; (b) PWAS phased array.

As a high-bandwidth strain sensor, the PWAS directly converts mechanical energy to electrical energy. The conversion constant is linearly dependent on the signal frequency. In the kilohertz range, signals of the order of hundreds of millivolts are easily obtained. No conditioning amplifiers are needed; the PWAS can be directly connected to a high-impedance measuring instrument such as a digitizing oscilloscope.

As a high-bandwidth strain exciter, the PWAS converts directly the electrical energy into mechanical energy. Thus, it can easily induce vibrations and waves in the substrate material. It acts very well as an embedded generator of waves and vibration. High-frequency waves and vibrations are easily excited with input signals as low as 10 V. These dual sensing and excitation characteristics of PWAS justify their name of "active sensors."

As a resonator, PWAS have the property that performs mechanical resonances under direct electrical excitation. Thus, very precise frequency standards can be created with a simple setup consisting of the PWAS and the signal generator. The resonant frequencies depend only on the wave speed (which is a material constant) and the geometric dimensions. Precise frequency values can be obtained through precise machining of the PWAS geometry.

As an embedded modal sensor, the PWAS is able to directly measure the high-frequency modal spectrum of a support structure. This is achieved with the electromechanical impedance method that reflects the mechanical impedance of the support structure into the real part of the electromechanical impedance measured at PWAS terminals. The high-frequency characteristics of this method, which have been proven to operate at hundreds of kilohertz and beyond, cannot be achieved with conventional modal measurement techniques. Thus, PWAS are the sensors of choice for high-frequency modal measurement and analysis.

This chapter describes the above functionalities of the PWAS devices, starting with the basic resonator equations and progressing through the other functionalities such as Lamb-wave transmitters and receivers (i.e., embedded ultrasonic transducers) and high-frequency modal sensors. The theoretical developments are performed in a step by step manner, with full presentation of the intermediate steps. This allows the readers the opportunity to duplicate the theoretical developments and extend them to fit their particular applications. To substantiate the theoretical developments and verify their predictions, several experimental tests are presented. Full description of the experimental setup and the performed measurements are provided. Again, the intention is to give readers the opportunity to duplicate the findings and extend them to suit their particular application needs.

11.2 Review of Elastic Waves and Structural Vibration

11.2.1 Waves

Waves are disturbances that travel, or propagate, from one region of space to another. In a generic description, the study of waves deals with the study of how disturbances propagate in an elastic medium. A generic disturbance, ϕ, propagating in the spatial direction, x, according to a propagation function, $f(\cdot)$, will have the form

$$\phi(x,t) = f(x - ct) \tag{11.2}$$

where c is the *wave speed*. All waves must satisfy the *wave equation* given by

$$c^2 \nabla^2 \phi = \ddot{\phi} \tag{11.3}$$

where ∇^2 is the *Laplace operator* and the dot above the variable signifies differentiation with respect to time, i.e., $\dot{\phi} = \partial \phi / \partial t$. In the simple case of Equation (11.2), the wave Equation (11.3) becomes

$$c^2 \phi'' = \ddot{\phi} \tag{11.4}$$

where the dash above the variable signifies differentiation with respect to space, i.e., $\phi' = \partial \phi / \partial x$. If the wave is *harmonic*, then Equation (11.2) takes the form

$$\phi(x,t) = \phi_0 e^{i(\gamma x - \omega t)} \tag{11.5}$$

where γ is the wave number and ω is the angular frequency. The wave number, γ, can be written in terms of the wave speed, c, and the angular frequency, ω, as

$$\gamma = \frac{\omega}{c} = \frac{2\pi}{\lambda} \tag{11.6}$$

The wave number can be also written in terms of wavelength, $\lambda = cT$, where T is the period of oscillation given by $T = 1/f$, where f is the oscillation frequency. The angular frequency, ω, and circular frequency, f, are related by the relation $\omega = 2\pi f$. Alternative forms of Equation (11.5) are

$$\phi(x,t) = \phi_0 e^{i\omega\left(\frac{x}{c} - t\right)} = \phi_0 e^{i2\pi\left(\frac{x}{\lambda} - \frac{t}{T}\right)} \tag{11.7}$$

TABLE 11.1

Waves in Elastic Solids

Wave Type	Particle Motion, Main Assumptions
Pressure waves (a.k.a. longitudinal; compressional; dilatational; P-wave)	Parallel to the direction of wave propagation, infinite 3-D medium
Axial waves	Parallel to the direction of wave propagation. Uniform over cross section for 1-D rods; uniform over thickness for 2-D plates
Shear waves (a.k.a. transverse; distortional; S-wave)	Perpendicular to the direction of wave propagation
Flexural waves (a.k.a. bending waves)	Elliptical. For beams, plane sections remain plane. For plates, straight normals remain straight
Rayleigh waves (a.k.a. surface acoustic waves, SAW)	Elliptical, amplitude decays quickly with depth
Lamb waves (a.k.a. guided plate waves)	Elliptical, free-surface conditions satisfied at the upper and lower plate surfaces

Different wave types are used in ultrasonic nondestructive evaluation (NDE) and structural health monitoring (SHM). Table 11.1 shows a brief listing of some of the wave types of interest to NDE and SHM and their applications. These wave types include pressure waves (a.k.a. P-waves, compressional waves, dilatational waves, or longitudinal waves); axial waves; shear waves (a.k.a. S-wave, transverse waves or distortional waves); Rayleigh waves (a.k.a. surface acoustic waves, SAW); and Lamb waves (a.k.a. guided plate waves). Other types of waves, which may be of importance to NDE and SHM but are not discussed here for the sake of brevity, include Love waves, Stonley waves, etc.

To achieve better understanding of elastic waves, visualization of the waveforms can be very useful. Putting the wave equations into mathematics software, the particle displacement can be calculated as function of space and time. Animations of different waves were calculated and posted on the Internet at http://www.me.sc.edu/research/lamss/default. htm under the research section. A quick review of the main wave types is given next. For each wave type, we will give a brief description of the wave speed and particle motion and a figure describing the particle motion. For a more extensive study, one can refer to the 'Waves' chapters in Giurgiutiu (2008) or the specialized works by Rayleigh (1887), Lamb (1917), Viktorov (1967), Achenbach (1999), Rose (1999), etc.

11.2.1.1 Pressure Waves

Pressure waves (P-waves) have the particle motion parallel to the direction of wave propagation (Figure 11.2). P-waves are one of the two wave types possible in an unbound

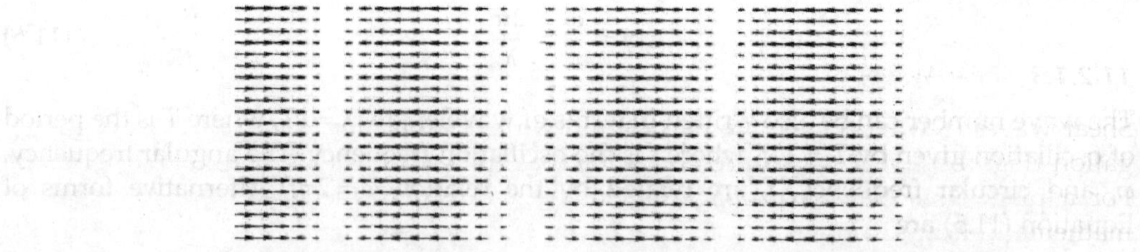

FIGURE 11.2

Simulation of particle motion for a plane-front P-wave. The same simulation illustrates an axial wave in a bar or plate having the displacement field constant across the plate thickness.

3-D solid medium. P-waves are also known as compressional, dilatational, or longitudinal waves. For a plane-front P-wave propagating in the x direction, the particle motion has the mathematical expression

$$u_x(x,t) = u_0 e^{i(\gamma x - \omega t)} \tag{11.8}$$

where $\gamma = \omega / c_P$ and c_P is the P-wave speed given by

$$c_P = \sqrt{\frac{\lambda + 2\mu}{\rho}} \tag{11.9}$$

The constants λ and μ are Lamé constants and ρ is the mass density. The Lamé constants are related to Young modulus, E, and Poisson's ratio, v, by the expressions

$$\lambda = \frac{vE}{(1-2v)(1+v)} \quad \text{and} \quad \mu = \frac{E}{2(1+v)} \tag{11.10}$$

A plot of Equation (11.8) is shown in Figure 11.2.

11.2.1.2 Axial Waves

Axial waves resemble P-waves because their particle motion is also parallel to the direction of wave propagation (Equation (11.8) and Figure 11.2). However, axial waves are just low frequency approximations of more complicated symmetric waves traveling in bars and plates. The main assumption of axial waves is that the displacement field is constant across the bar cross section or the plate thickness. For a 1-D slender bar, the axial wave speed has the simple expression

$$c = \sqrt{\frac{E}{\rho}} \tag{11.11}$$

For a 2-D plate, the axial wave speed, also known as the *longitudinal wave speed in a plate*, is given by

$$c_L = \sqrt{\frac{1}{1-v^2}\frac{E}{\rho}} \tag{11.12}$$

where v is the Poisson ratio.

11.2.1.3 Shear Waves

Shear waves (S-waves) have the particle motion perpendicular to the direction of propagation (Figure 11.3). S-waves are also known as transverse waves or distortional waves. For a plane-front S-wave propagating in the x direction, the particle motion has the mathematical expression

$$u_y(x,t) = u_0 e^{i(\gamma x - \omega t)} \tag{11.13}$$

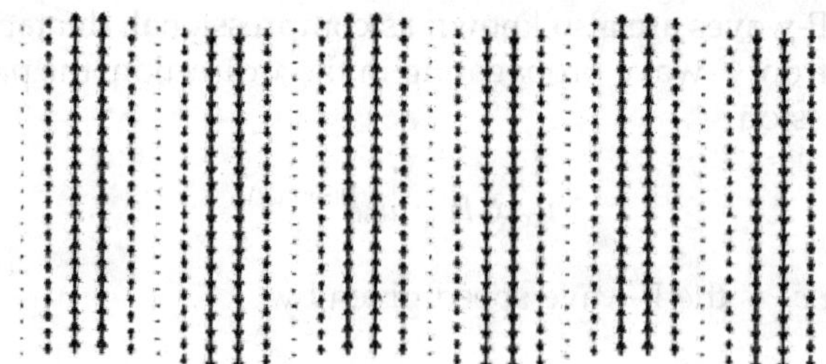

FIGURE 11.3
Simulation of a shear wave in a plate.

where $\gamma = \omega/c_S$ and c_S is the S-wave speed given by

$$c_S = \sqrt{\frac{\mu}{\rho}} \tag{11.14}$$

A plot of Equation (11.13) is shown in Figure 11.3.

11.2.1.4 *Flexural Waves*

Flexural waves appear due to the bending deformation of beams and plates in response to a transverse motion $u_y(x, t)$. The bending deformation generates a secondary in-plane motion, $u_x(x, t)$, which has a linear variation across the beam and plate thickness. In beams, the Euler–Bernoulli theory assumes plane sections to remain plane after the beam is bent. In plates, the Kirchhoff plate theory assumes straight normals to the plate to remain straight after the plate is bent. Both theories imply a linear distribution of axial displacement across the thickness. Based on these assumptions, the in-plane displacement, u_x, can be derived in terms of the transverse displacement, u_y, through simple geometric and kinematic arguments (Figure 11.4), i.e.,

$$u_x = -y u_y' \tag{11.15}$$

Assume a harmonic transverse displacement uniform across the beam or plate thickness, i.e.,

$$u_y(x, y, t) = u_0 e^{i(\gamma x - \omega t)} \tag{11.16}$$

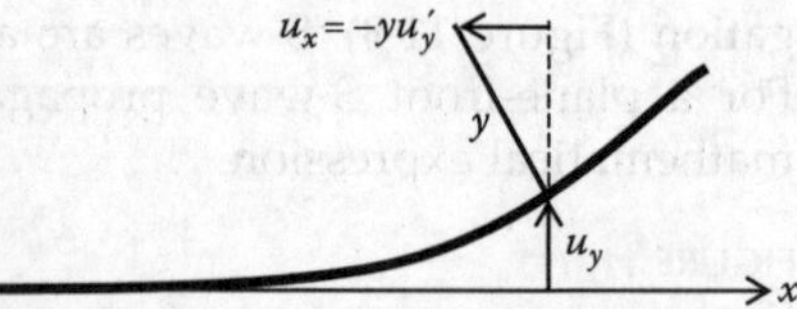

FIGURE 11.4
Definition of the axial displacement u_x in terms of the vertical displacement u_y for simple flexural motion.

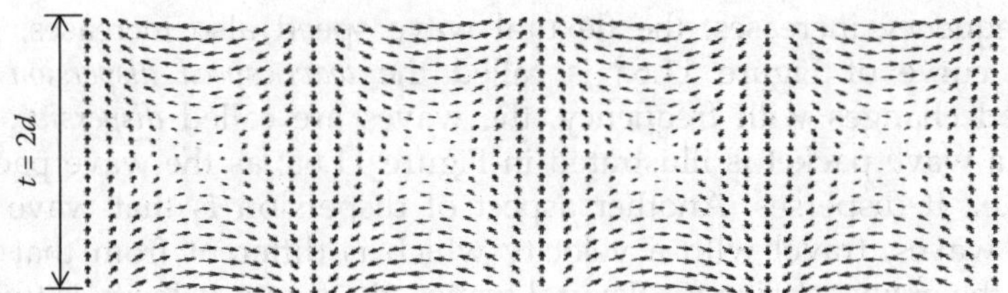

FIGURE 11.5
Simulation of flexural waves in a thin plate.

where $\gamma = \omega/c_F$. Hence, Equation (11.15) gives the in-plane displacement field across the beam or plate thickness as

$$u_x(x, y, t) = y\frac{\partial}{\partial x}u_0 e^{i(\gamma x - \omega t)} = i\gamma u_0 y e^{i(\gamma x - \omega t)} \tag{11.17}$$

The particle motion described by Equations (11.16) and (11.17) is shown in Figure 11.5. The particle motion is elliptical, with the ellipse aspect ratio varying from maximum at the surface to zero in the middle of the beam or plate thickness.

For a beam of flexural stiffness EI and mass per unit length m, the flexural wave speed is given by

$$c_F = a\sqrt{\omega} \tag{11.18}$$

where $a = (EI/m)^{1/4}$. If the beam is rectangular with thickness h and width b, then c_F takes the form

$$c_F = \left[\frac{Eh^2}{12\rho}\right]^{\frac{1}{4}}\sqrt{\omega} \tag{11.19}$$

The wave speed given by Equation (11.18) is frequency dependent. A plot of Equation (11.18) is given in Figure 11.6a. It is noticed that the flexural wave speed is zero when the frequency

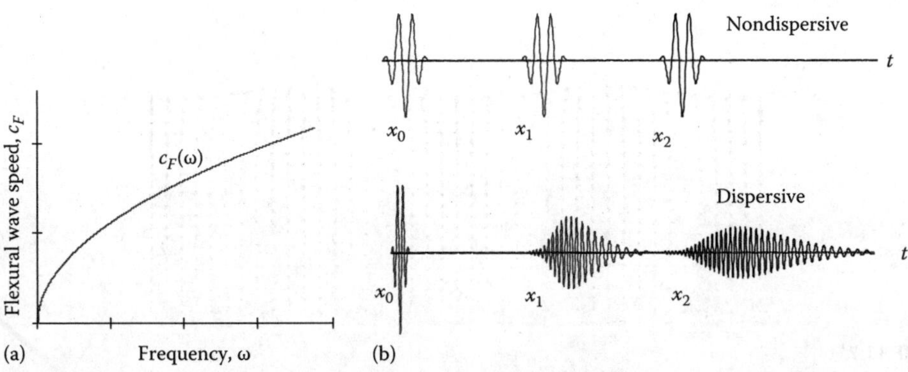

FIGURE 11.6
Dispersion of flexural waves: (a) dispersion curve; (b) dispersion effect on a traveling wave packet.

is zero. As frequency increases, the flexural wave speed also increases, following the $\sqrt{\omega}$ rule. The curve of Figure 11.6a is called the *wave speed dispersion curve*. When the wave speed changes with frequency, the waves are called *dispersive*. The effect of dispersion on a wave packet is illustrated in Figure 11.6a; as the wave packet travels, it spreads out, i.e., it disperses. Another aspect of dispersion is that wave packets, that are groups of waves, travel with a velocity which is different from that of individual waves. This is the *group velocity*; for flexural waves, the group velocity is twice as large as the wave speed, i.e.,

$$c_{gF} = 2a\sqrt{\omega} \tag{11.20}$$

For a plate of thickness h, the flexural wave speed is given by

$$c_F = \left[\frac{Eh^2}{12\rho(1 - \nu^2)}\right]^{\frac{1}{4}} \sqrt{\omega} \tag{11.21}$$

The difference between the beam and plate flexural speed expressions resides in the presence of the $(1 - \nu^2)$ term which represents the effect of the plane-strain condition imposed on the propagation of a straight crested flexural wave in an infinite plate.

11.2.1.5 Rayleigh Waves

Rayleigh waves, a.k.a. surface acoustic waves (SAW), have the property of propagating close to the body surface, with the motion amplitude decreasing rapidly with depth (Figure 11.7). The polarization of Rayleigh waves lies in a plane perpendicular to the surface. The effective depth of penetration is less than one wavelength. The wave velocity of Rayleigh waves is the solution of a cubic equation with up to three roots; for practical purposes, only one root is real giving the Rayleigh wave speed

$$c_R(\nu) = c_S \left(\frac{0.87 + 1.12\nu}{1 + \nu}\right) \tag{11.22}$$

where $c_S^2 = \mu/\rho$ is the shear (transverse) wave speed, μ is the Lamé constant, and ρ is the mass density.

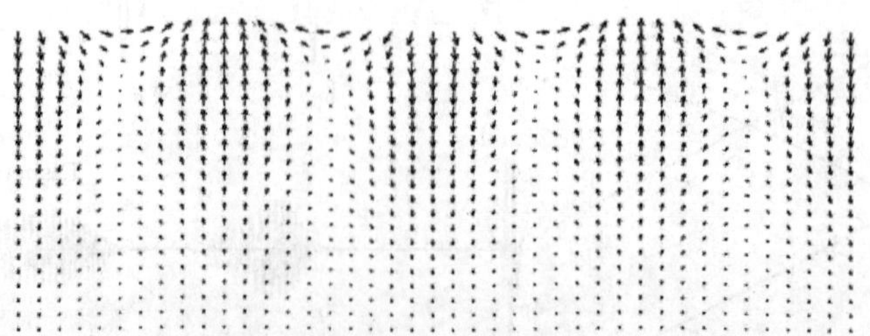

FIGURE 11.7
Simulation of Rayleigh wave in a semi-infinite medium.

The Rayleigh wave particle motion has two components, the horizontal component, u_x, and the vertical component, u_y. Both are functions of x, y, and t, i.e.,

$$\hat{u}_x(y) = Ai\left(\xi_R e^{-\alpha y} - \frac{\beta^2 + \xi_R^2}{2\xi} e^{-\beta y} \right)$$

$$\hat{u}_y(y) = A\left(-\alpha e^{-\alpha y} + i\frac{\beta^2 + \xi_R^2}{2\beta} e^{-\beta y} \right)$$

$$(11.23)$$

where $\xi_R = \omega/c_R$ is the Rayleigh wavenumber, $\alpha^2 = \xi^2 - \dfrac{\omega^2}{c_P^2}$, $\beta^2 = \xi^2 - \dfrac{\omega^2}{c_S^2}$, and c_P and c_S are given by Equations (11.9) and (11.14). The simulation of particle motion for a Rayleigh wave is shown in Figure 11.7.

11.2.1.6 Shear Horizontal Waves

Shear horizontal (SH) waves are a type of ultrasonic waves that are guided between two parallel free surfaces, such as the upper and lower surfaces of a plate, and have pure-shear particle motion polarized in the plane of the plate (Figure 11.8). SH waves are multimodal waves with each mode corresponding to a solution of the characteristic equation

$$\sin \eta d \cos \eta d = 0 \qquad (11.24)$$

where $\eta^2 = \dfrac{\omega^2}{c_S^2} - \xi^2$ and c_S is given by Equation (11.14). Solution of Equation (11.24) yields the symmetric (S) eigenvalues, $\xi_0^S, \xi_1^S, \xi_2^S, \ldots$ and the antisymmetric (A) eigenvalues, $\xi_0^A, \xi_1^A, \xi_2^A, \ldots$. The relationship $c = \omega/\xi$ yields the dispersive wave speed which is a function of the product fd between the frequency, $f = \omega/2\pi$, and the half thickness, d. A plot of the SH-wave dispersion curves is given in Figure 11.9a where the symmetric modes are designated $S_0, S_1, S_2, \ldots$ while the antisymmetric modes are designated $A_0, A_1, A_2, \ldots$. One notices that all the SH modes are dispersive except the first mode, S_0, which is not

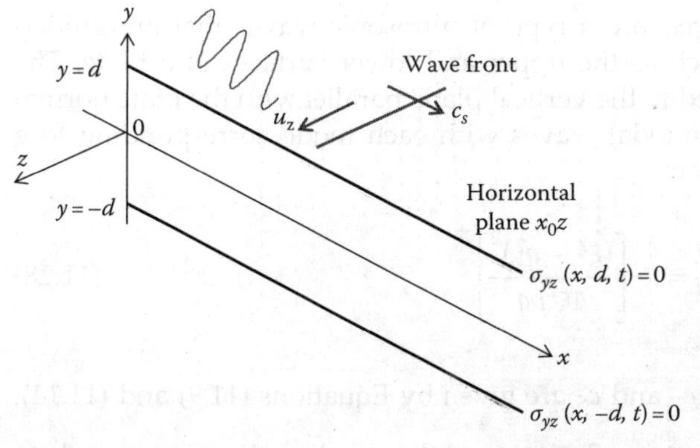

FIGURE 11.8
Axes definition and particle motion for SH waves.

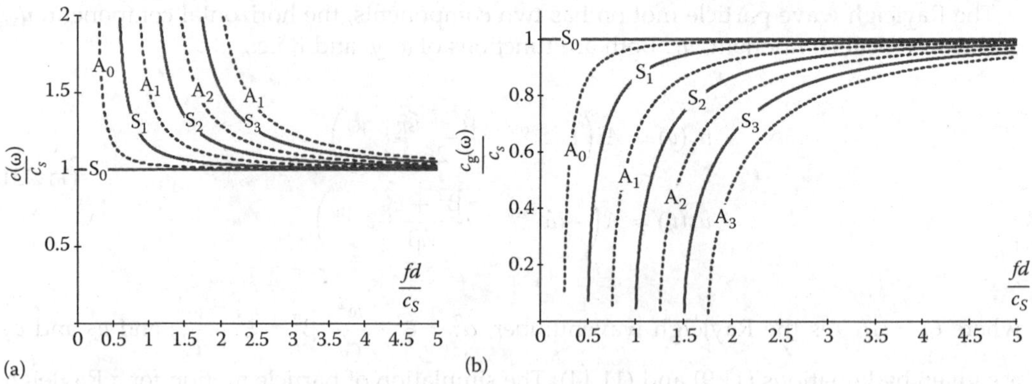

FIGURE 11.9
SH wave dispersion curves in aluminum plate (c_S = shear wave speed, d = half thickness of the plate): (a) wave speed (phase velocity); (b) group velocity.

dispersive. For this reason, the S_0 SH-wave mode is often sought after in ultrasonic NDE and SHM applications since nondispersive waves are much easier to analyze.

For each SH-wave mode, one can calculate the group velocity using the formula

$$c_g(\omega) = c_S \sqrt{1 - (\eta d)^2 \left(\frac{c_S}{\omega d}\right)^2} \tag{11.25}$$

The group velocity is important when examining the traveling SH-wave packets, as the case is with the pitch-catch and pulse-echo experiments. SH-wave particle motion takes place only in the z direction and has the form

$$u_z^S(x, y, t) = C_2 \cos \eta y \; e^{i(\xi x - \omega t)} \; \text{symmetric SH waves (S-modes)} \tag{11.26}$$

$$u_z^A(x, y, t) = C_1 \sin \eta y \; e^{i(\xi x - \omega t)} \; \text{antisymmetric SH waves (A-modes)} \tag{11.27}$$

11.2.1.7 Lamb Waves

Lamb waves, a.k.a. guided plate waves, are a type of ultrasonic waves that are guided between two parallel free surfaces such as the upper and lower surfaces of a plate. The Lamb-wave particle motion is polarized in the vertical plane parallel with the plate normal (Figure 11.10). Lamb waves are multimodal waves with each mode corresponding to a solution of the Rayleigh–Lamb equation

$$\frac{\tan pd}{\tan qd} = -\left[\frac{(\xi^2 - q^2)^2}{4\xi^2 pq}\right]^{\pm 1} \tag{11.28}$$

where $p^2 = \dfrac{\omega^2}{c_P^2} - \xi^2$, $q^2 = \dfrac{\omega^2}{c_S^2} - \xi^2$, and c_P and c_S are given by Equations (11.9) and (11.14). The positive sign corresponds to symmetric solutions and the negative sign corresponds to antisymmetric solutions. In expanded form, Equation (11.28) can be written as

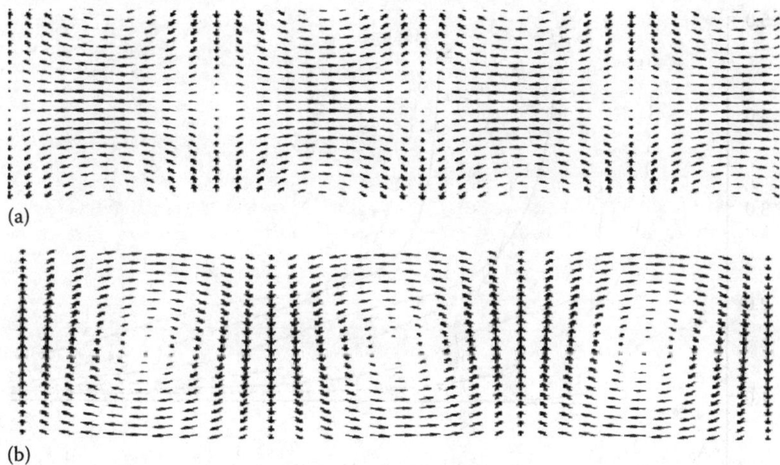

FIGURE 11.10
Simulation of Lamb waves particle motion: (a) symmetric S_0 mode; (b) antisymmetric A_0 mode.

$$D_S = \left(\xi^2 - q^2\right)^2 \cos pd \sin qd + 4\xi^2 pq \sin pd \cos qd = 0 \text{ (symmetric modes)} \tag{11.29}$$

$$D_A = \left(\xi^2 - q^2\right)^2 \sin pd \cos qd + 4\xi^2 pq \cos pd \sin qd = 0 \text{ (antisymmetric modes)} \tag{11.30}$$

Equations (11.28) through (11.30) are transcendental equations in which p and q depend on ξ. Numerical solution of Equation (11.28), or (11.29), (11.30), yields the symmetric (S) eigenvalues, $\xi_0^S, \xi_1^S, \xi_2^S, \ldots$ and the antisymmetric (A) eigenvalues, $\xi_0^A, \xi_1^A, \xi_2^A, \ldots$. The relationship $c = \omega/\xi$ yields the dispersive wave speed which is a function of the product fd between the frequency, $f = \omega/2\pi$, and the half thickness, d. At a given fd product, several Lamb modes may exist. At low fd values, only the lowest Lamb mode S_0 exists. A plot of the Lamb-wave dispersion curves is given in Figure 11.11a where the symmetric modes are designated $S_0, S_1, S_2, \ldots$ while the antisymmetric modes are designated $A_0, A_1, A_2, \ldots$.

For each Lamb-wave mode, one can calculate the group velocity using the formula

$$c_g = c^2 \left(c - fd \frac{\partial c}{\partial (fd)} \right)^{-1} \tag{11.31}$$

The group velocity is important when examining the traveling Lamb-wave packets, as the case is with the pitch-catch and pulse-echo experiments described later in this chapter.

Lamb-wave modes are illustrated graphically in Figure 11.12. At low values of the frequency–thickness product, the symmetric Lamb waves resemble the axial waves (Figure 11.12a) while the antisymmetric Lamb waves resemble the flexural waves (Figure 11.12b). In fact, it can be proved that, at low frequencies, the symmetric Lamb waves approach the behavior of the axial plate waves while the antisymmetric Lamb waves approach the behavior of the flexural plate waves. At higher frequencies, this is no longer true and the Lamb-wave modes can have quite a complicated behavior across the plate thickness. At very high values of the frequency–thickness product, the Lamb waves

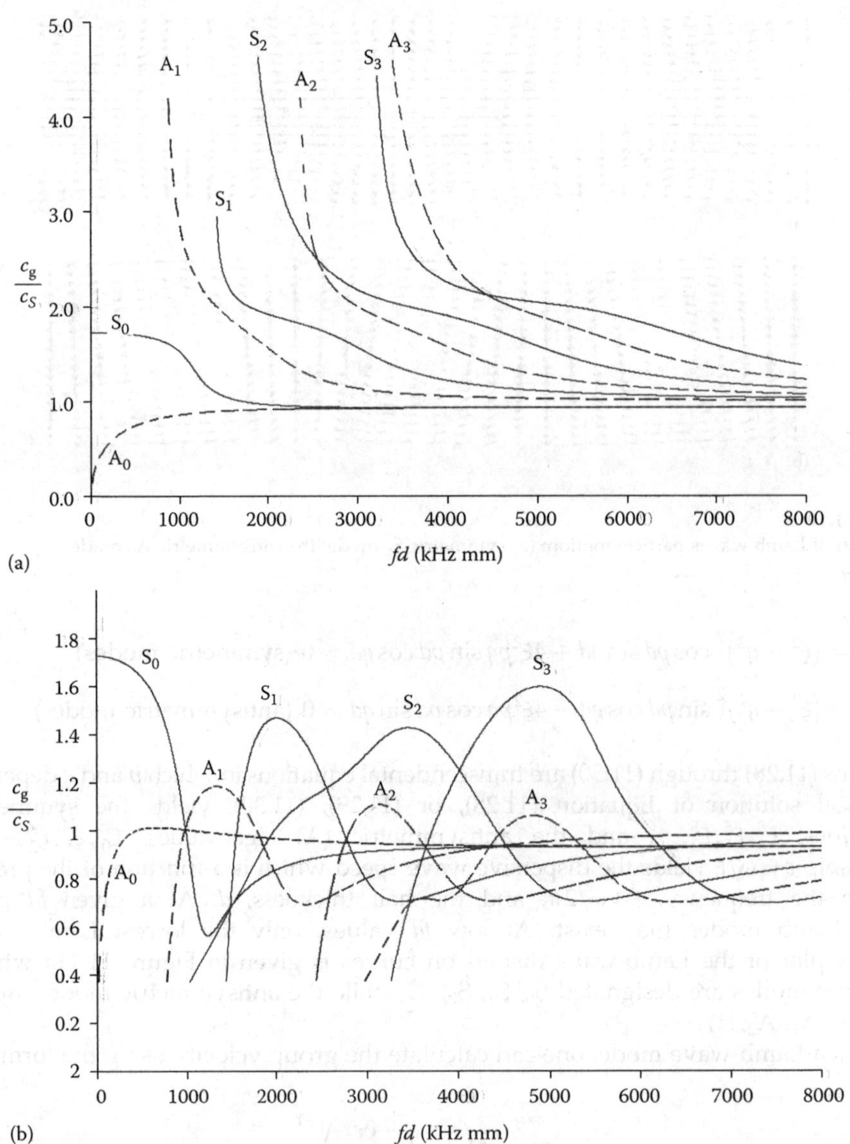

FIGURE 11.11

Lamb-wave dispersion curves in aluminum plate (c_S = shear wave speed, d = half thickness of the plate): (a) wave speed (phase velocity); (b) group velocity.

become concentrated close to the plate skin. In this case, Lamb-wave modes resemble the Rayleigh waves, and the wave speed approaches the Rayleigh wave speed, c_R (Figure 11.11).

The particle motion for straight-crested Lamb waves is given by

$$u_x^S(x, y, t) = A^S \left[-2\xi^2 q \cos qd \cos py + q(\xi^2 - q^2) \cos pd \cos qy\right] e^{i\xi x} e^{-i\omega t}$$
$$u_y^S(x, y, t) = A^S \left[-2i\xi pq \cos qd \sin py - i\xi(\xi^2 - q^2) \cos pd \sin qy\right] e^{i\xi x} e^{-i\omega t} \quad \text{(symmetric)} \quad (11.32)$$

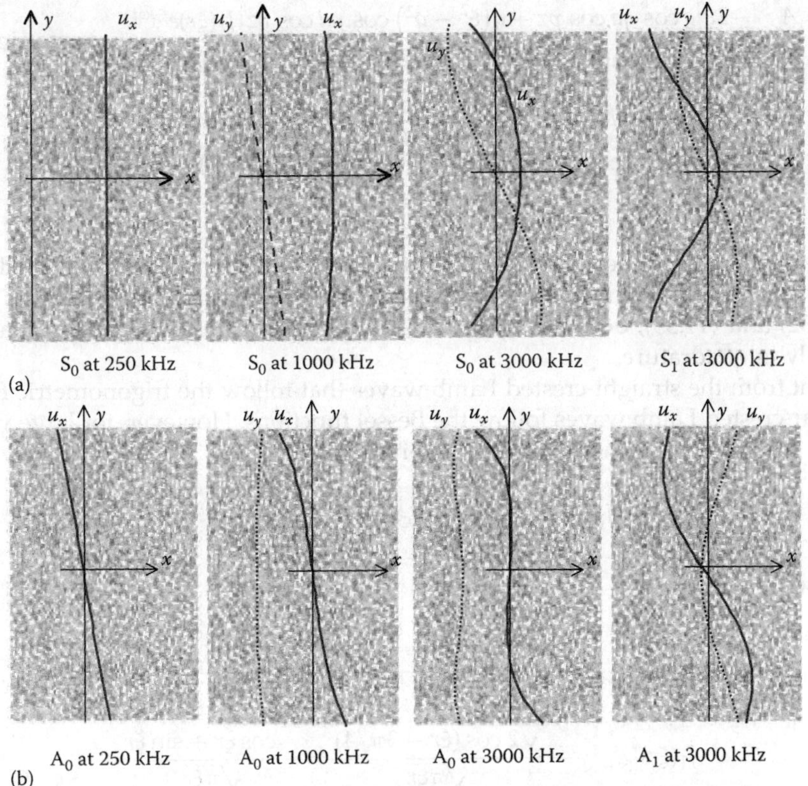

FIGURE 11.12
Displacement fields across the thickness for the Lamb-wave modes across the plate thickness at various frequencies: (a) symmetric mode S_0; (b) antisymmetric mode A_0.

$$u_x^A(x,y,t) = A^A\left[-2\xi^2 q \sin qd \sin py + q(\xi^2 - q^2)\sin pd \sin qy\right]e^{i\xi x}e^{-i\omega t}$$
$$u_y^A(x,y,t) = A^A\left[2i\xi pq \sin qd \cos py + i\xi(\xi^2 - q^2)\sin pd \cos qy\right]e^{i\xi x}e^{-i\omega t} \quad \text{(antisymmetric)}$$

$$(11.33)$$

where the mode amplitudes A^S and A^A have to be determined from the modes orthonormality conditions.

11.2.1.8 Circular-Crested Lamb Waves

Circular-crested Lamb waves are waves that emanate from a point and spread out in a circular pattern. Circular-crested waves are studies in cylindrical coordinates. Upon solution (Giurgiutiu, 2008, Section 6.4.2, p. 209), it is found that circular-crested Lamb waves have the same Rayleigh–Lamb characteristic equation as the straight-crested Lamb waves discussed in the previous section. Thus, Equation (11.28) can be used to determine the wave speed of both straight-crested and circular-crested Lamb waves. The particle motion of circular-crested Lamb waves is of the form

$$\begin{cases} u_r = A^{*S}\left[-2\xi^2 q \cos qd \cos pz + q(\xi^2 - q^2)\cos pd \cos qz\right]J_1(\xi r)e^{-i\omega t} \\ u_z = A^{*S}\left[2\xi pq \cos qd \sin pz + \xi(\xi^2 - q^2)\cos pd \sin qz\right]J_0(\xi r)e^{-i\omega t} \end{cases} \text{(symmetric)} \quad (11.34)$$

$$\begin{cases} u_r = A^{*A}\left[-2\xi^2 q \sin qd \sin pz + q(\xi^2 - q^2)\sin pd \sin qz\right]J_1(\xi r)e^{-i\omega t} \\ u_z = A^{*A}\left[2\xi pq \sin qd \cos pz + \xi(\xi^2 - q^2)\sin pd \cos qz\right]J_0(\xi r)e^{-i\omega t} \end{cases} \text{(antisymmetric)}$$

$$(11.35)$$

where the mode amplitudes A^{*S} and A^{*A} have to be determined from the modes orthonormality conditions. The factor i, which appears in the straight-crested waves in Equations (11.32) and (11.33), does not appear here because the Bessel functions J_0 and J_1 are intrinsically in quadrature.

Different from the straight-crested Lamb waves that follow the trigonometric functions, the circular-crested Lamb waves follow the Bessel functions. However, for large value of r, the Bessel functions have the asymptotic expression

$$J_\nu(z) \underset{z\to\infty}{\to} \sqrt{2/(\pi z)}\cos\left(z - \tfrac{1}{2}\nu\pi - \tfrac{1}{4}\pi\right) \qquad (11.36)$$

In our case,

$$J_0(\xi r)|_{\xi r \to \infty} \to \frac{\sqrt{2}\cos(\xi r - \pi/4)}{\sqrt{\pi \xi r}} = \frac{\cos \xi r + \sin \xi r}{\sqrt{\pi \xi r}}$$

$$(11.37)$$

$$J_1(\xi r)|_{\xi r \to \infty} \to \frac{\sqrt{2}\cos(\xi r - 3\pi/4)}{\sqrt{\pi \xi r}} = \frac{-\cos \xi r + \sin \xi r}{\sqrt{\pi \xi r}}$$

This means that the displacement pattern of circular-crested Lamb waves becomes periodic at large radial distances from the origin (i.e., at large values of ξr).

11.2.2 Structural Vibration

Structural vibrations are sustained structural oscillations that display preferred frequencies at which the oscillation amplitude increases significantly (i.e., resonance frequencies). The equations of motion for structural vibrations are identical to the wave equations for wave analysis. The main difference between wave analysis and vibration analysis lies in the propagation assumption; wave analysis is primarily concerned with propagating waves whereas vibration analysis treats stationary waves. On the one hand, the propagating waves of interests in NDE and SHM applications are manifested in terms of transitory wave packets that are sent into the structure and reflected at flaws, defects, boundaries, and other discontinuities. On the other hand, the structural vibrations are the effect of sustained excitation inducing stationary waves that establish themselves onto the structure in a steady-state subject to satisfying the boundary conditions.

Since structural vibrations are stationary waves, their analysis could start with wave propagation analysis and then proceed to study the conditions in which sustained harmonic waves can establish themselves in the structure subject to satisfying the boundary conditions. Such an approach is possible, but, in practice, it is only used in simple cases that aim at proving the connection between wave analysis and vibration analysis. The more common approach to vibration analysis relies on the physical concept of resonance and tries to identify the conditions under which a structure will vibrate with large amplitude while satisfying the boundary conditions. It is found that several solutions exist, called *natural modes of vibration* that happen at resonance frequencies. The analysis conducted in such a manner is called *modal analysis*. A brief review of these concepts with special attention to the cases relevant to using PWAS transducers for SHM applications will be considered next. For a more extensive study, one can refer to the 'Vibration' chapters in Giurgiutiu (2008) or specialized texts such as Timoshenko (1955), Leissa (1969), Harris (1996), etc.

11.2.2.1 Axial Vibration of Bars

Consider a uniform bar of length l, axial stiffness EA, and mass per unit length $m = \rho A$ undergoing axial motion in the longitudinal direction, $u(x, t)$. The equation of motion can be shown to be

$$c^2 u'' = \ddot{u} \tag{11.38}$$

where $c^2 = E/\rho$ is the axial wave speed as given by Equation (11.11). It can be shown that the general solution of Equation (11.38) is

$$u(x, t) = \hat{u}(x) e^{i\omega t} \quad \text{where} \quad \hat{u}(x) = A_1 \sin \gamma x + A_2 \cos \gamma x \tag{11.39}$$

The constants A_1 and A_2 are found from the boundary conditions. For free-free conditions, we have zero axial force at the bar ends, i.e., $N(0, t) = 0$, $N(l, t) = 0$. Substitution of the general solution (11.39) into the boundary conditions yields the *characteristic equation*

$$\sin \gamma l = 0 \tag{11.40}$$

Equation (11.40) has the roots $\gamma_j = j\pi/l$. Since $\gamma = \omega/c$, the corresponding natural frequencies are

$$\omega_j = j\frac{\pi}{l}\sqrt{\frac{E}{\rho}}, \quad j = 1, 2, 3, \ldots \tag{11.41}$$

To each natural frequency, there corresponds a natural mode of vibration given by

$$U_j(x) = A_j \cos \gamma_j x, \quad \text{where} \quad A_n = \sqrt{2/ml}, \quad j = 1, 2, 3, \ldots \tag{11.42}$$

The natural modes are orthonormal, i.e., satisfy the condition $\int_0^l m U_p(x) U_q(x) dx = \delta_{pq}$, where $\delta_{pq} = 1$ for $p = q$, and 0 otherwise. If the bar is excited at one of the natural frequencies ω_j, then it enters into the resonance state and vibrates in the natural mode $U_j(x)$. Off resonance, the bar vibrates with a combination of all its natural modes, i.e.,

$$u(x, t) = \sum_{j=1}^{\infty} \frac{f_j}{-\omega^2 + 2i\zeta_j\omega_j\omega + \omega_j^2} U_j(x) e^{i\omega t} \tag{11.43}$$

The quantity $f_j = \int_0^l \hat{f}(x)U_j(x)dx$ is the *modal excitation* (i.e., the projection of the excitation force onto the jth normal mode); the quantity ζ_j is the modal damping.

11.2.2.2 Flexural Vibration of Beams

Consider a uniform beam of length l, bending stiffness EI, and mass per unit length $m = \rho A$ undergoing flexural vibration of displacement $w(x, t)$. Under the Euler–Bernoulli theory of bending, the differential equation of motion can be written as

$$a^4 w'''' + \ddot{w} = 0 \tag{11.44}$$

where $a^4 = EI/m$. It can be shown that the general solution of Equation (11.44) is of the form

$$w(x, t) = \hat{w}(x)e^{i\omega t} \quad \text{where} \quad \hat{w}(x) = A_1 \sin \gamma x + A_2 \cos \gamma x + A_3 \sinh \gamma x + A_4 \cosh \gamma x \tag{11.45}$$

where $\gamma = (\sqrt{\omega})/a$. The constants $A_1, \ldots, A_4$ are found from the imposition of the boundary conditions. For a free-free beam, the boundary conditions are zero transverse force and zero bending moment at the beam ends, i.e., $M(0, t) = 0$, $M(l, t) = 0$, $V(0, t) = 0$, and $V(l, t) = 0$. Substitution of the general solution (11.45) into the boundary conditions yields the characteristic equation

$$\cos \gamma l \cosh \gamma l - 1 = 0 \tag{11.46}$$

Equation (11.46) is a transcendental equation. Numerical values of $(\gamma l)_j$ for $j \leq 5$ can be found in the literature, e.g., Giurgiutiu (2008), Table 3.5 p. 189; for $5 < j, (\gamma l)_j = (2j + 1)\pi/2$. Since $\gamma = (\sqrt{\omega})/a$, the corresponding natural frequencies are

$$\omega_j = a^2 \gamma_j^2, \quad j = 1, 2, 3, \ldots \tag{11.47}$$

To each natural frequency, there corresponds a natural mode of vibration given by

$$W_j(x) = A_j \left[\cosh \gamma_j x + \cos \gamma_j x - \beta_j \left(\sinh \gamma_j x + \sin \gamma_j x \right) \right] \tag{11.48}$$

where β is a modal parameter. Numerical values for β can be found in the literature, e.g., in Giurgiutiu (2008), Table 3.5 p. 189. For $5 < j$, $\sigma_{5 < j} = 1$. The amplitude A_j is determined through modal normalization, e.g., $A_j = 1 / \sqrt{\int_0^l m W_j^2(x)dx}$. The natural modes given in Equation (11.48) are orthonormal, i.e., satisfy the condition $\int_0^l m W_p(x)W_q(x)dx = \delta_{pq}$, where $\delta_{pq} = 1$ for $p = q$, and 0 otherwise. If the beam is excited at one of the natural frequencies ω_j, then it enters into the resonance state and vibrates in the natural mode $W_j(x)$. Off resonance, the beam vibrates with a combination of all its natural modes, i.e.,

$$w(x, t) = \sum_{j=1}^{\infty} \frac{f_j}{-\omega^2 + 2i\zeta_j \omega_j \omega + \omega_j^2} W_j(x)e^{i\omega t} \tag{11.49}$$

The quantity $f_j = \int_0^l \hat{f}(x)W_j(x)dx$ is the *modal excitation* (i.e., the projection of the excitation force onto the jth normal mode); the quantity ζ_j is the modal damping.

11.2.2.3 Axial Vibration of Rectangular Plates

Consider a uniform plate of length a, width b, thickness h, elastic modulus E, Poisson ratio v, and mass density ρ undergoing axial vibration with in-plane displacements $u_x(x,t)$ and $u_y(x,t)$. The equation of motion can be shown to be a coupled system of differential equations, i.e.,

$$c_L^2 \left[\frac{\partial^2 u_x}{\partial x^2} + \frac{1-v}{2}\frac{\partial^2 u_x}{\partial y^2} + \frac{1+v}{2}\frac{\partial^2 u_y}{\partial x \partial y} \right] = \ddot{u}_x$$

$$c_L^2 \left[\frac{1+v}{2}\frac{\partial^2 u_x}{\partial x \partial y} + \frac{1-v}{2}\frac{\partial^2 u_y}{\partial x^2} + \frac{\partial^2 u_y}{\partial y^2} \right] = \ddot{u}_y$$

$$(11.50)$$

where c_L is given by Equation (11.12). This system has a general solution representing plate vibration simultaneously in the x and y directions. The solution of this coupled system of differential equations is not immediate; this case will not be discussed here. However, numerical solutions can be obtained with the finite element method (FEM). For illustration, Figure 11.13 shows the axial mode shapes of a free–free square plate.

In the particular case of a plate with high aspect ratio ($a \gg b$), the vibration in the longitudinal direction $u_x(x,t)$ becomes separated from that in the transverse direction $u_y(x,t)$, at least at low frequencies. Under these conditions, the plate behaves more like a bar, as described in Section 11.2.2.1.

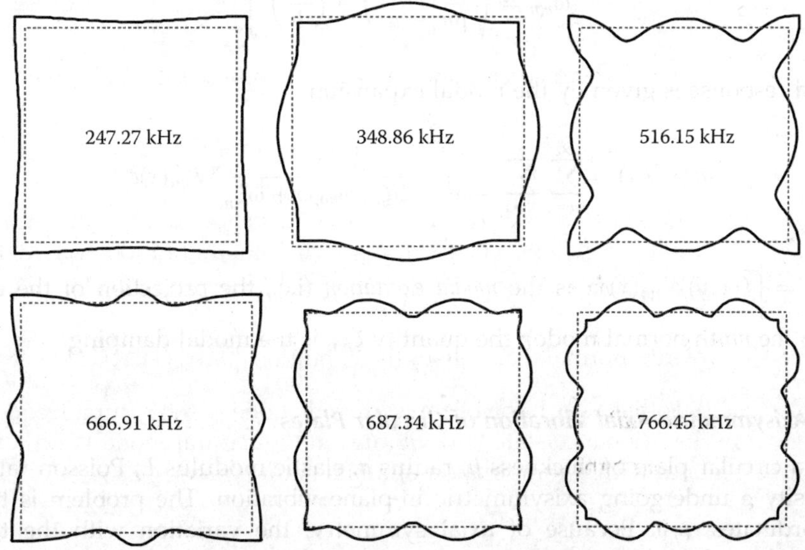

FIGURE 11.13
Axial mode shapes of a free-free square plate (7-mm square, 0.2 mm thick, $\rho = 7600$ kg/m^3, $E = 65$ GPa, $v = 0.35$).

11.2.2.4 Flexural Vibration of Rectangular Plates

Consider a uniform plate of length l, width b, thickness h, elastic modulus E, Poisson ratio v, and mass density ρ undergoing flexural vibration with out-of-plane displacement $u_z(x, t)$. The equation of motion can be shown to be

$$D\left(\frac{\partial^4 w}{\partial x^4} + 2\frac{\partial^4 w}{\partial x^2 \partial y^2} + \frac{\partial^4 w}{\partial y^4}\right) + \rho h \ddot{w} = 0 \tag{11.51}$$

where D is the *flexural plate stiffness* defined as

$$D = \frac{Eh^3}{12(1 - v^2)} \tag{11.52}$$

Equation (11.51) has the general solution of the form

$$w(x, y, t) = \hat{w}(x, y)e^{i\omega t} \tag{11.53}$$

For general boundary conditions, the expression of $\hat{w}(x, y)$ is complicated and will not be reproduced here. In the case of simply supported boundary conditions ($w = 0$, $M_x = 0$ for $x = 0$ $x = a$; $w = 0$, $M_y = 0$ for $y = 0$, $y = b$), the mode shapes take the remarkably simple form

$$W_{mn} = A_{mn} \sin\frac{m\pi x}{a} \sin\frac{n\pi y}{b} \tag{11.54}$$

where $A_{mn} = 1 \left/ \sqrt{\int_0^a \int_0^b \rho h W_{mn}^2(x)dxdy}\right.$. The corresponding natural frequencies are given by

$$\omega_{mn} = \sqrt{\frac{D}{\rho h}\left[\left(\frac{m\pi}{a}\right)^2 + \left(\frac{n\pi}{b}\right)^2\right]} \tag{11.55}$$

The forced response is given by the modal expansion

$$w(x, y, t) = \sum_{m=1}^{\infty} \sum_{n=1}^{\infty} \frac{f_{mn}}{-\omega^2 + 2i\zeta_{mn}\omega_{mn}\omega + \omega_{mn}^2} W_{mn}(x)e^{i\omega t} \tag{11.56}$$

where $f_{mn} = \int_0^l \hat{f}(x, y)W_{mn}(x)dx$ is the *modal excitation* (i.e., the projection of the excitation force onto the *mn*th normal mode); the quantity ζ_{mn} is the modal damping.

11.2.2.5 Axisymmetric Axial Vibration of Circular Plates

Consider a circular plate of thickness h, radius a, elastic modulus E, Poisson ratio v, and mass density ρ undergoing axisymmetric in-plane vibration. The problem is treated in polar coordinates, ρ, θ. Because of axial symmetry, the variation with the tangential coordinate θ and the tangential displacement is zero ($\partial/\partial\theta = 0$, $u_\theta = 0$). The radial displacement, $u_r(r, t)$, satisfies the differential equation

$$c_L^2 \left(\frac{\partial^2 u_r}{\partial r^2} + \frac{1}{r} \frac{\partial u_r}{\partial r} - \frac{u_r}{r^2} \right) = \ddot{u}_r \tag{11.57}$$

where c_L is given by Equation (11.12). Solution of Equation (11.57) is sought in the form

$$u_r(r,t) = \hat{u}(r)e^{i\omega t} \tag{11.58}$$

For free-edge boundary conditions, the *characteristic equation* is given in the form

$$z J_0(z) - (1 - v) J_1(z) = 0 \tag{11.59}$$

where $z = \gamma a$, $\gamma = \omega/c_L$, and c_L is given by Equation (11.12). The function $J_1(z)$ is the Bessel function of first kind and order 1. Equation (11.59) is transcendental in z and does not accept closed-form solution. Numerical solution of Equation (11.59) for $v = 0.30$ yields

$$z = 2.048652,\ 5.389361,\ 8.571860,\ 11.731771 \ \ldots \tag{11.60}$$

The corresponding resonance frequencies are given by

$$\omega_n = \frac{c_L}{a} z_n, \quad n = 1, 2, 3, \ldots \tag{11.61}$$

The corresponding mode shapes are given by

$$R_n(r) = A_n J_1(z_n r/a) \tag{11.62}$$

where $A_n = \sqrt{J_1^2(z_n) - J_0(z_n)J_2(z_n)}$. The mode shapes $R_n(r)$ form an orthonormal set that satisfies the orthonormality conditions

$$\rho h \int_0^a \int_0^{2\pi} R_i(r) R_j(r) r\, dr\, d\theta = m\delta_{ij} \tag{11.63}$$

where $m = \pi a^2 \rho h$ is the plate mass.

11.2.2.6 Axisymmetric Flexural Vibration of Circular Plates

Consider a circular plate of thickness h, radius r, elastic modulus E, Poisson ratio v, and mass density ρ undergoing axisymmetric in-plane vibration. The problem is treated in polar coordinates, ρ, θ. Because of axial symmetry, the variation with the tangential coordinate θ is zero ($\partial/\partial\theta = 0$). The out-of-plane displacement, $w(r,t)$, satisfies the differential equation

$$D\left(\frac{\partial^4 w}{\partial r^4} + \frac{2}{r} \frac{\partial^3 w}{\partial r^3} - \frac{1}{r^2} \frac{\partial^2 w}{\partial r^2} + \frac{1}{r^3} \frac{\partial w}{\partial r} \right) + \rho h \ddot{w} = 0 \tag{11.64}$$

where D is given by Equation (11.52). Solution of Equation (11.64) is sought in the form

$$w(r,t) = \hat{w}(r)e^{i\omega t} \tag{11.65}$$

For free-edge boundary conditions, the *characteristic equation* is

$$\frac{z^2 J_0(z) + (1-\nu)z^2 J_0'(z)}{z^2 I_0(z) + (1-\nu)z^2 I_0'(z)} = \frac{z^3 J_0'(z)}{z^3 I_0'(z)} \tag{11.66}$$

where $z = \gamma a$ and $\gamma^4 = \omega^2 \rho h/D$. The function $J_0(z)$ is the Bessel function of first kind and order zero whereas $I_0(z)$ is the modified Bessel function of first kind and order zero. Equation (11.66) is transcendental in z. Upon numerical solution, one finds

$$z = 3.01146, \, 6.20540, \, 9.37084, \ldots \tag{11.67}$$

The corresponding resonance frequencies are

$$\omega_j = z_j^2 \left(\frac{D}{\rho h a^4}\right)^{1/2}, \quad j = 1, 2, 3, \ldots \tag{11.68}$$

The corresponding mode shapes are

$$W_j(r) = A_j\left[J_0\left(z_j r/a\right) + C_j I_0\left(z_j r/a\right)\right] \tag{11.69}$$

where

$$A_j = \frac{1}{\sqrt{2}}\left\{\left[J_0(z_j) + C_j I_0(z_j)\right]^2 - \left[J_0'(z_j)\right]^2 - \left[C_j I_0'(z_j)\right]^2\right\}^{-1/2} \tag{11.70}$$

The mode shapes $W_j(r)$ form an orthonormal set that satisfies the orthonormality conditions

$$\rho h \int_0^a \int_0^{2\pi} W_p(r)W_m(r)r\,dr\,d\theta = m\delta_{pm} \tag{11.71}$$

where $m = \pi a^2 \rho h$ is the plate mass. Numerical values of eigenvalues z_j, mode shape parameters C_j, and mode shape amplitudes A_j can be found in Giurgiutiu (2008), Table 4.3, p. 127.

11.2.3 Section Summary

This section has presented a brief review of elastic waves and structural vibration principles to be used later in the chapter. It started with the basic pressure and shear waves in unbounded solid media and then addressed waves in bounded media such as rod and plates. The conventional axial and flexural wave plates were analyzed first. These simplified waves assume a certain displacement field across the thickness (constant for axial waves and linear for flexural waves). However, such simplified fields violate the stress-free boundary conditions on the lower and upper surfaces. When exact stress-free boundary conditions are imposed, the analysis yields the guided plate waves (SH waves and Lamb waves). These waves represent standing waves across the plate thickness and traveling waves along the plate length. Several guided plate wave modes exist, corresponding to the real roots of the characteristic equation. Guided waves are dispersive, i.e., the wave speed varies with frequency, as shown in Figures 11.9 and 11.11. Each wave mode has a different speed, a different wavelength, and a different standing wave pattern across the plate thickness. The standing mode pattern across the thickness varies with frequency. At low frequencies, the fundamental S_0 and A_0 Lamb-wave modes approach the conventional axial and flexural plate waves (Figures 11.11 and 11.12).

Figure 11.14 shows the wave-speed dispersion curves for S_0 and A_0 Lamb waves, axial waves, flexural waves, and Rayleigh waves in a 1-mm aluminum plate. Figure 11.14 shows that, at low frequency, the flexural waves and the A_0 Lamb-wave mode have similar wave speed curves. However, as the frequency increases, they become separated. This indicates that the conventional flexural waves are just low-frequency approximation of the A_0 Lamb-wave mode. Similarly, the conventional axial waves are a low-frequency approximation of the S_0 Lamb-wave mode. At low frequency, the wave speeds of the axial waves and the S_0 Lamb-wave mode are very close. However, at higher frequency, they differ substantially. At the other end of the spectrum, Rayleigh waves are a high-frequency approximation of the S_0 and A_0 Lamb waves. As indicated in Figure 11.14, as the frequency becomes very high, the S_0 and the A_0 wave speeds coalesce, and both have the same value, which is the Rayleigh wave speed. At high frequency, the particle motion of Lamb waves becomes restricted to the proximity of the free surfaces, and thus resembles that of the Rayleigh waves.

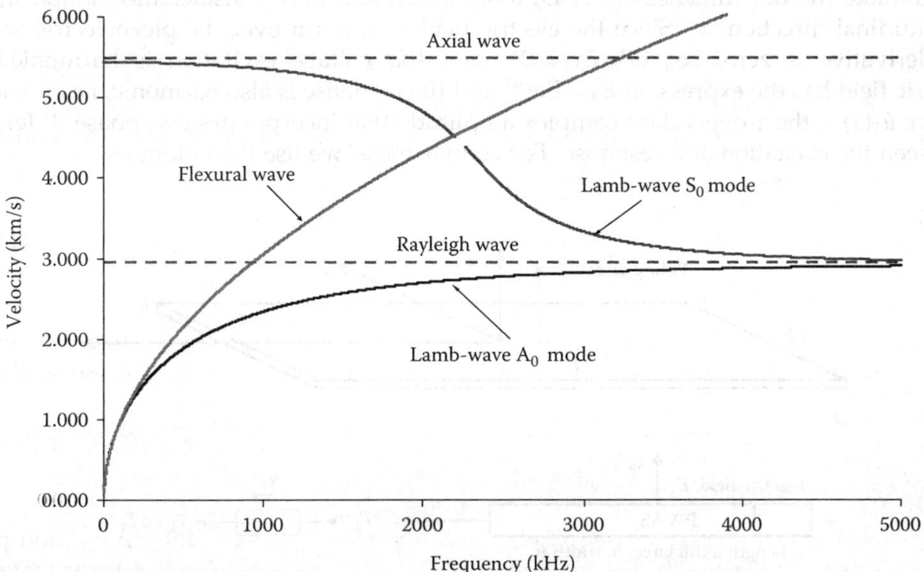

FIGURE 11.14
Frequency dependence of wave speed for axial, flexural, Lamb, and Rayleigh waves in 1 mm thick aluminum plate. Note that axial and flexural waves are only low-frequency approximations of the Lamb wave S_0 and A_0 modes. The Rayleigh wave is the high-frequency asymptote of the Lamb wave S_0 and A_0 modes.

The second part of this section has briefly reviewed the structural vibration. Axial and flexural vibration of bars, beams, and plates were considered. Closed-form solutions were given for the vibration of bars and beams. For plate vibration, closed-form solution was given only in a selected number of cases such as the axisymmetric axial and flexural vibrations of free circular plates, etc.

11.3 PWAS Resonators

11.3.1 1-D PWAS Resonators

This section addresses the behavior of a free piezoelectric wafer active sensor (PWAS). When excited by an alternating electric voltage, the free PWAS acts as an electromechanical resonator. The modeling of a free PWAS is useful for (1) understanding the electromechanical coupling between the mechanical vibration response and the complex electrical response of the sensor and (2) sensor screening and quality control prior to installation on the monitored structure.

Consider a piezoelectric wafer of length l, width b, and thickness t undergoing longitudinal expansion, u_1, induced by the thickness polarization electric field, E_3 (Figure 11.15). The electric field is produced by the application of a harmonic voltage $V(t) = \hat{V}e^{i\omega t}$ between the top and bottom surfaces (electrodes). The resulting electric field, $E_3 = V/t$, is assumed to be uniform over the piezoelectric wafer.

Assume that the length, width, and thickness have widely separated values ($t \ll b \ll l$) such that the length, width, and thickness motions are practically uncoupled. This allows us to make the one-dimensional (1-D) assumption and only consider the motion in the longitudinal direction, x_1. Since the electric field is uniform over the piezoelectric wafer, its derivative is zero, i.e., $\partial E_3/\partial x_1 = 0$. Since the voltage excitation is harmonic, the electric field has the expression $E_3 = \hat{E}_3 e^{i\omega t}$ and the response is also harmonic, i.e., $u = \hat{u}e^{i\omega t}$, where $\hat{u}_1(x)$ is the x-dependant complex amplitude that incorporates any phase difference between the excitation and response. For compactness, we use the notations

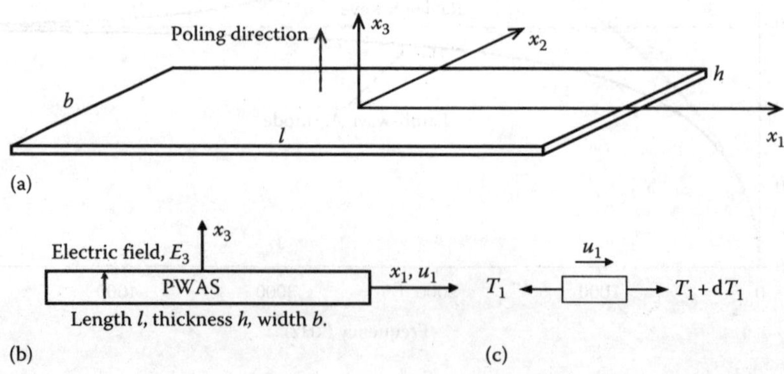

FIGURE 11.15
Schematic of a piezoelectric wafer active sensor and infinitesimal axial element: (a) 3-D view; (b) 1-D section; (c) infinitesimal element.

$$\frac{\partial}{\partial x_1}() = ()' \quad \text{and} \quad \frac{\partial}{\partial t}() = (^{\bullet})$$

Under the 1-D assumption, the general constitutive equations are reduced to the simpler expressions

$$S_1 = s_{11}^E T_1 + d_{31} E_3 \qquad \text{(strain)} \qquad (11.72)$$

$$D_3 = d_{31} T_1 + \varepsilon_{33}^T E_3 \qquad \text{(electric displacement)} \qquad (11.73)$$

where S_1 is the strain, T_1 is the stress, D_3 is the electrical displacement (charge per unit area), s_{11}^E is the mechanical compliance at zero field, ε_{33}^T is the dielectric constant at zero stress, and d_{31} is the induced-strain coefficient, i.e., mechanical strain per unit electric field. Typical values of these constants are given in Table 11.2.

Recall Newton's law of motion and the strain–displacement relation

$$T_1' = \rho \ddot{u}_1 \qquad \text{(Newton's law of motion)} \qquad (11.74)$$

$$S_1 = u_1' \qquad \text{(strain–displacement relation)} \qquad (11.75)$$

Differentiate Equation (11.72) with respect to x; since E_3 is constant with respect to x (i.e., $E_3' = 0$), the strain rate becomes

$$S_1' = s_{11}^E \cdot T_1' \qquad (11.76)$$

Substitution of Equations (11.74) and (11.75) into Equation (11.5) gives

$$u_1'' = s_{11}^E \rho \ddot{u}_1 \qquad (11.77)$$

TABLE 11.2

Properties of Piezoelectric Material APC-850

Property	Symbol	Value
Compliance, in plane	s_{11}^E	15.30×10^{-12} Pa^{-1}
Compliance, thickness wise	s_{33}^E	17.30×10^{-12} Pa^{-1}
Dielectric constant	ε_{33}^T	$\varepsilon_{33}^T = 1750\varepsilon_0$
Thickness wise induced-strain coefficient	d_{33}	400×10^{-12} m V^{-1}
In-plane induced-strain coefficient	d_{31}	-175×10^{-12} m V^{-1}
Coupling factor, parallel to electric field	k_{33}	0.72
Coupling factor, transverse to electric field	k_{31}	0.36
Coupling factor, transverse to electric field, polar motion	k_p	0.63
Poisson's ratio	ν	0.35
Density	ρ	7700 kg m^{-3}
Sound speed	c	2900 m s^{-1}

Note: $\varepsilon_0 = 8.85 \times 10^{-12}$ F m^{-1}.

Introduce the notation

$$c^2 = \frac{1}{\rho s_{11}^E} \quad \text{(wave speed)} \tag{11.78}$$

representing the longitudinal wave speed of the material. Substitution of Equation (11.78) into Equation (11.77) yields the 1-D wave equation

$$\ddot{u}_1 = c^2 u_1'' \tag{11.79}$$

The general solution of the wave equation is

$$u_1(x, t) = \hat{u}(x)e^{i\omega t} \tag{11.80}$$

where

$$\hat{u}_1(x) = C_1 \sin \gamma x + C_2 \cos \gamma x \tag{11.81}$$

and

$$\gamma = \frac{\omega}{c} \tag{11.82}$$

The quantity γ is the *wave number*. An associated quantity is the *wavelength*, $\lambda = cT = c/f$, where $f = \omega/2\pi$. The constants C_1 and C_2 are determined from the boundary conditions.

11.3.1.1 Mechanical Response

For a piezoelectric wafer of length l with the origin at its center, the interval of interest is $x \in \left(-\frac{1}{2}l, \frac{1}{2}l\right)$. Stress-free boundary conditions imply $T_1\left(-\frac{1}{2}l\right) = T_1\left(+\frac{1}{2}l\right) = 0$. Substitution into Equation (11.72) yields

$$S_1\left(-\tfrac{1}{2}l\right) = d_{31}E_3 \tag{11.83}$$

$$S_1\left(+\tfrac{1}{2}l\right) = d_{31}E_3 \tag{11.84}$$

Using the strain–displacement relation of Equation (11.75), we get

$$\hat{u}_1'\left(-\tfrac{1}{2}l\right) = \gamma\left(C_1 \cos \gamma \tfrac{1}{2}l + C_2 \sin \gamma \tfrac{1}{2}l\right) = d_{31}\hat{E}_3 \tag{11.85}$$

$$\hat{u}_1'\left(\tfrac{1}{2}l\right) = \gamma\left(C_1 \cos \gamma \tfrac{1}{2}l - C_2 \sin \gamma \tfrac{1}{2}l\right) = d_{31}\hat{E}_3 \tag{11.86}$$

Addition of the two equations yields an equation in only C_1, i.e.,

$$\gamma C_1 \cos \gamma \tfrac{1}{2}l = d_{31}\hat{E}_3 \tag{11.87}$$

Hence,

$$C_1 = \frac{d_{31}\hat{E}_3}{\gamma \cos \frac{1}{2}\gamma l} \tag{11.88}$$

Subtraction of the two equations yields an equation in only C_2, i.e.,

$$\gamma C_2 \sin \gamma \tfrac{1}{2} l = 0 \tag{11.89}$$

Assuming $\sin \tfrac{1}{2} \gamma l \neq 0$, we get

$$C_2 = 0 \tag{11.90}$$

An alternative derivation of C_1 and C_2 can be obtained by using Cramer's rule on the algebraic system

$$\begin{cases} C_1 \gamma \cos \tfrac{1}{2} \gamma l + C_2 \gamma \sin \tfrac{1}{2} \gamma l = d_{31} \hat{E}_3 \\ C_1 \gamma \cos \tfrac{1}{2} \gamma l - C_2 \gamma \sin \tfrac{1}{2} \gamma l = d_{31} \hat{E}_3 \end{cases} \tag{11.91}$$

Hence,

$$C_1 = \frac{\begin{vmatrix} d_{31}\hat{E}_3 & \gamma \sin \tfrac{1}{2} \gamma l \\ d_{31}\hat{E}_3 & -\gamma \sin \tfrac{1}{2} \gamma l \end{vmatrix}}{\begin{vmatrix} \gamma \cos \tfrac{1}{2} \gamma l & \gamma \sin \tfrac{1}{2} \gamma l \\ \gamma \cos \tfrac{1}{2} \gamma l & -\gamma \sin \tfrac{1}{2} \gamma l \end{vmatrix}} = \frac{d_{31}\hat{E}_3 \left(-2\gamma \sin \tfrac{1}{2} \gamma l \right)}{-\gamma^2 2 \sin \tfrac{1}{2} \gamma l \cos \tfrac{1}{2} \gamma l} = \frac{d_{31}\hat{E}_3}{\gamma \cos \tfrac{1}{2} \gamma l} \tag{11.92}$$

and

$$C_2 = \frac{\begin{vmatrix} \gamma \cos \tfrac{1}{2} \gamma l & d_{31}\hat{E}_3 \\ \gamma \cos \tfrac{1}{2} \gamma l & d_{31}\hat{E}_3 \end{vmatrix}}{\begin{vmatrix} \gamma \cos \tfrac{1}{2} \gamma l & \gamma \sin \tfrac{1}{2} \gamma l \\ \gamma \cos \tfrac{1}{2} \gamma l & -\gamma \sin \tfrac{1}{2} \gamma l \end{vmatrix}} = \frac{0}{-\gamma^2 2 \sin \tfrac{1}{2} \gamma l \cos \tfrac{1}{2} \gamma l} = 0 \tag{11.93}$$

In this derivation, we have assumed that the determinant appearing at the denominator is nonzero, i.e., $\Delta \neq 0$. Since $\Delta = -\gamma^2 2 \sin \tfrac{1}{2} \gamma l \cos \tfrac{1}{2} \gamma l = \gamma^2 \sin \gamma l$, this condition implies that

$$\sin \gamma l \neq 0 \tag{11.94}$$

Substitution of C_1 and C_2 into Equation (11.81) yields

$$\hat{u}_1(x) = \frac{d_{31}\hat{E}_3}{\gamma} \cdot \frac{\sin \gamma x}{\cos \tfrac{1}{2} \gamma l} \tag{11.95}$$

Using Equation (11.75), we get the strain as

$$\hat{S}_1(x) = d_{31}\hat{E}_3 \frac{\cos \gamma x}{\cos \tfrac{1}{2} \gamma l} \tag{11.96}$$

11.3.1.1.1 Solution in Terms of the Induced Strain, S_{ISA}, and Induced Displacement, u_{ISA}

Introduce the notations

$$S_{ISA} = d_{31}\hat{E}_3 \qquad \text{(Induced strain)} \qquad (11.97)$$

$$u_{ISA} = S_{ISA}l = \left(d_{31}\hat{E}_3\right)l \qquad \text{(Induced displacement)} \qquad (11.98)$$

where ISA signifies "induced-strain actuation." Hence, Equations (11.95) and (11.96) can be written as

$$\hat{u}(x) = \frac{1}{2}\frac{u_{ISA}}{\frac{1}{2}\gamma l}\frac{\sin\gamma x}{\cos\frac{1}{2}\gamma l} \qquad (11.99)$$

$$\hat{S}_1(x) = S_{ISA}\frac{\cos\gamma x}{\cos\frac{1}{2}\gamma l} \qquad (11.100)$$

Equations (11.99) and (11.100) show that, under dynamic conditions, the strain along the piezoelectric wafer is not constant. This is in contrast to the static case for which the strain is uniform along the length of the actuator. For $\cos\frac{1}{2}\gamma l \leq 1$, the maximum strain amplitude is observed in the middle and has the value

$$S_{\max} = \frac{S_{ISA}}{\cos\frac{1}{2}\gamma l} \qquad (11.101)$$

Introducing the notation $\phi = \frac{1}{2}\gamma l$, the displacement and strain equations can be rewritten as

$$\hat{u}(x) = \frac{1}{2}\frac{u_{ISA}}{\phi}\frac{\sin\gamma x}{\cos\phi} \qquad (11.102)$$

$$\hat{S}_1(x) = S_{ISA}\frac{\cos\gamma x}{\cos\phi} \qquad (11.103)$$

11.3.1.1.2 Tip Strain and Displacement

At the wafer ends, $x = \pm\frac{1}{2}l$, the tip strain and displacement are

$$\hat{u}\left(\pm\frac{1}{2}l\right) = \pm\frac{1}{2}\frac{u_{ISA}}{\frac{1}{2}\gamma l}\frac{\sin\frac{1}{2}\gamma l}{\cos\frac{1}{2}\gamma l} = \pm\frac{1}{2}\frac{u_{ISA}}{\frac{1}{2}\gamma l}\tan\frac{1}{2}\gamma l \qquad (11.104)$$

$$\hat{S}_1\left(\pm\frac{1}{2}\gamma l\right) = S_{ISA}\frac{\cos\frac{1}{2}\gamma l}{\cos\frac{1}{2}\gamma l} = S_{ISA} \qquad (11.105)$$

We notice that, at the tip, the strain takes exactly the same value as that of S_{ISA}.

11.3.1.2 Electrical Response

Consider a 1-D piezoelectric wafer active sensor under electric excitation (Figure 11.16). Recall Equation (11.73) representing the electrical displacement

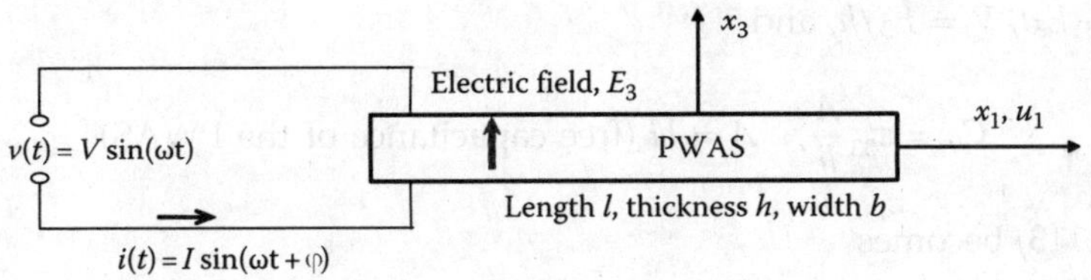

FIGURE 11.16
Schematic of a 1-D piezoelectric wafer active sensor under electric excitation.

$$D_3 = d_{31}T_1 + \varepsilon_{33}^T E_3 \tag{11.106}$$

Equation (11.72) yields the stress as function of strain and electric field, i.e.,

$$T_1 = \frac{1}{s_{11}}(S_1 - d_{31}E_3) \tag{11.107}$$

Hence, the electric displacement can be expressed as

$$D_3 = \frac{d_{31}}{s_{11}}(S_1 - d_{31}E_3) + \varepsilon_{33}^T E_3 \tag{11.108}$$

Upon substitution of the strain–displacement relation of Equation (11.75), we get

$$D_3 = \frac{d_{31}}{s_{11}}u_1' - \frac{d_{31}^2}{s_{11}}E_3 + \varepsilon_{33}^T E_3 \tag{11.109}$$

i.e.,

$$D_3 = \varepsilon_{33}^T E_3\left[1 - k_{31}^2\left(1 - \frac{u_1'}{d_{31}E_3}\right)\right] \tag{11.110}$$

where $\kappa_{13}^2 = d_{31}^2/(s_{11}\varepsilon_{33})$ is the electromechanical coupling coefficient. Integration of Equation (11.110) over the electrodes area $A = bl$ yields the total charge

$$Q = \int_A D_3 dx_1 dx_2 = \int_{-l/2}^{l/2}\int_0^b D_3 dx_1 dx_2 = \varepsilon_{33}^T E_3 b\left[l - k_{31}^2\left(1 - \frac{1}{d_{31}E_3}u_1\Big|_{-\frac{1}{2}l}^{\frac{1}{2}l}\right)\right] \tag{11.111}$$

Assuming harmonic time dependence, $Q = \hat{Q}e^{i\omega t}$, yields $\hat{Q}$ as function of $\hat{u}_1$ at both ends of the piezoelectric wafer

$$\hat{Q} = \varepsilon_{33}^T \hat{E}_3 b\left[l - k_{31}^2\left(l - \frac{1}{d_{31}\hat{E}_3}\left[\hat{u}_1\left(\tfrac{1}{2}l\right) - \hat{u}_1\left(-\tfrac{1}{2}l\right)\right]\right)\right] \tag{11.112}$$

Rearranging,

$$\hat{Q} = \varepsilon_{33}^T bl\left[1 - k_{31}^2 + k_{31}^2\left(\frac{\hat{u}_1\left(\tfrac{1}{2}l\right) - \hat{u}_1\left(-\tfrac{1}{2}l\right)}{d_{31}\hat{E}_3 l}\right)\right] \tag{11.113}$$

Use $\hat{u}_{ISA} = d_{31}\hat{E}_3 l$, $\hat{V} = \hat{E}_3/h$, and

$$C_0 = \varepsilon_{33}^T \frac{A}{h}, \quad A = bl \text{ (free capacitance of the PWAS)} \tag{11.114}$$

Equation (11.113) becomes

$$\hat{Q} = C_0 \hat{V} \left[1 - k_{31}^2 + k_{31}^2 \left(\frac{\hat{u}_1\left(\frac{1}{2}l\right) - \hat{u}_1\left(-\frac{1}{2}l\right)}{u_{ISA}} \right) \right] \tag{11.115}$$

The electric current is obtained as the time derivative of the electric charge, i.e.,

$$I = \dot{Q} = i\omega Q \tag{11.116}$$

Hence,

$$\hat{I} = i\omega C_0 \hat{V} \left[1 - k_{31}^2 + k_{31}^2 \left(\frac{\hat{u}_1\left(\frac{1}{2}l\right) - \hat{u}_1\left(-\frac{1}{2}l\right)}{u_{ISA}} \right) \right] \tag{11.117}$$

Alternatively, the electric current could have been obtained by first performing the derivative with respect to time and then the integration with respect to space, i.e.,

$$I = \int_A \frac{dD_3}{dt} dA = i\omega \int_A D_3 \, dA \tag{11.118}$$

The integral in Equation (11.118) is the same as the integral in Equation (11.112).

The admittance, Y, is defined as the ratio between the current and voltage, i.e.,

$$Y = \frac{\hat{I}}{\hat{V}} = i\omega C_0 \left[1 - k_{31}^2 + k_{31}^2 \left(\frac{\hat{u}_1\left(\frac{1}{2}l\right) - \hat{u}_1\left(-\frac{1}{2}l\right)}{u_{ISA}} \right) \right] \tag{11.119}$$

Recall Equation (11.99) giving the displacement solution

$$\hat{u}_1(x) = \frac{d_{31}\hat{E}_3}{\gamma} \frac{\sin \gamma x}{\cos \frac{1}{2}\gamma l}$$

Hence,

$$\frac{\hat{u}_1\left(\frac{1}{2}l\right) - \hat{u}_1\left(-\frac{1}{2}l\right)}{u_{ISA}} = \frac{1}{2}\frac{\sin\frac{1}{2}\gamma l - \left(-\sin\frac{1}{2}\gamma l\right)}{\frac{1}{2}\gamma l \cos\frac{1}{2}\gamma l} = \frac{1}{2}2\frac{\sin\frac{1}{2}\gamma l}{\frac{1}{2}\gamma l \cos\frac{1}{2}\gamma l} = \frac{\tan\frac{1}{2}\gamma l}{\frac{1}{2}\gamma l} \tag{11.120}$$

where the notation $\phi = \frac{1}{2}\gamma l$ was invoked. Substitution in Equation (11.119) yields

$$Y = i\omega C \left[1 - k_{31}^2 \left(1 - \frac{\tan\frac{1}{2}\gamma \ell}{\frac{1}{2}\gamma \ell} \right) \right] \tag{11.121}$$

This result agrees with Ikeda (1996). However, it may be more convenient, at times, to write it as

$$Y = i\omega C \left[1 - k_{31}^2 \left(1 - \frac{1}{\frac{1}{2}\gamma l \cot \frac{1}{2}\gamma l} \right) \right] \tag{11.122}$$

Note that the admittance is purely imaginary and consists of the capacitive admittance, $i\omega C$, modified by the effect of piezoelectric coupling between mechanical and electrical variables. This effect is apparent in the term containing the electromechanical coupling coefficient, k_{31}^2. The impedance, Z, is obtained as the ratio between the voltage and current, i.e.,

$$Z = \frac{\hat{V}}{\hat{I}} = Y^{-1} \tag{11.123}$$

Hence,

$$Z = \frac{1}{i\omega C} \left[1 - k_{31}^2 \left(1 - \frac{\tan \frac{1}{2}\gamma \ell}{\frac{1}{2}\gamma \ell} \right) \right]^{-1} \tag{11.124}$$

Recalling the notation $\phi = \frac{1}{2}\gamma l$, we can write

$$Y = i\omega C_0 \left[1 - k_{31}^2 \left(1 - \frac{\tan \phi}{\phi} \right) \right] \tag{11.125}$$

$$Z = \frac{1}{i\omega C_0} \left[1 - k_{31}^2 \left(1 - \frac{\tan \phi}{\phi} \right) \right]^{-1} \tag{11.126}$$

11.3.1.3 Resonances

In previous derivations, we made the assumption that $\sin \gamma l \neq 0$. This assumption implies that resonances do not happen. If resonances happened, they could be of two types:

1. Mechanical resonances
2. Electromechanical resonances

Mechanical resonances take place in the same conditions as in a conventional elastic bar. They happen under mechanical excitation that produces a mechanical response in the form of mechanical vibrations. Electromechanical resonances are specific to piezoelectric materials. They reflect the coupling between the mechanical and electrical variables. Electromechanical resonances happen under electric excitation that produces an electromechanical response, i.e., both a mechanical vibration and a change in the electric admittance and impedance. We will consider these two situations separately.

11.3.1.3.1 Mechanical Resonances

If a PWAS is excited mechanically with a frequency sweep, certain frequencies will be revealed at which the response is very large, i.e., the PWAS resonates. To study the mechanical resonances, assume that the material is not piezoelectric, i.e., $d_{31} = 0$. Hence, we recover the results for the classical mechanical resonances of an elastic bar. The stress-free boundary conditions become

$$\begin{cases} C_1\gamma\cos\tfrac{1}{2}\gamma l + C_2\gamma\sin\tfrac{1}{2}\gamma l = 0 \\ C_1\gamma\cos\tfrac{1}{2}\gamma l - C_2\gamma\sin\tfrac{1}{2}\gamma l = 0 \end{cases} \tag{11.127}$$

This homogenous system of equations accepts nontrivial solutions when the determinant of the coefficients is zero, i.e.,

$$\Delta = \begin{vmatrix} \gamma\cos\tfrac{1}{2}\gamma l & \gamma\sin\tfrac{1}{2}\gamma l \\ \gamma\cos\tfrac{1}{2}\gamma l & -\gamma\sin\tfrac{1}{2}\gamma l \end{vmatrix} = 0 \tag{11.128}$$

Upon expansion,

$$\Delta = -2\sin\tfrac{1}{2}\gamma l\cos\tfrac{1}{2}\gamma l = 0 \tag{11.129}$$

Equation (11.129) yields the characteristic equation

$$2\sin\tfrac{1}{2}\gamma l\cos\tfrac{1}{2}\gamma l = \sin\gamma l = 0 \tag{11.130}$$

One notes that the second part of Equation (11.130) is identical with Equation (11.40) discussed in the general review of axial vibration (Section 11.2.2.1). For the present discussion, it is more productive to retain Equation (11.130) in the form of the product

$$\sin\tfrac{1}{2}\gamma l\cos\tfrac{1}{2}\gamma l = 0 \tag{11.131}$$

Recall the notation $\phi = \tfrac{1}{2}\gamma l$. Hence, the characteristic Equation (11.131) can be written as

$$\sin\phi\cos\phi = 0 \tag{11.132}$$

This equation admits two sets of solutions, one corresponding to $\cos\phi = 0$ and the other corresponding to $\sin\phi = 0$. For reasons that will become apparent, the first condition leads to antisymmetric resonances and the second to symmetric resonances.

11.3.1.3.2 Antisymmetric Resonances (cos $\phi = 0$)

Antisymmetric resonances happen when $\cos\phi = 0$. This happens when ϕ is an odd multiple of $\pi/2$, i.e.,

$$\cos\phi = 0 \rightarrow \phi = (2n-1)\frac{\pi}{2} \tag{11.133}$$

This means that, at antisymmetric resonances, the variable ϕ can take one of the following values:

$$\phi^A = \frac{\pi}{2},\ \frac{3\pi}{2},\ \frac{5\pi}{2},\dots \tag{11.134}$$

These values are the antisymmetric eigenvalues of the system. The superscript A denotes antisymmetric resonances. Recall the notations $\phi = \tfrac{1}{2}\gamma l$, $\gamma = \omega/c$, and $\omega = 2\pi f$. Hence,

$$\gamma l = \frac{\omega l}{c} = \frac{2\pi f l}{c} = (2n-1)\pi \tag{11.135}$$

Therefore, the antisymmetric resonance frequencies are given by the formula

$$f_n^A = (2n - 1)\frac{c}{2l} \tag{11.136}$$

For each resonance frequency, Equation (11.127) accepts nontrivial values for the coefficients C_1 and C_2. However, they can only be determined up to a scaling factor. Using the notation $\phi = \frac{1}{2}\gamma l$, we write Equation (11.127) as

$$\begin{cases} C_1 \cos \phi + C_2 \sin \phi = 0 \\ C_1 \cos \phi - C_2 \sin \phi = 0 \end{cases} \tag{11.137}$$

Examination of Equation (11.137) reveals that, for $\cos \phi = 0$, the constant C_2 vanishes while the constant C_1 is indeterminate. We remove the indeterminacy by choosing $C_1 = 1$. Substituting $C_1 = 1$ and $C_2 = 0$ into Equation (11.81) yields the antisymmetric mode shapes

$$U_n^A(x) = \sin \gamma_n x \tag{11.138}$$

Using Equation (11.138), we write the mode shapes in the convenient form

$$U_n^A(x) = \sin (2n - 1)\pi\frac{x}{l} \tag{11.139}$$

For a graphical representation of these mode shapes, see the modes A_1, A_2, $A_3, \ldots$ in Table 11.3.

Another useful interpretation of these results is through the wavelength λ. Recall that the wavelength is defined as the distance traveled by the wave in a time period, i.e., $\lambda = cT$. Since the time period is the inverse of frequency, the wavelength is $\lambda = c/f$. Hence, Equation (11.135) gives

$$l_n^A = (2n - 1)\frac{\lambda}{2} \tag{11.140}$$

Equation (11.140) indicates that antisymmetric resonances will happen when the PWAS length is an odd multiple of the half wavelength of the elastic waves traveling in the PWAS. This relation is used in constructing PWAS that have to resonate at a given frequency.

11.3.1.3.3 Symmetric Resonances ($\sin \phi = 0$)

Symmetric resonances happen when $\sin \phi = 0$. This happens when ϕ is an even multiple of $\pi/2$, i.e.,

$$\sin \phi = 0 \rightarrow \phi = 2n\frac{\pi}{2} \tag{11.141}$$

This means that, at a symmetric resonance, the variable ϕ can take one of the following values

$$\phi^S = 2\frac{\pi}{2}, 4\frac{\pi}{2}, 6\frac{\pi}{2}, \ldots \tag{11.142}$$

TABLE 11.3

Resonance Mode Shapes of a 1-D PWAS of Length l

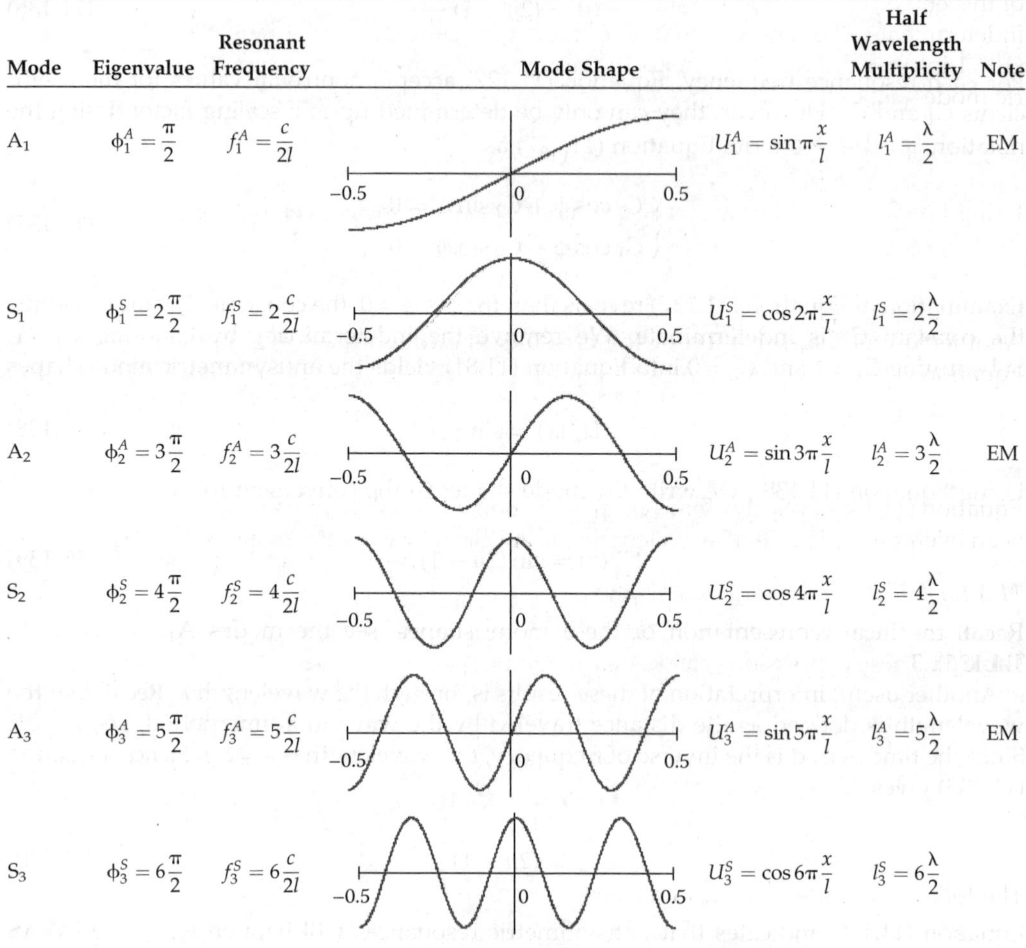

Mode	Eigenvalue	Resonant Frequency	Mode Shape	Half Wavelength Multiplicity	Note
A_1	$\phi_1^A = \dfrac{\pi}{2}$	$f_1^A = \dfrac{c}{2l}$		$U_1^A = \sin \pi \dfrac{x}{l}$ $l_1^A = \dfrac{\lambda}{2}$	EM
S_1	$\phi_1^S = 2\dfrac{\pi}{2}$	$f_1^S = 2\dfrac{c}{2l}$		$U_1^S = \cos 2\pi \dfrac{x}{l}$ $l_1^S = 2\dfrac{\lambda}{2}$	
A_2	$\phi_2^A = 3\dfrac{\pi}{2}$	$f_2^A = 3\dfrac{c}{2l}$		$U_2^A = \sin 3\pi \dfrac{x}{l}$ $l_2^A = 3\dfrac{\lambda}{2}$	EM
S_2	$\phi_2^S = 4\dfrac{\pi}{2}$	$f_2^S = 4\dfrac{c}{2l}$		$U_2^S = \cos 4\pi \dfrac{x}{l}$ $l_2^S = 4\dfrac{\lambda}{2}$	
A_3	$\phi_3^A = 5\dfrac{\pi}{2}$	$f_3^A = 5\dfrac{c}{2l}$		$U_3^A = \sin 5\pi \dfrac{x}{l}$ $l_3^A = 5\dfrac{\lambda}{2}$	EM
S_3	$\phi_3^S = 6\dfrac{\pi}{2}$	$f_3^S = 6\dfrac{c}{2l}$		$U_3^S = \cos 6\pi \dfrac{x}{l}$ $l_3^S = 6\dfrac{\lambda}{2}$	

These values are the symmetric eigenvalues of the system. The superscript S denotes symmetric resonances. Recall the notations $\phi = \frac{1}{2}\gamma l$, $\gamma = \omega/c$, and $\omega = 2\pi f$. Hence,

$$\gamma l = \frac{\omega l}{c} = \frac{2\pi f l}{c} = 2n\pi \tag{11.143}$$

Therefore, the symmetric resonance frequencies are given by the formula

$$f_n^S = 2n\frac{c}{2l} \tag{11.144}$$

Recall Equation (11.137)

$$\begin{cases} C_1 \cos \phi + C_2 \sin \phi = 0 \\ C_1 \cos \phi - C_2 \sin \phi = 0 \end{cases} \tag{11.145}$$

For each resonance frequency, Equation (11.137) accepts nontrivial values for the coefficients C_1 and C_2. However, they can only be determined up to a scaling factor. Examination of this equation reveals that, for $\sin\phi = 0$, the constant C_1 vanishes while the constant C_2 is indeterminate. We remove the indeterminacy by choosing $C_2 = 1$. Substituting $C_1 = 0$ and $C_2 = 1$ into Equation (11.81) yields the response at symmetric resonances, i.e., the symmetric mode shapes

$$U_n^S(x) = \cos\gamma_n x \tag{11.146}$$

Using Equation (11.143), we write the mode shapes in the convenient form

$$U_n^S(x) = \cos 2n\pi\frac{x}{l} \tag{11.147}$$

Another useful interpretation of these results is through the wavelength $\lambda = c/f$. Hence, Equation (11.143) becomes

$$l_n^S = 2n\frac{\lambda}{2} \tag{11.148}$$

Equation (11.148) indicates that symmetric resonances will happen when the PWAS length is an even multiple of half wavelengths of the elastic waves traveling in the PWAS.

11.3.1.3.4 *Electromechanical Resonances*

Recall the expressions for admittance and impedance given by Equations (11.125) and (11.126). These expressions can be rearranged as

$$Y = i\omega C_0\left[1 - k_{31}^2\left(1 - \frac{\tan\phi}{\phi}\right)\right] = i\omega C_0\left[\left(1 - k_{31}^2\right) + k_{31}^2\frac{\tan\phi}{\phi}\right] \tag{11.149}$$

$$Z = \frac{1}{i\omega C_0}\left[1 - k_{31}^2\left(1 - \frac{\tan\phi}{\phi}\right)\right]^{-1} = \frac{1}{i\omega C_0}\left[\left(1 - k_{31}^2\right) + k_{31}^2\frac{\tan\phi}{\phi}\right]^{-1} \tag{11.150}$$

The following conditions are considered:

- Resonance, when $Y \to \infty$, i.e., $Z = 0$
- Antiresonance, when $Y = 0$, i.e., $Z \to \infty$

Electrical resonance is associated with the situation in which a device is drawing very large currents when excited harmonically with a constant voltage at a given frequency. At resonance, the admittance becomes very large while the impedance goes to zero. As the admittance becomes very large, the current drawn under constant–voltage excitation also becomes very large since $I = Y \cdot V$. In piezoelectric devices, the mechanical response at electrical resonance also becomes very large. This happens because the electromechanical coupling of the piezoelectric material transfers energy from the electrical input into the mechanical response. For these reasons, the resonance of an electrically driven piezoelectric device must be seen as an electromechanical resonance. A piezoelectric wafer driven at electrical resonance may undergo mechanical deterioration and even break up.

Electrical antiresonance is associated with the situation in which a device draws almost no current. At antiresonance, the admittance goes to zero while the impedance becomes very large. Under constant-voltage excitation, this condition results in a very small current being drawn from the source. In a piezoelectric device, the mechanical response at electrical antiresonance is also very small. A piezoelectric wafer driven at the electrical antiresonance hardly moves at all. The resonance of an electrically driven piezoelectric device must be also seen as an electromechanical antiresonance.

The condition for electromechanical resonance is obtained by studying the poles of Y, i.e., the values of ϕ which make $Y \to \infty$. Equation (11.149) reveals that $Y \to \infty$ as $\tan \phi \to \infty$. This happens when $\cos \phi \to 0$, when ϕ is an odd multiple of $\pi/2$, i.e.,

$$\cos \phi = 0 \to \phi = (2n - 1)\frac{\pi}{2} \tag{11.151}$$

The angle ϕ can take the following values

$$\phi^{EM} = \frac{\pi}{2}, \frac{3\pi}{2}, \frac{5\pi}{2}, \ldots \tag{11.152}$$

These values are the electromechanical eigenvalues. They are marked by superscript EM. Since $\phi = \frac{1}{2}\gamma l$, Equation (11.152) implies that

$$\gamma l = \pi, 3\pi, 5\pi \tag{11.153}$$

Recall the definitions $\gamma = \omega/c$ and $\omega = 2\pi f$. Hence,

$$\gamma l = \frac{\omega l}{c} = \frac{2\pi f l}{c} = (2n - 1)\pi \tag{11.154}$$

Therefore, the electromechanical resonance frequencies are given by the formula

$$f_n^{EM} = (2n - 1)\frac{c}{2l} \tag{11.155}$$

It is remarkable that the electromechanical resonance frequencies do not depend on any of the electric or piezoelectric properties. They depend entirely on the speed of sound in the material and the geometric dimensions. In fact, they coincide with the antisymmetric resonance frequencies of Table 11.3. For each resonance frequency, f_n^{EM}, the angle ϕ_n is given by the formula

$$\phi_n^{EM} = \frac{\pi l}{c} f_n \tag{11.156}$$

The vibration modes corresponding to the first, second, and third electromechanical resonances are marked EM in Table 11.3.

11.3.1.3.5 Origin of the Electromechanical Resonance

The electromechanical resonance can be explained by analyzing the electromechanical response given by Equation (11.95). This equation states that

$$\hat{u}(x) = \frac{d_{31}\hat{E}_3}{\gamma} \frac{\sin \gamma x}{\cos \frac{1}{2}\gamma l} \tag{11.157}$$

The poles of Equation (11.95) correspond to frequency values where the mechanical response to electrical excitation becomes unbounded, i.e., when electromechanical resonance happens. The function $\cos\frac{1}{2}\gamma l$ in the denominator becomes zero when the argument $\phi = \frac{1}{2}\gamma l$ is an odd multiple of $\pi/2$, i.e.,

$$\phi^{EM} = \frac{\pi}{2}, \frac{3\pi}{2}, \frac{5\pi}{2}, \cdots \tag{11.158}$$

As $\cos\frac{1}{2}\gamma l$ becomes zero, the response becomes very large and resonance happens. Comparing Equations (11.158) and (11.134), we notice that these equations are the same, which means that the conditions for electromechanical resonance are the same as the conditions for antisymmetric mechanical resonance. This implies that electromechanical resonance is associated with the antisymmetric modes of mechanical resonance.

Explanation of why these particular modes are excited electrically resides in the electromechanical coupling taking place in the piezoelectric material. As indicated by the constitutive Equation (11.72), the electric field, E_3, couples directly with the strain, S_1, through the piezoelectric strain coefficient, d_{31}. The applied electric field is uniformly distributed over the piezoelectric wafer. A uniform distribution is a symmetric pattern. Since the strain is the space derivative of displacement, $S_1 = u_1'$, the strain distribution for an antisymmetric mode is also a symmetric pattern (Table 11.4). Since both the electric field and the strain have symmetric patterns, coupling between strain and electric field is promoted and resonances are excited. For this reason, the antisymmetric modes of mechanical resonance are also modes of electromechanical resonance.

By the same token, we can explain why the electromechanical resonance does not happen at the frequencies corresponding to the symmetric modes of mechanical resonance. The symmetrical modes have an antisymmetric strain pattern while the electric field

TABLE 11.4

Strain Distributions Corresponding to the First, Second, and Third Antisymmetric Modes of Mechanical Resonance

Mode	Eigenvalue	Resonant Frequency	Strain Distribution		Half Wavelength Multiplicity
A_1	$\phi_1^A = \dfrac{\pi}{2}$	$f_1^A = \dfrac{c}{2l}$		$S_1^A = \cos\pi\dfrac{x}{l}$	$l_1^A = \dfrac{\lambda}{2}$
A_2	$\phi_2^A = 3\dfrac{\pi}{2}$	$f_2^A = 3\dfrac{c}{2l}$		$S_2^A = \cos 3\pi\dfrac{x}{l}$	$l_2^A = 3\dfrac{\lambda}{2}$
A_3	$\phi_3^A = 5\dfrac{\pi}{2}$	$f_3^A = 5\dfrac{c}{2l}$		$S_3^A = \cos 5\pi\dfrac{x}{l}$	$l_3^A = 5\dfrac{\lambda}{2}$

Note: Note that the strain follows a symmetric pattern.

continues to have a symmetric pattern. When a symmetric electric field pattern tries to excite an antisymmetric strain pattern, the net result is zero since the effects on the left and right sides have opposite signs and cancel each other.

In conclusion, in a PWAS undergoing longitudinal vibrations, the electromechanical resonances happen at the poles of Y (zeros of Z) and correspond to the antisymmetric modes of mechanical resonances.

The condition for *electromechanical antiresonance* is obtained by studying the zeros of Y, i.e., the values of ϕ which make $Y = 0$. Since electromechanical antiresonances correspond to the zeros of the admittance (i.e., poles of the impedance), the current at antiresonance is zero, $I = 0$. Equation (11.149) indicates that $Y = 0$ happens when

$$\left(1 - k_{31}^2\right) + k_{31}^2 \frac{\tan \phi}{\phi} = 0 \rightarrow \frac{\tan \phi}{\phi} = -\frac{1 - k_{31}^2}{k_{31}^2} \text{ (antiresonance)} \tag{11.159}$$

This equation is a transcendental equation that does not have closed-form solutions. Its solutions are found numerically.

In practice, the numerical difference between the values of the electromechanical resonances and electromechanical antiresonances diminishes as we go to higher mode numbers. Table 11.5 shows the admittance and impedance poles calculated for $k_{31} = 0.36$ which is a common value in piezoelectric ceramics. The admittance poles, φ_Y, and impedance poles, φ_Z, differ significantly only for the first mode. By the fourth mode, the difference between them drops below 0.1%. This is important when trying to determine the resonance and antiresonance frequencies experimentally.

Equations (11.149) and (11.150) can be used to predict the frequency response of the admittance and impedance functions. To this purpose, we note that $\varphi = \frac{1}{2}\omega l/c$. As the excitation frequency varies and resonance and antiresonance frequencies are encountered, the admittance and impedance go through $+\infty$ to $-\infty$ transitions. Outside resonance, the admittance follows the linear function $i\omega C$ while the impedance follows the inverse function $1/(i\omega C)$.

11.3.1.3.6 Effect of Internal Damping

Materials under dynamic operation display internal heating due to several loss mechanisms. Such losses can be incorporated in the PWAS model by assuming complex compliance and dielectric constant, i.e.,

$$\bar{s}_{11} = s_{11}(1 - i\eta), \quad \bar{\varepsilon}_{33} = \varepsilon_{33}(1 - i\delta) \tag{11.160}$$

TABLE 11.5

Admittance and Impedance Poles for $k_{31} = 0.36$

Admittance poles, φ_Y (electromechanical resonances)	$\pi/2$	$3\pi/2$	$5\pi/2$	$7\pi/2$	$9\pi/2$	$11\pi/2$	...
Impedance poles, φ_Z (electromechanical resonances)	$1.0565\pi/2$	$3.021\pi/2$	$5.005\pi/2$	$7\pi/2$	$9\pi/2$	$11\pi/2$	...
Ratio φ_Z/φ_Y	1.0565	1.0066	1.0024	1.0012	1.0007	1.0005	...
Difference between φ_Z and φ_Y	5.6%	0.66%	0.24%	0.12%	0.07%	0.05%	

The values of η and δ vary with the piezoceramic formulation but are usually small (η, $\delta < 5\%$). The admittance and impedance become complex expressions

$$\overline{Y} = i\omega\overline{C}_0\left[1 - \overline{k}_{31}^2\left(1 - \frac{1}{\overline{\phi}\cot\overline{\phi}}\right)\right], \quad \overline{Z} = \frac{1}{i\omega\overline{C}_0}\left[1 - \overline{k}_{31}^2\left(1 - \frac{1}{\overline{\phi}\cot\overline{\phi}}\right)\right]^{-1} \tag{11.161}$$

where $\overline{k}_{13}^2 = d_{31}^2/(\overline{s}_{11}\overline{\varepsilon}_{33})$ is the complex coupling factor, and $\overline{C}_0 = (1 - i\delta)C_0$, $\overline{\phi} = \phi\sqrt{1 - i\eta}$. Similar expressions can be derived for the width and thickness vibrations with appropriate use of indices.

11.3.1.3.7 Admittance and Impedance Plots

Frequency plots of admittance and impedance are useful for the graphical determination of the resonance and antiresonance frequencies. Figure 11.17 presents the numerical simulation of admittance and impedance response for a piezoelectric active sensor ($l = 7$ mm, $b = 1.68$ mm, $t = 0.2$ mm, APC-850 piezoceramic). Light damping, $\delta = \eta = 1\%$, was assumed. As shown in Figure 11.17a, the admittance and impedance essentially behave like $Y = i\omega C$, and $Z = 1/i\omega C$ outside resonances and antiresonances. For example, the imaginary part of the admittance follows a straight-line pattern outside resonances while the real part is practically zero. At resonances and antiresonances, these basic patterns are modified by the addition of a pattern of behavior specific to resonance and antiresonance. These patterns include zigzags of the imaginary part and sharp peaks of the real part. The admittance shows zigzags of the imaginary part and peaks of the real part

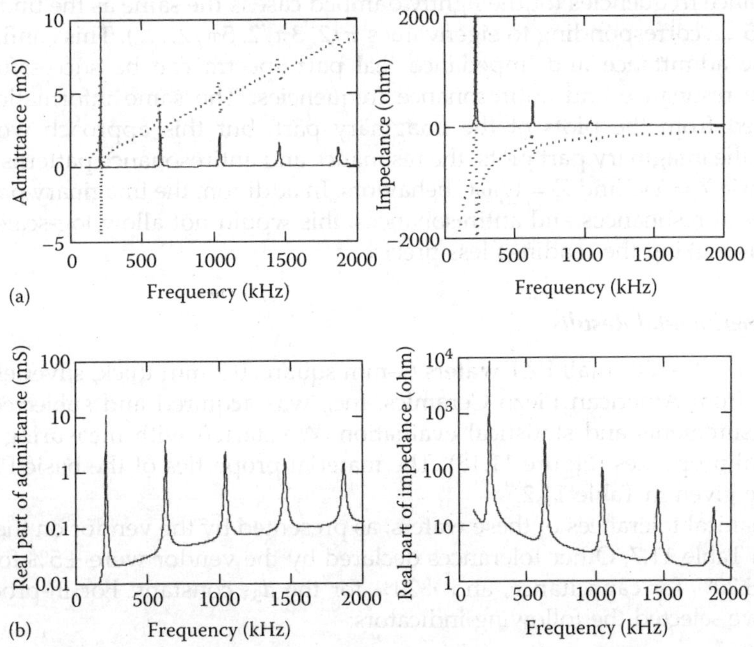

FIGURE 11.17
Simulated frequency response of admittance and impedance of a PWAS resonator ($l = 7$ mm, $b = 1.68$ mm, $t = 0.2$ mm, APC-850 piezoceramic, δ, $\eta = 1\%$): (a) complete plots showing both real (full line) and imaginary (dashed line) parts; (b) plots of real part only, log scale.

TABLE 11.6

Admittance and Impedance Poles Determined from the Numerical Simulation of a PWAS resonator ($l = 7$ mm, $b = 1.68$ mm, $t = 0.2$ mm, APC-850 piezoceramic, $\delta = \varepsilon = 1\%$)

Mode No.	1	2	3	4	5	6	...
Resonance frequencies (peaks of admittance real part spectrum), f_Y, kHz	207.17	621.57	1035.80	1450.20	1864.50	2278.90	...
Anti resonance frequencies (peaks of impedance real part spectrum), f_Z, kHz	218.94	625.70	1038.30	1452.00	1866.00	2280.00	...
f_Z/f_Y	1.0568	1.0066	1.0024	1.0012	1.0008	1.0005	...
f_{Yn}/f_{Y1}	1.000	3.000	5.000	7.000	9.000	11.000	

around the resonance frequencies (Figure 11.17a, left). The impedance shows the same behavior around the antiresonance frequencies (Figure 11.17a, right).

Figure 11.17b shows log-scale plots of the real parts of admittance and impedance. The log-scale plots are better for graphically identifying the resonance and antiresonance frequencies. Table 11.6 lists, to 4-digit accuracy, the frequencies read from these plots. Because the damping considered in this simulation was light, the values given in Table 11.6 compare well with the undamped data of Table 11.5. In fact, to 3-digit accuracy, the ratio of the resonance frequencies for the lightly damped case is the same as the undamped case (ratios 1, 3, 5, ... corresponding to eigenvalues $\pi/2, 3\pi/2, 5\pi/2, ...$). This confirms that the peaks of the admittance and impedance real part spectra can be successfully used to measure the resonance and antiresonance frequencies. The same information could be also extracted from the plots of the imaginary part, but this approach would be less practical. In the imaginary part plots, the resonance and antiresonance patterns are masked by the intrinsic $Y = i\omega C$ and $Z = 1/i\omega C$ behaviors. In addition, the imaginary parts undergo sign changes at resonances and antiresonances; this would not allow log-scale plots to be applied, thus making the readings less precise.

11.3.1.4 Experimental Results

A batch of 25 APC-850 small PZT wafers (7-mm square, 0.2 mm thick, silver electrodes on both sides), from American Piezo Ceramics, Inc., was acquired and subjected to experimental measurements and statistical evaluation. We started with measuring mechanical and electrical properties (Figure 11.18). The material properties of the basic PZT material APC-850 are given in Table 11.2.

The mechanical tolerances of these wafers, as presented by the vendor on their Web site, are given in Table 11.7. Other tolerances declared by the vendor were ±5% for resonance frequency, ±20% for capacitance, and ±20% for the d_{33} constant. For in-process quality assurance, we selected the following indicators:

1. Geometrical dimensions
2. Electrical capacitance
3. Electromechanical (E/M) impedance and admittance spectra

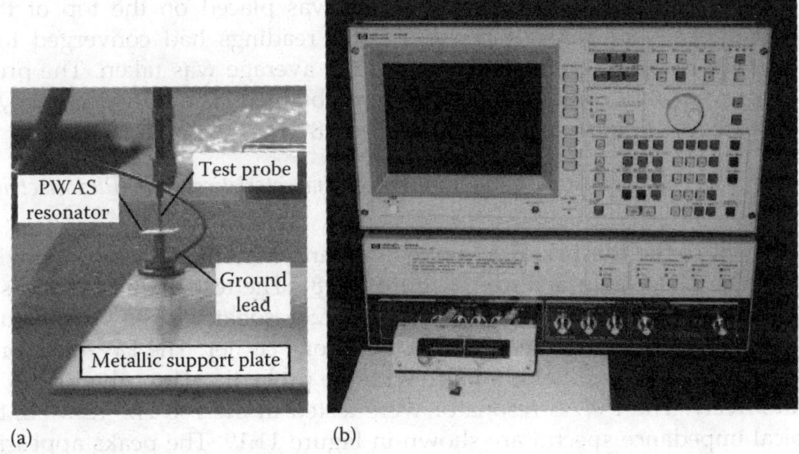

FIGURE 11.18
Dynamic measurement of PWAS resonator: (a) test jig ensuring unrestraint support; (b) HP 4194A impedance phase-gain analyzer.

TABLE 11.7

Manufacturing Tolerances for Small Piezoelectric Wafers from the Vendor APC, Inc. (www.americanpiezo.com)

Dimension	Units (mm)	Standard Tolerance
Length or width of plates	<13 mm	±0.13 mm
Thickness of all parts	0.20 to 0.49 mm	±0.025 mm

The geometric measurements were used to build up the investigator's confidence by showing an acceptable tolerance and a narrow spread. The electrical capacitance was used to verify electrical consistency of the fabrication process. It was found to be an important indicator but also somehow elusive. The E/M impedance and admittance spectra were found to be the most labor-intensive but also the most comprehensive and rewarding.

11.3.1.4.1 Geometric Measurements

Twenty-five nominally identical APC-850 wafers were measured with precision instrumentation consisting of a Mitutoyo Corp. CD 6″ CS digital caliper (0.01 mm precision) for length and width and a Mitutoyo Corp. MCD 1″ CE digital micrometer (0.001 mm precision) for thickness. Statistical analysis of the data obtained from these measurements showed good agreement with the normal (Gaussian) distribution. Mean and standard deviation values for length/width and thickness were 6.9478 mm ±0.5% and 0.2239 mm ±1.4%, respectively. The nominal values were 7 and 0.2 mm, respectively.

11.3.1.4.2 Electrical Capacitance Measurements

Electrical capacitance was measured with a BK Precision Tool Kit 27040 digital meter with a resolution of 1pF. Capacitance of the basic PWAS resonators was measured directly by putting the PWAS resonators on a flat metallic support plate. The negative probe was connected to the plate while the positive probe was placed on the top of the PWAS resonators. Readings were taken when the tester readings had converged to a stable value. At least six readings were recorded and the average was taken. The process was iteratively improved until consistent results were obtained. The statistical analysis of the direct capacitance test results gave $C = 3.276$ nF $\pm 3.8\%$.

11.3.1.4.3 Intrinsic E/M Impedance and Admittance Characteristics of the Piezoelectric Wafer Active Sensor

The measurement of the intrinsic E/M impedance and admittance was done with an HP 4194A impedance phase-gain analyzer (Figure 11.18). The test fixture for measuring the intrinsic E/M impedance and admittance of the PWAS resonator is shown in Figure 11.18a. We used a metallic plate with a lead connected at one corner. The PWAS resonator was centered on the bolt head and held in place with the probe tip. Thus, the PWAS resonator could vibrate freely. The PWAS resonator were tested in the 100 Hz–12 MHz frequency range. Typical impedance spectra are shown in Figure 11.19. The peaks appearing in the frequency range of up to 3000 KHz are associated with the *antiresonances of the in-plane modes*. These peaks are progressively smaller, with the fundamental resonance being the strongest. This is consistent with the fact that higher modes need higher energies to get excited and, under constant energy excitation, higher modes would have lower amplitudes. At around 11 MHz, a new solitary strong peak appears. This is associated with the *fundamental resonance of the thickness mode*.

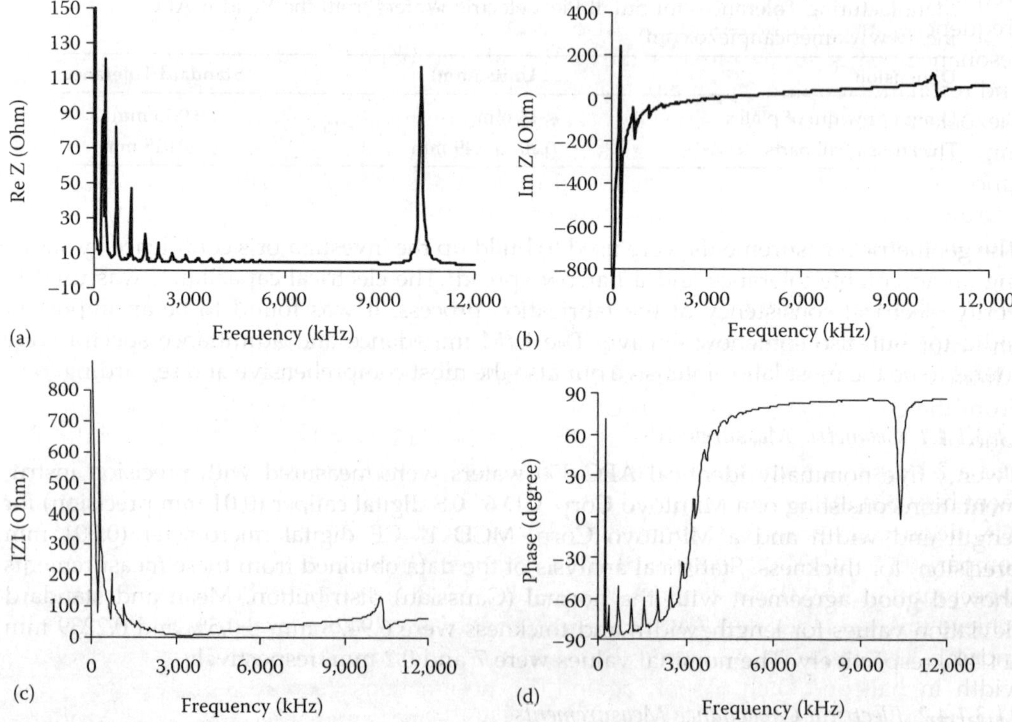

FIGURE 11.19
Intrinsic E/M impedance of a PWAS (7 mm², 0.2 mm thick, APC-850 piezoceramic): (a) real part; (b) imaginary part; (c) amplitude; (d) phase.

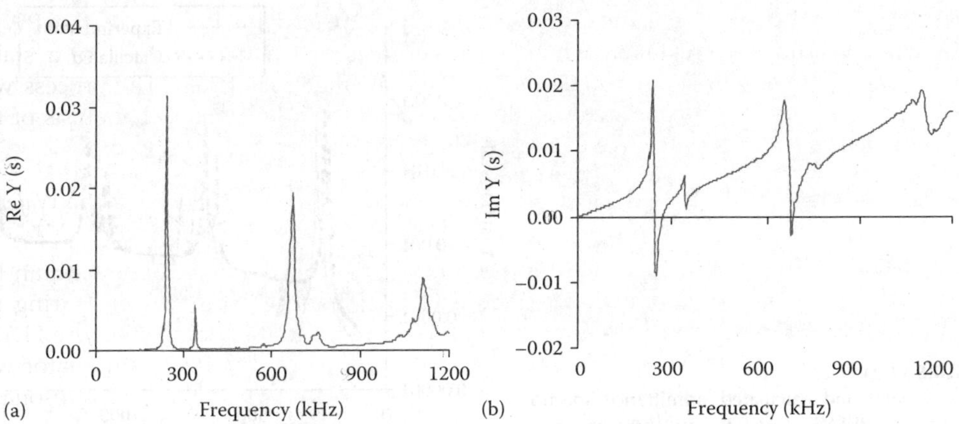

FIGURE 11.20
Intrinsic E/M admittance of a PWAS up to 1200 kHz (7 mm^2, 0.2 mm thick, APC-850 piezoceramic): (a) real part; (b) imaginary part.

The resonances of the in-plane modes were identified from the E/M admittance spectra. Figure 11.20 shows the spectra of the E/M admittance real and imaginary parts in the frequency range of up to 1200 kHz. The first, second, and third in-plane resonance frequencies are clearly observed. Also recorded were the values of the corresponding resonance peaks. Histograms of the statistical distribution of the resonance frequencies and resonance amplitudes for a batch of 25 PWAS resonators were collected and analyzed. Piezoelectric wafers of APC-850 piezoceramic had nominal dimensions 7 mm $\times$ 7 mm $\times$ 0.2 mm. The resonance frequency was found to have the value 251 kHz $\pm$ 1.2%. The admittance amplitude at resonance was found to be 67.152 mS $\pm$ 21%. These results indicate a narrow-band dispersion of the resonance frequency but a wider dispersion of the amplitude at resonance.

11.3.1.4.4 Comparison between Measured and Calculated E/M Admittance Spectra

The capability of Equations (11.149) and (11.150) to predict the E/M impedance and admittance response, and subsequently identify resonance frequencies, was investigated. From the beginning, it was observed that a rectangular PWAS resonator with an aspect ratio of 1:1 could not be well modeled by the 1-D analysis of Equations (11.149), (11.150), and a 2-D analysis would be required. However, for higher aspect ratios, a better agreement between the 1-D analysis and the experimental results would be expected. Hence, it was decided to progressively modify the aspect ratio and observe this effect on the comparison between the theoretical and experimental results. It was expected that, as the aspect ratio increased, the experimental results would converge to the 1-D predictions. To achieve this, we fabricated PWAS specimens with aspect ratios increasing from 1:1 to 2:1 and 4:1. These specimens were fabricated from a 7 mm^2 square wafer by cutting the width in half and then in half again. The designations "square," "half-width," and "quarter-width" were used. Superposed plots of the measured and calculated E/M admittance real parts for the frequency range of up to 1500 kHz are given in Figures 11.21 and 11.22. The in-plane resonance frequencies for each specimen were measured from the admittance plots. Also measured was the thickness resonance frequency obtained from

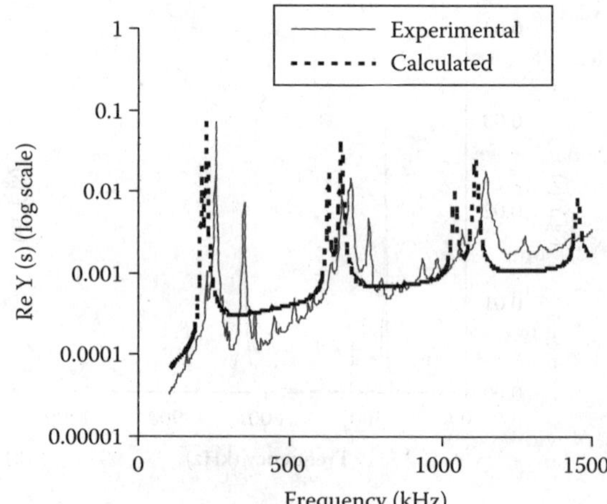

FIGURE 11.21
Experimental and calculated admittance spectra for square PWAS ($l = 6.99$ mm, $b = 6.56$ mm, $t = 0.215$ mm).

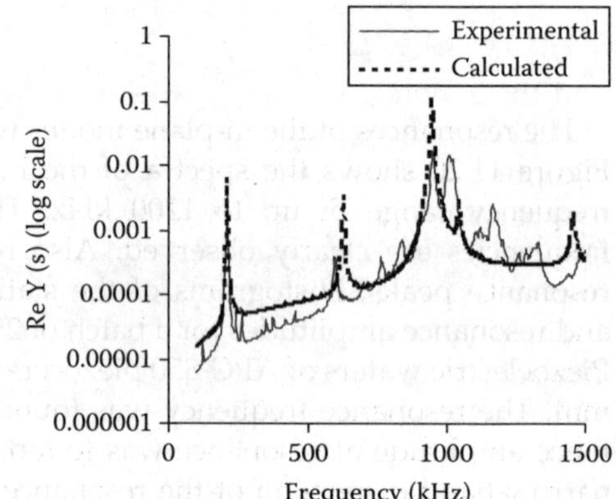

FIGURE 11.22
Experimental and calculated admittance spectra for quarter-width PWAS ($l = 6.99$ mm, $b = 1.64$ mm, $t = 0.215$ mm).

impedance plots. The measured and calculated results for various aspect ratios are given in Table 11.8 where L = length modes, W = width modes, and T = thickness modes.

The results for the square PWAS resonator are discussed first. The admittance real-part spectra are given in Figure 11.21 while the corresponding frequency values are given in the first major row of Table 11.8. The actual length and width values of this nominally square specimen were 6.99 and 6.56 mm, respectively. Since the length and the width were nearly identical, the corresponding lengthwise and widthwise resonance frequencies were close together, forming twin-peaks in the admittance plots. The 1-D analysis predicted that, in the 0–1500 kHz frequency band, seven peaks would exist (1L, 1W, 2L, 2W, 3L, 3W, 4L). The corresponding resonance frequencies (Table 11.8) were 207.6 kHz (1L), 221.6 kHz (1W), 623 kHz (2L), 665 kHz (2W), 1038 kHz (3L), 1108 kHz (3W), 1453 kHz (4L). It should be noted that the calculated frequencies for the L and W series are in the harmonic ratio 1:3:5:7 as predicted by Table 11.5. The 1L and 1W experimental results were found to be significantly different from the theoretical predictions, showing a high positive error (19% and 37%, respectively). These high positive errors are indicative of the 2-D stiffening effect, typical of in-plane vibrations of low-aspect ratio plates. These 2-D stiffening effects could not be captured by the 1-D theory. At higher modes, the 2-D stiffening effect diminishes,

TABLE 11.8

Results of Modal Characterization of Three Rectangular PWAS Resonators of Different Aspect Ratios

		Frequency (kHz)								
Square wafer	Exp.	257	352	670	702	1070	1150	1572	1608	10,565
6.99 mm ×	Calc.	207.6	221.6	623	665	1038	1108	1453	1551	10,488
6.56 mm ×		(1L)	(1W)	(2L)	(2W)	(3L)	(3W)	(4L)	(4W)	(1T)
0.215 mm	Error	23.8%	58.8%	7.5%	5.6%	3.1%	3.8%	8.2%	3.4%	0.7%
1/2 width wafer	Exp.	208	432	597	670	821	1153	1307	1491	10,567
6.99 mm ×	Calc.	207.6	439	621			1038	1318	1451	10,488
3.53 mm ×		(1L)	(1W)	(2L)			(3L)	(2W)	(4L)	(1T)
0.215 mm	Error	0.2%	−1.6%	−4.0%			11.1%	−0.8%	2.8%	0.7%
1/4 width wafer	Exp.	212		597	950	1020			1496	10,905
6.99 mm ×	Calc.	207.6		621	940	1038			1451	10,488
1.64 mm ×		(1L)		(2L)	(1W)	(3L)			(4L)	(1T)
0.215 mm	Error	2.1%		−4.0%	1.1%	−1.8%			3.1%	1.0%

Note: L, in-plane length vibration; W, in-plane width vibration; T, out-of-plane thickness vibration.

and the agreement between theory and experiment improves (7.3% and 5.6% for the 2L and 2W modes; 2.3% and 3.6% for the 3L and 3W modes). We conclude that, with the exception of fundamental modes, 1-D theory gives a reasonable approximation even at aspect ratios as low as 1:1. The agreement for the out-of-plane 1T thickness frequency (10, 565 kHz, 0.7% error) was also good since this mode is very little affected by the in-plane shape of the PZT wafer.

The results for the half-width PWAS with a nominal aspect ratio 2:1 are discussed next. The actual length and width values were 6.99 and 3.53 mm, respectively. The frequency values are given in the second major row of Table 11.8. In the 0–1500 kHz frequency band, the theoretical curve displays six peaks. The corresponding resonance frequencies are 207.6 kHz (1L), 621 kHz (2L), 1038 kHz (3L), 1451 kHz (4L) for lengthwise vibration and 439 kHz (1W), 1318 kHz (2W) for widthwise vibration. The experimental results are 208 kHz (1L), 597 kHz (2L), 1153 kHz (3L), 1491 kHz (4L) for lengthwise vibration and 432 kHz (1W), 1307 kHz (2W) for widthwise vibration. The experimental results agree very well for the 1L and 1W modes (<2% error), a little less for the 2L and 2W modes (4%–5% error), and reasonably well for the 3L mode (11% error). In addition to these clearly identifiable modes, several other peaks are present in the experimental curve. These modes are attributed to the edge roughness generated during the manufacturing process. This edge roughness produces secondary vibration effects. Overall, we conclude that a clear trend toward mode separation and a definite improvement in the first-mode prediction accuracy can be observed, although aspect ratio (2:1) of the half-width PWAS is still far from a proper 1-D situation.

Finally, we discuss the results for the quarter-width PWAS with a nominal aspect ratio of 4:1. The actual length and width values were 6.99 and 1.64 mm, respectively. The admittance real-part spectra are given in Figure 11.22 while the corresponding frequency values are given in the last major row of Table 11.8. In the 0–1500 kHz frequency band, the theoretical analysis predicts five resonance frequencies, as given in Table 11.8: 207.6 kHz (1L), 621 kHz (2L), 1038 kHz (3L), 1451 kHz (4L), 940 kHz (1W). The corresponding theoretical curve displays four peaks since the 1W and 3L peaks, having very close frequencies, have coalesced into a twin-peak. The experimental results, as presented in

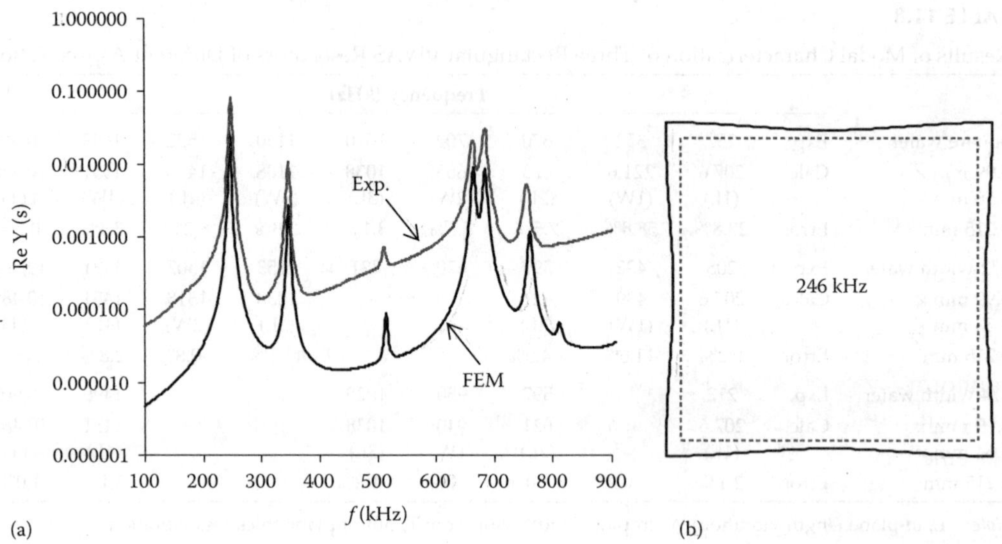

FIGURE 11.23
Analysis of square PWAS using multifield finite element method ($l = b = 7$ mm, $t = 0.215$ mm, APC-850 piezo-ceramics): (a) experimental and calculated admittance spectra; (b) mode shape of the first resonance mode at 246 kHz.

Figure 11.22 and Table 11.8, are in good agreement with the theory, especially for the 1L, 2L, 3L, 1W resonances (<4% error). The experimental curve shows an additional peak at 1167 kHz which is not predicted by the 1-D analysis but may have resulted due to edge roughness. The overall conclusion is that, for 4:1 aspect ratio, good agreement between experiment and 1-D theory is obtained.

11.3.1.4.5 Comparison with Finite Element Analysis Results

It has been seen in Section 11.3.1.4.4 that the 1-D analysis offered reasonably good agreement with experimental data for high-aspect ratio PWAS (e.g., 4:1 aspect ratio) but not for square PWAS (1:1 aspect ratio). In order to overcome the latter's shortcoming, a full analysis of the square PWAS was performed using finite element method (FEM). The multifield FEM approach was used, in which both the mechanical and the electrical properties are simultaneously taken into account. Figure 11.23a shows, in comparison the calculated and experimental admittance real parts. It is apparent that the agreement between calculated and experimental results is very good. The actual mode shape of the first resonance mode is shown in Figure 11.23b.

11.3.2 Circular PWAS Resonators

11.3.2.1 Modeling of a Circular PWAS Resonator

This section addresses the behavior of circular PWAS resonators, i.e., free piezo-electric wafer active sensors of circular form. Consider a circular PWAS resonator of radius a, and thickness h excited by the thickness polarization electric field, E_3 (Figure 11.24).

The electric field is produced by the application of a harmonic voltage $Ve^{i\omega t}$ between the top and bottom surfaces (electrodes). The resulting electric field, $E_3 = V/h$, is assumed

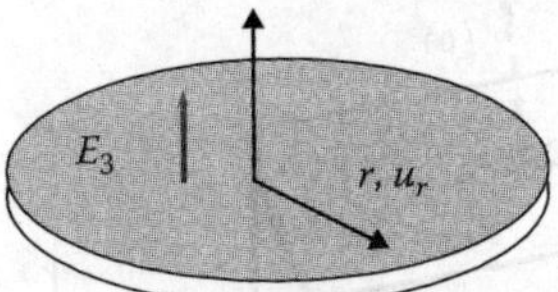

Radius a; thickness t

FIGURE 11.24
Circular piezoelectric wafer active sensor.

uniform over the electrodes area ($\partial E/\partial r = 0$). Because the electric field is uniform, the response will be assumed to be axially symmetric. This means that the circular PWAS resonator undergoes uniform radial and circumferential expansion. The term "uniform" implies that the derivative with respect to θ is zero, i.e., $\dfrac{\partial}{\partial \theta}$, and the circumferential displacement is also zero, i.e. $u_\theta = 0$. The only displacement is u_r. The strain–displacement relations in polar coordinates reduce to

$$
\begin{aligned}
S_{rr} &= \frac{\partial u_r}{\partial r} & S_{rr} &= \frac{du_r}{dr} \\[2mm]
S_{\theta\theta} &= \frac{1}{r}\frac{\partial u_\theta}{\partial \theta} + \frac{u_r}{r} & \rightarrow \quad S_{\theta\theta} &= \frac{u_r}{r} \\[2mm]
S_{r\theta} &= \frac{1}{2}\left(\frac{1}{r}\frac{\partial u_r}{\partial \theta} + \frac{\partial u_\theta}{\partial r} - \frac{u_\theta}{r}\right) & S_{r\theta} &= 0
\end{aligned}
\tag{11.162}
$$

Recall the strain constitutive equation in polar coordinates

$$
\begin{aligned}
S_{rr} &= s_{rr}^{E} T_{rr} + s_{r\theta}^{E} T_{\theta\theta} + d_{3r} E_3 \\[1mm]
S_{\theta\theta} &= s_{r\theta}^{E} T_{rr} + s_{\theta\theta}^{E} T_{\theta\theta} + d_{3\theta} E_3
\end{aligned}
\tag{11.163}
$$

Since the piezoelectric wafer is assumed in-plane isotropic, we will replace s_{rr}^{E}, $s_{\theta\theta}^{E}$, $s_{r\theta}^{E}$ by s_{11}^{E}, s_{11}^{E}, s_{12}^{E}, respectively, such that

$$
\begin{aligned}
S_{rr} &= s_{11}^{E} T_{rr} + s_{12}^{E} T_{\theta\theta} + d_{31} E_3 \\[1mm]
S_{\theta\theta} &= s_{12}^{E} T_{rr} + s_{11}^{E} T_{\theta\theta} + d_{31} E_3
\end{aligned}
\tag{11.164}
$$

Upon inversion,

$$
\begin{aligned}
T_{rr} &= \frac{1}{s_{11}^{E}(1 - v^2)}\left[(S_{rr} + v S_{\theta\theta}) - (1 + v)d_{31} E_3\right] \\[3mm]
T_{\theta\theta} &= \frac{1}{s_{11}^{E}(1 - v^2)}\left[(v S_{rr} + S_{\theta\theta}) - (1 + v)d_{31} E_3\right]
\end{aligned}
\tag{11.165}
$$

where v is the Poisson ratio defined as

$$
v = -\frac{s_{12}^{E}}{s_{11}^{E}} \qquad \text{(Poisson ratio)}
\tag{11.166}
$$

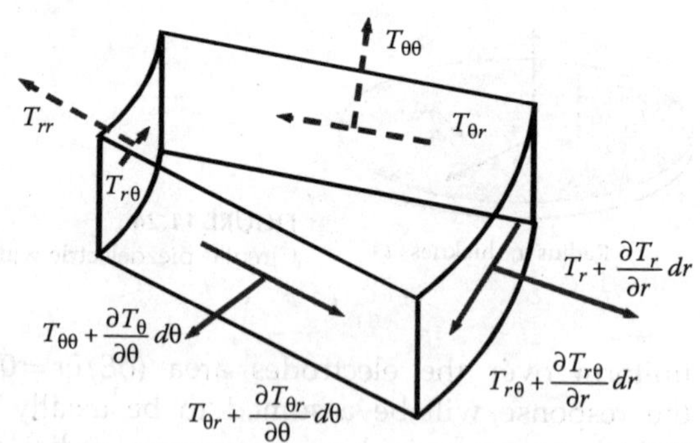

FIGURE 11.25
Stresses acting on an infinitesimal element of a circular piezoelectric wafer.

Substituting the strain–displacement relations of Equation (11.162) yields the stress expressed in terms of displacement, i.e.,

$$T_{rr} = \frac{1}{s_{11}^E(1-v^2)}\left[\left(\frac{du_r}{dr} + v\frac{u_r}{r}\right) - (1+v)d_{31}E_3\right]$$

$$T_{\theta\theta} = \frac{1}{s_{11}^E(1-v^2)}\left[\left(v\frac{du_r}{dr} + \frac{u_r}{r}\right) - (1+v)d_{31}E_3\right] \tag{11.167}$$

Newton's law of motion applied to the infinitesimal element of Figure 11.25 yields

$$\frac{dT_{rr}}{dr} + \frac{T_{rr} - T_{\theta\theta}}{r} = -\omega^2 \rho u_r \tag{11.168}$$

Substitution of the stresses T_{rr} and $T_{\theta\theta}$ yields

$$\frac{dT_{rr}}{dr} = \frac{1}{s_{11}^E(1-v^2)}\frac{d}{dr}\left\{\left[\left(\frac{du_r}{dr} + v\frac{u_r}{r}\right) - (1+v)d_{31}E_3\right]\right\}$$

$$= \frac{1}{s_{11}^E(1-v^2)}\left[\frac{d^2u_r}{dr^2} + v\left(\frac{1}{r}\frac{du_r}{dr} - \frac{u_r}{r^2}\right)\right] \tag{11.169}$$

and

$$\frac{T_{rrr} - T_{\theta\theta}}{r} = \frac{1}{s_{11}^E(1-v^2)}\frac{1}{r}\left(\frac{du_r}{dr} + v\frac{u_r}{r} - \frac{u_r}{r} - v\frac{du_r}{dr}\right)$$

$$= \frac{1}{s_{11}^E(1-v^2)}(1-v)\left(\frac{1}{r}\frac{du_r}{dr} - \frac{u_r}{r^2}\right) \tag{11.170}$$

Upon addition,

$$\frac{dT_{rr}}{dr} + \frac{T_{rr} - T_{\theta\theta}}{r} = \frac{1}{s_{11}^E(1-v^2)}\left(\frac{d^2u_r}{dr^2} + \frac{1}{r}\frac{du_r}{dr} - \frac{u_r}{r^2}\right) \tag{11.171}$$

Substitution into Equation (11.167) yields the equation of motion

$$\frac{1}{s_{11}^E(1-v^2)}\left(\frac{d^2u_r}{dr^2}+\frac{1}{r}\frac{du_r}{dr}-\frac{u_r}{r^2}\right)=-\omega^2\rho u_r \tag{11.172}$$

Introduce the notation

$$c_p=\sqrt{\frac{1}{\rho s_{11}^E(1-v^2)}} \qquad \text{(wave speed)} \tag{11.173}$$

which represents the wave speed of circular axial waves in the piezoelectric wafer. The equation of motion becomes the wave equation

$$\frac{d^2u_r}{dr^2}+\frac{1}{r}\frac{du_r}{dr}-\frac{u_r}{r^2}=-\frac{\omega^2}{c_p^2}u_r \tag{11.174}$$

Introducing the wave number, $\gamma=\dfrac{\omega}{c_p}$, and rearranging yields

$$r^2\frac{d^2u_r}{dr^2}+r\frac{du_r}{dr}+\left(r^2\gamma^2-1\right)u_r=0 \tag{11.175}$$

Using the change of variables

$$z=\gamma r\rightarrow r=\frac{1}{\gamma}z,\quad \frac{d}{dr}=\gamma\frac{d}{dz}, \tag{11.176}$$

Equation (11.172) becomes

$$z^2\frac{d^2u_r}{dz^2}+z\frac{du_r}{dz}+\left(z^2-1\right)u_r=0 \tag{11.177}$$

which is exactly the Bessel differential equation of order 1, accepting as solutions the Bessel function $J_1(z)$. Hence, the general solution of Equation (11.172) is

$$u_r(r)=AJ_1(\gamma r) \tag{11.178}$$

The constant A is determined from the boundary conditions. For stress-free boundary conditions at $r=a$, we write

$$T_{rr}(a)=0 \tag{11.179}$$

Using Equation (11.166), the boundary condition becomes

$$T_{rr}(a)=\frac{1}{s_{11}^E(1-v^2)}\left[\left(\frac{du_r}{dr}+v\frac{u_r}{r}\right)-(1+v)d_{31}E_3\right]=0 \tag{11.180}$$

i.e.,

$$\frac{du_r}{dr} + v\frac{u_r}{r} - (1+v)d_{31}E_3 = 0 \tag{11.181}$$

Substitution of Equation (11.178) into Equation (11.180) yields

$$A\gamma J_1'(\gamma a) + \frac{v}{a}AJ_1(\gamma a) = (1+v)d_{31}E_3 \tag{11.182}$$

Recall the Bessel functions identity

$$J_1'(z) = J_0(z) - \frac{1}{z}J_1(z) \tag{11.183}$$

Equation (11.182) becomes

$$A\left(\gamma J_0(\gamma a) - \frac{1}{a}J_1(\gamma a) + \frac{v}{a}J_1(\gamma a)\right) = (1+v)d_{31}E_3 \tag{11.184}$$

Hence,

$$A = \frac{(1+v)ad_{31}E_3}{(\gamma a)J_0(\gamma a) - (1-v)J_1(\gamma a)} \tag{11.185}$$

Substitution into the general solution (11.17) yields the displacement response

$$u_r(r) = d_{31}E_3a\frac{(1+v)J_1(\gamma r)}{(\gamma a)J_0(\gamma a) - (1-v)J_1(\gamma a)} \tag{11.186}$$

Introducing the notation

$$u_{ISA} = d_{31}\hat{E}_3a \qquad \text{(induced displacement)} \tag{11.187}$$

where ISA signifies induced-strain actuation, we can express Equation (11.186) as

$$u_r(r) = u_{ISA}\frac{(1+v)J_1(\gamma r)}{(\gamma a)J_0(\gamma a) - (1-v)J_1(\gamma a)} \tag{11.188}$$

Introducing $z = \gamma a$, we rewrite Equation (11.188) as

$$u_r(r) = u_{ISA}\frac{(1+v)J_1(\gamma r)}{zJ_0(z) - (1-v)J_1(z)} \tag{11.189}$$

11.3.2.2 Electrical Response

Consider a circular PWAS resonator under electric excitation (Figure 11.26). Recall the equation representing the electrical displacement as a function of stress and electric field

$$D_3 = d_{31}T_{rr} + d_{31}T_{\theta\theta} + \varepsilon_{33}^T E_3 \tag{11.190}$$

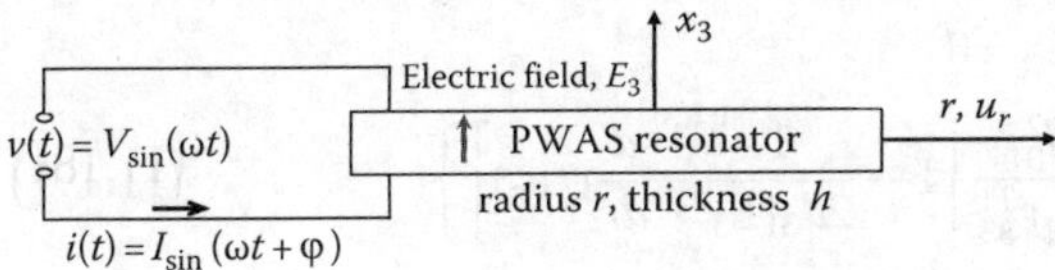

FIGURE 11.26
Schematic of a circular PWAS resonator under electric excitation.

This equation can be integrated to determine the electric charge, and hence the current. Before doing so, we need to express the stress as a function of applied electric field such that Equation (11.189) relates the electric displacement to the applied electric field only. To achieve this, we will use the displacement solution of Equation (11.186) and the stress-displacement relation of Equation (11.167). First, note that the stresses in Equation (11.189) can be grouped as

$$D_3 = d_{31}(T_{rr} + T_{\theta\theta}) + \varepsilon_{33}^T E_3 \tag{11.191}$$

Then, recall Equation (11.167), i.e.,

$$
\begin{aligned}
T_{rr} &= \frac{1}{s_{11}^E(1 - v^2)}\left[\left(\frac{du_r}{dr} + v\frac{u_r}{r}\right) - (1 + v)d_{31}E_3\right] \\
T_{\theta\theta} &= \frac{1}{s_{11}^E(1 - v^2)}\left[\left(v\frac{du_r}{dr} + \frac{u_r}{r}\right) - (1 + v)d_{31}E_3\right]
\end{aligned}
\tag{11.192}
$$

Upon addition,

$$T_{rr} + T_{\theta\theta} = \frac{1}{s_{11}^E(1 - v^2)}(1 + v)\left[\frac{du_r}{dr} + v\frac{u_r}{r} - 2d_{31}E_3\right] \tag{11.193}$$

The displacement terms in Equation (11.192) can be rewritten as

$$\frac{du_r}{dr} + \frac{u_r}{r} = \frac{1}{r}\left(r\frac{du_r}{dr} + u_r\right) = \frac{1}{r}\frac{d}{dr}(ru_r) \tag{11.194}$$

Hence,

$$
\begin{aligned}
T_{rr} + T_{\theta\theta} &= \frac{1}{s_{11}^E(1 - v^2)}(1 + v)\left[\frac{1}{r}\frac{d}{dr}(ru_r) - 2d_{31}E_3\right] \\
&= \frac{1}{s_{11}^E(1 - v)}\left[\frac{1}{r}\frac{d}{dr}(ru_r) - 2d_{31}E_3\right]
\end{aligned}
\tag{11.195}
$$

Substitution of Equation (11.194) into Equation (11.190) yields

$$D_3 = d_{31}\frac{1}{s_{11}^E(1 - v)}\left[\frac{1}{r}\frac{d}{dr}(ru_r) - 2d_{31}E_3\right] + \varepsilon_{33}^T E_3 \tag{11.196}$$

Upon rearrangement,

$$D_3 = \varepsilon_{33}^T E_3 \left\{ 1 - \frac{2}{(1-v)} \frac{d_{31}^2}{s_{11}^E \varepsilon_{33}^T} \left[1 - \frac{1}{2d_{31}E_3} \frac{1}{r} \frac{d}{dr}(ru_r) \right] \right\} \tag{11.197}$$

Introduce the planar coupling coefficient, k_p, defined as

$$k_p^2 = \frac{2}{(1-v)} \frac{d_{31}^2}{s_{11}^E \varepsilon_{33}^T} \tag{11.198}$$

Note that

$$k_p^2 = \frac{2}{(1-v)} k_{31}^2 \tag{11.199}$$

Hence,

$$D_3 = \varepsilon_{33}^T E_3 \left\{ \left(1 - k_p^2\right) + k_p^2 \left[\frac{1}{2d_{31}E_3} \frac{1}{r} \frac{d}{dr}(ru_r) \right] \right\} \tag{11.200}$$

The total charge Q is obtained by integration of the electric displacement D_3 over the electrodes area $A = \pi a^2$, i.e.,

$$Q = \int_A D_3 r\, dr\, d\theta = 2\pi \int_0^a D_3 r\, dr \tag{11.201}$$

Substitution of Equation (11.199) into Equation (11.200) yields

$$Q = 2\pi \int_0^a \varepsilon_{33}^T E_3 \left\{ \left(1 - k_p^2\right) + k_p^2 \left[\frac{1}{2d_{31}E_3} \frac{1}{r} \frac{d}{dr}(ru_r) \right] \right\} r\, dr \tag{11.202}$$

Hence,

$$\begin{aligned}
Q &= \pi \varepsilon_{33}^T E_3 \left\{ \left(1 - k_p^2\right) \int_0^a 2r\, dr + k_p^2 \frac{1}{d_{31}E_3} \int_0^a \left[\frac{1}{r} \frac{d}{dr}(ru_r) \right] r\, dr \right\} \\
&= \pi \varepsilon_{33}^T E_3 \left\{ \left(1 - k_p^2\right) r^2 \Big|_0^a + k_p^2 \frac{1}{d_{31}E_3} (ru_r) \Big|_0^a \right\} \\
&= \pi \varepsilon_{33}^T E_3 \left\{ \left(1 - k_p^2\right) a^2 + k_p^2 \frac{1}{d_{31}E_3} au_r(a) \right\}
\end{aligned} \tag{11.203}$$

i.e.,

$$Q = \pi a^2 \varepsilon_{33}^T E_3 \left\{ \left(1 - k_p^2\right) - k_p^2 \frac{u_r(a)}{d_{31}E_3 a} \right\} \tag{11.204}$$

Using $u_{ISA} = d_{31}E_3 a$, $V = E_3/h$, and

$$C_0 = \varepsilon_{33}^T \frac{A}{h}, \quad A = \pi a^2 \text{ (free capacitance of circular PWAS)} \tag{11.205}$$

$$Q = C_0 V\left[\left(1 - k_p^2\right) + k_p^2 \frac{u_r(a)}{u_{ISA}}\right] \tag{11.206}$$

The electric current is obtained as the time derivative of the electric charge, i.e.,

$$I = \dot{Q} = i\omega Q \tag{11.207}$$

Hence,

$$Q = i\omega C_0 V\left[\left(1 - k_p^2\right) + k_p^2 \frac{u_r(a)}{u_{ISA}}\right] \tag{11.208}$$

The admittance, Y, is defined as the ratio between the current and voltage, i.e.,

$$Y = \frac{I}{V} = i\omega C_0 \left[\left(1 - k_p^2\right) + k_p^2 \frac{u_r(a)}{u_{ISA}}\right] \tag{11.209}$$

Recall Equation (11.188) giving the displacement solution

$$u_r(r) = u_{ISA} \frac{(1 + \nu)J_1(\gamma r)}{(\gamma a)J_0(\gamma a) - (1 - \nu)J_1(\gamma a)} \tag{11.210}$$

Hence,

$$Y = i\omega C_0 \left[\left(1 - k_p^2\right) - k_p^2 \frac{(1 + \nu)J_1(\gamma a)}{(\gamma a)J_0(\gamma a) - (1 - \nu)J_1(\gamma a)}\right] \tag{11.211}$$

Recalling the notation $z = \gamma a$, we rewrite Equation (11.211) as

$$Y = i\omega C_0 \left[\left(1 - k_p^2\right) - k_p^2 \frac{(1 + \nu)J_1(z)}{zJ_0(z) - (1 - \nu)J_1(z)}\right] \tag{11.212}$$

This result agrees with Ikeda (1996). However, it may be more convenient at times to write it as

$$Y = i\omega C_0 \left[1 - k_p^2 \left(1 - \frac{(1 + \nu)J_1(\gamma a)}{(\gamma a)J_0(\gamma a) - (1 - \nu)J_1(\gamma a)}\right)\right] \tag{11.213}$$

Note that the admittance is purely imaginary and consists of the capacitive admittance, $i\omega C$, modified by the effect of piezoelectric coupling between mechanical and electrical variables. This effect is apparent in the term containing the electromechanical coupling coefficient, k_p^2. The impedance, Z, is obtained as the ratio between the voltage and current, i.e.,

$$Z = \frac{\hat{V}}{\hat{I}} = Y^{-1} \qquad (11.214)$$

Hence,

$$Y = i\omega C_0 \left[1 - k_p^2 \left(1 - \frac{(1+v)J_1(z)}{zJ_0(z) - (1-v)J_1(z)}\right)\right] \qquad (11.215)$$

Recalling the notation $z = \gamma a$, we rewrite

$$Y = i\omega C_0 \left[1 - k_p^2 \left(1 - \frac{(1+v)J_1(z)}{zJ_0(z) - (1-v)J_1(z)}\right)\right] \qquad (11.216)$$

$$Z = \frac{1}{i\omega C_0} \left[1 - k_p^2 \left(1 - \frac{(1+v)J_1(z)}{zJ_0(z) - (1-v)J_1(z)}\right)\right]^{-1} \qquad (11.217)$$

11.3.2.3 Resonances

If resonances happen, they could be of two types:

1. Electromechanical resonances
2. Mechanical resonances

Mechanical resonances take place in the same conditions as in a conventional elastic disc. They happen under mechanical excitation that produces a mechanical response in the form of mechanical vibrations. Electromechanical resonances are specific to piezoelectric materials. They reflect the coupling between the mechanical and electrical variables. Electromechanical resonances happen under electric excitation that produces an electromechanical response, i.e., both a mechanical vibration and a change in the electric admittance and impedance. We will consider these two situations separately.

11.3.2.3.1 Mechanical Resonances

If a PWAS is excited mechanically with a frequency sweep, certain frequencies will be revealed at which the response is very large, i.e., the PWAS resonates. To study the mechanical resonances, assume that the material is not piezoelectric, i.e., $d_{31} = 0$. Hence, we develop the analysis of the classical mechanical resonances of an elastic disc. Our discussion will be restricted to axisymmetric in-plane vibrations. Recall Equation (11.174) representing the wave equation for axisymmetric in-plane vibrations of an elastic disc

$$\frac{d^2 u_r}{dr^2} + \frac{1}{r}\frac{du_r}{dr} - \frac{u_r}{r^2} = -\frac{\omega^2}{c_p^2} u_r \qquad (11.218)$$

We have already shown that by introducing the wave number, $\gamma = \dfrac{\omega}{c_p}$, and the substitution of $z = \gamma r$, we recover the Bessel differential equation of order 1, as given by Equation (11.177)

$$z^2 \frac{d^2 u_r}{dz^2} + z \frac{du_r}{dz} + \left(z^2 - 1\right) u_r = 0 \tag{11.219}$$

Hence, the general solution is of the form given by Equation (11.178), i.e.,

$$u_r(r) = A J_1(\gamma r) \tag{11.220}$$

where the constant A must be determined from the boundary conditions. For stress-free boundary conditions at $r = a$, we have $T_{rr}(a) = 0$. Using Equation (11.166) and imposing no piezoelectric behavior, the stress-free boundary condition becomes

$$T_{rr}(a) = \frac{1}{s_{11}^E (1 - v^2)} \left(\frac{du_r}{dr} + v \frac{u_r}{r} \right) = 0 \tag{11.221}$$

Substitution of the solution $u_r = A J_1(\gamma r)$ yields

$$A \gamma J_1'(\gamma a) + \frac{v}{a} A J_1(\gamma a) = 0 \tag{11.222}$$

Using the identity $J_1'(z) = J_0(z) - \dfrac{1}{z} J_1(z)$, dividing through by A, and multiplying by a, we obtain the condition

$$(\gamma a) J_0(\gamma a) - (1 - v) J_1(\gamma a) = 0 \tag{11.223}$$

Equation (11.223) can be rearranged as

$$\frac{(\gamma a) J_0(\gamma a)}{J_1(\gamma a)} = (1 - v) \tag{11.224}$$

The left hand side of Equation (11.222) is also known as the modified quotient of Bessel functions, $\tilde{J}_1$, defined as

$$\tilde{J}_1(z) = \frac{z J_0(z)}{J_1(z)} \tag{11.225}$$

Note that Equation (11.223) depends on the Poisson ratio, v. Equation (11.223) indicates that nonzero values of A are only possible at particular values of (γa) which are the elastic system eigenvalues. Equation (11.223) is transcendental and does not accept closed-form solution. Numerical solution of Equation (11.223) for $v = 0.30$ yields

$$z = (\gamma a) = 2.048652;\ 5.389361;\ 8.571860;\ 11.731771 \ldots \tag{11.226}$$

It should be noted that, since these eigenvalues depend on v, the ratio between successive eigenvalues and the fundamental eigenvalues could be used for determining the Poisson ratio experimentally through a curve-fitting process.

For each eigenvalue, γa, we can determine the corresponding resonance frequency with the formula

$$f_n = \frac{1}{2\pi}\frac{c}{a}(\gamma a)_n \tag{11.227}$$

where $n = 1, 2, 3 \ldots$. The mode shapes are calculated using Equation (11.178) by putting the wave number γ corresponding to each eigenvalue. Thus, for the nth eigenvalue, $(\gamma a)_n$, the nth wave number is calculated with the formula

$$\gamma_n = \frac{1}{a}(\gamma a)_n \tag{11.228}$$

Thus, the nth mode shape is

$$R_n(r) = A_n J_1(z_n r/a) \tag{11.229}$$

The constant A_n is determined through modes normalization and depends on the normalization procedure used. A common normalization procedure, based on equal modal energy, yields

$$A_n = \sqrt{J_1^2(z_n) - J_0(z_n)J_2(z_n)} \tag{11.230}$$

Other normalization methods simply take $A_n = 1$. A graphical representation of the mode shapes is given in Table 11.9.

11.3.2.3.2 Electromechanical Resonances

Recall the expressions for admittance and impedance given by Equations (11.216) and (11.217). These expressions can be rearranged as

$$Y = i\omega C_0 \left[1 - k_p^2 \left(1 - \frac{(1+v)J_1(z)}{zJ_0(z) - (1-v)J_1(z)}\right)\right] \tag{11.231}$$

$$Z = \frac{1}{i\omega C_0}\left[1 - k_p^2 \left(1 - \frac{(1+v)J_1(z)}{zJ_0(z) - (1-v)J_1(z)}\right)\right]^{-1} \tag{11.232}$$

where $z = Ya$. The following conditions are considered:

- Resonance, when $Y \to \infty$, i.e., $Z = 0$
- Antiresonance, when $Y = 0$, i.e., $Z \to \infty$

Electrical resonance is associated with the situation in which a device is drawing very large currents when excited harmonically with a constant voltage at a given frequency. At resonance, the admittance becomes very large while the impedance goes to zero. As the admittance becomes very large, the current drawn under constant-voltage excitation also becomes very large since $I = YV$. In piezoelectric devices, the mechanical

TABLE 11.9

Resonance Mode Shapes of a circular PWAS resonator

Mode	Eigenvalue	Resonant Frequency	Mode Shape
R_1	$z_1 = \gamma_1 a = 2.048652$	$f_1 = \dfrac{1}{2\pi} z_1 \dfrac{c}{a}$	$R_1 = J_1(z_1 r/a)$
R_2	$z_2 = \gamma_2 a = 5.389361$	$f_2 = \dfrac{1}{2\pi} z_2 \dfrac{c}{a}$	$R_2 = J_1(z_2 r/a)$
R_3	$z_3 = \gamma_3 a = 8.571860$	$f_3 = \dfrac{1}{2\pi} z_3 \dfrac{c}{a}$	$R_3 = J_1(z_3 r/a)$

response at electrical resonance also becomes very large. This happens because the electromechanical coupling of the piezoelectric material transfers energy from the electrical input into the mechanical response. For these reasons, the resonance of an electrically driven piezoelectric device must be seen as an electromechanical resonance. A PWAS resonator driven at electrical resonance may undergo mechanical deterioration and even break up.

Electrical antiresonance is associated with the situation in which a device under constant-voltage excitation draws almost no current. At antiresonance, the admittance goes to zero while the impedance becomes very large. Under constant-voltage excitation, this condition results in very small current being drawn from the source. In a piezoelectric device, the mechanical response at electrical antiresonance is also very small. A PWAS resonator driven at the electrical antiresonance hardly moves at all. The antiresonance of an electrically driven piezoelectric device must be also seen as an electromechanical antiresonance.

The condition for electromechanical resonance is obtained by studying the poles of Y, i.e., the values of z which make $Y \to \infty$. The poles of Y are roots of the denominator. These are obtained by solving the equation

$$zJ_0(z) - (1 - \nu)J_1(z) = 0 \qquad \text{(resonance)} \qquad (11.233)$$

This equation is the same as Equation (11.223) used to determine the mechanical resonances. This is not surprising since in our analysis of mechanical resonances, we only considered the axisymmetric modes that couple well with a uniform electric field excitation. Hence, the frequencies of electromechanical resonance correspond identically to the frequencies of axisymmetric mechanical resonance given in Table 11.9.

The condition for electromechanical antiresonance is obtained by studying the zeros of Y, i.e., the values of z that make $Y = 0$. Since electromechanical antiresonances correspond to zeros of the admittance (i.e., poles of the impedance), the current at antiresonance is zero, $I = 0$. Equation (11.231) indicates that $Y = 0$ happens when

$$1 - k_p^2 \left(1 - \frac{(1 + \nu)J_1(z)}{zJ_0(z) - (1 - \nu)J_1(z)} \right) = 0 \tag{11.234}$$

Upon rearrangement,

$$1 = k_p^2 \left(1 - \frac{(1 + \nu)J_1(z)}{zJ_0(z) - (1 - \nu)J_1(z)} \right)$$
$$zJ_0(z) - (1 - \nu)J_1(z) = k_p^2(zJ_0(z) - (1 - \nu)J_1(z) - (1 + \nu)J_1(z)) \tag{11.235}$$

or

$$1 = k_p^2 \left(1 - \frac{(1 + \nu)J_1(z)}{zJ_0(z) - (1 - \nu)J_1(z)} \right)$$
$$zJ_0(z) - (1 - \nu)J_1(z) = k_p^2(zJ_0(z) - (1 - \nu)J_1(z) - (1 + \nu)J_1(z)) \tag{11.236}$$

Hence, the antiresonance condition is

$$\frac{zJ_0(z)}{J_1(z)} = \frac{1 - \nu - 2k_p^2}{\left(1 - k_p^2\right)} \qquad \text{(antiresonance)} \tag{11.237}$$

This equation is also transcendental and does not accept closed-form solutions. Its solutions are found numerically.

11.3.2.4 Experimental Results

Measured results and calculated predictions for circular PWAS resonators are given in Table 11.10. A superposed plot of the measured and calculated E/M admittance spectra is given in Figure 11.27. The E/M admittance of a circular PWAS resonator undergoing axisymmetric in-plane radial vibrations was modeled using Equation (11.232). Figure 11.27

TABLE 11.10

Results of the Dynamic Characterization of a Circular PWAS Resonator

	Frequency (kHz)				
Experimental	300 (1R)	784 (2R)	1247 (3R)	1697 (4R)	10,895 (1T)
Calculated	303 (1R)	796 (2R)	1267 (3R)	1733 (4R)	10,690 (1T)
Error	−1.0%	−1.5%	−1.6%	−2.1%	1.9%

Note: $a = 6.98$ mm, $t = 0.216$ mm, APC-850 Material; R, radial vibration; T, thickness vibration.

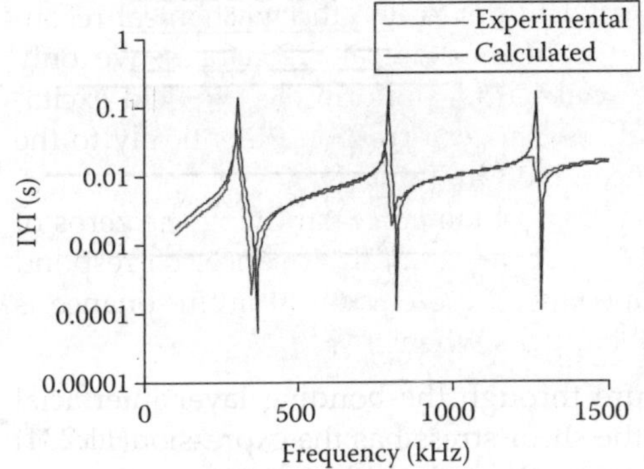

FIGURE 11.27
Experimental and calculated admittance spectra of a circular PWAS resonator ($a = 6.98$ mm, $t = 0.216$ mm, APC-850 material).

shows the predicted and measured results superposed on the same plot. Three resonance peaks are clearly visible. The corresponding frequencies (300, 784, and 1247 kHz, as indicated in Table 11.10) correspond to the first three in-plane radial modes. The fourth in-plane frequency (1697 kHz), which lies outside the 0–1500 kHz plotting range, was not plotted. However, its value appears in Table 11.10. During our experiments, we also noticed a very high frequency peak in the E/M impedance real-part response at 10,895 kHz. This value can be identified with the out-of-plane thickness vibration. Comparison of measured and calculated results listed in Table 11.10 for the circular disk case indicates very good agreement between theory and experiments (2.1% maximum error).

11.3.3 Section Summary

This section has discussed the behavior of circular PWAS resonators. It was shown that the vibration of circular PWAS resonators can be expressed in terms of Bessel functions. Subsequently, closed-form solutions for the impedance and admittance of circular PWAS resonators were derived. Comparison was made with experimental results obtained on 7-mm diameter circular PWAS resonators. The comparison between theoretical predictions and experimental observations for circular PWAS resonators was much better than linear PWAS resonators because (1) the closed-form solution was exact and (2) the experimental specimens had much more clear modes.

11.4 PWAS Attached to Structures

11.4.1 Shear-Layer Coupling between PWAS and Structure

The transmission of actuation and sensing between the PWAS and the structure is achieved through the adhesive layer. The adhesive layer acts as a shear layer in which the mechanical effects are transmitted through shear effects. Figure 11.28 shows a thin-wall structure of thickness t and elastic modulus E, with a PWAS of thickness t_a and elastic modulus E_a attached to its upper surface through a bonding layer of thickness t_b and shear modulus G_b. The PWAS length is l_a while the half-length is $a = l_a/2$. In addition, the definition $d = t/2$ is used. Upon application of an electric voltage, the PWAS experiences an induced strain,

$$\varepsilon_{ISA} = d_{31}\frac{V}{t_a}$$

(11.238)

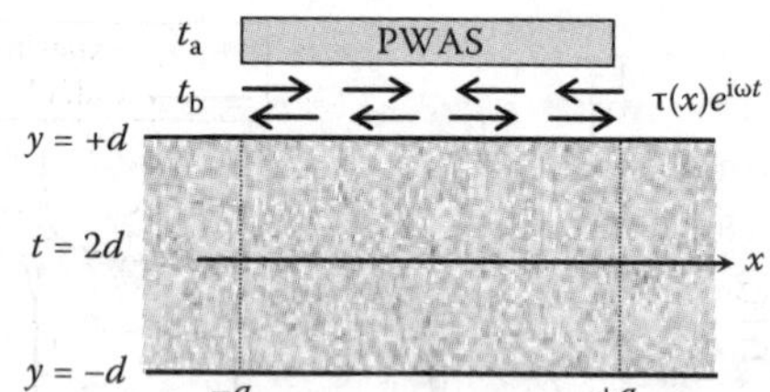

FIGURE 11.28
Interaction between the PWAS and the structure showing the bonding layer interfacial shear stress, $\tau(x)$.

The induced strain is transmitted to the structure through the bonding layer interfacial shear stress τ. For harmonic varying excitation, the shear stress has the expression $\tau(x)e^{i\omega t}$.

The PWAS expansion is transmitted to the structure through the bonding layer that acts predominantly in shear. Construction of the free body diagrams of the PWAS, bonding layer, and thin-wall structure over the infinitesimal length dx leads to the following equilibrium equations (Figure 11.28):

$$t_a \frac{d\sigma_a}{dx} - \tau = 0 \qquad \text{(PWAS)} \qquad (11.239)$$

$$t \frac{d\sigma}{dx} + \alpha\tau = 0 \qquad \text{(structure)} \qquad (11.240)$$

The coefficient α in Equation (11.240) depends on the stress, strain, and displacement distributions across the plate thickness. Under static and low-frequency dynamic conditions, one applies the usual hypothesis associated with simple axial and flexural motions, i.e., constant displacement for axial motion and linear displacement strain for flexural motion. In this case, $\alpha = 4$. For high-frequency motion, the displacement field across the plate thickness takes the more complicated forms associated with the Lamb wave modes. However, in the present section, we will restrict our analysis to the static and low-frequency dynamic conditions. These conditions are analyzed next; as a result, the value $\alpha = 4$ in Equation (11.240) will be derived. The analysis proceeds as follows. The shear stress τ applied to the upper surface is partitioned into symmetric and antisymmetric pairs, $(\tau/2, \tau/2)$ and $(\tau/2, -\tau/2)$, and applied to the upper and lower surfaces, respectively (Figure 11.29), such that

- At the upper surface $\tau_S\big|_{y=d} + \tau_A\big|_{y=d} = \dfrac{\tau}{2} + \dfrac{\tau}{2} = \tau$

- At the lower surface $\tau_S\big|_{y=-d} + \tau_A\big|_{y=-d} = \dfrac{\tau}{2} - \dfrac{\tau}{2} = 0$

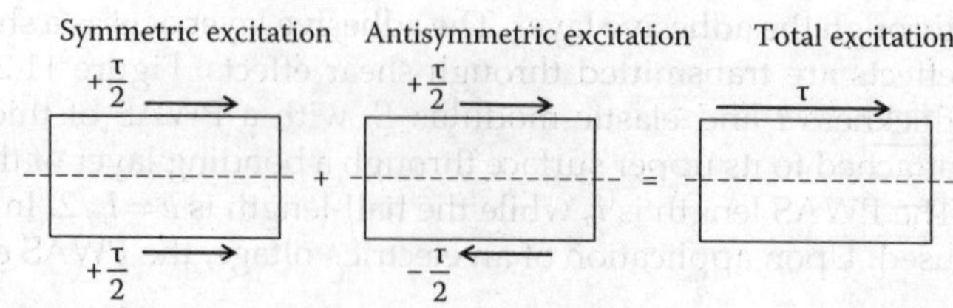

FIGURE 11.29
Symmetric and antisymmetric particle motion across the plate thickness.

Under static and low-frequency dynamic conditions, the following assumptions apply. For the symmetric case, uniform stress and strain distribution across the thickness is assumed. For the antisymmetric case, linear stress and strain distribution across the thickness is assumed. These two separate cases are analyzed next.

11.4.1.1 Symmetric Case

In the symmetric case, the stress and strain are assumed constant across the thickness (Figure 11.30). Since the stress is assumed uniform across the thickness, the stress distribution can be expressed as

$$\sigma(y) = \sigma_S \tag{11.241}$$

where σ_S is the value of the stress in the plate evaluated in the upper surface. The subscript S stands for "symmetric."

The stress resultants are evaluated by integration of the stress across the thickness. Since the stress distribution is symmetric, the only stress resultant is the axial force per unit width

$$N = \int_{-d}^{+d} \sigma(y)\,dy = \int_{-d}^{+d} \sigma_S\,dy = 2d\sigma_S \tag{11.242}$$

The force due to the shear stresses applied to the plate surface over the length dx is

$$dN_\tau = 2\frac{\tau}{2}dx = \tau\,dx \tag{11.243}$$

Hence, equilibrium of the infinitesimal element of Figure 11.30 gives

$$\cancel{N} + dN + dN_\tau - \cancel{N} = 0 \tag{11.244}$$

Substitution of Equations (11.242) and (11.243) into Equation (11.244) yields

$$t\frac{d\sigma_S}{dx} + \tau = 0 \tag{11.245}$$

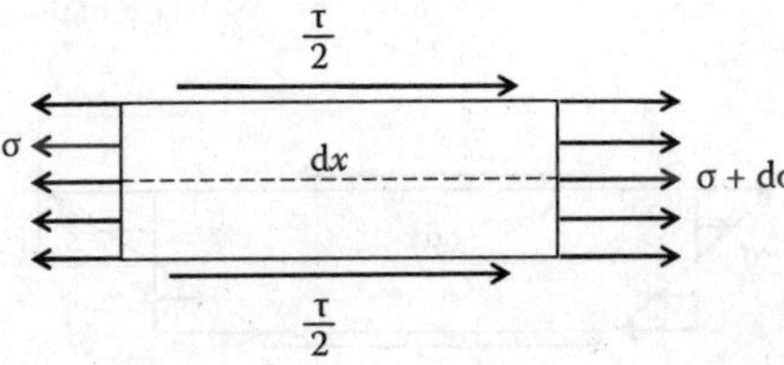

FIGURE 11.30
Symmetric stress distribution across the plate thickness.

In arriving at Equation (11.245), the definition $d = t/2$ was used. Equation (11.245) can be rewritten in the form

$$t\frac{d\sigma_S}{dx} + \alpha_S \tau = 0 \tag{11.246}$$

where $\alpha_S = 1$.

11.4.1.2 Antisymmetric Case

In the antisymmetric case, the stress and strain are assumed to vary linearly across the thickness (Figure 11.31). Since the stress is assumed linearly varying across the thickness, the stress distribution can be expressed as

$$\sigma(y) = \frac{y}{d}\sigma_A \tag{11.247}$$

where σ_A is the value of the stress in the plate evaluated in the upper surface. The subscript A stands for "antisymmetric."

The stress resultants are evaluated by integration of the stress across the thickness. Since the stress distribution is antisymmetric, the only stress resultant is the moment per unit width

$$M = \int\limits_{-d}^{+d} \sigma(y)y\,dy = \int\limits_{-d}^{+d} \frac{y}{d}\sigma_A y\,dy = \sigma_A \frac{1}{d}\int\limits_{-d}^{+d} y^2\,dy = \sigma_A \frac{2d^2}{3} \tag{11.248}$$

The moment due to the shear stresses applied to the plate surface over the length dx is

$$dM_\tau = 2d\frac{\tau}{2}dx = d\tau\,dx \tag{11.249}$$

Hence, equilibrium of the infinitesimal element of Figure 11.31 gives

$$\cancel{M} + dM + dM_\tau - \cancel{M} = 0 \tag{11.250}$$

Substitution of Equations (11.248) and (11.249) into Equation (11.250) yields

$$\frac{2d^2}{3}d\sigma_A + d\tau\,dx = 0 \tag{11.251}$$

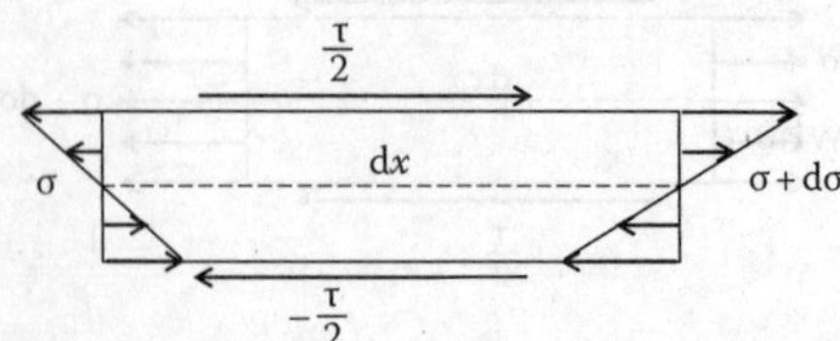

FIGURE 11.31
Antisymmetric stress distribution across the plate thickness.

Upon simplification and using the definition $d = t/2$, we get

$$t\frac{d\sigma_A}{dx} + 3\tau = 0 \qquad (11.252)$$

Equation (11.252) can be rewritten in the form

$$t\frac{d\sigma_A}{dx} + \alpha_A\tau = 0 \qquad (11.253)$$

where $\alpha_A = 3$.

11.4.1.3 Shear Lag Solution

The superposition $\sigma = \sigma_S + \sigma_A$ yields Equation (11.240) with $\alpha = \alpha_S + \alpha_A = 4$. (For Lamb-wave modes, which have complex stress and strain distributions, the value of α will actually vary from mode to mode and depend on the frequency-thickness product). The strain–displacement equations in the PWAS, bonding layer, and structure are

$$\varepsilon_a = \frac{du_a}{dx} \qquad \text{(PWAS)} \qquad (11.254)$$

$$\gamma = \frac{u_a - u}{t_b} \qquad \text{(bonding layer)} \qquad (11.255)$$

$$\varepsilon = \frac{du}{dx} \qquad \text{(structure)} \qquad (11.256)$$

whereas the stress–strain relations are

$$\sigma_a = E_a(\varepsilon_a - \varepsilon_{ISA}) \qquad \text{(PWAS)} \qquad (11.257)$$

$$\tau = G_b\gamma \qquad \text{(bonding layer)} \qquad (11.258)$$

$$\sigma = E\varepsilon \qquad \text{(structure)} \qquad (11.259)$$

In addition, recall the induced-strain Equation 11.238, i.e., $\varepsilon_{ISA} = d_{31}V/t_a$. Upon substitution, we obtain two second order coupled differential equations in ε_a and ε which, upon further differentiation, yield the following pair of decoupled fourth-order differential equations

$$\frac{d^4\varepsilon_a}{dx^4} - \Gamma^2\frac{d^2\varepsilon_a}{dx^2} \qquad \text{(PWAS)} \qquad (11.260)$$

$$\frac{d^4\varepsilon}{dx^4} - \Gamma^2\frac{d^2\varepsilon}{dx^2} \qquad \text{(structure)} \qquad (11.261)$$

where

$$\Gamma^2 = \frac{G_b}{E_a}\frac{1}{t_a t_b}\frac{\alpha + \psi}{\psi} \qquad \text{(shear lag parameter)} \qquad (11.262)$$

and

$$\psi = \frac{Et}{E_a t_a} \tag{11.263}$$

Solution of Equations (11.260) and (11.261) is obtained in the form

$$\varepsilon_a(x) = \frac{\alpha}{\alpha + \psi} \varepsilon_{ISA} \left(1 + \frac{\psi}{\alpha} \frac{\cosh \Gamma x}{\cosh \Gamma a} \right) \qquad \text{(PWAS actuation strain)} \tag{11.264}$$

$$\sigma_a(x) = -\frac{\psi}{\alpha + \psi} E_a \varepsilon_{ISA} \left(1 - \frac{\cosh \Gamma x}{\cosh \Gamma a} \right) \qquad \text{(PWAS stress)} \tag{11.265}$$

$$u_a(x) = \frac{\alpha}{\alpha + \psi} \varepsilon_{ISA} a \left(\frac{x}{a} + \frac{\psi}{\alpha} \frac{\sinh \Gamma x}{(\Gamma a)\cosh \Gamma a} \right) \qquad \text{(PWAS displacement)} \tag{11.266}$$

$$\tau(x) = \frac{t_a}{a} \frac{\psi}{\alpha + \psi} E_a \varepsilon_{ISA} \left(\Gamma a \frac{\sinh \Gamma x}{\cosh \Gamma a} \right) \qquad \text{(interfacial shear stress in bonding layer)} \tag{11.267}$$

$$\varepsilon(x) = \frac{\alpha}{\alpha + \psi} \varepsilon_{ISA} \left(1 - \frac{\cosh \Gamma x}{\cosh \Gamma a} \right) \qquad \text{(structure strain at the surface)} \tag{11.268}$$

$$\sigma(x) = \frac{\alpha}{\alpha + \psi} E \varepsilon_{ISA} \left(1 - \frac{\cosh \Gamma x}{\cosh \Gamma a} \right) \qquad \text{(structure stress)} \tag{11.269}$$

$$u(x) = \frac{\alpha}{\alpha + \psi} \varepsilon_{ISA} a \left(\frac{x}{a} - \frac{\sinh \Gamma x}{(\Gamma a)\cosh \Gamma a} \right) \qquad \text{(structure displacement at the surface)} \tag{11.270}$$

These equations apply for $|x| < a$. Outside the $|x| < a$ interval, the strain and stress variables in these equations are zero. Note that $\sigma_a \neq E_a \varepsilon_a$ because of Equation (11.257). The shear lag parameter, Γ, plays a very important role in determining the distribution of ε_a, ε, τ along the PWAS span $(-a, a)$. The effect of the PWAS is transmitted to the structure through the interfacial shear stress of the bonding layer. A small shear stress in the bonding layer produces a gradual transfer of strain from the PWAS to the structure whereas a large shear stress produces a rapid transfer. Since the PWAS ends are stress free, the build up of strain takes place at the ends and it is more rapid when the shear stress is more intense. For large values of Γa, the shear transfer process becomes concentrated toward the PWAS ends.

Example:

To illustrate these equations, we considered an APC-850 PWAS with $E_a = 63$ GPa, $t_a = 0.2$ mm, and $l_a = 7$ mm mounted on a thin-wall aluminum structure with $E = 70$ GPa and $t = 1$ mm. (The value $t = 2$ mm was also considered). The mounting is done with cynoacrylate adhesive ($G_b = 2$ GPa) of variable thickness, $t_b = 1$ μm, 10 μm, 100 μm. The piezoelectric constant of the APC-850 material used in the PWAS is $d_{31} = -175$ mm/kV. The applied voltage was $V = 10$ V. Figure 11.32a shows the strain distribution in the structure and PWAS while Figure 11.32b shows the shear stress distribution. For the range of values considered here, the value of Γa varied between 16 and 58.

Examination of Figure 11.32 reveals that the shear lag parameter, Γa, plays a very important role in determining the distribution of ε_a, ε, τ along the PWAS span $(-a, a)$. The effect of

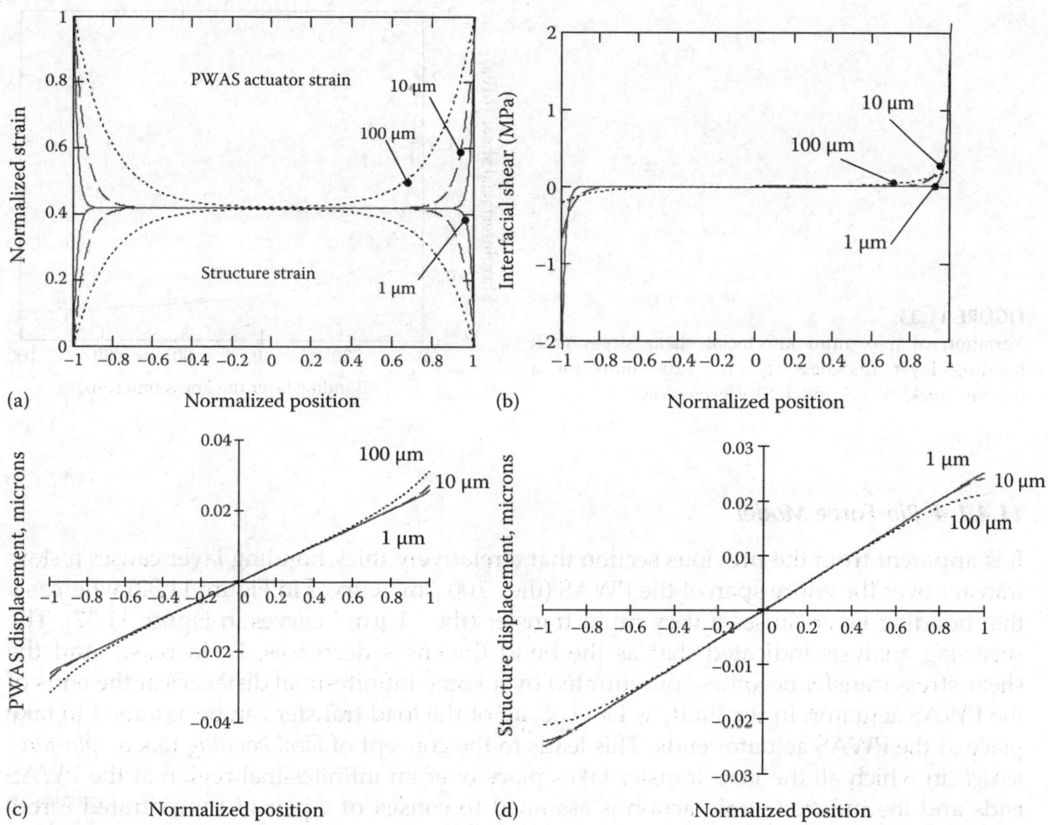

FIGURE 11.32
Variation of shear-lag transfer mechanism with bond thickness: (a) strain distribution in the PWAS and the structure; (b) interfacial shear stress distribution; (c) displacement distribution in the PWAS; (d) displacement distribution in the structure (bond thickness $t_b = 1$, 10, and 100 μm).

the PWAS is transmitted to the structure through the shear stress in the bonding layer. A small shear stress in the bonding layer produces a gradual transfer of strain from the PWAS to the structure whereas a large shear stress produces a rapid transfer. Since the PWAS ends are stress free, the build up of strain takes place at the ends and it is more rapid when the shear stress is more intense. As indicated by Equation (11.262), a relatively thick bonding layer produces a low Γa value, i.e., a slow transfer over the entire span of the PWAS (the "100 μm" curves in Figure 11.32) that a very thin bonding layer produces a very rapid transfer (the "1 μm" curves in Figure 11.32) that is confined to the ends.

Another aspect of interest is the maximum interfacial shear stress. As indicated by Figure 11.32b, the maximum interfacial shear stress takes place at the ends. A possible concern might be that the large shear stress values at the PWAS ends would exceed the bond strength and promote failure. A plot of the interfacial shear stress with bond thickness is presented in Figure 11.33 for a 0.2 mm thick PWAS under 10 V excitation. It is apparent that the maximum interfacial shear stress does not exceed 2.5 MPa, which is about an order of magnitude lower than the bonding layers shear strength. ∎

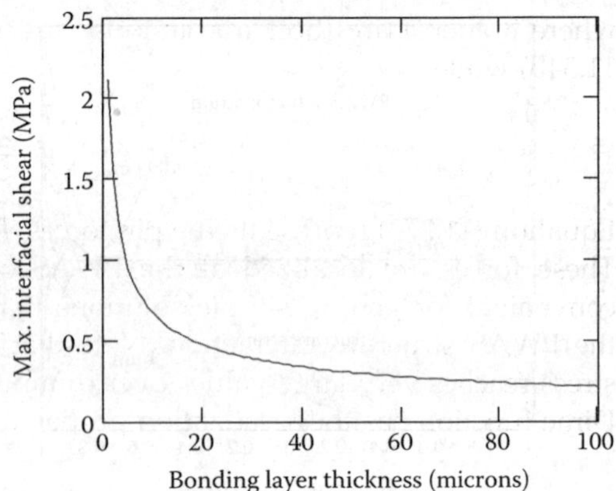

FIGURE 11.33
Variation of maximum interfacial shear stress with bonding layer thickness ($t_b = 1, \ldots, 100$ mm) for a 0.2 mm thick PWAS under 10 V excitation.

11.4.1.4 Pin-Force Model

It is apparent from the previous section that a relatively thick bonding layer causes a slow transfer over the entire span of the PWAS (the "100 μm" curves in Figure 11.32) whereas a thin bonding layer causes a very rapid transfer (the "1 μm" curves in Figure 11.32). The shear-lag analysis indicated that as the bond thickness decreases, Γa increases and the shear stress transfer becomes concentrated over some infinitesimal distances at the ends of the PWAS actuator. In the limit, as $\Gamma a \to \infty$, all of the load transfer can be assumed to take place at the PWAS actuator ends. This leads to the concept of *ideal bonding* (a.k.a., *pin-force model*) in which all the load transfer takes place over an infinitesimal region at the PWAS ends and the induced-strain action is assumed to consist of a pair of concentrated forces applied at the ends (Figure 11.34a), i.e.,

$$\tau(x) = a\tau_a[\delta(x - a) - \delta(x + a)] \qquad \text{(ideal-bonding shear stress)} \qquad (11.271)$$

$$F(x) = F_a[-H(x - a) + H(x + a)] \qquad \text{(shear force due to pin-end forces)} \qquad (11.272)$$

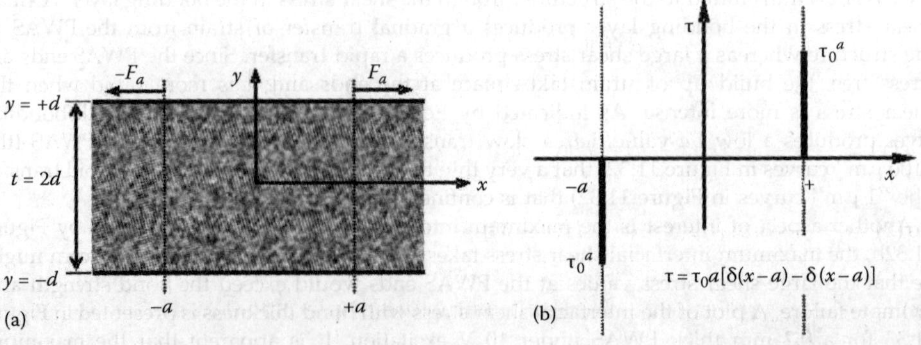

FIGURE 11.34
Pin-force model: (a) surface shear distribution; (b) direct strain induced in the structure at the upper structural surface.

where δ and H are the Dirac impulse function and the Heaviside step function (Figure 11.34b) while

$$F_a = a\tau_a \qquad \text{(pin-end forces)} \qquad (11.273)$$

Equation (11.273) represents the pin forces, $F_a = a\tau_a$, applied by the PWAS to the structure. These forces are localized at the PWAS ends (Figure 11.34a). The pin-force model is convenient for getting simple solutions that represent a first-order of approximation to the PWAS–structure interaction. Note that this extreme situation implies that the shear stress reaches very large values over diminishing areas at the PWAS ends. Recall that the Dirac function has the localization property

$$\int f(x)\delta(x - x_0)dx = f(x_0) \qquad \text{(Dirac function localization property)} \qquad (11.274)$$

and the Dirac function is the derivative of the Heaviside function, i.e.,

$$\delta(x) = H'(x) \qquad (11.275)$$

Under these assumptions, Equations (11.264) and (11.270) take the simple forms

$$\varepsilon_a(x) = \frac{\alpha}{\alpha + \psi}\varepsilon_{ISA}[H(x + a) - H(x - a)] \qquad \text{(PWAS actuation strain)} \qquad (11.276)$$

$$\sigma_a(x) = -\frac{\psi}{\alpha + \psi}\varepsilon_{ISA}[H(x + a) - H(x - a)] \qquad \text{(stress in the PWAS)} \qquad (11.277)$$

$$u_a(x) = \frac{\alpha}{\alpha + \psi}\varepsilon_{ISA}x[H(x + a) - H(x - a)] \qquad \text{(displacement in the PWAS)} \qquad (11.278)$$

$$\tau(x) = \frac{\psi}{\alpha + \psi}t_a E_a \varepsilon_{ISA}[-\delta(x + a) + \delta(x - a)] \qquad (11.279)$$
$$\text{(interfacial shear stress in bonding layer)}$$

$$F(x) = \frac{\psi}{\alpha + \psi}t_a E_a \varepsilon_{ISA}[-H(x + a) + H(x - a)] \qquad (11.280)$$
$$\text{(interfacial shear force in bonding layer)}$$

$$\varepsilon(x) = \frac{\alpha}{\alpha + \psi}\varepsilon_{ISA}[H(x + a) - H(x - a)] \qquad \text{(structure strain at the surface)} \qquad (11.281)$$

$$\sigma(x) = \frac{\alpha}{\alpha + \psi}E\varepsilon_{ISA}[H(x + a) - H(x - a)] \qquad \text{(structure stress at the surface)} \qquad (11.282)$$

$$u(x) = \frac{\alpha}{\alpha + \psi}\varepsilon_{ISA}x(H(x + a) - H(x - a)) \qquad \text{(structure displacement at the surface)} \qquad (11.283)$$

where $H(x)$ and $\delta(x)$ are the Heaviside step function and the Dirac delta function, respectively. Note that the strains in the PWAS and at the surface of the structure are equal, but the stresses are not, due to Equation (11.257).

The axial force and bending moment associated with the ideal-bonding hypothesis (pin-force model) are

$$N_a = F_a \qquad \text{(axial force)} \tag{11.284}$$

$$M_a = F_a d = F_a \frac{t}{2} \qquad \text{(bending moment)} \tag{11.285}$$

The axial force and bending moment described by Equations (11.284) and (11.285) represent the excitation induced by the PWAS into the plate under the ideal-bonding hypothesis.

11.4.1.5 Energy Transfer between the PWAS and the Structure

The transfer of energy between the PWAS and the structure will be studied for both the shear-lag model and the pin-force model.

11.4.1.5.1 Energy Transfer through the Shear-Lag Model

The energy retained in the actuator can be evaluated as

$$W_a = \int_{V_a} \frac{1}{2} \frac{\sigma_a^2}{E_a} dV = \frac{\psi^2}{(\alpha + \psi)^2} \left(\frac{1}{2} E_a \varepsilon_{ISA}^2 \right) \int_{V_a} \frac{1}{2} \left(1 - \frac{\cosh \Gamma x}{\cosh \Gamma a} \right)^2 dx$$

$$= \frac{\psi^2}{(\alpha + \psi)^2} \left(\frac{1}{2} E_a \varepsilon_{ISA}^2 l_a \right) I(\Gamma a) \tag{11.286}$$

The function $I(\Gamma a)$ is given by

$$I(\Gamma a) = 1 - \frac{3}{2} \frac{\sinh \Gamma a}{\Gamma a \cosh \Gamma a} + \frac{1}{2} \frac{1}{(\cosh \Gamma a)^2} \qquad \text{(bond efficiency)} \tag{11.287}$$

Note that $I(\Gamma a)$ approaches 1 for large Γa values (Figure 11.35). It represents a measure of the bond efficiency. For the values considered in our preliminary study, the values of $I(\Gamma a)$ were found to vary between 90% and 97%.

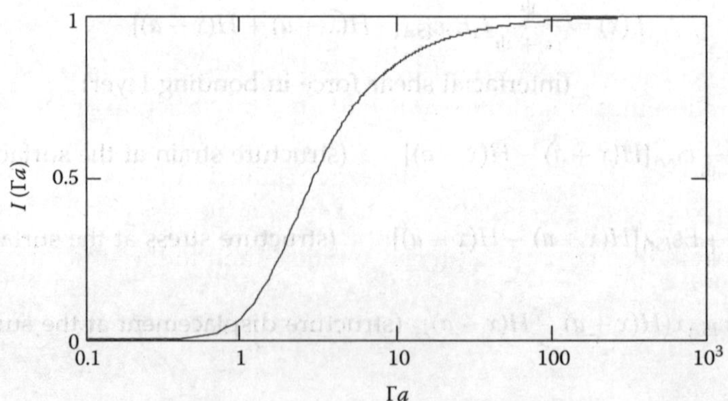

FIGURE 11.35
Bond energy efficiency variation with Γa.

The term $\frac{1}{2}E_a\varepsilon_{ISA}^2 l_a$ is a measure of the total induced-strain energy whereas the term $\psi^2/(\alpha+\psi)^2$ represents how much of this energy gets stored as elastic compression in the PWAS actuator and cannot be transmitted to the structure.

The energy transmitted to the structure can be evaluated either as the elastic energy in the structure or the work done by the interfacial shear stresses on the structural surface. The two paths are equivalent:

The elastic energy in the structure is given by

$$W_1 = \int_V \frac{1}{2}\Big[(\sigma\varepsilon)_{axial} + (\sigma\varepsilon)_{flexural}\Big]dV = \frac{\alpha}{(\alpha+\psi)^2}\left(\tfrac{1}{2}E\varepsilon_{ISA}^2\right)\int_{-a}^{a}\frac{1}{2}\left(1 - \frac{\cosh\Gamma x}{\cosh\Gamma a}\right)^2 dx$$

$$= \frac{\alpha\psi}{(\alpha+\psi)^2}\left(\tfrac{1}{2}E\varepsilon_{ISA}^2 l_a\right)I(\Gamma a) \tag{11.288}$$

The work done by the shear stresses at the structural surface is given by

$$W_2 = \int_{l_a}\left(\tfrac{1}{2}\tau u\right)dx = \int_{-a}^{a}\frac{1}{2}\frac{t_a}{a}\frac{\psi}{\alpha+\psi}E_a\varepsilon_{ISA}\left(\Gamma a\frac{\sinh\Gamma x}{\cosh\Gamma a}\right)\frac{\alpha}{\alpha+\psi}\varepsilon_{ISA}a\left(\frac{x}{a} - \frac{\sinh\Gamma x}{(\Gamma a)\cosh\Gamma a}\right)dx$$

$$= \frac{\alpha\psi}{(\alpha+\psi)^2}\left(\tfrac{1}{2}E\varepsilon_{ISA}^2 l_a\right)I(\Gamma a) \tag{11.289}$$

These two approaches give the same results, and the expression of the energy transmitted to the structure is

$$W = \frac{\alpha}{(\alpha+\psi)^2}\left(\tfrac{1}{2}E\varepsilon_{ISA}^2 l_a\right)I(\Gamma a) \qquad \text{(energy transmitted to the structure)} \tag{11.290}$$

We note that the integral $I(\Gamma a)$ and the term $\frac{1}{2}E_a\varepsilon_{ISA}^2 l_a$ of Equation (11.286) reappear in Equations (11.288) and (11.289). The term $\alpha/(\alpha+\psi)^2$ represents how much of the induced-strain energy gets transmitted into the structure.

11.4.1.5.2 Energy Transfer through the Pin-Force Model

Since the pin-force model is a constant-strain and constant-stress model, the evaluation of the energy stored in the actuator and transmitted to the structure can be readily done

$$W_a = \left(\tfrac{1}{2}E_a\varepsilon_{ISA}^2 l_a\right) \qquad \text{(energy retained in the actuator)} \tag{11.291}$$

$$W = \frac{\alpha\psi}{(\alpha+\psi)^2}\left(\tfrac{1}{2}E\varepsilon_{ISA}^2 l_a\right) \qquad \text{(energy transmitted to the structure)} \tag{11.292}$$

11.4.1.5.3 Conditions for Optimum Energy Transfer

Figure 11.36 represents a plot of the energy transferred into the structure versus parameter ψ in nondimensional form. It can be noticed that the energy transfer reaches a maximum as the parameter ψ equals the parameter α. Denote the apparent stiffness ratio, r, by

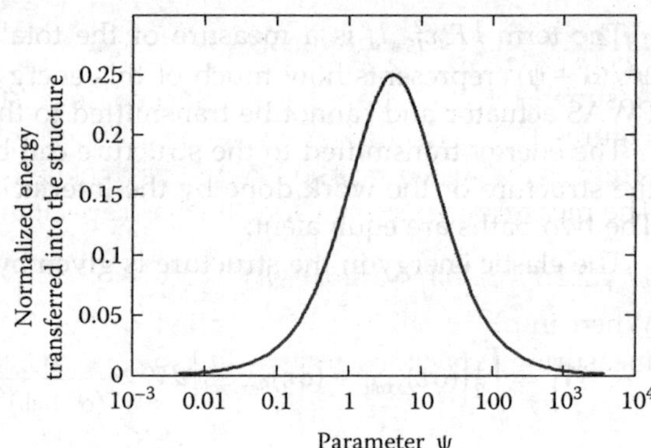

FIGURE 11.36
Energy transfer into the structure versus parameter ψ in nondimensional form. The plot is normalized by the factor $\frac{1}{2}E\varepsilon_{ISA}^2 l_a$.

$$r = \frac{\psi}{\alpha} = \frac{Et/\alpha}{E_a t_a} \qquad (11.293)$$

We notice that, under the quasi-static conditions considered here, maximum energy transfer is attained at $r = 1$. This represents the well known *stiffness matching principle* with the proviso that the stiffness of the PWAS, $E_a t_a$, needs to be matched with the *apparent stiffness* of the structure, Et/α. At the stiffness match condition, the energy in the structure and the energy in the PWAS coincide and are equal to the maximum extractable energy

$$W_{\max} = \frac{1}{4}\left(\tfrac{1}{2}E_a\varepsilon_{ISA}^2 l_a\right) \qquad \text{(maximum extractable energy)} \qquad (11.294)$$

The value $r = 1$ represents the *condition for maximum energy transfer* under quasi-static conditions. Under dynamic conditions, this condition may be modified by several factors

- The apparent structural stiffness varies with frequency due to inertia loads.
- Modal resonances will affect the apparent stiffness which will go through minima at every resonance.
- At ultrasonic frequencies, the apparent structural stiffness depends on the Lamb modes which have frequency-dependent displacement distribution across the thickness.

The stiffness match principle corresponds to the *impedance match principle* commonly used in electronic circuits design in order to attain maximum power transfer conditions.

11.4.2 Strain Sensing with PWAS

PWAS can be used for strain and stress measurements because they directly convert the mechanical energy into electrical energy. Since the resulting voltage is proportional to the strain rate, this type of measurement would be especially beneficial at high frequency. PWAS can be also used for low frequency and quasi-static measurements, but their effectiveness is not as good due to charge leakage. These aspects will be detailed in the following sections.

11.4.2.1 Static Measurements

Assume a PWAS transducer attached to a structural surface as previously shown in Figures 11.1 and 11.28. Assume ideal bonding such that the PWAS experiences the same strain as the structure. The PWAS has thickness h and top and bottom electrodes of area A. The measuring external capacitor has capacitance C_e (Figure 11.37).

11.4.2.1.1 Strain Measurements

When in-plane strain S_1 is applied to the PWAS, the voltage V is experienced at the measuring capacitor. Under 1-D assumptions, the relevant constitutive equations are

$$S_1 = s_{11}T_1 + d_{31}E_3 \tag{11.295}$$

$$D_3 = d_{31}T_1 + \varepsilon_{33}E_3 \tag{11.296}$$

The charge Q moved to the measuring capacitor is given by

$$Q = D_3 A \tag{11.297}$$

The relation between electric field E_3 and voltage V is

$$E_3 = -\frac{V}{h} \tag{11.298}$$

In the measuring capacitor, the relation between voltage V and capacitance C_e is

$$Q = C_e V \tag{11.299}$$

We want to find the voltage V as a function of applied strain S_1. To attain this, we do the following. Eliminate T_1 between Equations (11.295) and (11.296) to get

$$d_{31}S_1 - s_1 D_3 = \left(d_{31}^2 - s_{11}\varepsilon_{33}\right)E_3 \tag{11.300}$$

Combine Equations (11.297) and (11.299) to write

$$D_3 = \frac{Q}{A} = \frac{C_e}{A}V \tag{11.301}$$

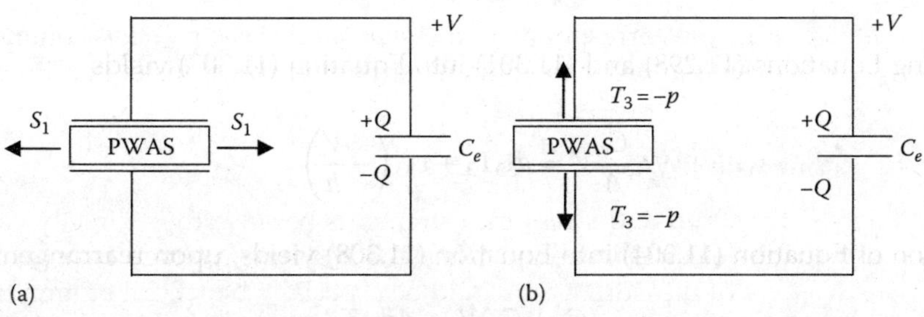

FIGURE 11.37
Schematic of a PWAS connected with measuring equipment of capacitance C_e: (a) measurement of static in-plane strain S_1; (b) measurement of static out-of-plane pressure ($T_3 = -p$).

Substitute Equations (11.298) and (11.301) into Equation (11.300) to obtain

$$d_{31}S_1 - s_{11}\frac{C_e}{A}V = (d_{31}^2 - s_{11}\varepsilon_{33})\left(-\frac{V}{h}\right) = (1 - k_{31}^2)s_{11}\varepsilon_{11}\frac{V}{h} \tag{11.302}$$

Upon rearrangement, Equation (11.302) becomes

$$d_{31}\frac{A}{s_{11}}S_1 = C_eV + (1 - k_{31}^2)\varepsilon_{33}\frac{A}{h}V = \left[C_e + (1 - k_{31}^2)C_0\right]V \tag{11.303}$$

where

$$C_0 = \varepsilon_{33}\frac{A}{h} \tag{11.304}$$

is the free capacitance of the PWAS, as previously defined by Equation (11.114). Upon solution, Equation (11.303) gives the expression of voltage as function of applied strain with the value of the measuring capacitor as a parameter, i.e.,

$$V(S_1; C_e) = \frac{1}{C_e + (1 - k_{31}^2)C_0}\frac{d_{31}A}{s_{11}}S_1 \tag{11.305}$$

Equation (11.305) indicates that a lower value of the measuring capacitor C_e will generally ensure a larger voltage V for a given strain S_1. This means that a measuring instrument with low input capacitance is desired. However, this argument is limited by the internal capacitance C_0 of the PWAS; decreasing C_e below C_0 would not yield any further benefits.

11.4.2.1.2 Stress Measurements

A similar argument with that of Equations (11.295) through (11.305) can be used to determine the measured voltage as a function of applied stress. This situation would be relevant if the PWAS were used to measure pressure, i.e., $T_3 = -p$. In this case, the relevant constitutive equations would be

$$S_3 = s_{33}T_3 + d_{33}E_3 \tag{11.306}$$

$$D_3 = d_{33}T_3 + \varepsilon_{33}E_3 \tag{11.307}$$

Substituting Equations (11.298) and (11.301) into Equation (11.307) yields

$$\frac{C_e}{A}V = d_{33}T_3 + \varepsilon_{33}\left(-\frac{V}{h}\right) \tag{11.308}$$

Substitution of Equation (11.304) into Equation (11.308) yields, upon rearrangement,

$$(C_e + C_0)V = Ad_{33}T_3 \tag{11.309}$$

Upon solution, we obtain the expression of voltage as a function of applied strain with the value of the measuring capacitor as a parameter, i.e.,

$$V(T_1; C_e) = \frac{A d_{33}}{C_e + C_0} T_3 = -\frac{A d_{33}}{C_e + C_0} p \tag{11.310}$$

Equation (11.310) indicates that a lower value of the measuring capacitor C_e will generally ensure a larger voltage V for a given strain T_1. This means that a measuring instrument with low input capacitance is desired. However, this argument is limited by the internal capacitance C_0 of the PWAS; decreasing C_e below C_0 would not yield any further benefits.

11.4.2.2 Dynamic Measurements

Assume again a PWAS transducer attached to a structural surface as previously shown in Figures 11.1 and 11.28. Assume ideal bonding such that the PWAS experiences the same strain as the structure. The PWAS has thickness h and top and bottom electrodes of area A. For dynamic measurement, we assume a measuring equipment of admittance Y_e and impedance Z_e ($Z_e = 1/Y_e$).

11.4.2.2.1 Strain Measurements

When in-plane dynamic strain $S_1(t)$ is applied to the PWAS, the voltage $V(t)$ is experienced at the measuring equipment (Figure 11.38). Under 1-D assumptions, the relevant constitutive equations are

$$S_1 = s_{11} T_1 + d_{31} E_3 \tag{11.311}$$

$$D_3 = d_{31} T_1 + \varepsilon_{33} E_3 \tag{11.312}$$

Taking time derivative of Equations (11.311) and (11.312), we write

$$\dot{S}_1 = s_{11} \dot{T}_1 + d_{31} \dot{E}_3 \tag{11.313}$$

$$\dot{D}_3 = d_{31} \dot{T}_1 + \varepsilon_{33} \dot{E}_3 \tag{11.314}$$

The current flowing through the circuit is given by the product between the time-derivative of the electric displacement $\dot{D}_3$ and the electrodes area A, i.e.,

$$I = \dot{D}_3 A \tag{11.315}$$

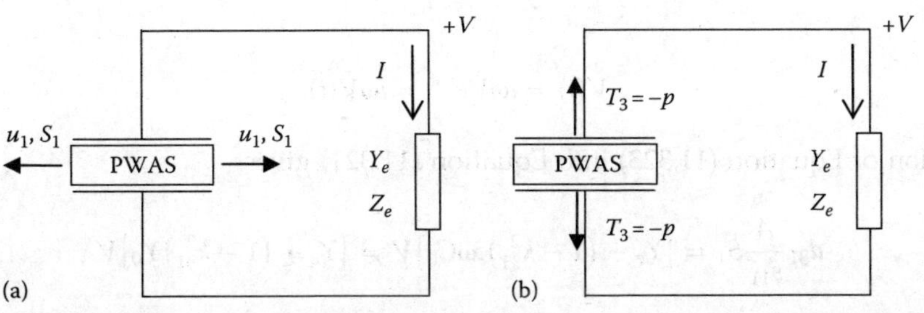

FIGURE 11.38
Schematic of a PWAS connected with measuring equipment of admittance Y_e and impedance Z_e ($Z_e = 1/Y_e$): (a) measurement of dynamic in-plane strain S_1; (b) measurement of dynamic out-of-plane pressure ($T_3 = -p$).

Take the time-derivative of Equation (11.298) to get

$$\dot{E}_3 = -\frac{\dot{V}}{h} \tag{11.316}$$

The relation between voltage V and the internal admittance of the measuring equipment is

$$I = Y_e V \tag{11.317}$$

We want to find the voltage V as a function of applied strain S_1. To attain this, we do the following. Eliminate $\dot{T}_1$ between Equations (11.313) and (11.314) to get

$$d_{31}\dot{S}_1 - s_1 \dot{D}_3 = \left(d_{31}^2 - s_{11}\varepsilon_{33}\right)\dot{E}_3 \tag{11.318}$$

Combine Equations (11.315) and (11.317) to write

$$\dot{D}_3 = \frac{I}{A} = \frac{Y_e}{A}V \tag{11.319}$$

Substitute Equations (11.316) and (11.319) into Equation (11.318) to obtain

$$d_{31}\dot{S}_1 - s_{11}\frac{Y_e}{A}V = \left(d_{31}^2 - s_{11}\varepsilon_{33}\right)\left(-\frac{\dot{V}}{h}\right) = \left(1 - k_{31}^2\right)s_{11}\varepsilon_{11}\frac{\dot{V}}{h} \tag{11.320}$$

Upon rearrangement, Equation (11.320) becomes

$$d_{31}\frac{A}{s_{11}}\dot{S}_1 = Y_e V + \left(1 - k_{31}^2\right)\varepsilon_{33}\frac{A}{h}\dot{V} = Y_e V + \left(1 - k_{31}^2\right)C_0 \dot{V} \tag{11.321}$$

where C_0 is the free capacitance of the PWAS, as previously defined by Equation (11.304). Note that Equation (11.321) is an ordinary differential equation in $V(t)$. Under harmonic assumptions, we write

$$S_1(t) = \hat{S}_1 e^{i\omega t}, \quad V(t) = \hat{V}e^{i\omega t} \tag{11.322}$$

Hence,

$$\dot{V}(t) = i\omega\hat{V}e^{i\omega t} = i\omega V(t) \tag{11.323}$$

Substitution of Equation (11.323) into Equation (11.321) gives

$$d_{31}\frac{A}{s_{11}}\dot{S}_1 = \left[Y_e + \left(1 - k_{31}^2\right)i\omega C_0\right]V = \left[Y_e + \left(1 - k_{31}^2\right)Y_0\right]V \tag{11.324}$$

where Y_0 is the admittance of the unconstrained PWAS transducer given by

$$Y_0 = i\omega C_0 \tag{11.325}$$

Upon solution, Equation (11.324) gives the expression of voltage as a function of applied strain rate with the value of the measuring admittance as a parameter, i.e.,

$$V(t, \dot{S}_1; Y_e) = \frac{1}{Y_e + (1 - k_{31}^2)Y_0} \frac{d_{31}A}{s_{11}} \dot{S}_1 \tag{11.326}$$

Equation (11.326) indicates that a lower value of the measuring admittance Y_e will generally ensure a larger voltage V for a given strain rate $\dot{S}_1$. This means that a measuring instrument with low input admittance (high impedance) is desired. However, this argument is limited by the internal admittance Y_0 of the PWAS; decreasing Y_e below Y_0 would not yield any further benefit.

Equation (11.322) can be used to express Equation (11.326) in terms of amplitudes $\hat{S}_1$ and $\hat{V}$. Note that

$$\dot{S}_1(t) = i\omega \hat{S}_1 e^{i\omega t} \tag{11.327}$$

Substitution of Equations (11.322) and (11.327) into Equation (11.326) yields an expression in terms of harmonic amplitudes, i.e.,

$$\hat{V}(\dot{S}_1; Y_e) = \frac{1}{Y_e + (1 - k_{31}^2)Y_0} \frac{d_{31}A}{s_{11}} i\omega \hat{S}_1 \tag{11.328}$$

For an ideal measurement instrument with $Y_e - i\omega C_e$, Equation (11.328) can be simplified to yield

$$\hat{V}(\dot{S}_1; Y_e) = \frac{1}{C_e + (1 - k_{31}^2)C_0} \frac{d_{31}A}{s_{11}} \hat{S}_1 \tag{11.329}$$

Note the similarity between Equations (11.329) and (11.305) used in the static case. However, practical measuring instruments do not have a purely capacitive input admittance since some resistive losses and inductive coupling are always present; similarly, actual PWAS transducers always have internal dissipation, especially at high frequencies. For these reasons, the simplifications that lead from Equations (11.328) to (11.329) may not be always possible.

11.4.2.2.2 Stress Measurements

A similar argument with that of Equations (11.312) through (11.329) can be used to determine the measured voltage as a function of applied stress. This situation would be relevant if the PWAS were used to measure transient or dynamic pressures, i.e., $T_3(t) = -p(t)$. In this case, the relevant constitutive equations, in time-derivative form, would be

$$\dot{S}_3 = s_{33}\dot{T}_3 + d_{33}\dot{E}_3 \tag{11.330}$$

$$\dot{D}_3 = d_{33}\dot{T}_3 + \varepsilon_{33}\dot{E}_3 \tag{11.331}$$

Substituting Equations (11.316) and (11.319) into Equation (11.331) yields

$$\frac{Y_e}{A}V = d_{33}\dot{T}_3 + \varepsilon_{33}\left(-\frac{\dot{V}}{h}\right) \tag{11.332}$$

Substitution of Equation (11.304) into Equation (11.332) yields, upon rearrangement,

$$Y_e V + C_0 \dot{V} = A d_{33} \dot{T}_3 \tag{11.333}$$

Equation (11.333) is an ordinary differential equation in $V(t)$. Under the harmonic assumptions of Equation (11.323) and using the definition of Y_0 given in Equation (11.325), we solve Equation (11.333) to obtain an expression of voltage as a function of applied stress rate with the value of the measuring admittance as a parameter, i.e.,

$$V(t, T_1; C_e) = \frac{A d_{33}}{Y_e + Y_0} \dot{T}_3(t) = -\frac{A d_{33}}{Y_e + Y_0} \dot{p}(t) \tag{11.334}$$

Equation (11.334) indicates that a lower value of the measuring admittance Y_e will generally ensure a larger voltage V for a given stress rate $\dot{T}_3$. This means that a measuring instrument with low input admittance (high impedance) is desired. However, this argument is limited by the internal admittance Y_0 of the PWAS; decreasing Y_e below Y_0 would not yield any further benefit.

Since harmonic assumptions apply, we can further simplify Equation (11.334) by using the relations

$$\dot{p}(t) = i\omega \hat{p} e^{i\omega t}, \quad \dot{T}_3(t) = i\omega \hat{T}_3 e^{i\omega t} \tag{11.335}$$

Substitution of Equations (11.323) and (11.335) into Equation (11.334) yields an expression in terms of harmonic amplitudes, i.e.,

$$\hat{V}(T_1; C_e) = \frac{A d_{33}}{Y_e + Y_0} i\omega \hat{T}_3 = -\frac{A d_{33}}{Y_e + Y_0} i\omega \hat{p} \tag{11.336}$$

For an ideal measurement instrument with $Y_e = i\omega C_e$, Equation (11.336) can be simplified to yield

$$\hat{V}(T_1; C_e) = \frac{A d_{33}}{C_e + C_0} \hat{T}_3 = -\frac{A d_{33}}{C_e + C_0} \hat{p} \tag{11.337}$$

Note the similarity between Equations (11.337) and (11.310) derived in the static case. However, practical measuring instruments do not have a purely capacitive input admittance since some resistive losses and inductive coupling are always present; similarly, actual PWAS transducers always have internal dissipation, especially at high frequencies. For these reasons, the simplifications that lead from Equations (11.336) to (11.337) may not be always possible.

11.4.2.3 Vibration Sensing Experiments on a Cantilever Beam

Experiments were conducted to demonstrate the PWAS vibration strain sensing capabilities on a stainless steel beam of length $L = 300$ mm, width $w = 19.2$ mm, thickness $t = 3.23$ mm, density $\rho = 8030$ kg/m^3, and Young's modulus $E = 195$ GPa. The theoretical natural frequencies of the beam were $f_1^{theory} = 28.6$ Hz, $f_2^{theory} = 179$ Hz, $f_3^{theory} = 501$ Hz. The beam was held in a fixture that ensured cantilever boundary conditions (Figure 11.39a). Three PWAS transducers, varying in piezoelectric material and thickness were used: (1) 28 μm thick PVDF-PWAS; (2) 110 μm thick PVDF-PWAS; (3) 200 μm PZT-PWAS. All three PWAS

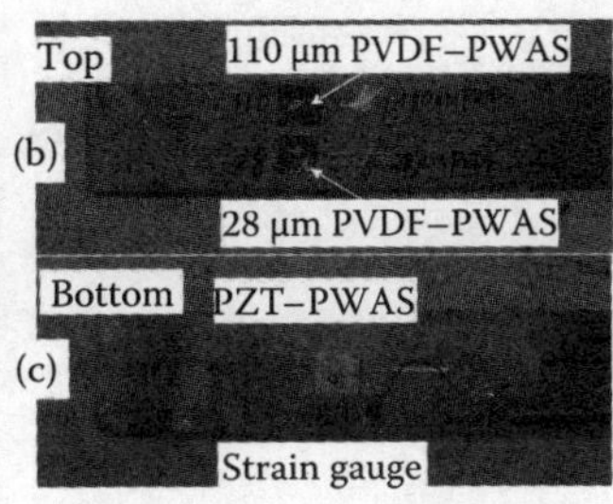

FIGURE 11.39

PZT-PWAS, PVDF-PWAS, and strain gauge on a cantilever beam: (a) experimental setup; (b) close-up view of the top surface showing 28 μm and 110 μm PVDF-PWAS transducers; (c) close-up view of the bottom surface showing the 200 μm PZT-PWAS and the strain gauge.

had the same in-plane geometry, viz. 7-mm × 7-mm squares. The piezoelectric properties of PZT and PVDF are given in Table 11.11. The PVDF-PWAS were cut out of piezo film sheets from Measurement Specialties, Sensor Products Div. A conventional resistance strain gauge was also used. All four transducers were placed at the same span position close to the beam root, two on top and two at the bottom (Figure 11.39b,c). A 4-channel Tektronix TDS5030B oscilloscope was used to collect the signals from the transducers. The strain gauge was connected to Channel 1 through a Vishay P3 strain indicator; the three PWAS were connected directly to the oscilloscope Channels 2 through 4.

During the experiments, the beam was displaced approximately 10 mm and suddenly released into free vibration. Figure 11.40 shows oscilloscope traces recorded during the experiment. Figure 11.41 shows the Fourier transform of these traces indicating the natural frequencies of the beam $f_1^{\text{exp}} = 29.7$ Hz, $f_2^{\text{exp}} = 181$ Hz, $f_3^{\text{exp}} = 501$ Hz. Examination of Figures 11.40 and 11.41 indicate that the PZT-PWAS was found to give the largest voltage but it was less responsive to the higher frequencies. The PVDF-PWAS was found to be more responsive to the high frequencies but it was found to give a lower voltage (Figure 11.41). The resistive strain gauge was only responsive to low frequency. Detailed data of the frequencies and peak amplitudes measured by the transducers are given in Table 11.12.

Using the calibration constants of the Vishay P3 strain indicator (0–2.5 V for strain excursion from −320 to +320 με) and the peak-to-peak 1.32 V value (CH 1 trace in Figure 11.41), we concluded that the peak-to-peak vibration strain experienced during the experiment was equal to 338 με. With this data, we used Equation (11.329) with an assumed 3 nF input capacitance for the oscilloscope to predict the peak-to-peak signal

TABLE 11.11

Properties of PZT and PVDF Compared with Those of $BaTiO_3$

Property	Units	PZT	$BaTiO_3$	PVDF
Density	10^3 kg/m^3	7.5	5.7	1.78
Relative permittivity	$\varepsilon/\varepsilon_0$	1200	1700	12
d_{31}	10^{-12} C/N	110	78	23
g_{31}	10^{-3} Vm/N	10	5	216
k_{31}	% at 1 kHz	30	21	12
Acoustic impedance	10^6 kg/m^{2}s	30	30	27

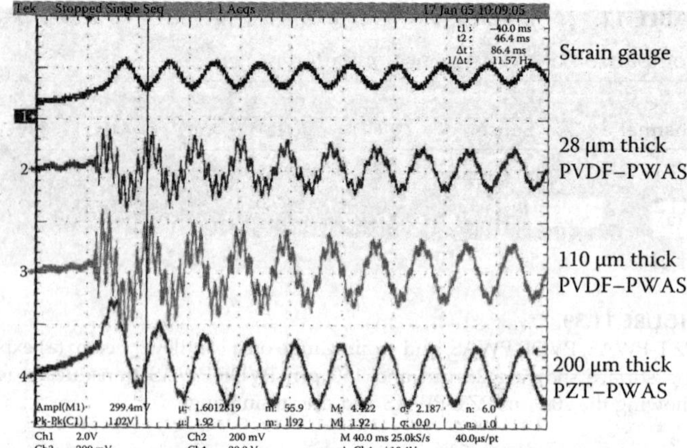

FIGURE 11.40
Vibration signal recorded by strain gauge, PVDF–PWAS and PZT–PWAS.

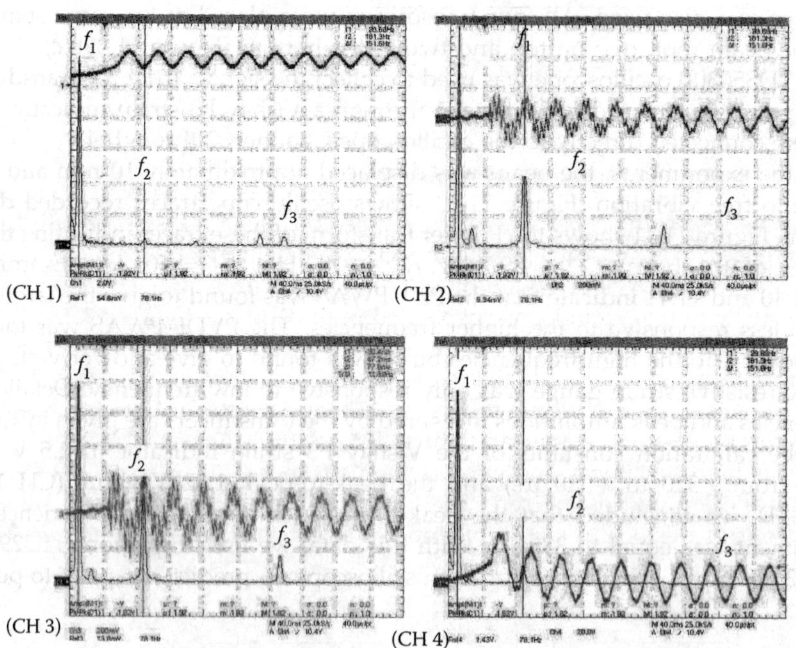

FIGURE 11.41
Vibration signal and spectrum of magnitude (Fourier transform) recorded by (CH 1) strain gauge; (CH 2) 28 μm thick PVDF –PWAS; (CH 3) 110 mm thick PVDF–PWAS; (CH 4) 200 μm thick PZT–PWAS.

that would be generated by the PWAS transducers. These predictions are shown in Table 11.13 in comparison with the experimental values. Good agreement is observed for the PZT–PWAS and the thin (28 mm) PVDF–PWAS; the prediction for the thicker (100 mm) PVDF–PWAS was not so good.

TABLE 11.12

Comparison of the Response of Different Sensors

Channel	Sensor	V_{PP} (V)	f_1 (Hz)	A_1 (mV)	f_2 (Hz)	A_2 (mV)	A_2/A_1 (%)	f_3 (Hz)	A_3 (mV)	A_3/A_1 (%)
CH 1	Strain gauge	1.32	29.69	327.4	182.8	12	3.66	509.4	20	6.1
CH 2	28 μm PVDF–PWAS	0.332	29.69	55.64	181.3	28.75	51.67	501.6	6.16	11.07
CH 3	110 mm PVDF–PWAS	0.508	29.69	82.5	181.3	45.65	55.33	501.6	11	13.33
CH 4	200 μm PZT–PWAS	30.8	29.69	8568	181.3	800	9.337	502.5	100	0.44

TABLE 11.13

Comparison of PWAS Response

	PZT–PWAS (V)	110 mm PVDF–PWAS (V)	28 μm PVDF–PWAS (V)
Theoretical	30	0.361	0.346
Experimental	30.8	0.508	0.332
Relative error (%)	2.67	40.7	−4.04

11.4.3 Waves Sensing Experiments on a Long Rod

PWAS strain wave sensing capabilities were studied on a long rod specimen in which axial waves were generated through an axial impact. A schematic of the experimental setup is shown in Figure 11.42. A 6061-T6 aluminum alloy rod with length $L = 2.5$ m (98 in.) and diameter $D = 6.35$ mm (0.25 in.) was suspended in three places by a monofilament line to ensure free–free conditions. The rod was impacted at one end with a 16 mm diameter steel ball. The impactor was centered on the rod to produce axial waves that bounced back and forth from the rod ends in multiple reflections. The rod was instrumented with one BLH semiconductor type strain gauge and three PWAS transducers. The BLH gauge was selected for its high sensitivity and good dynamic response; it was mounted at 58.4 mm (23 in.) from the left end of the rod. The PWAS transducers were 7-mm square PVDF patches cut out of piezo film sheets of three different thicknesses (28, 52, and 110 μm)

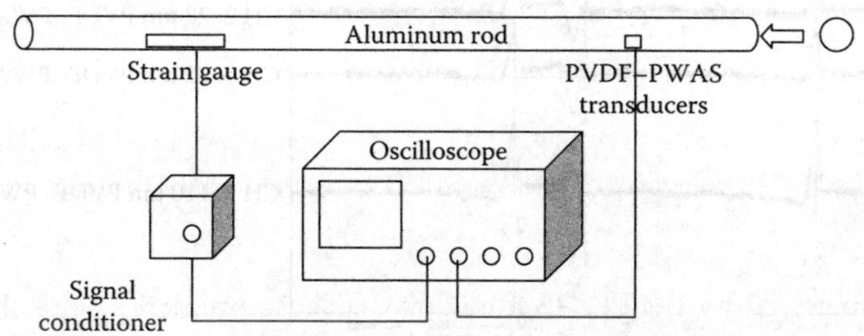

FIGURE 11.42
Experimental setup for rod impact wave propagation experiment.

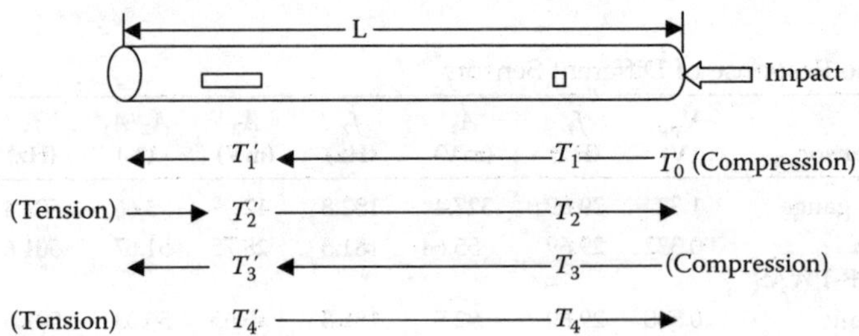

FIGURE 11.43
Stress wave propagation to and fro in the rod.

obtained from Measurement Specialties, Sensor Products Div.; these PVDF–PWAS were mounted at 50.8 mm (20 in.) from the right end of the rod. The BLH strain gauge was connected to Channel 1 (CH 1) of the oscilloscope through a signal conditioner; the PVDF–PWAS transducers were connected directly to CH 2, CH 3, CH 4.

Figure 11.43 shows the strain wave propagation events inside the rod. When the impactor hits the rod at time T_0, a compression wave is created traveling from right to left; it arrives at the PVDF–PWAS at time T_1 and then arrives at the BLH strain gauge at time T_1'. When a compression wave reaches the free end of the rod, it undergoes stress reversal and is reflected as a tension wave. Hence, the compression wave traveling from right to left will be reflected as a tension wave at the left free end and start traveling toward the right end of the rod. The tension wave will reach the BLH strain gauge at time T_2' and then the PVDF–PWAS at time T_2. This tension wave will undergo stress reversal at the right free end and be reflected as a compression wave traveling toward the left end. The distance traveled by the two adjacent positive or negative pulses is $2L$, where L is the length of the rod.

Figure 11.44 shows the signals recorded by the BLH strain gauge and the PVDF–PWAS transducers. The time base of 200 µs per division used in this capture allowed the

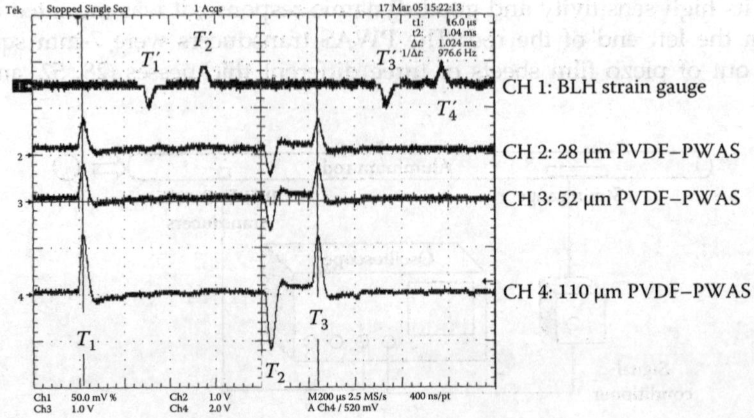

FIGURE 11.44
Impact responses for free-free boundary condition recorded by (CH 1) strain gauge; (CH 2) 28 µm thick PVDF–PWAS; (CH 3) 52 µm thick PVDF–PWAS; (CH 4) 110 µm thick PVDF–PWAS.

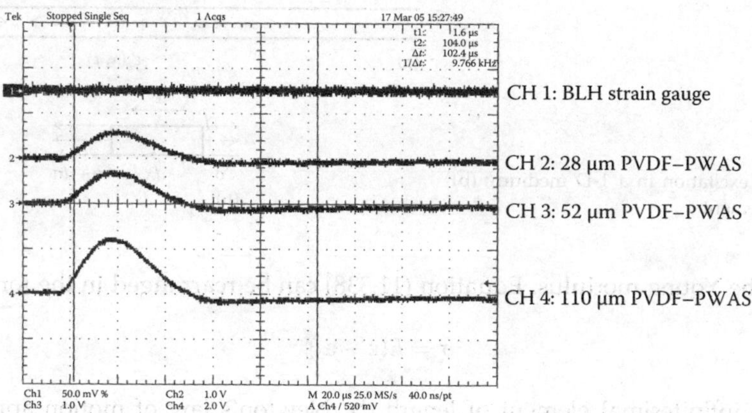

FIGURE 11.45
Impact responses recorded by (CH 1) strain gauge; (CH 2) 28 μm thick PVDF–PWAS; (CH 3) 52 μm thick PVDF–PWAS; (CH 4) 110 μm thick PVDF–PWAS.

recording of several transit cycles of the wave pulse. Three features are apparent: (1) compression peaks (T_1, T'_1, T_3, T'_3), (2) tension peaks (T_2, T'_2, T_4, T'_4), and (3) a dwell period while the wave traverses the length of the rod in which no signal is present. Note that the compression wave produces a positive peak on the PWAS transducer (e.g., T_1) but a negative peak on the BLH strain gauge (e.g., T'_1). The negative peak in the BLH gauge is correct because the compression wave has a negative strain and stress. The positive peak in the PWAS transducer is also correct because the piezoelectric coefficient d_{31} used in Equation (11.326) is negative. The speed of the pulse moving up and down the rod can be estimated from the time between peaks. The experimental wave speed is $c_{\exp} = 4860$ m/s. For 6061 rod material, the elastic modulus is $E = 69$ GPa and density is $\rho = 2700$ kg/m^3, thus giving a theoretical velocity $c = \sqrt{E/\rho} = 5055$ m/s. The relative error of the experimental results is a very reasonable 3.8%.

Figure 11.45 shows the signal traces captures with 20 μs per division such as to present in more detail the initial pulse; one observes that the PWAS which have a smaller internal admittance due to the greater thickness gives a higher signal (CH 4 is at 2 V/div whereas CH 2 is at 1 V/div and has smaller amplitude).

11.4.4 Structural Waves Excited by PWAS

11.4.4.1 Axial Waves Excited by PWAS

11.4.4.1.1 General Solution

Assume a 1-D medium in which an external cause induces an actuation strain, $\varepsilon_0(x, t)$, as shown in Figure 11.46. Such an actuation strain may be induced by surface mounted PWAS, applied symmetrically to the top and bottom surfaces. Hence, the total strain is given by

$$\varepsilon(x, t) = \frac{\sigma(x, t)}{E} + \varepsilon_e(x, t) \tag{11.338}$$

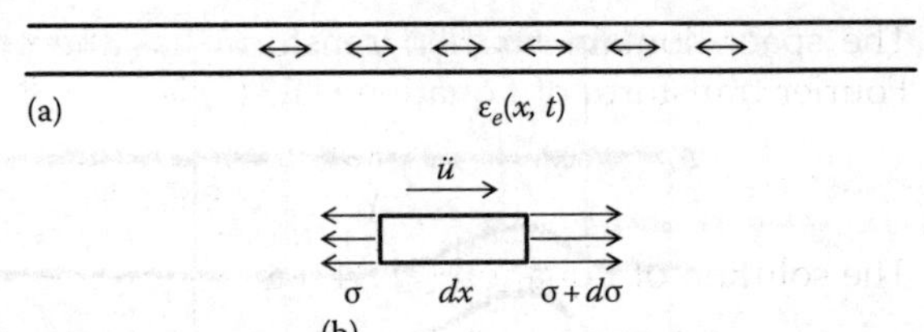

FIGURE 11.46
(a) axial strain excitation in a 1-D medium; (b) infinitesimal element.

where E is the Young modulus. Equation (11.338) can be rearranged in the form

$$\sigma = E(\varepsilon - \varepsilon_e) \tag{11.339}$$

Consider an infinitesimal element of length dx. Newton's law of motion applied to the infinitesimal element yields

$$A\,d\sigma = \rho A \ddot{u}\,dx \tag{11.340}$$

where $\dot{u} = \partial u / \partial t$. Hence,

$$\sigma' = \rho \ddot{u} \tag{11.341}$$

where $u' = \partial u / \partial x$. Substitution of Equation (11.339) into Equation (11.341) yields

$$E(\varepsilon - \varepsilon_e) = \rho \ddot{u} \tag{11.342}$$

Recall $\varepsilon = u'$. Substitution into Equation (11.342) and differentiation with respect to x yields

$$\varepsilon'' - \frac{1}{c^2}\ddot{\varepsilon} = \varepsilon_e'' \tag{11.343}$$

where $c^2 = E/\rho$. For harmonic excitation and subsequent motion, we have

$$\varepsilon_e(x, t) = \varepsilon_e(x)e^{-i\omega t}$$
$$\varepsilon(x, t) = \varepsilon(x)e^{-i\omega t} \tag{11.344}$$

Substitution of Equation (11.344) into Equation (11.343) gives

$$\varepsilon'' - \xi_0^2 \varepsilon = \varepsilon_e'' \tag{11.345}$$

where $\xi_0^2 = \omega^2 / c^2$ is the wave number of axial waves in the 1-D medium. Take the space-domain Fourier transform

$$f(x) = \int_{-\infty}^{+\infty} \tilde{f}(\xi)e^{i\xi x}\,d\xi$$

$$\tilde{f}(\xi) = \frac{1}{2\pi} \int_{-\infty}^{+\infty} f(x)e^{-i\xi x}\,dx \tag{11.346}$$

The space-domain Fourier transform has the property that $\tilde{f}' = i\xi\tilde{f}$. The space-domain Fourier transform of Equation (11.345) is

$$-\xi^2\tilde{\varepsilon} - \xi_0^2\tilde{\varepsilon} = -\xi^2\tilde{\varepsilon}_e \tag{11.347}$$

The solution of Equation (11.347) is

$$\tilde{\varepsilon} = \frac{\xi^2}{\xi^2 - \xi_0^2}\tilde{\varepsilon}_e \tag{11.348}$$

Equation (11.348) represents the solution in the Fourier domain. Taking the inverse space-domain Fourier transform yields the solution in the space domain

$$\varepsilon(x) = \frac{1}{2\pi}\int\limits_{-\infty}^{+\infty}\frac{\xi^2}{\xi^2 - \xi_0^2}\tilde{\varepsilon}_e(\xi)e^{i\xi x}d\xi \tag{11.349}$$

Of course, the exact expression of the solution $\varepsilon(x)$ depends on the particular form of the excitation, $\varepsilon_a(x)$, and its space-domain Fourier transform.

11.4.4.1.2 Solution for an Ideally Bonded PWAS

For an ideally bonded PWAS, the induced-strain ε_a is uniform over the PWAS length. Assume the PWAS is positioned symmetrically about the origin, i.e., it stretches from $x = -a$ to $x = +a$ (Figure 11.47a). The mathematical expression of this strain distribution is

$$\varepsilon_e(x, t) = \begin{cases} \varepsilon_a e^{-i\omega t}, & |x| < a \\ 0, & \text{otherwise} \end{cases} \tag{11.350}$$

The space-domain portion of Equation (11.350) is the rectangular pulse function (Figure 11.47b)

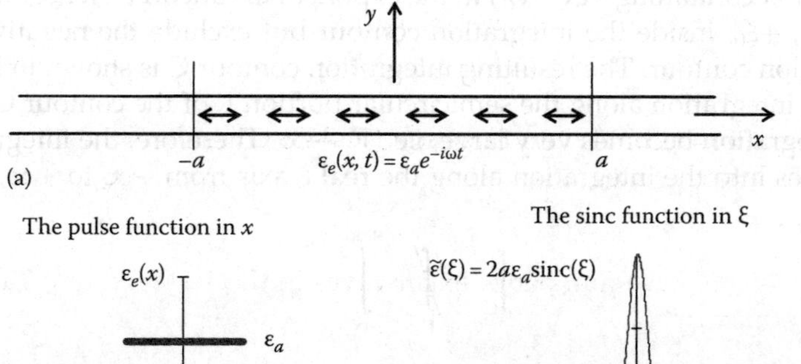

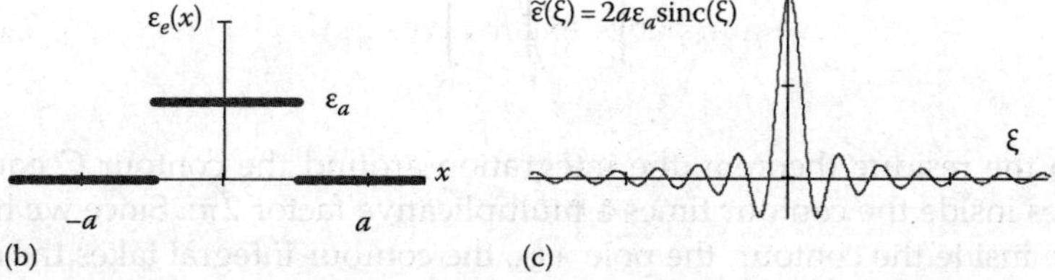

FIGURE 11.47

Axial strain excitation by ideally bonded PWAS: (a) infinitesimal element; (b) the pulse function in x; (c) Fourier transform, the sinc function in ξ.

$$\varepsilon_e(x) = \begin{cases} \varepsilon_a, & |x| < a \\ 0, & \text{otherwise} \end{cases} \qquad (11.351)$$

The space-domain Fourier transform of Equation (11.351) is the sinc function (Figure 11.47c).

$$\tilde{\varepsilon}_e = \varepsilon_a \frac{2}{\xi} \sin(\xi a) = 2a\varepsilon_a \frac{\sin(\xi a)}{\xi a} = 2a\varepsilon_a \text{sinc}(\xi a) \qquad (11.352)$$

To obtain the solution, recall Equation (11.345)

$$\varepsilon'' - \xi_0^2 \varepsilon = \varepsilon_a'' \qquad (11.353)$$

Upon taking the space-domain Fourier transform, we obtain

$$-\xi^2 \tilde{\varepsilon} - \xi_0^2 \tilde{\varepsilon} = -\xi^2 \varepsilon_a \frac{2}{\xi} \sin \xi a \qquad (11.354)$$

Hence, the Fourier-domain solution is

$$\tilde{\varepsilon} = \varepsilon_a \frac{2\xi}{\xi^2 - \xi_0^2} \sin \xi a \qquad (11.355)$$

Taking the inverse Fourier transform, we obtain the space-domain solution

$$\varepsilon(x) = \frac{2\varepsilon_a}{2\pi} \int_{-\infty}^{+\infty} \frac{\xi \sin \xi a}{\xi^2 - \xi_0^2} e^{i\xi x} d\xi \qquad (11.356)$$

The integral in Equation (11.356) can be resolved analytically using the residues theorem and a semicircular contour C in the complex ξ domain. We note that the integrant in Equation (11.356) has two poles, corresponding to the wave numbers $-\xi_0$ and $+\xi_0$. We will resolve Equation (11.356) for the forward traveling wave which exists for $x > 0$ and generates a solution containing $i(\xi x - \omega t)$ in the exponential function. Hence, we will retain the positive pole, $+\xi_0$, inside the integration contour but exclude the negative pole, $-\xi_0$, from the integration contour. The resulting integration contour C is shown in Figure 11.48. We note that the integration along the semicircular portion Γ of the contour C vanishes as the radius of integration becomes very large, i.e., $R \to \infty$. Therefore, the integration on the contour C resolves into the integration along the real ξ axis from $-\infty$ to $+\infty$, i.e.,

$$\oint_C = \int_\Gamma + \int_{-\infty}^{+\infty} \qquad (11.357)$$

According to the residue theorem, the integration around the contour C equals the sum of the residues inside the contour times a multiplicative factor $2\pi i$. Since we have retained only one pole inside the contour, the pole $+\xi_0$, the contour integral takes the expression

$$\oint_C = 2\pi i \text{Res}\big|_{\xi=\xi_0} \qquad (11.358)$$

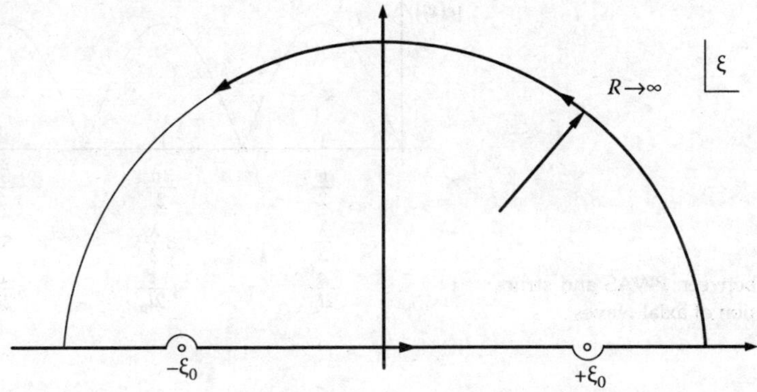

FIGURE 11.48
Contour for evaluating the strain solution under ideally bonded PWAS excitation. The residue at positive wave number is included while the residue for the negative wave number is excluded.

The residue is calculated as

$$\mathrm{Res}\left(\frac{\xi \sin \xi a}{\xi^2 - \xi_0^2} e^{i\xi x}\right) = \left(\frac{\xi \sin \xi a}{\xi + \xi_0} e^{i\xi x}\right)_{\xi = \xi_0} = \frac{\xi_0 \sin \xi_0 a}{\xi_0 + \xi_0} e^{i\xi_0 x} = \frac{1}{2}(\sin \xi_0 a)e^{i\xi_0 x} \tag{11.359}$$

Substitution of Equations (11.358) and (11.359) into Equation (11.356) gives

$$\varepsilon(x) = \frac{2\varepsilon_a}{2\pi} 2\pi i \frac{1}{2}(\sin \xi_0 a)e^{i\xi_0 x} \tag{11.360}$$

Equation (11.360) yields

$$\varepsilon(x) = i\varepsilon_a \sin \xi_0 a e^{i\xi_0 x} \tag{11.361}$$

Equation (11.361) is the space-domain solution of the problem. Substitution of Equation (11.361) into Equation (11.344) yields the complete solution which is the strain response in the structure due to a harmonically oscillating PWAS perfectly bonded to the structural surface. This solution has the form

$$\varepsilon(x) = i\varepsilon_a(\sin \xi_0 a)e^{i(\xi_0 x - \omega t)} \tag{11.362}$$

It is apparent that the response amplitude follows a sinusoidal variation with respect to the parameter $\xi_0 a$. A plot of this variation is presented in Figure 11.49. Response peaks are observed at odd integer multiples of $\pi/2$, i.e., when

$$\xi_0 a = (2n - 1)\frac{\pi}{2}, \quad n = 1, 2, 3, \ldots \tag{11.363}$$

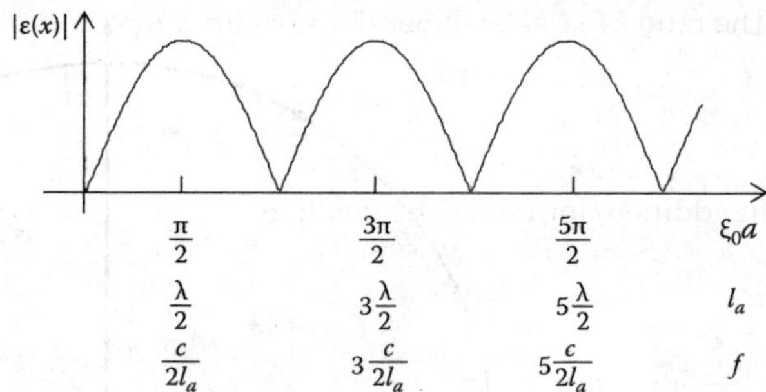

FIGURE 11.49
Optimum match between PWAS and structure for the excitation of axial waves.

Recalling that $\xi_0 = 2\pi/\lambda$, and the PWAS length is $l_a = 2a$, relation (11.363) implies that maximum excitation will happen when the PWAS length is an odd integer multiple of the half wavelength, i.e.,

$$l_a = (2n - 1)\frac{\lambda}{2}, \quad n = 1, 2, 3, \ldots \tag{11.364}$$

Thus, for a given wavelength, Equation (11.364) provides a method to construct optimum PWAS geometries that will optimally excite the structure with the prescribed wavelength. Of course, the first such geometry will be that in which the PWAS characteristic length is exactly half the wavelength. Higher matches are also possible at odd integer multiples of the half wavelength.

Since the wavelength depends on frequency, $\lambda = c/f$, it is also conceivable that, for a given PWAS, certain frequencies exist at which the excitation is optimal. These optimal excitation frequencies are given by

$$f_n = (2n - 1)\frac{c}{2l_a}, \quad n = 1, 2, 3, \ldots \tag{11.365}$$

11.4.4.2 Flexural Waves Excited by PWAS

Consider the general equation of flexural vibrations under external moment excitation, $M_e(x, t)$,

$$EIv'''' + \rho A\ddot{v} = M_e''(x, t) \tag{11.366}$$

where $\dot{v} = \partial v/\partial t$ and $v' = \partial v/\partial x$. Assume harmonic variation in the time domain of the form $e^{-i\omega t}$. Hence, Equation (11.366) becomes

$$EIv'''' - \omega^2 \rho Av = M_e''(x) \tag{11.367}$$

Divide by EI and obtain

$$v'''' - \omega^2 \frac{\rho A}{EI} v = \frac{M_e''}{EI} \tag{11.368}$$

The ratio M_e/EI has dimensions of curvature. We will call it the *excitation curvature*, κ_e, i.e.,

$$\kappa_e = \frac{M_e}{EI} \tag{11.369}$$

In addition, introduce the notation

$$\xi_F^4 = \omega^2 \frac{\rho A}{EI} = \frac{\omega^2}{\bar{a}^2}, \quad \text{where} \quad \bar{a}^2 = \frac{EI}{\rho A} = \frac{E\dfrac{bd^3}{3}}{\rho(bd)} = \frac{d^2}{3}\frac{E}{\rho} \tag{11.370}$$

Hence, Equation (11.368) becomes

$$v'''' - \xi_F^4 v = \kappa_e'' \tag{11.371}$$

Take the space-domain Fourier transform

$$f(x) = \int_{-\infty}^{+\infty} \tilde{f}(\xi)e^{i\xi x}d\xi$$

$$\tilde{f}(\xi) = \frac{1}{2\pi} \int_{-\infty}^{+\infty} f(x)e^{-i\xi x}dx \tag{11.372}$$

The space-domain Fourier transform has the property that $\tilde{f}' = i\xi\tilde{f}$. The space-domain Fourier transform of Equation (11.371) is

$$(\xi^4 - \xi_F^4)\tilde{v} = -\xi^2\tilde{\kappa}_e \tag{11.373}$$

The solution of Equation (11.373) is

$$\tilde{v} = \frac{-\xi^2}{\xi^4 - \xi_F^4}\tilde{\kappa}_e \tag{11.374}$$

Equation (11.374) represents the solution to the problem in the Fourier domain. Taking the inverse space-domain Fourier transform yields the solution in the space domain

$$v(x) = \frac{1}{2\pi} \int_{-\infty}^{+\infty} \frac{-\xi^2}{\xi^4 - \xi_F^4}\tilde{\kappa}_e e^{i\xi x}d\xi \tag{11.375}$$

The integral in Equation (11.375) can be resolved using the residues theorem and a semi-circular contour C in the complex ξ domain. We note that the integrant in Equation (11.375) has four poles, corresponding to the wave numbers $+\xi_F$, $-\xi_F$, $+i\xi_F$, and $-i\xi_F$. We will resolve Equation (11.375) for the forward traveling wave which exists for $x > 0$ and generates a solution containing $i(\xi_F x - \omega t)$ in the exponential function. Hence, we will retain the positive pole, $+\xi_F$, inside the integration contour but exclude the negative pole, $-\xi_F$, from the integration contour. The resulting integration contour C is shown in Figure 11.50.

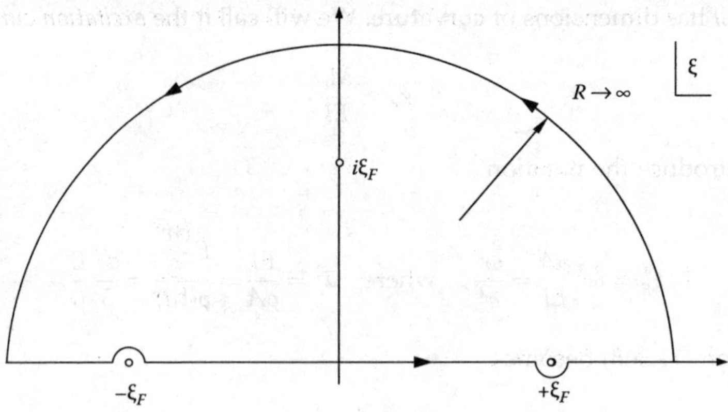

FIGURE 11.50
Contour for evaluating the flexural wave solution. The residue at positive wave number, $+\xi_F$, is included while the residue for the negative wave number, $-\xi_F$, is excluded. Also included is the residue at the imaginary wave number, $i\xi_F$.

This integration contour includes the real pole $+\xi_F$ and the imaginary pole $+i\xi_F$. We note that the integration along the semicircular portion Γ of the contour C vanishes as the radius of integration becomes very large, i.e., $R \to \infty$. Therefore, the integration on the contour C resolves into the integration along the real ξ axis from $-\infty$ to $+\infty$, i.e.,

$$\oint_C = \int_\Gamma + \int_{-\infty}^{+\infty} \tag{11.376}$$

According to the residue theorem, the integration around the contour C equals the sum of the residues inside the contour times a multiplicative factor $2\pi i$. Since we have retained only two poles inside the contour, $+\xi_0$ and $+i\xi_0$, the contour integral takes the expression

$$\oint_C = 2\pi i \left(\mathrm{Res}\big|_{\xi=+\xi_0} + \mathrm{Res}\big|_{\xi=+i\xi_0} \right) \tag{11.377}$$

The residue at $+\xi_F$ is calculated as

$$\mathrm{Res}\left(\frac{\xi^2 \tilde{\kappa}_e(\xi)}{\xi^4 - \xi_F^4} e^{i\xi x} \right)_{\xi=+\xi_F} = \left(\frac{\xi^2 \tilde{\kappa}_e(\xi)}{4\xi^3} e^{i\xi x} \right)_{\xi=+\xi_F} = \frac{\tilde{\kappa}_e(\xi_F)}{4\xi_F} e^{i\xi_F x} \tag{11.378}$$

The residue at $+i\xi_F$ is calculated as

$$\mathrm{Res}\left(\frac{\xi^2 \tilde{\kappa}_e(\xi)}{\xi^4 - \xi_F^4} e^{i\xi x} \right)_{\xi=+i\xi_F} = \left(\frac{\xi^2 \tilde{\kappa}_e(\xi)}{4\xi^3} e^{i\xi x} \right)_{\xi=+i\xi_F} = \frac{\tilde{\kappa}_e(i\xi_F)}{4i\xi_F} e^{-\xi_F x} \tag{11.379}$$

Substitution of Equations (11.376) and (11.379) into Equation (11.8) gives

$$v(x) = \frac{1}{2\pi} 2\pi i \left[\frac{\tilde{\kappa}_e(\xi_F)}{4\xi_F} e^{i\xi_F x} + \frac{\tilde{\kappa}_e(i\xi_F)}{4i\xi_F} e^{-\xi_F x} \right] \tag{11.380}$$

Adding the harmonic variation in the time domain of the form $e^{-i\omega t}$ yields the complete solution

$$v(x) = i\frac{\tilde{\kappa}_e(\xi_F)}{4\xi_F} e^{i(\xi_F x - \omega t)} + \frac{\tilde{\kappa}_e(i\xi_F)}{4\xi_F} e^{-\xi_F x} e^{-i\omega t} \tag{11.381}$$

Note that the first term in Equation (11.381) represents a propagating wave while the second term does not. In fact, the second term represents a vibration that is decaying fast with x. This term represents a local vibration that does not propagate. It is called an *evanescent wave*. Thus, we will retain only the propagating wave part of Equation (11.381), i.e.,

$$v(x, t) = i\frac{\tilde{\kappa}_e(\xi_F)}{4\xi_F} e^{i(\xi_F x - \omega t)} \tag{11.382}$$

The imaginary number i that multiplies Equation (11.382) indicates that the displacement $v(x, t)$ is in quadrature with the strain excitation, $\varepsilon_e(x, t)$.

The strain solution, ε_x, can be derived from the y-displacement solution, $v(x, t)$, as follows. First, derive the x-displacement

$$u_x = -yu_y' = -yi(i\xi_F)\frac{\tilde{\varepsilon}_e(\xi_F)}{4\xi_F} e^{i(\xi_F x - \omega t)} = y\frac{\tilde{\varepsilon}_e(\xi_F)}{4} e^{i(\xi_F x - \omega t)} \tag{11.383}$$

Upon space differentiation of Equation (11.383), obtain the strain ε_x as

$$\varepsilon_x = u_x' = y(i\xi_F)\frac{\tilde{\varepsilon}_e(\xi_F)}{4} e^{i(\xi_F x - \omega t)} \tag{11.384}$$

11.4.4.2.1 Solution for Ideally Bonded PWAS

For an ideally bonded PWAS, the excitation moment is represented by the rectangular pulse function

$$M_e(x) = \begin{cases} M_a, & |x| < a \\ 0, & \text{otherwise} \end{cases} \tag{11.385}$$

The space-domain Fourier transform of Equation (11.385) is the sinc function (Figure 11.51).

$$\tilde{M}_e(\xi) = M_a \frac{2}{\xi} \sin(\xi a) = 2aM_a \frac{\sin(\xi a)}{\xi a} = 2aM_a \text{sinc}(\xi a) \tag{11.386}$$

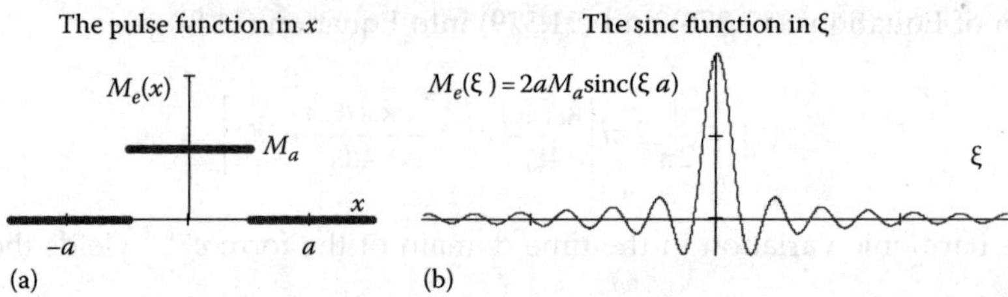

FIGURE 11.51
Moment excitation by ideally bonded PWAS: (a) the pulse function in x; (b) its Fourier transform, the sinc function in ξ.

Recall that $\kappa_e = \dfrac{M_e}{EI}$ and denote

$$\kappa_a = \frac{M_a}{EI} \tag{11.387}$$

Hence,

$$\kappa_e(x) = \begin{cases} \kappa_a, & |x| < a \\ 0, & \text{otherwise} \end{cases} \tag{11.388}$$

and

$$\tilde{\kappa}_e(\xi) = \kappa_a \frac{2}{\xi} \sin(\xi a) \tag{11.389}$$

Thus, Equation (11.384) becomes

$$\varepsilon_x = y(i\xi_F)\frac{\kappa_a \dfrac{2}{\xi_F}\sin\xi_F a}{4}e^{i(\xi_F x - \omega t)} = iy\frac{\kappa_a \sin\xi_F a}{2}e^{i(\xi_F x - \omega t)} \tag{11.390}$$

The strain at the material surface can be obtained from Equation (11.390) by taking $y = d$, i.e.,

$$\varepsilon_x\big|_{y=d} = id\frac{\kappa_a \sin\xi_F a}{2}e^{i(\xi_F x - \omega t)} \tag{11.391}$$

Recall the expression of actuation moment, M_a, in terms of the PWAS pin force, F_a, i.e.,

$$M_a = F_a d \tag{11.392}$$

Substitution of Equation (11.392) into Equation (11.387) yields

$$\kappa_a = \frac{F_a d}{EI}$$ (11.393)

Recall that

$$I = \frac{bd^3}{3} \quad \text{and} \quad A = b(2d)$$ (11.394)

Hence,

$$\kappa_a = \frac{F_a d}{E\dfrac{bd^3}{3}} = 3\frac{2}{d}\frac{F_a}{EA} = 3\frac{2}{d}\varepsilon_a$$ (11.395)

where

$$\varepsilon_a = \frac{F_a}{EA}$$ (11.396)

Substitution of Equation (11.395) into Equation (11.391) yields the following expression for the strain at the material surface of a thin plate undergoing flexural wave excitation under ideally bonded surface PWAS

$$\varepsilon_x\Big|_{y=d} = id\frac{1}{2}3\frac{2}{d}\varepsilon_a(\sin\xi_F a)e^{i(\xi_F x-\omega t)} = i3\varepsilon_a(\sin\xi_F a)e^{i(\xi_F x-\omega t)}$$ (11.397)

It is apparent that the response amplitude follows a sinusoidal variation with respect to the parameter $\xi_F a$. A plot of this variation is represented in Figure 11.52.

Response peaks are observed at odd integer multiples of $\pi/2$, i.e., when

$$\xi_F a = (2n-1)\frac{\pi}{2}, \quad n = 1,2,3,\ldots$$ (11.398)

Recalling that $\xi_F = 2\pi/\lambda_F$ and the PWAS length is $l_a = 2a$, relation (11.397) implies that maximum excitation of flexural waves will happen when the PWAS length is an odd integer multiple of the flexural half wavelength, i.e.,

$$l_a = (2n-1)\frac{\lambda_F}{2}, \quad n = 1,2,3,\ldots$$ (11.399)

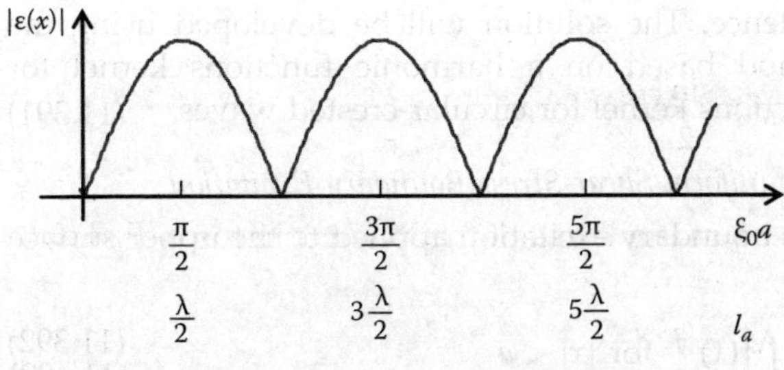

FIGURE 11.52
Optimum match between PWAS and structure for the excitation of flexural waves.

Thus, for a given wavelength, Equation (11.399) provides a method to construct PWAS geometries that will optimally excite the structure with flexural waves of prescribed wavelength. Of course, the first such geometry will be that in which the PWAS characteristic length is exactly half the flexural wavelength. Higher matches are also possible. These will occur at odd integer multiples of the half wavelength.

Since the wavelength depends on frequency and wave speed, $\lambda_F = c_F/f$, it is also conceivable that, for a given PWAS, certain frequencies exist at which the excitation is optimal. These optimal excitation frequencies are given by

$$f_n = (2n - 1)\frac{c_F(f)}{2l_a}, \quad n = 1, 2, 3, \ldots \tag{11.400}$$

Equation (11.400) indicates that the wave speed, c_F, is also a function of frequency. This is to be expected since flexural waves are dispersive. Recall that

$$c_F = \sqrt{\bar{a}\omega} = \sqrt{2\pi\bar{a}f} = \sqrt{\left(\sqrt{\frac{E}{3\rho}}\right)f} = \left(\frac{E}{3\rho}\right)^{\frac{1}{4}}(2\pi d)^{\frac{1}{2}}f^{\frac{1}{2}} \tag{11.401}$$

Hence, Equation (11.401) can be substituted into Equation (11.400) to obtain a solution of the optimal excitation frequencies in terms of the structural elastic, mass, and geometric properties, i.e.,

$$f_n = (2n - 1)^2 \frac{\pi}{2l_a^2}\left[\frac{Ed^2}{3p}\right]^{\frac{1}{2}}, \quad n = 1, 2, 3, \ldots \tag{11.401a}$$

11.4.4.3 Lamb Waves Excited by PWAS

In this section, we develop the theoretical foundation for selective Lamb-wave mode excitation with PWAS transducers. Figure 11.53 presents the coupling between PWAS and two Lamb-wave modes, S_0 and A_0. It is apparent from Figure 11.53 that maximum coupling between the PWAS and the Lamb-wave would occur when the PWAS length is an odd multiple of the half wavelength. Since different Lamb-wave modes have different wavelengths, which vary with frequency, the opportunity arises for selectively exciting various Lamb-wave modes at various frequencies, i.e., making the PWAS tuned to one or another Lamb-wave mode.

The analysis presented in this section considers surface shear distribution resulting from PWAS action and determines the Lamb-wave response of the structure. In this analysis, the choice of shear distribution is very important. As indicated by Equation (11.267) of Section 11.4.1.3, the shear stress distribution is nonuniform, with high values at the PWAS ends. Hence, we will have to assume the shear distribution of the shear lag model corrected for frequency dependence. The solution will be developed using the space-domain Fourier transform method based on a harmonic functions kernel for straight-crested waves and a Bessel functions kernel for circular-crested waves.

11.4.4.3.1 Lamb Wave Solution under Nonuniform Shear-Stress Boundary Excitation

Assume a generic harmonic shear-stress boundary excitation applied to the upper surface of the plate (Figure 11.54), i.e.,

$$\tau_a(x, h) = \begin{cases} \tau(x) & \text{for } |x| < a \\ 0 & \text{otherwise} \end{cases} \tag{11.402}$$

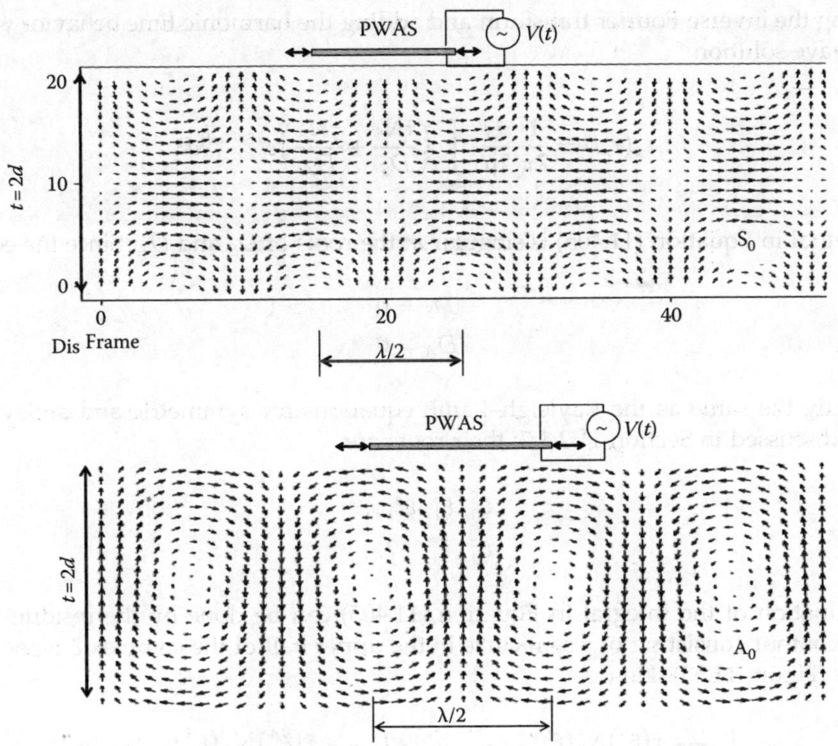

FIGURE 11.53
Typical structure of S_0 and A_0 Lamb-wave modes and the interaction of PWAS with the Lamb wave.

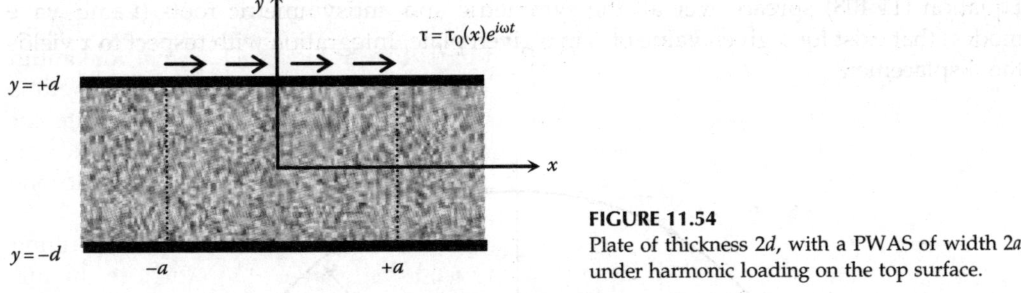

FIGURE 11.54
Plate of thickness $2d$, with a PWAS of width $2a$, under harmonic loading on the top surface.

Upon application of space-domain Fourier transform, one obtains the Fourier-domain solution (Giurgiutiu, 2008)

$$\tilde{\varepsilon}_x = -i\frac{\tilde{\tau}}{2\mu}\left(\frac{N_S}{D_S} + \frac{N_A}{D_A}\right) \tag{11.403}$$

where $\tilde{\tau}$ is the space-domain Fourier transform of $\tau_a(x,h)$ and

$$N_S = \xi q(\xi^2 + q^2)\cos pd \cos qd, \quad D_S = (\xi^2 - q^2)^2 \cos pd \sin qd + 4\xi^2 pq \sin pd \cos qd$$

$$N_A = \xi q(\xi^2 + q^2)\sin pd \sin qd, \quad D_A = (\xi^2 - q^2)^2 \sin pd \cos qd + 4\xi^2 pq \cos pd \sin qd$$

$$\tag{11.404}$$

Applying the inverse Fourier transform and adding the harmonic time behavior yields the strain wave solution

$$\varepsilon_x(x,t) = \frac{1}{2\pi}\frac{-i}{2\mu}\int_{-\infty}^{\infty}\left(\frac{\tilde{\tau}N_S}{D_S}+\frac{\tilde{\tau}N_A}{D_A}\right)e^{i(\xi x-\omega t)}d\xi \tag{11.405}$$

The integral in Equation (11.405) is singular at the roots of D_S and D_A. Since the equations

$$\begin{aligned} D_S &= 0 \\ D_A &= 0 \end{aligned} \tag{11.406}$$

are exactly the same as the Rayleigh–Lamb equations for symmetric and antisymmetric motion discussed in Section 11.2.1.7; their roots are

$$\begin{aligned} &\xi_0^S,\ \xi_1^S,\ \xi_2^S,\dots \\ &\xi_0^A,\ \xi_1^A,\ \xi_2^A,\dots \end{aligned} \tag{11.407}$$

The evaluation of the integral in Equation (11.405) can be done by the residue theorem using a contour consisting of a semicircle in the upper half of the complex ξ plane and the real axis (Figure 11.55). Hence,

$$\varepsilon_x(x,t) = \frac{1}{2\mu}\sum_{\xi^S}\frac{\tilde{\tau}(\xi^S)N_S(\xi^S)}{D_S'(\xi^S)}e^{i(\xi^S x-\omega t)}+\frac{1}{2\mu}\sum_{\xi^A}\frac{\tilde{\tau}(\xi^A)N_A(\xi^A)}{D_A'(\xi^A)}e^{i(\xi^A x-\omega t)} \tag{11.408}$$

where D_S' and D_A' are the derivatives of D_S and D_A with respect to ξ. The summations in Equation (11.408) spread over all the symmetric and antisymmetric roots (Lamb-wave modes) that exist for a given value of ω in a given plate. Integration with respect to x yields the displacement

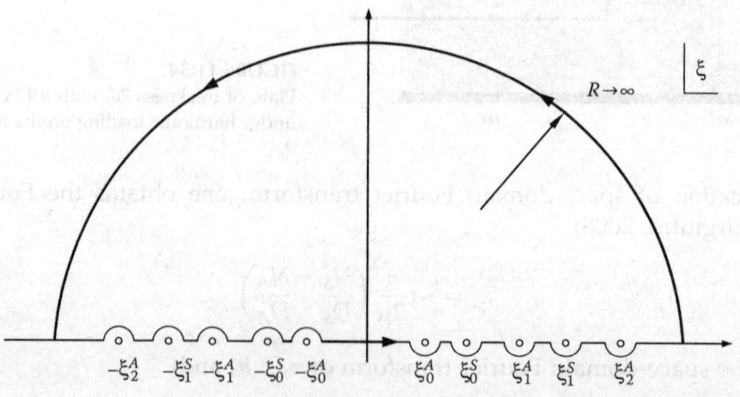

FIGURE 11.55
Contour for evaluating the surface strain under S_0 and A_0 mode. Residues at positive wave numbers are included; residues for the negative wave numbers are excluded.

$$u_x(x,t) = \frac{-i}{2\mu}\sum_{\xi^S}\frac{1}{\xi^S}\frac{\tilde{\tau}(\xi^S)N_S(\xi^S)}{D_S'(\xi^S)}e^{i(\xi^S x-\omega t)} + \frac{-i}{2\mu}\sum_{\xi^A}\frac{1}{\xi^A}\frac{\tilde{\tau}(\xi^A)N_A(\xi^A)}{D_A'(\xi^A)}e^{i(\xi^A x-\omega t)} \tag{11.409}$$

At low frequencies, only two Lamb-wave modes exist, S_0 and A_0, and the general solution has only two terms, i.e.,

$$\varepsilon_x(x,t) = \frac{1}{2\mu}\frac{\tilde{\tau}(\xi_0^S)N_S(\xi_0^S)}{D_S'(\xi_0^S)}e^{i(\xi_0^S x-\omega t)} + \frac{1}{2\mu}\frac{\tilde{\tau}(\xi_0^A)N_A(\xi_0^A)}{D_A'(\xi_0^A)}e^{i(\xi_0^A x-\omega t)} \quad \text{(Low frequency)} \tag{11.410}$$

$$u_r(x,t) = \frac{-i}{2\mu}\frac{1}{i\xi_0^S}\frac{\tilde{\tau}(\xi_0^S)N_S(\xi_0^S)}{D_S'(\xi_0^S)}e^{i(\xi_0^S x-\omega t)} + \frac{-i}{2\mu}\frac{1}{\xi_0^A}\frac{\tilde{\tau}(\xi_0^A)N_A(\xi_0^A)}{D_A'(\xi_0^A)}e^{i(\xi_0^A x-\omega t)} \quad \text{(Low frequency)}$$

$$\tag{11.411}$$

11.4.4.3.2 Ideal-Bonding Solution

For ideal bonding, the shear stress in the bonding layer is concentrated at the ends, i.e.,

$$\tau(x) = a\tau_0[\delta(x-a) - \delta(x+a)] \tag{11.412}$$

The Fourier transform of Equation (11.412) is

$$\tilde{\tau} = a\tau_0[-2i\sin\xi a] \tag{11.413}$$

Hence, Equations (11.410) and (11.411) become

$$\varepsilon_x(x,t) = -i\frac{a\tau_0}{\mu}\sum_{\xi^S}(\sin\xi^S a)\frac{N_S(\xi^S)}{D_S'(\xi^S)}e^{i(\xi^S x-\omega t)} - i\frac{a\tau_0}{\mu}\sum_{\xi^A}(\sin\xi^A a)\frac{N_A(\xi^A)}{D_A'(\xi^A)}e^{i(\xi^A x-\omega t)} \tag{11.414}$$

$$u_x(x,t) = -\frac{a^2\tau_0}{\mu}\sum_{\xi^S}\frac{\sin\xi^S a}{\xi^S a}\frac{N_S(\xi^S)}{D_S'(\xi^S)}e^{i(\xi^S x-\omega t)} - \frac{a^2\tau_0}{\mu}\sum_{\xi^A}\frac{\sin\xi^A a}{\xi^A a}\frac{N_A(\xi^A)}{D_A'(\xi^A)}e^{i(\xi^A x-\omega t)} \tag{11.415}$$

Equations (11.414) and (11.415) contain the $\sin\xi a$ function. The behavior of this function is such that it displays maxima when the PWAS length, $l_a = 2a$, equals an odd multiple of the half wavelength and minima when it equals an even multiple. A complex pattern of such maxima and minima is involved since several Lamb-wave modes, each with its own different wavelength, coexist at the same time. However, frequencies can be found when the response is dominated by certain modes that can be preferentially excited through mode tuning. An additional factor must be considered besides wavelength tuning, i.e., the mode amplitude at the top plate surface. This factor is contained in the values taken by the functions N/D'. Hence, it is conceivable that some higher modes may have little surface amplitude while others may have larger surface amplitudes at a given frequency. Thus, two important design factors have been identified:

1. The variation of $|\sin\xi a|$ with frequency for each Lamb-wave mode
2. The variation of the surface strain with frequency for each Lamb-wave mode

A plot of this solution in the 0–1000 kHz bandwidth is represented in Figure 11.56.

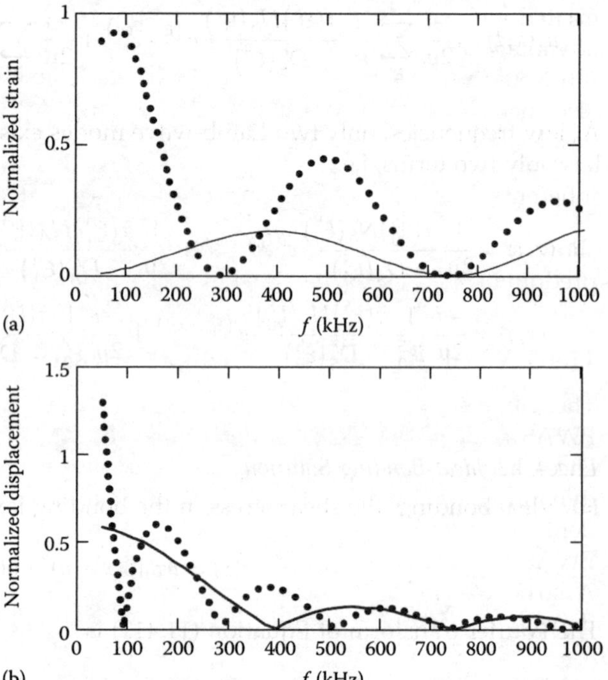

FIGURE 11.56
Predicted Lamb-wave response of a 1 mm aluminum plate under a 7 mm PWAS excitation: (a) strain response; (b) displacement response.

The $\sin\xi a$ contained in Equation (11.414) displays maxima when the PWAS length $l_a = 2a$ equals an odd multiple of the half wavelength and minima when it equals an even multiple of the half wavelength. A complex pattern of such maxima and minima evolves since several Lamb-wave modes coexist, each with its own different wavelength. At certain frequencies (e.g., 300 kHz in Figure 11.56a), the amplitude of the A_0 mode goes through zero while that of the S_0 remains strong, i.e., we have tuning of the S_0 mode and rejection of the A_0 mode. At other frequencies, the amplitude of the S_0 mode is quite small while that of A_0 mode is very large (e.g., 100 kHz in Figure 11.56a). Furthermore, frequencies also exist at which both A_0 and S_0 modes are rejected; for example, the 750 kHz frequency in Figure 11.56a. Experimental validation of these predictions together with a sweet spot for S_0 mode tuning at around 300 kHz will be given in Section 11.5.2.3. This sweet spot is especially important for embedded PWAS ultrasonics since, at this frequency, the S_0 mode has very little dispersion and hence can be successfully used in the pulse-echo mode. These observations illustrate the mode tuning opportunities offered by PWAS excitation of Lamb waves.

Another important fact to be noticed in Figure 11.56 is the difference between the strain response and the displacement response. As shown in Figure 11.56a, the strain response of the S_0 Lamb wave remains rather constant as the frequency increases whereas the corresponding displacement response, shown in Figure 11.56b, decreases rapidly with frequency. This is due to the fact that the strain response versus frequency varies as the sine function $\sin \xi^S a$ whereas the displacement response varies as the sinc function, $\dfrac{\sin \xi^S a}{\xi^S a}$. The sinc function has the maximum at the origin and then decreases rapidly as its argument increases. This observation indicates that the strain waves are more likely to be excited with significant amplitudes than the displacement waves. This fact highlights the advantage of using PWAS, which are strain-coupled devices, rather than conventional ultrasonic transducers that are displacement coupled devices. Similar observations can be also made with respect to the A_0 mode which is also shown in Figure 11.56. However, besides the $\sin \xi^A a$ dependency, the A_0 mode also shows a more complex dependency inherent in the $\dfrac{N_A(\xi^A)}{D'_A(\xi^A)}$ function. For this reason, the strain response peaks of the A_0 mode show a tendency to decrease with frequency, which indicates that the A_0 mode will predominantly be excitable only at lower frequencies.

11.4.5 Vibration of Structurally Constrained PWAS

The vibration of structurally constrained PWAS is different from the vibration of free PWAS resonators because, when applied to the structure, the PWAS transducer experiences a constraint from the structure in the form of a structural stiffness, k_{str}. Under dynamic conditions, the structural stiffness experienced by the PWAS is frequency dependent due to structural resonances. When the structure goes through a resonance under PWAS harmonic excitation, the effective dynamic stiffness goes through a minimum value. A structure can have many resonances, hence the PWAS will experience a frequency-dependent structural stiffness, $k_{str}(\omega)$.

11.4.5.1 1-D Analysis of Structurally Constrained PWAS

Consider a PWAS of length l_a, thickness t_a, and width b_a undergoing longitudinal expansion, u_1, induced by the thickness polarization electric field, E_3. The electric field is produced by the application of a harmonic voltage $V(t) = \hat{V}e^{i\omega t}$ between the top and bottom surfaces (electrodes). The resulting electric field, $E = V/t$, is assumed uniform with x_1 ($\partial E/\partial x_1 = 0$). The length, width, and thickness are assumed to have widely separated values ($t_a \ll b_a \ll l_a$) such that the length, width, and thickness motions are practically uncoupled.

The constitutive equations of the piezoelectric material are

$$S_1 = s_{11}^E T_1 + d_{31} E_3 \tag{11.416}$$

$$D_3 = d_{31} T_1 + \varepsilon_{33}^T E_3 \tag{11.417}$$

where S_1 is the strain, T_1 is the stress, D_3 is the electrical displacement (charge per unit area), s_{11}^E is the mechanical compliance at zero field, ε_3^T is the dielectric constant at zero stress, and d_{31} is the induced-strain coefficient, i.e., mechanical strain per unit electric field.

When the PWAS is bonded to the structure, the structure will constrain the PWAS motion with a structural stiffness, k_{str}. Hence, the present analysis will be concerned with the study of an elastically constrained PWAS, as shown in Figure 11.57. In our model, the

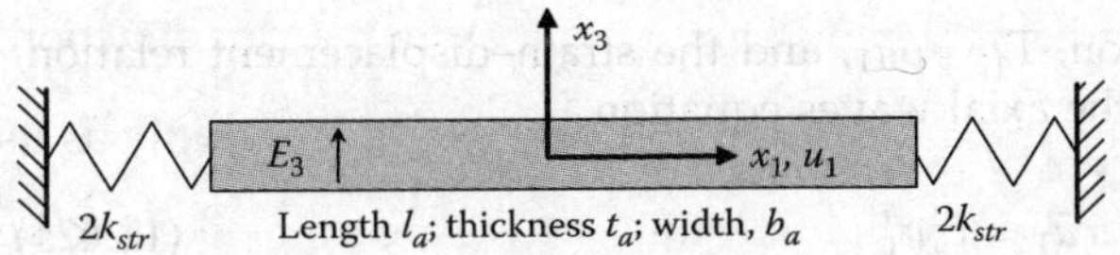

FIGURE 11.57
PWAS constrained by structural stiffness, k_{str}.

overall structural stiffness applied to the PWAS has been split into two equal components applied to the PWAS ends. The values of these components are $2k_{str}$ each such that

$$k_{total} = \left[(2k_{str})^{-1} + (2k_{str})^{-1} \right]^{-1} = k_{str} \tag{11.418}$$

The boundary conditions applied at the PWAS ends balance the resulting force, $T_1 b_a t_a$, with the spring reaction force, $2k_{str}u_1$, i.e.,

$$T_1 \left(\tfrac{1}{2}l_a\right) b_a t_a = -2k_{str}u_1 \left(\tfrac{1}{2}l_a\right)$$
$$T_1 \left(-\tfrac{1}{2}l_a\right) b_a t_a = 2k_{str}u_1 \left(-\tfrac{1}{2}l_a\right) \tag{11.419}$$

Note that the $+$ and $-$ signs in Equation (11.419) are chosen so as to be consistent with the sign convention. Recall the strain–displacement relation

$$S_1 = u_1' \tag{11.420}$$

Substitution of Equations (11.416) and (11.420) into Equation (11.419) gives

$$u_1' \left(\tfrac{1}{2}l_a\right) b_a t_a = -2k_{str} \frac{s_{11}^E}{bt} u_1 \left(\tfrac{1}{2}l_a\right) + d_{31}E_3$$
$$u_1' \left(-\tfrac{1}{2}l_a\right) b_a t_a = 2k_{str} \frac{s_{11}^E}{bt} u_1 \left(-\tfrac{1}{2}l_a\right) + d_{31}E_3 \tag{11.421}$$

Introducing the quasi-static PWAS stiffness

$$k_{PWAS} = \frac{b_a t_a}{s_{11}^E l_a} \tag{11.422}$$

and the frequency-dependent stiffness ratio

$$r(\omega) = \frac{k_{str}(\omega)}{k_{PWAS}} \tag{11.423}$$

allows us to rearrange Equation (11.421) in the form

$$u_1' \left(+\tfrac{1}{2}l_a\right) b_a t_a + \frac{r}{\tfrac{1}{2}l_a} \cdot u_1 \left(\tfrac{1}{2}l_a\right) = d_{31}E_3$$
$$u_1' \left(-\tfrac{1}{2}l_a\right) b_a t_a - \frac{r}{\tfrac{1}{2}l_a} \cdot u_1 \left(\tfrac{1}{2}l_a\right) = d_{31}E_3 \tag{11.424}$$

11.4.5.1.1 Mechanical Response

Substitution of Newton's law of motion, $T_1' = \rho \ddot{u}_1$, and the strain–displacement relation, $S_1 = u_1'$, into Equation (11.416) yields the axial waves equation

$$\ddot{u}_1 = c_a^2 u_1'' \tag{11.425}$$

where $\dot{u} = \partial u/\partial t$, $u' = \partial u/\partial x$, and $c_a^2 = 1/\rho s_{11}^E$ is the piezoelectric material wave speed. Assume harmonic excitation by the alternating electric field $E_3 = \hat{E}_3 e^{i\omega t}$. The general solution of Equation (11.425) is

$$u_1(x,t) = \hat{u}_1(x)e^{i\omega t} \tag{11.426}$$

where

$$\hat{u}_1(x) = (C_1 \sin \gamma x + C_2 \cos \gamma x) \tag{11.427}$$

The variable $\gamma = \omega/c_a$ is the wave number. The constants C_1 and C_2 are to be determined from the boundary conditions. Substitution of the general solution given by Equation (11.427) into the boundary conditions shown in Equation (11.424) yields the following linear system in C_1 and C_2

$$\begin{aligned}
\tfrac{1}{2}l_a \cdot \gamma\left(C_1 \cos\tfrac{1}{2}\gamma l_a - C_2 \sin\tfrac{1}{2}\gamma l_a\right) + r\left(C_1 \sin\tfrac{1}{2}\gamma l_a + C_2 \cos\tfrac{1}{2}\gamma l_a\right) = \tfrac{1}{2}l_a \cdot d_{31}\hat{E}_3 \\
\tfrac{1}{2}l_a \cdot \gamma\left(C_1 \cos\tfrac{1}{2}\gamma l_a + C_2 \sin\tfrac{1}{2}\gamma l_a\right) - r\left(-C_1 \sin\tfrac{1}{2}\gamma l_a + C_2 \cos\tfrac{1}{2}\gamma l_a\right) = \tfrac{1}{2}l_a \cdot d_{31}\hat{E}_3
\end{aligned} \tag{11.428}$$

Rearranging, we get

$$\begin{aligned}
\tfrac{1}{2}\gamma l_a\left(\cos\tfrac{1}{2}\gamma l_a + r\sin\tfrac{1}{2}\gamma l_a\right)C_1 + \left(-\tfrac{1}{2}\gamma l_a \sin\tfrac{1}{2}\gamma l_a + r\cos\tfrac{1}{2}\gamma l_a\right)C_2 = \tfrac{1}{2}l_a \cdot d_{31}\hat{E}_3 \\
\tfrac{1}{2}\gamma l_a\left(\cos\tfrac{1}{2}\gamma l_a + r\sin\tfrac{1}{2}\gamma l_a\right)C_1 + \left(+\tfrac{1}{2}\gamma l_a \sin\tfrac{1}{2}\gamma l_a - r\cos\tfrac{1}{2}\gamma l_a\right)C_2 = \tfrac{1}{2}l_a \cdot d_{31}\hat{E}_3
\end{aligned} \tag{11.429}$$

Recall the notations $u_{ISA} = d_{31}\hat{E}_3 l_a$ and $\phi = \tfrac{1}{2}\gamma l_a$. Upon substitution, we get the following linear system in C_1 and C_2

$$\begin{aligned}
(\phi \cos \phi + r \sin \phi)C_1 - (\phi \sin \phi - r \cos \phi)C_2 = \tfrac{1}{2}u_{ISA} \\
(\phi \cos \phi + r \sin \phi)C_1 + (\phi \sin \phi - r \cos \phi)C_2 = \tfrac{1}{2}u_{ISA}
\end{aligned} \tag{11.430}$$

Assume the system determinant is nonzero, i.e., $\Delta \neq 0$. Then, the solution of Equation (11.430) can be obtained as follows. Subtraction of the first equation from the second equation yields

$$2(\phi \sin \phi - r \cos \phi)C_2 = 0 \tag{11.431}$$

Hence, $C_2 = 0$. Now, add the two equations. The result is

$$2(\phi \cos \phi + r \sin \phi)C_1 = 2\frac{1}{2}u_{ISA} \tag{11.432}$$

Hence,

$$C_1 = \tfrac{1}{2}u_{ISA}\frac{1}{(\phi \cos \phi + r \sin \phi)}, \quad C_2 = 0 \tag{11.433}$$

Substituting C_1 and C_2 into Equation (11.427) yields the solution

$$\hat{u}_1(x) = \tfrac{1}{2}u_{ISA}\frac{\sin \gamma x}{\phi \cos \phi + r \sin \phi} \tag{11.434}$$

Substituting $\phi = \tfrac{1}{2}\gamma l_a$, we write

$$\hat{u}_1(x) = \tfrac{1}{2}u_{ISA}\frac{\sin \gamma x}{\tfrac{1}{2}\gamma l_a \cos \tfrac{1}{2}\gamma l_a + r \sin \tfrac{1}{2}\gamma l_a} \tag{11.435}$$

11.4.5.1.2 Electrical Response

Consider the constrained PWAS under harmonic electric excitation as shown in Figure 11.58. Recall Equation (11.417) representing the electrical displacement

$$D_3 = d_{31}T_1 + \varepsilon_{33}^T E_3 \tag{11.436}$$

Equation (11.416) yields the stress as a function of strain and electric field, i.e.,

$$T_1 = \frac{1}{s_{11}^E}(S_1 - d_{31}E_3) \tag{11.437}$$

Hence, the electric displacement can be expressed as

$$D_3 = \frac{d_{31}}{s_{11}}(S_1 - d_{31}E_3) + \varepsilon_{33}^T E_3 \tag{11.438}$$

Upon substitution of the strain–displacement relation of Equation (11.420), we get

$$D_3 = \frac{d_{31}}{s_{11}^E}u_1' - \frac{d_{31}^2}{s_{11}^E}E_3 + \varepsilon_{33}^T E_3 \tag{11.439}$$

i.e.,

$$D_3 = \varepsilon_{33}^T E_3 \left[1 - k_{31}^2\left(1 - \frac{u_1'}{d_{31}E_3}\right)\right] \tag{11.440}$$

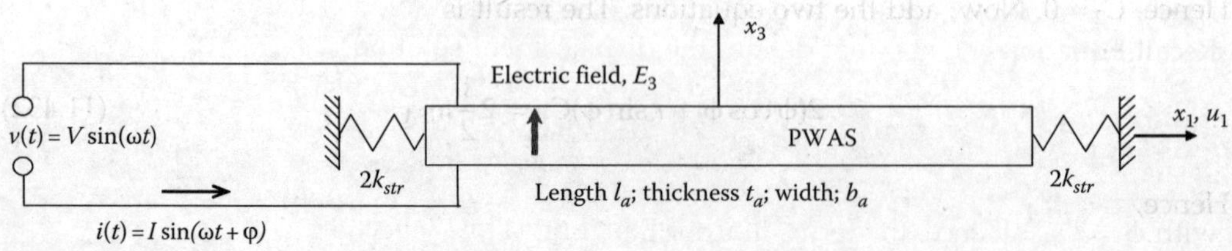

FIGURE 11.58
Schematic of a constraint PWAS under harmonic electric excitation.

where $\kappa_{13}^2 = d_{31}^2 / s_{11}^E \varepsilon_{33}^T$ is the electromechanical coupling coefficient. Integration of Equation (11.440) over the electrodes area $A_a = b_a l_a$ yields the total charge. Equation (11.417) can be reexpressed as

$$D_3 = \frac{d_{31}}{s_{11}^E}\left(u_1' - d_{31}E_3\right) + \varepsilon_{33}^T E_3 = \varepsilon_{33}^T E_3\left[1 + \kappa_{31}^2\left(\frac{u_1'}{d_{31}E_3} - 1\right)\right] \tag{11.441}$$

where $\kappa_{13}^2 = d_{31}^2 / \left(s_{11}^E \varepsilon_{33}\right)$ is the electromechanical coupling coefficient. Integration of Equation (11.441) over the volume of the piezoelectric wafer yields the total charge

$$Q = \int\limits_{-\frac{l_a}{2}}^{+\frac{l_a}{2}} \int\limits_{-\frac{b_a}{2}}^{+\frac{b_a}{2}} D_3\,dx\,dy = \varepsilon_{33}^T \frac{b_a l_a}{t_a} V\left[1 + \kappa_{31}^2\left(\frac{1}{l_a}\frac{1}{d_{31}E_3}u_1 \Big|_{-\frac{1}{2}l_a}^{\frac{1}{2}l_a} - 1\right)\right] \tag{11.442}$$

Since the time dependence is harmonic ($Q = \hat{Q}e^{i\omega t}$, etc.), we can write

$$\hat{Q} = C\hat{E}_3\left[1 + \kappa_{31}^2\left(\frac{1}{l_a}\frac{1}{d_{31}\hat{E}_3}\left[\hat{u}_1\left(\tfrac{1}{2}l_a\right) - \hat{u}_1\left(-\tfrac{1}{2}l_a\right)\right] - 1\right)\right] \tag{11.443}$$

where $C_0 = \varepsilon_{33}^T b_a l_a / t_a$ is the stress-free capacitance of the piezoelectric wafer active sensor. Using the definitions $u_{ISA} = d_{31}\hat{E}_3 l_a$ and $\hat{V} = \hat{E}_3/t_a$, we obtain

$$\hat{Q} = C_0\hat{V}\left[1 - k_{31}^2 + k_{31}^2\left(\frac{\hat{u}_1\left(\tfrac{1}{2}l_a\right) - \hat{u}_1\left(-\tfrac{1}{2}l_a\right)}{u_{ISA}}\right)\right] \tag{11.444}$$

The electric current is obtained as the time derivative of the electric charge, i.e.,

$$I = \dot{Q} = i\omega Q \tag{11.445}$$

Hence,

$$\hat{I} = i\omega C_0 \hat{V}\left[1 - k_{31}^2 + k_{31}^2\left(\frac{\hat{u}_1\left(\tfrac{1}{2}l_a\right) - \hat{u}_1\left(-\tfrac{1}{2}l_a\right)}{u_{ISA}}\right)\right] \tag{11.446}$$

The admittance, Y, is defined as the ratio between the current and voltage, i.e.,

$$Y = \frac{\hat{I}}{\hat{V}} = i\omega C_0\left[1 - k_{31}^2 + k_{31}^2\left(\frac{\hat{u}_1\left(\tfrac{1}{2}l_a\right) - \hat{u}_1\left(-\tfrac{1}{2}l_a\right)}{u_{ISA}}\right)\right] \tag{11.447}$$

Recall Equation (11.434) that defines the displacement solution

$$\hat{u}_1(x) = \tfrac{1}{2}u_{ISA}\frac{\sin\gamma x}{\phi\cos\phi + r\sin\phi} \tag{11.448}$$

with $\phi = \tfrac{1}{2}\gamma l_a$. Hence, the terms in $\hat{u}_1$ of Equation (11.447) become

$$\hat{u}_1\left(\tfrac{1}{2}l_a\right) - \hat{u}_1\left(-\tfrac{1}{2}l_a\right) = \frac{1}{2}\frac{\sin\tfrac{1}{2}\gamma l_a - \left(-\sin\tfrac{1}{2}\gamma l_a\right)}{\phi\cos\phi + r\sin\phi} = \frac{1}{2}2\frac{\sin\phi}{\phi\cos\phi + r\sin\phi} = \frac{1}{r + \phi\cot\phi} \tag{11.449}$$

Upon substitution, we get

$$Y = \frac{\hat{I}}{\hat{V}} = i\omega C_0 \left[1 - k_{31}^2 + k_{31}^2 \frac{1}{r + \phi \cot \phi} \right] \tag{11.450}$$

or

$$Y = \frac{\hat{I}}{\hat{V}} = i\omega C_0 \left[1 - k_{31}^2 \left(1 - \frac{1}{r + \phi \cot \phi} \right) \right] \tag{11.451}$$

where $r = r(\omega)$ is the frequency-dependent stiffness ratio given by Equation (11.423). Note that the admittance consists of a capacitive admittance, $i\omega C$, modified by the effect of piezoelectric coupling between mechanical and electrical variables. The impedance, Z, is obtained as the ratio between the voltage and current, i.e.,

$$Z = \frac{\hat{V}}{\hat{I}} = \frac{1}{i\omega C_0} \left[1 - k_{31}^2 \left(1 - \frac{1}{r + \phi \cot \phi} \right) \right]^{-1} \tag{11.452}$$

Thus, we have deduced the admittance and impedance expressions for a PWAS constrained by the structural substrate with a frequency-dependent stiffness ratio $r(\omega) = k_{str}(\omega)/k_{PWAS}$. In a later section, we will illustrate how the frequency-dependent structural stiffness could be calculated and experimentally verified.

In Equations (11.451) and (11.452), the structural stiffness ratio, r, is additive to the PWAS resonance term, $\phi \cot \phi$. When the PWAS is used in a frequency sweep, the apparent structural stiffness, k_{str}, will vary with frequency, going through zero at structural resonances and extreme values at structural antiresonances. For this reason, the structural stiffness is frequency dependent, i.e., $k_{str}(\omega)$. Subsequently, the stiffness ratio is also frequency dependent, i.e., $r(\omega)$. Rearrangement of Equations (11.451) and (11.452) and explicit indication of the terms that may depend on ω yield the general expressions

$$Y(\omega) = i\omega C_0 \left[1 - \kappa_{31}^2 \left(1 - \frac{1}{\phi(\omega) \cot \phi(\omega) + r(\omega)} \right) \right]$$

$$Z(\omega) = \frac{1}{i\omega C_0} \left[1 - \kappa_{31}^2 \left(1 - \frac{1}{\phi(\omega) \cot \phi(\omega) + r(\omega)} \right) \right]^{-1} \tag{11.453}$$

Equation (11.453) implies that both structural resonances and PWAS resonances will be reflected in the admittance and impedance frequency spectra.

11.4.5.2 Asymptotic Behavior

Analysis of the asymptotic behavior of Equation (11.453) allows us to recover situations for which the results are either known from previous investigations or can be easily determined through simple analysis. The asymptotic conditions to be considered here are

- Free piezoelectric wafer, i.e., $r = 0$
- Fully constrained (blocked) piezoelectric wafer, i.e., $r \rightarrow \infty$
- Constrained piezoelectric wafer under quasi-static conditions, i.e., $\phi = 0$

These asymptotic conditions will be considered in turn.

11.4.5.2.1 Free PWAS

A free piezoelectric wafer corresponds to $k_{str} = 0$, i.e., $r = 0$. In this case, the admittance and impedance expressions for a free piezoelectric wafer are recovered. Indeed, as the r term in the denominator of Equation (11.453) vanishes, the following expressions are obtained

$$Y_{free}(\omega) = i\omega C_0 \left[1 - k_{31}^2 \left(1 - \frac{1}{\phi(\omega)\cot\phi(\omega)}\right)\right]$$

$$Z_{free}(\omega) = \frac{1}{i\omega C_0}\left[1 - k_{31}^2 \left(1 - \frac{1}{\phi(\omega)\cot\phi(\omega)}\right)\right]^{-1}$$

(11.454)

These are exactly the previously determined expressions for the admittance and impedance of a free PWAS as given in Equation (11.161).

11.4.5.2.2 Fully Constrained (Blocked) PWAS

A fully constrained piezoelectric wafer has $k_{str} \to \infty$, i.e., $r \to \infty$. In this case, the fraction containing r in the denominator vanishes all together, and the admittance and impedance expressions become

$$Y_{blocked}(\omega) = i\omega C_0\left[1 - k_{31}^2\right]$$

$$Z_{blocked}(\omega) = \frac{1}{i\omega C_0}\left[1 - k_{31}^2\right]^{-1}$$

(11.455)

These are indeed the expressions for a "blocked" piezoelectric resonator cited in the specialized literature (Ikeda, 1996).

11.4.5.2.3 Constrained PWAS under Quasi-static Conditions

Quasi-static conditions are met when the frequency of oscillation is so low that the dynamic effects inside the piezoelectric wafer are negligible. This implies that the driving frequency is well below the first natural frequency of the piezoelectric wafer. Another way of looking at this is to say that the wavelength associated with this frequency is much larger than the wafer length, i.e., $\lambda \gg l_a$. In other words, the length of the wafer is so small that the elastic waves travel very quickly from one end to the other, and no stress and strain gradients are present due to the dynamic effects. In terms of the wave number, γ, this assumption reduces to $\gamma = \omega/c = 2\pi/\lambda$ and $\gamma l_a = (\omega/c)l_a = 2\pi l_a / \underset{\lambda \gg l_a}{\longrightarrow} 0$. The condition $\gamma l_a \to 0$ implies $\phi \to 0$. But for $\phi \to 0$, Equation (11.453) becomes

$$Y(\omega) = i\omega C_0\left[1 - \kappa_{31}^2 \frac{r(\omega)}{1 + r(\omega)}\right]$$

$$Z(\omega) = \frac{1}{i\omega C_0}\left[1 - \kappa_{31}^2 \frac{r(\omega)}{1 + r(\omega)}\right]^{-1}$$

(11.456)

This analysis of the asymptotic behavior illustrated how general are the admittance and impedance expressions of Equation (11.453). The simpler expressions contained in Equation (11.456) can be used for low-frequency structure-focused analysis in which the PWAS behavior can be considered quasi-static. However, for high-frequency analysis in which the PWAS resonances are also apparent, the complete expressions contained in Equation

(11.453) must be used. These comprehensive expressions cover the complete frequency spectrum and encompass both structure and PWAS dynamics.

11.4.5.3 Damping Effects

The damping effects can be either associated with the piezoelectric material or the elastic constrained. The damping effects in the piezoelectric material are covered through the adoption of complex compliance and dielectric constant expressions

$$\bar{s}_{11} = s_{11}(1 - i\eta), \quad \bar{\varepsilon}_{33} = \varepsilon_{33}(1 - i\delta) \tag{11.457}$$

The values of η and δ vary with the piezoceramic formulation but are usually small (η, $\delta < 5\%$).

The damping in the elastic constraint is similarly accounted for by assuming a complex stiffness expression, $\bar{k}_{str}$. As a result, the stiffness ratio will also take complex values, $\bar{r} = \bar{k}_{str}/\bar{k}_{PWAS}$. This frequency-dependent complex stiffness ratio reflects both the elastic constraint damping and the sensor dissipation mechanisms. Therefore, the admittance and impedance expressions of Equation (11.453) take the complex notation form

$$\bar{Y} = i\omega\bar{C}_0\left[1 - \bar{\kappa}_{31}^2\left(1 - \frac{1}{\bar{\phi}\cot\bar{\phi} + \bar{r}}\right)\right]$$

$$\bar{Z} = \frac{1}{i\omega\bar{C}_0}\left[1 - \bar{\kappa}_{31}^2\left(1 - \frac{1}{\bar{\phi}\cot\bar{\phi} + \bar{r}}\right)\right]^{-1} \tag{11.458}$$

It is worth noting that the elastic constraint can actually have a dynamic behavior of its own, e.g., of the form

$$\bar{k}_{str}(\omega) = \left[k(\omega) - \omega^2 m(\omega)\right] - i\omega c(\omega) \tag{11.459}$$

where $k(\omega)$, $m(\omega)$, and $c(\omega)$ are some frequency-dependent stiffness, mass, and damping coefficients.

11.4.5.4 Resonances

In order to determine resonances, we analyze the behavior of the determinant of the algebraic system (11.430), i.e.,

$$\Delta = \begin{vmatrix} (\phi\cos\phi + r\sin\phi) & -(\phi\sin\phi - r\cos\phi) \\ (\phi\cos\phi + r\sin\phi) & (\phi\sin\phi - r\cos\phi) \end{vmatrix} \tag{11.460}$$

i.e.,

$$\Delta = 2(\phi\cos\phi + r\sin\phi)(\phi\sin\phi - r\cos\phi) \tag{11.461}$$

This determinant Δ is zero when either the first parenthesis or the second parenthesis is zero. When the first parenthesis in Δ is zero, the denominator of Equation (11.434) vanishes and the response of the system to electrical excitation increases indefinitely. We identify this situation with an electromechanical resonance. When the second parenthesis in Δ vanishes, the denominator of the Equation (11.434) does not vanish and the electromechanical response does not increase indefinitely. We identify this situation with a purely mechanical resonance that cannot be excited electrically under the constant electric field distribution considered here. Therefore, the following two resonance conditions are identified:

$$\phi \cos \phi + r \sin \phi = 0 \qquad \text{(electromechanical resonance)} \qquad (11.462)$$

$$\phi \sin \phi + r \cos \phi = 0 \qquad \text{(mechanical resonance)} \qquad (11.463)$$

Using Equation (11.461), the condition that Δ is zero can be also expressed as follows

$$(\phi \cos \phi + r \sin \phi)(\phi \sin \phi - r \cos \phi) = 0 \qquad (11.464)$$

$$\phi^2 \cos \phi \sin \phi + r\phi (\sin^2 \phi - \cos^2 \phi) - r^2 \sin \phi \cos \phi = 0 \qquad (11.465)$$

$$(\phi^2 - r^2) \sin 2\phi - r\phi \cos 2\phi = 0 \qquad (11.466)$$

Assuming $\cos 2\phi \neq 0$, one writes the characteristic equation

$$\tan 2\phi(\omega) = \frac{r(\omega)\phi(\omega)}{\phi^2(\omega) - r^2(\omega)} \qquad (11.467)$$

It should be noted that all the terms in Equation (11.467) are frequency dependent. Solution of Equation (11.467) should give all the system resonances, both the mechanical resonances and the electromechanical resonances.

11.4.5.5 2-D Analysis of a Constrained Circular PWAS

To model a constrained circular PWAS, we start from the piezoelectric constitutive equations in cylindrical coordinates

$$S_{rr} = s_{11}^E T_{rr} + s_{12}^E T_{\theta\theta} + d_{31} E_z$$

$$S_{\theta\theta} = s_{12}^E T_{rr} + s_{11}^E T_{\theta\theta} + d_{31} E_z \qquad (11.468)$$

$$D_z = d_{31}(T_{rr} + T_{\theta\theta}) + \varepsilon_{33}^T E_z$$

where S_{rr} and $S_{\theta\theta}$ are the mechanical strains, T_{rr} and $T_{\theta\theta}$ the mechanical stresses, E_z the electrical field, D_z the electrical displacement, s_{11}^E and s_{12}^E the mechanical compliances at zero electric field ($E = 0$), ε_{33}^T the dielectric permittivity at zero mechanical stress ($T = 0$), and d_{31} the piezoelectric coupling between the electrical and mechanical variables. For axisymmetric motion, the problem is θ-independent and the space variation is in r only. Hence, $S_{rr} = \partial u_r / \partial r$ and $S_{\theta\theta} = u_r / r$. Applying Newton's 2nd law, one recovers, upon substitution, the wave equation in polar coordinates

$$\frac{\partial^2 u_r}{\partial r^2} + \frac{1}{r} \frac{\partial u_r}{\partial r} - \frac{u_r}{r^2} = \frac{1}{c_p^2} \frac{\partial^2 u_r}{\partial t^2} \qquad (11.469)$$

where

$$c_p = \sqrt{\frac{1}{\rho s_{11}^E \left(1 - v_a^2\right)}} \qquad \text{(wave speed)} \qquad (11.470)$$

is the wave speed in the PWAS for axially symmetric radial motion, with v_a being the Poisson's ratio of the piezoelectric material $\left(v_a = -s_{12}^E/s_{11}^E\right)$. Equation (11.469) admits a general solution in terms of Bessel functions of the first kind, J_1, in the form

$$u_r(r, t) = AJ_1\left(\frac{\omega r}{c}\right)e^{i\omega t} \qquad (11.471)$$

where the coefficient A is determined from the boundary conditions.

We assume that the PWAS circumference is elastically constrained by the dynamic structural stiffness $k_{str}(\omega)$ (Figure 11.59). At the boundary $r = r_a$, we have the boundary condition

$$T_{rr}(r_a)t_a = k_{str}(\omega)u_r(r_a) \qquad (11.472)$$

where t_a is the PWAS thickness. Hence, the radial stress can be expressed as

$$T_{rr}(r_a) = \frac{k_{str}(\omega)u_r(r_a)}{t_a} \qquad (11.473)$$

Substitution of Equation (11.473) into Equation (11.468) gives, upon rearrangement,

$$\frac{\partial u_r(r_a)}{\partial r} = \chi(\omega)(1 + v_a)\frac{u_r(r_a)}{r_a} - v_a\frac{u_r(r_a)}{r_a} + (1 + v_a)d_{31}E_z \qquad (11.474)$$

where

$$\chi(\omega) = k_{str}(\omega)/k_{PWAS} \qquad (11.475)$$

is the dynamic stiffness ratio and

$$k_{PWAS} = \frac{t_a}{r_a s_{11}^E\left(1 - v_a\right)} \qquad (11.476)$$

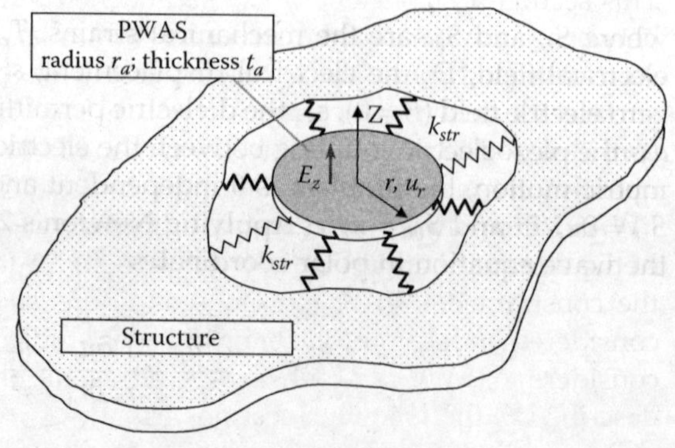

FIGURE 11.59
Circular piezoelectric wafer active sensor constrained by structural stiffness, $k_{str}(\omega)$.

is the static stiffness of the circular PWAS. Substitution of Equation (11.471) into Equation (11.474) yields

$$A = \frac{(1 + \nu_a)d_{31}E_0}{\frac{\omega}{c}J_0\left(\frac{\omega r_a}{c}\right) - \frac{(1 - \nu_a + \chi(\omega)(1 + \nu_a))}{r_a}J_1\left(\frac{\omega r_a}{c}\right)} \tag{11.477}$$

The electrical admittance is calculated as the ratio between the current and the voltage amplitudes, i.e., $Y = \hat{I}/\hat{V}$. The current is calculated by integrating the electric displacement D_3 over the PWAS area to obtain the total charge and then differentiating the result with respect to time while the voltage is calculated by multiplying the electric field by the PWAS thickness. Hence, the electrical admittance is expressed as

$$Y(\omega) = i\omega C_0\left(1 - k_p^2\right)\left[1 + \frac{k_p^2}{1 - k_p^2}\frac{(1 + \nu_a)J_1(\phi_a)}{\phi_a J_0(\phi_a) - (1 - \nu_a)J_1(\phi_a) - \chi(\omega)(1 + \nu_a)J_1(\phi_a)}\right] \tag{11.478}$$

where $\phi_a = \omega r_a/c$ and $k_p = \sqrt{2d_{31}^2/\left[s_{11}^E \cdot (1 - \nu_a)\varepsilon_{33}^T\right]}$ is the planar coupling factor. Then, the inverse relationship between impedance and admittance, $Z(\omega) = 1/Y(\omega)$, yields

$$Z(\omega) = \left\{i\omega C_0\left(1 - k_p^2\right)\left[1 + \frac{k_p^2}{1 - k_p^2}\frac{(1 + \nu_a)J_1(\phi_a)}{\phi_a J_0(\phi_a) - (1 - \nu_a)J_1(\phi_a) - \chi(\omega)(1 + \nu_a)J_1(\phi_a)}\right]\right\}^{-1} \tag{11.479}$$

Equation (11.479) predicts the E/M impedance spectrum as it would be measured by the impedance analyzer at the embedded PWAS terminals and it allows for direct comparison between calculated predictions and experimental results. The structural dynamics is reflected in Equation (11.479) through the dynamic stiffness factor, $\chi(\omega) = \dfrac{k_{str}(\omega)}{k_{PWAS}}$, which contains the dynamic stiffness of the structure, $k_{str}(\omega)$. This latter quantity results from the structural dynamics analysis.

11.4.6 Section Summary

This section has considered PWAS attached to the structure. First, the shear lag coupling between a PWAS and the structure substrate was discussed, and closed form solutions were obtained for some particular cases. Next, the excitation of structural waves with surface-attached PWAS transducers was studied. Axial, flexural, and Lamb waves were considered. The phenomenon of wave-mode tuning was observed and discussed. Subsequently, the vibration of structurally attached PWAS was discussed. The vibration of a PWAS constrained by an external stiffness was studied and closed form solutions in terms of the dynamic stiffness ratio were derived for the electrical impedance and admittance of the constrained PWAS. Extreme conditions such as "blocked" and "free" PWAS were considered and asymptotic behavior was studied. In this way, the foundation was laid for considering PWAS as both in-situ ultrasonic transducers and in-situ modal sensors, as described in the following sections.

11.5 PWAS Ultrasonic Transducers

For embedded NDE applications, PWAS can be used as embedded ultrasonic transducers. PWAS act as both Lamb-wave exciters and Lamb-wave detectors. PWAS couple their in-plane motion with the Lamb-waves' particle motion on the material surface. The in-plane PWAS motion is excited by the applied oscillatory voltage through the d_{31} piezoelectric coupling. Optimum excitation and detection happens when the PWAS length is an odd multiple of the half wavelength of particle Lamb-wave modes. The PWAS action as ultrasonic transducers is fundamentally different from that of conventional ultrasonic transducers. Conventional ultrasonic transducers act through surface tapping, applying vibrational pressure to the object's surface. PWAS, on the other hand, act through surface pinching and are strain-coupled with the object surface. This imparts to PWAS a much better efficiency in transmitting and receiving ultrasonic Lamb and Rayleigh waves than conventional ultrasonic transducers.

PWAS are capable of geometric tuning through matching between their characteristic direction and the half wavelength of the excited Lamb mode. Rectangular shaped PWAS with high length to width ratio can generate unidirectional Lamb waves through half wavelength tuning in the lengthwise direction. Circular PWAS excite omnidirectional Lamb waves that propagate in circular wave fronts. Unidirectional and omnidirectional Lamb wave propagation is illustrated in Figure 11.60. Omnidirectional Lamb waves are also generated by square PWAS, although their pattern is somewhat irregular in the PWAS proximity. At far enough distance, $r \gg a$, the wave front generated by square PWAS is practically identical with that generated by circular PWAS.

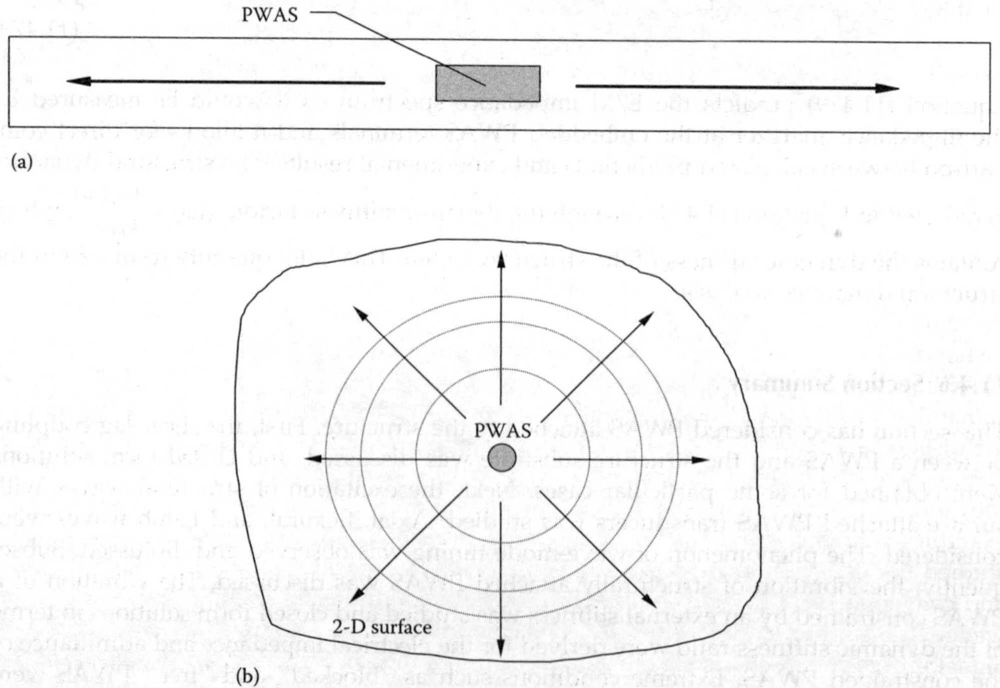

FIGURE 11.60
(a) Elastic waves generated by a PWAS in a 1-D structure; (b) circular-crested Lamb waves generated by a PWAS in a 2-D structure.

11.5.1 Guided Wave Methods for NDE and Damage Detection

Current ultrasonic inspection of thin wall structures (e.g., aircraft shells, storage tanks, large pipes, etc.) is a time consuming operation that requires meticulous through-the-thickness C-scans over large areas. One method to increase the efficiency of thin-wall structures inspection is to utilize guided waves (e.g., Lamb waves) instead of the conventional pressure waves. Guided waves propagate along the mid-surface of thin-wall plates and shallow shells and can travel relatively large distances with very little loss (Viktorov, 1967; Rose, 1999). Guided waves offer the advantage of large-area coverage with a minimum of installed sensors. In recent years, a large number of papers have been published on the use of Lamb waves for nondestructive evaluation and damage detection. Rose (2002) pointed out that the improved inspection potential of guided Lamb waves over other ultrasonic methods is due to their

- Variable mode structure and distributions
- Multimode character
- Sensitivity to different type of flaws
- Propagation for long distances
- Capability to follow curvature and reach hidden and/or buried parts

Guided Lamb waves method has opened new opportunities for cost-effective detection of damage in aircraft structures (Dalton et al. 2001). The use of Lamb waves to detect the corrosion in aluminum structures using the pitch-catch method was explored by Chahbaz et al. (1999). An A_1 Lamb wave mode traveling through a corroded zone ends up with a lower amplitude and longer time-of-flight (TOF) than in a pristine zone. Experiments were performed to detect corrosion around rivets and other fasteners. The pitch-catch method was also used by Grondel et al. (1999) to assess the damage progression in a riveted aircraft splice specimen during fatigue testing. The TOF changes were correlated with the appearance microcracks and macrocracks during fatigue testing. Alleyne et al. (2001) detected corrosion in pipes using the pulse-echo method using the mode conversion approach. Rose et al. (1995) used Lamb waves to detect disbond in adhesive joints with the pitch-catch method. In adhesive bonds, Lamb waves are able to leak from one side of the bond to the other side of the bond through "wave leakage."* Two situations were considered: (1) lap splice joint disbonds detected as a loss of reception signal due to the signal no longer leaking across the splice, and (2) tear-strap disbond or corrosion observed as an increase of the reception signal due to the signal no longer leaking away in the tear-strap. The use of Lamb waves for disbond detection was also reported by Singher et al. (1997), Mustafa and Chahbaz (1997), and Todd and Challis (1999). Rose (2002) outlined the inspection potential of ultrasonic guided wave for the detection of cracks, delaminations, and disbonds and gave examples utilizing conventional angle-probe ultrasonic transducers. Light et al. (2001) studied the detection of defects in thin steel plates using ultrasonic guided Lamb waves and conventional ultrasonic equipment. Conventional ultrasonic equipment consisting of wedge transmitter and bubbler receiver mounted on a scanning arm was utilized. Signal processing methods for the determination of the arrival times and the flaw location

---

* In contrast, conventional ultrasonic P-waves cannot cross the adhesive due to impedance mismatch.

were explored. Flaw localization results for simulated cracks (notches) with various sizes (from 2 to 3.5 in.) and different inclinations (from 0° to 45°) were presented. Further advancements in this direction were achieved through acousto-ultrasonics (Duke, 1988).

Traditional generation of guided Lamb waves has been through a wedge probe which impinges the plate obliquely with a tone-burst generated by a relatively large ultrasonic transducer. Snell's law ensures mode conversion at the interface, hence a combination of pressure and shear waves is simultaneously generated into the thin plate. They construct-ively and destructively interfere to generate Lamb waves. Modification of the wedge angle and excitation frequency allows the selective tuning of various Lamb-wave modes (Alleyne and Cawley, 1992; Rose et al., 1995). Another traditional way to selectively excite Lamb waves is through a comb probe in which the comb pitch is matched with the half wavelength of the targeted Lamb mode. Both the wedge and the comb probes are rela-tively large and expensive. If they were to be deployed in large numbers inside an aerospace structure as part of a structural health monitoring system, the cost and weight penalties would be exorbitant. Therefore, for structural health monitoring, a different type of Lamb wave transducer that is smaller, lighter, and cheaper than the conventional ultrasonic probes is required. Such a transducer could be deployed into the structure as sensor arrays which would be permanently wired and interrogated at will. Thus, the opportunity for embedded ultrasonic NDE would be created.

11.5.2 Pitch-Catch PWAS Experiments

11.5.2.1 Experimental Setup

To understand and calibrate the Lamb waves excitation and reception with PWAS, a set of experiments were conducted on a thin metallic plate with $t = 1.6$ mm. The plate was made of 2024-aluminum alloy. Its overall dimensions were 914×504 mm. The plate was instru-mented with eleven 7-mm $\times$ 7-mm PWAS positioned on a rectangular grid (Figure 11.61 and Table 11.14).

The sensors were connected with thin insulated wires to a 16-channel signal bus and two 8-pin connectors (Figure 11.61). An HP33120A arbitrary signal generator was used to generate a smoothed 300 kHz tone-burst excitation with a 10 Hz repetition rate. The signal was sent to active sensor #11 that generated a package of elastic waves that spread out into the entire plate. A Tektronix TDS210 two-channel digital oscilloscope, synchronized with the signal generator, was used to collect the signals captured at the remaining 10 active sensors. A digitally controlled switching unit and a LabVIEW data acquisition program were used. A Motorola MC68HC11 microcontroller was also tested as an embedded stand-alone option.

11.5.2.2 Excitation Signal

The excitation signal considered in our studies consisted of a smoothed tone-burst. The smoothed tone-burst was obtained from a pure tone-burst of frequency f filtered through a Hanning window. The Hanning window is described by the equation

$$x(t) = \frac{1}{2}[1 - \cos(2\pi t/T_H)], \quad t \in [0, T_H] \tag{11.480}$$

The number of counts, N_B, in the tone-burst matches the length of the Hanning window, i.e.,

$$T_H = N_B/f \tag{11.481}$$

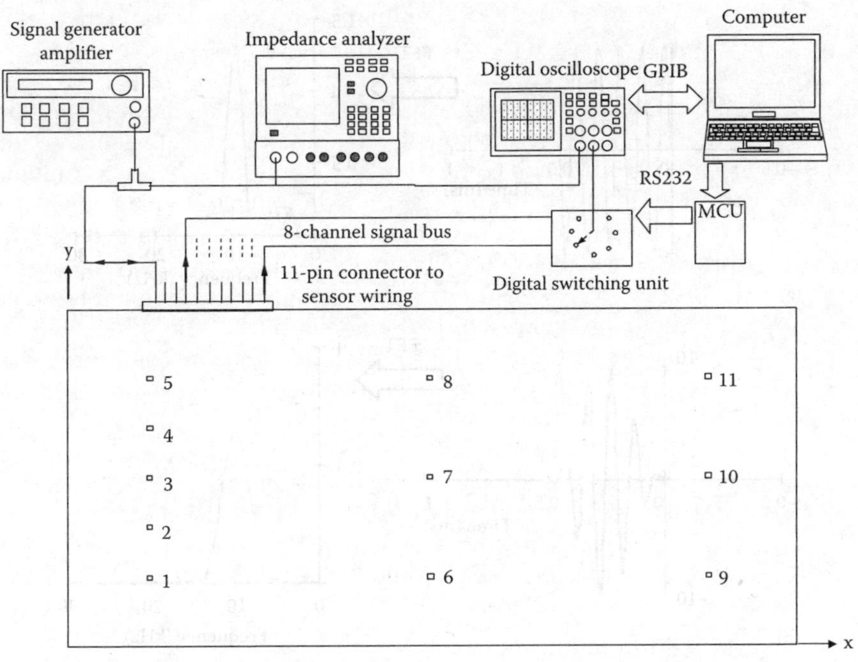

(a)

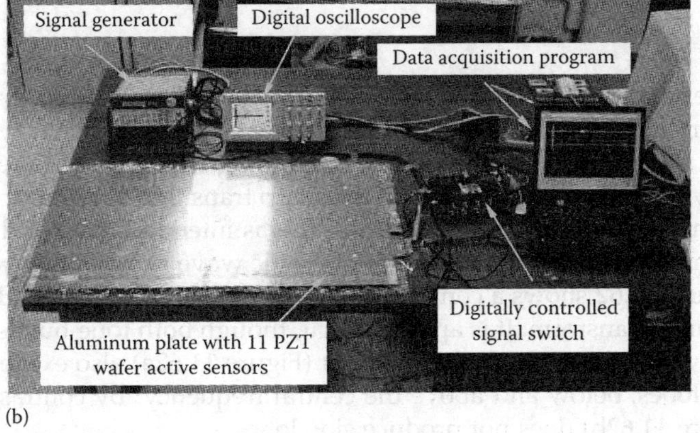

(b)

FIGURE 11.61
Schematic of experimental setup for PWAS wave propagation experiments on a rectangular plate: (a) schematic;
(b) photograph of actual experiment.

TABLE 11.14

Locations of PWAS Transducers on the Thin Rectangular Plate Specimen

Sensor #	1	2	3	4	5	6	7	8	9	10	11
x (mm)	100	100	100	100	100	450	450	450	800	800	800
y (mm)	100	175	250	325	400	100	250	400	100	250	400

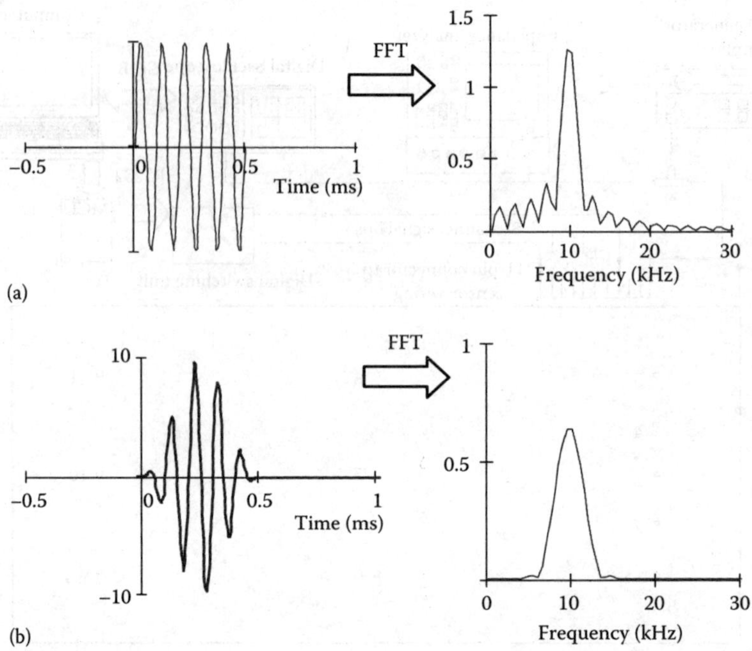

FIGURE 11.62
Example of 10 kHz, 5-count tone-burst excitation and its Fourier transform: (a) raw tone-burst; (b) smoothed tone-burst.

The tone burst excitation was chosen in order to excite coherent single-frequency waves. This aspect is very important especially when dealing with dispersive wave types (flexure, Rayleigh, Lamb, etc.). The Hanning window smoothing was applied to reduce the excitation of frequency side lobes associated with the sharp transition at the start and the end of a conventional tone burst. Through these means, it was intended that the dispersion effects would be minimized and the characteristics of elastic wave propagation would be readily understood. Figure 11.62 shows a comparison of raw and smoothed 10 kHz tone-bursts as well as their Fourier transform. It is apparent that, though both tone-bursts excite the same central frequency of 10 kHz, the raw tone-burst (Figure 11.62a) also excites a considerable number of side lobes, below and above the central frequency. By contrast, the smoothed tone-burst (Figure 11.62b) does not produce side lobes.

The smoothed tone-burst that resulted from this process was numerically synthesized and stored in PC memory as the excitation signal. This numerically generated excitation signal was used in the finite element analysis and in the experimental investigation.

11.5.2.3 Lamb Mode Tuning

During our experiments, we investigated the effect of excitation frequency on the excited wave amplitude. It was found that, at low frequencies (e.g., 10 kHz), the excitation of flexural wave was much stronger than that of axial waves. However, as frequency increased beyond 150 kHz, the excitation of flexural waves decreases while that of axial waves increased significantly. A "sweet spot" for axial wave (S_0) excitation was found around 300 kHz (Figure 11.63a). These trends validate the theoretical prediction of Lamb

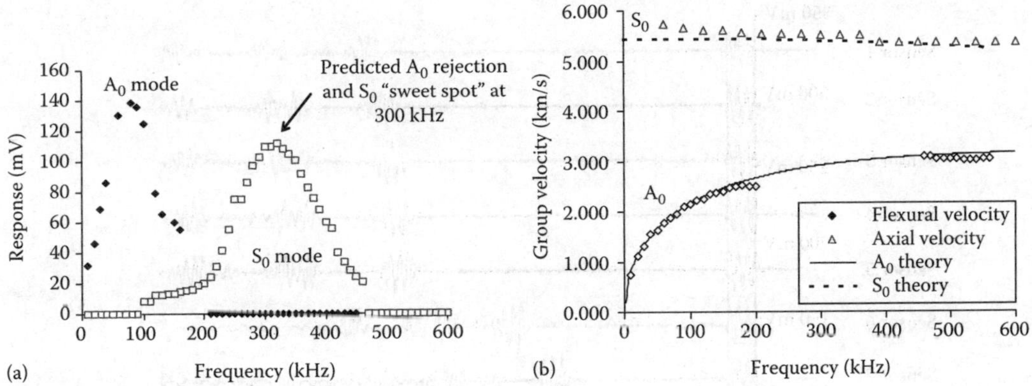

FIGURE 11.63
Pitch-catch PWAS wave propagation results: (a) frequency tuning studies identified a maximum wave response around 300 kHz; (b) group velocity dispersion curves for Lamb-wave A$_0$ and S$_0$ modes.

wave-PWAS tuning presented in Section 11.4.4. Thus, in this plate, the antisymmetric Lamb waves (A$_0$ mode) were best excited at around 100 kHz while the symmetric Lamb waves (S$_0$ mode) were best excited at around 300 kHz. This allowed us to produce S$_0$ Lamb waves only which have much less dispersion at this low frequency. Subsequently, for the pitch-catch and pulse-echo experiments described in the next sections, excitation at 300 kHz was adopted. The experimental investigation also reproduces the group-velocity dispersion curves for the axial (S$_0$) and flexural (A$_0$) modes (Figure 11.63b).

11.5.2.4 Pitch-Catch Results

Figure 11.64a shows a sample of the signals measured during this investigation. The first row shows the signal associated with sensor 11 (the transmitter). The "initial bang" generated by this active sensor as well as a number of wave packages received on the same sensor in the pulse-echo mode are present. The wave packages are reflection from the plate edges, and their TOF position is consistent with the distance from the sensor to the respective edges. The other rows of signals correspond to the receptor active sensors 1 through 8. Their TOF position is consistent with the distance between the transmitting and receiving active sensors. The consistency of the wave patterns is remarkable.

These raw signals were processed using a narrow-band signal correlation algorithm followed by an envelope detection method. As a result, the exact TOF for each wave package could be precisely identified. When the TOF was plotted against radial distance between the receiving active sensor and the transmitting active sensor, a remarkably good straight line fit (99.99% R^2 correlation) was obtained (Figure 11.64b). The slope of this straight line is the wave speed, 5.461 km/s. For the 1–6 mm aluminum alloy used in this experiment, the theoretical group velocity for S$_0$ mode is 5.440 km/s. The speed detection accuracy (0.3% error) is remarkable.

These systematic experiments gave conclusive results regarding the feasibility of exciting elastic waves in aircraft-grade metallic plates using small inexpensive and unobtrusive piezoelectric-wafer active sensors:

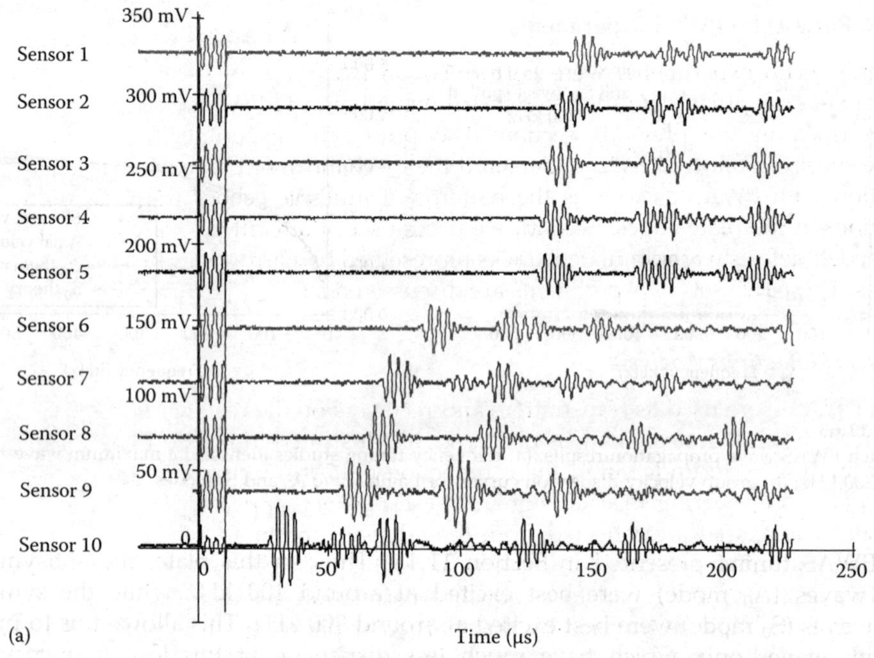

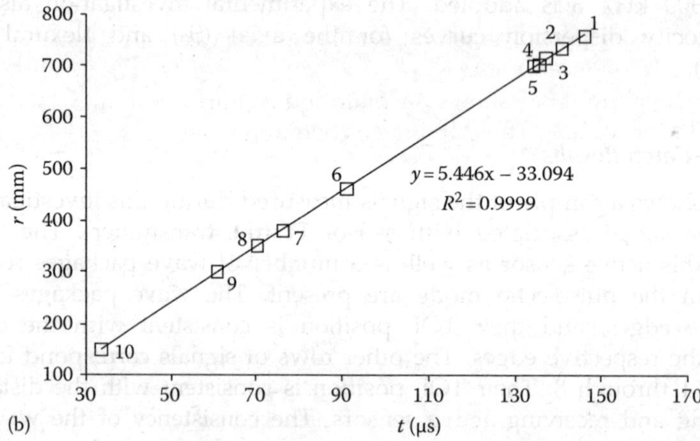

FIGURE 11.64
More pitch-catch PWAS wave propagation results: (a) excitation signal and echo signals on active sensor 11, and reception signals on active sensors 1 through 8; (b) correlation between radial distance and TOF.

1. Excitation and reception of high-frequency Lamb waves were verified over a large frequency range (10–600 kHz). For axial, S_0, waves, an excitation "sweet spot" was found at around 300 kHz (Figure 11.63a).

2. The elastic waves generated by this method had remarkable clarity and showed a 99.99% distance–time correlation. The group velocity correlated very well with the theoretical predictions (Figure 11.63b).

11.5.3 Pulse-Echo PWAS Experiments

The pulse-echo experiments were performed on two types of specimens. First, we used the simple rectangular plate specimen instrumented with 11 PWAS placed on a grid pattern, as described in the previous section. The pulse-echo experiments performed on this simple specimen allowed us to understand the mechanism of Lamb wave transmission and reception with PWAS as well as the patterns of multiple echoes resulting from multiple reflections at the plate edges. Second, we used a set of aircraft panels with seeded defects. The seeded defects were simulated cracks represented by narrow slits produced by the EDM process. These two sets of experiments are discussed next.

11.5.3.1 Reflections Analysis

When PWAS were used as transmitters and receivers on the rectangular plate specimen, a pattern of multiple reflections was observed, as illustrated in Figure 11.64a. The next step in our analysis is to understand these reflection patterns, represented by the subsequent wave packets appearing in each signal. These packets, appearing after the first packet, represent waves that are reflected from the edges of the plate. Understanding the wave reflection patterns is essential for implementing the pulse-echo method for damage detection. In our analysis, we had to establish a correlation between the TOF of each wave packet and the distance traveled by that particular packet (path length). The TOF determination was immediate but the determination of the actual traveled distance was more involved. Since elastic waves propagate in a circular wave front through the 2-D plate, they reflect at all the edges and continue to travel around until fully attenuated. AutoCAD drawing software was used to assist in the analysis and calculate the actual path length of the reflected waves (Figure 11.65).

The drawing in Figure 11.65 shows the plate and its mirror reflections with respect to the four edges and four corners. Thus, an image containing nine adjacent plates was obtained.

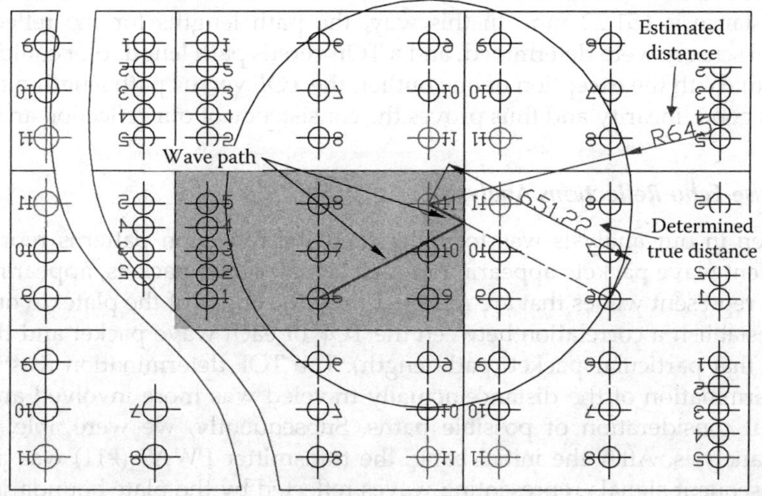

FIGURE 11.65
Method for finding the distance traveled by the reflected waves. The actual plate (shaded) is surrounded by eight mirror images to assist in the calculation of the reflection path length.

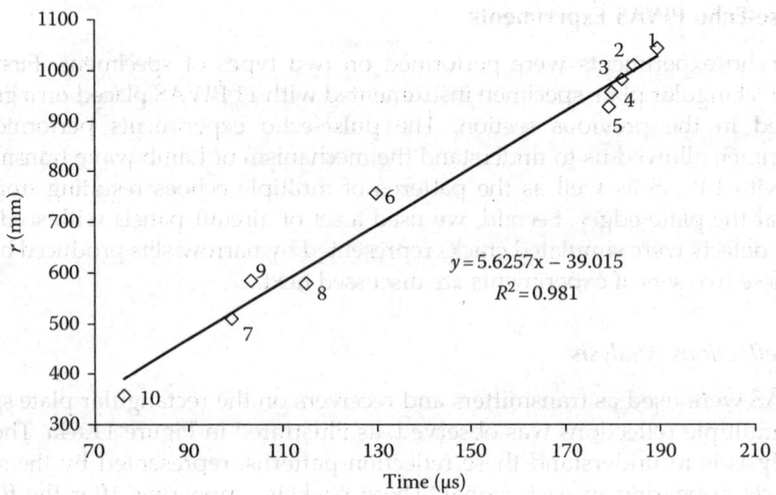

FIGURE 11.66
Correlations between path length and TOF for the wave reflection signals (second wave packet) captured on active sensors 1 through 10.

To identify which reflection path corresponds to a particular wave packet, we took its TOF and multiplied it by the wave speed to get the first estimate of the path length. For example, Figure 11.65 shows that for the reflected reception at sensor #6, the estimated distance is 645 mm. Then, a circle was drawn with its center at the transmitter sensor (#11) and with a radius equal to half of the estimated distance. The circle intersects with, or very close to, one of the reflected images of the reception sensor (#6). In this way, we identified which of the reflected images of the reception sensor #6 was the actual image to be processed. Next, the true distance (path length) between the identified sensor image and the transmitter sensor was determined using the distance feature of the AutoCAD software. In Figure 11.65, this true distance is 651.22 mm. In this way, the path lengths for the reflection wave packets of all sensors were determined, and a TOF versus path length plot could be created (Figure 11.66). With the exception of one outlier, the TOF versus path length plot of Figure 11.66 shows good linearity and thus proves the consistency of our reflection analysis.

11.5.3.2 Pulse-Echo Reflections Analysis

The next step in our analysis was to understand the reflection patterns represented by the subsequent wave packets appearing in each signal. These packets, appearing after the first packet, represent waves that are reflected from the edges of the plate. In our analysis, we had to establish a correlation between the TOF of each wave packet and the distance traveled by that particular packet (path length). The TOF determination was immediate, but the determination of the distance actually traveled was more involved and required more careful consideration of possible paths. Subsequently, we were able to perform pulse-echo analysis. After the initial bang, the transmitter PWAS (#11) was also able to capture subsequent signals representing waves reflected by the plate boundaries and sent back to the transmitter. These subsequent wave packets were recorded for evaluating the pulse-echo method. Figure 11.67a shows the signal recorded on PWAS #11. This signal contains the excitation signal (initial bang) and a number of wave packets received in the pulse-echo mode. The wave generated by the initial bang undergoes multiple reflections from the plate edges, as shown in Figure 11.67b. The values of the true path length for these reflections are given in Table 11.15. It should be noted that the path lengths for reflections R_1 and R_2 are very close. Hence, the echoes for these two reflections virtually superpose on the pulse-echo signal in Figure 11.67a. It is also important to notice that reflection R_4 has two possible paths, R_{4a} and R_{4b}. Both paths have the same length. Hence, the echoes corresponding to these two reflection paths arrive simultaneously and form a single echo signal in Figure 11.67a, with roughly double the intensity of the adjacent signals. Figure 11.67c shows the TOF of the echo wave packages plotted against their path lengths. The straight line fit has a very good correlation ($R^2 = 99.99\%$).

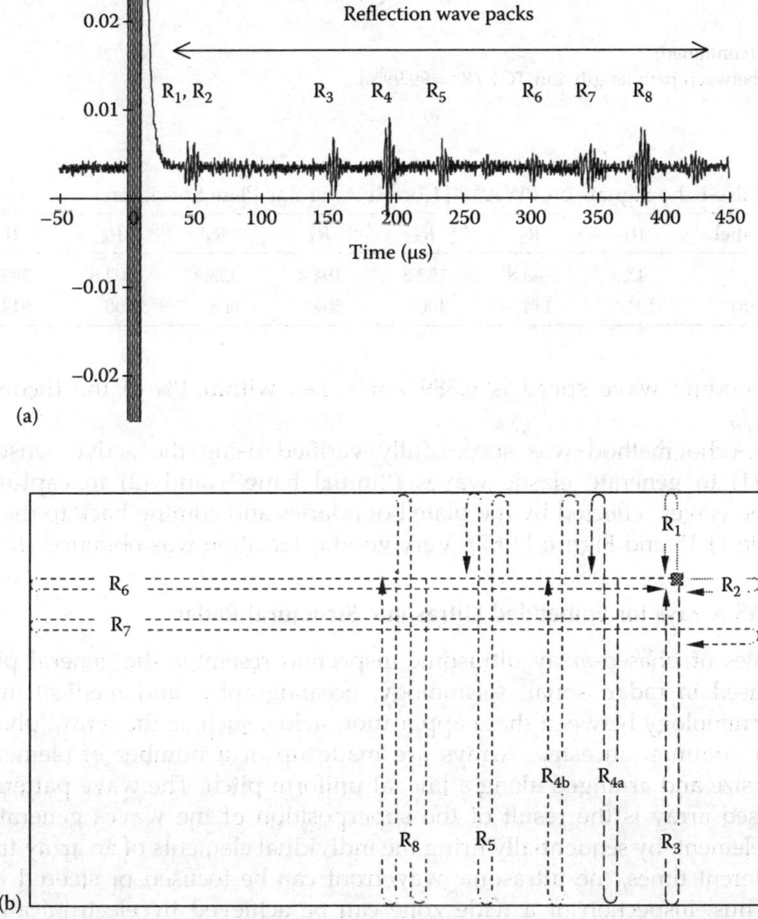

FIGURE 11.67
Pulse-echo method applied to active sensor #11: (a) the excitation signal and the echo signals on active sensor 11; (b) schematic of the wave paths for each wave packet; and

(continued)

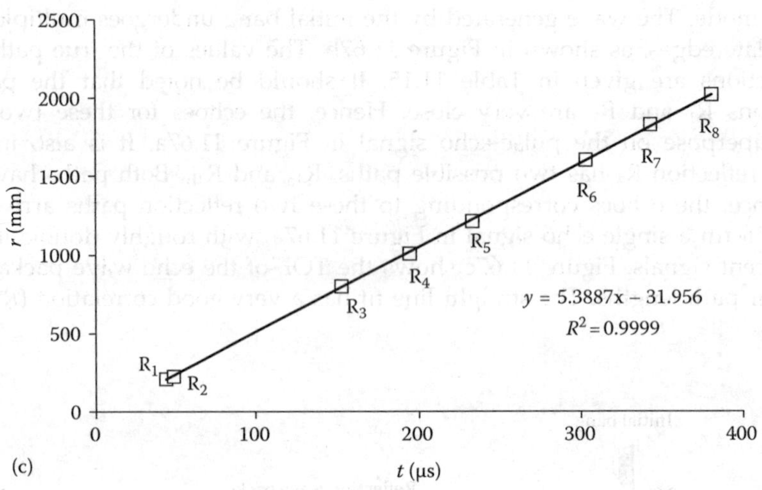

FIGURE 11.67 (continued)
(c) correlation between path length and TOF ($R^2 = 99.99\%$).

TABLE 11.15

Analysis of Pulse-Echo Signals on PWAS #11 on Rectangular Plate Specimen

Wave Packet Label	R_1	R_2	R_3	R_4	R_5	R_6	R_7	R_8
TOF (μs)	43.8	48.8	152.8	194.4	233.2	302.8	343.2	380.8
Path length (mm)	104	114	400	504	608	800	914	1008

The corresponding wave speed is 5.389 km/s, i.e., within 1% of the theoretical value of 5.440 km/s.

The pulse-echo method was successfully verified using the active sensor #11 in a dual role: (1) to generate elastic waves ("initial bang") and (2) to capture the echo signals of the waves reflected by the plate boundaries and coming back to the transmitter sensor (Table 11.15 and Figure 11.67). Very good correlation was obtained ($R^2 = 99.99\%$).

11.5.4 PWAS Arrays for Embedded Ultrasonic Structural Radar

The principles of phased-array ultrasonic inspection resemble the general phased-array principles used in radar, sonar, seismology, oceanography, and medical imaging. The common terminology between these application fields, such as the term "phased array," shows their common ancestry. Arrays are made up of a number of elements, usually identical in size and arranged along a line, at uniform pitch. The wave pattern generated by the phased array is the result of the superposition of the waves generated by each individual element. By sequentially firing the individual elements of an array transducer at slightly different times, the ultrasonic wavefront can be focused or steered in a specific direction. Thus, inspection of a wide zone can be achieved by electronically sweeping and/or refocusing without physically manipulating the transducer.

Once the beam steering and focusing have been established, the detection of internal flaws is done with the pulse-echo method. Conventional radar technology uses high-frequency electromagnetic waves to detect the presence and position of a target using

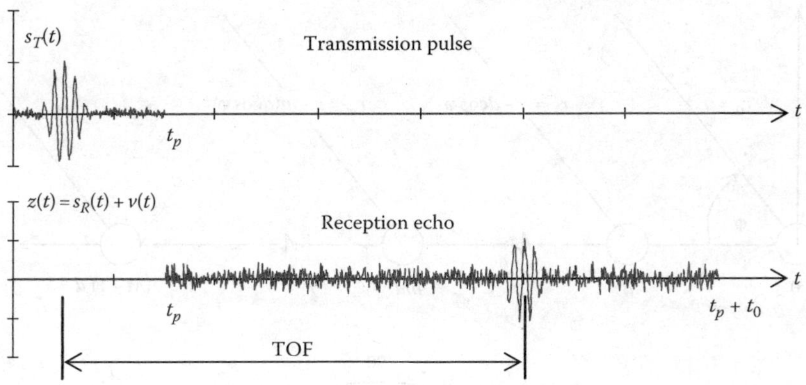

FIGURE 11.68
The pulse-echo method: (a) the transmitted pulse, $s_T(t)$; (b) the received echo, (t), consisting of the back-scattered signal, $s_R(t)$, and the noise, $v(t)$. The difference between the pulse transmission and echo reception is the TOF.

the pulse-echo principles (Figure 11.68). A pulse, consisting of a smooth-windowed tone-burst of duration t_p, is transmitted toward the target. The target reflects the signal and creates an echo that is detected by the radar. By analyzing the radar signal in the interval $(t_p, t_p + t_0)$, one identifies the delay, τ, representing the TOF taken by the wave to travel to the target and back. Knowing TOF and wave speed allows one to precisely determine the target position relative to the radar. A short signal (pulse) is sent, and the echoes resulting from wave backscatter at the internal material flaws are detected. Information regarding the geometric shape, size, and orientation as well as microstructure can be extracted from the patterns of ultrasonic backscatter echoes.

11.5.4.1 Embedded-Ultrasonic Structural Radar

The embedded-ultrasonic structural radar (EUSR) is a concept that utilizes the phased-array radar principles and ultrasonic guided waves (Lamb waves) to scan large surface areas of thin-wall structures for the detection of cracks and corrosion defects (Giurgiutiu et al., 2006). In the EUSR concept, the guided Lamb waves are generated with surface-mounted PWAS. The guided Lamb waves have the property of staying confined inside the walls of a thin-walled structure. Hence, they can travel over large distances with little attenuation. In addition, they can also travel inside curved walls with shallow curvature. Thus, the target location is described by the radial position, R, and the azimuth angle, ϕ. In conventional radars, the radar dish sweeps the horizon with a search beam that registers an echo when a target is detected. The phased-array radar replaces the rotating radar dish with an array of M active sensors that are electronically switched so as to generate a virtual sweep beam. The EUSR algorithm is adopted from the beamforming process currently used in phased-array radar applications. Consider a PWAS array as presented in Figure 11.69a. Each element in the PWAS array plays the role of both transmitter and receiver. A methodology is designed to change the role of each PWAS in a round robin fashion. The responses of the structure to all the excitation signals are collected. By applying the EUSR algorithm, an appropriate delay is applied to each signal in the data set to make them all focus on a direction denoted by angle ϕ. When

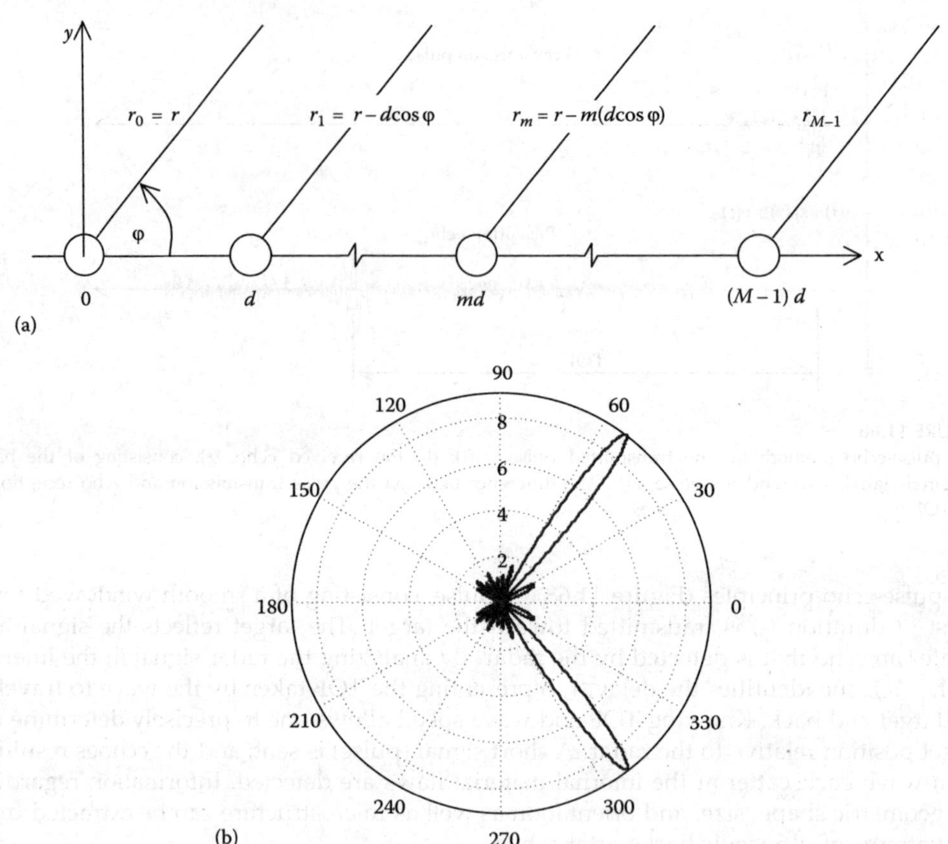

FIGURE 11.69

(a) Uniform linear array of M omnidirectional active sensors spaced at pitch d; (b) calculated beamforming pattern for a 9-sensor array (spacing $l = \lambda/2$) with 53° target azimuth.

this angle ϕ is changed from 0° to 180°, a virtual scanning beam is formed and a large area of the structure can be interrogated (Figure 11.69b).

Lamb waves can exist in a number of dispersive modes. However, through smoothed tone-burst excitation and frequency tuning, it is possible to confine the excitation to a particular Lamb-wave mode of carrier frequency F_c, wave speed c, and wave length $\lambda = c/F_c$. Hence, the smoothed tone-burst signal generated by one PWAS is assumed to be of the form

$$s_T(t) = s_0(t)\cos 2\pi F_c t, \quad 0 < t < t_p \tag{11.482}$$

where $s_0(t)$ is a short-duration smoothing window that is applied to the carrier signal of frequency F_c between 0 and t_p. As in conventional phased-array radar, we assume a uniform linear array of M active sensors (PWAS), with each PWAS acting as a point-wise omnidirectional transmitter and receiver. The PWAS in the array are spaced at a distance d which is assumed to be much smaller than the distance r to a generic, far-distance point, P. Since $d \ll r$, the rays joining the sensors with the point P can be assimilated with a parallel fascicle of azimuth ϕ (Figure 11.69b).

Because of the array spacing, the distance between one PWAS and the generic point P will be different from the distance between another PWAS and P. For the mth PWAS, the distance will be shortened by $m(d \cos \phi)$. If all the PWAS are fired simultaneously, the signal from the mth PWAS will arrive at P quicker by $\Delta_m(\phi) = m(d \cos \phi)/c$. If the PWAS are not fired simultaneously, but with some individual delays, δ_m, $m = 0, 1, \ldots, M-1$, then the total signal received at point P will be

$$s_P(t) = \frac{1}{\sqrt{r}} \sum_{m=0}^{M-1} s_T\left(t - \frac{r}{c} + \Delta_m(\phi) - \delta_m\right) \tag{11.483}$$

where $1/r$ represents the decrease in the wave amplitude due to the omnidirectional 2-D radiation and r/c is the delay due to the travel distance between the reference PWAS ($m = 0$) and the point P. (Wave-energy conservation, i.e., no dissipation, is assumed).

11.5.4.1.1 Transmitter Beamforming

Beamforming at angle ϕ_0 with an array of M omnidirectional sensors is based on the principles of constructive interference in the fascicle of parallel rays emanating from the array. The simplest way of achieving constructive interferences is to have $\delta_m = m\Delta(\phi)$ such that Equation (11.483) becomes

$$s_P(t) = M \frac{1}{\sqrt{r}} s_T\left(t - \frac{r}{c}\right) \tag{11.484}$$

i.e., an M times increase in the signal strength with respect to a simple sensor. This leads directly to the beamforming principle, i.e., if $\delta_m = m\dfrac{d}{c}\cos(\phi_0)$, and since $\Delta_m = m\dfrac{d}{c}\cos(\phi)$, then constructive interference (beamforming) takes place when $\cos(\phi) = \cos(\phi_0)$, i.e., at angles $\phi = \phi_0$ and $\phi = -\phi_0$. Thus, the forming of a beam at angles ϕ_0 and $-\phi_0$ is achieved through delays in the firing of the sensors in the array. Figure 11.69b shows the beamforming pattern for $\phi_0 = 53°$.

11.5.4.1.2 Receiver Beamforming

The receiver beamforming principles are reciprocal to those of the transmitter beamforming. If the point P is an omnidirectional source at azimuth ϕ_0, then the signals received at the mth sensor will arrive quicker by $m\Delta_0(\phi) = m(d \cos\phi_0)/c$. Hence, we can synchronize the signals received at all the sensors by delaying them by $\delta_m(\phi_0) = m\dfrac{d}{c}\cos(\phi_0)$.

11.5.4.1.3 Phased-Array Pulse-Echo

Assume that a target exists at azimuth ϕ_0 and distance R. The transmitter beamformer is sweeping the azimuth in increasing angles ϕ and receives an echo when $\phi = \phi_0$. The echo will be received on all sensors, but the signals will not be in synch. To synchronize the sensors signals, the delays $\delta_m(\phi_0) = m\dfrac{d}{c}\cos(\phi_0)$ are needed to be applied. The process is as follows. The signal sent by the transmitter beamformer is an M times boost of the original signal

$$s_P(t) = \frac{M}{\sqrt{R}} s_T\left(t - \frac{2R}{c}\right)$$

(11.485)

At the target, the signal is backscattered with a backscatter coefficient, A. Hence, the signal received at each sensor will be

$$\frac{AM}{R} s_T\left(t - \frac{2R}{c} + \Delta_m(\phi)\right)$$

(11.486)

The receiver beamformer assembles the signals from all the sensors with the appropriate time delays, i.e.,

$$s_R(t) = \frac{AM}{R} \sum_{m=0}^{M-1} s_T\left(t - \frac{2R}{c} + \Delta_m(\phi) - \delta_m\right)$$

(11.487)

Constructive interference between the received signals is achieved when $\delta_m = m\frac{d}{c}\cos(\phi_0)$. Thus, the assembled received signal will be again boosted M times with respect to the individual sensors, i.e.,

$$s_R(t) = \frac{AM^2}{R} \sum_{m=0}^{M-1} s_T\left(t - \frac{2R}{c}\right)$$

(11.488)

The time delay of the receive signal, $s_R(t)$, with respect to transmit signal, $s_T(t)$, is

$$\tau = \frac{2R}{c}$$

(11.489)

Measurement of the time delay τ observed in $s_R(t)$ allows one to calculate the target range, $R = c\tau/2$.

11.5.4.1.4 Practical Implementation of the EUSR Algorithm

The practical implementation of the signal generation and collection algorithms is described next. In a round-robin fashion, one active sensor at a time is activated as the transmitter. The reflected signals are received at all the sensors. The activated sensor acts in pulse-echo mode, i.e., as both transmitter and receiver, the other sensors act as passive sensors. Thus, an $M \times M$ matrix of signal primitives is generated (Table 11.16). The signal primitives are assembled into synthetic beamforming responses using the synthetic beam-former algorithm given by Equations (11.483) and (11.484). The delays, δ_j, are selected in such a way as to steer the interrogation beam at a certain angle, ϕ_0. The synthetic-beam sensor responses, $w_i(t)$, synthesized for a transmitter beam with angle ϕ_0, are assembled by the receiver beamformer into the total received signal, $s_R(t)$, using the same delay as for the transmitter beamformer. However, to apply this method directly, one needs to know

TABLE 11.16

$M \times M$ Matrix of Signal Primitives Generated in a Round-Robin Activation of the PWAS Elements of the Phased Array

| | | Transmitters | | | Synthetic Beamforming |
		T_0	T_1	T_{M-1}	Response
Receivers	R_0	$p_{0,0}(t)$	$p_{0,1}(t)$	$\cdots$ $p_{0,M-1}(t)$	$w_0(t)$
	R_1	$p_{1,0}(t)$	$p_{1,1}(t)$	$\cdots$ $p_{1,M-1}(t)$	$w_1(t)$
	R_2	$p_{2,0}(t)$	$p_{2,1}(t)$	$\cdots$ $p_{2,M-1}(t)$	$w_2(t)$
	$\cdots$	$\cdots$	$\cdots$	$\cdots$ $\cdots$	$\cdots$
	R_{M-1}	$p_{M-1,\,0}(t)$	$p_{M-1,1}(t)$	$\cdots$ $p_{M-1,\,M-1}(t)$	$w_{M-1}(t)$

the target angle ϕ_0. Since, in general applications, the target angle is not known, we need to use an inverse approach to determine it. Hence, we write the received signal as a function of the parameter ϕ_0, using the array unit delay for the direction ϕ_0 as $\Delta_0(\phi_0) = \dfrac{d}{c}\cos\phi_0$. (To accurately implement the time shifts when the time values fall in between the fixed values of the sampled time, we have used a spline interpolation algorithm).

A coarse estimate of the target direction is obtained by using an azimuth sweep technique in which the beam angle, ϕ_0, is modified until the maximum received energy is attained, i.e.,

$$\max E_R(\phi_0), \quad E_R(\phi_0) = \int\limits_{t_p}^{t_p+t_0} |s_R(t,\phi_0)|^2 dt \tag{11.490}$$

After a coarse estimate of the target direction is found ϕ_0, the actual round-trip TOF, τ_{TOF}, is calculated using an optimal estimator, e.g., the cross-correlation between the receiver and the transmitter signals

$$y(\tau) = \int\limits_{t_p}^{t_p+t_0} s_R(t)s_T(t-\tau)dt \tag{11.491}$$

Then, the estimated $\tau_{TOF} = 2R/c$ is obtained as the value of τ where $y(\tau)$ is maximum. Hence, the estimated target distance is

$$R_{\exp} = c\frac{\tau_{TOF}}{2} \tag{11.492}$$

This algorithm works best for targets in the far field for which the "parallel-rays" assumption holds. For targets in the near and intermediate field, a more sophisticated self-focusing algorithm that uses triangulation principles is used. This algorithm is an outgrowth of the passive-sensors target-localization methodologies. The self-focusing algorithm modifies the delay times used in each synthetic-beam response, $w_i(t)$. The total response is maximized by finding the focal point of individual responses, i.e., the common location of the defect that generated the echoes recorded at each sensor. For very close range targets, triangulation techniques are utilized.

11.5.4.2 EUSR System Design and Experimental Validation

The EUSR system consists of three major modules: (1) the PWAS array, (2) the DAQ module, and (3) the signal processing module. A system diagram is shown in Figure 11.70a. A proof-of-concept EUSR system was built in the Laboratory for Active Materials and Smart Structures (LAMSS) at the University of South Carolina to evaluate the feasibility and capability of the EUSR system.

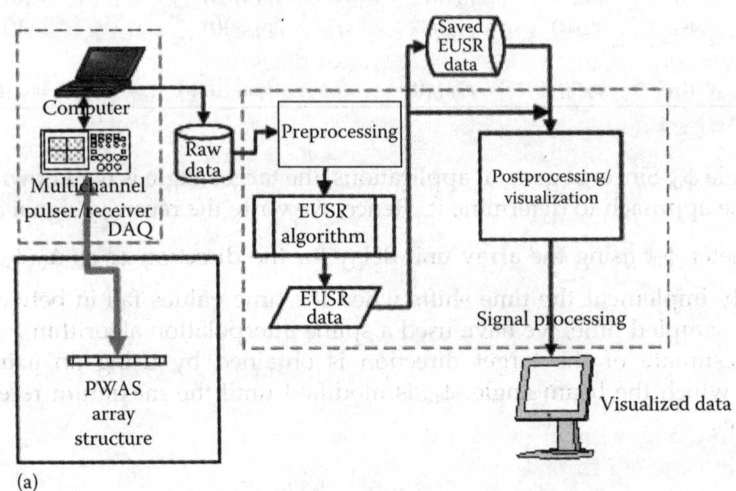

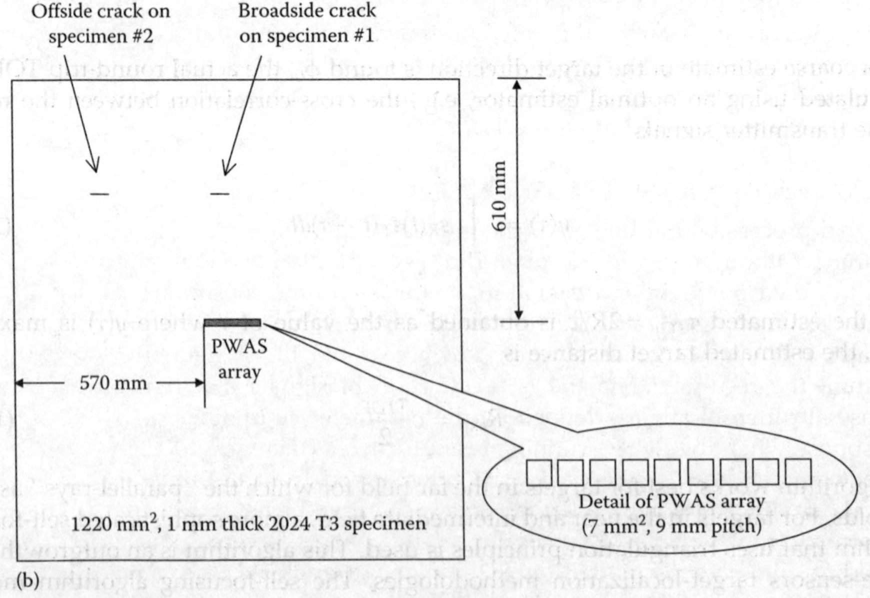

FIGURE 11.70
Proof-of-concept EUSR experiment: (a) data flow diagram; (b) thin plate specimens with broadside and offside cracks, and 9-element PWAS array at the center.

11.5.4.2.1 Experimental Setup

Three specimens were used in the experiments. These specimens were 1220-mm square panels of 1 mm thick 2024-T3 Al-clad aircraft grade sheet metal stock. One of the specimens (specimen #0) was pristine and used to obtain baseline data. The other two specimens were manufactured with simulated cracks. The cracks were placed on a line midway between the center of the plate and its upper edge (Figure 11.70b). The cracks were 19 mm long and 0.127 mm wide. On specimen #1, the crack was placed broadside with respect to the phase array at coordinates (0, 0.305 m), i.e., at $R = 305$ mm, $\phi_0 = 90°$. On the specimen #2, the crack was placed offside with respect to the phased array at coordinates (-0.305 m, 0.305 m) which corresponds to $R = 409$ mm, $\phi_0 = 136.3°$ with respect to the reference point of the PWAS array. The PWAS array was constructed using nine 7 mm sq., 0.2 mm thick piezoelectric wafers (American Piezo Ceramic Inc., APC-850) placed on a straight line in the center of the plate. The sensors were spaced at pitch $d = \lambda/2$, where $\lambda = c/f$ is the wavelength of the guided wave propagating in the thin-walled structure. Since the first optimum excitation frequency for S_0 mode was 300 kHz, and the corresponding wave speed was $c = 5.440$ km/s, the wavelength was $\lambda = 18$ mm. Hence, the spacing in the PWAS array was selected as $d = 9$ mm (Figure 11.70b).

The DAQ module consisted of an HP33120A arbitrary signal generator, a Tektronix TDS210 digital oscilloscope, and a portable PC with DAQ and GPIB interfaces. The HP33120A arbitrary signal generator was used to generate a 300 kHz Hanning-windowed tone-burst excitation with a 10 Hz repetition rate. Under the Hanning-windowed tone-burst excitation, one element in the PWAS array generated a Lamb waves package that spread out into the entire plate in an omnidirectional pattern (circular wave front). The Tektronix TDS210 digital oscilloscope, synchronized with the signal generator, collected the response signals from the PWAS array. One of the oscilloscope channels was connected to the transmitter PWAS while the other was switched among the remaining elements in the PWAS array by using a digitally controlled switching unit. A LabVIEW computer program was developed to digitally control the signal-switching, record the data from the digital oscilloscope, and generate the group of raw data files. Photographs of the experimental setup are presented in Figure 11.71.

11.5.4.2.2 Implementation of the EUSR Data Processing Algorithm

The signal-processing module reads the raw data files and processes them using the EUSR algorithm. Although the EUSR algorithm is computationally nonintensive, the large amount of data points in each signal made this step time consuming. Hence, we elected to save the resulting EUSR data on the PC for later retrieval and postprocessing. This approach also enables other programs to access the EUSR data. Based on the EUSR algorithm, the resulting data file is a collection of signals that represent the structure response at different angles, defined by the parameter ϕ. In other words, they represent the response when the EUSR scanning beam turned at incremental angles ϕ.

After being processed, the data were transformed from the time domain to the 2-D physical domain. Knowing the Lamb wave speed c, and using $r = ct$, the EUSR signal was transformed from voltage V versus time t to voltage V versus distance r. The signal detected at angle ϕ was plotted on a 2-D plane at angle ϕ. Since angle ϕ was stepped from 0° to180°, at constant increments, the plots covered a half space. These plots generate a 3-D surface, which is a direct mapping of the structure being interrogated, with the z value of the 3-D surface representing the detected signal at that (x, y) location (Figure 11.72a).

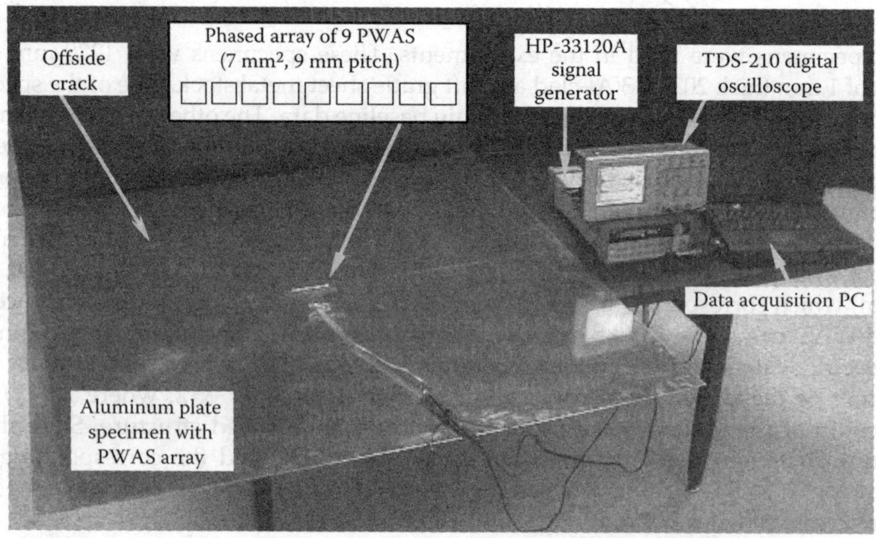

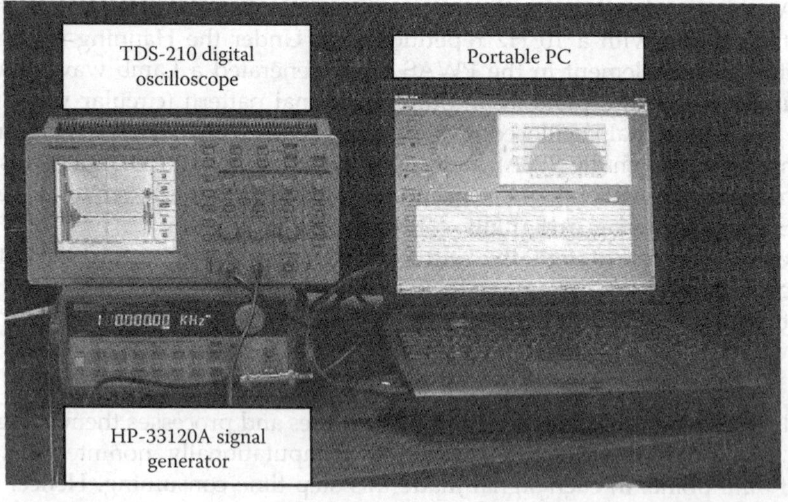

FIGURE 11.71
Instrumentation setup for EUSR experiment: (a) overall view showing the plate, active sensors, and instrumentation; (b) detail of the instrumentations and the data acquisition program.

If we present the z value on a color scale, then the 3-D surface is projected to the 2-D plane, and the color of each point on the plane represents the intensity of the reflections.

11.5.4.2.3 *Experimental Results*

Both the broadside crack (specimen #1) and the offside crack (specimen #2) were successfully detected. However, the offside crack presented a detection challenge since the direct

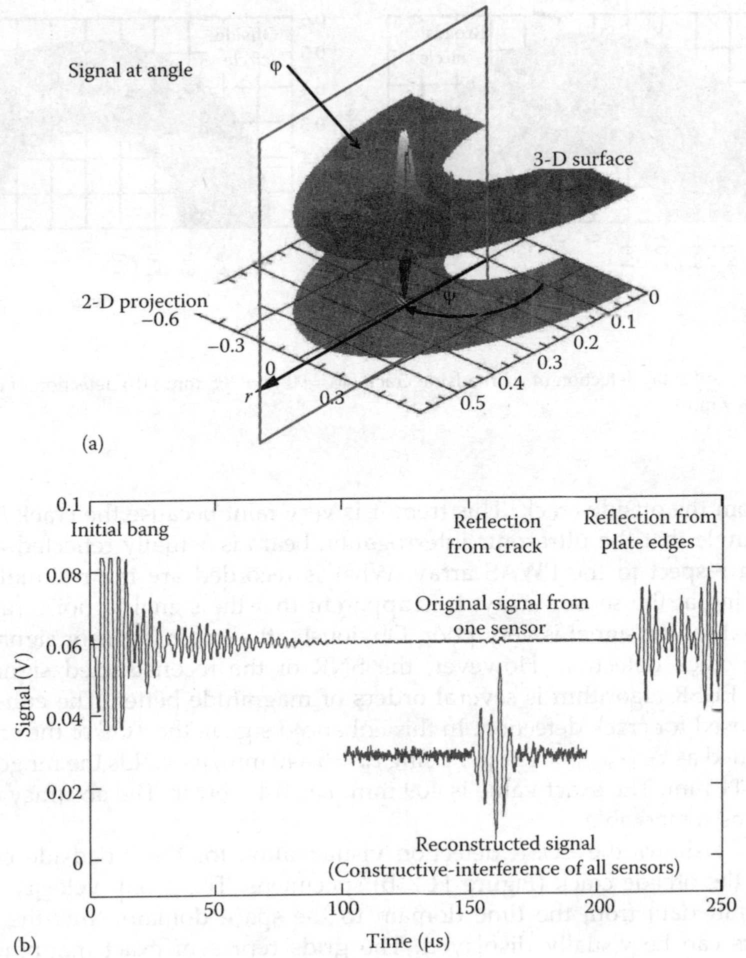

FIGURE 11.72
EUSR signal reconstruction examples: (a) 3-D visualization of EUSR signal reconstruction for the broadside crack specimen #1; (b) remarkable signal enhancement through the EUSR phased-array method illustrated on the offside crack specimen #2.

reflections from the crack were deflected away from the sensors array due to the inclination of the crack with respect to the beam axis. Only secondary backscatter signals, which were due to the discontinuity created by the crack tips in the elastic field, were recorded. These signals were very faint, and would not be detected by conventional methods. However, they could be detected through the constructive interference principles contained in the EUSR algorithm. An example of the EUSR signal reconstruction for the offside crack experiment is presented in Figure 11.72b. The original signal (top, blue) and the reconstructed signal (bottom, red) are presented. The reconstructed signal (bottom, red) was obtained by the constructive interference of all the original sensor signals with the appropriate time delays. (For display purposes, the signals in Figure 11.72a were separated vertically by the application of a DC bias). One notices that the original sensor signal (top, blue) displays a faint tremor at approximately 151 µs representing the backscatter

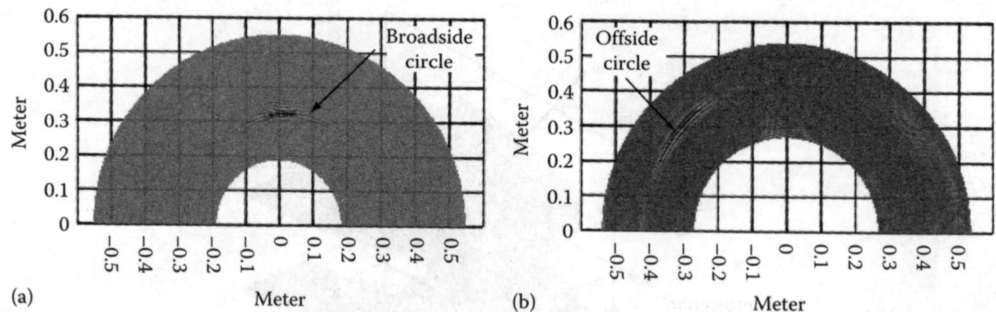

FIGURE 11.73

EUSR mapping results: (a) detection of a broadside crack ($\phi_0 = 90°$, $r = 305$ mm); (b) detection of an offside crack ($\phi_0 = 136°$, $r = 409$ mm).

reflection from the offside crack. This tremor is very faint because the crack is positioned at such an angle that the ultrasonic interrogation beam is actually reflected away and to the left with respect to the PWAS array. What is recorded are the residual backscatter signals arriving at the sensor array. It is apparent that the signal to noise ratio (SNR) of the individual sensor signal is very poor. Obviously, the original sensor signals could not be used for crack detection. However, the SNR of the reconstructed signal enhanced through the EUSR algorithm is several orders of magnitude better. The enhanced signal can now be used for crack detection. In this enhanced signal, the TOF of the crack signal is easily identified as $\tau_{TOF,offside} = 151$ μs. Using $c_g = 5.440$ mm/μs yields the range of the crack as $R_{offside} = 411$ mm. The exact value is 409 mm, i.e., 0.4% error. The accuracy of the EUSR method seems remarkable.

Figure 11.73 shows the EUSR detection visualization for the broadside crack (Figure 11.73a) and the offside crack (Figure 11.73b) specimens. The group velocity was used to map the EUSR data from the time domain to the space domain, thus the locations of the reflectors can be visually displayed. The grids represent exact mapping in meters. The shaded area represents the swept surface. The signal amplitude is presented on a color/grayscale intensity scale. The location of the crack is easily determined from the color/grayscale change. Figure 11.73a represents the results for the broadside specimen. The small area with darker color represents the high amplitude echo (reflected wave) generated when the scanning angular beam intercepted the crack. From the picture scale, we observe that this area is located at an angle of 90° and at approximately 0.3 m from the center of the plate. Careful analysis of the reconstructed signal yielded the exact $\tau_{TOF,broadside} = 112.4$ μs, corresponding to a radial position $R_{broadside} = 305.7$ mm. This differs a mere 0.2% from the actual position of the broadside crack on this specimen ($\phi_0 = 90°$, $r = 305$ mm). The dark area in the EUSR result predicted the simulated crack perfectly. Similarly, Figure 11.73b presents the results for the offside specimen. It is apparent that the offside crack is located just beyond the (-0.3 m, 0.3 m) coordinates, which compares very well with the actual values (-0.305 m, 0.305 m). The crack range $R_{offside} = 411$ mm, determined from the analysis of the reconstructed echo, showed 0.4% accuracy. Both Figures 11.72 and 11.73 prove that the detection sensitivity and accuracy of the EUSR method is very good. The implementation of these concepts in a graphical user interface (GUI) is represented in Figure 11.74. The angle sweep is performed automatically to produce the structure/defect imaging picture on the right. Manual sweep of the beam

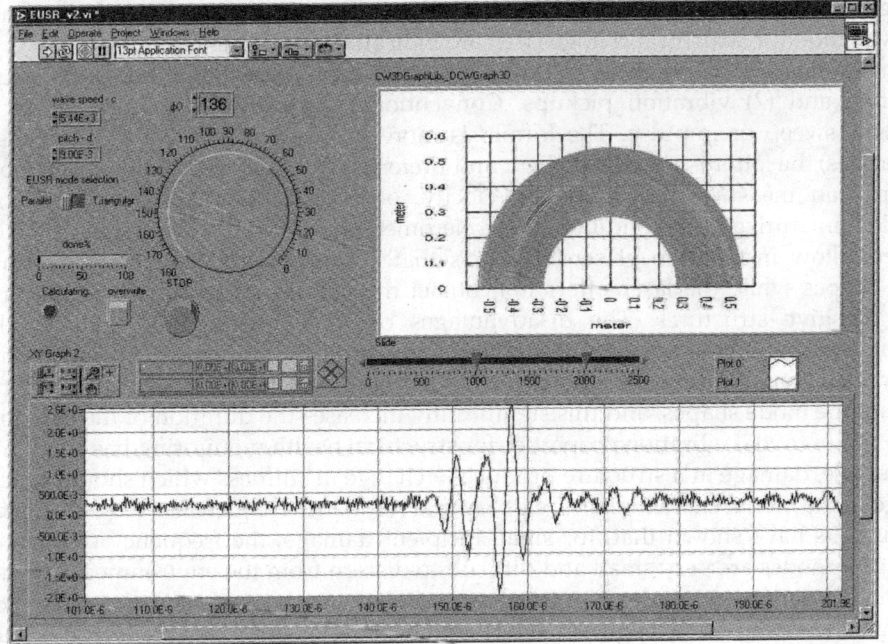

FIGURE 11.74

Graphical user interface (EUSR-GUI) front panel. The angle sweep is performed automatically to produce the structure/defect imaging picture on the right. Manual sweep of the beam angle can be also performed with the turn knob; the signal reconstructed at the particular beam angle (here, $\phi_0 = 136°$) is shown in the lower picture.

angle can be also performed with the turn knob; the signal reconstructed at the particular beam angle (here, $\phi_0 = 136°$) is shown in the lower picture.

11.5.5 Section Summary

This section has discussed the use of phased arrays of PWAS ultrasonic transducers. It was shown that the phased-array principles, initially derived for electromagnetic waves radar, can be also used with guided ultrasonic waves traveling inside plates and shells. Beam steering examples were theoretically derived and experimentally verified. The EUSR principles were presented together with several simple experiments. It was found that the EUSR principle has great potential for inspecting large structural areas from a single location with a minimal number of PWAS transducers arranged in a phased-array pattern.

11.6 PWAS Modal Sensors

11.6.1 Vibrations-Based Structural Health Monitoring

Modal analysis and dynamic structural identification have become an intrinsic part of current engineering practice. Structural frequencies, damping, and modes shapes identified through this process are subsequently used to predict dynamic response, avoid resonances, and even monitor structural changes that are indicative of incipient failure.

Conventional modal analysis testing relies on two essential components: (1) structural excitation and (2) vibration pickups. Conventional structural excitation can be either harmonic sweep or impulse. The former is more precise and can zoom in on resonant frequencies; the latter is more expedient and preferred for quick estimations. The vibration pickups can measure displacement, velocity, or acceleration. Advanced technologies include miniaturized self-conditioning accelerometers and laser velocimeters. The accelerometers allow installation of sensor arrays that accurately and efficiently measure the mode shapes while the laser offers noncontact measurements that are essential for low mass sensitive structures. The disadvantages of accelerometers are substantial cost, unavoidable bulkiness, and possible interference with the structural dynamics through their added mass. Laser velocimeters, on the other hand, need to scan the structure to measure the mode shapes, and this significantly increases the duration of the experiments.

The use of modal vibration properties for structural health monitoring has a rich history. In principle, damage in a structure produces a change in stiffness which should reflect in a change in the modal characteristics, e.g., natural resonance frequencies. However, practical experiments have shown that, for small incipient damage, the frequency changes in the low-order modes are very small and difficult to discern from the environmentally induced frequency changes. In order for frequency changes to be sensitive to small incipient damage, the modal wave length must be commensurable with the damage size. To achieve this desiderate, one has to go to very high frequencies that are beyond the capabilities of conventional modal analysis equipment.

11.6.2 1-D PWAS Modal Sensors

The advent of commercially available low-cost piezoceramics has opened new opportunities for high-frequency modal structural identification. Through their intrinsic electromechanical (E/M) coupling, the PWAS can act as both sensors and actuators. Additionally, their frequency bandwidth is orders of magnitude larger than that of conventional shakers and even impact hammers. Small piezoelectric wafers can be permanently attached to the structural surface. They could form sensor and actuator arrays that permit effective modal identification in a wide frequency band (Giurgiutiu and Zagrai, 2001).

11.6.2.1 Analytical Model

Consider a 1-D beam structure with a PWAS attached to its surface (Figure 11.75a). The beam has length l, axial stiffness EA, flexural stiffness EI, and mass per unit length $m = \rho A$. The PWAS has length l_a and lies between x_a and $x_a + l_a$. Upon activation, the PWAS expands by ε_{PWAS}. This generates a reaction force F_{PWAS} from the beam onto the PWAS and an equal and opposite force from the PWAS onto the beam (Figure 11.75b). This force excites the beam. At the neutral axis, the effect is felt as an axial force excitation, N_a, and a bending moment excitation, M_a. As the active sensor is electrically excited with a high-frequency harmonic signal, it will induce elastic waves into the beam structure. The elastic waves travel sideways into the beam structure and set it into oscillation.

In a steady-state regime, the structure oscillates at the PWAS excitation frequency. The reaction force per unit displacement (dynamic stiffness) presented by the structure to the PWAS will depend on the internal state of the structure, the excitation frequency, and the boundary conditions

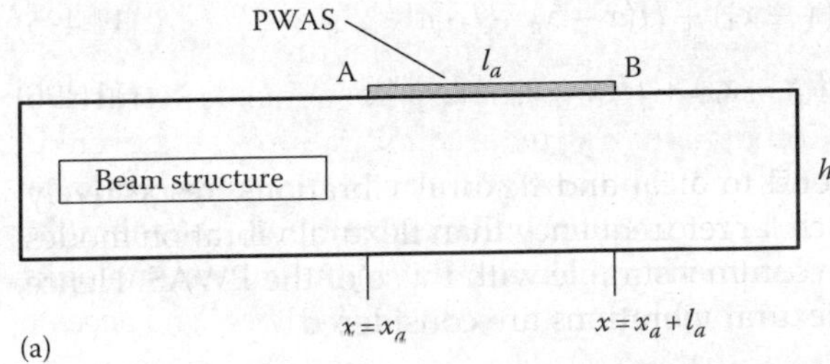

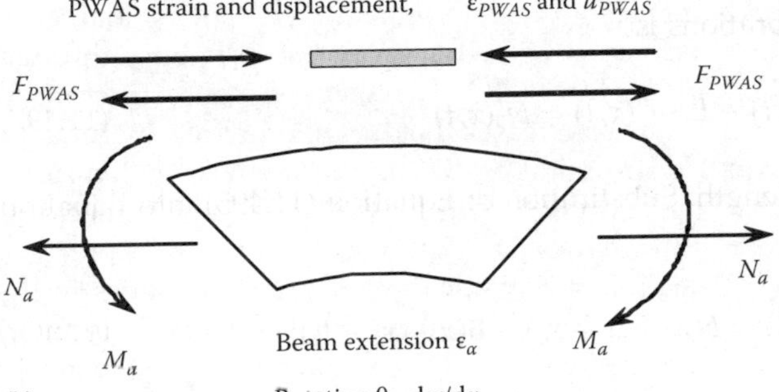

FIGURE 11.75
Interaction between a PWAS and a beam-like structural substrate: (a) geometry; (b) forces and moments.

$$k_{str}(\omega) = \frac{\hat{F}_{PWAS}(\omega)}{\hat{u}_{PWAS}(\omega)} \tag{11.493}$$

where $\hat{u}_{PWAS}(\omega)$ is the displacement amplitude at frequency ω, $\hat{F}_{PWAS}(\omega)$ is the reaction force, and $k_{str}(\omega)$ is the dynamic stiffness. The symbol $^\wedge$ signifies the complex amplitude of a time-varying function. Since the size of the PWAS is very small with respect to the size of the structure, formula (2) represents a point-wise structural stiffness.

11.6.2.2 Dynamics of the Structural Substrate

The response of the structural substrate to the PWAS excitation is deduced from the general theory of beam vibrations. However, the PWAS excitation departs from the typical textbook formulation since it acts as a pair of self-equilibrating axial forces and bending moments that are separated by a small finite distance, l_{PWAS}. This feature gives *gusto* to our analysis.

11.6.2.2.1 Definition of the Excitation Forces and Moments

The excitation forces and moments acting upon the beam structure are derived from the PWAS force, $F_{PWAS} = \hat{F}_{PWAS}e^{i\omega t}$, using the beam cross-section geometry (Figure 11.75b),

$$M_a = F_{PWAS}\frac{h}{2}, \quad N_a = F_{PWAS} \tag{11.494}$$

The space-wise distribution of excitation bending moment and axial force is expressed using the Heaviside function, $H(x - x_a)$, defined as $H(x - x_a) = 0$ for $x < x_a$, and $H(x - x_a) = 1$ for $x_a \le x$, i.e.,

$$N_e(x,t) = N_a[-H(x - x_a) + H(x - x_a - l_a)]e^{i\omega t} \tag{11.495}$$

$$M_e(x,t) = -M_a[-H(x - x_a) + H(x - x_a - l_a)]e^{i\omega t} \tag{11.496}$$

Equations (11.495) and (11.496) correspond to axial and flexural vibrations, respectively. Axial vibration modes are usually of much larger frequency than flexural vibration modes. However, their vibration frequencies are commensurable with those of the PWAS. Hence, in the present analysis, both axial and flexural vibrations are considered.

11.6.2.2.2 Axial Vibrations

The equation of motion for axial vibrations is

$$m\ddot{u}(x,t) - EAu''(x,t) = N_e'(x,t) \tag{11.497}$$

where $m = \rho A$ is the mass per unit length. Substitution of Equation (11.495) into Equation (11.497) yields

$$m\ddot{u}(x,t) - EAu''(x,t) = \hat{N}_a[-\delta(x - x_a) + \delta(x - x_a - l_a)]e^{i\omega t} \tag{11.498}$$

where δ is Dirac's function. Assume modal expansion

$$u(x,t) = \sum_{n=0}^{\infty} C_n U_n(x)e^{i\omega t} \tag{11.499}$$

where $U_n(x)$ are orthonormal mode shapes, i.e., $\int mU_mU_n dx = \delta_{mn}$, with $\delta_{mn} = 1$ for $m = n$ and 0 otherwise. C_n are the modal amplitudes. The mode shapes satisfy the free-vibration differential equation

$$EAU_n + \omega_n^2 mU_n = 0 \tag{11.500}$$

Hence, multiplication by $U_n(x)$ and integration over the length of the beam yields

$$C_n = \frac{1}{\omega_n^2 - \omega^2} \frac{\hat{N}_a}{\rho A}[-U(x_a) + U(x_a + l_a)] \tag{11.501}$$

Thus,

$$u(x,t) = \frac{\hat{N}_a}{m} \sum_{n=0}^{\infty} \frac{-U_n(x_a) + U_n(x_a + l_a)}{\omega_n^2 - \omega^2} U_n(x)e^{i\omega t} \tag{11.502}$$

11.6.2.2.3 Flexural Vibrations

For Euler–Bernoulli beams, the equation of motion under moment excitation is

$$m\ddot{w}(x,t) + EIw''''(x,t) = -M_e''(x,t) \tag{11.503}$$

Substitution of Equation (11.496) into Equation (11.503) yields

$$m\ddot{w}(x,t) + EIw''''(x,t) = \hat{M}_a[-\delta'(x - x_a) + \delta'(x - x_a - l_a)]e^{i\omega t} \tag{11.504}$$

where δ' is the first derivative of Dirac's function ($\delta' = H''$). Assume the modal expansion

$$w(x,t) = \sum_{n-N_1}^{N_2} C_n W_n(x)e^{i\omega t} \tag{11.505}$$

where $W_n(x)$ are the orthonormal bending mode shapes and N_1 and N_2 are the mode numbers enclosing the frequency band of interest. The mode shapes satisfy the free-vibration differential equation

$$EIW_n'''' = \omega_n^2 m W_n \tag{11.506}$$

Hence, multiplication by $W_n(x)$ and integration over the length of the beam yields

$$C_n = \frac{1}{\omega_n^2 - \omega^2}\frac{\hat{M}_a}{\rho A}\int_0^l W_n(x)[\delta'(x - x_a) - \delta'(x - x_a - l_a)]dx \tag{11.507}$$

Integration by parts and substitution into the modal expansion expression yields

$$C_n = -\frac{1}{(\omega_n^2 - \omega^2)}\frac{\hat{M}_a}{\rho A}\left[-W_n'(x_a) + W_n'(x_a + l_a)\right] \tag{11.508}$$

Hence,

$$w(x,t) = -\frac{\hat{M}_a}{\rho A}\sum_{n=1}^{\infty}\frac{-W_n'(x_a) + W_n'(x_a + l_a)}{\omega_n^2 - \omega^2}e^{i\omega t} \tag{11.509}$$

11.6.2.2.4 Calculation of Frequency Response Function and Dynamic Structural Stiffness

To obtain the dynamic structural stiffness, k_{str}, presented by the structure to the PWAS, we first calculate the elongation between the two points, A and B, connected to the PWAS ends. Kinematic analysis gives the horizontal displacement of a generic point P placed on the surface of the beam in terms of the axial and flexural displacements as

$$u_P(t) = u(x) - \frac{h}{2}w'(x) \tag{11.510}$$

where u and w are the axial and bending displacements measured at the neutral axis (Figure 11.76a). Letting P be A and B (Figure 11.76b), and taking the difference, yields

$$u_{PWAS}(t) = u_B(t) - u_A(t) = u(x_a,t) - u(x_a + l_a,t) - \frac{h}{2}[w'(x_a,t) - w'(x_a + l_a,t)] \tag{11.511}$$

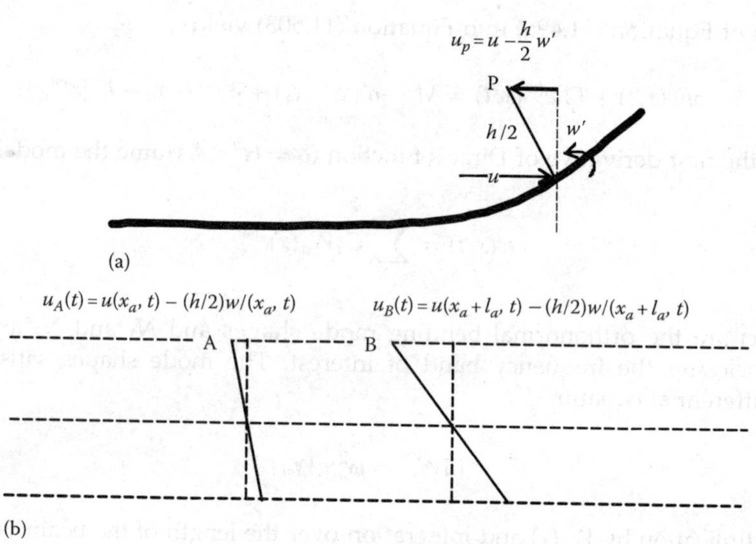

FIGURE 11.76
The total horizontal displacement, u_P, results from the superposition of axial displacement u and the rotation w': (a) single point, P; (b) two points, A and B.

Equations (11.494), (11.502), (11.509), (11.511) give

$$u_{PWAS} = \frac{F_{PWAS}}{\rho A} \left\{ \sum_{n_u} \frac{[U_{n_u}(x_a + l_a) - U_{n_u}(x_a)]^2}{\omega_{n_u}^2 - \omega^2} + \left(\frac{h}{2}\right)^2 \sum_{n_w} \frac{\left[W'_{n_w}(x_a + l_a) - W'_{n_w}(x_a)\right]^2}{\omega_{n_w}^2 - \omega^2} \right\}$$

$$(11.512)$$

where the axial and flexural vibrations frequencies and mode shapes were distinguished by the use of n_u, ω_{n_u}, and $U_{n_u}(x)$ and n_w, ω_{n_w}, and $W_{n_w}(x)$, respectively.

The frequency response function (FRF) for single input single output (SISO) excitation/response is the ratio between response and excitation. The FRF is usually denoted by $H(\omega)$. In the case of a PWAS driving a structure, the FRF is given by

$$H_{PWAS}(\omega) = \frac{u_{PWAS}}{F_{PWAS}}$$

$$(11.513)$$

In view of Equation (11.513), we can obtain the FRF by taking the $\hat{u}_{PWAS}$ given by Equation (11.512) and dividing it by $\hat{F}_{PWAS}$, i.e.,

$$H_{PWAS}(\omega) = \frac{u_{PWAS}}{F_{PWAS}} = \frac{1}{\rho A} \left\{ \sum_{n_u} \frac{[U_{n_u}(x_a + l_a) - U_{n_u}(x_a)]^2}{\omega_{n_u}^2 - \omega^2} \right.$$

$$\left. + \left(\frac{h}{2}\right)^2 \sum_{n_w} \frac{\left[W'_{n_w}(x_a + l_a) - W'_{n_w}(x_a)\right]^2}{\omega_{n_w}^2 - \omega^2} \right\}$$

$$(11.514)$$

This process yields the FRF of the structure under the excitation applied by the PWAS. This situation is similar to conventional modal testing with the proviso that the PWAS wafers are unobtrusive and permanently attached to the structure. For convenience, the axial and flexural components of the structural FRF can be expressed separately, i.e.,

$$H_u(\omega) = \frac{1}{\rho A} \sum_{n_u} \frac{[U_{n_u}(x_a + l_a) - U_{n_u}(x_a)]^2}{\omega_{n_u}^2 + 2i\zeta_{n_u}\omega_{n_u}\omega - \omega^2} \tag{11.515}$$

$$H_w(\omega) = \frac{1}{\rho A}\left(\frac{h}{2}\right)^2 \sum_{n_w} \frac{\left[W'_{n_w}(x_a + l_a) - W'_{n_w}(x_a)\right]^2}{\omega_{n_w}^2 + 2i\zeta_{n_w}\omega_{n_w}\omega - \omega^2} \tag{11.516}$$

Note that modal damping, ζ_n, has been introduced to account for the inherent dissipative losses encountered in practical experiments. The FRFs are additive such that the total FRF is simply

$$H(\omega) = H_u(\omega) + H_w(\omega) \tag{11.517}$$

The SISO FRF is the same as the dynamic structural compliance seen by the PWAS transducer. The dynamic structural stiffness is the inverse of the structural compliance, i.e.,

$$k_{str}(\omega) = \frac{F_{PWAS}}{u_{PWAS}} = \rho A \left\{ \sum_{n_u} \frac{[U_{n_u}(x_a + l_a) - U_{n_u}(x_a)]^2}{\omega_{n_u}^2 + 2i\zeta_{n_u}\omega_{n_u}\omega - \omega^2} \right.$$

$$\left. + \left(\frac{h}{2}\right)^2 \sum_{n_w} \frac{\left[W'_{n_w}(x_a + l_a) - W'_{n_w}(x_a)\right]^2}{\omega_{n_w}^2 + 2i\zeta_{n_w}\omega_{n_w}\omega - \omega^2} \right\}^{-1} \tag{11.518}$$

For free–free beams, the axial and flexural mode shapes can be calculated with the formulae given in Sections 11.2.2.1 and 11.2.2.2.

11.6.2.3 Numerical Simulations and Experimental Results

The analytical model was used to perform several numerical simulations that directly predict the E/M impedance and admittance signature at the PWAS terminals during structural identification. Subsequently, experiments were performed to verify these predictions. We consider a set of specimens consisting of small steel beams ($E = 200$ GPa, $\rho = 7750$ kg/m^3) of various thicknesses and widths fabricated in the laboratory. All beams were $l = 100$ mm long with various widths $b_1 = 8$ mm (narrow beams) and $b_2 = 19.6$ mm (wide beams). The nominal thickness of the specimen was $h_1 = 2.6$ mm; by gluing two specimens back-to-back, we were also able to create double thickness specimens $h_2 = 5.2$ mm. Thus, four beam types were used (Figure 11.77): (1) narrow-thin, (2) narrow-thick, (3) wide-thin, and (4) wide-thick.

The comparison of wide and narrow beams was aimed at identifying the width effects in the frequencies spectrum while the change from double to single thickness was aimed at simulating the effect of corrosion (for traditional structures) and disbonding/delamination

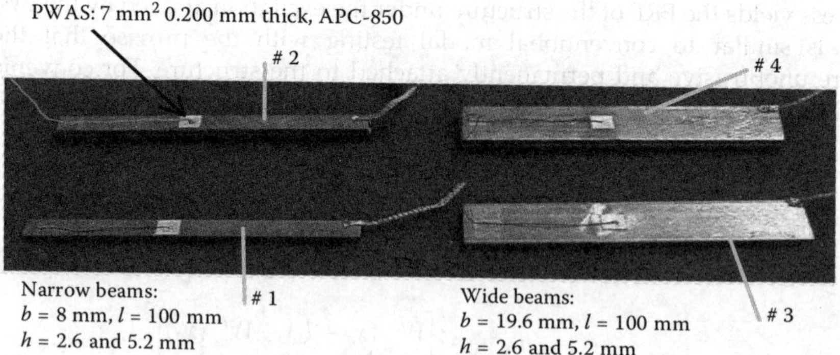

FIGURE 11.77
Steel-beam specimens to simulate 1-D structure.

on adhesively bonded and composite structures. All specimens were instrumented with thin 7-mm square PWAS ($l_a = 7$ mm, $b_a = 7$ mm, and $t_a = 0.22$ mm) placed at $x_a = 40$ mm from one end.

The numerical simulation was performed using the modal analysis theory discussed in the previous sections. Numerically exact expressions for the axial and flexural frequencies and mode shapes were used. Steel beams were analyzed. The damping coefficient was assumed $\zeta = 1\%$. The simulation was performed over a modal subspace that incorporates all modal frequencies in the frequency bandwidth of interest. The theoretical analysis indicates that these frequencies should be identical with the basic beam resonances, as predicted by classical vibration analysis. The "Calc." columns of Table 11.17 show the first six predicted resonances for axial and flexural vibrations. The experimental setup is shown in Figure 11.78.

During the experiments, recording of the E/M impedance real part spectrum with the HP 4194A Impedance Analyzer was performed in the 1–30 kHz range. To approximate the free–free boundary conditions, the beams were supported on common packing foam. The beam natural frequencies were identified from the E/M impedance spectrum.

The results are given in the "Exp." columns of Table 11.17. It should be noted that the errors are small and within the range that is normally accepted in experimental modal analysis. When the beam thickness was doubled, the frequencies also doubled. This is consistent with theoretical prediction. However, the error between theory and experiment seems larger for the double thickness beam, which may be caused by the compliance of the adhesive layer used in the construction of the double thickness beam. Even so, the confirmation of theoretical predictions by the experimental results is quite clear. The effect of beam width is demonstrated by comparison of "Narrow Beam" and "Wide Beam" columns in Table 11.17. The experimental results indicate clusters of frequencies that move to higher frequencies as the width of the beam is reduced. We associate these clusters with width vibrations that are not covered by the simple 1-D beam theory.

The width vibrations are also influenced by the thickness, i.e., they shift to higher frequencies as the thickness is increased. This is also noticeable in Table 11.17, which shows that the lowest cluster appeared for the single thickness wide specimen and the highest cluster for double thickness narrow beam. Another important aspect that needs to be noticed is that the first axial mode of vibrations is at 25 kHz. This explains the ~25 kHz

TABLE 11.17

Theoretical and Experimental Results for Wide and Narrow Beams with Single and Double Thickness

#	Beam #1 (Narrow Thin)			Beam #2 (Narrow Thick)			Beam #3 (Wide Thin)			Beam #4 (Wide Thick)		
	Calc., kHz	Exp., kHz	$\Delta\%$	Calc., kHz	Exp., kHz	$\Delta\%$	Calc., kHz	Exp., kHz	$\Delta\%$	Calc., kHz	Exp., kHz	$\Delta\%$
1	1.396	1.390	−0.4	2.847	2.812	−1.2	1.390	1.363	−1.9	2.790	2.777	−0.5
2	3.850	3.795	−1.4	7.847	7.453	−5.2	3.831	3.755	−2	7.689	7.435	−3.4
3	7.547	7.4025	−2	15.383	13.905	−10.6	7.510	7.380	−1.7	15.074	13.925	−8.2
4	12.475	12.140	−2.7		20.650		12.414	12.093	−2.6		21.825	
5	18.635	17.980	−3.6	25.430	21.787	−16.7	18.545	17.965	−3.2	24.918	22.163	−12.4
6		24.840						24.852				
7	26.035	26.317	1	26.035	26.157	0.5	26.022	26.085	0.2	25.944	26.100	0.6
		Cluster			Cluster			Cluster			Cluster	
		175 kHz			210 kHz			35 kHz			60 kHz	

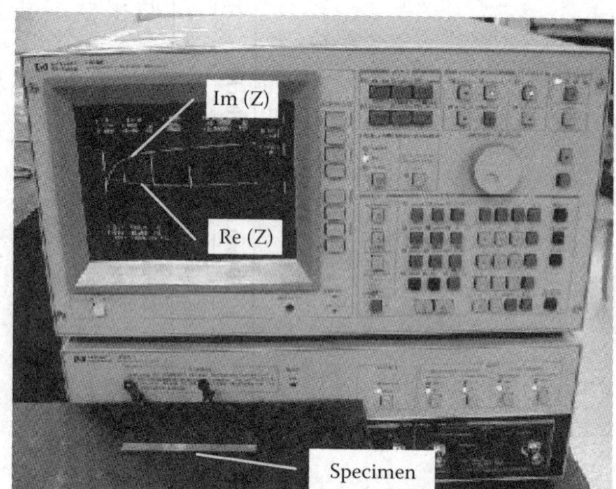

FIGURE 11.78

Experimental setup for dynamic identification of steel beams.

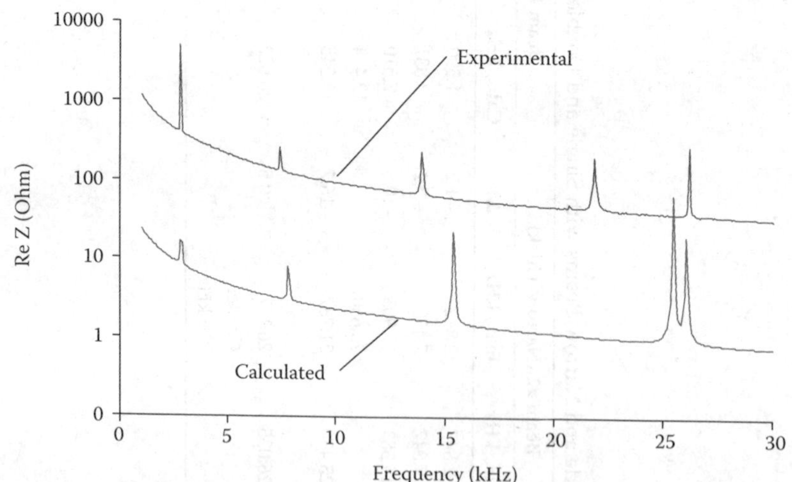

FIGURE 11.79

Experimental and calculated spectra of frequencies for double-thickness narrow beam (Beam #2).

double frequencies observed in the single-thickness beams, both narrow and wide. An example of the actual impedance spectra is given in Figure 11.79. The calculated and measured results are shown superposed. It was found that for the first four modes of single-thickness beams, the predicted and measured frequency values almost superpose. For the fifth mode, there is a slight difference.

These results prove that the predicted and measured results are in close agreement, well within the tolerance that is normally expected from experimental modal analysis. The results obtained for double-thickness beams are less precise due to inhomogeneity introduced by the layer of glue between single-thickness beams. This inhomogeneity was found to alter E/M impedance response of a beam.

11.6.2.4 Comparison with Conventional Methods

In order to highlight the advantages of our new technology, dynamic identification was also attempted with conventional modal analysis methods. A small steel beam with dimensions identical to the #1 specimen (single-thickness beam) was instrumented with two CEA-13-240UZ-120 strain gauges connected in half bridge configuration to a P-3500 strain indicator obtained from Measurements Group, Inc. The specimen was suspended in a free–free configuration and excited with a sharp impact. The resulting signal was collected with an HP 54601B digital oscilloscope and processed numerically on a PC. Standard signal analysis algorithms (FFT) were used to extract the frequency spectrum. The first natural frequency (1.387 kHz) was clearly displayed.

The second natural frequency (3.789 kHz) could also be identified, but with a much weaker amplitude. These results are consistent with theoretical values and the experimental results obtained with our new technology, as presented in Table 11.17. However, the impact excitation method was not able to excite the other higher frequencies depicted in Table 11.17, most probably, due to bandwidth limitations inherent in impact excitation approach. We also considered the use of another conventional modal analysis method, specifically the sweep excitation method. In principle, sweep excitation would be able to reach all the natural frequencies within the sweep bandwidth. However, the application of this method to our small specimen was not found feasible due to attachment difficulties and the fact that the kHz frequency range could not be easily achieved with conventional shakers. This demonstrates that, for the type of small rigid specimens as considered in our study, our proposed new technology has a niche of its own that cannot be filled by conventional modal analysis methods.

11.6.2.5 Noninvasive Characteristics of the PWAS Modal Sensors

The active sensors used in the experiments were very small and did not significantly disturb the dynamic properties of the structure under consideration. Table 11.18 presents the mass and stiffness for the sensor and structure. For comparison, the mass of the accelerometer is also included in Table 11.18.

The data in Table 11.18 illustrate numerically the noninvasive properties of PWAS. The mass and stiffness additions brought by sensor are within the 1% range (0.5% for mass and 1.5% for stiffness). In spite of its small dimensions, the PWAS was able to adequately perform the dynamic structural identification of the test specimens, as illustrated in Figure 11.79. The numerical results of this dynamic identification are given in Table 11.17. If the same identification was to be attempted with classical methodology, i.e., using an accelerometer and an impact hammer, the mass addition due to accelerometer would have been around 4.3%, which would greatly contaminate the results. This simple

TABLE 11.18

Numerical Illustration of the Noninvasive Properties of PWAS Transducers

Item	Mass (g)	Percentage of Structural Mass	Stiffness (M N/m)	Percentage of Structural Stiffness
PWAS	0.082	0.5%	15	1.5%
Structure	16.4	N/A	1000	N/A
Accelerometer: 352A10, PCB Piezotronics	0.7	4.3%	N/A	N/A

example underlines the fact that the use of PWAS is not only advantageous but also irreplaceable in certain situations. For small components such as in precision machinery and computer industry, the use of PWAS could be the only practical option for in-situ structural identification.

11.6.2.6 PWAS Self-Diagnostics

PWAS affixed to, or embedded into, a structure play a major role in the successful operation of a health monitoring and damage detection system. The integrity of the sensor and consistency of the sensor/structure interface are essential elements that can "make or break" an experiment. The general expectation is that once the PWAS have been placed on or into the structure, they will behave consistently throughout the duration of the health monitoring exercise. For real structures, the duration of the health monitoring process is extensive and can span several years. It will also encompass various service conditions and several loading cases. Therefore, in-situ self-diagnostics methods are mandatory. The PWAS should be scanned periodically to determine their integrity. They should be also scanned prior to any damage detection cycle. Self-diagnostic methods for assessing the sensor integrity are necessary to ensure that the health monitoring process is progressing as expected. For PWAS, a sensor self-diagnostic method is readily available through the E/M impedance technique. This self-diagnostic method works as follows.

Our preliminary tests have shown that the reactive (imaginary) part of the impedance (Im Z) can be a good indication of active sensor integrity. This is justified by the fact that the PWAS is predominantly a capacitive device and its impedance is dominated by its reactive part, $1/i\omega C$. Base-line signatures taken at the beginning of the health monitoring process can be compared with current reading in order to identify the defective PWAS. In this comparison, the imaginary part of the PWAS complex impedance should be used. Figure 11.80 compares the Im Z spectrum of a well-bonded PWAS with that of a disbonded (free) PWAS. For the disbonded PWAS, the appearance of the free-vibration resonance and the disappearance of structural resonances constitute unambiguous features that can be used for automated PWAS self-diagnostics.

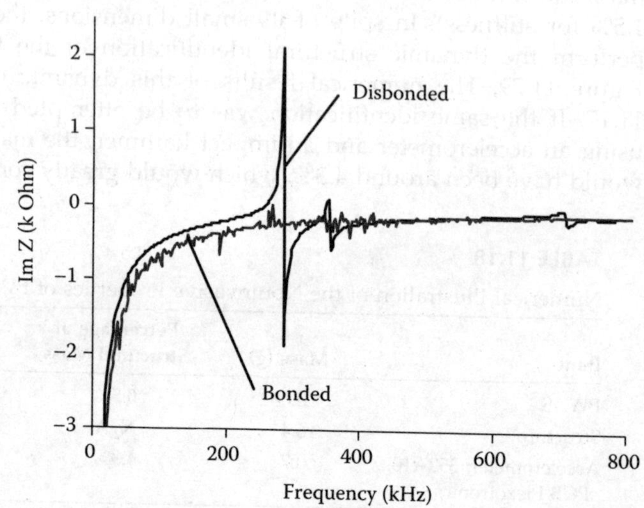

FIGURE 11.80
PWAS self-diagnostic using the imaginary part of the E/M impedance, when PWAS is disbonded, new free-vibration resonance features appear at ~267 kHz.

11.6.2.7 Typical Applications

The PWAS and the associated structural dynamics identification methodology based on the electromechanical impedance response are ideally suited for small rigid machinery parts that have natural frequencies in the kilohertz range. As an example, we considered the aircraft turbo-engine blade shown in Figure 11.81. Two PWAS were installed, one on the blade, the other on the root. E/M impedance spectrum was recorded, and natural frequencies could be easily identified (Figure 11.82).

11.6.2.8 Section Summary

The benefits and limitations of using embedded piezoelectric active sensors for structural identification at ultrasonic frequency are highlighted. An analytical model based on structural vibration theory and theory of piezoelectricity was developed and used to predict the electromechanical (E/M) impedance response as it would be measured at the piezoelectric

FIGURE 11.81
Aircraft turbo-engine blade equipped with PWAS.

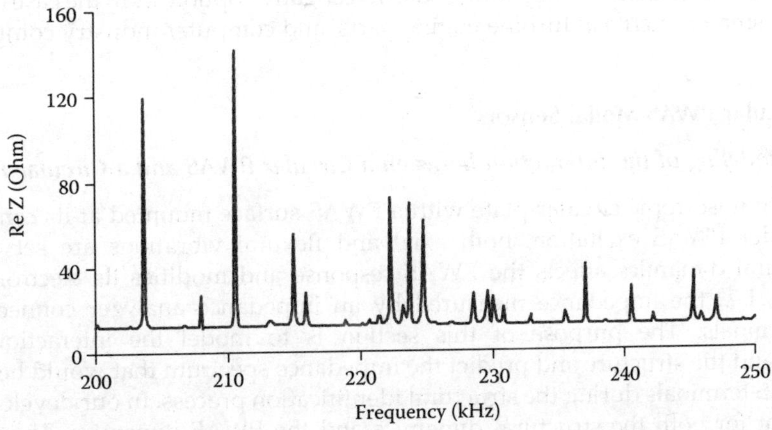

FIGURE 11.82
E/M impedance spectrum of aircraft turbo-engine blade.

active sensor's terminals. The model considers 1-D structures and accounts for both axial and flexural vibrations. Elastically constrained piezoelectric active sensors permanently bonded to the structure were considered. The derived mathematical expressions accounted for the dynamic response of both the sensor and the structure. Experiments were conducted on simple specimens in support of the theoretical investigation and on realistic turbine blade specimen to illustrate the method's potential. It was shown that the E/M impedance spectrum recorded by the piezoelectric active sensor accurately represents the mechanical response of a structure. It was further proved that the response of the structure is not modified by the presence of the sensor, thus validating the latter's noninvasive characteristics. It is shown that such sensors of negligible mass can be permanently applied to the structure creating a nonintrusive sensor array adequate for online automatic structural identification and health monitoring. The sensor calibration procedure was presented. Numerical estimation of the noninvasive properties of the proposed active sensors in comparison with conventional sensors is presented. Self-diagnostics capabilities of the proposed sensors were also investigated, and methods for automatic self-test were were discussed. As discussed in this chapter, the E/M impedance method, when using just one active sensor, can only detect structural resonances. The detection of structural mode shapes is also possible but requires the simultaneous use of several sensors, their number being in direct relationship to the desired modal resolution.

A limitation of the E/M impedance technique is that it is less effective at low frequencies than high frequencies. In our experiments, we found that below 5 kHz, the resonance peaks are buried in the overall electric response. These difficulties were alleviated by narrowband tuning. The numerical results reported for the 1–5 kHz range were obtained with this technique. However, below 1 kHz, the E/M impedance method is simply not recommended. Above 5 kHz, the E/M impedance method is definitely superior to other experimental modal analysis and dynamic identification methods.

In view of these advantages and disadvantages, it is felt that PWAS in conjunction with the E/M impedance technique have their niche as a structural identification methodology using self-sensing, permanently attached active sensors. Due to their perceived low cost (<$10), these active sensors can also be inexpensively configured as sensor arrays. The proposed method can be a useful and reliable tool for automatic online structural identification in the ultrasonic frequencies range. The use of PWAS can not only be advantageous but also, in certain situations, may be the sole investigative option, as in the case of precision machinery, small but critical turbine-engine parts, and computer industry components.

11.6.3 Circular PWAS Modal Sensors

11.6.3.1 Modeling of the Interaction between a Circular PWAS and a Circular Plate

Assume a thin isotropic circular plate with a PWAS surface mounted at its center (Figure 11.83). Under PWAS excitation, both axial and flexural vibrations are set in motion. The structural dynamics affects the PWAS response and modifies its electromechanical impedance, i.e., the impedance measured by an impedance analyzer connected to the PWAS terminals. The purpose of this section is to model the interaction between the PWAS and the structure and predict the impedance spectrum that would be measured at the PWAS terminals during the structural identification process. In our development, we will account for both the structural dynamics and the PWAS dynamics. The interaction between the PWAS and the structure is modeled as shown in Figure 11.84. The structure is assumed to present to the PWAS an effective structural dynamic stiffness, $k_{str}(\omega)$, which

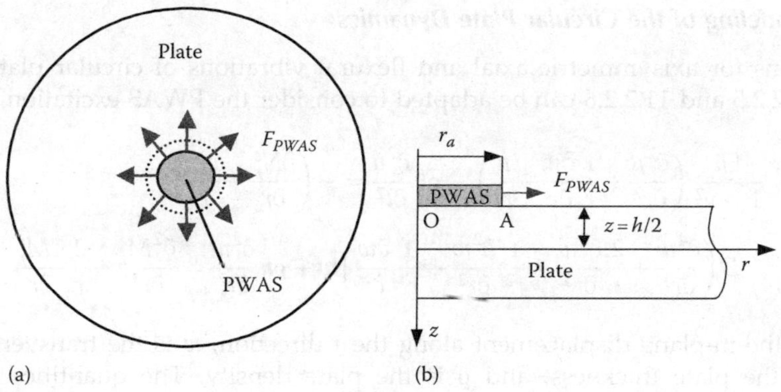

FIGURE 11.83
(a) Circular PWAS mounted on a circular plate; (b) cross-section schematics.

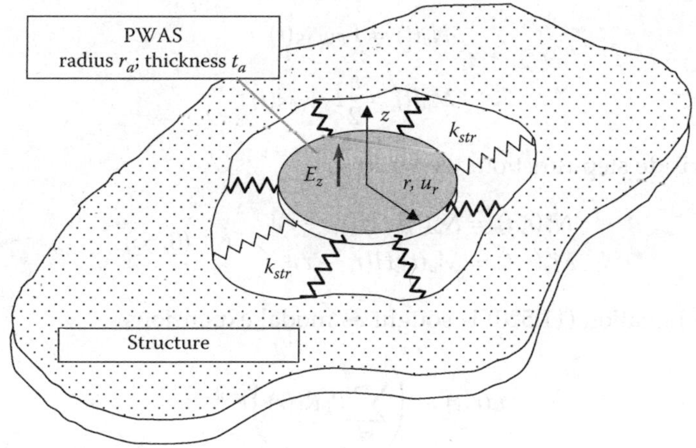

FIGURE 11.84
Circular PWAS constrained by structural stiffness, $k_{str}(\omega)$.

includes both axial and flexural modes. The problem is formulated in terms of interaction line force, F_{PWAS}, and the corresponding displacement, u_{PWAS}, measured at the PWAS circumference. Under the idea bonding assumption, the surface adhesion between the PWAS and the structure can be reduced to an effective boundary interaction between the radial displacement and the line force at the PWAS circumference (similar to the pin-force model in 1-D analysis of PWAS–structure interaction).

When the PWAS is excited with an oscillatory voltage, its volume expands in phase with the voltage in accordance with the piezoelectric effect. Expansion of the PWAS mounted on the surface of the plate induces a surface reaction from the plate in the form of the line force distributed around the PWAS circumference, $F_{PWAS}(t) = \hat{F}_{PWAS}e^{i\omega t}$. The reaction force, $F_{PWAS}(t)$, depends on the PWAS radial displacement, $u_{PWAS}(t)$, and the frequency-dependent dynamic stiffness, $k_{str}(\omega)$, presented by the structure to the PWAS

$$F_{PWAS}(t) = k_{str}(\omega)\, u_{PWAS}(t) \tag{11.519}$$

11.6.3.2 *Modeling of the Circular Plate Dynamics*

The equations for axisymmetric axial and flexural vibrations of circular plates given in Sections 11.2.2.5 and 11.2.2.6 can be adapted to consider the PWAS excitation, i.e.,

$$\frac{Eh}{1-v^2}\left(\frac{\partial^2 u}{\partial r^2}+\frac{1}{r}\frac{\partial u}{\partial r}-\frac{u}{r^2}\right)-\rho h\frac{\partial^2 u}{\partial t^2}=-\left(\frac{\partial N_r^e}{\partial r}+\frac{N_r^e}{r}\right)$$

$$D\left(\frac{\partial^4 w}{\partial r^4}+\frac{2}{r}\frac{\partial^3 w}{\partial r^3}-\frac{1}{r^2}\frac{\partial^2 w}{\partial r^2}+\frac{1}{r^3}\frac{\partial w}{\partial r}\right)w+\rho h\frac{\partial^2 w}{\partial t^2}=\frac{\partial^2 M_r^e}{\partial r^2}+\frac{2}{r}\frac{\partial M_r^e}{\partial r} \tag{11.520}$$

where u is the in-plane displacement along the r-direction, w is the transverse displacement, h is the plate thickness, and ρ is the plate density. The quantities N_r^e and M_r^e are excitation line forces and line moments acting over the whole surface of the plate. These excitation forces and moments originate in the PWAS force, $F_{PWAS}(t)$, acting at the surface of the plate at $r=r_a$. Resolving this force at the plate midplane, we get a line force and a line moment, i.e.,

$$N_a(t)=F_{PWAS}(t)$$

$$M_a(t)=\frac{h}{2}F_{PWAS}(t) \tag{11.521}$$

Using the Heaviside step function, we write

$$\begin{aligned}N_r^e(r,t)&=N_a(t)[-H(r_a-r)]\\ M_r^e(r,t)&=M_a(t)[H(r_a-r)]\end{aligned}\;,\quad r\in(0,\infty) \tag{11.522}$$

The solution of Equation (11.520) is sought as modal expansions

$$u(r,t)=\left(\sum_k P_k R_k(r)\right)e^{i\omega t}$$

$$w(r,t)=\left(\sum_m G_m W_m(r)\right)e^{i\omega t} \tag{11.523}$$

where $R_k(r)$ and $W_m(r)$ are the mode shapes given by Equations (11.62) and (11.69) for axial and flexural vibrations and P_k and G_m are the corresponding modal participation factors. Substitution of Equation (11.523) into Equation (11.520), multiplication by the mode shapes, and integration over the plate with the use of orthonormality conditions (11.63) and (11.71) yield the modal participation factors

$$P_k=\frac{2N_a}{\rho h a^2}\frac{\left[r_a R_k(r_a)-\int_0^a R_k(r)H(r_a-r)dr\right]}{(\omega_k^2-2i\varsigma_k\omega\omega_k+\omega^2)} \tag{11.524}$$

$$G_m=\frac{2M_a}{\rho h a^2}\frac{\left[3W_m(r_a)+r_a W_m'(r_a)\right]}{(\omega_m^2-2i\varsigma_m\omega\omega_m+\omega^2)}$$

where ζ_k and ζ_m are modal damping ratios.

11.6.3.3 Calculation of the Effective Structural Stiffness

The axial modes have in-plane motion whereas the flexural modes have out-of-plane flexural motion. We calculate the effective structural stiffness $k_{str}(\omega)$ from the structural response to PWAS excitation. The radial displacement at the rim of the PWAS can be expressed in the form

$$u_{PWAS}(r_a, t) = u(r_a, t) - \frac{h}{2} w(r_a, t) \tag{11.525}$$

Note that $u(r_a, t)$ and $w(r_a, t)$ represent displacements at the plate mid-surface while $u_{PWAS}(r_a, t)$ is measured at the plate upper surface (Figure 11.83). Using Equation (11.523), we write

$$u_{PWAS}(r_a, t) = \sum_k P_k R_k(r_a) e^{i\omega t} - \frac{h}{2} \sum_m G_m W'_m(r_a) e^{i\omega t} \tag{11.526}$$

Substitution of Equation (11.524) into Equation (11.526) yields the PWAS displacement in terms of the interface force, F_{PWAS}, and the axial and flexural dynamics of the plate. Discarding the time dependence, we write

$$FRF(\omega) \frac{\hat{u}_{PWAS}(\omega)}{F_{PWAS}(\omega)}$$

$$= \frac{1}{\rho a^2} \left[\frac{2}{h} \sum_k \frac{\left[r_a R_k(r_a) - \int_0^a R_k(r) H(r_a - r) dr \right] R_k(r_a)}{(\omega_k^2 - 2i\varsigma_k \omega\omega_k + \omega^2)} + \frac{h}{2} \sum_m \frac{\left[3W_m(r_a) + r_a W'_m(r_a) \right] W'_m(r_a)}{(\omega_m^2 - 2i\varsigma_m \omega\omega_m + \omega^2)} \right] \tag{11.527}$$

Recalling Equation (11.519), we write $k_{str}(\omega) = \hat{F}_{PWAS}(\omega)/\hat{u}_{PWAS}(\omega)$, i.e.,

$$k_{str}(\omega) = \rho a^2 \left[\frac{2}{h} \sum_k \frac{\left[r_a R_k(r_a) - \int_0^a R_k(r) H(r_a - r) dr \right] R_k(r_a)}{(\omega_k^2 - 2i\varsigma_k \omega\omega_k + \omega^2)} \right.$$

$$\left. + \frac{h}{2} \sum_m \frac{\left[3W_m(r_a) + r_a W'_m(r_a) \right] W'_m(r_a)}{(\omega_m^2 - 2i\varsigma_m \omega\omega_m + \omega^2)} \right]^{-1} \tag{11.528}$$

Upon inversion, one can obtain the frequency response function (FRF) of the structure when subjected to PWAS excitation, i.e., $FRF_{str}(\omega) = k_{str}^{-1}(\omega)$.

11.6.3.4 Model Validation through Numerical and Experimental Results

A series of experiments were conducted on thin-gauge aluminum plates to validate the theoretical results. The set of specimens consisted of five identical circular plates

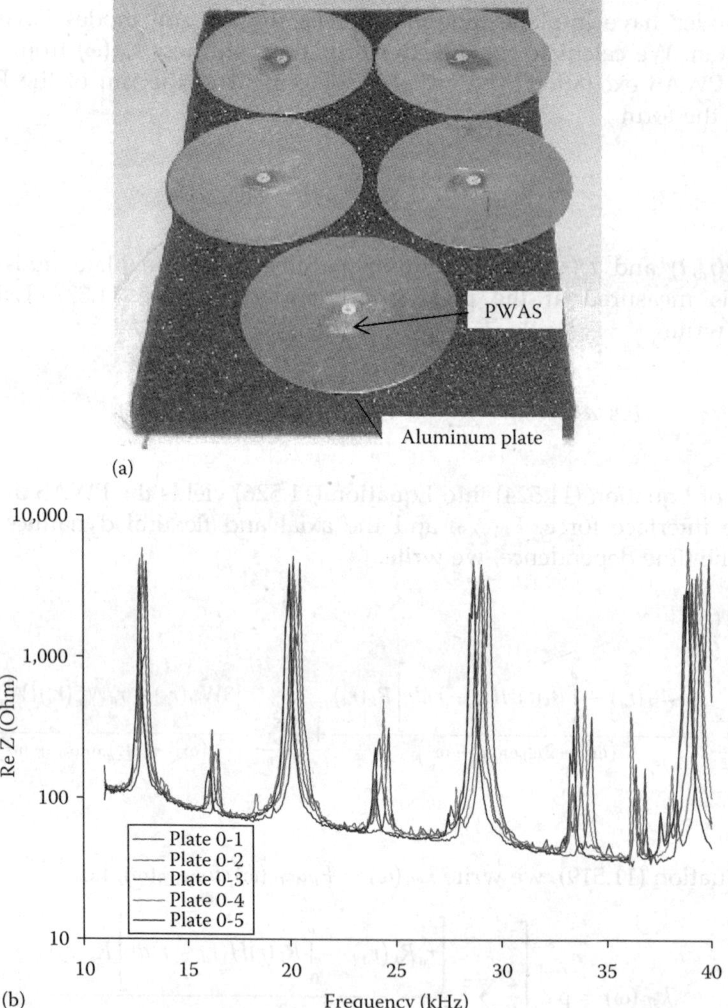

FIGURE 11.85

(a) Thin-gauge aluminum plate specimens with centrally located piezoelectric sensors: 100 mm diameter, 0.8 mm thickness; (b) E/M impedance spectra taken from pristine plates in the 11–40 kHz frequency band.

manufactured from aircraft-grade aluminum stock. The diameter of each plate was 100 mm and the thickness was approximately 0.8 mm. Each plate was instrumented at its center with a 7 mm diameter PWAS. During the experiments, the specimens were supported on packing foam to simulate free boundary conditions. Impedance readings were taken using an HP 4194A Impedance Analyzer. The collected spectra are shown superposed in Figure 11.85.

It is noticed that the spectra collected from the five specimens showed very little variation from specimen to specimen (1% standard deviation in the resonance frequencies as identified from the E/M impedance real part spectra Re (Z)). Plate resonance

TABLE 11.19

Statistical Summary for Resonance Peaks of First Four Axisymmetric Modes
of a Circular Plate as Measured with a PWAS Modal Sensor Using the E/M
Impedance Method

Average Frequency kHz	Frequency STD kHz (%)	Log_{10}-Average Amplitude Log_{10}-Ohms	Log_{10}-Amplitude STD Log_{10}-Ohms (%)
12.856	0.121 (1)	3.680	0.069 (1.8)
20.106	0.209 (1)	3.650	0.046 (1.2)
28.908	0.303 (1)	3.615	0.064 (1.7)
39.246	0.415 (1)	3.651	0.132 (3.6)

frequencies were identified from the E/M impedance real part spectra. Table 11.19 shows
the statistical data in terms of resonance frequencies and log_{10} amplitudes. It should be
noted that the resonance frequencies have very little variation (1% standard deviation)
while the log_{10} amplitudes vary more widely (1.2%–3.6% standard deviation).

To validate the theory, we compared the experimental Re(Z) spectrum with the theoretical Re(Z) and the theoretical FRF spectrum. The FRF spectrum was utilized to illustrate
the fact that the Re(Z) spectrum reflects the structural dynamics, i.e., the peaks of the Re(Z)
spectrum coincide with the FRF peaks which are the structural resonances. Figure 11.86a
shows the FRF spectrum calculated with Equation (11.527) for $r_a = 3.5$ mm, $a = 50$ mm,
$h = 0.8$ mm, $\varsigma_k = 0.07\%$, and $\varsigma_m = 0.4\%$. The frequency range was 0.5–40 kHz. Six flexural
resonances and one axial resonance were captured.

Figure 11.86b compares the experimental and theoretical Re(Z). The theoretical Re(Z)
was calculated with a modification of Equation (11.529). The modifications consisted of
introducing a multiplicative correction factor a/r_a in front of the stiffness ratio $\chi(\omega)$. This
correction factor was needed to account for the difference between the k_{str} distributed over
the radius of the plate and the k_{PWAS} distributed over the radius of the PWAS. The
modified formula is

$$Z(\omega) = \left\{ i\omega C\left(1 - k_p^2\right)\left[1 + \frac{k_p^2}{1 - k_p^2} \frac{(1 + v_a)J_1(\phi_a)}{\varphi_a J_0(\varphi_a) - (1 - v_a)J_1(\varphi_a) - \dfrac{a}{r_a}\chi(\omega)(1 + v_a)J_1(\varphi_a)}\right]\right\}^{-1}$$

$$(11.529)$$

As seen in Figure 11.86b, good matching between theoretical and experimental Re(Z)
spectra was obtained.

The numerical values of the predicted and measured frequencies are compared in Table
11.20. The match between the theoretical and experimental resonance values was very
good. It is noted that the error is consistently very low (<2%). For most of the modes, the
matching error is less than 1%. Two exceptions are noted:

1. The first flexural frequency has a matching error of −7.7%. This can be attributed
   to experimental error since the E/M impedance does not work as well at low
   frequencies as it works at high frequencies. At low frequencies, the response due to

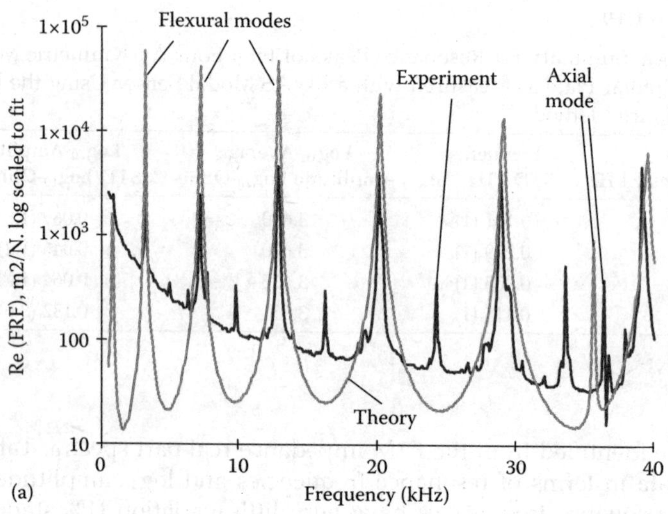

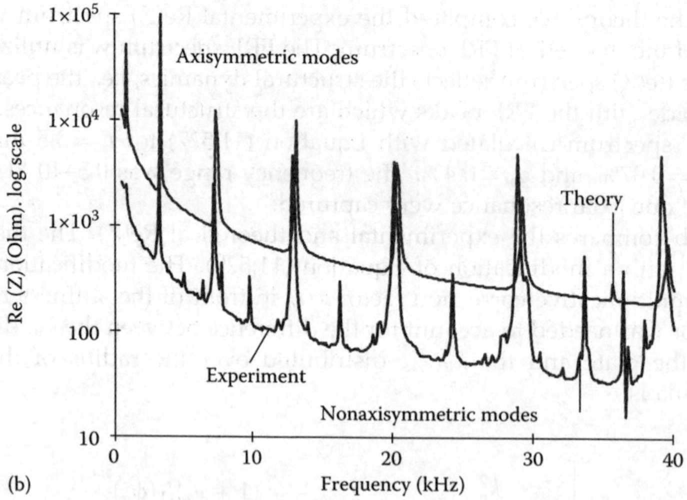

FIGURE 11.86

Experimental and calculated spectra for circular plate specimen: (a) FRF in 0.5–40 kHz frequency range; (b) E/M impedance in 0.5–40 kHz frequency range.

TABLE 11.20

Theoretical and Experimental Results for a Circular Plate with a PWAS Modal Sensor Installed in the Center

Frequency #	1	2	3	4	5	6	7	8
Mode	F	F	F	F	F	F	A	F
Calc., (kHz)	0.742	3.152	7.188	12.841	20.111	29.997	35.629	39.498
Exp. (kHz)	0.799	3.168	7.182	12.844	20.053	28.844	36.348	39.115
Error, Δ%	−7.708	−0.520	0.078	−0.023	0.288	0.528	1.978	0.97

Note: F, flexural mode; A, axial mode.

mechanical resonances is buried in the overall electric response. Hence, the peak of the first flexural mode is rather weak.

2. The first axial mode has a 2% error that can be attributed to slight imperfections in the plate contour.

Thus, we can conclude that Equation (11.529) permits direct comparison of experimental and theoretical E/M impedance data, which was the aim of our analysis.

Although the simulation gives a good matching with experimental results, the theoretical model presented here is limited to the analysis of purely axisymmetrical modes. In principle, the axisymmetric assumption is consistent with a geometry in which a PWAS of circular shape is placed exactly at the center of a circular plate. However, if the sensor is slightly misaligned, nonaxisymmetric modes will also be excited and appear in the spectrum. This effect is observable in Figure 11.86 where the low-amplitude peaks that appear at 15, 24, and 33 kHz on the experimental curves have no match on the theoretical curves. These small peaks are due to nonaxisymmetric modes that get parasitically excited due to slight misalignment in the placement of the sensor at the center of the plate.

11.6.3.5 Section Summary

A circular plate model, which accounts for sensor–structure interaction in 2-D geometry, was derived and validated through experimental testing. The model considers both the structural dynamics and the sensor dynamics. In the structural dynamics, both the axial and the flexural vibrations were considered. The structural dynamics was incorporated into the model through the pointwise dynamic stiffness presented by the structure to the PWAS. The analytical model predicts the electromechanical (E/M) impedance response as it would be measured at the PWAS terminals. The real part of the E/M impedance reflects with fidelity the natural frequencies of the structure on which the PWAS is mounted. Through experimental tests, we were able to validate that the model is capable to correctly predict the E/M impedance and the structural frequencies can be determined directly from the E/M impedance real part. Thus, it was verified that PWAS, in conjunction with the E/M impedance, act as self-excited high-frequency modal sensors that correctly sense the structural dynamics.

11.6.4 Damage Detection with PWAS Modal Sensors

This section will present the use of the high-frequency PWAS modal sensors for incipient damage detection using the electromechanical (E/M) impedance technique. The advantage of using PWAS for damage detection resides in their very high-frequency capability that exceeds by orders of magnitudes the frequency capability of conventional modal analysis sensors. Thus, PWAS are able to detect subtle changes in the high-frequency structural dynamics at local scales. Such local changes in the high-frequency structural dynamics are associated with the presence of incipient damage that would not be detected by conventional modal analysis sensors which operate at lower frequencies.

In this section, two sets of experiments are presented. The first set of experiments was performed on circular plates which have clean and reproducible spectra, as seen in the previous section. Thus, the effect of damage on the resonance spectrum was easily observed. This first set of experiments allowed a tractable development of the damage identification algorithms. It was found that two algorithm types could be used, one based on overall statistics of the entire spectrum, the other based on the spectral features. The second set of experiments was performed on realistic specimens which are representative of actual structures. Aircraft panel specimens were used. These specimens had seeded faults such as simulated cracks and corrosion. The E/M impedance method was used to detect the presence of seeded faults by comparing "pristine" and "with faults" spectra. Details of these two sets of experiments are given next.

11.6.4.1 Damage Detection Experiments on Circular Plates

Systematic experiments were performed on circular plates to assess the crack detection capabilities of the method. As shown schematically in Table 11.21 and Figure 11.87, five damage groups were considered: one group consisted of pristine plates (Group 0) and four groups consisted of plates with simulated cracks placed at increasing distance from the plate edge (Group 1 through 4). Each group contained five nominally "identical" specimens. Thus, the statistical spread within each group could also be assessed. In our study, a 10 mm circumferential slit was used to simulate an in-service crack. The simulated crack was placed at decreasing distance from the plate edge. The following radial positions of the crack from the PWAS were considered: 40, 25, 10, and 3 mm.

The experiments were conducted over three frequency bands: 10–40 kHz; 10–150 kHz, and 300–450 kHz. The data were processed by plotting the real part of the E/M impedance spectrum and determining a damage metric to quantify the difference between spectra. The data for the 10–40 kHz band are shown in Figure 11.87. As damage is introduced in the plate, resonant frequency shifts, peaks split, and the appearance of new resonances are noticed. As the damage becomes more severe, these changes become more profound. The most profound changes are noticed for Group 4. For the higher frequency bands, similar behavior was observed. Numerical values of the measured frequencies of the major resonances are given in Table 11.21.

11.6.4.1.1 Overall-Statistics Damage Metrics

The damage metric is a scalar quantity that results from the comparative processing of impedance spectra. The damage metric should reveal the difference between spectra due to damage presence. Ideally, the damage index would be a metric which captures only the spectral features that are directly modified by the damage presence while neglecting the variations due to normal operation conditions (i.e., statistical difference within a population of specimens and expected changes in temperature, pressure, ambient vibrations, etc.). To date, several damage metrics have been used to compare impedance spectra and assess the presence of damage. Among them, the most popular are the root mean square deviation (RMSD), the mean absolute percentage deviation (MAPD), and the correlation coefficient deviation (CCD). The mathematical expressions for these metrics, given in terms of the impedance real part Re (Z), are as follows

$$RMSD = \sqrt{\sum_N [\mathrm{Re}(Z_i) - \mathrm{Re}(Z_i^0)]^2 \Big/ \sum_N [\mathrm{Re}(Z_i^0)]^2} \qquad (11.530)$$

$$MAPD = \sum_N |[\mathrm{Re}(Z_i) - \mathrm{Re}(Z_i^0)]/\mathrm{Re}(Z_i^0)| \qquad (11.531)$$

TABLE 11.21

Damage Detection Experiments on Circular Plates Using PWAS Modal Sensors
and the Electromechanical (E/M) Impedance Method

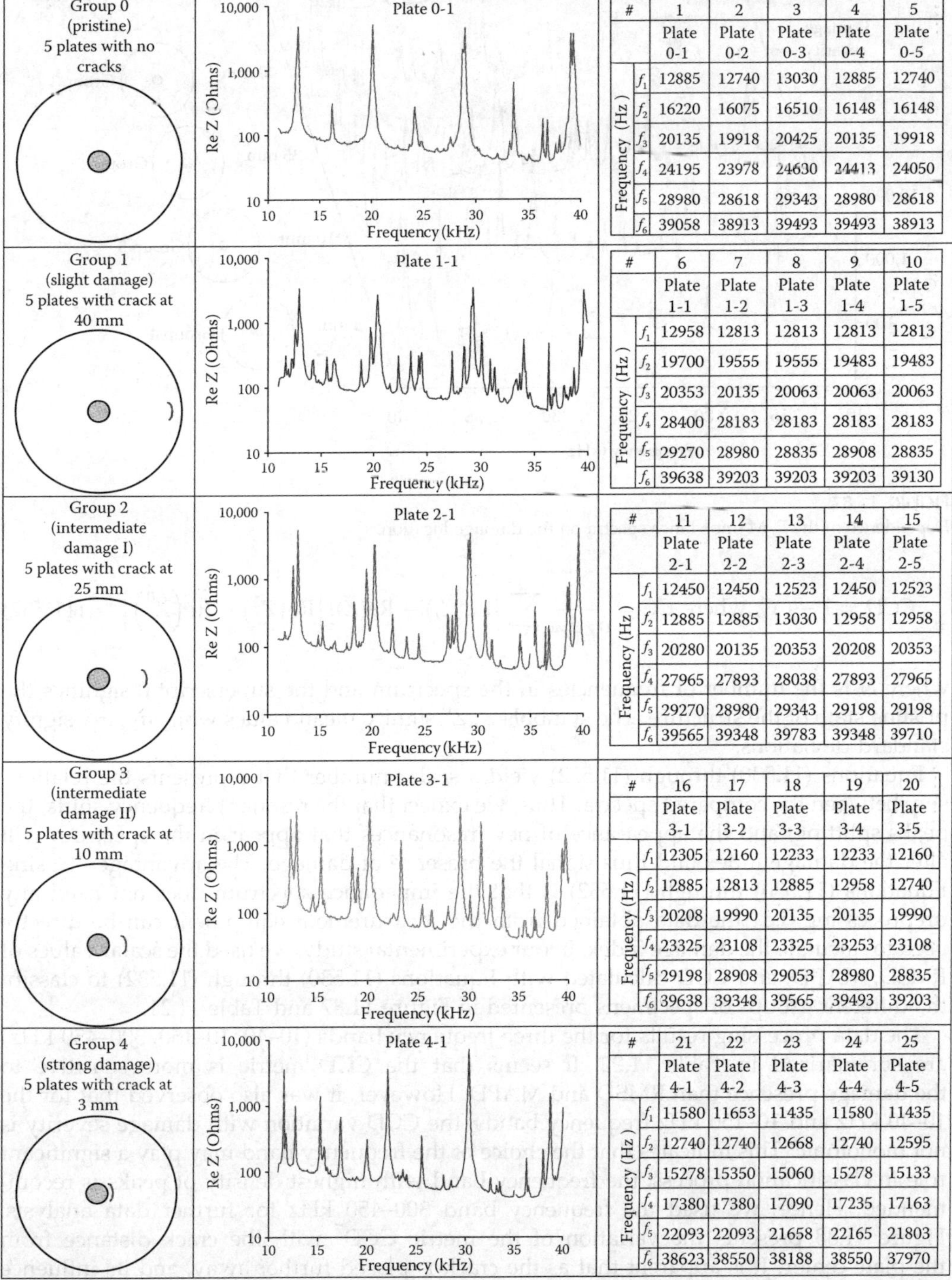

Group 0 (pristine) — 5 plates with no cracks — Plate 0-1

#	1	2	3	4	5
	Plate 0-1	Plate 0-2	Plate 0-3	Plate 0-4	Plate 0-5
f_1	12885	12740	13030	12885	12740
f_2	16220	16075	16510	16148	16148
f_3	20135	19918	20425	20135	19918
f_4	24195	23978	24630	24413	24050
f_5	28980	28618	29343	28980	28618
f_6	39058	38913	39493	39493	38913

Group 1 (slight damage) — 5 plates with crack at 40 mm — Plate 1-1

#	6	7	8	9	10
	Plate 1-1	Plate 1-2	Plate 1-3	Plate 1-4	Plate 1-5
f_1	12958	12813	12813	12813	12813
f_2	19700	19555	19555	19483	19483
f_3	20353	20135	20063	20063	20063
f_4	28400	28183	28183	28183	28183
f_5	29270	28980	28835	28908	28835
f_6	39638	39203	39203	39203	39130

Group 2 (intermediate damage I) — 5 plates with crack at 25 mm — Plate 2-1

#	11	12	13	14	15
	Plate 2-1	Plate 2-2	Plate 2-3	Plate 2-4	Plate 2-5
f_1	12450	12450	12523	12450	12523
f_2	12885	12885	13030	12958	12958
f_3	20280	20135	20353	20208	20353
f_4	27965	27893	28038	27893	27965
f_5	29270	28980	29343	29198	29198
f_6	39565	39348	39783	39348	39710

Group 3 (intermediate damage II) — 5 plates with crack at 10 mm — Plate 3-1

#	16	17	18	19	20
	Plate 3-1	Plate 3-2	Plate 3-3	Plate 3-4	Plate 3-5
f_1	12305	12160	12233	12233	12160
f_2	12885	12813	12885	12958	12740
f_3	20208	19990	20135	20135	19990
f_4	23325	23108	23325	23253	23108
f_5	29198	28908	29053	28980	28835
f_6	39638	39348	39565	39493	39203

Group 4 (strong damage) — 5 plates with crack at 3 mm — Plate 4-1

#	21	22	23	24	25
	Plate 4-1	Plate 4-2	Plate 4-3	Plate 4-4	Plate 4-5
f_1	11580	11653	11435	11580	11435
f_2	12740	12740	12668	12740	12595
f_3	15278	15350	15060	15278	15133
f_4	17380	17380	17090	17235	17163
f_5	22093	22093	21658	22165	21803
f_6	38623	38550	38188	38550	37970

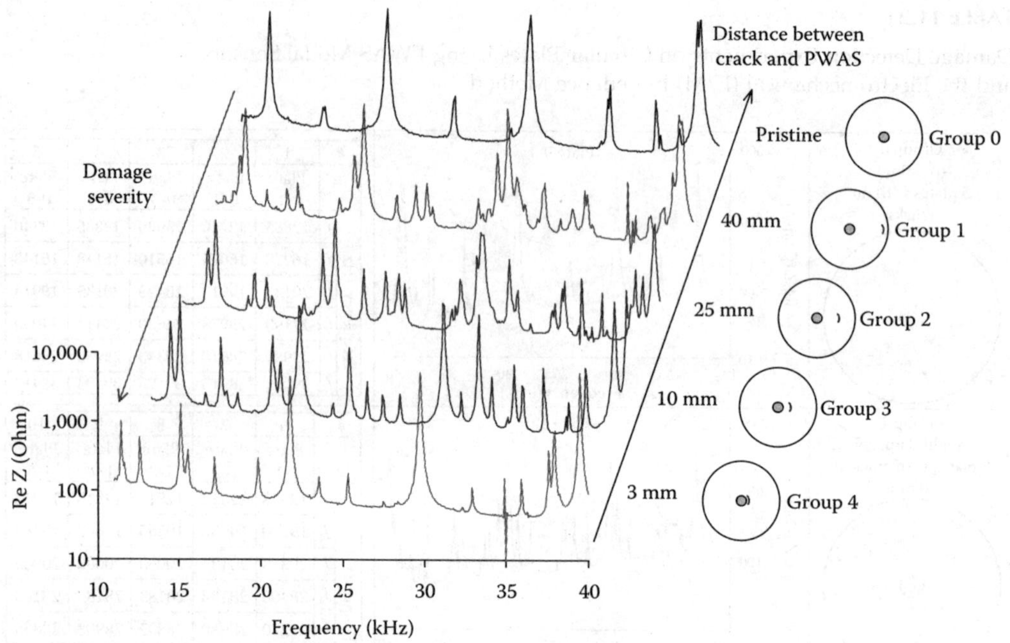

FIGURE 11.87
Dependence of the E/M impedance spectra on the damage location.

$$CCD = 1 - CC, \text{ where } CC = \frac{1}{\sigma_Z \sigma_{Z^0}} \sum_N \left[\text{Re}(Z_i) - \text{Re}(\overline{Z}) \right] \left[\text{Re}(Z_i^0) - \text{Re}(\overline{Z}^0) \right] \quad (11.532)$$

where N is the number of frequencies in the spectrum and the superscript 0 signifies the pristine state of the structure. The symbols $\overline{Z}$, $\overline{Z}^0$ signify mean values while σ_Z, σ_{Z^0} signify standard deviations.

Equations (11.530) through (11.532) yield a scalar number that represents the relationship between the compared spectra. Thus, we expect that the resonant frequency shifts, the peaks splitting, and the appearance of new resonances that appear in the spectrum will alter the damage index and thus signal the presence of damage. The advantage of using Equations (11.530) through (11.532) is that the impedance spectrum does not need any preprocessing, i.e., the data obtained from the measurement equipment can be directly used to calculate the damage index. In our experimental study, we used the scalar values of RMSD, MAPD, and CCD calculated with Equations (11.530) through (11.532) to classify the different groups of specimens presented in Figure 11.87 and Table 11.21.

The data processing results for the three frequency bands (10–40, 10–150, 300–450 kHz) are summarized in Table 11.22. It seems that the CCD metric is more sensitive to the damage presence than RMSD and MAPD. However, it was also observed that for the 10–40 kHz and 10–150 kHz frequency bands, the CCD variation with damage severity is not monotonic. This indicates that the choice of the frequency band may play a significant role in classification process; the frequency band with highest density of peaks is recommended. Hence, we used the frequency band 300–450 kHz for further data analysis. Figure 11.88 presents the variation of the metric CCD^7 with the crack distance from the plate center. It is apparent that as the crack is placed further away, and its influence

TABLE 11.22

Overall-Statistics Damage Metrics for Various Frequency Bands

Frequency Band	11–40 kHz				11–150 kHz				300–450 kHz			
Compared groups	0_1	0_2	0_3	0_4	0_1	0_2	0_3	0_4	0_1	0_2	0_3	0_4
RMSD %	122	116	94	108	144	161	109	118	93	96	102	107
MAPD %	107	89	102	180	241	259	170	183	189	115	142	242
CCD %	84	75	53	100	93	91	52	96	81	85	87	89

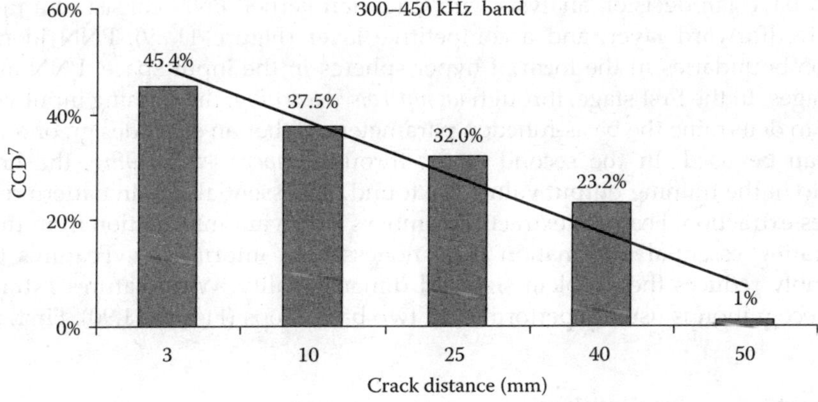

FIGURE 11.88

Monotonic variation of the CCD^7 damage metric with the crack radial position on a 50 mm radius plate in the 300–450 kHz band.

diminishes, the value of the CCD^7 metric also diminishes. In fact, the CCD^7 damage metric tends to decrease linearly with the crack position, which may be very useful in automated damage assessment.

11.6.4.1.2 Probabilistic Neural Networks for Damage Identification

Probabilistic neural networks (PNN) are an efficient tool for solving classification problems. PNN are hybrid multilayer networks that use basis functions and competitive selection concepts. In contrast to other neural network algorithms used for damage identification, PNN have a statistically derived activation function and utilize a Bayesian decision strategy for the classification problem. The kernel-based approach to probability density function (PDF) approximation is used. This method permits the construction of the PDF of any sample of data without any *a priori* probabilistic hypothesis. The PDF reconstruction is achieved by approximating each sample point with kernel function(s) to obtain a smooth continuous approximation of the probability distribution. In other words, using the kernel technique, it is possible to map a pattern space (data sample) into the feature space (classes). However, the result of such transformation should retain essential information presented in the data sample and be free of redundant information that may contaminate the feature space. In this study, we used the resonance frequencies as a data sample to classify spectra according to the damage severity. The classical multivariate Gaussian kernel was chosen for PNN implementation

$$p_A(x) = \frac{1}{n(2\pi)^{d/2}\sigma^d} \sum_{i=1}^{n} \exp\left(-\frac{(x - x_{Ai})^T (x - x_{Ai})}{2\sigma^2}\right) \tag{11.533}$$

where i is the pattern number, x_{Ai} is the ith training pattern from A category, n is the total number of training patterns, d is the dimensionality of measurement space, and σ is the spread parameter. Although the Gaussian kernel function was used in this work, its form is generally not limited to being Gaussian.

A MATLAB® implementation of the PNN algorithm is shown in Figure 11.89. PNN achieve a Bayesian decision analysis with Gaussian kernel. PNN consist of a radial-basis layer, a feedforward layer, and a competitive layer (Figure 11.89). PNN identify class separation boundaries in the form of hyper-spheres in the input space. PNN are trained in two stages. In the first stage, through *unsupervised learning*, the training input vectors, x_n, are used to determine the basis-function parameters. Either an exact design or an adaptive design can be used. In the second stage, through *supervised learning*, the linear-layer weights to fit the training output values are found. An essential step in pattern recognition is features extraction. Features extraction removes irrelevant information from the data set by separating essential information from nonessential information. Features extraction considerably reduces the problem size and dimensionality. With features extraction, the pattern recognition is usually performed in two basic steps (Figure 11.90). First, a number

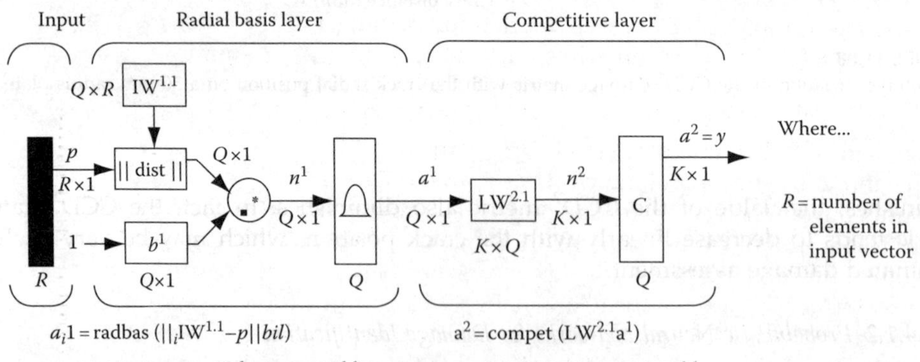

FIGURE 11.89
The general architecture of a probabilistic neural net implemented in the MATLAB package (http://www.math works.com/access/helpdesk/help/toolbox/nnet/nnet.shtml).

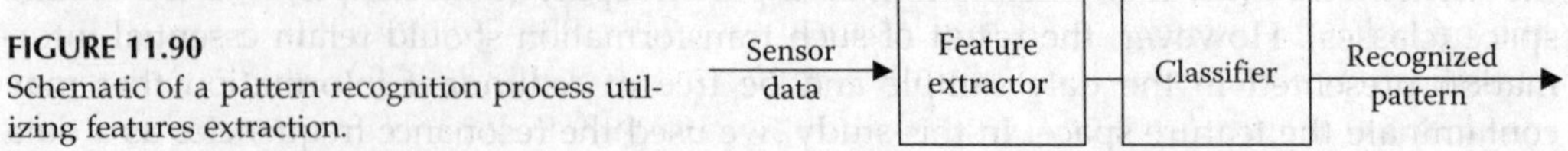

FIGURE 11.90
Schematic of a pattern recognition process utilizing features extraction.

of features are extracted and placed into a features vector. Second, the features vector is used to perform the classification. In spectral analysis, the features-vector approach recognizes that the number and characteristics of resonance peaks play a major role in describing the dynamic behavior of the structure.

The physics-based features extraction utilizes the deterministic relation between the peaks in the spectrum and the phenomenon of structural resonance. Hence, the spectrum is described through its essential dynamical features, i.e., resonance frequencies, resonance peaks, and modal damping coefficients. Through this process, the problem dimensionality may be reduced by at least one order of magnitude (from the 400-points spectrum to a few tens of features). However, this reduction may not be large at very high frequencies where lightly damped structures may have very high density of resonance peaks. To account for this, an incremental approach can be used; the complexity of the features vector is gradually increased so as to balance breadth of details with computational efficiency. For example, one can consider first a reduced-order features vector containing only the resonance frequencies and then gradually extend it to include the other features (relative amplitudes and damping factors).

11.6.4.1.3 Damage Detection in Circular Plates with Probabilistic Neural Networks

To construct the PNN input vectors, the spectra of Figure 11.90 were processed to extract the resonance frequencies for each group. Once the feature vectors are established, the classification problem can be approached in the features space. Although several features-based classification algorithms are available, in our study, we explored the PNN classification algorithm. Typical results are shown in Table 11.23. These results were obtained with a features vector of size six, corresponding to the six most predominant frequencies in the 10–40 kHz spectrum. Five analysis tests were conducted: I, II, III, IV, and V. In each test, one spectrum from each class was selected for training and the other four spectra in each class were used for validation. Thus, in total, 5 vectors were used for training (one from each class) and 20 vectors were used for validation (four from each class). The vectors used for training are designated with letter T in Table 11.23. The vectors used for validation are designated with letter V in Table 11.23.

The input to the neural network is marked with IN in Table 11.23. This input is constituted from the vectors that are presented to the neural network during the validation cycle. The output of the neural network is marked with OUT in Table 11.23. This output is constituted from the responses that the neural network gives to each input vector. There are five possible responses that the neural network can return. These responses are represented by the numbers: 0, 1, 2, 3, and 4. Each number corresponds to a group. Thus, if the neural network returns the response "3," it means that the neural network has identified the input as belonging to group 3. It is apparent from Table 11.23 that the neural network was able to correctly recognize all the presented spectra. Thus, clear distinction could be established between the spectra generated by the "pristine" case (Group 0) and the "damage" cases (Groups 1, 2, 3, and 4). In addition, clear distinction could also be determined between the spectra of various "damage" groups that correspond to various crack positions (Group 4, 3, 2, and 1 correspond to $r = 3$, 10, 25, and 40 mm, respectively). These examples have shown that the PNN approach, in conjunction with a sufficiently large features vector, can successfully identify the damage presence and its location.

It should be noted that if the number of features in the features vector is not sufficient, the neural network may give misclassifications. This situation was met when the PNN classification was performed with a four-frequency features vector. The four frequencies used in the features vector were the strongest resonance frequencies. In this case, good

TABLE 11.23

Synoptic Classification Table for Circular Plates Using 6-Frequency Feature Vectors

Test #	Plate #	Group 0 (No Damage)					Group 1 ($r=40$ mm)					Group 2 ($r=25$ mm)					Group 3 ($r=10$ mm)					Group 4 ($r=3$ mm)					
		1	2	3	4	5	6	7	8	9	10	11	12	13	14	15	16	17	18	19	20	21	22	23	24	25	
I		T	V	V	V	V	T	V	V	V	V	T	V	V	V	V	T	V	V	V	V	T	V	V	V	V	IN
		—	0	0	0	0	—	1	1	1	1	—	2	2	2	2	—	3	3	3	3	—	4	4	4	4	OUT
II		V	T	V	V	V	V	T	V	V	V	V	T	V	V	V	V	T	V	V	V	V	T	V	V	V	IN
		0	—	0	0	0	1	—	1	1	1	2	—	2	2	2	3	—	3	3	3	4	—	4	4	4	OUT
III		V	V	T	V	V	V	V	T	V	V	V	V	T	V	V	V	V	T	V	V	V	V	T	V	V	IN
		0	0	—	0	0	1	1	—	1	1	2	2	—	2	2	3	3	—	3	3	4	4	—	4	4	OUT
IV		V	V	V	T	V	V	V	V	T	V	V	V	V	T	V	V	V	V	T	V	V	V	V	T	V	IN
		0	0	0	—	0	1	1	1	—	1	2	2	2	—	2	3	3	3	—	3	4	4	4	—	4	OUT
V		V	V	V	V	T	V	V	V	V	T	V	V	V	V	T	V	V	V	V	T	V	V	V	V	T	IN
		0	0	0	0	—	1	1	1	1	—	2	2	2	2	—	3	3	3	3	—	4	4	4	4	—	OUT

Note: T, training vector; V, validation vector; IN, input to PNN; OUT, output from PNN.

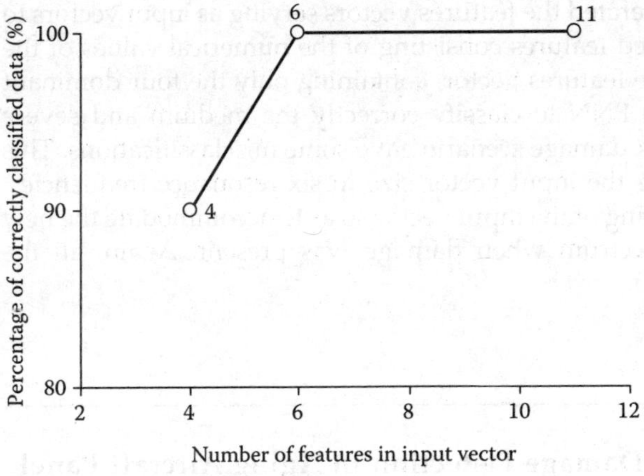

FIGURE 11.91
Number of features in input vector of PNN vs. percent of correctly classified data.

classification was attained for Groups 2, 3, and 4 but not for Group 1. Group 1 (slight damage) was misclassified with Group 0 (pristine). This misclassification problem could not be fixed by increasing the number of training vectors. However, it was fixed when the size of the features vector was increased from four to six.

Further in our study, we considered an extension of the features vector size by incorporating the new resonances that appeared in the damaged plates but were not present in the pristine plates. This feature plays an important role in distinguishing healthy structures from damaged structures, especially when the damage is incipient or located away from the sensor. To achieve this, we expanded the features vector to incorporate the new resonance that appeared in the damaged plates. In order to preserve dimensionality, the pristine plates vectors were zero filled as needed. Thus, the features vector size was expanded to 11. Then, one vector from each group was used for training and the rest for validation. The PNN algorithm was again able to correctly classify data regardless of the choice of the training vectors. Figure 11.91 presents the percent of correct classified data versus number of features in the input vectors of PNN. The good classification results obtained with PNN encouraged us to use this method for damage classification in aircraft structural specimens.

11.6.4.2 Section Summary

The theoretical and practical aspects of the application of the E/M impedance method to the SHM of thin-walled structures using PWAS were discussed. Damage detection experiments were perform on circular plates and aircraft panels. It was observed that the presence of damage significantly modifies the E/M impedance spectrum that features frequency shifts, peaks splitting, and appearance of new harmonics. The rate of changes in the spectrum increases with the severity of damage. To quantify these changes and classify the spectra according to the severity of damage, two approaches were used: (1) overall-statistics damage metrics and (2) PNN. In the overall-statistics damage metrics approach, RMSD, MAPD, and CCD were considered. Through the circular plates experiments, it was found that CCD^7 damage metric is a satisfactory classifier in the high frequency band where the resonance peaks density is high (300–450 kHz). In the PNN approach, the spectral data was first preprocessed with a features extraction algorithm.

This features extraction algorithm generated the features vectors serving as input vectors to the PNN. In our investigation, we used features consisting of the numerical values of the resonance frequencies. A reduced-size features vector, containing only the four dominant resonance frequencies, permitted the PNN to classify correctly the medium and severe damage scenarios. However, the weak damage scenario gave some misclassifications. This problem was overcome by increasing the input vector size to six resonance frequencies. Further, we studied the adaptive resizing of the input vector so as to accommodate the new frequencies that appeared in the spectrum when damage was present. Again, all the damage cases were properly classified.

11.7 Case Study: Multimethod Damage Detection in Aging Aircraft Panel Specimens

This section will present a realistic case study in which damage detection in aging aircraft panel specimens is performed with several methods. Realistic specimens representative of actual aircraft structures with aging-induced damage (cracks and corrosion) were designed and fabricated. Figure 11.92 presents a CAD drawing of these panels. The panels have a built-up construction typical of conventional aircraft structures. A lap splice joint, tear straps, and hat-shaped stringer/stiffeners are featured. The whole construction is made of 1 mm (0.040″) thick 2024-T3 Al-clad sheet assembled with 4.2 mm (0.166″) diameter countersunk rivets. Simulated cracks (EDM hairline cuts) and simulated corrosion damage (milled-out areas) were incorporated. Several specimens were constructed: (1) pristine, (2) with cracks, and (3) with corrosion. The specimens were instrumented with several PWAS, 7-mm square and 0.2 mm thick, made of APC-850 piezoceramic. These specimens are aircraft-type panels with structural details typical of metallic aircraft structures (rivets, splices, stiffeners, etc.). Their presence complicates the structural dynamics and makes the damage detection task more difficult. The specimens were made of 1 mm (0.040″) thick 2024-T3 Al-clad sheet assembled with 4.2 mm (0.166″) diameter countersunk rivets. Cracks were simulated with electric discharge machine (EDM). In our study, we investigated crack damage and considered two specimens: pristine (Panel 0) and damaged (Panel 1). The objective of the experiment was to detect a 12.7 mm (0.5″) simulated crack originating from a rivet hole (Figure 11.92).

The examples presented next will illustrate how crack damage in these specimens could be detected with two PWAS-based methods: (1) a wave propagation method based on the pulse-echo principle and (2) a high-frequency vibration using the electromechanical (E/M) impedance. In the first method, PWAS will act as an in-situ ultrasonic transducer performing transmission and reception of guided elastic waves traveling in the panels. Thus, the PWAS transducers will be able to detect the echoes reflected by the cracks. In the second method, PWAS will act as in-situ modal sensors evaluating the high-frequency structural vibration spectrum; being high-frequency, this spectrum is sensitive to the presence of small structural cracks.

11.7.1 Pulse-Echo Damage Detection in Aging Aircraft Panels

Wave propagation experiments were conducted on these realistic aircraft panel specimens using the PWAS in pulse-echo mode. The experimental apparatus was similar to that used

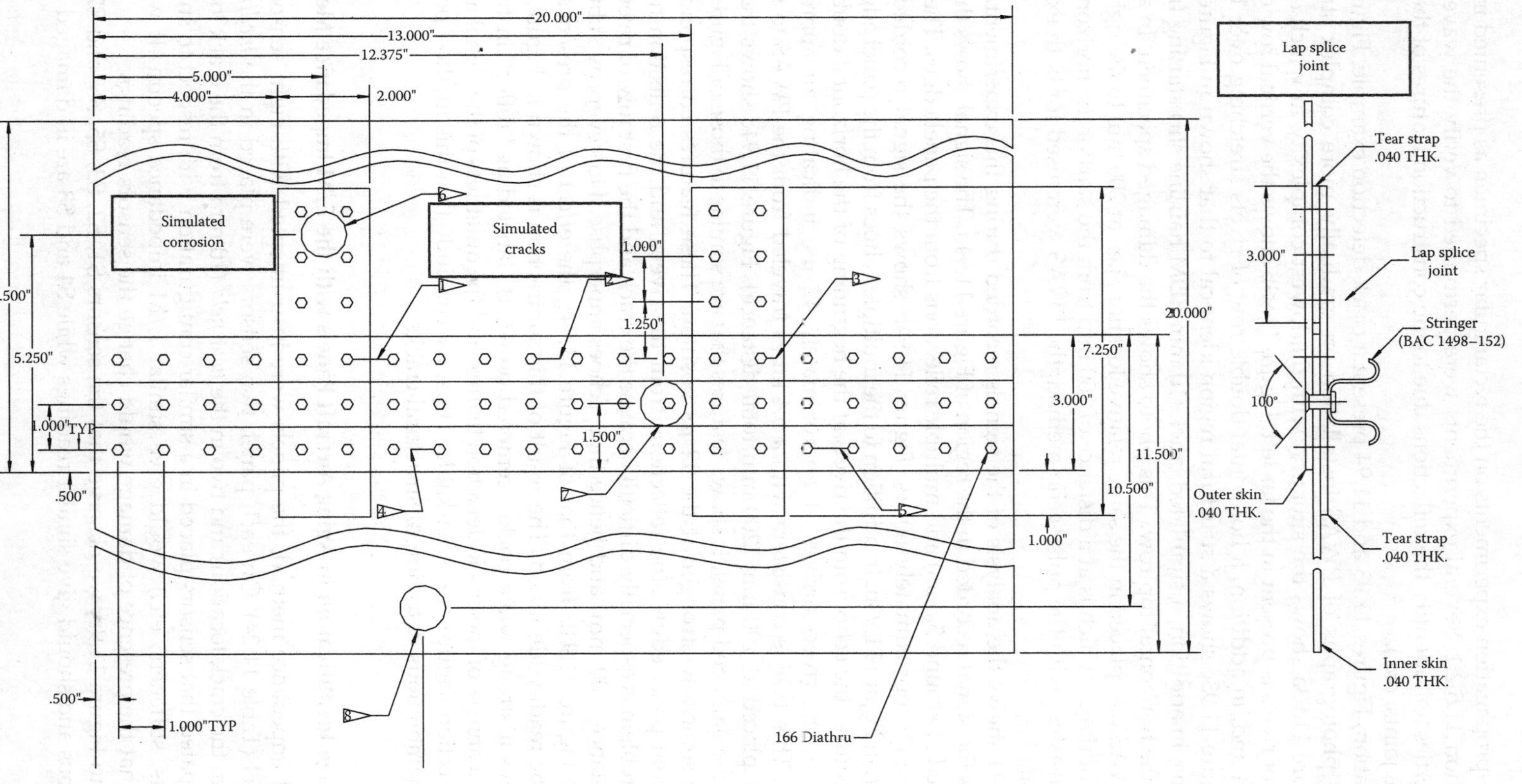

FIGURE 11.92
Realistic aging aircraft lap-splice joint panel with simulated cracks (EDM slits) and corrosion (milled-out areas).

for the wave-propagation experiments on the rectangular specimen as presented in Figure 11.61 of Section 11.5.2.1. Several experiments were performed to verify the wave propagation properties and identify the reflections due to the construction features of the panels (rivets, splice joints, etc.).

For illustration, Figures 11.93 and 11.94 present a crack detection example. Figure 11.93 shows three photographs of PWAS installation on gradually more complex structural regions. Figure 11.93a shows the situation with the lowest complexity in which only the vertical row of rivets is present in the far left. Figure 11.93b shows the vertical row of rivets in the far left and, in addition, a horizontal double row of rivets stretching over 100 mm distance. Figure 11.93c shows a structural region identical to that shown in Figure 11.93b but which has, in addition, a simulated crack (10 mm EDM hairline slit) starting from the first rivet in the horizontal top row. This photo shows the damaged specimen. In all three regions, a PWAS was placed in the same relative location, i.e., at 200 mm to the right of the vertical row of rivets, which is at a distance of 100 mm from the start of the horizontal row of rivets. Consistent with the pulse-echo method, the PWAS were used for both excitation and reception.

Figure 11.94 shows the analysis of the signals recorded during this experiment. Figure 11.94a shows the signal recorded in the region of Figure 11.94a. The signal shows the initial bang (centered at around 5.3 μs) and multiple reflections from the panel edges. The echoes start to arrive at approximately 60 μs. Figure 11.94b shows the signal recorded in the region shown in Figure 11.93b. In addition to the multiple echoes from the panel edges, this signal also features the echo from the rivets at the beginning of the horizontal double row. The echo from the rivets arrives at approximately 42 μs, indicating an approximate TOF $= 37$ μs. This TOF is consistent with a 5.4 km/s traveled from the PWAS to the first line of rivets placed at 100 mm (200 mm total distance). Figure 11.94c shows the signal recorded on the damaged panel. It shows features that are similar to those of Figure 11.93c, but they are somehow stronger at the 42 μs position. These features correspond to the reflections from panel edges, the reflections from the rivets, and the reflection from the crack. The problem is especially difficult because the crack and the first row of rivets are at the same distance (100 mm) and hence their echoes superpose. However, by subtracting the signal of Figure 11.94b from that of Figure 11.94c, the effect of the presence of the crack could be readily identified. The result of this subtraction is shown in Figure 11.94d, which features a strong wave packet centered on 42 μs, labeled as "reflection from the crack." The cleanness of the crack-detection feature and the quietness of the signal ahead of the crack-detection feature are remarkable. Thus, we conclude that this method permits a clean and unambiguous detection of structural cracks.

11.7.2 Damage Identification in Aging Aircraft Panels with the E/M Impedance Method

For the E/M impedance method, the panels were instrumented with eight sensors, four on each panel (Table 11.24). On each panel, two sensors were placed in the *medium field* (100 mm from the crack location) and two in the *near field* (10 mm from the crack location). It was anticipated that sensors placed in a similar configuration with respect to structural details (rivets, stiffeners, etc.) would give similar E/M impedance spectra. It was also anticipated that the presence of damage would change the sensors readings.

Referring to Figure 11.95, we observe that the sensors S1, S2, S3, S5, S6, and S7 are in pristine regions and should give similar readings while S4 and S8 are in damaged regions

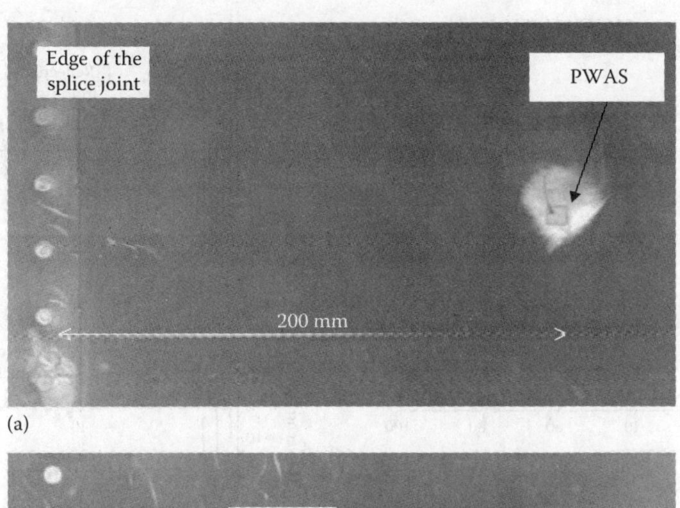

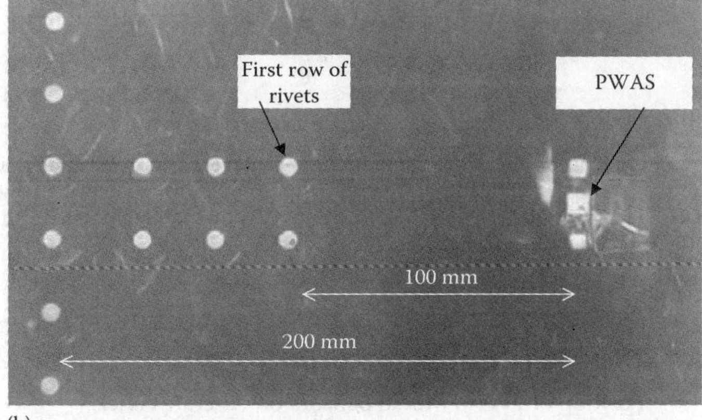

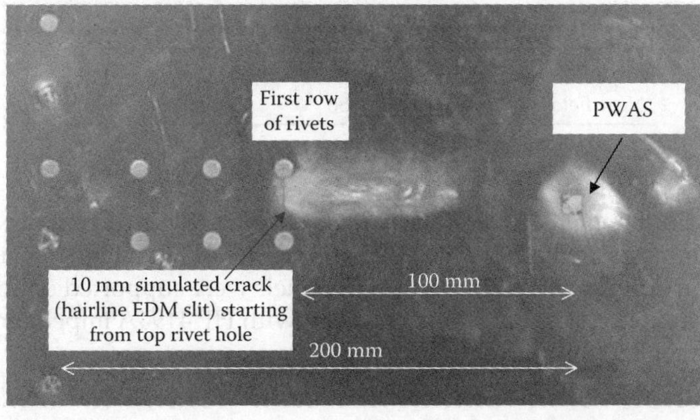

FIGURE 11.93
Crack detection experiment on aging aircraft panel: (a) pristine panel featuring a PWAS transducer placed in a rivet-free region; (b) pristine panel featuring a PWAS transducer placed at 100 mm from a row of rivets; (c) damaged panel featuring a 10 mm simulated crack (hairline EDM slit) starting from the top rivet.

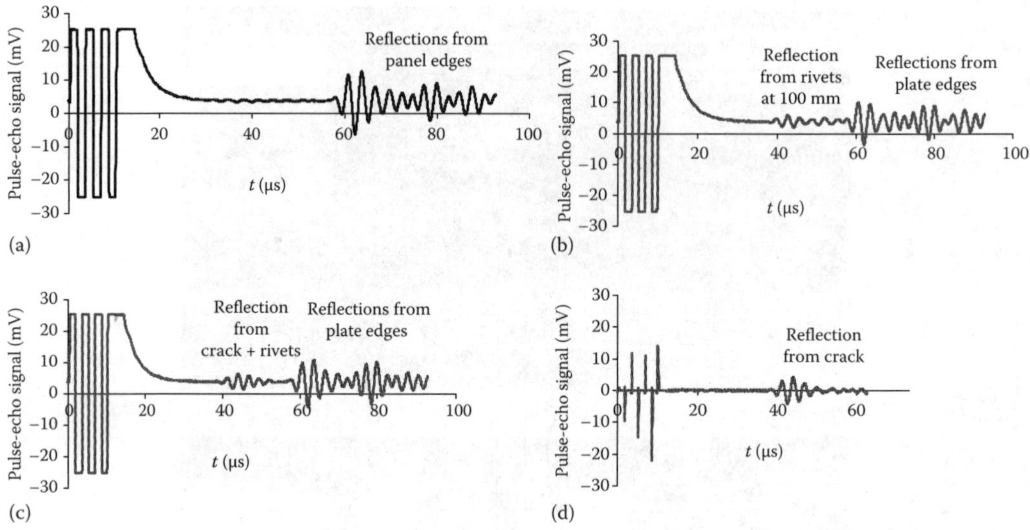

FIGURE 11.94
Analysis of the pulse-echo signals for crack detection: (a) signal showing only reflections from the panel edges; (b) signal recorded in the pristine panel featuring the reflection from the rivets and the panel edges; (c) signal recorded on the cracked panel featuring, in addition, the reflections due to the presence of the crack; (d) signal difference revealing the strong reflection from the crack.

TABLE 11.24

Position of PWAS Modal Sensors on Aircraft Panels

	Panel 0		Panel 1	
	Pristine	**Pristine**	**Pristine**	**Damaged**
Medium field	S1	S2	S3	S4
Near field	S5	S6	S7	S8

and should give different readings. The high frequency E/M impedance spectrum was collected for each sensor in the 200–550 kHz band which shows a high density of resonance peaks. During the experiment, both the aircraft panels were supported on foam to simulate free boundary conditions. The data were collected with HP 4194A impedance analyzer and then loaded into the PC through the GPIB interface.

11.7.2.1 Classification of Crack Damage in the PWAS Near-Field

The near-field PWAS were S5, S6, S7, and S8. All were placed in similar structural configurations. Hence, in the absence of damage, all should give similar impedance spectra. However, S8 is close to the simulated crack originating from the rivet hole. Hence, it was anticipated that S8 should give an impedance spectrum different from that of S5, S6, and S7. The change in the spectrum would be due to the presence of crack damage. Hence, we call S5, S6, and S7 "pristine" and S8 "damage." Figure 11.96a shows

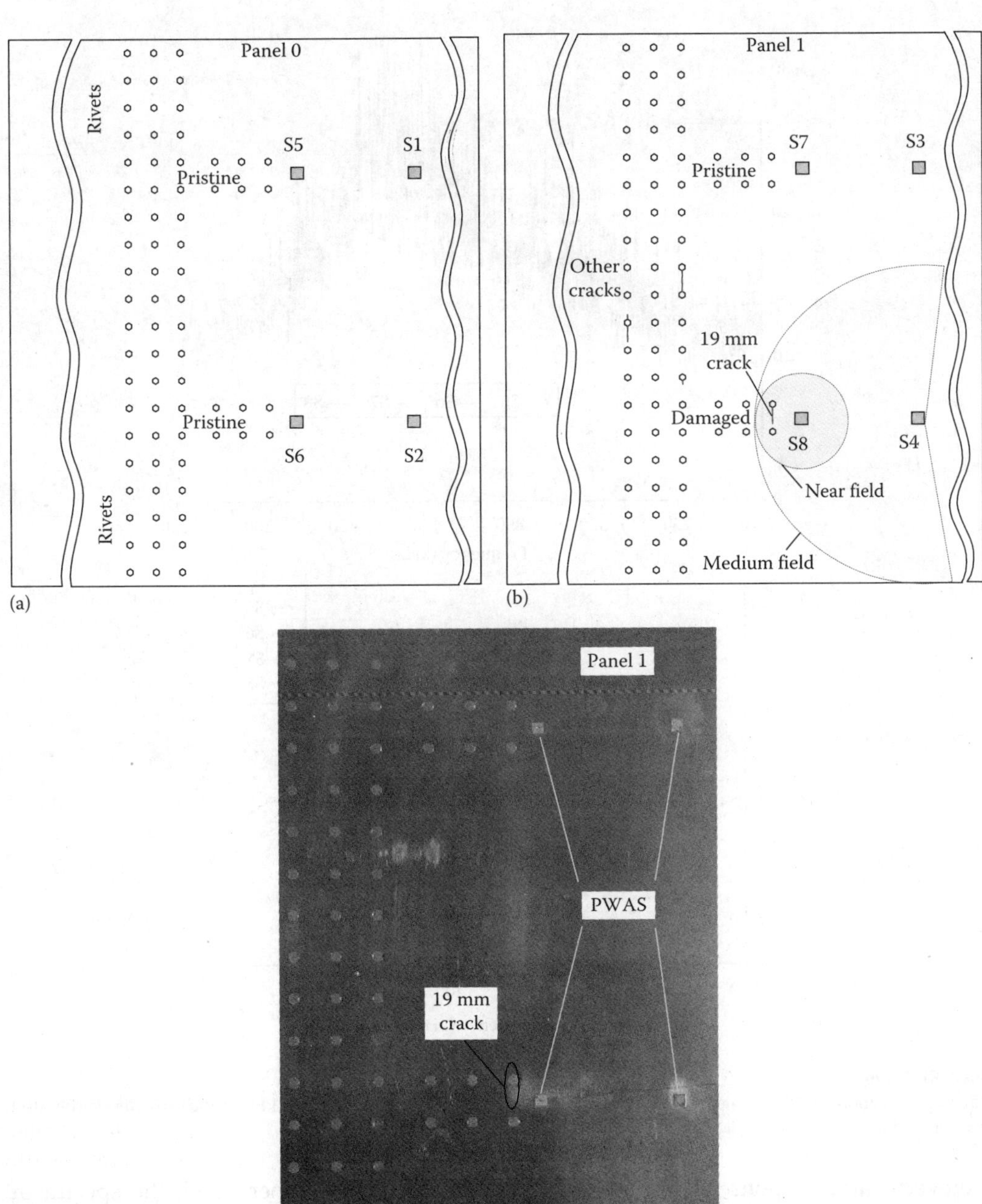

FIGURE 11.95
Schematics of the aging aircraft panel specimens and PWAS configuration: (a) panel 0, sensors S1, S2, S5, and S6; (b) panel 1, sensors S3, S4, S7, and S8; (c) actual photograph of panel 1.

the superposition of the spectra obtained from these sensors. Examination of these spectra reveals that sensor S8, placed next to the crack, has two distinct features that make it different from the other three spectra: (1) a higher-density of peaks and (2) an elevated

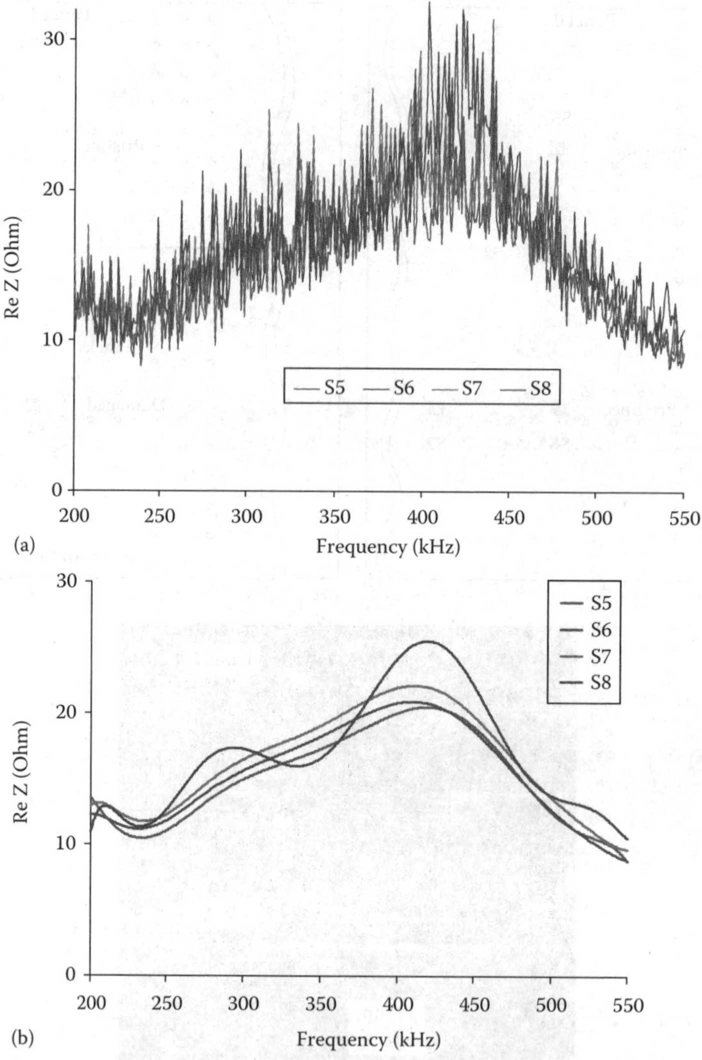

FIGURE 11.96
(a) Superposition of PWAS near field E/M impedance spectra in the 200–550 kHz band; (b) dereverberated response (DR).

dereverberated response in the 400–450 kHz range. On the other hand, the spectra of sensors S5, S6, and S7 do not show significant differences. To quantify these results, we used two methods:

1. Overall statistics metrics of the dereverberated response (DR)
2. PNN

Figure 11.96b shows the DR curves extracted from the Figure 11.96a spectra. It is clear that the three DR curves for the pristine scenario (S5, S6, and S7) are very similar. In contrast, the DR for the damage scenario, S8, is clearly different. To quantify these DR differences, we

TABLE 11.25

Response of RMSD, MAPD, and CCD Metrics to Near Field Damage

Class	Pristine vs. Pristine			Damage vs. Pristine		
Sensors	S5_S6	S5_S7	S6_S7	S5_S8	S6_S8	S7_S8
RMSD	4.09%	8.74%	5.71%	15.64%	14.10%	10.10%
		6.18%			13.28%	
MAPD	3.75%	8.43%	5.88%	13.26%	11.89%	8.05%
		6.02%			11.07%	
CCD	0.94%	0.63%	0.07%	5.70%	7.45%	6.77%
		0.55%			6.64%	

Note: S5, S6, and S7 = pristine; S8 = damage.

used the overall-statistics damage metrics defined by Equations (11.530) through (11.532), i.e., RMSD, MAPD, and CCD. The results of this analysis are presented in Table 11.25.

Two sets of results are presented: (1) pristine versus pristine and (2) damage versus pristine. The former is used to quantify the statistical differences between members of the same class, i.e., the pristine sensors S5, S6, and S7. The latter is used to quantify the differences between the damage sensor, S8, and any of the pristine sensors, S5, S6, or S7. Then, in each set, the mean value was calculated. Examination of Table 11.25 indicates that the RMSD and MAPD values for the damage case are almost double of that for the pristine case. This indicates good damage detection capability. However, the CCD values indicate an even better detection capability since the value for the damage case is an order of magnitude larger than the pristine case (6.64% vs. 0.55%). This confirms that CCD is potentially a very powerful damage detection metric. The mean CCD, MAPD, and RMSD values are represented graphically in Figure 11.97. The stronger detection capability of the CCD metric is again apparent.

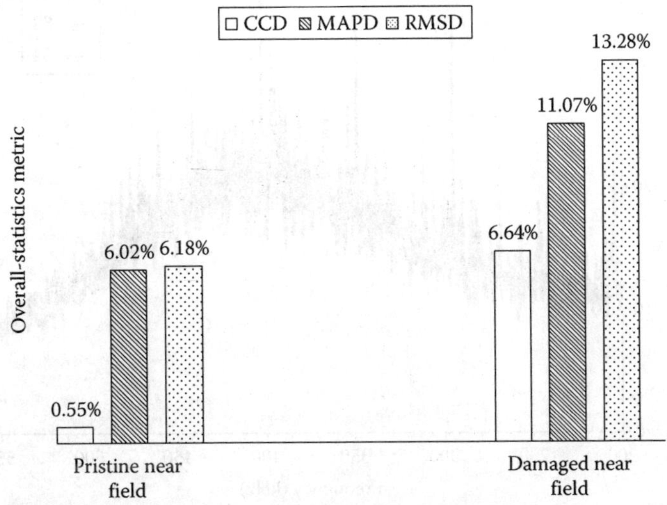

FIGURE 11.97
Overall-statistics damage metrics comparison as calculated on the near field spectra.

11.7.2.2 Classification of Crack Damage in the PWAS Medium-Field

The medium field experiment was designed to estimate the ability of PWAS to detect damage in a wider area. In this study, the medium field is called the area with a radius of about 100 mm where the detection of damage is still possible but the effect of damage is not manifested as drastically on the E/M impedance spectra as in the near field. The distance between PWAS and crack for the medium field experiment was eight times bigger than the near field experiment. The relative size of the near field and medium field of PWAS is depicted in Figure 11.95b.

The medium field PWAS were S1, S2, S3, and S4. They were located approximately 100 mm from the first rivets in the horizontal rows of rivets. Though placed at different locations, all four PWAS were placed in similar structural situations. Hence, in the absence of damage, they should give the same spectral readings. However, S4 is not exactly in the same situation due to the presence of the 12.7 mm simulated crack originating from the first rivet hole in Figure 11.95. Therefore, the S4 spectrum is expected to be slightly different. Since S4 is not in the crack near field, this difference is not expected to be as large as that observed in the case of S8 in the near-field experiments. To summarize, S1, S2, and S3 are in pristine situations while S4 is in a damage situation.

The E/M impedance spectra for S1, S2, S3, and S4 are presented in Figure 11.98. It could be noted that the spectrum of the damage sensor S4 displays some higher amplitudes of some of the spectral resonances in comparison with the spectra for S1, S2, and S3. On the other hand, no significant difference was observed between the S1, S2, and S3 spectra that are in the pristine class. However, the changes due to damage are much slighter than those observed in the near-field experiment, and no change in the dereverberated response could be observed. In other words the dereverberated responses observed in the pristine (S1, S2, and S3) and damage (S4) classes follow similar general patterns. For this reason, the analysis of the dereverberated response did not yield practical results.

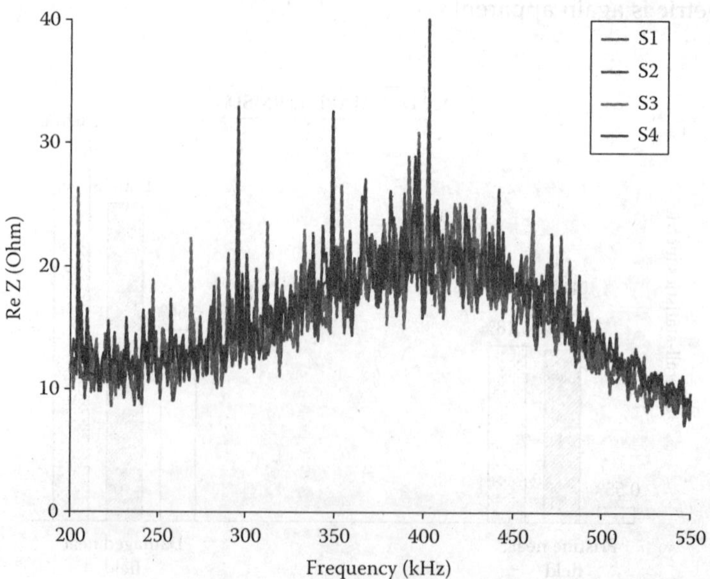

FIGURE 11.98

Superposition of PWAS medium field E/M impedance spectra in the 200–550 kHz band.

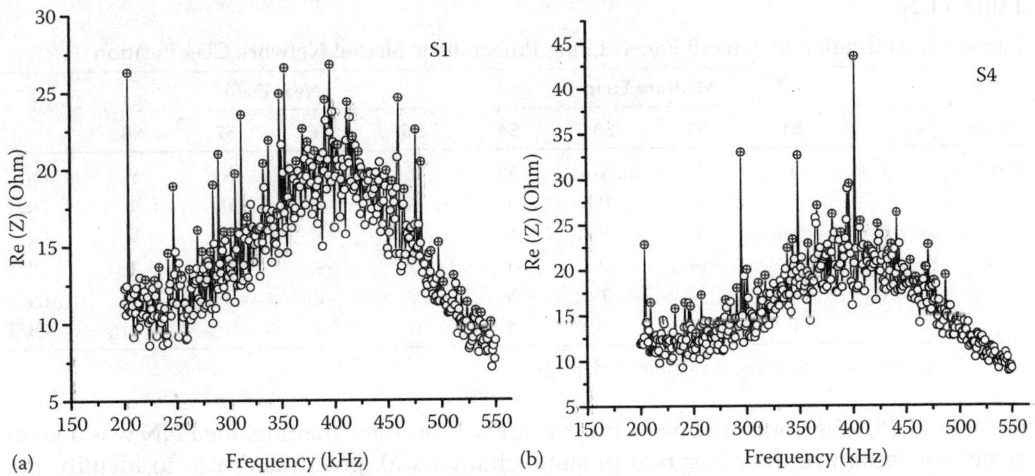

FIGURE 11.99
E/M impedance spectra for damage detection experiment in the PWAS medium field: (a) sensor S1 is the "pristine" case; (b) sensor S4 is the "damaged" case.

To compare the medium-field spectra, we used the PNN algorithm. As in the case of the circular plates, the spectral features considered in the analysis were the resonance frequencies. A features extraction algorithm was used to obtain the features vectors. The algorithm was based on a search window and amplitude thresholds. The 48 extracted features are shown graphically in Figure 11.99. We observe that Figure 11.99a shows the S1 results corresponding to a pristine scenario whereas Figure 11.99b shows the S4 results corresponding to a damage scenario. The resonance peaks picked by the feature extraction algorithm are marked with a cross in the data point. It can be seen that the resonance frequencies for the pristine and damage scenarios are different. For example, the pristine scenario features several peaks above 500 kHz while the damage scenario does not show any peak in this bandwidth. As a result of this process, we were able to construct four feature vectors, each 48 long, which served as inputs to the PNN. Of these, three vectors represented pristine condition (S1, S2, and S3) while the fourth vector represented a damage condition (S4).

The PNN was designed to classify data into two classes: pristine and damage. Since we have three pristine input vectors and one damage input vector, we decided to use one of three pristine input vectors for training and the other three vectors (two pristine and one damage) for validation. (No training was feasible for damage scenario). This created a dichotomous situation in which the PNN would recognize data when it belongs to the pristine class and reject it when it does not belong to the pristine class. Thus, the damage situation was recognized as "nonpristine." We would like to emphasize that this type of classification problem often occurs in practice where the response of structure in pristine condition is usually known and the response in damaged conditions is frequently unknown. The results of the PNN study indicated that the PNN was able to correctly classify data into the correspondent classes regardless of the choice of training vector for the "pristine" class. This is indicated in Table 11.26 which presents the damage identification via PNN classification in the PWAS medium field and near field using 48-frequencies features vectors extracted from the spectra presented in Figure 11.99. In Table 11.26, T = "training," V = "validation," 0 = "pristine," and 1 = "damage." As shown

TABLE 11.26

Damage Identification in Aircraft Panels Using Probabilistic Neural Network Classification

		Medium Field				Near Field				
Vector		S1	S2	S3	S4	S5	S6	S7	S8	
Test	I	T	V	V	V	T	V	V	V	IN
		—	0	0	1	—	0	0	1	OUT
	II	V	T	V	V	V	T	V	V	IN
		0	—	0	1	0	—	0	1	OUT
	III	V	V	T	V	V	V	T	V	IN
		0	0	—	1	0	0	—	1	OUT

Note: T, training; V, validation, 0, pristine; 1, damage.

in Table 11.26, when one pristine feature vector was used for training, the PNN was able to positively recognize the other two pristine situations (0, green) and also to identify the damage situation (1, red). The fact that the sensors S1 and S2 were consistently identified as pristine is not surprising since they were placed on a pristine panel. However, it is interesting to note that sensor S3, which was placed on a damaged panel, was also identified as "pristine." This is correct since sensor S3, though placed on a damaged panel, is placed in a pristine location that is reasonably away from the damage location. The sensor that was more in the vicinity of the damage was sensor S4. The PNN algorithm identified the sensor S3 output as corresponding to a pristine condition and that of sensor S4 as corresponding to a damage condition. This is, again, correct. This proves the detection localization capability of the PWAS-based electromechanical impedance method which is only sensitive to local damage due to its high-frequency operating principle.

After successfully using the PNN method to detect damage in the medium field, we backtracked and used it for damage classification in the near field. The raw spectra obtained for the near field were processed with the features extraction algorithm using the same method as for the medium field experiment. The PNN design, training, and validation were similar to the medium field. The PNN correctly classified data into "pristine" and "damage" for all training vector choices. We conclude that the classification problem for both medium field and near field problems on aging aircraft panels can be successfully solved with the PNN algorithm.

An additional note regarding the localization property of the E/M impedance method is worth making. As seen in Figure 11.95, Panel 1 had several other simulated cracks besides those discussed so far. However, these cracks were away from the PWAS, and hence outside their sensing range. For this reason, these other cracks did not noticeably influence the sensors, readings. The localization property of the E/M impedance method is important for finding the approximate location of the damage on the structure. Since the E/M imped-ance method is intended for structural health monitoring, an approximate location of the damage would be sufficient for a first alert. Further investigation of the exact damage location and its severity would be the object of detailed NDE investigations. Such detailed NDE investigations would be initiated once the vehicle was pulled out of service due to structural health deficiency as signaled by the structural health monitoring equipment.

11.7.3 Section Summary

Aging aircraft panels were used to study the ability of the pitch-catch and E/M impedance method to capture incipient damage represented by a crack growing from a rivet hole.

The experiments were conducted with four PWAS placed in the near field and another four placed in the medium field. Data was collected in the 200–550 kHz band. The overall statistics damage metrics could only be used for the near-field case when the damage spectrum showed a clear change in the dereverberated response. In this case, the CDD metric applied to the dereverberated response showed the biggest chance between the pristine and the damage situations. The PNN approach was able to correctly classify both near-field and medium-field spectra. The PNN inputs consisted of 48-long features vectors extracted from the spectrum with a peak-selection algorithm. The PNN had to select between four spectra (three pristine and one damaged). The procedure was to train the PNN on one of the pristine spectra and then present the remaining three to the PNN for validation. In all cases, correct classification was obtained and the damage spectrum was recognized as nonpristine. We conclude that PNN (with the appropriate choice of the spread constant) shows good potential for attacking and solving the complex classification problems associated with high-frequency spectrum-based damage detection encountered in actual applications.

The SHM example of aging aircraft specimens shows that permanently attached unobtrusive and minimally invasive PWAS, used in conjunction with E/M impedance method, can be successfully used to assess the presence of incipient damage through E/M impedance spectra classification. The E/M impedance method and the wave propagation method form complementary techniques and enabling technologies for in-situ structural health monitoring.

11.8 Summary and Conclusions

This chapter has presented the use of PWAS transducers. To facilitate understanding, a gradual presentation of this subject was used. First, the general wave types encountered in elastic solids were briefly reviewed. Then, attention was focused on the Lamb waves. A brief review of Lamb-wave equations and their solution for both symmetric and antisymmetric modes was provided. Lamb-wave modes and the associated dispersive wave speeds were examined. The circular-crested Lamb waves were also discussed. It was found that they have same Lamb-wave modes and dispersive wave speeds as the straight-crested Lamb wave. However, their radial behavior was found to be governed by the Bessel functions, while that of the straight-crested Lamb waves was governed by the conventional trigonometric functions.

Analysis of PWAS resonators was done next. Detailed analysis of the mechanical and electrical responses was presented and the associated resonances were identified. Closed form solutions of the electromechanical impedance and admittance were derived. Comparison with experimental data was presented.

The use of PWAS as ultrasonic transducers was then presented. The analysis started with the investigation of the shear layer coupling between the PWAS and the support structure. Closed form solutions were derived indicating the dependence of shear distribution on the shear layer stiffness. For very stiff shear layers (i.e., high modulus thin adhesives), the shear distribution is concentrated at the PWAS ends, thus giving rise to ideal bonding assumption and the pin-force model. The pin-force model assumes that transmission of force between the PWAS and the structure takes place in the form of concentrated pin forces at the PWAS ends. The energy transfer between the PWAS and the structure was also analyzed, and conditions for optimum energy transfer were identified.

The use of PWAS ultrasonic transducers for the generation and reception of elastic waves was examined. It was found that PWAS couple intimately with the surface motion patterns of the guided Lamb wave traveling in thin-walled structures. An analysis of the mechanism through which PWAS can excite and detect the Lamb waves used in embedded nondestructive evaluation. Examples of using PWAS for Lamb-wave nondestructive evaluation was given. The pitch-catch and pulse-echo methods were presented. The detection of cracks in realistic aircraft panels was demonstrated. A PWAS phased array was used to illustrate the generation of a virtual scanning beam that was able to detect the presence of isolated cracks in large plates. This approach and the associated algorithm were termed embedded ultrasonics structural radar (EUSR).

The use of PWAS as embedded modal sensors was described in the last part of this chapter. The analysis started with the examination of an elastically constrained PWAS. A complete analysis was performed and closed-form solutions for the electromechanical impedance and admittance as function of the dynamic stiffness ratio between the PWAS and the structure were derived. Both 1-D and 2-D analyses of this phenomenon were performed. It was demonstrated that the real part of the PWAS electromechanical impedance and admittance spectrum closely follows the mechanical resonance spectrum of the substrate structure. These theoretical developments were verified against experimental results, and very good agreement was obtained for 1-D and 2-D situations. Thus, PWAS were identified as true modal sensors that can directly identify the mechanical resonances of the structural substrate. Methods for structural damage detection based on the electromechanical impedance spectrum were examined. It was found that the electromechanical impedance method works best in the high kilohertz range. This high-frequency range is most sensitive to localized incipient damage because it can excite the high-frequency local modes. Examples of damage detection in circular plates and realistic aircraft panels with simulated cracks were presented. The classification of the high-frequency spectra resulting from the high-frequency impedance measurements was implemented with various algorithms, some based on simple overall statistics, others based on neural networks. Several examples based on actual experiments were provided.

The chapter culminates with a case study in which the detection of cracks in realistic aircraft panels is performed using the above technique. This multi-method approach highlights the advantages and disadvantages of the methods presented above and illustrate the versatility of PWAS transducers for damage detection in realistic structures.

11.9 Problems and Exercises

PROBLEM 11.1

Consider a PWAS transducer with length $l = 7$ mm, width $b = 1.65$ mm, thickness $t = 0.2$ mm, and material properties as given in Table 11.2.

1. Find the free capacitance, C_0, of the PWAS.

2. Calculate the voltage, V, measured at PWAS terminals by an instrument with input capacitance $C_e = 1$ pF when the applied strain is $S_1 = -1000$ μɛ.

3. Plot the variation of measured voltage with instrument capacitance $C_e = 0.1 \ldots 10$ nF when $S_1 = -1000$ μɛ (use log scale for C_e axis).

4. Plot variation of measured voltage with strain in the range $S_1 = 0 \ldots -1000$ $\mu\varepsilon$; make a carpet plot of V vs. S_1 for various values of C_e (0.1 nF, 1 nF, 10 nF).

5. Extend to dynamic strain: calculate the voltage V for dynamic strain of amplitude $\hat{S}_1 = 1000$ $\mu\varepsilon$ with frequency $f = 100$ kHz (take $C_e = 1$ pF).

PROBLEM 11.2

Consider a PWAS transducer with length $l = 7$ mm, width $b = 1.65$ mm, thickness $t = 0.2$ mm, and material properties as given in Table 11.2.

1. Calculate the approximate value of Young's modulus, Y^E, in GPa. What common material is this value close to?

2. Calculate the electric field, E_3, for an applied voltage $V = 100$ V. Express the electric field E_3 in kV/mm.

3. Calculate the strain S_1 for the simultaneous application of stress $T_1 = -1$ MPa and voltage $V = -100$ V.

4. Plot on the same chart the variation of strain S_1 with voltage V for $T_1 = 0$; -2.5, -5, -7.5, -10 MPa.

5. Plot on the same chart the variation of strain S_1 with stress T_1 for $V = 0$, 25, 50, 75, 100 V.

PROBLEM 11.3

Consider a PWAS transducer with length $l = 7$ mm, width $b = 1.65$ mm, thickness $t = 0.2$ mm, and piezoelectric material properties as given in Table 11.2. Assume internal damping ratio $\eta = 1\%$ and electric loss factor $\delta = 1\%$.

1. Calculate the complex compliance, $\bar{s}_{11}^E$, complex dielectric permittivity, $\bar{\varepsilon}_{33}^T$, complex coupling coefficient, $\bar{k}_{31}$, real and complex sound speeds, c and $\bar{c}$, and real and complex electrical capacitances, C_0 and $\bar{C}_0$.

2. Calculate the natural resonance frequencies of the first, second, and third modes of antisymmetric vibration in kilohertz. Plot on the same chart the first, second, and third modes of antisymmetric vibration.

3. Calculate the natural resonance frequencies of the first, second, and third modes of symmetric vibration in kilohertz. Plot on the same chart the first, second, and third modes of symmetric vibration.

4. Plot on separate charts the real part, the imaginary part, the amplitude, and the phase of the PWAS electromechanical admittance at 401 equally spaced points in the 10–1000 kHz frequency range. Identify on this chart some of the resonance frequencies determined earlier. State the mode type (symmetric or antisymmetric) and mode number (e.g., f_2^A for the second antisymmetric mode).

5. Plot on separate charts the real part, the imaginary part, the amplitude, and the phase of the PWAS complex impedance at 401 equally spaced points in the 100–1000 kHz frequency range. Identify on this chart the frequencies at which the real part of the impedance peaks. Explain how these frequencies are related, if at all, with the resonance frequencies.

PROBLEM 11.4

Consider a PWAS of length $l=7$ mm, width $b=1.65$ mm, thickness $t=0.2$ mm, and material properties as given in Table 11.2. The PWAS is bonded to a 1 mm thick aluminum strip with material properties as given in Table 11.27. The PWAS length is oriented along the strip.

1. Calculate the first three frequencies at which the PWAS will tune into the axial waves propagating into the aluminum strip.
2. Repeat the calculations considering the PWAS length oriented across the strip.

PROBLEM 11.5

Consider a PWAS of length $l=7$ mm, width $b=1.65$ mm, thickness $t=0.2$ mm, and material properties as given in Table 11.2. The PWAS is bonded to a 1 mm thick aluminum strip with material properties as given in Table 11.27. The PWAS length is oriented along the strip.

1. Calculate the first three frequencies at which the PWAS will tune into the flexural waves propagating into the aluminum strip.
2. Repeat the calculations considering the PWAS length oriented across the strip.
3. Discuss the difference between these results and those of axial tuning calculated in the previous problem. Comment on the results of (1) and (2).

PROBLEM 11.6

Consider a PWAS of width $b=1.65$ mm, thickness $t=0.2$ mm, and material properties as given in Table 11.2. The PWAS is bonded to a 1 mm thick aluminum strip with material properties as given in Table 11.27. The PWAS length is oriented along the strip.

1. Calculate the smallest PWAS length l that will tune into the flexural waves propagating into the aluminum strip at 10 kHz.
2. Calculate the smallest PWAS length l that will reject the flexural waves propagating into the aluminum strip at 10 kHz.
3. Calculate the PWAS length l that will tune into the axial waves propagating into the aluminum strip at 100 kHz.
4. Calculate the PWAS length l that will reject the axial waves propagating into the aluminum strip at 100 kHz.

TABLE 11.27

Typical Material Properties of Aluminum and Steel

	Aluminum (7075 T6)	Steel (AISI 4340 Normalized)
Modulus, E	70 GPa	200 GPa
Poisson's ratio	0.33	0.3
Density, ρ	2700 kg/m^3	7750 kg/m^3
Yield stress, Y	500 MPa	860 MPa

PROBLEM 11.7

Consider a steel beam of thickness $h_1 = 2.6$ mm, width $b_1 = 8$ mm, length $l = 100$ mm, and material properties given in Table 11.27. A 7-mm square PWAS ($l_a = 7$ mm, $b_a = 7$ mm, and $t_a = 0.22$ mm) is bonded to the beam surface at $x_a = 40$ mm from the left-hand end. The material properties of the PWAS are given in Table 11.2. Assume 1% mechanical damping and electric loss.

1. Use the 1-D analytical expressions deduced in this chapter to calculate the admittance and impedance response in the interval 1 to 30 kHz of the PWAS attached to the structure.

2. Plot superposed on the same chart (with appropriate scale factors) the admittance and impedance real parts and the frequency response function imaginary part. Comment on the significance of the peaks observed in these plots.

PROBLEM 11.8

Consider again the data in Problem 11.7 above, but let either double the thickness ($h_2 = 5.2$ mm) or wider width ($b_2 = 16$ mm) or both. Use the 1–D analytical expressions deduced in this chapter to recalculate the admittance and impedance response of the PWAS in the interval 1–30 kHz for these other combinations of thickness and width. Discuss your results.

PROBLEM 11.9

Consider a circular aluminum plate of thickness 0.8 mm, diameter 100 mm, and material properties given in Table 11.27. A 7 mm circular PWAS ($r_a = 7$ mm, $t_a = 0.2$ mm) is bonded in the center of the plate. The material properties of the PWAS are given in Table 11.2. Assume 1% mechanical damping and electric loss.

1. Use the 2–D analytical expressions deduced in this chapter to calculate the admittance and impedance response in the interval 1–40 kHz of the PWAS attached to the structure. (In these calculations, use the modified Equation (11.529) for impedance and its equivalent for admittance).

2. Plot superposed on the same chart (with appropriate scale factors) the admittance and impedance real parts and the frequency response function imaginary part. Comment on the significance of the peaks observed in these plots.

PROBLEM 11.10

Repeat the calculations in Problem 11.9 above, but exclude the axial vibration from the structural stiffness calculation. Discuss your results.

PROBLEM 11.11

Repeat the calculations in Problem 11.9 above, but exclude the flexural vibration from the structural stiffness calculation. Discuss your results.

PROBLEM 11.12

Consider a PWAS transducer of length $l = 7$ mm, width $b = 1.65$ mm, thickness $t = 0.2$ mm, and material properties as given in Table 11.2. The PWAS is bonded at one end of a

1-mm thick 1000-mm long aluminum strip with $E = 70$ GPa and $\rho = 2.7$ g/cc. The PWAS length is oriented along the strip. The PWAS is excited with a 3.5 counts tone-burst.

1. Calculate the first frequency at which the S_0 wave propagation is dominant.

2. Assuming that the frequency is adjusted to the value at which the S_0 wave propagation is predominant as calculated in part (1), calculate the time taken by the wave packet to travel to the other end of the strip specimen and come back. Sketch the wave pattern.

3. Repeat part (2) assuming that a through-the-thickness crack reflector is present at 400 mm from the PWAS. Sketch the wave pattern.

4. Superpose the effects of (2) and (3) assuming that the energy is equally partitioned between the waves reflecting from the end and reflecting from the crack. Sketch the wave pattern. Discuss.

PROBLEM 11.13

Consider a PWAS transducer of length $l = 7$ mm, width $b = 1.65$ mm, thickness $t = 0.2$ mm, and material properties as given in Table 11.27. The PWAS is bonded at one end of a 1 mm thick 1000 mm long aluminum strip with $E = 70$ GPa and $\rho = 2.7$ g/cc. The PWAS length is oriented along the strip. The PWAS is excited with a 3.5 counts tone-burst.

1. Calculate the first frequency at which the A_0 wave propagation is dominant.

2. Assuming that the frequency is adjusted to the value at which the A_0 wave propagation is dominant as calculated in part (1), calculate the time taken by the wave packet to travel to the other end of the strip specimen and come back. Sketch the wave pattern.

3. Repeat part (3) assuming that a through-the-thickness crack reflector is present at 400 mm from the PWAS. Sketch the wave pattern.

4. Superpose the effects of (2) and (3) assuming that the energy is equally partitioned between the waves reflecting from the end and reflecting from the crack. Sketch the wave pattern. Discuss.

References

Achenbach, J. D., *Wave Propagation in Solids*, North-Holland, Amsterdam, 1999.

Alleyne, D. N. and Cawley, P., Optimization of Lamb wave inspection techniques, *NDTE International*, 25(1), 11–22, 1992.

Alleyne, D. N., Pavlakovic, B., Lowe, M. J. S., and Cawley, P., Rapid, long range inspection of chemical plant pipework using guided waves, *Review of Progress in QNDE*, 20, 180–187, 2001.

Banks, H. T., Smith, R. C., and Wang, Y., *Smart Material Structures: Modeling, Estimation and Control*, Masson, John Wiley & Sons, Paris, 1996.

Chahbaz, A., Gauthier, J., Brassard, M., and Hay, D. R., Ultrasonic technique for hidden corrosion detection in aircraft wing skin, *Proceedings of the 3rd Joint Conference on Aging Aircraft*, Albuquerque, NM, 1999.

Chang, F.-K., Built-in damage diagnostics for composite structures, *Proceedings of the 10th International Conference on Composite Structures (ICCM-10)*, Whistler, BC, Canada, Vol. 5, pp. 283–289, August 14–18, 1995.

Chang, F.-K., Manufacturing and design of built-in diagnostics for composite structures, *52nd Meeting of the Society for Machinery Failure Prevention Technology*, Virginia Beach, VA, March 30–April 3, 1998.

Chang, F.-K., Structural health monitoring: Aerospace assessment, *Aero Mat 2001, 12th ASM Annual Advanced Aerospace Materials and Processes Conference*, Long Beach, CA, June 12–13, 2001.

Chaudhry, Z., Joseph, T., Sun, F., and Rogers, C., Local-area health monitoring of aircraft via piezoelectric actuator/sensor patches, *Proceedings, SPIE North American Conference on Smart Structures and Materials*, San Diego, CA, Vol. 2443, pp. 268–276, February 26–March 3, 1995.

Crawley, E. A. and deLuis, J., Use of piezoelectric actuators as elements of intelligent structures, *AIAA Journal*, 25(10), 1375–1385, 1987.

Culshaw, B., Pierce, S. G., and Staszekski, W. J., Condition monitoring in composite materials: An integrated system approach, *Proceedings of the Institute of Mechanical Engineers*, Vol. 212, Part I, pp. 189–202, Professional Engineering Publishing, London, U.K., 1998.

Dalton, R. P. Cawley, P., and Lowe, M. J. S., The potential of guided waves for monitoring large areas of metallic aircraft structure, *Journal of Nondestructive Evaluation*, 20, 29–46, 2001.

Diamanti, K., Hodgkinson, J. M., and Soutis, C., Damage detection of composite laminates using PZT generated Lamb waves, 1st European Workshop on Structural Health Monitoring, Paris, France, pp. 398–405, July 10–12, 2002.

Duke, J. C. Jr., *Acousto-Ultrasonics—Theory and Applications*, Plenum Press, New York, 1988.

Fuller, C. R., Snyder, S. D., Hansen, C. H., and Silcox, R. J. Active control of interior noise in model aircraft fuselages using piezoceramic actuators, *American Institute of Aeronautics and Astronautics Paper # 90-3922*, 1990.

Giurgiutiu, V., *Structural Health Monitoring with Piezoelectric Wafer Active Sensors*, Elsevier Academic Press, Amsterdam; Boston, 2008.

Giurgiutiu, V., Bao, J., and Zagrai, A. N., Structural health monitoring system utilizing guided lamb waves embedded ultrasonic radar, U.S. Patent Office, 2006.

Giurgiutiu, V. and Zagrai, A. N., Embedded self-sensing piezoelectric active sensors for on-line structural identification, *Transactions of ASME, Journal of Vibration and Acoustics*, 124, 116–125, January 2001.

Giurgiutiu, V., Zagrai, A. N., and Bao, J., Embedded active sensors for in-situ structural health monitoring of thin-wall structures, *ASME Journal of Pressure Vessel Technology*, 124(3), 293–302, August 2002.

Grondel, S., Moulin, E., and Delebarre, C., Lamb wave assessment of fatigue damage in aluminum plates, *Proceedings of the 3rd Joint Conference on Aging Aircraft, 1999*.

Harris, C. M., *Shock and Vibration Handbook*, McGraw Hill, 1996.

Ihn, J.-B., and Chang, F.-K., Built-in diagnostics for monitoring crack growth in aircraft structures, *Proceedings of the SPIE's 9th International Symposium on Smart Structures and Materials*, San Diego, CA, paper #4702-04, March 17–21, 2002.

Ikeda, T., *Fundamentals of Piezoelectricity*, Oxford University Press, Oxford, 1996.

Lamb, H., On waves in an elastic plate, *Proceedings of the Royal Society of London, Series A*, Vol. 93, pp. 114, 1917.

Leissa, A., *Vibration of Plates*, NASA SP-160, U.S. Gov't Printing Office, 1969.

Lester, H. C. and Lefebvre, S., Piezoelectric actuator models for active sound and vibration control of cylinders, *Journal of Smart Materials and Structures*, 4, 295–306, July 1993.

Liang, C., Sun, F. P., and Rogers, C. A., Coupled electro-mechanical analysis of adaptive material system-determination of the actuator power consumption and system energy transfer, *Journal of Intelligent Material Systems and Structures*, 5, 12–20, January 1994.

Light, G. M., Minachi, A., and Spinks, R. L., Development of a guided wave technique to detect defects in thin steel plates prior to stamping, *Proceeding of the 7th ASME NDE Topical Conference*, San Antonio, TX, NDE-Vol. 20, pp. 163–165, 2001.

Lin, X. and Yuan, F. G., Damage detection of a plate using migration technique, *Journal of Intelligent Material Systems and Structures*, 12(7), July 2001.

Mustafa, V., Chahbaz, A., Hay, D. R., Brassard, M., and Dubois, S., Imaging of disbond in adhesive joints with Lamb waves; *UT Online Journal* (on the NDT Network), 2(3), 1997. http://www.ndt.net/article/tektren2/tektren2.htm

Osmont, D., Dupont, M., Gouyon, R., Lemistre, M., and Balangeas, D., Damage and damaging impact monitoring by PZT sensors-based HUMS, *Proceedings of the SPIE Smart Structures and Materials 2000: Sensory Phenomena and Measurement Instrumentation for Smart Structures and Materials*, Newport Beach, CA, SPIE Vol. 3986, pp. 85–92, 2000.

Park, G. and Inman, D. J., Impedance-based structural health monitoring, monograph: Nondestructive testing and evaluation methods for infrastructure condition assessment, Woo, S. C. (Ed.), Kluwer Academic Publishers, New York, 2001.

Rayleigh, J. W. S., On waves propagated along the plane surface of an elastic solid, *Proceedings of the London Mathematical Society*, Vol. 17, pp. 4–11, 1887.

Rose, J. L., *Ultrasonic Waves in Solid Media*, Cambridge University Press, Cambridge, 1999.

Rose, J. L., A baseline and vision of ultrasonic guided wave inspection potential, *ASME Journal of Pressure Vessel Technology—Special Issue on Nondestructive Characterization of Structural Materials*, 124(3), 273–282, August 2002.

Rose, J. L., Rajana, K. M., and Hansch, M. K. T., Ultrasonic guided waves for NDE of adhesively bonded structures, *Journal of Adhesion*, 50, 71–82, 1995.

Singher, L., Segal, Y., and Shamir, J., Interaction of a guided wave with a nonuniform adhesion bond, *Ultrasonics*, 35, 385–391, 1997.

Sun, F. P., Liang, C., and Rogers, C. A., Experimental modal testing using piezoceramic patches as collocated sensors-actuators, *Proceeding of the 1994 SEM Spring Conference & Exhibits*, Baltimore, MI, June 6–8, 1994.

Timoshenko, S., P., *Vibration Problems in Engineering*, D. Van Nostrand Company Inc., New York, 1955.

Todd, P. D. and Challis, R. E., Quantitative classification of adhesive bondline dimensions using Lamb waves and artificial neural networks, *IEEE Transactions of Ultrasonics, Ferroelectrics, and Frequency Control*, 46(1), 167–181, January 1999.

Tzou, H. S. and Tseng, C. I., Distributed piezoelectric sensor/actuator design for dynamic measurement/control of distributed parametric systems: A piezoelectric finite element approach, *Journal of Sound and Vibration*, 138, 17–34, 1990.

Viktorov, I. A., *Rayleigh and Lamb Waves: Physical Theory and Applications*, Plenum Press, New York, 1967.

Wang, C. S. and Chang, F.-K., Built-in diagnostics for impact damage identification of composite structures, in *Structural Health Monitoring 2000*, Fu-Kuo Chang (Ed.), Technomic, Lancaster, PA, pp. 612–621, 2000.

12

Microcontrollers for Sensing, Actuation, and Process Control

12.1 Introduction

This chapter deals with microcontrollers and their use in sensing, actuation, and control mechatronics applications. The chapter starts with an introduction of microcontroller hardware and software, followed by the treatment of each of the specific functions the microcontroller can perform. Subsequently the sensing, actuation, and process control applications are discussed. This chapter contains worked out examples, flowcharts, and codes. Several experiments using the microcontroller functions are described. Problems and exercises that can be used for homework assignments are given as a separate section at the end of the chapter. However, due to space constraints, not everything that could be said about this subject could be inserted into this chapter. But a complete collection of solved tutorial examples, homework assignments, and laboratory exercises is posted on the publisher's Web site. The reader is encouraged to visit the publisher's Web site in order to grasp a full understanding of this subject.

12.1.1 Motivation

There is a second computing revolution going on, this time involving microcontrollers. These powerful bundles of computing power are now showing up in automobiles, machine control systems, and home appliances—any application, in fact, where you need to embed some "intelligence" inside a device. A deluge of computers, sensors, microcontrollers, and actuators has permeated the very fabric of present-day society. Microcontroller-based devices and appliances are to be found in all the crevices of our everyday life. The auto industry is putting tens of microcontrollers in a modern automobile and plans to increase this number multifold as new technologies are being introduced. Automotive customers need high performance to tackle increasingly complex applications such as emission control, maximized engine performance, and electronic transmission control. To deliver these requirements, a high-speed, cost-effective solution is needed with higher levels of system integration. Hybrid propulsion, 42V wiring, "steer-by-wire," "brake-by-wire," collision avoidance, and autopilot are being currently developed, and automobiles with such capabilities will hit the market in the near future. The designers of the smart commercial kitchen (SCK) systems intend to use embedded microcontroller technology to create appliance controls that can communicate to an online kitchen in a plug and play environment as well as access the Internet. In the medical field, an aging population is accelerating development of diagnostic equipment with higher electronic content. More semiconductor content like microcontrollers, microprocessors, programmable logic, and analog products are being designed into high-end medical devices, and the technical requirements are getting tougher. In response to this situation, an interdisciplinary engineering branch, that spans mechanical engineering, electronics, embedded microcontrollers/digital signal processing, controls, and information technology, has emerged under the name of mechatronics. At close examination, one cannot but notice that most of today's machinery, from the simplest bread maker and robotics toys, through automobiles and manufacturing facilities contain at least one mechatronics component, whether overt or covert.

12.1.2 Microprocessor and Microcontroller

Microprocessors and microcontrollers are essential to many of the products we use every day, such as TVs, cars, radios, home appliances, and of course computers. Microprocessors are the core of a computer, but they are used also in many other applications, such as in "embedded devices." The biggest chunk of the digital business comes from processors, but processors are only 2% of all semiconductors market. Thirty percent of the world's semiconductor sales come from microprocessors, DSP's, microcontrollers, and programmable peripheral chips. Processors contribute only 2% of the market, but they seem to generate 30% of the sales.

A *microcontroller* can be defined as a complete computer system on a chip including CPU, memory, clock oscillator, and I/O devices. When some of these elements (such as the I/O or memory) are missing, the integrated circuit would be called a *microprocessor*. The CPU in a personal computer (PC) is a microprocessor.

The small 8-bit chips (e.g., Intel 8051s and Motorola 6805s) are the best-selling type of microprocessors and microcontrollers. These small devices are embedded in a wide variety of electronic devices, ranging from small gadgets and home equipment control to car electronics. Current sales of the small microcontrollers are at the rate of over 3 billion per year. But they are not very expensive, so they are less than 15% of the total revenue. At the opposite end of the scale are the 32-bit microprocessors. This category includes PC processors like Pentium 5 and Athlon and dozens of embedded processors such as PowerPC, 68k, MIPS, and ARM chips. Most of the 32-bit microprocessors and microcontrollers are used in embedded systems, not PCs. PC processors are only 2% of all processors sales volume, but PC processors are 50% of all processor sales revenue, since they are much more expensive. This means that PC processor makes 15% of all the money made from every type of semiconductor from every company everywhere in the world. Overall, the average price for a microprocessor, microcontroller, or DSP is just over $6. Most are much cheaper than that, but the few expensive ones pull the average price up.

12.1.2.1 Embedded Processing

Simply (and naively stated) an *embedded controller* is a controller that is embedded in a greater system. A rigid definition is difficult if not impossible to formulate. You could say that an embedded controller is a controller (or computer) that is embedded into some device for some purpose other than to provide generic computing. Embedded controllers adhere to the philosophy of high integration. By including the features necessary for the task at hand, an embedded controller (processor) can be a powerful yet cost effective solution. An embedded controller might need external components before it is considered a true "computer." This is especially true regarding RAM.

Usually an embedded system is a system whereby the user is not given direct access to any level of code. An embedded system is treated as a black box, i.e., one presses a button and sees some action taking place. "Embedded" has a structured definition as "being supplied as a component part of a larger system." Most embedded computers do not "look like" computers. Generally, PCs are not defined as embedded systems; however if the case is removed and some changes are implemented, they can become embedded systems. For example, many embedded industrial systems are based on PC hardware; an x86 motherboard of different form factor and without some of the extra options like PCI slots.

12.1.2.2 *Microcontrollers, Microprocessors, and Digital Signal Processors*

Microprocessors are commonly associated with PCs. A common example would be a typical PC clone. The x86 microprocessor is considered as a general purpose device, since the machine is typically used for general purpose computing (computing, word processing, Web browsing, multimedia, etc.).

Microcontrollers are commonly associated with embedded applications. A microcontroller can be viewed as a PC on a chip. It incorporates all the necessary functions (including the memory) in one IC chip. You just need to apply the power (and possibly clock signal), and the device starts executing the program that was "burned" in its read-only memory. The original idea behind the microcontroller was to limit the capabilities of the CPU itself, allowing a complete computer (memory, I/O, interrupts, etc.) to fit the available IC volume. Generally, a microcontroller has the main CPU core and some memory (ROM/EPROM/EEPROM/FLASH, and RAM) and some accessory functions (like timers and I/O controllers) integrated into one IC chip.

Microcontrollers are typically used in applications where the amount of the processing power is not very important. In such applications, compact construction, small size, low power consumption, and low cost are more important. For example controlling a microwave oven is easily accomplished with the smallest of microcontrollers. Nowadays, there are many small electronic devices that are based on a microcontroller. A modern home can include easily tens or hundreds of microcontrollers, since almost every modern appliance has one or more microcontrollers inside it.

A special application that microcontrollers are well suited for is data logging. Place a microcontroller at the centre of a corn field or up in a balloon, and it will monitor and record environmental parameters (temperature, humidity, rain, etc.). Small size, low power consumption, and flexibility make microcontrollers ideal for unattended data monitoring and recording. The automotive market is possibly the most important single driving force in the microcontroller market, especially at its high end. Several microcontroller families are specifically developed for automotive applications. With the continuing process of high scale integration well underway, many standard architecture processors are turning up as microcontrollers.

Embedded controllers may come in many flavors and varieties. Depending on the power and features that are needed, you might choose a 4, 8, 16, or 32-bit microcontroller. On one end of the scale you find the small one-chip microcontrollers. At the other end, you may find high power microprocessors (such as the Motorola 68000 or National 32032 types) used as the CPU of multifunctional embedded controllers. In addition, specialized microcontrollers are available which include features specific for communications, keyboard handling, signal processing, video processing, and other tasks.

Embedded programming: Embedded programming implies programming a system in which resources are limited, and which may need to run without manual intervention. In such a situation, all program errors need to be handled internally. The simplest error handler would just restart the system if things go wrong. Many embedded systems use watchdog systems to handle errors like software crash. The watchdog just resets the system if software stops notifying watchdog timer that is running properly.

"Real-time" implies that time is a critical factor in the functioning of an embedded system. Real-time is associated with a system in which if a single event is missed or overruns its time slot, the whole system would fail. Since such a failure may have disastrous results, it cannot be allowed to happen. *Hard real-time* is usually accepted to mean that any time constraint that is missed would generate a failure. *Soft real-time* is usually accepted to mean that some missed deadlines are acceptable, or at least will not endanger life or property. For example, an automatically controlled airplane represents a situation of hard real-time; if it takes too long for the system to react, the plane will crash. But a PC game could be considered soft real-time; if the player does not get the next frame within so many milliseconds, then the player would be dissatisfied with the game, but nobody would get hurt and nothing would get damaged.

12.1.2.3 Microcontroller Nomenclature

Some common terms related to programmable microprocessors (MPU), microcontrollers (MCU), and digital signal processors (DSP) used in embedded electronic equipments are listed below:

- A/D (Analog to Digital Conversion)

  Converts an external analog signal (typically relative to voltage) and converts it to a digital representation. Microcontrollers that have this feature can be used for instrumentation, environmental data logging, or any application that lives in an analog world without using an external A/D converter IC for this.

- Brownout Protection

  The device is held in reset and will remain in reset when Vcc stays below the Brownout voltage. The device will resume execution (from reset) after Vcc has risen above the brownout voltage.

- CAN (Controller Area Network)

  CAN is a multiplexed wiring scheme that was developed jointly by Bosh and Intel for wiring in automobiles. The CAN specification seems to be the one that is being used in industrial control both in North American and Europe.

- CISC (Complex Instruction Set Computer)

  Almost all of today's microcontrollers are based on the CISC concept. The typical CISC microcontroller has well over 80 instructions, many of them very powerful and very specialized for specific control tasks.

- Clock Monitor

  A clock monitor can shut the microcontroller down if the input clock is too slow.

- CMOS (Complementary Metal Oxide Semiconductor)

  CMOS is the name of a common technique used to fabricate most (if not all) of the microcontrollers. CMOS-based microcontroller requires much less power than ones made with older fabrication techniques, which permits battery operation. CMOS chips can also be fully or near-fully static, which means that the clock can be slowed up (or even stopped) putting the chip in sleep mode. CMOS has a quite high immunity to noise (power fluctuations or spikes), although it does not like static electricity spikes (unless special protection is included in the chip).

- D/A (Digital to Analog) Converters

This feature takes a digital number and converts it to an analog output.

- EPROM (Erasable Programmable Read Only Memory)

Nonvolatile memory that can be completely erased and reprogrammed several times. Erasure is performed by exposure to UV light through a quartz window

- EEPROM (Electrically Erasable Programmable Read Only Memory)

Memory similar to EPROM, but electrically erasable.

- Field programming/reprogramming

Using nonvolatile memory as a place to store program memory allows the device to be reprogrammed in the field without removing the microcontroller from the system that it controls. Field-programmable microcontrollers are generally programmable by connecting the programming device to few pins on the microcontroller and doing the programming while the microcontroller is in place in the circuit board.

- FLASH

Nonvolatile memory has the high-speed access and in-circuit erasability of EEPROM but with higher density and lower cost. FLASH is better solution than regular EEPROM for handling large amounts of nonvolatile program memory. It is both faster than EEPROM and permits more erase/write cycles.

- Halt mode

In Halt mode, all activities are stopped (including timers and counters). The only way to wake up is by a reset or device interrupt (such as an I/O port). The power requirements of the device are minimal.

- Harvard Architecture

Microcontrollers based on the Harvard Architecture have separate data bus and an instruction bus. This means that data and instructions are stored into separate memories that are accessed separately.

- I2C bus (Inter-Integrated Circuit bus)

The I2C bus is a simple two wire serial interface developed by Philips. It was developed for 8-bit applications and is widely used in consumer electronics, automotive, and industrial applications. In addition to microcontrollers, several peripherals also exist that support the I2C bus.

- Idle mode

In Idle mode, all activities are stopped except on-board oscillator circuitry, watchdog logic, clock monitor, and idle timer. Power supply requirements on the microcontroller in this mode are typically around 30% of normal power requirements of the microprocessor. Idle mode is exited by a reset, or some other stimulus (such as timer interrupt, serial port, etc.). A special timer/counter (the idle timer) causes the chip to wake up at a regular interval to check if things are OK. The chip then goes back to sleep.

- Idle/Halt/Wakeup

  The device can be placed into Idle/HALT mode by software control. In both Halt and Idle conditions the state of the microcontroller is preserved. RAM is not cleared, and outputs are not changed. Normal operation is rested with wakeup (usually interrupt) or reset.

- Interrupts

  On receipt of an interrupt, the controller suspends its current operation, identifies the interrupting peripheral, then jumps (vectors) to the appropriate interrupt service routine. After this routine is executed, normal program execution continues. Most microcontrollers have at least one external interrupt, which can be edge selectable (rising or falling) or level triggered. Both systems (edge/level) have advantages. Edge is not time sensitive, but it is susceptible to glitches. Level must be held high (or low) for a specific duration (which can be a pain but is not susceptible to glitches).

- J1850

  J1850 is the SAE (Society of Automotive Engineers) multiplexed automotive wiring standard that is currently in use in North America.

- MICROWIRE/PLUS

  MICROWIRE/PLUS is a serial synchronous bidirectional communications interface. This is used on National Semiconductor Corporation's devices (microcontrollers, A/D converters, display drivers, EEPROM's, etc.).

- Monitor Program

  A monitor is a program installed in the microcontroller which provides basic development and debug capabilities.

- OTP (One Time Programmable)

  An OTP is a PROM (Programmable Read-Only-Memory) device. Once your program is written into the device with a standard EPROM programmer, it cannot be erased or modified. An OTP part usually uses standard EPROM, but the package has no window for erasing (package with window is expensive).

- PWM (Pulse Width Modulator)

  A controllable pulse train is generated at microcontroller output. This is often used as a DA conversion technique: A pulse train is generated and regulated with a low-pass filter to generate a voltage proportional to the duty cycle.

- Resident Program Loader

  Loads a program by Initializing program/data memory from either a serial or parallel port. Convenient for prototyping.

- RISC (Reduced Instruction Set Computer)

  RISC microcontroller has smaller number of commands than CISC microcontroller, and those commands are generally simple (do less work per command). By implementing fewer instructions, the chip designed is able to dedicate some of the precious silicon real-estate for performance enhancing features. Benefits are usually smaller chip and lower power consumption. The industry trend for microprocessor design is for RISC designs.

- RAM (Random Access Memory)

  Volatile memory contains temporary data that is lost when the power is turned off. RAM can be accessed by the CPU to read and write by direct addressing. RAM is widely used in PCs. In microcontrollers, RAM is not used as much due to its relative large physical size. Microcontrollers used in expanded mode will access additional RAM memory provided on a separate chip.

- ROM (Read Only Memory)

  Nonvolatile memory contains information permanently stored during the micro-controller manufacturing process, such as a built-in program.

- SCI (serial communications interface)

  SCI is an enhanced UART (universal asynchronous receiver/transmitter serial port).

- SISC (Specific Instruction Set Computer)

  SISC microcontroller instruction set is designed for a special application in mind. At the expense of the more general-purpose instructions that make the standard microprocessors (8088, 68000) so easy to use, the instruction set was designed for the specific purpose of control (powerful bit manipulation, easy and efficient I/O, and so on).

- Software protection

  Either by encryption or fuse protection, the programmed software is protected against unauthorized snooping (reverse engineering, modifications, piracy, etc.).

- SPI (serial peripheral interface)

  SPI is a synchronous serial port. This is commonly used by Motorola in their microcontroller. Many peripheral chips (like A/D converters) exist which can be connected to a SPI port.

- Synchronous serial port

  A synchronous serial port does not require start/stop bits and can operate at much higher clock rates than an asynchronous serial port. It is used to communicate with high-speed devices such as memory servers, display drivers, additional A/D ports, etc. It can also be used to implement a simple microcontroller network.

- UART (Universal Asynchronous Receiver Transmitter)

  UART is a serial port adapter for asynchronous serial communications.

- USART (Universal Synchronous/Asynchronous Receiver Transmitter)

  USART is a serial port adapter for either asynchronous or synchronous serial communications.

- Vectored Interrupts

  When an interrupt occurs, the hardware interrupt handler automatically branches to a specific address depending on what interrupt occurred.

- Von Neuman Architecture

  Microcontrollers based on the Von Neuman architecture have a single "data" bus that is used to fetch both instructions and data. Program instructions and data are stored in a common main memory. When such a controller addresses main memory, it first fetches an instruction, and then it fetches the data to support the instruction (if such data is needed).

- Watchdog Timer

  A watchdog timer provides a means of graceful recovery from a system problem. If the program fails to reset the watchdog at some predetermined interval, a hardware reset will be initiated.

12.1.3 Common Microcontroller Types

A list of manufacturers of microprocessors, microcontroller, and digital signal processors includes in alphabetical order:

- AB-Semicon—makers of Z180 Assembler code compatible microprocessors, http://www.ab-semicon.com/products.htm
- Ajile Systems—makers of Java-compatible micros, http://www.ajile.com/products.htm
- Altera—makers of the Excalibur processor, http://www.altera.com/products/devices/dev-index.jsp
- AMD (Advanced Micro Devices)—makers of Athlon processors, http://www.amd.com
- Analog Devices—makers of digital signal processor (DSP) chips, http://www.analog.com/dsp/products/index.html
- Basic Micro—makers of ATOM Microcontroller, http://www.basicmicro.com
- Atmel—makers of 89C1051/89C2051 (8051 architecture) chips and AVR RISC chips, http://www.atmel.com
- Basic-X—makers of the BasicX (BX-1) and BasicX-24 (BX-24) microcontrollers, http://www.basicx.com/
- Comfile (in Korea)—makers of PIC-based SBC and TinyPLC, http://www.comfile.co.kr
- Cybernetic Micro Systems—makers of the P-51 chip, http://www.controlchips.com/P51.htm
- Cygnal—makers of 8051-type micros, http://www.cygnal.com
- Cypress Semiconductor—makers of the PSoC microcontrollers, http://www.cypressmicro.com
- Dallas Semiconductor Maxim—makers of 8051-compatible chips, http://www.dalsemi.com/Microcontrollers.cfm
- Embedded Systems—makers of AVR Sprint Microcontroller, http://www.avrsprint.com/default.asp
- Fairchild Semiconductor—makers of the ACEx microcontrollers, http://www.fairchildsemi.com/products/micro
- Hitachi Semiconductor—makers of several micros, http://www.renesas.com/eng/products/mpumcu/index.html
- Infineon—makers of several micros, http://www.infineon.com/microcontrollers

- Intel—makers of x86, Pentium, and other micros, http://www.intel.com/design/embcontrol

- Maxim—makers of several microcontrollers, part of Dallas Semiconductors Maxim, http://dbserv.maxim-ic.com/Microcontrollers.cfm

- Microchip—makers of the PIC microcontrollers, http://www.microchip.com

- Micromint—makers of the Micromint 80C52 and PicStics microcontrollers, http://www.micromint.com/products/picstic.htm

- Mitsubishi—makers of several microcontrollers, http://www.mitsubishichips.com/startMCU.html

- Motorola makers of the 68HCxx microcontrollers, http://e-www.motorola.com

- National Semiconductor—makers of the COP8xx microcontrollers, http://www.national.com/appinfo/mcu

- NEC Electronics—makers of several microcontrollers, http://www.necelam.com/microcontrollers/index.asp

- OO-PIC—makers of the OO-PIC microcontroller, http://www.oopic.com

- Parallax Inc.—makers of the BASIC-Stamps and Javelin-Stamp Microcontroller, http://www.parallaxinc.com

- Philips—makers of several variants of the 8051 chip, http://www-us.semiconductors.philips.com/markets/mms/products/microcontrollers/index.html

- Protean Logic—makers of the TICkit microcontrollers, http://www.protean-logic.com

- Renesas Microcomputers—makers of Hitachi microcontrollers, http://www.renesas.com/eng/products/mpumcu/index.html

- Scenix (now Ubicom)—makers of the SX microcontrollers and IP2000 chips, http://www.ubicom.com/products/index.html

- ST Microelectronics—makers of STxx microcontrollers, http://us.st.com/stonline/profiles/mcu/index.shtml

- Sun—makers of SPARC and MAJC microprocessors, http://www.sun.com/processors/index.html

- Systronix—makers of J-Stamp Microcontroller and other embedded Java microcontrollers, http://www.systronix.com

- Transmeta—makers of the Crusoe embedded processor, http://www.transmeta.com

- Ubicom (formerly Scenix)—makers of the SX microcontrollers and IP2000 chips, http://www.ubicom.com

- Western Design Center—makers of the W65C02/65Cxx chips, http://www.westerndesigncenter.com

- VIA Technologies—makers of Cyrix and other microprocessors, http://www.viatech.com/en/viac3/cyrix_MII.jsp

- Wilke Technology—makers of the TINY-Tiger and the BASIC-Tiger, http://www.wilke-technology.com/

- Zilog—makers of the Z8/Z8plus and Z80 micros, http://www.zilog.com

Of these, some of the most popular microcontrollers are

- Motorola 6811
- Microchip PIC
- Basic STAMP
- Intel 8051 clones

The 8051 was one of the first microcontroller families, and remains one of the most commonly used. The devices are available from multiple sources, are cheap, have decent tools, and offer a nice upgrade path to larger and more capable parts. The 8051 family was created by Intel, but is now largely driven by other companies, including Atmel, Dallas Semiconductor, and Philips. Devices in the family include the 8031, 8032, 8051, 8052, 80151, 80251, and XA series.

The Microchip Technologies PIC has gained popularity in recent years. Year 2000 statistics ranked Microchip second among microcontroller producers in terms of total units shipped.

Motorola has dominated the microcontrollers market for more than 10 years, and is still ranked number one in terms of total units shipped. Its most popular product is the 68HC11 family, a 8-bit microcontroller with extensive peripheral capabilities including parallel and serial I/O, timer input and output functions, and A/D conversion. In the rest of this chapter, we will base our discussion of the microcontroller on the Motorola 68HC11 chip. Nevertheless, the general principles developed in conjunction with this microcontroller can be easily extended to other microcontroller types.

12.2 Microcontroller Architecture

A simple and generic compute representation is shown in Figure 12.1. It features a CPU (central processing unit) that runs a *program*. The CPU receives *inputs*, creates *outputs*, and exchanges information with the *memory*. The CPU operates in steps that are controlled by a

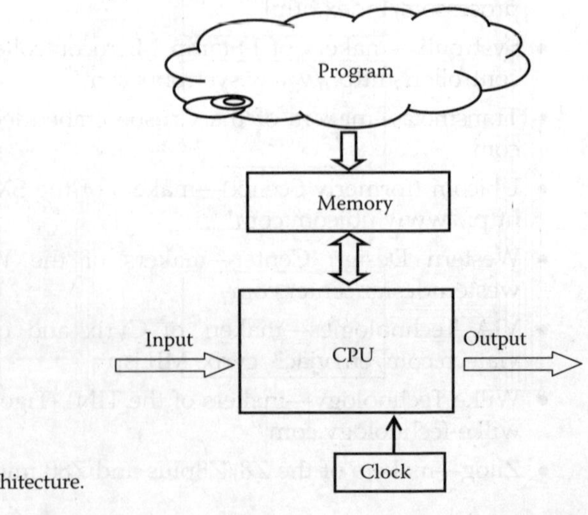

FIGURE 12.1
Generic computer architecture.

clock. The clock sets the tempo of the computer. At each clock signal, the computer executes a certain operation. The operations to be executed are provided by the program. The memory contains *data*. In many instances, the program is also stored in memory, however in a separate partition. And, needless to say, the computer needs to be energized by a power supply.

In PCs and mainframe computers, various elements of the computer system (CPU, memory, clock, input/output, etc.) reside in different chips, circuit boards, or even cases. In a microcontroller, all the computer system elements are closely packed into one integrated circuit chip. A microcontroller can be generically described as a "computer on a chip." In microcontroller parlance, the function of the CPU is performed by a microprocessor unit that is integrated with the other system element inside the microcontroller chip.

12.2.1 Basic Microcontroller Architecture

A microcontroller unit (MCU) with von Neumann architecture has the central processing unit (CPU) and the memory interconnected by three buses: a *data bus*, an *address bus*, and a *control bus* (Figure 12.2). The data bus contains *factual information*, either an instruction or data used to perform the instruction. The address bus contains *location information*. It tells where in memory the instruction or the data is placed. The control bus contains *timing and synchronization information*. It contains a timing signal, which synchronizes the operation of various components of the computer system. When a pulse is sent through the control bus, an operational step is executed in the computer system, and the system transitions from one state to the next state. Computers are sequential automata. They transition sequentially from one state to another state. Since the total number of states is finite, computers are finite state machines.

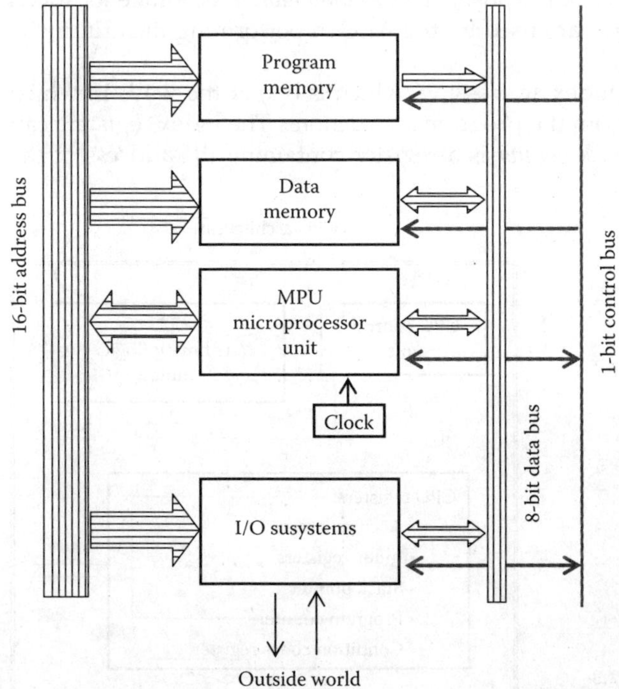

FIGURE 12.2
Microcontroller architecture.

The data bus and the address bus are multilane information corridors. The width of these multilane data corridors depends on the size of the information packets that have to travel on them. For example, in MC68HC11 microcontroller, the data information is contained in 8-bit (1 byte) words, hence the data bus has 8 lines in parallel. The address information is contained in 16-bit (2 byte) words, hence the address bus has 16 lines in parallel. The control bus is a single lane bus.

Every clock beat prompts the CPU to execute a step. The CPU generates a control prompt that synchronizes the rest of the microcontroller, i.e., the program memory, data memory, and I/O subsystems. The 16-bit addresses generated by the CPU are communicated to the program memory, data memory, and I/O subsystems. If the 16-bit address corresponds to a location in the program memory, the contents of that location is loaded on the 8-bit data bus and sent to CPU. If the 16-bit address corresponds to a location in the data memory, then, depending on the specific instruction, one of the two things can happen. Either the content of the data memory location is *read*, i.e., loaded on the data bus or the data memory location is *written*, i.e., the content of the data bus is unloaded and stored in that specific location. The situation with the I/O subsystems is similar. Information can be taken from an I/O subsystem and put on the data bus, and in this case one has an *input* of information. Or, information is taken from the data bus and put into the I/O subsystem, in which case one has an *output* of information.

12.2.2 Microcontroller CPU Architecture

The CPU has its own internal architecture (Figure 12.3). The CPU consists of the CPU control unit, the arithmetic/logic unit (ALU), and a number of registers. The *CPU control unit* generates the control and synchronization signal that goes from the CPU to the other parts of the microcontroller. The ALU performs arithmetic and logic operations on the data. The CPU registers serve a variety of purposes. The *accumulators* are storage locations intimately connected to the ALU. They are used by the ALU in performing the arithmetic and logic operations.

The *index registers* are used in the index addressing, which is one of the ways in which the CPU generates 16-bit addresses from the program instructions. The index registers can be also used as accumulators. The *stack pointer* is a register containing the address of the

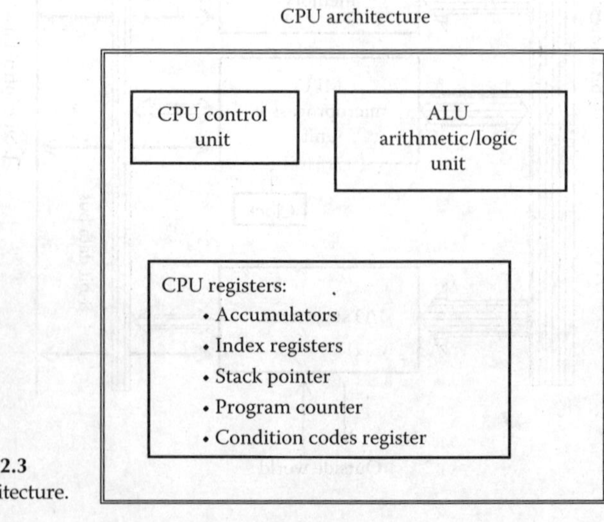

FIGURE 12.3
CPU architecture.

"top of the stack," a location in memory used for temporary storage. The *program counter* stores the memory address of the next instruction to be executed. The *condition codes register* (CCR) stores a number of bits, each with a special significance related to the results of the operation that has just been executed (e.g., if the result of the operation was zero, then a zero bit, Z, is set. If the result of the operation was a negative number, then the negative bit, N, is set, and so on).

12.2.3 The MC68HC11 Microcontroller

MC68HC11 microcontroller unit (MCU) family is an 8-bit MCU with advanced on-chip peripheral capabilities. It has a nominal bus speed of 2 MHz, which is quite appropriate for many mechatronics applications which interface with mechanical and electromechanical systems. The microcontroller has various types of on-chip memory types (Table 12.1). For example, the variant MC68HC11E1 includes

- 512 bytes of electrically erasable programmable ROM (EEPROM)
- 512 bytes of RAM

The major on-chip peripheral functions provided in the MC68HC11 family are

- A 16-bit, free-running timer system with three input-capture lines, five output-compare lines, and a real-time interrupt function (port A)
- A unidirectional 8-bit parallel output port (port B)
- A programmable bidirectional 8-bit parallel input/output port (port C)
- An asynchronous serial communication interface (SCI)
- A separate synchronous serial peripheral interface (SPI)
- An 8-bit pulse accumulator subsystem of the free-running timer that can count external events or measure external periods
- An 8-channel A/D converter with 8-bits of resolution (port E)

A block diagram of the MC68HC11 microcontroller is shown in Figure 12.4. This diagram shows the major subsystems listed above and the designation of the pins used by each subsystem. Note that the asynchronous SCI and the synchronous SPI are part of the 10-pin Port D.

TABLE 12.1

Memory Options for Various MC68HC11 Types

Device	ROM	EPROM	EEPROM	RAM
		Memory (Bytes)		
MC68HC11E0	—	—	—	512
MC68HC11E1	—	—	512	512
MC68HC11E9	12 K	—	512	512
MC68HC711E9	—	12 K	512	512
MC68HC11E20	20 K	—	512	768
MC68HC711E20	—	20 K	512	768
MC68HC811E2	—	—	2048	256

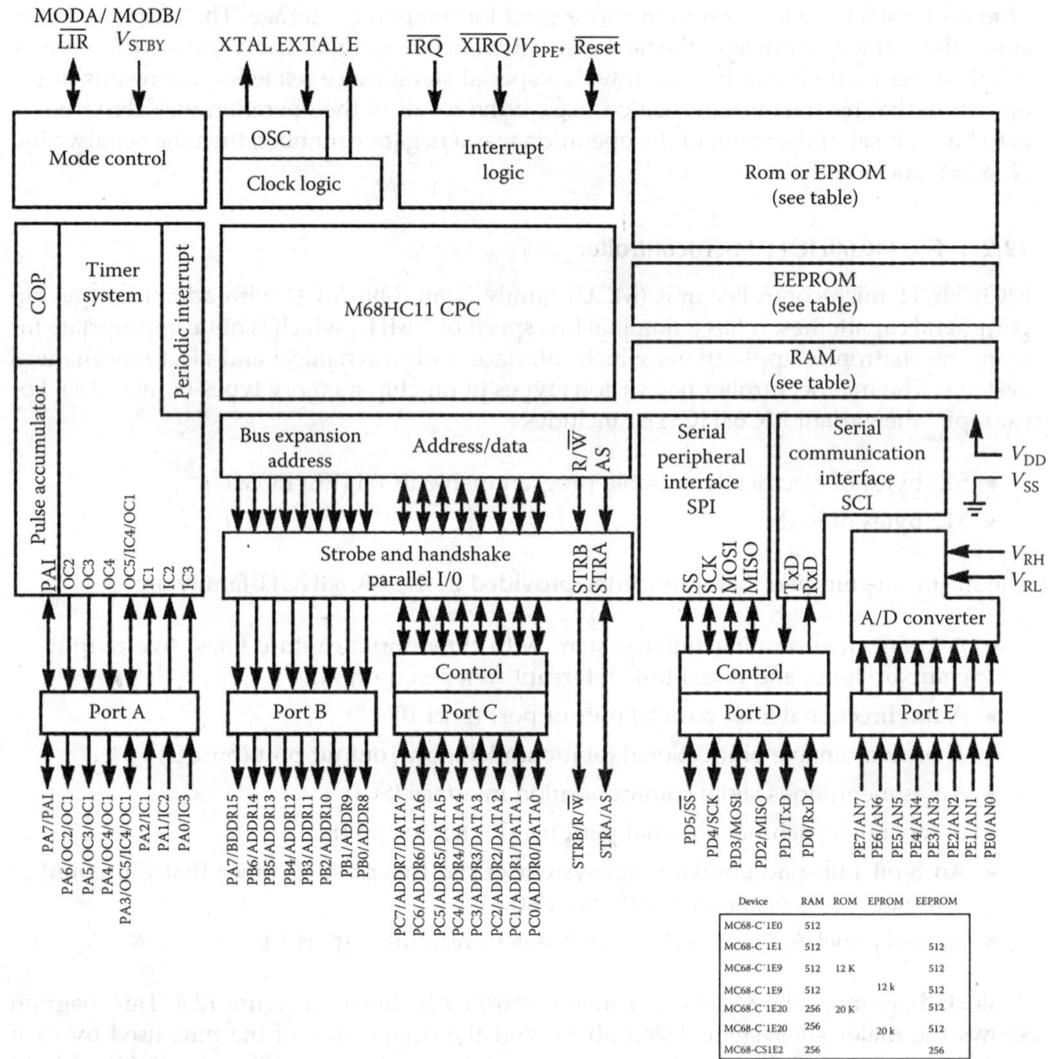

FIGURE 12.4
Block diagram of the MC68HC11 microcontroller family.

12.2.4 Microcontroller Packaging

The microcontroller can be packed in either square or rectangular package. Figure 12.5a shows the typical PLCC square packaging. PLCC stand for plastic leadless chip carrier. It allows quick connection of multipin microcontrollers. The MC68HC11 microcontroller in PLCC square packaging has 52 pins, as shown in Figure 12.5b. Examination of Figure 12.5b in combination with Figure 12.4 allows one to observe the layout of the individual pins of each I/O ports, as well as other pin-related functions. It is important to notice that the A/D converter pins are intercalated, i.e., their physical position is in the order PE0, PE4, PE1, PE5, PE2, PE6, PE3, PE7. This detail is important for the design of the actual wiring in microcontroller applications. It should also be noted that the port C and port B pins (PC0–PC7 and PB0–PB7) are also designated as part of the 16-bit address bus (A0–A15).

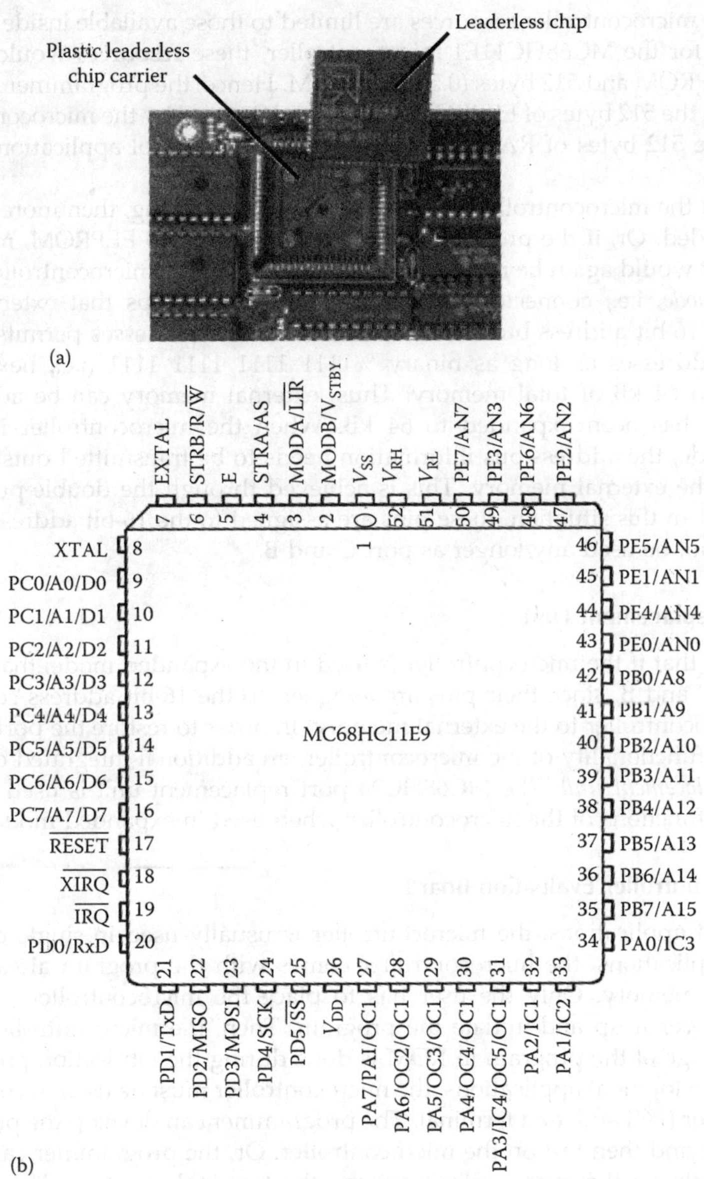

FIGURE 12.5
Microcontroller packaging: (a) plastic leadless chip carrier (PLCC) allows quick connection of multipin microcontrollers; (b) pins assignment of the MC68HC11E9 microcontroller in the 52-pin PLCC square package.

Thus, the port C and B pins are *double purpose pins*. This aspect is important when using the microcontroller in expanded mode, as will be discussed in the next section.

12.2.5 Single-Chip and Expanded Modes of Operation

When used in actual embedded applications, the microcontroller unit can operate as a self-sufficient single chip. This mode of operation is called *single-chip operation*. In single-chip operation, the microcontroller resources are limited to those available inside its packaging. For example, for the MC68HC11E1 microcontroller, these resources would be 512 bytes (0.5 kB) of EEPROM and 512 bytes (0.5 kB) of RAM. Hence, the programmer will have to fit its program in the 512 bytes of EEPROM, while the data used by the microcontroller should not exceed the 512 bytes of RAM. For some embedded control applications this may be sufficient.

However, if the microcontroller is to be used for data logging, then more data memory would be needed. Or, if the program is larger than the 0.5 kB EEPROM, more programming memory would again be needed. In such situations, the microcontroller can be used in *expanded mode*, i.e., connected with external memory chips that extend its internal memory. The 16-bit address bus that the microcontroller possesses permits the handling of memory addresses as long as binary %1111 1111 1111 1111 (i.e., hex $ffff), which corresponds to 64 kB of total memory. Thus, external memory can be added until the total memory has been expanded to 64 kB. When the microcontroller is operated in expanded mode, the address bus information needs to be transmitted outside the microcontroller to the external memory. This is achieved through the double-purpose pins of ports C and B. In this situation, these pins are assigned to the 16-bit address communication, and cannot be used any longer as port C and B.

12.2.6 Port Replacement Unit

We have seen that if the microcontroller is used in the expanded mode, than it looses the use of ports C and B, since their pins are assigned to the 16-bit address communication from the microcontroller to the external memory. In order to restore the port C and port B input/output functionality of the microcontroller, an additional integrated circuit is used, called *port replacement unit*. The MC68HC24 port replacement unit is used to restore the input/output functions of the microcontroller when used in expanded mode.

12.2.7 Microcontroller Evaluation Board

For embedded applications, the microcontroller is usually used in single chip mode. In embedded applications, the microcontroller comes with the program already "burned" into its ROM memory. Only the user has to place the microcontroller in its intended location to power it up and initiate the program. Then, the microcontroller will run by itself. The storage of the program in ROM is done during the fabrication process.

For code development applications, the microcontroller must be used in connection with a host computer (PC) and/or a terminal. The programmer can develop the program on the host computer and then test on the microcontroller. Or, the programmer can develop the program directly on the microcontroller using the terminal interface. The electronic circuitry and IC chips associated with this process are placed on an evaluation board (EVB). The EVB contains expanded memory chips and a port replacement unit, as well as IC chips for servicing the connection to the host computer and/or a terminal. The EVB is essential for program development, since it allows the software programmer to develop and test the microcontroller application software. Once the microcontroller application software is developed and tested, it will be "burned" into the ROM of the mass production microcontrollers. By using the EVB expanded system containing an MC68HC24 and a PC, the user can develop software intended for either single-chip mode or expanded mode microcontroller applications.

A variety of microcontroller evaluation boards are available. They range from the simplest to the most complex. In the following sections, we will discuss some of them in detail.

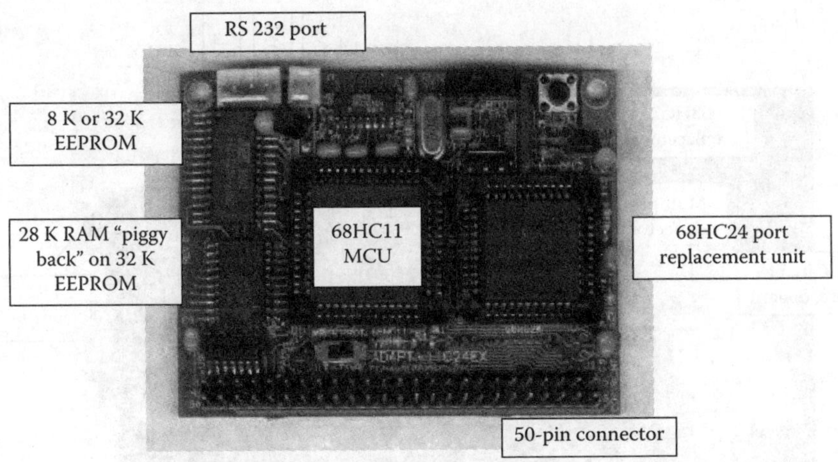

FIGURE 12.6
Adapt11C24DX microcontroller evaluation board. (Courtesy of Technological Arts, Inc. Available at: http://www.technologicalarts.com/myfiles/t1.html#EVBU.)

12.2.7.1 Adapt11C24DX Microcontroller Evaluation Board

The Adapt11C24DX microcontroller evaluation board from Technological Arts, Inc. is characterized by its compactness (Figure 12.6). On a 2.1-in by 2.8-in printed circuit board, this compact EVB accommodates the MCU, the port replacement unit, up to 32 kB of EEPROM and up to 28 kB of RAM "piggy back" on the 32 kB EEPROM. It has a 50-pin connector that can accommodate all the microcontroller port. Due to its compactness, it is also recommended in embedded applications in which expanded memory and the port replacement unit are required.

12.2.7.2 EVBplus2 Microcontroller Evaluation Board

The EVBplus2 microcontroller evaluation board from http://www.evbplus.com/ is a multi-functional product that incorporates several other features beyond the basic EVB functionality (Figure 12.7). In addition to the microcontroller, port replacement unit, and expanded memory, it has a series of other components. These extra components are provided for facilitating the EVB use for classroom instruction, project development, and product design. As presented in Figure 12.7, the EVBplus2 contains a convenient bread board for ad-hoc circuit design, a 4-digit 7-LED display, a row of 8 LEDs for port B monitoring, a liquid crystal (LC) display, several port interfaces, and other features.

12.3 Programming the Microcontrollers

Internally, the microcontrollers use a set of instructions consisting of *opcodes* and *operands*. These opcodes and operands are expressed as numeric codes. Each code is 8-bit long, i.e., it fits into a 1 byte memory location. The codes are usually expressed in 2 hex numbers. (Details of binary numbers and hex numbers are given later in this section.) The opcodes

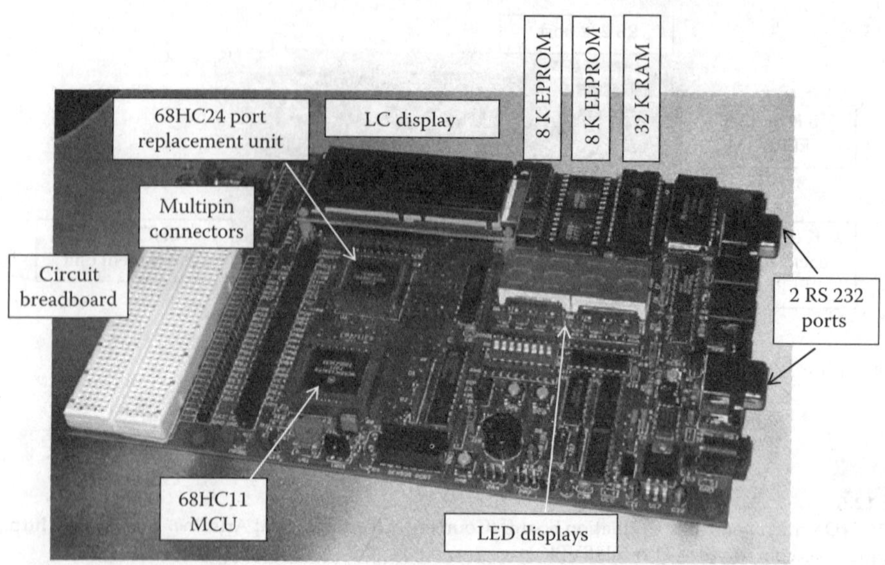

FIGURE 12.7
EVBplus2 evaluation board from http://www.evbplus.com/ for MC68HC11 microcontroller.

and operands expressed numerically constitute the *machine language*. Microcontrollers can be programmed directly in machine language, i.e., writing the actual opcodes and operands into the microcontroller memory. In this case, the program consists of a sequence of 2 hex numbers that are interpreted by the microcontroller directly. However, direct programming in machine language is tedious and nonproductive for the human programmer. It is rarely used, except for direct debugging of an existing program.

The MCU can be more easily programmed using a PC and the *assembly language*. The assembly language is made up of a set of *mnemonic* instructions that are directly related to the MCU opcodes and operands. The mnemonic instructions are easier to memorize for the human programmers, since they resemble actual spoken words. The assembly language is interpreted by an *assembler program*. The assembler program resides in a PC and helps the programmer write the program. One commonly used assembly program is the software miniIDE. The assembler program takes the program written in assembly language (the *source code*) and converts it into a program written in machine language (the *object code*). It also produces a *list file* that presents side by side the object code and the source code. The list file is used for program debugging.

A very useful software tool in programming the microcontroller is a microcontroller simulator. The author of this book has used with good results the THRSIM11 microcontroller simulator. The simulator allows the user to write assembly code and then assemble to code into the object code and the list file.

The next level of complexity is to program in a high-level language, such as BASIC, FORTRAN, C, or C++. The high-level languages use actual English words in their instructions, not just mnemonics as the assembly language. Thus, a high-level language program would be more portable between different types of microcontroller, while an assembly program would have to be specific to a certain microcontroller. (The set of mnemonic instructions used in the assembly program are directly related to the type of microcontroller being used.) The high-level language program is translated into the machine language program (object code) by a *compiler*. Compilers are programs larger and more complex than assemblers. Compilers are more expensive and require more memory storage. *Interpreters* are programs that translate the high-level language into the machine language line-by-line. As soon as a line has been translated, the interpreter program instructs the microcontroller to execute it. Thus, an interpreter program can be viewed as a more economic version of a compiler program.

Both compilers and interpreters have advantages and disadvantages vis-à-vis assemblers. One major advantage of a compiler (or interpreter) vis-à-vis an assembler is that they use high-level source code, which can be written without detail knowledge of the specific microcontroller instruction set. This makes the programming more versatile and portable. One major disadvantage of compilers/interpreters is that they usually produce longer instruction sets (object codes) than assemblers. This is understandable, since a programmer writing in assembly language can take advantage of the specific features of the particular microcontroller being used, whereas a high-level language programmer does not usually need to know what microcontroller is being used. Thus, convenience of programming is being traded for programming length and increased overhead. The object codes produced from high-level language source codes are usually more extensive and take longer time to execute than those produced from assembly language.

Another aspect of this problem is that high-level languages are not generally suited to those microcontroller applications that deal with controlling external devices or processes. These applications involve sending data and control information to output devices and receiving data and status information from input devices. Very often, the control and status information is made up of few binary bits that have specific hardware-related meanings. High-level languages are oriented towards processing decimal numbers; they are less efficient than assembly language in handling bits.

12.3.1 Binary and Hex Numbers

This section gives a brief introduction of binary and hex numbers. Readers familiar with these concepts may skip to the next section. The binary number system is a base-2 numbering system. In binary representation, any value is represented using a combination of 1s and 0s. For example: $19_{10} = 10011_2$, where the subscript 10 on the first number indicates that the number 19 is represented in decimal (base 10), whereas the subscript 2 on the second number indicates that 1110 is represented in binary (base 2).

The binary representation is also called "digital." "Digit" also means finger, and you can imagine a numbering representation in which you use your digits to generate a number containing 1s and 0s (if a finger is up, then it is a 1; if down, then a 0). The ability to represent numbers in terms of 1s and 0s is important, because it is the easiest and most unambiguous way to represent and communicate information. In a computer, a 1 is represented by a "high" voltage ($\sim$5 V) and a 0 by a "low" voltage ($\sim$0 V). The binary system is the backbone of all digital computers and other high-tech applications. Because binary representation of large numbers can be quite lengthy, a more compact representation, the hex number, is also used. The hex number is a number expressed in base 16, e.g., $19_{10} = 13_{16}$. More details about the binary and hex numbers will be given later in this section.

To quickly grasp the use of binary and hex arithmetic, use the binary and hex modes in your scientific pocket calculator. Many of the inexpensive scientific pocket calculators have the hex and binary representations as an option. This feature can be also found in the "calculator" program available as standard in the Windows operating system on most PCs. In order to avoid confusion and stay consistant with the assembly language rules, we will use $ prefix for hex numbers, % prefix for binary numbers, and no prefix for decimal.

12.3.1.1 Binary System

To understand how the binary system works, let us first examine how the conventional base-10 system works. The base-10, or decimal, system constructs numbers using increasing powers of 10. For example, the number 132 is constructed using three powers of 10: 10^0, 10^1, and 10^2. These numbers correspond to 1, 10, and 100. The number 135 is constructed as

$$132 = 1 \times 100 + 3 \times 10 + 2 \times 1 \quad \text{or} \quad 132 = 1 \times 10^2 + 3 \times 10^1 + 2 \times 10^0$$

The equivalent of number 132 in base 2 is %10000100. This is constructed as an addition of powers of 2, i.e.,

$$132 = 1 \times 128 + 0 \times 64 + 0 \times 32 + 0 \times 16 + 0 \times 8 + 1 \times 4 + 0 \times 2 + 0 \times 1$$

or

$$132 = 1 \times 2^7 + 0 \times 2^6 + 0 \times 2^5 + 0 \times 2^4 + 0 \times 2^3 + 1 \times 2^2 + 0 \times 2^1 + 0 \times 2^0$$

It can be seen that the only significant difference between the two systems is the base number.

Each one or zero in the binary representation is called a "bit." A collection of 8-bits is called a "byte" and, in a somewhat humorous note, a collection of 4-bits is called a "nibble." Two nibbles make a byte. The bit associated with the highest power of two is called the most significant bit (MSB), and the bit associated with the lowest power of two is the least significant bit (LSB). The 4-bit nibble forms the basis of hex numbers, which will be discussed a little later.

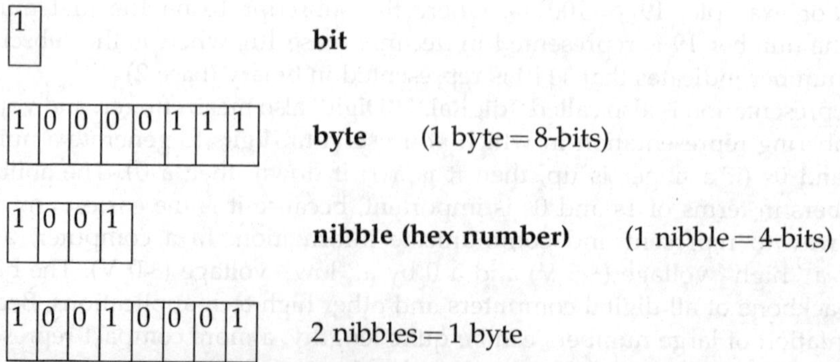

12.3.1.2 Decimal to Binary Conversion

Because most people are more comfortable using, and thinking in, the decimal system, it is important to know how to convert from the decimal to the binary system. This is most easily achieved through a series of divisions by two and by tracking the resulting remainders. Let us consider the binary conversion of the decimal number 132:

$$
\begin{array}{rcll}
132 \div 2 &=& 66 & \text{Remainder} \quad 0 \\
66 \div 2 &=& 33 & \text{Remainder} \quad 0 \\
33 \div 2 &=& 16 & \text{Remainder} \quad 1 \\
16 \div 2 &=& 8 & \text{Remainder} \quad 0 \\
8 \div 2 &=& 4 & \text{Remainder} \quad 0 \\
4 \div 2 &=& 2 & \text{Remainder} \quad 0 \\
2 \div 2 &=& 1 & \text{Remainder} \quad 0 \\
1 \div 2 &=& 0 & \text{Remainder} \quad 1
\end{array}
$$

$$132 = \%10000100$$

MSB   LSB

One notices that this process generates the binary digits from right to left. The binary digits are the remainders of each consecutive division by 2. The remainder resulting from the *last* division is the MSB, whereas the remainder resulting from the *first* division is the LSB. From this example, we see that the decimal number 132 is equal to the binary number %10000100.

The conversion from binary to decimal can be done in the same manner, by adding together power of 2 multiplied by the appropriate values of the bits (i.e., either 1 or 0).

More complicated conversions between binary and decimal systems are best handled with a pocket calculator. Many of the inexpensive scientific pocket calculators have the binary representation as an option. This feature can be also found in the "calculator" program available as standard in the Windows operating system on most PCs.

12.3.1.3 Hexadecimal (Hex) Numbers

As one might have already surmised, binary numbers quickly become long and hard to remember. For this reason, it is more convenient to convert the binary values into hexadecimal numbers (hex). Hexadecimal numbers are numbers in base 16. This requires six additional characters to represent the values 10, 11, 12, 13, 14, and 15. These values will be represented by the letters a, b, c, d, e, and f. The counting order in hex is 0, 1, 2, 3, 4, 5, 6, 7, 8, 9, a, b, c, d, e, f. The reason hex notations are used is that it allows for a one to one correspondence between the 4-bit binary nibble and a single hexadecimal value. If a binary number is broken down into 4-bit nibbles, then each nibble can be replaced with the corresponding hexadecimal number, and the compression is complete. Consider the number 132. The binary equivalent is 10000100. It can be broken down into two separate nibbles: 1000 and 0100. Convert each nibble into the corresponding hex value (8 and 4, respectively), and the hex equivalent of 132 is \$84. This number is much easier to handle! For example, the hex number \$a23e3 can be quite easily converted to the binary number 1010 0010 0011 1110 0011 since 1010 = \$a, 0010 = \$2, 0011 = \$3, 1110 = \$e, and 0011 = \$3.

Conversion from decimal to hex numbers and vice-versa can be also achieved directly using the powers of 16.

Example 12.1:

Consider again the number 132 used in Sections 12.3.1.1 and 12.3.1.2. In power of 16, this number can be represented as

$$132 = 8 \times 16 + 4 \times 1 \quad \text{or} \quad 132 = 8 \times 16^1 + 4 \times 16^0$$

Example 12.2:

To convert the number \$a23e3 to decimal, do the following:

$$\$a23e3 = \$a \times 16^4 + \$2 \times 16^3 + \$3 \times 16^2 + \$e \times 16^1 + \$3 \times 16^0$$
$$= 10 \times 16^4 + 2 \times 16^3 + 3 \times 16^2 + 14 \times 16^1 + 3 \times 16^0$$
$$= 10 \times 65{,}536 + 2 \times 4{,}096 + 3 \times 256 + 14 \times 16 + 3 \times 1 = 664{,}547$$

More complicated conversions among the binary, hex, and decimal systems are best handled with a pocket calculator. Many of the inexpensive scientific pocket calculators have the hex and binary representations as an option. This feature can be also found in the "calculator" program available as standard in the Windows operating system on most PC's.   ∎

Example 12.3:

$$1100\ 1010 = \$ca = 202$$

We leave to the reader to work out this example using a scientific pocket calculator, PC scientific calculator application, or manually.   ∎

12.3.1.4 Numerical Conversion Charts for Unsigned Hex and Binary Integers

Table 12.2 gives the conversion from decimal into hex and binary numbers over the range 0 through 15. In this range, there are 16 hex numbers, 0, 1, 2,..., 9, a, b, c, d, e, f. This is the range of the simplest hex number, which has only one hex digit. Over the same range,

TABLE 12.2

Numerical Conversion Chart for Unsigned 1-Digit Hex Numbers (4-Bit Binary Integers)

Decimal (Base 10)	1-Digit Hex (Base 16)	4-Bit Binary (Base 2)
0	0	0000
1	1	0001
2	2	0010
3	3	0011
4	4	0100
5	5	0101
6	6	0110
7	7	0111
8	8	1000
9	9	1001
10	a	1010
11	b	1011
12	c	1100
13	d	1101
14	e	1110
15	f	1111

TABLE 12.3

Numerical Conversion Chart for Unsigned 2-Digit
Hex Numbers (8-Bit Binary Integers)

Decimal (Base 10)	2-Digit Hex (Base 16)	8-Bit Binary (Base 2)
0	00	0000 0000
1	01	0000 0001
2	02	0000 0010
3	03	0000 0011
4	04	0000 0100
5	05	0000 0101
6	06	0000 0110
7	07	0000 0111
8	08	0000 1000
9	09	0000 1001
10	0a	0000 1010
11	0b	0000 1011
12	0c	0000 1100
13	0d	0000 1101
14	0e	0000 1110
15	0f	0000 1111
16	10	0001 0000
17	11	0001 0001
18	12	0001 0010
19	13	0001 0011
20	14	0001 0100
21	15	0001 0101
...	...	...
255	ff	1111 1111

the 4-bit binary numbers increase from 0000 to 1111. This is as far as one can go with one hex digit and four binary bits. Going further, one requires more digits and bits.

For hex numbers with two digits, a wider range can be covered. Table 12.3 gives the conversion from decimal to hex and binary over the range that can be covered by the 2-digit hex numbers. One notes that this range is from 0 through 255, i.e., from $00 through $ff hex. The corresponding binary numbers range from 0000 0000 through 1111 1111. In this range there are 256 numbers in total. This is as far as the 2-digit hex and 8-bit binary numbers are concerned. Higher numbers would require additional digits and bits.

One notes that the numbers in Tables 12.2 and 12.3 are all positive, i.e., they are *unsigned integers*. If one also wants to consider negative numbers, than one would have to deal with *signed integers*, which will be discussed in a forthcoming section.

12.3.2 Program Development and File Types

We have already seen that programming the microcontroller consists of writing the source code that is processed by a PC program into the object code (to be run into the microcontroller) and the list file that presents side by side the object code and the source code. Thus, a microcontroller programmer has to deal with three types of program files: *source file, object file, list file*.

1. The source file is written in assembly language and has extension .asm. When the source file is processed, the other two files are generated.

2. Object file can be run in the microcontroller. The object file for the MC68HC11 microcontroller is in ASCII-hex format. Its generic name is "S19 file." Its extension is .S19.

3. List file, extension .lst, contains side by side the original code in assembly language and the corresponding hex codes that resulted from the assembly process. The list file is used by the programmer to verify and debug the coding of the program.

Figure 12.8 presents the steps in developing a successful microcontroller software program. The .asm source files can be opened, viewed, edited, and saved in the THRSIM11 simulator. The simulator assembles the .asm code and generates a list file (*.lst). Next, the user steps through the program in the simulator and debugs it until it performs as intended. All this can be done remotely from the hardware on a PC. Once the microcontroller program has been debugged with the simulator, it can then be tested with hardware. The hardware used for testing the program is a MCU mounted on an EVB. The EVB hardware is connected to a lab PC which contains emulator software such as the MiniIDE. MiniIDE reads the .asm file from the lab PC or a removable media and transforms it into machine-language executable code (*.S19). This code is downloaded to the MCU on the EVB. After downloading the code into the MCU, the user makes the MCU run the code using the MiniIDE interface screens. The MiniIDE also generates a list file (.lst) that can be used locally for debugging.

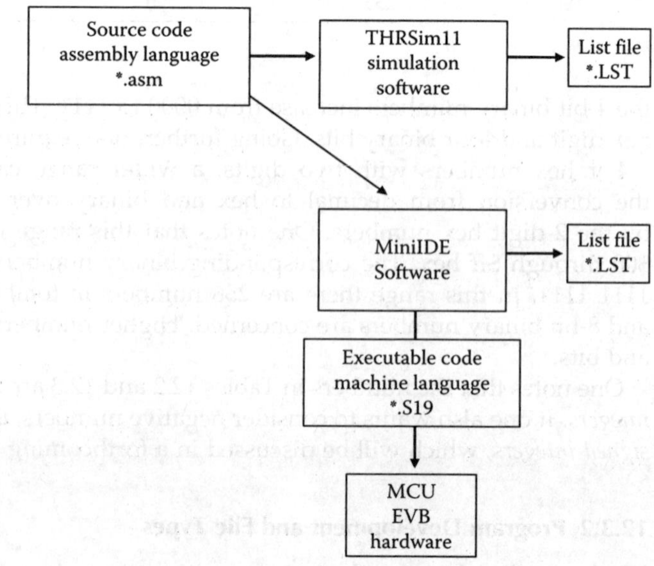

FIGURE 12.8
Flowchart of typical programming steps used in microcontroller software development.

Alternatively, all the three file types (.asm, .lst, .S19) can be also processed in a text editor, e.g., the Notepad application. Examples of .asm and .lst files are shown later in this section.

In many microcontroller applications, the existence of a fast and compact object code is essential for application success. In applications such as machine control, process control, and communications, the microcontroller is required to execute the program as fast as possible. These are called *real-time* applications, because the program is controlling a process that takes place at the same time as the program runs in the microcontroller. For such real-time applications, object codes generated from assembly language are much better suited than those generated from high-level languages.

12.3.3 Assembly Language

In this section we will discuss some basic features of the assembly language. An assembly language code consists of

1. Program statement lines
2. Comment lines

A program statement is a line that contains four fields in the following format:

[<LABEL>]   [<OPCODE   [<OPERANDS>]   [; <comments>]
                 MNEMONIC>]

or

[<LABEL>]   [<DIRECTIVE   [<OPERANDS>]   [; <comments>]
                 MNEMONIC>]

where [ ] indicates an optional field that may not be always required. The fields are separated by a tab or space. (Tab is recommended, since it ensures an orderly appearance to your code. For the same reason, when a field is not used, the tab or blank should still to be used, such that the fields of the same type stay aligned in same columns.) When writing < LABEL >, < OPCODE MNEMONIC > or < DIRECTIVE MNEMONIC >, and < OPERANDS > use upper case characters. When writing < comments > use lower case. The maximum size of LABEL is seven characters.

The < OPCODE MNEMONICS > correspond to the microcontroller opcodes. A brief listing of these opcode mnemonics are given in Section 12.3.4; a more extensive description can be found in the Motorola MC68HC11 programming reference guide and related literature.

The < DIRECTIVE MNEMONICS > are native to the assembly language. A list of directives is given in Table 12.4. The directives that you will use often are shown in *bold*.

The < OPERAND > contains a value, an expression, an address, or a label that the opcodes or the directives need. The operand could be up to 4 bytes long, separated by commas. Some opcodes or directives do not require operands (inherent mode). The

TABLE 12.4

Assembler Directives

Name of Assembler Directive	What It Does	Alias for
END	End program	
DB	Define bytes	FCB
DW	Define words	FDB
DS	Define storage	RMB
EQU	Equate	
FCB	Form constant byte	
FCC	Form constant characters	
FDB	Form double bytes	
ORG	Set origin	
RMB	Reserve memory bytes	
#INCLUDE	Include source file	
$INCLUDE	Include source file	#INCLUDE

TABLE 12.5

Assembler Symbols for Constants

Symbol	Meaning	Example
$<alphanumeric>	Hex number	$A1
<number>	Decimal number	20
%<number>	Binary number	%11001010
@<number>	Octal number	@73
'<string>', '<string>	ASCII string	'A' or 'A (the latter does not work with #INCLUDE)

TABLE 12.6

Assembler Symbols for Expressions

Symbol	Meaning	Example
−	Unary minus	−4
&	Binary AND	%11111111&%10000000
!	Binary OR	%11111111!%10000000
*	Multiplication	3*$2A
/	Division	$7E/3
+	Addition	1+2
−	Subtraction	3−1
( )	Parentheses used for grouping	3*(1+2)

constants used in the < OPERAND > can be hex, decimal, binary, or octal numbers. Table 12.5 shows prefix conventions for differentiating between different number systems: the hex numbers are prefixed by $, the binary numbers by %, etc. Decimal number do not use any prefix.

TABLE 12.7

Other Important Programming Conventions

Symbol	Meaning	Example
#	Immediate mode (IMM)	#$A3
<	Direct mode (DIR)	<$A3
;	Start of comment line and of comment inside a program statement	LDAA #$FF; Load accA
*	Alternate sign for start of comment line only	* This is a comment
,X	Index X mode (IND,X)	LDAA TFLG1,X
,Y	Index X mode (IND,Y)	LDAA TFLG2,Y

The < OPERAND > may contain arithmetic or logical expressions. The arithmetic or logical expressions used in the < OPERAND > can use any of the operators listed in Table 12.6.

Some important conventions used in the < OPERAND > are given in Table 12.7.

The < LABEL > is a very powerful concept that can greatly simplify the programmer's task. The < LABEL > consists of a string of alphanumeric characters that make up a name somehow meaningful to the programmer. The placement of the < LABEL > can be in one of the following positions:

1. In the first column and terminates with a tab or blank character

2. In any column and terminates with a colon (:)

There are three different usages of the < LABEL >:

1. To assign the name inserted in the < LABEL > to a location in a program. The < LABEL > will be assigned the address of that location

2. To assign the value of an expression or constant to the name inserted in the < LABEL > using the EQU (equate) or SET directives.

3. To define the name of a subroutine (macro). Essentially, this is the same as (1), since an address (the subroutine starting address) is assigned to the label.

When labels are assigned to certain addresses, one can tell the program to go to that address by referring to the label (case 1 above). Alternatively, one can use the contents of certain address by referring to its label, just like when using variables (case 2 above).

A comment is prefixed by semicolon (;). When the assembler detects a semicolon, it knows that the rest of the line is a comment and does not expect any executable instructions from it. A comment can be a separate line (comment line) or can be inserted in a program statement. A comment line can be also prefixed by an asterisk (*). The comments, either in the comment field or as a separate comment line, are of great benefit to the programmer in debugging, maintaining, or upgrading a program. A comment should be brief and specific and not just reiterate its operation. A comment that does not convey any new information need not be inserted. When writing a comment use lower case characters.

12.3.4 Microcontroller Commands

The 68HC11 microcontroller has 145 different commands. These commands can be grouped into several categories. The categories and the commands in those categories are listed below:

1. Arithmetic operations:
   a. Addition:

   > ABA, ABX, ABY, ADCA, ADCB, ADDA, ADDB, ADDD, INC, INCA, INCB, INS, INX, INY

   b. Subtraction:

   > SBA, SBCA, SBCB, SUBA, SUBB, SUBD, DEC, DECA, DECB, DES, DEX, DEY

   c. Multiplication: MUL
   d. Division: FDIV, IDIV

2. Logical operations: (note: logical operations are carried out on a bit by bit basis)
   a. Standard logical operations: ANDA, ANDB, EORA, EORB, ORAA, ORAB, COM (Boolean inverse), COMA, COMB
   b. Operations that shift the location of the bits in the register:

   > ASL, ASLA, ASLB, ASLD, ASR, ASRA, ASRB, LSL, LSLA, LSLB, LSLD, LSR, LSRA, LSRB, LSRD, ROL, ROLA, ROLB, ROR, RORA, RORB

   e. Operations that compare two numbers:

   > BITA, BITB, CBA, CMPA, CMPB, CPD, CPX, CPY

3. Branching commands: BCC, BCS, BEQ, BGE, BGT, BHI, BHS, BLE, BLO, BLS, BLT, BMI, BNE, BPL, BRA, BRCLR, BRN, BRSET, BSR, BVC, BVS, JMP, JSR, RTS, RTI, WAI

4. Memory/Register Functions
   a. Move data into/out of memory: LDAA, LDAB, LDD, LDS, LDX, LDY, STAA, STAB, STD, STS, STX, STY
   b. Change the values in memory/registers: BCLR, BSET, CLC, CLI, CLR, CLRA, CLRB, CLV, COM, COMA, COMB, NEG, NEGA, NEGB, SEC, SEI, SEV
   c. Transfer data from one register to another: TAB, TAP, TBA, TPA, TSX, TSY, TXS, TYS, XGDX, XGDY

5. Stack Pointer Functions: PSHA, PSHB, PSHX, PSHY, PULA, PULB, PULX, PULY
6. Misc.: NOP, SWI

Note: Boolean inversion commands: COM, COMA, COMB

12.3.5 Condition Codes Register

The *condition codes register* (CCR) stores a number of bits, each with a special significance related to the results of the operation that has just been executed (e.g., if the result of the operation was zero, then a zero bit, Z, is set. If the result of the operation was a negative number, then the negative bit, N, is set, and so on). The CCR bits (numbered b0 through b7) are labeled as follows

b7	b6	b5	b4	b3	b2	b1	b0
S	X	H	I	N	Z	V	C

The meaning of these bits is as follows

S = Stop bit

Allows user to turn the microcontroller stop function on or off.

X = XIRQ mask

Used to disable interrupts from the XIRQ.

H = Half carry bit

Indicates a carry from bit 3 during addition. Only updated by ABA, ADD, and ADC. It is used by the DAA in BCD operations (setting a hexadecimal number to decimal).

I = Interrupt mask

Global interrupt mask. Allow user to turn on/off interrupts.

N = Negative bit

Set to 1 when the result of an operation is 1 in the MSB.

Set to 0 when the result of an operation is 0 in the MSB.

Z = Zero bit

Set to 1 when the result of an operation is zero.

Set to 0 when the result of an operation is anything other than zero.

V = Overflow bit

Set to 1 when a 2's complement overflow has occurred due to a specific operation

C = Carry bit

Set to 1 when a carry or borrow has occurred in the MSB. In addition operations, it is set if there was a carry from MSB (lost carry into the ninth bit). In subtractions, it is set if a number with a larger absolute value is subtracted from a number with a smaller absolute value (free borrow from the ninth bit). It is also used to monitor multiplication and division operations.

12.3.6 Binary and Hex Arithmetic

12.3.6.1 Binary and Hex Addition

The rules for addition of binary numbers are straightforward: $0+0=0$, $0+1=1$, $1+1=10$, etc. In the last operation, a carry of 1 took place. An 8-bit microcontroller has the word length limited to 8-bits; hence, special effects will be observed when the result of the addition yields a number larger than what can be stored in an 8-bit word, as will be discussed later in this section.

Example 1:

Let us do the 8-bit binary addition 1001 0100 + 0100 1100.

SOLUTION

This operation is best achieved by putting the two binary numbers one under the other and adding the digits one by one:

        1001 0100+
        0100 1100
        1110 0000

The result is the binary number 1110 0000. Note that a carry occurred when adding the third bit. This carry was then propagated into the fourth bit and generated another carry in the fifth bit, and then into the sixth bit.

  The 8-bit word can be viewed as the concatenation of two 4-bit words. The part to the right is the lower 4-bit part, the part to the left is the higher 4-bit part. When a carry occurs between the lower and the higher 4-bit parts (i.e., between the fourth and the fifth bits), we have a *half carry*. The microcontroller signals the user that a half-carry event took place by setting the H bit in the CCR.   ■

Example 2:

Let us do the binary addition 1010 0101 + 0100 0100 + 0001 0100.

SOLUTION

Put the binary numbers one under the other and add the digits one by one:

        1010 0101+
        0100 0100
        0001 0100
        1111 1101

In 8-bit arithmetic, if the arithmetic operation results in unsigned numbers outside the range 0000 0000 through 1111 1111, then the bits are truncated to the lower 8-bits. This is called "lost carry."   ■

Example 3:

Let us do the binary addition 1010 0101 + 1100 0100.

SOLUTION

Put the binary numbers one under the other and add the digits one by one:

        1010 0101+
        1100 0100
        0001 0100
        10111 1101

Note that the exact solution is 1 0111 1101. However, the ninth bit cannot be stored in an 8-bit word. What will actually be stored is the lower 8-bit portion 0111 1101. Thus, the ninth bit resulting from the carry process was lost. This is the lost carry effect. However, there is also a bright side to this effect, as we will see in the next section: when subtracting, we can liberally borrow from the ninth bit.

When a carry even happens (whether a lost carry into the 9th or a free borrow from the ninth bit), the microcontroller signals the even to the user by setting a C bit into the condition codes register CCR.

Addition of hex numbers follows the same rules and will not be extensively discussed here. There is a direct mapping between the 8-bit binary number and the 2 hex hexadecimal number. The lower 4-bits correspond to the lower hex digit, and the higher 4-bits correspond to the higher hex digit. The half-carry event takes place when a carry takes place between the lower and the higher hex digits. The carry event happens in relation to the higher hex digit.   ■

12.3.6.2 Negative Numbers in the Computer (2's Complement Numbers)

Until now, we have discussed only positive numbers. These numbers were called "unsigned 8-bit integers." In an 8-bit byte, we can represent a set of 256 positive numbers in the range 0–255. However, in many operations it is necessary to also have negative numbers. For this purpose, we introduce "signed 8-bit integers." Since we are limited to 8-bit representation, we are also limited to a total of 256 numbers. However, half of them will be negative (-128 through -1) and half will be positive (0 through 127).

This representation of signed (positive and negative) numbers is called 8-bit 2's complement representation. In this representation, the eighth bit indicates the sign of the number ($0 = +$ve; $1 = -$ve). A numerical conversion chart for 2's complement signed 2-digit hex (8-bit binary) integers is given in Table 12.8.

The signed binary numbers must conform to the obvious laws of signed arithmetic. For example, in signed decimal arithmetic, $-3 + 3 = 0$. When performing signed binary arithmetic, the same cancellation law must be verified. This is assured when constructing the 2's complement negative binary numbers through the following rule:

To find the negative of a number in 8-bit 2's complement representation, simply subtract the number from zero, i.e., $-X = 0 - X$ using 8-bit binary arithmetic.

Example 1:

Use the above rule to represent the number -3 in 8-bit 2's complement.

SOLUTION

Subtract the 8-bit binary representation of 3 from the 8-bit binary representation of 0 using 8-bit arithmetic (8-bit arithmetic implies that you can liberally take from, or carry into the ninth bit, since only the first 8-bits count!).

```
BINARY        DECIMAL
0000 0000-       0-
0000 0011        3
1111 1101       -3
```

Note that, in this operation, a 1 was liberally borrowed from the ninth bit and used in the subtraction!

Verification: We have established that $-3 = 11111101$. Let us now verify that $3 + (-3) = 0$ using 8-bit arithmetic.

```
BINARY       DECIMAL
0000 0011−      3+
1111 1101      −3
0000 0000       0
```

Note that, in this operation, a carry of 1 was liberally lost in the ninth bit!   ∎

Example 2:

Given the binary number 0011 0101 = 106, find it's 2's complement.

SOLUTION

Subtract the number from 00000000, i.e.,

```
BINARY       DECIMAL
0000 0000−      0−
0110 1010      106
1001 0110     −106
```

Verification: 01101010 + 10010110 = (1)0000 0000. Since the ninth bit cannot be stored in an 8-bit word (i.e., is irrelevant), the answer is actually 0000 0000, as expected.

The rule outlined above can be applied to both binary and hex numbers. The direct mapping between the higher 4-bit part and the lower 4-bit part of a 8-bit binary word into the higher digit and the lower digit of the corresponding 2 hex word apply as before. The following example illustrates it.   ∎

Example 3:

Given the hex number $6a, find its 8-bit 2's complement.

SOLUTION

Subtract the number from 00_{16} using 8-bit arithmetic:

```
HEX   DECIMAL
$00      0−
$6a     106
$96    −106
```

Verification: $6a + $96 = $(1)00. Since the ninth binary bit is irrelevant, the answer is actually $00, as expected.   ∎

12.3.6.3 Overflow During Signed Numbers Arithmetic

We have determined that the signed numbers range is $80 . . . $7f, i.e., from −128 to +127. This means that the smallest signed number that can be accommodated into the 8-bit microcontroller is −128, and the greatest number 127. The $80 . . . $7f range has two subranges: (1) the negative range $80 . . . $ff, i.e., from −128 to −1, and (2) the positive range $00 . . . $7f, i.e., from 0 to 127.

If the result of an arithmetic operation in signed numbers gives a number which is outside the signed-numbers range, then *overflow* happens. In other words, the result of the addition does not make since and the premises should be corrected before proceeding. The overflow event is signaled by V bit in the condition codes register CCR. The following example illustrates this situation.

Example 4:

Add the signed hex numbers \$6a and \$1f and interpret the result.

SOLUTION
The operation is shown below

HEX	CORRECT DECIMAL	WRONG DECIMAL
\$6a	106+	106+
\$1f	31	31
\$89	−119	137

The hex column above shows that the addition of the two positive numbers gives the result \$89 which is outside the permissible positive range \$00...\$7f. Thus, the result \$89 cannot be stored as a positive number since it is outside the positive numbers range. The result looks like a negative number, and the N bit in the condition codes register CCR will be set to indicate this. However, the V bit will also be set to indicate that the result came from an operation that generated an overflow. Thus, the user is cautioned about using this result.

The "correct decimal" column indicates this fact, i.e., the addition of two positive numbers resulted in the negative number −119 because an overflow occurred. The "wrong decimal" column shows that the interpretation that makes \$89 equals to 137 is wrong, because this would apply to unsigned arithmetic, but here we deal with signed number. The reader should not be surprised if this situation seems confusing or paradoxical; the simple answer is that when overflow occurs, one is in forbidden territory and should reassess the problem premises.   ∎

12.3.6.4 Numerical Conversion Chart for 2's Complement Signed 2-Digit Hex (8-Bit Binary) Integers

To accommodate signed numbers arithmetic, the set of 256 hex numbers that can be written with only 2 hex digits is split into two equal sets. One set represents the positive numbers, whereas the other set represents the negative numbers. The positive numbers are from \$00 to \$7f (i.e., 0 through 127 decimal). The negative numbers are from \$ff to \$80 which correspond to −1 to −128. As it can be verified, each set has exactly 128 elements. The fact that \$ff is assigned to represent −1 resides in the fact that by subtracting 1 from \$00 one gets \$ff according to the 2's complement convention.

12.3.7 Logic Gates and Boolean Algebra

12.3.7.1 Logic Gates

A logic gate performs a logical operation on one or more logic inputs and produces a single logic output. Because the output is also a logic value, an output of one logic gate can connect to the input of one or more other logic gates. The logic normally performed is Boolean logic and is most commonly found in digital circuits. The outcome of a logic

TABLE 12.8

Numerical Conversion Chart for 2's Complement Signed
2-Digit Hex Numbers (8-Bit Binary Integers)

Decimal	2-Digit 2's Complement Signed Hex	8-Bit 2's Complement Signed Binary
+127	7f	0111 1111
…	…	…
+16	10	0001 0000
+15	0f	0000 1111
+14	0e	0000 1110
+13	0d	0000 1101
+12	0c	0000 1100
+11	0b	0000 1011
+10	0a	0000 1010
+9	09	0000 1001
+8	08	0000 1000
+7	07	0000 0111
+6	06	0000 0110
+5	05	0000 0101
+4	04	0000 0100
+3	03	0000 0011
+2	02	0000 0010
+1	01	0000 0001
0	00	0000 0000
−1	ff	1111 1111
−2	fe	1111 1110
−3	fd	1111 1101
−4	fc	1111 1100
−5	fb	1111 1011
−6	fa	1111 1010
−7	f9	1111 1001
−8	f8	1111 1000
−9	f7	1111 0111
−10	f6	1111 0110
−11	f5	1111 0101
−12	f4	1111 0100
−13	f3	1111 0011
−14	f2	1111 0010
−15	f1	1111 0001
−16	f0	1111 0000
…	…	…
−128	80	1000 0000

operation performed by a logic gate is expressed through a *truth table*. The truth table describes the outcome of the logic operation for each combination of input values.

Logic gates are primarily implemented electronically using diodes or transistors, but can also be constructed using electromagnetic relays, fluidics, optics, or even mechanical elements. In electronic logic, a logic level is represented by a voltage or current (which depends on the type of electronic logic in use). Each logic gate requires power so that it can source and sink currents to achieve the correct output voltage. In logic circuit diagrams the power is not shown, but in a full electronic schematic power connections are required. Electronic logic gates are implemented as integrated circuits (IC). Table 12.9 gives logic gate symbols, IC designation, graphic symbol, Boolean algebra operation, and 68HC11 command for some commonly used logic gates.

TABLE 12.9

Logic Gate Symbols, Integrated Circuit Designation, Graphic Symbol, Boolean Algebra Operation, 68HC11 Command, and Truth Table Outcomes

Circuit	IC Name	Symbol	Boolean Algebra Operation	68HC11 Command	Truth Table 0 0	0 1	1 0	1 1	A B
Buffer	7407		$X - A$	—	0	0	1	1	
NOT (Inverter)	7404		$X = \bar{A}$	COMA	1	1	0	0	
AND	7408		$X = A \cdot B$	ANDA	0	0	0	1	
OR	7432		$X = A + B$	ORAA	0	1	1	1	
NAND	7400		$X = \overline{A \cdot B}$	[a]	1	1	1	0	
NOR	7402		$X = \overline{A + B}$	[a]	1	0	0	0	
Exclusive OR XOR	7486		$X = A \cdot \bar{B} + \bar{A} \cdot B$ $= A \oplus B$	EORA	0	1	1	0	
Comparator			$X = A \cdot B + \bar{A} \cdot \bar{B}$	[b]	1	0	0	1	

Notes: [a] During programming, to achieve NAND, use two steps, first AND, then NOT. This means using first the commands ANDA and then COMA. (Similar for NOR.)

[b] During programming, to achieve this function, use two steps, first EORA, then NOT. This means using first the commands EORA and then COMA.

12.3.7.2 Boolean Algebra

In formulating mathematical expressions for logic circuits, it is important to have knowledge of Boolean algebra, which defines the rules for expressing and simplifying binary logic statements. The basic Boolean laws and identities are listed below. A bar over a symbol indicates the Boolean operation NOT, which corresponds to the inversion of a signal.

Fundamental laws

$$
\begin{array}{lll}
\textbf{OR} & \textbf{AND} & \textbf{NOT} \\
A + 0 = A & A \cdot 0 = 0 & \\
A + 1 = 1 & A \cdot 1 = A & \\
A + A = A & A \cdot A = A & \bar{\bar{A}} = A \ \text{(double inversion)} \\
A + \bar{A} = 1 & A \cdot \bar{A} = 0 &
\end{array}
\tag{12.1}
$$

Commutative laws

$$
\begin{aligned}
A + B &= B + A \\
A \cdot B &= B \cdot A
\end{aligned}
\tag{12.2}
$$

Associative laws

$$
\begin{aligned}
(A + B) + C &= A + (B + C) \\
(A \cdot B) \cdot C &= A \cdot (B \cdot C)
\end{aligned}
\tag{12.3}
$$

Distributive laws

$$
A + (B \cdot C) = (A + B) \cdot (A + C)
\tag{12.4}
$$

Other useful identities

$$
A + (A \cdot B) = A
\tag{12.5}
$$

$$
A \cdot (A + B) = A
\tag{12.6}
$$

$$
A + (\bar{A} \cdot B) = A + B
\tag{12.7}
$$

$$
(A + B) \cdot (A + \bar{B}) = A
\tag{12.8}
$$

$$
(A + B) \cdot (A + C) = A + (B \cdot C)
\tag{12.9}
$$

$$
A + B + (A \cdot \bar{B}) = A + B
\tag{12.10}
$$

$$
(A \cdot B) + (B \cdot C) + (\bar{B} \cdot C) = (A \cdot B) + C
\tag{12.11}
$$

$$
(A \cdot B) + (A \cdot C) + (\bar{B} \cdot C) = (A \cdot B) + (\bar{B} \cdot C)
\tag{12.12}
$$

DeMorgan's Laws

    DeMorgan's Laws are useful in rearranging and simplifying long Boolean expressions or in converting between AND and OR gates:

$$
\overline{A + B + C + \ldots} = \bar{A} \cdot \bar{B} \cdot \bar{C} \cdot \ldots
\tag{12.13}
$$

$$
\overline{A \cdot B \cdot C \cdot \ldots} = \bar{A} + \bar{B} + \bar{C} + \ldots
\tag{12.14}
$$

If we invert both sides of these equations and apply the double NOT law for Equation 12.1, we can write DeMorgan's Laws in the following form:

$$A + B + C + \cdots = \overline{\overline{A} \cdot \overline{B} \cdot \overline{C} \cdots}$$ (12.15)

$$A \cdot B \cdot C \cdots = \overline{\overline{A} + \overline{B} + \overline{C} + \cdots}$$ (12.16)

12.3.8 ASCII Code

ASCII stands for *American Standard Code for Information Interchange*. Historically, the ASCII format precedes the hex format. For this reason, it was constructed on the minimum number of bits concept and has only 7 bits. When used in modern computers and micro-controllers, the ASCII format is zero filled to the left with an extra bit to achieve the 8-bit format of the hex representation. The 7 bit ASCII code can encode most of the commonly used characters. Table 12.10 illustrates in compact form the ASCII logic. The columns are designated by the most significant digit (MSD), which has only 3 bits and runs from $0 to $7. The rows are designated by the least significant digit (LSD), which has 4 bits and runs from $0 to $f. The upper case letters of the alphabet are coded in columns 4 and 5, whereas the lower case letters are coded in columns 6 and 7. The 7-bit ASCII is loaded right justified into the 8-bit byte, i.e., the eighth bit has always the value 0. For reader's convenience, Table 12.11 gives a cross-reference list of the ASCII codes in sequential order.

TABLE 12.10

The ASCII Code Character Set Highlighting the MSD and LSD Connectivity

		3-Bit MSD							
		$0	$1	$2	$3	$4	$5	$6	$7
4-bit LSD	$0	NUL	DLE	SP	0	@	P	'	p
	$1	SOH	DC1	!	1	A	Q	a	q
	$2	STX	DC2	''	2	B	R	b	r
	$3	ETX	DC3	#	3	C	S	c	s
	$4	EOT	DC4	$	4	D	T	d	t
	$5	ENQ	NAK	%	5	E	U	e	u
	$6	ACK	SYN	&	6	F	V	f	v
	$7	BEL	ETB	'	7	G	W	g	w
	$8	BS	CAN	(	8	H	X	h	x
	$9	HT	EM	)	9	I	Y	i	y
	$a	LF	SUB	*	:	J	Z	j	z
	$b	VT	ESC	+	;	K	[	k	{
	$c	FF	FS	,	<	L	\	l		
	$d	CR	GS	-	=	M	]	m	}
	$e	SO	RS	.	>	N	^	n	~
	$f	SI	US	/	?	O	–	o	DEL

TABLE 12.11

The ASCII Code Character Set in Sequential Listing

Hex	ASCII	Hex	ASCII	Hex	ASCII	Hex	ASCII
$00	NUL	$20	SP space	$40	@	$60	' grave
$01	SOH	$21	!	$41	A	$61	a
$02	STX	$22	"	$42	B	$62	b
$03	ETX	$23	#	$43	C	$63	c
$04	EOT	$24	$	$44	D	$64	d
$05	ENQ	$25	%	$45	E	$65	e
$06	ACK	$26	&	$46	F	$66	f
$07	BEL beep	$27	' apost.	$47	G	$67	g
$08	BS back sp	$28	(	$48	H	$68	h
$09	HT tab	$29	)	$49	I	$69	i
$0A	LF linefeed	$2A	*	$4A	J	$6A	j
$0B	VT	$2B	+	$4B	K	$6B	k
$0C	FF	$2C	, comma	$4C	L	$6C	l
$0D	CR return	$2D	- dash	$4D	M	$6D	m
$0E	SO	$2E	. period	$4E	N	$6E	n
$0F	SI	$2F	/	$4F	O	$6F	o
$10	DLE	$30	0	$50	P	$70	p
$11	DC1	$31	1	$51	Q	$71	q
$12	DC2	$32	2	$52	R	$72	r
$13	DC3	$33	3	$53	S	$73	s
$14	DC4	$34	4	$54	T	$74	t
$15	NAK	$35	5	$55	U	$75	u
$16	SYN	$36	6	$56	V	$76	v
$17	ETB	$37	7	$57	W	$77	w
$18	CAN	$38	8	$58	X	$78	x
$19	EM	$39	9	$59	Y	$79	y
$1A	SUB	$3A	:	$5A	Z	$7A	z
$1B	ESC	$3B	;	$5B	[	$7B	{
$1C	FS	$3C	<	$5C	\	$7C		
$1D	GS	$3D	=	$5D	]	$7D	}
$1E	RS	$3E	>	$5E	∧	$7E	~
$1F	US	$3F	?	$5F	_ under	$7F	DEL delete

12.3.9 BCD Code

BCD stands for *binary coded decimal*; it represents a method of coding decimal numbers digit-by-digit. Each decimal digit is converted into the equivalent binary number. Decimal digits run from 0 through 9. The equivalent binary range is from 0000 through 1001. This shows that it takes four binary bits to store a decimal digit. Single precision BCD code uses 8-bits (1 byte). Hence, single precision BCD code can only store two decimal digits, i.e., decimal numbers from 0 to 99. The following examples are provided to illustrate how the BCD code works.

Example 1:   Conversion from Decimal to BCD

Consider the decimal number 91 (ninety one). The BCD code of ninety one is 1001 0001. The first binary group represent the decimal digit 9; the second group represents the decimal digit 1, i.e.,

$$
\begin{array}{cc}
9 & 1 \\
\downarrow & \downarrow \\
1001 & 0001
\end{array}
$$

Example 2:   Conversion from BCD to Decimal

Consider the BCD-coded number 0111 0010 The corresponding decimal number is 72 (seventy two), because the first binary number is 7 whereas the second binary number is 2, i.e.,

$$
\begin{array}{cc}
0111 & 0010 \\
\downarrow & \downarrow \\
7 & 2
\end{array}
$$

Example 3:   Difference between Hex and BCD Codification of a Decimal Number

Consider the decimal number 75 (seventy five). Its hex equivalent is $4b, which makes the binary number 0100 1011. However, the BCD code of 75 is 0111 0101, because decimal 7 becomes 0111 and decimal 5 becomes 0101. Furthermore, if we want to express the BCD code in the more compact hex representation, then 0111 0101 = $75. This indicates that the hex representation of a BCD-coded decimal number is substantially different from the hex conversion of a decimal number, i.e.,

$$
75 \xrightarrow[\text{HEX}]{} \$4b = 0100\ 1011
$$
$$
75 \xrightarrow[\text{BCD}]{} 0111\ 0101 = \$75
$$

12.3.10 Program Development Steps

There are several steps in developing a successful microcontroller program:

1. Developing the source file
2. Testing the source file with a microcontroller simulation software, such as the THRSim11
3. Converting the source file into an object file using emulator software (e.g., mini-IDE) and test running the object file into the microcontroller using an evaluation board connected to a keyboard and monitor screen through a PC
4. Loading the object file into a stand-alone microcontroller (i.e., burning it into the ROM) and running it independently

First, one develops the source file. The source file can be written in a higher-level language, which is friendlier to the human user (Assembly, C^{++}, etc.) than the machine language.

The source-file examples considered in this book use the Assembly language. The source file is developed starting with the program tasks list, then drawing the flowchart, and finally typing in the program code. The typing of the assembly code can be done with a text editor. However, a more efficient approach is to type the source code in a microcontroller simulator, such as THRSim11, such that simulation running of the source code can be done immediately.

12.3.11 Flowcharts

A flowchart is a schematic representation of an algorithm or a process. A typical flowchart will have the following kinds of symbols:

- Start and end symbols, represented as lozenges, ovals, or rounded rectangles, usually containing the word "Start" or "End," or another phrase signaling the start or end of a process.

- Arrows showing the "flow of control." An arrow coming from one symbol and ending at another symbol represents graphically how the control passes from the first symbol to the second symbol. All processes involved in the flowchart should flow from top to bottom and left to right.

- Processing steps represented as rectangle boxes. Such processing steps can be additions, subtractions, multiplications, divisions, or other operations that the microcontroller can do.

- Input or output steps represented as parallelograms.

- Decision steps represented as diamonds (rhombuses). These typically contain a Yes/No question or a True/False test. The diamond symbol is unique among the flowcharting symbols because it has two arrows coming out of it, one corresponding to Yes or True, and the other corresponding to No or False. The arrows should always be labeled.

An important characteristic of flowcharting is the capability to loop. A loop may, for example, consist of a connector where control first enters, then some processing steps, followed by a decision step with one arrow exiting the loop, and one going back to the connector. These are useful to represent an iterative process.

Simple flowcharts can be drawn easily using standard software tools, such as those available on the Drawing toolbar of the MS Word package. When the flowcharts become lengthy, extension over several pages may be required. Off-page connectors are often used to signify a connection to a (part of another) process held on another sheet or screen. It is important to remember to keep these connections logical in order.

Example 1:

Assume we have to make a flowchart to represent the process of adding two numbers, say, 3 and 2, using the microcontroller accumulators A (accA) and accumulator B (accB) and the microcontroller operation ABA ("add accumulator B to accumulator A"). Then, we would like to check if the sum is greater than or equal to, say, 7. If this is true, then exit and stop. Otherwise, go back, add again B to A, and check again. Let us see how this process can be written in a flowchart.

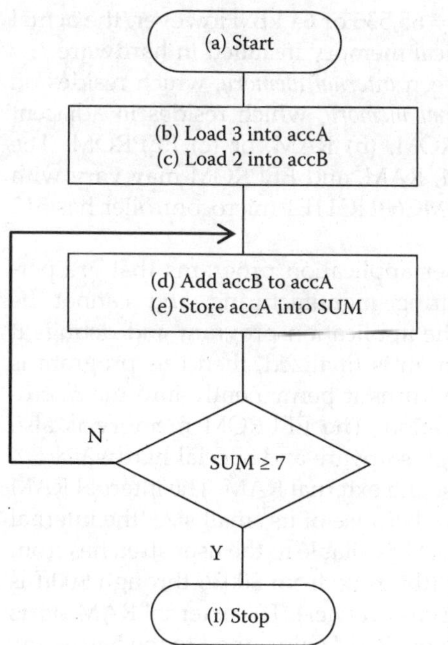

FIGURE 12.9
Flowchart Example 1.

SOLUTION

First, write down the programming steps of the process, i.e.,

(a) Start

(b) Load the first number in accA

(c) Load the second number in accB

(d) Add accB to accA

(e) Store the result in variable SUM

(f) Compare variable SUM with the threshold value 7and see if SUM $\geq$ 7

(g) If false (N), go back to (d) and add again

(h) If true (Y), exit to (i) and stop

(i) Stop

Next, draw a flowchart to represent these programming steps. The flowchart is shown in Figure 12.9. One notices the Start and Stop ovals, the two process boxes and the decision diamond. The first processing box contains steps (b) and (c), whereas the second processing box contains steps (d) and (e). The decision diamond covers steps (f), (g), and (h). In order to make room for the feedback arrow from the decision diamond, the four steps (a), (c), (d), and (e) where split between two boxes, such that the feedback arrow could enter the process just before step (d).

The program and flowchart shown above can be programmed into the microcontroller using the Assembly language. When doing this, several additional steps may need to be added, as will be shown in a future section. ∎

12.3.12 Microcontroller Memory Usage

The microcontroller *logical memory* is set to the size of the memory address, which is double precision, i.e., it runs from memory address \$0000 through memory address \$ffff. This means that the size of the logical memory is $10000 = 65,536$ or 64 kB. However, the actual memory available to the user depends on the physical memory installed in hardware.

The microcontroller *physical memory* is split between *internal memory*, which resides on the same chip with the microcontroller, and *external memory*, which resides in adjacent memory chips. The internal memory can be (a) ROM, (b) RAM, or (c) EEPROM. The external memory is usually RAM. The size of ROM, RAM, and EEPROM may vary with the microcontroller type and version. For example, MC68HC11E1 microcontroller has 512 bytes of RAM and 512 bytes of EEPROM.

The primary use for on-chip ROM is to hold user application programs that are permanently "burned" into the microcontroller during manufacturing and cannot be changed. In a typical scenario, the user develops the application program and debugs it using evaluation board hardware. If the user program is finalized, then this program is passed onto the microcontroller manufacturer who burns it permanently into the microcontroller chip during the microlithography fabrication. The EEPROM memory is also read-only, but it can be rewritten by the user through software and special hardware.

The RAM is split between internal RAM (512 bytes) and external RAM. The internal RAM covers logical memory locations $0000 through $01ff. Because of its small size, the internal RAM is usually used for storing data. The internal RAM available to the user stretches from $0000 through $0035 and from $0100 through $01ff (the zone from $0036 through $00ff is taken up by BUFFALO commands, which will be discussed later). The external RAM starts at logical memory location $c000. Its size depends on the RAM chip added to the hardware. Because it is large, external RAM is used to store program instructions.

12.3.13 Addressing Modes

The term *addressing mode* refers to the way the programming instructions address the microcontroller memory. Each addressing mode type uses different machine language instructions and may take different numbers of locations in memory and different number of cycles to execute. There are six addressing modes: (1) inherent, (2) immediate, (3) direct, (4) extended, (5) index, and (6) relative, as discussed next.

Inherent mode happens when the memory addressing is implied in the actual operation; this mode requires no programming action.

Immediate mode happens when the number contained in the operand will be immediately used to perform the operation. For example, the program line

LDAA   #fa

will load the number $fa into accA.

Direct and extended modes happen when the number contained in the operand is used as a memory address where the microcontroller goes to retrieve the required information or to deposit the information in processing. For example, the program line

LDAA $002c

will instruct the microcontroller to go to address $002c and retrieve the number that exists at that address. Similarly, the instruction line

STAB $001a

will instruct the microcontroller to go and deposit at address $001a the contents of accB.

The direct mode is used to address the internal memory of the microcontroller, i.e., the range $0000 through $0035. Since the lower memory addresses always start with 00, the direct mode addressing only needs the last 2 hex numbers in the address ($00 through $35). For example, the program line

LDAA $2c

will actually go to address $002c to find the information it needs; however, it will not store this whole 4-hex address in the program memory but only the useful lower 2-hex part.

The extended mode is used for addressing memory locations outside the internal RAM range. The extended mode uses the full 4-hex address. For this reason, the extended mode uses more memory to store the machine code than the direct mode. For example, the program line

LDAA $1004

will instruct the microcontroller to load into accA the content of memory at address $1004.

Index mode is another addressing mode that saves space in program memory. The index mode adds the operand to the value already existing in the index X or index Y registers, as selected. In this case, the operand acts as an offset. For example, the above instruction, which tells the microcontroller to load into accA the content of memory at address $1004, can be written in index mode as

LDAA $04,X

Assuming that index X contains the value $1000, then this program line will instruct the microcontroller to go the address $1000 + $0004 = $1004. The index mode saves program memory because it only needs 2 hex numbers to store the operant; however, it requires additional programming lines to ensure that the correct values are stored in the index registers. The index mode is useful when doing repetitive addressing in the same memory area, as for example in the $1000 zone.

Relative mode uses the operand as an offset relative to the present Program Counter value. They are used in branch and subroutine instructions.

In Assembly language, addressing modes are handled as follows:

1. Immediate addressing is designated by preceding the operand with the # sign.
2. Direct and extended addressing is automatically selected by the assembler when translating the assembly language into machine instructions. If the length of the address operand is 1 or 2 hex digits, then the direct mode is selected; if 3 or 4 hex digits, then the extended mode is used. If needed, the extended addressing mode can be forced by padding the address operand with leading zeros (e.g., $002c instead of $2c).
3. Indexed addressing is designated by a comma after the operand. The comma must be preceded by a 1 byte (2-hex) relative offset and followed by an X or Y to designate which index register to use.
4. Relative offsets for branch instructions are computed by the assembler. Therefore the valid operand for any branch instruction is the branch-if-true address, not the relative offset.

12.3.14 Sample Program in Assembly Language with MCU Commands

This section presents a sample program written in Assembly language using typical command of the MCU. This simple program is an example of addition. The program adds two variables, VAR0 and VAR1 and stores the result in the variable SUM, i.e., it performs the operation:

$$VAR0 + VAR1 \rightarrow SUM$$

Subsequently, the program checks if an overflow happened during the addition process. Then, the OVERFL flag is set accordingly ($ff if overflow occurred; $00 otherwise).

12.3.14.1 Program Description

The program uses accA to perform the arithmetic operations and accB to signal the presence of an overflow (i.e., accB is used as an internal flag). The data is taken from variables VAR0 and VAR1. Internal operations are performed using the accumulators A and B. The results are stored in the variables SUM and OVERFL. The following are the main steps of the program:

1. Initialize and define variables in lower memory starting with $0000. One memory byte is reserved for each variable in the order VAR0, VAR1, SUM, OVERFL
2. Load accB with zero, $00, using operation LDAB
3. Load VAR0 into accA using operation LDAA
4. Add VAR1 to accA using operation ADDA. The result of the addition stays in accA
5. Use operation BVC to branch to LABEL1 if no overflow occurres
6. This step is reached if an overflow has occurred during the addition process. Use operation COMB to invert accB from $00 to $ff
7. LABEL1 store the result of addition from accA into SUM using operation STAA
8. Store accB ($00 or $ff, depending on the logic just discussed) into the overflow flag OVERFL using operation STAB
9. Stop using operation SWI (software interrupt)

12.3.14.2 Flowchart

The flowchart of this program is given in Figure 12.10. The main steps discussed above can be easily identified in the flowchart. The following sections give the Assembly code (.asm) and the list file (.lst) of the program written using this flow chart. Note that the .asm file is much shorter than the .lst file, because the latter contains a number of standard definitions which were simply included in the .asm file using a single instruction, #INCLUDE 'A:\VAR-DEF.ASM'. This instruction included the standard file VAR-DEF.ASM during the compilation process.

It should be also noted that large number of comments are included in the .asm file. These comments are inserted using the * and ; separators (see Table 12.7 for details).

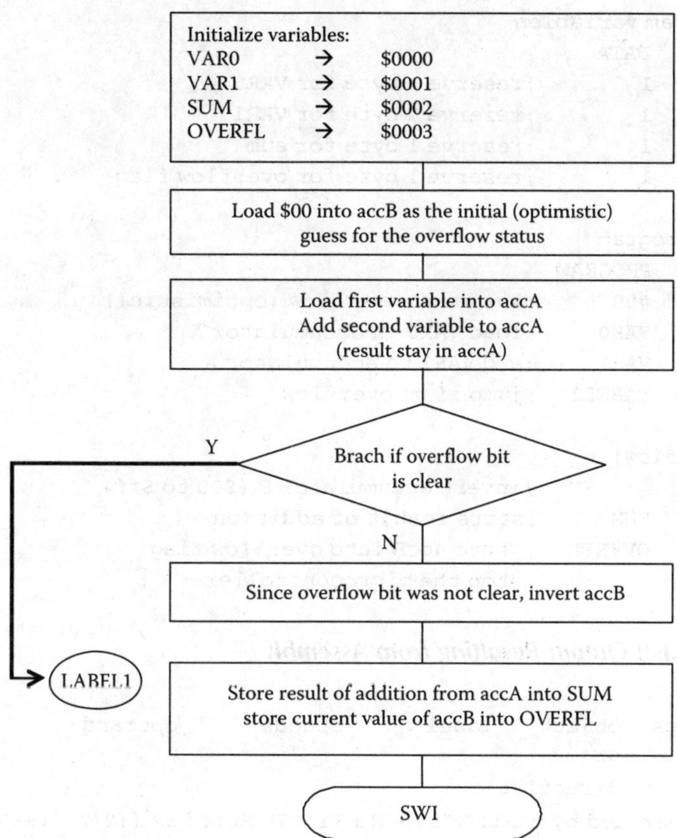

FIGURE 12.10
Flowchart for the sample program.

12.3.14.3 Assembly (.asm) Code

```
* DEMO.ASM
* This simple program adds the contents of
* VAR0 at address $0000 to the contents of
* VAR1 at address $0001 and stores the
* result in SUM at address $0002
*
*
* If an overflow occurs during the addition process,
* then the flag OVERFL (stored at address $0003)
* is set to $ff; else, it stays $00.
*
* Include definition of variables for MC68HC11
#INCLUDE        'A:\VAR_DEF.ASM'
```

```
* Define program variables
          ORG     DATA
VAR0      RMB     1              ;reserve 1 byte for VAR0
VAR1      RMB     1              ;reserve 1 byte for VAR1
SUM       RMB     1              ;reserve 1 byte for sum
OVERFL    RMB     1              ;reserve 1 byte for overflow flag

* Start main program
          ORG     PROGRAM
          LDAB    #00            ;assume no overflow (optimistic!)
          LDAA    VAR0           ;load VAR0 in accumulator A
          ADDA    VAR1           ;add VAR1 to accumulator A
          BVC     LABEL1         ;jump if no overflow

* We have overflow!
          COMB                   ;invert accumulator B ($00 to $ff)
LABEL1    STAA    SUM            ;store result of addition
          STAB    OVERFL         ;store accB into overflow flag
          SWI                    ;stop the microcontroller
```

12.3.14.4 List (.lst) Output Resulting from Assembly

```
list#    address    object    label       opcode       operand        comments
                     or
                     directive
DEMO.lst - generated by MiniIDE's ASM12 V1.07b Build 52 [12/29/1999,
  16:30:49]
 1:
              *124567890123456789012456789012345678901245678901234556789
 2:
 3:                             * DEMO.ASM
 4:                             * This simple program adds the contents of
 5:                             * VAR0 at address $0000 to the contents of
 6:                             * VAR1 at address $0001 and stores the
 7:                             * result in SUM at address $0002
 8:                             *
 9:                             *
10:                             * If an overflow occurs during the addition process,
11:                             * then the flag OVERFL (stored at address $0003)
12:                             * is set to $ff; else, it stays $00.
13:                             *
14:                             * Include definition of variables for MC68HC11
 1:                             * Define variables used by MC68HC11 microcontroller
 2:
 3:       0000       DATA      EQU          $0000          ;start of data
 4:       c000       PROGRAM   EQU          $C000          ;start of program
```

```
 5:      fffe     RESET     EQU          $FFFE      ;reset vector
 6:      1000     REGBAS    EQU          $1000      ;register base
 7:
 8:      0000     PORTA     EQU          $00
 9:      0002     PIOC      EQU          $02
10:      0003     PORTC     EQU          $03
11:      0004     PORTB     EQU          $04
12:      0005     PORTCL    EQU          $05
13:      0007     DDRC      EQU          $07
14:      0008     PORTD     EQU          $08
15:      0009     DDRD      EQU          $09
16:      000a     PORTE     EQU          $0a
17:      000b     CFORC     EQU          $0b
18:      000c     OC1M      EQU          $0c
19:      000d     OC1D      EQU          $0d
20:      000e     TCNT      EQU          $0e
21:      0010     TIC1      EQU          $10
22:      0012     TIC2      EQU          $12
23:      0014     TIC3      EQU          $14
24:      0016     TOC1      EQU          $16
25:      0018     TOC2      EQU          $18
26:      001a     TOC3      EQU          $1a
27:      001c     TOC4      EQU          $1c
28:      001e     TOC5      EQU          $1e
29:      0020     TCTL1     EQU          $20
30:      0021     TCTL2     EQU          $21
31:      0022     TMSK1     EQU          $22
32:      0023     TFLG1     EQU          $23
33:      0024     TMSK2     EQU          $24
34:      0025     TFLG2     EQU          $25
35:      0026     PACTL     EQU          $26
36:      0027     PACNT     EQU          $27
37:      0028     SPCR      EQU          $28
38:      0029     SPSR      EQU          $29
39:      002a     SPDR      EQU          $2a
40:      002b     BAUD      EQU          $2b
41:      002c     SCCR1     EQU          $2c
42:      002d     SCCR2     EQU          $2d
43:      002e     SCSR      EQU          $2e
44:      002f     SCDR      EQU          $2f
45:      0030     ADCTL     EQU          $30
46:      0031     ADR1      EQU          $31
47:      0032     ADR2      EQU          $32
48:      0033     ADR3      EQU          $33
```

```
49:       0034        ADR4     EQU        $34
50:       0039        OPTION   EQU        $39
51:       003a        COPRST   EQU        $3a
52:       003b        PPROG    EQU        $3b
53:       003c        HPRIO    EQU        $3c
54:       003d        INIT     EQU        $3d
55:       003e        TEST1    EQU        $3e
56:       003f        CONFIG   EQU        $3f


list#    address   object     label  opcode operand comments
                   or
                   directive


57:                 *123456789012345678901234567890123456789012345678901234567890123456789
15:                           #INCLUDE       'A:\VAR_DEF.ASM'
16:
17:                           * Define program variables
18:                                   ORG     DATA
19:                           VAR0    RMB     1          ;reserve 1 byte for VAR0
20:                           VAR1    RMB     1          ;reserve 1 byte for VAR1
21:                           SUM     RMB     1          ;reserve 1 byte for sum
22:                           OVERFL  RMB     1          ;reserve 1 byte for overflow
flag
23:
24:                           * Start main program
25:                                   ORG     PROGRAM
26:       c000    c6 00               LDAB    #00        ;assume no overflow
27:       c002    96 00               LDAA    VAR0       ;load VAR1 in accumulator A
28:       c004    9b 01               ADDA    VAR1       ;add VAR2 to accumulator A
29:       c006    28 01               BVC     LABEL1     ;jump if no overflow
30:                           * We have overflow!
31:       c008    53                  COMB               ;Invert accumulator B ($00 to
$FF)
32:       c009    97 02       LABEL1   STAA   SUM        ;store result of addition
33:       c00b    d7 03                STAB   OVERFL     ;store accB into overflow
flag
34:       c00d    3f                   SWI               ;stop the microcontroller


Symbols:
DATA      *0000
LABEL1    *c009
OVERFL    *0003
PROGRAM   *c000
SUM       *0002
VAR0      *0000
VAR1      *0001
```

12.3.15 Program DELAY

A simple program with multiple applications is the program DELAY. This program is very simple, but effective. It uses the repeated subtraction from an initial value until the result is equal or less than zero. Table 12.12 shows, in three columns, the program steps, the flowchart, and the assembly code for this program. The first column of Table 12.12 describes the six steps of the program: first, the variable DELAY is defined at location $0000. Then, its value is loaded in accA. Subsequently, the number 1 is subtracted from accA and the result is checked against zero. If the result is higher than zero, then the program returns and subtracts again. The return is done using a label, here LABEL1. The program exists when the result of the subtraction is not higher than zero.

The flowchart shown in the middle column of Table 12.12 is easy to follow. Each major operation takes place in a separate box. The branch is shown by a back arrow that returns to LABEL1. The assembly code shown in the last column of Table 12.12 has only five essential lines. The program uses the following operations: RMB, LDAA, SUBA, BHI, and SWI. The operation RMB reserves a memory byte for the variable DELAY. The operation LDAA loads accA with the value DELAY. The operation SUBA subtracts 1 from accA. The operation BHI branches if the result of the previous operation is higher than zero. The operation SWI is the software interrupt that stops the program.

These lines of codes are to be inserted in the program template and assembled together with the standard definition file using the instruction #INCLUDE 'A:\VAR_DEF.ASM' as shown in example program described in the previous section.

TABLE 12.12

Program DELAY

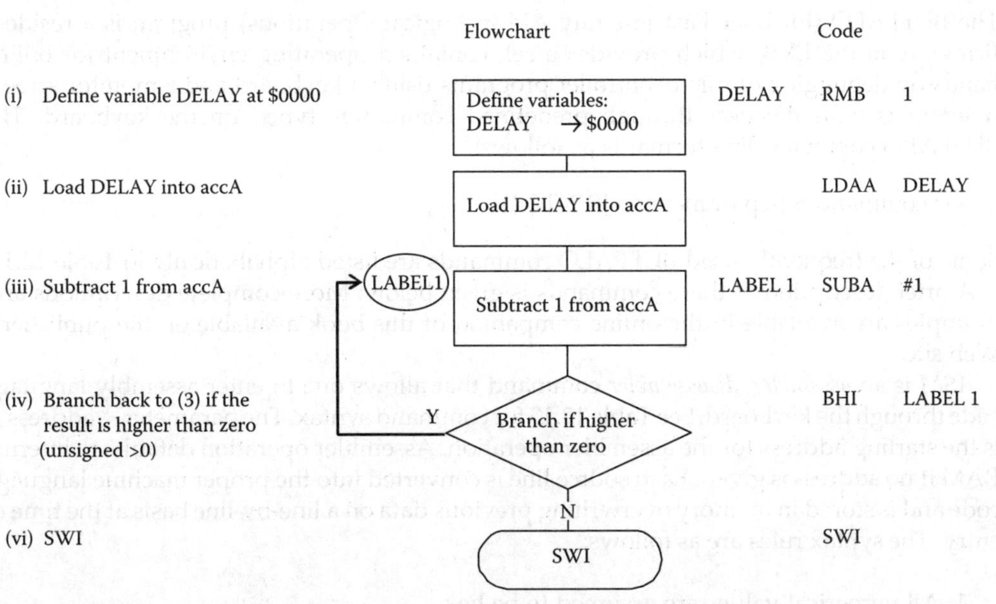

	Flowchart	Code
(i) Define variable DELAY at $0000		DELAY RMB 1
(ii) Load DELAY into accA		LDAA DELAY
(iii) Subtract 1 from accA		LABEL 1 SUBA #1
(iv) Branch back to (3) if the result is higher than zero (unsigned >0)		BHI LABEL 1
(vi) SWI		SWI

The time delay realized by running this program depends on the value of the variable DELAY. During execution, the operator will insert certain value in the location $0000 where the variable DELAY was defined. Then the program will loop until the initial value is used up by repeated subtraction. The time taken by each loop depends on the machine cycles needed for each operation in the loop. For example, the operation LDAA takes 3 machine cycles, the operation SUBA takes 2 machine cycles, and the operation BHI takes 3 machine cycles. Thus, each loop takes 5 machine cycles. For a 2 MHz microcontroller, the machine cycle duration is 0.5 μs. Hence, a pass through the loop takes 2.5 μs. Of course, the total delay will also depend on the initialization time, which is used only once before entering the loop.

The maximum delay that can be obtained with this program is obtained when the value of DELAY is maximum possible, i.e., $ff. This corresponds to 255. Multiplying 255 by 2.5 μs we get 637.5 μs. Considering that there is also some initialization time of say, 7 cycles, i.e., 3.5 μs, we conclude that an approximate value for this maximum possible delay is around ~640 μs. This is not a very large value. If longer delays are required, the first thing to do would be to use double precision DELAY, which would allow a maximum value $ffff = 65,535 cycles or 32,767 μs = ~32.8 ms. However, when using a double precision variable, one would have to use the appropriate double precision operation. If this delay is still insufficient, than one can use nesting of loops, in which case the outer loop would decrease by one every time the inner loop has performed a complete decrementation of the initial value. With nesting, the delay increases with the power law, where the exponent is the number of nests. For a simple nest of three loops inside each other, the simple precision $ff can produce a delay of $255^3 = 16,581,375$ machine cycles, which corresponds to a delay of ~8,290,687 μs = ~8.3 s.

12.3.16 BUFFALO Commands

The BUFFALO (Bit User Fast Friendly Aid to Logical Operations) program is a resident firmware in the EVB, which provides a self-contained operating environment for online hands-on debugging of microcontroller programs using a keyboard and a monitor screen. It interacts with the user through predefined commands typed on the keyboard. The BUFFALO command line format is as follows:

> < command > [<parameters>](ENTER)

Some of the frequently used BUFFALO commands are listed alphabetically in Table 12.13.

A brief description of these commands is given below. More complete descriptions and examples are available in the online companion of this book available on the publisher's Web site.

ASM is an *assembler/disassembler* command that allows one to enter assembly language code through the keyboard. See Table 12.13 for command syntax. The parameter < address > is the starting address for the assembler operation. Assembler operation defaults to internal RAM if no address is given. Each source line is converted into the proper machine language code and is stored in memory overwriting previous data on a line-by-line basis at the time of entry. The syntax rules are as follows:

1. All numerical values are assumed to be hex
2. Operands must be separated by one or more space or tab characters

TABLE 12.13

BUFFALO Commands for Online Debugging the Microcontroller Program

Command	Description
HELP	Display monitor commands
ASM [<address>]	Assembler/disassembler
BF <address1> <address2> <data>	Block fill memory with data
BR [-][<address>]	Set up breakpoint
CALL [<address>]	Execute subroutine
GO [<address>]	Execute program
G [<address>]	
HELP	Displays information about the BUFFALO commands
MD [<address1> [<address2>]]	Memory Display
MM [<address>]	Memory Modify
MOVE <address1> <address2> [<destination>]	Move memory to new location
OFFSET [-]<arg>	Offset for download
RD	Register Display
RM [p, y, x, a, b, c, s]	Register Modify
T [<n>]	Trace n instructions

Assembler/disassembler subcommands are

/	Assemble the current line and then disassemble the same address location.
^	Assemble the current line and then disassemble the previous sequential address location.
(ENTER)	Assemble the current line and then disassemble the next opcode address.
(CTRL)-J	Assemble the current line. If there is no new line to assemble, then disassemble the next sequential address location. Else, disassemble the next opcode address.
(CTRL)-A	Exit the assembler mode of operation.

BF is a command that fills a memory block with a given data value. Table 12.13 gives the command syntax. As a result, the memory block from < address1 > through < address2 > is filled with the hex value < data >.

BR is a command that allows the user to setup or modify a breakpoint during program execution. When a breakpoint is encountered during program execution, the program stops and memory content and program registers are displayed on the monitor screen. Table 12.13 gives the command syntax. The breakpoint is set at memory location < address >. The minus sign [−] is used to remove a previously set breakpoint at that location. Up to four separate breakpoints can be setup into the program execution.

CALL is a command to execute a subroutine starting at location < address >. See Table 12.13 for command syntax.

GO or just G is a command that starts execution of a program from location < address >. See Table 12.13 for command syntax.

HELP is a command that displays information about the BUFFALO commands.

MD is a command that displays a block of memory from < address1 > through < address2 >. See Table 12.13 for command syntax.

MM is a command that allows the user to examine and modify the contents memory at location < address >. See Table 12.13 for command syntax.

MOVE is a command that allows the user to move the contents of memory. See Table 12.13 for command syntax. The content of memory from < address1 > through < address2 > is moved to a new location starting with address < destination >.

OFFSET is a command used to make an offset during an operation.

RD is a command that displays the content of the MCU registers such as program counter (P), Y index (Y), X index (X), accA (A), accB (B), condition codes register (C), and stack pointer (S).

RM is a command that allows the user to display and modify the content of the MCU registers such as program counter (P), Y index (Y), X index (X), accA (A), accB (B), condition code register (C), and stack pointer (S). See Table 12.13 for command syntax. The command displays the contents of all the registers. If no parameters is entered, then the command allows the user to modify the first register, i.e., the program counter P. If a register is entered in the command, then the command allows the user to modify that specific register.

T is a command that allows the user to follow the program execution instruction by instruction. See Table 12.13 for command syntax. When executed, the command starts from the current value of the program counter P and goes on for < n > steps, where n is a single-precision hex number in the range 1 through ff. If no < n > is specified, then only one step is traced.

12.3.17 Debugging Tips

12.3.17.1 Microcontroller Problems

- Is the processor plugged into the PC serial port?
- Is the processor plugged into the power supply?
- Is the power supply turned on?
- Is the serial port plugged into the correct connector?

12.3.17.2 Hardware Problems

- Does the component have power?—Check all voltages
- Are the chips oriented correctly—notch in the correct direction?
- Do the chips straddle the gap in the center of the board?
- Make sure all chips have power (not just input and output lines).
- Verify the direction of diodes and electrolytic capacitors.
- Verify the power at intermediate locations—use 5 or 0 V from the supply instead of chip input to check various conditions.
- Verify that the PC ports are giving the expected output signals.
- Verify chip and transistor pins with the pin diagrams.
- Are there any "open" lines, no voltage connection instead of 0 V?
- Verify resistor codes and capacitor values.

12.3.17.3　Software Problems

- Is the correct program currently in memory?
- Is the correct starting location being used (G????).
- Verify the program with ASM.
- Use trace (T) to step through and verify branches, jumps, and data.
- Compare memory locations with expected information after the program stops.
- Insert SWI at a key location to allow verification of branch, memory, and accumulator values.
- Do branches and jumps have the correct offsets?
- Have RET and RTI commands been reversed somewhere?
- For serial communications, has TE or RE been set?
- For serial communications, has TDRE or RDRF been reset?
- For parallel port C, has 1007 been set for input or output?
- Has the interrupt mask been cleared (CLI)?
- Has the stack pointer changed substantially?

Use the BUFFALO commands to do step-by-step (Trace, T) and Break-Point (BR) execution of the program. Press F1 for details of the BUFFALO commands.

12.4　Parallel Communication with Microcontrollers

Parallel communication is a communication that occurs simultaneously on many lines—thus the word, parallel. It is used most often when the communicating devices are local to one another. On the 8-bit MC68HC11 microcontroller, the parallel ports have eight parallel lines. That is, each of the parallel ports has eight separate output lines, one line for each bit of data. Each port has two sides: the *physical side*, represented by the pins, and the *logical side*, represented by the *port register*. The port registers reside in the microcontroller memory. The information traffic through a port is directional. It can be either *output* or *input*. On some ports the information flow direction is fixed, on others the direction can be set through programming.

An *output port* sends information out. The voltage states on the port pins are set in correspondence with the logical states of the bits in the port register. A logical 1 on a register bit produces the voltage state of High (5 V) on the corresponding pin. A logical 0 on a register bit produces the voltage state of Low (0 V) on the corresponding pin. To generate a certain pattern of high and low voltage states on the port pins, one has to store the appropriate pattern of 1 and 0 states in the corresponding port register.

An *input port* receives information from outside. The voltage states at its physical pins are set by outside factors; the microcontroller sets the bits in the port register according to the voltage states of the corresponding pins in the port. A voltage state of High on a port pin produces a logical 1 in the corresponding bit in the port register. A voltage state of Low on a port pin produces a logical 0 in the corresponding bit in the port register.

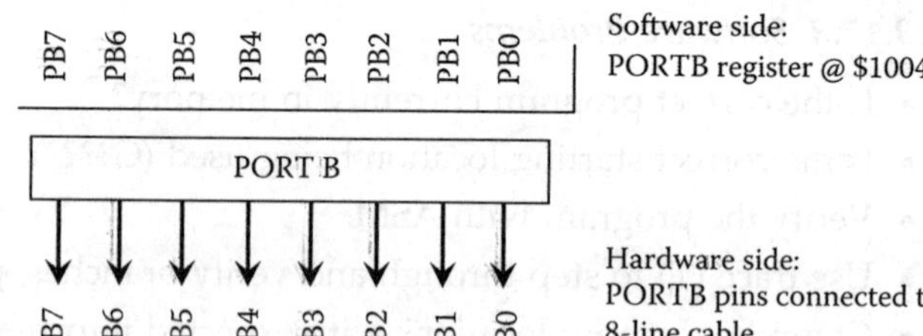

FIGURE 12.11
Port B pins and the corresponding PORTB bits.

A *programmable port* can have its pins set to either output or input. The port has a default direction, say input, and a *data direction register*, which serves the purpose of setting the direction of the pins traffic. A logical 0 in a data direction register bit leaves the corresponding physical pin in its default direction. A logical 1 in a data direction register pin changes the direction of the corresponding physical pin. If the default direction was, say input, the changed direction will be output.

Two multipurpose parallel ports are available on the MC68HC11 microcontroller: port B and port C. The other ports (A, D, and E) may also present themselves in a parallel configuration, but the user does not usually have direct access to these. The parallel ports B and C are configured differently. Parallel port B is restricted to output only. Parallel port C can be used for either input or output.

12.4.1 Port B

Port B is a general-purpose, 8-bit, fixed-direction output port (Figure 12.11). The port B register is call PORTB and is located at memory address $1004 (Table 12.14). Writings to PORTB register causes data to be latched and driven out of the port B pins.

The use of parallel port B is straightforward. Since parallel port B is an output-only port, there is only one thing to be done to make it work: store the output data into the PORTB register. A pattern of 0 and 1 stored to PORTB register will generate a pattern of 0 V and 5 V to be sent out through the port B pins. This single action specifies the voltage states on all the eight separate output lines.

Example:

Assume that we want to send a 5 V signal through pin PB1 of PORTB and 0 V signals through the rest of the port B pins. Since the pin PB1 is the second pin of port B, the pattern of 0 and 1 that needs to be stored in PORTB register will be %00000010. The programming sequence to achieve this action consists in loading this pattern in an accumulator, say accA, and then storing it to the PORTB register, i.e.,

```
    . . .
    LDAA    %00000010
    STAA    PORTB,X
    . . .
```

TABLE 12.14

Port B Register PORTB

$1004	PB7	PB6	PB5	PB4	PB3	PB2	PB1	PB0	PORTB

TABLE 12.15

Location of the Parallel Communication Port C Register, PORTC, and of Its Direction Control
Register, DDRC (the Bits Not Discussed in the Present Section are Shaded in Grey)

$1003	PC7	PC6	PC5	PC4	PC3	PC2	PC1	PC0	PORTC
$1007	DDC7	DDC6	DDC5	DDC4	DDC3	DDC2	DDC1	DDC0	DDRC
Reset	0	0	0	0	0	0	0	0	

Note on DDRC bits: 0, set corresponding port C pin to input; 1, set corresponding port C pin to output.

12.4.2 Port C

Port C is a general-purpose, 8-bit, bidirectional input/output port.

The default data flow direction through the port C pins is input. The data flow direction
through port C pins can be programmed with the DDRC register, the *data direction register
for port C* (Table 12.15). The parallel port C bits in DDRC do not have to be all set to the
same direction. Some can be set to output, others can be set to input.

Example:

For example, assume we wish to set the first four pins to output and the last four pins to input.
To set the first four pins of parallel port C (PC0, PC1, PC2, PC3) to output, we set the first
4-bits of DDRC to 1 (i.e., DDC0 = DDC1 = DDC2 = DDC3 = 1). To set the last four pins of
parallel port C (PC4, PC5, PC6, PC7) to input, we set the last 4-bits of DDRC to 0 (i.e.,
DDC4 = DDC5 = DDC6 = DDC7 = 0). To achieve this, we have to store the sequence
%00001111 to DDRC. The code to achieve this would be

```
    ...
    LDAA    %00001111
    STAA    DDRC,X
    ...
```

After setting the data flow direction of port C pins, we can proceed to send information
through the pins that were set for output, and to receive information from the pins that
were set for input. For example, if the first four pins were set to output and last four pins
were set to input, then we can send out information on the first four pins, and receive
information on the last four pins. This situation is shown in Figure 12.12.

In order to *use the output pins*, we would write data to the PORTC register. Say, we want
to send the signal 0 on pin PC3, 1 on pin PC2, 1 on pin PC1, and 0 on pin PC0. To achieve
this we would store in PORTC register a number that has the sequence 0110 in the
positions corresponding to bits PC3, PC2, PC1, and PC0. The other bits will not matter.
This can be achieved with the following code sequence:

```
    ...
    LDAA    #%00000110
    STAA    PORTC,X
    ...
```

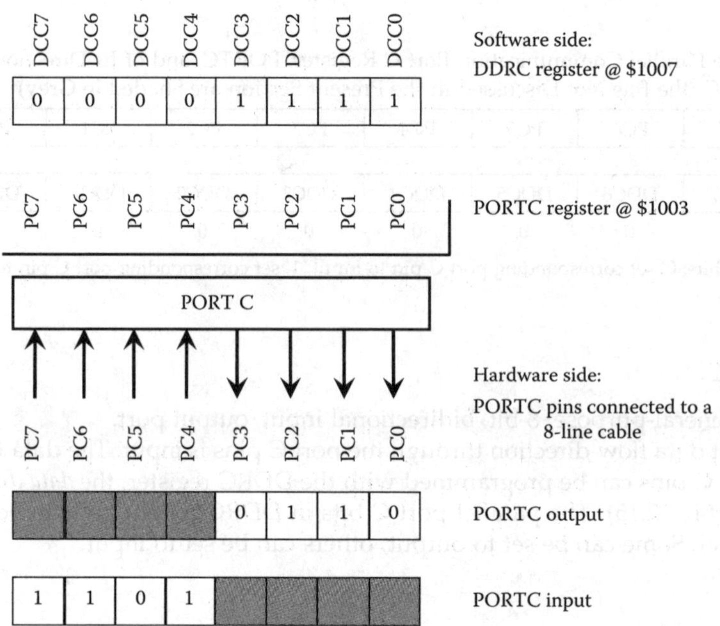

FIGURE 12.12
Port C pins and the corresponding PORTC and DDRC bits for setting the first port C pins to output and the last port C pins to input (the bits not active during a certain operation are shaded in grey).

In order to *use the input pins*, we would read data from the PORTC register. Say, the data coming to PORTC contains the signal 1 on pin PC7, 1 on pin PC6, 0 on pin PC5, and 1 on pin PC4. Hence, when reading (i.e., loading) from PORTC register, these bits will be set. The reading of PORTC register can be achieved with the following code sequence:

```
        ...

        LDAA        PORTC,X
        ...
```

The binary number that is read will be %11010110. This number contains the input sequence 1101 in its upper part. Note that the lower part of the read number, i.e., the last 4-bits, correspond to the pins that were set to output. The pattern loaded from these bits is 0110, i.e., exactly the pattern that was sent out as output in the previous step.

In conclusion, we see that, for parallel port C, three steps must be taken in order to complete the parallel communication. These three steps are

1. Set data direction for each pin (i.e., whether a pin is input or output) by using DDRC

2. Write the output data for each output pin by storing the appropriate bit pattern in PORTC

3. Read the input data from each input pin by storing the appropriate bit pattern to PORTC

The usage the of parallel port C in output is the same as the usage of the parallel port B. Changing the value of the bit changes the value of the output voltage. The usage of the parallel port C in input is by observing that the value in a PORTC bit set for input follows the voltage state of the corresponding physical pin. A value of 1 indicates a voltage is currently being received through that pin, a value of 0 indicates no voltage to the pin. Writing to an input pin has no effect.

> Do not apply an input signal into a pin specified for output! That will fry the micro-controller chip!

12.4.3 Square Wave

The parallel port B can be used to generate a square wave. A square wave is a signal that is high for some time, Δt_1, and then low for some time, Δt_2 (Figure 12.13). When the high time equals the low time we have a true *square wave*. If the high time differs from the low time, we have a rectangular wave. If the high time is shorter than the low time, the resulting wave is a *trail of pulses*.

We can use port B to generate a square wave. When using port B, the high-signal state is $+5$ V, while the low-signal state is 0 V. The square wave can be generated on any of the port B pins, or even on all of them. In this example we will send the square wave through the port B pin PB1, which is the second pin from the right. To generate the square wave, we will use two bit patterns:

- For the low-signal state on port B pin PB1 use %00000000
- For the high-signal state on port B pin PB1 use %00000010

The duration of each signal state is taken to be the same, Δt. To time the duration Δt, we will use a program variable, DELAY. The variable DELAY is used in a wait loop. The relationship between the value of the variable DELAY and the actual delay time, Δt, is determined through calibration. It can be estimated from the number of cycles taken by each instruction. The program will instruct the microcontroller to send a low signal, and then wait for the duration Δt. Next, the program will instruct the microcontroller to send a high signal, and again wait for the duration Δt. Finally, the program will loop back at the beginning and repeat the whole cycle.

To save program space, the wait segment of the program can be done in a subroutine. Two programs are presented below. One has the wait loop inserted twice in the program,

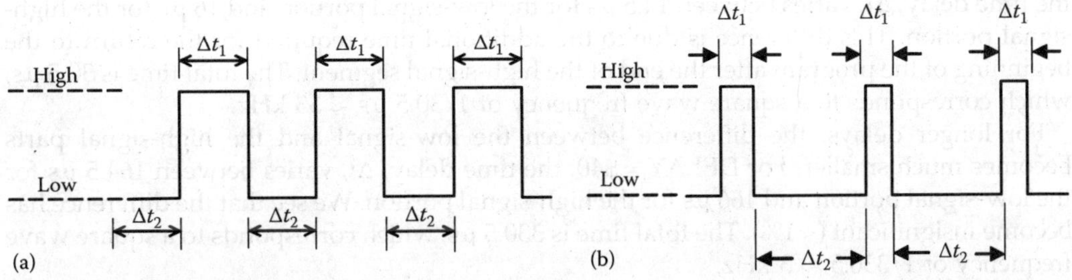

FIGURE 12.13
(a) Square wave, $\Delta t_1 = \Delta t_2 = \Delta t$; (b) trail of pulses, $\Delta t_1 < \Delta t_2$.

TABLE 12.16

Progam sq_wav1

Instructions	Flowchart	Code

(i)   Define variable DELAY at $0000
(ii)   Load REGBAS in reg. X
(iii)   Send low signal
(iv)   Load %00000000 in accA
(v)   Store accA in Port B
(vi)   Load DELAY into accB
(vii)   Decrease accB in a loop until zero
(viii)   Send high signal
(ix)   Load %00000010 in accA
(x)   Store accA in Port B
(xi)   Load DELAY into accB
(xii)   Decrease accB in a loop until zero
(xiii)   Branch back to (iii)
(xiv)   SWI
	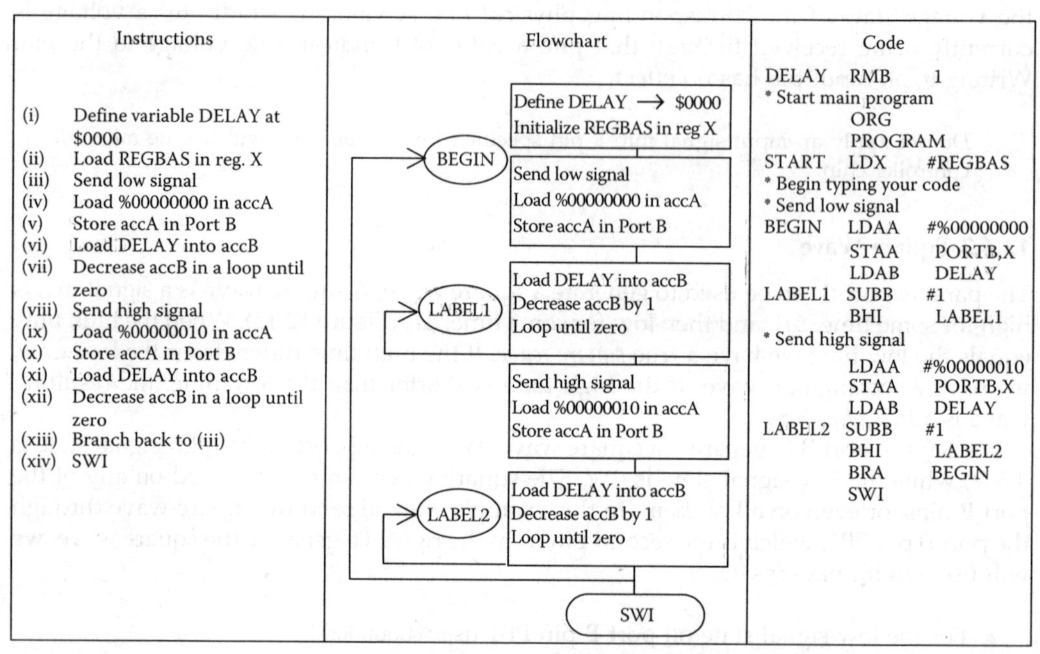	`DELAY   RMB      1`
		`* Start main program`
		`        ORG`
		`        PROGRAM`
		`START   LDX      #REGBAS`
		`* Begin typing your code`
		`* Send low signal`
		`BEGIN   LDAA     #%00000000`
		`        STAA     PORTB,X`
		`        LDAB     DELAY`
		`LABEL1  SUBB     #1`
		`        BHI      LABEL1`
		`* Send high signal`
		
		`        LDAA     #%00000010`
		`        STAA     PORTB,X`
		`        LDAB     DELAY`
		`LABEL2  SUBB     #1`
		`        BHI      LABEL2`
		`        BRA      BEGIN`
		`        SWI`

Note: REGBAS, the base of the register address, is equal to $1000 in most Motorola microcontrollers.

first, after the low-signal portion of the program and second, after the high-signal portion of the program. This program is longer, but simpler. The program is shown in Table 12.16. The program flowchart is shown to the right of the program instructions. Note the "send low signal" and "send high signal" blocks. Also note the two wait blocks, one after sending the low signal, the other after sending the high signal. The wait blocks are not detailed, since they have been covered in a previous example and can be directly reused from there. The essential code for this program is shown to the left of the flowchart. The file Sq_wav1.asm is the result of incorporating this code in the standard template.

The calibration of this program is shown in Table 12.17. It is noted that for DELAY = $04, the time delay, Δt, varies between 14.5 µs for the low-signal portion and 16 µs for the high-signal portion. This difference is due to the additional time required for the return to the beginning of the program after the end of the high-signal segment. The total time is 30.5 µs, which corresponds to a square wave frequency of 1/30.5 µs ≈ 33 kHz.

For longer delays, the difference between the low-signal and the high-signal parts becomes much smaller. For DELAY = $40, the time delay, Δt, varies between 164.5 µs for the low-signal portion and 166 µs for the high-signal portion. We see that the difference has become insignificant (<1%). The total time is 330.5 µs, which corresponds to a square wave frequency of 1/330.5 ≈ 3 kHz.

The other program has the wait loop placed separately at the end. Inside the main program, a jump to subroutine (JSR) is implemented every time the wait subroutine needs to be used. When the work of the subroutine is finished, a return from subroutine (RTS) is

TABLE 12.17

Calibration of the Wait Loop for the Program sq_wav1

| | | | Clock Cycles | | | Time (μs) | | | |
DELAY	State	N_1	N_2	ΔN	T_1	T_2	ΔT	$\Delta N/\Delta T$	$\Delta T/\Delta N$
$04	Initialize	2	11	9	1.0	5.5	4.5		
	L	11	40	29	5.5	20.0	14.5	2.0	0.5
	H	40	72	32	20.0	36.0	16.0	2.0	0.5
	L	72	101	29	36.0	50.5	14.5	2.0	0.5
	H	101	133	32	50.5	66.5	16.0	2.0	0.5
$40	Initialize	2	11	9	1.0	5.5	4.5		
	L	11	340	329	5.5	170.0	164.5	2.0	0.5
	H	340	672	332	170.0	336.0	166.0	2.0	0.5
	L	672	1001	329	336.0	500.5	164.5	2.0	0.5
	H	1001	1333	332	500.5	666.5	166.0	2.0	0.5

executed, and the logic flow returns in the main program. In this program, jumps to the
wait subroutine are implemented two times: after sending a low signal and after sending a
high signal. The program is shown in Table 12.18. The program flowchart is shown to the
left of the program instructions. Note the "send low signal" and "send high signal" blocks.
Also, note that the two wait blocks of the program Sq_wav1 have been replaced by a single
subroutine block that is repeatedly called. The essential code for this program is shown to

TABLE 12.18

Program sq_wav2

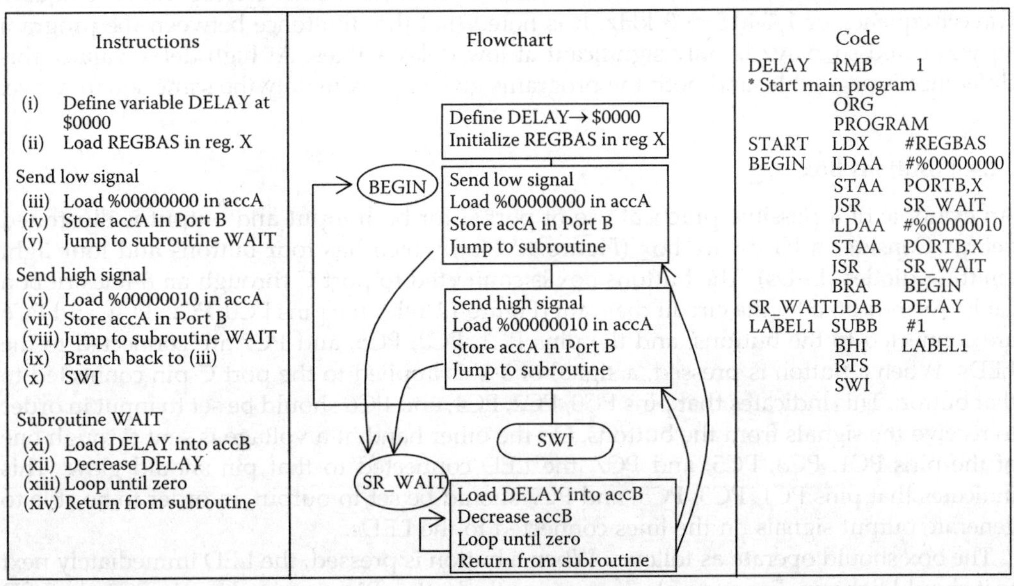

TABLE 12.19

Calibration of the Wait Loop for the Program sq_wav2

		Clock Cycles			Time (μs)				
DELAY	State	N_1	N_2	ΔN	T_1	T_2	ΔT	$\Delta N/\Delta T$	$\Delta T/\Delta N$
$04	Initialize	2	11	9	1.0	5.5	4.5		
	L	11	51	40	5.5	25.5	20.0	2.000	0.500
	H	51	94	43	25.5	47.0	21.5	2.000	0.500
	L	94	134	40	47.0	67.0	20.0	2.000	0.500
	H	134	177	43	67.0	88.5	21.5	2.000	0.500
$40	Initialize	2	11	9	0.0	5.5	5.5		
	L	11	351	340	5.5	175.5	170.0	2.000	0.500
	H	351	694	343	175.5	347.0	171.5	2.000	0.500
	L	694	1034	340	347.0	517.0	170.0	2.000	0.500
	H	1034	1377	343	517.0	688.5	171.5	2.000	0.500

the left of the flowchart. The file Sq_wav2.asm is the result of incorporating this code in the standard template.

The calibration of this program is shown in Table 12.19. It is noted that for DELAY $= \$04$, the time delay, Δt, varies between 20.0 μs for the low-signal portion and 21.5 μs for the high-signal portion. This difference is due to the additional time required for the return to the beginning of the program after the end of the high-signal segment. The total time is 40.5 μs, which corresponds to a square wave frequency of $1/40.5$ μs $\approx$ 25 kHz.

For longer delays, the difference between the low-signal and the high-signal parts becomes much smaller. For DELAY $= \$40$, the time delay, Δt, varies between 170.0 μs for the low-signal portion and 171.5 μs for the high-signal portion. We see that the difference has become insignificant ($<$1%). The total time is 340.5 μs, which corresponds to a square wave frequency of $1/340.5 \approx 3$ kHz. It is noted that the difference between the program sq_wav1 and sq_wav2 is only significant at low delay values. At high delay values, this difference is negligible, and both the programs give approximately the same square wave.

12.4.4 Buttons Box

An example of a possible practical use of port C for both input and output is illustrated below. Consider a hardware box (Figure 12.14a), which has four buttons and four light emitting diodes (LEDs). The buttons box is connected to port C through an 8-lines ribbon cable. As indicated by the circuit diagram (Figure 12.14b), the pins PC0, PC2, PC4, and PC6 are connected to the buttons, and the pins PC1, PC3, PC5, and PC7 are connected to the LEDs. When a button is pressed, a signal of 5 V is applied to the port C pin connected to that button. This indicates that pins PC0, PC2, PC4, and PC6 should be set to input in order to receive the signals from the buttons. On the other hand, if a voltage is sent through one of the pins PC1, PC3, PC5, and PC7, the LED connected to that pin should glow. This indicates that pins PC1, PC3, PC5, and PC7 should be set to output, in order to be able to generate output signals on the lines connected to the LEDs.

The box should operate as follows: When a button is pressed, the LED immediately next to it should light up. For example, if we press the button B1 connected to pin PC2, the LED L1 connected to pin PC3 should light up. This would happen, if a program would instruct

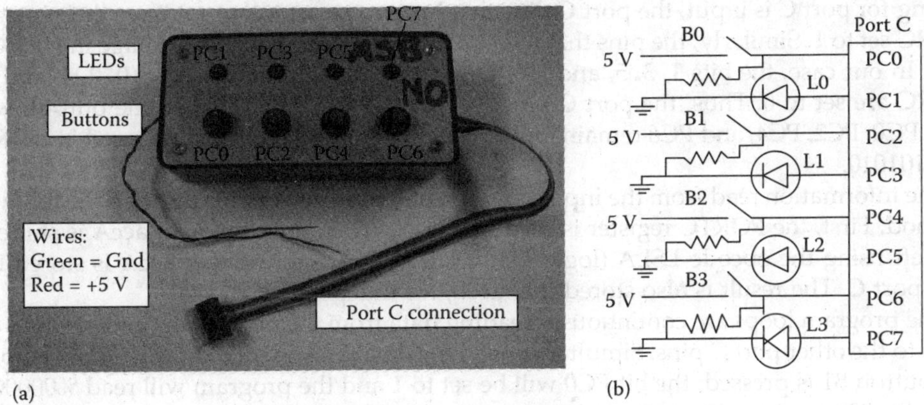

FIGURE 12.14
The buttons box containing four buttons and four LEDs: (a) photo; (b) circuit diagram.

the microcontroller to read all the port C input pins and then send out voltage signals through the port C output pins adjacent to input pins that have input signals coming through them.

The program to achieve this is shown in Table 12.20. The program flowchart is not provided, as it is simpler than the previous examples. The readers are encouraged to use the space provided for filling in their own flowchart of the program. The essential code for this program is shown to the right of the flowchart. The file Buttons_bx.asm is the result of incorporating this code in the standard template.

This program is an example of using port C for both input and output. In this program, port C is configured to have the pins PC0, PC2, PC4, and PC6 for input and the pins PC1, PC3, PC5, and PC7 for output. This is achieved using the DDRC register. Since the default

TABLE 12.20

Program Buttons_bx

Instructions	Flowchart	Code
	Please fill in yourself	START LDX #REGBAS
(i) Load REGBAS in reg. X		* Begin typing your code * Clear Port B
(ii) Clear port B by storing zero in it		LDAA #%00000000
Select odd Port C pins for output		STAA PORTB,X
(iii) Load %10101010 in accA		* Select PC0, PC2, PC4, PC6 for output
(iv) Store accA to DDRC		LDAA #%10101010 STAA DDRC,X
Main loop		* Main loop
(v) Load Port C to accA		BEGIN LDAA PORTC,X
(vi) Shift accA to the left		LSLA
(vii) Store accA to Port C and Port B		* STAA PORTC,X STAA PORTB,X
(viii) Branch back to (v)		BRA BEGIN
(ix) SWI		SWI

setting for port C is input, the port C pins that have to be set to output have their image in DDRC set to 1. Similarly, the pins that have to be set to input have their image in DDRC set to 0. In our case, the bits 1, 3, 5, and 7 of DDRC are set to 1, and the bits 0, 2, 4, and 6 of DDRC are set to 0. Thus, the port C pins PC1, PC3, PC5, and PC7 become output, while pins PC0, PC2, PC4, and PC6 remain input. The bit pattern that has to be stored in DDRC is %10101010.

The information read from the input pins is sent to the output pins through the following method. First, the PORTC register is read into accA. Then the content of accA is shifted to the left using the opcode LSLA (logical shift left accA). The resulting accA is stored back into port C. The result is also stored into accB, for independent display.

The program loops are continuously reading data from the port C pins and sending data back to the other port C pins. Simultaneously, the data is also sent to port B. For example, if the button B1 is pressed, the bit PC0 will be set to 1 and the program will read %00000001 from PORTC register. Thus, the value in accA will be %00000001. After applying the LSLA operation, the value in accA will become %00000010. When this value is stored back into PORTC, the pin PC1 will be set to 5 V, and the L0 LED will light up.

12.5 Serial Communication with Microcontrollers

12.5.1 Types of Serial Communication

Serial communications is a mode of communication in which the bits are sent through a single line, one bit at a time. To send an 8-bit word, the 8-bits must be streamed through the communication line in certain order. Since all the data are transferred on one line and the bits are transferred sequentially, the serial communication is much slower than the parallel communication.

The microcontroller MC68HC11 has two ways of performing serial communication:

- Synchronous, in which the serial communication is performed in synch with the microcontroller clock
- Asynchronous, in which no synchronization is required

These two ways of serial communication correspond to two independent serial communication subsystems. Both subsystems are connected to port D pins (Figure 12.15). The synchronous serial communication is used for communicating with peripherals. This is a high-speed synchronous communication with peripherals or other microcontrollers usually located on the same printed circuit board. For synchronous serial communication, the connected devices must be physically close together to minimize the unwanted interference and delays. In the MC68HC11 microcontroller, the synchronous serial communication is performed through the *synchronous SPI*.

The asynchronous serial communication is a slower interface that has the advantage of being able to connect at large distances. A common type of asynchronous SCI is the universal asynchronous receiver transmitter (UART) standard. In the MC68HC11 microcontroller, the asynchronous UART standard is implemented in the SCI. The SCI subsystem can be used to connect the microcontroller to a cathode ray oscilloscope (CRT),

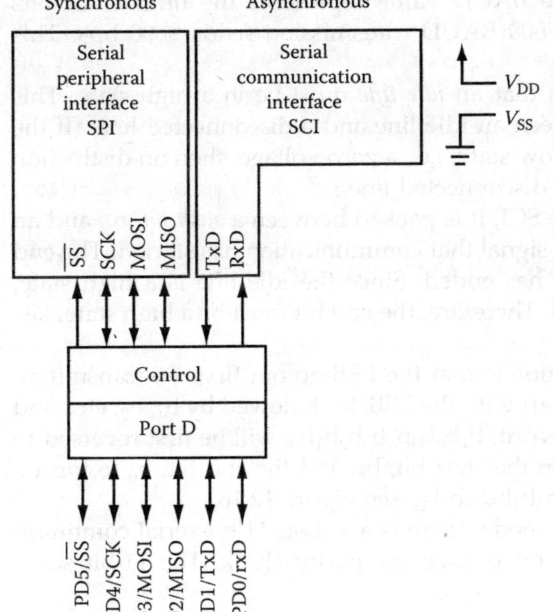

FIGURE 12.15
The two SCI subsystems on the MC68HC11 micro-controller, synchronous SPI, and asynchronous SCI.

keyboard, or PC. The SCI subsystem can be also used to connect several distributed microcontrollers to form a communication network over a wide area. When the microcontroller is installed in an EVB, the communication between the microcontroller and the PC is achieved through the SCI subsystem and a general RS232C serial connector cable. This important feature is used extensively as microcontroller programs are developed in labs and during projects using the EVB setup.

The asynchronous serial communication interface is more versatile than the synchronous SCI and has much more uses. In this chapter, we will focus our discussion on asynchronous SCI.

The reader could consult Sections 8 and 9 of the M68HC11 Reference Manual for more detailed information on the synchronous and asynchronous SCIs.

12.5.2 Serial Communication Basics

In serial communications, only one communication line is used. On such a line, a bit of data is specified by holding the voltage constant for a certain period of time, Δt, at either a high voltage state, V_{DD}, or a low voltage state, V_{SS}. The high voltage state corresponds to $V_{DD} \approx$ 5 V. The low voltage state corresponds to $V_{SS} \approx 0$ V. The V_{DD} state corresponds to a logical 1, while the V_{SS} state corresponds to a logical 0.

The duration Δt controls the speed of communication. The shorter the duration Δt, the faster is the serial communication. However, the duration Δt must be long enough for the receiving device to sense it and distinguish it from the ambient noise. For this reason, "noisy" communication lines cannot go at high speed.

The *baud rate* is the inverse of Δt. The BAUD value represents the number of bits transmitted per second. For example, a 9600 BAUD transmission sends 9600 bps. This corresponds to $\Delta t = 104$ μs ≈ 0.1 ms.

A basic rule of serial communication is that an *idle line* must be in a high state. This originates in the need to distinguish between an idle line and a disconnected line. (If the "idle line" condition were chosen to be a low state, i.e., a zero voltage, then no distinction could be made between an idle line and a disconnected line.)

When a binary word is sent through the SCI, it is packed between a *start bit*, b_s, and an *end bit*, b_e. The purpose of the start bit is to signal that communication has started. The end bit signals show that the communication has ended. Since the idle line is a high state, the start bit has to be a low state, i.e., $b_s = 0$. Therefore, the end bit must be a high state, i.e., $b_e = 1$.

Another basic rule of serial communication is that the LSB go out first. To transmit an 8-bit word, $b_7b_6b_5b_4b_3b_2b_1b_0$, one would start with the LSB, b_0, followed by b_1, b_2, etc., and will end with the MSB, b_7. Thus, an 8-bit word, $b_7b_6b_5b_4b_3b_2b_1b_0$, will be first reversed to $b_0b_1b_2b_3b_4b_5b_6b_7$ and then packed between the start bit, b_s, and the end bit, b_e to form a 10-bit serial transmission packet, $b_sb_0b_1b_2b_3b_4b_5b_6b_7b_e$ (see Figure 12.16).

Besides the 10-bit serial communication mode, there is another, 11-bit serial communication mode. In this case, the additional bit is used for parity check. The 11-bit serial communication mode is not detailed here.

In conclusion, we see that the data byte being sent is bracketed by 2-bits, the start bit ($b_s = 0$) and the end bit ($b_e = 1$). An idle line has a high voltage state (logical 1). Communication is open by dropping the line voltage to zero, i.e., by sending a start bit. The data bits are then sent from the LSB to the MSB. Then, the communication is closed by raising

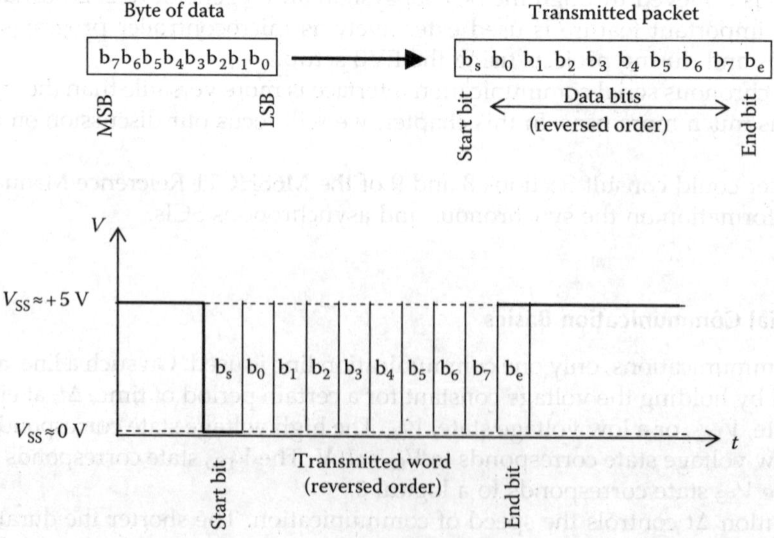

FIGURE 12.16
Principles of serial communication.

the line voltage back to high, i.e., by sending the end bit. During this process, all bits have the same duration in time, Δt. The duration Δt results from the BAUD rate which is expressed in bps. The rules of asynchronous serial communication are

1. Idle line has a logic state of 1 (voltage state $=$ High).
2. Each bit of information transmitted on a serial line has the same duration, Δt, which is the inverse of the BAUD rate.
3. For each byte of data, the communication is opened by a start bit, $b_s = 0$, and ended by an end bit, $b_e = 1$.
4. The byte of data is packed between the start bit and the end bit in reversed order, i.e., LSB is transmitted first.

12.5.3 ASCII Code Transmission through Serial Communication Interface

The serial communication data is usually sent in character format using the 7-bit ASCII code presented in Section 12.3.8. The ASCII code is a 7-bit code for encoding most of the commonly used characters (Table 12.10). The 7-bit ASCII is loaded right justified into the 8-bit byte, i.e., the eighth bit has always the value 0. A sequential listing of the ASCII codes is given in Table 12.11. An example of how the ASCII code is used to transmit an alphabet letter is given below.

Example:

Consider the transmission of the character M through the SCI. From Table 12.10, the ASCII code for M is

$$\text{ASCII(M)} = \$4d = \%01001101 = b_7 b_6 b_5 b_4 b_3 b_2 b_1 b_0$$

For serial transmission, the order is reversed to

$$b_0 b_1 b_2 b_3 b_4 b_5 b_6 b_7 = 10110010$$

After reversing the order, the result is packed between the start bit and the end bit, i.e., 0 and 1, respectively:

$$b_s b_0 b_1 b_2 b_3 b_4 b_5 b_6 b_7 b_e = 0101100101$$

This 10-bit sequence, 0101100101, will be transmitted through the SCI. If we tap the serial communication and view it on the oscilloscope screen, this sequence being transmitted will look like the image in Figure 12.17.   ■

12.5.4 Programming the Serial Communication Interface

The SCI is physically implemented through the port D pins PD1 and PD0 (see Figure 12.15). The pin PD1 is used for transmission (TxD), whereas the pin PD0 is used for reception (RxD). Since the transmission and reception lines are separated, simultaneous transmission and reception is physically possible. However, inside the microcontroller, the transmission and reception functions share the same data register. Thus transmission and

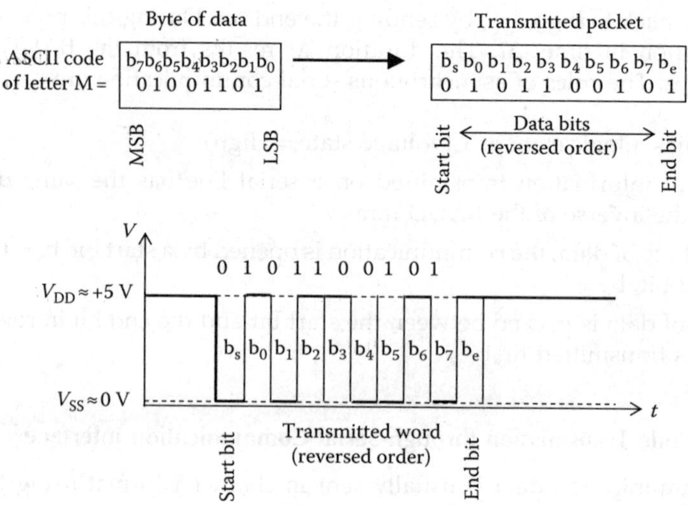

FIGURE 12.17
Transmission of the letter "M" through the SCI.

reception happen somehow sequentially, and not in parallel. These aspects will become clear further down this section.

12.5.5 SCI Registers and Pins

The microcontroller registers involved in the SCI discussed in this section are

- BAUD = register
- SCCR1 = serial communication control register #1
- SCCR2 = serial communication control register #2
- SCSR = serial communication status register
- SCDR = serial communication data register

These registers are located at address \$102b through \$102f, as shown in Table 12.21. Note that the serial communication data register is double buffered to accommodate both the received and the transmited data. In Table 12.21, the bits not discussed in the present section have been shaded in grey. The pins that are not shaded will be discussed in this section.

The microcontroller pins involved in the SCI discussed in this section are the port D pins PD1 and PD0.

12.5.6 Reception and Transmission Enable

The MC68HC11 microcontroller has the capacity to both receive and transmit data through the SCI. The selection of receive and/or transmit modes is done by setting to 1 the *reception enable* and/or *transmission enable* bits RE and TE in the *serial communication control register #2* (SCCR2) to 1. The SCCR2 register is located at the memory address \$102d. The RE and TE bits are the bit 2 and bit 3 of SCCR2. Simultaneous selection of

TABLE 12.21

Location of the Serial Communication Interface Registers

102b	TCLR	0	SCP1	SCP0	RCKB	SCR2	SCR1	SCR0	BAUD
Reset	0	0	0	0	0	U	U	U	
$102c	R8	T8	0	M	WAKE	0	0	0	SCCR1
Reset	U	U	0	0	0	0	0	0	
$102d	TIE	TCIE	RIE	ILIE	TE	RE	RWU	SBK	SCCR2
Reset	0	0	0	0	0	0	0	0	
$102e	TDRE	TC	RDRF	IDLE	OR	NF	FE	0	SCSR
Reset	1	1	0	0	0	0	0	0	
$102f	R7/T7	R6/T6	R5/T5	R4/T4	R3/T3	R2/T2	R1/T1	R0/T0	SCDR
Reset									

both receive and transmit modes is permitted, since the MC68HC11 microcontroller has separate lines for reception and transmission (RxD and TxD through port D pins PD0 and PD1, respectively).

12.5.7 Serial Communication Reception

Serial communication reception is performed with the receive mode enabled. The receive mode is enabled by setting the reception enable bit in SCCR2 register, i.e., by making $RE = 1$. In receive mode, the microcontroller waits for a serial communication to arrive. When a serial communication has arrived, the *receive-data-register-full* bit (RDRF) becomes set, i.e., $RDRF = 1$. This indicates that serial communication data has been received and is ready for pickup. The RDRF bit is bit 5 of SCSR at memory address $102e.

The received serial communication data can be retrieved from the SCDR at memory address $102f. Retrieving serial communication data is done by loading from SCDR into an accumulator.

12.5.7.1 Detection of Serial Communication Reception

One way of detecting whether the RDRF bit is set is by the polling method. In this method, the value of the RDRF bit is interrogated (polled) by comparing its value with an expected value. In our case, the expected value is $RDRF = 1$. Since the RDRF bit is bit 5 in the SCSR, the polling action will consist of observing when SCSR has the value 1 in bit 5. This can be achieved by the following simple polling sequence:

```
          ...
LABEL_1        LDAA        SCSR,X
               ANDA        #%00100000
               BEQ         LABEL_1
          ...
```

This sequence checks RDRF in a loop until it is found to be equal to 1. When this happens, the received serial communication data can be loaded from SCDR into an accumulator.

12.5.7.2 Retrieval of Received Serial Communication Data

When the reception of serial communication has been signaled by RDRF = 1, the received data can be retrieved from the SCDR register. This is achieved by loading from SCDR into an accumulator. For example, to load the received serial communication data from SCDR into accB, we can use the sequence:

```
          ...

          LDAB      SCDR,X
          ...
```

12.5.7.3 Clearing RDRF

The clearing of the RDRF bit does not require specific action from the user. The clearing sequence happens by itself in the normal course of event. The sequence of reading RDRF = 1 and loading data from SCDR will trigger the clearing of RDRF (i.e., will make RDRF = 0). In this way, the microcontroller becomes ready for the reception of the next serial communication data.

12.5.7.4 Program SCI_recept

This program is an example of SCI reception. The character T is being sent to SCDR when TDRE is set. The set of instructions, the flowchart, and the actual code are given in Table 12.22. The program flowchart is shown to the right of the program instructions. Note the initialization block, which contains reg. X initialization and SCI initialization. After

TABLE 12.22

Program Instructions, Flowchart, and the Actual Code for the Program SCI_recept

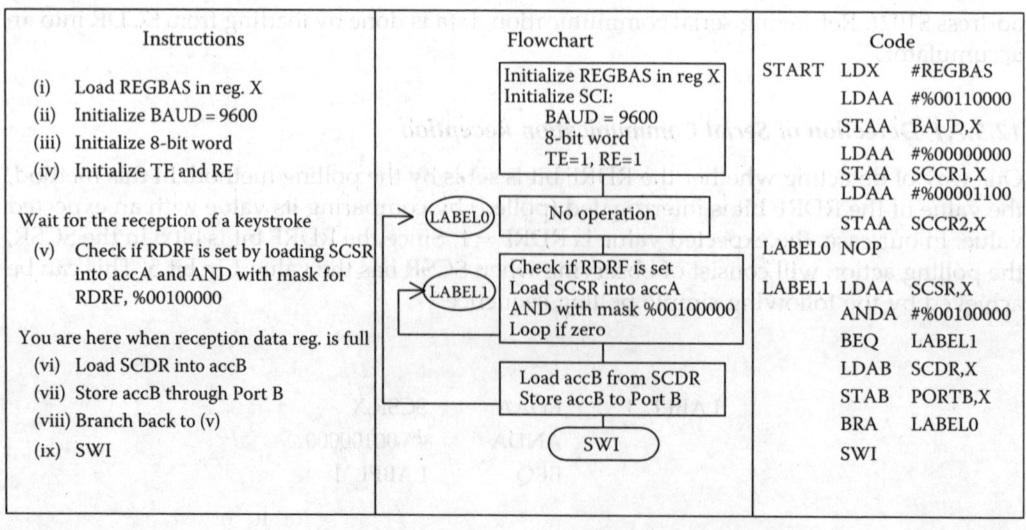

Instructions	Flowchart	Code

(i) Load REGBAS in reg. X
(ii) Initialize BAUD = 9600
(iii) Initialize 8-bit word
(iv) Initialize TE and RE
	Initialize REGBAS in reg X
	Initialize SCI:
	  BAUD = 9600
	  8-bit word
	  TE=1, RE=1
		START LDX #REGBAS
		        LDAA #%00110000
		        STAA BAUD,X
		        LDAA #%00000000
		        STAA SCCR1,X
		        LDAA #%00001100

Wait for the reception of a keystroke	LABEL0 No operation	STAA SCCR2,X LABEL0 NOP
(v) Check if RDRF is set by loading SCSR into accA and AND with mask for RDRF, %00100000	LABEL1 Check if RDRF is set Load SCSR into accA AND with mask %00100000 Loop if zero	LABEL1 LDAA SCSR,X ANDA #%00100000
You are here when reception data reg. is full		BEQ LABEL1
(vi) Load SCDR into accB (vii) Store accB through Port B	Load accB from SCDR Store accB to Port B	LDAB SCDR,X STAB PORTB,X
(viii) Branch back to (v) (ix) SWI	SWI	BRA LABEL0 SWI

initialization, the status of RDRF (reception data register full) flag is checked in a loop. When RDRF is set, the loop is exited and the content of SCDR is loaded into accB. This operation automatically resets RDRF. Now, the program loops back to the beginning and waits for another transmission.

The essential code for the program is shown to the right of the program flowchart. This essential code was incorporated into the standard asm template to generate the code file SCI_recept.asm.

12.5.8 Serial Communication Transmission

Serial communication transmission is performed with the transmit mode enabled. The transmit mode is enabled by setting the transmission enable bit in SCCR2 register, i.e., by making TE = 1. Serial communication transmission is performed by placing data into the serial-communication-data-register, SCDR. Placing the data into SCDR is done by storing it in SCDR. Once the data is stored in SCDR, the SCI takes over and ensures that the data is transmitted along the serial communication line. When data is stored in SCDR for transmission, the microcontroller packs the data between the start and stop bits and forms the serial communication transmit packet, which is then sent through the SCI.

Before new data can be placed in SCDR for transmission, one must ensure that the transmission of previously placed data has finished and that the transmission data register is empty and can accept new data. This verification is done with the help of the TDRE bit, where TDRE has the meaning "transmit data register empty." The TDRE bit is the bit 7 of the SCSR register at memory address \$102E. As long as TDRE = 0, the transmission of previous data is not finished, and the microcontroller is still transmitting data through the SCI. When the transmission is complete, the data register becomes empty and ready to receive new data to be transmitted. In this case TDRE = 1, which signals that new data can be placed in SCDR for transmission.

12.5.8.1 Detection of Serial Communication Readiness for Transmission

One way of detecting whether the TDRE bit is set is by the polling method. In this method, the value of the TDRE bit is interrogated (polled) by comparing its value with an expected value. In our case, the expected value is TDRE = 1. Since the TDRE bit is bit 7 in the serial communication status register, SCSR, the polling action will consist of observing when SCSR has the value 1 in bit 7. This can be achieved by the following simple polling sequence:

```
          ...

LABEL_1     LDAA      SCSR,X
            ANDA      #%10000000
            BEQ       LABEL_1

          ...
```

This sequence checks TDRE in a loop until it is found to be equal to 1. When this happens, the received serial communication data can be stored into SCDR for transmission.

12.5.8.2 Placing Data into SCDR to Be Transmitted through the Serial Communication Interface

When the serial communication is ready to receive new data to be transmitted (i.e., when TDRE = 1), the data to be transmitted can be placed into the SCDR register for transmission. This is achieved by storing the data from an accumulator into the SCDR. For example, to store data from accB into the serial communication data register SCDR we can use the sequence:

```
...
STAB        SCDR,X
...
```

12.5.8.3 Clearing TDRE

Clearing of the TDRE bit does not require specific action from the user. The clearing sequence happens by itself in the normal course of event. The clearing sequence for TDRE consists in reading TDRE = 1 followed by storing the data into SCDR. Subsequently, the microcontroller starts the serial communication transmission of the data that was placed in SCDR, and the bit TDRE will be kept at value TDRE = 0 until the transmission is complete.

12.5.8.4 Program SCI_transmit

This program is an example of SCI transmission. The character T is being sent to SCDR when TDRE (transmission data register empty) is set. The set of instructions, the flowchart, and the actual code are given in Table 12.23. The program flowchart is shown to the right of the program instructions. Note the initialization block, which contains reg. X initialization and SCI initialization. After initialization, the character T is loaded into accB. Then, the status of TDRE flag is checked in a loop. When TDRE is found set, the loop is exited and the contents of accB is stored in SCDR. This operation automatically resets TDRE. Now, the program loops back to the beginning and tries to send again.

TABLE 12.23

Program Instructions, Flowchart, and the Actual Code for the Program SCI_transmit

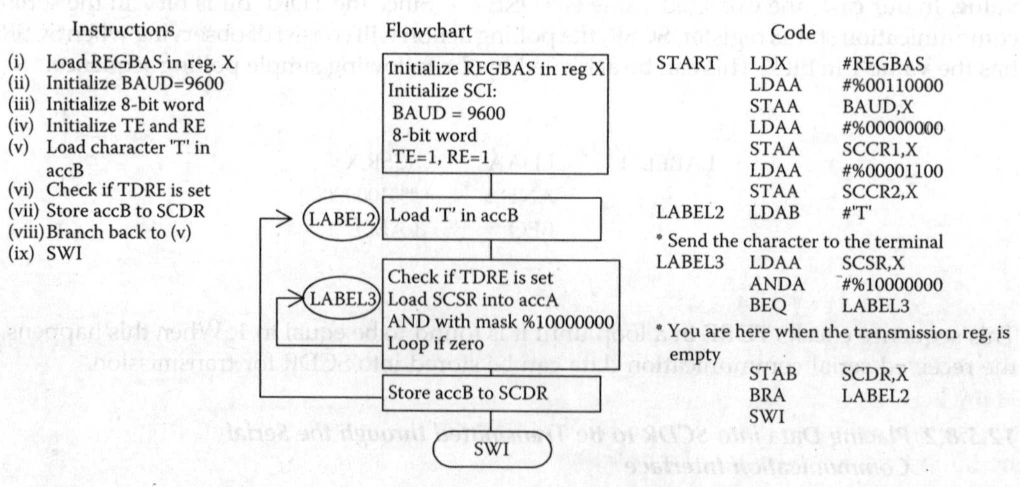

The essential code for the program is shown to the right of the program flowchart. This essential code was incorporated into the standard asm template to generate the file SCI_transmit.asm.

12.5.9 Interrogating Bits Using Masks

We have seen that the specific bits RDRF and TDRE in the SCSR register had to be interrogated in order to determine the action to be done next. One way of achieving this interrogation was to perform a logical AND between the contents of SCSR and an appropriate mask that reveals the bit to be interrogated. For RDRF (bit 5 in SCSR), the mask was %00100000. For TDRE (bit 7 in SCSR), the mask was %10000000. These are *hard masks*, because their values have to be set every time they are used. When these operations must be made many times inside a program, it might be more efficient to define variables that have the appropriate values. Such variables would be *soft masks*. This is attained by defining at the beginning of the program the following mask variables:

...

```
RDRF_MK      EQU      %00100000      mask for RDRF
TDRE_MK      EQU      %10000000      mask for TDRE
```
...

Subsequently, the interrogation sequence would simply be

- For RDRF:

```
             ...
LABEL_1      LDAA     SCSR,X
             ANDA     #RDRF_MK
             BEQ      LABEL_1
             ...
```

- For TDRE:

```
             ...
LABEL_1      LDAA     SCSR,X
             ANDA     #TDRF_MK
             BEQ      LABEL_1
             ...
```

After the logical AND is performed between the register and the appropriate mask, a BEQ instruction is used to loop back if the result is zero (i.e., if the interrogated bit is not yet set). If the interrogated bit is set, the result of the logical AND is 1, the BEQ is not satisfied, and program does not loop back, but rather passes forward.

However, there are also other ways of branching in correlation with the status of specific bits. For example, one can use the instruction BRCLR to branch if the masked bit is clear. Or one can use the instruction BRSET to branch if the masked bit is set. The reader should feel free to experiment with the use of other polling and branching methods besides those shown here.

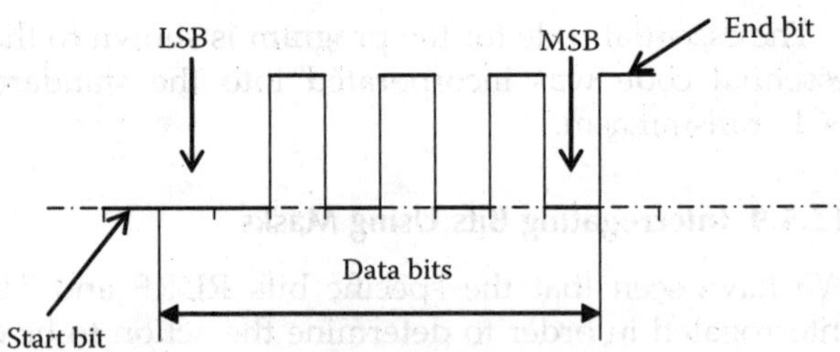

FIGURE 12.18
The representation on the oscilloscope screen of the letter T being transmitted through the SCI.

12.5.10 Readings of Serial Communication Data on the Oscilloscope Screen

One way of monitoring serial communication is to tap the traffic and display the data on an oscilloscope screen. The readings on the screen are high or low voltages which correspond to binary 1 or 0, respectively. Here is an example to show how to read the oscilloscope screen. Assume the capital letter "T" is being sent. The corresponding ASCII code is $54 = \%01010100$. Upon reversal and packing between the start and end bits, we get the 10-bit packet 0001010101. This packet is presented in Figure 12.18, which shows, from left to right, the start bit $b_s = 0$, the reverted code 00101010, and the end bit, $b_e = 1$.

To decipher the transmitted character from the oscilloscope screen, we first read the complete 10-bit transmission 0001010101. Then, we remove the first and last bits to get 00101010. To get the actual character being transmitted, we reverse the bits and obtain 01010100. Converting from binary to hex, we identify the number $54. Looking it up in Table 12.10, we observe that $54 corresponds to the capital letter T.

12.5.11 Serial Communication Echo

The echo function is very important in developing interactive programs with the microcontroller. We wish to create a program that will receive a keystroke from the terminal keyboard, and echo the same keystroke, continuously, to the terminal screen. The program will also let you see the binary pattern of keystrokes sent from the PC to the MCU. To achieve this, we will continuously check if the MCU has received, via SCI reception, a keystroke from the PC keyboard. When a keystroke has been received, the ASCII code will be sent back through the SCI transmission, through the parallel port B for display on an 8-LED unit. The 8-LED unit, affixed to the parallel port B, will display the bit pattern of the ASCII code. This way, we will check that the bit pattern corresponds to the ASCII table. We will also monitor the serial communication by tapping the serial communication signal and displaying it on the oscilloscope screen. In this way, we will be able to see the complete 10 bit pattern, and the time duration of the communication. We will perform a simple serial communication echo exercise. The instructions are as follows:

1. Fill in the ASCII codes for each key listed in Table 12.24.

2. Connect the LED board to port B. Attach a probe from channel 2 to parallel port C. Attach another probe from O-scope channel 1 to the serial port tap on the MCU board (jumper J6). Set the oscilloscope trigger to channel 2 and the oscilloscope mode to normal. Set display to "Both."

3. Load and run the code Echo.asm in the MCU. Press key "Q" on the keyboard. The character should appear on the screen continuously, while its ASCII code is sent out through the parallel port B and displayed on the LED board. Read the bit pattern displayed on the LED board and enter it in Table 12.24. Write the equivalent hex number. Compare the ASCII code and answer Y or N in the table.

TABLE 12.24

Keystrokes for Performing the Serial Communication Echo Exercise

	Keystroke	Q	0	3	8	D	E	F	^	!	&
	ASCII code from Table 12.10	$54									
LED board	8-bit pattern on the LED board	MSB→LSB 10011001									
	Equivalent hex values	$54									
	Agreement (Y/N)										
Oscilloscope screen	8-bit pattern on the oscilloscope screen										
	Equivalent hex values										
	Agreement (Y/N)										

4. Manipulate the time (horizontal) scale, amplitude (vertical) scale, and the trigger level, until you see both the triggering pulse and the serial communications transmission. Identify the start bit, the ASCII code for the character being transmitted, and the stop bit on the oscilloscope screen. Compare the ASCII code listed in Table 12.24 to the one displayed on the oscilloscope. Enter your answer in Table 12.24.

5. Use the oscilloscope to calculate the BAUD rate of the MCU.

   Duration of signal $\Delta t = \underline{\hspace{1cm}}$ ms; number of bits read in the signal $N = \underline{\hspace{1cm}}$; and calculated BAUD rate = # of bits/duration = $N/\Delta t = \underline{\hspace{1cm}}$ bps

6. Repeat for keys 0, 3, 8, D, E, F, ^, !.

Sketch the 10-bit pattern observed on the oscilloscope screen for character Q.

12.6 Microcontroller Timer Functions

12.6.1 Timer Functions

The timer functions are described in Chapter 10 of the Motorola M68HC11 Reference Manual. We distinguish the following main categories:

- Free running clock (timer counter)
- Input capture: detect voltage events (i.e., signal changes/transitions) on up to three input signal lines and records the time of their occurrence
- Output compare: generate voltage events (i.e., signal changes/transitions) on up to five output signal lines at independently present times

Timer functions allow the microcontroller to determine "time" by counting the number of machine cycles between events, the moment an event happens, or to generate an event at a preset time.

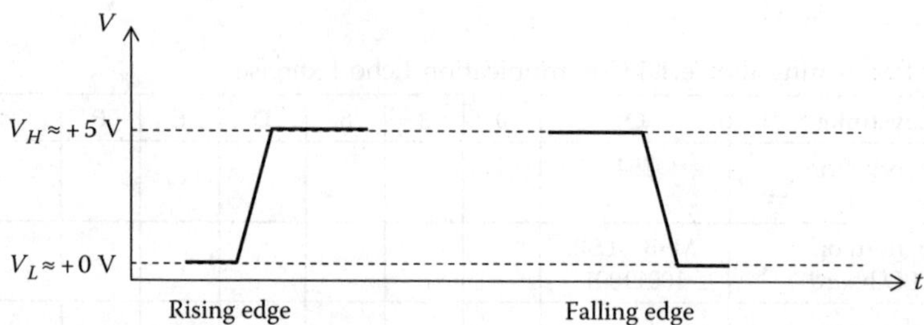

FIGURE 12.19
Voltage events corresponding to low–high (rising edge) and high–low (falling edge) voltage transitions.

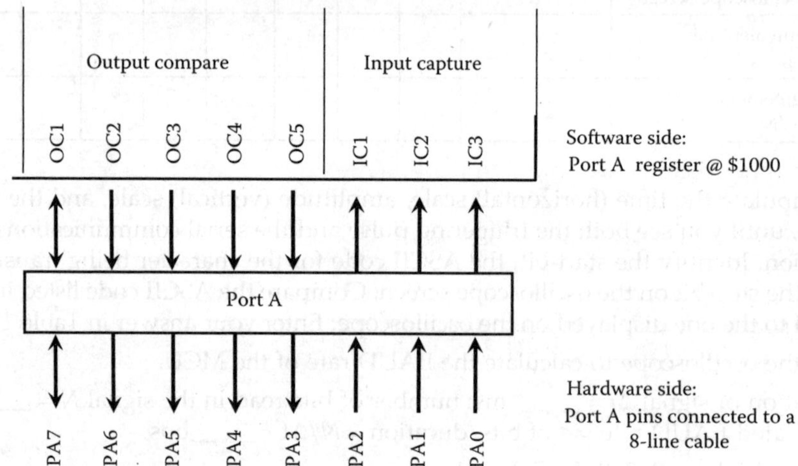

FIGURE 12.20
Port A pins and the corresponding input capture (IC) and output compare (OC) timer functions.

A voltage *event* is the change of voltage on a monitored electric line (Figure 12.19). The microcontroller monitors 8 electric lines, corresponding to the eight pins of port A (Figure 12.20). Such an event could be the voltage going from zero to a set value (*rising edge*), or from a set value to zero (*falling edge*). In timer mode, the microcontroller can be programmed to react to certain type of event, or to both. It can also be programmed to generate a certain type of event, or simply to flip the event type. The details of the timer functions are discussed next.

12.6.2 Timer Registers

The microcontroller registers involved in the timer functions discussed in this section are

- TCNT = timer counter register
- TIC1 = timer input capture register #1
- TIC2 = timer input capture register #2
- TIC3 = timer input capture register #3
- TOC1 = timer output compare register #1

- TOC2 = timer output compare register #2
- TOC3 = timer output compare register #3
- TOC4 = timer output compare register #4
- TOC5 = timer output compare register #5
- TCTL1 = timer control register #1
- TCTL2 = timer control register #2
- TFLG1 = timer flag register #1
- TFLG2 = timer flag register #2

The microcontroller pins involved in the timer functions discussed in this section are

- PORTA = Port A pins

12.6.3 Timer Counter

The timer counter is a free-running clock. It is incremented once every machine cycle. The values of the timer counter are stored in the timer counter register. The name of the *timer counter register* is TCNT (Table 12.25). It is a double-precision register, i.e., it takes up 2 bytes of memory. Its location is at addresses $100e and $100f. The higher part of the timer counter register (bits 8–15) is stored at $100e. The lower part (bits 0–7) is stored at $100f.

After the timer counter register reaches its maximum value, TCNT $-$ $ffff, no more increments are possible. Then, the next machine cycle causes the timer counter register to "overflow," i.e., to be reset from $ffff back to $0000. Thus, we see that the maximum number of cycles that TCNT can count is $\Delta T_{max} = \$ffff + \$0001 = \$10000 = 65,536$ cycles. At the microcontroller clock speed of 2 MHz, this corresponds to $t_{max} = 65,536/(2{*}10^6) \approx$ 32.768 ms. Using the timer functions of the microcontroller and the conversion factor of 0.5 μs/cycle, we can use the microcontroller for time measurement data acquisition.

12.6.4 Timer Overflow Flag

To let the user know that a timer overflow has happened, the microcontroller sets the *timer overflow flag*, TOF. The timer overflow flag is bit 7 in the timer flag register #2 (TFLG2) at address $1025 (Table 12.26). The value TOF $= 1$ implies that there has been a timer overflow; whereas TOF $= 0$ implies that there has not been a timer overflow. The timer

TABLE 12.25

Timer Counter Register, TCNT

$100e	Bit 15	Bit 14	Bit 13	Bit 12	Bit 11	Bit 10	Bit 9	Bit 8	TCNT
$100f	Bit 7	Bit 6	Bit 5	Bit 4	Bit 3	Bit 2	Bit 1	Bit 0	

TABLE 12.26

Timer Flag Register #2 (TFLG2) Showing the Position of the Timer Overflow Flag TOF

	Bit 7	Bit 6	Bit 5	Bit 4	Bit 3	Bit 2	Bit 1	Bit 0	
$1025	TOF	(RTIF)	(PAOVI)	(PAII)			(PR1)	(PR0)	TFLG2

Note: Bits not discussed in the text are in parentheses.

overflow flag can be used to count the number of times a timer overflow has happened. This can be achieved by setting a variable, say NOF, which is incremented each time we have a timer overflow, i.e., each time we record TOF $= 1$.

Example:

Assume we have to measure $\Delta t = 100$ and ms $= 100{,}000$ μs. This time duration is longer than $t_{max} = 32{,}768$ μs. This means that the microcontroller timer counter will overflow during the process. To understand how many timer overflows will take place, we write Δt as

$$t = 100{,}000 \text{ μs} = 3 \times 32{,}768 + 17{,}232 = 3 \times t_{max} + 1{,}696 \text{ μs} \tag{12.17}$$

This indicates that the timer counter TCNT will overflow at least three times during the timing of a 100 ms period. The software variable NOF which counts the number of times the timer counter has overflown, will take the value NOF $= 3$. Recalling the conversion rule "two cycles per microsecond" we can write this in terms of machine cycles as

$$\Delta T = 200{,}000 = 3 \times 65{,}536 + 3{,}392 = 3 \times \Delta T_{max} + 3{,}392 \text{ cycles} \tag{12.18}$$

In hex, this calculation is written as

$$\Delta T = \$30d40 = 3 \times \$10000 + \$00d40 \text{ cycles} \tag{12.19}$$

We notice that the writing of the total number of cycles, $\Delta T = \$30d40$ requires 5 hex digits. In an 8-bit processor, even a double precision timer register can only accommodate 4 hex digits. Therefore, the five-digit hex number resulting from this calculation could not be written as such. To overcome this problem, a fixed scaler that would reduce the size of the number to 4 hex digits can be used in programming the code for this calculation. For example, we could divide the total number by 100, and hence write

$$\Delta T\_100 = 200{,}000/100 = 3 \times 65{,}536/100 + 3{,}392/100$$
$$= 3 \times \Delta T_{max}\_100 + 33.92 \text{ hundred-cycles} \tag{12.20}$$

where $\Delta T_{max}\_100 = 655.36$ hundred-cycles. In hex, this will correspond to

$$\Delta T\_100 = \$30d40/\$64 = 3 \times \$10000/\$64 + \$00d40/\$64 = 3 \times \$028f + \$21 \tag{12.21}$$

We can write this as

$$\Delta T\_100 = 3 \times \Delta T_{max}\_100 + \$21 \text{ hundred-cycles} \tag{12.22}$$

where $\Delta T_{max}\_100 = \$028f$ hundred-cycles. However, we notice that the process of using a scaler in integer hex arithmetic leads to some accuracy loss, since

$$\$07d0 \approx 3 \times \$028f + \$21 = \$07ce \tag{12.23}$$

■

12.6.4.1 Detecting the Timer Overflow Flag

One way of detecting the TOF is by the polling method. In this method, the status of the TOF is interrogated (polled) by comparing its value with an expected value. In our case, the expected value is TOF $= 1$. Since the TOF is bit 7 in the timer flag register #2, TFLG2, the polling action will be to see if TFLG2 has a value 1 in bit 7. This can be achieved by the following simple polling sequence:

TABLE 12.27

Choices of Possible Instruction Sequences for Clearing the Timer Overflow Flag, TOF

Instruction Sequence	Opcode	Operand(s)	Address Mode	Bytes	Cycles	Total Sequence	
						Bytes	Cycles
1	LDAA	#%10000000	(IMM)	2	2		
	STAA	<TFLG2	(DIR)	2	3	4	5
2	BCLR	<TFLG2 %01111111	(DIR)	3	6	3	6
3	LDAA	#%10000000	(IMM)	2	2		
	STAA	TFLG2	(EXT)	3	4	5	6
4	LDAA	#%10000000	(IMM)	2	2		
	STAA	TFLG2,X	(IND,X)	2	4	4	6
5	BCLR	TFLG2,X %01111111	(IND,X)	3	7	3	7
6	LDAA	#%10000000	(IMM)	2	2		
	STAA	TFLG2,Y	(IND,Y)	3	5	5	7
7	BCLR	TFLG2,Y $7F	(IND,Y)	4	8	4	8

...

```
LABEL_0    LDAA    TFLG2,X
           ANDA    #%10000000
           BEQ     LABEL_0
```

...

12.6.4.2 Clearing the Timer Overflow Flag

To use TOF as a counting tool, one must clear the TOF after its presence has been recorded. The clearing of the TOF is done by writing the value 1 to it. This rather unusual procedure applies, in fact, to the clearing of all the timer flags (see Section 10.2.4 on page 387 in the Motorola Reference Manual). When clearing a flag, it is important that you do not interfere with the other bits in the register! Many instruction sequences can be used to clear timer flags. Table 12.27 illustrates a choice of seven different instruction sequences that could be used to clear the TOF status bit in TFLG2. In general, each sequence takes a different number of bytes of object code and a different number of cycles of execution time. The best choice of the appropriate sequence depends on a number of factors, including (but not limited to) whether the user wants minimum execution time or minimum program memory space. In most applications, the following simple sequence (row 4 of Table 12.27) suffices:

```
...

LDAA    #%10000000
STAA    TFLG2,X

...
```

12.6.5 Input Capture

The input capture function allows the microcontroller to record the value of the timer counter when the input voltage changes on the monitored external lines. As shown in Figure 12.20, there are three individual input capture channels: IC1, IC2, IC3. They correspond to

TABLE 12.28

Timer Control Register #2, TCTL2, Showing the EDGxB and EDGxA Control Bits for the Input Capture Function (x = 1, 2, 3)

	Bit 7	Bit 6	Bit 5	Bit 4	Bit 3	Bit 2	Bit 1	Bit 0	
\$1021	0	0	EDG1B	EDG1A	EDG2B	EDG2A	EDG3B	EDG3A	TCTL2

TABLE 12.29

Four Timer Input Capture Modes (x = 1, 2, 3 Corresponding to IC1, IC2, IC3)

EDGxB	EDGxA	Mode
0	0	Capture disabled
0	1	Capture on rising edges only
1	0	Capture on falling edges only
1	1	Capture on any edge (rising or falling)

three port A pins: PA2, PA1, PA0. They detect when a signal transition happens on one of the electrical lines connected to these pins. Up to three separate events can be monitored, one per each line. The type of signal transition that causes an input capture is determined by the edge bits, EDG1B, EDG1A; EDG2B, EDG2A; and EDG3B, EDG3A. The edge bits are located in the timer control register #2 (TCTL2) at address \$1021 (Table 12.28).

Because the EDGxB and EDGxA bits act together (x = 1, 2, 3 corresponding to IC1, IC2, IC3), their combination yields four different input capture modes: *disabled, on rising edge, on falling edge*, and *any edge* (i.e., both on rising and on falling edges). A summary of these four modes is given in Table 12.29.

When a signal transition is detected, the input capture function automatically records the value in the free running clock in a separate memory location. A flag is set to let the user know that an input capture has taken place. For each of the three input capture channels there are separate input capture registers (TIC1, TIC2, and TIC3) and input capture flags (IC1F, IC2F, and IC3F).

When an input capture is detected on a monitored channel, the value existing in TCNT is transferred in the appropriate TICx, where x = 1, 2, 3, depending on the input capture channels being selected. Note that one, two, or all three input capture channels can be used simultaneously, each with an independent setting of the input capture conditions. The input capture timer registers, TICx, are not affected by the microcontroller reset and cannot be overwritten.

For each of the three input capture channels, there is a separate input capture timer register. The input capture timer registers TIC1, TIC2, and TIC3 record the value of the free-running clock when the input capture was observed at that particular channel. Since the free-running clock generates a double precision number, each of the input capture registers must also be of 2 byte size. These 2 byte TICx registers (x = 1, 2, 3) are located at address \$1010–\$1015 (Table 12.30).

The input capture flags operate as follows. A value ICxF = 1 implies that there has been an input capture on that particular IC line; 0 implies that an input capture has not yet occurred. Each ICxF flag is cleared by writing a 1 to it in the appropriate location of the control register TFLG1. The clearing of the ICxF flags is achieved in the same way as the clearing of the TOF. To this purpose, any of the instruction sequences presented in

TABLE 12.30

Location of the Input Capture Registers TIC1, TIC2, and TIC3

	Bit 15	Bit 14	Bit 13	Bit 12	Bit 11	Bit 10	Bit 9	Bit 8	
$1010	Bit 15	Bit 14	Bit 13	Bit 12	Bit 11	Bit 10	Bit 9	Bit 8	TIC1
$1011	Bit 7	Bit 6	Bit 5	Bit 4	Bit 3	Bit 2	Bit 1	Bit 0	
$1012	Bit 15	Bit 14	Bit 13	Bit 12	Bit 11	Bit 10	Bit 9	Bit 8	TIC2
$1013	Bit 7	Bit 6	Bit 5	Bit 4	Bit 3	Bit 2	Bit 1	Bit 0	
$1014	Bit 15	Bit 14	Bit 13	Bit 12	Bit 11	Bit 10	Blt 9	Bit 8	TIC3
$1015	Bit 7	Bit 6	Bit 5	Bit 4	Bit 3	Bit 2	Bit 1	Bit 0	

TABLE 12.31

Location of Timer Input Capture Flags IC1F, IC2F, and IC3F in TFLG1register

	Bit 7	Bit 6	Bit 5	Bit 4	Bit 3	Bit 2	Bit 1	Bit 0	
$1023	(OC1F)	(OC2F)	(OC3F)	(OC4F)	(OC5F)	IC1F	IC2F	IC3F	TFLG1

Note: Bits not discussed in this section are in parentheses.

Table 12.27 can be used. The timer input capture flags ICxF are located in the timer flag register #1, TFLG1, at the memory address $1023 (Table 12.31).

The detection of timer input capture flags can be achieved with any of the methods previously discussed, e.g., polling. One simple polling sequence for polling, say IC1F, is

. . .

```
LABEL_0        LDAA        TFLG1,X
               ANDA        #%00000100
               BEQ         LABEL_0
```

. . .

In this sequence, IC1F was identified as bit 2 in TFLG1. Bit 2 is the third bit from the right. Hence, the number %00000100 was used for this interrogation.

In conclusion, the MC68HC11 microcontroller has three input capture channels IC1, IC2, and IC3. All act in the same way, with separate memory locations, EDG bits, and ICF flags.

12.6.6 Output Compare

Another timer function is the output compare. In essence, the output compare function schedules events to happen at a predetermined time. This function works as follows. A predetermined timer value is stored in an output compare register. The current value of the free running clock is compared with the value in the output compare register. When the value in the free running clock equals the value in an output compare register, a voltage event is generated on the output compare channel. The voltage event can be, depending on programming: (1) no action, (2) toggle the output line value, (3) set output line to low, and (4) set output line to high.

The microcontroller has five output compare channels: OC1, OC2, OC3, OC4, OC5. They correspond to five port A pins: PA7, PA6, PA5, PA4, PA3. Four of these output compare

TABLE 12.32

Timer Output Compare Registers TOC1, TOC2, TOC3, TOC4, TOC5

$1016	Bit 15	Bit 14	Bit 13	Bit 12	Bit 11	Bit 10	Bit 9	Bit 8	TOC1
$1017	Bit 7	Bit 6	Bit 5	Bit 4	Bit 3	Bit 2	Bit 1	Bit 0	

$1018	Bit 15	Bit 14	Bit 13	Bit 12	Bit 11	Bit 10	Bit 9	Bit 8	TOC2
$1019	Bit 7	Bit 6	Bit 5	Bit 4	Bit 3	Bit 2	Bit 1	Bit 0	

$101a	Bit 15	Bit 14	Bit 13	Bit 12	Bit 11	Bit 10	Bit 9	Bit 8	TOC3
$101b	Bit 7	Bit 6	Bit 5	Bit 4	Bit 3	Bit 2	Bit 1	Bit 0	

$101c	Bit 15	Bit 14	Bit 13	Bit 12	Bit 11	Bit 10	Bit 9	Bit 8	TOC4
$101d	Bit 7	Bit 6	Bit 5	Bit 4	Bit 3	Bit 2	Bit 1	Bit 0	

$101e	Bit 15	Bit 14	Bit 13	Bit 12	Bit 11	Bit 10	Bit 9	Bit 8	TOC5
$101f	Bit 7	Bit 6	Bit 5	Bit 4	Bit 3	Bit 2	Bit 1	Bit 0	

TABLE 12.33

Location of Timer Output Compare Flags OC1F, OC2F, OC3F, OC4F, OC5F in TFLG1register

	Bit 7	Bit 6	Bit 5	Bit 4	Bit 3	Bit 2	Bit 1	Bit 0	
$1023	OC1F	OC2F	OC3F	OC4F	OC5F	(IC1F)	(IC2F)	(IC3F)	TFLG1

Note: Bits not discussed in this section are in parentheses.

channels are similar (OC2, OC3, OC4, and OC5), and their functionality is programmed exactly in the same way. The fifth output compare channel, OC1 on PA7, is different, since the same physical pin is also used for the pulse accumulator function. In the subsequent discussion, we will focus our attention on the four commonly used output compare channels OC2, OC3, OC4, and OC5.

As the timer is a 2-byte register, each of the output compare registers is a 2-byte register. To set a value for output compare, simply store a 2-byte number into the appropriate output compare register. When a match is detected between the value in TCNT and the value existing in one of the output compare register, an output compare event takes place. The location of the five output compare registers TOC1, TOC2, TOC3, TOC4, and TOC5 is shown in Table 12.32.

When the output compare event takes place, an output compare flag is set. The output compare flags are OC1F, OC2F, OC3F, OC4F, and OC5F. They are located in the TFLG1 register at address $1023 (Table 12.33). A value OCxF = 1 indicates that output compare has occurred on output compare channel x (x = 1, 2, 3, 4, 5). A value of 0 indicates that output compare has not occurred yet. Each OCxF flag is cleared by writing a 1 to it in the appropriate location of the control register TFLG1. The clearing of the OCxF flags is achieved in the same way as the clearing of the TOF. To this purpose, any of the instruction sequences presented in Table 12.27 can be used.

The signal sent out by the microcontroller during an output compare event is controlled by 2-bits, the OMx and OLx bits, acting together (Table 12.34). Because the OMx and OLx bits act together (x = 2, 3, 4, 5), their combination yields four different output compare

TABLE 12.34

Timer Control Register #1, TCTL1, Showing the OMx and OLx Control Bits for the Output Compare Function (x = 2, 3, 4, 5)

	Bit 7	Bit 6	Bit 5	Bit 4	Bit 3	Bit 2	Bit 1	Bit 0	
$1020	OM2	OL2	OM3	OL3	OM4	OL4	OM5	OL5	TCTL1

TABLE 12.35

Encoding of the OMx and OLx Control Bits for the Output Compare Function of the Microcontroller Timer (x = 2, 3, 4, 5)

OMx	OLx	Action on Successful Compare on OCx
0	0	No action
0	1	Toggle OCx pin
1	0	Clear OCx pin
1	1	Set OCx pin

actions: (1) *no action*, (2) *toggle*, (3) *clear*, and (4) *set* (Table 12.35). These actions correspond to the following states of the output line voltage: (1) leave as it is, (2) toggle the output line voltage state, (3) set output line to low, and (4) set output line to high.

The detection of timer output compare flags can be achieved with any of the methods previously discussed, e.g., polling. One simple sequence for polling, say, OC3F, is

```
            ...
LABEL_0     LDAA      TFLG1,X
            ANDA      #%00100000
            BEQ       LABEL_0
            ...
```

In this sequence, OC3F was identified as bit 5 in TFLG1. Bit 5 is the sixth bit from the right. Hence, the number %00100000 was used for this interrogation.

12.6.7 Input Capture Example

This program is an example of timer input capture. The process is started when a keystroke is received. Then, the time is measured until a low–high transition is captured on line IC1. In this process, the initial time, T0, and the capture time, T1, as well as the number of overflows are recorded. After the input capture, the IC1F flag is reset; the overflow counter is zeroed; and the process is repeated.

The program instructions, flowchart, and assembly code are given in Table 12.36. The program flowchart is shown to the right of the program instructions. Note the variable definition block in which T0, T1, and NOF are defined. Next, the initialization block contains reg. X initialization, timer IC initialization, and SCI initialization. The program loop starts at label BEGIN with the overflow counter, NOF, being zeroed. First, the RDRF flag is checked in a loop to verify if a keystroke has been received. When keystroke was received, the time counter is read and stored in T0. Then, the program loops on LABEL1 until an input capture IC1 is recorded. In this loop, TOF is first checked to verify if timer overflow takes place. When timer overflow is detected, the overflow counter, NOF, is

TABLE 12.36

Instructions, Flowchart, and Code for the Use of the Input Capture Timer Caption

Instructions	Flowchart	Code

(i) Define variables:
  a. Origin of time, T0 = 2 bytes
  b. Capture time, T1 = 2 bytes
  c. Overflow counter, NOF1 = 1 byte
(ii) Initialize
  a. Initialize index X to REGBAS
  b. Initialize timer IC1 function: set EDG1A in TCTL2
  c. Initialize SCI
(iii) Zero overflow counter NOF1
(iv) Wait for a keystroke reception
(v) Store initial time
(vi) Wait for the input capture
  a. Check TOF
  b. Jump if no TOF; else
    i. Increment overflow counter
    ii. Reset TOF
  c. Check IC1F
  d. Loop back
(vii) After input capture
  a. Load and store t1
  b. Reset IC1F by writing 1 to it
(viii) Loop back to (iii) and do it again
	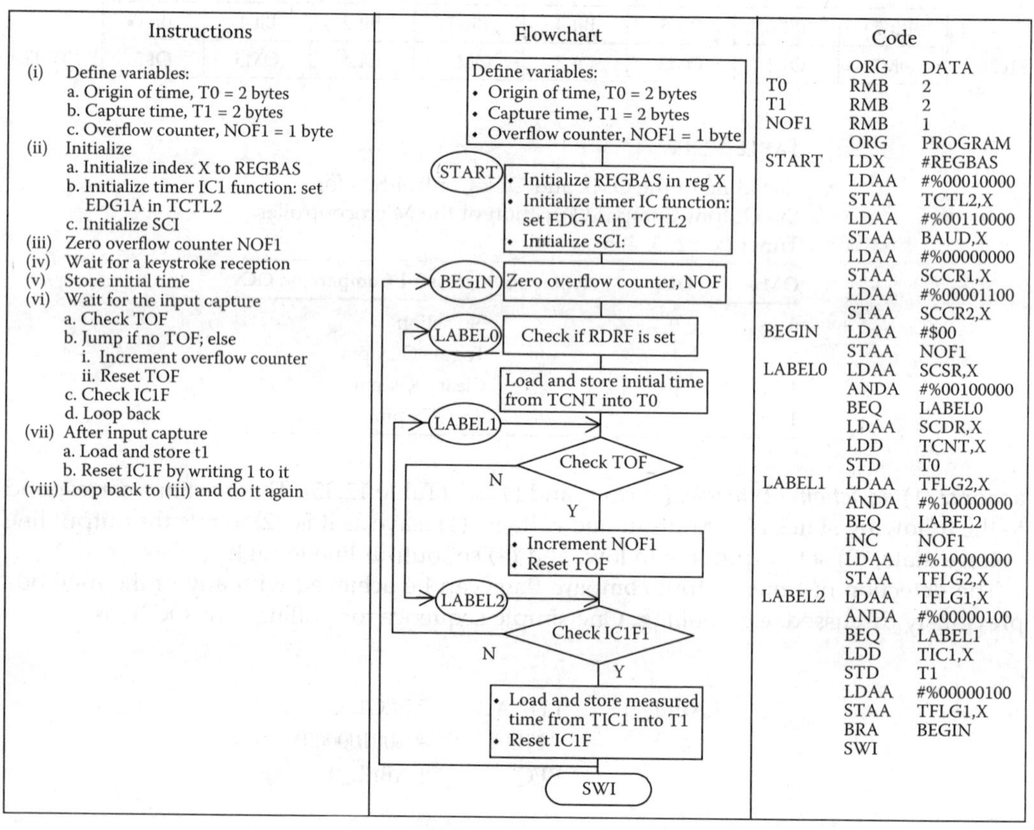	ORG DATA
		T0 RMB 2
		T1 RMB 2
		NOF1 RMB 1
		ORG PROGRAM
		START LDX #REGBAS
		LDAA #%00010000
		STAA TCTL2,X
		LDAA #%00110000
		STAA BAUD,X
		LDAA #%00000000
		STAA SCCR1,X
		LDAA #%00001100
		STAA SCCR2,X
		BEGIN LDAA #$00
		STAA NOF1
		LABEL0 LDAA SCSR,X
		ANDA #%00100000
		BEQ LABEL0
		LDAA SCDR,X
		LDD TCNT,X
		STD T0
		LABEL1 LDAA TFLG2,X
		ANDA #%10000000
		BEQ LABEL2
		INC NOF1
		LDAA #%10000000
		STAA TFLG2,X
		LABEL2 LDAA TFLG1,X
		ANDA #%00000100
		BEQ LABEL1
		LDD TIC1,X
		STD T1
		LDAA #%00000100
		STAA TFLG1,X
		BRA BEGIN
		SWI

incremented, and TOF is reset. Next, IC1F is checked to verify if input capture on IC1 took place. If input capture is not detected, the program returns to LABEL1. When input capture is detected, the program exits the loop, loads the IC1 timer from TIC1, and stores it in the capture time variable, T1. The program loops back to the beginning and waits for a new keystroke to restart the process. The essential code for the program is shown to the right of the program flowchart.

After the variables T0, T1, and NOF are determined, the total elapse time is found with the formula:

$$\Delta t = (T1 - T0 + \$10000*NOF)*0.5 \ \mu s \tag{12.24}$$

12.6.8 Output Compare Example

This program is an example of timer output compare. The program runs continuously and generates repeated toggles of the output compare pin OC3 (PA5) at equal intervals that are determined using the delay variable DT. Thus, a square wave as shown in Figure 12.21 is generated.

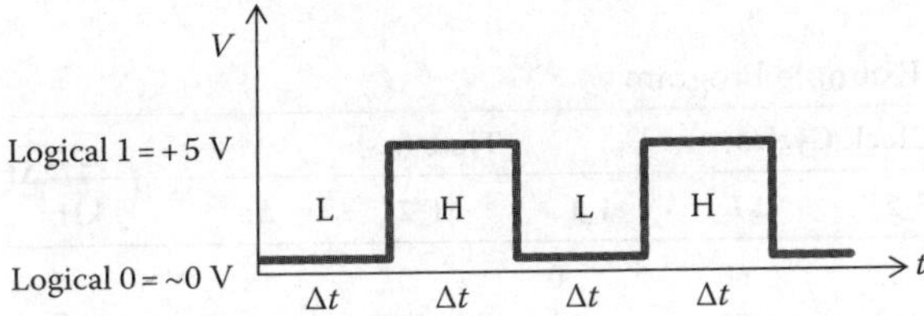

FIGURE 12.21
Square wave schematics showing the half-wave duration, Δt, and the low (L) and high (H) signal states.

TABLE 12.37

Instructions, Flowchart, and Code for the Generation of a Square Wave with the Output Compare Timer Function

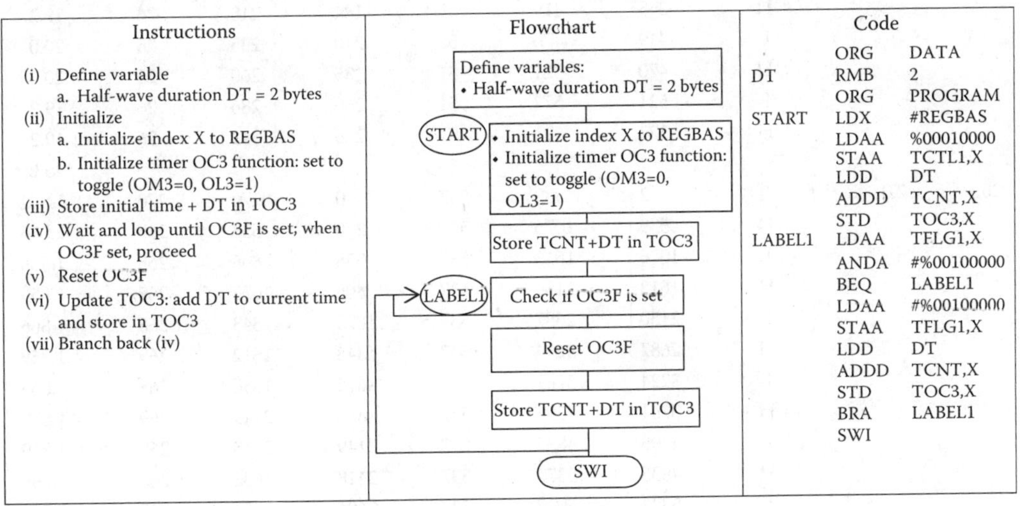

Instructions	Flowchart	Code

The table contents, combined:

Instructions

(i) Define variable
  a. Half-wave duration DT = 2 bytes
(ii) Initialize
  a. Initialize index X to REGBAS
  b. Initialize timer OC3 function: set to toggle (OM3=0, OL3=1)
(iii) Store initial time + DT in TOC3
(iv) Wait and loop until OC3F is set; when OC3F set, proceed
(v) Reset OC3F
(vi) Update TOC3: add DT to current time and store in TOC3
(vii) Branch back (iv)

Flowchart

- Define variables:
  - Half-wave duration DT = 2 bytes
- START
  - Initialize index X to REGBAS
  - Initialize timer OC3 function: set to toggle (OM3=0, OL3=1)
- Store TCNT+DT in TOC3
- LABEL1 → Check if OC3F is set
- Reset OC3F
- Store TCNT+DT in TOC3
- SWI

Code

```
        ORG     DATA
DT      RMB     2
        ORG     PROGRAM
START   LDX     #REGBAS
        LDAA    %00010000
        STAA    TCTL1,X
        LDD     DT
        ADDD    TCNT,X
        STD     TOC3,X
LABEL1  LDAA    TFLG1,X
        ANDA    #%00100000
        BEQ     LABEL1
        LDAA    #%00100000
        STAA    TFLG1,X
        LDD     DT
        ADDD    TCNT,X
        STD     TOC3,X
        BRA     LABEL1
        SWI
```

The program instructions are given in the first column of Table 12.37. The program flowchart is shown to the right of the program instructions. Note the variable definition block in which DT is defined. Next, the initialization block contains reg. X initialization and timer OC initialization. The program stores the current time (TCNT) plus the delay DT in the OC3 timer TOC3. A loop is entered in which the OC3 flag, OC3F, is checked. When OC3F is set, the loop is exited. The OC3F is reset, and the value of TOC3 is updated with the current time plus the delay. Then, the program loops back to the beginning and waits for a new keystroke to restart the process. The essential code for the program is shown to the right of the program flowchart.

12.6.8.1 Program Calibration

To calibrate the program, do the following sequence. Set breakpoints at $c013. Run through the program, and after each breakpoint stop, record the clock cycles and the simulated time in Table 12.38. Continue to do this until you have completed six L-H groups. These correspond to six square waves.

Calculate the half-wave time, Δt, and the equivalent frequency, $f = 1/2\Delta t$. (While doing this, omit the first row in your readings.) Note that the time interval oscillates between

TABLE 12.38

Simulated Reading of the Output Compare Example Program

DT (Hex)	Raw, f (kHz)	State	Clock Cycles			Time (μs)			$f = 1/2\Delta t$ kHz
			T_1	T_2	ΔT	t_1	t_2	Δt	
0020	31.25	L	2	62	60	0	31		
	kHz	H	62	113	51	31	56	25	20.0
		L	113	164	51	56	82	26	19.2
		H	164	215	51	82	108	26	19.2
		L	215	266	51	108	133	25	20.0
		H	266	317	51	133	158	25	20.0
		L	317	368	51	158	184	26	19.2
		H	368	419	51	184	210	26	19.2
		L	419	470	51	210	235	25	20.0
		H	470	521	51	235	260	25	20.0
		L	521	572	51	260	286	26	19.2
		H	572	623	51	286	312	26	19.2
									kHz
0200	2.0	L	2	539	537	0	270	270	1.852
		H	539	1076	537	270	538	268	1.866
		L	1076	1613	537	538	806	268	1.866
		H	1613	2150	537	806	1075	269	1.859
		L	2150	2687	537	1075	1343	268	1.866
		H	2687	3224	537	1343	1612	269	1.859
		L	3224	3761	537	1612	1880	268	1.866
		H	3761	4298	537	1880	2149	269	1.859
		L	4298	4835	537	2149	2418	269	1.859
		H	4835	5372	537	2418	2686	268	1.866
		L	5372	5909	537	2686	2954	268	1.866
		H	5909	6446	537	2954	3223	269	1.859
	Hz								Hz
2000	122	L	2	8216	8214	0	4108	4108	122
		H	8216	16430	8214	4108	8215	4107	122
		L	16430	24644	8214	8215	12322	4107	122
		H	24644	32858	8214	12322	16429	4107	122

25 and 26 μs, while the frequency oscillates between 20.0 and 19.2 kHz, i.e., 19.9 ± 0.2 kHz or $19.9 \pm 1\%$ kHz. These oscillations are an indication of the repeatability of your process, and are due to the displayed simulation time being limited to 1 μs, while one MCU cycle takes 0.5 μs.

Calculate the raw frequency corresponding to the delay \$0020 using the formula $f_{raw} = \dfrac{1}{2 \cdot DT \cdot 0.5 \text{ μs}} 10^3$ kHz. Notice that the raw frequency is 31.25 kHz, while the obtained frequency is 20 kHz. The difference is a programming overhead due to the extra cycles that are consumed in the program between two OC events. We expect that this overhead will become less and less significant as the delay increases, i.e., at lower frequencies.

Put the value \$0200 in DT. Repeat the procedure above and enter the clock cycles and simulate time values in Table 12.38. Note that the effective frequency oscillated between 1.859 and 1.866 kHz, i.e., 1.8625 ± 0.0035 kHz or $1.8625 \pm 0.2\%$. The repeatability of our simulation has greatly improved. Calculate the raw frequency corresponding to the delay \$0200 using the formula $f_{raw} = \dfrac{1}{2 \cdot DT \cdot 0.5 \ \mu s} 10^3$ kHz. Notice that the raw frequency is 2.0 kHz, while the effective frequency is 1.8625 kHz. The difference has been reduced considerably, (-6.9% error). This confirms the hypothesis that the programming overhead due to the extra cycles that are consumed in the program between two OC events is less significant at higher DT values, i.e., at lower frequencies.

Repeat for DT = \$2000. For this long delay, the frequency scale is switched from kilohertz to just hertz. In just a couple of cycles, you will note stable behavior and perfect frequency accuracy. This is due to the desired frequency (122 Hz) being much lower than the MCU clock frequency (2 MHz).

12.6.8.2 Generating a Desired Frequency

We will now practice a method for obtaining a desired frequency (Table 12.39). We will start with a low frequency, e.g., 100 Hz, since we have noticed that our square wave generation behaves much better at low frequencies. Then, we will increase the desired frequency, until a practical limit is obtained.

The operation of the program proceeds as follows:

1. Examine the row 100 Hz in Table 12.39. Notice that the corresponding half-period is 5000 µs. Hence, a raw estimation of the delay can be calculated with the formula, delay = half-period/0.5 and then converted to hex. The result is entered as *Raw delay* in Table 12.39.

TABLE 12.39

Generating a Desired Frequency with the Output Compare Timer Function Example

Desired f (Hz)	Half-Period (µs)	Raw Delay (Hex)	DT (Hex)	Time (µs)			$f = 1/\Delta t$ (Hz)	Error (%)
				t_1	t_2	Δt	Hz	
100	5000	2710	2710	5012	10024	5012	99.8	−0.2
200	2500	1388	1380	2510	5020	2510	199.2	−0.4
			1370	2502	5002	2500	200.0	0.0
500	1000	07D0	07d0	1012	2023	1011	495	−1.1
			07c0	1007	2010	1003	499	−0.3
1000	500	03E8	03e0	508	1015	507	986	−1.4
			03d0	503	1002	499	1002	0.2
2000	250	01F4	01f0	260	520	260	1923	−3.8
			01e0	256	506	250	2000	0.0
5000	100	00C8	00c8	112	223	111	4505	−9.9
			00c0	112	218	106	4717	−5.7
			00b8	108	210	102	4902	−2.0
			00b0	103	200	97	5155	3.1
			00b4	103	205	102	4902	−2.0

2. Enter DT = $2710 into the THRSim11, reset, and run. Note that the time of the breakpoint stop is 5012 μs. Run further, and note the second time as 10024 μs. These values are entered in Table 12.39. The resulting half-wave duration is 5012 μs, which corresponds to 99.8 Hz, i.e., a −0.2% error. We consider this good enough and select to stop.

3. Try to generate 200 Hz. According to Table 12.39, the corresponding half-period is 2500 μs, and the raw delay is $1388. Since we know that the MCU adds some extra cycles of its own, we select to round down the raw delay to the value $1380. Enter this value in the THRSim11, reset, and run. Record the times after first and second loop. These times are entered in Table 12.39 as 2510 μs and 5020 μs, respectively. The corresponding frequency is 199.2 kHz, i.e., −0.4% error. We improve on this error by making a minor adjustment to the DT from $1380 to $1370. The value $1370 gives us the desired frequency of 200 Hz exactly.

4. Try to generate 500 Hz. According to Table 12.39, the corresponding half-period is 1000 μs, and the raw delay is $07D0. Enter this value in the THRSim11, reset, and run. Record the times after first and second loop, and enter them in Table 12.39 as 1012 μs and 2023 μs, respectively. The corresponding frequency is 495 kHz, i.e., −1.1% error. We improve on this by making a minor adjustment to the DT, from $07d0 to $07c0. The value $07c0 gives us the frequency 499 Hz, which is only −0.3% in error. We select to stop here.

5. Try to generate 1000 Hz and 2000 Hz. Follow Table 12.39 for the values of the corresponding half-period, raw delay, and actual DT iterations. Notice how accuracy is being improved by small adjustments in second hex digit of the DT.

6. Try to generate 5000 Hz. According to Table 12.39, the corresponding half-period is 100 μs, and the raw delay is $00C8. This is a very small delay, and we expect some difficulties! Enter this value in the THRSim11, reset, and run. Record the times after first and second loop, and enter them in Table 12.39 as 112 μs and 223 μs, respectively. The corresponding frequency is 4505 kHz, i.e., −9.9% error. We try to improve on this by making minor adjustments to the DT, from $00c8 to $00c0, and then to $00b8, $00b0, and $00b4. However, the accuracy cannot be reduced below 2%. This indicates that we have reached a limit in our capability to fine-tune the frequency. This limit is due to the low value of the desired DT. We select to stop and conclude that 5000 Hz (5 kHz) is a practical limit of the square-wave frequency that could be generated with the MCU within 2% accuracy.

12.7 Analog/Digital Conversion with Microcontrollers

The A/D conversion is a process that allows the microcontroller to interface with a variety of devices such as potentiometers, strain-gages, accelerometers, cermets, thermocouples, LVDT's, and various other sensors. Thus, the microcontroller can accept analog inputs from external devices and convert them to digital values to be used in internal calculations.

12.7.1 Analog-to-Digital Conversion

An A/D converter takes an analog voltage, such as those produced by many electronic measuring devices and converts it to a digital value. Figure 12.22 shows an analog voltage

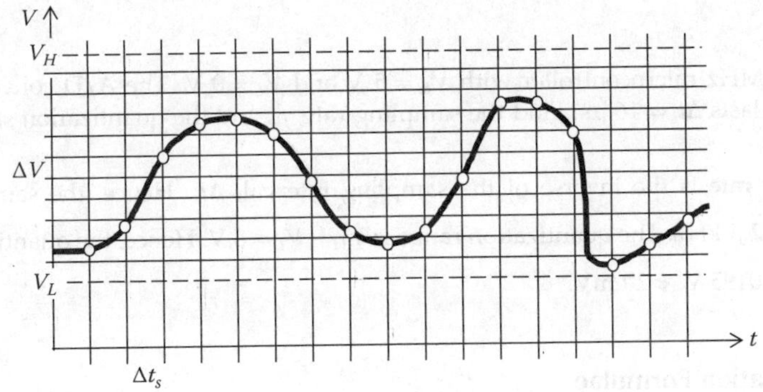

FIGURE 12.22
The A/D conversion process showing sampling, Δt_s, and discretization, ΔV.

that varies between a lower bound, V_L and a higher bound, V_H. The voltage is measured at discrete times, and the measured value has to be expressed as a number, i.e., it must be digitized. The A/D conversion process involves two steps:

- Sampling
- Digitization

Sampling is the process through which a continuously varying analog variable is measured (i.e., sampled) at fixed time intervals, Δt_s. The sampling rate, f_s, is the inverse of the sampling interval, Δt_s, i.e.,

$$f_s = \frac{1}{\Delta t_s} \tag{12.25}$$

In a microcontroller, the sampling rate, f_s, depends on the internal processes taking place during the A/D conversion. It is not equal to the clock speed, but to a subdivision of it. For example the MC68HC11 microcontroller with an internal clock speed of 2 MHz can achieve a sampling interval as low as 64 μs. Hence, the sampling rate can be as high as

$$f_s = \frac{1}{64 \ \mu s} = 15.625 \text{ kHz}.$$

Digitization is the process through which an analog value is converted to a digital value. Digitization is also called *quantization* since it involves the evaluation of an analog value using discrete quanta, i.e., incremental steps. The total number of steps used in the quantization process depends on the size of the binary number in which the quantized data will be stored. An 8-bit binary number has $2^8 = 256$ possible values: %00000000, %00000001, %00000010,..., %11111111. If quantization is implemented between a lower bound, V_L, and a higher bound, V_H, then the excursion $V_H - V_L$ will be divided into 2^8 levels. The step size from one level to next will be

$$\Delta V = \frac{V_H - V_L}{2^8} \tag{12.26}$$

Example:

Consider a 2 MHz microcontroller with $V_H = 5$ V and $V_L = 0$ V. The A/D conversion of one voltage value lasts $\Delta t_s = 16$ μs. Find the sampling rate, f_s, and the quantization step, ΔV.

SOLUTION

The sampling rate is the inverse of the sampling interval, Δt_s. Hence, the sampling rate is $f_s = \dfrac{1}{16 \text{ μs}} = 62.5$ kHz. The quantization range is $V_H - V_L = 5$ V. Hence, the quantization step is $\Delta V = 5/2^8 = 0.0195$ V ≈ 20 mV.　　　■

12.7.2 Quantization Formulae

The quantization formulae are used to assign a digital value to a given analog value. For example, in an 8-bit process, an analog voltage V will be converted to a digital voltage value $vv using the following *quantization formula*

$$\$vv = \frac{(V - V_L) \cdot \$ff}{V_H - V_L} \tag{12.27}$$

where $vv is the hex representation of the quantized digital voltage value. If the voltage V is equal to the lower bound, V_L, then the output of the quantization formula will be $00. On the other hand, if the voltage V equals the higher bound, V_H, then the output of the quantization formula will be $ff, i.e.,

If $V = V_L$ then $vv = $00

If $V = V_H$ then $vv = $ff

Example:

Consider an 8-bit quantization process with $V_H = 5$ V and $V_L = 0$ V. Assume that the analog voltage is $V = 2.457$ V. Find the corresponding digitized value $vv.

SOLUTION

$$\$vv = \frac{2457 \cdot \$ff}{5000} = \$7d \tag{12.28}$$

The reverse process is to recover an analog value from a digital value. Assuming that we know the result of the quantization process, $vv, we wish to estimate the actual value that was measured. To achieve this, we follow the quantization process in reverse. For 8-bit quantization, the following *dequantization formula* can be used:

$$V = V_L + \frac{(V_H - V_L) \cdot \$vv}{\$ff} \tag{12.29}$$

If the digitized voltage $vv is equal to zero, the actual analog voltage is estimated to be the lower bound, V_L. On the other hand, if the digitized voltage $vv equals $ff, then the analog voltage is estimated to be the higher bound, V_H, i.e.,

If $vv = $00 then $V = V_L$

If $vv = $ff then $V = V_H$　　　■

Example:

Consider an 8-bit quantization process with $V_H = 5$ V and $V_L = 0$ V. Assume that the digitized value is $vv = \$7d$. Find the corresponding analog voltage V

SOLUTION

$$V = 0 + \frac{5000 \cdot \$7d}{\$ff} = 2450 \text{ mV} = 2.450 \text{ V} \tag{12.30}$$

The small difference between the initial value, 2.457 V, and the recovered value, 2.450 V, represents the effect of the quantization error involved in any digitizing process. ■

12.7.3 Half-Bit versus One-Bit Quantization Error

In any A/D conversion process, the accuracy of conversion process depends on how much information is lost by missing 1-bit in the digitized number. Hence, the overall error of the A/D process can be estimated as one LSB, i.e., 1 LSB. To understand this concept, let us consider the simple, 2-bit quantization process shown in Figure 12.23. A 2-bit number can take four values: %00, %01, %10, %11. A rule must be devised to assign analog voltage values to each of these four digital values. Figure 12.23a shows the 1-*bit accuracy* assignment rule. For a 2-bit A/D process, this rule operates as follows:

$$\text{If} \quad 0 \le V < \frac{1}{4}V_0 \quad \text{then} \quad \%vv = \%00$$

$$\text{If} \quad \frac{1}{4}V \le V < \frac{2}{4}V_0 \quad \text{then} \quad \%vv = \%01$$

$$\text{If} \quad \frac{2}{4}V \le V < \frac{3}{4}V_0 \quad \text{then} \quad \%vv = \%10$$

$$\text{If} \quad \frac{3}{4}V \le V < V_0 \quad \text{then} \quad \%vv = \%11$$

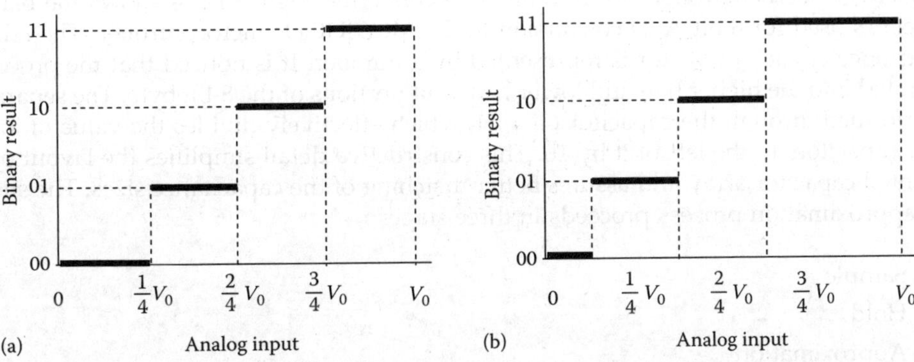

FIGURE 12.23
Various options for distributing the error in an A/D conversion process: (a) 1-bit accuracy; (b) half-bit accuracy.

It is apparent that the error involved in this rule is $-0/+1$ LSB. For an 8-bit A/D process the same concept would apply, and the error would be

$$err = -0/+1 \text{ LSB} = -0/+20 \text{ mV} \tag{12.31}$$

Figure 12.23b shows the *half-bit accuracy* assignment rule. For a 2-bit accuracy A/D process, this rule operates as follows:

$$\text{If} \quad 0 \leq V < \frac{1}{8}V_0 \quad \text{then} \quad \%vv = \%00$$

$$\text{If} \quad \frac{1}{8}V \leq V < \frac{3}{8}V_0 \quad \text{then} \quad \%vv = \%01$$

$$\text{If} \quad \frac{3}{8}V \leq V < \frac{5}{8}V_0 \quad \text{then} \quad \%vv = \%10$$

$$\text{If} \quad \frac{5}{8}V \leq V < V_0 \quad \text{then} \quad \%vv = \%11$$

It is apparent that, around the zero value, the error involved in the half-bit accuracy rule is $\pm 1/2$ LSB. This levels-assignment rule gives better accuracy at low voltage values. However, at high-voltage value, the accuracy is diminished, since the value %11 would be assigned starting with the value $5/8V_0$, which is lower than the value $3/4V_0$ of the previous rule. For an 8-bit A/D process the same concept would apply, and the error for the conversion of low-voltage values would be

$$err = \pm 1/2 \text{ LSB} = \pm 10 \text{ mV} \tag{12.32}$$

12.7.4 Physical Implementation of A/D Conversion

There are many types of A/D conversion techniques. One of them is the successive approximation technique used by the MC68HC11 microcontroller. Some other types are the *counter*, *integrative*, and *flash* A/D conversion methods. The successive approximation technique uses a bank of capacitors to successively evaluate the bits of the binary number using the charge redistribution principle. The capacitive charge-redistribution technique depends upon capacitance ratios rather than the absolute capacitance values. The MC68HC11 microcontroller has an 8-bit A/D converter. Figure 12.24 shows the bank of capacitors used for 8-bit A/D conversion in the MC68HC11 microcontroller. The relative capacitance of each capacitor is represented by a number. It is noticed that the process is separated into the higher half and lower half 4-bit portions of the 8-bit byte. The separation is performed through the capacitor $C_s = 1.1$, which effectively divides the value of lower-order capacitors to the left of it by 16. This constructive detail simplifies the layout of the weighted capacitor array and assures better matching of the capacitance sizes. The successive approximation process proceeds in three stages:

1. Sample
2. Hold
3. Approximation

The switching from one stage to the next is achieved through logically controlled analog switches shown as double-directional arrows in Figure 12.24.

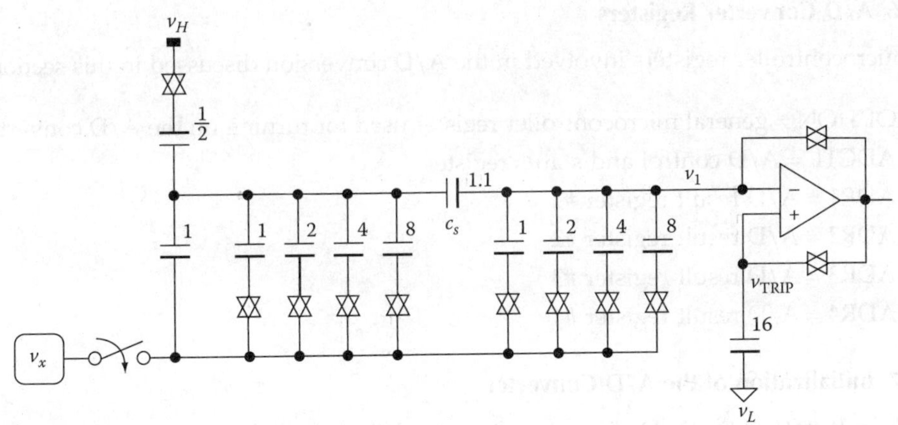

FIGURE 12.24
The bank of capacitors used for 8-bit A/D conversion in the MC68HC11 microcontroller.

12.7.5 A/D Conversion Channels, Port, and Reference Voltages

The A/D conversion can be performed on a single channel or simultaneously on four channels. The A/D conversion is implemented on port E. Since port E is an 8-bit port, two channel-selection options are available:

- Pins PE0, PE1, PE2, and PE3
- Pins PE4, PE5, PE6, and PE7

It should be noted that the physical wiring of the PE0 through PE7 pins is intercalated, as shown in Figure 12.25. The order of these pins is PE0, PE4, PE1, PE5, PE2, PE6, PE3, and PE7. This detail should be taken into consideration when wiring these pins for specific applications.

Also shown in Figure 12.25 are the reference voltage input pins V_H and V_L. By default, these values are assumed as $V_H = 5$ V and $V_L = 0$ V. Unless otherwise specified, these values will only be assumed in most applications.

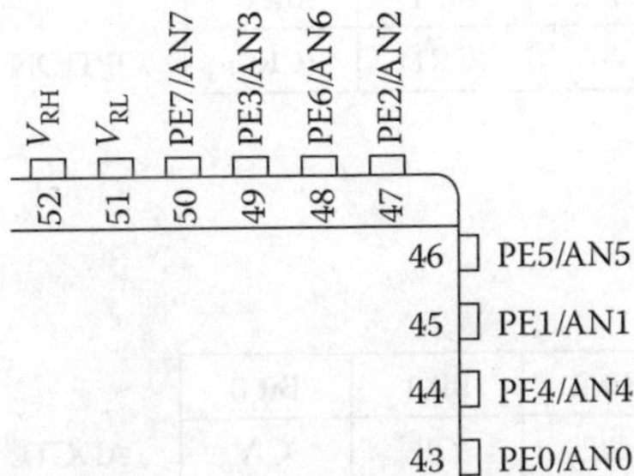

FIGURE 12.25

Physical pins associated with A/D conversion on the MC68HC11 microcontroller: input pins PE0 through PE7 and the reference voltage input pins V_H and V_L.

12.7.6 A/D Converter Registers

The microcontroller registers involved in the A/D conversion discussed in this section are

- OPTION = general microcontroller register used for turning on the A/D converter
- ADCTL = A/D control and status register
- ADR1 = A/D result register #1
- ADR2 = A/D result register #2
- ADR3 = A/D result register #3
- ADR4 = A/D result register #4

12.7.7 Initialization of the A/D Converter

The default state of the A/D converter after reset is *inactive*. To activate the A/D converter on the MC68HC11 microcontroller one has to set the A/D power up control bit, ADPU, to 1 in the OPTION control register at memory location $1039 (Table 12.40). When the ADPU is set to 1, the A/D charge pump and the comparator circuits are activated; after a short delay, they are ready to perform the A/D conversion.

12.7.8 Control and Status of the A/D Converter

The control of the A/D converter and the verification of its status is done with the bits of the ADCTL register at memory location $1030 (Table 12.41).

The *control* of the A/D converter is achieved through the control bits SCAN, MULT, CD, CC, CB, and CA. The *status* of the A/D converter is given by the flag bit CCF.

12.7.9 A/D Conversion Results

The results of the A/D conversions are stored in four separate registers, ADR1, ADR2, ADR3, and ADR4, at memory locations, $1031, $1032, $1033, and $1034, respectively (Table 12.42).

12.7.10 A/D Conversion Modes

The A/D converter can operate in four basic modes. These modes are achieved through combinations of the SCAN and MULT control bits. The four A/D conversion modes are

TABLE 12.40

The A/D Power Up Control Bit, ADPU, in the OPTION Control Register

	Bit 7	Bit 6	Bit 5	Bit 4	Bit 3	Bit 2	Bit 1	Bit 0	
$1039	ADPU	(CSEL)	(IRQE)	(DLY)	(CME)	—	(CR1)	(CR0)	OPTION

Note: Bits not discussed in the text are in parentheses.

TABLE 12.41

The Control and Status Bits of the A/D Converter

	Bit 7	Bit 6	Bit 5	Bit 4	Bit 3	Bit 2	Bit 1	Bit 0	
$1030	CCF	—	SCAN	MULT	CD	CC	CB	CA	ADCTL

TABLE 12.42

Location of the A/D Results Registers ADR1, ADR2, ADR3, and ADR4

$1031	Bit 7	Bit 6	Bit 5	Bit 4	Bit 3	Bit 2	Bit 1	Bit 0	ADR1
$1032	Bit 7	Bit 6	Bit 5	Bit 4	Bit 3	Bit 2	Bit 1	Bit 0	ADR2
$1033	Bit 7	Bit 6	Bit 5	Bit 4	Bit 3	Bit 2	Bit 1	Bit 0	ADR3
$1034	Bit 7	Bit 6	Bit 5	Bit 4	Bit 3	Bit 2	Bit 1	Bit 0	ADR4

- Single channel, single conversion (MULT = 0, SCAN = 0)
- Single channel, continuous conversion (MULT = 0, SCAN = 1)
- Multiple channel, single conversion (MULT = 1, Scan = 0)
- Multiple channel, continuous conversion (MULT = 1, SCAN = 1)

The bit MULT controls the number of channels associated with the A/D conversion process. When MULT = 0, only one channel is used for A/D conversion. In this case, the values recorded in ADR1, ADR2, ADR3, and ADR4 represent the result of four consecutive conversions being performed on the selected channel. The channel selection is done using the CD, CC, CB, and CA bits according to Table 12.43.

TABLE 12.43

A/D Channel Assignment Using the Bits CD, CC, CB, and CA for Single-Channel and Multi-Channel A/D Conversion

	MULT	CD	CC	CB	CA	Channel	ADR1	ADR2	ADR3	ADR4
Single-channel	0	0	0	0	0	PE0	X	X	X	X
(MULT = 0)	0	0	0	0	1	PE1	X	X	X	X
	0	0	0	1	0	PE2	X	X	X	X
	0	0	0	1	1	PE3	X	X	X	X
	0	0	1	0	0	PE4	X	X	X	X
	0	0	1	0	1	PE5	X	X	X	X
	0	0	1	1	0	PE6	X	X	X	X
	0	0	1	1	1	PE7	X	X	X	X
Multi-channel	1	0	0	NA	NA	PE0	X			
(MULT = 1)	1	0	0	NA	NA	PE1		X		
	1	0	0	NA	NA	PE2			X	
	1	0	0	NA	NA	PE3				X
	1	0	1	NA	NA	PE4	X			
	1	0	1	NA	NA	PE5		X		
	1	0	1	NA	NA	PE6			X	
	1	0	1	NA	NA	PE7				X

Note: NA indicates bits that are not active for this operating mode

When MULT = 1, four channels are simultaneously used for A/D conversion. The four channels can be either the lower or the upper group of Port E pins. The selection is done with the CD and CC bits (The CB and CA bits have no effect in this case). The combination CD = 0 and CC = 0 results in the selection of channels PE0, PE1, PE2, and PE4; whereas the combination CD = 0 and CC = 1 results in the selection of channels PE4, PE5, PE6, and PE7. The result of the conversion appears in the A/D result registers ADR1, ADR2, ADR3, and ADR4, in the order shown in Table 12.43.

When SCAN = 0, four A/D conversions are made, the ADR1, ADR2, ADR3, and ADR4 registers are filled, and the process stops. To signal the end of the A/D conversion process, the *conversion complete flag*, CCF, is set to 1. To restart the A/D conversion process, the program sequence has to start from the beginning, including the setting of the control bits in the ADCTL register. This will also clear the CCF flag. (Every time the ADCTL register is written, the flag bit CCF is automatically reset to 0).

When SCAN = 1, the A/D conversion is done continuously on the selected A/D channels, in a round-robin fashion. In this process, the result registers ADR1, ADR2, ADR3, and ADR4 are continuously updated. (The fact that the CCF flag gets set does not affect this process.)

12.7.11 A/D Conversion Programs

A program for performing A/D conversion must have four important segments:

1. Initialization of the A/D converter by setting the bit ADPU in the OPTION register to 1.

2. Selection of the appropriate A/D conversion mode and channels by setting the SCAN, MULT, CD, CC, CB, and CA bits in ADCTL. This will automatically clear the CCF flag bit.

3. Monitoring of the CCF flag which signals when the A/D conversion cycle has ended and the converted data is ready to be retrieved.

4. Retrieval of converted data from the A/D result registers ADR1, ADR2, ADR3, and ADR4.

To initialize the A/D converter, write the value %10000000 to the OPTION register. Then, write to ADCTL the appropriate value to configure SCAN, MULT, CD, CC, CB, and CA to the desired A/D conversion mode. This action automatically clears the CCF (bit 7 in ADCTL). The CCF will be set when the A/D conversion is complete and the four A/D result registers have been updated. (If SCAN = 1, CCF is set after the four A/D result registers are updated for the first time and remains set until a subsequent write to ADCTL.) Usual polling sequences can be used to monitor CCF.

It takes 128 machine cycles to complete the A/D conversion cycle consisting in updating four 8-bit result registers. This corresponds to a sampling time $\Delta t_s = 128 \times 0.5 \ \mu s/4 = 16 \ \mu s$. The corresponding sampling rate is $f_s = 1/(16 \ \mu s) = 62.5$ kHz. For small chip microcontroller, this is an impressive data acquisition rate.

12.7.11.1 *Program for Performing a Single A/D Conversion on a Single Channel*

In this section, we present a program for performing a single A/D conversion on a single channel. The channel selected in this example will be pin PE1. The program will do the following:

1. Initialize the A/D conversion
2. Set ADCTL to reflect the following control bits:
    a. CD = 0, CC = 0, CB = 0, and CA = 1 to select pin PE1
    b. SCAN = 0, i.e., no scanning
    c. MULT = 0, i.e., single channel
3. Check if the A/D conversion has finished. The conversion is finished when the flag CCF is set.
4. Load the results from the AD registers ADR1–ADR4 and store them into memory locations VAL1–VAL4.
5. Loop back to (2)

The program flowchart is shown in Table 12.44. Two flowchart levels are presented: the big-picture and the details. The big-picture is used to understand the overall architecture of the program. The details are used to explain some of the blocks. (Details are given only for those blocks which are somehow new and have not been used in previous programs.) The essential code for this program is shown to the right of the flowchart. The essential code was incorporated into the standard template asm to generate the code file Ex_AD_1.asm. When the program is run, the conversion values shown in Table 12.45 are obtained.

TABLE 12.44

Program AD_1

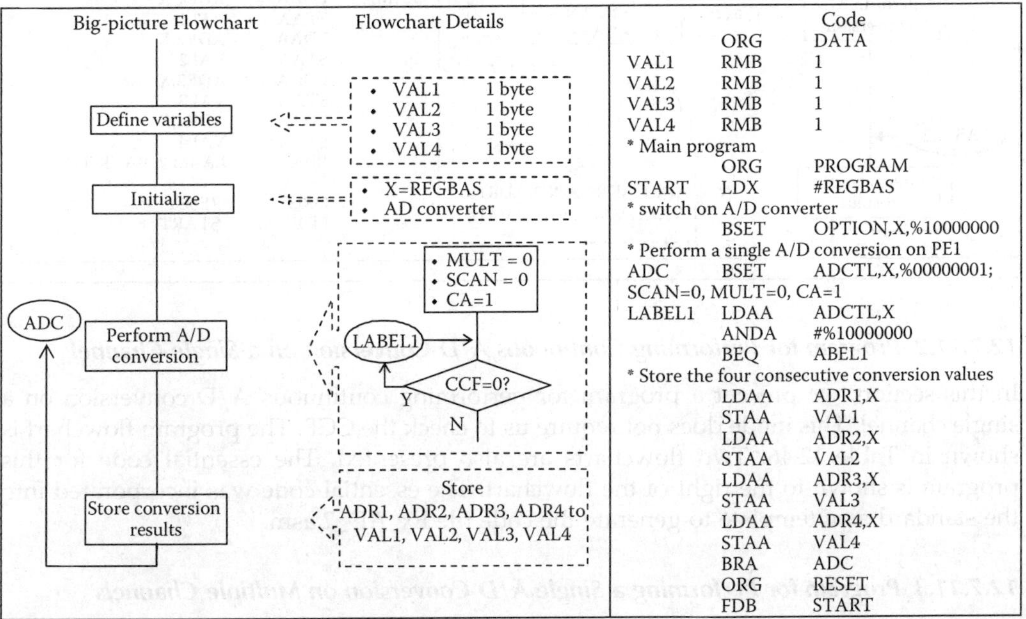


```
                                            Code
                         ORG        DATA
           VAL1          RMB        1
           VAL2          RMB        1
           VAL3          RMB        1
           VAL4          RMB        1
         * Main program
                         ORG        PROGRAM
           START         LDX        #REGBAS
         * switch on A/D converter
                         BSET       OPTION,X,%10000000
         * Perform a single A/D conversion on PE1
           ADC           BSET       ADCTL,X,%00000001;
           SCAN=0, MULT=0, CA=1
           LABEL1        LDAA       ADCTL,X
                         ANDA       #%10000000
                         BEQ        ABEL1
         * Store the four consecutive conversion values
                         LDAA       ADR1,X
                         STAA       VAL1
                         LDAA       ADR2,X
                         STAA       VAL2
                         LDAA       ADR3,X
                         STAA       VAL3
                         LDAA       ADR4,X
                         STAA       VAL4
                         BRA        ADC
                         ORG        RESET
                         FDB        START
```

TABLE 12.45

Conversion Values Obtained with the Microcontroller
A/D Converter on One Channel

Case	Voltage at Pin PE1 (mV)	A/D Converted Values (Hex)
1	1000	$33
2	1500	$4c
3	2000	$66
4	2500	$80
5	3500	$b3
6	5000	$ff

TABLE 12.46

Program AD_2

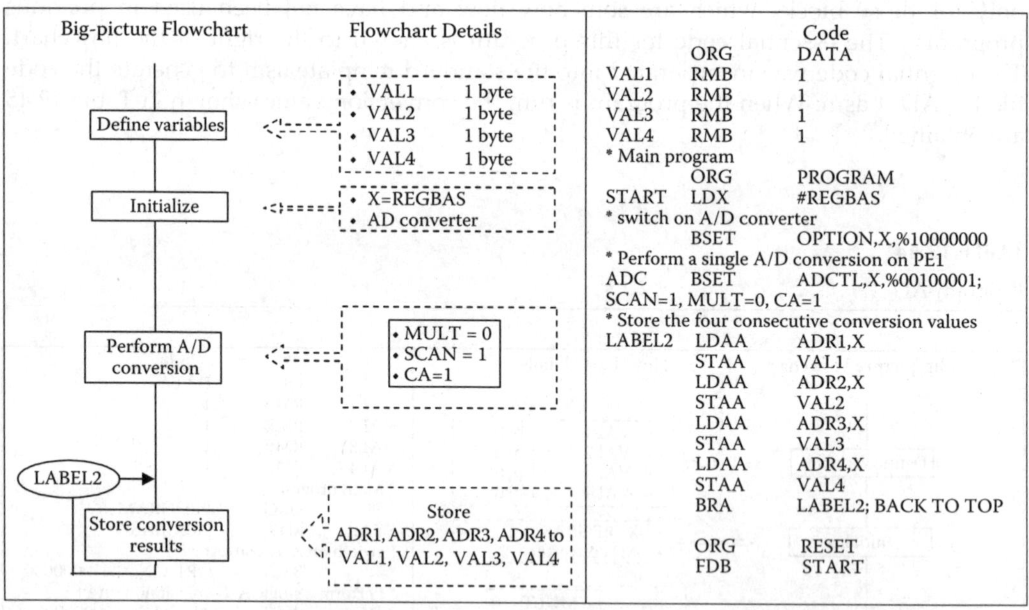


```
                          Code
          ORG     DATA
VAL1      RMB     1
VAL2      RMB     1
VAL3      RMB     1
VAL4      RMB     1
* Main program
          ORG     PROGRAM
START     LDX     #REGBAS
* switch on A/D converter
          BSET    OPTION,X,%10000000
* Perform a single A/D conversion on PE1
ADC       BSET    ADCTL,X,%00100001;
SCAN=1, MULT=0, CA=1
* Store the four consecutive conversion values
LABEL2    LDAA    ADR1,X
          STAA    VAL1
          LDAA    ADR2,X
          STAA    VAL2
          LDAA    ADR3,X
          STAA    VAL3
          LDAA    ADR4,X
          STAA    VAL4
          BRA     LABEL2; BACK TO TOP

          ORG     RESET
          FDB     START
```

12.7.11.2 Program for Performing Continuous A/D Conversion on a Single Channel

In this section, we present a program for performing continuous A/D conversion on a single channel. This mode does not require us to check the CCF. The program flowchart is shown in Table 12.46. Two flowcharts are also presented. The essential code for this program is shown to the right of the flowchart. The essential code was incorporated into the standard asm template to generate the code file Ex_AD_2.asm.

12.7.11.3 Program for Performing a Single A/D Conversion on Multiple Channels

In this section, we present a program for performing a single A/D conversion on multiple channels. The program flowchart is shown in Table 12.47. Two flowchart levels are

TABLE 12.47

Program AD_3

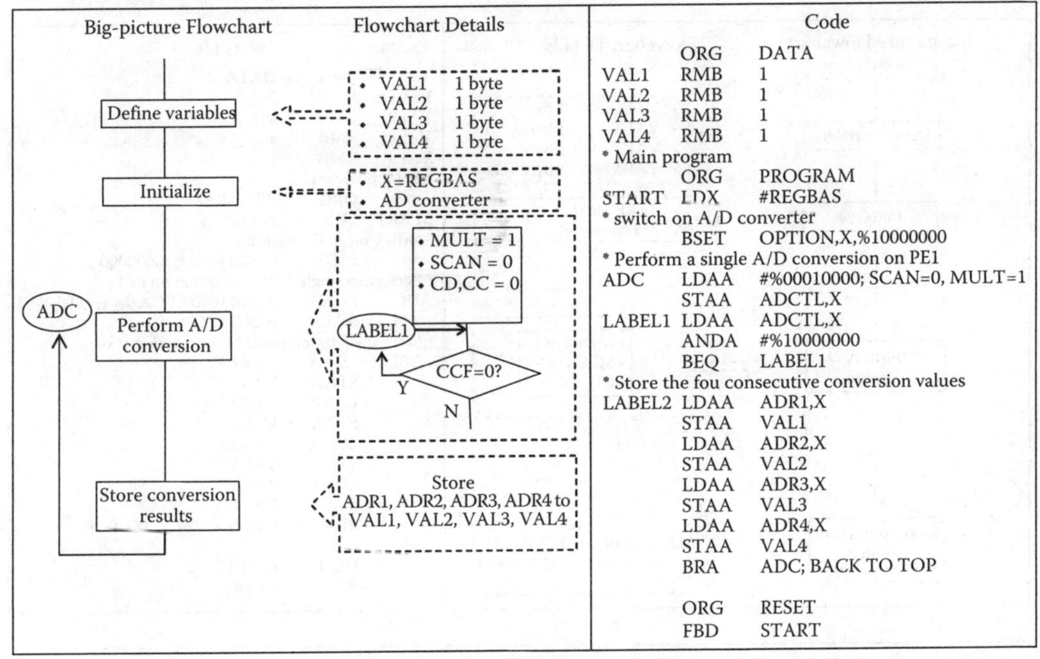

TABLE 12.48

Conversion Values Obtained with the Microcontroller
A/D Converter on Multiple Channels

Channel	Voltage (mV)	A/D Converted Values (Hex)
PE0	2500	$80
PE1	1000	$33
PE2	3500	$B3
PE3	2000	$66

presented: the big-picture and the details. The essential code for this program is shown to the right of the flowchart. The essential code was incorporated into the standard template asm to generate the code file Ex_AD_3.asm.

To test the program, enter 2500 mV in PE0, 1000 mV in PE1, 3500 mV in PE2, and 2000 mV in PE3. Run the program. The converted value $80, $33, $b3, and $66 appear in ADR1, ADR2, ADR3, and ADR4. Then, they are stored in the memory locations VAL1, VAL2, VAL3, and VAL4. The results are shown in Table 12.48.

12.7.11.4 Program for Performing Continuous A/D Conversion on Multiple Channels

We will now perform AD conversion on multiple channels, continuous conversion. This mode is selected by setting MULT = 1 and SCAN = 1 such that the conversion is performed continuously. The program flowchart is shown in Table 12.49. Two flowchart levels are

TABLE 12.49

Program AD_4

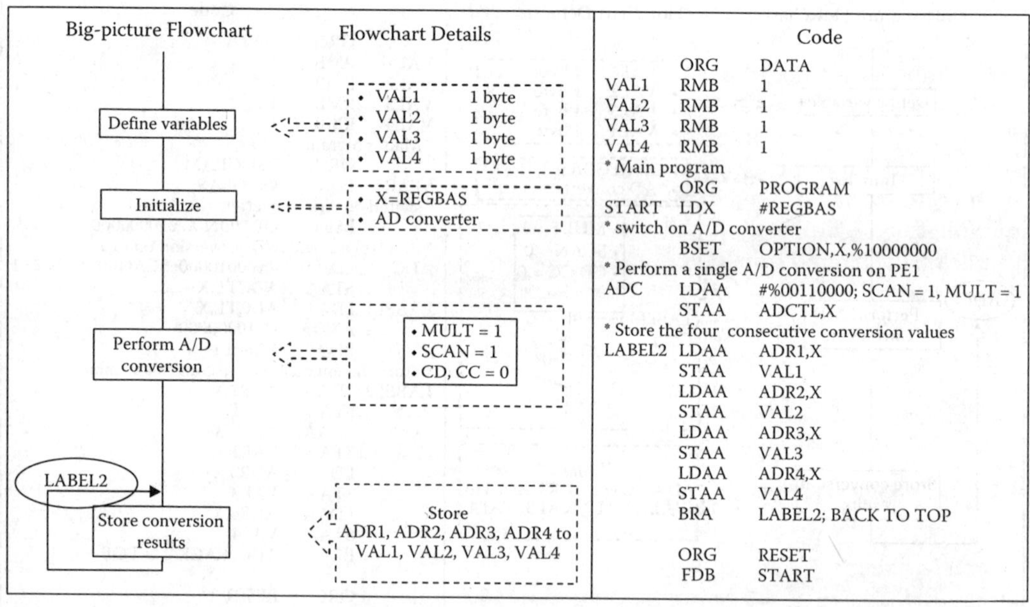


```
              ORG      DATA
VAL1     RMB      1
VAL2     RMB      1
VAL3     RMB      1
VAL4     RMB      1
* Main program
              ORG      PROGRAM
START    LDX      #REGBAS
* switch on A/D converter
              BSET      OPTION,X,%10000000
* Perform a single A/D conversion on PE1
ADC      LDAA     #%00110000; SCAN = 1, MULT = 1
              STAA     ADCTL,X
* Store the four  consecutive conversion values
LABEL2 LDAA     ADR1,X
              STAA     VAL1
              LDAA     ADR2,X
              STAA     VAL2
              LDAA     ADR3,X
              STAA     VAL3
              LDAA     ADR4,X
              STAA     VAL4
              BRA       LABEL2; BACK TO TOP

              ORG      RESET
              FDB      START
```

presented: the big-picture and the details. The essential code for this program is shown to the right of the flowchart. The essential code was incorporated into the standard template asm to generate the code file Ex_AD_4.asm.

12.8 Functional Modules

12.8.1 Functional Modules Overview

An essential part of the mechatronics education project is the construction of functional modules for teaching hands-on skills related to the interfacing of mechanical, electrical, and electronic components of a mechatronics system. Non-EE engineering students need hands-on experience to increase their ability and confidence in tackling electrical and electronics concepts, especially during the realization phase of a mechatronics project. To address this need, we will present here a suit of basic circuits in the form of functional modules: (1) voltage divider; (2) op-amp signal amplifiers; (3) Darlington transistor power amplifier; (4) MOSFET (field-effect transistor) power amplifiers; and (5) pulse-width modulation power control unit. Other basic circuits are presented in other parts of the book in connection with their direct applications. Optoelectronic sensors (emitter–detector pairs) and temperature sensors are treated in the section dedicated to sensing applications of the microcontroller. A linear power amplifier and a stepper motor drive unit are discussed in the section dedicated to actuation applications of the microcontroller. Also of interest, though not treated in this book, would be AC–DC converters and humidity sensors.

Accompanying each functional module are electrical and component schematics, applicable equations, and a complement of experimental results that can be used for calibration. Thus, the students will know what results to expect when using the functional module. These functional modules are intended as bolt-on building blocks with clearly defined inputs and outputs and an explanation of the underlying operational principles. To achieve this, the functional modules are housed in transparent casings, which allow the students to see the actual electric/electronic components of the circuitry, and to compare this image with the intricacy of the circuit diagram. The students are expected to use the functional modules as a learning tool. After understanding their functionality, they are expected to duplicate the circuitry on their own breadboards to be incorporated into their mechatronics class projects, as well as into other hands-on projects.

Our approach will help expand the student's understanding of hands-on mechatronics concepts. Though developed in the Department of Mechanical Engineering, these functional module concepts could be shared with other engineering departments in order to provide a useful teaching aid.

12.8.2 Voltage Divider

Voltage division allows voltage to be varied according to the values of the two resistors in series. The change of an input voltage is important in electronics and mechatronics because different loads (e.g., DC motors, operational-amplifier, and transistors) require certain voltages. Exceeding the maximum voltage of a load can cause damage beyond repair. Therefore, voltage division is useful in order to control the voltage that is supplied to a particular load.

A simple no-load circuit (Figure 12.26a) can be used to demonstrate the principles of voltage division. In a no-load circuit, the current through all components remains constant. According to Kirchhoff's voltage law (KVL), the sum of the voltage drops across each resistor is equal to the total voltage drop on the circuit. Additionally, resistors in series have a total resistance equal to the sum of the resistances. Inspection of the voltage divider circuit (Figure 12.26a) yields the theoretical output voltage

$$V_{out} = \frac{R_2}{R_1 + R_2} V_{in} = \frac{1}{1 + R_1/R_2} V_{in} \tag{12.33}$$

where V_{in} is the input voltage. We can observe that the output voltage is the input voltage times the ratio of the second resistor to the sum of the resistors.

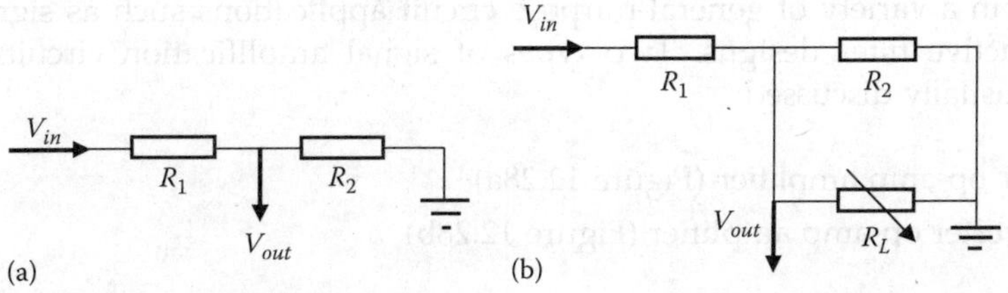

FIGURE 12.26
Voltage divider circuitry: (a) no load; (b) with load.

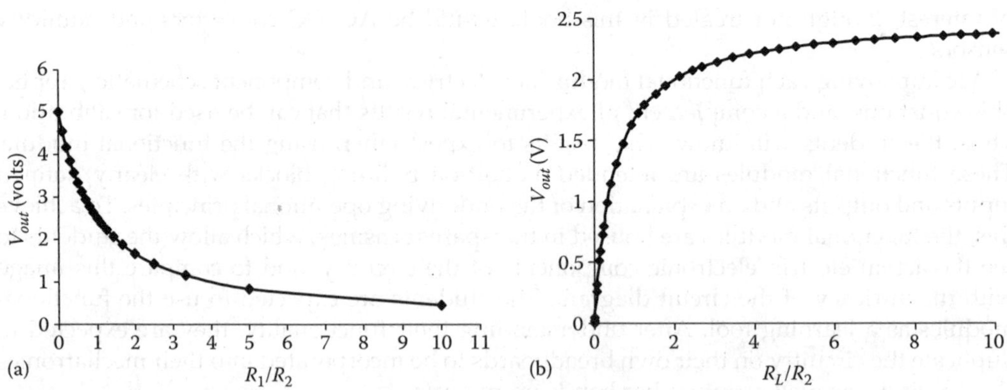

FIGURE 12.27
Voltage divider tests ($V_{in} = 5$ V): (a) no load and variable R_1/R_2 ratio; (b) with load and fixed $R_1 = R_2 = 25$ kΩ.

When a load is applied in parallel with the second resistor (Figure 12.26b), the current through the circuit is affected. The output voltage is found to also depend on the load resistor, R_L, i.e.,

$$V_{out} = \frac{R_2 R_L}{R_1 R_2 + R_1 R_L + R_2 R_L} V_{in} = \frac{R_L/R_1}{1 + R_L/R_2 + R_L/R_1} V_{in} \tag{12.34}$$

Experimental results using an input voltage of 5 V ($V_{in} = 5$ V) are presented in Figure 12.27. For zero load ($R_L = 0$), the output voltage decreases uniformly with an increase in the R_1/R_2 ratio as shown in Figure 12.27a. For example, if $R_1 = R_2 = 25$ kΩ, then the output voltage is half the input voltage, i.e., $V_{out} = 2.5$ V. With load R_L and fixed R_1/R_2 ratio (Figure 12.27b) the output voltage depends on how much current is drawn by the load. This plot confirms the results expected from Equation (12.34). One notices that for high load ($R_L \rightarrow 0$), the output voltage vanishes ($V_{out} \rightarrow 0$). As the load decreases ($R_L \rightarrow \infty$), the current drawn by R_L diminishes and the output voltage approaches the theoretical no-load value $V_{out} = \dfrac{1}{2} V_{in} = 2.5\,\text{V}$.

12.8.3 Op-Amp Signal Amplifier

An operational amplifier is a very-high-gain integrated circuit. The operational amplifier is almost the most useful single device in analog electronic circuitry. With only a handful of external components, it can be made to perform a variety of processing tasks. It is also quite affordable—most general-purpose amplifiers selling for under a dollar apiece. Op-amp is widely used in a variety of general-purpose circuit applications such as signal measurements and active filter designs. Two types of signal amplification circuits employing op-amp are usually discussed:

1. Inverter op-amp amplifier (Figure 12.28a)
2. Noninverter op-amp amplifier (Figure 12.28b)

Both circuit types amplify the input signal, as their names suggest, but the output polarity of the inverting op-amp is the inverse of the input signal polarity. A combined two-source configuration that covers both cases is shown in Figure 12.28c.

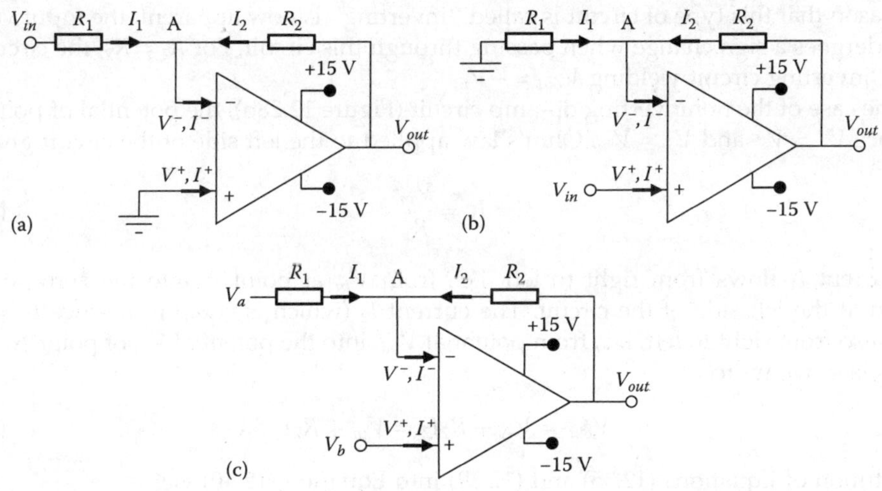

FIGURE 12.28
Op-amp circuits: (a) inverting op-amp circuit; (b) noninverting op-amp circuit; (c) combined two-source configuration.

The design of electric circuits involving op-amp is based on two simplifying principles. First, we assume that an infinite gain is achieved within the device; therefore, the differential input voltage is zero ($V^+ = V^-$) and that negligible current is drawn through either input terminals ($I^+ = I^- \approx 0$). Central to the performance of an op-amp circuit is a loop from the output voltage back to the negative input point. This feedback loop helps maintain stability and control gain of the op-amp circuit.

Analysis of the op-amp circuit yields the expected output voltage V_{out} and gain G. The analysis proceeds as follows. The condition of negligible current into the op-amp ($I^+ = I^- \approx 0$) yields

$$I_1 = I_2 = I \tag{12.35}$$

In the case of the inverting op-amp circuit (Figure 12.28a), the potential of point A is zero since $V^+ = V^-$ and $V^+ = 0$ (grounded). Ohm's law applied to the left side of the circuit gives

$$I_1 = \frac{V_{in}}{R_1} \tag{12.36}$$

The current I_1 flows from left to right, i.e., from V_{in} into the zero-potential point A. The current I_2 (which is equal to I_1 since $I_1 = I_2 = I$) flows also from left to right, i.e., from the zero-potential point A into an even lower potential. For this reason, the output potential V_{out} is negative. Using Ohm's law, we write

$$V_{out} = V_A - R_2 I_2 = -R_2 I_2 \tag{12.37}$$

Substitution of Equations (12.35) and (12.36) into Equation (12.37) yields

$$V_{out} = -\left(\frac{R_2}{R_1}\right) V_{in}, \quad G_{inv} = -\frac{R_1}{R_2} \tag{12.38}$$

The reason that this type of circuit is called "inverting" is now apparent: the input voltage V_{in} undergoes a sign change when passing through this circuit. For $R_1 = R_2$, the circuit is a simple inverting circuit yielding $V_{out} = -V_{in}$.

In the case of the noninverting op-amp circuit (Figure 12.28b), the potential of point A is V_{in} since $V^+ = V^-$ and $V^+ = V_{in}$. Ohm's law applied to the left side of the circuit gives

$$I_1 = \frac{V_{in}}{R_1} \tag{12.39}$$

The current I_1 flows from right to left, i.e., from V_{in} at point A into the zero-potential ground at the left side of the circuit. The current I_2 (which is equal to I_1 since $I_1 = I_2 = I$) flows also from right to left, i.e., from potential V_{out} into the potential V_{in} of point A. Using Ohm's law, we write

$$V_{out} = V_A + R_2 I_2 = V_{in} + R_2 I_2 \tag{12.40}$$

Substitution of Equations (12.35) and (12.39) into Equation (12.40) yields

$$V_{out} = \left(1 + \frac{R_2}{R_1}\right) V_{in}, \quad G_{noninv} = 1 + \frac{R_2}{R_1} \tag{12.41}$$

Combining Equations (12.38) and (12.41) we calculate the output voltage of the combined configuration (Figure 12.28c), i.e.,

$$V_{out} = -\frac{R_2}{R_1} V_a + \left(1 + \frac{R_2}{R_1}\right) V_b \tag{12.42}$$

For AC performance, Equations (12.38), (12.41), and (12.42) still apply, only that the resistances R_1 and R_2 are generally replaced by complex impedances $Z_1(\omega)$ and $Z_2(\omega)$, i.e.,

$$V_{out} = -\frac{Z_2(\omega)}{Z_1(\omega)} V_a + \left(1 + \frac{Z_2(\omega)}{Z_1(\omega)}\right) V_b \tag{12.43}$$

The practical realization of these concepts as an op-amp functional module is shown in the Figure 12.29. In this figure, the op-amp circuitry, which is laid out on a breadboard, is encased in the clear housing. Terminal posts are bolted into the plastic sides of the housing, each color coded to an uniform standards (i.e., "black" for ground, "yellow" for signal, "red" for 5 V input, etc.). Each post is connected on the inside of the housing to its corresponding location in the circuit. This allows the user to view and connect directly to the op-amp circuit through the terminal posts. In the circuits of op-amp functional module shown in Figure 12.29, the resistors R_1 and R_2 had the values $R_1 = 9.93$ kΩ and $R_2 = 216$ kΩ; hence, the expected gains were $G_{inv} = -21.75$ and $G_{noninv} = 22.08$, respectively.

The experimental results measured during the testing of this functional module are given in Figure 12.30. It is noticed that, upto a certain value, the output voltage has a linear dependence on the input voltage, as expected. However, above a certain level, one notices that the output voltage levels off, i.e., reaches a saturation voltage. The explanation for this behavior lies in the fact that the op-amp is an active device and requires external power. Therefore, the output voltage cannot exceed the voltage of the external power supply. In fact, the saturation voltage will be less than the voltage supplied due to the

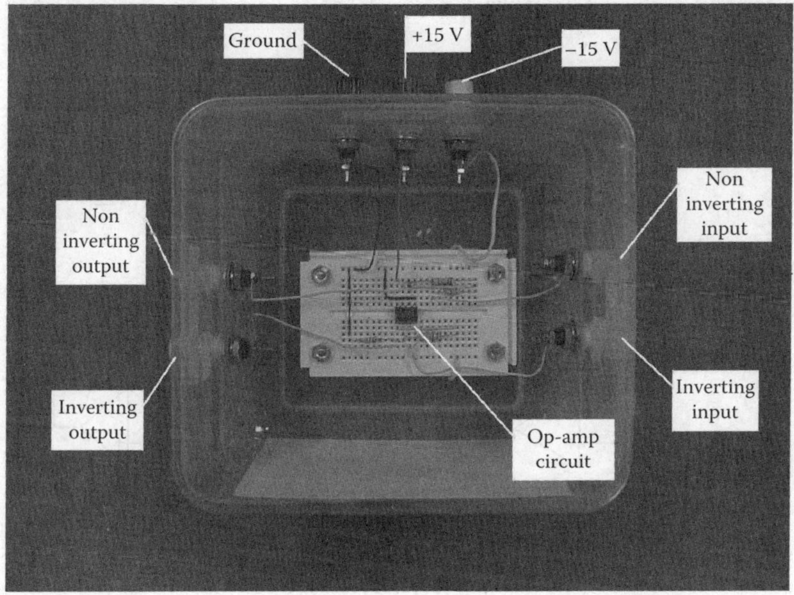

FIGURE 12.29
Op-amp functional module.

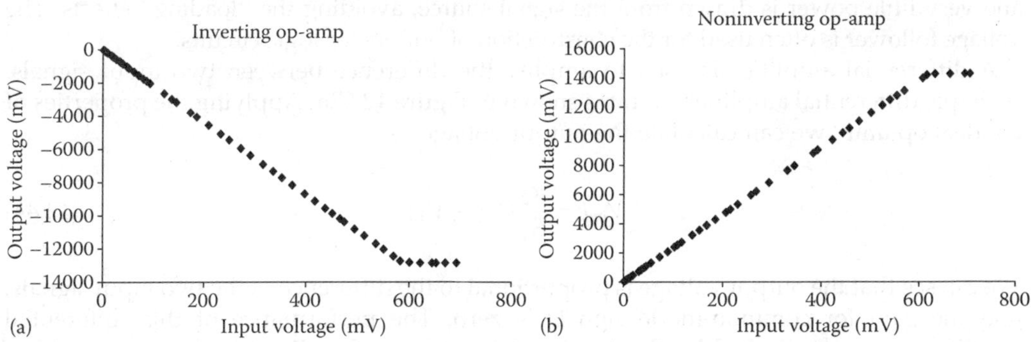

FIGURE 12.30
Tests results for the op-amp circuits ($R_1 = 9.93$ kΩ, $R_2 = 216$ kΩ): (a) inverting op-amp circuit; (b) noninverting op-amp circuit.

power required to run the op-amp, as well as source losses. Thus, the experimental results shown in Figure 12.30 present both the linear gain and the nonlinear saturation voltage.

Besides being used in the construction of inverting and noninverting amplifying circuits, the op-amp can be also used in building voltage followers and differential amplifiers. A typical voltage follower is shown in Figure 12.31. In this configuration the output of op-amp is connected back to the inverting input. As the op-amp has a very high open loop gain, the output voltage would be forced to come very close to the input voltage at noninverting input. The output voltage could be any voltage up to the power supplier voltage. The voltage follower is often used to electrically isolate the input from output, because the input imped-ance of the op-amp is very high, giving effective isolation of the output from the signal source

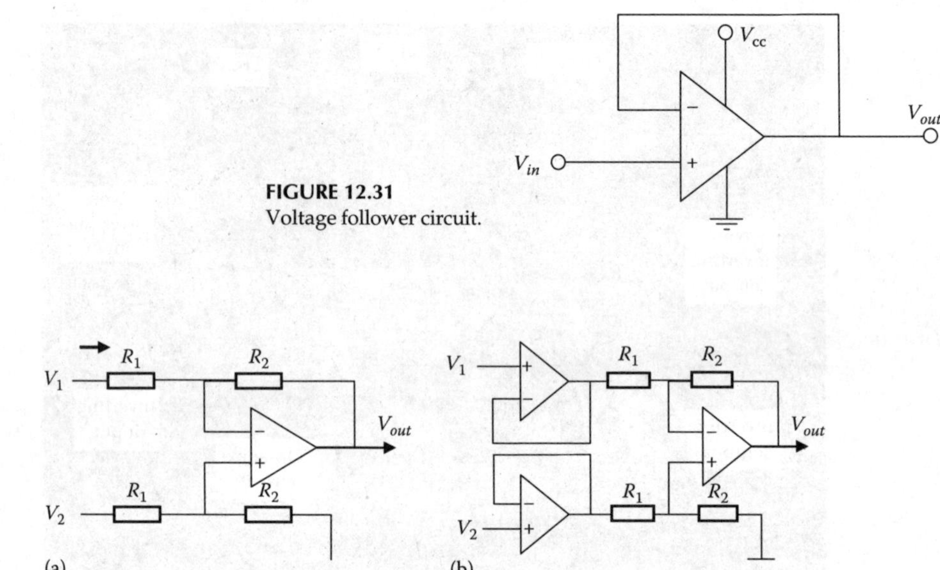

FIGURE 12.31
Voltage follower circuit.

FIGURE 12.32
Differential amplifier circuits: (a) simple differential amplifier; (b) instrumentation differential amplifier with input buffers.

and very little power is drawn from the signal source, avoiding the "loading" effects. The voltage follower is often used for the construction of buffers for logic circuits.

A differential amplifier is used to amplify the difference between two input signals. A simple differential amplifier circuit is shown in Figure 12.32a. Applying the properties of an ideal op-amp, we can calculate the output voltage as

$$V_{out} = \frac{R_2}{R_1}(V_2 - V_1) \tag{12.44}$$

We can see that the output voltage is proportional to the difference of the two input signals, and the gain for common-mode signals is zero. The performance of this differential amplifier is usually limited by the low-input impedance $R_1 + R_2$. For this reason, such a simple differential amplifier could not be directly connected to sensors that usually can only give a rather low current. To compensate for this deficiency, one can use two buffer amplifiers (voltage followers) added to the input. The resulting instrumentation differential amplifier has a very high-input resistance and can be successfully used in instrumentation applications (Figure 12.32b).

12.8.4 Charge Amplifier

The charge amplifier is an op-amp circuit that can be used to amplify electric charges such as those generated by a piezoelectric transducer. The advantage of the charge amplifier is that it prevents charge leakage, a plaguing for static and quasi-static operation of piezoelectric transducers. The charge amplifier principle was patented by W. P. Kistler in 1950 and gained practical significance in the 1960s. Basically, a charge amplifier consists of a high-gain capacitive voltage amplifier that ensures high insulation resistance.

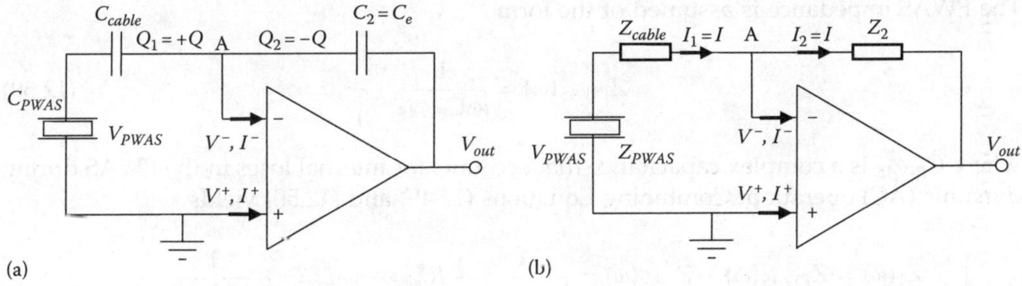

FIGURE 12.33
Schematic model of a charge amplifier consisting of an op-amp amplifier and passive components: (a) quasi-static operation; (b) dynamic operation.

An op-amp charge amplifier connected to a piezoelectric wafer active sensor (PWAS) is shown schematically in Figure 12.33. The cable effects (parasitic resistance, inductance, and capacitance) are included. To understand how this works, we should remember that the point A is at zero potential ($V^- = 0$), but almost perfectly insulated from the ground ($I^- \approx 0$).

For *static operation*, only capacitance aspects are relevant, i.e., the PWAS and cable capacitances, C_{PWAS} and C_{cable}, on the left of point A and the measuring capacitance, C_e, on the right of point A. The charge Q generated by the piezoelectric effect on the left of point A is mirrored with opposite sign on the right of point A, i.e.,

$$Q = C_1 V_{PWAS} = -C_2 V_{out} \tag{12.45}$$

Solution of Equation (12.45) yields

$$V_{out} = -\frac{C_1}{C_2} V_{PWAS} \tag{12.46}$$

But

$$C_1 = \left[C_{PWAS}^{-1} + C_{cable}^{-1} \right]^{-1}, \quad C_2 = C_e \tag{12.47}$$

Substitution of Equation (12.47) into Equation (12.46) yields the output voltage under quasi-static operation as

$$V_{out} = -\frac{\left[C_{PWAS}^{-1} + C_{cable}^{-1} \right]^{-1}}{C_e} V_{PWAS} \tag{12.48}$$

Equation (12.48) indicates that the output voltage is higher when the measuring capacitance is very small in comparison to the PWAS capacitance, provided the cable capacitance is kept in check.

For *dynamic operation*, the impedance of the cable and measuring capacitance must be taken into account. Assume that the cable presents resistance, inductance, and capacitance, whereas the measuring capacitance has only capacitance and internal resistance, i.e.,

$$Z_{cable}(\omega) = R_{cable} + i\omega L_{cable} + \frac{1}{i\omega C_{cable}}, \quad Z_e(\omega) = R_e + \frac{1}{i\omega C_e} \tag{12.49}$$

The PWAS impedance is assumed of the form

$$Z_{PWAS}(\omega) = \frac{1}{i\omega \bar{C}_{PWAS}}$$

(12.50)

where $\bar{C}_{PWAS}$ is a complex capacitance that accounts for internal loses in the PWAS during dynamic (AC) operation. Combining Equations (12.49) and (12.50) yields

$$Z_1(\omega) = Z_{PWAS}(\omega) + Z_{cable}(\omega) = \frac{1}{i\omega \bar{C}_{PWAS}} + R_{cable} + i\omega L_{cable} + \frac{1}{i\omega C_{cable}}$$

$$Z_2(\omega) = R_e + \frac{1}{i\omega C_e}$$

(12.51)

Substituting Equation (12.51) into Equation (12.43) with $V_a = V_{PWAS}$ and $V_b = 0$ yields the output voltage of the charge amplifier under dynamic operation, i.e.,

$$V_{out} = -\frac{Z_2(\omega)}{Z_1(\omega)} = -\frac{R_e + \dfrac{1}{i\omega C_e}}{\dfrac{1}{i\omega \bar{C}_{PWAS}} + R_{cable} + i\omega L_{cable} + \dfrac{1}{i\omega C_{cable}}} V_{in}$$

(12.52)

12.8.5 Darlington Transistor Power Amplifier

Transistors are composed of three parts: the base (B), the collector (C), and the emitter (E). As a power amplifier, the transistor utilizes a base current that controls the collector current. The gain factor, β, is the ratio between the collector current, I_C, and the base current, I_B, i.e.,

$$\beta = \frac{I_C}{I_B}$$

(12.53)

Since the emitter current, I_E, equals to the sum of the collector current and the base current, $I_E = I_C + I_B$, the emitter current can be written as

$$I_E = (\beta + 1)I_B$$

(12.54)

Hence, the transistor amplification is given by

$$G_{transistor} = \frac{I_E}{I_B} = \beta + 1$$

(12.55)

A Darlington transistor consists of two transistors arranged in cascade configuration known as a Darlington-pair. Figure 12.34a shows how the collector–emitter voltage from the first transistor triggers the second transistor. This cascade arrangement is used to increase the overall gain. To predict its behavior, we use Equation (12.54) repeatedly. The emitter current of the first transistor serves as the base current to the second transistor, I_{B2}.

$$I_{B2} = I_{E1}$$

(12.56)

Then, Equation (12.54) yields

$$I_{E1} = I_{C1} + I_{B1} = (\beta + 1)I_{B1}$$
$$I_{E2} = I_{C2} + I_{B2} = (\beta + 1)I_{B2}$$

(12.57)

Substituting and simplifying gives the formula for calculating the emitter current of the Darlington-pair transducer as

$$I_{E2} = (\beta + 1)^2 I_{B1}$$

(12.58)

The final gain shown in Equation (12.58) is $(\beta + 1)^2$.

$$G_{Darlington} = \frac{I_{E2}}{I_{B1}} = (\beta + 1)^2$$

(12.59)

This gain is a quadratic function of the individual gain, β. In practice, the gain, β, varies nonlinearly with the current amplitude. This nonlinearity is further accentuated by the quadratic law of Equation (12.59). In addition, the saturation which appears at high-current values further accentuates the nonlinear behavior of this configuration.

Experimental tests were performed with a functional module containing a TIP 120 Darlington transistor. The collected experimental data presented in Figure 12.34b indicates that, at the beginning, the gain is very small until the transistor reaches a threshold condition. At this point, the emitter current increases drastically until it reaches the saturation current, where it levels back off.

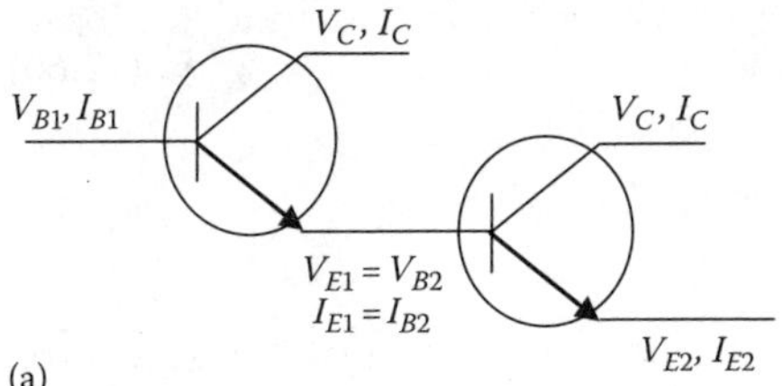

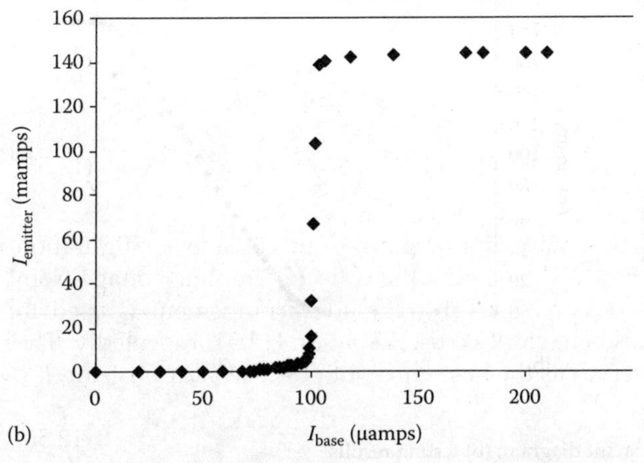

FIGURE 12.34
Darlington-pair transistor: (a) schematic showing how the two transistors are arranged in cascade; (b) testing results.

12.8.6 MOSFET Transistor Power Amplifiers

Metal-oxide-semiconductor field-effect transistors (MOSFET) are important devices that are used as amplifiers and logic switches. Due to their low-power consumption and easy fabrication process, MOSFETs nowadays are widely used in complex digital circuits such as memories and microprocessors. On the other hand, because of the wide selection of available types, BJT transistor discussed in the previous section remains a major device in analog circuit designs, especially those that require large output current and high frequency operations.

The MOSFET is composed of a channel of n-type or p-type semiconductor material and is accordingly called an NMOSFET or a PMOSFET. Electrons carrying charge will flow along these channels. The width of the channel, which determines how well the device conducts, is controlled by a gate electrode G. For the NMOSFET, one n+ region is connected to the drain electrode, D, which has a voltage applied to it. The other n+ region is connected to the source, S, which goes to the ground through a load resistance. Figure 12.35a shows a functional module circuit using this arrangement. When no voltage is applied to the transistor gate, G, there is no transfer of electrons from the drain to the source. This state is the normal condition for the transistor and is known as the cut-off state. When a voltage is applied to the gate, the electrons of the p-type substrate are attracted to the gate surface. These electrons form an n-channel, which allows current to flow from the drain to the source. The voltage that allows the n-channel to form is known as the threshold voltage, V_T. As the gate voltage increases and goes beyond the threshold voltage, the switch is turned on and the size of the n-channel increases, allowing more current to flow. This state is called the linear state, and the MOSFET operates like a transistor. The difference between the gate and threshold voltages is considered as the output of the transistor. This can be measured at the source terminal, i.e.,

$$V_{out} = V_G - V_T = V_s \tag{12.60}$$

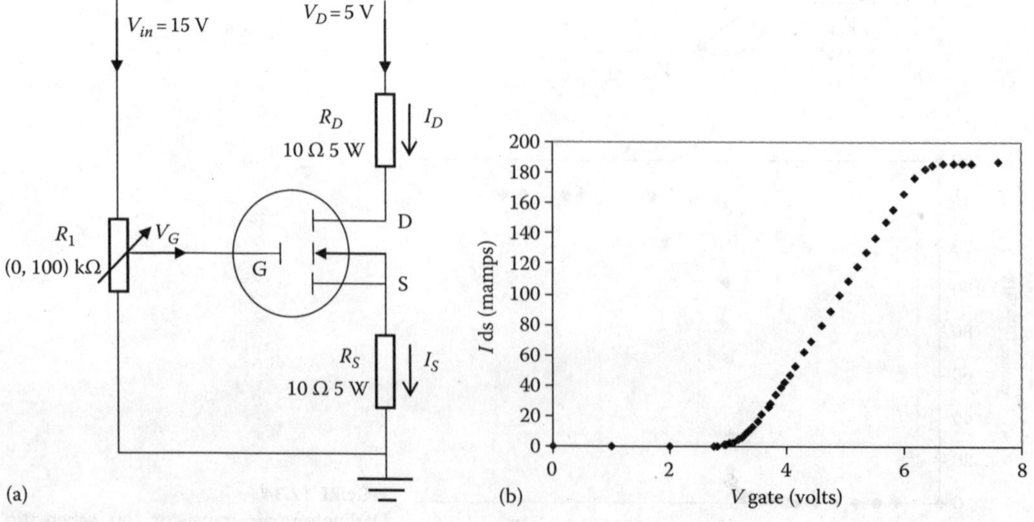

FIGURE 12.35
Field-effect transistor functional module: (a) circuit diagram; (b) testing results.

Assuming a static source resistance, R_s, Ohm's law can be used to find the drain to source current, I_{ds},

$$I_{ds} = (V_G - V_T)/R_s \tag{12.61}$$

However, the channel at the drain does not reach the same depth as the source; as the voltage increases further, it runs the risk of cutting the current flow. This "pinch off" takes place when the gate voltage is twice the threshold voltage, i.e., $V_G \geq 2V_T$. Above this value, the current no longer increases. In digital circuits the transistors are only operated in cut-off and triode mode. Experimental data collected with a MOSFET functional module (Figure 12.35b) confirms the predicted behavior. One notices that in a given range of gate voltage ($2.5\ \text{V} < V_{gate} < 6\ \text{V}$) the MOSFET current varies linearly with the applied voltage. Outside this range, the MOSFET is either off ($V_{gate} < 2.5\ \text{V}$) or on ($6\ \text{V} < V_{gate}$), thus working as a switch.

When comparing the behavior of the MOSFET circuit with that of the Darlington circuit, one notices similarities and differences. A similarity is that both can be used for on–off switching of the output current. However, the difference is that the MOSFET is switched by an applied voltage, V_{gate}, whereas the Darlington is switched by an applied current, I_{B1}.

12.8.7 Pulse-Width Modulation Power Control Unit

The Darlington and the MOSFET transistors act as fast on–off switches for current control. They are well suited for pulse-width modulation (PWM) control, in which the voltage applied to a device is rapidly switched on and off in a square wave fashion (Figure 12.36a). PWM control is used to modify the length of the high duration, t, of the square wave with respect to the total duration, T. The ratio of t to T is called "duty cycle." The effective voltage seen by the device is the applied voltage multiplied by the duty cycle. As the duty

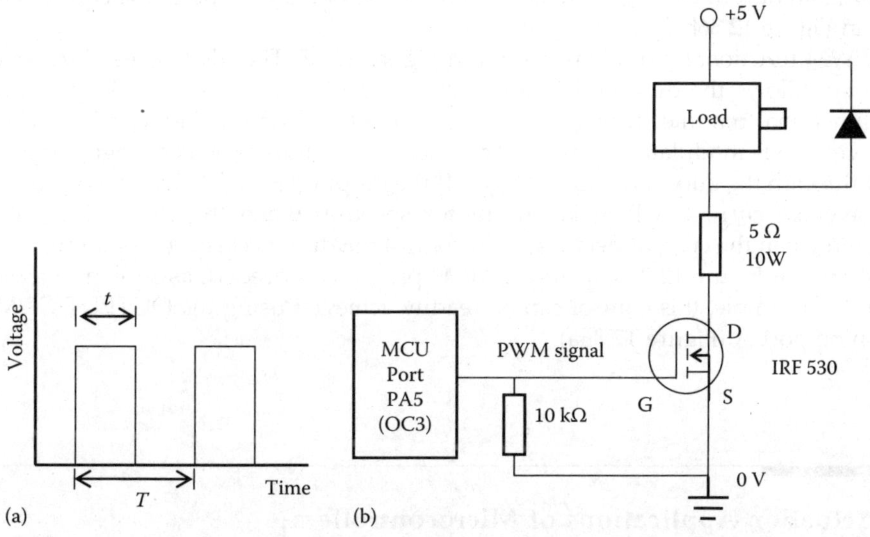

FIGURE 12.36
PWM control: (a) square wave schematic; (b) circuit diagram of the PWM functional module.

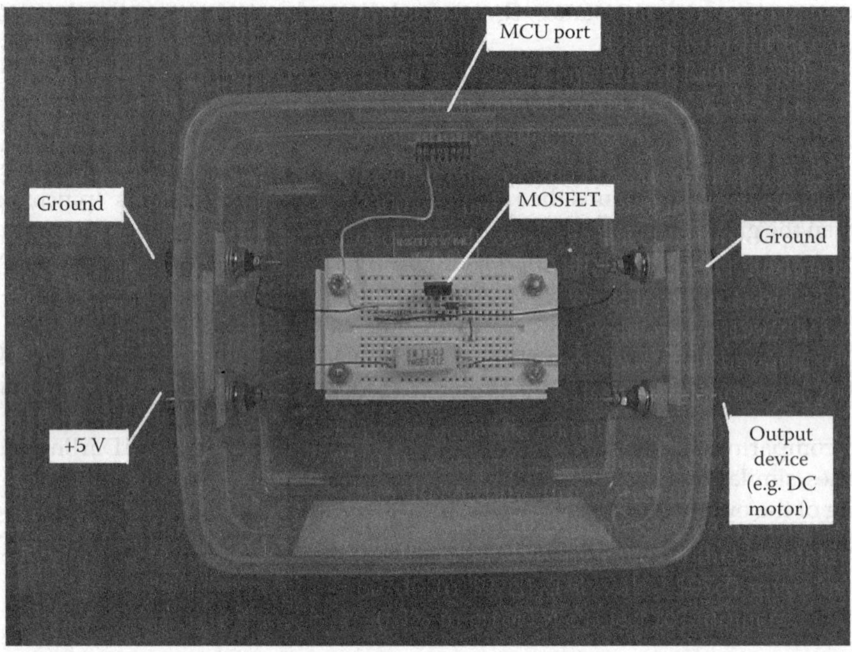

FIGURE 12.37
PWM functional module.

cycle is varied from zero to one, the effective voltage varies from zero to maximum. The PWM signal can be easily generated by a microcontroller. Thus, an effective scheme for controlling an output device, such as a DC motor, is obtained. The PWM control circuit is shown in Figure 12.36b.

The PWM functional module is shown in Figure 12.37. The MCU is used to generate a signal that follows the duty cycle shown in Figure 12.36a. This control signal enters the PWM functional module through the "MCU PORT" shown in Figure 12.37. The +5 V supply voltage is modulated by the duty cycle. As the duty cycle is changed, the average current through the output device changes. If the output device is a DC motor, this change in the average current will make the motor speedup when the current increases and slowdown when the current decreases. To control the duty cycle of the asymmetric square wave shown in Figure 12.36a, a simple MCU program is needed, as shown in a previous section. For example, this control can be readily achieved using the OUTPUT COMPARE function on port A (Figure 12.36a).

12.9 Actuation Applications of Microcontrollers

Embedded motor-control applications are a growing market expected to reach over 7 billion units. Motor control is a significant, but often ignored, segment of embedded applications. Motor control applications span everything from washing machines to fans, hand-held power tools, and automotive window lift to traction control systems. In most of these applications there is a move away from analog motor control to precision digital control of motors. Digital control of motors permits a much more efficient operation of the motor, resulting in longer life, lower power dissipation, and a lower overall system cost.

Both microcontrollers and DSPs are presently used in motor control; however, because of the real-time control algorithms that must be processed, majority of these applications are driven by microcontrollers.

12.9.1 DC Motors

Electric motors can be of many types: DC motors, AC motors, wound, brushless, synchronous, asynchronous, stepper, brushless permanent magnet servo, switched reluctance, etc. A review of various electric motor types can be found at the several specialized Web sites, such as the Applied Motion Products in http://www.applied-motion.com. In this section, we will focus our attention on the DC motor. DC motors can be considered as high-speed, low-torque motors. In many applications, DC motors are used in conjunction with a reducing gearbox that amplifies the torque but reduces the rotation speed.

12.9.1.1 Principles of Operation

Consider a simple DC motor consisting of a wound rotor and a permanent magnet stator (Figure 12.38). A voltage V is applied across the motor terminals. The current I flows through the motor from the positive terminal to the negative terminal. System consisting of brushes and a collector transfers the current from the stationary terminals of the rotor coils. Current flowing through the rotor coils interacts with the magnetic field of the permanent-magnet stator and generates an electromagnetic force, in accordance with the Ampere law. The electromagnetic force turns the rotor and sets the DC motor in action. In order to keep the DC motor rotating in the same direction, the collector switches (commutates) the rotor coils as the rotor turns. The rotation of the rotor makes its coils intersect the magnetic field lines. Thus, an electromotive force (emf) is induced in the coils in accordance with the Faraday law. The induced emf is called counter electromotive force (*back emf*), because it opposes the applied voltage. In Figure 12.38, the back emf is designated by the letter E. Also shown in Figure 12.38 is the internal resistance of the DC motor, R. This represents the resistance of the rotor windings, brushes, collector, etc. The internal resistance is the cause of power loss in the electrical motor. As current flows through the motor, electric energy gets converted into heat through the Joule effect. The resulting power must be dissipated,

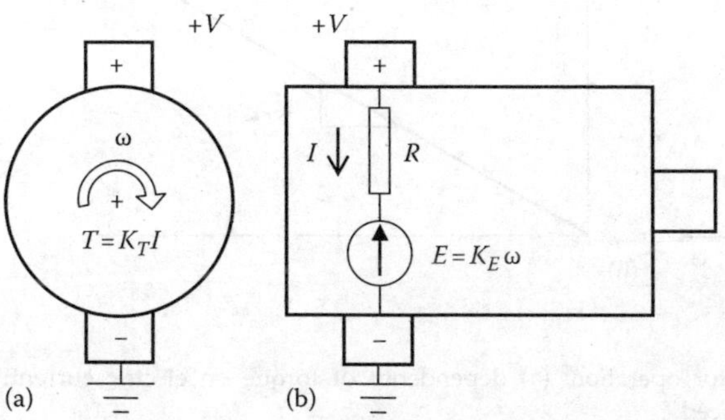

FIGURE 12.38
Schematic representation of an electric motor: (a) end view; (b) side view.

otherwise the DC motor overheats, which may result in burn out of the insulation and hence short-circuit.

12.9.1.2 DC Motor Equations

The torque developed by the DC motor originates in the electromagnetic force applied to rotor windings. Recall that the electromagnetic force on a conductor is proportional with the current flowing through the conductor. Hence, the torque generated by a DC motor is proportional with the current flowing through the rotor windings. The constant of proportionality is K_T, i.e.,

$$T = K_T I \qquad (12.62)$$

A schematic representation of Equation (12.62) is given in Figure 12.39. The amplitude of the back emf is proportional with the angular speed of the rotor. The constant of proportionality is K_E, i.e.,

$$E = K_E \omega \qquad (12.63)$$

The electric model of the DC motor is based on Ohm's law and Kirchhoff's laws, i.e.,

$$V = RI + E \qquad (12.64)$$

It can be shown that, in SI units, the two constants, K_T and K_E are equal and can be represented by just one constant, K, which is called the *motor constant*. The proof of this assertion starts with the principle of power conservation during the electromechanical transduction process. The electrical power input to the DC motor is

$$P_{in} = VI = (RI + E)I = RI^2 + EI \qquad (12.65)$$

Of this input power, a part is lost as heat (RI^2) and the rest is transduced into mechanical power. The electrical power that is transduced into mechanical power is the total input power minus the power dissipation. Using Equation (12.65), we see that the transduced electrical power is

$$P_{electrical} = EI \qquad (12.66)$$

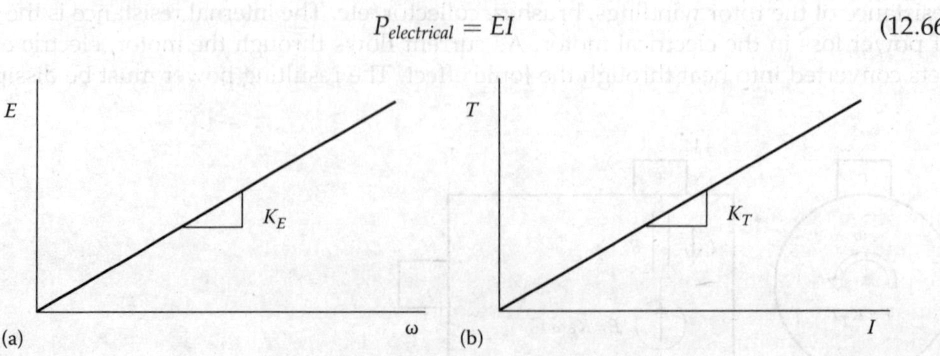

FIGURE 12.39
Schematic of the basic principles of the DC motor operation: (a) dependence of torque on electric current; (b) dependence of back emf on angular rotation speed.

The mechanical power output of the DC motor is

$$P_{mechanical} = T\omega \tag{12.67}$$

Equating the mechanical power with the transduced electrical power, we get

$$T\omega = EI \tag{12.68}$$

Substitution of Equations (12.62) and (12.63) into Equation (12.68) yields

$$K_T I \omega = K_E \omega I \tag{12.69}$$

Simplifying by $I\omega$, gives the relation:

$$K_T = K_E \tag{12.70}$$

Note that the SI unit for K_T are Nm/A, whereas the unit for K_E are V/(rad/s). It can be easily verified that the two are equivalent. Hence, we will use only one constant, K, and rewrite Equations (12.62) and (12.63) as

$$\begin{aligned} T &= KI \\ E &= K\omega \end{aligned} \tag{12.71}$$

Substitution of Equation (12.71) into Equation (12.64) yields the relationship between the applied voltage, the internal resistance, the torque, and the angular speed, i.e.,

$$V = \frac{R}{K}T + K\omega \tag{12.72}$$

Equation (12.72) represents the *electromechanical model* of the DC motor. It can be used to analyze various situations, such as the startup torque, the free-running speed, the torque capability for a given voltage and speed, the speed for a given voltage and torque load, etc. However, before deriving these characteristics, we need to consider another important element of the DC motor operation, i.e., the power dissipation constraint.

12.9.1.3 Power Dissipation Constraint on a DC Motor

The power dissipation capabilities of a DC motor design are essential during the DC motor operation. Since operation of the DC motor produces parasitic heat, adequate means must be provided to have this heat dissipated. Otherwise, the DC motor may overheat, which can produce burnout of the winding insulation and hence its short circuit. Since the heat inside a DC motor is generated in the internal resistance, R, the power that needs to be dissipated is

$$P_{diss} = RI^2 \tag{12.73}$$

Assuming that the maximum power dissipation capability of the cooling system is P_{diss}^{max}, we can calculate the maximum current that can be taken by the DC motor before burnout, i.e.,

$$I_{max} = \sqrt{\frac{P_{diss}^{max}}{R}} \tag{12.74}$$

Equation (12.74) sets an upper limit on all the current-dependent variables that appear in the DC motor analysis. The value I_{max} cannot be exceeded without permanent harm being produced to the DC motor, which may compromise its in-service operation. This value I_{max} will be considered in setting the upper bounds on our subsequent analysis. Associated with the maximum current, I_{max}, is the maximum torque, T_{max}, which can be developed by a DC motor before burn out. This is given by

$$T_{max} = KI_{max} = K\sqrt{\frac{P_{diss}^{max}}{R}} \tag{12.75}$$

12.9.1.4 Torque Characteristic of a DC Motor

Solving Equation (12.72) for torque gives the torque equation of a DC motor, i.e.,

$$T = \frac{K}{R}V - \frac{K^2}{R}\omega \tag{12.76}$$

For a given DC motor of known internal resistance, R, Equation (12.76) shows that the torque is directly proportional to the applied voltage, V. It also shows that the torque decreases linearly with the angular rotation speed, ω.

Operation under constant speed: Assume that we are running the motor at a given speed, and that the torque load increases. In order to hold the speed, we will have to increase the voltage. If the load increases further, the voltage will have to be further increased. This will continue until the maximum torque, T_{max}, given by Equation (12.75) is attained. Further increase of the voltage beyond this point is not recommended, since motor burnout will result.

Operation under constant voltage: Assume that the voltage, V, is constant. Equation (12.76) predicts that the torque, T, decreases as the angular rotation speed increases, ω. This decrease will continue, until a point of zero torque is attained.

12.9.1.5 Startup Torque

Equation (12.76) indicates that, for a given voltage, the torque increases as the angular rotation speed decreases. Hence, we expect that the torque will be greatest when the rotation speed is zero. This is the startup condition. The startup torque is given by

$$T_{startup} = \frac{K}{R}V \tag{12.77}$$

Equation (12.77) expresses one of the great advantages of the DC motor, when compared to other electric motors. The DC motor has a great capacity to startup from a stationary condition. This is especially beneficial when starting one's car on a cold winter morning!

In order to obtain greater startup torques, one can simply increase the voltage applied to the DC motor. However, the voltage should not be increased indefinitely, since the DC motor may burn. The maximum startup torque that a DC motor can sustain is limited by the power dissipation capabilities of the cooling system, i.e., it cannot exceed T_{max} which is given by the Equation (12.75). For this reason, the inequality $T_{startup} < T_{max}$ was included in Equation (12.77). Most DC motors are air cooled. The air is circulated by a fan placed at one end of the rotor. From this viewpoint, the startup condition is the most critical, since the fan does not rotate while the motor is at rest.

12.9.1.6 Speed Characteristic of a DC Motor

Solving Equation (12.72) for angular rotation speed gives the speed equation of a DC motor, i.e.,

$$\omega = \frac{1}{K}V - \frac{R}{K^2}T \tag{12.78}$$

Equation (12.78) shows that the angular rotation speed increases linearly with the applied voltage, V, and decreases linearly with the applied torque, T.

Operation under constant voltage: Assume that the voltage, V, is constant. Equation (12.78) predicts that the rotation speed will decrease as the torque is increased. Further increase of the torque will lead to further reduction of the speed, until the DC motor is stalled. However, the torque cannot be increased beyond the maximum torque, T_{max}, without inflicting damage to the DC motor. Hence, at higher constant voltages, the zero speed condition (motor stall) cannot be actually attained in safe way. Stalling the motor under high voltage input may result in the DC motor overheating and burning out.

Operation under constant torque: Assume that the DC motor is run under constant torque. Equation (12.78) indicates that the speed will increase as the applied voltage, V, increases.

12.9.1.7 Startup Voltage

Then, there will be a minimum voltage, $V_{startup}$, needed to start the DC motor. The value of $V_{startup}$ is obtained from Equation (12.78) by making $\omega = 0$, i.e.,

$$V_{startup} = \frac{R}{K}T \tag{12.79}$$

For higher torque values, a higher startup voltage will be required. This will continue until the maximum torque, T_{max}, given by Equation (12.75) is attained. Further increase of the voltage beyond this point is not recommended, since the motor will burnout.

12.9.1.8 Maximum Speed

Equation (12.78) indicates that the angular rotation speed increases as the torque decreases. For a given voltage, V, the angular rotation speed corresponding to zero torque is the *maximum rotation speed*, ω_{max}. Solving Equation (12.76) for $T = 0$ yields

$$\omega_{max} = \frac{V}{K} \tag{12.80}$$

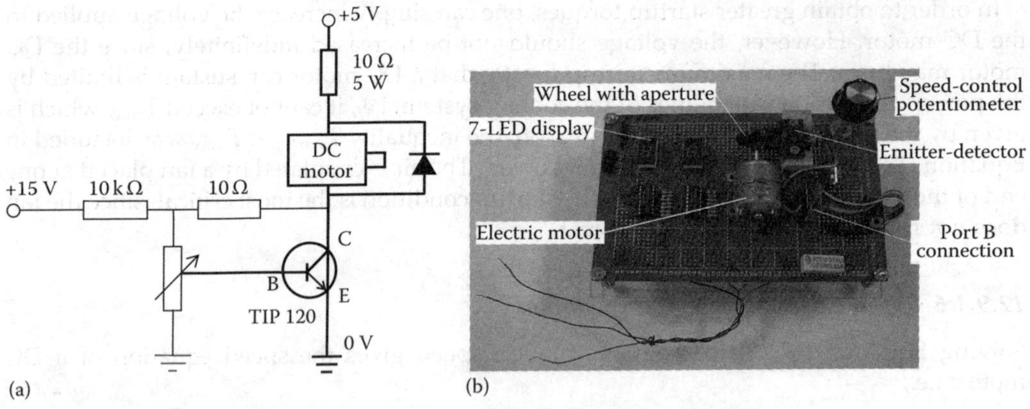

FIGURE 12.40
Linear control of a DC motor: (a) circuit diagram; (b) physical implementation.

In view of Equation (12.63), we can interpret Equation (12.80) to mean that the maximum angular rotation speed is the speed at which the induced emf, E, is balancing the applied voltage, V.

12.9.2 Analog Linear Control of a DC Motor

One of the simplest ways to control a DC motor is through a linear amplifier. Figure 12.40a shows the circuit diagram of a simple linear amplifier with feedback control used in conjunction with a potentiometer. The rotation of the potentiometer modifies its internal resistance and hence the voltage applied to the transistor base. The transistor used in this design is a TIP120 medium power transistor. The TIP120 transistor has a Darlington (two-stage cascade) construction. Hence, it is highly nonlinear. However, some linearity between the potentiometer position and the motor speed is obtained through the electric feedback connection implemented on the circuit board. The physical implementation of the circuit diagram is presented in Figure 12.40b.

12.9.3 Digital Linear Control of a DC Motor

Another way to achieve linear control of a DC motor is through a digital to analog converter. A digital to analog converter (DAC) converts a digital input into an analog output. Figure 12.41a shows a circuit used to achieve the digital linear control of a DC motor using a DAC0808LCN digital to analog converter. The DAC receives an 8-bit digital input from the microcontroller and outputs a current that is proportional to the value of the digital input. The current is converted to a voltage through a simple operational amplifier circuit. The voltage is applied to a second operational amplifier that controls the base of a TIP121 medium power transistor. The TIP121 transistor has a Darlington (two-stage cascade) construction. Hence, it is highly nonlinear. A simple electric feedback circuit is implemented to assure that the voltage at motor terminals is the same as the control voltage applied to one of the operational amplifier input terminals. Figure 12.41b shows the physical implementation of the circuit diagram depicted in Figure 12.41a.

An experimental setup for using the circuit of Figure 12.41a to control a DC motor with a MC68HC11 microcontroller is presented in Figure 12.42.

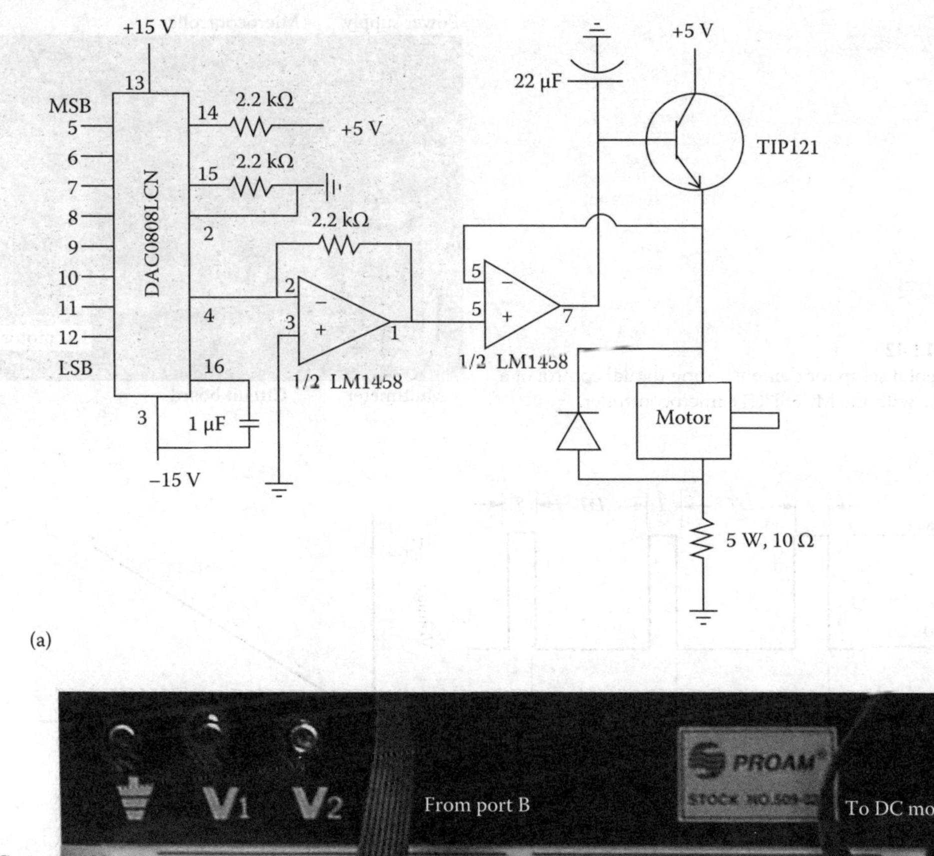

FIGURE 12.41
Digital linear control of a DC motor through an D/A converter: (a) circuit diagram; (b) physical implementation.

12.9.4 Pulse Width Modulation Control of a DC Motor

One of the most commonly used digital methods is pulse-width modulation (PWM). PWM uses a rectangular wave with the cycle period T. In each cycle, the rectangular wave is "on" for a duration t (Figure 12.43a). The on part of the rectangular wave corresponds to

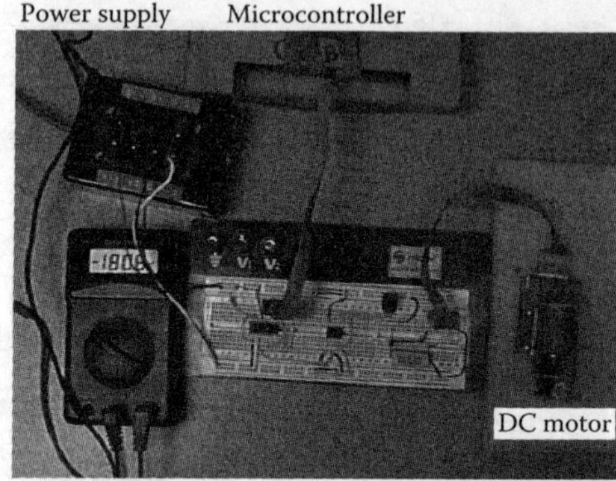

FIGURE 12.42

Experimental setup for demonstrating digital control of a DC motor with the MC68HC11 microcontroller.

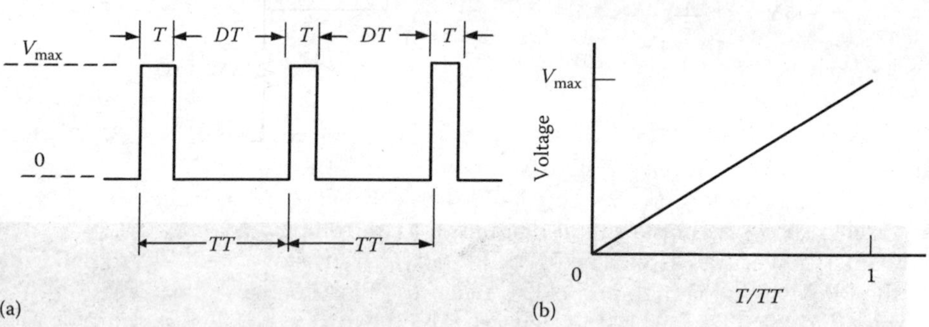

FIGURE 12.43

PWM control principle: (a) duty cycle; (b) equivalent voltage as function of the duty cycle.

the voltage applied to the DC motor. The ratio between the time on, t, and the cycle period, T, is called the *duty cycle*, i.e.,

$$duty\ cycle = \frac{t}{T} \tag{12.81}$$

The duty cycle is a fraction with values less or equal to one. It is usually expressed in percent. The duty cycle expresses the average value of the applied voltage. The average value can be easily calculated by integration over the cycle. If, say, the on-time is 20% of the period, then the duty cycle is 20%, and the average voltage over a period is 20%. This means, the DC motor experiences, on an average, only 20% of the maximum voltage. The relationship between the duty cycle and the effective voltage is plotted in Figure 12.43b.

The physical explanation of the PWM method is as follows. On one hand, the DC motor, due to its inertia, acts as a low pass filter. On the other hand, the PWM signal has high frequency. When applied to a low-pass filter, the high-frequency PWM signal gets rectified, and only the average value is retained. In simple terms, the PWM acts like a series of fast impulses applied to an inertial mass. The net effect is the integration of these impulses. If the frequency of the impulses is sufficiently fast, the ripple effects will be unnoticeable.

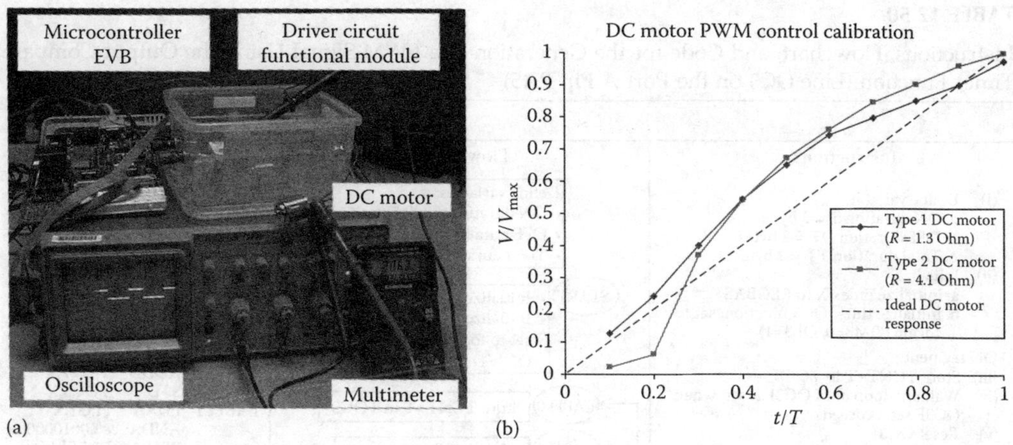

FIGURE 12.44
PWM control of a DC motor: (a) experimental setup; (b) experimental calibration curves for two DC motor types.

PWM excitation is especially effective for precise control, especially at low effective voltages. Since PWM applies a constant voltage, no linear amplification electronics is required. Only an electronic device capable of fast switching high currents is needed. Figure 12.44a shows a simple circuit diagram for controlling a DC motor through an IRF 530 MOSFET power transistor. The MOSFET power transistor acts like a gate: it opens on when the PWM signal is high and closes when the PWM signal is low. The PWM signal that drives the MOSFET transistor is generated by the microcontroller MC68HC11 through the output compare line OC3 on the port A pin PA5. The program instructions, flowchart, and assembly code to produce the PWM signal is reproduced in Table 12.50. Figure 12.44b and c presents the experimental setup for the implementation of the PWM DC motor control principles. The PWM signal for a duty cycle of ~35% is shown on the oscilloscope screen. Figure 12.44d gives the experimental results and the ideal response. Two DC motors were tested. Type 1 had an internal resistance of 1.3 Ω. Type 2 has an internal resistance of 4.1 Ω. In addition, the Type 1 motor was new, while the Type 2 motor was quite worn out. It is observed that the experimental results differ significantly from the ideal linear response. These nonlinear effects must be taken into consideration when implementing an actual design using the PWM method. The nonlinear effects can be compensated using closed-loop process control, as illustrated in Section 12.9.

12.9.5 Stepper Motors

12.9.5.1 Stepper Motor Construction

Stepper motors have coils wound on the stator and a permanent magnet with multiple poles as a rotor. The stepper motors can achieve angular motion in finite steps. Figure 12.45 shows the internal construction of a typical permanent magnet stepper motor. The cutout view of Figure 12.45a shows that the stator consists of two stacked coils. The coils engulf the permanent magnet rotor. A set or permalloy "teeth" are used to turn the magnetic polarity of the coils through 90° such that the rotor sees a sequence of N–S poles around its circumference. The sequential excitation of these teeth makes the permanent magnet rotor turn. Details are given in the next section.

TABLE 12.50

Instructions, Flowchart, and Code for the Generation of a PWM Signal Using the Output Compare Timer Function (Line OC3 on the Port A Pin PA5)

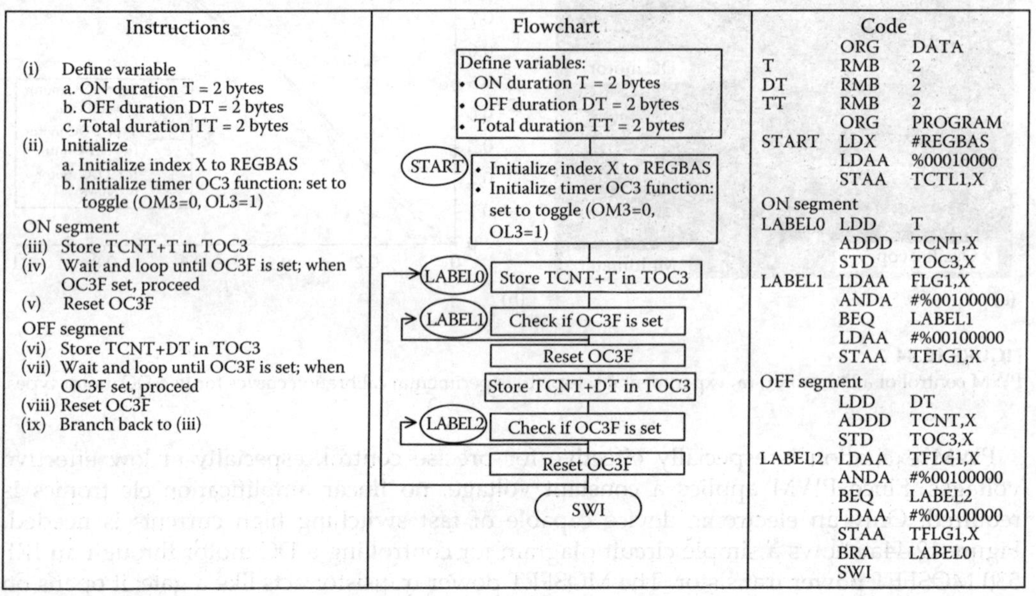


```
            ORG    DATA
T           RMB    2
DT          RMB    2
TT          RMB    2
            ORG    PROGRAM
START       LDX    #REGBAS
            LDAA   %00010000
            STAA   TCTL1,X
ON segment
LABEL0      LDD    T
            ADDD   TCNT,X
            STD    TOC3,X
LABEL1      LDAA   FLG1,X
            ANDA   #%00100000
            BEQ    LABEL1
            LDAA   #%00100000
            STAA   TFLG1,X
OFF segment
            LDD    DT
            ADDD   TCNT,X
            STD    TOC3,X
LABEL2      LDAA   TFLG1,X
            ANDA   #%00100000
            BEQ    LABEL2
            LDAA   #%00100000
            STAA   TFLG1,X
            BRA    LABEL0
            SWI
```

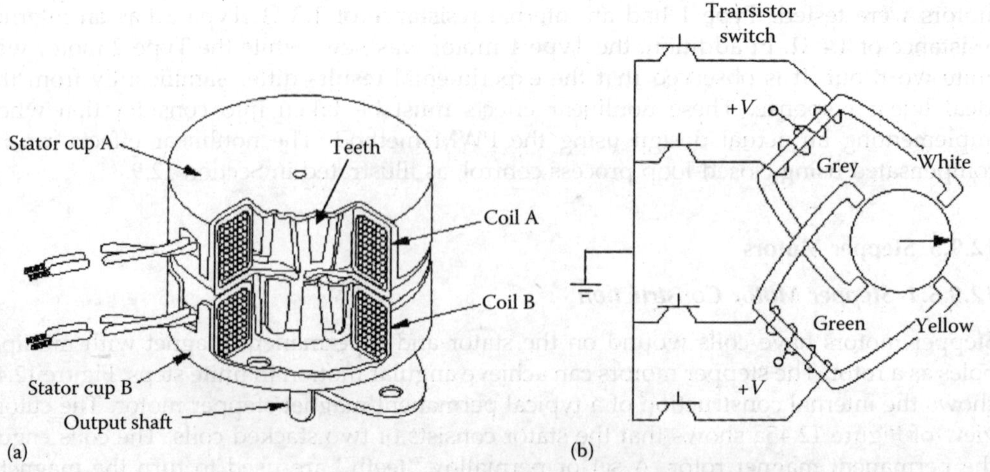

FIGURE 12.45
Stepper motor internal construction: (a) cut out view; (b) wiring diagram of a stepper motor showing the stator coils and the permanent magnet rotor. The coils have bifilar winding and middle taps.

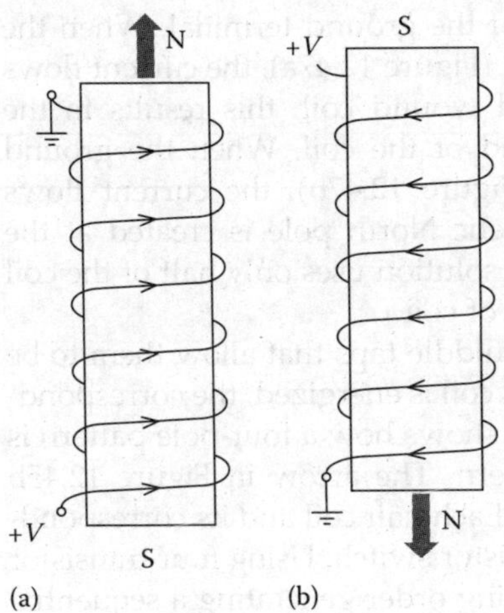

FIGURE 12.46
Bipolar energizing requires the switching of two terminals, the voltage and the ground: (a) north pole upward; (b) north pole downward.

12.9.5.2 Unipolar and Bipolar Energizing of Stepper Motor Coils

Stepper motor coils have a bifilar winding with middle taps, as indicated in Figure 12.45b. The use of middle taps allows the coils to be energized either in unipolar or in bipolar mode. Unipolar and bipolar energizing modes have their advantages and disadvantages, as discussed in the following paragraphs.

Bipolar energizing is illustrated in Figure 12.46. This energizing pattern uses the complete coil and hence results in a higher force density per unit mass of coil. However, the bipolar energizing requires the switching of two terminals, the voltage terminal and the ground terminal. This may be more cumbersome in applications since it requires two separate circuits, one for voltage and another for ground, and it may produce intense arcing due to the need to discharge the magnetic energy stored in the coil.

Unipolar energizing (Figure 12.47) requires the switching of only one terminal. Usually, the switched terminal is the ground terminal, because it is easier to switch the ground than to switch the voltage terminal. (Switching of the ground terminal produces less arcing than switching of the voltage terminal.) Voltage is applied to a central tap of the coil, and

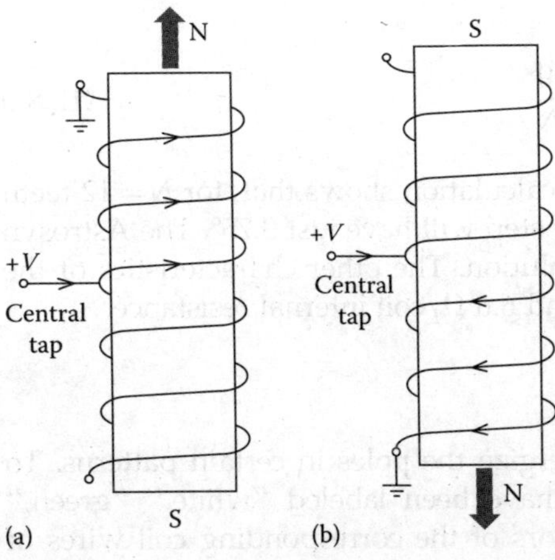

FIGURE 12.47
Unipolar energizing requires the switching of only one terminal: (a) north pole upward; (b) north pole downward.

current direction is changed by changing the location of the ground terminal. When the ground terminal is connected at the upper end of the coil (Figure 12.47a), the current flows upward through the upper half coil. For a right-hand wound coil, this results in the magnetic North pole being generated at the upper end of the coil. When the ground terminal is connected at the lower end of the coil (Figure 12.47b), the current flows downward through the lower half coil, and the magnetic North pole is created at the lower end of the coil. However, this unipolar energizing solution uses only half of the coil and hence results in a lower force density per unit mass of coil.

Figure 12.45 shows that the stepper motor coils have middle taps that allow them to be energized in unipolar mode. In simple terms, when a half coil is energized, the corresponding armature end becomes the North pole. Figure 12.45b shows how a four-pole pattern is created using the two coils arranged in a crossed pattern. The arrow in Figure 12.45b indicates the orientation of the rotor permanent magnet. Each half coil and its corresponding pole can be energized independently through a transistor switch. Using four transistor switches shown in Figure 12.45b one can create a switching order generating a sequential excitation of the four poles to make the permanent magnet rotor turn.

12.9.5.3 One Phase versus Two Phase Excitation

If one pole is energized, then the stepper motor has one phase energized. If two poles are energized, then the stepper motor has two phases energized. If two adjacent phases are energized at the same time, the rotor assumes a midpoint position. Energizing two phases confers a stronger holding force to the stepper motor. For this reason, we will call this situation *full step*. The intermediate situation in which only one phase is excited is called *half step*.

12.9.5.4 Stepper Motor Step Size

The representation in Figure 12.45b assumes a simplified four-pole configuration. In this case, the turning angle between two steps is 90°, while the turning angle between a half step and a full step is 45°. In practice, stepper motor construction uses intercalated teeth to achieve finer stepping resolution. If, say, 12 teeth were assigned to each pole, then a total of 48 intercalated teeth will be obtained. These intercalated teeth are apparent in Figure 12.45a. In this situation, the full circle is divided into 48 segments and the resulting step size is given by the formula

$$Step\ size = \frac{360°}{4N} \tag{12.82}$$

where N is the number of teeth of each pole. This calculation shows that, for $N = 12$ teeth per pole, the full step will have 7.5°, while the half step will have just 3.75°. The Astrosyn 23BB = H051-21 stepper motor has a 7.5°/step resolution. The other characteristics of the Astrosyn 23BB = H051-21 stepper motor are 5 V and 6.6 Ω/coil internal resistance.

12.9.5.5 Energizing Patterns of the Stepper Motor

To make the stepper motor rotate, one has to energize the poles in certain patterns. To facilitate discussion, the poles in Figure 12.45b have been labeled "white," "green," "gray," and "yellow" in accordance with the colors of the corresponding coil wires of the Astrosyn 23BB = H051-21stepper motor. Thus, each color corresponds to a certain half coil. When a color is energized, North (N) is produced on the corresponding pole of the stator.

Table 12.51 shows eight energizing sequences, S1 through S8, of the coils. When a coil is energized, a "1" is placed in the appropriate column of the table. Thus, sequence S1 has a "1" placed in the "white" column, indicating that the "white" pole is energized. Hence, the rotor needle points toward the white pole. For sequence S2, both white and yellow

TABLE 12.51

Stepper Motor Energizing Patterns and Their 2-Hex Equivalent Value

Sequence		Energizing Pattern				8-Bit 2-Hex Code	Phase Type	Angle (rad)	Step Position	Step Type
		White	Green	Grey	Yellow					
S1		1	0	0	0	$08	1-phase	$\frac{\pi}{4}$	$\frac{1}{2}$	Half step
S2		1	0	0	1	$09	2-phase	$\frac{2\pi}{4}$	1	Full step
S3		0	0	0	1	$01	1-phase	$\frac{3\pi}{4}$	$1\frac{1}{2}$	Half step
S4		0	1	0	1	$05	2-phase	$\frac{4\pi}{4}$	2	Full step
S5		0	1	0	0	$04	1-phase	$\frac{5\pi}{4}$	$2\frac{1}{2}$	Half step
S6		0	1	1	0	$06	2-phase	$\frac{6\pi}{4}$	3	Full step
S7		0	0	1	0	$02	1-phase	$\frac{7\pi}{4}$	$3\frac{1}{2}$	Half step
S8		1	0	1	0	$0A	2-phase	0	0	Full step

poles are energized, and the rotor needle points between the white and the yellow poles being equally attracted toward both. It is apparent that the sequence S1 corresponds to one-phase excitation whereas the sequence S2 corresponds to two-phase excitation. Sequence S1 is a *half step* whereas sequence S2 is a *full step*. For this simplified four-pole configuration, each half step turns the rotor through $\pi/4$ radians, whereas each full step turns it through $\pi/2$ radians. The odd sequences S1, S3, S5, and S7 are half-step sequences, whereas the even sequences S2, S4, S6, and S8 are full-step sequences. The sequencing patterns can be sent in coarse fashion (i.e., only the full-steps) or in a fine fashion (both the full and the half steps). Half-step patterns are intercalated between the full-step patterns. The combined use of half steps and full steps gives better resolution. However, the half-step holding force is weaker than the full-step holding force, since only half of the coils are energized at that moment.

12.9.5.6 Inertia Effects in Stepper Motor Excitation: Program Long Delay

As mentioned in 12.9.5.5, the stepper motor can be made to rotate by exciting its poles in certain sequences. At each sequence, the stepper motor is in a quasi-static state by pointing its permanent magnet rotor towards the pole that is currently excited. As the stepper motor is made to rotate by switching excitation between poles, it actually goes from a quasi-static position into another quasi-static position. In other words, the stepper motor "steps" through quasi-static positions having a break-dance like motion. The faster the pole switching is done, the faster is the stepper motor average speed. At low switching speeds, the stepping motion can be easily perceived with the naked eye. As the switching becomes faster, the perception speed of the human eye is exceeded; then, the steps blend into an apparently continuous motion.

However, this process has its limitations. When the pole excitation is switched to the next sequence, the stepper motor has to move from the current quasi-static state to the next quasi-static state. This mechanical motion is accompanied by mechanical inertia, i.e., the acceleration of rotational motion is proportional to the applied electromagnetic force. For a given stepper motor, the electromagnetic force exercised by the poles depends on the rated current which in turn depends on the applied rated voltage. Hence the acceleration of the rotor cannot exceed the ratio between the electromagnetic force moment and the rotor inertia. In other words, if the electric sequences are switched too fast, the stepper motor stalls and no longer moves. This physical fact sets an upper limit on how fast the stepper motor can rotate. This limit is dictated by the motor inertia and electromagnetic forces and not by the electronic switching process. When the electronic switching is programmed in the microcontroller, an upper limit must be imposed on how fast the switching is done; otherwise the stepper motor will stall.

For the Astrosyn 23BB = H051-21stepper motor, this upper speed is limited to 250 rpm. This corresponds to 200 steps/s, i.e., a minimum switching delay of 5 ms. The 250 rpm maximum speed of the stepper motor is rather low when compared to typical DC motor speeds, which range from 2,000 to 10,000 rpm.

During microcontroller programming of stepper motor excitation, a delay of minimum 5 ms must be realized. Such a delay is rather long for microcontroller operation, since it is equivalent to 10,000 internal cycles of a 2 MHz microcontroller. There are many ways in which a long delay can be obtained. Program *Long Delay* described below uses the microcontroller timer functions.

Program *Long Delay* uses a single precision variable, DELAY, to achieve a delay ranging from the minimum of 10,000 ($2710) cycles to the maximum of $ffff (65,535) cycles. Since

TABLE 12.52

Program Long Delay Using the Subroutine DLAY_SR

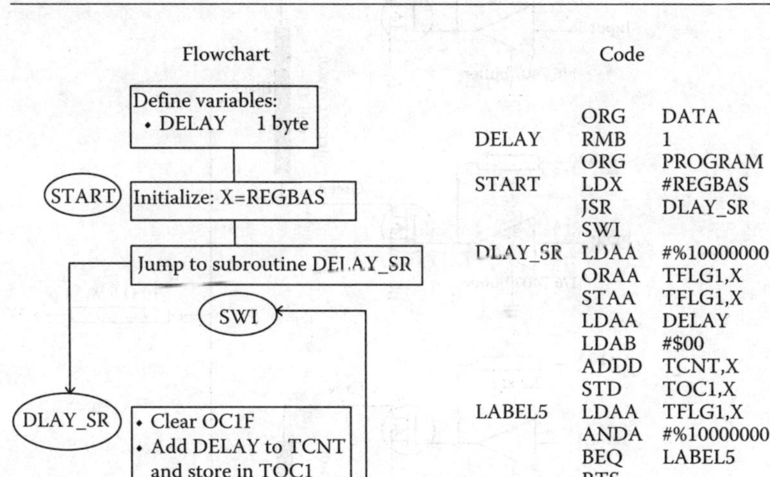

we use a single precision variable, DELAY, we need to round the numbers $2710 and $ffff to two significant hex numbers. Therefore we get $2800 and $ff00. (To maintain delay integrity, the first number was rounded up, the second number was rounded down.) Hence, the single precision variable, DELAY, will take values between $28 and $ff. The flowchart and the code for this program are given in Table 12.52. Please note that the program is just a shell in which the value of DELAY is entered manually. Subsequently, the program calls the subroutine DLAY_SR which does the delay work using the output compare register TOC1. When setting the TOC1 register, which is in double precision, the single precision variable DELAY is loaded in register A whereas the number $00 is loaded in register B. Then double precision addition of the current time TCNT with the content of the double precision register D yields the correct new wake-up time since the register D is the result of registers A and B being concatenated (D $\equiv$ AB).

12.9.5.7 Stepper Motor Drive Electronics

A stepper motor is controlled by sending binary patterns to its driver board. The binary patterns must correspond to the energizing patterns shown in Table 12.51. This is easily achieved with a microcontroller by programming the equivalent 8-bit 2 hex numbers and sending them out through an output port, e.g., parallel port B. Note that the 4-bit patterns corresponding to the four coils were zero filled to the left to generate 8-bit 2 hex numbers compatible with our microcontroller. Eight distinct binary patterns are recognized by the stepper motor driver board. They are the sequences S1–S8 given in Table 12.51. The stepper

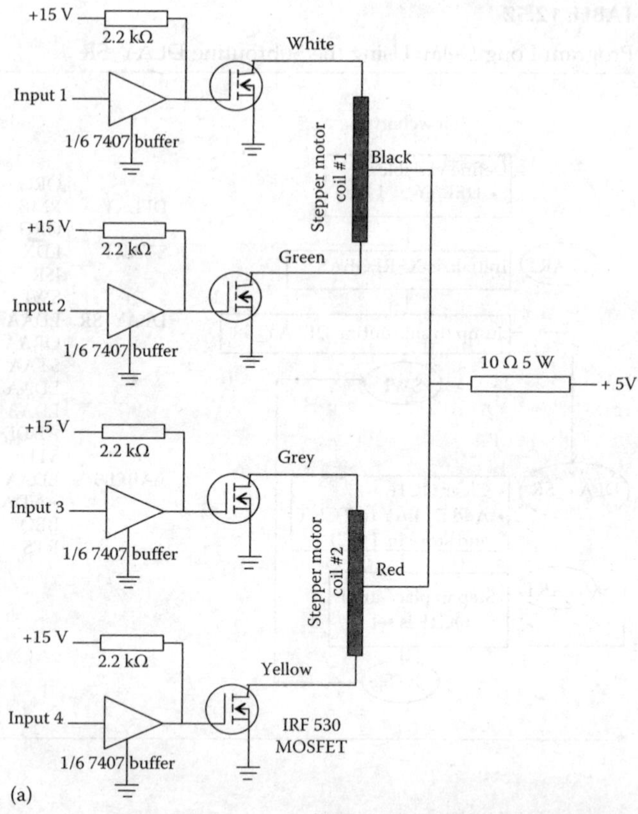

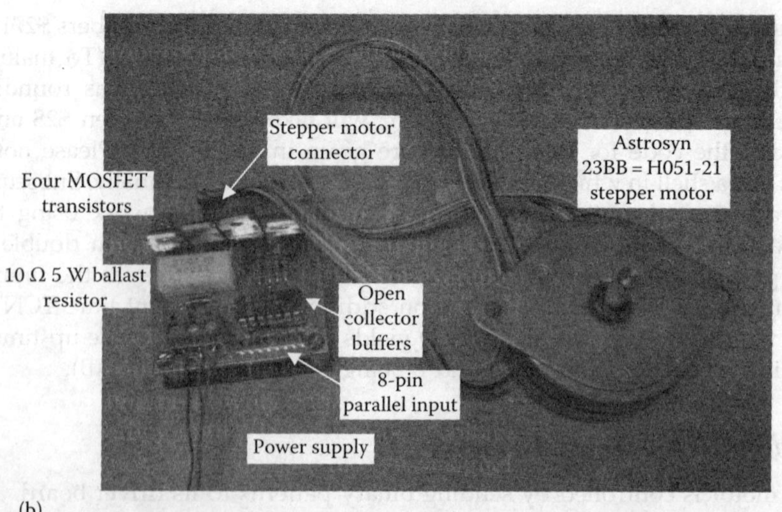

FIGURE 12.48
Stepper motor control with microcontroller: (a) circuit diagram of the stepper motor drive electronics; (b) physical implementation.

TABLE 12.53

Stepper Motor Speed Definitions

	Step	Start
Half speed forward	+1	Anywhere
Full speed forward'	+2	Even pattern
Half-speed backward	−1	Anywhere
Full speed backward'	−2	Even pattern

motor has four distinct coils, and a 1 in the energizing pattern signifies that the corresponding coil is energized. Note that in the odd sequences (S1, S3, S5, S7) only 1-bit in the energizing pattern is set. This indicates that only one out of the four stepper motor coils (1 phase) is energized. On the other hand, the even sequences (S2, S4, S6, S8) have 2-bits set, i.e., two coils (2 phases) are energized. A schematic of the electronic circuit used to drive it (stepper motor driver board) is given in Figure 12.48.

The sequences with two coils energized are labeled *full step*, while those with only one coil energized are labeled *half step*. When these patterns are hit in increasing order one after the other, the motion is called "half speed forward." If only the odd sequences are hit, i.e., every second pattern is hit, the motion is 'full speed forward'. If the patterns are hit in decreasing order, the motions are respectively, "half-speed backward" and "full speed backward" (Table 12.53).

12.9.5.8 Microcontroller Control of a Stepper Motor: Program Step

To control a stepper motor, the microcontroller is programmed to send a sequence of hex numbers through the parallel port B to the stepper motor driver board. The color-coded wire assignment to port B pins is shown in Table 12.54. Only the lower 4-bits of port B are used (PB0 through PB3), because the stepper motor has only four wires to be energized. The upper 4-bits (PB4 through PB7) are not used. They are set to zero, as already indicated by the left zero-filled hex numbers of Table 12.51.

The microcontroller program has to sequentially access memory locations using a memory pointer:

- The variable STEP is used to control the step size and the stepping direction, as indicated in Table 12.53. When running the program, one has to enter into memory location STEP one of the following options: STEP = $01 or $02, or $ff or $fe (corresponding to +1, +2, −1, −2, respectively). Two's complement 8-bit signed convention is used to generate the negative values (−1 = $ff, −2 = $fe). Variable STEP is stored in memory location $0000.

TABLE 12.54

Stepper Motor Color-Coded Wire Assignment to Port B Pins

MSB							LSB
PB7	PB6	PB5	PB4	PB3	PB2	PB1	PB0
N/A	N/A	N/A	N/A	I1 (white)	I2 (green)	I3 (gray)	I4 (yellow)

- The memory locations to be accessed have the variable names S1, S2, S3, S4, S5, S6, S7, and S8 and correspond to addresses $0001 through $0008.

- The memory pointer has the variable name POINTER and is stored at address $0009

- The memory pointer is used in index Y addressing. The memory pointer is entered in index register Y, and the offset is taken as zero ($00). Thus, through Y, the pointer controls directly the address to be accessed.

- The program automatically steps through the set of eight sequences S1–S8 and generates the next pointer value, by adding the variable STEP. As shown in Table 12.53, the values of STEP are +/−1 and +/−2. The positive values correspond to forward motion. The negative values correspond to backward motion.

- The program takes care not to send the pointer outside the eight-sequence range. Hence, when going forward, whenever the highest sequence, S8, is hit (i.e., we hit the roof) the program resets the pointer to the lowest sequence S1 (i.e., to the floor). When going backward, whenever the floor is hit, the program resets the pointer to the roof.

In the initialization phase, the program enters the values of S1–S8 given in Table 12.51 in the appropriate memory locations. Then sets STEP = $00 and POINTER = $00.

The program instructions, flowchart, and code are shown in Table 12.55. Two flowchart levels are presented: the big-picture and the details. The big-picture is used to understand the overall architecture of the program. The details are used to explain some of the blocks. (Details are given only for those blocks which are somehow new and have not been used in previous programs.) Note that the main program is only intended to demonstrate the use of the subroutine STEP_SR. For this reason, the main program is a mere call to the STEP_SR subroutine. The essential code for this program is shown to the right of the flowchart. The essential code was incorporated into the standard template asm to generate the code file Ex_Step.asm.

12.9.5.9 Example Program for Stepper Motor Control

As an example, consider the writing of a program to make the stepper motor go backward and forward with different speeds in full steps and half steps. The stepper motor should be controlled via the microcontroller through keystrokes sent through the keyboard and echoed on the screen.

Please note that the selective use of full steps and half steps constitutes a form of *coarse speed control*, since full-step jumps are twice as fast as half-step increments. Whereas *fine speed control* (increase speed, decrease speed) can be attained by modifying the delay inserted between steps. The shorter the delay is the faster the speed. However, when doing this, one should not reduce the delay below the minimum value mentioned in Section 12.9.5.6, otherwise the stepper motor will stall. The program will be constructed as to recognize *five keystroke commands*:

1. Move forward   >
2. Move backward   <
3. Increase speed (decrease delay)   +
4. Decrease speed (increase delay)   −
5. Stop program   S

TABLE 12.55

Program Step Using the Subroutine STEP_SR

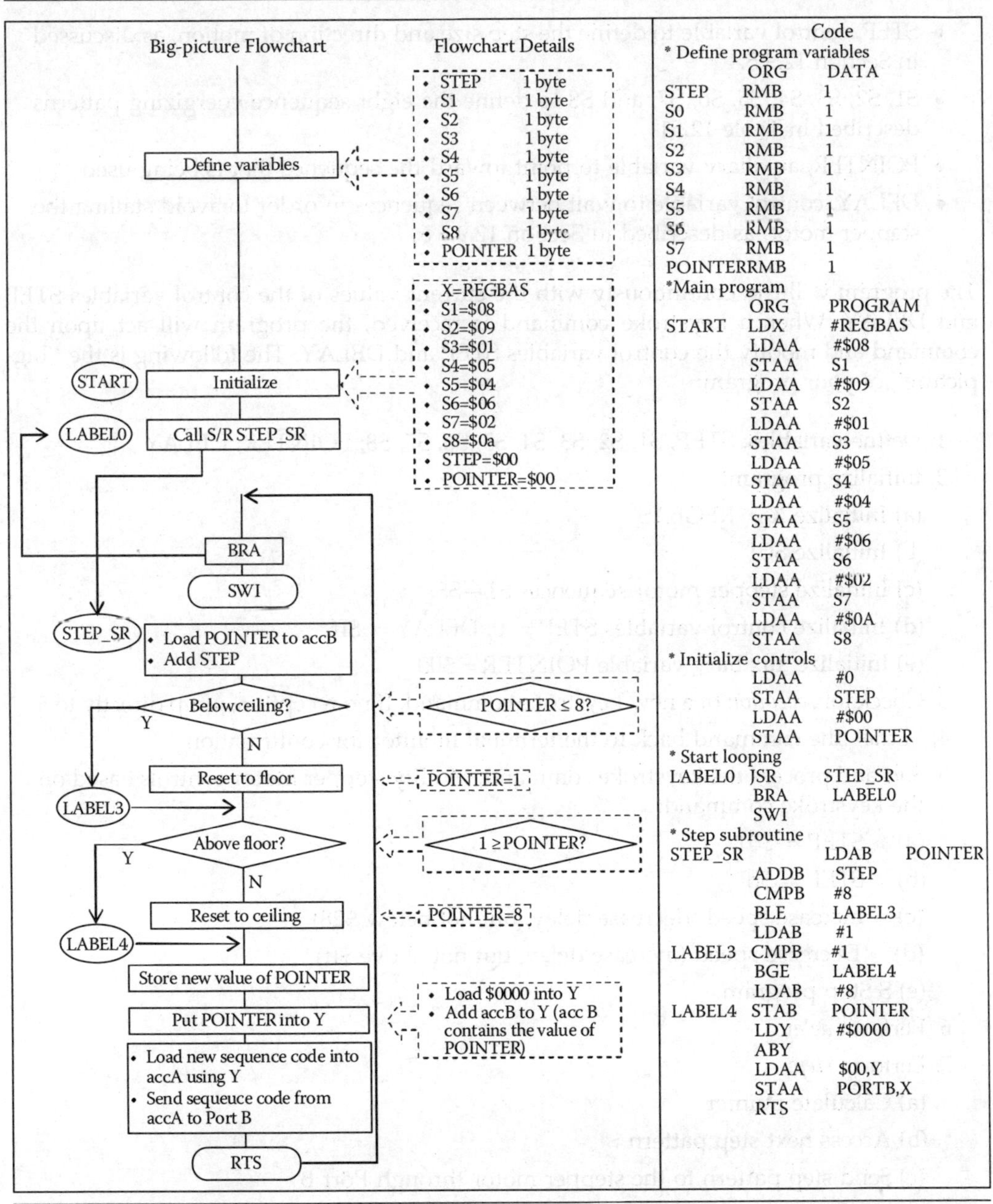

This program will take a little more thought. The programming will need to be done at several layers of complexity, using program segments and subroutines. Three main subroutine segments are envisaged: subroutine *Sort*, subroutine *Step*, subroutine *Long Delay*. In addition, a *Receive-Echo* segment will be used at the beginning of the program to receive the keystroke and echo it to the screen, as described in Sections 12.5.7 and 12.5.8. The program will use the following variables:

- STEP, control variable to define the step size and direction of motion, as discussed in Section 12.9.5.4
- S1, S2, S3, S4, S5, S6, S7, and S8 to define the eight sequence energizing patterns described in Table 12.51
- POINTER, auxiliary variable to point toward the sequence that is being used
- DELAY, control variable to wait between sequences in order to avoid stalling the stepper motor, as described in Section 12.9.5.6

The program will run continuously with the current values of the control variables STEP and DELAY. When a keystroke command is received, the program will act upon the command and modify the control variables STEP and DELAY. The following is the "big-picture" of your program:

1. Define variables: STEP, S1, S2, S3, S4, S5, S6, S7, S8; POINTER, DELAY
2. Initialize program:
   (a) Initialize X = REGBAS
   (b) Initialize SCI
   (c) Initialize stepper motor sequences S1—S8;
   (d) Initialize control variables STEP = 0, DELAY = $ff
   (e) Initialize auxiliary variable POINTER = $00
3. Check for *reception* of a new keystroke command. If no reception, jump directly to 5
4. "*Echo*" the command back to the terminal monitor for confirmation
5. *Sort* and process the keystroke command. Modify stepper motor controls based on the keystroke command:
   (a) > STEP = $01
   (b) < STEP = $ff
   (c) + Increase speed (decrease delay, but not below $28)
   (d) − Decrease speed (increase delay, but not above $ff)
   (e) S Stop program
6. Perform *delay*
7. Perform *step*:
   (a) Calculate pointer
   (b) Access next step pattern
   (c) Send step pattern to the stepper motor through Port B
8. Loop back to 3

A flowchart for the program can be drawn in a layered fashion. Subroutines should be used to perform *delay* and *step* actions based on the information presented in Sections 12.9.5.6 and 12.9.5.8. For performing the sort operation, please see the program code presented in Table 12.56.

TABLE 12.56

Program Sort

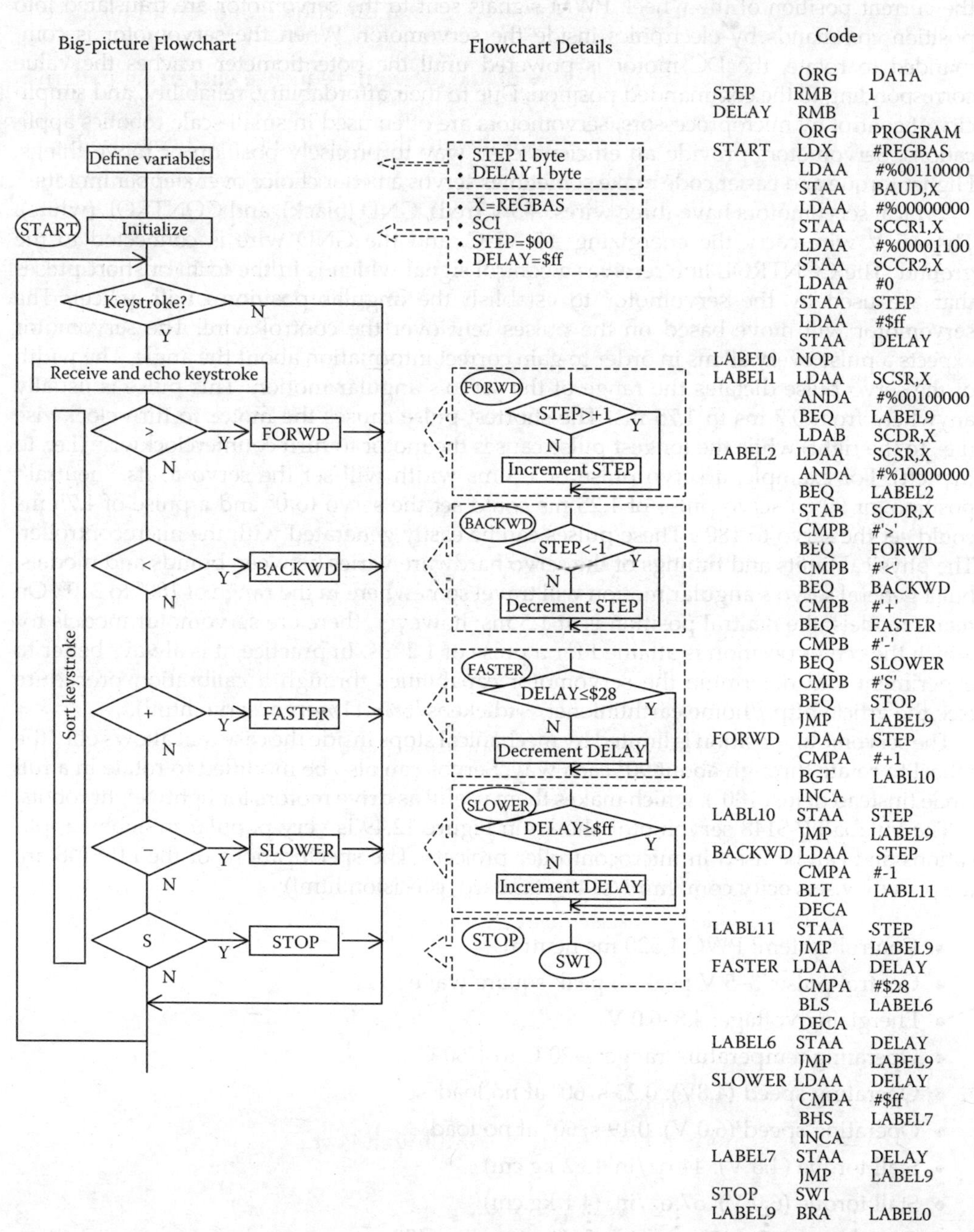


```
           ORG    DATA
STEP       RMB    1
DELAY      RMB    1
           ORG    PROGRAM
START      LDX    #REGBAS
           LDAA   #%00110000
           STAA   BAUD,X
           LDAA   #%00000000
           STAA   SCCR1,X
           LDAA   #%00001100
           STAA   SCCR2,X
           LDAA   #0
           STAA   STEP
           LDAA   #$ff
           STAA   DELAY
LABEL0     NOP
LABEL1     LDAA   SCSR,X
           ANDA   #%00100000
           BEQ    LABEL9
           LDAB   SCDR,X
LABEL2     LDAA   SCSR,X
           ANDA   #%10000000
           BEQ    LABEL2
           STAB   SCDR,X
           CMPB   #'>'
           BEQ    FORWD
           CMPB   #'<'
           BEQ    BACKWD
           CMPB   #'+'
           BEQ    FASTER
           CMPB   #'-'
           BEQ    SLOWER
           CMPB   #'S'
           BEQ    STOP
           JMP    LABEL9
FORWD      LDAA   STEP
           CMPA   #+1
           BGT    LABL10
           INCA
LABL10     STAA   STEP
           JMP    LABEL9
BACKWD     LDAA   STEP
           CMPA   #-1
           BLT    LABL11
           DECA
LABL11     STAA   -STEP
           JMP    LABEL9
FASTER     LDAA   DELAY
           CMPA   #$28
           BLS    LABEL6
           DECA
LABEL6     STAA   DELAY
           JMP    LABEL9
SLOWER     LDAA   DELAY
           CMPA   #$ff
           BHS    LABEL7
           INCA
LABEL7     STAA   DELAY
           JMP    LABEL9
STOP       SWI
LABEL9     BRA    LABEL0
```

12.9.6 Servomotors

A servomotor is comprised of a DC motor mechanically linked to a potentiometer that reads the current position of the wheel. PWM signals sent to the servomotor are translated into position commands by electronics inside the servomotor. When the servomotor is commanded to rotate, the DC motor is powered until the potentiometer reaches the value corresponding to the commanded position. Due to their affordability, reliability, and simplicity of control by microprocessors, servomotors are often used in small-scale robotics applications. Servomotors provide an efficient, easy way to precisely position or move things. Higher torque and easier code make sometimes servos a better choice over stepper motors.

Typical servomotors have three wires: +5 V (red), GND (black), and CONTROL (white). The +5 V wire carry the energizing +5 VDC, and the GND wire is connected to the ground. The CONTROL line receives a control signal which is in the form of short pulses that are used by the servomotor to establish the angular position of its wheel. The servomotor will move based on the pulses sent over the control wire. The servomotor expects a pulse every 20 ms in order to gain correct information about the angle. The width of the servo pulse dictates the range of the servo's angular motion. This pulse is usually anywhere from 0.7 ms to 1.75 ms. The shortest pulse causes the motor to turn clockwise (i.e., to the right) while the longest pulse causes the motor to turn counterclockwise (i.e., to the left). For example, a servo pulse of 1.5 ms width will set the servo to its "neutral" position, or 90°; a servo pulse of 1.25 ms could set the servo to 0° and a pulse of 1.75 ms could set the servo to 180°. These pulses can be easily generated with the microcontroller. The physical limits and timings of the servo hardware varies between brands and models, but a general servo's angular motion will travel somewhere in the range of 180° to 210°. On many models, the neutral position is at 1.5 ms; however, there are servomotor models for which the center position is attained for a pulse of 1.2 ms. In practice, it is always better to experiment and determine the servomotor capabilities through a calibration procedure (see the article http://home.earthlink.net/~tdickens/68hc11/servo/servo.html).

The servomotor position is limited by mechanical stops inside the case that allows only the wheel to rotate through about 90° each way. Servos can also be modified to rotate in a full circle (instead of just 180°), which makes them useful as drive motors for lightweight robots.

The Futaba FP-S148 servomotor shown in Figure 12.49 is very popular in hobby applications and can be used in microcontroller projects. The specifications of the FP-S148 are (http://www.servocity.com/html/s148_standard_precision.html):

- Control system: PWC 1.520 ms neutral
- Control pulse: 3–5 V peak to peak square wave
- Energizing voltage: 4.8–6.0 V
- Operating temperature range: −20°C to +60°C
- Operating speed (4.8V): 0.23 s/60° at no load
- Operating speed (6.0 V): 0.19 s/60° at no load
- Stall torque (4.8 V): 44 oz/in. (3.2 kg.cm)
- Stall torque (6.0 V): 57 oz/in. (4.1 kg.cm)
- Operating angle: 45° one side pulse traveling 400 us
- 360 modifiable: Yes
- Direction: Counterclockwise/pulse traveling 1.520–1.900 ms
- Current drain (4.8 V): 7.2 mA/idle

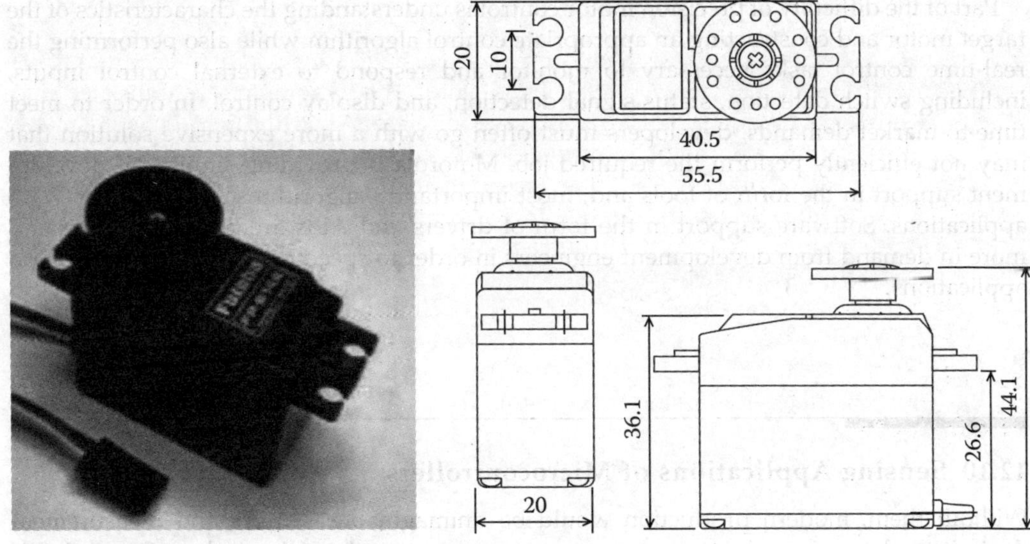

FIGURE 12.49
Futaba FP-S148 servomotor.

- Current drain (6.0 V): 8 mA/idle
- Motor type: 3 pole ferrite DC motor
- Potentiometer drive: Indirect drive
- Bearing type: Dual bronze bushings
- Gear type: All nylon gears
- Connector wire length: 12″
- Dimensions: $1.6'' \times 0.8'' \times 1.4''$ ($41 \times 20 \times 36$ mm^3)
- Weight: 1.6 oz. (44.4 g)

FP-S148 has a sintered-bronze brushing that reduces friction and wear. This metal bearing is of particular importance when the servo is modified to execute continuous rotation such as for use as a drive wheel on a mini robot. Metallic bushes resist much better to side loads than plastic bushes, which are known to wear out prematurely.

12.9.7 Dedicated Motor Drivers based on Microcontrollers and DSPs

Motorola offers a product mix for motor driving that includes the 8-bit 68HC08 core and Motorola's DSP56800 DSP. The 68HC08 solutions are targeted toward low-cost systems that utilize sensor control and use a UART-based system bus. They include a 15-bit PWM module specific to motor control, as well as integrated fault pins and a 10-bit A/D (accurate to ± 4 LSB). The DSP56800 solutions are for higher-performance systems that utilize sensorless control and use a CAN communications bus. It sports a 12-bit A/D (accurate to ± 1 LSB), an integrated quadrature encoder and a JTAG interface. The DSPs also have the same motor control PWM module as the 68HC08, insuring code compatibility between 68HC08 and DSP56800 solutions. Some products will be designed to meet automotive quality standards.

Part of the difficulty of precision motor control is understanding the characteristics of the target motor and constructing an appropriate control algorithm while also performing the real-time control tasks necessary to monitor and respond to external control inputs, including switch detection, status signal detection, and display control. In order to meet time-to-market demands, developers must often go with a more expensive solution that may not efficiently perform the required job. Motorola is providing significant development support in the form of tools and, most importantly, algorithms as well as software applications. Software support in the form of drivers and APIs are becoming more and more in demand from development engineers in order to speed development of common applications.

12.10 Sensing Applications of Microcontrollers

Without them, modern production would be unimaginable: sensors and measurement devices to determine pressures, temperatures, flows, and levels, and to forward this information to sophisticated control technology are part of every day life. In refineries and chemical plants and in small breweries and large water treatment facilities all around the world, various sensors and ancillary devices are playing essential roles in process control automation. There are a large number of sensing applications that the microcontroller can do. Their specific details depend on the type of sensors being used. In general, we distinguish two basic sensor types:

- Analog sensors that create a voltage output proportional to the amplitude of a physical quantity
- Digital sensors, that generate a binary number proportional with the amplitude of a physical quantity

12.10.1 Digital Sensors: The Emitter Detector

Depending on the physical stimulus, sensors types include optoelectronic sensors, pressure sensors, temperature sensors, humidity sensors, etc. The simplest sensor imaginable is just a switch. The switch is a simple binary sensor. It has two states, ON and OFF. For example, a switch can be used as a presence sensor. If something is present, say someone steps on a carpet that has a switch under it, then the switch comes ON. The output of a switch sensor is a voltage that is either LOW or HIGH. Because of its simplicity, the switch can be viewed as both a digital sensor and an analog sensor, depending on how its output is read.

A common electronic switch is the transistor. Depending on its type, it can be a voltage or a current switch. A linear transistor is a current switch; when its base receives a signal, then the linear transistor opens and allows a current to run from the collector to the emitter. Phototransistors perform in a similar way, but they respond to a photonic excitation of their base. The photonic excitation of common phototransistors is outside the visible spectrum, such as to not be affected by normal office lighting. Most of the time, phototransistors operate in the infrared spectrum, i.e., with infrared light.

One wide usage of phototransistors is in the *emitter-detector switches*. The emitter-detector is an optoelectronic sensor. It senses the presence of something passing through

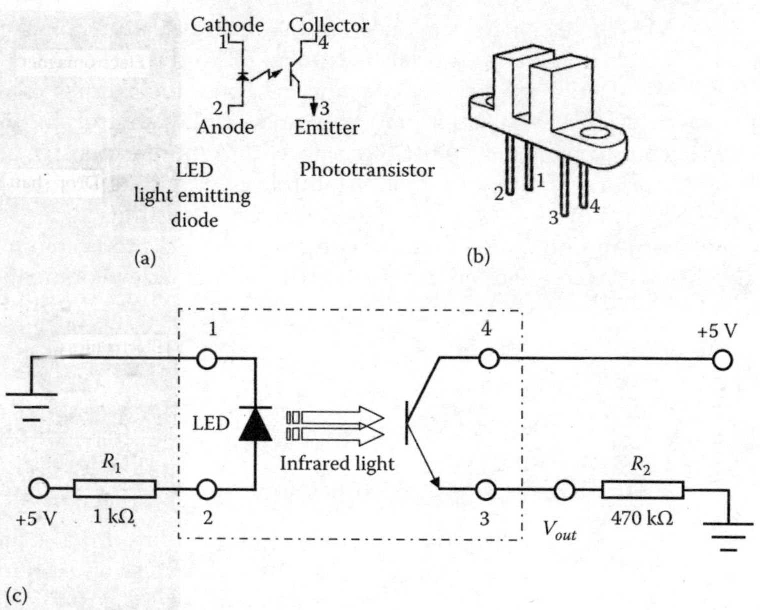

FIGURE 12.50

Emitter–detector (ED) optoelectronic sensor: (a) schematic representation of the ED sensor; (b) physical presentation; (c) implementation circuit.

its aperture. The schematic diagram of an emitter-detector sensor is shown in Figure 12.50. It consists of a light emitting diode (LED), that emits in the infrared band, and a phototransistor (Figure 12.50a). The LED releases an infrared signal across the device's gap. When the infrared signal reaches the phototransistor, this opens and lets a current pass. This is equivalent to the switch being ON. Otherwise, no current can pass and the switch is OFF. The implementation of the emitter–detector sensor in a microcontroller application is performed using the circuit shown in Figure 12.50c. On the left, we see a voltage source and a resistor in series with the LED. The voltage source can be +5 V. The resistor should be such that the current through the LED does not exceed the LED allowable current. In Figure 12.50c, this resistance has the value $R_1 = 1$ kΩ. On the right of Figure 12.50c we see the phototransistor. It is wire in series with the resistance $R_1 = 470$ kΩ. When current flows through the phototransistor (as triggered by the presence of infrared light), the voltage across the resistance R_2 is high. The output of the emitter–detector sensor, V_{out}, can be measured as shown in Figure 12.50c.

Some examples of the use of the emitter–detector optoelectronic sensor in conjunction with a microcontroller are given next. Two cases will be examined, the drop tower experiment and the digital tachometer.

12.10.2 Drop Tower Experiment

The drop tower experiment is an illustration of the use of the 68HC11 microcontroller in conjunction with an optoelectronic sensor. In this experiment, we measure the time of travel of an object drop from a given height. A 1.5 in. bolt will be held by an electromagnet at the top of a drop shaft (Figure 12.51).

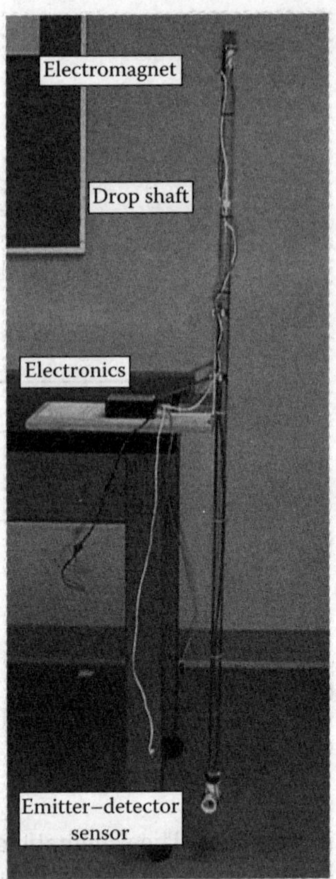

FIGURE 12.51
Bolt drop experimental setup.

An electromagnet is connected to the MCU. This magnet is controlled by any of the 8 pins on PORTB. It is turned on by sending 5 V through parallel PORTB. When a keystroke occurs, the MCU commands the electromagnet to release the bolt. Simultaneously, the time origin, T0, is recorded. The bolt travels down the shaft. After falling a specified distance ($L = 48.75$ in.), the bolt crosses an infrared emitter–detector sensor and interrupts its beam causing its output voltage to fall from high (+5 V) to low ($\sim$0 V). This event is marked as the time value T1. When the bolt exits the emitter–detector sensor, its output voltage comes back to high (+5 V). This event is marked as the time value T2. The output from the emitter–detector sensor is wired into the input capture pins IC1 and IC2 of the MCU. The microprocessor can be programmed to perform the following functions:

(a) Control the electromagnet

(b) Take the initial time when the process starts

(c) Measure the time when the free-falling bolt reaches the emitter–detector

(d) Measure the time the bolt passes through the emitter–detector

The electromagnet control is done using the keystroke command discussed in the serial communication section and the port B output discussed in the parallel communication section. The measuring of the times T0, T1, and T2 is done with a microcontroller program that is an extension of the program presented in the timer section. The only difference between that program and the program needed here is the number of input capture lines used. In the timer section, we presented a program to perform input capture timing on one input capture line. Here, we will need a program doing this on two lines, one for T1 and the other for T2.

12.10.3 Digital Tachometer

The digital tachometer is used to measure the rotation speed of a DC motor, as shown in Figure 12.52. The experimental setup consists of several parts. A DC electric motor is controlled by a potentiometer through an electronic driver. By turning the knob of the potentiometer, the speed of the DC motor can be increased and decreased. A disc with an aperture is mounted in the DC motor shaft. The disc passes through an emitter–detector sensor. As the DC motor spins, the aperture opens and closes the passage of light in the emitter–detector sensor. The emitter–detector sensor sends a High (5 V) signal when the aperture in the disc allows the beam of light to pass through. Hence, a pulse signal is generated at the emitter–detector sensor output. The period of the pulse signal corresponds to one full rotation of the disc. By measuring the period of the pulse signal one can determine the speed of rotation in RPM of the DC motor. The period of rotation, T, can be determined as the difference between two consecutive passages of the aperture through the emitter–detector sensor, i.e., the difference between two pulse fronts, t_1 and t_2. This will be achieved by using a microcontroller. The microcontroller receives the pulse signal from the emitter–detector sensor at one of the input-capture timer lines. The pulse signal is processed by the microcontroller to determine the values of t_1 and t_2. Then, the microcontroller calculates the time difference, T, and hence the motor RPM. The experimental setup also incorporates a 2-digit 7-LED display. This display is used to present the rotation speed values expressed in hundreds of RPM. The display is connected to the microcontroller port B. The microcontroller calculates the rotation speed values in hundreds of RPM and then converts them in binary-coded decimal (BCD) format. The BCD code is sent out through port B into the electronic driver of the 2-digit 7-LED display.

The program used in the microcontroller has to perform two main tasks. The first task is to measure the times T1 and T2. Since the microcontroller internal speed is considerably faster than the DC motor, it is expected that timer overflows will occur between two consecutive readings of T1 and T2. These timer overflows have to be also properly measured. The second task is to determine the RPM in BCD format and send it to the 2-digit 7-LED display through port B. This part involves calculations which are handled in the microcontroller. Some of the challenges involved in these calculations are associated with the fact that the actual times t_1 and t_2, expressed in microcontroller machine cycle, exceed the size of the microcontroller word, even when double precision is used. Hence, the arithmetic must be performed using scale factors to reduce the word size. Two separate programs, EX_RPM_1 and EX_RPM_2, each associated with one of the two tasks, are presented next. We leave to the reader the challenge of joining these two programs into a single program that would run the complete experiment.

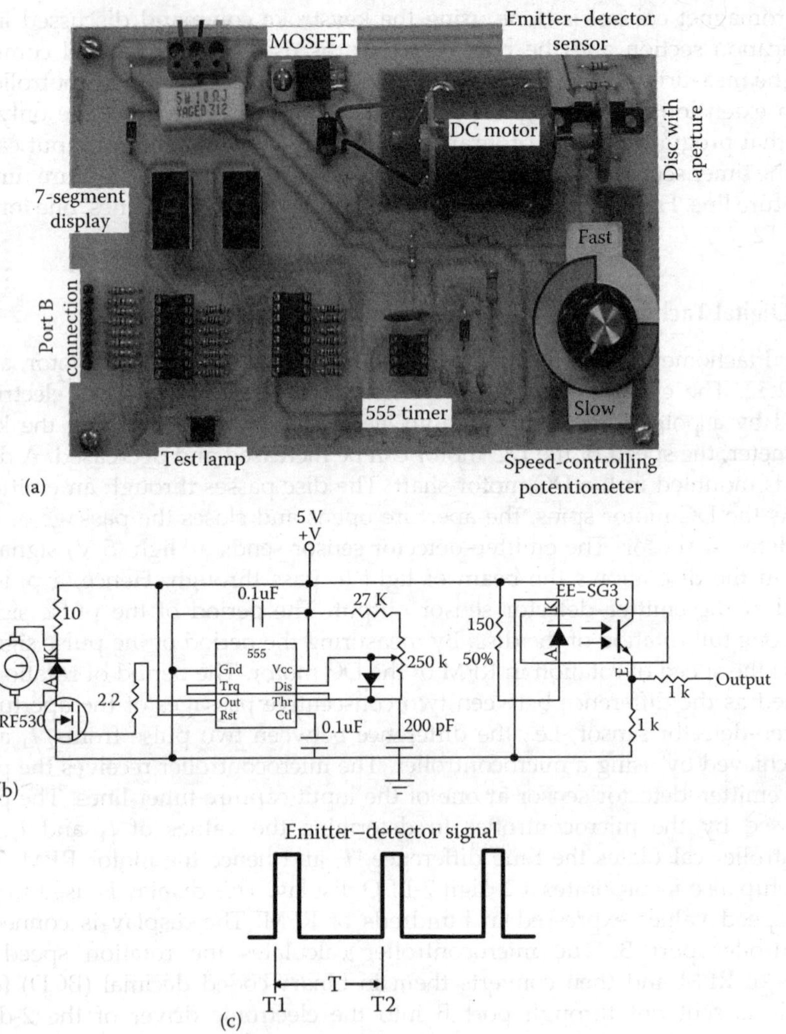

FIGURE 12.52
DC motor tachometer experiment: (a) experimental setup; (b) circuit diagram; (c) the pulse signal generated by the emitter–detector signal.

12.10.3.1 Program Ex_RPM_1

The program EX_RPM_1 captures the times t_1 and t_2 required in the digital tachometer RPM experiment. The output of the emitter–detector sensor is attached to the input capture pin IC1 on Port A. Two times are measured: T1, i.e., the first time when a falling edge transition is encountered on pin IC1; and T2, i.e., the second time when a falling edge transition is encountered on the same pin IC1. The values T1 and T2 are measured in machine cycles. Between T1 and T2, timer overflows may also happen. The number of timer overflows is captured in the variable NOF. From T1, T2, and number of overflows

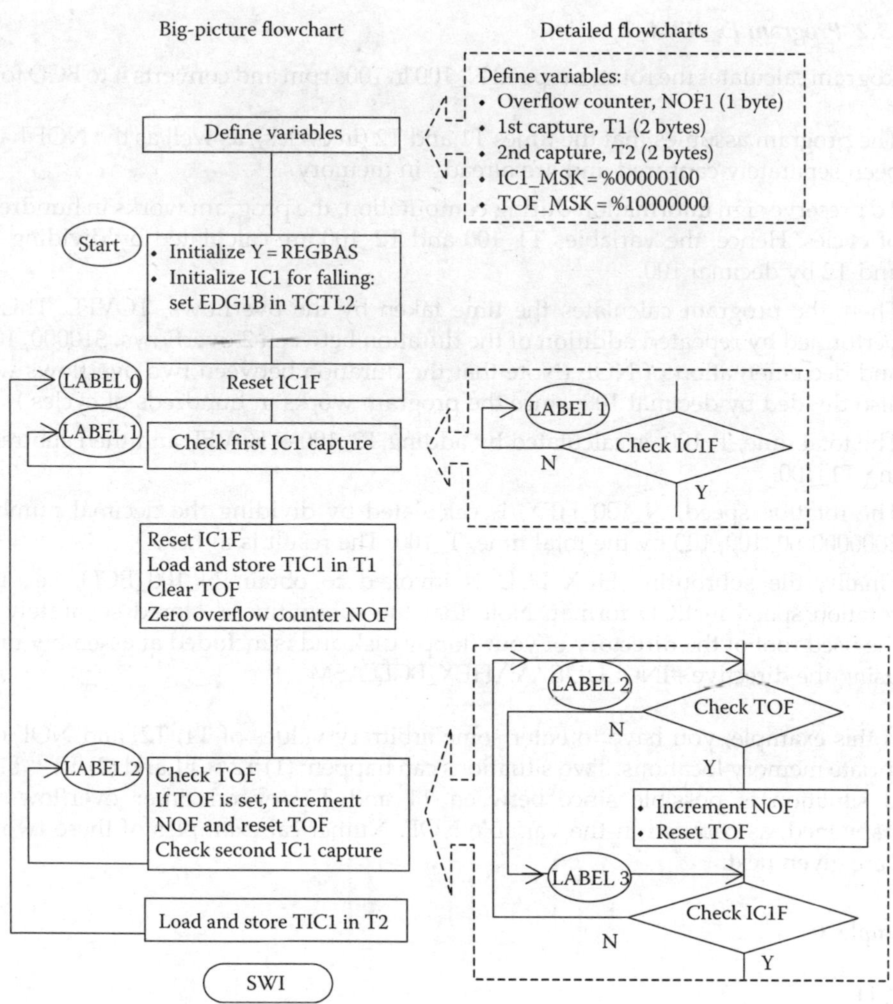

FIGURE 12.53
Flowchart for program EX_RPM_1.

(NOF), one can calculate the time between two consecutive passages of the disc aperture through the emitter–detector sensor. Thus, the difference between T2 and T1 plus the time taken by the overflows will represent the period of rotation of the disk, T. This information will be used in program EX_RPM_2 to calculate the rotation speed.

The program flowchart is shown in Figure 12.53. Two flowchart levels are presented: the big-picture and the details. The big-picture is used to understand the overall architecture of the program. The details are used to explain some of the blocks. (Details are given only for those blocks which are somehow new and have not been used in previous programs.) This flowchart is used to generate the code file Ex_RPM_1.asm using the standard asm template.

12.10.3.2 Program Ex_RPM_2

This program calculates the rotation speed N_100 in 100s rpm and converts it to BCD format.

- The program assumes that the times T1 and T2 (in cycles), as well as the NOF have been separately captured and are already in memory.

- To preserve sign information during computation, the program works in hundreds of cycles. Hence, the variables T1_100 and T2_100 are calculated by dividing T1 and T2 by decimal 100.

- Then, the program calculates the time taken by the overflows, TOVFL. This is performed by repeated addition of the duration between 2 overflows, $10000/100, and decrementation of NOF (Note that the duration between two overflows was also divided by decimal 100, since the program works in hundreds of cycles.)

- The total time, T_100, is calculated by adding T2_100 + TOVFL and then subtracting T1_100.

- The rotation speed, N_100_HEX, is calculated by dividing the decimal number 2000000*60/100/100 by the total time, T_100. The result is in hex.

- Finally, the subroutine HEX_BCD is invoked to obtain N_100_BCD, i.e., the rotation speed in BCD format. Note that the subroutine is stored separately in the used root of the directory of your floppy disk and is included at assembly time using the directive #INCLUDE 'A:\HEX_BCD.ASM'

To run this example, you have to enter some arbitrary values of T1, T2, and NOF in the appropriate memory locations. Two situations can happen: (1) T2 > T1 and (2) T2 < T1. The second situation is possible since between T1 and T2 several timer overflows may have happened, as counted in the variable NOF. Numerical examples of these two situations are given next:

Example 1:

T2 ≥ T1

 Data: NOF = 1, T1 = $0006, T2 = $110c

 Results: T1_100 = $0000, T2_100 = $002b, TOVFL = $028f, T_100 = $02ba,
     N_100_HEX = $0011, N_100_BCD = 17

The rotation speed is 1700 rpm.   ■

Example 2:

T2 ≤ T1

 This example gives the same time duration as Example 1, but the individual times were shifted to straddle the time-change line.

 Data: NOF = 2, T1 = $fffd, T2 = $1103

 Results: T1_100 = $028f, T2_100 = $002b, TOVFL = $051e, T_100 = $02ba,
     N_100_HEX = $0011, N_100_BCD = 17

The rotation speed is 1700 rpm.

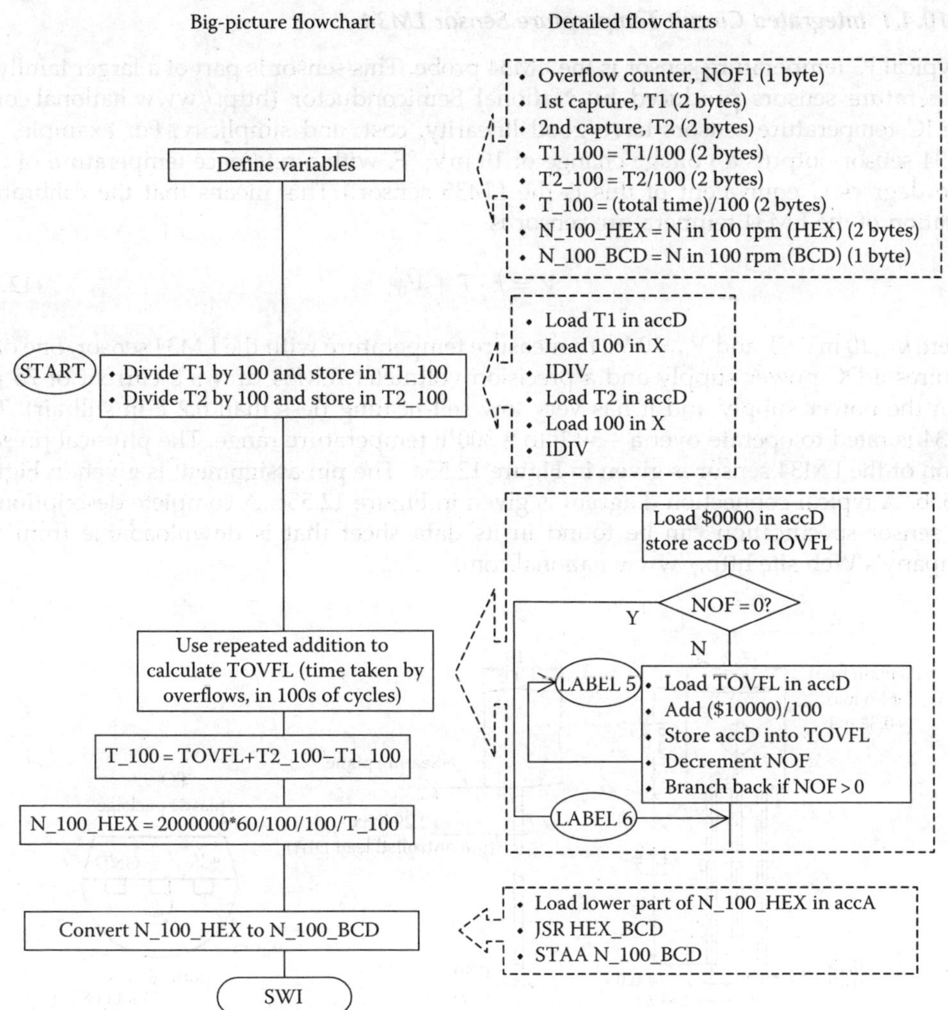

FIGURE 12.54
Flowchart for program EX_RPM_2.

The program flowchart is given in Figure 12.54. Two flowchart levels are presented: the big-picture and the details. The big-picture is used to understand the overall architecture of the program. The details are used to explain some of the blocks. (Details are given only for those blocks which are somehow new and have not been used in previous programs.) This flowchart was use to generate the code file Ex_RPM_2.asm using the standard asm template. ∎

12.10.4 Analog Sensors: Temperature Sensor

Temperature is one of the most common physical variables that are being measured in applications. From controlling the temperature of molten metal in a foundry to controlling liquid nitrogen in a cryogenics laboratory, the measurement, evaluation, and control of temperature are critical to many branches of industry. The principles of operation of a common temperature sensor and their interfacing with the microcontroller are presented in the following section.

12.10.4.1 *Integrated Circuit Temperature Sensor LM34*

A typical IC temperature sensor is the LM34 probe. This sensor is part of a larger family of temperature sensors produced by National Semiconductor (http://www.national.com). The IC temperature sensors have good linearity, cost, and simplicity. For example, the LM34 sensor outputs a voltage change of 10 mV/°F, with a reference temperature of 0°F. (The degrees C equivalent of this is the LM35 sensor.) This means that the calibration equation of the LM34 temperature sensor is

$$V = k \cdot T + V_0 \tag{12.83}$$

where $k = 10$ mV/°F and $V_0 = 0$ V. To measure temperature with the LM34 sensor, one only requires a DC power supply and a precision voltmeter. LM34 draws a current of 75 μA from the power supply and it has very low self-heating (less than 0.2°F in still air). The LM34 is rated to operate over a −50°F to +300°F temperature range. The physical presentation of the LM34 sensor is given in Figure 12.55a. The pin assignment is given in Figure 12.55b. A typical connection diagram is given in Figure 12.55c. A complete description of the sensor specification can be found in its data sheet that is downloadable from the company's Web site http://www.national.com.

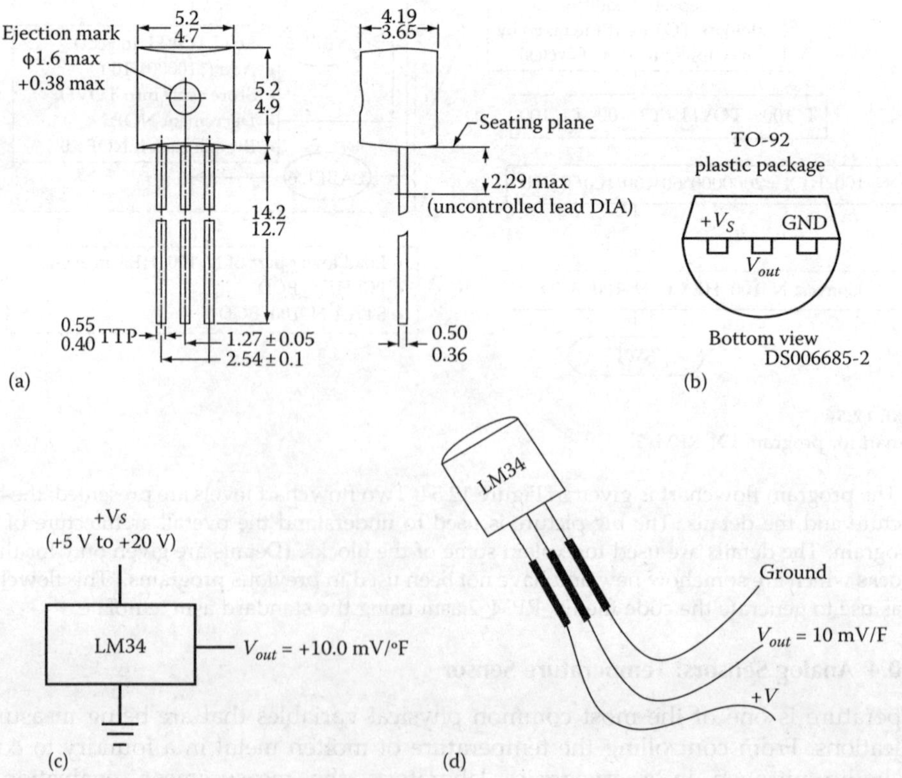

FIGURE 12.55
Temperature sensor LM34 from National Semiconductors (http://www.national.com): (a) physical dimensions; (b) pin assignment; (c) typical connection diagram; (d) physical connections.

By replacing the LM34 with another precision integrated-circuit temperature sensor LM35, we can easily get an output voltage proportional to the centigrade temperature. The LM35 sensor has a linear +10.0 mV/°C scale factor and a temperature range from −55°C to +150°C. In fact LM34 and LM35 are among the same series of temperature sensors so that they can be easily exchanged in different applications. The wiring for LM 35 is the same as that of LM34. Please refer to the manufacturer's datasheets for LM34 and LM35 to obtain more details about packaging and features.

12.10.4.2 Temperature Sensor Functional Module

A functional module was constructed to illustrate the use of the LM34 temperature sensor (Figure 12.56). The temperature sensor functional module consists of two parts: (1) the function module box and (2) the probe head. The functional module of Figure 12.56 shows the correct wiring connections. The supply is +5 V DC. The LM34 temperature sensor is mounted on the probe head. For LM34 in metal can package (TO-46), the small tab on the sensor indicates the position of "+Vs" leading pin. One should be careful to ensure that the sensor is properly mounted on the probe head (Figure 12.56 inset). The probe head is connected with flexible ribbon cable to the functional module. In this way, the probe can be placed where the temperature needs to be measured.

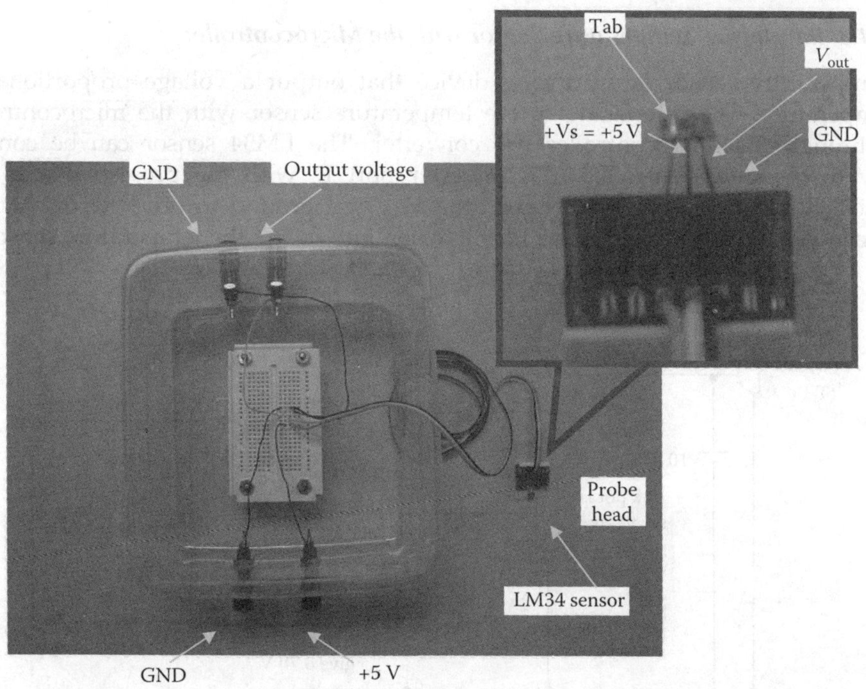

FIGURE 12.56
LM34 temperature sensor functional module hardware. The inset shows a close up of the sensing probe head with the correct connection of the LM34 chip.

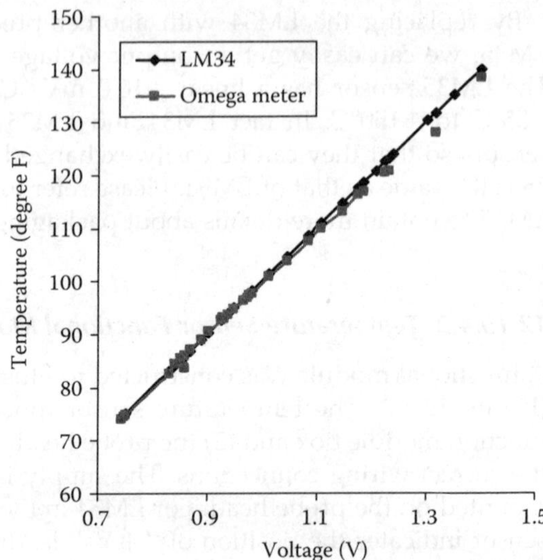

FIGURE 12.57
Calibration curve for the LM34 temperature sensor functional module.

The functional module was tested in parallel to a conventional Omega meter with a temperature probe. Thus, a rough calibration curve could be obtained. The experimental results are presented in Figure 12.57. One notices that good agreement exists between the LM34 temperature sensor and the reference Omega meter.

12.10.4.3 *Interfacing Temperature Sensor with the Microcontroller*

The temperature sensor is an analog device that output a voltage proportional with the temperature. The interfacing of the temperature sensor with the microcontroller is done through the microcontroller AD converter. The LM34 sensor can be connected directly to the microcontroller AD converter port E, with the proviso that a 10 µF capacitor should be connected between the V_{out} and ground in order to discharge the high frequency noise spikes (Figure 12.58). When interfacing the temperature sensor with

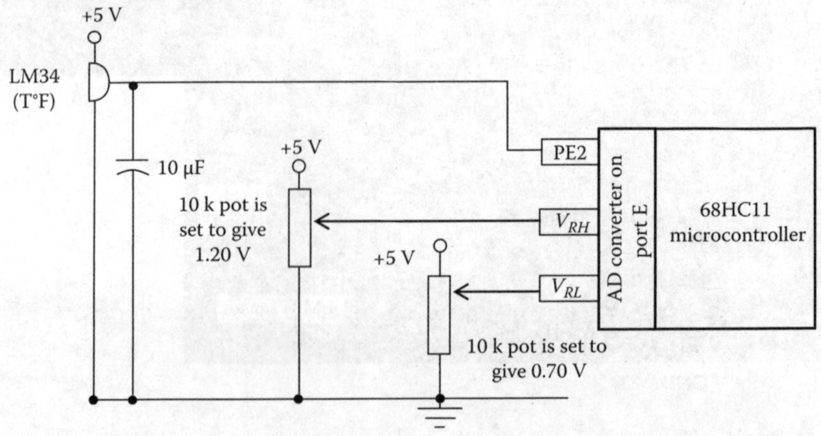

FIGURE 12.58
Schematic diagram of the interfacing of the LM34 temperature sensor with the MC68HC11 microcontroller.

the microcontroller, the reference voltages V_L and V_H must be set to such values as to encompass as well as possible the temperature range of interest. For example, if we desire to measure temperatures from 70°F (room temperature) to, say, 120°F, then, for best resolution, $V_{RL} = 700$ mV $= 0.7$ V and $V_{RH} = 1200$ mV $= 1.2$ V. This will give a resolution of

$$\Delta T = \frac{\Delta V}{k} \tag{12.84}$$

where

$$\Delta V = \frac{V_H - V_L}{\$ff + 1} = \frac{1200 - 700}{256} \approx 2 \text{ mV} \tag{12.85}$$

Substitution of Equation (12.85) into Equation (12.84) yields the temperature resolution as

$$\Delta T = \frac{\Delta V}{k} = \frac{2 \text{ mV}}{10 \text{ mV/°F}} = 0.2°F \tag{12.86}$$

The setting of the V_{RL} and V_{RH} values is done with adjustable potentiometers. Exact adjustment of the desired values will be done during the calibration stage of the experimental setup. However, if accuracy is not very important, one can stay with the nominal range of 0–5 V of the AD converter, in which case a resolution of $\Delta T = 2°F$ will be obtained.

12.11 Microcontroller Process Control

Industrial automation, process control, and instrumentation are essential elements of successful manufacturing. The beauty of microcontrollers is to be capable of embedding within the equipment all the processing power needed to perform sophisticated control functions.

Process control involves the use of *sensors* and *actuators*. The sensors collect data about the process and feed it into the *microcontroller*. The microcontroller processes the sensors data and compares it with the desired values according to the manufacturing process specifications. Based on the results of this comparison, the microcontroller takes corrective actions, if needed. The corrective actions are taken through the actuators.

The simplest example of process control with a microcontroller is the temperature control in a room or in an enclosure. Modern thermostats have microcontrollers incorporated in their design. Assume that the set point is at 70°F. A temperature sensor measures the room temperature and compares it with the set point. If the room temperature is below the set point, the microcontroller will order the heaters to come into action. If the temperature is above the set point, the microcontroller with start the air conditioning system. If the temperature is at the set point, no action will be taken.

In general, process control involves maintaining desired process conditions. Heating or cooling objects to a certain temperature, holding a constant pressure in a steam pipe, or setting a flow rate of material into a vat in order to maintain a constant liquid level are examples of continuous process control. The condition to be controlled is termed the *process variable*. Temperature, pressure, flow rate, and liquid level may be process variables.

Industrial output devices are the control elements, i.e., the actuators. Motors, valves, heaters, pumps, and solenoids are examples of actuators. They can control the energy input into the process and thus the outcome of the process. Measurement of the process outcomes is done with sensors.

12.11.1 Open-Loop versus Closed-Loop Control

Generally speaking, a process control can be classified into two types: open loop and closed loop. In open-loop control, no automatic check is made on the process to see whether corrective action is necessary. In closed-loop control, the process is continuously measured and evaluated, and the input parameters are controlled to maintain the process optimal.

12.11.1.1 Open-Loop Control

The simplest form of process control is open-loop control. Figure 12.59 shows the flow diagram of a basic open-loop system. On the left, we see the input to the system: energy is applied to the process, e.g., through an actuator. The type of actuator used will depend on the process. For a temperature control process, the actuator will be a heater or an air conditioning unit. For a pumping process, the actuator will be a pump. As shown in Figure 12.59, the energy applied to the process is metered through a "valve." The valve is adjusted to a *calibrated setting*. The calibrated setting determines how much of the actuator energy should be applied to the system. Since the process uses this energy to change its output, changing the actuator setting will change the energy level in the process and the resulting output. The fundamental concept of open-loop control is that the actuator's setting is based on an understanding of the process. This understanding includes knowing the effects of the energy on the process and an initial evaluation of any variables disturbing the process. This understanding of the process and the relevant equations make up the *system model*. If all of the variables that may affect the outcome of the process are steady, the output of the process will also be steady. Based on this understanding, the output "should" be correct for a given calibrated setting. In reality, many processes undergo perturbations due to changes in the outside conditions, modification in the prime material being processed, or just wear. Hence, the calibrated setting chosen at the beginning when the process was first started (i.e., when the plant was commissioned) may be no longer giving the optimal output in the present conditions. In this case, an external observer will have to reevaluate the process and take corrective actions. The corrective actions may be in the form of removing the disturbing parameters that affect the process and thus resetting the process to the initial (as new) conditions. Or, the corrective actions may be in the form of determining a new calibrated setting, at which the process will fare better under the new conditions. This later situation is viewed as *system recalibration*.

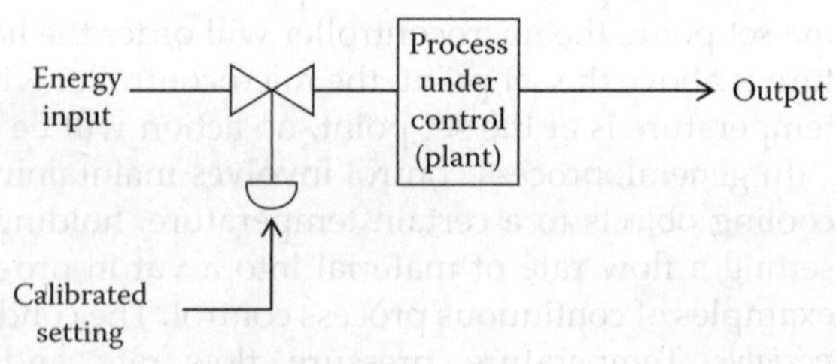

FIGURE 12.59
Schematic diagram of open loop control.

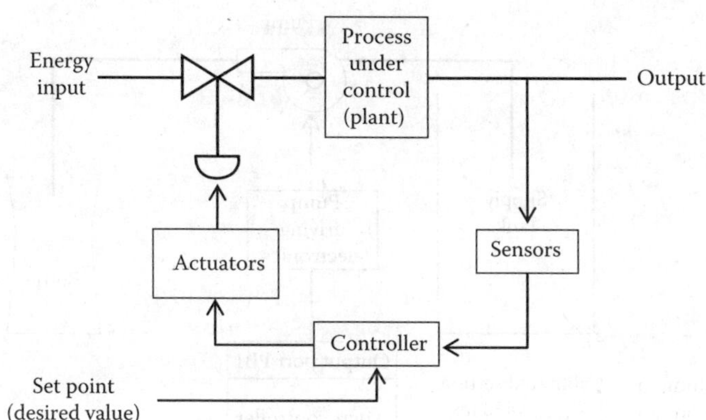

FIGURE 12.60
Schematic diagram of closed loop control.

12.11.1.2 Closed-Loop Control

We have seen in the previous section that open-loop control is alright as long as the system is steady and its behavior is exactly as it were at the time of the initial calibration, i.e., when it was commissioned. However, in reality, there are many systems that do not meet such requirements. Many systems undergo rapid changes in either the external variables or in their own behavior. In such situations, open-loop control will not be satisfactory, since it would often require recalibration and adjustments of the calibration setting.

To overcome the open-loop control shortcomings, closed-loop control can be utilized. Closed-loop control is based on the idea of *feedback*, i.e., information about the system performance is fed back to the system input, and corrective actions are taken in real time. A schematic of a closed-loop process control is given in Figure 12.60. It is noticed that a *controller* has been inserted. The role of the controller is to process information received from the *sensors* and to take adequate actions through the use of *actuators* in order to maintain the desired operational *set point*. We see that closed-loop control incorporates a continuous evaluation (measurement) of the output and a modification of the actuators settings based on the feedback information. The room temperature control process mentioned at the beginning of this section is a typical example of closed-loop control. Another example of closed-loop control would be a situation in which a reservoir is filled with liquid; a sensor will communicate to the controller when the reservoir is full and the pump has to be shut off.

12.11.1.3 Use of Microcontrollers for Process Control

Microcontrollers have proven to be a dependable, cost-effective way of adding automation to simple control schemes. Microcontrollers can be used to implement both open-loop and closed-loop control.

12.11.1.4 Use of Microcontrollers for Open-Loop Process Control

The microcontroller implementation of *open-loop control* requires that the process can be adequately modeled, and the process outcomes (process variables) can be unambiguously defined in terms of the process inputs. Two examples are provided below to illustrate this principle.

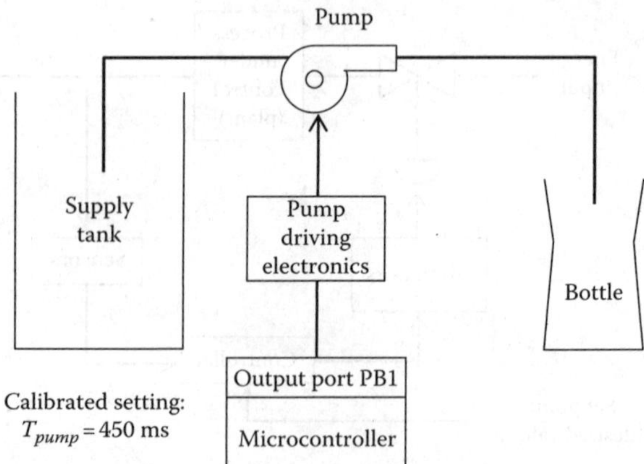

FIGURE 12.61
Schematic diagram of the setup for filling a bottle under microcontroller supervision in open-loop control.

Example: Open-Loop Control of a Bottle Filling Process

Consider the situation in which the microcontroller has to supervise the filling of bottles from a tank. In open-loop control, this operation would proceed as shown in Figure 12.61. A pump draws the liquid from a supply tank and feeds it into the bottle that has to be filled. Previous calibration has established that it takes 450 ms to fill each bottle. Hence, the pump has to be run for $T_{pump} = 450$ ms for each bottle. This timing can be achieved through a simple delay loop that is easily handled by the microcontroller (see Section 12.3). The microcontroller would control the pump action through the pump driving electronics by sending a signal through an output port, say PB1. While this signal is an ON signal the pump operates. When the time T_{pump} has elapsed, the microcontroller sends an OFF signal and the pump will stop. After the next bottle is set in position, the microcontroller would repeat the bottle filling cycle. Of course, the automation can be extended to include position sensors to inform the microcontroller about when the empty bottle is in place, as well as microcontroller outputs to inform the conveyer belt that the bottle is filled and can be taken away. An additional feature would be to extend the system from just one bottle size to several bottle sizes. Thus, the calibrated setting, T_{pump}, would be specific to the size of bottle being filled. A simple lookup table can be used to select the appropriate calibrated setting for the appropriate bottle size.

However, the important point about the open-loop process control described here is that *the accuracy of filling the bottle depends on having a reliable and reproducible model of the relationship between the calibrated setting, T_{pump}, and the amount of fluid being actually filled in the bottle.* If the pump wears out or if the liquid properties change such that the pump discharge is changed, then the existing value of the calibrated setting will produce inadequate filling of the bottle (either overfill or underfill, as the case may be). In this situation, operator action will be required to recalibrate the system and enter a new value for T_{pump} into the microcontroller. ∎

Example: Open-Loop Control of a Heating Process

Another open-loop control examined here is that of heating an enclosure. Consider the open-loop control presented in Figure 12.62. The heating element is placed in the enclosure. The heater driving electronics sends electrical energy to the heater. The microcontroller controls the heater through the heater driving electronics. For this purpose, the microcontroller may be using a PWM rectangular wave signal sent through one of the output pins, say PB3. The PWM

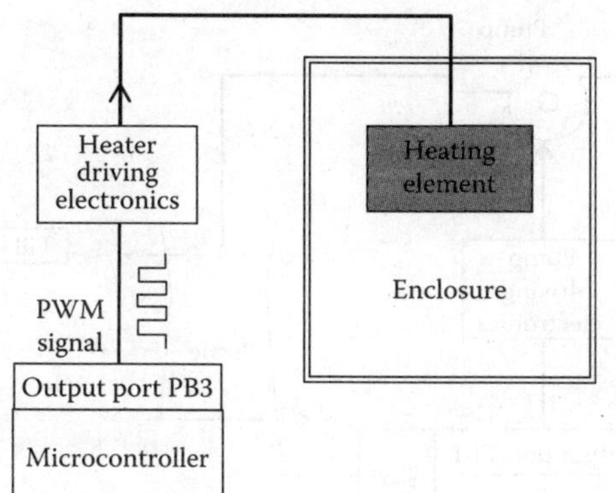

FIGURE 12.62
Schematic diagram of the setup for heating an enclosure under microcontroller supervision in open-loop control.

signal duty cycle (i.e., the ratio between ON time versus total time) controls how much energy is being actually sent to the heater. Assume that previous calibration has determined that a 45% duty cycle would produce the required heating under laboratory conditions. Thus, the input to the microcontroller is the calibrated setting $Duty\_cycle_{heater} = 45\%$. Of course, the microcontroller functions can be extended to turn on the heater at a certain moment of the day–night cycle and turn it off at a different moment. However, the main point about this example of an open-loop process control is that it essentially depends on the reliability of the calibrated setting. If the temperature in the environment surrounding the enclosure changes, or if the heater driving electronics wear out, then the enclosure will not have the expected temperature. In this case, recalibration of the system will be required, and a new calibrated setting will have to be determined. ∎

12.11.1.5 Use of Microcontrollers in Closed-Loop Control

In the previous examples we have seen the advantages and disadvantages of open-loop control. One of the major advantages is simplicity. A major disadvantage is the need for recalibration. This disadvantage can be eliminated by shifting from open-loop to closed-loop control. In closed-loop control, the microcontroller capabilities are better utilized, and the opportunities for large-scale automation are more fully exploited. Let us reconsider the previous two examples from the point of view of implementing closed-loop control.

Example: Closed-Loop Control of a Bottle Filling Process

Let us consider again the automation process in which the microcontroller supervises the filling of bottles from a tank. The open-loop control was presented in Figure 12.61. How can we automate this process in closed-loop control? There are several ways in which this can be done. One way would be to introduce a flow sensor in the bottle supply line to measure how much liquid has passed through. Thus, knowing the amount of liquid that has passed through the supply line, one can determine when the bottle is full and hence can stop the process. Another way would be to have a fill sensor determining directly when the bottle has filled up, and then sending the appropriate signal to the microcontroller to stop the pump. This latter solution is more attractive, because it addresses the problem directly. Hence, we will continue our discussion based on this implementation of the latter solution.

In closed-loop control, the bottle filling operation would proceed as shown in Figure 12.63. A pump draws the liquid from a supply tank and feeds it into the bottle. The bottle filling station is equipped with a fill sensor that detects when the bottle is full. (This sensor could be as

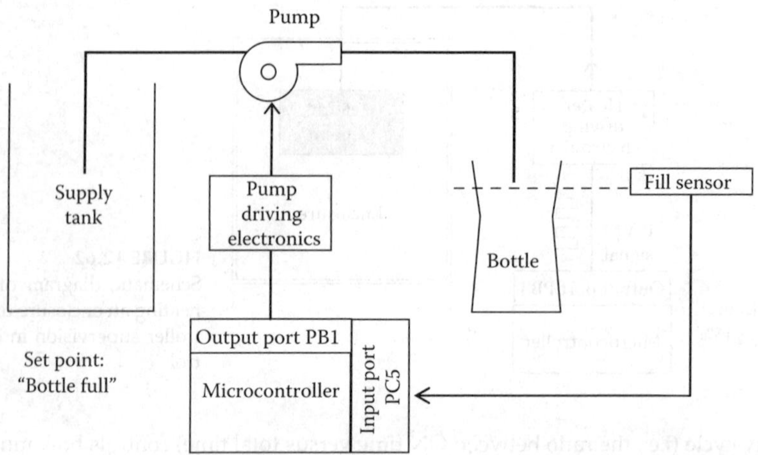

FIGURE 12.63
Schematic diagram of the setup for filling a bottle under microcontroller supervision in closed-loop control.

simple as an optoelectronic switch, or a capacitive switch, or a float, etc., depending on the application.) When the bottle is full, the fill sensor sends a signal to the microcontroller. In the implementation depicted in Figure 12.63, the fill sensor will send an ON signal to pin PC3 of the microcontroller input port C. Hence, the microcontroller will stop the pump by sending an OFF signal to the pump driving electronics. After the next bottle is set in position, the filling cycle is repeated. Of course, the automation can be extended to include position sensors to tell the microcontroller when the empty bottle is in place, as well as microcontroller outputs to tell the conveyer belt that the bottle is filled and can be take away. However, the important point about this closed-loop process control is that the accuracy of filling the bottle does not depend on having a dependable model of the relationship between a calibrated setting and the amount of fluid being actually filled in the bottle. If the pump wears out or if the liquid properties change such that the pump discharge changes, then the closed-loop process control will adjust itself automatically. In the particular solution presented in Figure 12.63, the process adjustment is done without even knowing what the relationship between the pump running time and the pump discharge is. This type of closed-loop control is called *model-less control* since it does not require a model of the process in order to implement the control algorithm. ∎

Example: Closed-Loop Control of a Heating Process

Let us consider again the automation process in which the microcontroller supervises the heating of an enclosure. The open-loop control was presented in Figure 12.62. How can this process be automated using closed-loop control? There are several ways in which this can be done. One way would be to introduce a power sensor in the heater supply line to measure how much energy is passing through it in unit time. This will tell us the power flow. By knowing the power passing through the heater supply line, one can evaluate if the heater driving electronics is operating properly. If necessary, one will inform the microcontroller that corrective actions are needed in the duty cycle in order to restore the electric power to its desired value. However, this solution will only partially take care of the problem. It will not detect if the heater delivery per unit electric power has changed or not.

Another solution for closed-loop control would be to install a temperature sensor in the enclosure. The temperature sensor will create an analog temperature signal proportional to the temperature inside the enclosure. The analog temperature signal will be sent to the

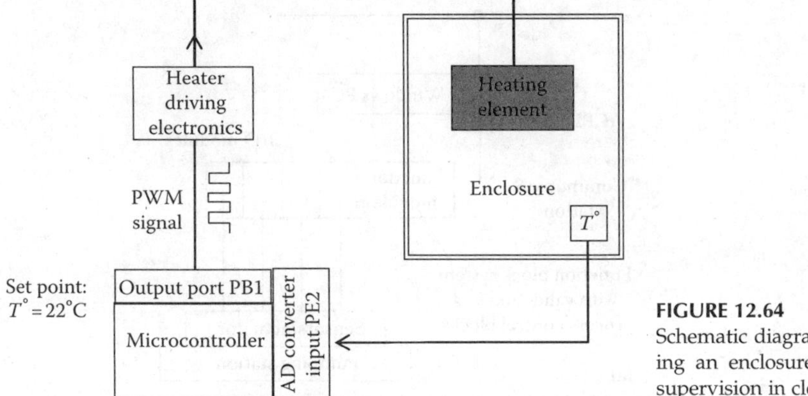

FIGURE 12.64
Schematic diagram of the setup for heating an enclosure under microcontroller supervision in closed-loop control.

microcontroller. In the implementation depicted in Figure 12.64, the analog temperature signal will be received at the AD converter pin AN2 which is the PE2 pin on port E. The AD converter will transform the analog signal into digital format. The microcontroller CPU will interpret the digitized temperature signal. i.e., it will compare the measured temperature with the set point temperature ($T = 22°C$ in Figure 12.64). Hence, the microcontroller will adjust the duty cycle to meet the new demands. If the measured temperature is less than the set point temperature, the microcontroller will increase the duty cycle value. If the measured temperature is above the set point temperature, the microcontroller will decrease the duty cycle value. If the measured temperature agrees with the set point temperature, the microcontroller will take no action. ∎

12.11.2 Hierarchical Process Control

Complicated processes may require complicated controls. One way of achieving the control of complicated processes is through *hierarchical process control*. In hierarchical process control, the control is set up at multiple levels, in a pyramidal structure. At the bottom of the pyramid lies the control of various independent actuators that perform local functions (pumps, heaters, etc.). At the top of the pyramid is the overall control of the process. At intermediate levels is the control of the sections or subprocesses that make up the complete process. Thus, a hierarchical tree-like structure is achieved, with branches and subbranches.

Example: Hierarchical Process Control of a Pump Station (Miniplant Simulation)

Let us consider a hierarchical control example developed by the Process Control Engineering Chair of the RWTH University of Technology Aachen, Germany (http://www.plt.rwth-aachen. de/english/vorstellung/vorstellung.html), the developers of the ACPLT© (Aachener Prozessleittechnik) process control simulation systems. The example under consideration is the *miniplant simulation*. The miniplant simulates a pump station such as the one that can be found in big waste water treatment plants. In the real plant, the process technology differs between heavy-polluted and light-polluted waste water. Heavy organically polluted waste water runs to a shelter basin. Light-polluted water goes to a waste water basin. To prevent the killing of all the bacteria in the treatment plant, the two types of polluted water must be mixed continuously. In the miniplant, this process is simulated by using differently tempered water.

Figure 12.65a shows the system control diagram of the miniplant. It consists of a function block library with different validation blocks which realize simple (e.g., variance or gradient checking) and complex (e.g., data reconciliation) validation methods. The simple validation blocks are

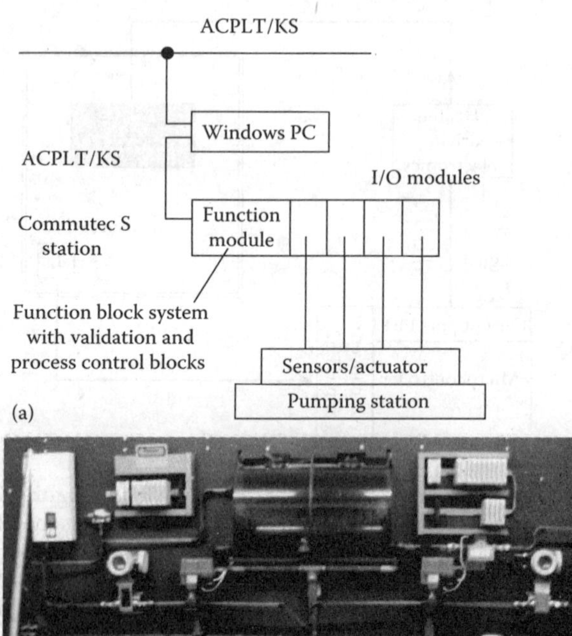

FIGURE 12.65
The microprocessor control of a simulated waste water plant (miniplant): (a) system control diagram; (b) photograph of the experimental setup. (From http://www.plt.rwth-aachen.de/english/vorstellung/vorstellung.html.)

realized inside the microcontroller-based function module of "Commutec S" made by Endress + Hauser (http://www.endress.com/). The more complex ones are realized in a function block system on a PC with Windows NT. A hierarchical control architecture was used to control the miniplant simulator. At the lowest level, standardized control units are used to control the actuators. In this application, also in the first group control level, standardized control units could be used. To develop the other superior control units, a design pattern was utilized. To control the miniplant, supervision information is integrated into the process control strategies. The actuator control units and validation blocks are implemented (inclusive ACPLT/OV and ACPLT/FB) in the microprocessor based control module Commutec S (Endress + Hauser). The group control units and higher lever validation blocks are implemented in a remote component, in this case a PC with Windows NT operating system. The communication, inclusive the distributed instruction oriented control mechanisms are realized with ACPLT/KS. Figure 12.65b presents a photograph of the experimental setup of the miniplant simulator. ■

12.12 Problems and Exercises

This section contains a selection of test problems and exercises. A complete collection of solved tutorial examples, homework assignments, and laboratory exercises is posted on the publisher's Web site. The reader is encouraged to visit the publisher's Web site in order grasp a full understanding of this subject.

PROBLEM 12.1

Write the range of 8-bit *unsigned* binary numbers in the form: from ... to ... (express your answer in decimal, hex, and binary)

PROBLEM 12.2

Write the range of 8-bit *signed* binary numbers in the form: from ... to ... (express your answer in decimal, hex, and binary)

PROBLEM 12.3

Predict if carry/borrow or overflow happen when the following operations are performed:

	Result	Did hex carry/borrow happen? (Y/N)	Did overflow happen? (Y/N)
$EA + $F5 =			
$7A + $3F =			
$00 − $07 =			

PROBLEM 12.4

Convert the *signed* binary numbers in the table below to hex and decimal

Signed Binary	Hex	Decimal
%01011101		
%11100001		
%01011011		
%11011110		

PROBLEM 12.5

Describe the meaning of the operation "STAA PORTB, X"

PROBLEM 12.6

Describe the meaning of the following characters when typed in an Assembly language program:

%	
*	
$	
,X	

PROBLEM 12.7

Perform the logical operations shown in the table below. Give your answers in both binary and hex.

	Prediction	
	Binary Operation	Hex Result
%10010010 **XOR** %01110100		
%01101111 **AND** %01011101		
%10100110 **NOR** %01010101		

PROBLEM 12.8

How many bits of data are being exchanged at one time when using (a) parallel communications; (b) serial communication?

PROBLEM 12.9

Which port(s) is (are) used for (a) parallel communication; (b) serial communication?

PROBLEM 12.10

How can one tell that serial communication data has been received?

PROBLEM 12.11

What has to occur first before you send out a serial communication transmission?

PROBLEM 12.12

What is the clock speed of the MCU in MHz? What is the duration of a machine cycle in microsecond?

PROBLEM 12.13

Draw a flowchart and write an assembly program to generate a rectangular pulse wave similar to the one shown in the picture below. Send the pulse continuously through pin PC7 of PORT C. Other pins of PORT C should be configured as input. The pulse should have the following form:

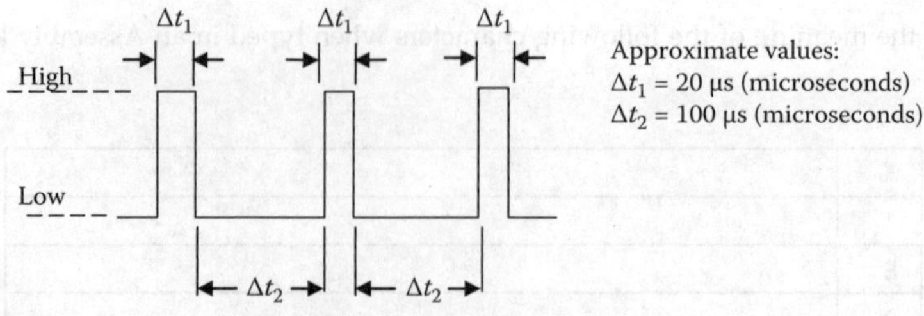

PROBLEM 12.14

Draw a flowchart and write a program to do the following:

1. Initialize serial communication
2. Wait until a keystroke (one character) is received through the serial communication
3. Read the received data into accB
4. Send the character ASCII code through serial communication as an "echo" to the monitor
5. Send the character ASCII code through Port C for LED display
6. Branch back such that you can repeat the sequence of reading the character and making the "echo" and sending it to port C.

PROBLEM 12.15

Explain the meaning of the opcodes BGT

PROBLEM 12.16

In what register is the IC3F flag bit located? How does one clear the IC3F flag bit?

PROBLEM 12.17

What binary number needs to be loaded into the timer register TCTL1 such that, upon a successful output compare operation, the following happens to the pins in port A: pin PA5 toggles, PA4 is set to one, and PA3 is set to zero?

PROBLEM 12.18

What binary number needs to be loaded into the timer register TCTL2 to set the pins in port A as follows: PA2 detects rising edges, PA1 detects falling edges, and PA0 detects both rising and falling edges?

PROBLEM 12.19

How many machine cycles does it take to execute each of the following lines of code?

LDX	#$1000
LDAA	#$FF
STAA	DDRC,X
LDAA	PORTB,X
STAA	PORTC,X

PROBLEM 12.20

In what register is the OC3F bit flag located? What event sets the OC3F bit flag?

PROBLEM 12.21

In what register is the IC1F bit flag located? What event sets the IC1F bit flag when EDG1B = 0 and EDG1A = 1?

PROBLEM 12.22

In what register is the TOF bit flag located? What event sets the TOF bit flag?

PROBLEM 12.23

How long (in ms) does it take to change the timer counter register TCNT from \$1000 to \$4000?

PROBLEM 12.24

The following serial-communication signal was captured on the oscilloscope screen. The time division knob is set to 0.005 s/div. Do the following:

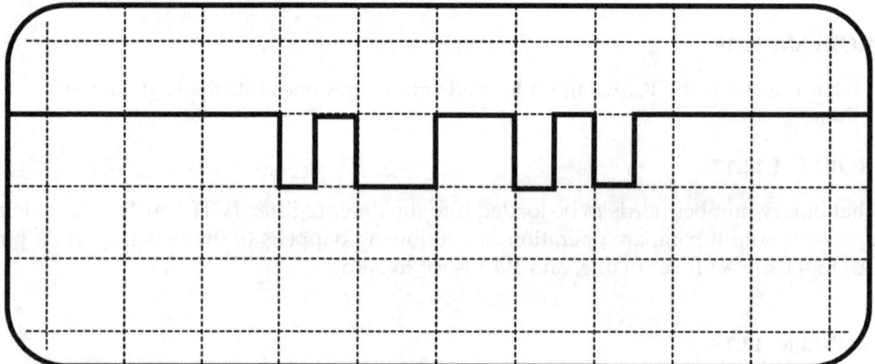

1. Identify and mark on the drawing the start bit and the end bit of the serial communication word
2. What is the *ASCII code* of the character transmitted through the SCI?
3. What is the character corresponding to this ASCII code?
4. How long (in s) does it take to send the complete character?
5. What was the Baud Rate used in this serial communication?

PROBLEM 12.25

An A/D converter is supplied with reference voltages $V_L = 0$ V, $V_H = 5$ V. What hex number will you get from the microprocessor when the input voltage is 3.0V?

PROBLEM 12.26

Fill in the rest of the table below to run the lab stepper motor in half steps forward (Note: the initial step is already filled in)

4-bit pattern				2-hex code	Sequence symbol	Position (rad)
0	1	0	1			

PROBLEM 12.27

Draw a flowchart and write a program to perform the tasks below:

1. Wait for a keystroke
2. Echo the keystroke to the monitor screen when received
3. If the transmitted character was "b"
   - Generate 50 cycles of a square wave on physical pin OC2; the half wavelength must be approximately 50 μs
   - After the wave is transmitted, go back to 1
4. If the transmitted character is "e," stop the program
5. If the transmitted character is neither "b," nor "e," go back to 1

PROBLEM 12.28

A DC electric motor supplied with $V = 24\,\text{V}$ has an internal resistance $R = 2.0\,\Omega$ and an electromechanical constant $K = 39.8\ \text{mV·s/rad}$. Find:

1. Maximum free running speed, in rad/s
2. Maximum free running speed, in RPM
3. Startup torque, T_S
4. Current associated with the startup torque, I_S
5. Power dissipation associated with the startup torque, P_S

References

Anon, *MC68HC11 Reference Manual*, Rev. 6, 4/2002, # M68HC11RM/D, Motorola Inc., 2000, http://e-www.motorola.com

Anon, *MC68HC11E Programming Reference Guide*, Rev. 1, 6/2002, # M68HC11ERG/AD, Motorola, Inc., 2002, http://e-www.motorola.com

Bolton, W., *Mechatronics: Electronic Control Systems in Mechanical and Electrical Engineering*, 2nd edn., Addison Wesley Longman, New York, 1999.

Driscoll, F. F., Coughlin, R. F., and Villanucci, R. S., *Data Acquisition and Process Control with the M68HC11 Microcontroller*, 2nd edn., Prentice-Hall, Upper Saddle River, NJ, 2000.

Haskell, R. E., *Design of Embedded Systems Using 68HC12/11 Microcontrollers*, Prentice-Hall, Upper Saddle River, NJ, 2000.

Histand, M. B. and Alciatore, D. G., *Introduction to Mechatronics and Measurement Systems*, WCB McGraw-Hill, Boston, MA, 1999.

Kheir, M., *The M68HC11 Microcontroller, Applications in Control, Instrumentation, and Communication* Prentice-Hall, Englewood Cliffs, NJ, 1997.

Morton, T. D., *Embedded Microcontrollers*, Prentice-Hall, Englewood Cliffs, NJ, 2001.

Spasov, P., *Microcontroller Technology, The 68C11*, 3rd edn., Prentice-Hall, Englewood Cliffs, NJ, 1999.

Tocci, R. J. and Ambrosio, F. J., *Microprocessors and Microcomputers, Hardware and Software*, 5th edn., Prentice-Hall, Upper Saddle River, NJ, 2000.

13

Fundamentals of Microfabrication

13.1 Introduction and Basic Processes

Various matured fabrication technologies and processes for distinct mechatronic systems and their components are available. To fabricate conventional, mini-, and microscale electromechanical motion devices and ICs, distinct technologies, processes, and materials are used. Affordable, high-yield, and effective conventional technologies have been used for many decades to fabricate various high-performance electromechanical motion devices, while complementary metal oxide semiconductor (CMOS) technology ensured high-yield for power electronics and ICs [1–7]. Power electronics, ICs, actuators, and sensors are profoundly different. However, the majority of microelectromechanical systems (MEMS) can be fabricated enhancing CMOS technologies [8–10]. MEMS imply not simply integration of electromechanical and electronic components, but utilization of enhanced microelectronics technologies and application of enabling materials to guarantee high-yield mass-produced cost-efficient system fabrications and packaging. This chapter is aimed to introduce the reader to the basics of MEMS fabrication. Different techniques should be utilized taking into the account various system components and specification. We will center on the fabrication basics for microelectromechanical devices which can be integrated with ICs forming MEMS [11]. Core micromachining and microelectronics processes will be introduced.

Micromachining, as a core fabrication technology, can be used to fabricate complex structures and devices which could be within micrometers in size. Integrated MEMS are composed from functional actuators, sensors, radiating energy devices, energy sources, ICs, etc. These MEMS can be fabricated utilizing different microfabrication technologies, e.g., CMOS, micromachining (*bulk* and *surface*), or high-aspect ratio processes. In contrast, the microelectronic components are fabricated using CMOS and biCMOS processes. The micromachining processes were developed as a well-defined sequence of steps to selectively etch away parts of the silicon wafer or other materials and deposit various structural and sacrificial layers of different materials forming mechanical, electromechanical, electro-opto-mechanical, electromagnetic, electrostatic, piezoelectric, and other devices and their components.

For integrated MEMS, the main goal is to integrate microelectronics with micromachined electromechanical devices in order to design high-performance systems. To guarantee high-performance, affordability, reliability, and manufacturability, well-developed CMOS-based batch-fabrication processes have been modified and enhanced. MEMS are packaged, and usually sealed, to protect them from harsh environments, prevent mechanical damage, minimize stresses, vibrations, contamination, and electromagnetic interference. It is impossible to specify *generic* MEMS, fabrication, and packaging solutions due to distinct device physics and unlimited number of possible application-dependent solutions.

Various fabrication techniques, processes, and materials must be applied to attain the desired performance, reliability, and cost. *Bulk* and *surface* micromachining, as well as high-aspect-ratio technologies such as Lithography–Galvanoforming–Molding (abbreviated as LIGA because it stands for Lithografie–Galvanik–Abformung in German), are the most developed fabrication methods. Silicon is the primary material which is used by the microelectronic industry. A single crystal ingot (solid cylinder hundreds millimeters in diameter and length) of very high purity silicon is grown, sawed to the desired thickness, and polished using chemical and mechanical polishing techniques. Electromagnetic and mechanical wafer properties depend on the orientation of the crystal growth, concentration and type of doped impurities. The major steps for ICs fabrication are diffusion, oxidation, photolithography, masking, deposition, etching, metallization, doping, planarization, packaging, and bonding. Extending conventional CMOS processing, additional processes and enabling materials are used to fabricate MEMS. There are a number of micromachining techniques which can be used in order to pattern and deposit thin films, as well as to shape silicon and other materials forming the needed microstructures and devices.

Photolithography (lithography) is the process used to transfer the mask pattern (desired pattern, surface topography, and geometry) to a layer of radiation- or light-sensitive material (photoresist). Then, the pattern is transferred to the films or substrates through development, etching, and other processes. The major steps in lithography are the masks production (pattern/topography generation) and transfer of the pattern. Important photoresist characteristics are resolution, sensitivity, etch resistance, thermal stability, adhesion, viscosity, flash point, and toxicity. The photoresist processing includes dehydration baking and priming, coating, soft baking, exposure, development, inspection, postbake (UV hardening), etc. The specified pattern is transferred to the wafer through etching, after which the photoresist is stripped by strong acid solutions (e.g., H_2SO_4), acid-oxidant solutions ($H_2SO_4 + Cr_2O_3$), organic solvents, alkaline strippers, oxygen plasma, or gaseous chemical reactants. Using wet and dry stripping, the photoresist must be removed without damaging silicon structures. In *surface* micromachining, an alternative solution to etching (in order to transfer the patterns from photoresist to thin films) is the lift-off. In lift-off, the photoresist is first patterned, the thin film to be patterned is deposited, and then the photoresist is dissolved. The photoresist acts as a sacrificial material under thin film regions to be removed.

High-resolution photolithography defines two-dimensional (planar) shapes, and, indirectly leads to three-dimensional features. Hence, microdevice and their structures (plates, stator, rotor, bearing, coils, and cavities.) geometry are defined photographically. First, a mask is produced on a glass plate. The silicon wafer is then coated with a polymer which is sensitive to ultraviolet light. This photoresistive layer is called photoresist. Ultraviolet light is shone through the mask onto the photoresist. The positive photoresist becomes softened, and the exposed layer can be removed. There are two types of photoresist, e.g., positive and negative. Where the ultraviolet light strikes the positive photoresist, it weakens the polymer. As the image is developed, the photoresist is rinsed where the light struck it. In contrast, if the ultraviolet light strikes negative photoresist, it strengthens the polymer. Therefore, a negative image of the mask results. When the photoresist is removed, the patterned oxide appears. Different chemical processes are involved to remove the oxide and other materials where they are exposed through the openings in the photoresist.

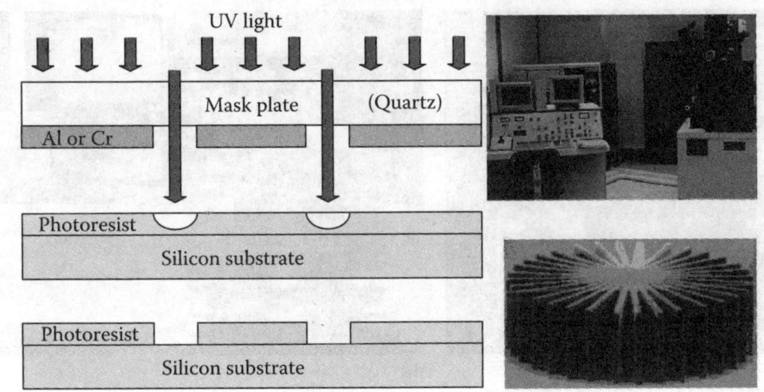

FIGURE 13.1
Photolithography process, photolithography system, and fabricated microstructure.

Alternatively, electron beam lithography can be used. Various computer aided design (CAD) tools are applied to support photolithography. The photolithography process and a photolithography system are illustrated in Figure 13.1.

A high-resolution positive image is needed, and different photolithography processes are developed. Deep UV lithography processes were developed to ensure the feature sizes of microstructures to be $\sim$0.1 μm. Different exposure wavelengths λ are used (e.g., 435, 365, 248, 193, and 157 nm) which are mainly predefined by the photoresist chemistry. Using the Rayleigh model for image resolution, one finds the expressions for image resolution i_R and the depth of focus d_F as $i_R = k_i \dfrac{\lambda}{N_A}$, $d_F = k_d \dfrac{\lambda}{N_A^2}$, where k_i and k_d are the lithographic process constants; λ is the exposure wavelength; N_A is the numerical aperture coefficient (for high numerical aperture we have N_A is $\sim$0.5).

The so-called g- and i-line IBM lithography processes (with wavelengths of 435 and 365 nm, respectively) allow one to attain $\sim$0.35 μm features. The deep ultraviolet light sources (mercury source or excimer lasers) with 248 nm wavelength enable $\sim$0.25 μm resolution. The changes to low exposure wavelength (λ is 193 or 157 nm) possess challenges due to photoresist robustness and stability. The resins photoresists, which are widely used for 365 and 248 nm lithography processes, are novolac and polyhydroxystyrene. These photoresists have absorption depths of 30–50 nm at 193 nm wavelength. Therefore, resins cannot be used as the single-layer photoresists at 193 nm wavelength. Methacrylates are semitransparent at 193 nm, and these polymers can serve as 193 nm lithography single-layer photoresists. Acid-catalyzed conversion of *t*-butyl methacrylate into methacrylic acid provides the chemical underpinning for several versions of these photoresists.

The computerized integrated steppers (step-and-repeat projection aligners) transfer the image of the microstructure or ICs from a master photomask to a specified area on the wafer surface. The substrate is then moved (stepped), and the image can be exposed once again to another area on the wafer. This process is repeated until the entire wafer is exposed. Different steppers are applied with different capabilities and features. The 1600DSA Ultratech computerized stepper system is illustrated in Figure 13.2a. The Karl Suss MA-6 mask aligners (for 6″ wafers) is commonly used to expose photoresist-coated substrates to ultraviolet light through photo masks, see Figure 13.2b.

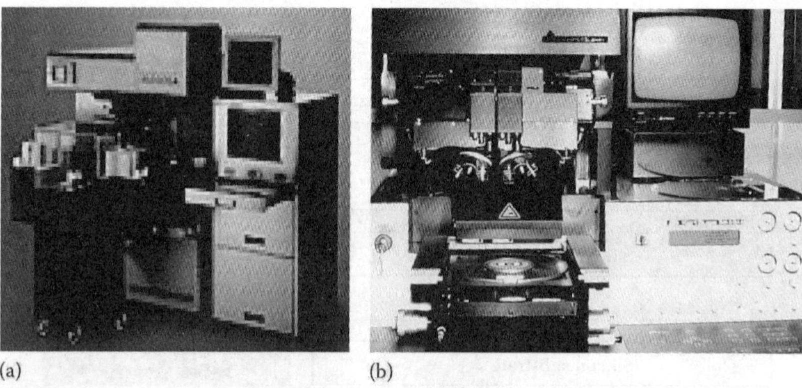

(a)   (b)

FIGURE 13.2
(a) Ultratech computerized stepper system; (b) Karl Suss MA-6 mask aligners.

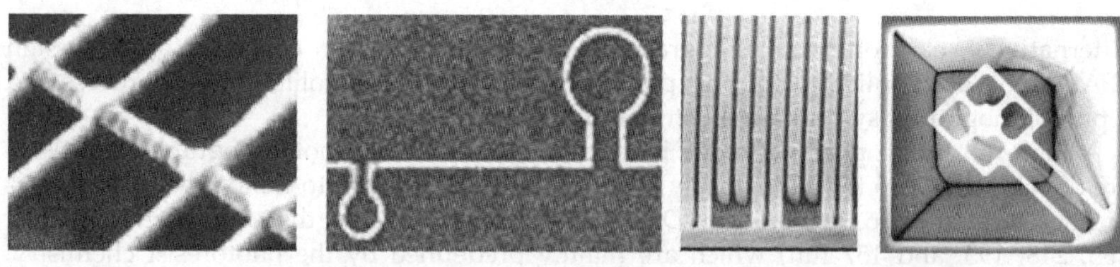

FIGURE 13.3
Copper–cobalt wires, gold current loops (0.5 μm and 1 μm diameter), gold antenna for the surface acoustic wave device, and microdevice with a suspended structure.

Some examples of the fabricated structures and devices are reported in Figure 13.3. In particular, we document the SEM images of Cu–Co wires (Co may be magnetized if needed after the fabrication process), 75 nm width gold loops (antenna or windings), electroplated 100 nm width Au for the surface acoustic wave device, and suspended structure for the microdevice with cavity. These structures were fabricated utilizing a sequence of steps, such as, photolithography, deposition, and etching.

Different microelectromechanical motion devices and microstructures can be designed, and silicon wafers with different crystal orientations can be used. Reactive ion etching (dry etching process) is commonly applied. Deep trenches and pits of desired shapes can be etched in a variety of materials including silicon, silicon oxide, and silicon nitride. A combination of dry and wet etching can be sequentially integrated. Metal and alloy thin films can be patterned using the lift-off stenciling technique. For example, a thin film of the assisting material (silicon oxide) is deposited first, and a layer of photoresist is deposited over and patterned. The silicon oxide is then etched to undercut the photoresist. The metal (or any other material) thin film is then deposited on the silicon wafer by the evaporation process. The metal pattern is stenciled through the gaps in the photoresist, which is then removed, lifting off the unwanted metal. The sacrificial layer is then stripped off, leaving the desired metal film pattern.

The *isotropic* and *anisotropic* wet etching, as well as the concentration-dependent etching, are widely used in bulk silicon micromachining. The microstructures are formed by etching away the bulk of silicon. *Surface* micromachining usually implies forming the structure in layers, formed by thin films, on the surface of the silicon wafer or other substrate. Hence, *surface* micromachining implies the use of thin films of two or more different materials, e.g., one uses the structural (polysilicon and metals.) and sacrificial layers. Both sacrificial and structural layers are deposited. Then, the sacrificial material is etched away to release the structure. A variety of different structures with different geometry can be fabricated. Wet chemical etching and dry etching (reactive ion etching, plasma etching, etc.) are used to etch various materials, such as semiconductors, conductors, alloys, and insulators (silicon oxide and silicon nitride). Wet and dry etchants, commonly used in micromachining and ICs, are listed in Tables 13.1 and 13.2 [8,9,11].

The *surface* micromachining is a well-developed, affordable, and high-yield technology. An illustrative *surface* micromachining process is represented in Figure 13.4.

For dry etching processes, such as physical etching, chemical plasma etching, and the combinations of physical and chemical etching, the recipes are available. Dry etching processes are based on the plasmas. The most important features of dry etching processes are feature size, control, wall profile, selectivity (the ratio of etch rates of the layer/material to be etched and to the layer/material to be kept), uniformities, defects, throughput, and radiation damage to dielectrics.

The basic microfabrication processes with the links to the manufacturer sites are reported on the MEMS Clearinghouse http://mems.isi.edu/ as well as other Web sites.

TABLE 13.1

Wet Etchants

Materials	Etchant and Expected Etch Rate
Polysilicon	6 mL HF, 100 mL HNO_3, 40 mL H_2O, 8000 Å/min, smooth edges
	1 mL HF, 26 mL HNO_3, 33 mL CH_3COOH, 1500 Å/min
Phosphorous-doped silicon dioxide (PSG)	Buffered hydrofluoric acid (BHF)
	28 mL HF, 170 mL H_2O, and 113 g NH_4F, 5000 Å/min
	1 mL BHF and 7 mL H_2O, 800 Å/min
Silicon nitride (Si_3N_4)	Hydrofluoric acid (HF)
	140 Å/min CVD at 1100°C
	750 Å/min CVD at 900°C
	1000 Å/min, CVD at 800°C
Silicon dioxide (SiO_2)	Buffered hydrofluoric acid (BHF)
	28 mL HF, 170 mL H_2O, and 113 g NH_4F, 1000–2500 Å/min
	1 mL BHF and 7 mL H_2O, 700–900 Å/min
Aluminum (Al)	4 mL H_3PO_4, 1 mL HNO_3, 4 mL CH_3COOH, 1 mL H_2O, 350 Å/min
	16–19 mL H_3PO_4, 1 mL HNO_3, 0–4 mL H_2O, 1500–2400 Å/min
Gold (Au)	3 mL HCl, 1 mL HNO_3, 25–50 μm/min
	4 g KI, 1 g I2, 40 mL H_2O, 0.5–1 μm/min
Chromium (Cr)	1 mL HCl, 1 mL glycerin, 800 Å/min, (need depassivation)
	1 mL HCl, 9 mL saturated $CeSO_4$ solution, 800 Å/min (need depassivation)
	1 mL (1 g NaOH in 2 mL H_2O), 3 mL (1 g $K_3Fe(CN)_6$ in 3 mL H_2O), 250–100 Å/min (photoresist mask)
Tungsten (W)	34 g KH_2PO_4, 13.4 g KOH, 33 g $K_3Fe(CN)_6$, and H_2O to make 1 L, 1600 Å/min (photoresist mask)

TABLE 13.2

Dry Etchants

Materials	Etchant (Gas) and Expected Etch Rate
Silicon dioxide (SiO_2)	$CF_4 + H_2$, C_2F_6, C_3F_8, or CHF_3, 500–800 Å/min
Phosphorous-doped silicon dioxide (PSG)	$SF_6 + Cl_2$, 1000–5000 Å/min
Silicon (single-crystal and polycrystalline)	CF_4, CF_4O_2, CF_3Cl, SF_6Cl, $Cl_2 + H_2$, $C_2ClF_5O_2$, SF_6O_2, SiF_4O_2, NF_3, $C_2Cl_3F_5$, or CCl_4He
Silicon nitride (Si_3N_4)	CF_4O_2, $CF_4 + H_2$, C_2F_6, or C_3F_8, SF_6He
Polysilicon	Cl_2, 500–900 Å/min
Aluminum (Al)	BCl_3, CCl_4, $SiCl_4$, BCl_3Cl_2, CCl_4Cl_2, or $SiCl_4Cl_2$
Gold (Au)	$C_2Cl_2F_4$ or Cl_2
Tungsten (W)	CF_4, CF_4O_2, C_2F_6, or SF_6
Al, Al-Si, Al-Cu	$BCl_3 + Cl_2$, 500 Å/min

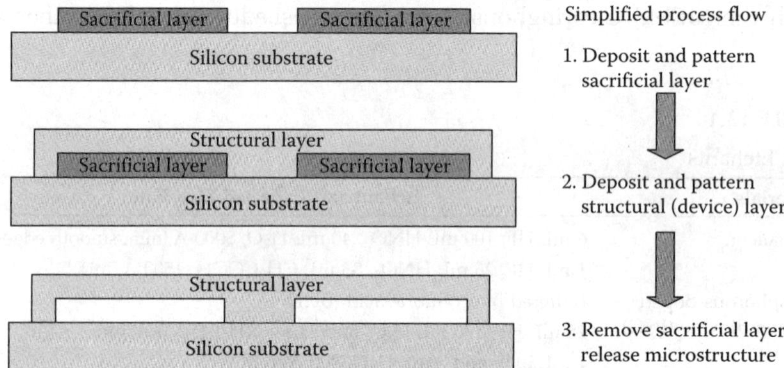

FIGURE 13.4
Surface micromachining process.

The author cannot assume responsibility for the validity, practicality, and fitness of any techniques, processes, sequential steps, data, chemicals, materials, as well as for the consequences of their use. The microelectronic, semiconductor, and MEMS manufacturers develop and utilize proprietary techniques, processes, steps, and materials. Only for introduction and educational purposes, the references to the MEMS Clearinghouse and university Web sites can be appropriate.

13.2 Microfabrication and Micromachining of ICs, Microstructures, and Microdevices

CMOS, high-aspect-ratio, and *surface* micromachining technologies are key factors for development, implementation, and commercialization of ICs and MEMS [1–11].

For MEMS, by utilizing the micromachining technology, one fabricates microscale structures and devices controlled by ICs. The low-cost high-yield CMOS technology enables fabrication of billions of transistors on a single chip, and currently, the minimal feature size is within $\sim$50 nm. This implies the *effective* FET size to be $\sim 1 \times 1$ μm. The size of each micromirror in the Texas Instrument DLP $\sim 1000 \times 1000$ mirror matrix is $\sim 10 \times 10$ μm. Signal processing, signal conditioning, interfacing, and control are performed by ICs. Microstructures and microdevices, integrated with ICs, comprise MEMS which have been widely used in actuation, sensing, and communication applications.

Microfabrication technologies are categorized as *bulk, surface,* and high-aspect-ratio. The major processes in fabrication of ICs and microstructures/microdevices are diffusion, deposition, patterning, lithography, etching, metallization, planarization, assembling, and packaging. Thin-film fabrication processes were developed and used for polysilicon, silicon dioxide, silicon nitride, and other different materials for years. In ICs, these thin films are used to build the active and passive circuitry components, as well as establish interconnect. Doping modifies the properties of the media. It was documented that the lithography processes are used to transfer the pattern from the mask (which defines the surface topography and geometry of ICs and microstructures) to the substrate or film surface, which is then selectively etched away to remove unwanted thin films, media, and regions to complete the pattern transfer. The number of masks depends on the design complexity, fabrication technology, processes, desired ICs and microstructure geometry. After testing, the wafers are diced, and devices are encapsulated (packaged) as final ICs, microdevices, or MEMS.

For MEMS, the material compatibility issues are important. The thermal conductivity and thermal expansion are studied, and the appropriate materials with the closest possible coefficients of thermal conductivity and expansion may be chosen. However, the desired electromagnetic and mechanical properties (permeability, resistivity, strength, and elasticity) should be ensured in order to guarantee specified overall performance and capabilities. Even simplest microstructures must be analyzed from mechanical, electromagnetic, and thermal prospectives emphasizing design, performance, and fabrication trade-offs. The designer tries to match the electromagnetic, mechanical, and thermal features with availability, feasibility, and affordability of the fabrication processes. The mechanical properties of different materials are documented in Table 13.3. Conductors,

TABLE 13.3

Mechanical Properties of Some Materials

Materials	Strength 10^9 N/m^2	Hardness kg/mm^2	Young's Modulus 10^9 Pa	Density g/cm^3	Thermal Conductivity W/cm-K	Thermal Expansion 10^{-6} 1/K
Si	7	850	190	2.3	1.6	2.6
SiC	21	2500	710	3.1	3.5	3.3
Diamond	54	6700	980	3.5	19	1
SiO$_2$	8.3	800	70	2.5	0.015	0.54
Si$_3$N$_4$	15	3500	380	3.1	0.2	0.8
Fe	13	420	200	7.8	0.8	12
Al	0.2	135	70	2.8	2.4	25
Mo	2	280	340	10	1.4	5.2
W	3.9	500	400	19	1.8	4.6

TABLE 13.4

Material Constants in SI Units

Material	Silver	Copper	Gold	Al	Tungsten	Zinc	Nickel	Iron
$\sigma \times 10^7$	6.2	5.8	4.1	3.8	1.82	1.67	1.45	1
ρ	1.6×10^{-8}	1.7×10^{-8}	2.4×10^{-8}	2.6×10^{-8}	5.5×10^{-8}	6×10^{-8}	6.9×10^{-8}	1×10^{-7}
μ_r	0.99998	0.999999	0.99999	1.00002	1.00008	500 nonlinear *BH*	600 nonlinear *BH*	900–280,000 nonlinear *BH*
$t_e\ 10^{-6}\ \mathrm{K}^{-1}$	19	16.5–17	14	23–24	4.6	30–31	13–13.5	11.5–12

Material	Si	SiO_2	Si_3N_4	SiC	GaAs	Ge
ε_r	11.8	3.8–3.9	7.6	6.5	11–13	16
$t_e \times 10^{-6}$	2.5–2.65	0.5–0.51	2.7	3–3.7	5.7–6.9	2.2–5
$\sigma \times 10^7$	1.5	0.01	N/a	2.2	0.5	0.7

insulators, magnetic, and other materials with the specified properties must be deposited. Some bulk material constants (conductivity σ, resistivity ρ at 20°C, relative permeability μ_r, thermal expansion t_e (per degree at 20°C), and dielectric constant—relative permittivity ε_r) are given in Table 13.4.

Remark. Constants, rates, and other data reported in this chapter are strongly affected and vary as functions of size, fabrication processes, operating envelope, environmental conditions, etc. The mechanical and electrical properties of various bulk and thin film materials, which are commonly used in ICs and MEMS fabrication, are given in the MEMS Clearinghouse Web site. One can use the Material Database on http://www.memsnet. org/material/.

Figure 13.5 illustrates the microtoroid and the linear microactuator. The windings and ferromagnetic cores can be fabricated using different materials and processes. Iron, nickel, cobalt, and their alloys (aluminum–nickel–cobalt) are ferromagnetic materials. In some application, one also may utilize ferrimagnetic materials and ferrites, such as iron oxide magnetite Fe_3O_4 or nickel ferrite $NiFe_2O_4$, which have low conductivity and μ_r.

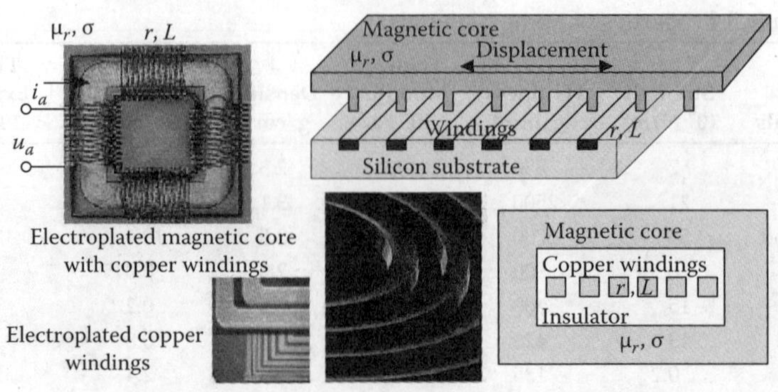

FIGURE 13.5
Ferromagnetic microtoroid and actuator with circular windings.

Various processes are utilized to fabricate MEMS. Oxidation of silicon wafers is used for passivation of the silicon surface (the formation of a chemically, electronically, and electromechanically stable surface), and making dielectric films.

Doping processes are used to selectively dope the substrate to produce either n- or p-type regions. These doped regions are used to fabricate passive and active circuitry components, form etch-stop-layers (important feature in *bulk* and *surface* micromachining), and produce conductive doped-silicon regions. Diffusion is achieved by placing wafers in a high-temperature furnace and passing a carrier gas which contains the desired dopant. For silicon, boron is the most common p-type dopant (acceptors), and arsenic and phosphorous are n-type dopants (donors). The dopant sources may be solid, liquid, and gaseous. Nitrogen is usually used as the carrier gas. Two major steps in diffusion are predeposition (impurity atoms are transported from the source to the wafer surface and diffused into the wafer, the number of atoms that enter the wafer surface is limited by the solid solubility of the dopant in the wafer), and drive-in (deposited wafer is heated in a diffusion furnace with an oxidizing or inert gas to redistribute the dopant in the wafer to reach a desired doping depth and uniformity). After deposition the wafer has a thin highly doped oxide layer on the silicon, and this oxide layer is removed by a hydrofluoric acid.

Metallization is the formation of metal films for interconnect, establishing contact, and protection. Metal thin films can be deposited on the surface by vacuum evaporation, sputtering, chemical vapor deposition, and electroplating.

Low-pressure chemical vapor deposition (LPCVD) is used to deposit metals, alloys, dielectrics, polysilicon, and other semiconductor, conductor, and insulator materials and compounds. The chemical reactants for the desired thin film are introduced into the CVD chamber in the vapor phase. The reactant gases then pyrochemically react at the heated surface of the wafer to form the desired thin film. Epitaxial growth, as the CVD process, allows one to grow a single crystalline layer on a single crystalline substrate. Homoepitaxy is the growth of the same type of material on the substrate (e.g., p+ silicon etch-stop-layer on an n-type substrate). Silicon homoepitaxy is used in *bulk* micromachining to form the etch-stop-layers. Heteroepitaxy is the growth of one material on a substrate which is a different type of medium. Plasma-enhanced chemical vapor deposition (PECVD) uses RF-induced plasma to provide additional energy to the reaction. The major advantage of PECVD is that it allows one to deposit thin films at low temperature.

Assembling and packaging of MEMS includes microstructure and die inspection, separation, attachment, wire bonding, and encapsulation. For robust packaging, one strives to match the thermal expansion coefficients for microstructures in order to minimize mechanical stresses. The connections and package must suit actuation and sensing features guaranteeing high-quality sealing, robustness, protection, interface, etc. Application-specific considerations may include operation in harsh adversarial environments (contaminates, electromagnetic interference, humidity, noise, radiation, shocks, temperature, and vibration), hybrid multichip packages, direct exposure to outside stimuli (light, gas, pressure, vibration, temperature, and radiation) or environment, etc. In MEMS, different bonding techniques are used to assemble individually micromachined (fabricated) structures to form microdevices, as well as integrate microstructures/ microdevices with controlling ICs. Silicon-direct bonding is used to bond a pair of silicon wafers together directly (face to face), while anodic bonding is used to bond silicon to glass. In silicon direct bonding, the polished sides of two silicon wafers are connected face to face and then annealed at high temperature. During annealing, the bonds are formed between the wafers. Anodic (electrostatic) bonding is used to bond silicon to glass.

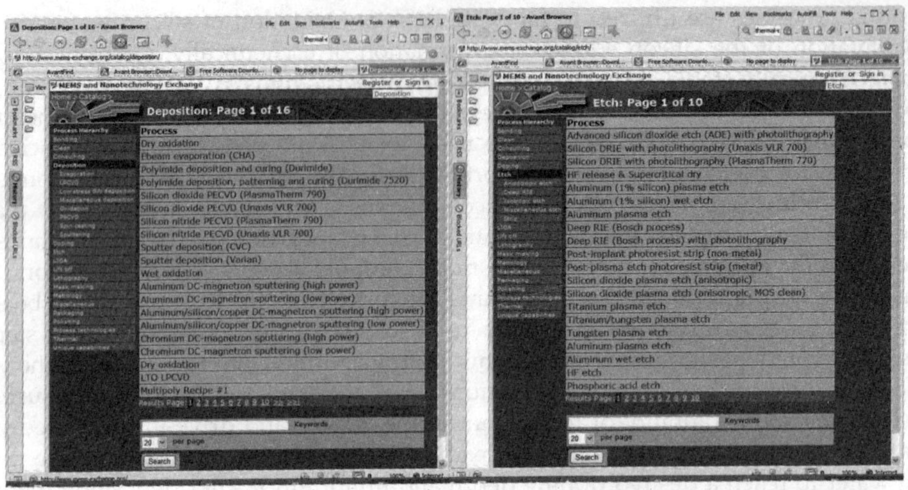

FIGURE 13.6
Deposition and etching processes, MEMS Clearinghouse. (From http://www.mems-exchange.org/catalog/deposition/ and http://www.mems-exchange.org/catalog/etch/)

The glass can be in the form of a plate or wafer, or as a thin film between two silicon wafers. Anodic bonding is performed at lower temperatures (450°C or less).

Specialized books on fabrication processes are available complimenting other recourses. For example, Figure 13.6 provides various deposition and etching processes, techniques, and data reported at the MEMS Clearinghouse.

The fabrication processes, techniques, and materials depend upon the available facilities. High-yield proprietary integrated CMOS-based MEMS-centered fabrication technologies are developed by leading MEMS manufacturers. There are various options in fabrication of microelectronics, microstructures, and microdevices. The general information, fabrication technologies, techniques, and processes are available at http://microlab.berkeley.edu/ (Microlab Laboratory, University of California at Berkeley), http://www-snf.stanford.edu/ (Stanford Nanofabrication Facility), and other. These recourses provide useful sources largely from educational standpoints because the high-technology industry has developed proprietary technique and processes to guarantee affordability, high-yield, specificity, and other features. The fitness, applicability, and reliability depend upon the equipment and infrastructure, as well as MEMSs themselves.

13.3 Bulk and Surface Micromachining, and Application of Microfabrication

Bulk and *surface* micromachining, as affordable and high-yield processes, are based on the modified CMOS processes. *Bulk* micromachining of silicon uses wet and dry etching techniques in conjunction with etch masks and etch-stop-layers to make microstructures from silicon. Microstructures are fabricated by etching specified areas of the silicon releasing the desired three-dimensional microstructures. The *anisotropic* and *isotropic* wet etching, as well as concentration-dependent etching, are widely used in *bulk* micromachining. The orientation-dependent *anisotropic* etchants (usually potassium hydroxide, KOH; sodium hydroxide, NaOH; H_2N_4; and ethylene-diamine-pyrocatecol, EDP) etch different crystallographic directions at different etch rates. Certain crystallographic planes (stopplanes) etch very slowly. Through *anisotropic* etching, various three-dimensional structures (cons, pyramids, cubes, and channels) are fabricated. The *anisotropic* etchants etch the (100) and (110) silicon crystal planes faster than the (111) crystal planes. For example, the etch rate could reach ~500:1 for (100) versus (111) orientations, respectively. Silicon dioxide, silicon nitride, and metal thin films (chromium and gold) provide good etch masks for typical silicon *anisotropic* etchants. These films are used to *mask* areas of silicon that must be protected from etching and to define the regions to be etched. Heavily boron-doped silicon (p^+ etch-stop) is effective in stopping the etch process. In contrast, the *isotropic* etching etches all directions in the silicon wafer at the same (or close) etch rate. Therefore, hemisphere and cylinder structures can be made. The widely implemented dry etching process is reactive ion etching. In this process, ions are accelerated toward the material to be etched, and the etching reaction is enhanced in the ion travel direction. Deep trenches and pits (up to a few tens of microns) of the specified shape with vertical walls can be etched in a variety of commonly used materials, e.g., silicon, polysilicon, silicon oxide, and silicon nitride. Compared with the *anisotropic* wet etching, dry etching is not limited by the crystal planes in the silicon. Figure 13.7 illustrates the Si structure with etched 2 μm holes, as well as 400 μm deep and 20 μm width grooves in 110-silicon.

Surface micromachining allows one to fabricate complex three-dimensional microstructures and devices. *Surface* micromachining is based on the use of sacrificial (temporary) layers which are used to maintain subsequent layers and then, removed to reveal (release) the fabricated microstructures. For example, on the surface of a silicon wafer, thin layers of structural and sacrificial materials are deposited and patterned. Then, the sacrificial material is removed, and a microelectromechanical structure or device is formed. Figure 13.8 documents an illustrative process sequence of the *surface* micromachining.

As the sacrificial layer, silicon dioxide (SiO_2), phosphorous-doped silicon dioxide (PSG), and silicon nitride (Si_3N_4) are widely used. The structural layers are typically formed with polysilicon, metals, and alloys. The sacrificial layer is removed. After fabrication of the surface microstructures and microdevices, the silicon wafer can be wet bulk etched to form cavities below the surface components, which allows one to ensure a wider range of desired motion for the device. The wet etching can be performed using hydrofluoric acid (HF), buffered hydrofluoric acid (BHF), KOH, EDP, tetramethylammonium hydroxide (TMAH), and NaOH.

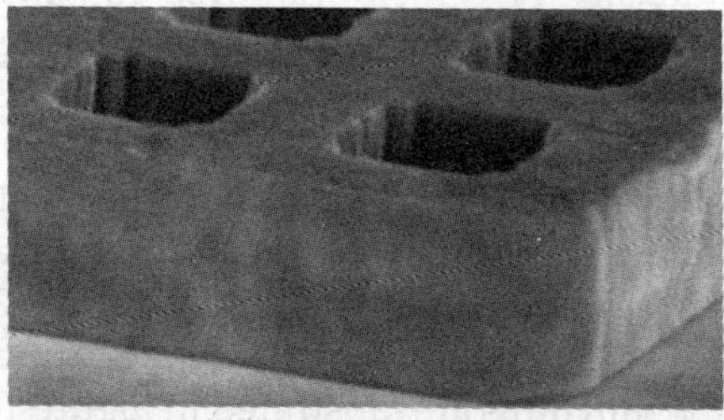

FIGURE 13.7
Si with 2 μm etched holes and deep grooves in 110-silicon.

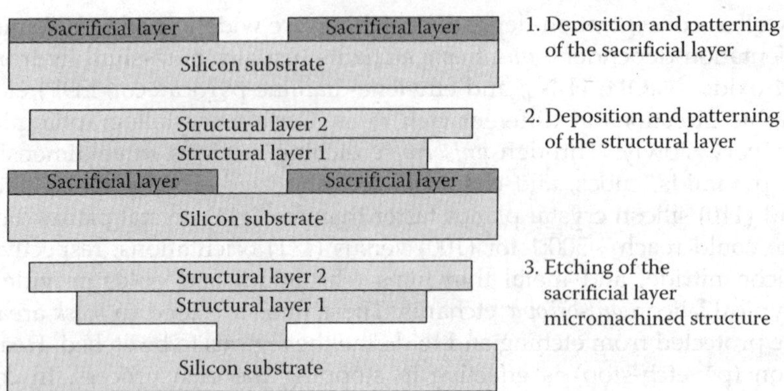

FIGURE 13.8
Surface micromachining.

Surface micromachining technology has been widely used in commercial microdevices (microtransducers, rotational/translational microservos, accelerometers, and gyroscopes.) and microstructures (beams, diaphragms, membranes, gears, electrostatic actuators, membranes, and mirrors). The most important attractive features of the *surface* micromachining technology are the small microstructure dimensions and the ability to integrate electromechanical, electronic, optical, and other components on the same chip.

Example process—The fabrication of microstructures and devices start as the synthesis, design, and analysis tasks are completed to meet the specifications and requirements. We report the simplified processes flow to fabricate the thin membrane. The polysilicon membrane can be fabricated by oxidizing a silicon substrate, patterning the silicon dioxide, deposition and patterning of polysilicon over the silicon dioxide, and removal of the silicon dioxide. To attain the actuation features, the Al thin film should be deposited and then, planar coils can be fabricated to achieve the displacement (as voltage is applied to coils in the external stationary magnetic field, established by a permanent magnet, the force is developed leading to actuation). The major steps and processes are given in Table 13.5 and illustrated in Figure 13.9.

In electromagnetic microstructures and microdevices, metals, alloys, ferromagnetic materials, magnets, and wires (windings) must be deposited. Using electron-beam lithography, the process of fabrication of micromagnets on the silicon substrate is illustrated in Figure 13.10. Development removes the photoresist which has been exposed to the electron beam. A ferromagnetic metal or alloy is then deposited, followed by lift-off of the unwanted material. This process allows one to make microstructures within ~50 nm dimension features. However, the volumetric density of ~100 × 100 × 100 nm magnet will be extremely low, making it virtually inapplicable for actuation and sensing the *engineered* silicon-technology systems. However, there is evidence that ~50 × 50 × 50 nm magnetic assemblies result in actuation and sensing in *natural* (living) systems.

The trade-offs in performance, capabilities, fabricationability, yield, and other features must be studied researching different materials. For example, nickel–iron alloys can lead to the saturational magnetization ~2 T [11,12], and the relative permeability can be ~2000. Narrow (soft) and medium *B–H* characteristics are achieved. The representative saturation magnetization and the *B–H* curves for $Ni_{x\%}Fe_{100-x\%}$ thin films are illustrated in Figure 13.11. The reader recalls that magnetic characteristics significantly depend upon dimensionality, thickness, and processes.

TABLE 13.5

Major Steps and Processes

Step (Process)	Brief Description
Step 1 (silicon dioxide growth)	Silicon dioxide is grown thermally on a silicon substrate. Growth can be performed in a water vapor ambient at $\sim$1000°C for 1 h. The silicon surfaces will be covered by $\sim$0.5 μm of silicon dioxide (thermal oxide thickness is limited to a few microns due to the diffusion of water vapor through silicon dioxide). Silicon dioxide can be deposited without modifying the surface of the substrate, but this process is slow to minimize the thin film stress. Silicon nitride may also be deposited, and its thickness is limited to $\sim$4 μm.
Step 2 (photoresist)	A photoresist is applied to the surface of the silicon dioxide. This can be done by spin coating the photoresist suspended in a solvent. The result after spinning and driving-off the solvent is a photoresist with thickness $\sim$2 μm. The photoresist is then soft baked to drive off the solvents inside.
Step 3 (photolithography exposure, and development)	The photoresist is exposed to ultraviolet light patterned by a photolithography mask (photomask). This photomask blocks the light and defines the pattern to guarantee the desired surface topography. Photomasks are usually made using fused silica. On one surface of the glass (or quartz), an opaque layer is patterned (usually hundreds of Å thick chromium layer). A photomask is generated based upon the desired form of the polysilicon membrane. The surface topography is specified by the mask. The photoresist is developed next. The exposed areas are removed in the developer. In a positive photoresist, the light will decrease the molecular weight of the photoresist, and the developer selectively removes (etches) the lower molecular weight material.
Step 4 (etch silicon dioxide)	The silicon dioxide is etched. The remaining photoresist will be used as a *hard mask* which protects sections of the silicon dioxide. The photoresist is removed by wet etching (hydrofluoric acid, sulfuric acid, and hydrogen peroxide) or dry etching (oxygen plasma). The result is a silicon dioxide thin film on the silicon substrate.
Step 5 (deposit polysilicon)	Polysilicon thin film is deposited over the silicon dioxide. For example, polysilicon can be deposited in the LPCVD system at $\sim$600°C in a silane (SiH_4) ambient. The typical deposition rate is $\sim$70 Å/min to minimize the internal stress and prevent bending and buckling (polysilicon thin film must be stress free or have a tensile internal stress). The thickness of the thin film is up to 4 μm.
Step 6 (photoresist)	Photoresist is applied to the polysilicon thin film, and planarization must be performed. The patterned silicon dioxide thin film changes the topology of the substrate surface. It is difficult to apply a uniform coat of photoresist over a surface with different heights. Hence, photoresist film has a different thickness, nonuniformity, and corners and edges of the patterns may not be covered. For 1 μm (or less) height, this problem is not significant; but, for thicker films and multiple layers, replanarization is required.
Step 7 (photolithography exposure, and development)	A photomask containing the desired topography (form) of the polysilicon membrane is aligned to the silicon dioxide membrane. Alignment accuracy (tolerances) can be achieved within the micrometer range, and the accuracy depends upon the size features of the microstructure and equipment used.
Step 8 (etch polysilicon)	The polysilicon thin film is etched with the photoresist protecting the specified polysilicon membrane. It is difficult to find a wet etch for polysilicon which does not attack the photoresist. Therefore, dry etching through plasma etching can be applied. Selectivity of the plasma between polysilicon and silicon dioxide is not a concern because the silicon dioxide will be removed later. Therefore, the polysilicon can be overetched by etching it longer than needed. This results in higher yield.

(continued)

TABLE 13.5 (continued)

Major Steps and Processes

Step (Process)	Brief Description
Step 9 (photoresist removal)	The photoresist protecting the polysilicon membrane is removed.
Step 10 (deposit Co)	Aluminum thin film is deposited and planar coils are fabricated utilizing the well-defined processes of microelectronics. For deposition, electroplating, evaporation, sputtering and other techniques can be applied.
Step 11 (remove silicon dioxide: release the thin film membrane)	The silicon dioxide is removed by wet etching (hydrofluoric or buffered hydrofluoric acids) because plasma etching cannot easily remove the silicon dioxide in the confined space under the polysilicon thin film. Hydrofluoric acid does not attack pure silicon. Hence, the polysilicon membrane and silicon substrate will not be etched. After the silicon dioxide is removed, the movable polysilicon membrane is formed (released). This membrane can bend down and stick to the surface of the substrate during drying after the wet etch. To avoid this, a rough polysilicon, which does not stick, can be used. Other solution is to fabricate the polysilicon membrane with the internal stress attaining that the polysilicon membrane it bended (curved) up during drying.

As reported in this book, the piezoelectric effect has been used in translational and rotational microtransducers. By applying voltage to the piezoelectric film, the film expands or contracts. Typical piezoelectric thin films used in microactuators are zinc oxide (ZnO), lead zirconate titanate (PZT), polyvinylidene difluoride (PVDF), and lead magnesium niobate (PMN). The linear microactuators fabricated using shape memory alloys have found limited applications compared with the piezoelectric devices.

Various AC and DC, as well as axial- and radial-topology permanent magnet electromechanical motion devices were covered. These microtransducers have stationary and rotating members (stator and rotor) and radiating energy microstructures. Surface micromachining technology was used to fabricate rotational micromachines with the rotor outer radius ~100 μm, air gap 1 μm, and bearing clearance 0.2 μm [11,13–16]. The cross section of the synchronous microdevice fabricated on the silicon substrate with polysilicon stator with deposited windings, polysilicon rotor with deposited permanent-magnets and bearing is illustrated in Figure 13.12a. Figure 13.12b illustrates the axial topology micromotor which significantly reduces the fabrication complexity. The stator is made on the substrate with deposited windings. The rotor with permanent-magnet thin films rotates due to the electromagnetic torque developed.

How to deposit material which traditionally was not widely used in microelectronics? We focus our attention on the deposition through electroplating. Electrochemistry has been utilized for copper, zinc, cobalt, nickel, and other metals. When aqueous solutions are electrolyzed using metal electrodes, an electrode will be oxidized if its oxidation potential is greater than that for water. As examples, nickel and copper are oxidized more readily than water, and the reactions are

For nickel Ni (s) $\rightarrow$ Ni^{2+} (aq) $+$ 2e^{-}, $E_{ox} = 0.28$ V,
$$2H_2O \text{ (l)} \rightarrow 4H^+ \text{ (aq)} + O_2 \text{ (g)} + 4e^-, \quad E_{ox} = -1.23 \text{ V};$$

For copper Cu (s) $\rightarrow$ Cu^{2+} (aq) $+$ 2e^{-}, $E_{ox} = -0.34$ V,
$$2H_2O \text{ (l)} \rightarrow 4H^+ \text{ (aq)} + O_2 \text{ (g)} + 4e^-, \quad E_{ox} = -1.23 \text{ V},$$

where s, aq, l, and g denote the solid, aqueous, liquid, and gas states.

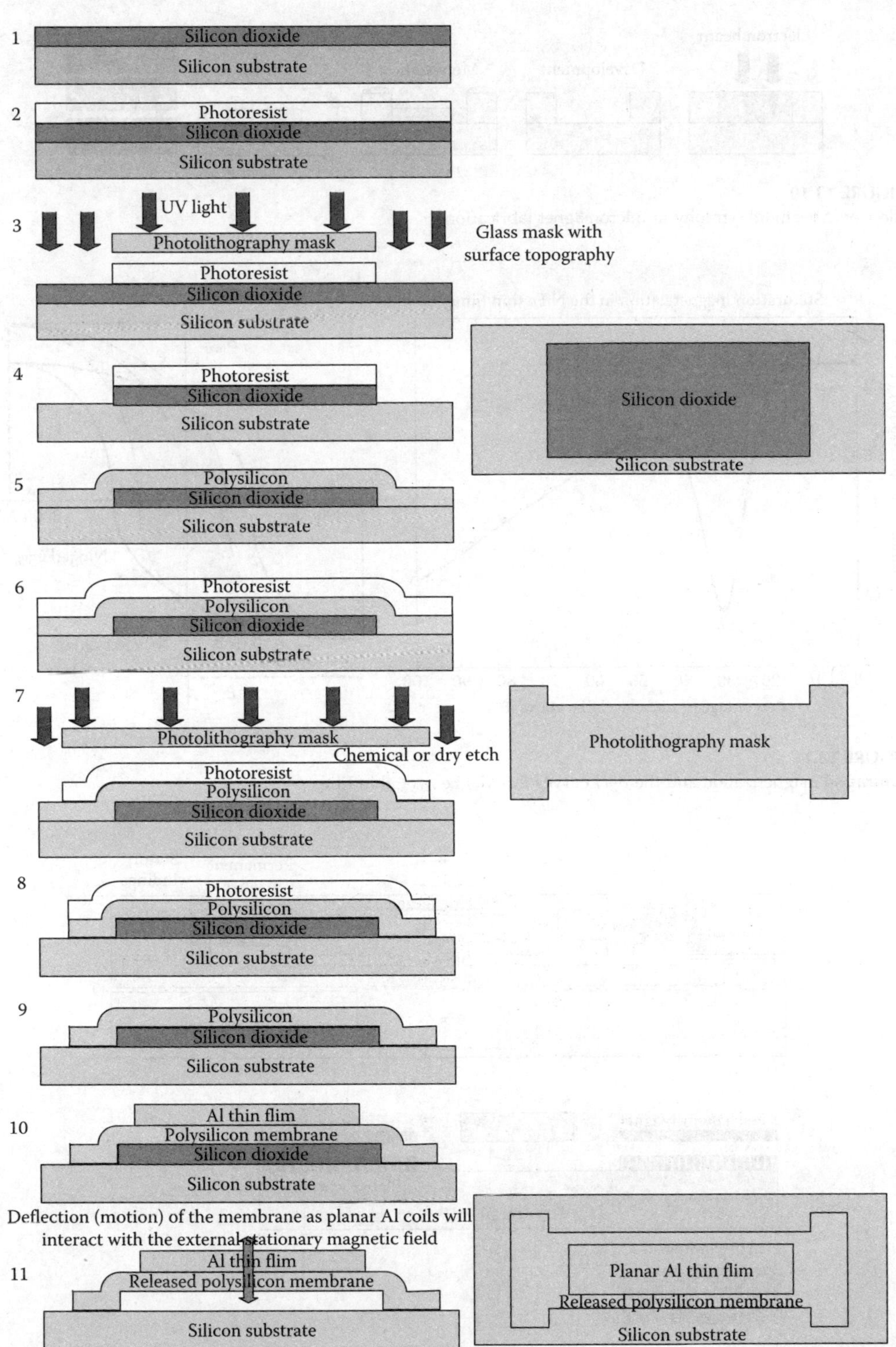

FIGURE 13.9
Micromachining fabrication of the polysilicon thin film membrane with Al film.

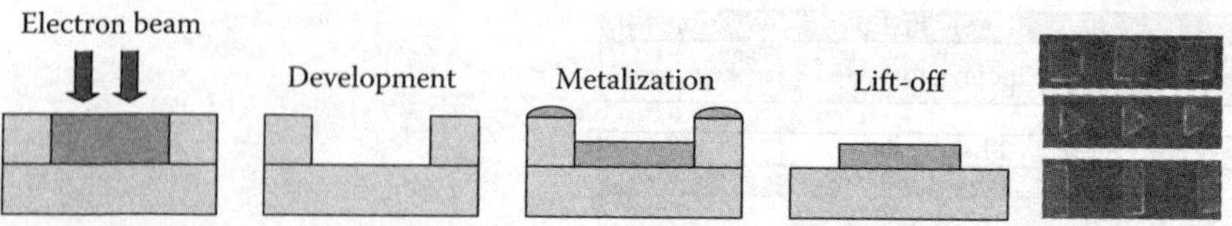

FIGURE 13.10
Electronic beam lithography in micromagnet fabrications.

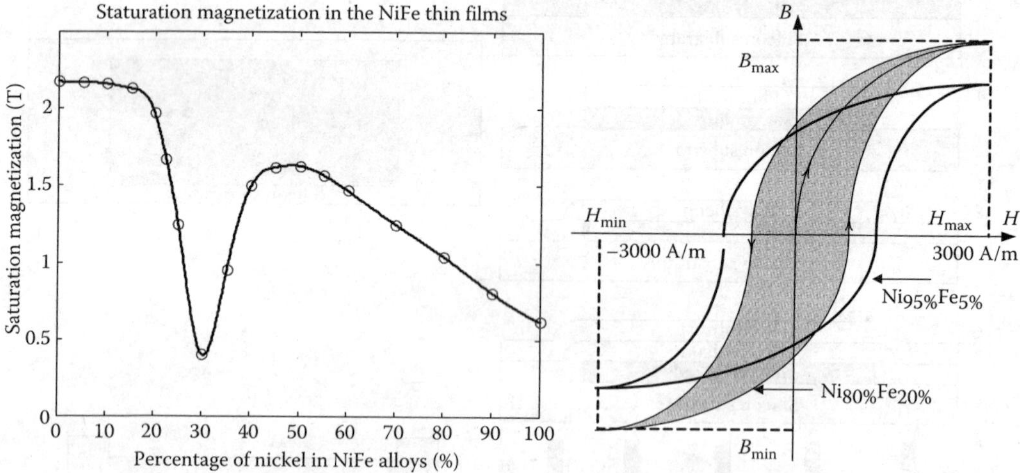

FIGURE 13.11
Saturation magnetization and the *B–H* curves for $Ni_{x\%}Fe_{100-x\%}$ thin films.

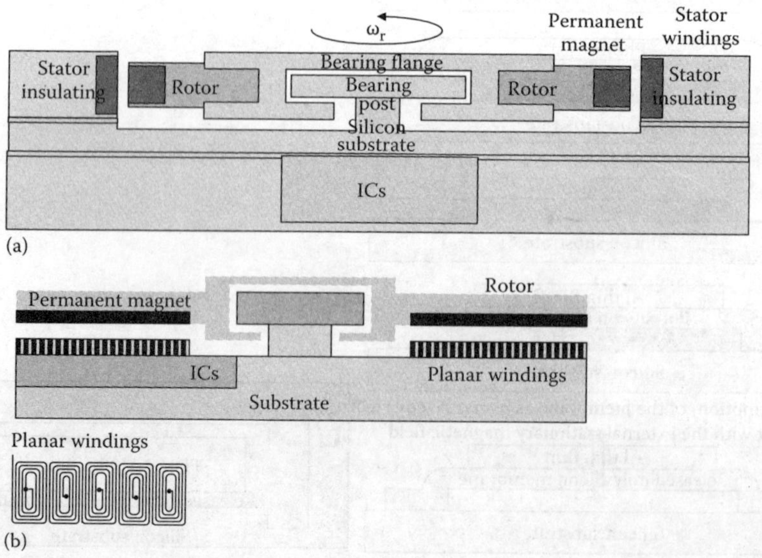

FIGURE 13.12
(a) Cross section of a permanent-magnet brushless slotless micromachine with driving/controlling ICs; (b) axial electromagnetic micromotor with driving/controlling ICs.

If the anode is made from nickel in an electrolytic cell, nickel metal is oxidized as the anode reaction. If $Ni^{2+}(aq)$ is the solution, it is reduced at the cathode in the preference to reduction of water. As current flows, nickel dissolves from the anode and deposits on the cathode. The reactions are

$$Ni(s) \rightarrow Ni^{2+}(aq) + 2e^- \text{ (anode)} \quad \text{and} \quad Ni^{2+}(aq) + 2e^- \rightarrow Ni(s) \text{ (cathode)}.$$

Most metals (chromium, iron, cobalt, nickel, copper, zinc, silver, and gold) used to fabricate MEMS are the transition metals. Electrodeposition of metals is made by immersing a conductive surface in a solution containing ions of the metal to be deposited. The surface is electrically connected to an external power supply, and current is fed through the surface into the solution. The reaction of the metal ions ($Metal^{x+}$) with x electrons (xe^-) to form metal (Metal) is

$$Metal^{x+} + xe^- = Metal.$$

The hydrated Cu ions reaction is $Cu^{2+} \rightarrow Cu(H_2O)_6^{2+}$, and the cathode reactions are

$$Cu^{2+} + 2e^- \rightarrow Cu, \quad Cu^{2+} + e^- \rightarrow Cu^+, \quad Cu^+ + e^- \rightarrow Cu, \quad 2Cu^+ \rightarrow Cu^{2+} + Cu,$$

$$H^+ + e^- \rightarrow \tfrac{1}{2} H_2.$$

Ferromagnetic cores in microstructures and microtransducers must be made. For example, the electroplated $Ni_{x\%}Fe_{100-x\%}$ thin films, such as *permalloy* $Ni_{80\%}Fe_{20\%}$, can be deposited to form the ferromagnetic core, inductors, transformers, switches, etc. The basic processes and sequential steps used are similar to the processes for the copper electrodeposition, and the electroplating is done in the electroplating bath. The windings (coils) must be insulated from the cores. Therefore, the insulation layers must be deposited. The sketched fabrication process with sequential steps to make the electromagnetic microtransducer with movable and stationary members is illustrated in Figure 13.13. On the silicon substrate, the chromium–copper–chromium (Cr–Cu–Cr) seed layer is deposited by evaporation, forming a seed layer for electroplating. The insulation layer (polyimide Dupont PI-2611) is spun on the top of the seed layer to form the electroplating molds. Several coats can be done to obtain the desired thickness of the polyimide molds (one coat results in ~10 μm insulation layer thickness). After coating, the polyimide is cured (at ~300°C) in nitrogen for 1 h. A thin aluminum layer is deposited on top of the cured polyimide to form a hard mask for dry etching. Molds for the lower conductors are patterned and plasma etched until the seed layer is exposed. After etching the aluminum (hard mask) and chromium (top Cr–Cu–Cr seed layer), the molds are filled with the electroplated copper through the copper electroplating process. One coat of polyimide insulates the lower conductors and the core (thus, ensuring insulation). The seed layer is deposited, mesh-patterned, coated with polyimide, and hard-cured. The aluminum layers (hard mask for dry etching) are deposited, and the mold for the ferromagnetic cores is patterned and etched until the seed layer is exposed. After etching aluminum (hard mask) and chromium (top Cr–Cu–Cr seed layer), the mold is filled with electroplated $Ni_{x\%}Fe_{100-x\%}$. The desired composition and thickness of the $Ni_{x\%}Fe_{100-x\%}$ thin films can be achieved. One coat of the insulation layer (polyimide) is spin-cast and cured to insulate the core and upper conductors. The via holes are patterned in the sputtered aluminum layer (hard mask) and etched through the polyimide layer using oxygen plasma. The vias are filled with the electroplated copper (electroplating process).

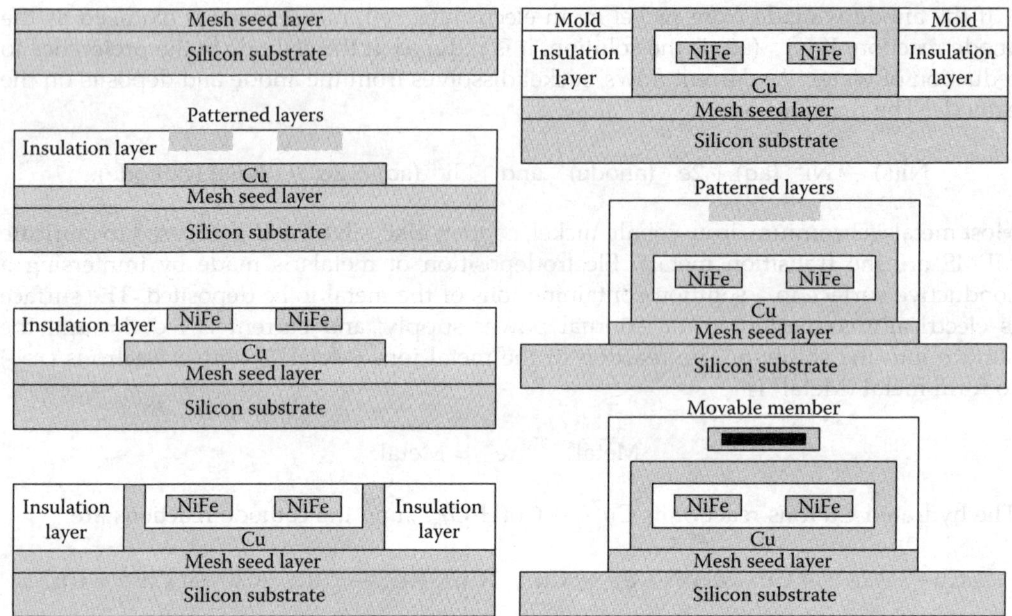

FIGURE 13.13
Basic fabrication sequential steps for the microtransducer fabrication.

The copper–chromium seed layer is deposited, and the molds for the upper conductors are formed using a thick photoresist. The molds are filled with the electroplated copper and then removed. Then, the gap for the movable member is made using the conventional processes. After removing the seed layer, the passivation layer (polyimide) is coated and cured to protect the top conductors. The polyimide is masked and etched to the silicon substrate. The bottom mesh seed layer is wet etched, and the microtransducer is diced and sealed.

Electroplating and micromolding processes are used to deposit NiFe alloys ($Ni_{x\%}Fe_{100-x\%}$ thin films). In conventional electromechanical motion devices, the $Ni_{80\%}Fe_{20\%}$ alloy is called *permalloy*, while $Ni_{50\%}Fe_{50\%}$ is called *orthonol*. To deposit $Ni_{x\%}Fe_{100-x\%}$ thin films, the silicon wafer is covered with a seed layer, for example, Cr–Cu–Cr. The photoresist layer ($\sim$20 μm Shipley STR-1110) is deposited on the seed layer and patterned. Then, the electrodeposition of $Ni_{x\%}Fe_{100-x\%}$ is performed at the specified temperature (usually $\sim$25°C) using a two-electrode system. The current density is from 1 to 30 mA/cm^2. The temperature and pH should be maintained within the recommended values. High pH causes highly stressed NiFe thin films, and the low pH reduces leveling and causes chemical dissolving of the iron anodes resulting in disruption of the bath equilibrium and nonuniformity. High temperature leads to hazy deposits, while low temperature causes high current density burning. Hence, many issues must be accounted for. The "pulse-width-modulation" (with varied waveforms, different forward and reverse magnitudes, and controlled duty cycle) can be used applying commercial or in-house made power supplies. Denoting the duty cycle length as T, the forward and reverse pulse lengths are denoted as T_f and T_r. The pulse length T is usually in the range from 5 to 20 μs, and the duty cycle (ratio T_f/T_r) can be varied from 0.1 to 1. The ratio T_f/T_r affects the percentage of Ni in the $Ni_{x\%}Fe_{100-x\%}$ thin films. Hence, the composition of $Ni_{x\%}Fe_{100-x\%}$ can be regulated based upon the desired properties. By varying the ratio T_f/T_r, the changes of the Ni are

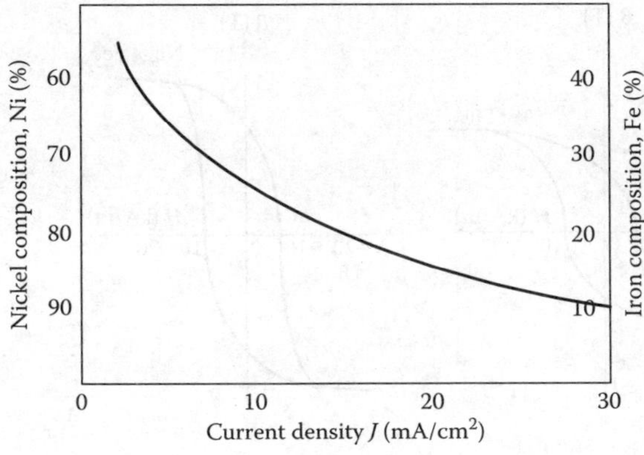

FIGURE 13.14

Nickel and iron compositions in $Ni_{x\%}Fe_{100-x\%}$ thin films as the functions of the current density.

relatively modest (from 85% to 79%). Therefore, other parameters vary to attain the desired composition. The nickel (and iron) composition is a function of the current density, and Figure 13.14 illustrates the nickel (iron) composition in the $Ni_{x\%}Fe_{100-x\%}$ thin films.

The $Ni_{80\%}Fe_{20\%}$ thin films of different thickness (which is a function of the electrodeposition time) are usually made at the current density ~15 mA/cm^2. This range of the current density can be used to fabricate *permalloy* thin films with 500 nm to 50 μm thickness. The rms value of the thin film roughness could reach ~10 nm for the 50 μm thickness. To guarantee good surface quality, the current density should be kept at the specified range, and usually to change the composition of the $Ni_{x\%}Fe_{100-x\%}$ thin films, the reverse current is controlled. The electroplating bath may contain various chemicals. The air agitation and saccharin are added to reduce internal stress and to keep the Fe composition stable. The deposition rate varies linearly as a function of the current density (the Faraday law is obeyed), and the electrodeposition rate is 100–150 nm-cm^2/min-mA. The *permalloy* thin films density is 9 g/cm^3 as for the bulk *permalloy*. The magnetic properties of the $Ni_{80\%}Fe_{20\%}$ (*permalloy*) thin films are of interest. The field coercivity H_c is a function of the thickness. For example, H_c is ~600 A/m for 150 nm thickness, and ~30 A/m for 600 nm films. Two possible solutions to electroplate $Ni_{80\%}Fe_{20\%}$ (deposited at 25°C) and $Ni_{50\%}Fe_{50\%}$ (deposited at 55°C) are (1) $Ni_{80\%}Fe_{20\%}$: $NiSO_4$–$6H_2O$ (200 g/L/), $FeSO_4$–$7H_2O$ (9 g/L), $NiCl_2$–$6H_2O$ (5 g/L), H_3BO_3 (27 g/L), saccharine (3 g/L), and pH (2.5–3.5); (2) $Ni_{50\%}Fe_{50\%}$: $NiSO_4$–$6H_2O$ (170 g/L), $FeSO_4$–$7H_2O$ (80 g/L), $NiCl_2$–$6H_2O$ (138 g/L), H_3BO_3 (50 g/L), saccharine (3 g/L), and pH (3.5–4.5).

In general, *permalloy* thin films have near-optimal magnetic properties in the following composition of nickel and iron: 80.5% of Ni and 19.5% of Fe. Thin films with minimal magnetostriction usually have optimal coercivity and permeability properties. For $Ni_{80.5\%}Fe_{19.5\%}$ thin films, the magnetostriction has zero crossing, the coercivity is 20 A/m or higher (coercivity is a nonlinear function of the film thickness), and μ_r is from 600 to 2000. Varying the composition of Fe and Ni, the characteristics of $Ni_{x\%}Fe_{100-x\%}$ will be changed. The composition of the $Ni_{x\%}Fe_{100-x\%}$ thin films is controlled by changing the current density, T_f/T_r ratio, bath temperature (varying the temperature, the composition of Ni can be varied from 75% to 92%), reverse current (varying the reverse current in the range from 0 to 1 A, the composition of Ni can be changed from 72% to 90%), air agitation of the solution, frequency (from 0.1 to 1 Hz), and forward and reverse pulses waveforms. The *B–H* curves for three different $Ni_{x\%}Fe_{100-x\%}$ thin films are illustrated in

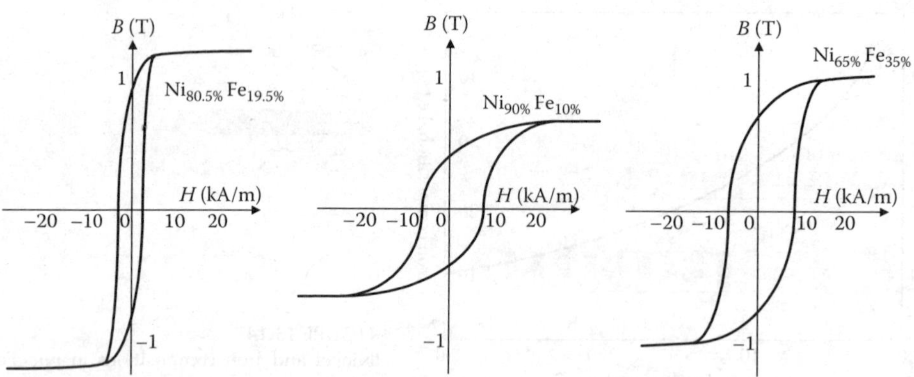

FIGURE 13.15
B–H curves for $Ni_{x\%}Fe_{100-x\%}$ thin films.

Figure 13.15. The $Ni_{80.5\%}Fe_{19.5\%}$ thin films have the saturation flux density 1.2 T, coercivity 20 (for *permalloy*) to 500 A/m, remanence B_r 0.26 T A/m, and μ_r from 600 to 2000.

Other electroplated permanent magnets (NiFeMo, NiCo, CoNiMnP, etc.) and micromachined polymer magnets exhibit good magnetic properties and can be used as the alternative solution to the $Ni_{x\%}Fe_{100-x\%}$ thin films. It was reported in the literature that [11,12] (1) $Ni_{79\%}Fe_{17\%}Mo_{4\%}$ thin films have the flux density $\sim$0.7 T, coercivity $\sim$5 A/m, and relative permeability $\sim$3500; (2) $Ni_{85\%}Fe_{14\%}Mo_{1\%}$ thin films have the flux density $\sim$1 T, coercivity from $\sim$10 to 300 A/m, and relative permeability from 3,000 to 20,000; (3) $Ni_{50\%}Co_{50\%}$ thin films have the flux density $\sim$1 T, coercivity from 1200 to 1500 A/m, and relative permeability $\sim$125 ($Ni_{79\%}Co_{21\%}$ thin films have much lower μ_r, which is $\sim$20). In general, high flux density, low coercivity, and high permeability lead to high-performance devices. The magnetic characteristics are significantly affected by thickness, and fabrication processes.

Polymer magnets (magnetically hard ceramic ferrite powder embedded in epoxy resin) can be used. The polymer magnets have thicknesses ranging from hundreds of micrometers to several millimeters. The micromachined polymer permanent-magnet disk with 80% strontium ferrite concentration (4 mm diameter and 90 μm thickness), magnetized normal to the thin-film plane (in the thickness direction), has the intrinsic coercivity $H_{ci} \sim 300,000$ A/m and a residual induction $B_r \sim 0.06$ T [11,12].

The electromagnetic microactuator (permanent magnet on the flexible beam and spiral planar windings controlled by ICs) is documented in Figure 13.16. In addition, the microstructure with eight cantilever beams and eight-by-eight fiber switching matrix are illustrated. The spiral planar coils can be made on the one-sided laminated copper layer using photolithography and wet etching in the ferric chloride solution. The x μm thick N-turn winding will establish the magnetic dipole moment $\mathbf{m}$ (the number of turns is a function of the footprint area, width, thickness, spacing, outer-inner radii, geometry, fabrication techniques, and material) which interacts with the stationary field $\mathbf{B}$ developed by a permanent magnet. The fabrication processes, briefly described in Figure 13.16, are covered in Ref. [11].

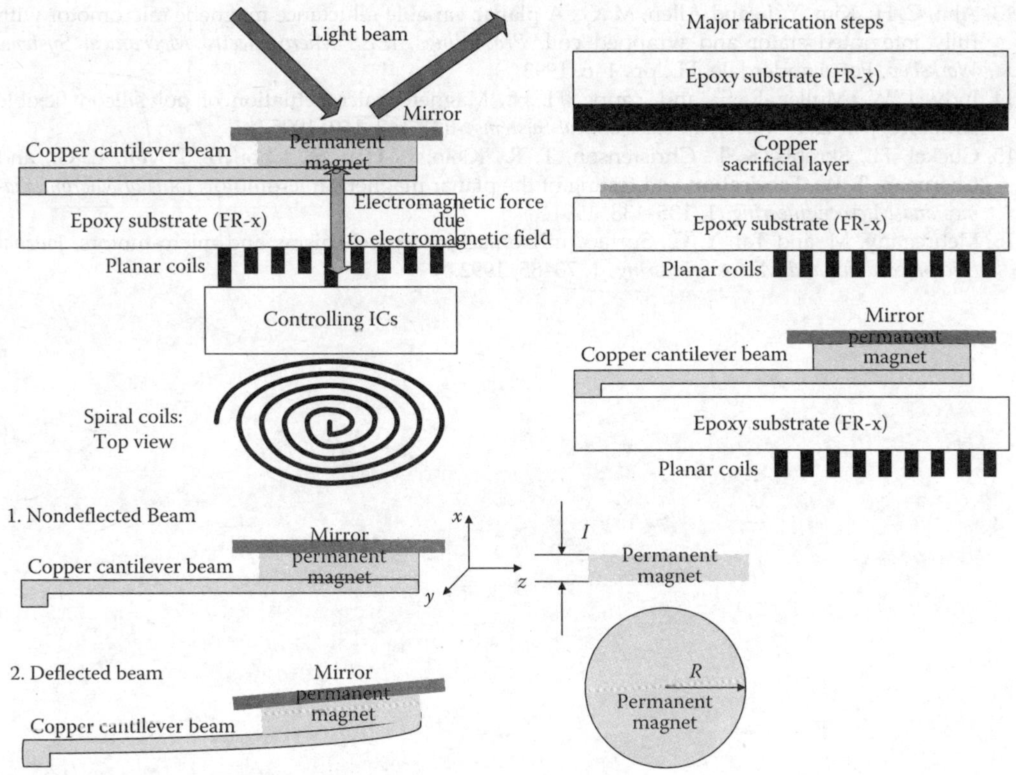

FIGURE 13.16

Electromagnetic microactuator with driving/controlling ICs and the suspended beam with permanent magnet and mirror.

References

1. Campbell, S. A., *The Science and Engineering of Microelectronic Fabrication*, Oxford University Press, New York, 2001.
2. Colclaser, R., *Microelectronics—Processing and Device Design*, John Wiley, New York, 1980.
3. Ghandi, S., *VLSI Fabrication Principles*, John Wiley, New York, 1983.
4. Guckel, H., Surface micromachined physical sensors, *Sensors and Materials*, 4(5), 251–264, 1993.
5. Petersen, K. E., Silicon as a mechanical material, *IEEE Proceedings*, 70(5), 420–457, 1982.
6. Beadle, W., Tsai, J., and Plummer, R., (Eds.), *Quick Reference Manual for Silicon IC Technology*, John Wiley, New York, 1985.
7. Wolfe, S. and Tauber, R., *Silicon Processing for the VLSI Era*, Lattice Press, Sunset Beach, CA, 1986.
8. Kovach, G. T. A., *Micromachined Transducers Sourcebook*, WCB McGraw-Hill, Boston, MA, 1998.
9. Madou, M., *Fundamentals of Microfabrication*, CRC Press, Boca Raton, FL, 2002.
10. Gad-el-Hak, M., (Ed.) *The MEMS Handbook*, CRC Press, Boca Raton, FL, 2001.
11. Lyshevski, S. E., *Nano- and Microelectromechanical Systems: Fundamentals of Nano- and Microengineering*, CRC Press, Boca Raton, FL, 2005.
12. Judy, J. W., Microelectromechanical systems (MEMS): Fabrication, design and applications, *Journal of Smart Materials and Structures*, 10, 1115–1134, 2001.

13. Ahn, C. H., Kim, Y. J., and Allen, M. G., A planar variable reluctance magnetic micromotor with fully integrated stator and wrapped coil, *Proceedings, IEEE Micro Electro Mechanical Systems Workshop*, Fort Lauderdale, FL, pp. 1–6, 1993.
14. Judy, J. W., Muller, R. S., and Zappe, H. H., Magnetic microactuation of polysilicon flexible structure, *Journal of Microelectromechanical Systems*, 4(4), 162–169, 1995.
15. Guckel, H., Skrobis, K. J., Christenson, T. R., Klein, J., Han, S., Choi, B., Lovell, E. G., and Chapman, T. W., Fabrication and testing of the planar magnetic micromotor, *Journal Micromechanics and Microengineering*, 1, 135–138, 1991.
16. Mehregany, M. and Tai, Y. C., Surface micromachined mechanisms and micro-motors, *Journal Micromechanics and Microengineering*, 1, 73–85, 1992.

Index

M